REVIEWS in MINERALOGY
Volume 38

URANIUM:
Mineralogy, Geochemistry and the Environment

Editors: Peter C. Burns[1] & Robert Finch[2]

[1]Department of Civil Engineering and Geological Sciences
University of Notre Dame
Notre Dame, Indiana

[2]Argonne National Laboratory
9700 South Cass Avenue
Argonne, Illinois

COVER PHOTOGRAPH:
Photomicrograph of a polished section showing uraninite, UO_{2+x} (opaque), ianthinite $U^{4+}_2(UO_2)O_4(OH)_6(H_2O)_9$ (purple to dark yellow), and uranophane, $Ca(UO_2)_2(SiO_3OH)_2(H_2O)_5$ (pleochroic mottled clear to blue-green). Originally purple ianthinite, remnants of which are apparent nearest the uraninite, has partly oxidized to yellow schoepite, a possible effect of sample preparation. Ianthinite and uraninite show an apparent epitaxial relationship, as do the overlying uranophane and ianthinite/schoepite, suggesting the potential importance of crystal structure on the paragenesis of uranium minerals. The paragenetic sequence pictured here, uraninite $\Rightarrow$ ianthinite $\pm$ schoepite $\Rightarrow$ uranophane, is observed spatially at micrometer to kilometer scales, as well as temporally, as uraninite alters through oxidization and dissolution to uranyl oxyhydroxides (ianthinite and schoepite), eventually forming uranyl silicates (uranophane) over time. Transmitted light, crossed polars; field: 0.5 mm wide.

Series Editor: Paul H. Ribbe
Virginia Polytechnic Institute and State University
Blacksburg, Virginia

MINERALOGICAL SOCIETY of AMERICA
Washington, DC

COPYRIGHT 1999

MINERALOGICAL SOCIETY OF AMERICA

The appearance of the code at the bottom of the first page of each chapter in this volume indicates the copyright owner's consent that copies of the article can be made for personal use or internal use or for the personal use or internal use of specific clients, provided the original publication is cited. The consent is given on the condition, however, that the copier pay the stated per-copy fee through the Copyright Clearance Center, Inc. for copying beyond that permitted by Sections 107 or 108 of the U.S. Copyright Law. This consent does not extend to other types of copying for general distribution, for advertising or promotional purposes, for creating new collective works, or for resale. For permission to reprint entire articles in these cases and the like, consult the Administrator of the Mineralogical Society of America as to the royalty due to the Society.

REVIEWS IN MINERALOGY

(Formerly: SHORT COURSE NOTES)
ISSN 0275-0279

Volume 38

URANIUM:
Mineralogy, Geochemistry and the Environment

ISBN 0-939950-50-2

ADDITIONAL COPIES of this volume as well as others in this series may be obtained at moderate cost from:

THE MINERALOGICAL SOCIETY OF AMERICA
1015 EIGHTEENTH STREET, NW, SUITE 601
WASHINGTON, DC 20036 U.S.A.

— URANIUM —
Mineralogy, Geochemistry and the Environment

FOREWORD

This volume was written and compiled by 24 authors and edited by Peter Burns and Robert Finch in preparation for a short course by the same title, sponsored by the Mineralogical Society of America, October 22 and 23, 1999 in Golden, Colorado, prior to MSA's joint annual meeting with the Geological Society of America. This is number 38 in the *Reviews in Mineralogy* series. As Series Editor, I apologize to our Eastern European authors for being unable to use appropriate diacritical marks on consonants in certain author and place names.

Paul H. Ribbe
Virginia Tech
August 26, 1999

PREFACE

Research emphasis in traditional mineralogy has often focused on detailed studies of a few hundred common rock-forming minerals. However, scanning the contents of a current issue of *American Mineralogist* or *Canadian Mineralogist*, or the titles of recent *Reviews in Mineralogy* volumes reveals that the emphasis of mineralogical research has undergone considerable change recently. Less-common, low-temperature minerals are receiving ever increasing attention, often owing to their importance to the environment. A tremendous challenge lies ahead for mineralogists and geochemists: the occurrences, structures, stabilities, and paragenesis of perhaps a thousand low-temperature minerals require detailed study if geoscientists are to be properly equipped to tackle environmental problems today and in the future. In many low-temperature environments mineral assemblages are extremely complex, with more than 10 species common in many cm-size samples. This *Reviews in Mineralogy* volume provides detailed reviews of various aspects of the mineralogy and geochemistry of uranium; hopefully the reader will benefit from this presentation, and perhaps more importantly, the reader may develop a sense of the tremendous amount of work that remains to be done, not only concerning uranium in natural systems, but for low-temperature mineralogy and geochemistry in general.

The low crustal abundance of uranium belies its mineralogical and geochemical significance: more than five percent of minerals known today contain uranium as an essential constituent. Uranium is a geochemical and geochronological indicator, and the U-Pb decay series has long been one of the most important systems for dating rocks and minerals. Uranium is an important energy source, and the uranium nuclear fuel cycle has generated a great deal of interest in uranium mineralogy and geochemistry since the first controlled nuclear fission reaction nearly sixty years ago. Current interest in uranium mineralogy and geochemistry stems in large part from the utilization of uranium as a natural resource. Environmental issues such as coping with uranium mine and mill tailings and other uranium-contaminated sites, as well as permanent disposal of highly radioactive uranium-based nuclear fuels in deep geologic repositories, have all refocused attention on uranium.

PREFACE

More than twenty years have passed since the 1978 Mineralogical Association of Canada's *Short Course in Uranium Deposits*. A realignment of research focus has clearly occurred since then, from exploration and exploitation to environmental remediation and geological "forecasting" of potential future impacts of decisions made today. The past decades have produced numerous remarkable advances in our understanding of uranium mineralogy and geochemistry, as well as technological and theoretical advances in analytical techniques which have revolutionized research of trace-elements, including uranium. It was these advances that provided us the impetus to develop this volume.

We have attempted to produce a volume that incorporates most important aspects of uranium in natural systems, while providing some insight into important applications of uranium mineralogy and geochemistry to environmental problems. The result is a blend of perspectives and themes: historical (Chapter 1), crystal structures (Chapter 2), systematic mineralogy and paragenesis (Chapters 3 and 7), the genesis of uranium ore deposits (Chapters 4 and 6), the geochemical behavior of uranium and other actinides in natural fluids (Chapter 5), environmental aspects of uranium such as microbial effects, groundwater contamination and disposal of nuclear waste (Chapters 8, 9 and 10), and various analytical techniques applied to uranium-bearing phases (Chapters 11-14).

We thank the numerous authors for making this volume possible; it has been a pleasure working with you. We are indebted to the many scholars who volunteered much of their time to provide detailed reviews of each chapter: they substantially improved the content and clarity of the chapters. We are especially indebted to Paul Ribbe, Series Editor for *Reviews in Mineralogy*, who made publication of the volume possible. We are grateful to the University of Notre Dame for financial support that has permitted many students to attend the Short Course at reduced cost. Finally, we thank our wives for putting up with many late nights over the past several months.

Peter C. Burns
University of Notre Dame

Robert Finch
Argonne National Laboratory
August 26, 1999

— URANIUM —
Mineralogy, Geochemistry and the Environment

Ch	p		
1	1	Radioactivity and the 20th Century	Rodney C. Ewing
2	23	The Crystal Chemistry of Uranium	Peter C. Burns
3	91	Systematics and Paragenesis of Uranium Minerals	Robert Finch & Takashi Murakami
4	181	Stable Isotope Geochemistry of Uranium Deposits	Mostafa Fayek & T. Kurtis Kyser
5	221	Environmental Aqueous Geochemistry of Actinides	William M. Murphy & Everett L. Shock
6	255	Uranium Ore Deposits: Products of the Radioactive Earth	Jane Plant, Peter R. Simpson, Barry Smith & Brian F. Windley
7	321	Mineralogy and Geochemistry of Natural Fission Reactors in Gabon	Janusz Janeczek
8	393	Geomicrobiology of Uranium	Yohey Suzuki & Jillian F. Banfield
9	433	Uranium Contamination in the Subsurface: Characterization and Remediation	Abdessalam Abdelouas, Werner Lutze & H. Eric Nuttall
10	475	Uranium Mineralogy and the Geologic Disposal of Spent Nuclear Fuel	David Wronkiewicz & Edgar Buck
11	499	Spectroscopic Techniques Applied to Uranium in Minerals	John M. Hanchar
12	521	Infrared Spectroscopy and Thermal Analysis of the Uranyl Minerals	Jirí Cejka
13	623	Analytical Methods for the Determination of Uranium in Geological and Environmental Materials	Stephen F. Wolf
14	653	Identification of Uranium-bearing Minerals and Inorganic Phases by X-ray Powder Diffraction	Frances C. Hill

TABLE OF CONTENTS

1   Radioactivity and the 20th Century
Rodney C. Ewing

INTRODUCTION	1
THE BEGINNING OF THE "ATOMIC ERA" : HISTORICAL NOTES	3
Henri Becquerel and the discovery of radioactivity	4
Marie and Pierre Curie	5
Ernest Rutherford and Frederick Soddy	6
Bertram Borden Boltwood	8
THE END OF THE CENTURY:	
THE NUCLEAR FUEL CYCLE INTERSECTS GEOCHEMICAL CYCLES	10
Geochemical cycles of radionuclides	11
Radiation exposure	14
SUMMARY	18
ACKNOWLEDGMENTS	19
REFERENCES	19

2   The Crystal Chemistry of Uranium
Peter C. Burns

INTRODUCTION	23
STRUCTURES OF U^{4+} MINERALS	24
BONDING IN U^{6+} POLYHEDRA	26
GEOMETRIES OF U^{6+} POLYHEDRA	27
Molecular-orbital calculations	27
EXAFS data	27
X-ray-diffraction data	28
Bond-valence parameters for U^{6+}	29
CONSTRAINTS ON POLYMERIZATION OF URANYL POLYHEDRA	30
HIERARCHY OF URANYL MINERAL STRUCTURES	32
The role of $(OH)^-$ and H_2O in uranyl mineral structures	33
Structures containing sheets of polyhedra	33
Anion topologies with squares	36
Anion topologies with triangles and pentagons	41
Anion topologies with triangles, squares and pentagons	49
Structures containing infinite chains	78
Structures containing finite clusters	82
Structures consisting of frameworks	85
ACKNOWLEDGMENTS	86
REFERENCES	86

Table of Contents

3   Systematics and Paragenesis of Uranium Minerals
Robert Finch & Takashi Murakami

INTRODUCTION	91
URANIUM MINERALS SYSTEMATICS	92
Minerals containing reduced uranium	92
Uranium(IV) silicates	97
Uranium(IV) niobates, tantalates and titanates	98
Uranium(IV) phosphates: rhabdophane group	105
Other uranium(IV) minerals	106
Uranyl minerals	107
Uranyl oxyhydroxides	108
Pb-uranyl oxyhydroxides	113
Uranyl carbonates	115
Uranyl silicates	120
Uranyl phosphates and arsenates	124
Uranyl vanadates, molybdates, tungstates	132
Uranyl sulfates, selenites, tellurites	134
ALTERATION OF REDUCED URANIUM MINERALS	137
Uraninite alteration under reducing conditions	138
Uraninite alteration under oxidizing conditions	140
Replacement of uraninite by uranyl minerals	144
ALTERATION OF URANYL MINERALS	146
Thermodynamic background	146
Measured and estimated thermodynamic parameters	146
Dehydration and the role of H_2O	152
Groundwater alteration	154
THE ROLE OF RADIOGENIC Pb IN U-MINERAL PARAGENESIS	156
PARAGENESIS OF THE URANYL PHOSPHATES	162
ACKNOWLEDGMENTS	166
REFERENCES	166

4   Stable Isotope Geochemistry of Uranium Deposits
Mostafa Fayek & T. Kurtis Kyser

INTRODUCTION	181
ISOTOPE FRACTIONATION: PRINCIPLES AND PROCESSES	181
Isotope exchange	182
Fractionation factors	183
Isotopic composition of systems and isotopic equilibrium	183
Causes of isotopic variation	184
Stable isotope geothermometry	185
ISOTOPIC COMPOSITION OF HYDROTHERMAL SYSTEMS	186
Open vs. closed system	187
Oxygen and hydrogen isotopes	188
Carbon isotopes	189
Sulfur isotopes	190

STABLE ISOTOPIC COMPOSITION OF PRINCIPAL FLUID RESERVOIRS 192
    Seawater 193
    Meteoric water.. 193
    Magmatic fluids and juvenile waters.. 195
    Metamorphic fluids 196
    Connate and formation waters, and basinal brines............................. 196
STABLE ISOTOPIC STUDIES OF URANIUM DEPOSITS................................. 197
    Unconformity-type uranium deposits 197
    Quartz-pebble conglomerate deposits.. 204
    Sandstone-type deposits 208
    Vein-type uranium deposits.. 213
CONCLUSIONS... 214
ACKNOWLEDGMENTS.. 214
REFERENCES.. 214

5   Environmental Aqueous Geochemistry of Actinides
William M. Murphy & Everett L. Shock

INTRODUCTION 221
GENERALITIES OF ENVIRONMENTAL AQUEOUS
   ACTINIDE GEOCHEMISTRY 222
    Complications in environmental aqueous actinide geochemistry........... 223
    Aqueous actinides in environmental and geochemical systems............. 225
        Uranium and other actinides in hot water.. 226
        The effects of pH and oxidation state on the speciation of aqueous U 226
CALCULATION OF STANDARD STATE PROPERTIES FOR AQUEOUS
   SPECIES AS FUNCTIONS OF TEMPERATURE AND PRESSURE 231
    Estimating the temperature dependence for
       equilibrium association constants 232
    Estimating values of log β at 25°C and 1 bar 236
        Inorganic complexes 236
        Organic complexes 238
        Activity coefficients for aqueous actinide ions and complexes......... 243
EXAMPLES OF SPECIATION CALCULATIONS FOR Am(III)
   IN GEOLOGICAL AND ENVIRONMENTAL FLUIDS.................................. 243
FUTURE DIRECTIONS FOR RESEARCH... 246
ACKNOWLEDGMENTS.. 246
APPENDIX.. 247
REFERENCES.. 248

6   Uranium Ore Deposits—Products of the Radioactive Earth
Jane A. Plant, Peter R. Simpson, Barry Smith, Brian F. Windley

INTRODUCTION 255
SOME CHARACTERISTICS OF URANIUM
   RADIOACTIVITY AND EARTH PROCESSES 256
SEPARATION OF RADIOTHERMAL CONTINENTAL CRUST 262

PRIMARY ENRICHMENT OF URANIUM IN FELSIC IGNEOUS ROCKS ... 264
Tonalite-trondhjemite granitoids ... 264
"I-type" calc-alkaline orogenic granitoids ... 264
Late and post-orogenic evolved I-type and A-type granites ... 266
CRUSTAL EVOLUTION ... 267
DISTRIBUTION OF RADIOACTIVITY IN THE CONTINENTAL CRUST ... 270
CLASSIFICATIONS OF URANIUM DEPOSITS ... 273
URANIUM DEPOSITS OF IGNEOUS PLUTONIC AND VOLCANIC ASSOCIATIONS ... 274
Changes associated with water-rock interaction in granites ... 276
Magmatic uranium deposits ... 280
Metasomatite uranium deposits ... 283
Breccia-complex uranium deposits ... 284
Volcanic uranium deposits ... 286
METAMORPHIC URANIUM DEPOSITS ... 290
Synmetamorphic uranium deposits ... 290
Uranium vein deposits in metamorphic rocks ... 291
URANIUM DEPOSITS OF SEDIMENT AND SEDIMENTARY BASIN ASSOCIATION ... 292
Quartz-pebble conglomerate uranium deposits ... 293
Unconformity-related uranium deposits ... 296
Sandstone uranium deposits ... 300
Sediment-hosted vein-type uranium deposits ... 302
Collapse-breccia pipe uranium deposits ... 303
Lignite uranium deposits ... 303
Surficial uranium deposits ... 304
Phosphorite uranium deposits ... 305
Uraniferous black shale ... 305
CONCLUSIONS ... 305
ACKNOWLEDGMENTS ... 307
REFERENCES ... 307

7 Mineralogy and Geochemistry of Natural Fission Reactors in Gabon
Janusz Janeczek

INTRODUCTION: NATURAL FISSION REACTORS ... 321
Geological setting ... 323
The origin of U deposits in the Franceville basin ... 324
Reactor zones ... 325
Physics of natural fission reactors ... 329
MINERALOGY OF NATURAL FISSION REACTORS ... 333
Uranium minerals ... 333
Uraninite ... 333
Coffinite ... 343
Uranyl minerals ... 345
Major associated minerals ... 347
Clay minerals ... 347
Galena ... 349

Accessory minerals.. 350
   Native elements.. 350
   Ru-, Rh-, and Pd-bearing minerals 350
   Pyrite and other sulfides.. 352
   Oxides 353
   Calcite... 354
   Phosphates 355
   Crandallite group 355
   Zircon... 356
Anomalous isotopic compositions of minerals in the reactor zones 357
URANIFEROUS ORGANIC MATTER.. 361
GEOCHEMISTRY OF NATURAL FISSION REACTORS 365
Actinides, fissiogenic and non-fissiogenic elements 365
Actinides... 368
   Uranium... 368
   Thorium 372
   Plutonium 372
   Neptunium 373
Actinide daughters: Bi and Pb 374
Fission products.. 375
   Alkali and alkaline earth elements (Rb, Sr, Cs, and Ba)................................ 375
   Rare Earth Elements.. 376
   Technetium.. 378
   Platinum group elements (PGE): Ru, Rh and Pd.. 379
   Low-yield mass region for fission products: Ag, Cd and Te 380
   Molybdenum... 380
Conditions for migration and retention of radionuclides
   in natural fission reactors 381
CONCLUDING REMARKS... 385
ACKNOWLEDGMENTS.. 387
REFERENCES.. 387

8   Geomicrobiology of Uranium
Yohey Suzuki & Jillian F. Banfield

INTRODUCTION... 393
BIOLOGICAL CLASSIFICATION OF METAL IONS 394
THE MICROBIAL CELL—SITES FOR METAL BINDING...................................... 394
   The cytoplasm and cytoplasmic membrane 395
   Cell walls and associated surfaces 396
   Extracellular polymers 398
MICROBIAL URANIUM ACCUMULATION... 399
   Metabolism-dependent metal uptake 399
   Metabolism-independent uptake of uranium.. 401
   Models for uranium biosorption 412
URANIUM TOXICITY, TOLERANCE, AND BIOAVAILABILITY 414
   Toxicity mechanisms in microorganisms... 415
   Factors affecting bioavailability and toxicity of uranium to microorganisms.... 415

Resistance and tolerance to uranium toxicity 415
MICROBIAL ECOLOGY IN CONTAMINATED SITES 417
   Microbial ecology in surficial uranium-contaminated sites and groundwaters 418
IMPLICATIONS FOR UNDERSTANDING GEOLOGICAL PROCESSES 419
   Microbial roles in the formation of uranium ore deposits 419
   Fossilization of microbial cells 420
   Subsurface transport of uranium 421
   Microbial impacts on the surface uranium cycle 421
BIOREMEDIATION OF URANIUM—
   AND OTHER ACTINIDES—CONTAMINATED SITES 422
IMPLICATIONS FOR MIGRATION OF OTHER ACTINIDES 422
SUGGESTIONS FOR FUTURE WORK 422
ACKNOWLEDGMENTS 424
REFERENCES 424

9   Uranium Contamination in the Subsurface: Characterization and Remediation
Abdesselam Abdelouas, Werner Lutze, H. Eric Nuttall

INTRODUCTION 433
LEACHING OF CONTAMINANTS FROM U MILL TAILINGS 435
ENVIRONMENTAL IMPACT AND HEALTH RISKS
   OF U MILL TAILINGS 438
   Remediation work in the western United States: The UMTRA program 438
   Remediation work in Germany by WISMUT 440
   Mill tailings in Canada: A case study 441
TECHNIQUES FOR U REMEDIATION 445
   Innovative or emerging processes 445
   Conventional technologies: Pump-and-treat 447
MICROBIAL REDUCTION OF URANIUM 449
   Reduction of uranium 449
   Reduction of uranium by microorganisms 450
   A laboratory study on bioremediation of uranium in groundwater
      from a mill tailings site 454
   Results 459
SUMMARY AND CONCLUSIONS 467
ACKNOWLEDGMENTS 468
REFERENCES 468

10   Uranium Mineralogy and the Geologic Disposal of Spent Nuclear Fuel
David J. Wronkiewicz & Edgar C. Buck

INTRODUCTION 475
THE STRUCTURE AND COMPOSITION OF SPENT NUCLEAR FUEL 476
THE REPOSITORY ENVIRONMENT 478

Table of Contents

An unsaturated setting 478
Effects of radiation on the repository environment 479
OXIDATION AND DISSOLUTION PROCESSES FOR URANIUM SOLIDS 480
EFFECTS OF SECONDARY URANIUM MINERALS 484
Influence on the dissolution of uranium solids 484
Comparison of corrosion products between spent fuel and UO_2 485
Comparisons with natural analogues 488
Influence of alteration phases on the migration of radionuclides 490
CONCLUSIONS 493
ACKNOWLEDGMENTS 494
REFERENCES 494

11 Spectroscopic Techniques Applied to Uranium in Minerals
John M. Hanchar

INTRODUCTION 499
OPTICAL SPECTROSCOPIC METHODS 499
Absorption spectroscopy 500
Luminescence spectroscopy 505
X-RAY ABSORPTION SPECTROSCOPY 509
XAS spectroscopy of uranium in minerals 510
X-RAY PHOTOELECTRON SPECTROSCOPY 515
X-ray photoelectron spectroscopy of uranium in minerals 515
CONCLUDING COMMENTS 516
ACKNOWLEDGMENTS 516
REFERENCES 517

12 Infrared Spectroscopy and Thermal Analysis of the Uranyl Minerals
Jiří Čejka

INFRARED SPECTROSCOPY OF THE URANYL MINERALS 521
Introduction 521
Uranyl, UO_2^{2+} 523
Uranyl vibrational modes (fundamental vibrations) 523
Uranyl stretching vibrations, v_1 UO_2^{2+} and v_3 UO_2^{2+} 524
Uranyl bending vibration v_2 (δ) UO_2^{2+} 527
Estimating the uranyl (U-O_t) bond length from its stretching vibrations 527
Equatorial ligand effect on v_3 UO_2^{2+} 528
Anion vibrational modes (fundamental vibrations) 528
Molecular water, hydroxyl, hydroxonium and $H_3O^+ \cdot nH_2O$ vibrational modes
(fundamental vibrations), hydrogen bonding and deuteroanalogues 530
Infrared spectra of:
Uranyl oxide hydrates, alkali, alkaline-earth and other uranyl oxide hydrates 535
Uranyl silicates 543
Uranyl phosphates and arsenates 548

Uranyl vanadates 557
   Uranyl molybdates 560
   Uranyl sulfates... 562
   Uranyl carbonates 567
   Uranyl selenites and tellurites.. 576
   Uranyl wolframates 578
THERMAL ANALYSIS OF THE URANYL MINERALS.............................. 578
  Introduction.. 578
  Thermal analysis of:
   Uranyl oxide hydrates, alkali, alkaline earth and other hydrates......... 582
   Uranyl silicates 587
   Uranyl phosphates and arsenates... 590
   Uranyl vanadates 598
   Uranyl molybdates 600
   Uranyl sulfates... 601
   Uranyl carbonates 604
   Uranyl selenites and tellurites.. 607
   Uranyl tungstates (wolframates).. 608
ACKNOWLEDGMENTS... 608
REFERENCES.. 608

13 Analytical Methods for Determination of Uranium in Geological and Environmental Materials
Stephen F. Wolf

INTRODUCTION... 623
SAMPLE DISSOLUTION METHODS... 623
  Fluxed decomposition 624
  Acid dissolution 624
CLASSICAL WET CHEMICAL TECHNIQUES... 624
  Gravimetric determination.. 625
  Volumetric determination... 625
  Colorimetric determination 626
URANIUM SEPARATIONS... 626
  Coprecipitation... 626
  Liquid-liquid extraction.. 626
  Ion-exchange.. 626
NUCLEAR METHODS... 627
  Radiometric techniques 627
    α-Spectrometry 627
    γ-Spectrometry 629
  Neutron activation analysis 632
    Fission track analysis 633
    Instrumental neutron activation analysis 633
    Radiochemical neutron activation analysis 634
ATOMIC SPECTROMETRIC TECHNIQUES... 634
  Atomic absorption spectrometry.. 635

Atomic emission spectrometry	636
Flame emission spectrometry	637
Inductively coupled plasma-atomic emission spectrometry	637
Atomic fluorescence spectrometry	638
Fluorometry	638
Phosphorimetry	639
X-ray fluorescence	639
MASS SPECTROMETRY	640
Solid state mass spectrometry	641
Spark source mass spectrometry	641
Glow discharge mass spectrometry	641
Secondary ion mass spectrometry	641
Laser ablation mass spectrometry	641
Resonant ionization mass spectrometry	642
Thermal ionization mass spectrometry	642
Inductively coupled plasma mass spectrometry	642
Calibration methods	645
High resolution-inductively coupled plasma mass spectrometry	646
Isotope ratio determination	646
SUMMARY	647
ACKNOWLEDGMENTS	648
REFERENCES	648

14   Identification of Uranium-bearing Minerals and Inorganic Phases by X-ray Powder Diffraction

Frances C. Hill

INTRODUCTION	653
A BRIEF REVIEW OF X-RAY DIFFRACTION	653
POWDER DIFFRACTION DATA FOR URANIUM-BEARING PHASES	654
COMPARISON OF EXPERIMENTAL AND CALCULATED POWDER X-RAY DIFFRACTOGRAMS	656
SUMMARY	659
ACKNOWLEDGMENTS	659
APPENDIX TABLES: Formulas, Names, Space Groups, Lattice Parameters, and d-Spacings of the Five Most Prominent Powder Diffraction Peaks	662
REFERENCES	673

1   Radioactivity and the 20th Century

Rodney C. Ewing

Departments of Nuclear Engineering & Radiological Sciences,
Materials Science & Engineering, and Geological Sciences
University of Michigan
Ann Arbor, Michigan 48109

"We stand to-day toward radio-activity exactly as our ancestor stood towards fire before he had learnt to make it."   H.G. Wells (1914) The World Set Free

INTRODUCTION

If a single word is used to describe the Twentieth Century, that word must be "Atomic" or "Nuclear." The developments in nuclear science during the twentieth century began with the discovery of X-rays by Röntgen in 1895, followed quickly with the discovery of radioactivity by Becquerel in 1896. The excitement and potential of these new discoveries quickly found their way into the popular culture with books such as *The Shape of Things to Come* and *The World Set Free* by H.G. Wells. The latter was dedicated to Frederick Soddy's book, *Interpretation of Radium*.

From the perspective of the end of this century, we can summarize the one-hundred-year legacy of the "Atomic Era" in terms of its impact: electrical power generation, weapons production and testing, nuclear waste management, and medical applications and benefits:

1. In 1995, 437 nuclear power plants generated approximately 17 percent of the world's electric power in 31 nations (Starr 1997; NEI 1997). Five new nuclear power plants became operational in 1996, and at the beginning of 1997, 45 nuclear power plants were under construction (NEI 1998). The United States has the largest number of operating nuclear power plants, 109, and the largest nuclear power generating capacity, 100.7 Gigawatts in 1996. Although 20% of the electricity in the U.S. is produced from nuclear power plants, other countries, such as France (~75%) and Belgium (~55%) use nuclear power for a much larger proportion of their electrical power production. In the United States, no nuclear reactor has been ordered by a utility company since 1978; and by the year 2000, 64 U.S. reactors will be more than 20 years old (Ahearne 1993). In 1996, world-wide use of nuclear power plants to generate electricity continued to increase reaching a new high of 22 percent.

2. In the United States, approximately 35,000 metric tons of spent (used) nuclear fuel from commercial power generation are presently stored by electric utilities at 70 sites throughout the country. Spent nuclear fuel is generated at a rate of 2,000 to 3,000 metric tons per year. The spent fuel is destined for direct disposal in the proposed repository at Yucca Mountain, one of the largest construction projects in the history of the Department of Energy. Approximately three billion dollars have been spent on investigations of the Yucca Mountain site, and the total cost and operation of the nuclear waste repository will probably exceed ten billion dollars. The originally scheduled opening date of 1998 has slipped to 2010.

3. Although the disposal of nuclear waste remains an abiding environmental issue, nuclear power plants have significantly reduced the emissions of CO_2 by reducing the amounts of fossil fuel consumption. In 1996, the production of energy from nuclear power plants was equivalent to a reduction of 147 million metric tons of

carbon emissions (Howard 1997). This is more than double the reduction called for by President Clinton's 1993 Global Climate Action Plan (NEI 1998). Whether the use of nuclear power plants can significantly reduce the CO_2 content of the atmosphere remains to be demonstrated, but the debate on global warming and consideration of possible solutions has certainly renewed interest in nuclear power.

4. Medical applications of X-rays and radioisotopes in diagnosis and treatment have extended the life expectancy of a burgeoning population (there are over 700,000 treatments of cancer each year with radiation in the U.S.). Within a decade of the discovery of radium, it was used in the treatment of cancer. Powerful radiation sources, such as ^{60}Co, Van de Graaff generators, betatrons, linear accelerators and cyclotrons are used for external beam radiation therapy, and man-made radionuclides, such as ^{99m}Tc, are used in medical imaging, diagnosis and treatment.

5. Weapons production in the United States has created approximately 400,000 cubic meters of high-level nuclear waste (Krauskopf 1988). Most of these wastes are presently held in underground storage tanks located mainly at Hanford, Washington, and Savannah River, South Carolina. The total activity of this waste exceeds one billion Curies.[1] The Department of Energy's budget request for nuclear waste management, remediation and disposal is approximately six billion dollars for the year 2000.

6. As a result of the first and second Strategic Arms Reductions Treaties, the dismantling of thousands of nuclear weapons in the United States and Russia have now made approximately 100 metric tons of weapons-grade plutonium available for geologic disposal or "burn-up" as a mixed oxide fuel in reactors.

7. Despite the very positive news that the United States and Russia have made available 100 metric tons of weapons-grade plutonium for disposal or disposition, the present world-wide inventory of fissile ^{239}Pu exceeds 1,300 metric tons and is growing at a rate of 50 to 70 metric tons per year. The major proportion of the plutonium produced continues to be from nuclear power plants (Stoll 1998; Ewing 1999).

8. A Comprehensive Test Ban Treaty (CTBT) has now been signed by 152 nations as an extension of the Non-Proliferation Treaty to permanently ban nuclear explosions. The Senate of the United States has yet to ratify the CTBT (Drell et al. 1999).

The nuclear legacy also includes the political and moral burden of two atomic bombs dropped on Hiroshima and Nagasaki in August of 1945, releasing energies equal to 12.5 and 22 kilotons of TNT, respectively. A recent survey of U.S. journalists and scholars ranks the atomic bombing of Hiroshima and Nagasaki as the top news story of this century. Many date the beginning of the Atomic Era from December 2, 1942, when the first nuclear chain reaction was sustained in the Chicago Pile Number One (CP-1) under the stands at Stagg Field at the University of Chicago, and this beginning quickly culminated in its military purpose. Nuclear energy was first used for electrical power generation on December 20, 1951. The Experimental Breeder Reactor Number 1 (EBR-1) in Idaho produced this electrical power, initially enough to light four bulbs. Although the early vision of nuclear power production anticipated 1,000 nuclear power plants by the year

---

[1] Radioactivity is expressed in Becquerel (Bq) = disintegration/sec. One curie (Ci), which is approximately equal to the activity of a single gram of radium, is equal to 3.7×10^{10} Bq.

2000, the promise of nuclear power generation was considerably damaged by the accidents at Three Mile Island (TMI) in 1978 and Chernobyl in 1986.

Despite the legacy of TMI and Chernobyl, nuclear power generation holds great promise. One of the most striking properties of nuclear fission is the magnitude of the energy released per fission, approximately 200 MeV, as compared to other nuclear reactions, such as an α-decay event with α-particle energies of approximately 5 MeV. The energy released from the complete fission of a single kilogram of ^{235}U is on the order of 10^{20} ergs or 10^{10} kilocalories, roughly equivalent to the explosion of 20,000 tons of TNT (Kaplan 1962). The possibility of breeding fissile ^{239}Pu from the more abundant ^{238}U isotope only deepens the well of the potential of nuclear power (Weinberg 1985). This potential stands as part of the solution to the projected electrical power needs of a growing population, the increased per capita energy consumption as the standard of living improves, and the limitations on non-renewable energy resources (Starr 1997). During the past 100 years the world's population has grown by more than 4 billion and the consumption of energy and raw materials has increased by a factor of ten. World energy needs are projected to double during the next several decades (Brown and Flavin 1999). Thus, today there is renewed interest and public debate on the future role of nuclear power. U.S. Senator Domenici (1998) has initiated the discussion of a new government policy toward the development of nuclear power and the treatment and disposal of nuclear waste.

Always at the center of the history of radioactivity and the present discussions of nuclear power and waste disposal is that remarkable, radioactive element, uranium, with its progeny of decay products, fission products and mostly man-made transuranium elements. Of the two primary isotopes of uranium, ^{235}U is fissile, and ^{238}U is the source of fissile ^{239}Pu (via neutron capture reactions and subsequent β-decays). This short chapter cannot be considered a history of radioactivity or the "Atomic Era;" there are already many extensive and informative histories (Smyth 1945; Badash 1979; Rhodes 1986). This chapter is not a contribution to the coming public debate on nuclear power; most of these words have yet to be written. This chapter is only meant to place the phenomenon of radioactivity in a broader perspective. A perspective that includes brief historical notes and a discussion relevant to the fears and concerns that the public has concerning radioactivity. The special, small subject of this brief chapter is to remind the geoscience community of what is often lost in discussions of "radio-activity" or "uranic rays": The mineralogy and geochemistry of uranium is an essential part of the discussion of *environmental radioactivity*. Environmental radioactivity, natural and anthropogenic, is now and will continue to be a major cause for public concern. The research of mineralogists and geochemists should contribute to the proper understanding of the extent and types of radioactivity in the environment. I think that it is the need for this perspective that makes the chapters presented in this volume relevant and important not only to scientists, but to the wider public.

THE BEGINNING OF THE "ATOMIC ERA" : HISTORICAL NOTES

The centennial celebrations of the discovery of radioactivity have just passed, and there are a number of commemorative papers that describe this history (Badash 1979; Genet 1995; Adloff and MacCordick 1995; Frame 1996; Adloff 1997; Halliday 1997). As part of the commemoration in 1998, the Muséum d'Histoire Naturelle presented an exhibit featuring the work of the early French scientists on radioactivity. Much of the exhibit is summarized in a beautifully illustrated catalogue (Schubnel 1996). The following summary draws heavily on these more extended histories.

The discovery of uranium predates the discovery of radioactivity, and the related radiation effects in minerals, by nearly one hundred years. The element uranium was discovered in 1789 by M.H. Klaproth, a German chemist and mineralogist at the University of Berlin, as part of his investigation of the pitchblende from Erzgebirge of Saxony. The dark metallic powder that Klaproth named uranium later proved to be UO_2. The French chemist, E. Péligot, isolated the first elemental uranium in 1841. The effects of radiation (Ewing 1994), e.g. metamictization due to decay of uranium, were recognized by the Swedish chemist Jacob Berzelius in 1815 and the Norwegian mineralogist W.C. Brøgger in 1893. However, the explanation for these phenomena, e.g. stored energy and optical isotropy, required the discovery of radioactivity.

Henri Becquerel and the discovery of radioactivity

The impetus for the search for radioactivity came with the discovery of X-rays by Wilhelm Conrad Röntgen in 1895 (for which he received the first Nobel Prize in physics) (Seliger 1995; Adams 1996). This was a rare instance in which the man-made phenomenon preceded the discovery of the natural phenomenon, radioactivity. Within a few weeks of Röntgen's discovery of X-rays, the news of this extraordinary phenomenon had arrived in Paris and was the subject of the weekly meeting of the French Academy of Sciences. One question, among many, was the source of the X-rays. Henri Poincaré maintained, as did Röntgen, that the X-rays originated from the walls of the Hittorf-Crookes tube. The issue was further complicated by the belief that the X-rays were related to the visible light produced by fluorescence. Henri Becquerel, Professor of Physics in the Museum of Natural History in Paris, was well equipped to pursue this idea, as his main field of research was fluorescence and phosphorescence. This field of study had a long history in the family of Henri Becquerel (Genet 1995). His great grandfather, Antonie César Becquerel (1788-1878), was the first to hold the Chair of Physics in the Muséum d'Histoire Naturelle, and Henri's father, Alexandre Edmond Becquerel (1820-1891), studied phosphorescence of minerals and had invented the "phosphoroscope." This remarkable family of scientists ended with the death of Henri's son, Jean Becquerel (1878-1953), a man much involved in subjects related to modern physics, such as radioactivity, nuclear energy and special relativity.

Henri Becquerel's long-standing interest in phosphorescence led to his discovery of radioactivity. In fact, he maintained, "It was perfectly appropriate that the discovery of radioactivity should have been made in our laboratory, and if my father had lived in 1896, it is he who would have made it." Antonine Henri had already noted that tetravalent uranium compounds were not phosphorescent; whereas, uranyl salts were luminescent. Thus, it is not surprising that Henri Becquerel used crystals of potassium uranyl sulfate stimulated by sunlight as a source of fluorescence. The crystals were placed on a photographic plate wrapped in aluminum foil and exposed to sunlight to stimulate the fluorescence. The developed film showed the dark exposure caused by the uranium salts, and Becquerel attributed the exposure on the film to radiation stimulated by sunlight. These results were reported to the Academy of Science on February 24th, 1896. Becquerel then recalled a wrapped plate with uranium salts in a closed drawer unexposed to sun, developed this plate and discovered the dark exposure caused by the uranium salts, of even greater intensity than the exposure when the salts were exposed to several hours of sunlight. He initially attributed this to delayed phosphorescence also stimulated by sunlight. Despite the misinterpretation of the origin of this image, the key observation was made on February 29th, 1896 and reported the next Monday. Becquerel immediately made other important observations concerning "uranic rays": the intensity of the radiation decreased with the increase of thickness of copper plates placed between the uranium salts and the film; the radiation rendered air conducting and discharged an electroscope; phosphorescent compounds without uranium did not expose the film, but nonphosphorescent uranyl sulfate did expose the photographic film. The importance of this phenomenon went mostly unrecognized by the scientific community. X-rays had been announced in the popular press before the first paper by Röntgen had been published, and thousands of papers followed, but radioactivity, emanating from uranium, produced no similar sensation. Becquerel's investigation and publications on radioactivity decreased, and he had temporarily abandoned this subject by 1898. Surprisingly, there was only a limited systematic effort to search for other substances that shared the peculiar properties of uranium. Two years later, Gerhardt S. Schmidt, also an expert on phosphorescence, discovered the radioactivity of thorium.

Marie and Pierre Curie

Fortunately, the lull in research on radioactivity ended with the arrival of Marie and Pierre Curie in Paris in 1898. Marie Curie, a doctoral student in physics, selected radioactivity as the subject of her dissertation. Marie Curie soon recognized that the intensity of the radiation was greater in the unprocessed uranium ore than in the purified uranium phases and concluded that there must be an additional source of radioactivity. Today, few appreciate the essentially mineralogical nature of her investigations (Curie 1903). Using a device to measure the electric current generated by ionized air (coupled to piezoelectric quartz, a property discovered by Pierre Curie and his brother, Jacques), she investigated numerous minerals. This device, a prototype of an ionization chamber, was much more sensitive to radiation than photographic film. Marie Curie's research revealed radioactivity from pitchblende, "chalcolite" (= torbernite), autunite, monazite, thorite, "orangite," fergusonite, and "cleveite." Although she attributed the source of the radioactivity to uranium and thorium, she established that some phases such as "chalcolite" and autunite were more radioactive than metallic uranium. Marie Curie wrote in her thesis,

> "Based on these observations it became highly probable that the high activity of pitchblende, chalcolite and autunite was related to the presence of small quantities of a highly radioactive substance, which was different from U, Th or any other known element."

Using a tedious chemical separation process involving successive fractional crystallizations, she isolated two new elements, polonium and radium, from the pitchblende ores of Joachimstahl. The Curies used standard chemical techniques for the separations, but took advantage of the radioactivity of the chemically fractionated compounds to follow the increasing concentrations of the radioactive species by measuring their total activities. This was certainly the birth of radiochemistry as it continues to be practiced today. Marie Curie published three notes on radioactivity during 1898 (Adloff and MacCordick 1995). In the first, she announced the discovery of radioactivity from thorium (the same observation had been published just two months earlier by Schmidt). In the second, she announced, with Pierre Curie as the first author and Gustave Bémont as the third author, the discovery of a new substance, polonium, which was 400 times more "radio-active" than uranium. Bémont was a chemist who would also be associated with the discovery of radium, but after this discovery his name disappears from future publications related to radioactivity. The third paper announced the discovery of a "new strongly radioactive" substance, radium, with an activity 900 times greater than uranium. Of interest to geoscientists, Marie Curie also noted low levels of activity in two potassium compounds, presumably due to ^{40}K (Adloff and MacCordick 1995). The Curies and Becquerel shared the Nobel Prize for Physics in 1903 for Becquerel's discovery of "spontaneous radioactivity" and the Curies' research on the "radiation phenomena." There was no mention of the discoveries of polonium and radium in the citation for this Nobel Prize. In 1911, Marie Curie received the Nobel Prize in Chemistry for the discovery of radium and polonium. The intervening decade had been filled with much controversy concerning whether polonium and radium were elements. In 1906, Lord Kelvin had argued that radium was a molecular combination of lead and helium, both associated with radium.

Ernest Rutherford and Frederick Soddy

At Cambridge, Ernest Rutherford was encouraged by J.J. Thomson (who had consulted on this matter with Lord Kelvin) to divert his interest from radiowaves in order to investigate Becquerel's newly discovered "uranic rays." Rutherford began his work in 1898, and his first publication on this subject appeared the following year (Rutherford, 1899). He identified two types of rays, α and β, the latter the more penetrating. In 1899, Sir William Crookes, the namesake and inventor of the cathode ray tube, began his studies of radioactivity by purchasing a half-ton of Joachimstahl pitchblende. Crookes purified the uranium to use as a standard and measured the radioactivity using a photographic plate (a much less sensitive technique than the ionization chamber used by Marie Curie). Surprisingly, the purified uranium did not expose the photographic plate. The pure uranium is an α-emitter, and the α-particles did not penetrate the film wrapping. Crookes assumed that another substance must carry the radioactivity, naming this substance UrX (Ur was the symbol used for uranium at that time). Crookes established that UrX was not polonium (also an α-emitter) and chemically different from radium. The UrX remained in the extraction and was in fact ^{234}Th (a result of the α-decay of ^{238}U with a half-life of 24.1 days) that decays to ^{234}Pa (a β-γ emitter with a half-life of 1.17 minutes). These were the two types of radiation that had been identified by Rutherford (1899). More remarkably, with time the original activity returned to U (Fig. 1). By 1900, Becquerel had returned to the subject of radioactivity and made the same observation as Crookes and noted that all radioactive materials had the same amount of activity (once normalized to the uranium concentration). Whatever the substance that caused the radioactivity, once separated from uranium, the activity was removed from the uranium only in the first moment, and then the radioactivity returned to the uranium over time.

The recognition of the radioactivity of thorium led to further important observations. The activity of thorium varied in an erratic way seemingly related to the movement of air past the sample. When shut in a box, the activity remained constant. Rutherford considered this "emanation" as a vapor of thorium. Rutherford, then at McGill, teamed with Frederick Soddy. Soddy, a skilled analytical chemist, examined the emanation from thorium and found it to be chemically inert, similar to the rare gases discovered just a few years before. Rutherford and Soddy also purified thorium in the same way

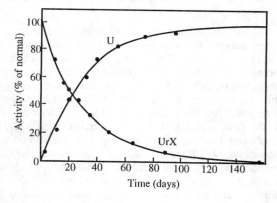

Figure 1. The decay of uranium X (UrX) activity and the concomitant recovery of the activity in the purified uranium. This effect was noted by both Becquerel and Crookes (after Kaplan, 1962).

Crookes had separated the radioactivity, UrX, from uranium. The separated filtrate had a high activity, more than 2,500 times that of Th. This was designated as ThX and later was identified to be ^{224}Ra ($t_{1/2}$ = 3.66 days), a decay product of ^{228}Ra ($t_{1/2}$ = 5.76 yr) (Fig. 2). These experiments were performed in December of 1901. When Rutherford and Soddy returned to their work after the Christmas vacation, they found that the thorium had

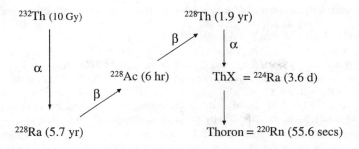

Figure 2. Partial decay scheme for thorium as determined by the early work of Rutherford and Soddy (1902c).

recovered its activity to its previously normal value, and the separated residues, ThX, had lost almost all of their original activity. These results were completely consistent with the results of Becquerel on the recovery of the activity of uranium and the loss of activity of UrX. Because of the shorter half-lives in the ^{232}Th decay series, Rutherford and Soddy (1902b) were able to perform the complete experiment. A thorium salt was treated with ammonia, resulting in the precipitation of thorium hydroxide. The residual solution, which now contained the radioactive ThX, was dried. The activities of the thorium hydroxide and ThX-bearing residue were measured over time (Fig. 3). The decrease in the activity of the ThX-residue was described by the exponential function:

$$A_{time}/A_{initial} = e^{-\lambda t} \tag{1}$$

where λ is the decay constant. The increase in activity of thorium hydroxide was similarly described by an exponential equation where λ has the same value as in Equation 1:

$$A_{time}/A_{initial} = 1 - e^{-\lambda t} \tag{2}$$

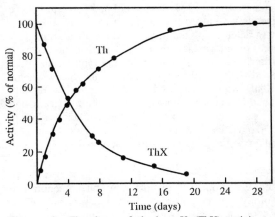

Figure 3. The decay of thorium X (ThX) activity as measured in the residue after the separation process and the concomitant recovery of the activity in the thorium hydroxide used to separate the thorium. This effect was first described by Rutherford and Soddy (1902b) (after Kaplan, 1962).

These results were only approximately correct because the chemical separation of ^{232}Th also included ^{228}Th and ^{228}Ac; the separation of ThX (^{224}Ra) also included ^{228}Ra (Fig. 2). There was a slight initial increase in the activity of ThX due to the in-growth of daughters such as ^{212}Pb (Rutherford and Soddy 1902b).

This simple experiment provided the evidence for important aspects of the process of radioactive decay: (1) one chemical element is transmuted to another chemical element; (2) the process can be described by a decay constant; (3) radioactive decay must involve changes at the atomic level because of the change in chemical properties of the decay products.

Prior to their classic series of papers on radioactive decay (Rutherford and Soddy 1902a,b,c), Rutherford had already noted an emanation from the decay of radioactive thorium which he named "thoron" (later identified as ^{220}Rn with a half-life of 55 seconds). Soddy, a chemist, determined that the thorium was transmuting to a chemically inert gas. As Rutherford and Soddy traced the radioactive decay of uranium, radium and thorium, they noted that the new decay products were radioactive and could be distinguished by their different half-lives. Chemically identical substances with different half-lives were called "isotopes", as later named by Soddy. In 1903, Soddy was among the first to note the tremendous amounts of energy released by radioactive decay (Rhodes 1986),

> "It may therefore be stated that the total energy of radiation during the disintegration of one gram of radium cannot be less than 10^8 gram-calories, and may be between 10^9 and 10^{10} gram-calories.... . The union of hydrogen and oxygen liberates approximately 4×10^3 gram-calories per gram of water produced, and this reaction sets free more energy for a given weight than any other chemical change known. The energy of radioactive change must therefore be at least twenty-thousand times, and may be a million times, as great as the energy of any molecular change."

Rutherford, still at McGill, with Howard T. Barnes began to examine the nature, magnitude and source of the heat associated with radioactive decay. They determined that the heat liberated was proportional to the number of α-particles emitted. In 1904 during a visit to the Royal Institution in England, Rutherford described this heating effect and concluded that Kelvin's calculations of the age of the Earth were only minimum estimates of geologic time. Rutherford lectured with Kelvin in the audience (Burchfield 1975),

> "The discovery of the radio-active elements, which in their disintegration liberate enormous amounts of energy, thus increases the possible limit of the duration of life on this planet, and allows the time claimed by the geologist and biologist for the process of evolution."

In addition to providing an additional source of heat for a cooling Earth and therefore substantially extending its age, the discovery of radioactivity provided a means of determining absolute ages of minerals and rocks. By 1904, Rutherford had come to believe that the helium locked in uranium minerals was a result of radioactive decay; however, there continued to be much controversy in determining the decay products of uranium. By that year, there was a considerable body of information on the radioactive elements. A number of elements, such as uranium, thorium, radium, actinium and polonium, had been identified chemically. Rutherford and Soddy had established the parts of the decay scheme for thorium: Th, ThX, ThEm, ThA and ThB. Rutherford had defined the series for radium: Ra, RaEm, RaA, RaB, RaC. Becquerel and Crookes had both described the transformation of uranium into UrX. These observations were the basis for the Rutherford-Soddy transformation theory in which one radioactive element was transmuted to a new radioactive element of different chemical properties and with a different half-life.

Bertram Borden Boltwood

The discovery of radioactivity and the properties of this new phenomenon had immediate and important effects on Geology. Bertram Borden Boltwood was one of the first Americans to note the importance of the discovery of radium (announced in the January, 1899, issue of *Scientific American* within a month of the paper by the Curies in *Comptes Rendus*). Boltwood realized that an elegant way to confirm the Rutherford-Soddy transformation theory was to show that the ratios of the amounts of radium and uranium, or other elements in the decay series, in minerals were constant. For minerals of great enough age, the radioelements in a decay series should have arrived at a secular equilibrium, and their ratios should be constant. Boltwood's simple experiment was elegant (Badash 1979). He dissolved five uranium ores in acid, and the gaseous emanations of radium (the radon) were collected. The gas was introduced into an air-tight electroscope. The rate of fall of the gold-leaf in the electroscope changed in proportion to the level of ionization of the air produced by radioactivity. The movement of the gold leaf was determined by reading a fine scale in the eyepiece of an attached microscope. The experiment assumed that the amount of radium was proportional to the gaseous emanation from the radium and that the degree of ionization was proportional to the activity. For the five samples, the measured ratios were 211, 212, 181, 207.2 and 218. The low value of 181 was attributed to the formation of gelatinous silica on the dissolution of uranophane that prevented the complete release of the emanations from the radium. The key to Boltwood's success, where others such as R.J. Strutt (the son of Lord Rayleigh) had been less successful, was in measuring the gaseous emanations on complete dissolution, rather than measuring the amount of radium directly. Based on his experiments, Boltwood concluded, "the quantities of radium present in uranium minerals which have been examined are apparently directly proportional to the quantities of uranium contained in the minerals" and that "radium is formed by the breaking down of the uranium atom." Others, including Robert A. Millikan, provided confirmation of Boltwood's experiments and conclusions. Boltwood continued his work, examining a wide variety of minerals. Much effort was devoted to determining the uranium content of minerals, a task that was particularly difficult for minerals such as monazite. His frustration shows in his words in a letter to Rutherford (Badash 1979),

> "When I have established the presence of the indicated quantities of uranium in monazite, which I mean to do even if it takes years to accomplish it, I am going to drop this line of research right there and rest on the assumption that the radium-uranium ratio is constant in ALL MINERALS, leaving it for others, who may so desire to attempt to PROVE the opposite."

By 1903, Ramsay and Soddy had discovered helium in radium and many believed that helium was the final decay product of a number of radioactive elements. Based on the helium content of uranium ores, Rutherford estimated their ages to be at least 400 million years. Robert Strutt had already calculated ages of two billion years. Strutt continued the work on uranium-helium dating and by 1910 had used zircon in dating that produced ages of 321 million years and titanite ("sphene") with ages of 700 million years. At the same time, Boltwood pursued uranium-lead dating. As early as 1905, Boltwood had observed that lead is invariably associated with uranium ores. Boltwood had found lead in all primary uranium minerals, and although hard pressed to detect the lead, had established its presence in all secondary uranium minerals. In general, greater concentrations of lead and He were found in the older primary uranium minerals. Rutherford had speculated that the emission of eight alpha particles, if they were indeed helium atoms, would reduce the atomic weight of uranium from 238.5 to 206.5, close to the atomic weight of lead. The initial motivation for Boltwood's studies was to use minerals of known geological ages in order to establish decay constants, but by the time he published his results in 1907, he had shown a good correlation between the uranium-lead ratios and the geological ages of forty-three minerals. Again, the elegance of the approach warrants a brief description (Badash 1979). At secular equilibrium, the activity ($A = N\lambda$, where N is the number of atoms) is equal for all of the radionuclides in the decay series. Since the decay constant is inversely proportional to the half-life ($\lambda = ln2/t_{1/2}$), the ratio of decays to half-life is also a constant for each decay series. Careful measurements by Rutherford and Boltwood established that the half-life of radium was about 2,600 years (the present value for the half-life of ^{226}Ra is 1,599 years) and that 10^{-10} grams of radium formed from a gram of uranium each year. At secular equilibrium the number of atoms of each member of the decay series changing to the next is constant. Thus, the conversion of U to Pb as a function of time can be used as the basis for age-dating using the simple formula (Pb/U is the atomic ratio),

$$\text{Age} = (\text{Pb/U}) (10^{10} \text{ years}) \tag{3}$$

Boltwood's mineral ages ranged between 410 million and 2.2 billion years. Neither Rutherford nor Boltwood pursued their age-dating studies, and the uranium-helium method continued to be regarded as the most useful and accurate. One of Strutt's students, Arthur Holmes, revived the uranium-lead dating technique and applied radioactive age-dating to the geologic time scale. Holmes began his work in 1911, the same year that Rutherford turned his attention to the structure of the atom and used alpha-particle irradiations in the famous scattering experiment with a thin foil of gold that established that the atom contained most of its mass concentrated in the small volume of the nucleus. This was the first time that the interior of the atom had been probed, and this marks the beginning of the field of nuclear physics. Holmes published the *Age of the Earth* in 1913 in which he combined the geological time scale with the results of radiometric dating. The debate over the age of the Earth continued for another 20 years until the publication of a series of reports by the National Research Council (NRC 1929, 1931) to which Arthur Holmes was a major, contributing author.

THE END OF THE CENTURY:
THE NUCLEAR FUEL CYCLE INTERSECTS GEOCHEMICAL CYCLES

At the end of the 20th century, radioactivity continues to be one of the most important phenomena and tools in the geosciences. Radioactive decay is now understood to provide much of the heat that drives the movement of material from the interior of the Earth to its crust. In the early part of this century, this newly discovered source of heat allowed scientists to extend the age of the Earth beyond Lord Kelvin's early calculations and much shorter estimates of tens of millions of years. The radioactive decay of nuclides from parents to daughters that have distinctly different chemical properties had its first important application in determining the absolute ages of rocks, but at the end of this century, this same phenomenon provides methods for tracing and understanding geologic and extraterrestrial processes (Fig. 4). These applications are not discussed in this chapter; however, a number of examples of the uses of radionuclides are summarized in Table 1.

For most of the 20th century radioactivity has been a tool for geoscientists, but at the end of this century and into the next, radioactivity in the environment has become a public concern. Environmental radioactivity is not a new subject, a book of this title, *Environmental Radioactivity*, was published in 1963 by Merril Eisenbud (the newest edition in 1987). Very quickly after the discovery of radioactivity one of the surprising discoveries by scientists was how common radioactivity is in the Earth's crust. By the second half of this century, scientists had acquired enough data to develop a good understanding of the sources of radiation in the environment (Adams and Lowder 1964). There is now a substantial knowledge of the sources, types, amounts and distribution of radioactivity in the environment (Eisenbud 1987). One of the most important aspects of this knowledge is to distinguish between natural and anthropogenic sources of radiation (Eisenbud and Paschoa 1989). This is the basis for determining pathways for human exposure and provides at least a qualitative basis for evaluating health effects and establishing regulations.

The development and evaluation of strategies for the long-term disposal of radioactive wastes rely heavily on an understanding of the behavior of naturally occurring radionuclides (Chapman 1992). This is the basis, in part, of "natural analogue" studies (Miller et al. 1994) at uranium deposits, such as Cigar Lake in Canada (Cramer and

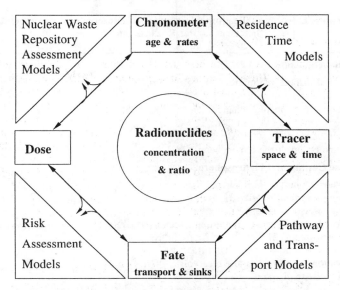

Figure 4. Relationship of radionuclides to four important aspects of the geosphere and exposure estimates (after Santschi and Honeyman 1989).

Smellie 1994), Poços de Caldas in Brazil (Chapman et al. 1990) and the Oklo natural reactors in Gabon (Gauthier-Lafaye et al. 1996; Janeczek, this volume; Jensen and Ewing, submitted). Thus, radioactive waste management is essentially the intersection of the nuclear fuel cycle with the geochemical cycles of radionuclides. Substantive evaluation of the environmental impact of any element or contaminant requires an understanding of the intersection of the short-term anthropogenic and technogenic cycles of mining, extraction, manufacturing, use and disposal with the long-term geochemical cycle. Such an approach is an extension (over time) of the relatively new methods used in industrial ecology to analyze global flows of materials and energy (Frosch and Gallopoulos 1989; Lifset 1997).

Geochemical cycles of radionuclides

Uranium is the most abundant of the naturally occurring actinides (Ac, Th, Pa, and U) with values ranging from 1.2 µg/g in sedimentary rocks to 120 µg/g in phosphate rocks (Langmuir 1997). The lower members of the actinide series are derived from the decay of three, long-lived parent nuclides ^{238}U, ^{235}U and ^{232}Th. The higher actinides (e.g. ^{237}Np and ^{239}Pu) occur naturally at very low levels (10^{-12} to 10^{-11} ppm) in the crust due to production by neutron capture reactions with ^{238}U (Peppard et al. 1951; Levine and Seaborg 1951) because actinides in the neptunium series (e.g. ^{241}Pu and ^{237}Np) that were initially present during the formation of the Earth have since decayed. Present day concentrations of higher actinides in the environment are essentially due to anthropogenic production from the nuclear fuel cycle. The annual productions of plutonium (50 to 70 metric tons/year) and neptunium (3 to 4 metric tons/year) are mainly due to nuclear reactors used for commercial power production. The global inventory of plutonium (>1,300 metric tons) has accumulated since its discovery in 1940 (Stoll 1998).

The general features of the geochemical cycle of uranium are well known (Langmuir 1978; Gascoyne 1982; Langmuir 1997) and many parts of this cycle are covered in detail in this volume. The most abundant uranium phase is uraninite, UO_{2+x}. In near surface, low temperature environments, U(IV) is highly immobile; however, oxidation to the higher

Table 1. Major applications of natural and anthropogenic radionuclides in the earth sciences (after Rozanski and Froehlich 1996).

Studies of the Atmosphere:
  dispersion, transport and mixing: ^{3}H, ^{85}Kr, ^{222}Rn, ^{14}C
  transport of water vapor: ^{3}H
  stratosphere-troposphere exchange: ^{3}H, ^{14}C, ^{85}Kr, ^{7}Be, ^{10}Be
  sources and sinks of CO_2 and CH_4: ^{3}H, ^{14}C
  atmospheric deposition: ^{36}Cl, ^{7}Be, ^{10}Be, ^{90}Sr, ^{137}Cs
  age and evolution of the atmosphere: ^{129}Xe

Studies of the Continental Hydrosphere:
  groundwater recharge: ^{3}H, ^{36}Cl
  dispersion of surface waters: ^{3}H
  aeration processes: ^{85}Kr
  interaction between surface and ground waters: ^{3}H, ^{222}Rn, ^{14}C
  groundwater dating: ^{3}H, ^{14}C, ^{85}Kr, ^{39}Ar, ^{36}Cl, ^{81}Kr
  rock-water interactions: ^{238}U, ^{234}U, ^{226}Ra, ^{228}Ra
  sedimentation rates in lakes and reservoirs: ^{137}Cs, ^{210}Pb, ^{230}Th
  radioactive waste disposal: ^{36}Cl, ^{129}I

Studies of the Oceans:
  circulation and mixing process: ^{3}H, ^{14}C, ^{85}Kr
  age of water masses: ^{3}H, ^{14}C, ^{39}Ar, ^{85}Kr
  transfer of anthropogenic CO_2 to the ocean: ^{14}C
  dating of ocean sediments: ^{14}C, ^{40}K
  dating of sea level changes: ^{14}C, ^{234}U, ^{230}Th

Studies of the Lithosphere:
  dating of rocks and minerals: ^{40}K/^{40}Ar, ^{40}Ar/^{39}Ar, ^{87}Rb/^{87}Sr, ^{176}Lu/^{176}Hf, ^{147}Sm/^{143}Nd, ^{187}Re/^{187}Os, ^{138}La/^{138}Ba, ^{138}La/^{138}Ce, ^{40}K/^{40}Ca, radionuclides in the decay series of ^{235}U, ^{238}U and ^{232}Th
  dating of carbonate deposits: ^{14}C, ^{234}U, ^{230}Th, ^{231}Pa
  dating of lacustrine deposits: ^{137}Cs, ^{210}Pb, ^{14}C, ^{234}U, ^{230}Th
  dating of surface exposures: ^{10}Be, ^{14}C, ^{26}Al, ^{36}Cl
  dating of meteorite surfaces: ^{80}Kr/^{81}Kr, ^{3}He/^{3}H, ^{41}K/^{40}K
  soil erosion: ^{137}Cs, ^{210}Pb, ^{10}Be
  mineral exploration: radionuclides in the decay series of ^{235}U, ^{238}U, and ^{232}Th
  earthquake monitoring: ^{222}Rn
  paleoseismicity and volcanic eruptions: ^{36}Cl, ^{26}Al, ^{10}Be

Studies of Climate Change: ^{230}Th, ^{231}Pa, ^{234}U, ^{210}Pb

Studies of the Early Solar System: extinct nuclides
  ^{129}I/^{129}Xe, ^{107}Pd/^{107}Ag, ^{53}Mn/^{53}Cr, ^{146}Sm/^{142}Nd, ^{41}Ca/^{41}K, ^{60}Fe/^{60}Ni, ^{26}Al/^{26}Mg, ^{182}Hf/^{182}W

---

valence, U(VI), results in the formation of the highly mobile uranyl ion, UO_2^{2+}. Depending on pH and solution composition a wide variety of uranyl complexes may form. Carbonate and organic complexes can be particularly important in determining uranium speciation and concentration in solution. Additionally, it has long been recognized that bacteria can play an important role in determining the oxidation state of uranium, and thus its mobility (Lovley et al. 1991). Bacteria may be used to enhance acid leaching of uranium from ores (Tuovinen and Kelly 1974) or for the bioremediation of uranium contaminated waters (Lovley and Phillips 1992). The other actinides follow uranium very closely in their geochemical behaviors; however, important differences result from a wider variety of oxidation states under natural aqueous conditions (e.g. Pu has four oxidation states: III, IV, V, VI and Np has two: IV and V, forming actinyl complexes such as PuO_2^+, PuO_2^{2+} and NpO_2^+). The mobility of each actinide depends essentially on a limited number of processes: precipitation/dissolution, co-precipitation, colloid formation and transport, complexation and sorption (Silva and Nitsche 1995). These processes not only control global distributions of the elements, but also control smaller scale redistributions of actinides, such as Pu, within uranium deposits (Curtis et al. 1999; Janeczek 1999).

For the purposes of this discussion it is important to know the contribution of naturally occurring actinides (mainly uranium and thorium) to the global inventory of radioactivity. Santschi and Honeyman (1989) have calculated the total activity of uranium and thorium isotopes in the upper 100 meters of the lithosphere (that is the portion of the elemental inventory readily exchangeable with air and water). The total activity is on the order of 10^{22} Bq, and an equal amount of activity is associated with the decay of ^{40}K (Fig. 5). For comparison, the global inventory of ^{239}Pu is 10^{18} Bq. Although the inventories and average concentrations are highest for natural radionuclides, global transfer rates due to anthropogenic processes have become comparable to transport rates due to natural processes. Although the magnitudes of the inventories of natural and anthropogenic radionuclides and their associated activities are of qualitative interest, they provide only limited insight into exposure and health affects. In order to develop the complete picture,

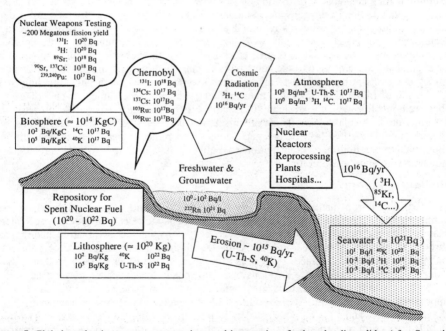

Figure 5. Global production rates, concentrations and inventories of selected radionuclides (after Santschi and Honeyman 1989). The sources of anthropogenic radioactivity are in boxes with the bold outline. Note: some of the anthropogenic radionuclides have short half-lives (e.g. ^{131}I = 8 days; ^{90}Sr = 28.78 yr; ^{137}Cs = 30.07 yr). The total estimated activity for the nuclear waste repository is based on 70,000 metric tons or equivalent of spent nuclear fuel. The range of values reflects variations in fuel burn-up and the proportion of high level waste from reprocessing that are included in the inventory. After 10,000 years the total activity in the repository will decrease to less than 0.01 percent of the initial value, mainly due to the decay of short-lived fission products.

radionuclide inventories must be analyzed in terms of their concentrations, distributions, mobility and half-lives, and finally placed in the broader context of total levels of radioactivity and exposure.

There are over 1,000 naturally occurring and man-made radionuclides found in the environment in concentrations that range from 10^{-15} to 10^{-20} molar up to activity levels of 10^3 Bq/kg (Santschi and Honeyman 1989). The radionuclides may have extremely short or long half-lives (<10^{-6} seconds to >10^9 years), and they may be geochemically mobile (e.g. ^{99}Tc) or relatively immobile (e.g. ^{239}Pu; however, see Kersting et al. (1999) for a recently documented example of colloidal transport of plutonium). The pathways that are responsible for human exposure depend on geochemical and biochemical processes, as well as the life-style of the exposed individual or group. Santschi and Honeyman (1989) have emphasized the connection between the dose response calculations and the geochemical cycles (Fig. 4). Ironically, the use of radionuclides in age-dating and as tracers very much facilitates the application of these concepts to studies of radionuclide pathways for exposure. A general knowledge of the residence time combined with the half-life and decay products of each radionuclide is an effective method of evaluating the likelihood of transport and pathways for human exposure. Global production rates and inventories for selected radionuclides are shown in Figure 5. The comparison of the natural fluxes and residence times in the geosphere to those that are related to a nuclear waste repository can also be used as a basis for establishing regulations for public safety and serve as a means of communicating concepts of dose and risk to the public (Miller and Smith 1993). That is, if the radionuclide flux in the geosphere is much greater than that induced by the presence of a repository, then the repository's radiological significance is not great. There are certainly technogenic radionuclides, such as ^{99}Tc, that are not present in any significant concentrations in the geosphere (Curtis et al. 1999), but the behavior of these nuclides can be evaluated by comparison to elements with similar chemistries. An ideal basis for public policy decisions in nuclear waste management should be a thorough understanding of the generation of radionuclides by the nuclear fuel cycle, the geochemical cycles of each of these radionuclides and the evaluation of pathways for human exposure. At present, such information is available, but generally dispersed between the relevant disciplines. The unique opportunity that lays before geochemists and mineralogists is to bridge this gap, thus making an important contribution to public policy decisions on the protection of human health and the environment.

Radiation exposure

Exposures from natural radioactivity from terrestrial and cosmic origins have been summarized by Eisenbud and Paschoa (1989). The cosmic rays from outer space interact mainly with the upper atmosphere, and the heavy ions interact with atmospheric gases through a number of nuclear reactions that result in a measurable flux of penetrating radiation at the Earth's surface. Typical exposures are on the order of 0.25 to 0.50 mSv/yr [2] depending on elevation. At elevations of 3,000 meters the exposure may be as high as 1.25 mSv, and during a transcontinental flight a passenger receives a dose of approximately 0.025 mSv. ^{14}C is produced by the interaction of cosmic rays with ^{14}N in the atmosphere, and ^{14}C ($t_{1/2}$ = 5,715 yr) is present in living tissue at concentrations of 0.28 Bq per gram of carbon (= 6.0 pCi per gram). This results in a dose of approximately 7 μSv/yr to the skeleton. Tritium ($t_{1/2}$ = 12.32 yr) is also produced by cosmic ray interactions with the atmosphere and exists mainly in the form of water vapor and

---

[2] The SI unit of equivalent radiation dose, a severest (Sv), is the product of the absorbed dose and a radiation weighting factor. One Sv = 100 em. A em is an alternate unit of dose equivalent equal to the quantity of radiation with the same biological effect as x-rays delivering 100 ergs per gram of tissue.

precipitates as rain and snow. Tritium occurs in surface waters at concentrations of 0.37 Bq per liter; thus, the dose from tritium is much less than that from cosmogenic ^{14}C. The global inventory of tritium is approximately 26 MCi (= 10^{18} Bq). Both ^{14}C and tritium decay by β⁻ emission. Cosmogenically produced ^{7}Be decays by electron capture ($t_{1/2}$ = 53.28 days). Air filters become radioactive due to both ^{7}Be and radon decay products that are sorbed onto airborne particles.

Of the approximately 350 naturally occurring nuclides, 70 are radioactive. These radionuclides occur as radioactive decay products in the three long-lived decay chains of ^{238}U, ^{235}U and ^{232}Th. Crustal abundances of uranium are in the order of 1 to 4 ppm in normal rocks and sediments to hundreds of ppm in phosphate rich deposits. The decay chains include more than 40 radionuclides with half-lives between 10^{-6} and 10^{8} years. Two important radionuclides in the decay chains are ^{226}Ra ($t_{1/2}$ = 1,599 yr) which is a bone-seeking nuclide and ^{222}Rn ($t_{1/2}$ = 3.8 days), a noble gas. The principal vector for exposure from radionuclides in the decay chains is ^{226}Ra which is highly soluble and whose decay leads to ^{222}Rn which further decays to radioactive isotopes of polonium, bismuth, lead and thallium. As an inert gas, radon readily diffuses through the atmosphere, and its principal pathway to the body is through the lungs. Average exposures are approximately 1.4 mS/yr. The exposure by radon dominates all other sources of radioactivity (see Table 2). Other nuclides in the decay chains reach the body through food (^{210}Pb and ^{210}Po), smoking (^{210}Po) or through the atmospheric discharge of fly ash from burning coal. Radionuclides that are not included in the decay chains are long-lived radionuclides that are still present in the Earth (e.g. ^{40}K, $t_{1/2}$ = 1.27×10^{9} yr; ^{87}Rb, $t_{1/2}$ = 4.88×10^{10} yr). Typical concentrations of ^{40}K in the body are 30 Bq per gram of potassium with an equivalent internal dose of 0.2 mSv/yr (external exposure from rocks is on the order of 0.1 mSv/yr).

Table 2. Average amounts of ionizing radiation received yearly by a member of the U.S. population (after NRC 1995).

Source	Dose	
	mSv/yr	%
Natural		
radon	2.0	55
cosmic	0.27	8
terrestrial	0.28	8
internal	0.39	11
Total Natural	**3.0**	**82**
Anthropogenic		
medical X-ray diagnosis	0.39	11
nuclear medicine	0.14	4
consumer products	0.10	3
occupational	<0.01	<0.3
nuclear fuel cycle	<0.01	<0.03
miscellaneous	<0.01	<0.03
Total Anthropogenic	**0.63**	**18**
Total Natural and Anthropogenic	**3.6**	**100**

In summary, typical exposures from natural sources are 2 mS/yr from radon and 1 mSv/yr from all other natural sources. The health effects of radon exposure are still much debated (see Cohen 1995, 1998; associated papers in *Health Physics*, the website for the

Environmental Protection Agency: http://www.epa.gov/iaq/radon/; the most recent report of the National Research Council 1999), and there have been recent advances in the evaluation of low radiation dose effects on cells (Miller et al. 1999).

The anthropogenic sources of radiation exposure are limited (Table 2). Medical and dental procedures are on average less than 1 mSv/yr and from consumer products the exposure is on the order of µSv/yr.

The nuclear fuel cycle and weapons testing are usually the principal radiation exposure concerns in the minds of the public. Uranium mining is estimated to have created tailings piles which contain 1.9×10^3 TBq (50,000 Ci) of ^{226}Ra. The principal exposure vector is through ^{222}Rn gas, and although ^{222}Rn has a half-life less than 4 days, the longer-lived ^{226}Ra ($t_{1/2}$ = 1,599 yr) and its parent ^{230}Th ($t_{1/2}$ = 75,400 yr) support the continuous production of ^{222}Rn.

The uranium extracted by mining fuels some 1,000 nuclear reactors, nearly one-half are used for electrical power generation. The airborne releases from these reactors are dominated by short-lived noble gases (e.g. ^{133}Xe, $t_{1/2}$ = 5 days; ^{135}Xe, $t_{1/2}$ = 9 hours; ^{138}Xe, $t_{1/2}$ = 14 minutes; ^{88}Kr, $t_{1/2}$ = 3 hours) and lesser amounts of ^{131}I ($t_{1/2}$ = 8 days); however, three radionuclides with relatively long half-lives are also released: ^{3}H ($t_{1/2}$ = 12.32 yr), ^{85}Kr ($t_{1/2}$ = 10.7 yr) and ^{14}C ($t_{1/2}$ = 5,715 yr). In 1981, exposures to individuals living near nuclear power plants were estimated to be in the range of 10^{-7} to 5×10^{-4} mSv/yr (Eisenbud 1987). These doses are very low, comparable or less than the exposures resulting from release of uranium, thorium and their daughter products in the airborne effluents of coal-fired power plants (McBride et al. 1978).

The largest source of radioactivity in the fuel cycle is from used or spent nuclear fuel. Approximately 20% of the electrical power is generated by approximately 100 nuclear power plants in the U.S. Each plant generates approximately 30 metric tons of spent nuclear fuel per year. By the year 2020 this inventory will have grown to nearly 80,000 metric tons with a total activity of over 35,000 MCi (10^{21} Bq). In the United States, this fuel will be disposed of directly in a geologic repository without reprocessing to reclaim fissile radionuclides (e.g. ^{235}U and ^{239}Pu). Most of the activity in spent fuel is associated with short-lived fission products, such as ^{90}Sr ($t_{1/2}$ = 28.78 yr) and ^{137}Cs ($t_{1/2}$ = 30.07 yr); however, over time the radiotoxicity of the fuel is mainly associated with the much longer-lived actinides, such as ^{239}Pu ($t_{1/2}$ = 24,100 yr) and ^{237}Np ($t_{1/2}$ = 2.14×10^6 yr). Although a typical burn-up (30 to 40 MWd/kg U) only converts 1% of the UO_2 fuel to transuranic elements (approximately 3% of the UO_2 is converted to fission products), the global inventory of transuranics is large. Since the creation of the first man-made atoms of plutonium by bombarding UO_2 with deuterons by Glenn Seaborg and colleagues at Berkeley in 1940, the world-wide inventory has grown to approximately 1,300 metric tons of ^{239}Pu (Stoll 1998). This is a large number in comparison to the bare critical mass of weapons grade plutonium that is only 15 kg. In addition, other actinides are generated: 3,500 kg/yr of ^{237}Np ($t_{1/2}$ = 2.14×10^6 yr), 2,740 kg/yr of ^{241}Am ($t_{1/2}$ = 432.7 yr).

In a relatively short period of time the total activity of the spent fuel decreases substantially by a factor of more than 100 in 100 years. Most of the radioactivity is associated with fission products that have half-lives of approximately 30 years. After only 7 half-lives, over 99 percent of the radionuclides have decayed. After 1,000 years, the radioactivity is dominated by longer-lived actinides and their daughter nuclides. Figure 6 compares the radiotoxicity due to inhalation of spent nuclear fuel with a burnup of 38MWd/kgU to the radiotoxicity of the quantity of uranium ore that was originally mined to produce the fuel (Hedin 1997). The major part of the long-term risk is directly related to

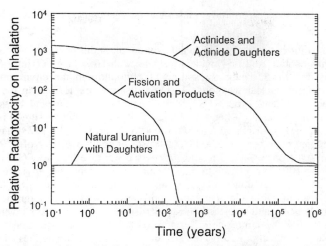

Figure 6. Relative radiotoxicity on inhalation of spent nuclear fuel with a burnup of 38 MWd/kg U. The radiotoxicity values are relative to the radiotoxicity (horizontal line) of the quantity of uranium ore that was originally mined to produce the fuel (eight tons of natural uranium yields one ton of enriched uranium, 3.5% ^{235}U) (after Hedin 1997).

the fate of three actinides in the geosphere, ^{239}Pu and ^{240}Pu out to periods of 100,000 years and ^{237}Np over longer geologic periods (10^6 yr). The total radiotoxicity reaches a value equivalent to that of the ore originally mined to produce the nuclear fuel after several million years.

Nuclear accidents present a more immediate pathway for exposure. The 1979 accident at Three Mile Island released considerable activity, mainly in the form of isotopes of the noble gases (10^7 Ci), such as ^{133}Xe and ^{85}Kr, and much smaller amounts of ^{131}I (15 to 24 Ci) (Sugarman 1979). The 1986 accident at Chernobyl released considerable amounts of activity from the core: 3% of the transuranium elements, 13% of the ^{137}Cs, 20% of the ^{131}I and all of the noble gases. This amounted to 12×10^{18} Bq during the first 10 days after the accident (IAEA 1996). The activity was spread over much of the northern hemisphere. Over 100 workers, many handling active fuel without proper protection, were diagnosed with "acute radiation syndrome" and 28 died within the first three months. Some 200,000 people involved in the initial clean up of the plant received average total body radiation doses on the order of 100 mSv (IAEA 1996).

The production and testing of nuclear weapons has been another source of anthropogenic radioactivity (DOE 1996). Weapons production generated radioactive waste at every stage: uranium mining, milling and refining, isotope separation and enrichment, fuel and target fabrication, reactor operation, chemical separations, component fabrication and now, thankfully, the dismantlement of nuclear weapons. The greatest volumes of radioactive waste are associated with the uranium mining, milling and refining. The greatest activities are associated with the high-level radioactive liquids and sludges resulting from the chemical separation processes used to reclaim the fissile nuclides. Within the DOE complex there are approximately 380,000 m^2 (100 million gallons) of HLW with a total activity of 960 million Ci. Over 99 percent of the present activity is associated with radionuclides with half-lives less than 50 years, as the reprocessing has removed a substantial proportion of the actinides. In addition to the HLW, there are 60,000 m^3 of waste contaminated with transuranic elements, TRU waste (defined as containing more than 100 nanocuries of alpha-emitting transuranic isotopes with half-lives > 20 years, per gram of waste).

Weapons testing has been a significant source of anthropogenic radioactivity. The United States has conducted 1,054 nuclear tests since the first at the Trinity Site in New Mexico on July 16, 1945. The main source of plutonium in the environment is from atmospheric testing of nuclear weapons. All U.S. weapons explosions in the atmosphere occurred between 1945 and 1963. The United States, Great Britain and Russia joined in terminating atmospheric testing when the Limited Test Ban Treaty was signed in September of 1963. From the 400 atmospheric tests prior to 1963, Hardy et al. (1973) have estimated that worldwide there have been 325,000 Ci (10^{16} Bq) or about 5,200 kg of ^{239}Pu and ^{240}Pu dispersed as fall-out from weapons testing. A revised estimate by Facer (1980) suggested that approximately 10^5 Ci of plutonium have been dispersed into the environment, of this 10^3 to 10^4 Ci are probably concentrated in surface soils around the U.S. test sites. In addition to the transuranium elements, the short-term activities are significant. Typical fission product elements and total activities per megaton of fission are given in Table 3 (Eisenbud 1987). For the atmospheric tests prior to 1963, the estimated total yield in megatons is approximately 220. The most significant radionuclides in terms of concentrations in the human body are ^{131}I ($t_{1/2}$ = 8.02 days) and ^{89}Sr ($t_{1/2}$ = 50.52 days) and ^{90}Sr ($t_{1/2}$ = 28.78 yrs). For ^{131}I, it has been estimated that in excess of 14 billion Ci (5×10^{20} Bq) were released into the atmosphere (Eisenbud 1987). Placing these numbers into perspective, atmospheric testing has put 100 to 1,000 times more radioactive material into the atmosphere than the Chernobyl accident (IAEA 1996); however, this amount of activity has increased the current average annual radioactive dose to the population by only a fraction of one percent (DOE 1997).

Table 3. Approximate yields of the principal nuclides per megaton of fission (after Eisenbud 1987).

Nuclide	Half-Life	Mci *
^{89}Sr	50 days	20.0
^{90}Sr	29 years	0.1
^{95}Zr	64 days	25.0
^{103}Ru	39 days	18.5
^{106}Ru	1 year	0.29
^{131}I	8 days	125.0
^{137}Cs	30 years	0.16
^{131}Ce	10 minutes	39.0
^{144}Ce	285 days	3.7

* One Curie = 3.7×10^{10} Bq.

SUMMARY

Science in the Twentieth Century has been dominated by the development and applications of the nuclear sciences. In the geosciences, radionuclides have been used to determine the age, fundamental character and processes of the Earth. At the end of the Twentieth Century, public and scientific concern for the Earth has led to a new, interdisciplinary field that is focused on environmental radioactivity. Thus, there is now an essential need to place the phenomenon of radioactivity, natural and anthropogenic, into a context that allows policy makers and the public to evaluate fundamental questions related to nuclear power and nuclear waste disposal. The most useful conceptual context is to consider the geochemical and biochemical cycles of each radionuclide and their intersection with the nuclear fuel cycle. This volume consists of chapters that provide a fundamental understanding of the relevant mineralogical and geochemical processes for uranium. The same approach can be used for other radionuclides. The perspective provided by a knowledge of the mineralogy and geochemical cycles of radionuclides provides the basis for understanding the modes and fluxes of radionuclide transport, the effective sinks and reservoirs, and the sources and extent of background exposures. This perspective should be essential in developing a regulatory framework that is reasonable and at the same time protects human health and the environment.

ACKNOWLEDGMENTS

This essay draws on numerous collaborations over the past decade related to the mineralogy and geochemistry of uranium. I have benefited greatly from work and discussions with each of these colleagues: Jordi Bruno, Peter Burns, Ignasi Casas, Fanrong Chen, Sue Clark, Mostafa Fayek, Bob Finch, François Gauthier-Lafaye, Frank Hawthorne, Janusz Janeczek, Keld A. Jensen, Werner Lutze, Mark Miller, Takashi Murakami, and Donggao Zhao. I gratefully acknowledge support from Lars Werme (SKB) and Bill Luth (BES-DOE) that allowed me to explore previously dormant areas of research.

REFERENCES

Adams GD 1996 Wilhelm Conrad Roentgen: The American Association of Physicists in Medicine: 1995 Radiology Centennial Hartman Oration. Health Physics 70:636-638
Adams JAS, Lowder WM 1964 Environmental Radiation Sources. The University of Chicago Press, Chicago, 1069 p
Adloff JP, MacCordick HJ 1995 The dawn of radiochemistry. Radiochim Acta 70/71:13-22
Adloff JP 1997 The "X" Compounds a Breakthrough in the Early History of Radioactivity. Radiochim Acta 77:1-7
Ahearne JF 1993 The future of nuclear power. Am Scientist 81:24-35
Badash L 1979 Radioactivity in America Growth and Decay of a Science. The Johns Hopkins Univ Press, Baltimore, Maryland, 327 p
Brown L, Flavin C 1999 It's getting late to switch to a viable world economy. International Herald Tribune, January 19, 1999, p 6
Burchfield JD 1975 Lord Kelvin and the Age of the Earth. MacMillian, London, 260 p
Chapman NA 1992 Natural radioactivity and radioactive waste disposal. J Volcan Geotherm Res 50:197-206
Chapman NA, McKinley IG, Shea ME, Smellie JAT 1990 The Poços de Caldas Project: Summary and implications for radioactive waste management. SKB Technical Report TR 90-24, 147 p
Cohen BL 1995 Test of the linear no-threshold theory of radiation carcinogenesis for inhaled radon decay products. Health Phys 68:157-174
Cohen BL 1998 Response to Lubin's proposed explanations of our discrepancy. Health Phys 75:18-22.
Cramer JJ, Smellie JAT (Eds) 1994 Final Report of the AECL/SKB Cigar Lake Analog Study. AECL-10851, 393 p
Curie MS 1903 Recherches sur les substances radioactives. PhD Thèse, Serie A., No. 445, No d'ordre, 1127
Curtis D, Fabryka-Martin J, Dixon P, Cramer J 1999 Nature's uncommon elements: Plutonium and technetium. Geochim Cosmochim Acta 63:275-285
Domenici "Pete" V 1998 Nuclear initiatives: Steps on a long journey. Nuclear News, August, 32-33
Drell S, Jeanloz R, Peurifoy R 1999 Maintaining a nuclear deterrent under the test ban treaty. Science 283:1119-1120
U.S. Department of Energy, Office of Environmental Management 1996 Closing the Circle on the Splitting of the Atom. DOE/EM-0266, 106 p
U.S. Department of Energy, Office of Environmental Management 1997 Linking Legacies–Connecting the Cold War Nuclear Weapons Productions Processes to Their Environmental Consequences, DOE/EM-0319, 231 p (plus an excellent poster summarizing the U.S. nuclear weapons complex)
Eisenbud M 1987 Environmental Radioactivity From Natural, Industrial, and Military Sources. Academic Press, New York, 475 p

Eisenbud M, Paschoa AS 1989 Environmental radioactivity. Nuclear Instr Methods in Phys Res A280:470-482
Ewing RC 1994 The metamict state: 1993—the centennial. Nucl Inst Meth Phys Res B91:22-29
Ewing RC 1999 Nuclear waste forms for actinides. Proc Nat Acad Sci 96:3432-3439
Facer G 1980 Transuranic Elements in the Environment. WC Hanson (Ed) U.S. Department of Energy, DOE/TIC-22800, 86-91
Frame PW 1996 Radioactivity: Conception to birth: The Health Physics Society 1995 Radiology Centennial Hartman Oration. Health Phys Soc J 70:614-620
Frosch RA, Gallopoulos NE 1989 Strategies for manufacturing. Sci Am 144-152
Gascoyne M 1982 Geochemistry of the actinides and their daughters. *In* Uranium Series Disequilibrium Applications to Environmental Problems. M Ivanovich, RS Harmon (Eds) Clarendon Press, Oxford, 33-55
Gauthier-Lafaye F, Holliger P, Blanc P-L 1996 Natural fission reactors in the Franceville basin, Gabon: A review of the conditions and results of a "critical event" in a geologic system. Geochim Cosmochim Acta 60:4831-4852.
Genet M 1995 The discovery of uranic rays: A short step for Henri Becquerel but a giant step for science. Radiochim Acta 70/71:3-12
Halliday AN 1997 Radioactivity, the discovery of time and the earliest history of the Earth. Contemp Phys 38:103-114
Hardy EP, Krey PW, Volchok HL 1973 Global inventories and distribution of fallout plutonium. Nature 241:444-447
Hedin A 1997 Spent Nuclear Fuel—How Dangerous Is It? Technical Report 97-13 of the Swedish Nuclear Fuel and Waste Management Co., 60 p
Howard AS 1997 Nuclear energy and global climate change. An address to The Energy Council Global Energy and Environmental Issues Conference, New Orleans, Louisiana, December 6, 1997
International Atomic Energy Agency 1996 Ten Years After Chernobyl: What Do We Really Know, 5-23
Jensen KA, Ewing RC (submitted) The Okélobondo natural fission reactor, southeast Gabon: Geology, mineralogy and retardation of nuclear reaction products. Geol Soc Am Bull
Kaplan I 1962 Nuclear Physics. Addison-Wesley Publishing Co., Reading, Massachusetts, 770 p
Kersting AB, Efurd DW, Finnegan DL, Rokop DJ, Smith DK, Thompson JL 1999 Migration of plutonium in ground water at the Nevada test site. Nature 397, 56-59.
Krauskopf KB 1988 Radioactive Waste Disposal and Geology. Chapman and Hall, New York, 145 p
Langmuir D 1978 Uranium solution-mineral equilibria at low temperatures with applications to sedimentary ore deposits. Geochim Cosmochim Acta 42:547-569
Langmuir D 1997 Aqueous Environmental Geochemistry. Prentice Hall, Upper Saddle River, NJ, 600 p
Lifset R 1997 A metaphor, a field and a journal. J Industrial Ecology 1:1-3
Levine CA, Seaborg GT 1951 The occurrence of plutonium in nature. J Am Chem Soc 73:3278-3283.
Lovley DR, Phillips EJP, Gorby YA, Landau ER 1991 Microbial reduction of uranium. Nature 350:413-416
Lovley DR, Phillips EJP 1992 Reduction of uranium by *Desulfovibrio desulfuricans*. Appl Env Microbiology 58:850-856.
McBride JP, Moore, RE, Witherspoon JP, Blanco RE 1978 Radiological impact of airborne effluents of coal and nuclear power plants. Science 202:1045-1050
Miller WM, Smith GM 1993 Fluxes of Elements and Radionuclides from the Geosphere. Report by Intera Environmental Division for the Swedish Institute for Radiation Protection, IE3438-1, 80 p
Miller W, Alexander M, Chapman N, Smellie J 1994 Natural Analogue Studies in the Geological Disposal of Radioactive Wastes Elsevier, Amsterdam, 395 p
Miller RC, Randers-Pehrson G, Geard CR, Hall EJ, Brenner DJ 1999 The oncogenic transforming potential of the passage of single α-particles through mammalian cell nuclei. Proc Nat Acad Sci 96:19-22
National Research Council 1929 Bulletin No. 51, 203 p
National Research Council 1931 Bulletin No. 80, 144 p
National Research Council 1995 Technical Bases for Yucca Mountain Standards. National Academy Press, Washington, DC, 205 p
National Research Council 1999 National Academy Press, Washington, DC, *Health Effects of Exposure to Radon: BEIR VI*. National Academy Press, Washington, DC, 516 p
Nuclear Energy Institute 1997 Nuclear Energy's Clean Air Benefits on a Worldwide Scale. Nuclear Energy Institute, Washington, DC, 46 p
Nuclear Energy Institute 1998 Fact Sheet: January 1998
Peppard DF, Studier MH, Gergel MV, Mason GW, Sullivan JC, Mech JF 1951 Isolation of microgram quantities of naturally-occurring plutonium and examination of its isotopic composition. J Am Chem Soc 73:2529-2531

Rhodes R 1986 The Making of the Atomic Bomb. Simon & Schuster, New York, 886 p
Rozanski K, Froehlich K 1996 Radioactivity and earth sciences: Understanding the natural environment. IAEA Bulletin 2:9-15
Rutherford E 1899 Uranium radiation and the electrical conduction produced by it. Philos Mag Ser 5, 47:109-163
Rutherford E, Soddy F 1902a The radioactivity of thorium compounds. I. An investigation of the radioactive emanation. J Chem Soc Trans 81:312-350
Rutherford E, Soddy F 1902b The radioactivity of thorium compounds. II. The cause and nature of radioactivity. J Chem Soc Trans 81:837-860
Rutherford E, Soddy F 1902c The cause and nature of radioactivity. Part II. Section IX. Phil Mag J Sci 6:569-585
Santschi PH, Honeyman BD 1989 Radionuclides in aquatic environments. Rad Phys Chem 34:213-240
Schubnel H-J, Editor 1996 Histoire Naturelle de la Radioactivité. Rev Gemmologie a.f.g., Paris, 132 p
Seliger HH 1995 Wilhelm Conrad Röntgen and the glimmer of light. Phys Today, November, 25-31
Silva RJ, Nitsche H 1995 Actinide environmental chemistry. Radiochim Acta 70/71:377-396
Smyth HD 1945 Atomic Energy for Military Purposes: The Official Report on the Development of the Atomic Bomb under the Auspices of the United States Government, 1940-1945. Princeton Univ Press, Princeton, New Jersey, 264 p
Starr C 1997 The future of nuclear power. Nucl News, March 1997, p 58-60
Stoll W 1998 What are the options for disposition of excess weapons plutonium? Mater Res Soc Bull 23:6-16
Sugarman R 1979 Nuclear power and the public risk. Spectrum 16:59-79
Tuovinen OH, Kelly DP 1974 Studies on the growth of *Thiobacillus ferrooxidans* II. Toxicity of uranium to growing cultures and tolerance conferred by mutation, other metal cations and EDTA. Arch Microbiol 95:153-164
Weinberg AM 1985 Continuing the Nuclear Dialogue—Selected Essays. American Nuclear Society, La Grange Park, Illinois, 204 p
Wells HG 1933 The Shape of Things to Come. Macmillan, New York, 431 p
Wells HG 1914 The World Set Free: a Story of Mankind. E.P. Dutton, New York, 308 p

2
The Crystal Chemistry of Uranium

Peter C. Burns

Department of Civil Engineering and Geological Sciences
University of Notre Dame
Notre Dame, Indiana 46556

INTRODUCTION

There are nearly 200 mineral species that contain uranium as a necessary structural constituent (Fleischer and Mandarino 1995), despite the low crustal abundance of U (~2.7 ppm, Taylor 1964). Owing to interest in U as an energy resource, and more recently, in its role in environmental problems associated with the disposal of radioactive waste and the remediation of contaminated sites, U minerals have been the subject of an increased amount of attention. In addition, the many striking colors of U minerals have attracted considerable interest amongst the amateur mineralogist community, who were responsible for the initial recognition of numerous new U species.

Most U minerals can be placed in two categories: reduced species that contain most U as U^{4+}, and oxidized species that contain U as U^{6+}. There are a few mixed-valence U^{4+}-U^{6+} minerals, and at least one U mineral contains U^{5+} (Burns and Finch 1999). There are many more U^{6+} species than U^{4+} species, and the paragenesis and structures of U^{6+} minerals are exceedingly complex. Generally, the structures of U^{4+} minerals contain regular coordination polyhedra about the U^{4+} cation, and tend to have high symmetry. It is common for U^{4+} minerals to be isostructural with non-U analogues, and U^{4+} often substitutes for other cations in a variety of mineral structures. In contrast, the structures of U^{6+} minerals are diverse owing to the unusual crystal chemistry of U^{6+}, and U^{6+} minerals are seldom isostructural with other minerals.

Despite the importance of U minerals, and the considerable attention that they have garnered, our understanding of their structures and crystal chemistry lags well behind many other mineral groups. To date, the structures have been determined and refined for only ~1/3 of the known U species. The lack of structural information largely reflects experimental difficulties; U minerals are extremely high absorbers of X-rays, and many do not form crystals of a size suitable for structure analysis by conventional means. Some undergo dehydration or other transformations that cause crystals to alter. However, with the recent introduction of CCD-based detectors for X-rays (Burns 1998a), and the use of synchrotron X-ray sources for the determination of the structures of microcrystals (Smith 1995), it is likely that many more U mineral structures will be determined over the next few years.

In this chapter I focus on the crystal chemistry of U and the details and relationships among the crystal structures of minerals that contain essential U. I have restricted coverage to those minerals for which structure solutions and refinements are available. I do not specifically cover non-mineral structures, except where comparison to mineral structures is appropriate; Burns et al. (1996) provide a more detailed coverage of the structures of non-mineral U inorganic phases. U minerals are discussed in two groups corresponding to reduced and oxidized species. In older literature reduced U species were often referred to as "primary", owing to their predominance in U deposits, whereas the oxidized species were designated "secondary" because they often form due to the oxidation of pre-existing reduced U minerals. This nomenclature is avoided here because oxidized U minerals can precipitate directly from solution without a reduced-species precursor. The details of U mineral systematics and paragenesis are provided by Finch and Murakami (Chapter 4, this volume).

Smith (1984) presented a classification of U^{4+} and U^{6+} minerals, with groupings based upon the identities of the principal anionic groups in the structures, and this work represents an important step forward in the understanding of U minerals. New U mineral structures are being described at the rate of a couple per year, and now ~65 U mineral structures are reasonably well characterized. Burns et al. (1996) considered the structures of the U^{6+} minerals, as well as numerous inorganic U^{6+} phases, and proposed a detailed structural hierarchy that is based upon those cation polyhedra of higher bond valence; this hierarchy is adhered to in this chapter.

STRUCTURES OF U^{4+} MINERALS

Many U deposits are dominated by reduced U minerals. However, there are relatively few U^{4+} mineral species, as compared to U^{6+} minerals, owing at least in part to the more regular crystal chemistry of U^{4+}. These minerals are often more complicated than their simple structures would suggest, owing to the tendency for U^{4+} to partially oxidize. Minerals that originally crystallized with all U as U^{4+} often contain both U^{4+} and U^{6+}, and possibly even U^{5+}. In addition, substitution of elements such as Th^{4+} and REE^{3+} for U^{4+} is common. Crystallographic information for U^{4+} minerals with known structures is presented in Table 1.

Table 1. Crystallographic parameters for U^{4+} minerals with known structures

Name	Formula	S. G.	a (Å)	b (Å)	c (Å)	$\beta(°)$	Ref.
uraninite	$(U^{4+}_{1-x-y-z-v}U^{6+}_{x}REE^{3+}_{y}M^{2+}_{z}[\ ]^{4+}_{v})O_{2+x-0.5y-z-2v}$	$Fm3m$	5.470–5.443				1
coffinite	$USiO_4$	$I4_1/amd$	6.979		6.253		2
brannerite	$(U,Ca,Ce)(Ti,Fe)_2O_6$	$C2/m$	9.79	3.72	6.87	118.4	3
uranmicrolite	$(U,Ca,Ce)_2(Nb,Ta)_2O_6(OH,F)$		10.40				
uranpyrochlore	$(U,Ca,Ce)_2(Ta,Nb)_2O_6(OH,F)$	$Fd3m$	10.44				

References: (1) Janeczek and Ewing (1992), (2) Fuchs and Gebert (1958), (3) Szymanski and Scott (1982).

The U^{4+} cation typically occurs in mineral structures in regular coordination polyhedra that contain from six to eight ligands. The effective ionic radius for $^{[6]}U^{4+}$ and $^{[8]}U^{4+}$ provided by Shannon (1976) are 0.89 and 1.00 Å, respectively, giving predicted U^{4+}-O bond lengths in the range 2.25 to 2.36 Å.

Uraninite, nominally UO_{2+x}, is the most common U^{4+} mineral species, and is the main ore mineral in many U deposits. Uraninite possesses the fluorite structure (Fig. 1a); the U^{4+} cation is coordinated by eight O atoms in a cubic arrangement, and each O atom bonds to four U^{4+} cations. The chemical and structural aspects of uraninite have received considerable attention owing to its economic importance, and to its close structural similarity to UO_2 nuclear fuel. Natural-analogue studies (e.g. Finch and Ewing 1992; Pearcy et al. *1994*) have taken advantage of these similarities and used the behavior of uraninite in natural systems as an analogue for the long-term behavior of spent nuclear fuel in a geological repository.

Uraninite is probably always at least partially oxidized in nature, with the formula

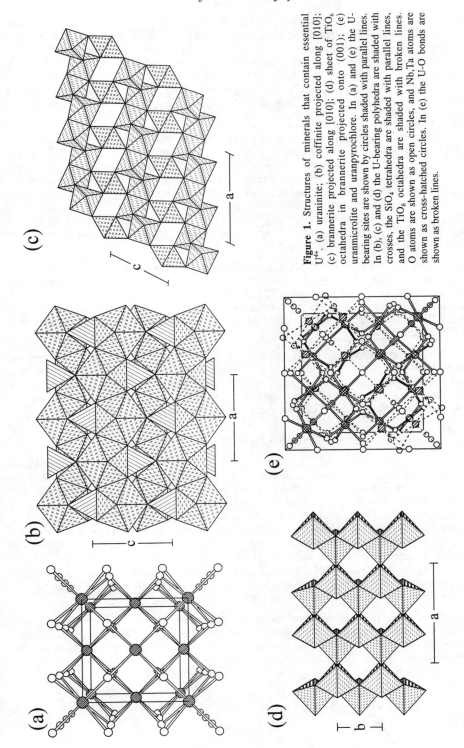

Figure 1. Structures of minerals that contain essential U^{4+}. (a) uraninite; (b) coffinite projected along [010]; (c) brannerite projected along [010]; (d) sheet of TiO_6 octahedra in brannerite projected onto (001); (e) U-uranmicrolite and uranpyrochlore. In (a) and (e) the U-bearing sites are shown by circles shaded with parallel lines. In (b), (c) and (d) the U-bearing polyhedra are shaded with crosses, the SiO_4 tetrahedra are shaded with parallel lines, and the TiO_6 octahedra are shaded with broken lines. O atoms are shown as open circles, and Nb,Ta atoms are shown as cross-hatched circles. In (e) the U-O bonds are shown as broken lines.

UO_{2+x} with X in the range 0.0 to 0.25 (Janeczek and Ewing 1992). The structure contains many defects because of oxidation and cationic substitution, as well as radiation damage. Geologically old uraninite contains up to ~20 wt % of PbO due to radioactive breakdown of U (Janeczek and Ewing 1992). Uraninite has also been found that contains ~11 wt % CaO (Xu et al. 1981) and ~12 wt % REE_2O_3 (Frondel 1958). In light of these complicating factors, Janeczek and Ewing (1992) proposed the following structural formula for uraninite: $(U^{4+}_{1-x-y-z-v}U^{6+}_xREE^{3+}_yM^{2+}_zO^{4-}_v)O_{2+x-0.5y-z-2v}$. Additional information on uraninite can be found in Chapter 3.

Coffinite is only known to occur as intergrowths of microscopic crystals that are too small for structure analysis using conventional techniques (Hansley and Fitzpatrick 1989). The structure was reported for a synthetic crystal by Fuchs and Gebert (1958), and although the structure is not precisely refined, it is apparent that coffinite is isostructural with zircon. Thus, coffinite is an orthosilicate with U^{4+} coordinated by eight O atoms in the form of a distorted cube-like polyhedron (Fig. 1b). There have been several reports that the substitution $4(OH)^- \leftrightarrow (SiO_4)^{4-}$ may occur in the structure, and that $P \leftrightarrow Si$ substitution may provide a charge-balancing mechanism for REE incorporation into the structure (Hansley and Fitzpatrick 1989).

Brannerite crystals are invariably metamict, but recover some of their crystallinity when heated. The structure of synthetic brannerite was determined by Szymanski and Scott (1982), and contains both U^{4+} and Ti^{4+} in octahedral coordination. The TiO_6 octahedra share corners and edges to form sheets that are parallel to (001) (Fig. 1c,d), and the $U^{4+}O_6$ octahedra link adjacent sheets of $Ti^{4+}\phi_6$ octahedra by sharing corners with $Ti^{4+}\phi_6$ octahedra.

Both uranmicrolite and uranpyrochlore crystallize with the pyrochlore structure, and in each case the A site is dominated by U^{4+}. The structure involves a three-dimensional framework of corner-sharing $(Ti,Ta,Nb)O_6$ octahedra (Fig. 1e), with the A site located in cavities where it is coordinated by eight O atoms in a distorted-cubic coordination. It is possible that some of the U^{4+} in these minerals has been oxidized to U^{6+}.

BONDING IN U^{6+} POLYHEDRA

The crystal structures of U^{6+} minerals are diverse owing to their unusual chemistries and to the unique nature of the coordination polyhedra about the U^{6+} cation. In minerals, the U^{6+} cation is invariably present as part of a nearly linear $(U^{6+}O_2)^{2+}$ uranyl ion (Ur) (Evans 1963). The U^{6+}-O bond lengths in the ion are ~1.8 Å, corresponding to very strong covalent bonds. The uranyl ion has a formal valence of 2+, and as such it must be coordinated by ligands in a crystal structure. The uranyl ion occurs coordinated by four, five, or six ligands that are arranged at the equatorial corners of $Ur\phi_4$ square bipyramids, $Ur\phi_5$ pentagonal bipyramids, and $Ur\phi_6$ hexagonal bipyramids, respectively (ϕ: O^{2-}, OH^-, H_2O) (Fig. 2). The O atoms of the uranyl ions (hereafter designated O_{Ur}) are located at the apices of the bipyramids. The U^{6+}-ϕ_{eq} (ϕ_{eq}: equatorial ϕ) bonds are longer than the U^{6+}-O_{Ur} bonds, and are substantially weaker (see below).

Craw et al. (1995) investigated the nature of bonding in the uranyl ion using *ab initio* molecular orbital calculations. They showed that the U^{6+}-O_{Ur} bonding mechanism is primarily through donation of electrons from the p orbitals of the O_{Ur} atoms into the empty d and f orbitals of the U^{6+} cation.

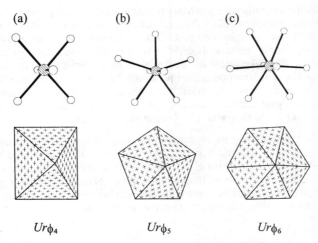

Figure 2. The different types of coordination polyhedra observed in U^{6+} minerals. In each case the polyhedron contains a nearly linear $(UO_2)^{2+}$ uranyl ion that is oriented roughly perpendicular to the plane of projection. The uranyl ion is coordinated by four to six additional anions arranged at the equatorial positions of bipyramids, with the O_{Ur} atoms forming the apices of the bipyramds. (a) $Ur\phi_4$ square bipyramid, (b) $Ur\phi_5$ pentagonal bipyramid, (c) $Ur\phi_6$ hexagonal bipyramid.

GEOMETRIES OF U^{6+} POLYHEDRA

Molecular-orbital calculations

Molecular-orbital calculations provide an *ab initio* approach to predicting the equilibrium geometries of uranyl polyhedra. Most calculations reported to date involve only the uranyl ion; equatorial anions of the bipyramids that are typically associated with uranyl ions in crystal structures are excluded from consideration. The calculations provide predicted geometries for the uranyl ion in vacuum. In crystal structures, the O_{Ur} atoms are not only bonded to the U^{6+} cation; they also almost always are weakly bonded to low-valence cations, and they commonly accept H bonds. Therefore, exact correspondence between the geometries of the uranyl ion calculated in vacuum with those observed in crystal structures is not expected. Craw et al. (1995) used the Hartree-Fock formalism to obtain an optimized U^{6+}-O bond length of 1.663 Å for the uranyl ion. Extending the model to include correlation of dynamic electrons with approximations gave bond lengths ranging from 1.700 to 1.783 Å. Craw et al. (1995) also reported Hartree-Fock calculations for uranyl nitrate and uranyl sulfate complexes, which gave optimized U^{6+}-O_{Ur} bond-lengths of 1.72 and 1.74 Å, respectively. Van Wezenbeek et al. (1991) used nonrelativistic Hartree-Fock-Slater calculations, which gave a predicted U^{6+}-O_{Ur} bond length of 1.67 Å. Modification of the calculations to include corrections for relativistic effects associated with the electrons resulted in an increase of the U^{6+}-O_{Ur} bond length to 1.70 Å. Within the Hartree-Fock limit, Pyykkö et al. (1994) obtained U^{6+}-O_{Ur} bond lengths that ranged from 1.66 to 1.72 Å, depending upon the parameterization of the calculation. Pyykkö et al. (1994) also provided Hartree-Fock calculations for the cluster $Ur(OH)_4$; optimized bond lengths were 1.784 Å for the uranyl ion, and 2.204 Å for the equatorial U^{6+}-OH bonds of the square bipyramid.

EXAFS data

Extended X-ray Absorption Fine Structure (EXAFS) spectra for solutions and solids containing U readily provide information pertaining to the local environment about the U^{6+} cation. A limited number of EXAFS spectra have been collected for U^{6+} minerals, but it is apparent that the method can provide U-O bond lengths within the uranyl ion, as well as average U^{6+}-ϕ_{eq} bond lengths. EXAFS spectra are useful for probing the local environment about the U^{6+} cation in the absence of material of suitable quality for structure analysis using diffraction techniques, or when U is a minor constituent of the phase. It is difficult to obtain accurate bond lengths for materials with high X-ray absorption using X-ray diffraction, thus EXAFS spectra also provide useful confirmation of polyhedral geometries in minerals for which crystal structures are known.

Typical U^{6+}-O_{Ur} bond lengths obtained from EXAFS spectra for minerals (or their synthetic analogues) are 1.78 and 1.79 Å (Charpin et al. 1985; Moll et al. 1994, 1995; Allen et al. 1995, 1996; Thompson et al. 1997). The method provides average U^{6+}-ϕ_{eq} bond lengths that are in accord with typical values obtained from X-ray diffraction studies (Moll et al. 1994), thus permitting distinction between $Ur\phi_4$ square bipyramids, $Ur\phi_5$ pentagonal bipyramids, and $Ur\phi_6$ hexagonal bipyramids.

X-ray-diffraction data

X-ray diffraction using single crystals is the most common method for determining the structures of uranyl minerals. However, such studies are difficult owing both to the high absorption of X-rays by the crystals, and to the common lack of crystals of a suitable quality for structural analysis. Currently, the structures have been determined and refined for only ~65 of the approximately 180 known uranyl minerals.

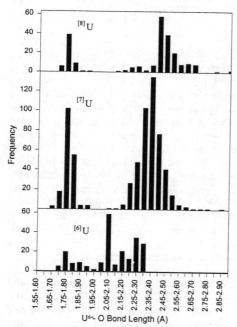

Figure 3. The distribution of U^{6+}-ϕ bond lengths in well-refined U^{6+} structures, grouped according to the number of coordinating anions. The modes at ~1.8 Å correspond to U^{6+}-O_{Ur} bonds, those at ~2.2-2.5 Å correspond to U^{6+}-ϕ_{eq} bonds. From Burns et al. (1997a).

Bond lengths. Burns et al. (1997a) provide a detailed evaluation of uranyl polyhedral geometries that occur in ~100 structures (minerals and inorganic phases) that were well-refined using X-ray-diffraction data. Bond-length distributions in these structures are shown in Figure 3. In the case of $Ur\phi_5$ pentagonal bipyramids and $Ur\phi_6$ hexagonal bipyramids, the distribution is strongly bimodal, with the mode at ~1.8 Å corresponding to U^{6+}-O_{Ur} bonds, and that at ~2.3-2.5 Å corresponding to the U^{6+}-ϕ_{eq} bonds. All $[7]U^{6+}$ and $[8]U^{6+}$ (where the superscript in square braces indicates the number of anions that bond to the U^{6+} cation, including the O_{Ur} atoms) polyhedra in minerals contain uranyl ions. The average $[7]U^{6+}$-O_{Ur} and $[7]U^{6+}$-ϕ_{eq} bond lengths for all 93 $Ur\phi_5$ polyhedra reported by Burns et al. (1997a) are 1.79 (σ = 0.04) and 2.37 (σ = 0.09) Å, respectively. In 28 $Ur\phi_6$ polyhedra, the average $[8]U^{6+}$-O_{Ur} and $[8]U^{6+}$-ϕ_{eq} bond lengths are 1.78 (σ = 0.03) and 2.47 (σ = 0.12) Å, respectively.

The bimodal distribution of bond

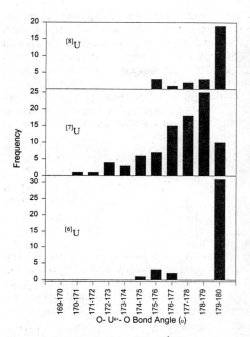

Figure 4. The distribution of O_{Ur}-U^{6+}-O_{Ur} bond angles in well-refined U^{6+} structures, grouped according to the number of coordinating anions. From Burns et al. (1997a).

lengths in $Ur\phi_4$ square bipyramids is less pronounced than for other uranyl polyhedra (Fig. 3). Burns et al. (1997a) noted that not all structures that contain $^{[6]}U^{6+}$ involve a uranyl ion; rather, several structures contain U^{6+} in octahedral or distorted-octahedral coordination. However, it is interesting to note that where $^{[6]}U^{6+}$ is present in a uranyl mineral, the $^{[6]}U^{6+}$ polyhedron always contains a uranyl ion; $U^{6+}\phi_6$ octahedra have only been observed in synthetic inorganic compounds. Burns et al. (1997a) observed that in the 16 $^{[6]}U^{6+}$ polyhedra that contain typical uranyl ions (i.e. U^{6+}-O_{Ur} ~1.8 Å), the average $^{[6]}U^{6+}$-O_{Ur} and $^{[6]}U^{6+}$-ϕ_{eq} bond lengths are 1.79 ($\sigma = 0.03$) and 2.28 ($\sigma = 0.05$) Å, respectively.

Burns et al. (1997a) demonstrated that $^{[6]}U^{6+}$ polyhedra display a trend from a $Ur\phi_4$ square pyramidal geometry with a typical U^{6+}-O_{Ur} bond length of ~1.8 Å, to a holosymmetric octahedron. Pyykkö and Zhao (1991) reported quasirelativistic *ab initio* calculations for $(UO_6)^{6-}$ clusters with $Ur\phi_4$ and octahedral geometries, as well as for geometries that were intermediate. The calculated trend was similar to that observed in crystal structures, and indicated that there is not a substantial energy barrier along the pathway transitional between the two types of coordination polyhedra, a finding that is in accord with the distribution of bonds in $^{[6]}U^{6+}$ polyhedra in well-refined structures.

Uranyl-ion linearity: The distribution of O_{Ur}-U^{6+}-O_{Ur} bond angles was evaluated for ~100 well-refined structures by Burns et al. (1997a) and is shown in Figure 4. The uranyl ion is close to linear in all polyhedra, with most uranyl-ion bond angles in $Ur\phi_4$ square bipyramids and $Ur\phi_6$ hexagonal bipyramids in the range 179° to 180°. However, in the case of $Ur\phi_5$ pentagonal bipyramids, there is a tendency for the uranyl-ion bond angle to be slightly distorted (Fig. 4). Burns et al. (1997a) attributed this to the distribution of anions in the equatorial plane of each polyhedron.

Bond-valence parameters for U^{6+}

The bond-valence approach (Brown 1981) is commonly used to evaluate the results of the analysis of a crystal structure. It provides a means to check bond lengths, the valence states of ions within the structure, a method to identify the cations and anions present, as well as providing the key to unraveling the details of H bonding in a structure where X-ray data does not provide H positions.

There have been several bond-valence parameters suggested for U^{6+} (Brown and Wu 1976; Zachariasen 1978; Brown and Altermatt 1985; Brese and O'Keeffe 1991; Burns et al. 1997a); they are listed in Table 2. Depending upon which set of bond-valence

Table 2. Bond-valence parameters for uranyl polyhedra.

Polyhedra	R_{ij}	b	Ref.
$Ur\phi_4, Ur\phi_5, Ur\phi_6$	2.049	N=4.3	1
$Ur\phi_4, Ur\phi_5, Ur\phi_6$	2.083	0.35	2
$Ur\phi_4, Ur\phi_5, Ur\phi_6$	2.075	0.37	3
$Ur\phi_4, Ur\phi_5, Ur\phi_6$	2.075	0.37	4
$Ur\phi_4$	2.074	0.554	5
$Ur\phi_5$	2.045	0.510	5
$Ur\phi_6$	2.042	0.506	5
$Ur\phi_4, Ur\phi_5, Ur\phi_6$	2.051	0.519	5

References: (1) Brown and Wu (1976), (2) Zachariasen (1978), (3) Brown and Altermatt (1985), (4) Brese and O'Keeffe (1991), (5) Burns et al. (1997a).

parameters is used, it is not uncommon for apparently well-refined uranyl structures to give bond-valence sums incident at the U^{6+} positions that differ substantially from the formal valence. A comparison of the bond-valence sums incident at U^{6+} positions in well-refined structures using the parameters of Brown and Wu (1976) and Brese and O'Keeffe (1991) is provided in Figure 5a. This may be compared to the sums obtained for the same structures with the coordination-specific parameters provided by Burns et al. (1997a) shown in Figure 5b. These parameters provide bond-valence sums that cluster about the formal valence of 6 vu, and display narrower distributions than other available parameters. Burns et al. (1997a) also provided coordination-independent bond-valence parameters for U^{6+}; the distribution of sums is shown in Figure 5c. Although these parameters do not perform as well as the coordination-specific parameters, their application is more straightforward.

Burns et al. (1997a) demonstrated that the bond-valence approach can also be used to identify the valence state of the U cation in well-refined crystal structures. The valence states U^{4+}, U^{5+}, and U^{6+} are readily distinguished with the aid of the coordination-independent parameters.

CONSTRAINTS ON POLYMERIZATION OF URANYL POLYHEDRA

Burns et al. (1997a) examined the nature of polyhedral polymerization in more than one hundred mineral and synthetic structures containing uranyl polyhedra, and made the following observations:

(1) $Ur\phi_4$ polyhedra often only share corners with other uranyl polyhedra, and where $Ur\phi_4$ polyhedra share edges with other uranyl polyhedra, it is almost invariably with $Ur\phi_5$ polyhedra.

(2) No structure contains either a $Ur\phi_5$ or a $Ur\phi_6$ polyhedron that only shares a corner with another uranyl polyhedron; where polymerization occurs, it involves the sharing of edges between polyhedra.

(3) $Ur\phi_5$ polyhedra most often share edges with other $Ur\phi_5$ polyhedra, although edge-sharing with both $Ur\phi_4$ or $Ur\phi_6$ polyhedra is common.

(4) $Ur\phi_6$ polyhedra often share edges with either $Ur\phi_5$ or $Ur\phi_6$ polyhedra, but never with $Ur\phi_4$ polyhedra.

(5) The most important factor in determining the mode of polymerization between $Ur\phi_n$ polyhedra and other cation polyhedra is cation charge.

(6) Those cation polyhedra (excluding U^{6+} polyhedra) with high cation charge (6+) and low coordination number (<6) do not commonly share edges with

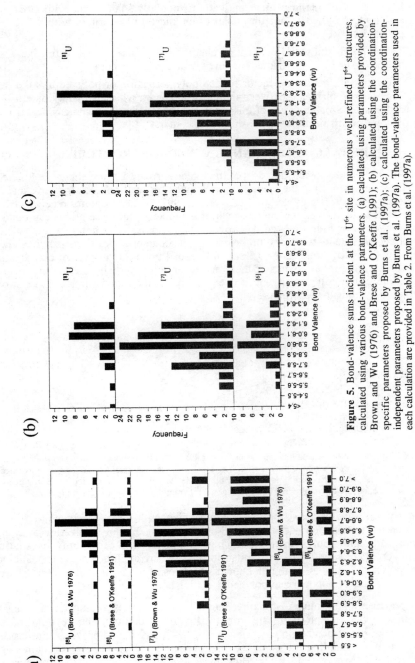

Figure 5. Bond-valence sums incident at the U^{6+} site in numerous well-refined U^{6+} structures, calculated using various bond-valence parameters. (a) calculated using parameters provided by Brown and Wu (1976) and Brese and O'Keeffe (1991); (b) calculated using the coordination-specific parameters proposed by Burns et al. (1997a); (c) calculated using the coordination-independent parameters proposed by Burns et al. (1997a). The bond-valence parameters used in each calculation are provided in Table 2. From Burns et al. (1997a).

$Ur\phi_n$ polyhedra.

(7) Polyhedra containing pentavalent cations regularly share edges with $Ur\phi_5$ and $Ur\phi_6$ polyhedra, but they also often share only corners with $Ur\phi_n$ polyhedra.

(8) Cation polyhedra with lower cation charge (≤ 4) almost always share edges with $Ur\phi_5$ or $Ur\phi_6$ polyhedra, but never with $Ur\phi_4$ polyhedra.

(9) An important geometric factor for the polymerization of polyhedra is the edge-length mismatch between the ideal polyhedra which is mainly due to cation size. Small degrees of mismatch favor edge-sharing; whereas, larger degrees of mismatch favor the sharing of corners.

HIERARCHY OF URANYL MINERAL STRUCTURES

The typical bond valences incident at the ligands around U^{6+} are shown in Figure 6. These were calculated using the average bond lengths for well-refined structures and the bond-valence parameters provided by Burns et al. (1997a). The O_{Ur} atoms receive ~1.7 vu from the bond to the U^{6+} cation, and as such, it is unlikely that these atoms can bond to additional cations of high valence. However, O_{Ur} atoms usually bond to low-valence cations or act as H-bond acceptors. In the case of the equatorial positions, bond valences associated with U^{6+}-ϕ_{eq} bonds are ~0.5 vu, and depend upon the number of coordinating anions. The bonding requirements of these anions are only partially met by the bond to the U^{6+} cation in the polyhedron, and it is necessary for them to form additional strong bonds to other cations in the structure.

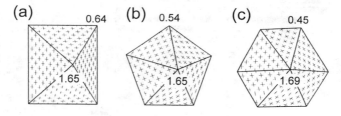

Figure 6. Typical bond valences incident upon the corners of uranyl polyhedra owing to the bond to the U^{6+} cation at the center of the polyhedron. The bond valences were calculated using the average polyhedral geometries and coordination-specific bond-valence parameters provided by Burns et al. (1997a). (a) $Ur\phi_4$ square bipyramid; (b) $Ur\phi_5$ pentagonal bipyramid; (c) $Ur\phi_6$ hexagonal bipyramid.

The strongly unequal distribution of bond valences in uranyl polyhedra (Fig. 6) dictates that polymerization between uranyl polyhedra, or the polymerization of uranyl polyhedra with other polyhedra of high bond valence, will occur only by sharing equatorial corners or edges. This two-dimensional polymerization favors the formation of sheets of polyhedra; indeed, of the 65 known structures of uranyl minerals, 50 are based upon infinite sheets of polymerized polyhedra of high bond valence.

The organization of mineral structures into hierarchies serves foremost as a means to order our knowledge and facilitate comparison of structures, the recognition of underlying controls of bond topology (Hawthorne 1983, 1990), and provides a necessary step towards understanding the complex relationships between mineral structures and their occurrences (Moore 1965, 1973; Hawthorne 1985). Burns et al. (1996) proposed a hierarchy of 180 inorganic structures that contain U^{6+}, 56 of which are minerals. The higher bond valence. The structures were subdivided into those that contained sheets of polyhedra (106, 43 minerals), chains of polyhedra (19, 5 minerals), finite clusters of polyhedra (22, 5 minerals), isolated polyhedra (7, no minerals), and frameworks (26, 3 minerals). The hierarchy of structures proposed by Burns et al. (1996) is used here to facilitate the following discussion of uranyl mineral structures.

The role of (OH)⁻ and H₂O in uranyl mineral structures

Most uranyl minerals are hydrated, and as such, both (OH)⁻ and H_2O play significant and varied structural roles. Most of the uranyl mineral structures that are known have been determined using X-ray diffraction methods, thus it is uncommon for the details of the H positions to be known. However, it is often possible to predict the positions of H atoms, and the resulting H bonds, based upon crystal-chemical criteria.

In oxide and oxysalt minerals, (OH)⁻ and H_2O groups that are included in the structural unit play a major role in the degree of polymerization of that unit because of the asymmetric nature of H-bonding systems (Hawthorne 1992). It is also common to find H_2O as an interstitial component, where it either bonds to an interstitial cation that is not part of the structural unit, or it is held in the structure by H bonds only (Hawthorne 1992). The H_2O groups play important roles in satisfying bond-valence requirements in structures because the H bonds associated with the H_2O group can effectively transmit bond valence between structural units, and between structural units and interstitial cations (Hawthorne 1992).

First, consider the structural unit that is composed of polyhedra of higher bond valence. In uranyl minerals the structural unit always contains uranyl polyhedra, and it also often contains other cation polyhedra of higher bond valence. It is common for (OH)⁻ to occur at equatorial positions of uranyl polyhedra in mineral structures, but the presence of (OH)⁻ in a uranyl ion has never been demonstrated. The bond valence incident upon equatorial anions of uranyl polyhedra due to the bond to the central U^{6+} cation is ~0.5 vu (Fig. 6), thus the presence of (OH)⁻ at the equatorial position of the polyhedron does not prevent the sharing of that polyhedral element with other cation polyhedra of higher bond valence. This is because the (OH)⁻ anion can accept ~1.2 vu from bonds to cations (other than H). It is common for (OH)⁻ groups to occur within sheets of uranyl polyhedra, where they are shared by either two or three adjacent uranyl polyhedra. In these situations, it is normal for the H bond to be accepted by an interlayer constituent, usually an H_2O group.

It is uncommon for H_2O groups to be bonded to U^{6+} in minerals, although there are a few examples. However, H_2O groups in uranyl mineral structures commonly occur bonded to a single interstitial cation. This situation occurs in all types of structures and is usually found in the interlayers of structures that are based upon sheets of polyhedra. The H_2O groups that are bonded to interstitial cations commonly accept H bonds from the structural unit or from other interstitial H_2O groups, and the H bonds are donated either to other interstitial H_2O groups or to anions of the structural unit. Those H_2O groups that are held in the structure by H bonding alone often accept two H bonds from other interstitial H_2O groups or from the structural unit. H_2O groups in interstitial positions commonly donate H bonds that are accepted by O_{Ur} anions of adjacent structural units.

Structures containing sheets of polyhedra

Fifty of the 65 known uranyl mineral structures are based upon sheets of polyhedra of higher bond valence. Because of the myriad of structural variations, it was necessary to develop a method to compare and classify different sheets. Burns et al. (1996) proposed that the topological arrangement of anions within each sheet is a convenient basis for the classification and comparison of these sheets. The *sheet anion topology* is derived as

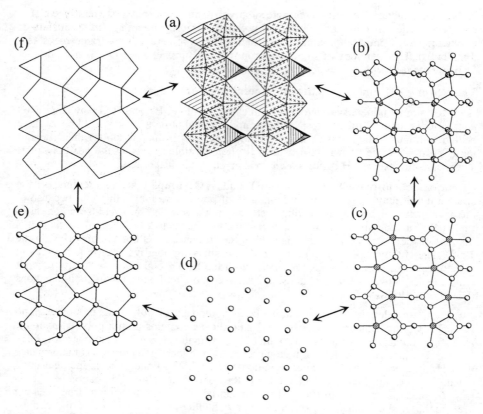

Figure 7. The development of a sheet anion topology following the method proposed by Burns et al. (1996): (a) a typical sheet of polyhedra using the polyhedral representation; (b) the ball-and-stick representation of the sheet; (c) each anion that is not bonded to at least two cations within the sheet, and is not an equatorial anion of a bipyramid or pyramid within the sheet, is removed from further consideration; (d) cations are removed, along with all cation-anion bonds, leaving an array of unconnected anions; (e) anions are joined by lines, with only those anions that may be realistically considered as part of the same coordination polyhedron being connected (i.e. those anions separated by less than about 3.5 Å); (f) anions are removed from further consideration, leaving only a series of lines that represent the sheet anion topology. From Burns et al. (1996).

follows (Fig. 7): (1) Each anion that is not bonded to at least two anions within the sheet, and is not an equatorial anion of a bipyramid or pyramid within the sheet, is not considered a constituent of the sheet anion topology (Fig. 7c). (2) Cations are ignored, as well as all cation-anion bonds, leaving an array of unconnected anions (Fig. 7d). (3) Anions are joined by lines, such that only anions that may be realistically considered as part of the same coordination polyhedron are connected (i.e. those anions separated by less than about 3.5 Å) (Fig. 7e). (4) Anions are removed from further consideration, leaving only a series of lines that represent the sheet anion topology.

The sheet anion topology is a two-dimensional tiling of space, and contains triangles, squares, pentagons and hexagons. The sheet anion topology does not contain information about the cation distribution in the sheet from which it was derived; sheets with substantially different cation distributions and types of polyhedra may have the same sheet anion topology. The concept of the sheet anion topology is especially well demonstrated by considering the two structures illustrated in Figure 8. The sheet shown in Figure 8a occurs in several mineral structures, such as becquerelite, $Ca[(UO_2)_3O_2(OH)_3]_2(H_2O)_8$, and compreignacite, $K_2[(UO_2)_3O_2(OH)_3]_2(H_2O)_7$, and contains only $Ur\phi_5$ pentagonal bipyramids that share corners and edges. The corresponding sheet anion topology (Fig. 8b) contains only pentagons and triangles. The sheets in the structure of $Mg[(UO_2)(SO_4)_2](H_2O)_{11}$ (Fig. 8c) contain both $Ur\phi_5$ pentagonal bipyramids and sulfate tetrahedra, and bear little resemblance to the sheet shown in Figure 8a. However, the topological arrangement of anions is identical in both sheets, hence both are based upon the same sheet anion topology.

The sheet anion topologies (for all but curite) of the sheets in minerals that contain only uranyl polyhedra (either square, pentagonal, or hexagonal bipyramids) can be obtained as stacking sequences of the four distinct chains shown in Figure 9 (Miller et al. 1996). The **P** chain is composed of edge-sharing pentagons, the **R** chain contains rhombs, and the **H** chain contains edge-sharing hexagons. The arrowhead chains (**U** and **D**), which have a directional aspect, contain both pentagons and triangles, arranged such that each triangle shares an edge with a pentagon, and the opposite corner with another pentagon in the chain. The development of a sheet anion topology as a chain-stacking sequence is illustrated in Figure 10. This approach also permits the prediction of sheet anion topologies that are as yet unknown from crystal structures (Miller et al. 1996).

In the discussion that follows, uranyl mineral structures that are based upon sheets of polyhedra of higher bond valence are discussed, arranged on the basis of the types of polygons within the corresponding sheet anion topologies. The minerals are arranged in Tables 3 to 7 according to the details of the sheet anion

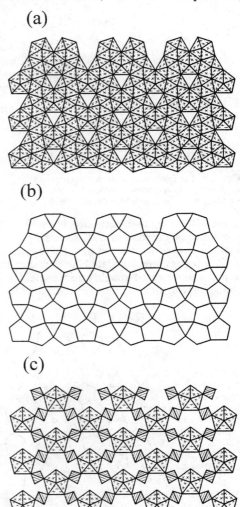

Figure 8. An example of two substantially different sheets of polyhedra that have identical underlying sheet anion topologies: (a) the sheet that occurs in the structures of becquerelite, billietite, compreignacite, richetite, and protasite; (b) the sheet anion topology that corresponds to the sheets shown in (a) and (c); (c) the sheet that occurs in the structure of $Mg[(UO_2)(SO_4)_2](H_2O)_{11}$. From Burns et al. (1996).

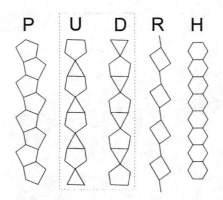

Figure 9. The four chain types that are required to develop sheet anion topologies as chain-stacking sequences. From Miller et al. (1996).

topologies that correspond to the sheet found in each structure. Within the tables, structures with the same sheet anion topologies are listed together, and structures that have different sheet anion topologies are separated by double lines. Structures that are based upon the same sheet anion topology, but that contain different sheets (i.e. different populations of the same anion topology), are separated by broken lines.

Anion topologies with squares

Only one anion topology is possible that contains squares arranged such that the anions of common edges of adjacent squares are all coincident (Fig. 11a). Three distinct populations of this anion topology are known among uranyl phases (Burns et al. 1996), but only the population shown in Figure 11b occurs in mineral structures. This sheet, referred to as the autunite-type sheet, contains equal numbers of $Ur\phi_4$ square bipyramids and $X^{5+}O_4$ tetrahedra (X^{5+} = P, As), and is the basis of at least seven mineral structures (Table 3). One half of the squares of the anion topology are populated; one quarter contain $Ur\phi_4$ square bipyramids with the uranyl ions oriented roughly perpendicular to the sheet, and one quarter are $X^{5+}O_4$ tetrahedra with all four anions contained within the sheet.

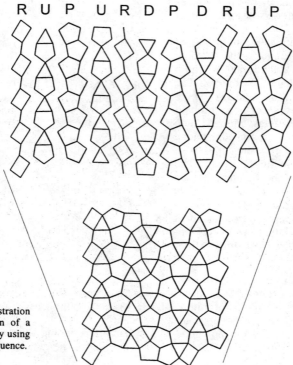

Figure 10. Demonstration of the construction of a sheet anion topology using a chain-stacking sequence.

(a)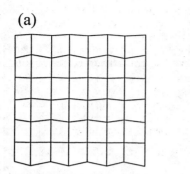

(b)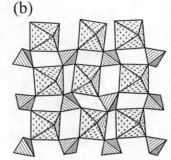

Figure 11. Sheets based upon the anion topology that contains only squares: (a) the anion topology; (b) the autunite-type sheet that occurs in the structures of saleeite, metazeunerite, metatorbernite, threadgoldite, meta-uranocircite, abernathyite, and meta-autunite. The uranyl polyhedra are shaded with crosses and $X^{5+}O_4$ tetrahedra are shaded with parallel lines.

Table 3. Uranyl mineral structures containing sheets based on anion topologies with squares

Name	Formula	S.G.	a (Å)	b (Å)	c (Å)	α (°)	β (°)	γ (°)	Ref.
saléeite	$Mg[(UO_2)(PO_4)]_2(H_2O)_{10}$	$P2_1/c$	6.951	19.947	9.896		135.17		1
metazeunerite	$Cu[(UO_2)(AsO_4)]_2(H_2O)_8$	$P4_2/nmc$	7.105		17.704				2
metatorbernite	$Cu_{0.92}[(UO_2)(PO_4)]_2(H_2O)_8$	$P4/n$	6.972		17.277				3
threadgoldite	$Al[(UO_2)(PO_4)]_2(OH)(H_2O)_8$	$C2/c$	20.168	9.847	19.719		110.71		4
meta-uranocircite	$Ba[(UO_2)(PO_4)]_2(H_2O)_6$	$P2_1/a$	9.789	9.882	16.868			89.95	5
abernathyite	$K[(UO_2)(AsO_4)](H_2O)_3$	$P4/ncc$	7.176		18.126				6
meta-autunite	$Ca[(UO_2)(PO_4)]_2(H_2O)_6$	$P4/nmm$	6.96		8.40				7

References: (1) Miller and Taylor (1986), (2) Hanic (1960), (3) Stergiou et al. (1993), (4) Khosrawan-Sazedj (1982b), (5) Khosrawan-Sazedj (1982a), (6) Ross and Evans (1964), (7) Makarov and Ivanov (1960).

With the exception of the O_{Ur} atoms, each anion contained within the autunite-type sheet is bonded to one X^{5+} and one U^{6+} cation. The bond valence incident upon each anion due to the bond to U^{6+} is ~0.64 vu (cf. Fig. 6), and the contribution from the X^{5+} cation is ~5/4 vu. Thus, the total bond valence incident upon each sheet anion from bonds within the sheet is ~1.89 vu, which approximately meets the bond-valence requirements of the O^{2-} anions, and indicates that the anions do not form strong bonds to interlayer constituents. Thus, most of the linkage between the autunite-type sheets and the interlayers must be through the O_{Ur} atoms, which can accept ~0.35 vu from the interlayer (cf. Fig. 6).

The $(UO_2)(X^{5+}O_4)$ sheets in meta-uranocircite, threadgoldite, metatorbernite, meta-autunite, and saleeite all contain X^{5+} = P, whereas the sheets in metazeunerite and abernathyite contain X^{5+} = As. In metazeunerite and metatorbernite the sheets are undistorted; the sheets in the other structures are variably distorted from ideal (Fig. 12).

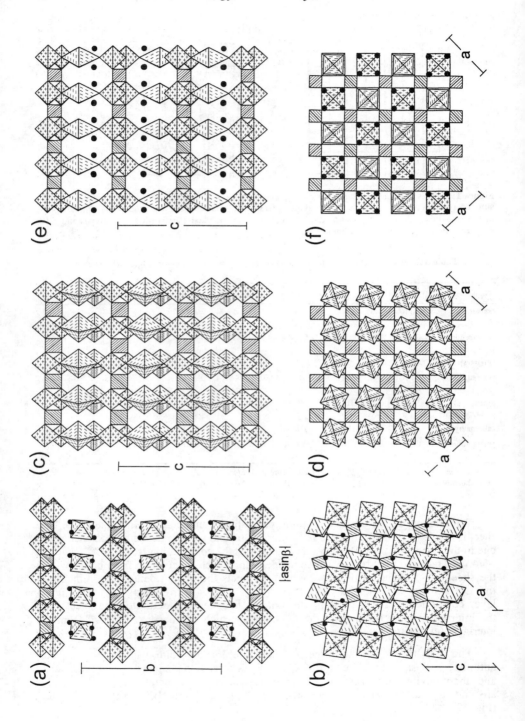

Figure 12 (opposite and next pages). Structures of minerals based upon the autunite-type sheet: (a) saleeite projected along [001]; (b) saleeite along [010]; (c) metazeunerite along [110]; (d) metazeunerite along [001]; (e) metatorbernite along [110]; (f) metatorbernite along [001]; (g) threadgoldite along [001]; (h) threadgoldite along [100]; (i) meta-uranocircite along [100]; (j) meta-uranocircite along [001]; (k) abernathyite along [110]; (l) abernathyite along [001]. The uranyl polyhedra are shaded with crosses and the $X^{5+}O_4$ tetrahedra are shaded with parallel lines. Interlayer polyhedra are shaded with broken parallel lines or herring-bone patterns. Interlayer H_2O groups are shown by black circles.

The structures of saleeite (Miller and Taylor 1986), metazeunerite (Hanic 1960), metatorbernite (Stergiou et al. 1993), and threadgoldite (Khosrawan-Sazedj 1982b) have octahedrally coordinated cations in the interlayer (Fig. 12). In the structures of both saleeite and threadgoldite, the interlayer octahedra are linked to the sheets by H bonds only, whereas the interlayer octahedra in metazeunerite and metatorbernite share corners with $Ur\phi_4$ square bipyramids in the sheets.

The interlayer in saleeite contains Mg cations that are coordinated by six H_2O groups in an octahedral arrangement (Fig. 12a). The interlayer also contains two symmetrically distinct H_2O groups that are H bonded into the structure. H atoms were not located in the structure solution (Miller and Taylor 1986), but an extensive H bonding network undoubtedly exists that binds the interlayer to the sheets on either side.

The structures of both metazeunerite and metatorbernite contain Cu^{2+} cations in the interlayer that are octahedrally coordinated by four H_2O groups and two O_{Ur} atoms, with one O_{Ur} atom from each adjacent sheet (Fig. 12c-f). The $Cu^{2+}\phi_6$ octahedra are (4+2) distorted owing to the Jahn-Teller effect, and in both structures the apical ligands, which correspond to the two long $Cu^{2+}-\phi$ bonds in each octahedron, are O_{Ur} atoms. In the case of metatorbernite, the interlayer also contains an H_2O group that is not bonded to Cu^{2+}, but is held in the structure by H bonds only. Projections of the interlayers of both structures onto the sheets (Fig. 12d,f) shows that metazeunerite contains twice as many interlayer octahedra as metatorbernite; this is because the octahedra in metazeunerite are only 50% occupied. Where the octahedral cation is locally absent, the H_2O groups are H bonded into the structure.

The interlayer of the structure of threadgoldite contains Al cations that are octahedrally coordinated by two (OH)⁻ groups and four H_2O groups (Fig. 12g). Symmetrically equivalent $Al\phi_6$ octahedra share edges that connect the two (OH)⁻ groups, resulting in dimers of composition $Al_2(OH)_2(H_2O)_8$ (Fig. 12h). These dimers are connected to the sheets of polyhedra by H bonding only. There are also four symmetrically distinct H_2O groups that are H bonded into the interlayer.

The structure of meta-uranocircite (Khosrawan-Sazedj 1982a) contains Ba cations and H_2O groups in the interlayer (Fig. 12i). Two Ba cations are coordinated by a total of nine ligands, five of which are interlayer H_2O groups. The remaining ligands correspond to two O_{Ur}, with one from each adjacent sheet, and two sheet anions that are bonded to both U^{6+} and P. The $Ba\phi_9$ polyhedra share an edge to form a dimer, as shown in Figure 12j. The structure also contains two symmetrically distinct H_2O groups that are held in the structure by H bonds only.

The structure of abernathyite (Ross and Evans 1964) contains both K cations and H_2O groups in the interlayer (Fig. 12k,l), and both constituents occur at the same structural site, with K replacing H_2O in $1/4$ of the sites. Each K cation is coordinated by five ligands with bond lengths less than 3.25 Å, one of these is an O_{Ur} atom, one is a sheet O atom that bonds to both As and U^{6+}, and three are H_2O groups. There must be a complex network of H bonds, although the positions of the H atoms were not located in

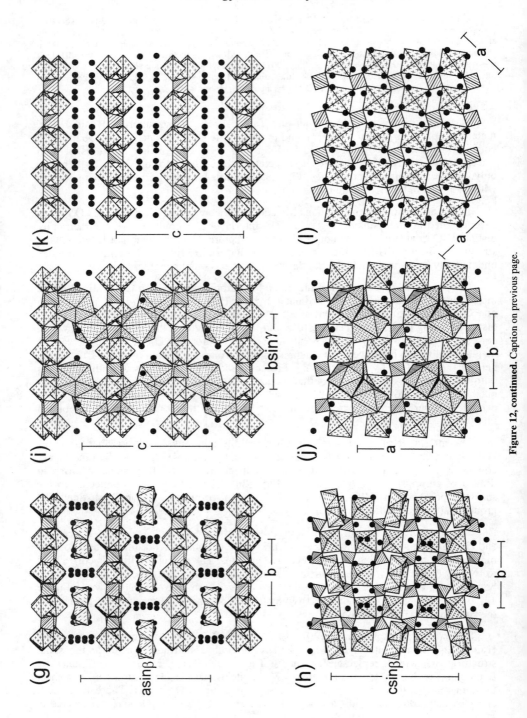

Figure 12, continued. Caption on previous page.

the structure determination.

A partial structure for meta-autunite was reported by Marakov and Ivanov (1960). The R factor remained high, and the structure contains some unrealistic atomic separations. The solution indicated that the structure contains the autunite-type sheet and that the interlayer involves some positional disorder and partially occupied sites. However, further refinement of the structure is hindered by the lack of crystals of suitable quality; crystals invariably show signs of dehydration and mottled extinction between crossed polarizers.

Anion topologies with triangles and pentagons

Nine uranyl mineral structures are based upon sheets with anion topologies that contain only triangles and pentagons (Table 4).

Table 4. structures containing sheets based upon anion topologies containing triangles and pentagons

Name	Formula	S. G.	a (Å)	b (Å)	c (Å)	α (°)	β (°)	γ (°)	Ref.
protasite	$Ba[(UO_2)_3O_3(OH)_2](H_2O)_3$	Pn	12.2949	7.2206	6.9558		90.40		1
billietite	$Ba[(UO_2)_3O_2(OH)_3]_2(H_2O)_4$	$Pbn2_1$	12.0720	30.167	7.1455				1
becquerelite	$Ca[(UO_2)_3O_2(OH)_3]_2(H_2O)_8$	$Pn2_1a$	13.8378	12.3781	14.9238				1
compreignacite	$K_2[(UO_2)_3O_2(OH)_3]_2(H_2O)_7$	$Pnmn$	14.8591	7.1747	12.1871				2
richetite	$M_xPb_{8.57}[(UO_2)_{18}O_{18}(OH)_{12}]_2(H_2O)_{41}$	$P1$	20.9391	12.1000	16.3450	103.87	115.37	90.27	3
masuyite	$Pb[(UO_2)_3O_3(OH)_2](H_2O)_3$	Pn	12.241	7.008	6.983		90.40		4
fourmarierite	$Pb[(UO_2)_4O_3(OH)_4](H_2O)_4$	$Bb2_1m$	13.986	16.400	14.293				5
schoepite	$[(UO_2)_8O_2(OH)_{12}](H_2O)_{12}$	$P2_1ca$	14.337	16.813	14.731				6
vandendriesscheite	$Pb_{1.57}[(UO_2)_{10}O_6(OH)_{11}](H_2O)_{11}$	$Pbca$	14.1165	41.478	14.5347				7

References: (1) Pagoaga et al. (1987), (2) Burns (1998c), (3) Burns (1998d), Burns and Hanchar (1999), (5) Piret (1985), (6) Finch et al. (1996), (7) Burns (1997).

Protasite anion topology. Two distinct populations of the protasite anion topology (Fig. 13b) are known, although only one has been found in minerals (Burns et al. 1996). If all pentagons of the protasite anion topology are populated with uranyl ions, $Ur\phi_5$ pentagonal bipyramids result, and this sheet occurs in the structures of protasite (Pagoaga et al. 1987), becquerelite (Pagoaga et al. 1987), billietite (Pagoaga et al. 1987), compreignacite (Burns 1998c), richetite (Burns 1998d), and masuyite (Burns and Hanchar 1999) (Fig. 13a). Topologically identical sheets also occur in the structures of a synthetic Cs uranyl oxide hydrate phase (Hill and Burns 1999), as well as the mixed-valence α-U_3O_8 phase (Loopstra 1977).

The protasite anion topology shown in Figure 13b does not distinguish between O^{2-} and $(OH)^-$ anions. However, the distributions of anions within the sheets based upon this anion topology are not identical, an observation that is significant because it pertains to the charge on the sheet, as well as to how the sheets connect to interlayer constituents by H bonding. The distribution of O^{2-} and $(OH)^-$ in the anion topologies of the sheets in the structures of protasite, becquerelite, billietite, compreignacite, richetite, and masuyite are shown in Figure 14, where the location of $(OH)^-$ groups is indicated by open circles in the anion topology. Note that the sheets in becquerelite, billietite and compreignacite have identical anion distributions (Fig. 14a), with the formula $[(UO_2)_3O_2(OH)_3]^{-1}$. The corners of all triangles in the protasite anion topology correspond to $(OH)^-$ groups. The

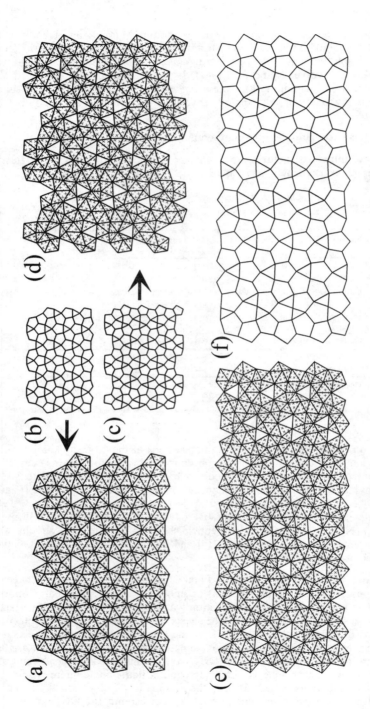

Figure 13. Sheets based upon anion topologies that contain triangles and pentagons. (a) the protasite-type sheet that occurs in the structures of protasite, billietite, becquerelite, compreignacite, richetite, and masuyite; (b) the protasite anion topology; (c) the fourmarierite anion topology; (d) the fourmarierite-type sheet that occurs in the structures of fourmarierite and schoepite; (e) the vandendriesscheite sheet; (f) the vandendriesscheite anion topology. Uranyl polyhedra are shaded with crosses.

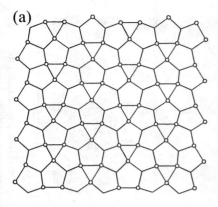

Becquerelite
Billietite
Compreignacite

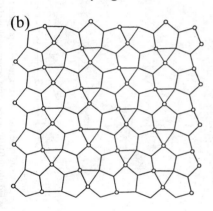

Protasite
Masuyite

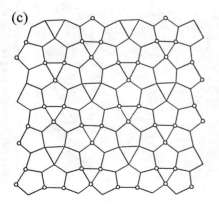

Richetite

Figure 14 (left). Anion topologies that correspond to the protasite-type sheets that occur in minerals. The positions of (OH)⁻ groups in the anion topologies are indicated by open circles. The topologies correspond to (a) becquerelite, billietite, compreignacite; (b) protasite and masuyite; (c) richetite.

distribution of (OH)⁻ groups in the protasite and masuyite sheets are different in that only two vertices of each triangle are occupied by (OH)⁻ groups (Fig. 14b), resulting in the formula $[(UO_2)_3O_3(OH)_2]^{-2}$. The formula of the richetite sheet is also $[(UO_2)_3O_3(OH)_2]^{-2}$, but there are two types of triangles in the anion topology; those that have (OH)⁻ groups at all corners, and those that contain no (OH)⁻ groups (Fig. 14c).

Projections of the structures of becquerelite, protasite, and billietite are provided in Figure 15. In becquerelite, the interlayer contains Ca that is coordinated by eight ligands; four are H₂O groups and four are O_{Ur} atoms of adjacent sheets of uranyl polyhedra, arranged such that the Ca polyhedra shares three vertices with one sheet and only one vertex with the opposing sheet (Fig. 15a). The structure of becquerelite contains four symmetrically distinct H₂O groups that are held in the structure by H bonding only. Projection of the interlayer constituents onto the sheet of polyhedra (Fig. 15b) shows that the Ca polyhedra and H-bonded H₂O groups are arranged in alternating strips that are parallel to the b axis.

The interlayer of the protasite structure (Figs. 15c,d) contains Ba cations that are coordinated by ten ligands; three are H₂O groups, one is an equatorial O atom of uranyl polyhedra in an adjacent sheet, and six are O_{Ur} atoms, three each from adjacent sheets. In billietite (Fig. 15e), the Ba cation in the interlayer is coordinated by at least seven ligands; one is an H₂O group and six are O_{Ur} atoms, three from each adjacent sheet. Chemical analysis for billietite indicate the presence of additional H₂O groups, but these were not located in the structural analysis (Pagoaga et al. 1987). On the basis of a spectroscopic study, Cejka et al. (1998) concluded that billietite contains 8 H₂O groups. By analogy with protasite, the Ba polyhedron

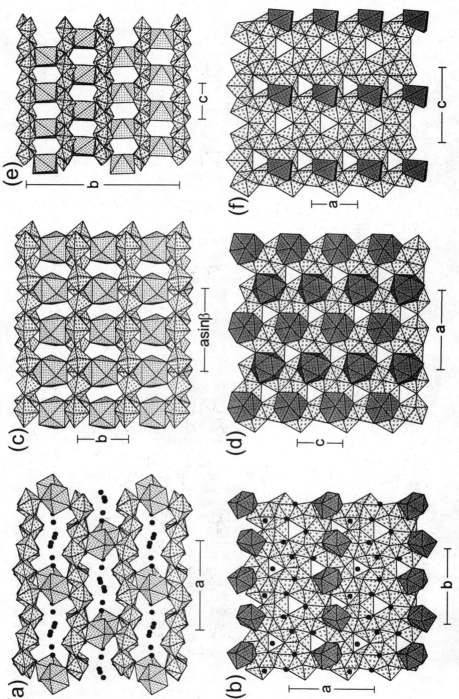

Figure 15. The structures of minerals that contain the protasite-type sheet. (a) becquerelite projected along [010]; (b) becquerelite projected along [001]; (c) protasite projected along [001]; (d) protasite projected along [010]; (e) billietite projected along [100]; (f) billietite projected along [010].

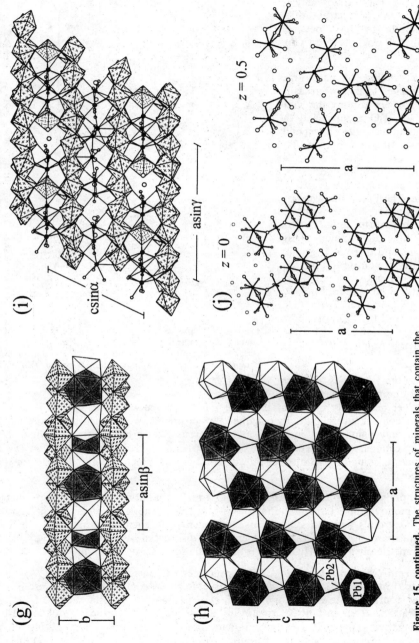

Figure 15, continued. The structures of minerals that contain the protasite-type sheet. (g) masuyite projected along [001]; (h) masuyite projected along [010]; (i) richetite projected along [100]; (j) interlayers of the richetite structure.

may contain additional H$_2$O groups, and, by analogy with becquerelite, H$_2$O groups may be H bonded into the structure in strips parallel to [100], between the strips of Ba polyhedra (Fig. 15f).

The structure of masuyite is closely related to that of protasite. The interlayers contain the same number of H$_2$O groups, and the Pb(1) site in masuyite corresponds to the Ba site in protasite. The structure differs in the presence of the Pb(2) site, which does not have a corresponding site in protasite. The Pb(2) site was only 12% occupied in the crystal studied by Burns and Hanchar (1999). It is coordinated by six O$_{Ur}$ atoms and three H$_2$O groups. The fully occupied Pb(1)φ$_{10}$ polyhedra do not share polyhedral elements with other Pb(1)φ$_{10}$ polyhedra, but each shares three faces with Pb(2)φ$_9$ polyhedra, resulting in sheets of Pbφ$_n$ polyhedra parallel to (010) (Fig. 15g,h).

Richetite possesses an extraordinarily complex structure that

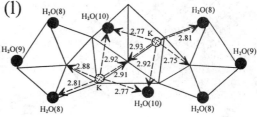

Figure 15, continued. The structures of minerals that contain the protasite-type sheet: (k) compreignacite projected along [100]; (l) enlargement of the section of the compreignacite structure enclosed in a box in (k). The legend is as in Figure 12.

has not yet been fully characterized. The interlayers between the sheets of uranyl polyhedra contain several partially occupied Pb^{2+} positions, two octahedrally coordinated sites that may contain Mg and Fe^{2+}, and numerous H$_2$O groups, some of which are bonded to cations, and some that are held in the structure by H bonding only (Fig. 15i,j). The structure contains two distinct interlayers, both of which involve disorder of the atomic positions.

Compreignacite is the only known mineral that is based upon protasite-type sheets that has monovalent cations in the interlayer (Fig. 15k). The K cation is coordinated by three H$_2$O groups and four O$_{Ur}$ atoms, two each from each adjacent sheet. Two K polyhedra share an edge to form a dimer that is disordered over two sites (Fig. 15k,l). One symmetrically distinct H$_2$O group is held in the structure by H bonding only.

Fourmarierite anion topology. The structures of fourmarierite (Piret 1985) and schoepite (Finch et al. 1996) are the only structures that contain sheets that are based upon the fourmarierite anion topology shown in Figure 13c. The sheets of uranyl polyhedra that occur in both fourmarierite and schoepite (Fig. 13d) are obtained by populating every pentagon in the anion topology with a uranyl ion, giving $Ur\phi_5$ pentagonal bipyramids and the formulae [(UO$_2$)$_4$O$_3$(OH)$_4$]$^{-2}$ and [(UO$_2$)$_4$O(OH)$_6$], respectively. The distribution of (OH)$^-$ groups in the anion topology of each structure is

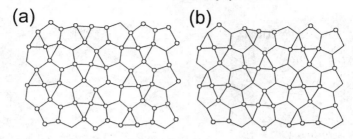

Figure 16. Sheet anion topologies corresponding to the sheets in (a) schoepite and (b) fourmarierite. The positions of (OH)⁻ groups are shown by circles.

shown in Figure 16. The fourmarierite sheet has a charge of −2 per formula unit, so that interlayer cations are required for charge balance. In the case of schoepite, the sheet of uranyl polyhedra is electrostatically neutral, so the interlayer must also be neutral.

The structures of fourmarierite and schoepite are shown in Figure 17. There are two Pb^{2+} cations in the interlayer of fourmarierite. The Pb(1) cation bonds to three H_2O groups and four O_{Ur} atoms, whereas the Pb(2) cation is bonded to three H_2O groups and six O_{Ur} atoms that are in adjacent sheets of uranyl polyhedra. The Pb(1)ϕ_7 and Pb(2)ϕ_9 polyhedra share an edge to form a dimer (Fig. 17b). In addition, the interlayer of fourmarierite contains four symmetrically distinct H_2O groups that are H bonded into the structure. The interlayer in schoepite contains only H_2O groups that are H bonded into the structure, and which provide the only linkages between the sheets of uranyl polyhedra (Fig. 17c,d).

Vandendriesscheite anion topology. The complex sheet of $Ur\phi_5$ pentagonal bipyramids shown in Figure 13e is the basis of the structure of vandendriesscheite (Burns 1997). The sheet anion topology is given in Figure 13f, and contains only pentagons and triangles. The structure contains two symmetrically distinct Pb^{2+} cations in the interlayer (Fig. 18); both are bonded to three H_2O groups, and Pb(1) and Pb(2) also bond to six and five O_{Ur} atoms of adjacent sheets of uranyl polyhedra, respectively. The Pb(1)ϕ_9 polyhedra are isolated in the interlayer, and two Pb(2)ϕ_8 polyhedra sharing a face to form a dimer (Fig. 18b). However, adjacent Pb(2) positions are separated by ~3 Å, and site-occupancy refinement indicates that both are not typically occupied locally. In addition to the $Pb^{2+}\phi_n$ polyhedra, there are six symmetrically distinct H_2O groups that are held in the interlayer of the structure by H bonds alone.

Anion topologies as chain-stacking sequences. The protasite, fourmarierite, and vandendriesscheite anion topologies all contain only triangles and pentagons. It is straightforward to construct each of these anion topologies as stacking sequences of **P** and arrowhead chains (**U** and **D**), as demonstrated in Figure 19. Only **P** and **D** chains are required to develop the protasite anion topology, with the repeat sequence **PDPD**… The fourmarierite anion topology requires **P**, **D** and **U** chains, with the sequence **DUPUDPDU**… The vandendriesscheite anion topology contains sections that are identical to the **PDPD**… or **PUPU**… repeats of the protasite anion topology, with the join between such sections involving the **DU** sequence of the fourmarierite anion topology. Thus, the vandendriesscheite anion topology can be considered to be a structural intermediate between the protasite and fourmarierite anion topologies.

Figure 17 (next page). Structures of minerals that contain the fourmarierite-type sheet: (a) fourmarierite projected along [100]; (b) fourmarierite projected along [001]; (c) schoepite projected along [100]; (d) schoepite projected along [001]. The legend is as in Figure 12.

48 URANIUM: Mineralogy, Geochemistry and the Environment

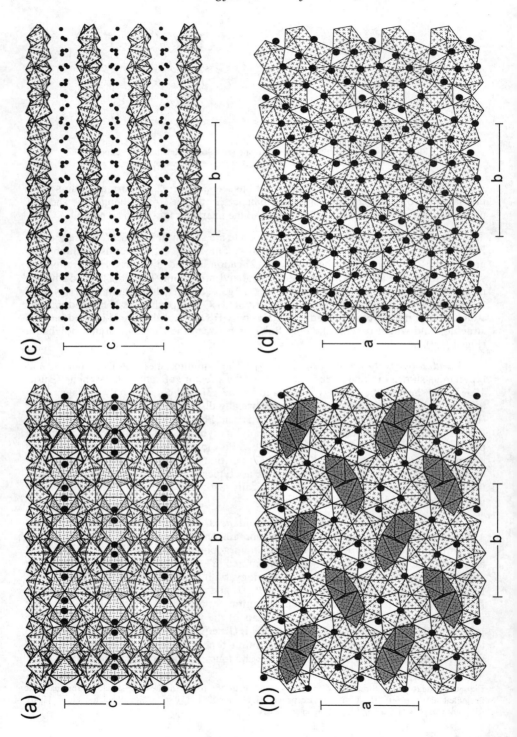

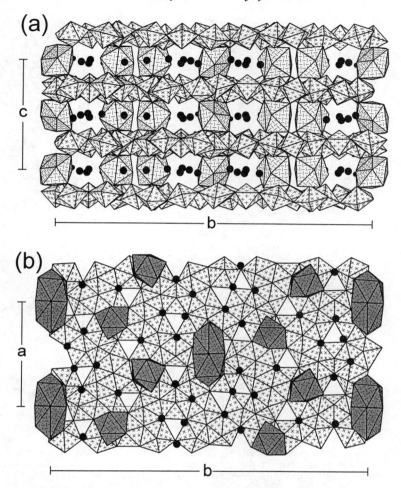

Figure 18. The structure of vandendriesscheite: (a) projected along [100]; (b) projected along [001]. The legend is as in Figure 12.

Anion topologies with triangles, squares and pentagons

Uranyl minerals that possess structures that are based upon sheets with underlying anion topologies composed of triangles, squares and pentagons are listed in Table 5.

Uranophane anion topology. The uranophane anion topology shown in Figure 20a is the basis of the sheets that occur in 16 structures (Burns et al. 1996), eight of which are minerals (Table 5). This anion topology contains chains of edge-sharing pentagons (P chains) that are connected through chains of alternating edge-sharing triangles and squares. The topology can also be obtained using a chain-stacking sequence of arrowhead (U and D) and R chains, with adjacent arrowhead chains pointing in opposite directions, as shown in Figure 21.

With the exception of ulrichite, the sheets based upon the uranophane anion topology that occur in minerals contain uranyl ions in the pentagons of the anion topology,

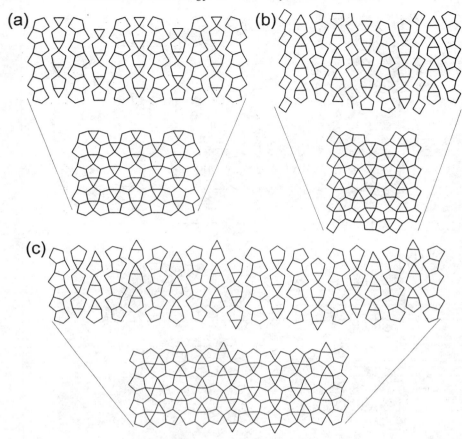

Figure 19. Development of anion topologies as chain-stacking sequences: (a) development of the protasite anion topology; (b) development of the fourmarierite anion topology; (c) demonstration that the vandendriesscheite anion topology is intermediate between those of protasite and fourmarierite.

resulting in chains of edge-sharing $Ur\phi_5$ pentagonal bipyramids (Fig. 20). Either the triangles or the squares of the anion topology are also populated by cations, but never both in the same sheet. Most sheets contain silicate tetrahedra positioned such that one tetrahedral face corresponds to a triangle in the underlying anion topology. Two types of uranyl silicate sheets that are based upon the uranophane anion topology occur in six minerals, and the distinction between the two is the orientation of the silicate tetrahedra. Consider the chains of squares and triangles that is vertical in the anion topology shown in Figure 20a. In the α-uranophane-type sheet, which occurs in five minerals, the silicate tetrahedra are positioned such that the apical (non-sheet) anions of adjacent tetrahedra in the triangle-square chain point alternately up and down (Fig. 20b). In the β-uranophane-type sheet, the tetrahedra in the triangle-square chain alternate in pairs, such that there are two adjacent tetrahedra with apical anions pointing up, then two pointing down, and so on (Fig. 20c).

Structures that contain the α-uranophane-type sheet are shown in Figure 22. In α-uranophane, boltwoodite, sklodowskite, and cuprosklodowskite the sheet contains an

Table 5. Uranyl mineral structures containing sheets based upon anion topologies with triangles, squares and pentagons

Name	Formula	S. G.	a (Å)	b (Å)	c (Å)	α (°)	β (°)	γ (°)	Ref.
α-uranophane	$Ca[(UO_2)(SiO_3OH)]_2(H_2O)_5$	$P2_1$	15.909	7.002	6.665		97.27		1
boltwoodite	$(K,Na)[(UO_2)(SiO_3OH)](H_2O)_{1.5}$	$P2_1/m$	7.0772	7.0597	6.6479		104.98		2
cupro-sklodowskite	$Cu[(UO_2)(SiO_3OH)]_2(H_2O)_6$	$P\bar{1}$	7.052	9.267	6.655	109.23	89.84	110.01	3
sklodowskite	$Mg[(UO_2)(SiO_3OH)]_2(H_2O)_6$	$C2/m$	17.382	7.047	6.610		105.9		4
kasolite	$Pb[(UO_2)(SiO_4)](H_2O)$	$P2_1/c$	6.704	6.932	13.252		104.22		5
β-uranophane	$Ca[(UO_2)(SiO_3OH)]_2(H_2O)_5$	$P2_1/a$	13.966	15.443	6.632		91.38		6
ulrichite	$Cu[Ca(UO_2)(PO_4)_2](H_2O)_4$	$C2/m$	12.79	6.85	13.02		91.03		7
schmitterite	$[(UO_2)(TeO_3)]$	$Pca2_1$	10.161	5.363	7.862				8
francevillite	$Ba_{0.96}Pb_{0.04}[(UO_2)_2(V_2O_8)](H_2O)_5$	$Pcan$	10.419	8.510	16.763				9
curienite	$Pb[(UO_2)_2(V_2O_8)](H_2O)_5$	$Pcan$	10.40	8.45	16.34				10
sengierite	$Cu_2[(UO_2)_2(V_2O_8)](OH)_2(H_2O)_6$	$P2_1/a$	10.599	8.093	10.085		103.42		11
billietite	$Ba[(UO_2)_3O_2(OH)_3]_2(H_2O)_4$	$Pbn2_1$	12.072	30.167	7.1455				12
ianthinite	$[U^{4+}_2(UO_2)_4O_6(OH)_4(H_2O)_4](H_2O)_5$	$P2_1cn$	7.178	11.473	30.39				13
wyartite	$Ca[U^{5+}(UO_2)_2(CO_3)O_4(OH)](H_2O)_7$	$P2_12_12_1$	11.2706	7.1055	20.807				14
iriginite	$[(UO_2)(MoO_3OH)_2(H_2O)](H_2O)$	$Pca2_1$	12.77	6.715	11.53				15
sayrite	$Pb_2[(UO_2)_5O_6(OH)_2](H_2O)_4$	$P2_1/c$	10.704	6.960	14.533		116.81		16
curite	$Pb_3[(UO_2)_8O_8(OH)_6](H_2O)_3$	$Pnam$	12.551	13.003	8.390				17
vandenbrandeite	$[(UO_2)Cu(OH)_4]$	$P\bar{1}$	7.855	5.449	6.089	91.44	101.90	89.2	18
zippeite	$K[(UO_2)_2(SO_4)(OH)_3](H_2O)$	$C2/c$	8.755	13.987	17.730		104.13		19
wölsendorfite	$Pb_{6.2}Ba_{0.4}[(UO_2)_{14}O_{19}(OH)_4](H_2O)_{12}$	$Cmcm$	14.131	13.885	55.969				20

References: (1) Ginderow (1988), (2) Burns (1998b), (3) Rosenzweig and Ryan (1975), (4) Ryan and Rosenzweig (1977), (5) Rosenzweig and Ryan (1977a), (6) Viswanathan and Harneit (1986), (7) Birch et al. (1988), (8) Meunier and Galy (1973), (9) Mereiter (1986d), (10) Borène and Cesbron (1971), (11) Piret et al. (1980), (12) Pagoaga et al. (1987), (13) Burns et al. (1997b), (14) Burns and Finch (1999), (15) Serezhkin et al. (1973), (16) Piret et al. (1983), (17) Taylor et al. (1981), (18) Rosenzweig and Ryan (1977b), (19) Vochten et al. (1995), (20) Burns (1999a).

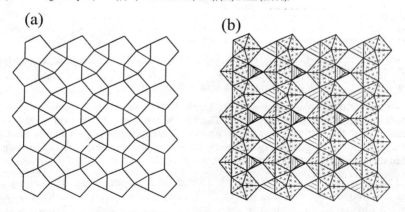

Figure 20. Sheets that are based on the uranophane anion topology: (a) the uranophane anion topology; (b) the α-uranophane-type sheet that occurs in α-uranophane, boltwoodite, cuproskolodowskite, sklodowskite, and kasolite; (b) the sheet that occurs in β-uranophane; [next page] (c) the sheet that occurs in schmitterite; (d) the sheet that occurs in ulrichite. Uranyl polyhedra are shaded with crosses and tetrahedra are shaded with parallel lines.

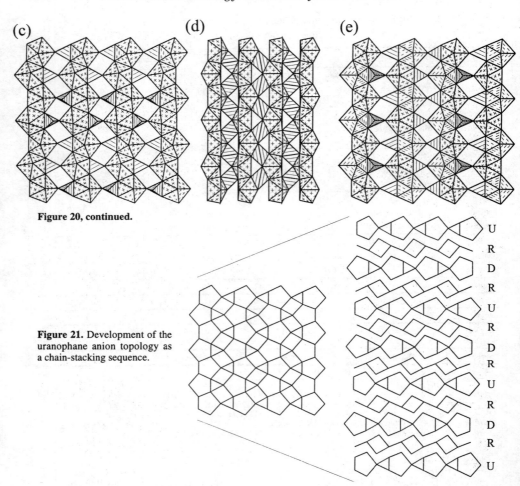

Figure 20, continued.

Figure 21. Development of the uranophane anion topology as a chain-stacking sequence.

acid silicate group, with the (OH)⁻ group located at the apical (non-sheet) position of the tetrahedron. The (OH)⁻ group typically donates a H bond that is accepted by an H_2O group in the interlayer, providing additional linkage between sheets.

The structure of α-uranophane (Ginderow 1988) contains one symmetrically distinct Ca position in the interlayer (Fig. 22a). The Ca coordination polyhedron includes two O_{Ur} atoms, one from each adjacent sheet, one (OH)⁻ group that is part of the acid silicate group, and four H_2O groups. In addition, the interlayer contains one H_2O group that is held in the structure by H bonds only.

The structures of sklodowskite (Ryan and Rosenzweig 1977) and cuprosklodowskite (Rosenzweig and Ryan 1975) contain octahedrally coordinated cations in their interlayers; in sklodowskite the octahedron is occupied by Mg, whereas $Cu^{2+}\phi_6$ octahedra occur in cuprosklodowskite. In addition, the structural connectivity of these two minerals is quite different (Fig. 22c,e). In sklodowskite, the $Mg\phi_6$ octahedron involves four H_2O groups as well as two (OH)⁻ groups that belong to acid silicate groups in sheets on either side. In cuprosklodowskite, the $Cu^{2+}\phi_6$ octahedron contains four H_2O

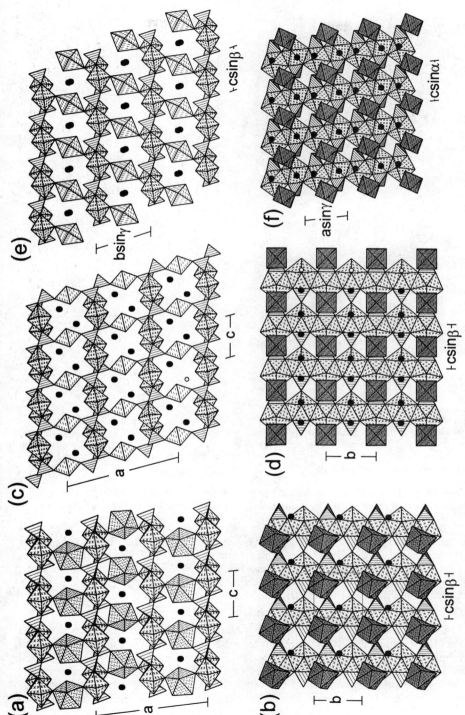

Figure 22. Structures of minerals that contain the α-uranophane-type sheet. (a) α-uranophane projected along [010]; (b) α-uranophane along [100]; (c) sklodowskite along [010]; (d) sklodowskite along [100]; (e) cuprosklodowskite along [100]; and (f) cuprosklodowskite along [010].

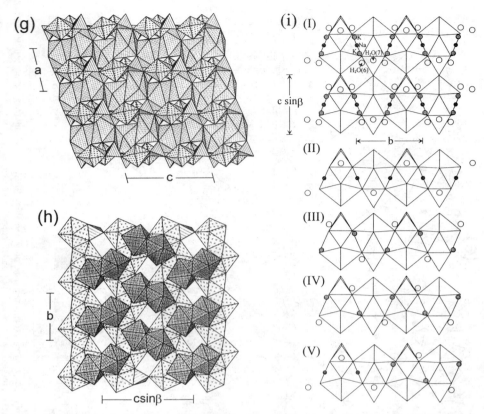

Figure 22, continued. Structures of minerals that contain the α-uranophane-type sheet: (g) kasolite projected along [010]; (h) kasolite along [100]; and (i) boltwoodite along [100]. Various possible local interlayer configurations for boltwoodite are illustrated (I to V). The legend is as in Figure 12 except that in (i) the K and Na cations are shown as circles shaded with dots and parallel lines, respectively, and the H$_2$O groups are shown as open circles.

groups, as in sklodowskite, but the Cu^{2+}φ$_6$ octahedron also contains two O$_{Ur}$ atoms. Both structures also contain an H$_2$O group that is held in the interlayer by H bonds only.

The structure of kasolite (Rosenzweig and Ryan 1977a) (Fig. 22g,h) is unusual in that the uranyl silicate sheet does not contain an acid silicate group. Instead, the apical (non sheet) anions of the silicate tetrahedra also bond to two Pb^{2+} cations in the interlayer, thus satisfying their bond-valence requirements. Each Pb^{2+} cation is bonded to two O$_{Ur}$ atoms, two apical O atoms of silicate tetrahedra, two O atoms that are equatorial ligands of Urφ$_5$ polyhedra and Si tetrahedra, and one H$_2$O group. The Pb^{2+}φ$_7$ polyhedra share edges with symmetrically equivalent Pb^{2+}φ$_7$ polyhedra, forming dimers (Fig. 22h).

Stohl and Smith (1981) provided the structure for end-member boltwoodite, and Burns (1998b) studied a crystal with substantial Na substitution in the interlayer. Stohl and Smith (1981) concluded that boltwoodite contains an (H$_3$O)$^+$ (hydronium) ion, based upon a structure refinement for a twinned crystal. Burns (1998b) obtained a higher-quality refinement, and concluded that the structure contains an acid silicate group, rather than a hydronium ion. The structure has not been done for end-member sodium boltwoodite, but X-ray powder diffraction studies indicate that the structure has a larger unit cell than boltwoodite (Vochten et al. 1997).

It is possible that a complete solid-solution series occurs between boltwoodite and sodium boltwoodite. The K and Na cations in the interlayer of the boltwoodite studied by Burns (1998b) are in different positions, with significantly different coordination environments about each cation. In addition, the K cations are distributed over two symmetrically equivalent sites. The Na cation is octahedrally coordinated by four O_{Ur} atoms and two H_2O groups, whereas the K cation is coordinated by four O_{Ur} atoms, the (OH)⁻ group that is part of the acid silicate group, as well as two H_2O groups. Possible local arrangements of interlayer species are shown in Figure 22i.

β-uranophane is a polymorph of α-uranophane, and the details of their structural connectivities differ substantially (Viswanathan and Harneit 1986). The uranyl silicate sheet in β-uranophane differs from that in α-uranophane in the orientation of the silicate tetrahedra (Fig. 20c), which results in a different interlayer. In β-uranophane, the Ca cation is coordinated by three O_{Ur} atoms of adjacent sheets, one (OH)⁻ group of an acid silicate group, and four H_2O groups, resulting in a Caφ$_8$ polyhedron with one more ligand than in α-uranophane. As for α-uranophane, there is an H_2O group in the interlayer that is held in the structure by H bonds only.

In the structure of β-uranophane, the Caφ$_8$ polyhedra occur in two rows that are parallel to the c axis (Fig. 23b). In one such row, the Caφ$_8$ polyhedra include (OH)⁻ groups of acid silicate groups, as well as one O_{Ur} atom in each polyhedron from the sheet below, and two O_{Ur} atoms in each polyhedron from the sheet above. In the adjacent rows of Caφ$_8$ polyhedra on either side within the same interlayer, the Ca cations bond to two O_{Ur} atoms from the sheet below, as well as one O_{Ur} atom and one (OH)⁻ group from the sheet above. In contrast, all Caφ$_7$ polyhedra in the interlayer of α-uranophane (Fig. 22b) include one (OH)⁻ group from the acid silicate group, and one O_{Ur} atom from the sheet on one side, and one O_{Ur} atom from the sheet on the other side.

The structure of schmitterite (Meunier and Galy 1973) contains the sheet shown in Figure 20d. The pentagons of the underlying anion topology are populated by uranyl

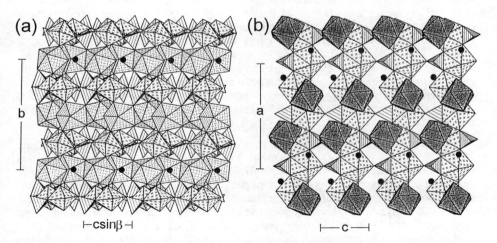

Figure 23. The structure of β-uranophane: (a) projected along [100]; (b) projected along [010]. The legend is as in Figure 12.

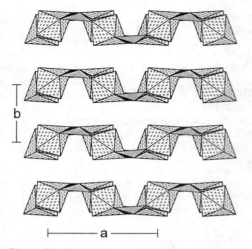

Figure 24. The structure of schmitterite projected along [001]. The legend is as in Figure 12.

ions, resulting in $Ur\phi_5$ pentagonal bipyramids. The squares of the anion topology contain $Te^{4+}\phi_4$ polyhedra. This is the only sheet that occurs in a mineral that is based upon the uranophane anion topology with the square sites occupied, although several such sheets occur in synthetic compounds (Burns et al. 1996). Owing to the lone pair of electrons on the Te^{4+} cation, all four anions of the coordination polyhedron occur on the same side of the cation (Fig. 24). The sheets have the composition $(UO_2)(TeO_3)$ and are neutral, and the structure does not contain any interlayer constituents. Adjacent sheets are held together by van der Waals forces and weak Te^{4+}-O interactions over distances exceeding 3.1 Å.

The structure of ulrichite (Birch et al. 1988) contains sheets that are based upon the uranophane anion topology, but the structure is unusual in that only one half of the pentagons in the anion topology are populated with uranyl ions; the others contain $Ca\phi_8$ polyhedra (Fig. 20e). The triangles of the anion topology contain PO_4 tetrahedra. There are two types of PO_4 tetrahedra; one type shares an edge with a $Ur\phi_5$ pentagonal bipyramid and an edge with the $Ca\phi_8$ polyhedron in the adjacent chain of edge-sharing pentagons, so that all four tetrahedral ligands of this type of PO_4 tetrahedron are bonded within the sheet. The edge shared with the $Ca\phi_8$ polyhedron is perpendicular to the sheet, and the tetrahedron projects as a triangle. The other type of PO_4 tetrahedron shares an edge with a $Ca\phi_8$ polyhedron and a corner with an adjacent $Ur\phi_5$ pentagonal bipyramid. The two types of PO_4 tetrahedra occur in equal numbers and are distributed such that only one type occurs in each chain of edge-sharing triangles and squares in the underlying anion topology. In the arrowhead chains of pentagons and triangles in the anion topology, the pentagons are alternately populated with $Ur\phi_5$ pentagonal bipyramids and $Ca\phi_8$ polyhedra (Fig. 20e). There are two types of arrowhead chains. In one, the apices of all tetrahedra point in the same direction, whereas in the other they alternate up and down. The interlayer of the ulrichite structure contains Cu^{2+} cations and H_2O groups and may involve considerable disorder, as the reported structure includes unrealistic atomic separations in the interlayer, and the Cu^{2+} coordination polyhedra involve unlikely configurations. The structure of ulrichite was done using film methods (Birch et al. 1988) and a precise refinement requires quantitative intensity data.

Francevillite anion topology. The francevillite anion topology is shown in Figure 25b. In this anion topology, edge-sharing dimers of pentagons share corners, creating both triangles and squares. The francevillite-type sheet (Fig. 25a) is the only known derivative of this anion topology. The pentagons of the anion topology are populated with uranyl ions, giving $Ur\phi_5$ pentagonal bipyramids, and the squares correspond to the base of VO_5 square pyramids. The sheet occurs in the structures of francevillite, curienite, and sengierite (Table 5), as well as three synthetic phases (Burns et al. 1996). A similar sheet containing NbO_5 square pyramids rather than VO_5 square pyramids is also known from a synthetic phase (Burns et al. 1996).

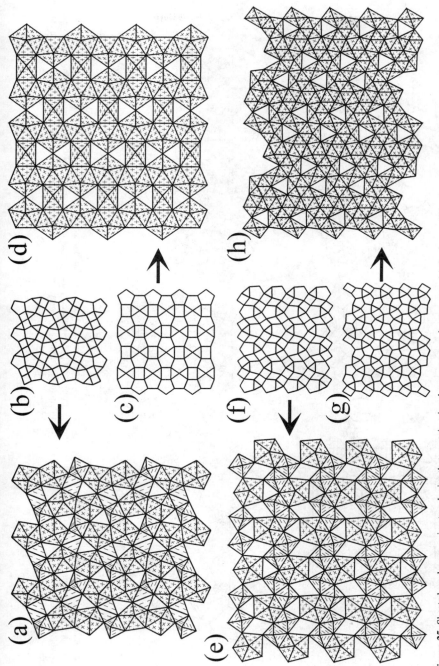

Figure 25. Sheets based upon anion topologies that contain triangles, squares and pentagons: (a) the francevillite-type sheet that occurs in francevillite, curienite, and sengierite; (b) the francevillite anion topology; (c) the β-U_3O_8 anion-topology; (d) the ianthinite sheet; (e) the iriginite anion topology; (f) the iriginite sheet; (g) the sayrite anion topology; (h) the sayrite sheet. The legend is as in Figure 12 except that non-U^{6+} pentagonal bipyramids, octahedra, and square pyramids are shaded with broken parallel lines.

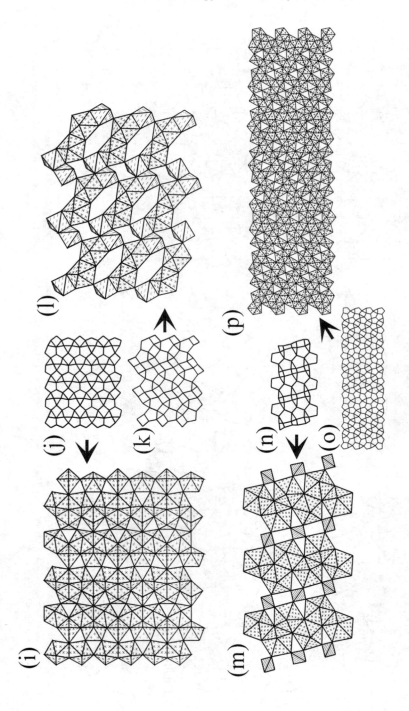

Figure 25, continued. Sheets based upon anion topologies that contain triangles, squares and pentagons. (i) the curite sheet; (j) the curite anion topology; (k) the vandenbrandeite anion topology; (l) the vandenbrandeite sheet; (m) the zippeite sheet; (n) the zippeite anion topology; (o) the wölsendorfite anion topology; (p) the wölsendorfite sheet. The legend is as in Figure 12 except that non-U^{6+} pentagonal bipyramids, octahedra, and square pyramids are shaded with broken parallel lines.

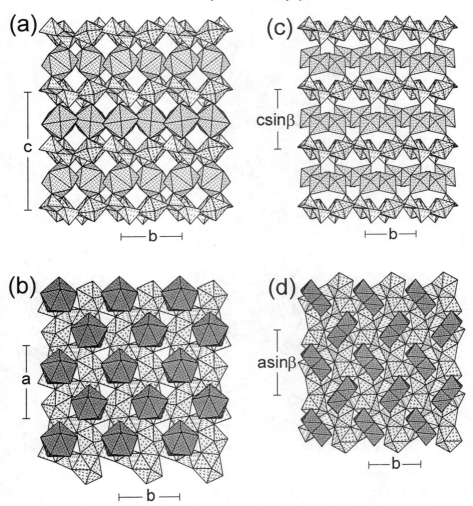

Figure 26. Structures that contain the francevillite-type sheet. (a) francevillite and curienite projected along [100]; (b) francevillite and curienite projected along [001]; (c) sengierite projected along [100]; (d) sengierite projected along [001]. The legend is as in Figure 12.

Francevillite (Mereiter, 1986d) and curienite (Borène and Cesbron, 1971) are isostructural minerals that respectively contain Ba and Pb in their interlayers (Fig. 26a,b). The structure solution for curienite (Borène and Cesbron, 1971) failed to reveal the position of one H_2O group, but it is most likely part of the $Pb^{2+}\phi_9$ polyhedron, as is the case in francevillite. Each divalent cation polyhedron contains nine ligands; of these, two are O_{Ur} atoms, one each from each adjacent sheet, two are O atoms that are apical ligands of VO_5 square bipyramids of each adjacent sheet, and five are H_2O groups.

The structure of sengierite (Piret et al., 1980) is shown in Figure 26c,d. The interlayer contains $Cu^{2+}\phi_6$ polyhedra that are (4+2) distorted owing to the Jahn-Teller effect. Two octahedra share an equatorial edge to form a dimer; the shared ligands are (OH)⁻ groups. One apical ligand of each $Cu^{2+}\phi_6$ octahedron is the apical ligand of the VO_5

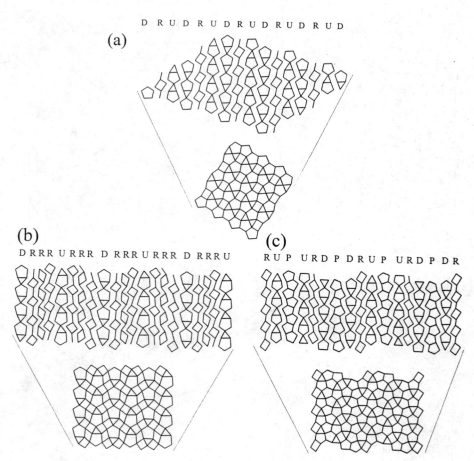

Figure 27. Development of the (a) β-U$_3$O$_8$; (b) iriginite; (c) sayrite anion topologies as chain-stacking sequences.

square pyramid of an adjacent sheet, such that the dimer of Cu^{2+}φ$_6$ octahedra provides a linkage between adjacent uranyl vanadate sheets.

β-U$_3$O$_8$ anion topology. The β-U$_3$O$_8$ sheet anion topology is shown in Figure 25c. This topology is conveniently described by a chain-stacking sequence involving **U**, **D** and **R** chains that are oriented diagonally in Figure 25c, with the repeat sequence **DRUDRU**... (Fig. 27a). Ianthinite is a mixed-valence U^{4+}-U^{6+} mineral which contains a sheet (Fig. 25d) that is based upon the β-U$_3$O$_8$ anion topology (Burns et al., 1997b). Each of the pentagons in the anion topology are populated with a uranyl ion, giving a chain of edge-sharing Urφ$_5$ pentagonal bipyramids. The square sites contain U^{4+}φ$_6$ octahedra, and the octahedra share edges with two adjacent Urφ$_5$ pentagonal bipyramids. The structure contains only H$_2$O groups in the interlayer (Fig. 28a), and some of the H$_2$O sites are partially occupied. Sheets based upon the β-U$_3$O$_8$ anion topology also occur in the structure of β-U$_3$O$_8$.

Recently, Burns and Finch (1999) reported the structure of wyartite and proposed

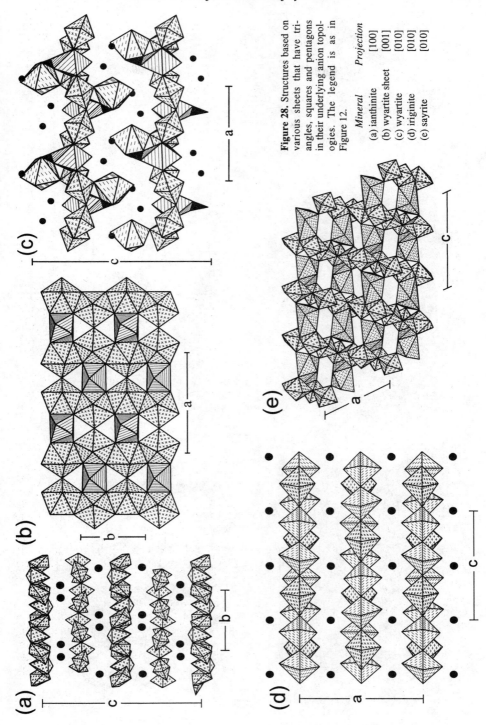

Figure 28. Structures based on various sheets that have triangles, squares and pentagons in their underlying anion topologies. The legend is as in Figure 12.

Mineral	Projection
(a) ianthinite	[100]
(b) wyartite sheet	[001]
(c) wyartite	[010]
(d) iriginite	[010]
(e) sayrite	[010]

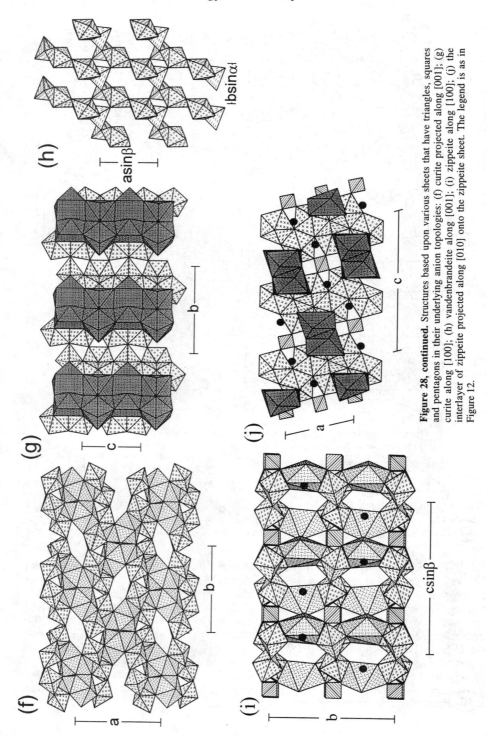

Figure 28, continued. Structures based upon various sheets that have triangles, squares and pentagons in their underlying anion topologies: (f) curite projected along [001]; (g) curite along [100]; (h) vandenbrandeite along [001]; (i) zippeite along [100]; (j) the interlayer of zippeite projected along [010] onto the zippeite sheet. The legend is as in Figure 12.

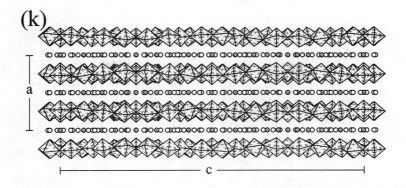

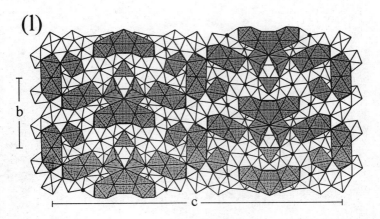

Figure 28, continued. Structures based upon various sheets that have triangles, squares and pentagons in their underlying anion topologies: (k) wölsendorfite projected along [010]; (l) the interlayer of wölsendorfite projected along [100] onto the wölsendorfite sheet. The legend is as in Figure 12.

that it is a mixed-valence U^{5+}-U^{6+} mineral. Wyartite is the first mineral reported to contain essential U^{5+}. The crystallographic evidence for U^{5+} is convincing (Burns and Finch 1999), but independent confirmation is desirable, although difficult to obtain owing to the scarcity of material.

The structure of wyartite contains three symmetrically distinct U positions; two are U^{6+} with typical uranyl pentagonal bipyramidal coordination polyhedra. The U^{5+} is also in pentagonal bipyramidal coordination, but a uranyl ion with U-O bond lengths ~1.8 Å is lacking. The uranyl polyhedra share equatorial edges forming chains that are cross-linked by sharing edges with $U^{5+}\phi_7$ polyhedra, forming sheets parallel to (001) (Fig. 28b). The carbonate groups are attached to the sheet of uranium polyhedra by sharing an edge with the $U^{5+}\phi_7$ polyhedron (Fig. 28b). Owing to the presence of the carbonate group, the wyartite sheet has a unique topology and a new sheet anion topology. However, it is closely related to the β-U_3O_8 anion topology (Fig. 25c), which results if the carbonate groups are omitted. The interlayer contains a $Ca\phi_7$ polyhedron and two H_2O groups that are hydrogen bonded into the structure (Fig. 28c).

Iriginite anion topology. The structure of iriginite (Serezhkin et al. 1973) contains sheets that are based upon the iriginite anion topology illustrated in Figure 25f. This anion topology contains alternating up (**U**) and down (**D**) arrowhead chains, oriented vertically in Figure 25f, that are separated by a combination of three **R** chains, giving the repeat sequence **URRRDRRRU**... (Fig. 27b). The sheet in the structure of iriginite is derived by populating each pentagon in the anion topology with a uranyl ion, resulting in $Ur\phi_5$ pentagonal bipyramids, and two-thirds of the squares correspond to $Mo^{6+}\phi_6$ octahedra, which share edges to form dimers. One third of the squares and all of the triangles of the anion topology are empty. The interlayer contains one symmetrically distinct H_2O group, and adjacent sheets are connected through H bonds (Fig. 28d). No other mineral is known that contains a sheet that is based upon the iriginite anion topology, but the sheets in the synthetic compound $[Ca(UO_2)MoO_4O_{14}]$ (Lee and Jaulmeds 1987) are based on the iriginite anion topology, with half of the pentagons in the anion topology containing uranyl ions and half containing Ca (Burns et al. 1996).

Sayrite anion topology. The sayrite anion topology is shown in Figure 25g. This complex anion topology contains both up (**U**) and down (**D**) arrowhead chains, as well as **P** and **R** chains. The chains are all oriented vertically in Figure 25g, and are arranged such that each **P** chain is flanked by two arrowhead chains with the same sense of direction, giving **UPU** or **DPD** sequences. These two sequences alternate in the anion topology, and are separated by **R** chains, giving the sequence **RUPURDPDRUPU**... (Fig. 27c). The sayrite-type sheet results when every pentagon and square of the anion topology are populated by uranyl ions, giving $Ur\phi_5$ pentagonal bipyramids and $Ur\phi_4$ square bipyramids (Fig. 25h). These sheets occur in the structure of sayrite (Piret et al. 1983) (Fig. 28e), as well as the synthetic phase $K_2[(UO_2)_5O_8](UO_2)$ (Kovba 1972, Burns et al. 1996).

The interlayer in the structure of sayrite contains one symmetrically distinct Pb^{2+} cation. It is coordinated by five O_{Ur} atoms of adjacent sheets of uranyl polyhedra, as well as one equatorial anion of the uranyl polyhedra of an adjacent sheet and two H_2O groups that are located in the interlayer. The $Pb^{2+}\phi_8$ polyhedra are isolated from each other in the interlayer, contrary to the trend observed in most other Pb^{2+} uranyl oxide hydrate minerals in which the $Pb\phi_n$ polyhedra occur as dimers in the interlayers.

Curite anion topology. The structure of curite contains sheets that are based upon the curite anion topology shown in Figure 25j. This unusually complex anion topology cannot be described as a simple chain-stacking sequence using only the chains previously introduced. A new chain is required that contains pentagons, triangles and squares, arranged such that each pentagon shares an edge with a triangle, and the opposite corner of the triangle is shared with a distorted square. These groups of polygons are arranged to form a chain with gaps between the squares and pentagons, as shown in Figure 29a. The chain has a directional sense owing to the presence of an arrowhead (a pentagon and a triangle sharing an edge), and is designated U^m and D^m, for up and down (modified) pointing chains, respectively. The curite anion topology can be characterized by the chain-stacking sequence U^mDU^mD... as illustrated in Figure 29b.

The curite sheet (Fig. 25i) is obtained by populating all pentagons and squares of the anion topology with uranyl ions, giving a sheet that is composed of edge- and corner-sharing $Ur\phi_4$ square bipyramids and $Ur\phi_5$ pentagonal bipyramids (Taylor et al. 1981). Curite is the only mineral that contains this sheet of uranyl polyhedra, although an identical sheet was recently found in a synthetic Sr analogue of curite, $Sr_{2.84}[(UO_2)_4O_4(OH)_3]_2(H_2O)_2$ (Burns and Hill 1999). The structure of curite contains two symmetrically distinct Pb^{2+} cations in the interlayer. Both are coordinated by nine

Figure 29. Development of the curite anion topology as a chain-stacking sequence: (a) the new chain required for the curite anion topology; (b) the chain-stacking sequence.

ligands. Each $Pb^{2+}\phi_9$ polyhedron contains seven O_{Ur} atoms of adjacent sheets, one O atom that is an equatorial ligand of three uranyl polyhedra within the sheet, and one H_2O group that is located in the interlayer. The $Pb^{2+}\phi_9$ polyhedra share faces, forming chains of polyhedra that are parallel to the c axis (Fig. 28f,g).

Vandenbraneite anion topology. The vandenbrandeite anion topology is shown in Figure 25k, and the sheet, which only occurs in vandenbrandeite, is shown in Figure 25l. The anion topology contains arrowhead and **R** chains, and corresponds to the chain-stacking sequence **DRURRRDRURRR**..., as shown in Figure 30. In the vandenbrandeite sheet (Rosenzweig and Ryan 1977b), the pentagons of the anion topology are populated with uranyl anions, giving $Ur\phi_5$ pentagonal bipyramids, and the squares are populated with Cu^{2+} cations, such that each square corresponds to the base of a $Cu^{2+}\phi_5$ square pyramid. The $Cu^{2+}\phi_5$ square pyramids occur as edge-sharing dimers, with the apical ligand of the adjacent pyramids in the dimers pointing in opposite directions. The sheets are linked directly through the apical ligands of the $Cu^{2+}\phi_5$ square pyramids, which are O_{Ur} atoms of adjacent sheets (Fig. 28h).

Zippeite anion topology. The structure of zippeite (Vochten et al. 1995) contains the unusual uranyl sulfate sheet shown in Figure 25m. The corresponding sheet anion topology (Fig. 25n) contains a

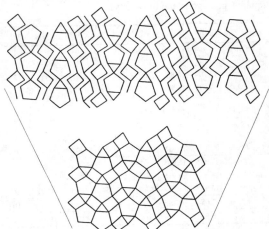

Figure 30. Development of the vandenbrandeite anion topology as a chain-stacking sequence.

chain of edge-sharing pentagons that is two pentagons wide, such that groups of three pentagons share a common vertex. The chains of pentagons are connected by sharing edges with a chain of edge-sharing squares. The sheets are derived by populating each pentagon in the anion topology with a uranyl ion, giving $Ur\phi_5$ pentagonal bipyramids, and every second square along the chain of edge-sharing chains with a SO_4 tetrahedron, such that all four tetrahedral vertices are shared within the sheet. Note that SO_4 tetrahedra and uranyl polyhedra polymerize with each other only by corner-sharing, as the sharing of polyhedral edges would result in excessive repulsion between the two hexavalent cations.

The structure of zippeite contains both K cations and H_2O groups in the interlayer (Fig. 28I,j). The K is distributed over two sites, such that each is 50% occupied, with the two K sites separated by 1.15 Å (Vochten et al. 1995). Each K cation is coordinated by eight ligands, one of which is a H_2O group located in the interlayer.

Wölsendorfite anion topology. The wölsendorfite anion topology (Fig. 25o), which only occurs in wölsendorfite, is the most complex anion topology known from uranyl minerals and synthetic phases. The anion topology contains R, D, P, and U chains, and may be generated as a complex chain-stacking sequence (Burns 1999a). However, on the basis of this approach, Burns (1999) was able to show that the wölsendorfite anion topology contains slabs of the α-U_3O_8 (protasite) and β-U_3O_8 anion topologies, thus it follows that the wölsendorfite sheet is structurally intermediate between the α-U_3O_8 and β-U_3O_8-type sheets.

The wölsendorfite sheet may be derived from the anion topology by populating each square and pentagon with uranyl ions, resulting in uranyl square and pentagonal bipyramids, respectively (Fig. 25p). The sheets are linked through a complex interlayer containing eight symmetrically distinct $Pb\phi_n$ polyhedra as well as two symmetrically distinct H_2O groups that are held in the structure by H bonding only (Fig. 28k,l). The $Pb\phi_n$ polyhedra are polymerized by the sharing of vertices, edges and faces, forming complex chains parallel to [001].

Table 6. Uranyl mineral structures containing sheets based upon anion topologies containing triangles, squares, pentagons and hexagons

Name	Formula	S. G.	a (Å)	b (Å)	c (Å)	α (°)	β (°)	γ (°)	Ref.
phosph-uranylite	$KCa(H_3O)_3(UO_2)[(UO_2)_3(PO_4)_2O_2]_2(H_2O)_8$	$Cmcm$	15.899	13.740	17.300				1
upalite	$Al[(UO_2)_3(PO_4)_2O(OH)](H_2O)_7$	$P2_1/a$	13.704	16.82	9.332		111.5		2
françoisite-(Nd)	$Nd[(UO_2)_3(PO_4)_2O(OH)](H_2O)_6$	$P2_1/c$	9.298	15.605	13.668		112.77		3
dewindtite	$Pb_3[H(UO_2)_3O_2(PO_4)_2]_2(H_2O)_{12}$	$Bmmb$	16.031	17.264	13.605				4
vanmeers-scheite	$U(OH)_4[(UO_2)_3(PO_4)_2(OH)_2](H_2O)_4$	$P2_1mn$	17.06	16.76	7.023				5
dumontite	$Pb_2[(UO_2)_3(PO_4)_2O_2](H_2O)_5$	$P2_1/m$	8.118	16.819	6.983		109.03		6
phurcalite	$Ca_2[(UO_2)_3O_2(PO_4)_2](H_2O)_7$	$Pbca$	17.415	16.035	13.598				7
phuralumite	$Al_2[(UO_2)_3(PO_4)_2(OH)_2](OH)_4(H_2O)_{10}$	$P2_1/a$	13.836	20.918	9.428		112.44		8
althupite	$AlTh(UO_2)[(UO_2)_3(PO_4)_2O(OH)]_2(OH)_3(H_2O)_{15}$	$P\bar{1}$	10.953	18.567	13.504	72.64	68.20	84.21	9
guilleminite	$Ba[(UO_2)_3(SeO_3)_2O_2](H_2O)_3$	$P2_1nm$	7.084	7.293	16.881				10
johannite	$Cu[(UO_2)_2(SO_4)_2(OH)_2](H_2O)_8$	$P\bar{1}$	8.903	9.499	6.812	109.87	112.01	100.40	11
roubaultite	$[Cu_2(UO_2)_3(CO_3)_2O_2(OH)_2](H_2O)_4$	$P\bar{1}$	7.767	6.924	7.850	92.16	90.89	93.48	12

References: (1) Demartin et al. (1991), (2) Piret and Declercq (1983), (3) Piret et al. (1988), (4) Piret et al. (1990), (5) Piret and Deliens (1982), (6) Piret and Piret-Meunier (1988), (7) Atencio et al. (1991), (8) Piret et al. (1979), (9) Piret and Deliens (1987), (10) Cooper and Hawthorne (1995), (11) Mereiter (1982c), (12) Ginderow and Cesbron (1985).

Anion topologies with triangles, squares, pentagons and hexagons

Uranyl minerals with structures that contain sheets of polyhedra that are based upon anion topologies that contain triangles, squares, pentagons and hexagons are listed in Table 6 (above).

Phosphuranylite anion topology. The phosphuranylite anion topology shown in Figure 31a is the basis for sheets that occur in 11 minerals (Table 6), as well as one synthetic phase (Burns et al. 1996). The anion topology contains pairs of edge-sharing pentagons that share edges with hexagons to form chains that have twice as many pentagons as hexagons, and which have alternating hexagons and pairs of pentagons along the chain length. Adjacent chains of pentagons and hexagons are offset so that hexagons of a given chain are adjacent to pentagons on either side. The chains of pentagons and hexagons are separated by a chain of edge-sharing squares and triangles.

There are five distinct populations of the phosphuranylite anion topology, each of which is shown in Figure 31. Note that in each case only the triangles, pentagons and hexagons are populated; no structure is known in which the squares of this anion topology are occupied. The sheets shown in Figures 31b,c,d are derived from the phosphuranylite anion topology by populating each pentagon and hexagon with a uranyl ion, giving $Ur\phi_5$ pentagonal bipyramids and $Ur\phi_6$ hexagonal bipyramids, respectively, and the triangles are the faces of tetrahedra. In each of these sheets, each tetrahedron shares an edge with a $Ur\phi_6$ hexagonal bipyramid, and the opposite corner with a $Ur\phi_5$ pentagonal bipyramid. The apical ligand of each tetrahedron is only one-connected within the sheet. Each sheet has the composition $(UO_2)_3(PO_4)\phi_2$ and sheets differ only in the orientation of the tetrahedra. Consider the chains of edge-sharing triangles and squares in the anion topology (vertical in Fig. 31a): in the phosphuranylite-type sheet (Fig. 31b), tetrahedra point up (u) and down (d) with the sequence uudduu..., and pairs of tetrahedra that share edges with a common $Ur\phi_6$ hexagonal bipyramid both point in the same direction. The phosphuranylite-type sheet occurs in the structures of phosphuranylite (Demartin et al. 1991), upalite (Piret and Declercq 1983), françoisite-(Nd) (Piret et al. 1988), and dewindtite (Piret et al. 1990).

In the vanmeersscheite-type sheet (Fig. 31c), going along the chain of edge-sharing triangles and squares, tetrahedra alternate up and down with the sequence udud..., and two tetrahedra that share edges with a common $Ur\phi_6$ hexagonal bipyramid point in the same direction. This sheet occurs in both vanmeersscheite (Piret and Deliens 1982) and dumontite (Piret and Piret-Meunier 1988). The phurcalite-type sheet (Fig. 31d) contains tetrahedra that are oriented in the sequence uudduu... along chains of edge-sharing triangles and squares, but in this case two tetrahedra that share their edges with a common $Ur\phi_6$ hexagonal bipyramid point in opposite directions. This sheet occurs in phurcalite (Atencio et al. 1991), phuralumite (Piret et al. 1979) and althupite (Piret and Deliens 1987).

The structure of phosphuranylite (Fig. 32a,b) is remarkable in that it contains all three types of uranyl coordination polyhedra that occur in minerals. $Ur\phi_5$ pentagonal bipyramids and $Ur\phi_6$ hexagonal bipyramids occur in the uranyl phosphate sheet, and a $Ur\phi_4$ square bipyramid occurs within the interlayer. The interlayer uranyl ion is oriented approximately parallel to the uranyl phosphate sheet, and apical ligands of phosphate tetrahedra of both adjacent sheets act as equatorial ligands of the square bipyramid. As such, the structure may be regarded as a framework of polyhedra of higher bond valence, but it is included here as a sheet structure because of the occurrence of the uranyl phosphate sheet in other structures. The interlayer of phosphuranylite also involves partially occupied sites that contain Ca and K cations. The Ca cations are coordinated by

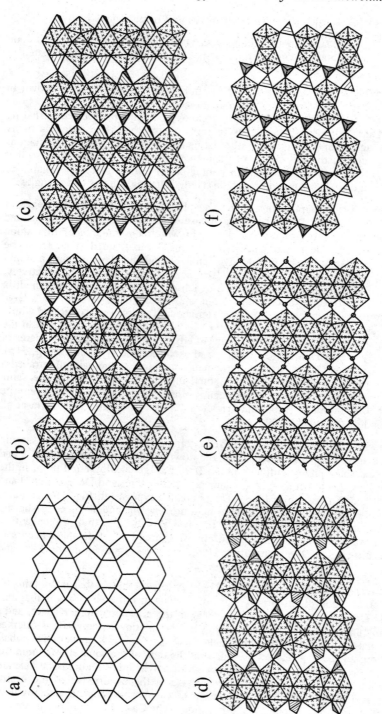

Figure 31. Sheets based upon the phosphuranylite anion topology that contains triangles, squares, pentagons and hexagons: (a) the phosphuranylite anion topology; (b) the phosphuranylite-type sheet that occurs in phosphuranylite, upalite, françoisite-(Nd), and dewindtite; (b) the vanmeersscheite-type sheet that occurs in vanmeersscheite and dumontite; (c) the phurcalite-type sheet that occurs in phurcalite, phuralumite and althupite; (d) the guilleminite sheet; (e) the johannite sheet. The legend is as in Figure 12.

two O_{Ur} atoms and six H_2O groups, with symmetrically equivalent pairs of Ca polyhedra sharing a face to form a dimer (Fig. 32b). The separation of the Ca positions in the dimer is only ~2.9 Å, thus both sites cannot be occupied locally. The K cation is coordinated by an H_2O group as well as eight O_{Ur} atoms; two are from the interlayer uranyl ion (one on either side of the $K\phi_9$ polyhedron) and six are provided by adjacent uranyl phosphate sheets. The interlayer also contains a single H_2O group that is held in place by H bonds only.

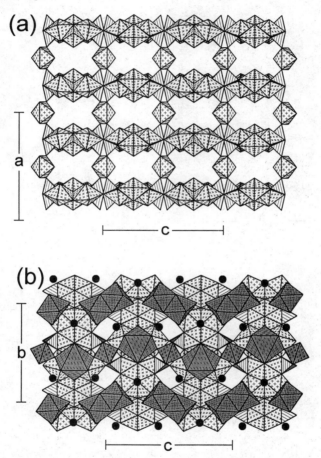

Figure 32. Structures that contain the phosphuranylite-type sheet: (a) phosphuranylite projected along [010] and (b) phosphuranylite along [100]. The legend is as in Figure 12.

The structure of upalite (Fig. 32c,d) contains the phosphuranylite-type sheet (Fig. 31b), and the interlayer contains two symmetrically distinct Al positions, each of which is octahedrally coordinated. One Al position is coordinated by four H_2O groups and two O atoms that are apical ligands of phosphate tetrahedra in both adjacent sheets, thus the $Al\phi_6$ octahedron connects adjacent sheets. All ligands of the other $Al\phi_6$ octahedron are H_2O groups, and this octahedron is bonded in position within the interlayer by H bonds only. There are also two symmetrically distinct H_2O groups in the interlayer that are held by H bonds only.

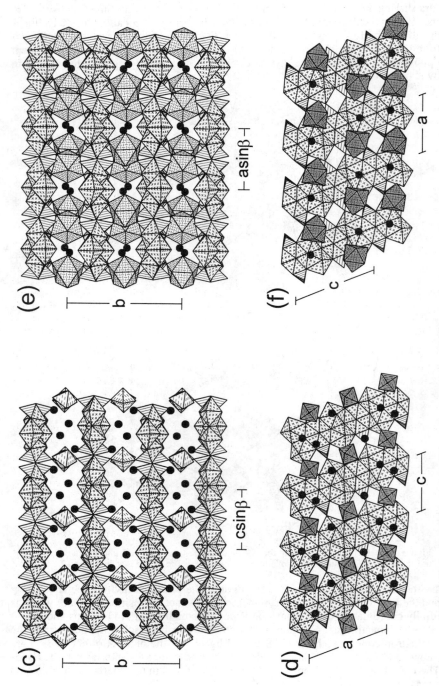

Figure 32, continued. Structures that contain the phosphuranylite-type sheet. (c) upalite projected along [010]; (d) upalite along [100]; (e) françoisite along [001]; (f) françoisite along [010]. The legend is as in Figure 12.

Françoisite is a REE-bearing uranyl phosphate that contains the phosphuranylite-type sheet. Chemical analysis showed that it contains Nd, Y, La, Ce, Pr, Sm, and Dy, and the structure determination indicates that each of these occur at a single site in the structure (Piret et al. 1988). The structure is illustrated in Figures 32e,f. The interlayer REE site is coordinated by nine ligands: one O_{Ur} atom from each adjacent sheet, one apical O atom of the phosphate tetrahedra of each adjacent sheet, and five H_2O groups. In addition, there is one symmetrically distinct H_2O group that is held in the interlayer by H bonds only.

Dewindtite contains the phosphuranylite-type sheet, with two symmetrically distinct Pb sites in the interlayer. Each Pb^{2+} cation is coordinated by eight ligands that correspond to O_{Ur} anions of adjacent sheets and interlayer H_2O groups. The Pb cations and H_2O groups in the interlayer are disordered. In addition to the H_2O groups that are bonded to Pb, the interlayer contains three symmetrically distinct H_2O groups that are H bonded into the structure.

The structures of vanmeersscheite and dumontite both contain the vanmeersscheite-type sheet of uranyl and phosphate polyhedra shown in Figure 31c. The structure of vanmeersscheite is imperfectly known (Piret and Deliens 1982) and requires further refinement. The structure appears to contain octahedrally coordinated U^{6+} in the interlayer. The refined bond lengths of the distorted octahedron are not consistent with the presence of a uranyl ion. There is also a U^{6+} cation in the interlayer of the structure of phosphuranylite, but in that case a uranyl ion is unambiguous. In vanmeersscheite, the precision of the structure refinement may be inadequate to demonstrate the presence of a uranyl ion, or positional disorder may obscure the uranyl ion. If future structure refinements demonstrate the absence of a uranyl ion in the interlayer of vanmeersscheite, this will be the first U^{6+} mineral structure in which U^{6+} does not form part of a uranyl ion.

There are two symmetrically distinct U^{6+} positions in the interlayer of vanmeersscheite, and both are coordinated by six ligands. Two ligands of each $U^{6+}\phi_6$ polyhedron are apical ligands of phosphate tetrahedra of adjacent sheets, providing linkage of the sheets (Fig. 33a). The remaining ligands of the $U^{6+}\phi_6$ polyhedron are either (OH)⁻ or H_2O.

The structure of dumontite contains one symmetrically distinct Pb^{2+} cation in the interlayer. The Pb^{2+} cation is coordinated by three O_{Ur} atoms of adjacent sheets, two apical atoms of phosphate tetrahedra in adjacent sheets, and three H_2O groups. Two $Pb^{2+}\phi_8$ polyhedra share an edge to form a dimer, with the ligands of the shared edge corresponding to the apical atoms of the phosphate tetrahedra of sheets on both sides (Fig. 33c). The Pb dimers share a common H_2O group to form chains of $Pb^{2+}\phi_8$ polyhedra that are parallel to the b direction (Fig. 33d).

The structures of phurcalite, phuralumite, and althupite each contain the phurcalite-type sheet of uranyl and phosphate polyhedra shown in Figure 31d. However, the interlayers of these structures are different. In phurcalite, there are two symmetrically distinct Ca cations located in the interlayer, as shown in Figure 34a,b. One of these is coordinated by seven ligands: two are O_{Ur} atoms, one from each adjacent uranyl phosphate sheet, two are apical O atoms of phosphate tetrahedra, both of which occur in the same sheet, and three are H_2O groups that are located in the interlayer. The other Ca position is coordinated by eight ligands; two are O_{Ur} atoms that are both part of the same sheet, one is an apical ligand of a phosphate tetrahedron, and four are H_2O groups. Two $Ca\phi_n$ polyhedra share an edge, forming a dimer that bridges between adjacent uranyl phosphate sheets (Fig. 34b). The structure of phurcalite contains a single H_2O group that is held in position in the interlayer by H bonds only.

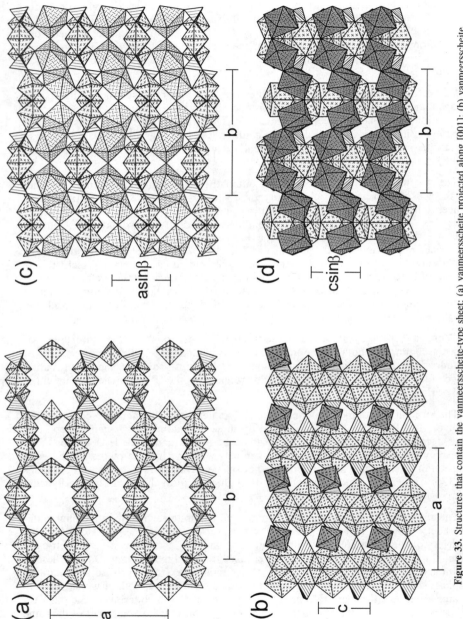

Figure 33. Structures that contain the vanmeersscheite-type sheet: (a) vanmeersscheite projected along [001]; (b) vanmeersscheite projected along [010]; (c) dumontite projected along [001]; (d) dumontite projected along [100]. The legend is as in Figure 12.

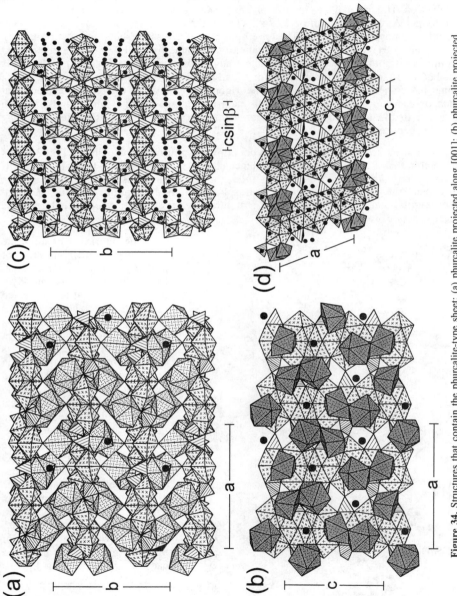

Figure 34. Structures that contain the phurcalite-type sheet: (a) phurcalite projected along [001]; (b) phurcalite projected along [010]; (c) phuralumite projected along [100]; (d) phuralumite projected along [010]. The legend is as in Figure 12.

The structure of phuralumite contains Alϕ_n polyhedra in the interlayer, as well as nine H$_2$O groups that are held in the structure by H bonds only. The Al cations are in two different coordinations; one is essentially a trigonal bipyramid and the other is an octahedron. The trigonal bipyramid contains four Al-ϕ bond lengths that are ~1.77 Å, and one longer Al-ϕ bond length of 2.19 Å. Two of the ligands are O atoms that are apical ligands of phosphate tetrahedra that are both part of the same sheet, and the remaining ligands are (OH)$^-$ groups. The Alϕ_6 octahedron contains five (OH)$^-$ groups and one H$_2$O group, and is not directly connected to the uranyl phosphate sheets (Fig. 34c). Rather, this octahedron shares an edge with a symmetrically equivalent octahedron, and the resulting dimer of octahedra shares edges with the trigonal bipyramid on the other side, such that each octahedron shares one edge with each of the two trigonal bipyramids that are attached to the adjacent sheets of uranyl and phosphate polyhedra, thus providing linkages between the sheets.

The structure of althupite, which contains the phurcalite-type sheet (Fig. 31d), is unusual in that it contains three high-valence cations in the interlayer: U^{6+}, Th, and Al. If these cations are included in the structural unit, an open framework structure results. However, the sheet of uranyl and phosphate polyhedra dominates the structural connectivity, and is known from other structures, so the structure is classified here as a sheet structure.

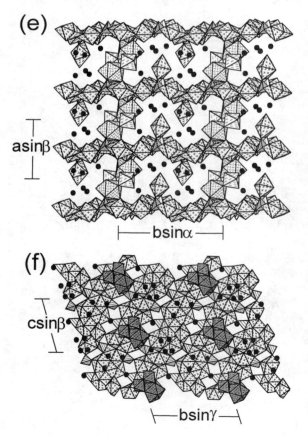

The interlayer of althupite contains three different types of polyhedra. It is rather unusual for U^{6+} to occur in the interlayer of a structure that is dominated by sheets of polyhedra, and this is the only known case of a $Ur\phi_5$ pentagonal bipyramid in the interlayer. The uranyl ion is oriented roughly parallel to the plane of the adjacent sheet, and two equatorial ligands of the pentagonal bipyramid are apical ligands of phosphate tetrahedra in one adjacent sheet (Fig. 34e,f). The remaining three equatorial ligands are H$_2$O groups. The Th cation is co-ordinated by nine ligands, three of which are shared with an adjacent uranyl phosphate sheet; these correspond to one O$_{Ur}$ atom and two apical ligands of phosphate tetrahedra. Three each of the remaining ligands

Figure 34, continued. Structures that contain the phurcalite-type sheet: (e) althupite projected along [001]; (f) althupite projected along [100]. The legend is as in Figure 12.

are (OH)⁻ and H_2O groups. The Al cation is octahedrally coordinated by four (OH)⁻ and two H_2O groups. Two symmetrically equivalent $Al\phi_6$ octahedra share an edge, and the resulting dimer shares two edges on either side with adjacent $Th\phi_9$ polyhedra (Fig. 34e,f). Thus, a combination of two $Th\phi_9$ and two $Al\phi_6$ polyhedra provides linkage between adjacent sheets of uranyl and phosphate polyhedra. There are seven symmetrically distinct H_2O groups in the interlayer that are held in position by H bonds only; the structure must include a very complex network of H bonds.

The structure of guilleminite (Cooper and Hawthorne 1995) contains the sheet shown in Figure 31e. The sheet, which is unique to guilleminite, is obtained from the phosphuranylite anion topology (Fig. 31a) by populating each pentagon and hexagon with a uranyl ion, giving $Ur\phi_5$ pentagonal bipyramids and $Ur\phi_6$ hexagonal bipyramids, respectively, and populating triangles with $Se^{4+}O_3$ triangles. The $Se^{4+}O_3$ triangles are distorted, with the Se^{4+} cation displaced above the plane of the anions, owing to the presence of a lone pair of electrons on the Se^{4+} cation. The interlayer contains a single symmetrically distinct Ba cation, as well as two symmetrically distinct H_2O groups (Fig. 35a,b). Each Ba cation is coordinated by 10 ligands, three of which are interlayer H_2O groups. The remaining ligands are each O_{Ur} atoms of adjacent sheets of polyhedra; three in one sheet and four in the other.

The structure of johannite (Mereiter 1982c) is based upon the unusually open sheet shown in Figure 31f. This sheet is obtained from the phosphuranylite anion topology (Fig. 31a) by populating each pentagon with a uranyl ion, giving $Ur\phi_5$ pentagonal bipyramids, and each triangle of the anion topology is the face of a SO_4 tetrahedron. The hexagons of the anion topology remain empty. An identical sheet is found in the synthetic phase $Sr[(UO_2)_2(CrO_4)_2(OH)_2](H_2O)_8$ (Serezhkin et al. 1982; Burns et al. 1996), but with chromate tetrahedra replacing sulfate tetrahedra in johannite. Note that the SO_4 tetrahedra in johannite only share corners with the uranyl polyhedra; population of the hexagons of the anion topology with uranyl ions would necessitate the sharing of a tetrahedral edge with a uranyl polyhedron, which is presumably unfavorable owing to the repulsion that would occur between two hexavalent cations.

The interlayer of the structure of johannite (Fig. 35c,d) contains a single symmetrically distinct $Cu^{2+}\phi_6$ octahedron that shows the usual (4+2) distortion owing to the Jahn-Teller effect. The equatorial ligands of the octahedron are H_2O groups. The apical ligands of the $Cu^{2+}\phi_6$ octahedron are O_{Ur} atoms from both adjacent sheets. Thus, cross linking between the sheets is both through the Cu^{2+} octahedra and H bonds. The structure contains two additional symmetrically distinct H_2O groups that are held in the interlayer by H bonds only.

Roubaltite anion topology. The structure of roubaultite (Ginderow and Cesbron 1985) contains the sheet shown in Figure 36b, which has the anion topology shown in Figure 36a. The roubaultite anion topology is similar to the phosphuranylite anion topology in that it contains the same chain of edge-sharing pentagons and hexagons. However, the two anion topologies differ in the way the chains are aligned and attached. In the roubaultite anion topology, chains of pentagons and hexagons are separated by a chain of edge-sharing squares that share corners with pentagons. This configuration results in three triangles that share edges with each other, share edges with adjacent pentagons and hexagons, and share corners with adjacent squares.

The roubaultite sheet is obtained by populating each pentagon and hexagon of the anion topology with uranyl ions, giving $Ur\phi_5$ pentagonal bipyramids and $Ur\phi_6$ hexagonal

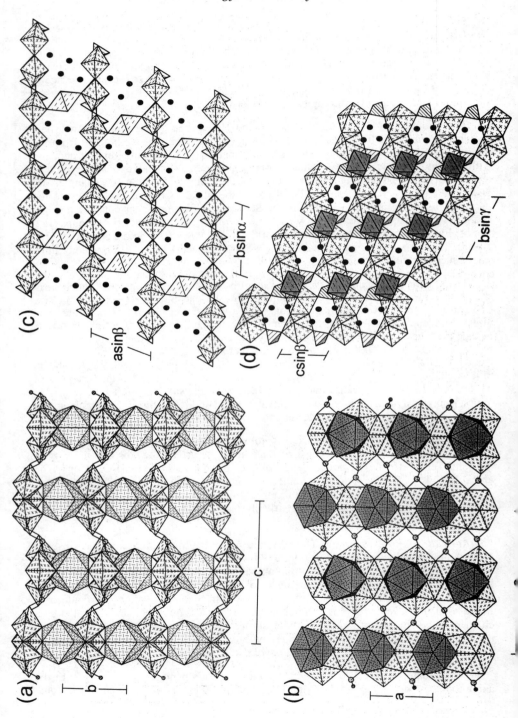

Figure 35 (opposite page). Structures based upon the guillimenite and johannite sheets: (a) guilleminite projected along [100]; (b) guilleminite projected along [010]; (c) johannite projected along [001]; (d) johannite projected along [100]. The legend is as in Figure 12.

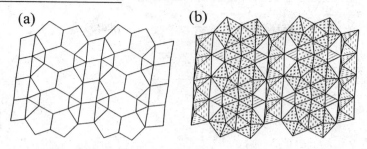

Figure 36. The roubaultite anion topology (a) and sheet (b). The legend is as in Figure 25.

bipyramids. The triangles that share an edge with the hexagons are populated by CO_3 triangles, and each square contains a $Cu^{2+}\phi_6$ octahedron that is (4+2) distorted owing to the Jahn-Teller effect. The $Cu^{2+}\phi_6$ octahedron contains two H_2O groups, located at equatorial positions of the distorted octahedron, that are only one-connected within the sheet. These H_2O groups donate H bonds that are accepted by O_{Ur} atoms of adjacent sheets of polyhedra (Fig. 37). There are no interlayer constituents in the roubaultite structure.

Other anion topologies containing hexagons

The structures of umohoite and rutherfordine (Table 7) contain sheets that have anion topologies containing hexagons.

α-UO₃ anion topology. The structure of synthetic α-UO_3 contains only $Ur\phi_6$ hexagonal bipyramids (Loopstra and Cordfunke 1966), which share edges to form sheets that have the anion topology shown in Figure 38b. About 10 synthetic structures are known that have sheets based upon this anion topology (Burns et al. 1996). However, umohoite (Makarov and Anikina 1963) (Table 7) is the only mineral that has been demonstrated to have sheets of edge-sharing hexagonal bi-pyramids, and the corres-ponding sheet is shown in Figure 38a. In this sheet, only

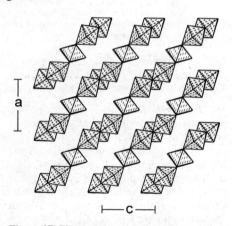

Figure 37. The structure of roulbaultite projected along [010]. The legend is as in Figure 25.

Table 7. Miscellaneous uranyl mineral structures containing sheets based upon anion topologies containing hexagons

Name	Formula	S. G.	a (Å)	b (Å)	c (Å)	α (°)	β (°)	γ (°)	Ref.
umohoite	[$(UO_2)(MoO_4)$]-$(H_2O)_4$	$P2_1/c$	6.32	7.50	57.8		94		1
rutherfordine	[$(UO_2)(CO_3)$]	$Pmmn$	4.845	9.205	4.296				2

References: (1) Makarov and Anikina (1963), (2) Finch et al. (1999).

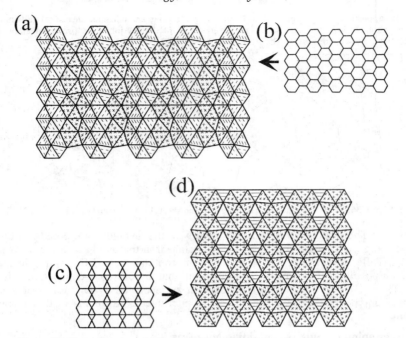

Figure 38. Sheets based upon anion topologies that contain hexagons: (a) the umohoite sheet; (b) the α-UO$_3$ anion topology; (c) the rutherfordine anion topology; (d) the rutherfordine sheet. The legend is as in Figure 12 except that CO$_3$ triangles are shaded with parallel lines.

half of the hexagons of the anion topology are populated by uranyl ions; the other half are populated by Mo^{6+}, giving Mo^{6+}O$_8$ hexagonal bipyramids. The structure contains only H$_2$O groups in the interlayer (Fig. 39a), and adjacent sheets are linked by H bonds.

Rutherfordine anion topology. The structure of rutherfordine (Finch et al. 1999) contains sheets with the anion topology shown in Figure 38c. This anion topology contains chains of edge-sharing hexagons that share corners, creating pairs of edge-sharing triangles. The rutherfordine-type sheet is obtained by populating all hexagons in the anion topology with uranyl ions, giving $Ur\phi_6$ hexagonal bipyramids, and one half of the triangles are populated by CO$_3$ groups (Fig. 38d). An identical sheet occurs in the structure of synthetic [(UO$_2$)(SeO$_3$)] (Loopstra and Brandenburg 1978), with the CO$_3$ groups in rutherfordine replaced with distorted Se^{4+}O$_3$ triangles. The sheets in rutherfordine are held together by van der Waals bonds, although it has been noted that many samples of rutherfordine contain significant amounts of H (Cejka and Urbanec 1988).

Structures containing infinite chains

Burns et al. (1996) listed 19 uranyl phases that have structures based upon infinite chains of polymerized polyhedra of higher bond valence. Eight types of chains were recognized. Only six uranyl minerals are known that have structures based upon chains (Table 8), and there are five distinct chain types amoung these six minerals (Fig. 40).

The chain shown in Figure 40a, which occurs in moctezumite (Swihart et al. 1993), contains $Ur\phi_5$ pentagonal bipyramids as well as Te^{4+}O$_3$ triangles that are strongly

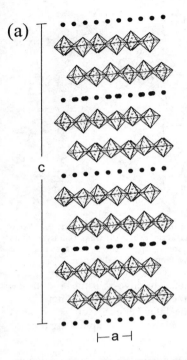
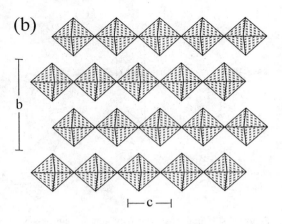

Figure 39. The structures of umohoite and rutherfordine: (a) umohoite projected along [010]; (b) rutherfordine projected along [100]. The legend is as in Figure 38.

Table 8. Uranyl mineral structures based upon infinite chains

Name	Formula	S. G.	a (Å)	b (Å)	c (Å)	α (°)	β (°)	γ (°)	Ref.
moctezumite	PbTe$_2$O$_3$[(UO$_2$)O$_3$]	$P2_1/c$	7.813	7.061	13.775		93.71		1
derriksite	Cu$_4$[(UO$_2$)(SeO$_3$)$_2$](OH)$_6$	$Pn2_1m$	5.570	19.088	5.965				2
demesmaekerite	Pb$_2$Cu$_5$[(UO$_2$)(SeO$_3$)$_3$]$_2$(OH)$_6$(H$_2$O)$_2$	$P\bar{1}$	11.955	10.039	5.639	89.78	100.36	91.34	3
walpurgite	Bi$_4$O$_4$[(UO$_2$)(AsO$_4$)$_2$](H$_2$O)$_2$	$P\bar{1}$	7.135	10.426	5.494	101.47	110.82	88.20	4
ortho-walpurgite	Bi$_4$O$_4$[(UO$_2$)(AsO$_4$)$_2$](H$_2$O)$_2$	$Pbcm$	5.492	13.324	20.685				5
deloryite	Cu[(UO$_2$)(MoO$_8$)](OH)$_6$	$C2/m$	19.91	6.116	5.520		104.18		6
parsonsite	Pb$_2$[(UO$_2$)(PO$_4$)$_2$]	$P\bar{1}$	6.842	10.383	6.670	101.265	98.174	86.378	7

References: (1) Swihart et al. (1993), (2) Ginderow and Cesbron (1983a), (3) Ginderow and Cesbron (1983b), (4) Mereiter (1982b), (5) Krause et al. (1995), (6) Pushcharovsky et al. (1996), (7) Burns (1999b).

distorted owing to the presence of a lone pair of electrons on the Te^{4+} cation. $Ur\phi_5$ pentagonal bipyramids share equatorial edges to form chains, and Te^{4+}O$_3$ triangles share edges and corners with the chains of $Ur\phi_5$. The chains are oriented parallel to the b axis in the structure of moctezumite (Fig. 41a). The chains are linked by sharing polyhedral elements with a single symmetrically distinct Pb$^{2+}\phi_6$ polyhedron.

The structure of derriksite (Ginderow and Cesbron 1983a) contains the chain shown in Figure 40b. The chain contains both $Ur\phi_4$ square bipyramids and Se^{4+}O$_3$ triangles, and the latter are strongly distorted owing to the presence of a lone pair of electrons on the Se^{4+} cation. The chain involves only corner-sharing of polyhedra, with Se^{4+}O$_3$ triangles bridging between adjacent $Ur\phi_4$ square bipyramids that have their uranyl ions oriented perpendicular to the chain length. The chains are parallel to the c axis in the structure, and

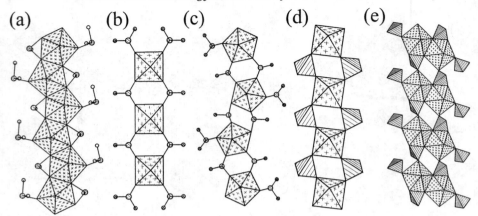

Figure 40. Infinite chains of polyhedra of higher bond valence that occur in the structures of uranyl minerals: (a) moctezumite chain; (b) derriksite chain; (c) demesmaekerite chain; (d) walpurgite-type chain that occurs in walpurgite, orthowalpurgite and deloyrite; (e) the parsonsite chain. The legend is as in Figure 12 except that Se^{4+} cations are shown as open circles, and O atoms are shown as shaded circles.

the non-chain O atoms of the $Se^{4+}O_3$ triangles are linked to edge-sharing brucite-like sheets of $Cu^{2+}\phi_6$ octahedra (Fig. 41b,c). There are three symmetrically distinct $Cu^{2+}\phi_6$ octahedra, each of which is (4+2) distorted owing to the Jahn-Teller effect. The ligands shared between the $Se^{4+}O_3$ triangles and the $Cu^{2+}\phi_6$ octahedra are apical ligands of elongated $Cu^{2+}\phi_6$ octahedra.

The chain shown in Figure 40c is only known to occur in the structure of demesmaekerite (Ginderow and Cesbron 1983b). The chain contains $Ur\phi_5$ pentagonal bipyramids, as well as $Se^{4+}O_3$ triangles that are distorted by the presence of a lone pair of electrons on the Se^{4+} cation. Adjacent $Ur\phi_5$ pentagonal bipyramids have their uranyl ions oriented approximately perpendicular to the chain length, and $Ur\phi_5$ polyhedra are linked by sharing corners with $Se^{4+}O_3$ triangles, such that each $Ur\phi_5$ pentagonal bipyramid is linked to four SeO_3 triangles. The fifth equatorial ligand of the $Ur\phi_5$ pentagonal bipyramid is shared with a $Se^{4+}\phi_3$ triangle that is one-connected to the chain. The structure also contains three symmetrically distinct $Cu^{2+}\phi_6$ octahedra. Two of these are (4+2) distorted due to the Jahn-Teller effect, whereas the other $Cu^{2+}\phi_6$ octahedron adopts an unusual (2+4) octahedral distortion that involves compression of the apical ligands. The $Cu^{2+}\phi_6$ octahedra link by sharing edges to form zigzag chains that are parallel to the c axis; adjacent chains of $Cu^{2+}\phi_6$ octahedra link by corner-sharing to form sheets parallel to (010). The structure contains a single symmetrically distinct $Pb^{2+}\phi_6$ polyhedron that is linked to the sheet of $Cu^{2+}\phi_6$ octahedra. The $Pb\phi_6$-$Cu^{2+}\phi_6$ sheets are cross-linked by the uranyl selenite chains (Fig. 41d).

The structures of walpurgite (Mereiter 1982b) and orthowalpurgite (Krause et al. 1995) are based upon the chain shown in Figure 40d. The chain contains $Ur\phi_4$ square bipyramids and AsO_4 tetrahedra, and each $Ur\phi_4$ square bipyramid shares all four corners with AsO_4 tetrahedra that provide the linkages between adjacent uranyl polyhedra along the chain length. The uranyl ions of the $Ur\phi_4$ square bipyramids are oriented roughly perpendicular to the chain length. The structures of both orthowalpurgite and walpurgite

---

Figure 41 (opposite page). Uranyl minerals that contain infinite chains: (a) moctezumite projected along [100]; (b) derriksite along [001]; (c) derriksite along [010]; (d) demesmaekerite along [100]; (e) walpurgite along [100]; (f) orthowalpurgite along [010]. The legend is as in Figure 40.

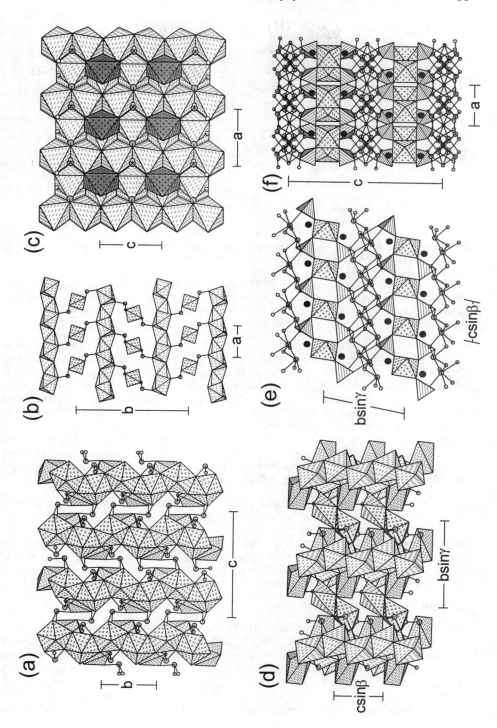

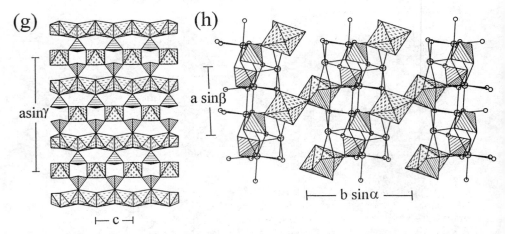

Figure 41, continued. Uranyl minerals that contain infinite chains: (g) deloryite projected along [010]; (h) parsonsite projected along [001]. The legend is as in Figure 40.

contain two symmetrically distinct Bi polyhedra, which are coordinated by six or seven ligands. The Biϕ_n polyhedra link to form sheets that are in turn linked by the uranyl arsenate chains (Fig. 41e,f). The two structures are polymorphs, and differ mainly in the alignment of adjacent uranyl arsenate chains.

The structure of deloryite (Pushcharovsky et al. 1996) contains the chain shown in Figure 40d, with $Ur\phi_4$ square bipyramids and $Mo^{6+}O_4$ tetrahedra. Note, however, that Pushcharovsky et al. (1996) argued that an unusually long Mo-O separation of 2.58 Å may correspond to a bond; if this is the case, the Mo^{6+} cation is in a distorted square-pyramidal coordination, rather than a tetrahedron. The structure contains two symmetrically distinct $Cu^{2+}\phi_6$ octahedra, both of which show usual (4+2) distortions owing to the Jahn-Teller effect. One $Cu\phi_6$ octahedron contains five (OH)⁻ groups, whereas the other contains only four. The remaining ligands are O atoms bonded to Mo^{6+} cations. The $Cu\phi_6$ octahedra share edges to form brucite-like sheets similar to those found in derriksite. The sheets of octahedra are linked to the uranyl molybdate chains by sharing apical O atoms of the (4+2)-distorted octahedra with MoO_4 tetrahedra (Fig. 41g).

The structure of parsonsite (Burns 1999b) contains the chain shown in Figure 40e, composed of uranyl pentagonal bipyramids and phosphate tetrahedra. The chain contains dimers of uranyl pentagonal bipyramids that share an edge. Each uranyl polyhedron also shares an edge with a phosphate tetrahedron, and the dimers of uranyl polyhedra are linked by phosphate-O_{eq} vertex-sharing to form a chain. There are two symmetrically distinct Pb^{2+} cations that are coordinated by nine and six O atoms, respectively. Layers of Pb polyhedra parallel to (010) link the uranyl phosphate chains (Fig. 41h).

Structures containing finite clusters

At least 23 uranyl phases have structures that are based upon finite clusters of polyhedra of higher bond valence (Burns et al. 1996), but only five of these are minerals (Table 9). In comparison to structures based upon sheets of polyhedra, uranyl minerals are not commonly based upon finite clusters, and when they are, the cluster is invariably $(UO_2)(CO_3)_3$ (Fig. 42). This cluster is composed of a $Ur\phi_6$ hexagonal bipyramid that shares three equatorial edges with CO_3 triangles. The ligands that are shared between the uranyl and carbonate polyhedra are herein designated O_{U-C}, and the unshared ligand of the carbonate triangle is designated O_C.

Table 9. Uranyl mineral structures based upon finite clusters

Name	Formula	S. G.	a (Å)	b (Å)	c (Å)	α (°)	β (°)	γ (°)	Ref.
liebigite	$Ca_2[(UO_2)(CO_3)_3](H_2O)_{11}$	$Bba2$	16.699	17.557	13.697				1
schröckingerite	$NaCa_3[(UO)_2(CO_3)_3](SO_4)F(H_2O)_{10}$	$P\bar{1}$	9.634	9.635	14.391	91.41	92.33	120.26	2
bayleyite	$Mg_2[(UO_2)(CO_3)_3](H_2O)_{18}$	$P2_1/a$	26.560	15.256	6.505		92.90		3
swartzite	$CaMg[(UO_2)(CO_3)_3](H_2O)_{12}$	$P2_1/m$	11.080	14.634	6.439		99.43		4
andersonite	$Na_2Ca[(UO_2)(CO_3)_3](H_2O)_5$	$R\bar{3}m$	17.904		23.753				5

References: (1) Mereiter (1982a), (2) Mereiter (1986a), (3) Mayer and Mereiter (1986), (4) Mereiter (1986b), (5) Mereiter (1986c).

The $(UO_2)(CO_3)_3$ complex is known to occur in U-rich carbonate-rich waters with high pH (Langmuir 1978), and minerals precipitating from these solutions will often incorporate the complex as a $(UO_2)(CO_3)_3$ cluster in the structure.

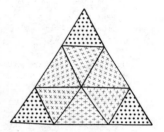

Figure 42. The $(UO_2)(CO_3)_3$ cluster that occurs isolated in uranyl mineral structures. The uranyl polyhedron is shaded with crosses and the CO_3 triangles are stippled.

The structure of liebigite (Mereiter 1982a) contains $(UO_2)(CO_3)_3$ clusters (Fig. 43a) and three symmetrically distinct Ca polyhedra. The Ca(1) cation is coordinated by seven ligands, whereas the Ca(2) and Ca(3) cations are each coordinated by eight ligands. Each $Ca\phi_n$ polyhedron includes four H_2O groups, and the remaining ligands are shared with $(UO_2)(CO_3)_3$ clusters. The $Ca(1)\phi_7$ polyhedron includes one O_C and two O_{U-C} atoms. Only one of the O_C atoms of a given $(UO_2)(CO_3)_3$ cluster is shared with a $Ca\phi_7$ polyhedron. In addition to the H_2O groups that are bonded to the Ca cations, there are three symmetrically distinct H_2O groups that are held in the structure by H bonds only. The $Ca\phi_7$ polyhedra do not share polyhedral elements with other $Ca\phi_7$ polyhedra but are linked only though $(UO_2)(CO_3)_3$ clusters and H bonds, resulting in a loose framework of polyhedra (Fig. 43a).

Schröchingerite (Mereiter 1986a) is a structurally and chemically complex mineral. In addition to the $(UO_2)(CO_3)_3$ cluster, it also contains a $Na\phi_6$ octahedron, a SO_4 tetrahedron, and three symmetrically distinct $Ca\phi_8$ polyhedra. The structure is shown in Figures 43b,c, from which it is apparent that the polyhedra link to form sheets that are parallel to (001), with H_2O groups located between the sheets, where they are held in position by H bonds. The details of the heteropolyhedral sheet are shown in Figure 43b. Each $(UO_2)(CO_3)_3$ cluster shares three equatorial edges of the $Ur\phi_6$ hexagonal bipyramid with $Ca\phi_8$ polyhedra, and the O_C ligands are also shared with a $Ca\phi_8$ polyhedron and a $Na\phi_6$ octahedron. Each $Ca\phi_8$ polyhedron shares an edge with a $(UO_2)(CO_3)_3$ cluster, an O_C ligand with another $(UO_2)(CO_3)_3$ cluster, and an edge with each of two $Ca\phi_8$ polyhedra and an $Na\phi_6$ octahedron. The structure contains a F atom that is shared between three $Ca\phi_8$ polyhedra. The SO_4 tetrahedron is located above the F anion, and shares three of its anions with three pairs of $Ca\phi_8$ polyhedra. The fourth tetrahedral anion of the SO_4 tetrahedron is directed into the interlayer, and accepts H bonds from interlayer H_2O groups. The O_{Ur} atoms are only one-connected within the sheet, and also accept H bonds from interlayer H_2O groups.

The structure of bayleyite (Mayer and Mereiter 1986) is shown in Figure 43d. The structure contains three symmetrically distinct $Mg(H_2O)_6$ octahedra, as well as the

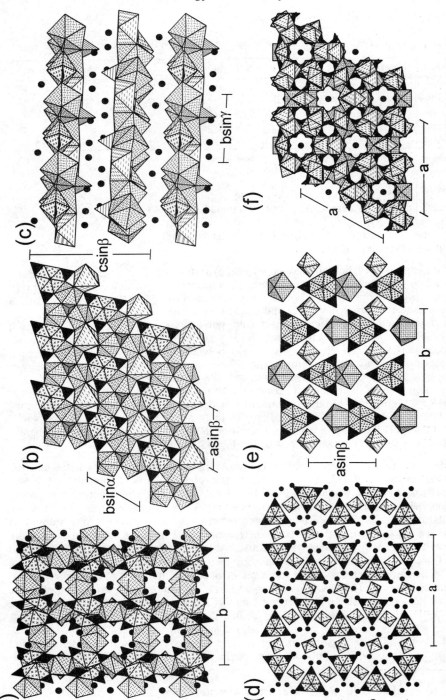

Figure 43. Uranyl mineral structures that contain isolated clusters of polyhedra of higher bond valence: (a) liebigite projected along [001]; (b) schröckingerite projected along [100]; (c) schröckingerite projected along [001]; (d) bayleyite projected along [001]; (e) swartzite projected along [001]; (f) andersonite projected along [001]. The legend is as in Figure 25 except that CO_3 triangles are black.

$(UO_2)(CO_3)_3$ cluster and six symmetrically distinct H_2O groups that are H bonded into the structure. The $(UO_2)(CO_3)_3$ clusters are linked to the $Mg(H_2O)_6$ octahedra by H bonding only.

The structure of swartzite (Mereiter 1986b) is illustrated in Figure 43e. In addition to the $(UO_2)(CO_3)_3$ cluster, swartzite contains a $Ca\phi_8$ polyhedron and a $Mg(H_2O)_6$ octahedron. The Ca polyhedron shares an edge with the $(UO_2)(CO_3)_3$ cluster, and the other six ligands are H_2O groups. A complex network of H bonds provide linkages between the $Mg(H_2O)_6$ octahedra and the adjacent $(UO_2)(CO_3)_3$ clusters.

The structure of andersonite (Mereiter 1986c), a complex hexagonal uranyl carbonate, is shown in Figure 43f. In addition to the $(UO_2)(CO_3)_3$ cluster, the structure contains two symmetrically distinct $Na\phi_6$ octahedra, one $Ca\phi_7$ polyhedron, and one H_2O group that is H bonded into the structure. The $Na\phi_6$ octahedra and $Ca\phi_7$ polyhedra link to each other, as well as to the $(UO_2)(CO_3)_3$ cluster, forming a complex heteropolyhedral framework structure. The structure contains a channel along [001] that contains disordered H_2O groups (Coda et al. 1981).

Structures consisting of frameworks

Uranyl minerals that have structures based upon frameworks of polyhedra of higher bond valence are uncommon (Table 10). Of those uranyl minerals with framework structures, soddyite is the most common. The structure of soddyite (Demartin et al. 1992) contains $Ur\phi_5$ pentagonal bipyramids that share equatorial edges to form chains. The chains are cross-linked by sharing edges with SiO_4 tetrahedra such that each tetrahedron shares two of its edges with adjacent chains (Fig. 44a). The unshared corners of the $Ur\phi_5$ pentagonal bipyramids are H_2O groups, making soddyite one of the few uranyl minerals that involves a bond between U^{6+} and H_2O. The synthetic phase $(UO_2)_2(GeO_4)(H_2O)_2$ (Legros and Jeannin 1975) is isostructural with soddyite.

The structure of weeksite (Baturin and Sidorenko 1985) (Fig. 44b) contains chains of edge-sharing $Ur\phi_5$ pentagonal bipyramids that share edges with SiO_4 tetrahedra. These chains are linked through disordered SiO_4 tetrahedra to form complex sheets, which are in turn linked into a framework through SiO_4 tetrahedra.

The structure of cliffordite (Branstätter 1981b) (Fig. 44c) is the only known framework structure that contains $Ur\phi_6$ hexagonal bipyramids. There are two symmetrically distinct $Ur\phi_6$ hexagonal bipyramids, and in each case the O_{Ur} atoms are only one-connected. Each equatorial anion of the uranyl polyhedra is bonded to one of two symmetrically distinct Te^{4+} cations. The coordination polyhedra about the Te^{4+} cations are highly distorted owing to the presence of a lone pair of electrons. Each Te^{4+} is coordinated by five anions in an approximate distorted square-pyramidal geometry.

Table 10. Uranyl mineral structures based on frameworks of polyhedra

Name	Formula	S. G.	a (Å)	b (Å)	c (Å)	α (°)	β (°)	γ (°)	Ref.
soddyite	$(UO_2)_2(SiO_4)(H_2O)_2$	$Fddd$	8.334	11.212	18.668				1
weeksite	$(K_{0.62}Na_{0.38})_2(UO_2)_2(Si_5O_{13})$-$(H_2O)_3$	$Cmmm$	7.092	17.888	7.113				2
cliffordite	$(UO_2)(Te_3O_7)$	$Pa\bar{3}$	11.335						3

References: (1) Demartin et al. (1992), (2) Baturin and Sidorenko (1985), (3) Branstatter (1981b).

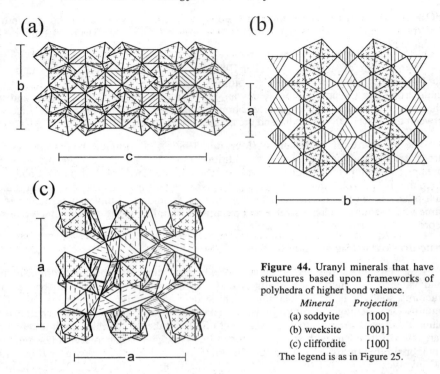

Figure 44. Uranyl minerals that have structures based upon frameworks of polyhedra of higher bond valence.

Mineral	Projection
(a) soddyite	[100]
(b) weeksite	[001]
(c) cliffordite	[100]

The legend is as in Figure 25.

ACKNOWLEDGMENTS

My research efforts on the structures of uranium minerals are supported by the Environmental Management Sciences Program of the United States Department of Energy (DE-FG07-97ER14820) and the National Science Foundation (NSF-EAR98-04723). The manuscript was improved following careful reviews by Robert Finch and Jiří Cejka. I am indebted to the many colleagues and students with whom I have had the privilege of studying U minerals: Christopher Cahill, Rod Ewing, Bob Finch, Alex Garza, Bob Gault, Rebecca Glatz, John Hanchar, Fran Hill, Jennie Jackson, Erin Keppel, Sergey Krivovichev, Yaping Li, Mark Miller, Michel Pipard, Andy Roberts, and Sarah Scott.

REFERENCES

Allen PG, Shuh DK, Bucher JJ, Edelstein NM, Reich T, Denecke MA, Nitsche H 1996 EXAFS determinations of uranium structures: The uranyl ion complexed with tartaric, citric, and malic acids. Inorg Chem 35:784-787

Allen PG, Bucher JJ, Clark DL, Edelstein NM, Ekberg SA, Gohdes JW, Hudson EA, Kaltsoyannis N, Lukens WW, Neu MP, Palmer PD, Reich T, Shuh DK, Tait CD, Zwick BD 1995 Multinuclear NMR, raman, EXAFS, and X-ray diffraction studies of uranyl carbonate complexes in near-neutral aqueous solution. X-ray structure of $[C(NH_2)_3]_6[(UO_2)_3(CO_3)_6]\cdot 6.5H_2O$. Inorg Chem 34:4797-4807

Atencio D, Neumann R, Silva AJGC, Mascarenhas YP 1991 Phurcalite from Perus, São Paulo, Brazil, and redetermination of its crystal structure. Can Mineral 29:95-105

Baturin SV, Sidorenko GA 1985 Crystal-structure of weeksite $(K_{.62}Na_{.38})_2(UO_2)_2[Si_5O_{13}]\cdot 3H_2O$. Dokl Akad Nauk SSSR 282:1132-1136 (in Russian)

Birch WD, Mumme WG, Segnit ER 1988 Ulrichite: A new copper calcium uranium phosphate from Lake Boga, Victoria, Australia. Aust Mineral 3:125-131

Borène J, Cesbron F 1970 Structure cristalline de l'uranyl-vanadate de nickel tétrahydraté

Ni(UO$_2$)$_2$(VO$_4$)$_2$·4H$_2$O. Bull Soç fr Minéral Cristallogr 93:426-432

Borène J, Cesbron F 1971 Structure cristalline de la curiénite Pb(UO$_2$)$_2$(VO$_4$)$_2$·5H$_2$O. Bull Soç fr Minéral Cristallogr 94:8-14

Branstätter F 1981 Non-stoichiometric, hydrothermally synthesized cliffordite. Tschermaks Mineral Petrogr Mitt 29:1-8

Brese NE, O'Keeffe M 1991 Bond-valence parameters for solids. Acta Crystallogr B47:192-197

Brown ID 1981 The bond-valence method: an empirical approach to chemical structure and bonding. *In* Structure and Bonding in Crystals II. M O'Keeffe, A Navrotsky (Eds) Academic Press, New York, p 1-30

Brown ID, Altermatt D 1985 Bond-valence parameters obtained from a systematic analysis of the inorganic crystal structure database. Acta Crystallogr B41:244-247

Brown ID, Wu Kang Kun 1976 Empirical parameters for calculating cation-oxygen bond valences. Acta Crystallogr B32:1957-1959

Burns PC, Finch RJ 1999 Wyartite: crystallographic evidence for the first pentavalent-uranium mineral. Am Mineral (in press)

Burns PC, Hanchar JM 1999 The structure of masuyite, Pb[(UO$_2$)$_3$O$_3$(OH)$_2$](H$_2$O)$_3$, and its relationship to protasite. Can Mineral (in press)

Burns PC 1999a A new complex sheet of uranyl polyhedra in the structure of wölsendorfite. Am Mineral (in press)

Burns PC 1999b A new uranyl phosphate chain in the structure of parsonsite. Am Mineral (submitted)

Burns PC, Hill FC 1999 Implications of the synthesis and structure of the Sr analogue of curite. Can Mineral (accepted)

Burns PC 1998a CCD area detectors of X-rays applied to the analysis of mineral structures. Can Mineral 36:847-853

Burns PC 1998b The structure of boltwoodite and implications of solid-solution towards sodium boltwoodite. Can Mineral 36:1069-1075

Burns PC 1998c The structure of compreignacite, K$_2$[(UO$_2$)$_3$O$_2$(OH)$_3$]$_2$(H$_2$O)$_7$. Can Mineral 36:1061-1067

Burns PC 1998d The structure of richetite, a rare lead uranyl oxide hydrate. Can Mineral 36:187-199

Burns PC 1997 A new uranyl oxide hydrate sheet in the structure of vandendriesscheite: Implications for mineral paragenesis and the corrosion of spent nuclear fuel. Am Mineral 82:1176-1186

Burns PC, Ewing RC, Hawthorne FC 1997a The crystal chemistry of hexavalent uranium: Polyhedral geometries, bond-valence parameters, and polymerization of polyhedra. Can Mineral 35:1551-1570

Burns PC, Finch RJ, Hawthorne FC, Miller ML, Ewing RC 1997b The crystal structure of ianthinite, [U$^{4+}_2$(UO$_2$)$_4$O$_6$(OH)$_4$(H$_2$O)$_4$](H$_2$O)$_5$: A possible phase for Pu^{4+} incorporation during the oxidation of spent nuclear fuel. J Nucl Mater 249:199-206

Burns PC, Miller ML, Ewing RC 1996 U^{6+} minerals and inorganic phases: a comparison and hierarchy of structures. Can Mineral 34:845-880.

Cejka J, Sejkora J, Skála R, Cejka J, Novotná M, Ederová J 1998 Contibution to the crystal chemistry of synthetic becquerelite, billietite and protasite. Neues Jahrb Mineral Abh 174:159-180

Cejka J, Urbanec Z 1988 Contibution to the hydrothermal origin of rutherfordine. Casopis Národního Muzea, rada prírodovedná 150:1-10

Charpin P, Dejean A, Folcher P, Rigny P, Navaza P 1985 EXAFS sur des composés de coordination de l'uranium en phase solide et en solution. J Chim Phys 82:925-932

Coda A, Della Giusta A, Tazzoli V 1981 The structure of synthetic andersonite, Na$_2$Ca[UO$_2$(CO$_3$)$_3$]·xH$_2$O ($x \cong 5.6$). Acta Crystallogr B37:1496-1500

Cooper MA, Hawthorne FC 1995 The crystal structure of guilleminite, a hydrated Ba-U-Se sheet structure. Can Mineral 33:1103-1109

Craw JS, Vincent MA, Hillier IH, Wallwork AL 1995 *Ab initio* quantum chemical calculations on uranyl UO$_2^{2+}$, plutonyl PuO$_2^{2+}$, and their nitrates and sulfates. J Phys Chem 99:10181-10185

Demartin F, Diella V, Donzelli S, Gramaccioli CM, Pilati T 1991 The importance of accurate crystal structure determination of uranium minerals. I. Phosphuranylite KCa(H$_3$O)$_3$(UO$_2$)$_7$(PO$_4$)$_4$O$_4$·8H$_2$O. Acta Crystallogr B47:439-446

Demartin F, Gramaccioli CM, Pilati T 1992 The importance of accurate crystal structure determination of uranium minerals. II. Soddyite (UO$_2$)$_2$(SiO$_4$)·2H$_2$O. Acta Crystallogr C48:1-4

Evans HT Jr. 1963 Uranyl ion coordination. Science 141:154-7

Finch RJ, Cooper MA, Hawthorne FC, Ewing RC 1999 Refinement of the crystal structure of rutherfordine. Can Mineral (in press)

Finch RJ, Cooper MA, Hawthorne FC, Ewing RC 1996 The crystal structure of schoepite, [(UO$_2$)$_8$O$_2$(OH)$_{12}$](H$_2$O)$_{12}$. Can Mineral 34:1071-1088

Finch RJ, Ewing RC 1992 The corrosion of uraninite under oxidizing conditions. J Nucl Mater 190:133-156

Fleischer M, Mandarino JA 1995 Glossary of mineral species. Mineralogical Record Inc, Tucson, Arizona

Frondel C 1958 Systematic mineralogy of uranium and thorium. U S Geol Surv Bull 1064

Fuchs LH, Gebert E 1958 X-ray studies of synthetic coffinite, thorite and uranothorites. Am Mineral 43:243-248

Ginderow D 1988 Structure de l'uranophane alpha, $Ca(UO_2)_2(SiO_3OH)_2 \cdot 5H_2O$. Acta Crystallogr C44:421-424

Ginderow D, Cesbron F 1983a Structure de la derriksite, $Cu_4(UO_2)(SeO_3)_2(OH)_6$. Acta Crystallogr C39:1605-1607

Ginderow D, Cesbron F 1983b Structure de la demesmaekerite, $Pb_2Cu_5(SeO_3)_6(UO_2)_2(OH)_6 \cdot 2H_2O$. Acta Crystallogr C39:824-827

Ginderow D, Cesbron F 1985 Structure de la roubaultite, $Cu_2(UO_2)_3(CO_3)_2O_2(OH)_2 \cdot 4H_2O$. Acta Crystallogr C41:654-657

Hanic F 1960 The crystal structure of meta-zeunerite $Cu(UO_2)_2(AsO_4)_2 \cdot 8H_2O$. Czech J Phys 10:169-181

Hansley PL, Fitzpatrick JJ 1989 Compositional and crystallographic data on REE-bearing coffinite from the Grants uranium region, northwestern New Mexico. Am Mineral 74:263-270

Hawthorne FC 1983 Graphical enumeration of polyhedral clusters. Acta Crystallogr A39:724-736

Hawthorne FC 1985 Towards a structural classification of minerals: the $^{vi}M^{iv}T_2\phi_n$ minerals. Am Mineral 70, 455-473

Hawthorne FC 1992 The role of OH and H_2O in oxide and oxysalt minerals. Z Kristallogr 201:183-206.

Hawthorne FC 1990 Structural hierarchy in $M^{[6]}T^{[4]}\phi_n$ minerals. Z Kristallogr 192:1-52

Hill FC, Burns PC 1999 Structure of a synthetic Cs uranyl oxide hydrate and its relationship to compreignacite. Can Mineral (in press)

Janeczek J, Ewing RC 1992 Structural formula of uraninite. J Nucl Mater 190:128-132

Khosrawan-Sazedj F 1982a The crystal structure of meta-uranocircite II, $Ba(UO_2)_2(PO_4)_2 \cdot 6H_2O$. Tschermaks Mineral Petrogr Mitt 29:193-204

Khosrawan-Sazedj F 1982b On the space group of threadgoldite. Tschermaks mineral petrogr Mitt 30: 111-115

Kovba LM 1972 Crystal structure of $K_2U_7O_{22}$. J St Chem 13:235-238

Krause W, Effenberger H, Brandstätter F 1995 Orthowalpurgite, $(UO_2)Bi_4O_4(AsO_4)_2 \cdot 2H_2O$, a new mineral from the Black Forest, Germany. Eur J Mineral 7:1313-1324

Langmuir D 1978 Uranium solution-mineral equilibria at low temperatures with applications to sedimentary ore deposits. Geochim Cosmochim Acta 42:547-569

Lee MR, Jaulmes S 1987 Nouvelle serie d'oxydes derives de la structure de α-U_3O_8: $M^{II}UMo_4O_{16}$. J Solid State Chem 67:364-368

Legros JP, Jeannin Y 1975 Coordination de l'uranium par l'ion germanate. II. Structure du germanate d'uranyle dihydraté $(UO_2)_2GeO_4(H_2O)_2$. Acta Crystallogr B31:1140-1143

Loopstra BO 1977 On the structure of α-U_3O_8. J Inorg Nucl Chem 39:1713-1714

Loopstra BO, Brandenburg NP 1978 Uranyl selenite and uranyl tellurite. Acta Crystallogr B34:1335-1337

Loopstra BO, Cordfunke EHP 1966 On the structure of α-UO_3. Rec Trav Chim Pays-Bas 85:135-142.

Makarov YS, Ivanov VI 1960 The crystal structure of meta-autunite, $Ca(UO_2)_2(PO_4)_2 \cdot 6H_2O$. Dokl Akad Nauk SSSR 132:601-603

Makarov YS, Anikina LI 1963 Crystal structure of umohoite $(UMoO_6(H_2O)_2) \cdot 2H_2O$. Geochemistry 1963:14-21

Mayer H, Mereiter K 1986 Synthetic bayleyite, $Mg_2[UO_2(CO_3)_3] \cdot 18H_2O$: Thermochemistry, crystallography and crystal structure. Tschermaks Mineral Petrogr Mitt 35:133-146.

Mereiter K 1982a The crystal structure of liebigite, $Ca_2UO_2(CO_3)_3 \cdot \sim 11H_2O$. Tschermaks Mineral Petrogr Mitt 30:277-288

Mereiter K 1982b The crystal structure of walpurgite, $(UO_2)Bi_4O_4(AsO_4)_2 \cdot 2H_2O$. Tschermaks Mineral Petrogr Mitt 30:129-139

Mereiter K 1982c Die Kristallstruktur des Johannits, $Cu(UO_2)_2(OH)_2(SO_4)_2 \cdot 8H_2O$. Tschermaks Mineral Petrogr Mitt 30:47-57

Mereiter K 1986a Crystal structure and crystallographic properties of a schröckingerite from Joachims-thal. Tschermaks Mineral Petrogr Mitt 35:1-18

Mereiter K 1986b Synthetic swartzite, $CaMg[UO_2(CO_3)_3] \cdot 12H_2O$, and its strontium analogue, $SrMg[UO_2(CO_3)_3] \cdot 12H_2O$: Crystallography and crystal-structures. N Jahrb Mineral Monatsh 1986:481-492

Mereiter K 1986c Neue kristallographische Daten ueber das Uranmineral Andersonit. Anz Oesterr Akad Wiss Math-Naturwiss Kl 3:39-41

Mereiter K 1986d Crystal structure refinements of two francevillites, $(Ba,Pb)[(UO_2)_2V_2O_8] \cdot 5H_2O$. Neues Jahrb Mineral Monatsh 1986:552-560

Meunier G, Galy J 1973 Structure cristalline de la schmitterite synthétique $UTeO_5$. Acta Crystallogr

B29:1251-1255

Miller ML, Finch RJ, Burns PC, Ewing RC 1996 Description and classification of uranium oxide hydrate sheet topologies. J Mater Res 11:3048-3056

Miller SA, Taylor JC 1986 The crystal structure of saleeite, $Mg[UO_2PO_4]_2 \cdot 10H_2O$. Z Kristallogr 177: 247-253

Moll H, Matz W, Schuster G, Brendler E, Bernhard G, Nitsche H 1995 Synthesis and characterization of uranyl orthosilicate $(UO_2)_2SiO_4 \cdot 2H_2O$. J Nucl Mater 227:40-49

Moll H, Reich T, Matz W, Bernhard G, Nitsche H, Shuh DK, Bucher JJ, Kaltsoyannis N, Edelstein NM, Scholz A 1994 Structural analysis of $(UO_2)_2SiO_4 \cdot 2H_2O$ by XRD and EXAFS. Forschungszentrum Rossendorf, Inst Radiochem Ann Report 1994:75-78

Moore PB 1965 A structural classification of Fe-Mn orthophosphate hydrates. Am Mineral 50, 2052-2062

Moore PB 1973 Pegmatite phosphates: descriptive mineralogy and crystal chemistry. Mineral Rec 4:103-130

Pagoaga MK, Appleman DE, Stewart JM 1987 Crystal structures and crystal chemistry of the uranyl oxide hydrates becquerelite, billietite, and protasite. Am Mineral 72:1230-1238.

Pearcy EC, Prikryl JD, Murphy WM, Leslie BW 1994 Alteration of uraninite from the Nopal I deposit, Peña Blanca District, Chihuahua, Mexico, compared to degradation of spent nuclear fuel in the proposed U.S. high-level nuclear waste repository at Yucca Mountain, Nevada. Appl Geochem 9:713-732

Piret P 1985 Structure cristalline de la fourmariérite, $Pb(UO_2)_4O_3(OH)_4 \cdot 4H_2O$. Bull Minéral 108:659-665.

Piret P, Declercq J-P 1983 Structure cristalline de l'upalite $Al[(UO_2)_3O(OH)(PO_4)_2] \cdot 7H_2O$. Un exemple de macle mimétique. Bull Minéral 106:383-389

Piret P, Declercq J-P, Wauters-Stoop D 1980 Structure cristalline de la sengiérite. Bull Minéral 103: 176-178.

Piret P, Deliens M 1982 La vanmeersscheite $U(UO_2)_3(PO_4)_2(OH)_6 \cdot 4H_2O$ et la méta-vanmeersscheite $U(UO_2)_3(PO_4)_2(OH)_6 \cdot 2H_2O$, nouveaux minéraux. Bull Minéral 105:125-128

Piret P, Deliens M 1987 Les phosphates d'uranyle et d'aluminium de Kobokobo IX. L'althupite $AlTh(UO_2)[(UO_2)_3O(OH)(PO_4)_2]_2(OH)_3 \cdot 15H_2O$, nouveau minéral; propriétés et structure cristalline. Bull Minéral 110:65-7

Piret P, Deliens M, Piret-Meunier, J 1988 La françoisite-(Nd), nouveau phosphate d'uranyle et de terres rares; propriétés et structure cristalline. Bull Minéral 111:443-449

Piret P, Deliens M, Piret-Meunier, J, Germain G 1983 La sayrite, $Pb_2[(UO_2)_5O_6(OH)_2] \cdot 4H_2O$, nouveau minéral; propriétés et structure cristalline. Bull Minéral 106:299-304

Piret P, Piret-Meunier J 1988 Nouvelle détermination de la structure cristalline de la dumontite $Pb_2[(UO_2)_3O_2(PO_4)_2] \cdot 5H_2O$. Bull Minéral 111:439-442

Piret P, Piret-Meunier J, Declercq J-P 1979 Structure of phuralumite. Acta Crystallogr B35:1880-1882

Piret P, Piret-Meunier J, Deliens M 1990 Composition chimique et structure cristalline de ia dewindtite $Pb_3[H(UO_2)_3O_2(PO_4)_2]_2 \cdot 12H_2O$. Eur J Mineral 2:399-405

Pushcharovsky DYu, Rastsvetaeva RK, Sarp H 1996 Crystal structure of deloryite, $Cu_4(UO_2)[Mo_2O_8](OH)_6$. J Alloys Comp 239:23-26

Pyykkö P, Zhao Y 1991 The large range of uranyl bond lengths: Ab initio calculations on simple uranium-oxygen clusters. Inorg Chem 30:3787-3788

Pyykkö P, Li J, Runeberg N 1994 Quasirelativistic pseudopotential study of species isoelectronic to uranyl and the equatorial coordination of uranyl. J Phys Chem 98:4809-4813

Rosenzweig A, Ryan RR 1975 Refinement of the crystal structure of cuprosklodowskite, $Cu[(UO_2)_2(SiO_3OH)_2] \cdot 6H_2O$. Am Mineral 60:448-453

Rosenzweig A, Ryan RR 1977a Kasolite, $Pb(UO_2)(SiO_4) \cdot H_2O$. Cryst Struct Commun 6:617-621

Rosenzweig A, Ryan RR 1977b Vandenbrandeite $CuUO_2(OH)_4$. Cryst Struct Commun 6:53-56

Ross M, Evans HT Jr 1964 Studies of the torbernite minerals (I): The crystal structure of abernathyite and the structurally related compounds $NH_4(UO_2AsO_4) \cdot 3H_2O$ and $K(H_3O)(UO_2AsO_4)_2 \cdot 6H_2O$. Am Mineral 49:1578-1602

Ryan RR, Rosenzweig A 1977 Sklodowskite, $MgO \cdot 2UO_3 \cdot 2SiO_2 \cdot 7H_2O$. Cryst Struct Commun 6:611-615.

Serezhkin VN, Boiko NV, Trunov VK 1982 Crystal-structure of $Sr[UO_2(OH)CrO_4]_2 \cdot 8H_2O$. J St Chem 23:270-273

Serezhkin VN, Chuvaev VF, Kovba LM, Trunov VK 1973 Structure of synthetic iriginite. Dokl Akad Nauk SSSR 210:873-876 (in Russian)

Shannon RD 1976 Revised effective ionic radii and systematic studies of interatomic distances in halides and chalcogenides. Acta Crystallogr A32:751-767

Smith DK 1984 Uranium mineralogy. In Uranium Geochemistry, Mineralogy, Geology, Exploration and Resources. F. Ippolito, B. DeVero, G. Capaldi (Eds) Institution of Mining and Metallurgy, London, 43-71

Smith JV 1995 Synchrotron X-ray sources: Instrumental characteristics: New applications in microanalysis, tomography, absorption spectroscopy and diffraction. Analyst 120:1231-1245

Stergiou AC, Rentzeperis PJ, Sklavounos S 1993 Refinement of the crystal structure of metatorbernite. Z Kristallogr 205:1-7

Stohl FV, Smith DK 1981 The crystal chemistry of the uranyl silicate minerals. Am Mineral 66:610-625.

Swihart GH, Gupta PKS, Schlemper EO, Back ME, Gaines RV 1993 The crystal structure of moctezumite [PbUO$_2$](TeO$_3$)$_2$. Am Mineral 78:835-839

Szymanski JT, Scott JD 1982 A crystal structure refinement of synthetic brannerite UTi$_2$O$_6$ and its bearing on rate of alkaline-carbonate leaching of brannerite in ore. Can Mineral 20:271-280

Taylor JC, Stuart WI, Mumme IA 1981 The crystal structure of curite. J Inorg Nucl Chem 43:2419-2423.

Taylor, SR 1964 The abundance of chemical elements in the continental crust – a new table. Geochim Cosmochim Acta 28, 1273-1285

Thompson HA, Brown GE Jr, Parks GA 1997 XAFS spectroscopic study of uranyl coordination in solids and aqueous solution. Am Mineral 82:483-496

van Wezenbeek EM, Baerends EJ, Snijders JG 1991 Relativistic bond lengthening of UO$_2^{2+}$ and UO$_2$. Theor Chim Acta 81:139-155

Viswanathan K, Harneit O 1986 Refined crystal structure of β-uranophane Ca(UO$_2$)$_2$(SiO$_3$OH)$_2$·5H$_2$O. Am Mineral 71:1489-1493

Vochten R, Blaton N, Peeters O, Van Springel K, Van Haverbeke L 1997 A new method of synthesis of boltwoodite and of formation of sodium boltwoodite, uranophane, sklodowskite and kasolite from boltwoodite. Can Mineral 35:735-741

Vochten R, Van Haverbeke L, Van Springel K, Blaton N, Peeters OM 1995 The structure and physico-chemical characteristics of synthetic zippeite. Can Mineral 33:1091-1101

Zachariasen WH 1978 Bond lengths in oxygen and halogen compounds of d and f elements. J Less-Common Met 62:1-7

3
Systematics and Paragenesis of Uranium Minerals

Robert Finch
Argonne National Laboratory
9700 South Cass Avenue
Argonne, Illinois 60439

Takashi Murakami
Mineralogical Institute
University of Tokyo
Bunkyo-ku, Tokyo 113, Japan

INTRODUCTION

Approximately five percent of all known minerals contain U as an essential structural constituent (Mandarino 1999). Uranium minerals display a remarkable structural and chemical diversity. The chemical diversity, especially at the Earth's surface, results from different chemical conditions under which U minerals are formed. U minerals are therefore excellent indicators of geochemical environments, which are closely related to geochemical element cycles. For example, detrital uraninite and the absence of uranyl minerals at the Earth's surface during the Precambrian are evidence for an anoxic atmosphere before about 2 Ga (Holland 1984, 1994; Fareeduddin 1990; Rasmussen and Buick 1999).

The oxidation and dissolution of U minerals contributes U to geochemical fluids, both hydrothermal and meteoric. Under reducing conditions, U transport is likely to be measured in fractions of a centimeter, although F and Cl complexes can stabilize U(IV) in solution (Keppler and Wyllie 1990). Where conditions are sufficiently oxidizing to stabilize the uranyl ion, UO_2^{2+}, and its complexes, U can migrate many kilometers from its source in altered rocks, until changes in solution chemistry lead to precipitation of U minerals. Where oxidized U contacts more reducing conditions, U can be reduced to form uraninite, coffinite, or brannerite. The precipitation of U(VI) minerals can occur in a wide variety of environments, resulting in an impressive variety of uranyl minerals. Because uraninite dissolution can be rapid in oxidizing, aqueous environments, the oxidative dissolution of uraninite caused by weathering commonly leads to the development of a complex array of uranyl minerals in close association with uraninite. Understanding the conditions of U mineral formation and alteration is an important part of understanding the geochemical behavior of U.

Renewed interest in the paragenesis and structures of uranyl minerals has arisen lately due, in part, to their roles as alteration products of uraninite under oxidizing conditions (Frondel 1958; Garrels and Christ 1959; Finch and Ewing 1992b). But uranyl compounds are also important corrosion products of the UO_2 in spent nuclear fuel (Finch and Ewing 1991; Forsyth and Werme 1992; Johnson and Werme 1994; Wronkiewicz et al. 1992, 1996; Buck et al. 1998; Finn et al. 1998; Finch et al. 1999b), and they may control groundwater concentrations of U in contaminated soils (Buck et al. 1996; Morris et al. 1996). Studies of the natural occurrences of uranyl minerals can be used to test the extrapolation of results from short-term experiments to periods relevant to high-level nuclear-waste disposal (Ewing 1993) and to assess models that predict the long-term behavior of spent nuclear fuel buried in a geologic repository (Bruno et al. 1995).

URANIUM MINERALS SYSTEMATICS

The most complete descriptions of U minerals to date were provided by Frondel (1958); however, during the intervening 40 years there has been a dramatic increase in our understanding of U mineralogy and crystal chemistry, and many new species have been described. Smith (1984) provided an extensive review of U mineralogy, including summaries of structures, occurrences and mineral descriptions for U minerals described since Frondel (1958). In this volume, detailed descriptions of U-mineral structures are provided by Burns (this volume). Here, we focus on U-mineral paragenesis and chemistries. Some detailed descriptions are provided for U minerals reported since Smith's (1984) paper. Minerals containing reduced U are discussed first, followed by uranyl minerals, in which U occurs as U^{6+}. Minerals are further divided chemically according to the major anionic component (e.g. silicate, phosphate, etc.), with some chemical groups listed together because of structural similarities. Tables list minerals in alphabetical order within each chemical group. The name, formula, and references are provided, together with comments pertaining to recent work reported for these minerals.

Minerals containing reduced U

Most U in nature occurs in accessory minerals in which it may be a major (Table 1) or minor component (Table 2), but only a few of these minerals are found with U concentrations sufficient to be economically important. By far the most important U mineral, both in terms of abundance and economic value, is the nominally simple oxide, uraninite. The U silicate, coffinite, is of secondary economic importance, with most major coffinite-bearing deposits restricted to low-temperature deposits, such as the sandstone-hosted deposits in the west central USA (Finch 1996; Plant et al., Chapter 6, this volume). However, coffinite is increasingly recognized as an important alteration product of uraninite under reducing conditions and has been identified in many diverse U deposits (Janeczek 1991, 1992c; Fayek and Kyser 1997; Fayek et al. 1997). The U titanate brannerite is perhaps the next most abundant U(IV) mineral, occurring in quantities sufficient to be economically mined in a few localities. Most of the remaining minerals listed in Tables 1 and 2 do not form economic ore deposits; they are, nevertheless, important hosts for U in the rocks in which they reside, and the alteration of these minerals by both hydrothermal and meteoric waters is the source of dissolved U in surface waters, groundwaters and hydrothermal fluids, from which many U deposits are derived (Plant et al. this volume).

Uraninite. Uraninite is a common accessory mineral in pegmatites and peraluminous granites, and is probably the most important source of dissolved U in groundwaters emanating from weathered granite terrains (Frondel 1958; Förster 1999 and references therein; Plant et al. this volume). Uraninite is isometric (fluorite structure type, $Fm3m$) with nominal composition UO_{2+x} (Z = 4); however, pure UO_2 is not known in nature, being always at least partly oxidized (x < 0.25-0.3) and containing additional elements. In addition to radiogenic Pb (and other radiogenic daughter products) produced by decay of ^{238}U and ^{235}U, uraninite commonly contains Th, REE, Ca, and other elements. The commonly extreme non-stoichiometry exhibited by uraninite is apparent from the structural formula: $(U^{4+}_{1-x-y-z}U^{6+}_xREE^{3+}_yM^{2+}_z\square_v)O_{2+x-0.5y-z-2v}$ (Janeczek and Ewing 1992b). Synthetic UO_2 is brick red, becoming black upon slight oxidation. Uraninite is commonly black with an iron-black metallic luster, although various dark shades of brown and green have also been reported for more weathered material (Frondel 1958). The unit-cell parameter, a, for synthetic $UO_{2.03}$ is 5.4682 Å, decreasing as U^{4+} oxidizes to U^{6+} up to $UO_{2.25}$, for which a equals 5.440 Å (Smith et al. 1982). The unit-cell parameter for synthetic UO_{2+x} decreases linearly with increasing values of x, whereas unit cell parameters reported for uraninite are highly variable, depending in a complex way on composition (Brooker and Nuffield 1952;

Table 1. Minerals in which reduced U is an essential component

Name	Formula	Comments
Uraninite*	$(U^{4+}_{1-x-y-z}U^{6+}_xREE^{3+}_yM^{2+}_z)O_{2+x-_y-z}$	Fluorite structure type; s.s. with ThO_2; Frondel (1958); Janeczek & Ewing (1992b); Kotzer & Kyser (1993); Pourcelot et al. (1998); Förster (1999)
Coffinite	$USiO_4 \cdot nH_2O$	Zircon structure type; minor P; REE; As?; Frondel (1958); Speer (1982); Smits (1989); Janeczek (1991
Brannerite*	$(U,Ca,Y,Ce)(Ti,Fe)_2O_6$	metamict; brannerite-thorutite structure type; Frondel (1958); Szymanski & Scott (1982); Smith (1984); Singh et al. (1990); Gaines et al. (1997)
Orthobrannerite	$(U^{6+},U^{4+})(Ti,Fe)_2O_6(OH)$	metamict; Singh et al. (1990); Gaines et al. (1997). Compare synth. $UTiO_5$ (Marshall & Hoekstra 1965; Miyake et al. 1994), $UMoO_5$ (D'yachenko et al 1996), and UVO_5 (Dickens et al. 1992).
Ianthinite*	$U^{4+}(U^{6+}O_2)_4O(OH)_6(H_2O)_9$	β-U_3O_8 structure derivative; Finch & Ewing (1994); Burns et al. (1997b)
Ishikawaite*	$(U,Ca,Y,Ce)(Nb,Ta)O_4$	Samarskite group; Wolframite structure type; Gaines et al. (1997); Hanson et al. (1999)
Lermontovite	$U(PO_4)(OH)(H_2O)$ (?)	ill defined; compare vyacheslavite; Smith (1984); Gaines et al. (1997)
Moluranite	$H_4U(UO_2)_3(MoO_4)_7(H_2O)_{18}$	amorphous; Gaines et al. (1997)
Mourite	$UMo_5O_{12}(OH)_{10}$ (?)	IR suggests molecular H_2O; Gaines et al. (1997)
Ningyoite	$(U,Ca,Ce,Fe)_2(PO_4)_2 \cdot 1\text{-}2H_2O$	Rhabdophane grp.; compare synthetic $U(HPO_4)_2(H_2O)_4$ [G&92]; Gaines et al. (1997)
Petscheckite	$UFe^{2+}(Nb,Ta)_2O_8$	hypothetical formula for unaltered material; Smith (1984); Gaines et al. (1997)
Sedovite	$U(MoO_4)_2$	Gaines et al. (1997)
Uranomicrolite	$(U,Ca,Ce)_2(Ta,Nb)_2O_6(OH,F)$	Pyrochlore grp.; microlite subgroup (Ta > Nb); Frondel (1958); Lumpkin & Ewing (1992a)
Uranopolycrase	$(U,Y)(Ti,Nb)_2O_6$	Columbite structure type; Aurisicchio et al. (1993); Gaines et al. (1997)
Uranopyrochlore	$(U,Ca,Ce)_2(Nb,Ta)_2O_6(OH,F)$	Pyrochlore grp.; pyrochlore subgroup (Nb > Ta); Frondel (1958); Lumpkin & Ewing (1995)
Vyacheslavite	$U(PO_4)(OH)(H_2O)_{2.5}$	Rhabdophane grp.; compare lermontovite; Belova et al. (1984)
Wyartite*	$CaU^{5+}(UO_2)O(OH)_4(CO_3)(H_2O)_7$	mixed U^{5+}-U^{6+}; Clark (1960); Burns & Finch (1999)
Wyartite II*	$CaU^{5+}(UO_2)O(OH)_4(CO_3)(H_2O)_3$	mixed U^{5+}-U^{6+}; dehydration product of wyartite; Clark (1960); unpublished data
Unnamed*	$U(HAsO_4)_2(H_2O)_4$	possible synthetic analogue: PDF 38-644; Chernorukov et al. (1985); Ondrus et al. (1997c)

* New or supplemental data or not listed in Dana's New Mineralogy (Gaines et al. 1997).
**Minerals containing both U^{4+} and U^{6+} are also listed in tables according to the predominant oxyanion.

Table 2. Miscellaneous minerals that commonly contain substantial U^{4+} as a minor component

Name	Formula	Comments; Ref.
Betafite*	$(Ca,Na,U)_2(Ti,Nb,Ta)_2O_6(OH)$	Pyrochlore grp.; betafite subgroup; Lumpkin & Ewing (1996); Gaines et al. (1997)
Brabantite*	$Ca(Th,U)(PO_4)_2$	Monazite group; Gaines et al. (1997); Förster (1998a,b)
Brockite	$(Ca,Th,REE)(PO_4)\cdot H_2O$	Rhabdophane group; Gaines et al. (1997)
Cheralite	$(Ca,Ce,Th)(P,Si)O_4$	Monazite group; Gaines et al. (1997)
Davidite	$(Ce,La)(Y,U,Fe^{2+})(Ti,Fe^{3+})_{20}(O,OH)_{30}$	Crichtonite group; Frondel (1958); Gaines et al. (1997)
Ekanite	$Ca_2(Th,U)Si_8O_{20}$	cf. iraqite; steacyite; thornasite; umbozerite; unnamed ("thorsite"); Gaines et al. (1997)
Euxenite-(Y)	$(Y,Ca,Ce,U,Th)(Nb,Ta,Ti)_2O_6$	metamict; polycrase structure type; Gaines et al. (1997)
Grayite	$(Th,Ca,Pb)PO_4\cdot H_2O$	Rhabdophane grp.; ill defined; questionable; Gaines et al. (1997)
Huttonite	$(Th,U)SiO_4$	Monazite structure type; Speer (1982); Boatner & Sales (1988); Gaines et al. (1997)
Iraqite-(La)	$K(Ca,Na)_4(La,Ce,Th)_2(Si,Al)_{16}O_{40}$	cf. steacyite; ekanite; thornasite; umbozerite; unnamed ("thorsite"); Gaines et al. (1997)
Kobeite-(Y)	$(Y,U)(Ti,Nb)_2(O,OH)_6$	Columbite structure derivative; (= polycrase?); 8; Gaines et al. (1997)
Mckelveyite-(Y)	$Ba_3Na(Ca,U)Y(CO_3)_6(H_2O)_3$	Gaines et al. (1997)
Monazite-(REE)*	$(La-Sm)PO_4$	Monazite grp.; REE = La; Ce; Nd; Sm; 7; 9; Gaines et al. (1997)
Plumbobetafite*	$(Pb,U,Ca)(Ti,Nb)_2O_6(OH,F)$	Pyrochlore grp.; betafite subgroup; Lumpkin & Ewing (1996); Gaines et al. (1997)
Plumbomicrolite*	$(Pb,U,Ca)_2Ta_2O_6(OH)$	Pyrochlore grp.; microlite subgroup; Lumpkin & Ewing (1992a); Gaines et al. (1997)
Plumbopyrochlore*	$(Pb,U,Ca)_{2-x}Nb_2O_6(OH)$	Pyrochlore grp.; pyrochlore subgroup; Lumpkin & Ewing (1995); Gaines et al. (1997)
Samarskite-(Y)*	$(Y,REE,U,Fe^{3+},Fe^{2+})(Nb,Ta)O_4$	Samarskite group; Wolframite structure type; Warner & Ewing (1993); Hanson et al. (1999)
Steacyite	$K_x(Ca,Na)_2ThSi_8O_{20}$ ($x = 0.6 - 0.8$)	cf. ekanite; iraqite; thornasite; umbozerite; unnamed; Gaines et al. (1997)
Thorianite	$(Th,U)O_2$	Fluorite structure type; s.s. with UO_2 & CeO_2; Gaines et al. (1997)
Thorite	$(Th,U)SiO_4$	Zircon structure type; Frondel (1958); Smits (1989); Speer (1982); Lumpkin & Ewing (1992b)
Thornasite	$(Na,K)ThSi_{11}(O,F,OH)_{25}(H_2O)_8$	cf. ekanite; iraqite; steacyite; umbozerite; unnamed ("thorsite"); Gaines et al. (1997)
Thorogummite	$(Th,U)(SiO_4)_{1-x}(OH)_{4x}$	Zircon structure type; Gaines et al. (1997)
Thorutite	$(Th,U,Ca)Ti_2(O,OH)_6$	Brannerite structure type; Gaines et al. (1997)
Tristramite	$(Ca,U^{4+},Fe^{3+})(PO_4,SO_4)(H_2O)_2$	Rhabdophane group; Atkins et al. (1983)
Umbozerite	$Na_3Sr_4ThSi_8(O,OH)_{24}$	amorphous; cf. ekanite; iraqite; thornasite; steacyite; unnamed ("thorsite"); Gaines et al. (1997)
Yttrobetafite-(Y)	$(Y,U,Ce)(Ti,Nb,Ta)_2O_6(OH)$	Pyrochlore grp.; betafite subgroup; Gaines et al. (1997)
Yttrocolumbite-(Y)	$(Y,U,Fe^{2+})(Nb,Ta)O_4$ or $(Y,U,Fe^{2+})(Nb,Ta)_2O_6$	ill defined; compare yttrotantalite; Gaines et al. (1997)
Yttrocrasite-(Y)	$(Y,Th,Ca,U)(Ti,Fe^{3+})_2(O,OH)_6$	Pyrochlore structure type; compare thorutite; Gaines et al. (1997)
Yttropyrochlore-(Y)	$(Y,Na,Ca,U)_{1-2}(Nb,Ta,Ti)_2(O,OH)_7$	Pyrochlore grp.; ill defined; Gaines et al. (1997)
Yttrotantalite-(Y)	$(Y,U,Fe^{2+})(Ta,Nb)O_4$ or $(Y,U,Fe^{2+})(Ta,Nb)_2O_6$	Samarskite group (?); Gaines et al. (1997)
Unnamed*	$Ca_{3.5}(Th,U)_{1.5}Si_3O_{12}(OH)$	Apatite structure type; Jamveit et al. (1997)
Unnamed*	$Th_2(Ca,Ba)[Si_8O_{22}](OH)_2(H_2O)_n$	"thorsite" cf. Ekanite; iraqite; steacyite; thornasite; umbozerite; Lazebnik et al (1994)

* New or supplemental data or not listed in Dana's New Mineralogy (Gaines et al. 1997).

Berman 1957; Frondel 1958). There is no simple correlation between the oxidation state of U in uraninite and unit-cell size (Janeczek and Ewing 1992a).

The impurity content of uraninite depends strongly upon the environment of deposition, as well as the conditions under which dissolved U may have been transported. Three types of uraninite can be roughly defined in terms of their geneses (McMillan 1978; Plant et al. this volume): (1) igneous, magmatic, and metamorphic, including pegmatitic uraninite; (2) hydrothermal (e.g. vein type and unconformity-related deposits); (3) low temperature (sedimentary-hosted deposits). The chemistry of unaltered uraninite can be a reasonably reliable indicator of its origin (Frondel 1958). Magmatic uraninite commonly contains Th and REE, whereas these elements are largely absent from hydrothermal and low-temperature sedimentary uraninite (Frondel 1958). These compositional differences reflect differences in the aqueous chemistries of U, Th and REE; U may be readily oxidized and transported as the UO_2^{2+} ion and its complexes, whereas Th and REE tend to be less mobile (Langmuir 1978). These different solution behaviors fractionate U from Th and REE in many aqueous environments, especially where redox conditions favor UO_2^{2+} formation.

Impurities can provide insight into the genesis of uraninite and uraninite-fluid interactions, and may also affect uraninite stability. The most important impurities in uraninite are Pb, Th, Ca, Y and lanthanides.

Radiogenic Pb can reach quite high levels in ancient uraninite, with reports of 15-20 wt % PbO in some analyses, although 7-10 wt % is more common (Berman 1957; Frondel 1958; Janeczek and Ewing 1995). In many uraninite specimens, especially those of sedimentary and hydrothermal origin, Pb is the most abundant cation after U. Lead is incompatible in the UO_2 structure, however, and the structural role of Pb in uraninite has long perplexed researchers. Berman (1957) suggested that Pb exsolves as PbO (massicot) at the unit-cell scale, such that it is not detectable by XRD; however, high-resolution transmission electron microscopy (HRTEM) has not supported this hypothesis (Janeczek et al. 1993). It appears that Pb may replace some U and occupy interstitial sites within the uraninite structure (Janeczek and Ewing 1992b). X-ray powder diffraction data indicate that unit-cell volumes of Pb-rich uraninite are larger than those of Pb-poor uraninite (Janeczek and Ewing 1992c, 1995), suggesting that Pb accumulation can induce significant strain. Because Pb is incompatible in the uraninite structure, it is eventually lost from uraninite. Lead is relatively immobile in most groundwaters (Mann and Deutscher 1980), and under reducing conditions, Pb released from uraninite commonly forms galena, provided the activity of S is sufficient (Janeczek and Ewing 1995).

The decay of U to Pb also influences the average oxidation state of uraninite. This is because U in uraninite is predominantly U^{4+}, leading to Pb^{4+} as the ionic species produced by decay (ignoring potential redox steps for intermediate daughters). But Pb^{4+} is a strong oxidizer and is unstable in the presence of U^{4+}; thus U^{4+} oxidizes to U^{6+} (or two U^{4+} may each oxidize to U^{5+}) and Pb^{4+} is reduced to Pb^{2+}. This process has been called "auto-oxidation" (Frondel 1958) and may lead to relatively high U^{6+}/U^{4+} ratios in uraninite. The valence of Pb^{2+} in uraninite has been verified by X-ray photon spectroscopy (XPS) (Sunder et al. 1994, 1996). These same authors report U^{6+}/U^{4+} ratios of approximately 0.02 to as high as ~0.75 in uraninite from the Cigar Lake U deposit in northern Saskatchewan, despite reducing to anoxic conditions at the depth of the ore deposit (>400 m). Radioactive decay of U in "old" uraninite can therefore destabilize uraninite by two mechanisms: (1) auto-oxidation, which leads to U^{6+}/U^{4+} ratios at which the uraninite structure becomes unstable; and (2) accumulation of Pb^{2+} to levels that cannot be accommodated by the uraninite structure. These two processes occur simultaneously, leading to Pb loss and, commonly, recrystallization of uraninite under reducing conditions (Janeczek and Ewing 1992b, Kotzer and Kyser 1993, Janeczek and Ewing 1995). The effects of Pb on the stability of uraninite and other U minerals are discussed further in another section.

After radiogenic Pb, the most important impurity elements in most uraninite occurrences are probably Th, Ca, and REE (Janeczek and Ewing 1992b). Synthetic UO_2 and ThO_2 are isostructural, and form a complete solid solution, with the lattice parameter varying linearly with Th content (Frondel 1958). Thorium contents of magmatic uraninite, however, rarely reach levels above approximately 10 or 12 wt % ThO_2 (Frondel 1958; Grandstaff 1976; Förster 1999). Most Th in Ca-poor peraluminous granites resides in monazite-group minerals (Friederich and Cuney 1989), suggesting a greater affinity of Th for phosphate minerals than U. Reduced U shows a strong tendency to crystallize as uraninite (Förster 1998a). Although Podor et al. (1995) demonstrated that U + Ca can substitute without limit for La in synthetic monazite-(La), even U-rich granites with low molar Th/U ratios tend to crystallize monazite-group minerals with low U contents and Th well in excess of U (Förster 1998a,b). In contrast to uraninite, thorianite is relatively rare. Crystallizing in some Th-rich pegmatites, thorianite is most commonly found as detrital grains, sometimes forming economic Th deposits (Frondel 1958).

Calcium contents reported in magmatic uraninite tend to be rather low, up to perhaps 0.5 wt % CaO (~12 mol %) (Frondel 1956; Förster 1999). The ionic radius of Ca^{2+} (1.12 Å) makes it reasonably compatible with the uraninite structure, provided a charge-balance mechanism is available. High-temperature studies of UO_2-CaO solid solutions suggest that approximately 47 mol % CaO may be incorporated into the UO_2 structure above approximately 1500 to 1800°C (Alberman et al. 1951; Pialoux and Touzelin 1998). The high-temperature solid solution is disordered, whereas below approximately 1250°C an ordered solid solution, cubic $U_{1-x}Ca_xO_{2-x}$, exists between $x \cong 0.25$ and $x \cong 0.15$-0.05 (x decreasing with temperature) (Pialoux and Touzelin 1998). Maximum reported Ca contents in natural magmatic uraninite are close to the lower end of this solid-solution range. The degree to which these results apply to natural uraninite is uncertain; however, Ca^{2+} may help charge balance U^{5+} or U^{6+} in uraninite, a factor not addressed in the studies by Alberman et al. (1951) and Pialoux and Touzelin (1998). Comparatively high concentrations of Ca reported for hydrothermal and low-temperature uraninite (Frondel 1958; Janeczek and Ewing 1992b may reflect Ca^{2+} incorporated for charge-balance during uraninite formation, possibly compensating U^{6+} (Janeczek and Ewing 1992b; Finch and Ewing 1992b); however, high Ca contents are most commonly reported for fine-grained uraninite and may include mineral inclusions such as calcite (Janeczek and Ewing 1995; Janeczek this volume). Calcium may also be an important charge-balancing species replacing Pb^{2+} during uraninite alteration, especially under conditions where U^{6+} (or U^{5+}) are not completely reduced. A Ca-Pb exchange reaction may be something like

$$(U^{4+}_{1-2y}U^{6+}_y Pb^{2+}_y)O_{2\,(cr)} + yCa^{2+} + yHS^- \Rightarrow (U^{4+}_{1-2y}U^{6+}_y Ca_y)O_{2\,(cr)} + yPbS_{(cr)} + yH^+$$

Support for the potential importance of this reaction is found in commonly negative correlations between Pb and Ca in uraninite altered under anoxic or reducing conditions (Berman 1957; Frondel 1958; Janeczek and Ewing 1992c and references therein; Fayek et al. 1997). The above reaction may be most important where volume- and grain-boundary-enhanced diffusion dominate, because dissolution and reprecipitation under reducing conditions should favor reduction of U^{6+}.

Yttrium and lanthanides (REE) tend to be relatively minor substituents in uraninite, with typical concentrations being a few tens of ppm to a few tenths of a weight percent

REE_2O_3. Although concentrations of 10-15 wt % REE_2O_3 have been reported for some pegmatitic uraninites (Berman 1957; Frondel 1958), such high values may reflect spurious mineral inclusions, such as monazite. Most recent analyses by microprobe techniques indicate that REE_2O_3 concentrations in uraninite rarely exceed 2-4 wt % (Janeczek and Ewing 1992b; Foord et al. 1997; Förster 1999). Concentrations of REE in uraninite can vary widely, depending on local environment (Pagel et al 1987; Foord et al. 1997; Förster 1999), and, as for Th, concentrations of P in mineralizing fluids probably play a crucial role in determining the degree to which REE are incorporated into uraninite. Relative REE trends in uraninite are variable, with some uraninite displaying light REE (LREE) enrichments relative to Y and heavy REE (HREE). Others display relative enrichments in HREE, whereas yet others show no REE fractionation or even middle REE enrichment (Pagel et al. 1987). Positive Ce anomalies are rare (though not unknown), whereas negative Eu anomalies are rather common (Pagel et al. 1987); both observations are consistent with expected redox influences on Ce and Eu under reducing conditions where uraninite is stable. Substitution of REE^{3+} for U^{4+} in synthetic UO_2 increases the lattice parameter (Stalbauer et al. 1974), but the influence of minor REE substitution on the lattice parameter of natural uraninite has not been examined.

Other elemental impurities reported in uraninite include Si, P, Al, Fe, Mg, Na and K (Berman 1957; Frondel 1958; Finch and Ewing 1992b; Pearcy et al. 1994; Janeczek at al. 1996; Foord et al. 1997). The exact roles of these elements in uraninite remain uncertain, but most are not compatible with the uraninite structure. Unaltered magmatic uraninite commonly contains few if any of these elements (Berman 1957; Frondel 1958; Förster 1999), although Foord et al. (1997) report more than 2 wt % K_2O and up to 0.5 wt % Na_2O in pegmatitic uraninite (exposed to alkali metasomatism). The highest concentrations of impurities are most commonly reported for exceptionally fine-grained uraninite ("pitchblende"). Silicon and P may be relatively high in partly altered uraninite due to replacement of uraninite by coffinite (Janeczek 1991; Janeczek and Ewing 1992a,c). It is not always clear how much Si and P are due to inclusions of coffinite in uraninite, but reports of relatively high Si contents in uraninite are not uncommon, even where coffinite is not identified (Finch and Ewing 1992b; Pearcy et al. 1994; Foord et al. 1997). Uraninite precipitated from low-temperature groundwaters (T < 100°C) in sedimentary-hosted U deposits can be very fine grained, and analyses reported for such fine-grained uraninite commonly include elements from mineral inclusions along grain and sub-grain boundaries (Janeczek et al. 1996; Janeczek and Ewing 1995). Detailed TEM examinations of fine-grained uraninite commonly reveal sub-micron-scale Si-rich inclusions along grain boundaries. The most common mineral inclusion is probably coffinite (Janeczek and Ewing 1991, 1992c; Fayek et al. 1997), but other minerals, especially clay minerals such as kaolinite (Pearcy et al. 1994) and chlorite (Janeczek and Ewing 1995) may be intimately intergrown with fine-grained uraninite. It is increasingly apparent that elements such as Al, Si and P do not substitute in the uraninite structure to any significant degree, so that uraninite analyses that include these elements probable indicate spurious mineral inclusions.

Uranium(IV) silicates

Coffinite. After uraninite, coffinite is the most important ore mineral for U. Coffinite is a tetragonal orthosilicate isostructural with zircon, $ZrSiO_4$, hafnon, $HfSiO_4$, thorite, $ThSiO_4$, and synthetic $PuSiO_4$ and $NpSiO_4$ (Speer 1982). Uranium occupies triangular-dodecahedral sites coordinated by isolated Si tetrahedra (see Burns this volume). Chemical analyses of coffinite commonly indicate H_2O, and the formula was first reported as $U(SiO_4)_{1-x}(OH)_{4x}$ in allusion to thorogummite (Stieff et al. 1956); however, compositional and IR data indicate that H_2O is molecular and the correct formula is $USiO_4 \cdot nH_2O$ (Speer

1982; Lumpkin and Chakoumakos 1988; Smits 1989; Janeczek 1991) (Table 1). Uranium-Th silicates form two complete series of anhydrous and hydrated compounds, for which the general formulas are $(U,Th)SiO_4$ and $(Th,U)SiO_4 \cdot nH_2O$ ($n < 4$). These two formulas encompass three mineral species (Smits 1989): coffinite, the hydrated and anhydrous end member with U > Th; thorite, the anhydrous Th end member with Th > U; and thorogummite, the hydrated Th end member with Th > U (Tables 1 and 2). All three minerals are tetragonal and thorogummite may be redundant as a separate species. Coffinite samples tend to be so fine grained and impure that accurate analyses can be exceedingly difficult to obtain. Metamictization and aqueous alteration may also be significant, especially for thorite and thorogummite (Lumpkin and Chakoumakos 1988). Coffinite may be less amenable to substitutions than uraninite, and many impurities reported in coffinite analyses may be present as mineral inclusions. There are, however, some chemical substitutions that are well documented.

Phosphorous and REE are among the most common impurity elements in most reported coffinite analyses. Coffinite may contain substantial amounts of REE and P, suggesting some solid solution with xenotime, YPO_4 (Hansley and Fitzpatrick 1989; Janeczek and Ewing 1996), with which coffinite is isostructural. Calcium-rich coffinite from Bangombé, Gabon, has P well in excess of REE, which was explained by fine-scale (< 1 μm) inclusions of amorphous material with a composition similar to that of ningyoite (Belova et al. 1980); however, Janeczek and Ewing (1996) found no evidence for such inclusions and proposed a limited solid solution between coffinite and ningyoite, according to the substitutions,

$$2\ Ca^{2+} + 0.8\ P^{5+} + 0.2\ ^{IV}\square = U^{4+} + Si^{4+} \text{ and } 2\ Ca^{2+} + P^{5+} + (OH)^- = U^{4+} + Si^{4+},$$

where $^{IV}\square$ represents tetrahedral-site vacancies in the coffinite structure. It is unclear whether the extra Ca^{2+} cation is proposed to occupy interstitial sites in coffinite or what the structural role of $(OH)^-$ might be (Janeczek and Ewing 1996 only refer to "hydroxylation"). Perhaps the second substitution should be written, $Ca^{2+} + P^{5+} + (OH)^- = U^{4+} + Si^{4+} + O^{2-}$, although it has already been noted that hydroxyl substitution in coffinite is relatively minor (Speer 1982; Janeczek 1991). Of course, because of structural differences, a complete solid solution between coffinite and ningyoite is impossible, whereas substantial solid solution between coffinite and xenotime seems likely.

Coffinite is the major U-bearing mineral in many sandstone-hosted U deposits that extend from western South Dakota to eastern Arizona in the United States (Finch 1996). In these mostly low-temperature U deposits, coffinite most commonly occurs intimately intermixed with organic material, such as lignite. Coffinite is also a common alteration product of uraninite in Si-rich, reducing environments (Janeczek 1991; Janeczek and Ewing 1992a). Coffinite occurs in placer deposits in the Dominion Reef of the Witwatersrand, South Africa, where it replaces detrital uraninite grains (Smits 1989). Janeczek (1991) describes the crystal chemistry and paragenesis of coffinite from Jáchymov, Czech Republic. Plant et al. (this volume) provide more detail on the origin of coffinite-bearing U deposits.

U(IV) niobates, tantalates and titanates

A large number of complex Ta, Nb and Ti oxides are known that contain U in various amounts (Table 2). For the most part these minerals occur as accessory minerals in granitic rocks and granite pegmatites. Several are important Ta and Nb ore minerals and may be mined for REEs. A few contain U as an essential constituent (Table 1), and the U is usually oxidized to some degree (Smith 1984). Nearly all contain some U and Th in solid solution, and are therefore important actinide hosts in many granitic rocks, as well as being important sources for dissolved U in hydrothermal and meteoric waters with which they interact. Many of these minerals have long defied detailed understanding because, due to their abilities to incorporate radioactive elements, they are commonly metamict; specimens may also be strongly altered. Metamict minerals offer a special challenge to mineralogists trying to glean structural information about their crystalline precursors. Attempts to elucidate compositional and structural details of metamict minerals may involve annealing (heating) mineral specimens in order to recrystallize the original structure. Although annealing is commonly successful, it is not always known whether the recrystallized compounds represent the original minerals. Redox conditions during annealing may change oxidation states of some elements (e.g. Fe or U), and, because of possible post-formation alteration, it not always clear what oxidation state some elements were in at the time of crystallization (Sugitani et al. 1984; Warner and Ewing 1993). Such experimental difficulties are exacerbated by the nearly ubiquitous alteration of these minerals (Ewing 1975; Lumpkin and Ewing 1992a, 1995, 1996; Warner and Ewing 1993), so that annealed samples may include spurious compounds formed from aqueous alteration products.

These minerals share a common structural feature: Nb, Ta, and Ti occupy octahedral sites, and the octahedra share corners or edges (or both) to form the structural framework; these are designated as B sites in the structural formulas. Additional cations occupy the so-called A sites, which are (ideally) either six-coordinated (octahedral) or eight coordinated (distorted cube), depending on the size of the A-site cation. U and Th occupy the A sites in these minerals. Eight structurally related groups are listed in Table 3, which gives the name of each mineral or mineral group as a function of the dominant B-site cation. Minerals in these groups may have prefixes that identify the dominant A-site cation (e.g. uranopyrochlore). There are also Sb^{5+} analogues for several of these minerals, including sibiconite, bindhemite, triphuyite, and others (Gaines et al. 1997), but they are not discussed in detail here. Recent data for the pyrochlore-group mineral roméite are reported by Brugger et al. (1997).

Structurally, these minerals may be divided into two groups. The structures of the ixiolite, samarskite (wolframite), and columbite groups consist of approximately hexagonally close packed O atoms. The A and B sites are both octahedrally coordinated. Octahedra share edges to form chains along [001] and layers parallel to (100) (notation after Warner and Ewing 1993). The A and B octahedral layers alternate along [100]. All three groups have structures that are derivatives of the α-PbO_2 structure (Graham and Thornber 1974; Warner and Ewing 1993) (Fig. 1). The structure of brannerite is distinct from the above groups; however, it is related, with both A and B sites being octahedrally coordinated (Szymanski and Scott 1982; Burns this volume). The second group is closely similar to the first, with octahedrally coordinated B cations; however, the A sites occupy distorted cubic sites (ideally 8-coordinated, but distortions are known). The larger A site in this second group accommodates cations such as Ca^{2+} and Na^+, in addition to actinides and REE. Fergusonite and aeschynite are structurally similar, differing primarily in how their alternate (100) layers stack (Warner and Ewing 1993). The structures of pyrochlore and zirconolite share some basic features with those of fergusonite and aeschynite: octahedral chains and large A sites; however, the pyrochlore structure is a derivative of the CaF_2 structure (Chakoumakos 1984) and zirconolite is a derivative of the pyrochlore structure (Mazzi and Munno 1983; Bayliss et al. 1989). The most important U-bearing Ta-Nb-Ti oxides are discussed further below.

Brannerite AB_2O_6. After uraninite and coffinite, brannerite is the most important ore mineral for U. Nominally $U^{4+}Ti_2O_6$, the U in brannerite is nearly always partly oxidized. Brannerite is typically metamict and requires annealing to produce an XRD pattern. As noted, the structure of brannerite is related to, but distinct from other AB_2O_6

MO$_2$ (ixiolite) ABO$_4$ (samarskite) AB$_2$O$_6$ (columbite)

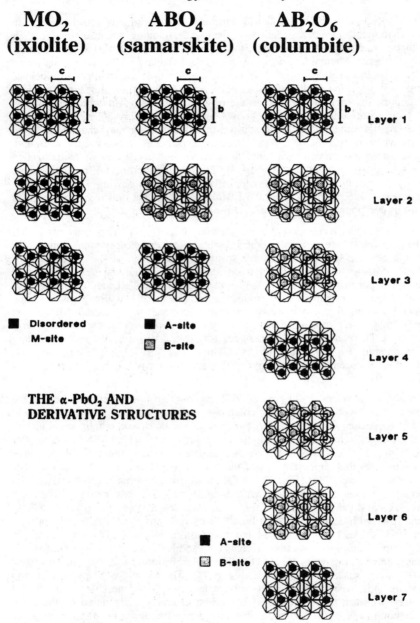

Figure 1. Depiction of the α-PbO and derivative structures (above) and the mixed coordinate structures (next page) in Nb-Ta-Ti oxides. Both A and B sites in the α-PbO and derivative structures are octahedral, whereas the mixed coordination structures have sheets of eight-coordinated A sites alternating with sheets of zigzag chains of octahedra (B sites). The a cell dimension is tripled in the AB$_2$O$_6$ structure types (columbite and polycrase) relative to the ABO$_4$ and MO$_2$ structure types due to cation ordering. Both AB$_2$O$_6$ structure types and the α-PbO-type MO$_2$ structure are orthorhombic, but distortions caused by ordering result in the ABO$_4$ structure types being monoclinic (modified slightly from Warner and Ewing 1993).

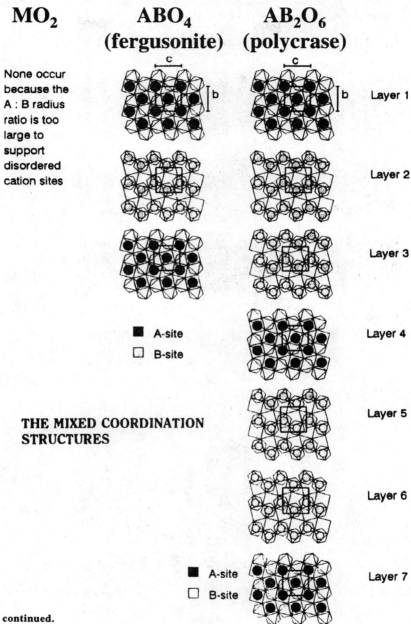

Figure 1, continued.

oxides, with layers of edge-sharing Ti octahedra and layers of distorted U octahedra (see Burns this volume). The structure of orthobrannerite remains unknown and crystals are essentially metamict. It is not simply a polymorph of brannerite, as it reportedly contains substantial U^{6+} (Table 1). Singh et al. (1990) report that a metamict brannerite heated to 900°C in air revealed $U^{6+}TiO_5$ and rutile in addition to brannerite. When heated to 1000°C,

Table 3. Structural comparisons among some Nb, Ta, Ti oxides

Representative mineral	Formula	A-site*	Dominant B-site cation			Structure type
			Ta	Nb	Ti	
Ixiolite	MO_2	octahedral	ixiolite	ashanite	zirkelite	α-PbO_2
Wolframite	ABO_4	octahedral	X	samarskite	X	α-PbO_2 derivative
Columbite	AB_2O_6	octahedral	tantalite	columbite	X	α-PbO_2 derivative
Brannerite	AB_2O_6	octahedral	X	X	brannerite	$ThTi_2O_6$
Fergusonite	ABO_4	distorted cube	formanite	fergusonite	X	interlayered α-PbO_2 derivative
Polycrase	AB_2O_6	distorted cube	rynersonite	vigezzite	polycrase, aeschynite	interlayered α-PbO_2 derivative
Pyrochlore	$A_{1-2}B_2O_6(O,OH,F)$	distorted cube	microlite	pyrochlore	betafite	distorted CaF_2 derivative
Zirconolite	$AA'B_2O_7$	distorted cube	X	(unamed)[#]	zirconolite	pyrochlore derivative

* U substitution occurs at the A site.   [#] See Williams & Gieré (1998).

the UTiO$_5$ and rutile disappeared and only brannerite was detected in the XRD powder pattern. Singh et al. (1990) proposed that UTiO$_5$ and rutile are formed during metamictization and alteration of brannerite, caused by oxidation of U^{4+} in brannerite according to the reaction, 2UTi$_2$O$_6$ + O$_2$ $\Rightarrow$ 2UTiO$_5$ + 2TiO$_2$. Smith (1984) suggested that orthobrannerite is related to synthetic UTiO$_5$, which tends to be substoichiometric (UTiO$_{5-x}$, x 0.15) and contains U^{5+} (Miyake et al. 1994). Though readily synthesized (Bobo 1964; Marshall and Hoekstra 1965), the structure of UTiO$_5$ is apparently unknown. However, it is almost certainly related to U^{6+}Mo^{4+}O$_5$ (D'yachenko et al 1996) and U^{6+}V^{4+}O$_5$ (Chevalier and Gasperin 1970; Dickens et al. 1992), both of which are orthorhombic and possess the uranophane-type sheet anion topology (cf. Miller et al. 1996; Burns et al. 1996, Burns this volume). These two synthetic compounds lack UO$_2^{2+}$ ions (Bobo 1964; D'yachenko et al 1996), and if orthobrannerite turns out to be isostructural with UMoO$_5$ or UVO$_5$, it would be the only mineral known to contain U^{6+} without forming UO$_2^{2+}$ ions.

Brannerite is a common accessory mineral in numerous uraninite and coffinite U deposits and has been identified at both unconformity-type and hydrothermal-vein U deposits (Finch 1996). In some deposits, brannerite may form following adsorption of U onto Ti oxides (McCready and Parnell 1997, 1998). The paragenesis of brannerite is distinct from that of most Ta, Nb, and Ti oxides, which form primarily in magmatic systems. This may explain why there are no known Ta- or Nb-analogues of brannerite (Table 3), although coupled cation substitutions such as REE^{3+} + Ta^{5+} = U^{4+} + Ti^{4+} might be expected. Thorutite is the Th-analogue of brannerite (Smith 1984); it too is commonly metamict.

Columbite group AB_2O_6. The columbite group of minerals comprises a large number of structurally related orthorhombic AB$_2$O$_6$ compounds (B = Ta, Nb). The columbite subgroup is Nb-dominant, and the tantalite subgroup Ta-dominant; there are no known minerals in which Ti is dominant (Table 3), although there seems no reason not to expect one, provided charge balance can be maintained. Most commonly occurring as accessory minerals in granite pegmatites (Gaines et al. 1997), columbite-group minerals contain U (and Th) in various amounts and are commonly metamict (Table 3), but none has been described with U as an essential constituent. The relatively small octahedral A site is commonly occupied by Mg^{2+} (magnesiocolumbite) and transition-metal cations, such as Fe^{2+} (ferrocolumbite) and Mn^{2+} (manganocolumbite), and U and Th substitution tend to relatively minor. The structures of brannerite and thorutite might be considered as actinide analogues of the columbite structure, but distortions caused by U and Th in octahedral sites results in these two structure types being significantly different.

Polycrase group AB_2O_6. The polycrase structure type is comparable to that of the columbite group, except that the A site is a distorted cube (Fig. 1). Ti exceeds (Ta+Nb) in the B site of polycrase-(Y); Nb is dominant in the B site of euxenite-(Y) and vigezzite; and tantaeuxenite-(Y) and rynersonite are Ta-dominant in the B site (Table 3). Charge balance is maintained by Ca^{2+} replacing REE^{3+} at the A site. Euxenite-(Y) and tantaeuxenite-(Y) are compositionally intermediate between polycrase-(Y) and vigezzite and rynersonite, the latter two being Ca-dominant at the A site. This group occurs primarily as accessory minerals in granites and granite pegmatites (Gaines et al. 1997). All are commonly metamict (Table 3) owing to U (and Th) substitution at the A site, especially, it seems, in the Y-dominant minerals. However, only uranopolycrase contains U as an essential constituent (Table 1). Uranopolycrase was originally described from a pegmatite near Campo Village on Elba Island, Italy, and occurs as opaque, reddish brown, elongate [100] orthorhombic crystals that display good (100) cleavage (Aurisicchio et al. 1993). Uranopolycrase is "almost completely metamict" and heating to 900°C produces a sharp X-ray diffraction pattern indicating orthorhombic symmetry (*Pbcn*) (Aurisicchio et al. 1993). Uranopolycrase occurs in a pegmatitic vein, where it is associated with uranmicrolite, euxenite-(Y), manganocolumbite, and titanowodginite. The recent structure determination of a crystalline polycrase-(Y) crystal from Malawi (Johnsen et al. 1999) verifies the iso-structural relationship between polycrase and uranopolycrase, and with synthetic $Y(Nb_{0.5}Ti_{0.5})_2O_6$ (von Weitzel and Schröcke 1980).

Samarskite group, ABO_4. Minerals of the samarskite group have Nb > Ta in the B site. The structures of the samarskite-group minerals are similar to that of wolframite, $MnWO_4$, a derivative of the α-PbO_2 structure (Graham and Thornber 1974; Warner and Ewing 1993). The samarskite group has the ideal formula $A^{3+}B^{5+}O_4$ (Warner and Ewing 1993), and the A sites are in octahedral coordination (Table 3). Hanson et al. (1999) reported detailed analyses of ishikawaite specimens and showed that this mineral is properly classified as a member of the samarskite group. There are three known members of the samarskite group of minerals, defined on the dominant A-site cation: samarskite-(Y) is Y dominant, ishikawaite is (U+Th) dominant, and calciosamarskite is Ca dominant (Hanson et al. 1999); however all contain substantial U and Th. An XRD study of a metamict samarskite was conducted by Keller and Wagner (1983), who reported the radial distribution function.

Fergusonite group ABO_4. The fergusonite group consists of REE-bearing Ta and Nb oxides, many of which are metamict and, therefore, commonly poorly characterized, with most available structural information derived from studies of heated material or synthetic analogues. The structure of the fergusonite group is comparable to that of samarskite group (wolframite structure type) but with large A sites (distorted cubes) (Fig. 1). Most of these minerals are monoclinic, although orthorhombic and tetragonal unit cells arise from cation ordering (Gaines et al. 1997). Fergusonite-(Nd) is reportedly tetragonal (Gong 1991b), and partially metamict fergusonite-(Y) may consist of both monoclinic and tetragonal forms (Gong 1991a,b). A Nd-dominant polymorph of fergusonite was described from a REE deposit in Proterozoic dolomite at Bayan Obo, Inner Mongolia, China, where it occurs as small (0.02-0.25 mm) irregular grains and prisms in association with fergusonite-(Ce) and fergusonite-(Nd), aeschynite-(Nd), monazite, bastnaesite, riebeckite, ferroan dolomite, ilmenite, biotite, magnetite and pyrite (Weijun et al. 1983). It is proposed to be a monoclinic polymorph that the authors call "β-fergusonite-(Nd);" however, its definition as a new mineral is hampered by the fact that it is nearly metamict and is considered a questionable species (Nickel and Nichols 1992). Fergusonite-(Nd) contains approximately 1.1 wt % ThO_2 and 1.5 wt % UO_2. The B site in the fergusonite group is dominated by Nb, whereas Ta dominates in formanite; no Ti analogue is known (Table 3).

Pyrochlore group, $A_{1-2}B_2O_6(O,OH,F)$. The pyrochlore group is a particularly important group of Nb-Ta-Ti oxides that can contain substantial U. The structure of ideal pyrochlore, $A_{1-2}B_2O_6(O,OH,F)$, is a defect derivative of the fluorite structure type (Chakoumakos 1984, 1986). The structure is essentially a framework of B site octahedra with Ta, Nd, and Ti, and which can also contain Fe, Sn, W, and Sb (Fleischer and Mandarino 1999); Sb^{5+} can even dominate at the B site, as for roméite (Brugger et al. 1997). The A site is eight coordinated (distorted cube) and may contain alkalis, alkaline earths, REE and actinides. Charge balance is maintained in pyrochlore through cation substitutions at either A or B sites as well as through anionic substitutions. Three pyrochlore subgroups are defined, depending on the predominant cation in the B site. Niobium exceeds Ta in the pyrochlore subgroup, whereas Ta exceeds Nb in the microlite subgroup. Both the pyrochlore and microlite subgroups have (Ta + Nb) > 2Ti, whereas the betafite subgroup is characterized by 2Ti > (Ta + Nb) (Table 3). The compositions of most pyrochlores cluster near the pyrochlore-betafite join, with Ta less than 10 mol %, and near the microlite end member, with 30 mol % Nb or less and less than 15 mol % Ti. U substitutes at the A site, and metamict pyrochlores are common. Although virtually all these minerals contain some U, only two minerals of the pyrochlore group contain U as an essential constituent: uranmicrolite and uranopyrochlore (Table 1). Pyrochlore has been studied as a potential actinide-bearing waste form and is a constituent of Synroc® and related crystalline ceramics being developed for nuclear waste disposal (Ringwood et al. 1988; Lumpkin et al. 1994). Numerous defect structures can be derived from the pyrochlore structure type (Lumpkin and Ewing 1988). Additional information on the pyrochlore structure can be found in Lumpkin and Ewing (1988), in Gaines et al. (1997) and by Burns (Chapter 2, this volume).

Zirconolite group, $A_2B_2O_7$. The structure of zirconolite, $CaZrTi_2O_7$, can be described as a derivative of the pyrochlore structure, with octahedrally coordinated B sites and A sites in distorted cubes. Zirconolite is monoclinic and has two distinct A sites, designated A (Ca) and A' (Zr) in Table 3. Zirconollite is Ti dominant at the B site, and Nb-dominant zirconolite minerals were identified from carbonatites in Kovdor (Williams and Gieré 1996); however, no Ta-dominant zirconolite-group minerals are known (Table 3). As for most other Ta-Nb-Ti oxides, U substitutes at the large cation sites, primarily for Ca at the A site in zirconolite; however, no U dominant zirconolite-group minerals are known. Nevertheless, U and Th substitution in zirconolite can be sufficient to induce substantial structural damage, and metamict zirconolites are not uncommon. In addition to its being an important accessory mineral in a wide variety of rocks, zirconolite has been studied as a potential actinide-bearing nuclear waste form (Vance et al. 1994; Lumpkin et al. 1994; Hart et al. 1996; Putnam et al. 1999; Woodfield et al. 1999). Gieré et al. (1998) suggest that redox conditions strongly influence cation substitutions in zirconolite; notably, ferric-ferrous ratios in crystallizing fluids can affect charge-balance. Williams and Gieré (1996) review numerous zirconolite occurrences and report chemical analyses. Mazzi and Munno (1983) clarified distinctions among zirconolite and zirkelite, and elucidated their structural relationships to the pyrochlore-group. Bayliss et al. (1989) further explain polytypism and "polytypoids" among zirconolite and related minerals.

Ixiolite and other α-PbO_2 structure types, MO_2. Ixiolite, $(Ta,Mn,Nb)O_2$, ashanite, $(Nb,Ta,Fe,Mn,V)_4O_8$, and zirkelite, $(Ti,Ca,Zr)O_{2-x}$, are structurally related to α-PbO_2, with octahedrally coordinated cation sites (Mazzi and Munno 1983; Warner and Ewing 1993). All three minerals may contain minor U (and Th). The papers by Mazzi and Munno (1983) and Bayliss et al. (1989) provide additional information on zirkelite and its relationship to zirconolite and the pyrochlore group.

Petschekite and liandratite. Both of these minerals are reportedly metamict and known from only one locality, a pegmatite in Madagascar (Mücke and Strunz 1978). Petschekite crystallizes when annealed, and Smith (1984) noted that the structure appears related to synthetic UTa_2O_8, a derivative of the U_3O_8 structure (Gasperin 1960). If so, petschekite may be more closely related to orthobrannerite than to other Nb,Ta oxides. Petschekite reported alters to a series of partly oxidized ($Fe^{2+} \rightarrow Fe^{3+}$), hydrous, Fe-depleted compounds (Mücke and Strunz 1978). Liandratite is the fully oxidized alteration product of petschekite and contains U^{6+}. It occurs as thin 1-2 mm-thick yellow to yellow-brown glassy coatings on the surfaces of petschekite crystals. Both minerals remain poorly described, and no data have been reported since their initial descriptions (Gaines et al. 1997).

U(IV) phosphates: rhabdophane group

Ningyoite. The rhabdophane group of phosphates are hexagonal phosphates with ideal formula $APO_4 \cdot H_2O$ (Bowels and Morgan 1984) or $APO_4 \cdot nH_2O$ ($0.5 \leq n \leq 1$) (Hikichi et al. 1989). The most common minerals of this group contain REE in the A site, although four are known to contain actinides: ningyoite (A = U, Ca, Fe), grayite (A = Th, Pb, Ca), tristramite (A = Ca, U, Fe^{3+}), and brockite (A = Ca, Th, REE). Charge balance is maintained by substitutions of divalent cations at the A site or, possibly, by OH substitution for apical O atoms of PO_4 tetrahedra; Sharmová and Sharm (1994) suggested the formula $Ca_{2-x}U_x[P(O,OH)_4]_2 \cdot nH_2O$ for ningyoite (note that the formula for grayite in Dana's New Mineralogy (Gaines et al. 1997) is not charge balanced). Rhabdophane-group minerals have acicular habits, reflecting the ring-like structure of isolated PO_4 tetrahedra and A sites that form large channels parallel to the c axis; H_2O groups occupy sites within the channels. Ningyoite is orthorhombic, pseudo-hexagonal, due to doubling of the b cell edge. The structure of ningyoite has not been determined. Ningyoite is the most important U mineral at the Ningyo-toge mine in Japan, where it was first discovered (Muto et al. 1959), occurring as microcrystalline crusts and within cracks, rarely forming micrometer-sized crystals. Ningyoite has also been found in the U ore district of northern Bohemia, Czech Republic (Sharmová and Scharm 1994), where it reportedly forms a continuous solid-solution series with brockite: $Ca_{2-y}(U_{1-x}Th_x)_y[P(O,OH)_4]_2 \cdot nH_2O$. Sharmová and Scharm (1994) noted that many ningyoite crystals are P deficient, which they suggest is due to CO_3 and SO_4 replacing PO_4 groups. As noted above, Janeczek and Ewing (1996) proposed a limited solid solution between ningyoite and coffinite.

Vyacheslavite and lermontovite. Lermontovite occurs in reduced hydrothermal deposits in the Kola Peninsula, Russia, where it is associated with Tl minerals ("Tl ochre"), molybdenum sulfates and marcasite (Gaines et al. 1997). Lermontovite is remarkably similar to vyacheslavite, which has nearly the same nominal formula, being slightly more hydrated (Table 1). Both minerals display similar physical and optical properties, although their optical orientations and crystal habits differ. Vyacheslavite was described from an undisclosed location in Uzbekistan, where it occurs on quartz associated with pyrite (Belova et al. 1984).

Other U(IV) minerals

Numerous minerals contain U as a minor substituent and some are listed in Table 2. U^{4+} and Th^{4+} may substitute freely for one another, and Th-bearing minerals are especially important among those minerals that contain U. Many of these are included in discussions of various mineral groups above (e.g. thorianite and thorite). The structurally related REE phosphates of the *monazite group* and *xenotime* are also important U- and Th-bearing minerals in many granitic rocks, with U and Th substituting for primarily LREE in monazite and HREE in xenotime, through appropriate charge-balance substitutions (Förster 1998a,b; Hanchar et al. 1999; Finch et al. 1999c), and, as noted, coffinite may form a limited solid solution with xenotime (Janeczek and Ewing 1996). We will not attempt to completely describe all minerals with minor U in detail. They are described in Dana's New Mineralogy (Gaines et al. 1997) and additional references are provided for some minerals listed in Table 2. Additional mineralogical information is also available in Fleischer's Glossary of Mineral Species (Mandarino 1999) and Mineral Reference Manual (Nickel and Nichols 1992). Of particular note, however, are a group of chemically similar Th silicates discussed below.

Ekanite, steacyite, thornasite, iraqite-(La) and related Th silicates: Four of these minerals are structurally related, based on rings of silicate tetrahedra. Thorium is an essential constituent in ekanite, steacyite, and thornasite, whereas it is a minor substituent in iraqite, substituting for REE (Table 2). Ekanite and steacyite have often been confused (the structure of steacyite was even reported as that of ekanite; Gaines et al. 1997). The structure of ekanite (and possibly thornasite) is based on two-dimensional infinite sheets of puckered four-member rings of SiO_4 tetrahedra; that of steacyite and iraqite-(La) are based on isolated four-member rings of tetrahedra. Thorium and other cations occupy interstitial sites coordinated by SiO_4, O atoms, and in some cases, OH and H_2O groups. These minerals are commonly metamict and may contain substantial amounts of U^{4+} in substitution for Th^{4+} (Diella and Mannucci 1986). Ekanite is known to form gem-quality material (Gauthier and Fumey 1988), although it is not advisable to wear such radioactive jewelry for long! These minerals occur as accessory minerals in a variety of felsic rocks and granite pegmatites. Numerous occurrences are described for ekanite and steacyite (Gaines et al. 1997); thornasite is known only from Mount St.-Hillarie, Quebec (Ansell and Chao 1987), and iraqite-(La) occurs in granite near a dolomite contact at Shakhi-Rash Mt. in northern Iraq (Livingstone et al. 1976).

An unnamed Th-silicate, $Th_2(Ca,Ba)(Si_9O_{22})(OH)_2 \cdot nH_2O$, conditionally called "thorsite" (Lazebnik et al. 1985,1994), but closely resembling metamict ekanite and steacyite in many respects, was described from calcite carbonatites in the Murun massif, where it is associated with thorite, quartz, K-feldspar, aegerine, titanite, tinaksite, apatite, and dalyite. The translucent yellow grains are highly radioactive and X-ray amorphous, presumably metamict; relict prism faces on grains suggest originally tetragonal symmetry. X-ray powder data for heated material (900°C) resemble those of huttonite. The authors note that heating metamict ekanite (though only to 650°C) also leads to crystallization of huttonite. "Thorsite" has not been accepted as a valid mineral name, and similarities to ekanite and steacyite suggest that "thorsite" may not be a distinct species (Table 2).

An unnamed U-bearing Ca-Th silicate, $Ca_{3.5}(Th,U)_{1.5}(SiO_4)_3(OH)$, was described by Jamtveit et al. (1997) from a regionally metamorphosed shale-limestone xenolith in the Skrim plutonic complex, Oslo Rift, southern Norway, where it apparently forms during prograde reactions of hydrothermal fluids on apatite (T = 820-870°C). This mineral apparently has the apatite type structure and occurs with other silicate apatites, along with wollastonite, melilite, phlogopite, titanian grossular, kalsilite, nepheline, perovskite, and other minerals.

Uranyl minerals

The most oxidized state for U in nature is U^{6+}, and in oxidizing, aqueous environments, U^{6+} always bonds strongly to two O atoms, forming the approximately linear uranyl ion, UO_2^{2+}. In the absence of fluoride, the (hydrated) uranyl ion is the dominant aqueous species in most waters below a pH of approximately 5. At higher pH, the uranyl ion hydrolyzes, forming a number of aqueous hydroxide complexes, according to the general hydrolysis reaction,

$$UO_2^{2+} + yH_2O \Rightarrow UO_2(OH)_y + yH^+ ,$$

and in more U-rich solutions, polymeric U complexes become increasingly important:

$$xUO_2^{2+} + yH_2O \Rightarrow (UO_2)_x(OH)_y^{2x-y} + yH^+$$

These hydroxy complexes are moderately weak solution complexes, and in most groundwaters dissolved carbonate combines with UO_2^{2+} to form uranyl carbonate solution complexes (Langmuir 1978; Grenthe et al. 1992; Clark et al. 1995; also see Murphy and Shock this volume). The uranyl carbonate complexes are quite stable in most groundwaters, and most dissolved U in near-surface groundwaters is probably present as uranyl carbonate complexes (Langmuir 1978; Clark et al. 1995). Where sulfide minerals are undergoing oxidation and dissolution in the presence of U minerals, uranyl sulfate complexes can be important. These, too, are stable solution complexes, and dissolved sulfate can be an important factor for the transport of U in some low pH groundwaters. Most other oxyanions that form complexes with UO_2^{2+} form relatively insoluble uranyl oxysalt minerals. Among these are uranyl silicates, phosphates, vanadates, arsenates, and molybdates.

The paragenesis of many uranyl minerals can be understood in terms of local groundwater chemistry, the relative solubilities of minerals, and the stabilities of relevant solution complexes. Most uranyl carbonates and sulfates are soluble in dilute groundwaters, precipitating where evaporation is significant. Uranyl oxyhydroxides are substantially less soluble than most uranyl carbonates (with the notable exception of rutherfordine), and can precipitate in abundance if solution complexes other than OH$^-$ are absent. Most uranyl silicates, phosphates, vanadates and arsenates are relatively insoluble, but require dissolved Si, P, V^{5+} and As, which may be derived from a variety of sources. Dissolved silica is a common constituent of many natural waters and, not surprisingly, the uranyl silicates, uranophane and ß-uranophane, are the most common uranyl minerals in nature (Frondel 1958; Smith 1984). Phosphate is another common constituent in many groundwaters, and uranyl phosphates are also common; however, these minerals present some interesting problems for those attempting to understand their paragenesis (Murakami et al. 1997; see below). Figure 2 summarizes uranyl mineral paragenesis for most important uranyl mineral groups from weathered U deposits in the Colorado Plateau, USA, as originally set forth by Garrels and Christ (1959) more that forty years ago. Their conclusions remain valid today and are relevant to many uranyl-mineral occurrences worldwide.

Uranyl oxyhydroxides

The uranyl oxyhydroxides (Table 4) can be represented by the general formula

$$M_n[(UO_2)_xO_y(OH)_z](H_2O)_m$$

where M represents divalent cations, commonly Ca^{2+}, Pb^{2+}, Ba^{2+}, and Sr^{2+}, although K$^+$-bearing phases are also known. The compositions of known uranyl oxyhydroxides are plotted in Figure 3, which shows mole fractions of MO as a function of total H$_2$O (the mole fraction of UO$_3$ is [1 − MO + H$_2$O]). Because some H$_2$O in uranyl oxyhydroxides occurs as H$_2$O groups in interlayer positions with the M cations, there is a general decrease in total H$_2$O as MO increases. This trend also reflects decreasing OH$^-$ in the structural sheets that compensates increased interlayer cation occupancies.

The uranyl oxyhydroxides form in U-rich aqueous solutions and develop early during the oxidation and corrosion of uraninite-bearing ore deposits, most commonly at or near the surface of corroded uraninite. Alteration of uranyl oxyhydroxides is ubiquitous, and the question of their long-term stability under various environmental conditions is pertinent to understanding the often complex assemblages of uranyl minerals found at many U deposits. The formation and alteration of uranyl oxyhydroxides can, in part, determine reaction paths and uranyl-minerals paragenesis at weathered U deposits, which help to control dispersion and fixation of U in many dilute groundwaters.

Significant progress has been made in our knowledge of this group of minerals in the 15 years since the review by Smith (1984). This has come with the advent of improved analytical methods, most notably the introduction of charge-coupled device (CCD) detectors for X-ray diffraction (Burns 1998b). The use of CCD detectors permits accurate structure determinations of very small crystals and of minerals with large unit cells, both of which are common problems among the uranyl minerals (see Burns 1998b for a discussion of the use of CCD detectors in X-ray structure analysis). Two new uranyl oxyhydroxides have been named since the review by Smith (1984), although Ondrus et al. (1997c) report several new unnamed species, which are listed in Table 4. One new mineral is the Pb-uranyl oxyhydroxide sayrite, discussed with other Pb minerals below. The other is

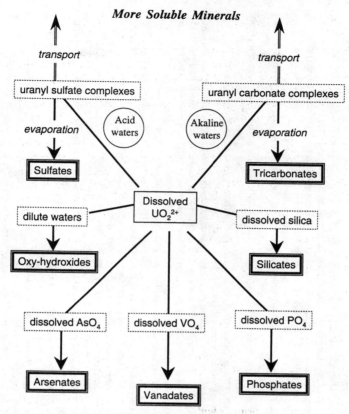

Figure 2. Schematic representation of the paragenesis of several important uranyl-mineral groups. Dissolved UO_2^{2+} is derived from the oxidative dissolution of U-bearing minerals and the arrows indicate interactions with additional dissolved species. Arrows pointing downwards indicate precipitation, and those pointing upwards indicate transport. Minerals are shown with more soluble species above those with generally lower solubilities; for qualitative comparison only (modified after Garrels and Christ 1959).

protasite, a rare hydrated Ba-uranyl oxyhydroxide from the Shinkolobwe mine, where it forms bright orange pseudohexagonal plates flattened on {010} with good {010} cleavage (Pagoaga et al 1986). The structure of protasite is monoclinic (*Pn*), though almost dimensionallyorthorhombic ($\beta = 90.4°$), with edge- and corner-sharing U polyhedra forming structural sheets parallel to (010) (Pagoaga et al. 1987). Protasite, which closely resembles fourmarierite in color and habit, is associated with Pb-uranyl oxyhydroxides and uranophane. The synthetic equivalent of protasite has been known for some time (Protas 1959), as has the compositionally similar Ca-analogue, although the Ca-analogue of protasite is unknown as a mineral.

At the time of Smith's (1984) review, the uranyl peroxides, studtite and metastudtite, were considered rare; however, both minerals have now been identified from a large number of localities, including the Shinkolobwe mine in Shaba, southern Democratic Republic of Congo, the Kobokobo pegmatite in western Democratic Republic of Congo, the Menzenschwand U mine, and U deposits in Lodève, France, sometimes in large

Table 4. Uranyl oxy-hydroxides

Name	Formula	Comments
Agrinierite*	$(K_2,Ca,Sr)(UO_2)_3O_2(OH)_3(H_2O)_4$	structure similar to protasite, Cahill & Burns (1999)
Bauranoite*	$BaO \cdot UO_3 \cdot 4\text{-}5H_2O$	ill defined; Belova et al. (1985)
Becquerelite*	$Ca(UO_2)_6O_4(OH)_6(H_2O)_8$	Pagoaga et al. (1987), Cejka et al. (1998a)
Billietite*	$Ba(UO_2)_6O_4(OH)_6(H_2O)_8$	Pagoaga et al. (1987), Cejka et al. (1998a); structure refinement, Finch et al. (in prep.)
Calciouranoite*	$(Ca,Ba,Pb,K_2,Na_2)U_2O_7 \cdot 5H_2O$	amorphous; ill defined; compare wölsendorfite, bauranoite, clarkeite; Belova et al. (1985)
Clarkeite*	$(Na,Ca)(UO_2)(O,OH)(H_2O)_n$, ($n = 0\text{-}1$)	cf. calciouranoite, bauranoite, wölsendorfite; Gaines et al. (1997); compositional zoning, Finch & Ewing (1997); compare synthetic $K_5[(UO_2)_{10}O_8(OH)_9](H_2O)$ of Burns & Hill (l99i)
Compreignacite*	$K_2(UO_2)_6O_4(OH)_6(H_2O)_7$	Gaines et al. (1997); Ondrus et al. (1997b); similar to becquerelite, Burns (1998b)
Curite*	$Pb_{1.5+x}(UO_2)_4O_{4+2x}(OH)_{3-2x}(H_2O)$ $(0 < x < 0.15)$	Gaines et al. (1997); Cejka et al. (1998a); variable Pb content, Li & Burns (1999a); Burns & Hill (1999a)
"Dehydrated schoepite"*	$(UO_2)O_{0.15-x}(OH)_{1.5+2x}$ $(0 < x < 0.25)$	dehydration of schoepite & metaschoepite; forms series with $\alpha\text{-}UO_2(OH)_2$; Christ & Clark (1960); Finch et al. (1998)
Fourmarierite*	$Pb(UO_2)_4O_3(OH)_4(H_2O)_4$	Deliens (1977a); Piret & Deliens (1985); Gaines et al. (1997); variable Pb content, Li & Burns (in prep.)
Ianthinite*	$U^{4+}(UO_2)_4O_6(OH)_6(H_2O)_9$	mixed-valence U species; Burns et al. (1997b)
Masuyite*	$Pb[(UO_2)_3O_3(OH)_2](H_2O)_3$	recent structure suggests possibility for Pb:U variability (Burns & Hanchar 1999). Formula is that of "grooved masuyite"; 4PbO·9UO$_3$·10H$_2$O also reported as "type masuyite" with distinct X-ray powder pattern (Deliens & Piret 1996); synthetic analogue is 3PbO·8UO$_3$·10H$_2$O (Protas 1959; Noe-Spirlet & Sobry 1974). Christ & Clark (1960); Deliens (1977a); Deliens et al. (1984) Finch & Ewing (1992); Ondrus et al. (1997b).
Metacalciouranoite*	$(Ca,Ba,Pb,K,Na)_2U \cdot UO_3 \cdot 2H_2O$	amorphous; ill defined; compare wölsendorfite, bauranoite, clarkeite; Belova et al. (1985)
Metaschoepite*	$(UO_2)_8O_2(OH)_{12}(H_2O)_{10}$	uncertain formula; Christ & Clark (1960), Debets & Loopstra (1963), Ondrus et al. (1997b); Finch et al. (1998)
Metastudtite	$UO_4 \cdot 2H_2O$	Peroxide; Deliens & Piret (1983b); Gaines et al. (1997)
Meta-vandendriesscheite	$PbO \cdot 7UO_3 \cdot (12\text{-}x)H_2O$ (?)	ill defined; structure probably similar to vandendriesscheite; Christ & Clark (1960)
Paraschoepite*	$UO_3 \cdot nH_2O$ (?) ($n \approx 2$)	ill defined; questionable species; Finch et al. (1992, 1997, 1998)
Protasite	$Ba(UO_2)_3O_3(OH)_2(H_2O)_3$	very rare; compare masuyite; Pagoaga et al. (1987); Cejka et al. (1998a)
Rameauite	$K_2Ca(UO_2)_6O_4(OH)_6(H_2O)_6$	Gaines et al. (1997)
Richetite*	$M_x Pb_{8.57}[(UO_2)_{36}O_{36}(OH)_{24}](H_2O)_{41}$	Variable H$_2$O? M is probably a transition-metal; e.g. Fe^{2+}, Burns (1998a); Piret & Deliens (1984); Ondrus et al. (1997b)
Sayrite	$Pb_2(UO_2)_5O_6(OH)_2(H_2O)_4$	Piret et al. (1983); Gaines et al. (1997)
Schoepite*	$(UO_2)_8O_2(OH)_{12}(H_2O)_{12}$	Finch et al. (1996); Finch et al. (1997); Finch et al. (1998)
Studtite*	$UO_4 \cdot 4H_2O$	Peroxide; Cejka et al. (1996a)
Uranosphaerite*	$Bi_2O_3 \cdot 2UO_3 \cdot 3H_2O$	ill defined; Kolitschn (1997); Gaines et al. (1997)
Vandenbrandeite*	$Cu(UO_2)(OH)_4$	Cejka (1994); Gaines et al. (1997)
Vandendriesscheite*	$Pb_{1.57}(UO_2)_{10}O_6(OH)_{11}(H_2O)_{11}$	Christ & Clark (1960); Deliens (1977a); Gaines et al. (1977a); Gaines et al. (1997); Burns (1997)
Wölsendorfite*	$(Ca,Ba)_x Pb_{7-x}[(UO_2)_{14}O_{19}(OH)_4(H_2O)_{12}$ ($x \le 1$?)	structure of barian wölsendorfite by Burns (1999e); $Pb_{6.16}Ba_{0.36}(UO_2)_{14}O_{19}(OH)_4(H_2O)_{12}$; compare calciouranoite, bauranoite, clarkeite. Deliens (1977a); Beddoe-Stephens & Secher (1977); Belova et al. (1985); Gaines et al. (1997)
Unnamed*	Hydrated uranyl hydroxide	Mineral "A" of Frondel (1956); hydrothermal origin; minor Pb, K, Th, Ca, Sr; Korzeb et al. (1997); Foord et al. (1997)
Unnamed*	Hydrated uranyl hydroxide	Amber brown crystals, with minor Pb; synthetic analogue: PDF 15-569; Threadgold (1960); Ondrus et al. (1997c)
Unnamed*	Hydrated uranyl oxide hydroxide	olive yellow; compare dehydrated schoepite; Ondrus et al. (1997c)

* New or supplemental data or not listed in Dana's New Mineralogy (Gaines et al. 1997).

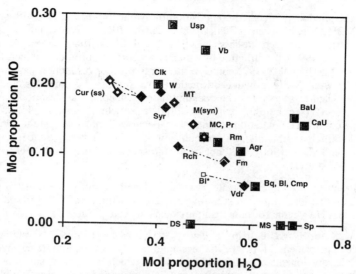

Figure 3. Compositions of the uranyl oxyhydroxides as a function of the molecular proportions of MO and H$_2$O (M = Ca, Pb, Ba, Sr, K$_2$). The Pb- uranyl oxyhydroxides are indicated by black-filled diamonds, richetite and wölsendorfite, which may contain cations besides Pb, are indicated with slightly lighter shadings. Pb-free uranyl oxyhydroxides are represented as lightly filled squares. Labels are as follows. Sp: schoepite; MS: metaschoepite; DS: "dehydrated schoepite;" Vdr: vandendriesscheite; Bq: becquerelite; Bl: billietite (Bl*: partly dehydrated billietite of Pagoaga et al. 1987); Cmp: compreignacite; Rch: richetite; Fm: fourmarierite; Agr: agrinierite; Rm: rameauite; MC: masuyite (Burns and Hanchar 1999; "gooved masuyite" of Deliens and Piret 1996); Pr: protasite; M(syn) synthetic analogue of masuyite (Protas 1959); MT: "type masuyite" of Deliens and Piret (1996); Syr: sayrite; W: wölsendorfite; Clk: clarkeite; Cur (ss): curite (solid solution indicated by shaded line at left; Li and Burns 1999; third diamond at right from curite of Taylor et al. 1981); CaU: calciouranoite; BaU: Bauranoite; Vb: vandenbrandeite; Usp: uranosphaerite.

quantities (e.g. at Menzenschwand). The structures of these two minerals are unknown, but the unit-cell parameters and space group of studtite suggest it is topologically similar to α-UO$_2$(OH)$_2$ (Taylor 1971), with alternate rows of U atoms missing along [001]. By placing a U atom at (0,0,0) in space groups $C2/m$ (studtite) and $Immm$ (metastudtite), X-ray powder diffraction patterns can be calculated that match well with data published for natural studtite (Debets 1963; Cejka et al. 1996a) and metastudtite (Debets 1963; Deliens and Piret 1983b). X-ray examinations of natural studtite commonly reveal mixtures of studtite and meta-studtite as well as other minerals, such as lepersonnite and oxyhydroxidesfourmarierite, the latter commonly occurring as thread-like inclusions within studtite and metastudtite crystals (Deliens and Piret 1983b).

The structure of ianthinite was reported by Burns et al. (1997b), and the formula differs slightly from that originally reported (Bignand 1955; Guillemin and Protas 1959) by being more reduced (U^{4+}:U^{6+} = 2:4 rather than U^{4+}:U^{6+} = 1:5). The structure determination by Burns et al. (1997b) was not of optimal quality, with an agreement index (R) of 9.7 %, and additional refinement may help clarify the details of the structure. Ianthinite is rarely found, but because it oxidizes readily in air to schoepite, it may be a common precursor to schoepite and metaschoepite (Deliens et al. 1984). Pearcy et al. (1994) identified ianthinite as an early oxidation product of uraninite at the Nopal I mine near Peña Blanca, Mexico.

A recent structure determination of schoepite (Finch et al. 1996) indicates that the composition is equivalent to that determined by Billiet and de Jong (1935), 4UO$_3$·9H$_2$O, rather than the previously accepted composition of synthetic preparations with slightly less water, $UO_3 \cdot 2H_2O$ (Christ and Clark 1960; Christ 1965). The synthetic compound apparently corresponds to metaschoepite (Christ and Clark 1960; Debets and Loopstra 1963; Finch et al. 1998), the structure of which has reportedly been determined (M Weller, pers. comm.). Finch et al. (1997) compare the X-ray powder diffraction pattern calculated for schoepite with X-ray powder diffraction data for several minerals and synthetic compounds related to schoepite. The dehydration of schoepite was described by Finch et al. (1992, 1998) (see below), who proposed a structural formula for "dehydrated schoepite", a common dehydration product of schoepite and metaschoepite in nature (Christ and Clark 1960). Paraschoepite, an inadequately described mineral (Schoep and Stradiot 1947; Christ and Clark 1960), may not be a valid species (Finch et al. 1997, 1998) and requires further study.

The structure of compreignacite was determined recently, demonstrating that it contains a sheet similar to the structure of becquerelite (Burns 1998c), and indicating that there is one less H_2O in the formula than previously accepted (Table 4).

New data for clarkeite help to clarify the formula. Clarkeite crystallizes during metasomatic replacement of pegmatitic uraninite by late-stage, oxidizing hydrothermal fluids. Samples are zoned compositionally: clarkeite, which is Na-rich, surrounds a K-rich core (commonly with remnant uraninite) and is surrounded by more Ca-rich material. Clarkeite is hexagonal ($R\bar{3}m$). The structure of clarkeite is based on anionic sheets with composition $[(UO_2)(O,OH)_2]$, and these are bonded to each other through interlayer cations and water molecules. The ideal formula for clarkeite is $Na[(UO_2)O(OH)](H_2O)_{0.1}$, although it commonly contains additional elements, including K, Ca, and Th. The structure of a hexagonal synthetic K-uranyl oxyhydroxide, $K_5[(UO_2)_{10}O_8(OH)_9](H_2O)$, recently reported by Burns and Hill (1999b), may be structurally related to clarkeite and the "K-rich phase" that forms near the cores of clarkeite (Finch and Ewing 1997).

New data on mineral "A" described by Frondel (1956) were reported by Foord et al. (1997), and some details on its paragenesis at the Ruggles and Palermo pegmatites are discussed by Korzeb et al. (1997). This material may be a mixture of metaschoepite and "dehydrated schoepite," and more data are needed to clarify the structural details of mineral "A."

An extensive study natural and synthetic samples of becquerelite, billietite and protasite was reported by Cejka et al. (1998b), who demonstrated that billietite contains eigth interlayer H_2O groups, rather than the four reported by Pagoaga et al. (1987). This is consistent with a recent structure refinement for billietite (Finch et al. in prep.).

Although a large number of uranyl oxyhydroxide minerals are known, none contains Na or Mg. Such phases have been synthesized, and reports of Na-uranyl oxyhydroxides are especially common in the literature (e.g. Diaz Arocas and Grambow 1997); these display close structural similarities to schoepite and metaschoepite (Diaz Arocas and Grambow 1997). Their lack of occurrence in nature probably reflects their high solubilities. Several Mg-uranyl oxyhydroxides have been synthesized, and all are structurally related to becquerelite and protasite: $Mg[(UO_2)_6O_6(OH)_4]_2 \cdot 4H_2O$ and $Mg_3[(UO_2)_6O_5(OH)_5]_2 \cdot 19H_2O$ (Vdovina et al. 1984) and $Mg[(UO_2)_6O_4(OH)_6] \cdot 10H_2O$ (Vochten et al. 1991). Vochten et al. (1991) also synthesized two Ni^{2+} and Mn^{2+} uranyl oxyhydroxides and measured their solubilities along with that of the Mg compound. They concluded that the Mg compound was approximately as soluble as synthetic becquerelite (Vochten and Van Haverbeke 1990), whereas the Ni and Mn compounds were more so. Vochten et al. (1991) concluded that Mg, Ni, and Mn uranyl oxyhydroxides should be stable in natural groundwaters and may occur as minerals. None has been found yet.

A synthetic Ca-uranyl oxyhydroxide was described from experiments with U-doped cements that has the empirical formula $CaO \cdot 4UO_3 \cdot 7H_2O$, and which appears to be structurally related to becquerelite (Moroni and Glasser 1996; Skakle et al 1997). The Ca:U ratio (1:4) is the same as the Mg:U ratio of the more Mg-rich uranyl oxyhydroxide synthesized by Vdovina et al. (1984), and is similar to the Ba:U ratio of protasite (1:3); however, no Ca-uranyl oxyhydroxide with a similar Ca:U ratio is known as a mineral.

Pb-uranyl oxyhydroxides

Structure determinations for several Pb-uranyl oxyhydroxides, vandendriesscheite, fourmarierite, masuyite, richetite, and wölsendorfite, have greatly improved our understanding of this enigmatic group of minerals. Masuyite and, especially, vandendriesscheite are common at oxidized U deposits; whereas richetite is exceedingly rare.

Vandendriesscheite. Vandendriesscheite is one of the most common Pb-bearing uranyl oxyhydroxides, occurring at numerous uraninite deposits exposed to weathering environments (Frondel 1956, 1958). Christ and Clark (1960) reported that, like schoepite, vandendriesscheite undergoes spontaneous, irreversible dehydration with structural modifications, although they did not determine water contents. Oriented inclusions of extraneous phases, such as masuyite, within vandendriesscheite crystals may complicate interpretations of XRD data. The recent structure determination for vandendriesscheite revises the formula slightly (Burns 1997), although it is not clear how the crystal examined is related to meta-vandendriesscheite as described by Christ and Clark (1960) (Table 4).

Fourmarierite. Fourmarierite has been confused in the past with vandendriesscheite and wölsendorfite (Frondel 1958; Deliens 1977a). These minerals display similar X-ray powder diffraction patterns. The structure determination of fourmarierite clarifies the formula (Piret 1983), and additional detail on its structure is provided by Burns (this volume).

Richetite. Richetite's dark color prompted Frondel (1958) to suggest that some U is reduced; however, XPS analyses verified that U in richetite is fully oxidized (Piret and Deliens 1984). This is consistent with the recent structure determination, which verifies that richetite is related to becquerelite, but with a complex interlayer containing cations in addition to Pb, possibly Fe (Burns 1998a); this may explain its dark color. The new structural formula indicates that the richetite crystal studied by Burns (1998a) contained significantly more H_2O than material described by Piret and Deliens (1984). The reason for this apparent discrepancy is uncertain, but richetite may exhibit variable hydration states without significant structural differences (cf. curite). The fact that richetite is rare may be a reflection of unusual conditions required for its formation. Richetite, like wölsendorfite, which also contains interlayer cations in addition to Pb, does not form a series with other Pb-uranyl oxyhydroxides but requires a unique paragenesis. The precise composition of richetite remains uncertain, and further work is needed to clarify the formula (Burns 1998a). Richetite has been reported from the Shinkolobwe mine in Shaba, Democtratic Republic of Congo (Deliens et al. 1981) and at Jáchymov, Czech Republic (Ondrus et al. 1997b); specimens from the two localities display subtle differences in unit-cell parameters. Its color (dark brown to tan), strong pleochroism, and triclinic symmetry help distinguish richetite from other Pb-uranyl oxyhydroxides. X-ray powder diffraction data for richetite are reported by Piret and Deliens (1984) and by Ondrus et al. (1997b). Burns (this volume) discusses the structure of richetite in more detail.

Masuyite. The composition of masuyite has long been uncertain. Piret et al. (1983) noted that masuyite from Shinkolobwe is heterogeneous, with variations in Pb:U ratios and XRD powder spectra. Deliens and Piret (1996) report two compositions for masuyite, $PbO \cdot 3UO_3 \cdot 4H_2O$ ("grooved" masuyite) and $4PbO \cdot 9UO_3 \cdot 10H_2O$ ("type" masuyite). These bracket the composition of the synthetic analogue, $3PbO \cdot 8UO_3 \cdot 10H_2O$ (Protas 1959; Noe-Spirlet and Sobry 1974). Finch and Ewing (1992a, 1992b) reported masuyite inclusions within vandendriesscheite crystals with the composition $PbO \cdot 3UO_3 \cdot 4H_2O$. Burns and Hanchar (1999) report the crystal structure for masuyite with this same composition, demonstrating that it is isostructural with protasite, although the presence of an additional cation site in the interlayer may permit compositional variability.

Sayrite. The only new Pb-uranyl oxide hydrate described since Smith's (1984) review, sayrite, a very rare Pb-uranyl oxyhydroxide, occurs at the Shinkolobwe U mine in close association with other Pb-uranyl oxyhydroxides, masuyite and richetite, as well as becquerelite and uranophane (Piret et al. 1983). Sayrite (monoclinic, space group $P2_1/c$) forms reddish orange to yellow-orange prismatic crystals (optically biaxial negative) flattened on {-102} and elongated along [010], with perfect (-102) cleavage parallel to the structural sheets. Compositionally, it lies between masuyite and curite (Fig. 3).

Curite. A structural examination of thirteen curite crystals showed that curite displays a limited range of interlayer Pb occupancies but no evidence for OH^- ions in interlayer positions (Li and Burns 1999), as proposed by Mereiter (1979) and Taylor et al. (1981). Charge balance for different Pb occupancies is maintained through $OH^- \Leftrightarrow O^{2-}$ substitution in the structural sheets, and curite can be represented by the general formula $Pb_{1.5+x}[(UO_2)_4O_{4+x}(OH)_{3-2x}](H_2O)$ ($0 < x < 0.15$) (Li and Burns 1999a). Cejka et al. (1998a) provide a comprehensive review of analytical data for curite and report new chemical analyses, X-ray powder-diffraction data, IR data and thermal analyses. Burns and Hill (1999a) report the structure of a synthetic Sr-bearing analogue of curite.

Wölsendorfite and related minerals. The structure of a barian wölsendorfite was recently reported by Burns (1999e), demonstrating that wölsendorfite is structurally related to other uranyl oxyhydroxides of the protasite group but with a structural sheet that has a considerably more complex anion topology. Belova et al. (1985) report chemical and X-ray diffraction data for four compositionally similar uranyl oxyhydroxide minerals wölsendorfite, calciouranoite, metacalciouranoite, and bauranoite. They consider these four to comprise a structurally related group with solid solution between minerals rich in Ca, Ba and Pb. Except for wölsendorfite, these minerals are known only from unspecified occurrences in the former Soviet Union. Calciouranoite and bauranoite are not generally well crystallized (Rogova et al. 1973, 1974), and X-ray powder-diffraction data originally reported for these two minerals are for samples heated to 900°C. Powder patterns of heated material closely resemble synthetic anhydrous Ca and Ba uranates structurally related to clarkeite (Loostra and Rietveld 1969) but structurally distinct from wölsendorfite and other uranyl oxyhydroxides. X-ray diffraction patterns reported by Belova et al. (1985) show considerable variation (with some unindexed lines), and assignment of group status to these minerals is probably premature. Fayek et al. (1997) reported finding an unidentified uranyl oxyhydroxide at the Cigar Lake U deposit, where uranyl minerals reportedly occur as colloform bands along the outer edges of fine-grained botryoidal uraninite and single uraninite grains, as well as occurring in small veins that cut across uraninite and coffinite. Isobe et al. (1992b) reported Ca-free wölsendorfite from the Koongarra U deposit in Northern Territories, Australia.

Uranyl carbonates

Uranium carbonates (Tables 5 and 6) may precipitate where evaporation is significant or where the fugacity of CO_2 is greater than atmospheric (Garrels and Christ 1959; Hostetler and Garrels 1962; Finch and Ewing 1992b; Finch 1997a). Uranyl di- and tri-carbonates tend to form only where evaporation is high. Most of these minerals are ephemeral, dissolving readily when re-exposed to fresh water. Uranyl carbonate complexes in solution are quite stable and are probably the most important solution complexes responsible for U migration in oxidizing environments (Langmuir 1978; Clark et al. 1995). They are important in near-neutral to alkaline waters (pH > 7), and precipitation of uranyl tricarbonate minerals usually reflects evaporation of alkaline waters (Fig. 2). On the other hand, the monocarbonates rutherfordine, joliotite, blatonite, urancalcarite and wyartite probably precipitate from fresh water where (at least local) pCO_2 values are elevated relative to atmospheric values ($pCO_2 > 10^{-3.5}$ atm). Elevated pCO_2 most commonly results from organic respiration, transpiration and decomposition. Such waters tend to have relatively acid pH values (pH < 6 or 7). Unlike the uranyl di- and tri-carbonates, the uranyl monocarbonates are relatively insoluble. The solubility of rutherfordine, for example, is comparable to that of schoepite (Grenthe et al. 1992), and rutherfordine may persist in some environments for several tens to hundreds of thousands of years (Finch et al. 1996). Understanding the structural and thermodynamic stabilities of U carbonates is particularly germane to our ability to manage U-contaminated sites and our understanding of potential radionuclide behavior around a high-level waste repository (Clark et al. 1995; Finch 1997a) because of the potentially high mobility of U in carbonate-bearing groundwaters (Hostetler and Garrels 1962; Langmuir 1978; Grenthe et al. 1992). Uranyl carbonates may be important in actinide-contaminated soils and certain high-level nuclear-waste repositories (e.g. the proposed repositoy at Yucca Mountain) because they may be sinks for ^{14}C (Murphy 1995), transuranic actinides (Burns et al. 1997a; Wolf et al. 1997) and possibly certain fission-products (Finch and Ewing 1991; Wronkiewicz et al. 1996).

Only three uranyl carbonates with no cations besides U and C are known, rutherfordine, joliotite and blatonite, and all are monocarbonates. Rutherfordine and joliotite were discussed by Smith (1984). A recent structure refinement of rutherfordine indicates that there are two polytypes of rutherfordine with space groups $Pmmn$ (Christ et al. 1955) and $Imm2$ (Finch et al. 1999). Thermal and IR analyses of synthetic and natural uranyl monocarbonates, including rutherfordine, indicate variable amounts of OH and H_2O in some of these phases, with minor modifications to the structure (Cejka et al. 1988). Blatonite, $UO_2CO_3 \cdot H_2O$, was recently described from the Jomac mine in San Juan County, Utah (Vochten and Deliens 1998) Blatonite occurs as fibrous bundles of finely acicular, canary-yellow crystals within gypsum seams in siltstones of the Triassic Shinarump Conglomerate and is associated with the uranyl minerals boltwoodite, cocoininoite, metazeunerite, and rutherfordine, as well as azurite, malachite, brochantite, smithsonite and carbonate-cyanotrichite. Blatonite is hexagonal or trigonal; Vochten and Deliens (1998) report XRD and IR data. Walenta (1976) studied the dehydration of joliotite, $UO_2CO_3 \cdot 2H_2O$, and found that heating increased the refractive indices and caused diffraction peaks to become more diffuse; however, data from the dehydration of joliotite do not resemble those of blatonite, leading Vochten and Deliens (1998) to conclude that blatonite is not a simple dehydration product of joliotite and therefore cannot be described as "meta-joliotite."

The structure of wyartite, $Ca(CO_3)U^{5+}(UO_2)_2O_4(OH)(H_2O)_7$, was reported by Burns and Finch (1999), who demonstrated that it contains U^{5+}, rather than U^{4+} as originally reported (Guillemin and Protas 1959), making wyartite the first mineral known to contain pentavalent U. This may have important implications for the stability of the UO_2^+ ion in

Table 5. Miscellaneous uranyl carbonates

Name	Formula	Comments
Albrechtschraufite*	$MgCa_4(UO_2)_2(CO_3)_3F_2(H_2O)_{17}$	compare schröckingerite; Mereiter (1984); Ondrus et al. (1997a)
Astrocyanite-(Ce)	$Cu_2(REE)_2(UO_2)(CO_3)_5(OH)_2(H_2O)_{1.5}$	compare shabaite; Piret & Deliens (1990); Gaines et al. (1997)
Blatonite*	$UO_2CO_3 \cdot H_2O$	not produced by dehydration of joliotite; Vochten & Deliens (1999)
Bijvoetite-(Y)*	$(REE)_2(UO_2)_4(CO_3)_4(OH)_6(H_2O)_{11}$	Deliens & Piret (1982a); Gaines et al. (1997); Li & Burns (1999b)
Fontanite	$Ca(UO_2)_3(CO_3)_4(H_2O)_3$	Deliens & Piret (1992); Gaines et al. (1997)
Joliotite	$UO_2CO_3 \cdot nH_2O$ ($n \approx 2$)	Gaines et al. (1997)
Kamotoite-(Y)	$Y_2(UO_2)_4(CO_3)_3O_4(H_2O)_{14.5}$	Gaines et al. (1997)
Lepersonnite-(Gd)	$Ca(REE)_2(UO_2)_{24}(CO_3)_8Si_4O_{12}(H_2O)_{60}$	Deliens & Piret (1982a); Gaines et al. (1997)
Roubaultite	$Cu^{2+}{}_2(UO_2)_3(CO_3)_2O_2(OH)_2(H_2O)_4$	Ginderow & Cesbron (1985); B&97; Gaines et al. (1997)
Rutherfordine*	UO_2CO_3	two polytypes (*Pmmn* & *Imm2*); Christ et al. (1955), Frondel (1958); Čejka & Urbanec (1988); Gaines et al. (1997); Finch et al. (1999); Vochten & Blaton (1999)
Schröckingerite	$NaCa_3(UO_2)(CO_3)_3(SO_4)F(H_2O)_{10}$	compare albrechtschraufite; Urbanec & Čejka (1979b); Mereiter (1986); Gaines et al. (1997)
Shabaite-(Nd)	$Ca(REE)_2(UO_2)(CO_3)_4(OH)_2(H_2O)_6$	compare astrocyanite; Deliens & Piret (1990a), Gaines et al. (1997)
Sharpite	$Ca(UO_2)_6(CO_3)_5(OH)_4(H_2O)_6$	Čejka et al. (1984); Gaines et al. (1997)
Urancalcarite	$Ca_2(UO_2)_3(CO_3)_3(OH)_6(H_2O)_3$	oxidation product of wyartite (?); Deliens & Piret (1984b); Gaines et al. (1997)
Voglite*	$Ca_2Cu(UO_2)_2(CO_3)_4(H_2O)_6$ or $Ca_2Cu(UO_2)_2(CO_3)_3(OH)_2(H_2O)_4$ [PD79]	Jachymov: $Ca_2Cu(UO_2)_2(CO_3)_2O_3$; Ondrus et al. (1997c); Urbanec & Čejka (1979b); Piret & Deliens (1979); Gaines et al. (1997)
Wyartite*	$CaU^{5+}(CO_3)(UO_2)O(OH)_4(H_2O)_7$	Gaines et al. (1997); Smith (1984); mixed $U^{5+}-U^{6+}$; structure determination reveals first known U^{5+} mineral; Burns & Finch 1999
"Wyartite II"*	$CaU^{5+}(CO_3)(UO_2)O(OH)_4(H_2O)_3$	Clark (1960); mixed $U^{5+}-U^{6+}$; dehydration product of wyartite, structure similar to wyartite, unpublished data
Unnamed*	$Ca_2Cu(UO_2)_2(CO_3)_2O_3(H_2O)_3$ or $Ca_2Cu(UO_2)_2(CO_3)_2(OH)_6$	compare voglite; Ondrus et al. (1997c)
Unnamed*	$Ca_3Cu(UO_2)_4(CO_3)_6(OH)_8(H_2O)_4$	"pseudo-voglite"; Ondrus et al. (1997c)
Unnamed*	hydrated Ca-Cu-uranyl carbonate	Ondrus et al. (1997c)

* New or supplemental data or not listed in Dana's New Mineralogy (Gaines et al. 1997).

Table 6. Uranyl dicarbonates ($UO_2:CO_3 = 1:2$) and tricarbonates ($UO_2:CO_3 = 1:3$)

Name	Formula	Comments
Andersonite*	$Na_2Ca(UO_2)(CO_3)_3(H_2O)_6$	synthetic intermediate with liebigite also known; Čejka et al. (1987); Urbanec & Čejka (1979b); Vochten et al. (1994); Gaines et al. (1997)
Bayleyite	$Mg_2(UO_2)(CO_3)_3(H_2O)_{18}$	Frondel (1958); Gaines et al. (1997)
Grimselite	$K_3Na(UO_2)(CO_3)_3(H_2O)$	Gaines et al. (1997)
Liebigite*	$Ca_2(UO_2)(CO_3)_3(H_2O)_{11}$	synthetic intermediate with andersonite also known; Urbanec & Čejka (1979b); Vochten et al. (1994); Gaines et al. (1997)
Metazellerite*	$Ca(UO_2)(CO_3)_2(H_2O)_n$ (n < 5)	dicarbonate; dehydration product of zellerite; Gaines et al. (1997); Ondrus et al. (1997b)
Rabbittite	$Ca_3Mg_3(UO_2)_2(CO_3)_6(OH)_4(H_2O)_{18}$	Frondel (1958); Gaines et al. (1997); Ondrus et al. (1997b)
Swartzite	$CaMg(UO_2)(CO_3)_3(H_2O)_{12}$	Gaines et al. (1997)
Unnamed*	$Na_4(UO_2)(CO_3)_3$	may be structurally distinct from the synthetic analogue; Douglass (1956); Ondrus et al. (1997c)
Widenmannite*	$Pb_2(UO_2)(CO_3)_3$	Walenta (1976); Gaines et al. (1997)
Zellerite*	$Ca(UO_2)(CO_3)_2(H_2O)_5$	dicarbonate; Gaines et al. (1997); Ondrus et al. (1997b)
Znucalite*	$CaZn_{12}(UO_2)(CO_3)_3(OH)_{22}(H_2O)_4$ or $CaZn_{11}(UO_2)(CO_3)_3(OH)_{20}(H_2O)_4$	possible compositional variability; Ondrus et al. (1990); Chiappero & Sarp (1993); Ondrus et al. (1997b); Gaines et al. (1997)

* New or supplemental data or not listed in Dana's New Mineralogy (Gaines et al. 1997).

anoxic U-rich waters (Langmuir 1978), because U^{5+} is considered unstable in aqueous solutions, disproportionating to U^{4+} plus U^{6+}, especially in acid media. Whether wyartite forms from thermodynamically stable UO_2^+-bearing waters, or whether UO_2^+ in these solution persists metastably is not known; however, the crystal structure apparently stabilizes U^{5+} in the mineral. The crystal structure of a dehydration product of wyartite, wyartite II, $Ca(CO_3)U^{5+}(UO_2)_2O_4(OH)(H_2O)_{3-4}$, originally described by Clark (1960), has also been determined (unpublished data), and it is also consistent with U^{5+} in structural sites.

Deliens and Piret (1984b) described urancalcarite from the Shinkolobwe mine in Shaba, Democratic Republic of Congo, where it is found on uraninite, and is associated with uranophane, masuyite and wyartite. Urancalcarite is orthorhombic (*Pbnm* or *Pbn*2_1), forming bright yellow crystals that occur as radiating aggregates to 4 mm in diameter. Individual crystals form fibers elongated on [001] and flattened on (100), and display no cleavage. Unit-cell parameters of urancalcarite resemble those of the mixed-valence uranyl carbonates wyartite and wyartite II. Similarities between the mixed-valence carbonates and fully oxidized urancalcarite resemble similarities between the mixed-valence uranyl oxyhydroxide, ianthinite, and fully oxidized schoepite. The structural sheets in these mixed-valence minerals posses the β-U_3O_8 anion topology (cf. Miller et al. 1996; Burns et al. 1996) and have similar dimensions in the plane of the structural sheets (Table 7). Schoepite and urancalcarite have similar cell dimensions in the plane of their structural sheets, suggesting that urancalcarite has structural sheets similar to those in schoepite (Table 7). Ianthinite oxidizes to schoepite, with little or no lattice strain (Schoep and Stradiot 1947), the implication being that wyartite (or wyartite II) may oxidize to urancalcarite. The U:Ca:C ratios in urancalcarite (3:1) are identical to those in wyartite and wyartite II, so their compositions are similar. Whether urancalcarite might form by the oxidation of pre-existing wyartite or wyartite II is uncertain; however, both minerals commonly occur together (Deliens and Piret 1984b).

Fontanite, $Ca(UO_2)_3(CO_3)_4(H_2O)_3$, is orthorhombic (*Pmnm*, *Pmn*2_1, or *P*2_1*nm*: *a* 15.337; *b* 17.051; *c* 6.931 Å), and occurs as bright yellow, elongated [001], platy (010), lath-shaped crystals in pelitic rocks of Permian age, in the weathered zone of the Rabejac U deposit in Lodève, Hérault, France, where it is associated with billietite and uranophane (Deliens and Piret 1992). The unit-cell parameters are similar to those of urancalcarite (*a* 15.42; *b* 16.08; *c* 6.970 Å), although the composition is not.

Znucalite is a uranyl tricarbonate that occurs with gypsum, hydrozincite, aragonite, sphalerite, galena, pyrite and other minerals as a weathering product from uraninite-bearing mine tailings (Ondrus et al. 1990), and with hydrozincite, calcite and gypsum on the surfaces of mine workings and adits (Ondrus et al. 1997b). Znucalite forms porous, fine-grained coatings and aggregates of minute scaly white to yellow-green crystals that are reportedly triclinic (Ondrus et al. 1997b). At Jáchymov, Czech Republic, znucalite is noted to be absent from weathered sulfide-bearing veins, where pH values can drop below ~4. Chiappero and Sarp (1993) report that znucalite from the French U deposits Mas d'Alary and Lodève is orthorhombic (space group unknown). At the French occurrences, znucalite occurs as submillimetric sperules of elongate [001], platy (010) crystals with perfect {010} cleavage, and is biaxial negative. The composition of the crystals from France, $CaZn_{11}(UO_2)(CO_3)_3(OH)_{20} \cdot 4H_2O$, apparently differs slightly from that reported by Ondrus et al. (1990): $CaZn_{12}(UO_2)(CO_3)_3(OH)_{22} \cdot 4H_2O$. The UO_2:CO_3 ratio (1:3) suggests that the structure of znucalite contains uranyl tricarbonate ions $(UO_2)(CO_3)_3$, with water and cations occupying structural interstices, as for most other uranyl tricarbonates. It seems reasonable that znucalite may display some compositional variability without significantly changing the underlying structural framework.

Table 7. Unit-cell parameters of selected orthorhombic Ca-uranyl carbonates (Å)

	uranyl carbonates		uranyl oxy-hydroxides	
	wyartite	urancalcarite*	ianthinite	schoepite
a	11.2706	15.42	11.473	16.813
b	7.1055	6.97	7.178	14.337
c	20.807	16.08	30.39	14.731
V (Å^3)	1666.3	1728.2	2502.7	3550.9
sp. grp.	$P2_12_12_1$	$Pmcn$ or $P2_1cn$	$P2_1ca$	$P2_1ca$

*Note: Unit-cell parameters of urancalcarite ianthinite and schoepite are transposed $a\ c\ b$, to compare with wyartite; c is always perpendicular to the plane of the sheets.

There are five recently described uranyl carbonates that contain essential REE. The rare REE-uranyl carbonate, bijvoetite, occurs close to uraninite with sklodowskite, uranophane, lepersonnite, oursinite, becquerelite, curite, studtite and rutherfordine in the lower part of the oxidation zone at Shinkolobwe mine (Deliens and Piret 1982). Bijvoetite is orthorhombic (C-centered), forming elongated [110] yellow {001} tablets with good {001} cleavage. Bijvoetite is pleochroic (colorless to dark yellow) and, unlike most known uranyl minerals, is optically biaxial positive. Lepersonnite, a REE-uranyl carbonate-silicate, orthorhombic ($Pnnm$ or $Pnn2$), forms long (~1 mm), slender, bright yellow crystals in close proximity to uraninite in the lower part of the oxidation zone at Shinkolobwe mine. Associated uranyl minerals include sklodowskite, uranophane, bijvoetite, oursinite, becquerelite, curite, rutherfordine and studtite—with which lepersonnite has sometimes been confused (Deliens and Piret 1982). Three REE bearing carbonates have been reported from the Kamoto-Est Cu-Co deposit west of Kalwezi, near Kivu in southern Shaba province, Democratic Republic of Congo. Astrocyanite-(Ce) is an oxidation product of uraninite, associated with uranophane, kamototite-(Y), françoisite-(Nd), shabaite-(Nd), masuyite, and schuilingite-(Nd), $PbCu^{2+}(REE)(CO_3)_3(OH)(H_2O)_{1.5}$ (Deliens and Piret 1990b). Named for its habit and striking color, the hexagonal ($P6/mmm$ or subgroup) crystals occur as blue to blue-green, millimeter-sized rosettes of micaceous to platy {001} tablets, rarely occasionally as isolated tabular blue crystals, translucent to opaque, optically uniaxial negative, and strongly pleochroic (blue to colorless). Another REE carbonate from Kamoto, shabaite-(Nd) occurs with uraninite, kamotoite-(Y), uranophane and schuilingite-(Nd). Shabaite is monoclinic($P2$, Pm or $P2_1/m$), optically biaxial negative, and commonly forms rosettes (to 5 mm) of pale yellow, elongated [100] micaceous flakes {010} (Deliens and Piret 1990a). The third REE carbonate from Kamoto, kamotoite-(Y), forms bladed yellow crystals and fine-grained crusts on uraninite; it is monoclinic ($P2_1/n$), optically biaxial negative, strongly pleochroic (colorless to yellow) and has two good cleavages, {001} and {-101}, commonly occurring as crusts or yellow blades on uraninite (Piret and Deliens 1986).

Ondrus et al. (1997b) report voglite with two distinct compositions from Jáchymov, Czech Republic, along with several compositionally similar Ca-Cu-uranyl carbonates (Table 5). X-ray data show similarities to published data but with characteristic differences. Clearly additional work is required to clarify the composition of this mineral (or minerals), as reported stoichiometries display a range of U:C ratios (Frondel 1958; Piret and Deliens 1979; Ondrus et al. 1997b) (Table 5).

Ondrus et al. (1997c) also report an unnamed anhydrous Na-uranyl tri-carbonate,

$Na_4(UO_2)(CO_3)_3$ from Jáchymov, where it forms minute earthy aggregates (~100 μm) on surfaces. The mineral is apparently structurally distinct from the synthetic phase with the same composition (Douglass 1956).

Roubaultite was originally described as a cupric uranyl oxyhydroxide (Cesbron et al. 1970), but a structural analysis demonstrated that it is a uranyl carbonate (Ginderow and Cesbron 1985), revising the formula from that reported by Smith (1984). The occurrence and paragenesis of roubaultite are described by Smith (1984) and by Deliens et al. (1981, 1990).

Elton and Hooper (1992) describe new occurrences for andersonite and schröckingerite from the Geevor mine, Cornwall, England, where these two tri-carbonates form on the walls of the 17-level of the mine. Schöckingerite occurs as greenish-yellow globules; andersonite, which is less common, forms bright yellow, pseudocubic crystals to 3 mm. Other U minerals at this mine include johannite and a zippeite-like mineral.

Uranyl silicates

The uranyl silicates (Table 8) are moderately insoluble in most natural groundwaters. Because of the ubiquity of dissolved Si in most groundwaters, uranyl silicates are the most abundant group of uranyl minerals. Uranophane, the most common uranyl mineral and possibly the most common U mineral after uraninite, precipitates from near neutral to alkaline groundwaters that contain dissolved Si and Ca. When exposed to dilute meteoric waters (low carbonate and pH below ~7), uranophane may be replaced by soddyite (Deliens 1977b; Finch 1994), a phenomenon also observed in experimental dissolution studies of uranophane in deionized water (Casas et al. 1994). Sklodowskite is replaced by kasolite where radiogenic Pb that accumulates within the sklodowskite reaches sufficient levels to be exsolved and reprecipitated as kasolite (Isobe et al. 1992). This appears to be independent of changes in water compositions, but is rather a result of the durability of sklodowskite, which persists long enough to accumulate radiogenic Pb in substantial amounts. Replacement of other uranyl silicates is not well documented, a reflection, perhaps, of their durability. Uranophane and soddyite can persist in some environments for more than one hundred thousand years (Finch et al. 1996).

Several especially Si-rich minerals are known, the most important being the weeksite group with a U:Si ratio of 2:5 (Rastsvetaeva et al. 1997; Burns 1999b) (Table 8). These minerals are relatively rare and occur in arid environments, which may reflect evaporation of Si-rich waters of relatively high pH, in which dissolved polymeric silicate species may reach quite elevated concentrations (Stumm and Morgan 1981; Dent Glasser and Lachowski 1980). Unlike the uranyl carbonates and sulfates, however, weeksite-group minerals apparently do not dissolve so readily upon re-exposure to fresh water. Like uranophane, when exposed to carbonate-free waters, these minerals might lose some Ca and Si preferentially, altering to more U-rich minerals such as uranophane or soddyite. In more alkaline, carbonate-rich waters, they may lose U preferentially, altering to amorphous or microcrystalline silica.

Uranosilite, with a U:Si ratio (1:7) far lower than any other known uranyl silicates, is found at the Menzenschwand U deposit, where it occurs on quartz and is intimately intergrown with studtite and uranophane (Walenta 1974, 1983). Uranosilite forms finely acicular, yellowish white orthorhombic ($P22_12_1$, $Pmmb$ or $Pmcb$) crystals, that are optically biaxial negative. Heating causes no change in the X-ray powder diffraction pattern but lowers the refractive indices, which is explained by the loss of a small amount of "zeolitic water" (Walenta 1983). Uranium(VI) in uranosilite almost certainly occurs as the uranyl ion, and the structure of uranosilite may be that of a silicate framework with UO_2^{2+}

Table 8. Uranyl silicates

Name	Formula	Comments
Soddyite	$(UO_2)_2SiO_4(H_2O)_2$	Gaines et al. (1997); Vochten et al. (1995a); Moll et al. (1995)
Uranosilite	$(UO_2)Si_7O_{15}(H_2O)_n$ $(0 < n < 1)$	Rare; Walenta (1983)
Ursilite	$(Mg,Ca)_4(UO_2)_4(Si_2O_5)_5(OH)_6(H_2O)_{15}$	probably equivalent to haiweeite; Chernikov et al. (1957); Smith (1984)
Lepersonnite-(Gd)	$Ca(Gd,Dy)_2(UO_2)_{24}(CO_3)_8Si_4O_{12}(H_2O)_{60}$	Gaines et al. (1997)
"Pilbarite"	$PbO \cdot ThO_2 \cdot UO_3 \cdot 2SiO_2 \cdot 4H_2O$ (?)	species not verified; Frondel (1958)
Unnamed *	$2CaO \cdot 2UO_3 \cdot 6SiO_2 \cdot 10H_2O$ "$Ca_2(UO_2)_2(Si_2O_5)_3 \cdot 10H_2O$"	Ondrus et al. (1997c); proposed formula for synthetic analogue is unbalanced; Moroni & Glasser (1995); Skakle et al. (1997)
Uranophane Group	$M^{m+}[(UO_2)(SiO_3OH)]_m(H_2O)_n$	
Boltwoodite*	$K(UO_2)(SiO_3OH)(H_2O)_{1.5}$	Stohl & Smith (1981); Pu Congjian (1990); Gaines et al. (1997); Vochten et al. (1997c); Na-K solid solution, Burns (1998c); Na/K-Cs ion exchange, Burns (1998a)
Cuprosklodowskite	$Cu(UO_2)_2(SiO_3OH)_2(H_2O)_6$	Frondel (1958); Smith (1984)
Kasolite	$Pb(UO_2)(SiO_4)(H_2O)$	Frondel (1958); Stohl & Smith (1981); Gaines et al. (1997)
Na-boltwoodite*	$(Na,K)(UO_2)(SiO_3OH)(H_2O)_{1.5}$	Stohl & Smith (1981); Gaines et al. (1997); Burns (1998c); Vochten et al. (1997c)
Oursinite*	$(Co,Mg)(UO_2)_2(SiO_3OH)_2(H_2O)_5$	rare; Deliens & Piret (1983a); Gaines et al. (1997)
Sklodowskite*	$Mg(UO_2)_2(SiO_3OH)_2(H_2O)_6$	Frondel (1958); Gaines et al. (1997)
Swamboite	$U^{6+}(UO_2)_6(SiO_3OH)_6(H_2O)_{30}$	Deliens & Piret (1981a); Gaines et al. (1997)
Uranophane-α	$Ca(UO_2)_2(SiO_3OH)_2(H_2O)_5$	most common uranyl mineral; Frondel (1958); Ginderow (1988); Gaines et al. (1997)
Uranophane-β*	$Ca(UO_2)_2(SiO_3OH)_2(H_2O)_5$	common mineral but never synthesized; Viswanathan & Harnett (1986); Cesbron et al. (1993); Gaines et al. (1997); Palenzona & Selmi (1998)
Weeksite Group	$M^{m+}(UO_2)_2[Si_5O_{13}](H_2O)_4$	
Haiweeite*	$Ca(UO_2)_2[Si_5O_{12}(OH)_2](H_2O)_{4.5}$	Stohl & Smith (1981); structure reported by Rastsvetaeva et al. (1997); Burns (1999f)
Metahaiweeite	$Ca(UO_2)_2[Si_5O_{12}(OH)_2](H_2O)_n$ $(n < 5)$	ill defined; structural formula inferred; Gaines et al. (1997)
Mg-haiweeite*	$Mg(UO_2)_2[Si_5O_{12}(OH)_2](H_2O)_n$	Structural formula inferred; Smith (1984)
Weeksite*	$K_{1-x}Na_x(UO_2)_2(Si_5O_{13})(H_2O)_4$ $(x = 0.4)$	Stohl & Smith (1981); Baturin & Sidorenko (1985), Vochten et al. (1997a); Jackson & Burns (1999)

* New or supplemental data or not listed in Dana's New Mineralogy (Gaines et al. 1997).

ions occupying structural interstices, similar to the framework uranyl tellurite cliffordite, and suggesting the structural formula $(UO_2)Si_7O_{15}·nH_2O$. The conditions necessary for the precipitation of uranosilite are unknown, although the fact that it occurs with studtite suggests that it may be stable only in highly oxidizing environments and in waters with low carbonate concentrations. No data are available on its stability or solubility, although it is relatively insoluble in weak hydrochloric acid (Walenta 1983).

Slender yellow crystals of swamboite occur at the Swambo U mine in Shaba, Democratic Republic of Congo, where it is associated with soddyite and curite. Swamboite is monoclinic ($P2_1/a$). Optical properties, crystal habit and unit-cell data demonstrate its close structural relationship to uranophane, but the ideal formula, $U^{6+}_{0.33}H^+_2(UO_2)_2(SiO_4)_2(H_2O)_{10}$, derived by analogy with the uranophane-group, is difficult to reconcile with the uranophane structure. The structural roles and chemical forms of the additional U^{6+} (presumably as the UO_2^{2+} ion) and H^+ ions (possibly as part of acid $(SiO_3OH)^{3-}$ groups) are uncertain, although the large cell, with Z = 18, suggests a complex ordering of interlayer constituents. Compositionally similar to soddyite, with which it is associated, swamboite and soddyite must precipitate under similar conditions, and the coexistence of these two uranyl silicates suggests that slight changes in groundwater chemistry may be sufficient to shift from the stability regime of one mineral to the other. Soddyite has a higher U:Si ratio (2:1) than swamboite (1.167:1), suggesting that swamboite is stable relative to soddyite in waters with higher activities of orthosilicic acid.

Deliens and Piret (1983a) report oursinite from the Shinkolobwe mine, Shaba, Democratic Republic of Congo, where it occurs with soddyite, kasolite, schoepite and curite. Oursinite occurs as pale yellow acicular [001] crystals that form radial aggregates. Crystal habit, optical properties and X-ray diffraction data indicate that oursinite is structurally related to the uranophane-group minerals, with a unit cell that resembles that of sklodowskite. The Co and Ni in oursinite are derived from the oxidation of primary Co- and Ni-sulfides in the host rocks at Shinkolobwe.

New structure determinations show that neither polymorph of uranophane contains H_3O^+ (hydroxonium), as previously proposed by Stohl and Smith (1981). Charge balance is maintained by acid silica tetrahedra instead (Ginderow 1988, Viswanathan and Harnett 1986). The structure of boltwoodite contains the alpha-uranophane type sheet, with K and Na in the interlayer. Burns (1998d) demonstrated solid solution between boltwoodite and Na-boltwoodite. The structure is apparently compatible with a range of cation occupancies within the interlayer. Vochten et al. (1997c) synthesized Na-boltwoodite, demonstrating by spectroscopy that the structure contains acid $(SiO_3OH)^{3-}$ groups, rather than H_3O^+ ions as previously reported, consistent with the structure determination (Burns 1998d). The existence of H_3O^+ in uranyl mineral structures is commonly proposed but has not been convincingly demonstrated. Vochten et al. (1997c) used the synthetic Na-boltwoodite as starting material for the synthesis of several structurally related uranophane-group silicates. Burns (1999a) obtained Cs-substituted boltwoodite by immersing natural boltwoodite single crystals in Cs-rich solutions. Burns (1999a) demonstrated that the Cs replaced K and Na in the interlayer by cation exchange, rather than by dissolution and re-precipitation, a testament to the stability of the structural unit in boltwoodite. The demonstration of Cs exchange in boltwoodite has significant potential implications for the retention of Cs released from corroded spent nuclear fuel, because, under conditions relevant to the proposed high-level waste repository at Yucca Mountain, Na-boltwoodite is a major corrosion product of spent UO_2 fuel (Finch et al. 1999b).

Haiweeite has been poorly understood, with symmetry variously reported as monoclinic or orthorhombic. Based on single-crystal XRD data for a crystal of haiweeite from Teofilo Otoni, Minas Gerais, Brazil, Rastsvetaeva et al. (1997) report the structural formula for haiweeite as $Ca(UO_2)_2[Si_5O_{12}(OH)_2]\cdot 4.5H_2O$, which was confirmed recently by a more precise structure refinement (Burns 1999b). This is slightly different from the formula that had been accepted, $Ca(UO_2)_2Si_6O_{15}\cdot 5H_2O$ (Cejka and Urbanec 1990; Nickel and Nichols 1992; Mandarino 1999). Smith (1984) suggested that "ursilite," $(Mg,Ca)_4(UO_2)_4(Si_2O_5)_{5.5}(OH)_5\cdot 13H_2O$ (Chernikov et al. 1957) is equivalent to haiweeite (or Mg-haiweeite); however, "ursilite" is not considered a valid species according to Mandarino (1999). Weeksite contains the same structural sheets as haiweeite, but the sheets are bonded directly to each other, forming a framework structure (Baturin and Sidorenko 1985). Low valence cations and water molecules are located in channels within the framework.

Ondrus et al. (1997c) reported an unnamed uranyl silicate from Jáchymov, Czech Republic, which forms thin coatings of tabular yellow crystals, and is associated with liebigite, voglite, gypsum, rösselite, and an unnamed hydrated Cu-Ca-uranyl carbonate. The phase described by Ondrus et al. (1997c) displays an X-ray powder diffraction pattern similar to that of a Ca-uranyl silicate synthesized from U-loaded cement (Moroni and Glasser 1995; Skakle et al. 1997). The natural and synthetic compounds both appear to be distinct from haiweeite. The formula suggested by Skakle et al (1997) for the uranyl silicate, $Ca_2(UO_2)_2(Si_2O_5)_3\cdot 10H_2O$, cannot be correct, as it is not charge-balanced. Alternative formulas may be $Ca_2(UO_2)_2Si_6O_{16}\cdot 10H_2O$ [cf. "ursilite" and the synthetic K-U silicate of Plesko et al. (1992) or, $Ca_2(UO_2)_2(Si_4O_{10})_2\cdot nH_2O$, by analogy with a synthetic K-Na-uranyl silicate described by Burns et al. (1999)].

Several studies have been reported on synthetic uranyl silicates. The synthesis of minerals and related compounds can provide important insights into the conditions of mineral formation and their structural relationships. Alpha-uranophane has been synthesized (Nguyen et al. 1992; Cesbron et al. 1993); however no successful synthesis of _-uranophane has been reported, despite concentrated efforts to do so (Cesbron et al. 1993). This is a curious dilemma because β-uranophane is the more common of the two known uranophane polymorphs at several U deposits (Frondel 1958; Palenzona and Selmi 1998), including the Nopal I U deposit in Peña Blanca, northern Mexico (Pearcy et al. 1994). Hydrothermal reaction of synthetic U-bearing borosilicate glass resulted in the formation of a previously unknown compound, $KNa_3(UO_2)_2(Si_4O_{10})_2(H_2O)_4$ (Burns et al. 1999). This novel uranyl silicate may be relevant to hydrothermal U(VI) occurrences, such as near Spruce Pine, North Carolina, Ruggles mine, Grafton, New Hampshire, and near Rajputana, India (Frondel and Meyrowitz 1956; Korzeb et al. 1997). This compound has a U:Si ratio of 1:4, placing it intermediate between haiweeite-group minerals (2:5) and uranosilite (1:7).

Burns et al (1999) discuss how the crystal structure of the synthetic $KNa_3(UO_2)_2(Si_4O_{10})_2\cdot 4H_2O$ provides insight into the structural connectivities of uranyl silicates. Si tetrahedra become increasingly polymerized as Si:U ratios increase, with linkages between Si and U polyhedra depending on the U:Si ratio. In soddyite (U:Si = 2:1) each Si tetrahedron shares two of its edges with U polyhedra, whereas in structures with U:Si = 1:1 only one edge of each Si tetrahedron is shared with a U polyhedron, and each tetrahedron links to another U polyhedron by sharing vertices. In structures with a 2:5 ratio, some Si tetrahedra share a single edge with a U polyhedron, whereas others share none. In synthetic $KNa_3(UO_2)_2(Si_4O_{10})_2\cdot 4H_2O$ (U:Si = 1:4), U polyhedra and Si tetrahedra share only vertices; however, all four equatorial vertices of the U polyhedra are shared with Si tetrahedra in adjacent sheets.

During a study to examine phase relationships among uranyl silicates and uranyl oxyhydroxides at elevated temperatures, Plesko et al. (1992) synthesized several uranyl silicates in sealed tubes at 200 to 300°C and 30 MPa, including weeksite and a K-uranyl silicate with a U:Si ratio close to that of the synthetic K-Na-uranyl silicate of Burns et al. (1999): $K_2O \cdot 4UO_3 \cdot 15SiO_2$. However, Plesko et al. (1992) performed their syntheses in a Na-free system. The phase synthesized by Plesko et al. (1992) appears structurally distinct from, and contains less alkali than, synthetic $KNa_3(UO_2)_2(Si_4O_{10})_2 \cdot 4H_2O$. Plesko et al. (1992) found that, under the conditions of their experiments, the only uranyl silicates in the system $K_2O-UO_3-SiO_2-H_2O$ are soddyite, boltwoodite, weeksite and $K_2O \cdot 4UO_3 \cdot 15SiO_2$.

Uranyl phosphates and arsenates

Uranyl phosphates and arsenates (Tables 9-13) constitute by far the most diverse group of uranyl minerals, with approximately 70 species described. Dissolved phosphate is a common constituent of many groundwaters, and the uranyl phosphates have a correspondingly wide distribution in nature. Their genesis, however, appears to be somewhat unique and is discussed in more detail in a separate section below. Uranyl arsenates precipitate where dissolved AsO_4 is available, which is most commonly where arsenide minerals and As-bearing sulfide minerals are being oxidized. Thus, uranyl arsenates commonly occur in the same localities as uranyl sulfates. However, unlike the sulfates, uranyl arsenates are quite insoluble in most natural waters. In fact, they are structurally related to uranyl phosphates and display virtually identical physical properties, including low solubilities. Many phosphates and arsenates show substantial substitution of P and As in structural sites (see Burns this volume), and complete solid solution may be possible between some end members. A large number of uranyl phosphates and arsenates has been described since Smith's (1984) review. One particularly interesting occurrence is the uraniferous quartz-albite-muscovite pegmatite at Kobokobo, near Kivu, in western Democratic Republic of Congo, where at least ten uranyl phosphates occur, most of which also contain Al (Deliens and Piret 1980). Kobokobo is the type locality for nine of these: kamitugaite, triangulite, althupite, moreauite, ranunculite, threadgoldite, phuralumite, vanmeerscheite and metavanmeerscheite (Deliens et al. 1981, 1984, 1990). Furongite also occurs at Kobokobo (Deliens and Piret 1985). Uranyl phosphates not discussed by Smith (1984) are described below within their respective structural groups.

Uranyl phosphates and arsenates can be divided into at least three structurally and chemically related groups. The autunite and meta-autunite groups are tetragonal or pseudotetragonal with U:P and U:As ratios of 1:1 (Tables 9 and 10). The phosphuranylite group (Table 11) is based on a structural sheet with U:P = 3:2. A small but interesting group are uranyl phosphates and arsenates with U:P and U:As ratios of 1:2 (Table 12), most of which are triclinic and may be structurally related to the arsenates walpurgite and orthowalpurgite (Mereiter 1982; Krause et al. 1995). Several phosphates and arsenates that are not so readily categorized are listed in Table 13.

Autunite and meta-autunite groups (Tables 9 and 10). These two mineral groups are the most numerous of the phosphates and arsenates, with approximately 40 species known. They are common in a wide variety of deposits (Frondel 1958; Smith 1984), and probably control U concentrations in many groundwaters. Four new autunite-group minerals and the redefinition of one described before 1984 have been reported since Smith's (1984) review.

Vochtenite is found at the Basset mine, southeast of Camborne, Cornwall, England, where it occurs as aggregates of approximately square, monoclinic (space group unknown) brown crystals with one perfect {010} cleavage; optical properties, crystal habit, and the $UO_2:PO_4$ ratio clearly indicate that vochtenite is a member of the autunite-group (Zwaan et al. 1989). Vochtenite is associated with chalcopyrite, chalcocite, and cassiterite, and minor

Table 9. Autunite group of phosphates and arsenates (U:P = 1:1) $M^{m+}[(UO_2)(PO_4)]_m(H_2O)_n$ (n = 10–12)

Name	Formula	Comments
Autunite	$Ca[(UO_2)(PO_4)]_2(H_2O)_{10-12}$	Frondel (1958); Gaines et al. (1997)
Sabugalite	$Al[(UO_2)_4(HPO_4)(PO_4)_3](H_2O)_{16}$	Monoclinic; compare threadgoldite, uranospathite, ranunculite; Frondel (1958); Gaines et al. (1997)
Uranospathite	$AlH[(UO_2)(PO_4)]_4(H_2O)_{40}$	Compare threadgoldite, sabugalite, ranunculite; Walenta (1978); Smith (1984); Gaines et al. (1997)
Saléeite*	$Mg[(UO_2)(PO_4)]_2(H_2O)_{10}$	Murakami (1996a); Vochten & Van Springel K (1996); Murakami et al. (1997); Gaines et al. (1997)
Torbernite	$Cu[(UO_2)(PO_4)]_2(H_2O)_8$	Gaines et al. (1997)
Uranocircite	$Ba[(UO_2)(PO_4)]_2(H_2O)_8$	Gaines et al. (1997)
Arsenuranospathite	$AlH[(UO_2)_4(AsO_4)_4(H_2O)_{40}$	Smith (1984); Gaines et al. (1997)
Heinrichite	$Ba[(UO_2)(AsO_4)]_2(H_2O)_{10-12}$	Smith (1984); Gaines et al. (1997)
Kahlerite	$Fe^{2+}[(UO_2)(AsO_4)]_2(H_2O)_{10-12}$	only qualitative chemical analysis reported; Frondel (1958); Gaines et al. (1997)
Kirchheimerite	$Co[(UO_2)(AsO_4)]_2(H_2O)_{12}$	Smith (1984); Gaines et al. (1997)
Novácekite	$Mg[(UO_2)(AsO_4)]_2(H_2O)_{9-12}$	Frondel (1958); Gaines et al. (1997)
Trögerite	$(UO_2)_3(AsO_4)_2(H_2O)_{12}$ or possibly $[H(UO_2)(AsO_4)]_2(H_2O)_8$	ill defined; cf. synthetic "H-uranospinite"; Frondel (1958); Smith (1984); Gaines et al. (1997)
Uranospinite	$Ca[(UO_2)(AsO_4)]_2(H_2O)_{10}$	Ca-free analogue synthesized: "H-uranospinite"; Frondel (1958); Gaines et al. (1997)
Zeunerite	$Cu[(UO_2)(AsO_4)]_2(H_2O)_{16}$	Frondel (1958); Gaines et al. (1997)

* New or supplemental data or not listed in Dana's New Mineralogy (Gaines et al. 1997).

Table 10. Meta-autunite group of phosphates and arsenates (U:P = 1:1) $M^{m+}[(UO_2)(PO_4)]_m(H_2O)_n$ $(2 < n < 8)$

Name	Formula	Comments
Bassetite	$Fe[(UO_2)(PO_4)]_2(H_2O)_8$	Vochten et al. (1984); Gaines et al. (1997)
Chernikovite	$(H_3O)[(UO_2)(PO_4)]_2(H_2O)_8$	"hydrogen autunite"; Gaines et al. (1997)
Lehnerite	$Mn[(UO_2)(PO_4)]_2(H_2O)_8$	Mücke (1988); Gaines et al. (1997)
Meta-autunite	$Ca[(UO_2)(PO_4)]_2(H_2O)_6$	Gaines et al. (1997)
Meta-autunite II*	$Ca[(UO_2)(PO_4)]_2(H_2O)_2$	questionable species; Smith (1984)
Meta-ankoleite	$K_2[(UO_2)(PO_4)]_2(H_2O)_6$	Gaines et al. (1997)
Metatorbernite*	$Cu[(UO_2)(PO_4)]_2(H_2O)_6$	Stergiou et al. (1993); Calos & Kennard (1996); Gaines et al. (1997)
Meta-uranocircite	$Ba[(UO_2)(PO_4)]_2(H_2O)_8$	Gaines et al. (1997)
Meta-uranocircite II	$Ba[(UO_2)(PO_4)]_2(H_2O)_6$	questionable species; Gaines et al. (1997)
Na-autunite	$(Na_2,Ca)[(UO_2)(PO_4)]_2(H_2O)_8$	Smith (1984); Gaines et al. (1997)
Przhevalskite	$Pb[(UO_2)(PO_4)]_2(H_2O)_4$	ill defined; Gaines et al. (1997)
Ranunculite	$AlH(OH)_3[(UO_2)(PO_4)](H_2O)_4$ or possibly $Al(OH)_2[(UO_2)(PO_4)](H_2O)_5$	compare sabugalite, uranospathite; Gaines et al. (1997)
Threadgoldite	$Al(OH)[(UO_2)(PO_4)]_2(H_2O)_8$	Smith (1984); K-S82; Gaines et al. (1997)
Uramphite	$(NH_4)_2[(UO_2)(PO_4)]_2(H_2O)_{4-6}$ (?)	from oxidized coal deposits (Russia); Gaines et al. (1997)
Vochtenite	$(Fe^{2+},Mg)Fe^{3+}[(UO_2)(PO_4)]_4(OH)(H_2O)_{12-13}$	Zwaan et al. (1990); Gaines et al. (1997)
Abernathyite	$K_2[(UO_2)(AsO_4)]_2(H_2O)_8$	Frondel (1958); Gaines et al. (1997)
Metaheinrichite	$Ba[(UO_2)(AsO_4)]_2(H_2O)_8$	Frondel (1958); Gaines et al. (1997)
Metakahlerite	$Fe^{2+}[(UO_2)(AsO_4)]_2(H_2O)_8$	Mn^{2+} and Fe^{3+} analogues have been synthesized; Gaines et al. (1997)
Metalodevite	$Zn[(UO_2)(AsO_4)]_2(H_2O)_{10}$	Smith (1984); Gaines et al. (1997)
Metanováčekite	$Mg[(UO_2)(AsO_4)]_2(H_2O)_{4-8}$	polymorph of seelite ?; Frondel (1958); Smith (1984); Bachet et al. (1991); Gaines et al. (1997)
Na-uranospinite	$(Na_2,Ca)[(UO_2)(AsO_4)]_2(H_2O)_5$	Frondel (1958); Gaines et al. (1997)
Seelite	$Mg(UO_2)(AsO_3)_x(AsO_4)_{1-x}(H_2O)_7$ (x = 0.7)	Arsenate-arsenite; polymorph of metanováčekite?; Bachet et al. 1991; Bariand et al. (1993); Gaines et al. (1997)
Meta-uranospinite*	$Ca[(UO_2)(AsO_4)]_2(H_2O)_8$	two polymorphs (18 & 17 Å) ?; Frondel (1958); Gaines et al. (1997); Ondrus et al. (1997b)
Metazeunerite	$Cu[(UO_2)(AsO_4)]_2(H_2O)_8$	Frondel (1958); Gaines et al. (1997)
Unnamed*	$Ni[(UO_2)(AsO_4)]_2(H_2O)_{6-8}$	XRD of synthetic analogue: PDF 12-586; Ondrus et al. (1997c)
Unnamed*	$(H_3O)(UO_2)_2(AsO_4)_2(H_2O)_8$ or possibly $(UO_2)_2(HAsO_4)_2(AsO_4)(H_2O)_{10}$	"hydronium uranospinite" (compare trögerite); Ondrus et al. (1997c)

* New or supplemental data or not listed in Dana's New Mineralogy (Gaines et al. 1997).

bassetite.

Lehnerite is found in the Hagendorf pegmatite, Oberpfalz, Germany, where it occurs in close association with altered zweiselite, $(Fe^{2+},Mn^{2+})_2(PO_4)F$, and rockbridgeite, $(Fe^{2+},Mn^{2+})Fe^{3+}_4(PO_4)_3(OH)_5$, as well as with morinite, carlhintzeite, pachnolite and fluellite (Mücke 1988). Lehnerite is monoclinic ($P2_1/n$), forming bronze-yellow to yellow, micaceous, pseudotetragonal plates (to 1 mm), flattened on (010) and layered perpendicular to [010]; it is optically biaxial negative with perfect {010} cleavage. The optical properties of lehnerite crystals vary from core to rim, with the optic axial angle ($2V$) decreasing and extinction becoming more oblique toward crystal edges. The outermost edges of crystals have axial planes perpendicular to the axial-plane orientation at the core, with intermediate zones being uniaxial. These changes in optical properties are explained by variations in H_2O contents (Mücke 1988).

Seelite is known from both the Rabejac U deposit in Lodève and the Talmessi mine in central Iran, where it forms bright yellow, pleochroic (colorless to yellow), tabular, elongate [010] monoclinic ($C2/m$) crystals, flattened on (100) (Bariand et al. 1993). The formula, originally reported as $Mg[UO_2)(AsO_4)]_2(H_2O)_4$ for a crystal from Talmessi (Bachet et al. 1991), but revised to $Mg[UO_2)(AsO_3)_x(AsO_4)_{1-x}]_2(H_2O)_7$ ($x \cong 0.7$), based on the chemistry and structure of a crystal from Lodève, indicates that it is the first known mixed-valence uranyl arsenite-arsenate and a member of the meta-autunite group. Lambor (1994) suggested that seelite might be the tetragonal polymorph of metanovacekite. Seelite is the As analogue of saléeite (Table 9).

Sodium meta-autunite, $Na_2(UO_2)_2(PO_4)_2 \cdot 6\text{-}8H_2O$, can be reversibly hydrated to form fully hydrated sodium autunite, $Na_2(UO_2)_2(PO_4)_2 \cdot 10\text{-}16H_2O$ (Tschernikov and Organova 1994). Both minerals are tetragonal ($P4/nmm$). Sodium meta-autunite forms rapidly in air from sodium autunite.

Chernikovite, the new mineral name for "hydrogen autunite," occurs as thin, transparent, micaceous plates, elongated [010], with perfect (001) cleavage. Chernikovite is pale yellow to lemon green and fluoresces an intense yellow green in ultraviolet light; it is weakly pleochroic. Chernikovite is known from a number of localities (Atencio 1988), including Perus (Sao Paolo) Brazil, where it occurs as oriented inclusions within autunite and meta-autunite. These uranyl phosphates are found in fractures within pegmatitic granites at the Perus occurrence. Finch and Ewing (1992b) tentatively identified chernikovite from the Shinkolobwe mine, where oriented flakes of chernikovite appear to have formed epitaxially on pre-existing curite. Finch and Ewing (1992b) suggest that chernikovite is a precursor to later-formed uranyl phosphates, and that it may replace uranyl oxyhydroxides in P-bearing groundwaters where P concentrations are lower than required to precipitate uranyl phosphates directly from solution. If true, chernikovite may be an important, though potentially short-lived, mineral in the paragenesis of U phosphate minerals.

Phosphuranylite group (Table 11). This group of minerals includes 16 mostly orthorhombic uranyl phosphates, with U:P equal to 3:2. The U:P ratio refers to that of the structural sheets, as there are exceptions: a few minerals in this group contain U in interlayer positions, such as phosphuranylite. The phosphuranylite group contains structural sheets with composition $[(UO_2)_3(O,OH)_2(PO_4)_2]$. Unit-cell dimensions within the plane of the structural sheets in phosphuranylite-group minerals are ~7 and ~17.3 Å (or multiples thereof). In addition to the phosphates, a few minerals without P are structurally related to the phosphuranylite group, including the uranyl selenites guilleminite (Cooper and Hawthorne 1995), haynesite (Cejka et al. 1999), and piretite (Vochten et al. 1996) (see Table 18, below) and probably the arsenate hügelite (Table 11). The structural unit in

Table 11. Phosphuranylite group (U:P = 3:2) $(M^{p+})_m[(UO_2)_3(O,OH)_2(PO_4)_2]_{(p-m)/2}(H_2O)_n$

Name	Formula	Comments
Althupite	ThAl(OH)(UO$_2$)[(UO$_2$)$_3$O$_2$(PO$_4$)$_2$]$_2$(H$_2$O)$_{17}$	Gaines et al. (1997)
Bergenite	(Ba,Ca)$_2$[(UO$_2$)$_3$O$_2$(PO$_4$)$_2$](H$_2$O)$_{6.5}$	Gaines et al. (1997)
Dewindtite	Pb$_3$[(UO$_2$)$_3$O(OH)(PO$_4$)$_2$]$_2$(H$_2$O)$_{12}$	Piret et al. (1990); Gaines et al. (1997)
Dumontite	Pb$_2$[(UO$_2$)$_3$O$_2$(PO$_4$)$_2$](H$_2$O)$_5$	Gaines et al. (1997)
Hügelite*	Pb$_2$[(UO$_2$)$_3$O$_2$(AsO$_4$)$_2$](H$_2$O)$_5$	As-analogue of dumontite (formula inferred); Gaines et al. (1997)
Françoisite-(Nd)	(REE)[(UO$_2$)$_3$O(OH)(PO$_4$)$_2$](H$_2$O)$_6$	Gaines et al. (1997)
"Kivuite"*	(Th,Ca,Pb)(H$_3$O)$_2$(UO$_2$)$_4$(PO$_4$)$_2$(OH)$_8$(H$_2$O)$_5$ (?)	ill defined, discredited species; Nickel & Nichols (1992); Gaines et al. (1997)
Mundite	Al(OH)[(UO$_2$)$_3$(OH)$_2$(PO$_4$)$_2$](H$_2$O)$_{5.5}$	Al[(UO$_2$)$_3$O(OH)(PO$_4$)$_2$](H$_2$O)$_{6.5}$ DP81b, Gaines et al. (1997)
Phosphuranylite*	Ca(UO$_2$)[(UO$_2$)$_3$(OH)$_2$(PO$_4$)$_2$]$_2$(H$_2$O)$_{12}$ or KCa(H$_3$O)$_3$(UO$_2$)[(UO$_2$)$_3$O$_2$(PO$_4$)$_2$]$_2$(H$_2$O)$_8$	D&91, PP91, Gaines et al. (1997)
Phuralumite	Al$_2$(UO$_2$)$_3$O$_2$(PO$_4$)$_2$(OH)$_2$(H$_2$O)$_{12}$	Gaines et al. (1997)
Phurcalite	Ca$_2$[(UO$_2$)$_3$O$_2$(PO$_4$)$_2$](H$_2$O)$_7$	Gaines et al. (1997)
Renardite	Pb(UO$_2$)$_4$(PO$_4$)$_2$(OH)$_4$(H$_2$O)$_7$ possibly Pb(UO$_2$)[(UO$_2$)$_3$O$_2$(PO$_4$)$_2$]$_2$(H$_2$O)$_9$	Possible mixture (dewindtite + phosphuranylite); Deliens et al. (1990),Gaines et al. (1997)
Upalite	Al[(UO$_2$)$_3$O(OH)(PO$_4$)$_2$](H$_2$O)$_7$	Gaines et al. (1997)
Vanmeersscheite	U^{6+}(UO$_2$)$_3$(PO$_4$)$_2$(OH)$_6$(H$_2$O)$_4$	Piret & Deliens (1982b); Gaines et al. (1997)
Meta-vanmeersscheite	U^{6+}(UO$_2$)$_3$(PO$_4$)$_2$(OH)$_6$(H$_2$O)$_2$	Piret & Deliens (1982b); Gaines et al. (1997)
Yingjiangite	K$_2$Ca(UO$_2$)$_7$(PO$_4$)$_4$(OH)$_6$(H$_2$O)$_6$ or K$_2$Ca(UO$_2$)[(UO$_2$)$_3$O(OH)(PO$_4$)$_2$]$_2$(H$_2$O)$_8$	Similar to phosphuranylite; K ⇔ H$_2$O exchange?; Chen et al. (1990); Gaines et al. (1997)

*New or supplemental data or not listed in Dana's New Mineralogy (Gaines et al. 1997).

phosphuranylite-type structures is composed of sheets with a strong linear component, reflected in the physical properties: cleavage parallel to the layers (17.3 × 7 plane), and crystals are commonly elongated along [100] (in the 17.3 Å direction).

Phosphuranylite is one of a few minerals with U in interlayer positions. Demartin et al. (1991) demonstrated that phosphuranylite from Sardinia contains K in interlayer positions, $KCa(H_3O)_3(UO_2)[(UO_2)_3O_2(PO_4)_2]_2(H_2O)_8$, which is slightly different from the composition reported for a crystal from the Shinkolobwe mine: $Ca(UO_2)(UO_2)_3(OH)_2(PO_4)_2(H_2O)_{12}$ (Piret and Piret-Meunier 1991). In order to explain charge balance, Demartin et al. (1991) proposed that hydroxonium ions (H_3O^+) also occur in the interlayer; however, the evidence for H_3O^+ is not convincing, especially given the apparent disorder of interlayer constituents in this mineral (Piret and Piret-Meunier 1991; Demartin et al. 1991).

Althupite was described from the pegmatite at Kobokobo, Kivu, Democratic Republic of Congo, where it occurs with beryl and columbite as thin, transparent yellow tablets (to 0.1 mm); triclinic ($P\bar{1}$), elongated [001] and flattened on (001), optically biaxial negative and pleochroic (pale yellow to dark yellow); the structure has been determined (Piret and Deliens 1987).

Vanmeerscheite and meta-vanmeerscheite both occur with studtite at the Kobokobo pegmatite, forming orthorhombic ($P2_1mn$) yellow plates, elongated on [001] with good {010} cleavage (subordinate {100} cleavage); it is optically biaxial negative (Piret and Deliens 1982). Both minerals are believed to contain phosphuranylite-type structural sheets parallel to (010) (Piret and Deliens 1982).

Another member of the phosphuranylite group, yingjiangite occurs as golden yellow to yellow, compact microcrystalline aggregates (Chen et al. 1990). The IR spectrum indicates both OH and H_2O groups in the structure, and X-ray powder diffraction demonstrates a close structural similarity to phosphuranylite; yingjiangite is orthorhombic ($C222_1$), optically biaxial negative, length slow and pleochroic (colorless to yellow). Described as an alteration product in an oxidized zone that contains uraninite and uranothorite from Yingjiang County, Yunnan Province, China, yingjiangite is associated with studtite, calcurmolite, tengchongite, and autunite. Zhang et al. (1992) reported chemical analyses of material from Xiazhuang U deposit, Guangdong Province, China, confirming the formula $(K_2,Ca)(UO_2)_7(PO_4)_4(OH)_6 \cdot 6H_2O$, and orthorhombic symmetry (*Bmmb*) originally reported (Chen et al. 1990). Yingjiangite bears close structural and chemical similarities to phosphuranylite, possibly with K^+ replacing some interlayer H_2O (or H_3O^+).

Françoisite-(Nd) was described from the Kamoto-est Cu-Co-deposit near Kolwezi, Democratic Republic of Congo, where it occurs as yellow, tabular (010), elongated [001] crystals (monoclinic, $P2_1/c$), and is associated with uraninite, schoepite, uranophane, curite, kamotoite-(Y) and schuilingite-(Nd) (Piret et al. 1988). The structure determination verifies françoisite is a member of the phosphuranylite group (Piret et al. 1988).

Single-crystal structure determinations demonstrate that dewindtite (Piret et al. 1990) and dumontite (Piret and Piret-Meunier 1988) are both members of the phosphuranylite group. The close chemical and structural similarity of dumontite to hügelite (Table 11) suggests that hügelite is also a phosphuranylite-group mineral, with possible formula $Pb_2[(UO_2)_3O_2(AsO_4)_2] \cdot 5H_2O$. A crystal structure determination and electron microprobe analysis of phurcalite from São Paulo, Brazil, showed this mineral to be a member of the phosphuranylite group (Atencio et al. 1991). Braithwaite et al. (1989) report phurcalite from Dartmoor in southwest England and demonstrate that "nisaite," described from Nisa,

Portugal, is equivalent to phurcalite (nisaite was never accepted by the IMA as a valid species), and Singh (1999) report phurcalite from Putholi in Rajasthan, India. A recent review of phurcalite occurrences is provided by Walenta (1993a) who also describes a new occurrence of phurcalite from the Schmiedestollen-Halde, Wittechen, in central Schwarzwald, Germany. In another paper, Walenta (1993b) reports on established and discredited mineral species in the German Schwarzwald, a compilation of interest for more than the discussion of U minerals. Steen (1998) describes U and Pb mineralization near Sulzburg in the south Schwarzwald, Germany.

Walpurgite group (Table 12). These are compositionally and structurally similar minerals with U:P or U:As ratios of 1:2. Although they display many similarities, many of their structures are still unknown, and their classification here as a single mineral group remains to be demonstrated.

Krause et al. (1995) reported orthowalpurgite from a mine dump at Schmiedestollen, Wittichen, Germany, where it occurs as transparent yellow, orthorhombic (*Pbcm*), tabular, elongate [100] crystals (to 0.3 mm) flattened on {010}, also forming fan-shaped aggregates to 1 mm. Orthowalpurgite is associated with preisingerite, $Bi_3(AsO_4)_2O(OH)$, quartz, and anatase; it is isochemical with walpurgite.

Birch et al. (1988) described ulrichite from a granite quarry near Lake Boga, Victoria, Australia, where it occurs as radiating sprays of apple-green to lime-green acicular crystals (to 1 mm) and as flat prisms (monoclinic *C2/m*), optically biaxial (probably negative). Ulrichite is found in miarolitic cavities in a pegmatoidal granite along with chalcosiderite, turquoise, torbernite, saléeite, as well as fluorite, cyrilovite, libethenite, apatite, and an unidentified Fe phosphate.

Furongite was originally described from a U deposit in western Hunan, China, where it occurs in strongly weathered carbonaceous shales in association with autunite, limonite, halloysite, opal, evansite and variscite (Hunan 230 Laboratory 1976). Furongite was subsequently described from the Kobokobo quartz-beryl pegmatite, where it occurs with other Al-uranyl phosphates, principally triangulite and moreauite (Deliens and Piret 1985b). Furongite, which is triclinic (P1), occurs as bright yellow masses of optically biaxial negative tabular crystals with three perfect cleavages, and apparently displays some compositional variability, with slight changes in Al, and large differences in H_2O contents (Table 13). The formula is close to $Al_2(UO_2)(PO_4)_2(OH)_2 \cdot 8H_2O$ (Gaines et al. 1997; Mandarino 1999), with Kobokobo material being less hydrated (~1.5 H_2O instead of 8); H_2O loss begins at 38°C. The composition and space group resemble those of walpurgite-type minerals (Table 12). However, Shen and Peng (1981) reportedly determined the structure (but did not publish their results or data), describing furongite as a sheet-type structure with "a lot of water" between the structural sheets; the structure reportedly supports the more complex formula (Table 13).

Some uranyl phosphates are not readily categorized, having compositions and structures distinct from the phosphuranylite and walpurgite, autunite and meta-autunite groups (Table 13). Kamitugaite is a Pb-Al-uranyl phosphate-arsenate (P > As) from the pegmatite at Kobokobo, Kivu, Democratic Republic of Congo, where it occurs as thin yellow plates (to 0.5 mm) and tufts on the surfaces of quartz grains, triclinic, with two cleavages, (010) and (001). It is optically biaxial negative, as for most other uranyl phosphates of the (meta)autunite and phosphuranylite groups; however, cleavage and unit-cell data suggest that it is not structurally related to any known uranyl phosphates or arsenates (Deliens and Piret 1984a). Kamitugaite is associated with other uranyl phosphates such as triangulite, threadgoldite and dumontite, as well as the uranyl peroxide studtite.

Table 12. Walpurgite group of phosphates and arsenates (U:P = 1:2) $M^{2+}[(UO_2)(PO_4)_2](H_2O)_n$

Name	Formula	Comments
Hallimondite	$Pb[(UO_2)(AsO_4)_2]$	Smith (1984); Gaines et al. (1997)
Orthowalpurgite	$(BiO)_4[(UO_2)(AsO_4)_2](H_2O)_2$	Krause (1995); Gaines et al. (1997)
Parsonsite	$Pb_2[(UO_2)(PO_4)_2](H_2O)_2$	Frondel (1958); Gaines et al. (1997); structure determination indicates presence of uranyl phosphate chains, Burns (1999c)
Pseudo-autunite	$Ca_2[(UO_2)_2(PO_4)_4](H_2O)_9$	discredited; inadequate data; Gaines et al. (1997)
Ulrichite	$CaCu^{2+}[(UO_2)_2(PO_4)_2](H_2O)_4$	Frondel (1958); Gaines et al. (1997)
Unnamed	$(BiO)_4[(UO_2)(PO_4)_2](H_2O)_2$	phosphate analogue of walpurgite ?; Smith (1984); Ondrus et al. (1997c)
Walpurgite	$(BiO)_4[(UO_2)(AsO_4)_2](H_2O)_2$	Mereiter (1982); Gaines et al. (1997)

*New or supplemental data or not listed in Dana's New Mineralogy (Gaines et al. 1997).

Table 13. Miscellaneous uranyl phosphates and arsenates

Name	Formula	Comments
Arsenuranylite	$Ca(UO_2)_4(AsO_4)_2(OH)_4(H_2O)_6$ possibly $Ca(UO_2)[(UO_2)_3O_2(AsO_4)_2](H_2O)_8$	Possible phosphuranylite structure type; Smith (1984); Gaines et al. (1997)
Asselbornite	$(Pb,Ba)(BiO)_4(UO_2)_6(AsO_4)_2(OH)_{12}(H_2O)_3$	Sarp et al. (1983); Gaines et al. (1997)
Coconinoite	$Fe^{3+}{}_2Al_2(UO_2)_2(PO_4)_4(SO_4)(OH)_2(H_2O)_{20}$	Compare xiangjiangite; $SO_4 \Leftrightarrow PO_4$?; Gaines et al. (1997)
Furongite*	Approximately $Al_2(OH)_2[(UO_2)(PO_4)_2](H_2O)_8$	Akita, Japan: $Al_{13}(UO_2)_7(PO_4)_{13}(OH)_{14}(H_2O)_{58}$; Shen & Peng (1981); Kivu, DRC: $Al_2(UO_2)_7(PO_4)_3(OH)(H_2O)_{13.5}$, Deliens & Piret (1985); Deliens et al. (1990); Gaines et al. (1997). Compare with walpurgite group (Table 12)
Kamitugaite	$PbAl(UO_2)_5[(P,As)O_4)_2(OH)_9(H_2O)_{9.5}$	Deliens & Piret (1984a), Gaines et al. (1997)
Moreauite	$Al_3(UO_2)(PO_4)_3(OH)_2(H_2O)_{13}$	Gaines et al. (1997)
Triangulite	$Al_3(OH)_5[(UO_2)(PO_4)]_4(H_2O)_5$	Deliens & Piret (1982c); Gaines et al. (1997)
Xiangjiangite	$(Fe^{3+},Al)(UO_2)_4(PO_4)_2(SO_4)_2(OH)(H_2O)_{22}$	Compare coconinoite; $SO_4 \Leftrightarrow PO_4$?; Gaines et al. (1997)

*New or supplemental data or not listed in Dana's New Mineralogy (Gaines et al. 1997).

Triangulite, $Al_3[(UO_2)(PO_4)]_2(OH)_5 \cdot 5H_2O$, occurs in the oxidized zone of the Kobokobo pegmatite (Deliens and Piret 1982b). Its structure is triclinic and is apparently distinct from known uranyl phosphates. It bears a compositional similarity to the synthetic compound, $Ba_3[(UO_2)(PO_4)(PO_3OH)]_2 \cdot xH_2O$ (Guesdon and Raveau 1998), which does not possess autunite-type structural sheets. Triangulite occurs as bright yellow triangular crystals and as mm-sized nodules and is most commonly associated with phosphuranylite, meta-autunite and ranunculite. Crystals are flattened on (010), and elongated [001], and display two good cleavages: (010) and (001).

Deliens and Piret (1985) described moreauite from the Kobokobo pegmatite, where it is associated with furongite, ranunculite and phosphosiderite. Moreauite crystals (monoclinic, space group $P2_1/c$) are greenish yellow and form books of flattened plates, on (001), with perfect {001} cleavage (optically biaxial negative). Crystals of moreauite are commonly partially dehydrated, and the reported formula corresponds to fully hydrated moreauite. The structure of moreauite is unknown; however, the composition and unit-cell data suggest that it is not structurally related to either the autunite and meta-autunite groups, or to the phosphuranylite group.

Sarp et al. (1983) identified asselbornite from Schneeberg, Saxony, Germany, where it occurs within quartz-rich gangue minerals and is associated with β-uranophane, uranospinite, and uranosphaerite. Asselbornite forms translucent brown to lemon-yellow cubic (*I*-centered) crystals up to 0.3 mm across. Martin and Massanek (1995) reported orange crystals of asselbornite associated with zeunerite and uranophane in the Schneeberg area of Saxony, Germany.

Recent chemical analyses of a coconinoite specimen from an unspecified locality within the Kizylkhum Formation demonstrate the predominance of Al: $Al_4(UO_2)_2(PO_4)_4(SO_4)(OH)_2 \cdot 18H_2O$ (Belova et al. 1993), which contrasts with the original description of coconinoite from the Colorado Plateau, for which Al and Fe occur in approximately equal molar proportions: $Fe^{3+}_2Al_2(UO_2)_2(PO_4)_4(SO_4)(OH)_2 \cdot 20H_2O$ (Young et al. 1966). Combined electron and X-ray diffraction analyses confirm that coconinoite is monoclinic (*C2/c* or *Cc*) as originally reported (Belova et al. 1993).

Uranyl vanadates, molybdates, tungstates

Uranyl vanadates (Table 14) comprise perhaps the most insoluble of uranyl minerals (Garrels and Christ 1959; Langmuir 1978; Smith 1984). Uranyl vanadates are important U ores in the Colorado Plateau of the US (Finch 1996). Uranyl vanadates are so stable that it is likely that they will form wherever dissolved U comes in contact with waters containing dissolved vanadate ions. The uranyl vanadates occur where reduced U minerals (e.g. uraninite, coffinite or brannerite) and reduced V minerals (e.g. montroseite) are undergoing oxidation. Vanadium(V) may also be derived from rocks that contain reduced V, such as organic-rich shales and other clay-rich rocks. The stabilities of the uranyl vanadates is manifested in their potential longevity. Some of the carnotite-group minerals have reported ages of 350 Ka (Kaufman and Ku 1989) No new uranyl vanadates have been described since the review by Smith (1984), although Paulis (1992) described curienite from Abertamy near Jáchymov, Czech Republic, where it occurs on a phyllite fragment penetrated by a quartz-sphalerite-galena-uraninite vein.

Uranyl molybdates (Table 15) are locally important as ore minerals, and are also common as accessories in roll-front deposits and other deposits where uraninite and Mo-bearing minerals are being weathered. Molybdenum in these minerals occurs as Mo^{6+}, and U may be of mixed valence, U^{4+} and U^{6+}. The paragenesis of many of these minerals remains uncertain and many are poorly described. Uranyl molybdates are potentially more

Table 14. Uranyl vanadates

Name	Formula	Comments
Carnotite	$K_2(UO_2)_2(V_2O_8)(H_2O)_3$	Carnotite group; Frondel (1958); Gaines et al. (1997)
Curiénite	$Pb(UO_2)_2(V_2O_8)(H_2O)_5$	Carnotite group; Smith (1984); Paulis (1992); Gaines et al. (1997)
Ferghanite*	$(UO_2)_3(V_2O_8)(H_2O)_6$ (?)	Not verified; possible carnotite group mineral; Frondel (1958)
Francevillite	$(Ba,Pb)(UO_2)_2(V_2O_8)(H_2O)_5$	Carnotite group; Gaines et al. (1997)
Fritzscheite	$Mn^{2+}(UO_2)_2(V_2O_8)(H_2O)$	Carnotite group; Frondel (1958); Gaines et al. (1997)
Margaritasite	$(Cs,K)_2(UO_2)_2(V_2O_8)(H_2O)_n$ (n ~ 1-3)	Carnotite group; Gaines et al. (1997)
Rauvite*	$Ca(UO_2)_2V_{10}O_{28}(H_2O)_{16}$	ill defined; possibly isostructural w/ pascoeite; Frondel (1958); Smith (1984)
Sengierite	$Cu_2(UO_2)_2(V_2O_8)(OH)_2(H_2O)_5$	Carnotite group; Gaines et al. (1997)
Strelkinite	$Na_2(UO_2)_2(V_2O_8)(H_2O)_6$	Carnotite group; Gaines et al. (1997)
Tyuyamunite	$Ca(UO_2)_2(V_2O_8)(H_2O)_8$	Carnotite group; Frondel (1958); Gaines et al. (1997)
Metatyuyamunite	$Ca(UO_2)_2(V_2O_8)(H_2O)_3$	Carnotite group; Frondel (1958); Gaines et al. (1997)
Vanuralite*	$Al(UO_2)_2(V_2O_8)(OH)(H_2O)_{11}$	Carnotite group; Gaines et al. (1997)
Metavanuralite	$Al(UO_2)_2(V_2O_8)(OH)(H_2O)_8$	Carnotite group; Gaines et al. (1997)
Vanuranylite	$(H_3O,Ba,Ca,K)_{1.6}(UO_2)_2(V_2O_8)(H_2O)_4$	ill defined; possible carnotite-group mineral; Gaines et al. (1997)
Uvanite	$(UO_2)_2V_6O_{17}\cdot 15H_2O$	ill defined; Frondel (1958); Gaines et al. (1997)

* New or supplemental data or not listed in Dana's New Mineralogy (Gaines et al. 1997).

Table 15. Uranyl molybdates, tungstates, niobate-tantalate

Name	Formula	Comments
Calcurmolite	$Ca(UO_2)_3(MoO_4)_3(OH)_2(H_2O)_{11}$	Gaines et al. (1997)
Cousinite	$Mg(UO_2)_2(MoO_4)_2(OH)_2(H_2O)_5$	Gaines et al. (1997)
Deloryite	$Cu^{2+}_4(UO_2)(MoO_4)_2(OH)_6$	Sarp & Chiappero (1992); Gaines et al. (1997)
Iriginite	$(UO_2)Mo_2O_7(H_2O)_3$	alteration product of umohoite; Gaines et al. (1997)
Moluranite	$H_4U^{4+}(UO_2)_3(MoO_4)_7(H_2O)_{18}$	Gaines et al. (1997)
Tengchongite	$Ca(UO_2)_6(MoO_4)_2O_5(H_2O)_{12}$	Gaines et al. (1997)
Umohoite	$(UO_2)MoO_4(H_2O)_4$	minor Mg, Ni, Ca; Frondel (1958); Gaines et al. (1997)
Uranotungstite	$(Ba,Pb,Fe^{2+})(UO_2)_2(WO_4)(OH)_4(H_2O)_{12}$	Walenta (1985); Gaines et al. (1997)
Liandratite	$U^{6+}(Nb,Ta)_2O_8$	Oxidation product of of petschekite (Table 1); possible α-U_3O_8 structure derivative; compare with UTa_2O_8; Gasperin (1960); Smith (1984); Gaines et al. (1997)

* New or supplemental data or not listed in Dana's New Mineralogy (Gaines et al. 1997).

abundant than currently appreciated due to confusion (by color) with reduced U minerals. U-molybdates commonly coexist as fine-grained masses (Smith 1984; Gaines 1997).

Two new uranyl molybdates have been described since Smith (1984). Deloryite (Table 15) is monoclinic ($C2$, Cm or $C2/m$) and occurs as dark green rosettes of tabular (010) elongate [001] crystals, transparent to opaque, and displaying three cleavages (Sarp and Chiappero 1992). Deloryite was described from the Cap Garonne mine near Le Pradet, Var, France, where it occurs with metazeunerite, atacamite, paratacamite, malachite, tourmaline, and barite on quartz gangue. Unit-cell dimensions of deloryite are similar to those of the Cu-uranyl selenite, derricksite, suggesting a structural relationship.

Chen et al. (1986) described the hydrated uranyl molybdate, tengchongite, from Tengchong County, Yunan province, China, where it occurs in a migmatitic gneiss. Tengchongite is orthorhombic ($A2_122$) [a 15.616; b 13.043; c 17.716 Å] and forms yellow (biaxial negative) irregular grains, flattened on {001} with perfect (001) cleavage. Closely associated with studtite, and calcurmolite, tengchongite has a higher U:Mo ratio (3:1) than other known uranyl molybdates. Notably, the IR spectrum indicates that the structure contains H_2O groups but no OH^- ions. Both structurally and compositionally, tengchongite resembles an orthorhombic Cs- and Ba-bearing uranyl molybdate identified from corrosion experiments on spent UO_2 fuel (Buck et al. 1997).

A single-crystal structure determination for synthetic iriginite (Serezhkin etal. 1973), combined with more recent TEM data (Vishnev et al. 1991), confirm orthorhombic symmetry ($Pca2_1$), and indicate the structural formula $(UO_2)[Mo_2O_7(H_2O)_2](H_2O)$, similar to the composition reported for natural iriginite, $(UO_2)Mo_2O_7 \cdot H_2O$ (Epstein 1959) but with more H_2O.

Uranyl tungstates (Table 15). There is only one known uranyl tungstate, uranotungstite, which is known from two localities in the Black Forest: Menzenschwand (Southern Black Forest) and the Clara mine near Oberwolfach (Central Black Forest) (Walenta 1985). Specimens from the two localities have different compositions. Material from Menzenschwand contains approximately equal molar proportions of Fe, Ba and Pb, whereas uranotungstite from the Clara mine has no Pb and only a trace of Ba. Uranotungstite is orthorhombic (optically biaxial negative) and forms spherulitic clusters of yellow, orange or brownish, lath-shaped crystals. Crystals have a perfect {010} cleavage and are pleochroic (colorless to yellow) perpendicular to the cleavage. Molon (1990) provides a synopsis of mineral paragenesis at the Clara mine, including that of uranotungstite.

Uranyl sulfates, selenites, tellurites

Uranyl sulfates (Table 16) are important only where sulfides are being oxidized, providing dissolved sulfate to groundwater that can complex with UO_2^{2+} to form stable uranyl sulfate complexes in solution. Evaporation is required to precipitate uranyl sulfates. Waters from which uranyl sulfates precipitate tend to be somewhat acidic, with pH values below approximately 6 (Garrels and Christ 1959; Ondrus et al. 1997a) (Fig. 2). Like uranyl di- and tri-carbonates, most uranyl sulfates are ephemeral and redissolve upon exposure tofresh water. Uranyl sulfates occur where uranyl carbonates are absent (and vice versa), a reflection of the different pH ranges over which uranyl sulfate and uranyl carbonate complexes are important (Ondrus et al. 1997a). Synthesis experiments on uranyl sulfates suggest that a "slow decrease in solution pH" is required for the precipitation of zippeite (-group minerals), otherwise uranyl oxyhydroxides tend to form, decreasing the activities of dissolved U species (Frondel et al. 1976; Brindley and Bastanov 1982).

Table 16. Uranyl sulfates

Name	Formula	Comments
Coconinoite*	$Fe^{3+}{}_2Al_2(UO_2)_6(PO_4)_4(SO_4)(OH)_2(H_2O)_{20}$	compare xiangjiangite; Smith (1984); Gaines et al. (1997)
Deliensite*	$Fe(UO_2)_2(SO_4)_2(OH)_2(H_2O)_3$	Vochten et al. (1997b)
Jáchymovite*	$(UO_2)_8(SO_4)(OH)_{14}(H_2O)_{13}$	compare uranopilite; C&96b
Johannite	$Cu(UO_2)_2(SO_4)_2(OH)_2(H_2O)_{6-8}$	Frondel (1958); Gaines et al. (1997); Ondrus et al. (1997b)
Rabejacite	$Ca(UO_2)_4(SO_4)_2(OH)_6(H_2O)_6$	DP93; Gaines et al. (1997); Ondrus et al. (1997b)
Schrökingerite	$NaCa_3(UO_2)(CO_3)_3(SO_4)F(H_2O)_{10}$	Frondel (1958); Gaines et al. (1997)
Uranopilite*	$(UO_2)_6(SO_4)(OH)_{10}(H_2O)_{12}$	compare jáchymovite; Gaines et al. (1997); Ondrus et al. (1997b)
Meta-uranopilite	$(UO_2)_6(SO_4)(OH)_{10}(H_2O)_5$	reversible dehydration of uranopilite; Gaines et al. (1997); Ondrus et al. (1997b)
Xiangjiangite	$(Fe^{3+},Al)(UO_2)_4(PO_4)_2(SO_4)_2(OH)(H_2O)_{22}$	compare coconinoite; Gaines et al. (1997)
Unnamed*	Hydrated Ca-uranyl sulfate	"pseudo-johannite", Ondrus et al. (1997c)
Unnamed*	$(Fe,Mg)_2(UO_2)_4(SO_4)_2(OH)_8(H_2O)_3$	"ferro-zippeite", Ondrus et al. (1997c)
Unnamed*	Hydrated Mg-Fe-K-uranyl-sulfate	"pseudo-Mg-zippeite", Ondrus et al. (1997c)
Unnamed	Hydrated Pb-uranyl sulfate	few data; Ondrus et al. (1997c)
Zippeite Group	$M^{2+}{}_2[(UO_2)_2(SO_4)(OH)_3]_{10}(H_2O)_n$ or $M^{p+}{}_{ml}[(UO_2)_2(SO_4)(OH)_3]_{(p-m)}(H_2O)_n$	
Zippeite*	$K_4(UO_2)_6(SO_4)_3(OH)_{10}(H_2O)_4$	synthetic analogue, $K[(UO_2)_2(SO_4)(OH)_3](H_2O)$ (Vochten et al. 1995b); Frondel (1958); Frondel et al. (1976)
Co-zippeite	$Co_2(UO_2)_6(SO_4)_3(OH)_{10}(H_2O)_{16}$	Frondel et al. (1976); Gaines et al. (1997)
Mg-zippeite	$Mg_2(UO_2)_6(SO_4)_3(OH)_{10}(H_2O)_{16}$	Frondel et al. (1976); Gaines et al. (1997); Ondrus et al. (1997b)
Na-zippeite	$Na_4(UO_2)_6(SO_4)_3(OH)_{10}(H_2O)_4$	Frondel et al. (1976); Gaines et al. (1997); Ondrus et al. (1997b)
Ni-zippeite	$Ni_2(UO_2)_6(SO_4)_3(OH)_{10}(H_2O)_{16}$	Frondel et al. (1976); Gaines et al. (1997)
Zn-zippeite	$Zn_2(UO_2)_6(SO_4)_3(OH)_{10}(H_2O)_{16}$	Frondel et al. (1976); Gaines et al. (1997)

* New or supplemental data or not listed in Dana's New Mineralogy (Gaines et al. 1997).

Several new uranyl sulfates have been described since Smith's (1984) review. Jáchymovite was described from Jáchymov, Czech Republic, where it occurs as translucent yellow acicular crystals (monoclinic $P2_1$ or $P2_1/m$), with good {010} cleavage, forming coatings on uraninite-bearing dolomite veins, and is associated with uraninite, uranopilite, and gypsum (Cejka et al. 1996b). The formula for jáchymovite, $(UO_2)_8(SO_4)(OH)_{14}\cdot 13H_2O$, is similar to that of uranopilite, $(UO_2)_6(SO_4)(OH)_{10}\cdot(12\text{-}13)H_2O$ (Table 16), and X-ray diffraction data are probably necessary to distinguish these minerals with confidence. Jensen et al. (1997) identified uranopilite (tentatively, by microprobe analysis) from the Bangombé natural fission reactor in Gabon, where it is a recent alteration product of uraninite.

Rabejacite, found at both Rabejac and Mas de d'Alary in Lodève, Herault, France, forms bright yellow to amber-yellow spherulitic aggregates and individual tablets flattened on {001} (to 0.1 mm) (Deliens and Piret 1993). Rabejacite is orthorhombic (space group unknown: a 8.73; b 17.09; c 15.72 Å), optically biaxial negative, and strongly pleochroic (pale yellow to sulfur yellow). Rabejac is associated with gypsum, and other uranyl minerals derived from the oxidative dissolution of uraninite, including fontanite, billietite, and uranophane. Deliensite was also discovered at the Mas d'Alary U deposit in Lodève, Hérault, France, where it occurs as spherical aggregates of submillimeter-sized pale yellow to grayish white, tabular, orthorhombic ($Pnnm$ or $Pnn2$) crystals (Vochten et al. 1997b). Deliensite is associated with uraninite, gypsum and pyrite, and results from the oxidative dissolution of uraninite and primary sulfides.

The composition and unit-cell parameters of zippeite-group minerals may need revision in light of the crystal-structure study of synthetic zippeite by Vochten et al. (1995b) (Table 16). They showed that the structural formula for synthetic zippeite, $K(UO_2)_2(SO_4)(OH)_3\cdot H_2O$, is significantly different from the previously accepted formula, $K_4(UO_2)_6(SO_4)_3(OH)_{10}\cdot 16H_2O$ (Frondel et al. 1976). Minerals of the zippeite group are thought to be orthorhombic; however, the synthetic crystal examined by Vochten et al. (1995b) was monoclinic. Re-examination of this mineral group is in order, a potentially daunting challenge given the extremely small grain sizes and pulverulent habits common to most zippeite-group minerals.

Uranyl selenites (Table 17) occur where Se-bearing sulfide minerals are undergoing oxidation and dissolution. Selenium in all known uranyl minerals occurs as Se(IV), in the form of the selenite ion, SeO_3^{2-}, although there seems no reason not to expect the existence of Se(VI) minerals under sufficiently oxidizing conditions. These would probably be as soluble (and as ephemeral) as uranyl sulfates, and might be confused with them. No data are available on the solubilities of these minerals, only six of which are known. Two-thirds of the uranyl selenites are from the Musonoi deposit in southern Shaba, Democratic Republic of Congo, where Se is derived from the oxidation of seleniferous digenite, Cu_9S_5 (Deliens et al. 1981). Two new uranyl selenites have been described since Smith's (1984) review.

A structure determination of guilleminite by Copper and Hawthorne (1995) revises the structural formula slightly and demonstrates that the structural unit of guilleminite is similar to that of the phosphuranylite-group minerals.

Piretite is found at the Shinkolobwe mine in the Democratic Republic of Congo, where it occurs as lemon-yellow tablets, orthorhombic ($Pmn2_1$ or $Pmnm$), with good {001} cleavage, and is optically biaxial negative (Vochten et al. 1996). Piretite forms crusts on the surfaces of uraninite crystals and is also associated with masuyite (or similar Pb-uranyl oxyhydroxide minerals). The composition, physical properties, and unit-cell parameters of piretite indicate a close structural relationship with guilleminite (Cooper and Hawthorne

Table 17. Uranyl selenites and tellurites

Name	Formula	Comments
Demesmaekerite	$Cu_5Pb_2(UO_2)_2(SeO_3)_6(OH)_6(H_2O)_2$	Gaines et al. (1997)
Derricksite	$Cu_4(UO_2)(SeO_3)_2(OH)_2$	Gaines et al. (1997)
Guilleminite*	$Ba[(UO_2)_3O_2(SeO_3)_2](H_2O)_3$	Phosphuranylite structure type; Gaines et al. (1997); Cejka et al. (1995, 1999)
Haynesite	$(UO_2)_3(OH)_2(SeO_3)_3(H_2O)_5$	Phosphuranylite structure type; Gaines et al. (1997); Cejka et al. (1999)
Marthosite	$Cu(UO_2)_3(OH)_2(SeO_3)_3(H_2O)_7$	Gaines et al. (1997)
Piretite*	$Ca(UO_2)_3(OH)_2(SeO_3)_2(H_2O)_4$	Possible phosphuranylite structure type; Vochten et al. (1996)
Cliffordite	UTe_3O_9	Smith (1984); Gaines et al. (1997)
Moctezumite	$Pb(UO_2)(TeO_3)_2$	Swihart et al. (1993); Gaines et al. (1997)
Schmitterite	$(UO_2)TeO_3$	Smith (1984); Gaines et al. (1997)

* New or supplemental data or not listed in Dana's New Mineralogy (Gaines et al. 1997).

1995) and suggests that piretite has a phosphuranylite-type structural sheet.

Haynesite was described from the Repete mine near Blanding, Utah, where it occurs in mudstones and sandstones with boltwoodite and andersonite, as well as gypsum and calcite (Deliens and Piret 1991). Haynesite is orthorhombic (*Pnc2* or *Pncm*) and occurs as amber-yellow, elongate [001] tablets or acicular crystals in rosettes; individual crystals display perfect {010} cleavage. The chemical formula, unit-cell parameters, and physical properties suggest that haynesite contains structural sheets similar to those found in guilleminite and the phosphuranylite group of uranyl phosphates, and a preliminary structure determination suggests that this is true (Burns, pers. comm.). Infrared data for haynesite are reported by Cejka et al. (1999) and compared with other uranyl selenites (also see Cejka, Chapter 12 this volume).

Uranyl tellurites (Table 17). The three known uranyl tellurites all occur at the San Miguel and Moctezuma mines in Moctezuma, SON, Mexico, although schmitterite is also known from the Shinkolobwe mine in southern Democratic Republic of Congo. They occur where sulfides are undergoing oxidation, which probably provide Te. All uranyl tellurites are anhydrous. The structures of all three are known, the most recent being reported for moctezumite by Swihart et al. (1993) (also see Burns et al. 1996, and Burns this volume). Few details are known about the conditions of their formation, although moctezumite reportedly alters pseudomorphously to schmitterite, and both schmitterite and moctezumite are more soluble in dilute HCl than cliffordite is (Gaines 1965). Thermodynamic data have been reported recently for synthetic cliffordite (Mishra et al. 1998) and synthetic schmitterite (Mishra et al. 1998; Singh et al. 1999).

ALTERATION OF REDUCED URANIUM MINERALS

Except for coffinite and brannerite, most U^{4+} minerals occur as accessory minerals in granitic igneous rocks, pegmatites, and aluminous metamorphic host rocks. Alteration of U-bearing accessory minerals is a major factor affecting U concentrations in groundwaters emanating from exposed granite terranes. Most U-bearing accessory minerals are formed under reducing conditions, crystallizing from deep-seated Si-rich magmas or hydrothermal fluids. Where uplifted and exposed to percolating meteoric waters, these minerals are commonly destabilized, and U^{4+} may be oxidized, lost from the crystalline host, and enter into aqueous solution as the uranyl ion or its complexes. Because U^{4+} is readily oxidized at near-atmospheric O fugacities, and because the uranyl ion, UO_2^{2+}, can form many stable solution complexes, U is potentially mobile in oxidizing aqueous solutions, whereas other elements, such as Th and REE, commonly incorporated into U^{4+}-bearing accessory minerals during crystallization are generally less amenable to aqueous transport. Differences in aqueous mobilities of U and its erstwhile companion elements readily explain why U that is mobilized from Th- and REE-bearing minerals is so commonly segregated from these other, less-mobile elements. Uraninite deposited from low-temperature U-bearing groundwaters are nearly always devoid of Th and REE (Berman 1957; Frondel 1958).

Initial alteration of most U^{4+} minerals probably occurs as U^{4+} oxidizes to U^{5+} or U^{6+} without significantly affecting the structure. Accumulation of radiogenic Pb, He, and U^{6+} (the latter also caused by "auto-oxidation") may contribute to destabilization of some U^{4+} minerals, although the degree to which this is important has rarely been addressed. Of course, the importance of these effects in a mineral depends on the mineral's age and concentrations of radioactive elements. On the other hand, the effects of radioactive decay in U- (and Th-) bearing minerals caused by alpha-recoil damage commonly exceeds most detectable effects of compositional changes. Notably, the UO_2 structure is remarkably resilient to alpha-decay damage, due to rapid annealing kinetics (Stout et al. 1988; Janeczek and Ewing 1991), which is one reason why UO_2 was chosen as a nuclear fuel. Many minerals are significantly less resilient to radiation damage, and metamictization (radiation-induced amorphization), partial or complete, can have a profound influence on the geochemical durability of most radioactive minerals.

Geochemical alteration of pyrochlore-group minerals was described in a series of papers by Lumpkin and Ewing (1992a, 1995, 1996). They have shown that microlite, pyrochlore and betafite are subject to cation and anion exchange and incongruent dissolution when exposed to a wide range of aqueous fluids in a variety of environments: from deep hydrothermal (to ~650°C and 5 kbar) to near-surface weathering environments. The A-site cation mobility is limited by ionic charge and radius, with the high field-strength cations, REE^{3+}, Th^{4+} and U^{4+}, being least mobile. Actinides are essentially retained in microlite, pyrochlore and most betafite samples for periods up to ~1.4 Ga. Preferential loss of some cations from betafite shifts the composition towards the stability field of liandratite, uranopyrochlore and rutile (or anatase). Betafite may undergo extensive incongruent dissolution and recrystallization, and exhibit U loss under "extreme conditions." Alteration of the pyrochlore group minerals depends in part on grain size, radiation damage and micro-fracturing, as well as fluid flow and duration of exposure to fluids. Pyrochlore-group minerals exposed to lateritic weathering experience cation exchange at a rate that exceeds mineral dissolution, which is primarily controlled by the stability of the Nb-Ta-Ti octahedral framework. Electron energy-loss spectroscopy (EELS) was used to identify Ce oxidation states in a partly altered pyrochlore sample (Xu and Wang 1999). Cerium and other REE occur as trivalent ions in unaltered regions, whereas Ce is oxidized to Ce^{4+} in a neighboring altered region of the pyrochlore. The oxidation of Ce was reportedly accompanied by losses of REE, U, and radiogenic Pb during alteration.

Uraninite alteration under reducing conditions

Uraninite is by far the most important U mineral in terms of abundance, wide-spread occurrence, and economic value. Coffinite and brannerite are of secondary importance, both geochemically and economically (Finch 1996). Because of its importance to U geochemistry, a rather detailed discussion of uraninite alteration is provided here.

Ancient uraninite commonly consists of two cubic phases with slight but distinct differences in unit-cell parameters (Janeczek and Ewing 1992a; Janeczek and Ewing 1991a); an effect that appears to be caused by segregation of Pb-rich and Pb-poor uraninite, with the larger unit cell corresponding to the Pb-rich phase (Janeczek 1993). In strongly reducing environments ancient uraninite may recrystallize as fine-grained Pb-depleted, Si-enriched material from precursor Pb-rich, Si-poor coarsely crystalline material (Janeczek and Ewing 1992b; Kotzer and Kyser 1993; Fayek et al. 1997). Recrystallization of ancient uraninite in reducing environments may be most affected by the combination of two effects of radioactive decay of U: (1) destabilization of uraninite by decreased U content and increased Pb content (upwards of 20 wt % PbO has been reported in some specimens of uraninite), and (2) increased $U^{6+}:U^{4+}$ ratios caused by "auto-oxidation," which increasingly destabilizes uraninite under reducing conditions. The rate at which synthetic UO_{2+x} dissolves exceeds that of UO_2 in reducing waters above pH 5 (Janeczek and Ewing 1992c). Variations in Si contents within uraninite demonstrate the important role of groundwaters during uraninite alteration under reducing conditions, and incipient coffinite is common in many uraninite specimens that have recrystallized in reducing, Si-rich groundwaters (Janeczek and Ewing 1995; Fayek et al. 1998).

Uraninite is sparingly soluble at normal pHs in dilute, reducing groundwaters (Langmuir 1978; Parks and Pohl 1988). The solubility of uraninite increases with temperature and dissolved F, Cl and CO_2 (Giblin and Appelyard 1987; Keppler and Wyllie 1990), and dissolved U^{4+} concentrations in Na-K-Ca-Cl brines are an order of magnitude higher than in fresh waters (Giblin and Appelyard 1987). U^{4+}-fluoride complexes are stable below pH 4, contributing to potentially significant migration of U in reducing groundwaters. The influence that impurity elements in uraninite may have on uraninite dissolution under reducing conditions is uncertain. Grandstaff (1976) reports that increased Th concentrations increase the dissolution rate, whereas Posey-Dowty et al. (1987) found no effect. Janeczek and Ewing (1992c) examined uraninite from Cigar Lake in Saskatchewan, Canada, and Oklo, Gabon, and found evidence for extensive dissolution and replacement of uraninite. Replacement by U-free minerals such as illite, chalcopyrite and apatite suggests that the amount of U that migrated was substantial. Janeczek and Ewing (1992c) attribute much of this alteration to hydrothermal interactions, although they note that radiolysis (the radiolytically induced decomposition of water) may have been a factor by increasing redox conditions at the uraninite-water interface, particularly at Oklo (Dubessy et al. 1988; Meere and Banks 1997; Savary and Pagel 1997; Janeczek this volume). Kotzer and Kyser (1993) showed that most uraninite at the Cigar Lake deposit has been recrystallized, possibly through migration and reprecipitation. Janeczek (this volume) notes that uraninite of secondary origin from the Francevillian FA sandstone at Bangombé often can be distinguished from primary uraninite within the nearby natural reactor only on the basis of $^{235}U/^{238}U$ ratios. A recent experimental study comparing the dissolution of synthetic UO_2 and natural uraninite suggests that "the mobilization of U in reducing environments can occur by only slight changes [in] the surrounding conditions, even if the [uraninite] is chiefly in its reduced form" (Casas et al. 1998).

Coffinitization of uraninite. The alteration of uraninite to coffinite is common (Smits 1989; Janeczek 1991; Janeczek and Ewing 1992a,b), although it is not always well understood whether, when uraninite initially precipitated in Si-rich groundwaters, coffinite was not thermodynamically stable, or whether coffinite formation is kinetically inhibited. The factors that determine whether uraninite or coffinite form are poorly understood. Uraninite certainly crystallizes from Si-rich groundwaters under reducing conditions, as well as from higher temperature Si-rich hydrothermal and magmatic fluids. Perhaps certain impurity cations stabilize uraninite relative to coffinite in some Si-rich groundwaters.

Because uraninite always contains some oxidized U, whereas coffinite may not, we can see that coffinite is more stable than uraninite in strongly reducing waters according to the simplified reaction:

$$UO_{2+x} + SiO_{2(aq)} \Rightarrow USiO_4 + 1/2\ xO_2$$

The alteration of uraninite to coffinite may arise from an increase in dissolved silica, destabilizing existing uraninite. In addition, radiation-induced changes to the composition of uraninite can destabilize uraninite, which, in reducing groundwaters with adequate dissolved silica, combined with slow release of U from uraninite, favors crystallization of coffinite. Many elements commonly contained in uraninite are not compatible in the coffinite structure. For example Pb commonly precipitates as galena where S^{2-} activities are sufficient.

Radiogenic helium. Coarsely crystalline uraninite may display high concentrations of voids, due to accumulation of radiogenic He (Stout et al. 1988; Finch 1994; L. Thomas pers. comm.) (Fig. 4a). Voids are concentrated along sub-grain boundaries and dislocations, where present, but are also distributed throughout homogenous crystals. The role of increased strain due to the accumulation of radiogenic He is uncertain, but this may contribute to destabilization and subsequent recrystallization of some uraninite, as well as to the alteration of other radioactive minerals. Finch (1994) suggested that He voids in uraninite exposed to oxidizing groundwaters may also accelerate oxidative corrosion by increasing surface area and providing pathways for groundwater penetration (Fig. 4b).

Uraninite alteration under oxidizing conditions

The oxidative alteration of uraninite was described by Frondel (1958) as follows: first, partial oxidation of U^{4+} to U^{6+} without decomposition of the uraninite structure, although the uraninite commonly displays a change in color from black to dark brown and a change in luster from sub-metallic to "dull or pitch-like." At this stage and as alteration progresses, there is commonly "some degree of hydration, and U and especially Th may be [lost to groundwater]" relative to Pb (Frondel 1958). Further alteration, which Frondel (1958) considered as a "second type," resulted in the complete conversion of uraninite into uranyl minerals, the compositions of which depended primarily on the composition of local groundwaters.

By analogy with synthetic UO_2, it has been proposed by several authors that the early stages of oxidation and alteration of uraninite proceeds through intermediate U oxides such as U_3O_7 or U_3O_8 (Voultsidis and Clasen 1978; Posey-Dowty et al. 1987; Waber et al. 1990; Bruno et al. 1991; Sunder et al. 1996; Trocellier et al. 1998); however, crystalline compounds with these compositions are unknown in nature. It has been proposed that these oxides may occur as thin (~1 μm) layers on uraninite, but their existence has not been demonstrated (Janeczek et al. 1993). The formation of hydrated, X-ray amorphous U^{6+} hydroxides from uraninite has been postulated (Garrels and Christ 1959; Finch 1994), sometimes being given names such as "hydronasturan" and "urghite" (see Cejka and Urbanec 1990), but these are poorly described and not considered valid species.

The initial oxidation and subsequent dissolution of uraninite has not been studied in the same detail as the oxidation, dissolution and corrosion of synthetic UO_2 and UO_2 nuclear fuels (Forsyth and Werme 1992; Sunder et al. 1992; Gray et al. 1993; Einziger et al. 1992; Bruno et al 1992; Matzke 1992; Wronkiewicz et al. 1992, 1996). Because of structural similarities of uraninite and UO_2, the oxidation and corrosion of uraninite has often been assumed to be identical to that of synthetic UO_2 (Miller 1958; Parks and Pohl 1990). This is probably true to a limited degree, but elements besides U

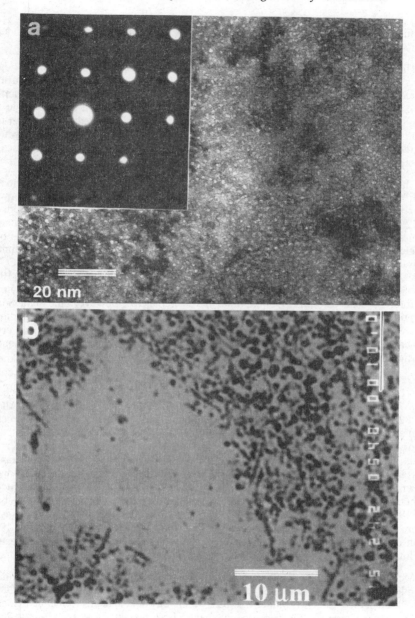

Figure 4. (a) Bright field TEM image of uraninite from Shinkolobwe mine, Shaba, Dem. Rep. Congo displaying mottled contrast attributed to inclusions of radiogenic-He produced during radioactive decay of U. Inset: SAED pattern along [001] illustrating cubic UO_2-type pattern. There is no indication of super-cell or ordering as might be expected for U_4O_9 or U_3O_7 (TEM photo courtesy of Larry Thomas). (b) Voids in altered uraninite grain possibly reflecting preferential dissolution of radiogenic-He bubbles (from Finch 1994).

and O can affect the alteration behavior of uraninite (Grandstaff 1976), of irradiated (spent) UO_2 fuels (Einziger et al. 1992; Thomas et al. 1993) and of doped U oxides (Anderson 1953; Anderson et al. 1954; Thomas et al. 1993).

The slow diffusion of oxygen into the UO_2 structure necessitates that most oxidation studies of UO_2 be performed at elevated temperatures (~150 to > 800°C). (The diffusion coefficient of oxygen in UO_2 at 25°C is on the order of 10^{-23} to 10^{-25} m²·s⁻¹; Grambow 1989). Even slow kinetics of oxygen diffusion may be unimportant over geologic time spans, as interstitial sites in ancient uraninite probably become saturated with oxygen in all but the most reducing environments (Janeczek and Ewing 1992b). The question of whether U oxides that form in nature, especially during weathering, resemble those synthesized at elevated temperatures has been the subject of some debate (e.g. Janeczek et al. 1993).

The mechanisms by which synthetic UO_2 oxidizes in water may differ from those by which it oxidizes in dry air (Grambow 1989). For example, Grambow (1989) reports that the activation energy for weight gain in water is approximately one-third of that in dry air (21-42 kJ·mol⁻¹ in water vs. ~109 kJ mol⁻¹ in air). XPS has shown that the earliest stages of aqueous oxidation of synthetic UO_{2+x}, prior to dissolution, probably mimic those of air oxidation (Sunder et al. 1992; Sunder et al. 1997). Whether this is true for uraninite is uncertain, but recent studies indicate that the oxidation pathway of Gd-doped UO_2 and spent UO_2 fuels differ from that of pure UO_2 (Einziger et al. 1992; Thomas et al. 1993). Inferring the oxidation and corrosion behavior of uraninite from that of pure UO_2 must, therefore, be done carefully. Despite potential discrepancies, however, the oxidation of synthetic UO_2 is a useful analogue for the oxidation behavior of uraninite, and studies on the low temperature oxidation of UO_2 in aqueous solutions provide insight into crystal chemical and micro-structural changes occurring in uraninite during oxidative aqueous corrosion (Finch and Ewing 1992b; Wronkiewicz et al. 1996).

Oxidation of synthetic UO_2 in air. The initial oxidation of synthetic UO_2 to U_4O_9 ($UO_{2.25}$) in air above ~350°C has been studied extensively (Willis 1963; Belbeoch et al. 1967; Contamin et al.1972; Naito 1974; Willis 1978; Allen and Tempest 1982; Allen et al. 1982; Allen 1985; Allen et al. 1987; St A Hubbard, and. Griffiths 1987). The structural aspects of oxidation were summarized by Willis (1987). Excess oxygen enters unoccupied U-equivalent interstices in UO_2, displacing two adjacent oxygens to opposite interstices (Willis 1978). At the oxidation state $UO_{2.25}$, there is one excess oxygen atom per uraninite unit cell. In β-U_4O_{9-y}, oxygen occupies a different interstitial site in each of four adjacent cells in the three crystallographic directions giving rise to a $4a \times 4a \times 4a$ superstructure. Synthetic UO_{2+x} consists of two coexisting phases, representing a miscibility gap (Belbeoch et al. 1967; Naito 1974; Smith et al. 1982), and oxidation proceeds in a "step-wise" fashion as domains in the structure transform from $UO_{2.04}$ to U_4O_{9-y} ($y < 0.2$; Smith et al. 1982). The influence of these two phases on the electronic properties of UO_{2+x} strongly affects the dissolution of synthetic UO_{2+x} in oxidizing waters (Johnson and Shoesmith 1988).

McEachern and Taylor (1998) provide an especially thorough review of the oxidation of synthetic UO_2 below 400°C. The formation of crystalline U_4O_9 from synthetic UO_2 may be inhibited below ~250°C (Grambow 1989), which may be due to slow kinetics associated with long-range ordering of interstitial oxygens (Janeczek et al. 1993). Below 250°C, synthetic UO_2 oxidizes to a phase that is a tetragonal distortion of the fluorite structure with a composition near $UO_{2.33}$: α-U_3O_7 or β-U_3O_7. Evidence for the transient formation of β-U_4O_9 does exist, but the small change in stoichiometry between $UO_{2.25}$ and $UO_{2.33}$ may explain the rapid transformation of cubic U_4O_9 to tetragonal U_3O_7, perhaps before complete conversion of all domains of UO_{2+x} to U_4O_9. Further oxidation may crystallize orthorhombic U_3O_8 (commonly non-stoichiometric) when time or temperature (or both) are sufficient. No stable phases between tetragonal U_3O_7 and orthorhombic U_3O_8 exist at ambient pressures. Continued oxidation of U_3O_8 to UO_3 does not occur, although the reasons for this are not fully understood (McEachern and Taylor 1998). At elevated temperatures (~800°C) UO_3 decomposes to non-stoichiometric U_3O_8 through O loss.

Oxidation of doped UO_2 and spent UO_2 fuels in air. The dry air oxidation of spent UO_2 fuel differs from that of unirradiated UO_2 in several respects. The primary difference between the two materials is that spent fuel contains impurities, such as fission products and transuranic elements, and the effects of radiation-induced changes on the structure of spent fuel can be important (Matzke 1992). The oxidation of spent fuel below 150°C produces disordered cubic U_4O_{9+x} (γ-U_4O_9) (Thomas et al. 1989). The fluorite-type structure is maintained to at least an O:U ratio of 2.4 (Einziger et al. 1992; Thomas et al. 1993). The oxidation of spent UO_2 fuel to γ-U_4O_9 proceeds along a reaction front that advances from grain boundaries into the centers of grains without significant strain. The smaller unit cell volume of γ-U_4O_9 causes shrinkage cracks at grain boundaries and embrittlement of the fuel. Further oxidation in dry air leads to crystallization of a hexagonal polymorph of U_3O_8, although the onset of formation is delayed to longer times or higher temperatures compared with formation of orthorhombic α-U_3O_8 during oxidation of undoped UO_2 (McKeachern et al. 1998) There is no evidence for tetragonal U_3O_7 at any stage in the oxidation sequence of spent fuels (Thomas et al. 1989; Thomas et al. 1993) or Gd-doped UO_2 (Thomas et al. 1993).

Understanding the disparate oxidation behaviors of UO_2 and doped-UO_2 (or spent UO_2 fuel) depends on understanding the role of the non-uranium cations in the UO_2 structure. Many of these cations have valences below 4+ (e.g. Sr^{2+}, Gd^{3+}), others do not oxidize readily (Th^{4+}); these elements can stabilize the UO_2 structure (Grandstaff 1976). The persistence of cubic U_4O_{9+x} to O:U ratios close to 2.4 probably explains the nucleation and growth of U_3O_8 without intermediate formation of tetragonal U_3O_7 (Thomas et al. 1993) since the O:U ratio of $UO_{2.4}$ exceeds that of U_3O_7. The reason that hexagonal U_3O_8 forms during dry-air oxidation of spent fuel, rather than orthorhombic α-U_3O_8, which forms from pure UO_{2+x}, is not understood (Thomas et al. 1993). Additional comparisons between synthetic UO_2 and spent UO_2 fuels are discussed by Janeczek et al. (1995) and by Janeczek (this volume).

Oxidation of uraninite. The dry-air oxidation of uraninite at elevated temperatures (150-300°C) is similar to that described for spent UO_2 fuel and doped synthetic UO_2. Cubic symmetry is maintained up to 300°C during air oxidation of untreated uraninite. No tetragonal distortions are evident, nor is crystalline U_3O_8 detected (Janeczek and Ewing 1991a ; Janeczek et al. 1993). By contrast, uraninite that was pre-annealed at 1200°C in H_2 gas for 24 hours forms cubic U_4O_9, followed by crystallization of orthorhombic α-U_3O_8 when oxidized in air at 300°C (Janeczek et al. 1993). In neither case, however, is tetragonal U_3O_7 observed. As for spent fuel and doped UO_2, the formation of hexagonal U_3O_8 is apparently inhibited during dry-air oxidation of uraninite. Thus, as for spent UO_2 fuels and doped synthetic UO_2, formation of tetragonal U_3O_7 and orthorhombic U_3O_8 are not formed during the dry-air oxidation of unannealed uraninite.

The initial oxidation of uraninite may result in a reduced unit cell parameter, as noted for oxidized spent fuel, although the effect of other cations may offset this to varying degrees (Janeczek 1993). The upper limit on the amount of U^{5+} or U^{6+} that can be accommodated in uraninite is unknown and probably depends strongly on the original composition. Reported U^{6+}:U^{4+} ratios for uraninite commonly exceed the highest expected for U_4O_9, without loss of cubic symmetry (Frondel 1958; Berman 1957; Janeczek and Ewing 1992b). Despite such high U^{6+}:U^{4+} ratios, interstitial O probably does not exceed

1/8 of U-equivalent sites in uraninite (Janeczek and Ewing 1992b), equal to the interstitial O occupancy in synthetic U_4O_9, and the actual number of O interstitials in uraninite may be less if interstitial sites are also occupied by cations such as Pb, Ca or REEs. Recalculation of reported uraninite formulas by accounting for impurity elements generally gives O:U ≤ 2.25, as expected (Janeczek et al. 1993).

The effects described above for the oxidation of uraninite provide no clear indication of how oxidation affects the physical properties of uraninite and what role these changes play in the dissolution of uraninite in oxidizing groundwaters. Reduced unit-cell parameters caused by oxidation could result in the formation of gaps between uraninite grains, increasing surface areas and providing pathways for groundwater penetration. There is no clear evidence, however, for a transformation of the type: $UO_{2+x} \Rightarrow U_4O_{9+x}$ in natural uraninite. This may be because most uraninite is already partially oxidized at the time of formation, and may also reflect how the accumulation of radiogenic Pb (and other large cations) can offset unit-cell shrinkage caused by partial oxidation. Where uraninite with two different unit-cell sizes have been reported within a single sample, this has been attributed to the segregation of radiogenic Pb, rather than oxidation (Janeczek and Ewing 1992c; Janeczek et al. 1993).

Replacement of uraninite by uranyl minerals

When in contact with oxidizing water, U oxides above U_4O_9 cannot form from synthetic UO_2, because the kinetics of dissolution in oxidizing waters are orders of magnitude faster than those of dry-air oxidation and dissolution dominates (Posey-Dowty et al. 1987; Shoesmith and Sunder 1991; Shoesmith et al. 1998). This is consistent with the observation that cubic uraninite is the mineral most often found in contact with the corrosion rind formed by oxidation in water (Frondel 1958; Finch and Ewing 1992b). Rapid oxidative dissolution of uraninite can lead to quite elevated concentrations of dissolved U, and the precipitation of uranyl oxyhydroxides appears to be kinetically favored over precipitation of more complex uranyl minerals (Finch and Ewing 1992b; Finch 1994). The reasons for kinetic constraints on the precipitation of many uranyl minerals are uncertain. Precipitation of uranyl oxyhydroxides may require little atomic rearrangement of uranyl solution complexes (Evans 1963) compared with atomic rearrangements required to precipitate uranyl silicates or phosphates. Many uranyl tricarbonate solution complexes might be expected to precipitate as minerals without much atomic rearrangement, because their structures contain the uranyl tricarbonate ion, $(UO_2)(CO_3)_3^{4-}$. However, U concentrations required to precipitate most uranyl carbonates (and sulfates) are much higher than for uranyl oxyhydroxides (with the exception of rutherfordine). Uranyl oxyhydroxides may therefore have the lowest solubilities among those minerals for which precipitation kinetics do not present significant barriers to nucleation. Given that the early-formed uranyl minerals are kinetic (i.e. metastable) products, it is not surprising that they are pervasively altered by continued interaction with groundwaters from which they precipitated.

Thus, the uranyl oxyhydroxides comprise an especially important group of uranyl minerals because they commonly form early during the oxidative dissolution of reduced U minerals. The uranyl oxyhydroxides that form earliest during the oxidation of uraninite are ianthinite, schoepite, becquerelite, vandendriesscheite and fourmarierite (Snellling 1980; Finch and Ewing 1992b; Pearcy et al. 1994; Finch 1994). Schoepite and becquerelite are also the first phases to form from Si-saturated waters during corrosion experiments on synthetic UO_2 at 90°C (Wronkiewicz 1992, 1996).

Finch and Ewing (1992b) described a simplified reaction for the early-stage oxidative corrosion of uraninite at Shinkolobwe (written with all U retained in schoepite and vandendriesscheite):

$U^{4+}_{0.48}U^{6+}_{0.39}(Y,Ce)_{0.02}Pb_{0.09}Ca_{0.02}O_{2.27}$
+ 0.24 O_2
+ 1.57 H_2O
+ 0.10 H^+

0.03375 $[(UO_2)_8O_2(OH)_{12}](H_2O)_{12}$
+ 0.06 $Pb_{1.5}[(UO_2)_{10}O_3(OH)_{11}](H_2O)_{11}$
+ 0.02 Ca^{2+}
+ 0.02 $(Y,Ce)^{3+}$

The minerals on the right, schoepite and vandendriesscheite, are common corrosion products formed early in contact with dissolving uraninite. In Ca-bearing water, becquerelite will also form, at the expense of some or all schoepite, depending on the activity ratio $\{Ca^{2+}\}/\{H^+\}^2$ (cf. Fig. 7, below). If pO_2 levels drop, due to consumption of oxygen by uraninite oxidation or from organic influences, the mixed valence oxyhydroxide, ianthinite, may replace schoepite in the above reaction. Little is known about the conditions necessary for the formation of ianthinite, but it may be a common initial oxidation product of uraninite (Finch and Ewing 1994; Pearcy et al. 1994; Burns et al. 1997b), subsequently oxidizing in air to schoepite. The above reaction represents a somewhat simplified view of the natural system, being written without dissolved silica, carbonate or other complexing ligands. However, even in silica saturated solutions, uranyl silicates do not play an important role during the initial stages of uraninite corrosion (Frondel 1956; Wronkiewicz 1992, 1996; Pearcy et al. 1994). Slightly alkaline carbonate waters tend to solubilize U, although a significant amount of U remains in solids near the dissolving uraninite in most weathered uraninite deposits (Frondel 1956; Frondel 1958; Finch and Ewing 1992b).

Continued interaction between groundwaters and the early-formed uranyl oxyhydroxides results in their replacement by uranyl silicates (and carbonates) and increasingly Pb-rich uranyl oxyhydroxides (Frondel 1956; Finch and Ewing 1991; Finch and Ewing 1992b). The most common uranyl minerals to persist after uraninite has been essentially replaced at the Shinkolobwe mine are soddyite and curite, where pseudomorphic replacement of uraninite by approximately equal volumes of curite plus soddyite is common (Schoep 1930; Finch 1994). At Koongarra, where Mg-chlorites are abundant, uraninite alteromorphs are commonly composed of curite and sklodowskite (Isobe et al. 1992). In the somewhat drier environment at the Nopal I U mine near Peña Blanca, Mexico, alteromorphs after Pb-poor Tertiary-age uraninite are commonly composed of uranophane, soddyite, and minor weeksite (Pearcy et al. 1994). Although kasolite is a persistent uranyl mineral that withstands prolonged groundwater interaction, kasolite is not known to replace uraninite, but instead tends to fill veins that may transect uraninite alteromorphs. Kasolite commonly fills veins within uraninite early during alteration (Isobe et al. 1992; Finch 1994), probably owing to the availability of Pb from dissolving galena located along uraninite grain and subgrain boundaries. These kasolite veins can persist even after the surrounding uraninite has been entirely replaced by uranyl minerals. They do not, however, comprise a substantial volume of the alteromorphs (Isobe et al. 1992; Finch 1994).

Elton and Hooper (1995) described a rich assemblage of supergene U, Pb, and Cu minerals at a coastal exposure near Low Warren, Cornwall, England. These minerals were produced by the action of surface waters and sea spray on uraninite, chalcocite and chalcopyrite. Elton and Hooper (1995) identified three zones of mineralization in terms of the major alteration minerals found in each.

1. Rutherfordine, boltwoodite, and vandendriesscheite closely associated with the uraninite, along with minor trögerite and an unidentified Ca-REE uranyl

carbonate.
2. Central part of the exposure with few U minerals, but there are several Cu-bearing products of chalcocite alteration.
3. Kasolite, wölsendorfite, widenmannite, dewindtite, and an unidentified "basic" Pb uranyl carbonate.

ALTERATION OF URANYL MINERALS

Thermodynamic background

Uraninite can be a remarkably heterogeneous mineral, with a composition and microstructure that can vary widely among samples from different localities, or even among specimens from within a single occurrence (Janeczek and Ewing 1992b,c, Janeczek and Ewing 1995). Thermodynamics of uraninite may be difficult to generalize, and the synthetic pure oxides, UO_2 and U_4O_9, are necessary surrogates in geochemical codes. The influences that impurity elements may have on thermodynamic stabilities are poorly understood. Elements such as Ca^{2+} and Th^{4+}, as well as U that is more oxidized than U^{4+}, may increase the stability range of uraninite to higher oxidation potentials relative to pure oxides; elements incompatible in the uraninite structure, notably radiogenic Pb^{2+}, probably decrease uraninite stability. Accumulation of radiogenic Pb may affect thermodynamic stabilities of all U minerals of sufficient age. A recent study comparing solubilities of several uraninite samples and synthetic UO_2, suggests that synthetic UO_2 and natural uraninite have comparable solubilities in reducing groundwater, and in the near-neutral pH range (5 to 7) variability among uraninite samples can exceed differences between uraninite and synthetic UO_2 (Casas et al. 1998). The thermodynamics of coffinite are also poorly known, and impurities such as REE, P and Ca, will influence its stability (Hansley and Fitpatrick 1989; Janeczek and Ewing 1996). For the most part, thermodynamic stabilities of other U^{4+}-bearing minerals remain poorly constrained. This is also true of most uranyl minerals. The solubilities of uranyl minerals, measured in terms of total dissolved U, span several orders of magnitude: from as low as 10^{-9} to 10^{-8} mol·L^{-1} for some uranyl phosphates, arsenates and vanadates, to as high as perhaps 10^{-3} to 10^{-2} mol·L^{-1} for uranyl carbonates and sulfates (Langmuir 1978). Despite their importance in controlling U concentrations in U-rich waters, reliable thermodynamic data are available for fewer than ten of the more than 160 uranyl minerals known (Robie et al. 1979; Hemingway 1982; Vochten and Van Haverbeke 1990; Grenthe et al. 1992; Nguyen et al. 1992; Casas et al. 1994, 1997).

The lack of thermodynamic data for uranyl minerals is not due to a lack of effort. Many studies have been done on the solubilities and other thermodynamic properties of U minerals. However, natural samples are rarely pure; minerals are often fine-grained and intimately intergrown at even a sub-micron scale. Variations in composition, including hydration state, are common and can lead to large differences in measured parameters. Studies of the thermodynamics of synthetic phases are useful, but for many of these, accurate thermodynamic measurements are hampered by kinetic effects, and mixed or amorphous phases are common in laboratory studies. Synthetic phases do not always correspond to known minerals, and, due in part to inadequate knowledge about the compositions of many uranyl minerals, the degree to which synthetic phases are representative of minerals may be uncertain.

Measured and estimated thermodynamic parameters

The desire to understand the thermodynamics of U minerals focused early on trying to describe geochemical conditions favorable for concentration of ·U in economic deposits.

Hostetler and Garrels (1962) were perhaps the first to predict mineral-solution equilibria for low-temperature U-rich waters. Their study provided important insights into the occurrences of U deposits and the conditions under which U is transported in and precipitated from groundwaters. Langmuir (1978) published a comprehensive study of U thermodynamics, and applied his results successfully to understanding a wide variety of U-mineral occurrences. Out of necessity, many of Langmuir's data were estimated. The United States Geological Survey recognized the need for thermodynamic data for the great number of U minerals, leading to the work by Hemingway (1982), which focussed on the formation of coffinite and uraninite deposits. Hemingway (1982) used the method of Chen (1975) to estimate many of the Gibbs free energies of formation values for U many minerals; however, Chen's method works best if a large number of thermodynamic data can be used for the calculation, and such was not the case for U minerals at that time. Despite some inaccurate estimates, data reported by Langmuir (1978) and Hemingway (1982) were remarkably successful at helping to understand U-mineral occurrences. Estimates of Gibbs free energies of formation for some anhydrous phosphates were published by Van Genderen and Van der Weijden (1984), who used the method developed by Tardy and Garrels (1976, 1977); however, Van Genderen and Van der Weijden (1984) ignored the contribution of structurally bound H_2O in uranyl phosphates.

During the 1980s, the emphasis on U geochemistry began to shift. Demand for nuclear energy diminished (particularly in the US), due in part to public perception and escalating costs of building commercial nuclear reactors. There was also the increasingly pressing concern for a permanent solution to the problem of the growing volume of highly radioactive nuclear waste worldwide. Because of its remarkable durability as a reactor fuel and the apparent persistence of uraninite in nature, many countries consider direct geologic disposal of spent UO_2 fuel to be a potentially safe and cost-effective means of solving the nuclear-waste problem. Research efforts in U geochemistry began to shift away from U exploration and towards understanding the chemical durability of the UO_2 component of spent nuclear fuel, which houses many radioactive and toxic radionuclides generated during fission reactions towards predicting future geochemical behavior of U in and around a geologic repository for nuclear waste (see Wronkiewicz and Buck, and Janeczek, this volume). Towards this end, the Nuclear Energy Agency (NEA) under the direction of the Organization for Economic Co-operation and Development (OECD) in Europe, published a compilation and critical review of existing thermodynamic data for U (Grenthe et al. 1992).

Published a decade after the papers by Hemingway (1982) and Langmuir (1978), Grenthe et al. (1992) provided a valuable up-to-date resource for research in U geochemistry. Unfortunately, of the approximately 200 U minerals known only four were accepted in the NEA compilation, all of them synthetic analogues: uraninite (synthetic UO_{2+x}, x = 0, 0.25, 0.33), coffinite (synthetic $USiO_4$), metaschoepite (synthetic $UO_3 \cdot 2H_2O$) and rutherfordine (synthetic UO_2CO_3).

Concurrent with the publication of the NEA database, several studies of the thermodynamics of U solids were published. Vochten and van Haverbeke (1990) reported the solubilities of three synthetic uranyl oxyhydroxides: becquerelite, billietite and $PbU_2O_7 \cdot 2H_2O$ (the latter is isostructural with wölsendorfite). Sandino and Bruno (1992) determined the solubility of $(UO_2)_3(PO_4)_2 \cdot 4H_2O$ (a phase unknown in nature) and reported stability constants for several U(VI)-phosphate solution complexes. Nguyen et al. (1992) published experimentally determined Gibbs free energies of formation for four synthetic uranyl silicates, uranophane, soddyite, Na-boltwoodite, and Na-weeskite. Sandino and Grambow (1995) studied the solubility of synthetic becquerelite and compreignacite, reporting a solubility constant for becquerelite quite close to that determined by Vochten and van Haverbeke (1990). A subsequent study of the solubility of a natural becquerelite crystal determined a solubility constant that is lower than that reported for synthetic becquerelite by approximately 13 orders of magnitude, suggesting that the stability of becquerelite may be greater than previously thought (Casas et al. 1997). Gibbs free energies of formation were reported recently for the synthetic analogues of cliffordite (Mishra et al. 1998) and schmitterite (Mishra et al. 1998; Singh et al. 1999).

The paucity of thermodynamic data for even common U minerals prompted Finch (1994) to estimate the Gibbs free energies of formation for some uranyl oxyhydroxides in order to help explain the paragenesis of uranyl minerals at the Shinkolobwe mine in southern Democratic Republic of Congo. He used a method similar to that developed by Tardy and Garrels (1976, 1977) and used by Van Genderen and Van der Weijden (1984): summing free energy contributions of fictive oxide components to the total Gibbs free energy of formation of the mineral of interest. Activity-activity (mineral stability) diagrams were used to compare predicted and observed mineral relationships, providing an independent check on the reliabilities of estimates. Due to the success of this method, it was expanded to include uranyl silicates and uranyl carbonates (Finch and Ewing 1995). Finch (1997a) published estimated Gibbs free energies of formation for several uranyl oxyhydroxides, uranyl silicates and uranyl carbonates, and used these values to construct mineral stability diagrams for two geochemically important aqueous systems: $CaO-CO_2-UO_3-H_2O$ and $CaO-SiO_2-UO_3-H_2O$. A brief description of the method reported by Finch (1997a) follows.

Given the $\Delta G_f°$ values for stoichiometrically simple oxides and hydroxides, one may estimate the contribution of the constituent oxides to the total $\Delta G_f°$ value of each mineral; that is, the ΔG_f^* of each oxide *in the mineral structure*. The $\Delta G_f°$ value of a mineral is the arithmetic sum of the oxide contributions: $\Delta G_f° = \Sigma \Delta G_f^*{}_i$. Finch (1997a) estimated $\Delta G_f^*{}_i$ values from a small number of synthetic hydrated uranyl minerals. Chen et al. (1999) expanded on this by deriving $\Delta G_f^*{}_i$ values by regression analyses of a large number of uranyl compounds, both hydrated and anhydrous. The large data set used by Chen et al. (1999) appears to be an improvement over the limited set used by Finch (1997a), and Tables 18 and 19 list the data derived by Chen et al. (1999). The values for the hypothetical oxides in Table 18 are used to estimate $\Delta G_f°$ for the uranyl minerals in Table 19 by adding $\Delta G_f^*{}_i$ contributions from the constituent oxides in their stoichiometric proportions.

Estimated $\Delta G°_f$ values are used to construct activity-activity (stability) diagrams, and the predicted stability fields can be compared with observed mineral occurrences and reaction pathways. With some exceptions, natural occurrences agree well with the mineral stability fields estimated for the systems $SiO_2-CaO-UO_3-H_2O$ and $CO_2-CaO-UO_3-H_2O$, providing some confidence in the estimated thermodynamic values. Activity-activity diagrams are sensitive to small differences in $\Delta G°_f$ values, and mineral compositions must be known accurately, including structurally bound H_2O (Finch 1997a). Estimated $\Delta G°_f$ values may not be reliable for a few minerals (e.g. liebigite, zellerite, uranosilite) for two reasons: (1) the structures of the minerals in question are not closely similar to those used to estimate the ΔG_f^* values of the component oxides, or (2) the minerals in question may be exceptionally fine grained, leading to large surface energies that increase effective mineral solubilities (Finch 1997a).

As an illustration of the difficulties encountered when constructing stability diagrams from experimental data, Figure 5 compares two stability diagrams for the $SiO_2-CaO-UO_3-H_2O$ system. The reported solubility constant for synthetic becquerelite (Vochten and Van Haverbeke 1990; Sandino and Grambow 1995): log K_{So} 42, is more than ten orders of magnitude greater than that reported for a natural becquerelite crystal (Casas et al. 1997): log K_{So} 29. The difference this makes to a stability diagram is clearly seen in Figure 5. If

Finch & Murakami: *Systematics, Paragenesis of U Minerals*

Table 18. Molar contributions of structural components to $\Delta G^0_{f,298}$ and $\Delta H^0_{f,298}$ of U(VI) phases reported by Chen et al. (1999) (kJ·mol^{-1})

Component	UO_3	$Li_2O_{(l)}$	$Na_2O_{(l)}$	$K_2O_{(l)}$	$Rb_2O_{(l)}$	$Cs_2O_{(l)}$	$CaO_{(l)}$
$\Delta G_f{}^*_i$	-1161.05	-692.14	-686.54	-637.45	-639.05	-644.35	-715.77
$\Delta H_f{}^*_i$	-1233.75	-737.75	-736.3	-686.95	-688.95	-694.25	-726.57

Component	$BaO_{(l)}$	$SiO_{2(IV)}$	$SO_{3(IV)}$	$CO_{2(III)}$	$N_2O_{5(III)}$	$P_2O_{5(IV)}$	$H_2O_{(S)}$	$H_2O_{(H)}$
$\Delta G_f{}^*_i$	-725.91	-853.96	-538.87	-400.61	21.95	-1638.25	-237.94	-241.1
$\Delta H_f{}^*_i$	-761.98		-624.17	-455.59	-78.99	-1802.37	-299.93	-295.58

Table 19. $\Delta G^\circ_{f,298}$ and $\Delta H^\circ_{f,298}$ for uranyl minerals indicated in Figs. 5, 6, and 7 (kJ·mol^{-1}).

Mineral	$\Delta G^0_{f,298}$	$\Delta H^0_{f,298}$	M/C*	Ref.
Schoepite	-13299.4	-14908.7	C	1
Metaschoepite	-13092.0	-14608.8	M	2
Becquerelite	-10324.7		C	1
Rutherfordine	-1563.0	1689.6	M	3
Urancalcarite	-6036.7		C	1
Sharpite	-11607.6		C	1
Fontanite	-6524.7		C	1
Zellerite	-3879.9		C	1
Liebigite	-6226.0	-7301.6	M	4
Uranosilite	-7126.1		C	1
Haiweeite	-9367.2		C	1
Ursilite	-20377.4		C	1
Soddyite	-3658.0		M	5
Uranophane	-6210.6		M	5

*M/C designates measured or calculated values. References: 1. Chen et al. (1999); 2. O'Hare et al. (188); 3. Sergeyeva et al. (172); 4. Alwan & Williams (1980); 5 Nguyen et al. (1992). Chen et al. re-evaluated the data of Nguyen et al. (1992) and determined GFE values of -3655.7 and -6192.3 kJ·mol^{-1} for soddyite and uranophane, respectively. The ΔG_f° values for dissolved species, calcite and CO_2, used to construct Figs. 5, 6, and 7 are from Grenthe et al. (1992).

we assume that the solubility of synthetic becquerelite best represents the solubility of becquerelite in nature (Fig. 5a), then becquerelite is unstable in all but those groundwaters with vanishingly small dissolved silica concentrations ($<10^{-6}$ mol L^{-1} H_4SiO_4); a conclusion that seems contrary to observation (Finch and Ewing 1992b; Finch 1994; Finch et al. 1995, 1996). On the other hand, the exceptionally large stability field indicated for the natural becquerelite crystal suggests that becquerelite should predominate in most natural waters (Fig. 5b), which also seems contrary to observation (however, note that most other compounds in Fig. 5 are synthetic). Of course, simple comparisons of thermodynamic stabilities with natural occurrences ignore potential kinetic effects on mineral relationships. Nevertheless, because of the apparent discrepancies between observation and the stability diagrams shown in Figure 5, we will use the estimated Gibbs free energy of formation for becquerelite derived by Chen et al. (1999), which corresponds to a solubility product intermediate between the experimental values: log K_{so} 36.

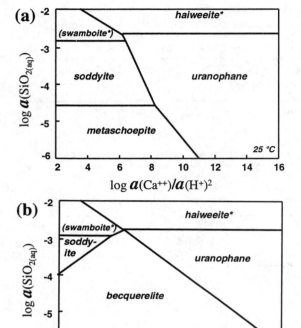

Figure 5. Activity-activity diagrams for the system SiO_2-CaO-UO_3-H_2O, calculated with $\Delta G°_f$ values derived from from solubility studies of (a) synthetic becquerelite powders (Vochten and Van Haverbeke 1990; Sandino and Grambow 1995), and (b) a natural becquerelite crystal (Casas et al. 1997). All data are derived from experimental studies except minerals marked with an asterisk (*), for which stability fields were calculated from estimated $\Delta G°_f$ values in Table 19.

Two systems are analyzed in detail: SiO_2-UO_3-CaO-H_2O and CO_2-UO_3-CaO-H_2O. The SiO_2– and CO_2–bearing system are important in nature, and will help to understand the paragenesis of uranyl minerals. The Ca-con-taining systems are considered because Ca is virtually ubiquitous in near-surface groundwaters, and many Ca-bearing uranyl minerals are known, representing a variety of near-surface environ-ments.

System CO_2-CaO-UO_3-H_2O. Figure 6 illustrates the stability fields for several uranyl carbonate minerals. The concentration of Ca in many groundwaters is controlled by equilibrium with calcite and dissolved $CO_{2(g)}$: $CaCO_3 + 2H+ \Leftrightarrow Ca^{2+} + H_2O + CO_{2(g)}$. This equilibrium is indicated as a diagonal dotted line in Figure 6. The calcite equilibrium line passes through the fields for becquerelite, schoepite and rutherfordine. These are the most common minerals among those represented in Figure 6, and rutherfordine is by far the most common uranyl carbonate (Frondel 1958; Smith 1984). Chen et al. (1999) showed that many natural groundwater compositions plot within the stability field of becquerelite, consistent with it being the most common uranyl oxyhydroxide mineral in nature (Frondel 1958; Smith 1984). Sharpite and urancalcarite are rare and occur in deposits where the predominant carbonate mineral in the host rocks is magnesian calcite or dolomite (Deliens et al. 1981); however, if Figure 6 is accurate, sharpite may be more common than currently thought, likely to form in calcite-bearing rocks in high pCO_2 groundwaters, such as saturated soils. Fontanite, which coexists with becquerelite and uranophane, occurs in pelitic silts and shales (Deliens and Piret 1992).

Two uranyl carbonates not shown in Figure 6 are liebigite and zellerite (Table 19). Stability fields calculated for these two minerals replace most of the area shown in Figure 6, contrary to observation. The Gibbs free energy of formation for liebigite was determined by Alwan and Williams (1980) for synthetic material. Liebigite and zellerite are commonly found as efflorescences on mine walls, surface outcrops, and elsewhere that evaporation is high. These minerals also tend to be extremely fine grained, suggesting that high surface free energies may enhance their effective solubilities.

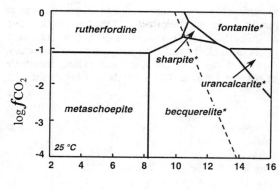

Figure 6. Activity-activity diagram for the system CO_2-CaO-UO_3-H_2O, calculated using measured and estimated $\Delta G°_f$ values in Table 19. Data are derived from experimental studies except minerals marked with an asterisk (*), for which stability fields were calculated from estimated Gibbs free energies of formation (Table 19). Diagonal dotted line represents calcite equilibrium.

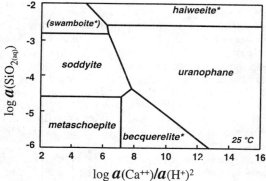

Figure 7. Activity-activity diagram for the system SiO_2-CaO-UO_3-H_2O, calculated using measured and estimated $\Delta G°_f$ values from Table 19. Data are derived from experimental studies except minerals marked with an asterisk (*), for which stability fields were calculated from estimated $\Delta G°_f$ values (Table 19). Swamboite field is approximate only, shown for completeness.

System SiO_2-CaO-UO_3-H_2O. Figure 7 illustrates the stability fields of minerals common in Si-bearing groundwaters. The most common Pb-free U(VI) minerals are schoepite and metaschoepite, becquerelite, soddyite, uranophane and rutherfordine. Schoepite forms early in the alteration paragenesis of uraninite oxidation products (Finch and Ewing 1992b; Finch 1994). Though common, schoepite is not usually abundant at most oxidized U deposits, because it is commonly replaced by uranyl silicates and carbonates, especially uranophane, soddyite and rutherfordine (Frondel 1956; Finch and Ewing 1992a,b; Finch 1994; Pearcy et al. 1994). Although the direct replacement of schoepite by becquerelite is not readily confirmed, this reaction has been reported from several experimental studies (Vochten and Van Haverbeke 1990; Sandino and Grambow 1995; Sowder et al. 1996). Although well-formed crystals are known, schoepite, metaschoepite and "dehydrated schoepite" commonly form fine-grained masses, effectively increasing their solubilities, and schoepite dehydrates spontaneously, becoming polycrystalline (Finch et al. 1998, also see below). The lower hydrates also have higher solubilities in water at ~25°C (O'Hare et al. 1988). Uranophane is the most common U(VI) mineral in nature (Smith 1984) and is the stable uranyl mineral in contact with groundwaters whose compositions are controlled by calcite and silica equilibria (Langmuir 1978). These observations are consistent with the large uranophane stability field in Figure 7. Soddyite is another common mineral in oxidized U deposits, where it replaces schoepite and, less commonly, uranophane (Finch 1994). Haiweeite is relatively rare, and is usually associated with volcaniclastic rocks in arid environments.

The rare mineral swamboite commonly occurs with soddyite and uranophane (Deliens et al. 1984), suggesting a genetic relationship. The swamboite stability field indicated in

Figure 7 is only approximate, and was estimated by assuming that the additional U atom in the swamboite formula (Table 8), presumably in an interlayer site, makes a lower free energy contribution than does U in the structural unit. Swamboite is probably stable near the upper limit for the activity H_4SiO_4 in natural waters. The extremely rare minerals, calciouranoite and metacalciouranoite are not shown in Figure 7, and should only be stabile in very Si-poor waters at high values of pH and dissolved Ca, a conclusion consistent with their occurrence (Rogova et al. 1973, 1974).

Stability fields estimated for two uranyl silicates, uranosilite and "ursilite," are excluded from Figure 7, as they would replace the entire field of uranophane illustrated in Figure 7, contrary to observation. Like the carbonates liebigite and zellerite, uranosilite and "ursilite" always form as fine-grained masses, and surface-free energy must contribute to their solubilities (ursilite also contains Mg and may be synonymous with haiweeite, Smith 1984). The structure of the rare mineral uranosilite is unknown but is probably unique among the uranyl silicates (Table 8).

Structural considerations. A fundamental assumption of this method to estimate Gibbs free energies is that each oxide component occurs in every structure with the same relative "fit." That is, we ignore possible structural distortions that may arise in individual minerals, due to, for example, changing the size of an interstitial cation. An example of such a distortion is evident by comparing the nearly isostructural minerals becquerelite and billietite. The ionic radii of Ca^{2+} (1.12 Å) and Ba^{2+} (1.42) (both for 8-coordination) differ by nearly 25%. This difference has two effects. (1) the structural sheets in becquerelite are more corrugated than those in billietite (Pagoaga et al. 1987), and (2) uranyl coordination in one-half of the structural sheets of billietite differs slightly from that in becquerelite. Corrugated structural sheets in becquerelite are required to accommodate the smaller Ca^{2+} ion, and this may contribute strain energy to the becquerelite structure that is not a factor in billietite. However, $\Delta G°_f{}^*$ for UO_3 is the same for both minerals (i.e. an average value). Structural strain may become increasingly severe as the number of interlayer Ca^{2+} ions increase progressively in synthetic $Ca(UO_2)_3O_4 \cdot 5H2O$ ("Ca-protasite") and calciouranoite. In fact, additional cations in calciouranoite may help reduce structural strain and stabilize this mineral in nature (Finch 1994). That the larger Ba^{2+} cation induces less structural strain is evident from the nearly flat structural sheets in protasite, which may explain why protasite occurs as a mineral (albeit rare) but not its Ca analogue.

Dehydration and the role of H_2O

Virtually all minerals that contain the uranyl ion, $UO_2{}^{2+}$, also contain substantial amounts of structurally bound H_2O, as well as OH^- ions. Many uranyl minerals are weathering products, formed at low-temperatures in near-surface aqueous environments, and structurally bound molecular H_2O strongly influences the structural and thermodynamic stabilities of these minerals. The hydronium ion, H_3O^+, may occur in some uranyl minerals, but it has never really been verified and is an unusual constituent in mineral structures (Hawthorne 1992). Structurally bound H_2O groups most commonly occur in interstitial (interlayer) sites in mineral structures, where they may be bonded to an interlayer cation, or they may occupy sites in which H_2O groups act as an H-bond "bridge" only. The ease with which H_2O groups may be removed from interstitial sites depends on bonding environments, and H_2O groups that are H-bonded only appear to be lost most readily (Finch et al. 1998). Numerous uranyl minerals display significant structural changes caused by the loss of structurally bound H_2O groups. Such structural changes are the reason that many uranyl minerals dehydrate irreversibly, although a few uranyl sulfates, vanadates, and uranyl phosphates are known to dehydrate reversibly (see Cejka, Chapter 12, this volume). No uranyl oxyhydroxides are known to dehydrate reversibly. This is a significant fact, since these minerals are commonly the earliest to form during corrosion of uraninite in weathering environments.

Finch (1997b) described the structural role of H_2O in uranyl minerals. The structures of most uranyl minerals are based on sheets of uranyl polyhedra polymerized in the two dimensions perpendicular to the approximately linear uranyl ion. These structural sheets are most commonly bonded to each other through divalent cations and molecular H_2O. The uranyl O atom is strongly bonded to the U^{6+} cation, with a bond valence (b.v.) of ~1.8 valence units (v.u.) and, therefore, contributes only ~0.2 v.u. to interlayer cations. In minerals based on sheet structures, the cations in interlayer sites can bond to no more than six neighboring uranyl O atoms. This contributes a maximum of ~1.2 v.u. to the central cation, leaving a deficit of ~0.8 v.u. This deficit is accommodated in most of these structures through cation-H_2O bonds within the interlayer, each of which contributes ~0.2 v.u. Interlayer divalent cations are commonly coordinated by four H_2O groups, thereby satisfying the cations' valence requirements. Interlayer H_2O groups that are not bonded to a cation are not "excess water" but contribute ~0.2 v.u. from H-bonds to those uranyl O atoms not bonded to an interlayer cation. These interlayer H_2O groups also act as H-bond acceptors for OH groups in the structural sheets. In addition, interlayer H_2O groups may act as bond modifiers by distributing the bond valence from central cations to multiple uranyl O atoms through an array of H-bonds, with each H-bond contributing ~ 0.2 v.u.

The removal of structurally bound H_2O groups from interlayer sites requires significant re-adjustment of local bonding arrangements and commonly results in phase transformations or even complete structural decomposition. For these reasons, most dehydration reactions among the uranyl minerals are irreversible, even at near-ambient temperatures. Furthermore, the relatively low energies required to remove H_2O groups from the interlayer sites in many of these structures results in narrow temperature ranges over which each mineral is stable with respect to other uranyl phases with more or less H_2O.

Perhaps the most dramatic and best understood effect of dehydration is that due to the loss of H_2O from schoepite (Finch et al. 1998). Schoepite, $[(UO_2)_8O_2(OH)_{12}](H_2O)_{12}$, transforms slowly in air at ambient temperature (Fig. 8) to metaschoepite, $[(UO_2)_8O_2(OH)_{12}](H_2O)_{10}$. The transformation is characterized by a two-percent decrease in

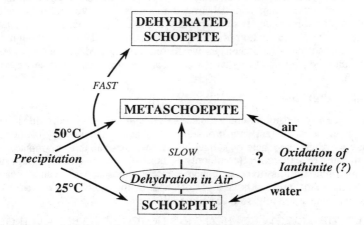

Figure 8. Representation of the phase relationships among schoepite, metaschoepite, and dehydrated schoepite, inferred from natural occurrences and experimental studies (from Finch et al. 1998).

the a cell dimension. There may be a slight decrease in the b cell dimension, but there is no significant change in the c cell dimension. The observed unit-cell changes may be due to the loss of one-sixth of the interlayer H_2O groups in schoepite, and this must result in changes to H-bonding arrangements. Differences in unit cell volumes induce strain to crystals for which the transformation to metaschoepite is incomplete, and stored strain energy may be sufficient to rapidly drive the transformation of schoepite to "dehydrated schoepite" when exposed to an external stress (e.g. heat or mechanical pressure). The complete transformation of schoepite to "dehydrated schoepite" is

$$[(UO_2)_8O_2(OH)_{12}](H_2O)_{12} \Rightarrow 8\ [(UO_2)O_{0.25}(OH)_{1.5}] + 12 H_2O$$

"Dehydrated schoepite" is a defect structure-derivative of α-$UO_2(OH)_2$. Crystals that undergo dehydration change from translucent yellow schoepite to opaque yellow, polycrystalline "dehydrated schoepite" ± metaschoepite. The complete transformation occurs in three steps: (1) loss of interlayer H_2O from schoepite, causing collapse of the layers; (2) atomic rearrangement within the sheets from a schoepite-type arrangement to a configuration which may be similar to that of metaschoepite; (3) a second rearrangement to the defect α-$UO_2(OH)_2$-type sheet. Finch et al. (1998) proposed that the formula of "dehydrated schoepite" be written $(UO_2)O_{0.25-x}(OH)_{1.5+2x}$ ($0 \leq x \leq 0.15$) to reflect the observed non-stoichiometry. Dehydration of schoepite and metaschoepite to "dehydrated schoepite" are irreversible (Christ and Clark 1960). Upon re-exposure to water, "dehydrated schoepite" does not hydrate, but vacancies and O atoms in the structural sheets may be replaced by OH groups (increasing x).

Suzuki et al. (1998) examined dehydration behaviors of saleeite and metatorbernite. These two uranyl phosphates each lose H_2O groups two at a time as temperature is increased. This reduces their d_{200} spacings by ~0.1 nm for each pair of H_2O groups lost. Both saleeite and metatorbernite rehydrate at room temperature; however, the d_{200} spacings of the rehydrated minerals differ slightly from their original values (Suzuki et al. 1998).

Some important physical effects of dehydration of uranyl minerals include the expansion of gaps between grain boundaries (due a reduction in molar volume) and reduced grain sizes (due to structural changes). These phenomena can increase available pathways for the penetration of groundwater into corrosion rinds formed on altered uraninite and increase reactive surface areas of exposed minerals. Dehydration of uranyl oxyhydroxides is irreversible, and dehydrated minerals dissolve when recontacted by groundwater, replaced by minerals such as soddyite and uranophane. Dehydrated schoepite may even inhibit the reprecipitation of schoepite (Finch et al. 1992). Although common, schoepite is not normally abundant in weathered uraninite deposits (Frondel 1958). Becquerelite, and the Pb-uranyl oxyhydroxides, vandendriesscheite and curite, tend to be more common.

Groundwater alteration

Precipitated early during uraninite alteration, the uranyl oxyhydroxides themselves alter as they continue to interact with groundwaters. Their alteration may include complete dissolution; e.g. where carbonate or sulfate complexes are available, or replacement; e.g. by uranyl silicates or carbonates (commonly rutherfordine). At uraninite deposits where carbonates are abundant, dissolved carbonate concentrations and pH can increase during interaction with the host rocks. Above a pH of about 8, the dissolution of schoepite in the presence of bicarbonate can release U to solution:

$$[(UO_2)_8O_2(OH)_{12}](H_2O)_{12} + 24\ HCO_3^{2-} \Rightarrow 8\ UO_2(CO_3)_3^{2-} + 32\ H^+ + 14\ H_2O$$

where pH remains relatively low (< 6), schoepite may be replaced by rutherfordine if groundwater pCO_2 increases due to biological respiration and decomposition (Fig. 6):

$$[(UO_2)_8O_2(OH)_{12}](H_2O)_{12} + 8\,CO_{2\,(g)} \Rightarrow 8\,UO_2CO_3 + 16\,H_2O$$

The alteration of becquerelite and many other uranyl oxyhydroxides is essentially identical to that of schoepite, with UO_2^{2+} being solubilized in carbonate groundwaters, except where pCO_2 exceeds atmospheric levels, under which conditions, rutherfordine or one of several uranyl carbonates illustrated in Figure 6 may become stable. Becquerelite is more stable than schoepite at higher $\{Ca^{2+}\}/\{H^+\}^2$ values (Figs. 6 and 7) and so, perhaps, is more resistant to dissolution. In dilute, low pH, non-complexing waters (e.g. fresh rain water), becquerelite may dissolve incongruently by losing Ca preferentially to U (Casas et al. 1994), possibly to form schoepite:

$$4\,Ca[(UO_2)_6O_4(OH)_6](H_2O)_8 + 8\,H^+ + H_2O \Rightarrow 4\,Ca^{2+} + 3\,[(UO_2)_8O_2(OH)_{12}](H_2O)_{12}$$

If the activity of dissolved silica is sufficient, UO_2^{2+} can complex with silicic acid to precipitate as uranyl silicates. Whether soddyite or uranophane forms depends on the activity ratio, $\{Ca^{2+}\}/\{H^+\}^2$ (Fig. 7). Low pH, Ca-poor waters favor formation of soddyite (Fig. 7); for example, replacing schoepite:

$$[(UO_2)_8O_2(OH)_{12}](H_2O)_{12} + 4\,H_4SiO_4 \Rightarrow 4\,(UO_2)_2SiO_4(H_2O)_2 + 12\,H_2O$$

or becquerelite

$$Ca[(UO_2)_6O_4(OH)_6](H_2O)_8 + 2\,H^+ \Rightarrow 3\,(UO_2)_2SiO_4(H_2O)_2 + Ca^{2+} + 6\,H_2O$$

More alkaline, Ca-bearing waters favor formation of uranophane, assuming carbonate concentrations remain relatively low (Fig. 7):

$$[(UO_2)_8O_2(OH)_{12}](H_2O)_{12} + 4\,Ca^{2+} + 8\,H_4SiO_4 \Rightarrow$$
$$4\,Ca(UO_2)_2(SiO_3OH)_2(H_2O)_5 + 8\,H^+ + 12\,H_2O$$

and

$$Ca[(UO_2)_6O_4(OH)_6](H_2O)_8 + 2\,Ca^{2+} + 6\,H_4SiO_4 \Rightarrow$$
$$3\,Ca(UO_2)_2(SiO_3OH)_2(H_2O)_5 + 8\,H^+ + H_2O$$

The above reactions constitute some of the more important replacement reactions that occur as groundwaters interact with corrosion rinds of uranyl oxyhydroxides. The alteration of the Pb-uranyl oxyhydroxides is slightly different from that of the Pb-free minerals, and we will consider their alteration in the next section.

Comparable reactions are relevant for the replacement of uranyl oxyhydroxides by minerals such as uranyl phosphates, arsenates and vanadates. However, in the presence of these ions, the uranyl oxyhydroxides tend to be rare or non-existent (Garrels and Christ 1959; Langmuir 1978).

The alteration of uranyl silicates is less commonly reported than that of the uranyl oxyhydroxides. Uranyl silicates are less soluble than oxy-hydoxides in most natural waters and tend to be less vulnerable to groundwater attack. However, as for becquerelite, uranophane may dissolve incongruently in fresh waters, releasing Ca^{2+} and silica, and precipitating soddyite. The replacement of uranophane by soddyite has been observed in natural samples from Shinkolobwe (Deliens 1977b; Finch 1994), as well as experimentally (Casas et al. 1994):

$$Ca(UO_2)_2(SiO_3OH)_2(H_2O)_5 + 2\,H^+ \Rightarrow (UO_2)_2SiO_4(H_2O)_2 + Ca^{2+} + H_4SiO_4 + 3\,H_2O$$

To our knowledge, the reverse reaction has not been reported.

The interaction of uranyl silicates with strongly complexing solutions, such as alkaline bicarbonate waters, might result in the preferential release of U, which would increase the Si:U ratio in the residual solid, and might help explain compositional uncertainties for minerals such as "ursilite."

The following sequence summarizes observed U mineral paragenesis at many oxidized uraninite deposits:

1. Dissolution of uraninite and precipitation of uranyl oxyhydroxides: becquerelite, schoepite, (ianthinite) and vandendriesscheite. These minerals tend to first precipitate within voids and fractures in uraninite as it dissolves. Deliens (1977b) noted that the earliest-formed uranyl oxyhydroxides tend to be very fine grained.

2. Replacement of earlier-formed uranyl oxyhydroxides by uranyl silicates and replacement of Pb-poor minerals by Pb-enriched uranyl minerals. In weakly complexing waters, some U is transported short distances to precipitate on the outer surfaces of corroding uraninite as a coarsely crystalline rind of uranyl oxyhydroxides schoepite and becquerelite (± ianthinite). Monocarbonates such as rutherfordine may also precipitate where pCO$_2$ values are sufficient.

3. The continued replacement of both coarse and fine grained uranyl oxyhydroxides repeats the alteration sequence. Uranyl phosphates commonly form relatively late.

The dehydration and alteration of uranyl oxyhydroxides notwithstanding, schoepite and becquerelite, may persist for many thousands of years. Uranium-series activity ratios for several uranyl minerals from the Shinkolobwe mine in southern Democratic Republic of Congo indicate that these minerals did not experience significant preferential loss of U since their formation more than 100,000 years ago. The minerals examined included rutherfordine, schoepite, becquerelite, and uranophane. No correlation was found between mineral species and mineral age (Finch et al. 1995, 1996). Finch et al. (1996) concluded that the oxidative dissolution of primary uraninite maintains locally high dissolved U, keeping waters supersaturated with respect to most uranyl minerals and providing an inexhaustible source of dissolved U^{6+} for new mineral precipitation and growth. These results suggest that, as long as uraninite persists in an oxidizing environment, the assemblage of secondary uranyl minerals is determined by local groundwater chemistry (including transitory changes), but not necessarily towards formation of uranyl minerals with lower solubilities.

THE ROLE OF RADIOGENIC Pb IN U MINERAL PARAGENESIS

Because U is radioactive and ultimately decays to Pb, the mineralogy of U is intimately tied to that of Pb. As radiogenic Pb accumulates in a U mineral, U content decreases concomitantly. Because the crystal chemistries of U and Pb are so different, the accumulation of Pb and attrition of U can combine to destabilize structures of U minerals that are old enough to contain substantial radiogenic Pb. What we mean by "substantial" depends, of course, on the mineral, its age, and the amount of U initially present.

Lead is incompatible in the uraninite structure (Berman 1957; Janeczek and Ewing 1995), where it may occupy interstitial sites. X-ray studies indicate that unit-cell volumes of Pb-rich uraninite are larger than those of Pb-poor uraninite (Janeczek and Ewing 1992c, 1995), suggesting that Pb accumulation can induce significant strain. Lead is relatively immobile in most groundwaters (Mann and Deutscher 1980), and under reducing conditions, Pb released from U minerals commonly forms galena, provided the activity of S is sufficient. Recrystallization of uraninite under reducing conditions may proceed according to a reaction such as,

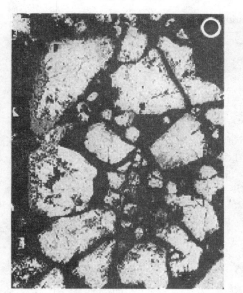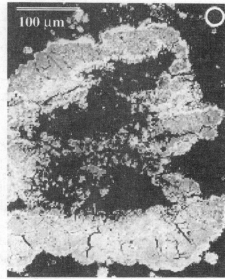

Figure 9. BSE images of fragmented uraninite crystals dissolved at their cores, a texture that may be related to strain caused by accumulation of radiogenic Pb and precipitation of galena. (from Janeczek and Ewing 1995; reprinted by permission of Pergamon Press).

$$(U^{4+}_{1-2y}U^{6+}_{y}Pb^{2+}_{y})O_{2\,(cr)} + yHS^{-} \Rightarrow (1-y)UO_{2\,(cr)} + yPbS_{(cr)} + yOH^{-} + 1/2\,yO_{2\,(aq)}$$

An interesting consequence of this reaction is that the volume of uraninite steadily decreases without any loss of U to groundwater! Galena has a larger molar volume than uraninite (31.490 cm³·mol⁻¹ and ~24.62 cm³·mol⁻¹, respectively), and precipitation of galena within uraninite could induce substantial strain, helping to fragment crystals and providing pathways for groundwater infiltration. Brecciation of uraninite and dissolution of their cores may reflect this phenomenon (Janeczek and Ewing 1992c, 1995) (Fig. 9).

Under oxidizing conditions, Pb can combine with UO_2^{2+} to form one or more of the nearly 25 known Pb-uranyl minerals (Table 20). Virtually every chemical group of uranyl minerals is represented by at least one species in which Pb is an essential constituent, with eight or more being oxyhydroxides. The Pb-uranyl oxyhydroxides commonly form directly from precursor uraninite; however, many Pb-uranyl minerals do not, and their genesis is probably related to the accumulation of radiogenic Pd in nominally Pb-free minerals after formation. For example, kasolite formed at depth in the Koongarra U deposit does not replace uraninite, as originally believed (Snelling 1980). Instead, accumulation of radiogenic Pb in sklodowskite causes continual recrystallization of Pb-free sklodowskite (Isobe et al. 1992). Lead lost from the original sklodowskite precipitates as kasolite within sklodowskite veins (Fig. 10). Accumulation of radiogenic Pb and U loss combine to destabilize the structure of clarkeite, which eventually recrystallizes to wölsendorfite or curite (Finch and Ewing 1997). Studtite crystals commonly have Pb-rich cores, which appear as thin "threads" of Pb-uranyl oxyhydroxides, such as fourmarierite, that are evident from XRD data (Deliens and Piret 1983b). Asselbornite is compositionally zoned, with Pb concentrations highest at the cores of crystals (Sarp et al. 1983). Crystals of vandendriesscheite, schoepite, and becquerelite commonly contain inclusions of Pb-uranyl oxyhydroxides such as masuyite and fourmarierite (Finch and Ewing 1992b; Finch 1994)

Table 20. Pb-bearing U minerals

Mineral	Formula	Notes
Plumbobetafite	$(Pb,U,Ca)(Ti,Nb)_2O_6(OH,F)$	
Plumbomicrolite	$(Pb,U,Ca)_2Ta_2O_6(OH)$	
Plumbopyrochlore	$(Pb,U,Ca)_{2-x}Nb_2O_6(OH)$	
Uraninite	$(U,Pb)O_2$	To ~20 wt % PbO
Wölsendorfite	$xCaO \cdot (6-x)PbO \cdot 12UO_3 \cdot 12H_2O$ ($x = 0 - 1$)	
Sayrite	$Pb_2(UO_2)_5O_6(OH)_2(H_2O)_4$	
Curite	$Pb_3(UO_2)_8O_8(OH)_6(H_2O)_2$	
Masuyite	$Pb[(UO_2)_3O_3(OH)_2](H_2O)_3$ also $4PbO \cdot 9UO_3 \cdot 12H_2O$ and $3PbO \cdot 8UO_3 \cdot 10H_2O$	"grooved masuyite" "type masuyite" and a synthetic analogue
Fourmarierite	$Pb(UO_2)_4O_3(OH)_4(H_2O)_4$	
Richetite	$M_xPb_{8.57}[(UO_2)_{18}O_{18}(OH)_{12}]_2(H_2O)_{41}$	
Vandendriesscheite	$Pb_{1.57}(UO_2)_{10}O_6(OH)_{11}(H_2O)_{11}$	
Meta-vandendriesscheite	$PbO \cdot 7UO_3 \cdot (12-x)H_2O$	uncertain formula
Calciouranoite	$(Ca,Ba,Pb,K,Na)O \cdot UO_3 \cdot 5H_2O$	minor Pb
Meta-calciouranoite	$(Ca,Ba,Pb,K,Na)O \cdot UO_3 \cdot 2H_2O$	minor Pb
Kasolite	$Pb(UO_2)(SiO_4)(H_2O)$	
"Pilbarite"	$PbO \cdot ThO_2 \cdot UO_3 \cdot 2SiO_2 \cdot 4H_2O$ (?)	doubtul species
Widenmannite	$Pb_2(UO_2)(CO_3)_3$	
Demesmaekerite	$Cu_5Pb_2(UO_2)_2(SeO_3)_6(OH)_6(H_2O)_2$	
Parsonsite	$Pb_2(UO_2)(PO_4)_2(H_2O)_2$	
Dewindtite	$Pb_3[(UO_2)_3O(OH)(PO_4)_2]_2(H_2O)_{12}$	
Dumontite	$Pb_2[(UO_2)_3O_2(PO_4)_2](H_2O)_5$	
Przhevaskite	$Pb[(UO_2)(PO_4)]_2(H_2O)_2$	
"Renardite"	$Pb[(UO_2)_4(OH)_4(PO_4)_2](H_2O)_7$	mixture
Kamitugaite	$PbAl(UO_2)_5([P,As]O_4)_2(OH)_9(H_2O)_{9.5}$	
Hügelite	$Pb_2[(UO_2)_3O_2(AsO_4)_2](H_2O)_5$	
Hallimondite	$Pb(UO_2)(AsO_4)_2$	
Asselbornite	$(Pb,Ba)(BiO)_4(UO_2)_6(AsO_4)_2(OH)_{12}(H_2O)_3$	zoned: Pb-rich cores
"Kivuite"	$(Th,Ca,Pb)(UO_2)_4(HPO_4)_2(OH)_8(H_2O)_7$	uncertain species
Curiénite	$Pb(UO_2)_2(V_2O_8)(H_2O)_5$	
Francevillite	$(Ba,Pb)(UO_2)_2(V_2O_8)(H_2O)_5$	minor Pb
Moctezumite	$Pb(UO_2)(TeO_3)_2$	
Uranotungstite	$(Ba,Pb,Fe^{2+})(UO_2)_2(WO_4)(OH)_4(H_2O)_{12}$	minor Pb

(Fig. 11). Schoepite and metaschoepite may contain epitaxial intergrowths of fourmarierite or vandendriesscheite that are evident in X-ray precession photographs of schoepite and metaschoepite single crystals, which could be misinterpreted as being representative of the host mineral, and may explain conflicting unit-cell data for minerals such as paraschoepite, masuyite and vandendriesscheite (Christ and Clark 1960). Confusion about compositions of Pb-bearing minerals such as calciouranoite and bauranoite (Belova et al. 1993), clarkeite (Finch and Ewing 1997), renardite (Deliens et al. 1990) and masuyite (Deliens and Piret 1996) may also reflect continuous recrystallization of Pb-rich minerals from Pb-poor minerals at a fine scale. Recent detailed analyses (Deliens et al. 1990; Deliens and Piret 1996) and single-crystal structure determinations (Burns 1997, 1998a, 1999e; Burns and Hanchar 1999) have greatly improved our understanding of the structural and paragenetic relationships among Pb-uranyl oxyhydroxides, although many questions still remain.

Figure 10. BSE image of acicular kasolite crystals (K) at the margin of a sklodowskite (S) vein in a sample from Koongarra. Kasolite replaces the sklodowskite. Width of vein is approximately 2 mm. (from Isobe et al. 1992; reprinted by permission of North-Holland, Elsevier).

Another factor in the development of the complex (and often confusing) mineralogy of Pb-uranyl minerals is alteration by groundwater. Due to the different mobilities of Pb and U in most groundwaters, Pb-bearing uranyl minerals tend to dissolve incongruently. This is especially notable during alteration of Pb-uranyl oxyhydroxides.

Frondel (1956) first explained the enrichment of Pb in corrosion rinds by the preferential loss of U to groundwater. Vandendriesscheite and fourmarierite are commonly the earliest Pb-uranyl oxyhydroxides to precipitate when Pb-bearing uraninite corrodes, forming more Pb-enriched minerals as they interact with Si- and carbonate-bearing groundwaters (Frondel 1956; Deliens 1977a; Finch and Ewing 1992a,b; Finch 1994). Preferential removal of U increases the residual Pb content within the dissolving vandendriesscheite, and polycrystalline masuyite, sayrite and curite may precipitate as inclusions within vandendriesscheite crystals (Finch and Ewing 1992a,b; Finch 1994). Inclusions of masuyite are common where vandendriesscheite is replaced by uranyl silicates (Fig. 12). Cryptocrystalline corrosion rinds that are veined by uranophane also show increased Pb concentrations adjacent to uranophane veins (Fig. 13); Ca from becquerelite and U from becquerelite and vandendriesscheite are incorporated into uranophane, whereas Pb is not. Because Pb is not removed by groundwater, it accumulates adjacent to uranyl-silicate veins as fine-grained Pb-enriched uranyl oxyhydroxides. Thus,

Figure 11. (a) BSE image of a polished section through an altered vandendriesscheite grain showing polycrystalline masuyite inclusion within optically continuous vandendriesscheite. The vandendriesscheite has partially dissolved and been replaced by rutherfordine (black); masuyite has also dissolved at the upper left of the image (from Finch 1994). (b) BSE image of polycrystalline becquerelite with 1-10 μm inclusions of an undetermined Pb-uranyl oxyhydroxide.

the precipitation of Pb-rich uranyl oxyhydroxides does not necessarily require high concentration of dissolved Pb.

Lead is not incorporated into rutherfordine, soddyite or uranophane, minerals that commonly replace early-formed Pb uranyl oxyhydroxides. The only known Pb-uranyl silicate, kasolite, has a Pb:U ratio (1:1), higher than that of any of the Pb-uranyl oxyhydroxides. The same is true for the uranyl carbonate widenmannite (Pb:U = 2:1). Rather than being associated with the alteration of vandendriesscheite, kasolite commonly occurs within fractures inherited from precursor uraninite (Isobe et al. 1992; Finch 1994) where Pb is derived from dissolving galena, or kasolite crystallizes from nominally Pb-free uranyl silicates that have accumulated radiognic Pb (Isobe et al. 1992). Kasolite commonly coexists with curite, a late-stage weathering product of uraninite.

The incongruent dissolution of vandendriesscheite to masuyite by carbonate groundwaters releases some U to groundwater but retains some in the residual solid (Finch and Ewing 1992a,b):

$$Pb_{1.5}(UO_2)_{10}O_6(OH)_{11}(H_2O)_{11(cr)} + 11\,HCO_3^-{}_{(aq)} \Rightarrow$$
$$1.5\,Pb[(UO_2)_3O_3(OH)_2](H_2O)_{3(cr)} + 5.5\,UO_2(CO_3)_2{}^{2-} + 16.5\,H_2O$$

The replacement of Pb-uranyl oxyhydroxides by uranyl silicates is common (Finch and Ewing 1992a,b). In the presence of sufficient dissolved silica and Ca^{2+}, vandendriesscheite alters incongruously to masuyite plus uranophane (Finch and Ewing 1992a):

$$\text{vandendriesscheite} + Ca^{2+} + H_4SiO_{4\,(aq)} \Rightarrow \text{uranophane} + \text{masuyite} + 4\,H^+ + H_2O$$

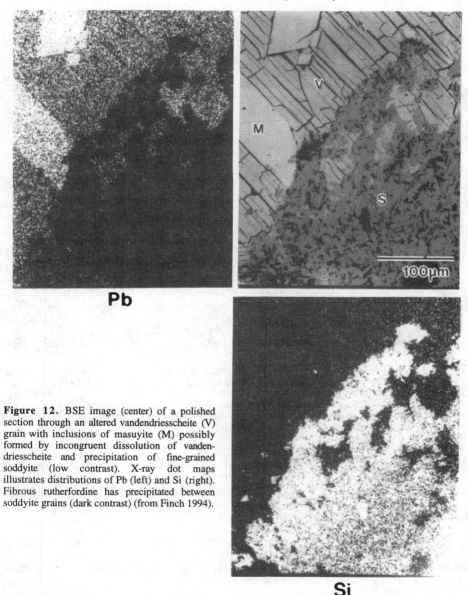

Figure 12. BSE image (center) of a polished section through an altered vandendriesscheite (V) grain with inclusions of masuyite (M) possibly formed by incongruent dissolution of vandendriesscheite and precipitation of fine-grained soddyite (low contrast). X-ray dot maps illustrates distributions of Pb (left) and Si (right). Fibrous rutherfordine has precipitated between soddyite grains (dark contrast) (from Finch 1994).

Or, without sufficient dissolved Ca, soddyite precipitates (Fig. 12):

vandendriesscheite + $H_4SiO_{4(aq)}$ ⇒ soddyite + masuyite + H_2O

These reactions are written with U retained in the solids, which, given the relative insolubility of uranyl silicates (Langmuir 1978; Grenthe et al. 1992), is a reasonable simplification (Both reactions are illustrative and not strictly balanced.). These three reactions demonstrate how Pb helps reduce the mobility of U during the interaction of

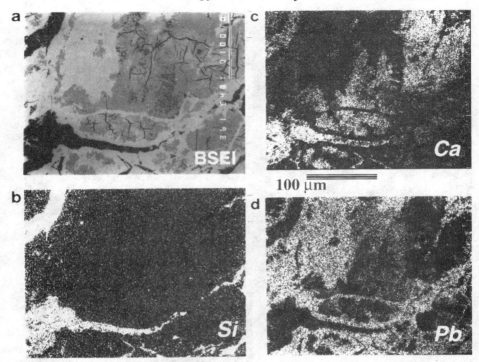

Figure 13. (a) BSE image of a polished section through an altered region of a fine-grained uraninite corrosion rind showing Pb enrichment and Ca depletion associated with alteration by Si-bearing groundwater. Becquerelite dissolves during alteration, with Ca and U incorporated into uranophane. Central area of BSE image (a) (dark gray contrast) is intermixed vandendriesscheite and becquerelite. Darkest contrast veins contain uranophane. Regions of brightest contrast correspond to Pb-rich regions, probably masuyite or fourmarierite. (b) X-ray dot maps illustrating the distribution of Si, (c) Ca, and (d) Pb. (from Finch 1994).

uranyl oxyhydroxides in carbonate- and Si-bearing groundwaters. The earliest minerals to form tend to have relatively low Pb:U ratios (Frondel 1956, Isobe et al. 1992, Finch and Ewing 1992b, Finch 1994). Continued interaction with groundwater forms increasingly enriched Pb-bearing minerals, such as curite or wölsendorfite (Isobe et al. 1992, Finch and Ewing 1992a,b, Finch 1994) (Fig. 14).

Curite is commonly one of the last remaining minerals after the complete oxidation, dissolution and replacement of uraninite (Isobe et al. 1992; Finch 1994). Uranyl phosphates and curite are so commonly associated that a genetic relationship has long been supposed (Frondel 1956; Frondel 1958; Deliens 1977b). The direct replacement of curite by uranyl phosphates has been observed at the Shinkolobwe mine (Fig. 15), and Finch and Ewing (1992b) proposed that curite may serve as a "substrate" for the nucleation of certain phosphates, similar to the role of Fe and Mn oxyhydroxides (Murakami et al. 1997; see below), except that curite contains abundant U required for uranyl phosphate formation.

PARAGENESIS OF THE URANYL PHOSPHATES

Uranyl phosphates help control U concentrations in many natural waters. They generally have solubilities below those of the uranyl silicates and are associated with a wide

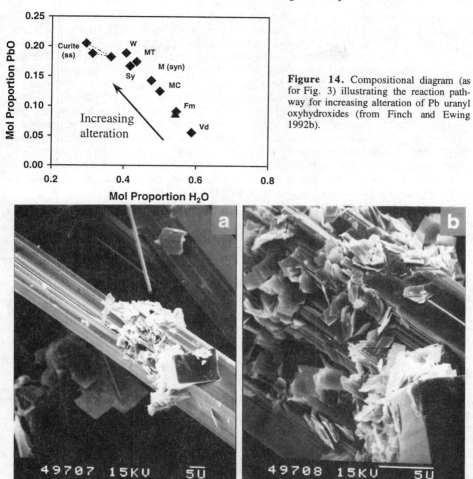

Figure 14. Compositional diagram (as for Fig. 3) illustrating the reaction pathway for increasing alteration of Pb uranyl oxyhydroxides (from Finch and Ewing 1992b).

Figure 15. SEM images of curite crystal partly replaced by an unidentified uranyl phosphate, possibly chernikovite. (b) Magnified view of (a) (from Finch 1994).

range of weathered U deposits. Uranyl phosphates are known to precipitate from groundwater with U concentrations in the range 10^{-8} to 10^{-9} mol/kg (Dall'aglio et al. 1974), values that approach the solubility of uraninite in some reducing environments. Uranyl phosphates may occur well removed from any U source (Weeks and Thompson 1954; Frondel 1956). In groundwaters where $\log\{[PO_4^{3-}]_T/[CO_3^{2-}]_T\} > -3.5$, uranyl phosphate complexes predominate over uranyl carbonate complexes (Sandino and Bruno 1992). Apatite controls the phosphate concentrations in many natural waters, keeping phosphate activities below 10^{-7} mol/kg^{-1} above pH = 7 (Stumm and Morgan 1981), but synthesis experiments show that the phosphate concentrations necessary to precipitate uranyl phosphates can be quite high (on the order of 10^{-2} mol/kg) (Markovic and Pavkovic 1983; Sandino 1991). Uranyl phosphates are typically most stable below approximately pH 5, where apatite solubility tends to increase (Stumm and Morgan 1981).

Perhaps one of the most studied occurrences of uranyl phosphates is the Koongarra U deposit in the Northern Territory, Australia. The alteration of uraninite and the genesis of uranyl phosphates at Koongarra were examined extensively by Snelling (1980, 1992) and Isobe et al. (1992, 1994). A variety of uranyl phosphates has been identified among the Koongarra U minerals, including, saléeite, dewindtite, sabugalite, and torbernite. Saléeite is the predominant uranyl phosphate in the weathered zone at Koongarra. Three mineralogical zones are defined at Koongarra, based on the predominant types of U minerals that occur within them: U oxide zone, uranyl silicate zone, and uranyl phosphate zone (Snelling 1980). Dissolved U is transported roughly from the U oxide zone at depth to the silicate zone, also at depth, then upward to the phosphate zone. The uranyl phosphate zone is in the most oxidized weathered zone near the surface. Macroscopic saléeite crystals can be seen with the unaided eye as yellow-green, platy crystals within millimeter-wide veins, suggesting that saléeite is the thermodynamically stable mineral in contact with the groundwater that flowed through those veins. In addition, microscopic saléeite crystals, tens to hundreds of μm across, commonly replace sklodowskite and apatite (Fig. 16).

Most groundwaters in and around the Koongarra deposit are considered undersaturated with respect to saléeite (Payne et al. 1992, Murakami et al. 1997). Autunite, the Ca analogue of saléeite, is absent even on the surfaces of dissolving uranophane and apatite crystals, both of which should supply abundant Ca to solution. The solubility

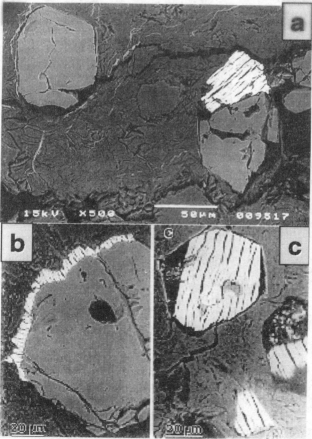

Figure 16. BSE images of polished sections showing (a) apatite crystal on right partly replaced by saléeite (brightest contrast); (b) incipient saléeite on the surface of an apatite crystal; (c) saléeite pseudomorph after apatite with remnant apatite inclusions (gray) (from Murakami et al. 1996a; Murakami et al. 1997).

constant for autunite, log K_{So} = -19.43 (Grenthe et al. 1992), is greater than that of saléeite, log K_{So} = -22.3 (Magelhaes et al. 1985) (both for 25°C), and Murakami et al. (1997) showed from thermodynamic calculations that saléeite should precipitate instead of autunite even at apatite surfaces. Their calculations, combined with observed textural relationships between saléeite and sklodowskite and between saléeite and apatite (Fig. 16), indicate that microscopic saléeite crystals precipitate directly on the surfaces of sklodowskite and apatite due to local saturation, with Mg, U and P derived from dissolving sklodowskite and apatite in addition to the availability of those elements in groundwater. The two types of saléeite occurrences (vein-filling and surface-controlled precipitation) strongly suggest that uranyl phosphate formation at Koongarra not only reflects macroscopic thermodynamic equilibrium with percolating groundwaters, but is also influenced by kinetic factors. This is probably true for uranyl phosphate precipitation in many natural environments, helping to explain the existence of uranyl phosphates in groundwaters with especially low U concentrations (Dall'aglio et al. 1974).

Phosphate is a ubiquitous anion in near-surface environments. Dissolved phosphate concentrations in most river waters are on the order of 20 ppb, and dissolved U concentrations are typically less than 0.3 ppb (Holland 1978), suggesting that most rivers are well below saturation with respect to uranyl phosphates. Sorption of U onto Fe and Mn oxyhydroxides and clay minerals is an important mechanism influencing U migration in natural waters (e.g. Guthrie 1989; Ivanovich et al. 1994; McKinley et al. 1995; Buck et al. 1996, 1999). Groundwater concentrations of Si may also influence U adsorption if dissolved Si competes with U for surface sites on goethite, as demonstrated experimentally by Gabriel et al. (1998). The chemical forms of U adsorbed on various minerals have been examined by several authors. For example, Waite et al. (1994) modeled the adsorption of U as mononuclear uranyl complexes on ferrihydrite surfaces, and extended X-ray-absorption fine-structure spectroscopy (EXAFS) shows that U may be sorbed either as outer-sphere uranyl complexes on smectite (Dent et al. 1992) or as inner-sphere complexes, which may influence subsequent reduction of U (Giblin 1982; Giaquinta et al. 1997). Glinka et al. (1997) found that U is sorbed weakly onto colloidal silica as outer-sphere complexes. Uranium sorption on Fe and Mn oxyhydroxides has long been known to be an important process. A combined experimental and modeling study by Bruno et al. (1995) suggested that U may co-precipitate on Fe(III) oxyhydroxides surfaces as schoepite or "dehydrated schoepite." Murakami et al. (1997) examined a sample from Koongarra by high-resolution transmission electron microscopy (HRTEM) and found microcrystals of saléeite, 10-50 nm across, within a microvein of goethite or hematite a few μm wide. The sample was from a region where the groundwater is far below saturation with respect to saléeite (Mg = 13 ppm, P < 5 ppb, U = 30 ppb). Murakami et al. (1997) explained the crystallization of saléeite microcrystals ("microcrystallization") as follows. Ferrihydrite is formed early during weathering of ferrous minerals (Murakami et al. 1996b). Dissolved P from groundwater is incorporated onto or into ferrihydrite by adsorption, co-precipitation, or both. Phosphorous in ferrihydrite is then released during the transformation of ferrihydrite to goethite and hematite (Ostwald ripening). This localized source of P and locally available U (which can strongly sorb onto fine-grained ferric oxyhydroxides) permits precipitation of saléeite microcrystals directly on the surfaces of goethite and hematite. Buck et al. (1996) identified microcrystals of meta-autunite in U-contaminated soils at Fernald, Ohio, USA, and Sato et al. (1997) reported microcrystals of torbernite on Fe-oxyhydroxide nodules at Koongarra. Microcrystallization is therefore a potentially important mechanism by which U can be immobilized in natural environments for long periods.

ACKNOWLEDGMENTS

We are grateful to Peter Burns for substantial help editing the manuscript and for many informative discussions about U mineralogy. We thank Christopher Cahill, Jennifer Jackson, and Sarah Scott for critically reading the manuscript in record time. Helpful comments and suggestions about Ta-Nb-Ti oxides were provided by Reto Gieré, who also generously provided data on zirconolites. TM is grateful to Toshihiko Ohnuki, Hiroshi Isobe, and Tsutomu Sato for many discussions about U mineralogy and geochemistry. Leslie Finch helped with typing references and several sections. Much of the content here derives from our long and fruitful collaborations with Rodney Ewing, whose guidance, encouragement, and insight provided the seeds from which this chapter grew. Finally, and perhaps most importantly, we thank our wives, Leslie and Emiko, whose support and patience were invaluable to achieving this end. Financial support for this work was provided to RJF by the U.S. Department of Energy, Environmental Management and Sciences Program (DE-FG07-97ER14820), and to TM by the Ministry of Education, Science and Culture [Japan] (11440160).

> This manuscript has been created by the University of Chicago as Operator of Argonne National Laboratory ("Argonne") under Contract No. W-31-109-ENG-38 with the U.S. Department of Energy. The U.S. Government retains for itself, and others acting on its behalf, a paid-up, nonexclusive, irrevocable worldwide license in said article to reproduce, prepare derivative works, distribute copies to the public, and perform publicly and display publicly, by or on behalf of the Government.

REFERENCES

Alberman KB, Blakely RC, Anderson JS 1951 The oxides of uranium—Part II, Binary system UO_2-CaO. J Chem Soc 1951:1352-1356

Allen GC 1985 An angle-resolved X-ray photoelectron spectroscopic study of air oxidized UO_2 pellet surfaces. Phil Mag B51:465-473

Allen GC, Tempest PA 1982 Linear ordering of oxygen clusters in hyperstoichiometric uranium dioxide. J Chem Soc, Dalton Trans 1982:2169-2173

Allen GC, Tucker PM, Tyler JW 1982 Oxidation of uranium dioxide at 298 K studied by using X-ray photoelectron spectroscopy. J Phys Chem 86:224-228

Allen GC, Tempest PA, Tyler JW 1987 Oxidation of polycrystalline UO_2 studied by using X-ray photoelectron spectroscopy. J Chem Soc, Faraday Trans 1 83:925-935

Alwan AK, Williams PA 1980) The aqueous chemistry of uranium minerals. Part 2. Minerals of the liebigite group. Mineral Mag 43:665-667

Anderson JS 1953 Recent work on the chemistry of uranium oxides. Bull Soc Chim France. 1953:781-788

Anderson JS, Edgington DN, Roberts LEJ, Wait E 1954 Oxides of uranium. IV. System UO_2-ThO_2-O. J Chem Soc (London) 3:3324-3331

Ansell VE, Chao GY 1987 Thornasite, a new hydrous sodium thorium silicate from Mount St-Hillaire, Quebec. Can Mineral 25:181-183

Atkin D, Basham I, Bowles JFW 1983 Tristramite, a new calcium uranium phosphate of the ekanite group. Mineral Mag 47:393-396

Atencio D 1988 Chernikovite, a new mineral name for $(H_3O)_2(UO_2)_2(PO_4)_2 \cdot 6H_2O$ superceding "hydrogen autunite." Mineral Rec 19:249-252

Atencio D, Newman R, Silva AJGC, Mascarenhas YP 1991) Phurcalite from Perus, São Paulo, Brazil, and redetermination of its crystal structure. Can Mineral 29:95-105

Aurisicchio C, Orlandi P, Pasero M, Perchiazzi N 1993 Uranopolycrase, the uranium-dominant analogue of polycrase-(Y), a new mineral from Elba Island, Italy, and its crystal structure. Eur J Mineral 5:1161-1165

Bachet B, Brassey C, Cousson A 1991) Structure of $Mg[(UO_2)(AsO_4)]_2 \cdot 4H_2O$. Acta Crystalogr C47:2013-2015 (in French)

Bariand P, Bachet B, Brassey C, Medenbach O, Deliens M 1993 Seelite, a new uranium mineral from the Talmessi mine, Iran, and Rabejac, France. Mineral Record 24:463-467

Baturin SV, Sidorenko GA 1985 Crystal structure of weeksite, $(K_{0.62}Na_{0.38})_2(UO_2)_2[Si_5O_{13}] \cdot 3H_2O$. Doklady Akad Nauk SSSR 282:1132-1136 (in Russian)

Bayliss P, Mazzi F, Munno R, White TJ 1989 Mineral nomenclature: Zirconolite. Mineral Mag 53:565–569
Beddoe-Stephens B, Secher K 1982 Barian wölsendorfite from east Greenland. Mineral Mag 46:130-132
Belbeoch B, Boivineau JC, Perio P 1967 Changements de structure de l'oxyde U_4O_9. J Phys Chem Solids 28:1267
Belova LN, Gorshkov AI, Doinikova OA, Mokhov AV, Trubkin NV, Sivtsov AV 1993 New data on coconinoite. Doklady Akad Nauk 329:772-775
Belova LN, Gorshkov AI, Ivanova, Sivtsov AV 1980 Nature of the so-called phosporous-bearing coffinite. Doklady Akad Nauk SSSR 255:428-430 (English translation in Doklady Earth Sci Sec 255:156-158)
Belova LN, Gorshkov AI, Ivanova, Sivtsov AV, Lizorkina, Boronikhin 1984 Vyacheslavite. Zap Vses Mineral Obshch 113:360 (in Russian)
Belova LN, Ryzhov BI, Federov OV, Lyubomilova GV 1985 Typomorphism and isomorphism of the wölsendorfite group of uranium hydroxides. Izv Akad Nauk SSSR Ser Geol 2:65-72
Berman RM 1957 The role of lead and excess oxygen in uraninite. Am Mineral 42:705-731
Bignand C 1955 Sur les propriétés et les synthèses de quelques minéraux uranifères. Bull Soc fr Minér Crist 78:1-26
Billiet V, de Jong WF 1935 Schoepiet en Becquereliet. Natuur Tijd Ned-Indië 17:157-162
Birch WD, Mumme WG, Segnit ER 1988 Ulrichite-a new copper calcium uranium phosphate from Lake Boga, Victoria, Australia. Australian Mineral 3:125-134
Boatner LA, Sales BC 1988 Monazite. In Radioactive Waste Forms for the Future. Lutze W, Ewing RC (Eds) North-Holland, Amsterdam, p 495-564
Bobo JC 1964 Physical and chemical properties in systems between U-O and a metallic element. Rev Chim Minerale 1:3-37
Bowels JFW, Morgan DJ 1984 The composition of rhabdophane. Mineral Mag 48:146-148
Brasseur H 1949) Étude de la billietite. Bull Acad Royale Belg Cl Sci 35:793-804
Bruno J, Casas I, Cera E, Finch RJ, Ewing RC, Werme LO 1995 The assessment of the long-term evolution of the spent nuclear fuel matrix by kinetic, thermodynamic and spectroscopic studies of uranium minerals. Mater Res Soc Symp Proc 353:633-639
Bruno J, Casas I, Puigdomènech I 1991 The kinetics of dissolution of UO_2 under reducing conditions and the influence of an oxidized surface layer (UO_{2+x}): Application of a continuous flow-through reactor. Geochim Cosmochim Acta 55:647-658
Blanc P-L 1996 Oklo—Natural Analogue for a Radioactive Waste Repository. Vol. 1. Acquirements of the Project. Nuclear Science and Technology Report EUR 16857/1 EN, 123 p
Braithwaite RSW, Paar WH, Chisholm JE 1989 Phurcalite from Dartmoor, Southwest England, and its identity with 'nisaite' from Portugal. Mineral Mag 53:583–589
Brindley GW, Bastanov M 1982 Interaction of uranyl ions with synthetic zeolites of type A and formation of compreignacite-like and becquerelite-like products. Clays Clay Minerals 30:135-142
Brooker EJ Nuffield EW 1952 Studies of radioactive compounds: IV—pitchblende from Lake Athabasca, Canada. Am Mineral 37:363–385
Brugger J, Gieré R, Graeser S, Meisser N 1997 The crystal chemistry of roméite. Contrib Mineral Petrol 127:136-146
Bruno J, Casas I, Cera E, Ewing RC, Finch RJ, Werme LO 1995 The assessment of the long-term evolution of the spent nuclear fuel matrix by kinetic, thermodynamic and spectroscopic studies of uranium minerals. Mater Res Soc Symp Proc 353:633-639
Bruno J, Casas I, Sandino A 1992 Static and dynamic SIMFUEL dissolution studies under oxic conditions. J Nucl Mater 190:61-69
Bruno J, de Pablo J, Duro L, Figuerola E 1995 Experimental study and modeling of the U(VI)-Fe(OH)$_3$ surface precipitation/coprecipitation equilibria. Geochim Cosmochim Acta 59:4113-4123
Buck EC, Bates JK 1999) Microanalysis of colloids and suspended particles from nuclear waste glass alteration, Appl Geochem 14:635-653
Buck EC, Wronkiewicz DJ, Finn PA, Bates JK 1997 A new uranyl oxide hydrates phase derived from spent fuel alteration. J Nucl Mater 249:70-76
Buck EC, Brown NR, Dietz NL 1996 Contaminant uranium phases and leaching at the Fernald site in Ohio. Environ Sci Technol 30:81-88
Buck EC, Finch RJ, Bates JK 1998 Retention of neptunium in uranyl alteration phases formed during spent fuel corrosion. Mater Res Soc Symp Proc 506:83-91
Burns PC 1997 A new uranyl oxide hydrate sheet in vandendriesscheite: Implications for mineral paragenesis and the corrosion of spent nuclear fuel. Am Mineral 82:1176-1186
Burns PC 1998a The structure of richetite, a rare lead uranyl oxide hydrate. Can Mineral 36:187-199
Burns PC 1998b CCD area detectors of X-rays applied to the analysis of mineral structures. Can Mineral 36:847-853

Burns PC 1998c The structure of compreignacite, $K_2[(UO_2)_3O_2(OH)_3]_2(H_2O)_7$. Can Mineral 36:1061-1067

Burns PC 1998d The structure of boltwoodite and implications for the solid-solution with sodium boltwoodite. Can Mineral 36:1069-1075

Burns PC 1999a Cs boltwoodite obtained by ion exchange from single crystals: Implications for radionuclide release in a nuclear repository. J Nucl Mater 265:218-223

Burns PC 1999b A new uranyl silicate sheet in the structure of haiweeite and comparison to other uranyl silicates. Can Mineral (submitted).

Burns PC 1999c A new uranyl phosphate chain in the structure of parsonsite. Am Mineral (submitted).

Burns PC 1999d The crystal chemistry of uranium. In Rev Mineral 38 (this volume)

Burns PC 1999e Wölsendorfite: a masterpiece of structural complexity. GAC-MAC Abstracts 24:16

Burns PC 1999f A new uranyl silicate sheet in the structure of haiweeite and comparison to other uranyl silicates. Can Mineral (submitted)

Burns PC, Ewing RC, Miller ML 1997a Incorporation mechanisms of actinide elements into the structures of U^{6+} phases formed during the oxidation of spent nuclear fuel. J Nucl Mater 245:1-9

Burns PC, Finch RJ 1999) Wyartite, crystallographic evidence for the first pentavalent uranium mineral. Am Mineral 84 (in press)

Burns PC, Finch RJ, Hawthorne FC, Miller ML, Ewing RC 1997b The crystal structure of ianthinite, $[U_2^{4+}(UO_2)_4O_6(OH)_4(H_2O)_4](H_2O)_5$: A possible phase for Pu^{4+} incorporation during the oxidation of spent nuclear fuel. J Nucl Mater 249:199-206

Burns PC, Hanchar JM 1999) The structure of masuyite, $Pb[(UO_2)_3O_3(OH)_2](H_2O)_3$. Can Mineral (in press)

Burns PC, Hawthorne FC, Ewing RC 1997c The crystal chemistry of hexavalent uranium: Polyhedral geometries, bond-valence parameters, and polyhedral polymerization. Can Mineral 35:1551-1570

Burns PC, Hill FC 1999a Implications of the synthesis and structure of the Sr analogue of curite. Can Mineral 37 (in press)

Burns PC, Hill FC 1999b A new uranyl sheet in $K_5[(UO_2)_{10}O_8(OH)_9](H_2O)$. Implications for understanding sheet anion-topologies. Can Mineral (submitted)

Burns PC, Miller ML, Ewing RC 1996 U^{6+} minerals and inorganic phases: A comparison and hierarchy of crystal structures. Can Mineral 34:845-880

Burns PC, Olson RA, Finch RJ, Hanchar JM, Thibault Y. 1999) $KNa_3(UO_2)_2(Si_4O_{10})_2(H_2O)_4$, a new compound formed during vapor hydration of an actinide-bearing borosilicate waste glass. J Nucl Mater (submitted)

Cahill CL, Burns PC 1999) The structure of agrinierite: a K-Ca-Sr uranyl oxide hydrate sheet of the alpha-U_3O_8 type. Geol Soc America Annual Meeting, Abstracts

Calos NJ, Kennard CHL 1996 Crystal structure of copper bis(uranyl phosphate) octahydrate (metatorbernite), $Cu(UO_2PO_4)_2·8(H_2O)$. Z Zristallogr 211:701-702

Casas I, Bruno J, Cera E, Finch RJ, Ewing RC 1994 Kinetic and thermodynamic studies of uranium minerals. Assessment of the long-term evolution of spent nuclear fuel. SKB Technical Report 94-16, Stockholm, 73 p

Casas I, Bruno J, Cera E, Finch RJ, Ewing RC 1997 Characterization and dissolution kinetics of becquerelite from Shinkolobwe, Zaire. Geochim Cosmochim Acta 61:3879-3884

Casas I, de Pablo J, Giménez J, Torrero ME, Bruno J, Cera E, Finch RJ, Ewing RC 1998 The role of pe, pH, and carbonate on the solubility of UO_2 and uraninite under nominally reducing conditions. Geochim Cosmochim Acta 62:2223-2231

Cejka J 1994 Vandenbrandeite, $CuUO_2(OH)_4$: Thermal analysis and infrared spectrum. N Jb Mineral Mh 1994:112-120

Cejka J, Cejka J, Skála R, Sejkora J, Muck A 1998 New data on curite from Shinkolobwe, Zaire. N Jb Miner Mh 1998:385-402

Cejka J, Mrázek Z, Urbanec Z 1984 New data on sharpite, a calcium uranyl carbonate. N Jb Miner Mh, H3:109-117

Cejka J, Sejkora J, Deliens M 1996a New data on studtite, $UO_4·4H_2O$ from Shinkolobwe, Shaba, Zaire. N Jb Miner Mh 3:125-134

Cejka J, Sejkora J, Deliens M 1999 To the infrared spectrum of haynesite, a hydrated uranyl selenite, and its comparison with other uranyl selenites. Neues Jb Miner Mh H6:241-252

Cejka J, Sejkora J, Mrázek Z, Urbanec Z, Jarchovsky T 1996b Jáchymovite, $(UO_2)_8(SO_4)(OH)_{14}·13H_2O$, a new uranyl mineral from Jáchymov, the Krusné Hory Mts., Czech Republic, and its comparison with uranopilite. N Jb Mineral Abh 170:155-170

Cejka J, Sejkora J, Skála R, Cejka J, Novotná M, Ederová J 1998 Contribution to the crystal chemistry of synthetic becquerelite, billietite and protasite. N Jb Miner Abh 174:159-180

Cejka J, Urbanec Z 1988 Contribution to the hydrothermal origin of rutherfordine, UO_2CO_3. Casopis Národního Muzea 157:1-10

Cejka J, Urbanec Z 1990 Secondary uranium minerals. Trans Czechoslovak Academy of Sciences, Math

Natur History Series 100, Academia, Prague. 93 p

Cejka J, Urbanec Z, Cejka J Jr 1987 Contribution to the crystal chemistry of andersonite. N Jb Mineral Mh 11:488-501

Cesbron F 1970 La roubaultite, $Cu_2(UO_2)_3(CO_3)_2O_2(OH)_2(H_2O)_4$, une nouvelle espèce minérale. Bull Soc fr Minér Crist 93:550-554

Cesbron F, Ildefonse P, Sichere M-C 1993 New mineralogical data on uranophane and β-uranophane; synthesis of uranophane. Mineral Mag 57:301-308

Chakoumakos BC 1984 Systematics of the pyrochlore structure type, ideal $A_2B_2X_6Y$. J Solid State Chem 53:120-129

Chakoumakos BC 1986 Pyrochlore. In McGraw-Hill Yearbook of Science and Technology 1987. Parker SP (Ed) McGraw-Hill, New York, p 393-395

Chen C-H 1975 A method of estimation of standard free energies of formation of silicate minerals at 298.15 K. Am J Sci 275:801-817

Chen F, Ewing RC, Clark SB 1999 The Gibbs free energies and enthalpies of formation of U^{6+} phases: An empirical method of prediction. Am Mineral 84:650-664

Chen Z, Luo K, Tan F, Zhang Y, Gu X 1986 Tengchongite, a new mineral of hydrated calcuim uranyl molybdate. Kexue Tongbao 31:396-401

Chen Z, Huang Z, Gu X 1990 A new uranium mineral-Yingjiangite. Acta Mineral Sinica 10:102-105

Chernikov AA, Krutetskaya OV, Sidelnikova VD 1957 Ursilite—a new silicate of uranium. Atomnaya Energ, Voprosy Geol Urana (Suppl) 6:73-77 (English translation In The Geology of Uranium. Consultants Bureau, Inc., New York, 1958)

Chernorukov N, Korshunov E, and Voenova L 1985 New uranium compound. Uranium(IV) hydrogen-arsenate tetrahydrate. Radiokhimiya 5:676-679 (in Russian)

Chevalier R, Gasperin M 1970 Crystal structure of UVO_5. Bull Soc fr Minér Crist 93:18-22

Chiappero P-J, Sarp H 1993 New data and the second occurrence of znucalite. Archs Sci Genève 46:291-301 (in French, English abstract)

Christ CL 1965 Phase transformations and crystal chemistry of schoepite. Am Mineral 50:235-239

Christ CL, Clark JR 1960 Crystal chemical studies of some uranyl oxide hydrates. Am Mineral 45:1026-1061

Christ CL, Clark JR, Evans HT Jr 1955 Crystal structure of rutherfordine, UO_2CO_3. Science 121:472-473

Clark DL, Hobart DE, Neu MP 1995 Actinide cabonate complexes and their importance in actinide environmental chemistry. Chem Rev 95:25-48

Clark JR 1960 X-ray study of alteration in the uranium mineral wyartite. Am Mineral 45, 200-208

Contamin P, Backmann JJ, Marin JF 1972 Autodiffusion de l'oxygene dans le 'dioxide d'uranium surstochiometrique. J Nucl Mater 42:54-64

Cooper MA, Hawthorne FC 1995 The crystal structure of guilleminite, a hydrated Ba-U-Se sheet structure. Can Mineral 33:1103-1109

Dall'aglio M, Gragnani R, Locardi E 1974 Geochemical factors controlling the formation of the secondary minerals of uranium. In Formation of Uranium Ore Deposits. Int'l Atomic Energy Agency, Vienna, p 33-48

Debets PC 1963 X-ray diffraction data on hydrated uranium peroxide. J Inorg Nucl Chem 25:727-730

Debets PC, Loopstra BO 1963 On the uranates of ammonium—II. X-ray investigation of the compounds in the system NH_3-UO_3-H_2O. J Inorg Nucl Chem 25:945-953

Deliens M 1977a Review of the hydrated oxides of U and Pb, with new X-ray powder data. Mineral Mag 41:51-57

Deliens M 1977b Associations de minéraux secondaires d'uranium à Shinkolobwe (région du Shaba, Zaïre). Bull Soc fr Minér Crist 100:32-38

Deliens M, Piret P 1980 New aluminum and uranyl phosphates from Kobokobo, Kivu, Zaire. Rocks Minerals 55:169-171

Deliens M, Piret P 1981a La swamboite, nouveau silicate d'uranium hydraté du Shaba, Zaïre. Can Mineral 19:553-557

Deliens M, Piret P 1981b Les phosphates d'uranyl et d'aluminum de Kobokobo; V, la mundite, nouveau minéral. Bull Minéral 104:669-671

Deliens M, Piret P 1982a Bijvoetite et lepersonnite, carbonates hydratés d'uranyle et de terres rares de Shinkolobwe, Zaïre. Can Mineral 20:231-238

Deliens M, Piret P 1982b Les phosphates d'uranyl et d'aluminum de Kobokobo; VI, la triangulite, $Al_3(OH)_5[(UO_2)(PO_4)]_4(H_2O)_5$, nouveau minéral. Bull Minér 105:611-614

Deliens M, Piret P 1983a L'oursinite, $(Co_{0.86}Mg_{0.10}Ni_{0.04})O\cdot2UO_3\cdot2SiO_2\cdot6H_2O$, nouveau minéral de Shinkolobwe, Zaïre. Bull Minér 106:305-308

Deliens M, Piret P 1983b Metastudtite, $UO_4\cdot2H_2O$, a new mineral from Shinkolobwe, Shaba, Zaire. Am Mineral 68:456-458

Deliens M, Piret P 1984a Kamitugaite, $PbAl(UO_2)_5[(P,As)O_4]_2(OH)_9 \cdot 9.5H_2O$, a new mineral from Kobokobo, Kivu, Zaire. Bull Minéral 107:15-19 (in French)

Deliens M, Piret P 1984b L'urancalcarite, $Ca(UO_2)_3CO_3(OH)_6 \cdot 3H_2O$, nouveau minéral de Shinkolobwe, Zaïre. Bull Minéral 107:21-24

Deliens M, Piret P 1985a Les phosphates d'uranyle et d'aluminium de Kobokobo VII. La moreauite, $Al_3UO_2(PO_4)_3(OH)_2 \cdot 13H_2O$, nouveau minéral. Bull Minéral 108:9-13

Deliens M, Piret P 1985b Les phosphates d'uranyl et d'aluminium de Kobokobo. VIII. La furongite. Ann Soc Geol Gelg 108:365-368

Deliens M, Piret P 1990a La shabaite-(Nd), $CaREE_2(UO_2)(CO_3)_4(OH)_6 \cdot 6H_2O$, nouveau espèce minérale de Kamoto, Shaba, Zaïre. Eur J Mineral 1:85-88

Deliens M, Piret P 1990b L'astrocyanite-(Ce), $Cu_2(TR)_2(UO_2)(CO_3)_5(OH)_2 \cdot 1.5H_2O$, nouvelle espèce minérale de Kamoto, Shaba, Zaïre. Eur J Mineral 2:407-411

Deliens M, Piret P 1991 La haynesite, sélénite hydraté d'uranyl, nouvelle espèce minérale de la mine Repete, comté de San Juan, Utah. Can Mineral 29:561-564

Deliens M, Piret P 1992 La fontanite, carbonate hydraté d'uranyle et de calcium, nouvelle espèce minérale de Rabejac, Hérault, France. Eur J Mineral 4:1271-1274.

Deliens M, Piret P 1993 Rabejacite, $Ca(UO_2)_4(SO_4)_2(OH)_6 \cdot 6H_2O$, a new uranyl and calcium sulfate from Lodève, Hérault, France. Eur J Mineral 5:873-877

Deliens M, Piret P 1996 Les masuyites de Shinkolobwe (Shaba, Zaïre) constituent un group formé de deux variétés distinct par leur composition chimique et leuras propriétés radiocristallographiques. Bull Mus Royal Central Africa, Tervuren, Belgium, Sci Terres 66:187-192

Deliens M, Piret P, Comblain G 1981 Les minéraux secondaires du Zaïre. Editions du Musée royal de l'Afrique centrale, Tervuren, Belgium, 113 p

Deliens M, Piret P, Comblain G 1984 Les minéraux secondaires du Zaïre. Complement. Institute Editions du Musée royal de l' Afrique centrale, Tervuren, Belgium, 37 p

Deliens M, Piret P, van der Meersche E 1990 Les minéraux secondaires du Zaïre. Deuxième complement. Editions du Musée royal de l' Afrique centrale, Tervuren, Belgium, 39 p

Demartin F, Diella V, Donzelli S, Gramaccioli CM, Pilati T 1991 The importance of accurate crystal structure determination of uranium minerals. I. Phosphuranylite $KCa(H_3O)_3(UO_2)_7(PO_4)_4O_4 \cdot 8H_2O$. Acta Crystallogr B47:439-446

Demartin F, Diella V, Donzelli S, Gramaccioli CM, Pilati T 1992 The importance of accurate crystal structure determination of uranium minerals. II. Soddyite $(UO_2)_2(SiO_4) \cdot 2H_2O$. Acta Crystallogr C48:1-4

Dent AJ, Ramsay JDF, Swanton SW 1992 An EXAFS study of uranyl ion in solution and sorbed onto silica and montmorillonite clay colloids. J Colloid Interface Sci 150:45-60

Dent Glasser LS, Lachowski EE 1980 Silicate species in solution. Part 1. Experimental Observations. J Chem Soc (London) Dalton Trans 1980:393-398

Diaz Arocas P, Grambow B 1998 Solid-liquid phase equilibria of U(VI) in NaCl solutions. Geochim Cosmochim Acta 62:245-263

Diella V, Mannucci G 1986 A uranium-rich ekanite, $(Th_{0.75},U_{0.21})(Ca_{2.01},Fe_{0.04}Mn_{0.01})Si_{7.99}O_{20}$, from Pitigliano, Italy. Rend Soc Ital Mineral Petrol 41:3-6

Dickens PG, Stuttard GP, Ball RGJ, Powell AV, Hull S, Patat S 1992 Powder neutron-diffraction study of the mixed uranium vanadium oxides, $Cs_2(UO_2)_2(V_2O_8$ and UVO_5. J Mater Chem 2(2):161-166

Douglass RM 1956 Tetrasodium uranyl tricarbonate, $Na_4(UO_2)(CO_3)_3$. Anal Chem 28:1635

Dubessy J, Pagel M, Beny JM, Christensen, H, Hickel B, Kosztolanyi C, Poty B 1988 Radiolysis evidenced by H_2-O_2 and H_2-bearing fluid inclusions in three uranium deposits. Geochim Cosmochim Acta 52:1155-1167

D'yachenko OGD, Taachenko VV, Tali R, Kovba LM, Marinder B-O, Sudberg M 1996 Structure of $UMoO_5$ studied by single-crystal X-ray diffraction and high-resolution transmission miocroscopy. Acta Crystallogr B52:961-965

Einziger RE, Thomas LE, Buchanan HC, Stout RB 1992 Oxidation of spent fuel in air at 175 to 195°C. J Nucl Mater 190:53-60

Elton NJ, Hooper, JJ 1992 Andersonite and schröckingerite from Geevor mine, Cornwall: two species new to Britain. Mineral Mag 56:124-125

Elton NJ, Hooper JJ 1995 Supergene U-Pb-Cu mineralization at Loe Warren, St Just, Cornwall, England. J Russel Soc 6:17-26

Evans HT Jr 1963 Uranyl ion coordination. Science 141:154-158.

Ewing RC 1975 Alteration of metamict, rare earth, AB_2O_6-type Nb-Ta-Ti oxides. Geochim Cosmochim Acta 39:521-530

Ewing RC 1993 The long-term performance of nuclear waste forms: natural materials—three case studies. Mater Res Soc Symp Proc 294:559-568

Fareeduddin JAS 1990 Significance of the occurrence of detrital pyrite and uraninite on the Kalasapura conglomerate, Karnataka. Indian Minerals 44:335-340

Fayek M, Janeczek J, Ewing RC 1997 Mineral chemistry and oxygen isotopic analyses of uraninite, pitchblende and uranium alteration minerals from the Cigar Lake deposit, Saskatchewan, Canada. Appl Geochem 12:549-565

Fayek M, Kyser TK 1997 Characterization of multiple fluid-flow events and rare-earth-element mobility associated with formation of unconformity-type uranium deposits in the Athabasca basin, Saskatchewan. Can Mineral 35:627-658

Finch RJ 1994 Paragenesis and crystal chemistry of the uranyl oxide hydrates. Ph.D. Thesis, University of New Mexico, 257 p

Finch RJ 1997a Thermodynamic stabilities of U(VI) minerals: Estimated and observed relationships. Mater Res Soc Symp Proc 465:1185-1192

Finch RJ 1997b The role of H_2O in minerals containing the uranyl ion. Eos 78:S328

Finch RJ, Cooper MA, Hawthorne FC, Ewing RC 1996 The crystal structure of schoepite, $[(UO_2)_8O_2(OH)_{12}](H_2O)_{12}$. Can Mineral 34:1071-1088

Finch RJ, Cooper MA, Hawthorne FC, Ewing RC 1999a Refinement of the crystal structure of rutherfordine. Can Mineral 37 (in press)

Finch RJ, Ewing RC 1991 Uraninite alteration in an oxidizing environment and its relevance to the disposal of spent nuclear fuel. SKB Technical Report 91-15, SKB, Stockholm, 107 p

Finch RJ, Ewing RC, 1992a Alteration of uranyl oxide hydrates in Si-rich groundwaters: Implications for uranium solubility. Mater Res Soc Symp Proc 257:465-472

Finch RJ, Ewing RC 1992b The corrosion of uraninite under oxidizing conditions. J Nucl Mater 190:133-156

Finch RJ, Ewing RC 1994 Formation, oxidation and alteration of ianthinite. Mater Res Soc Symp Proc 333:625-630

Finch RJ, Ewing RC 1995 Estimating the Gibbs free energies of some uranium(VI) minerals: calculated and observed mineral relationships. Joint Ann Meeting Geol Assoc Canada, Mineral Assoc Canada, Victoria BC, Canada, May, 1995, Abstracts Vol 20

Finch RJ, Ewing RC 1997 Clarkeite: New chemical and structural data. Am Mineral 82:607-619

Finch RJ, Finn PA, Buck EC, Bates JK 1999b Oxidative corrosion of spent UO_2 fuel in vapor and dripping groundwater at 90°C. Mater Res Soc Symp Proc 556 (in press)

Finch RJ, Hanchar JM, Hoskin PWO, Burns PC 1999c Rare-earth elements in zircon. 2. Single crystal X-ray study of xenotime substitution. Am Mineral (submitted)

Finch RJ, Hawthorne FC, Ewing RC 1998 Crystallographic relations among schoepite, metaschoepite and dehydrated schoepite. Can Mineral 36:831-845

Finch RJ, Hawthorne FC, Miller ML, Ewing RC 1997 Distinguishing among schoepite, $[(UO_2)_8O_2(OH)_{12}](H_2O)_{12}$, and related minerals by X-ray powder diffraction. Powder Diff 12:230-238

Finch RJ, Miller ML, Ewing RC 1992 Weathering of natural uranyl oxide hydrates: schoepite polytypes and dehydration effects. Radiochim Acta 58/59:433-443

Finch RJ, Suksi J, Rasilainen K, Ewing RC 1995 The long-term stability of becquerelite. Mater Res Soc Symp Proc 353:647-652

Finch RJ, Suksi J, Rasilainen K, Ewing RC 1996 Uranium series ages of secondary uranium minerals with applications to the long-term storage of spent nuclear fuel. Mater Res Soc Symp Proc 412:823-830

Finch WI 1996 Uranium Provinces of North Amrerica—Their definition, Distribution, and Models. U S Geol Surv Bull 2141, 18 p

Finn PA, Finch RJ, Buck EC, Bates JK 1998 Corrosion mechanisms of spent fuel under oxidizing conditions. Mater Res Soc Symp Proc 506:123-131

Finn PA, Hoh JC, Wolf SF, Surchik MT, Buck EC, Bates JK 1997 Spent fuel reaction: The behavior of the ε-phase over 3.1 years. Mater Res Soc Symp Proc 465:527-534

Foord EE, Korzeb SL, Lichte FE, Fitpatrick JJ 1997 Additional studies on mixed uranyl oxide-hydroxide hydrate alteration products of uraninite from Palermo and Ruggles granitic pegmatities, Grafton County, New Hampshire. Can Mineral 35:145-151

Förster H-J 1998a The chemical composition of REE-Y-Th-U-rich accessory minerals in peraluminous granites of the Erzgebirge-Fichtelgebirge region, Germany. Part I: The monazite-(Ce)-brabantite solid solution series. Am Mineral 83:259-272

Förster H-J 1998b The chemical composition of REE-Y-Th-U-rich accessory minerals in peraluminous granites of the Erzgebirge-Fichtelgebirge region, Germany. Part II: Xenotime. Am Mineral 83:1302-1315

Förster, H-J 1999 The chemical composition of uraninite in Variscan granites of the Erzgebirge, Germany. Mineral Mag 63:239-252

Forsyth RS, Werme LO 1992 Spent fuel corrosion and dissolution. J Nucl Mater 190:3-19

Frondel C 1956 The mineralogical composition of gummite. Am Mineral 41:539-568

Frondel C 1958 Systematic Mineralogy of Uranium and Thorium. US Geol Surv Bull 1064, 400 p
Frondel C, Ito J, Honca RM, Weeks AM 1976 Mineralogy of the zippeite group. Can Mineral 14:429-436
Frondel C, Meyrowitz R 1956 Studies of uranium minerals (XIX) rutherfordine, diderichite and clarkeite. Am Mineral 41:127-133
Gabriel U, Gaudet J-P, Spadini L, Charlet L 1998 Reactive transport of uranyl in a goethite column: an experimental and modelling study. Chem Geol 151:107-128
Gaines RV, Skinner HCW, Foord EE, Mason B, Rosenzweig A, King VT, Dowty E 1997 Dana's New Mineralogy. 8th Edition. Wiley & Sons, New York, 1819 p
Gaines RV 1965 Moctezumite, a new lead uranyl tellurate. Am Mineral 50:1158-1163
Garrels RM, Christ CL 1959 Behavior of uranium minerals during oxidation. *In* Geochemistry and Mineralogy of the Colorado Plateau Uranium Ores. Garrels RM, Larsen ES (Eds) US Geol Surv Prof Paper 320:81-89
Gasperin M 1960 Contribution à l'étude de quelques oxydes doubles que forme le tantale avec l'uranium et la calcium. Bull Soc fr Minéral Cristallogr 83:1-21
Gauthier JP, Fumey P 1988 Un gemme metamict; l'ekanite. Rev Gemmol Assoc France Gemmol 94:3-7
Giaquinta DM, Soderholm L, Yuchs SE, Waserman SR 1997 The speciation of uranium in a smectite clay: evidence for catalysed uranyl reduction. Radiochim Acta 76:113-121
Giblin AM 1982 The role of clay adsorption in genesis of uranium ores. *In* Uranium in the Pine Creek Geosyncline. Ferguson J, Goleby AB (Eds) Int'l Atomic Energy Agency, Vienna, p. 521-529
Giblin AM, Appleyard EC 1987 Uranium mobility in non-oxidizing brines: field and experimental evidence. Appl Geochem 2:285-295
Gieré R, Williams CT, Lumpkin GR 1998 Chemical characteristic of zirconolite. Schweiz mineral petrogr Mitt 78:443-459
Ginderow D 1988 Structure de l'uranophane alpha, $Ca(UO_2)_2(SiO_3OH)_2 \cdot 5H_2O$. Acta Crystallogr C44:421-424
Ginderow D, Cesbron F 1983a Structure de demesmaekerite. Acta Crystallogr C39:824-827
Ginderow D, Cesbron F 1983b Structure de derriksite. Acta Crystallogr C39:1605-1607
Ginderow D, Cesbron F 1985 Structure de la roubaultite, $Cu_2(UO_2)_3(CO_3)_2O_2(OH)_2 \cdot 4H_2O$. Acta Crystallogr C41:654-657
Glinka YD, Jaroniec M, Rozenbaum VM 1997 Adsorption energy evaluation from luminescence spectra of uranyl ions UO_2^{2+} adsorbed on disperse silica surfaces. J Colloid Interface Sci 194:455-469.
Gong W 1991a A study on high-temperature phase transition of metamict fergusonite group minerals. Acta Mineral Sinica 11:1-8 (in Chinese with English abstract)
Gong W 1991b Chemistry and evolution of fergusonite group minerals, Bayan, Obo, Inner Mongolia. Acta Mineral Sinica 11:200-208 (English translation in: Chinese J Geochem 10:266-276
Graham J, Thornber MR 1974 The crystal chemistry of complex niobium and tantalum oxides. I. Structural classification of MO_2 phases. Am Mineral 59:1026-1039
Grambow B 1989 Spent fuel dissolution and oxidation: an evaluation of literature data. SKB Technical Report 89-13, SKB, Stockholm, 42 p
Grandstaff DE 1976 A kinetic study of the dissolution of uraninite. Econ Geol 71:1493
Gray WJ, Thomas LE, Einziger RE 1993 Effects of air oxidation on the dissolution rate of LWR spent fuel. Mater Res Soc Symp Proc 294:47-54
Grenthe I, Fuger J, Konings RJM, Lemire RJ, Muller AB, Nguyen Trung C, Wanner H 1992 Chemical Thermodynamics. North-Holland, Amsterdam, 656 p
Guesdon A, Raveau B 1998 A new uranium(VI) monophosphate with a layered structure: $Ba_3[UO_2(PO_4)(PO_3(OH))]_2 \cdot xH_2O$ (x 0.4). Chem Mater 10:3471-3474
Guillemin C, Protas J 1959 Ianthinite et wyartite. Bull Soc fr Minér Crist 82:80-86
Guthrie VA 1989 Fission-track analysis of uranium distribution in granitic rocks. Chem Geol 77:87–103
Hanchar JM, Finch RJ, Watson EB, Hoskin PWO, Cherniak D, Mariano A 1999 Rare-earth elements in zircon. 1. Synthesis and rare-earth element and phosphorous doping. Am Mineral (submitted)
Hansley PL, Fitzpatrick JJ 1989 Compositional and crystallographic data on REE-bearing coffinite from the Grants Uranium region, northwestern New Mexico. Am Mineral 74:263-270
Hanson SL, Simmons WB, Falster AU, Foord EE, Lichte FE 1999 Proposed nomenclature for samarskite-group minerals: New data on ishikawaite and calciosamarskite. Mineral Mag 63:27-36
Hart KP, Lumpkin GR, Gieré R, Williams CT, McGlinn PJ, Payne TE 1996 Naturally-occurring zirconolites analogues for the long-term encapsulation of actinides in Synroc. Radiochim Acta 74:309-312.
Hawthorne FC 1992 The role of OH and H_2O in oxide and oxysalt minerals. Z Kristallogr 201:183-206
Hemingway BS 1982 Thermodynamic properties of selected uranium compounds and aqueous species at 298.15 K and 1 bar and at higher temperatures—Preliminary models for the origin of coffinite deposits. U S Geol Surv Open-file Report 82-619, 89 p

Hikichi Y, Murayama K, Ohsato H, Nomura T 1989 Thermal changes of rare earth phosphate minerals. J Mineral Soc Japan 19:117-126 (Japanese with English abstract)

Hodgeson NA, Le Bas MJ 1992 The geochemistry and cryptic zonation of pyrochlore from San Vincente, Cape Verde Islands. Mineral Mag 56:201-214

Hoekstra HR, Siegel S 1973 The uranium trioxide-water system. J Inorg. Nucl. Chem. 35:761-779

Holland HD 1978 The Chemistry and the Atmosphere and Oceans. Wiley-Interscience, NY, 351 p

Holland HD 1984 The Chemical Evolution of the Atmosphere and Oceans. Princeton Univ Press, Princeton, New Jersey, 582 p

Holland HD 1994 Early Proterozoic atmospheric change. In Begtson S (Ed) Early Life on Earth: Nobel Symposium 84, New York. Columbia Univ Press, New York, p 237-380

Hostetler PB, Garrels RM 1962 Transportation and precipitation of uranium and vanadium at low temperatures, with special reference to sandstone-type uranium deposits. Econ Geol 57:137 p

Hunan 230 Laboratory 1976 Furongite—a new uranium mineral found in China. Acta Geologica Sinica 22:203-204 (in Chinese, abstracted in Am Mineral 63:198, 1978)

Isobe H, Murakami T, Ewing RC 1992 Alteration of uranium minerals in the Koongarra deposit, Australia: Unweathered zone. J Nucl Mater 190:174-187

Isobe H, Ewing RC, Murakami T 1994 Formation of secondary uranium minerals in the Koongarra deposit, Australia. Mater Res Soc Symp Proc 333:653-660

Jackson J, Burns PC 1999 The structure of weeksite: a potassium uranyl silicate hydrate. Geol Soc America Annual Meeting. Abstracts

Janeczek J 1991 Composition and origin of coffinite from Jachymov, Czechoslovakia. N Jb Miner Mh 9:385-395

Janeczek J 1999 Natural fission reactors. In Rev Mineral 38 (this volume)

Janeczek J, Ewing RC 1991 X-ray powder diffraction study of annealed uraninite. J Nucl Mater 185:66-77

Janeczek J, Ewing RC 1992a Coffinitization—a mechanism for the alteration of UO_2 under reducing conditions. Mater Res Soc Symp Proc 257:497-504

Janeczek J, Ewing RC 1992b Structural formula of uraninite. J Nucl Mater 190:128-132

Janeczek J, Ewing RC 1992c Dissolution and alteration of uraninite under reducing conditions. J Nucl Mater 190:157-173

Janeczek J, Ewing RC 1995 Mechanisms of lead release from uraninite in the natural fission reactors in Gabon. Geochim Cosmochim Acta 59:1917-1931

Janeczek J, Ewing RC 1996 Phosphatian coffinite with rare earth elements and Ce-rich françoisite-(Nd) from sandstone beneath a natural fission reactor at Bangombé, Gabon. Mineral Mag 60:665-669

Janeczek J, Ewing RC, Oversby VM, Werme LO 1996 Uraninite and UO_2 in spent nuclear fuel: a comparison. J Nucl Mater 238:121-130

Janeczek J, Ewing RC, Thomas LE 1993 Oxidation of uraninite: Does tetragonal U_3O_7 occur in nature? J Nucl Mater 207:177-191

Jamtveit B, Dahlgren S, Austrheim H 1997 High-grade metamorphism of calcareous rocks from the Oslo Rift, southern Norway. Am Mineral 82:1241-1254

Jensen KA, Ewing RC, Gauthier-Lafaye F 1997 Uraninite: A 2 Ga spent nuclear fuel from the natural fission reactor at Bangombé in Gabon, West Africa. Mater Res Soc Symp Proc 465:1209-1218

Johnsen O, Stahl K, Petersen OV, Micheelsen HI 1999 Structure refinement of natural non-metamict polycrase-(Y) from Zomba-Malosa complex, Malawi. N Jb Mineral Mh 1999:1-10

Johnson LH, Shoesmith, DW 1988 Spent fuel. In Radioactive Wasteforms for the Future. Lutze W, Ewing RC (Eds) Elsevier, Amsterdam, p 635-698

Johnson LH, Werme LO 1994 Materials characteristic and dissolution behavior of spent nuclear fuel. Mater Res Soc Bull XIX(12):24-27

Kaufman A, Ku T-L 1989 The U-series ages of carnotites and implications regarding their formation. Geochim Cosmochim Acta 53:2675-2681

Keller L, Wagner CNJ 1983 Diffraction analysis of metamict samarskite. Am Mineral 68:459-465

Keppler H, Wyllie PJ 1990 Role of fluids in transport and fractionation of uranium and thorium in magmatic provinces. Nature 348:531-533

Kolitschn U 1997 Uranosphärite und weitere neue Mineralien von der Grube Clara im Schwarzwald. Mineralien Welt 8:18-26

Korzeb SL, Foord EE, Lichte FE 1997 The chemical evolution and paragenesis of uranium minerals from Ruggles and Palermo granitic pegmatites, New Hampshire. Can Mineral 35:135-144

Kotzer TG, Kyser TK 1993 O, U, and Pb isotopic and chemical variations in uraninite: Implications for determining the temporal and fluid history of ancient terranes. Am Mineral 78:1262-1274

Krause W, Effenberger H, Brandstätter F 1995 Orthowalpurgite, $(UO_2)Bi_4O_4(AsO_4)_2 \cdot 2H_2O$, a new mineral from the Black Forest, Germany. Eur J Mineral 7:1313-1324

Lambor 1994 New mineral names. Am Mineral 79:1012

Langmuir D 1978 Uranium solution-minerals equilibria at low temperatures with applications to sedimentary ore deposits. Geochim Cosmochim Acta 42:547-569

Lazebnik KA et al. 1985 Typomorphism and Geochemical Features of Endogenic Minerals of Yakutia. Yakutsk 132-142

Lazebnik KA, Kayakina NV, Mathotko VF 1994 A new thorium silicate from carbonates of the Syrenevyi Kamen' deposit. Doklady Akad Nauk 334:735-738

Li Y, Burns PC 1999a A single-crystal diffraction study of the crystal chemistry of curite. GAC-MAC Abstracts 24:71

Li Y, Burns PC 1999b A new REE-bearing uranyl carbonate sheet in the structure of bijvoetite. Geol Soc Am Annual Mtg. Abstracts (in press)

Livingstone A, Atkin D, Hutchison D, Al-Hermezi HM 1976 Iraqite, a new rare earth mineral of the ekanite group. Mineral Mag 40:441-445

Loopstra BO, Rietveld HM 1969 The structures of some alkaline-earth metal uranates. Acta Crystallogr B25:787-791

Lumpkin GR, Chakoumakos BC 1988 Chemistry and radiation effects of thorite-group minerals from the Harding pegmatite, Taos County, New Mexico. Am Mineral 73:1405-1419

Lumpkin GR, Ewing RC 1988 Alpha-decay damage in minerals of the pyrochlore group. Phys Chem Mineral 16:2-20

Lumpkin GR, Ewing RC 1992a Geochemical alteration of pyrochlore group minerals: Microlite subgroup. Am Mineral 77:179-188

Lumpkin GR, Ewing RC 1992b Chemistry and radiation effects of thorite-group minerals of the Harding pegmatite, Taos County, New Mexico. Am Mineral 73:1405-1419.

Lumpkin GR, Ewing RC 1995 Geochemical alteration of pyrochlore group minerals: Pyrochlore subgroup. Am Mineral 80:732-743

Lumpkin GR, Ewing RC 1996 Geochemical alteration of pyrochlore group minerals: Betafite subgroup. Am Mineral 81:1237-1248

Lumpkin GR, Hart KP, McGlinn PJ, Payne TE, Gieré R, Williams CT 1994 Retention of actinides in natural pyrochlores and zirconolites. Radiochim Acta 66/67:469-474

Magalhaes MCF, Pedrosa de Jesus JD, Williams PA 1985 The chemistry of uranium dispersion in groundwaters at the Pinhal do Souto mine, Portugal. Inorg Chim Acta 109:71-78

Mandarino J 1999 Fleischer's Glossary of Mineral Species. Mineralogical Record, Tucson, Arizona

Mann AW, Deutscher RL 1980 Solution chemistry of lead and zinc in water containing carbonate, sulfate and chloride ions. Chem Geol 29:293-311

Markovic M, Pavkovic N 1983 Solubility and equilibrium constants of uranyl(2+) in phosphate solutions. Inorg Chem 22:978–982

Marshall RH, Hoekstra HR 1965 Preparation and properties of $TiUO_5$. J Inorg Nucl Chem 27:1947-1950

Martin M, Massanek A 1995 Asselbornite aus Schneeberg/Sachsen. Lapis 20:34-35

Matzke Hj 1992 Radiation damage-enhanced dissolution of UO_2 in water. J Nucl Mater 190:101-106

Mazzi F, Munno R 1983 Calciobetafite (new mineral of the pyrochlore group) and related minerals from Campi Flegrei, Italy; crystal structures of polymignyte and zirkelite: cmparison with pyrochlore and zirconolite. Am Mineral 68:262-276

McCready AJ, Pernell J 1997 Uraniferous bitumens from the Orcadian basin, Scotland: analogues for Witwatersrand uranium mineralization. In Mineral Deposits. Papunen H (Ed). Blakema, Rotterdam. p 83-86

McCready AJ, Pernell J 1998 A Phanerozoic analogue for Witwatersrand-type uranium mineralization: uranium-titanium-bitumen nodules in Devonian conglomerate/sandstone, Orkney, Scotland. Trans Inst Min Metall B, Appl Earth Sci 107:B89-B97

McEachern RJ, Dorn DC, Wood DD 1998 The effect of rare-earth fission products on the rate of U_3O_8 formation on UO_2. J Nucl Mater 252:145-149

McEachern RJ, Taylor P 1998 A review of the oxidation of uranium dioxide at temperatures below 400°C. J Nucl Mater 254:87-121

McKinley JP, Zachara JM, Smith SC, Turner GD 1995 The influence of uranyl hydrolysis and multiple site-binding reactions on absorption of U(VI) to montmorillonite. Clays Clay Mineral 43:586-598

McMillan RH 1978 Genetic aspects and classification of important Canadian uranium deposits. In Uranium Deposits: Their Mineralogy and Origin. Kimberly MM (Ed) Mineral Soc Canada Short Course Handbook 3:187-204

Meere PA, Banks DA 1997 The effects of water radiolysis on local redox conditions in the Oklo, Gabon, natural fission reactors 10 and 16. Geochim Cosmochim Acta 61:4479-4494

Mereiter K 1979 The crystal structure of curite, $[Pb_{6.56}(H_2O,OH)_4][(UO_2)_8O_6(OH)_8]_2$. Tschermaks mineral petrogr Mitt 26:279-292

Mereiter K 1982 The crystal structure of walpurgite. Tschermaks mineral petro Mitt 30:129-139

Mereiter K 1984 The crystal structure of albrechtschaufite, $MgCa_4F_2(UO_2)(CO_3)_3 \cdot 17(H_2O)$. Acta Crystallogr A40, Supplement C-247 (abstract)

Mereiter K 1986 Crystal structure and crystallographic properties of schröckingerite from Joachimsthal. Tschermaks mineral petrogr Mitt 35:1-18

Miller LJ 1958 The chemical environment of pitchblende. Econ Geol 53:521-544

Miller ML, Finch RJ, Burns PC, Ewing RC 1996 Description and classification of uranium oxide hydrate sheet topologies. J Mater Res 11:3048-3056

Mishra R, Namboodiri PN, Tripathi SN, Bharadwaj SR, Dharwadkar SR 1998 Vaporization behaviour and Gibbs' energy of formation of $UTeO_5$ and UTe_3O_9 by transpiration. J Nucl Mater 256:139-144

Miyake C, Sugiyama D, Mizuno M 1994 Oxidation states of U and Ti in U-Ti-O ternary mixed oxides. J Alloys Compounds 213/214:516-519

Moll H, Matz W, Schuster G, Brendler E, Bernhard G, Nitsche H 1995 Synthesis and characterization of uranyl orthosilicate, $(UO_2)_2SiO_4 \cdot 2H_2O$. J Nucl Mater 227:40-49

Molon J 1990 Through the 'scope: minerals of the Clara mine, Wolfach, Germany—an update. Rocks and Minerals 65:448-454

Moroni LP, Glasser FP 1995 Reactions between cement components and U(VI) oxide. Waste Management 15:243:254

Morris DE, Allen PG, Berg JM, Chisolm-Brause CJ, Conradson SD, Hess NJ, Musgrave JA, Tait CD 1996 Speciation of uranium in Fernald soils by molecular spectroscopic methods: Characterization of untreated soils. Environ Sci Technol 30:2322-2331

Mücke A 1988 Lehnerite, $Mn[UO_2/PO_4]_2 \cdot 8H_2O$, a new mineral from the Hagendorf pegmatite, Oberpfalz. Aufschluss 39:209-217 (in German; abstract in Am Mineral 75:221

Mücke A, Strunz H 1978 Petscheckite and liandratite, two new pegmatite minerals from Madagascar. Am Mineral 63:941

Murakami T, Isobe H, Ohnuki T, Sato T, Yanase N, Kiyoshige J 1996a Mechanism of saléeite formation at the Koongarra secondary uranium deposit. Mater Res Soc Symp Proc 412:809-816

Murakami T, Isobe H, Sato T, Ohnuki T 1996b Weathering of chlorite in a quartz-chlorite schist: I. Mineralogical and chemical changes. Clays Clay Minerals 44:244-256

Murakami T, Ohnuki T, Isobe H, Sato T 1997 Mobility of uranium during weathering. Am Mineral 82:888-899

Murphy WM 1995 Natural analogs for Yucca Mountain. Radwaste Mag 2:44-50

Murphy WM, Shock EL 1999 Environmental aqueous geochemistry of actinides. Rev Mineral 38 (this volume)

Naito K 1974 Phase transitions of U_4O_9. J Nucl Mater 51-126

Nguyen SN, Silva RJ, Weed HC, Andrews JE Jr 1992 Standard Gibbs free energies of formation at the temperature 303.15 K of four uranyl silicates: soddyite, uranophane, sodium boltwoodite and sodium weeksite. J Chem Therm 24:359-376

Nickel EH, Nichols MC 1992 Mineral Reference Manual. Van Norstrand-Reinhold, New York, 250 p

Noe-Spirlet MR, Sobry R 1974 Les uranates hydratés de forment pas une série continue. Bull Soc Royale Sci Liège 43:164-171

Oliver NHS, Pearson PJ, Holcombe RJ, Ord A 1999 Mary Kathleen metamorphic-hydrothermal uranium—rare-earth element deposit: ore genesis and numerical model of coupled deformation and fluid flow. Australian J Earth Sci 46:467-483

O'Hare PAG, Lewis BM, Nguyen SN 1988 Thermochemistry of uranium compounds XVII. Standard molar enthalpy of formation at 298.15 K of dehydrated schoepite $UO_3 \cdot 0.9H_2O$. Thermodynamics of (schoepite + dehydrated schoepite + water). J Chem Therm 20:1287-1296

Ondrus P, Veselovsky, F, Hlousek J 1997a A review of mineral associations and paragenetic groups of secondary minerals of the Jáchymov (Joachimsthal) ore district. J Czech Geol Soc 42:109-114

Ondrus Minerals of the Jáchymov (Joachimsthal) ore district. J Czech Geol Soc 42:3-76

Ondrus O, Veslovsky F, Rybka R 1990 Znucalite, $Zn_{12}(UO_2)Ca(CO_3)_3(OH)_{22} \cdot 4H_2O$, a new mineral from Pribram, Czecholslovakia. N Jb Mineral Mh 1990:393-400

Ondrus P, Veselovsky F, Skála R, Císarová I, Hlousek J, Fryda J, Vavrín I, Cejka J, Gabasová A 1997c New naturally occurring phases of secondary origin from Jáchymov (Joachimsthal). J Czech Geol Soc 42:77-108

Pagel M, Pinte G, Rotach-Toulhoat N 1987 The rare earth elements in natural uranium oxides. In Monograph Series on Mineral Deposits 27. Gebruder Bortraeger, Berlin-Stuttgart. p 81-85.

Pagoaga MK, Appleman DE, Stewart JM 1986 A new barium uranyl oxide hydrate mineral, protasite. Mineral Mag 50:125-128

Pagoaga MK, Appleman DE, Stewart JM 1987 Crystal structure and crystal chemistry of the uranyl oxide hydrates becquerelite, billietite, and protasite. Am Mineral 72:1230-1238

Palenzona A, Selmi P 1998 Uranophane-beta ed altri minerali del ghiacciaio della Brenva. Rivista

Mineralogica Italiana 22:58-60
Paulis P 1992 Curiénite from Abertamy near Jáchymov. Casopis Mineral Geol 37:55-56 (in Czech)
Parks, GA Pohl DC 1988 Hydrothermal solubility of uraninite. Geochim Cosmochim Acta 52:863-875
Pearcy EC, Prikryl JD, Murphy WM, Leslie BW 1994 Alteration of uraninite from the Nopal I deposi Peña Blanca District, Chihuahua, Mexico, compared to degradation of spent nuclear fuel in th proposed US high-level nuclear waste repository at Yucca Mountain, Nevada. Appl Geochem 9:71: 732
Pialoux A, Touzelin B 1998 Étude du système U-Ca-O par diffractométrie de rayons X à haute températur J Nucl Mater 255:14-25
Piret P 1984 Structure cristalline de la fourmariérite, $Pb(UO_2)_4O_3(OH)_4 \cdot 4(H_2O)$. Bull Minéral 108:659-665
Piret P, Deliens M 1979 New crystal data for $Ca-Cu-UO_2$-hydrate carbonate: voglite. J Appl Crystallo; 12:616
Piret P, Deliens M 1982 Vanmeersscheite and metavanmeersscheite. Bull Minéral 105:125-128
Piret P, Deliens M. 1984 New data for richetite $PbO \cdot 4UO_3 \cdot 4H_2O$. Bull Minéral 107:581-585
Piret P, Deliens M 1986 La kamotoïte-(Y), un nouveau carbonate d'uranyl et de terres rares de Kamot Shaba, Zaïre. Bull Minér 109:643-647
Piret P, Deliens M 1987 Les phosphates d'uranyl et d'aluminium de Kobokobo. L'althupit $AlTh(UO_2)[UO_2)_3O_2(OH)(PO_4)_2]_2(OH)_3 \cdot 15H_2O$, nouveau minéral. Propriétés et structure cristalline. Bt Minéral 110:65-72
Piret P, Deliens M 1990 L'astrocyanite, $Cu_2(REE)_2(UO_2)(CO_3)_5(OH)_2(H_2O)_{1.5}$, nouvelle espèce minérale Kamoto, Shaba, Zaïre. Eur J Miner 2:407-411
Piret P, Deliens M, Piret-Meunier J 1988 La françoisite-(Nd), nouveau phosphate d'uranyl et terres rare propriétés et structure cristalline. Bull Minéral 111:443-449
Piret P, Deliens M, Piret-Meunier J, Germain G 1983 La sayrite, $Pb_2[(UO_2)_5O_6(OH)_2] \cdot 4H_2O$, nouvea minéral; propriétés et structure cristalline. Bull Minéral 106, 299-304
Piret P, Piret-Meunier J 1988 Nouvelle détermination de la structure de la dumonti $Pb_2[(UO_2)_3O_2(PO_4)_2] \cdot 5H2O$. Bull Minéral 111:439-442
Piret P, Piret-Meunier J 1991 Composition chimique et structure cristalline de la phosphuranyli $Ca(UO_2)(UO_2)_3(OH)_2(PO_4)_2 \cdot 12H_2O$. Eur J Mineral 3:69-77
Piret P, Piret-Meunier J, Deliens M 1990 Composition chimique et structure cristalline de la dewindti $Pb_3[H(UO_2)_3O_2(PO_4)_2] \cdot 12H_2O$. Eur J Mineral 2:399-405
Plesko EP, Scheetz BE, White WB 1992 Infrared vibrational characterization and synthesis of a family hydrous alkali uranyl silicates and hydrous uranyl silicate minerals. Am Mineral 77:431-437
Podor R, Cuney M, Nguyen Trung C 1995 Experimental study of the solid solution between monazite-(L and $(Ca_{0.5}U_{0.5})PO_4$ at 780°C and 200 MPa. Am Mineral 80:1261-1268
Posey-Dowty J, Axtmann E, Crerar D, Borcsik M, Ronk A, Woods W 1987 Dissolution rate of uranini and uranium role-front ores. Econ Geol 82:184-194
Pourcelot L, Gauthier-Lafaye F 1998 Mineralogical, chemical and O-isotopic data on uraninites fro natural fission reactors (Gabon): effects of weathering conditions. CR Acad Sci Paris, Sci terre planét 326:485-492
Protas J 1959 Contribution à l'étude des oxydes d'uranium hydratés. Bull Soc franç Minéral Cristallo 82:239-272
Pu Congjian 1990 Boltwoodite discovered for the first time in China. Acta Mineral Sinica 10:157-160 (Chinese)
Putnam RL, Navrotsky A, Woodfield BF, Boerio-Goates J, Shapiro JL 1999 Thermodynamics of formati for zirconolite ($CaZrTi_2O_7$). J Chem Thermodynamics 31:229-243
Rasmussen B, Buick R 1999 Redox state of the Archean Atmosphere: evidence from detrital heavy miner: in ca. 3250-2750 Ma sandstones from the Pilbara Craton, Australia. Geology 27:115-118
Rastsvetaeva RK, Arakcheeva AV, Pushcharovsky DY, Atencio D, Menezes Filho LAD 1997 New silic band in the haiweeite structure. Crystallogr Reports 42:927-933
Ringwood AE, Kesson SE Reeve KD, Levins DM Ramm EJ 1988 Synroc. In Radioactive Waste Fon for the Future. Lutze W, Ewing RC (Eds) p 233-334
Robie RA, Hemingway BS, Fisher JR 1979 Thermodynamic properties of minerals and related substanc at 298.15 K and 1 bar (10^5 Pascals) pressure and at higher temperatures. US Geol. Surv. Bulletin 14: 456 p
Rogova VP, Belova LN, Kiziyarov GP, Kuznetsova NN 1973 Bauranoite and metacalciouranoite—n minerals of the group of hydrous uranium oxides. Zap Vses Mineral Obshch 102:75-81 (in Russian)
Rogova VP, Belova LN, Kiziyrov GP, Kuznetsova NN 1974 Calciouranoite, a new hydroxide of uraniu Zap Vses Mineral Obshch 103:103-109 (In Russian; abstract in Am Mineral 60:161, 1975)
Sandino A, Bruno J 1992 The solubility of $(UO_2)_3(PO_4)_2 \cdot 4H_2O(s)$ and the formation of U(VI) phosph complexes: Their influence in uranium speciation in natural waters. Geochim Cosmochim A

56:4135-4145

Sandino MCA, Grambow B 1995 Solubility equilibria in the U(VI)-Ca-K-Cl-H$_2$O system: Transformation of schoepite into becquerelite and compreignacite. Radiochim Acta 66/67:37-43

Sarp H, Bertrand J, Deferne J 1983 Asselbornite, (Pb,Ba)(BiO)$_4$(UO$_2$)$_6$(AsO$_4$)$_2$(OH)$_{12}$(H$_2$O)$_3$, a uranium, bismuth and barium arsenate. N Jb Mineral Mh 1983:417-423

Sarp H, Chiappero PJ 1992 Deloryite, Cu$_4$(UO$_2$)(MoO$_4$)$_2$(OH)$_6$, a new mineral from the Cap Garonne mine near Le Pradet, Var, France. N Jb Miner Mh 1992:58-64

Sato T, Murakami T, Yanase N, Isobe H, Payne TE, Airey PL 1997 Iron nodules scavenging uranium from groundwater. Environ Sci Technol 31:2854-2858

Savary V, Pagel M 1997 The effects of water radiolysis on local redox conditions in the Oklo, Gabon, natural fission reactors 10 and 16. Geochim Cosmochim Acta 61:4479-4494

Schoep A 1930 Les minéraux du gîte uranifère du Katanga. Musée Congo Belge Ann Ser 1, Tome 1, Fasc 2, 43 p

Serezhkin VN, Tshubaev VF, Kovba LM, Trnov VK 1973 The structure of synthetic iriginite. Doklady Akad Nauk SSSR Ser Chem 210:873-876 (in Russian)

Sergeyeva EI, Nikitin AA. Khodakovkiy IL, Naumov GB 1972 Experimental investigation of equilibria in the system UO$_3$-CO$_2$-H$_2$O in 25-200°C temperature interval. Geokhimiya 11:1340 (in Russian) [English translation in Geochem Int'l 9:900]

Sharmová M, Scharm B 1994 Rabdophane-group minerals in the uranium ore district of northern Bohemia (Czech Republic). 39:267-280

Shen J, Peng Z 1981 The crystal structure of furongite. Acta Crystallogr A37 supplement C-186

Shoesmith DW, Sunder S 1991 An electrochemistry-based miodel for the dissolution of UO$_2$. AECL-10488, Atomic Energy of Canada, Whiteshell Laboratories, 97 p

Shoesmith DW, Sunder S, Tait JC 1998 Validation of an electrochemical model for the oxidative dissolution of used CANDU fuel. J Nucl Mater 257:89-98

Singh Z, Dash S, Krishnan K, Prasad R, Venugopal V 1999 Standard Gibbs energy of formation of UTeO$_5$(s) by the electrochemical method. J Chem Thermodynamics 31:197-204

Singh KDP, Bhargava LR, Ali MA, Swarnkar BM 1990 An unusual brannerite from Tai, Arunachal Pradesh, India. Explor Res Atomic Minerals [India] 1:117-122

Singh Y 1999 Phurcalite: a rare secondary calcium uranium phosphate mineral from Putholi, Chittaurgarh district, Rajasthan. J Geol Soc India 53:355-357.

Skakle JMS, Moroni LP, Glasser FP 1997 X-ray diffraction data for two new calcium uranium(VI) hydrates. Powd Diff 12:81-86

Smith DK Jr 1984 Uranium mineralogy. In Uranium Geochemistry, Mineralogy, Geology, Exploration and Resources. DeVivo B, Ippolito F, Capaldi G, Simpson PR (Eds) Institute of Mining and Metallurgy, London, p 43-88

Smith DK Jr, Scheetz BE, Anderson CAF, Smith KL 1982 Phase relations in the uranium-oxygen-water system and its significance on the stability of nuclear waste forms. Uranium 1:79-111

Smits G 1989 (U,Th)-bearing silicates in reefs of the Witwatersrand, South Africa. Can Mineral 27:643-656

Snelling AA 1980 Uraninite and its alteration products, Koongarra uranium deposit. In Uranium in the Pine Creek Geosyncline. Ferguson J, Goleby AB (Eds) Int'l Atomic Energy Agency, Vienna, p 487-498

Snelling AA 1992 Geologic setting, Alligator Rivers Analogue Project Final Report 2, ISBN 0-642-59928-9. Australian Nuclear Science and Technology Organisation, Sydney, 118 p

Speer A 1982 Actinide orthosilicates. Rev Mineral 5:113-135

St A Hubbard HV, Griffiths TR 1987 An investigation of defect structures in single-crystal UO$_{2+x}$ by optical absorption spectroscopy. J Chem Soc, Faraday Trans 2 83:1215-1227

Stalbauer E, Wichmann V, Lott V, Keller C 1974 Relationships of the ternary La-U-O system. J Solid State Chem 10:341-350

Steen H 1998 Ein kleines Vorkommen von Uran- und Bleimineralien im Schweizergrund bie Sulzburg, Südschwarzwald. Der Erzgräber 7:1-3

Stieff LR, Stern TW, Sherwood AM 1956 Coffinite, a uranous silicate with hydroxyl substitution—a new mineral. Am Mineral 41:675-688

Stergiou AC, Rentzeperis PJ, Sklavounos S 1993 Refinement of the crystal structure of metatorbernite. Z Kristallogr 205:1-7

Stohl FV, Smith DK Jr 1981 The crystal chemistry of the uranyl silicate minerals. Am Mineral 66:610-625

Stout PJ, Lumpkin GR, Ewing RC, Eyal Y 1988 An annealing study of alpha-decay damage in natural UO$_2$ and ThO$_2$. Mater Res Soc Symp Proc 112:495-504

Stumm W, Morgan JJ 1981 Aquatic Chemistry, 2nd Edn. Wiley-Interscience, New York, 780 p

Sugitani Y, Suzuki Y, Nagashima K 1984 Recovery of the original samarskite structure by heating in a reducing atmosphere. Am Mineral 69:377-379

Sunder S, Cramer JJ, Miller NH 1996 Geochemistry of the Cigar Lake deposit: XPS studies. Radiochim Acta 74:303-307

Sunder S, Miller NH Duclos AM 1994 XPS and XRD studies of samples from the natural fission reactors in the Oklo uranium deposits. Mater Res Soc Symp Proc 333:631-638

Sunder S, Shoesmith DW, Christensen H, Miller NH 1992 Oxidation of UO_2 fuel by the products of gamma radiolysis of water. J Nucl Mater 190:78-86

Suzuki Y, Murakami T, Kogure T, Isobe H, Sato T 1998 Crystal chemistry and microstructures of uranyl phosphates. Mater Res Soc Symp Proc 506:839-846

Swihart GH, Sen Gupta PK, Schlemperer EO, Back ME, Gaines RV 1993 The crystal structure of moctezumite [$PbUO_2$](TeO_3)$_2$. Am Mineral 18:835-839

Szymanski JT, Scott JD 1982 A crystal structure refinement of synthetic brannerite UTi_2O_6 and its bearing on rate of alkaline-carbonate leaching of brannerite in ore. Can Mineral 20:271-280

Tardy Y, Garrels RM 1976 Prediction of Gibbs free energies of formation: I. Relationship between Gibbs energies of formation of hydroxides, oxides and aqueous ions. Geochim Cosmochim Acta 40:1051-1056

Tardy Y, Garrels RM 1977 Prediction of Gibbs free energies of formation: II. Monovalent and divalent metal silicates. Geochim Cosmochim Acta 41:87-92

Taylor JC 1971 The structure of the α form of uranyl hydroxide. Acta Crystallogr B27:1088-1091

Taylor JC, Stuart WI, Mumme IA 1981 The crystal structure of curite. J Inorg Nucl Chem 43:2419-2423

Thomas LE, Einziger RE, Woodley RE 1989 Microstructural examination of oxidized spent PWR fuel by transmission electron microscopy. J Nucl Mater 166:243-251

Thomas LE, Einziger RE, Buchanan HC 1993 Effects of fission products on air oxidation of LWR spent fuel. J Nucl Mater 201:310-319

Threadgold IM 1960 The mineral composition of some uranium ores from the south Alligator River area, Northern Territory. Mineralog Invest CSIRO Technical Paper 2

Tschernikov AA, Organova NI 1994 Sodium autunite and sodium meta-autunite. Doklady Akad Nauk 338:368-371

Trocellier P, Cachoir C, Guilbert S 1998 A simple thermodynamic model to describe the control of the dissolution of uranium dioxide in granite groundwater by secondary phase formation. J Nucl Mater 256:197-206

van Genderen LCG, van der Weijden CH 1984 Prediction of Gibbs free energies of formation and stability constants of some secondary uranium minerals containing the uranyl group. Uranium 1:249-256

Vance ER, Ball CJ, Day RA, Smith KL, Blackford MG, Begg BD, Angel P 1994 Actinide and rare earth incorporation into zirconolite. J Alloys Comp 213/214:406-409

Vdovina OL, Serezhkina LB, Serezhkin VN, Boiko NV 1984 Interaction of magnesium and uranyl chromates in aqueous solution. Radiokhimiya 25:345-349

Vishnev AI, Gorshkov AI, Federov OV 1991 Iriginite according to the data of electron microscopy and electronography. Izvestia Akad Nauk SSSR Ser Geol 1991:143-149

Viswanathan K, Harnett O 1986 Refined crystal structure of beta-uranophane, $Ca(UO_2)_2(SiO_3OH)_2 \cdot 5H_2O$. Am Mineral 71:1489-1493

Vochten R, Blaton N 1999 Synthesis of rutherfordine and its stability in water and alkaline solutions. N Jb Miner Mh 1999:372-384

Vochten R, Blaton N, Peeters O 1997a Synthesis of sodium weeksite and its transformation into weeksite. N Jb Miner Mh 1997:569-576

Vochten R, Blaton N, Peeters O 1997b Deliensite, $Fe(UO_2)_2(SO_4)_2(OH)_2 \cdot 3$ H_2O, a new ferrous uranyl sulfate hydroxyl hydrate from Mas d'Alary, Lodève, Hérault, France. Can Mineral 35:1021-1025

Vochten R, Blaton N, Peeters O, Deliens M 1996 Piretite, $Ca(UO_2)_3(SeO_3)_2(OH)_2 \cdot 4H_2O$, a new calcium uranyl selenite from Shinkolobwe, Shaba, Zaire. Can Mineral 34:1317-1322

Vochten R, Blaton N, Peeters O, van Springel K, van Haverbeke L 1997c A new method of synthesis of boltwoodite and of formation of sodium boltwoodite, uranophane, sklodowskite and kasolite from boltwoodite. Can Mineral 35:735-741

Vochten R, de Grave E, Pelsmaekers J 1984 Mineralogical study of bassetite in relation to its oxidation. Am Mineral 69:967-978

Vochten R, Deliens M 1998 Blatonite, $UO_2CO_3 \cdot H_2O$, a new uranyl carbonate monohydrate from San Juan County Utah. Can Mineral 36:1077-1081

Vochten R, van Haverbeke L 1990 Transformation of schoepite into the uranyl oxide hydrates: becquerelite, billietite and wölsendorfite. Mineral Petrol 43:65-72

Vochten R, van Haverbeke L, Sobry R 1991 Transformation of schoepite into the uranyl oxide hydrates: of the bivalent cations Mg^{2+}, Mn^{2+} and Ni^{2+}. J Mater Chem 1:637-642

Vochten R, van Haverbeke L, van Springel K 1995 Soddyite: Synthesis under elevated temperature and pressure, and study of some physicochemical characteristics. N Jb Miner Mh H 10:470-480

Vochten R, van Haverbeke L, van Springel K, Blaton N, Peeters OM 1994 The structure and physicochemical characteristics of a synthetic phase compositionally intermediate between liebigite and andersonite. Can Mineral 32:553-561

Vochten R, van Haverbeke K, van Springel K, Blaton N, Peeters OM 1995 The structure and physicochemical characteristics of synthetic zippeite. Can Mineral 33:1091-1101

Vochten R, van Springel K 1996 A natural ferrous substituted saléeite from Arcu su Linnarbu, Capoterra, Cagliari, Sardinia. Mineral Mag 60:647-651

Von Weitzel H, Schröcke H 1980 Kristallstrukturfeinerungen von Euxenit, $T(Nb_{0.5}Ti_{0.5})_2O_6$, und M-Fergusonit, $YNbO_4$. Z Kristallogr 152:69-82

Voultsidis V, Clasen D 1978 Probleme und grenzbereiche der uranmineralogie. Erzmetall 31:8-13

Waber N, Schorscher HD, Peters T 1990 Mineralogy, petrology and geochemistry of the Poços de Caldas analogue study sites, Minas Gerais, Brazil. I: Osamu Utsumi uranium mine. SKB Technical Report 90-11, SKB, Stockholm, 514 p

Waite TD, Davis JA, Payne TE, Waychunas GA, Xu N 1994 Uranium(VI) adsorption to ferrihydrite: Application of a surface complexation model. Geochim Cosmochim Acta 58:5465-5478

Walenta K 1974 On studtite and its composition. Am Mineral 59:166-171

Walenta K 1976 Widenmannit und Joliotite, zwei neue Uranylcarbonatmineralien aus den Schwartzwald. Schweiz mineral petrogr Mitt 56:167-185

Walenta K 1983 Uranosilite, a new mineral from the uranium deposit at Menzenschwand (southern Black Forest). N Jb Mineral Mh 1983:259-269

Walenta K 1985 Uranotungstite, a new secondary uranium mineral from the Black Forest, West Germany. Tschermaks mineral petrol Mitt 25-34 (in German)

Walenta K 1993a Phurcalit von Wittichen. Der Erzgräber 7:1-3

Walenta K 1993b Gesicherte und fragliche Mineralfunde von Vorkommen im Schwarzwald. Der Erzgräber 7:54-64

Warner JK, Ewing RC 1993 Crystal chemistry of samarskite. Am Mineral 78:419-424

Weeks AD, Thompson ME 1954 Identification and occurrence of uranium and vanadium minerals from the Colorado plateaus. U S Geol Surv Bull 1009-B, 62 p

Williams CT, Gieré R 1996 Zirconolite: a review of localities worldwide, and a compilation of its chemical compositions. Bull Nat Hist Museum Geol Ser 52:1-24

Willis BTM 1963 Positions of the oxygen atoms in $UO_{2.13}$. Nature 197:755-756

Willis BTM 1978 The defect structure of hyperstoichiometric uranium dioxide. Acta Crystallogr A34:88

Willis BTM 1987 Crystallographic studies of anion-excess uranium oxides. J Chem Soc Faraday Trans 2:1073-1081

Weijun S, Fengjun M, Shijie Z 1983 Fergusonite-beta-Nd. Sci Geol Sinica 78-81 (in Chinese)

Wolf SF, Bates JK, Buck EC, Dietz NL, Fortner JA, Brown NR 1997 Physical and chemical characterization of actinides in soil from Johnston Atoll. Environ Sci Technol 31:467–471

Woodfield BF, Boerio-Goates J, Shapiro JL, Putnam RL, Navrotsky A 1999 Molar heat capacity and thermodynamic functions of zirconolite $CaZrTi_2O_7$. J Chem Thermodynamics 31:245-253

Wronkiewicz DJ, Bates JK, Gerding TJ, Veleckis E, Tani BS 1992 Uranium release and secondary phase formation during unsaturated testing of UO_2 at 90°C. J Nucl Mater 190:107-127

Wronkiewicz DJ, Bates JK, Wolf SF, Buck EC 1996 Ten-year results from unsaturated drip tests with UO_2 at 90°C: Implications for the corrosion of spent nuclear fuel. J Nucl Mater 238:78-95

Wronkiewicz DJ, Buck EC 1999 Uranium mineralogy and the geologic disposal of spent nuclear fuel. Rev Mineral 38 (this volume)

Xu HF, Wang YF 1999 Electron energy-loss spectroscopy (EELS) study of oxidation states of Ce and U in pyrochlore and uraninite—natural analogues for Pu- and U-bearing waste forms. J Nucl Mater 265:117-123

Young EJ, Weeks AD, Meyrowitz R 1966 Coconinoite, a new uranium mineral from Utah and Arizona. Am Mineral 51:651-663

Zhang J, Wan A, Gong W 1992 New data on yingjiangnite. Acta Petrol Mineral 11:178-184 (in Chinese)

Zhangru C, Keding L, Falan T, Yi Z, Xiaofa G 1986 Tengchongite, a new mineral of hydrated calcium uranyl molybdate. Kexue Tongbao 31:396-401

Zwaan PC, Arps CES, de Grave E 1989 Vochteinte, $(Fe^{2+},Mg)Fe^{3+}[UO_2/PO_4]_4(OH)\cdot 12\text{-}13H_2O$, a new uranyl phosphate mineral from Wheal Basset, Redruth, Cornwall, England. Mineral Mag 53:473-478

4 Stable Isotope Geochemistry of Uranium Deposits

Mostafa Fayek
Department of Earth and Space Sciences
University of California, Los Angeles
3806 Geology Building, Box 951567
Los Angeles, California 90095

T. Kurtis Kyser
Department of Geological Sciences and Geological Engineering
Miller Hall, Queen's University
Kingston, Ontario K7L 3N6 Canada

INTRODUCTION

The application of stable isotope geochemistry to mineral deposit research is based on understanding the mechanisms and magnitudes of isotopic fractionation that are associated with various geological processes. Knowledge of equilibrium isotopic fraction-ation factors between compounds as a function of temperature and other geochemical parameters and the kinetic isotope effects accompanying conversion of one substance to another under a variety of conditions is essential to the study of mineral deposits.

Stable isotope studies, in conjunction with other types of geochemical, geological, and mineralogical studies including paragenesis and phase relations of minerals, trace elements, fluid inclusions, and structural settings of mineral deposits can provide constraints on ore-forming processes. The physicochemical characteristics (e.g. H, O, S, C isotopic composition, pH, fO_2, T, and P) of fluids associated with mineral deposits can indicate the origin of mineral-precipitating fluids and constituents, the geometry of plumbing systems, the duration of hydrothermal activity, the mechanism of metal transport and deposition, and evolutionary processes. These data, when examined in a geological context, can be used to develop a genetic model for mineral deposition and help refine exploration strategies.

The study of fluids associated with U deposits relies heavily on H and O isotopic data because H_2O is the solvent of mineral constituents in U deposits. The H and O isotopic composition of the mineral-precipitating fluid reflects the origin of the fluids and provides detailed information on water-rock interaction. Sulfur and C isotopes provide information on the chemical aspects of mineral precipitation. Therefore, this chapter outlines the principles and methods of application of light stable isotope geochemistry to the study of U deposits. The problems associated with the methodologies will also be discussed. Secondly, this chapter summarizes stable isotopic studies of selected U deposits from different deposit types.

ISOTOPE FRACTIONATION: PRINCIPLES AND PROCESSES

In general, the O, H, C, and S isotopic composition of fluids associated with mineral deposits can be linked ultimately to one or more principal sources: seawater, meteoric waters, mantle fluids (juvenile waters), formation waters, magmatic waters, metamorphic waters, and their mixtures in the crust (Sheppard 1986). Therefore, the isotopic composition of minerals and rocks is derived from the modification of these reservoirs through isotopic fractionation and mixing of fluids of isotopically different components (Taylor 1987). Isotope fractionation is the partitioning of isotopes of the same element between two or more phases, and is produced through isotope exchange reactions and kinetic processes. These processes depend on differences in reaction rates between molecules containing the heavy isotopes and those containing the light isotopes of the same element (Hoefs 1997). The following is a brief summary of the principles of isotopic fractionation and exchange. A more detailed introduction can be found in Bigeleisen and Mayer (1947), Urey (1947), Melander (1960), Bigeleisen (1965), Bottinga and Javoy (1973), Javoy (1977), Richet et al. (1977), Hulston (1978), O'Neil (1986), and Kyser (1987).

Isotope exchange

Variations in the atomic mass of an element result in differences in chemical and physical properties. These differences are called "isotope effects." The electronic structure of an element essentially determines its chemical behavior, whereas the nucleus is more-or-less responsible for its physical properties (Hoefs 1997). Although isotopes of the same element contain the same number and arrangement of electrons, small differences exist in physicochemical properties due to isotope mass differences (Hoefs 1997). These mass differences are most pronounced for the lightest elements. Differences in the physicochemical properties of isotopes arise from quantum mechanical effects, such that bonds formed between light isotopes are weaker than bonds between heavy isotopes. Therefore, during a chemical reaction involving isotopic exchange, molecules with the light isotope will, in general, react more readily than those with the heavy isotope.

Isotope exchange reactions can be written:

$$aA_1 + bB_2 = aA_2 + bB_1 \tag{1}$$

where the subscripts 1 and 2 represent the light and heavy isotope, respectively, contained in species A and B. Therefore, for this reaction the equilibrium constant (K) is expressed by:

$$K = (A_2/A_1)^a / (B_2/B_1)^b \tag{2}$$

Isotopic equilibrium constants may be expressed in terms of partition functions (Q) of the various species:

$$K = (QA_2/QA_1)/(QB_2/QB_1) \tag{3}$$

where the partition function is defined by:

$$Q = \Sigma_i (g_i \exp(-E_i/kT)) \tag{4}$$

The summation is over the energy level, E_i, of the molecules, g_i is the degeneracy of the ith level of E_i, k is the Boltzmann constant, and T is the temperature.

The partition function of a molecule can be separated into translational, rotational, and vibrational energy modes. The difference in translational and rotational energy between molecules appearing on both sides of the exchange reaction equation is very small (except for H) leaving vibrational energy as the dominant effect on isotopic exchange (O'Neil 1986; Kyser 1987; Hoef 1997).

The dependence of the equilibrium constant K on temperature is important to studies of U deposits and geological processes in general (Eq. 3) because if a mineral system is shown to be in isotopic equilibrium, the equilibrium constant can be used to calculate the temperature of mineral precipitation. For ideal gases at low temperatures (generally below room temperature), ln K ~ 1/T and at higher temperatures, ln K ~ 1/T², where T is the absolute temperature (Bigeleisen and Mayer 1947; Bottinga and Javoy 1973). When solid materials are considered, the evaluation of isotope partition function ratios is more complicated because it is necessary to consider lattice vibrations as well as the internal vibrations for each molecule (Hoefs 1997).

Fractionation factors

The fractionation factor is defined as the ratio of isotopes (R) in one chemical compound, A, divided by the corresponding ratio in chemical compound, B:

$$\alpha_{A-B} = R_A/R_B \tag{5}$$

If the isotopes are randomly distributed over all positions in compounds A and B, then α is related to the equilibrium constant K by:

$$\alpha = K^{1/n} \tag{6}$$

where n is the number of atoms exchanged (Hoefs 1997). Factors that can influence the sign and magnitude of α are temperature, chemical composition, crystal structure, and to a lesser extent, pressure (O'Neil 1986). Recent studies have shown that pressure can effect H isotope fractionation by as much as 7.5‰ at pressures of 2000 bars and ~400°C (Driesner 1997, 1998; Horita et al. 1998). However, the effect of pressure on O isotope fraction is small (0.2‰) and will hardly be detectable in natural samples (Driesner 1997; Polyakov 1998).

In isotope geochemistry, the difference of absolute ratios between two substances is measured, rather than the absolute ratios in each phase. As such, stable isotope abundances are normally reported as delta (δ) values in units of ‰ (per mil) relative to a standard:

$$\delta_A = (R_A/R_S - 1) \times 10^3 \tag{7}$$

where R_S is the absolute ratio in the standard. The standard values used were adopted by the International Atomic and Energy Agency (IAEA), Vienna and are Vienna Standard Mean Ocean Water (VSMOW) for H and O, belemnite from the Cretaceous Pee Dee formation (VPDB) for C (sometimes for O in carbonates), and troilite from the Canyon Diablo iron meteorite (CDT) for S. For example, a $\delta^{18}O$ value of +10‰ for quartz means that the quartz has an absolute $^{18}O/^{16}O$ ratio that is 1% (or 10‰) greater than VSMOW. Similarly, biotite with a δD value of -70‰ means that it has a D/H ratio that is 7% lower than that ratio in VSMOW. For two compounds A and B, the δ-values and fractionation factor α are related by:

$$\delta_A - \delta_B = \Delta_{A-B} \sim 10^3 \ln \alpha_{A-B} \tag{8}$$

This approximation is good for differences in δ-values of less than 10 (O'Neil 1986; Kyser 1987; Hoefs 1997). Assuming that atomic vibrations can be described by simple harmonic oscillators, the fractionation factor, α, between two phases A and B is related to temperature, T, by:

$$\ln \alpha_{A-B} \sim X + Y/T^2 \text{ at high T and,} \tag{9}$$

$$\ln \alpha_{A-B} \sim X + Y/T \text{ at low T} \tag{10}$$

where X and Y are constants (Kyser 1987). These equations describe the differences expected in the ratios of an element in two phases at equilibrium for a given temperature.

Isotopic composition of systems and isotopic equilibrium

In this chapter the "system" refers to the mineral-precipitating fluid and the dissolved constituents. The isotopic composition of a system is determined by:

$$\delta_{SYS} = X_A\delta_A + X_B\delta_B + \cdots\cdots + X_N\delta_N \tag{11}$$

where X and δ are the atom fractions and isotopic composition, respectively, of its individual components A and B.

Isotopic equilibrium in hydrothermal systems (i.e. the system has remained closed to the isotopes of interest) is difficult to assess because these systems have varied mineral assemblages that consist of only a few phases (Deines 1977). Changes in the isotopic composition of components in a system largely occur through continuous equilibrium and fractional (Rayleigh) equilibrium processes. Continuous equilibrium processes include phase separation during confined boiling and mineral precipitation, chemical reaction (e.g. redox), and fluid mixing. Examples of fractional equilibrium include magmatic degassing, boiling, and mineral precipitation from a finite reservoir with no subsequent recrystallization or exchange (Taylor 1987).

In the case of U deposits, the main factors that cause U mineral precipitation are an increase in pH and oxidation-reduction reactions. Thus, isotopic equilibrium is likely achieved through continuous equilibrium processes. However, boiling may be an important U precipitating mechanism in relatively shallow, open hydrothermal systems, where fluids percolate through open fractures (Romberger 1985). In this environment, isotopic equilibrium is likely attained through fractional equilibrium processes.

Causes of isotopic variation

The major cause of H and O isotopic variations in most systems are the partial conversions of liquid water to and from water vapor (i.e. evaporation and condensation), fluid mixing, and water/rock interaction (Ohmoto 1986). For example, mixing between H_2O and CO_2 during decarbonation reactions can cause shifts in the $\delta^{18}O$ value of the C-O-H fluid system (Fig. 1).

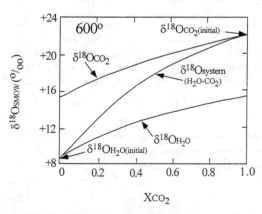

Figure 1. Isotopic variation of H_2O in a C-O-H fluid when mixed with CO_2 at 600°C as a function of X_{CO_2} (from Taylor and O'Neil 1977; reprinted with permission of Springer-Verlag, from *Contrib. Mineral. Petrol.*, 63, p. 14, Fig. 3).

In contrast, the major causes of C and S isotopic variation in nature are redox reactions involving the isotopic species. This occurs because C and S occur in nature in several different valence states (i.e. C^{4+} in organic compounds, C^0 in graphite, C^{4+} in carbonates, S^{2-} in sulfides, S^0 in native S, S^{4+} in SO_2, and S^{6+} in sulfate minerals), and many mechanisms, both biological and non-biogenic, control the variation in redox state of the S and C species (Ohmoto 1986). In addition, large kinetic effects (10 to 70‰) are associated with many redox and bacterial reactions, and large fractionation factors exist between the oxidized and reduced species (O'Neil 1986; Ohmoto 1986; Kyser 1987). These kinetic effects occur because molecules containing isotopes of elements with higher atomic mass such as ^{34}S or ^{32}S are more stable and have higher disassociation energies than those containing isotopes of lighter elements (i.e. H or deuterium). Therefore, it is easier to break bonds such as $H-^{18}O$ than to break bonds such as ^{34}S-^{18}O (O'Neil 1986).

Chemical and isotopic variability of a fluid can occur from boiling. The change in the isotopic composition of fluids due to boiling depends on the extent of vapor loss, magnitude of the fractionation factors between vapor phase, liquid phases, and other compounds, open- vs. closed-system behavior, and isotopic exchange kinetics (Taylor 1987). Boiling at moderate temperatures (~350°C) can result in loss of H to the vapor phase (Drummond and Ohmoto 1985), causing an increase in the oxidation state (fO_2) of the fluid (Taylor 1987). Therefore, H_2S may be oxidized, increasing the SO_4/H_2S ratio, reducing the ligand capacity and precipitating sulfides (Taylor 1987). However, if metals are in solution as chloride complexes, an increase in fO_2 may increase the solubility of chloride-metal complexes and therefore limit the sulfide precipitation (Taylor 1987).

Stable isotope geothermometry

The principles of stable isotope geothermometry have been discussed in detail by O'Neil (1986) and Kyser (1987). The application of stable isotope geothermometry to mineral deposits requires that (1) the fractionation factors between mineral pairs or minerals and fluids are well-calibrated and are strongly temperature dependent, (2) the phases or species were in isotopic equilibrium at the time of deposition, and (3) they have retained their isotopic composition since formation.

Stable isotope data, in conjunction with fluid inclusion data, can provide P and T estimates for mineral precipitation. Stable isotope geothermometry may give the true temperature of mineral precipitation, whereas homogenization temperatures of fluid inclusions generally give minimum temperatures (Ohmoto 1986). The P and T conditions of mineral precipitation lie along an isochore of the fluid inclusion, which is defined by the fluid composition and P-V-T properties (Fig. 2). The temperature calculated from stable isotope data (i.e. fractionation factor between two coexisting minerals) and the appropriate stable isotope geothermometer, plots on the isochore and thus defines the pressure of mineral precipitation. The pressure can then be used to estimate the depth of mineral deposition. However, uncertainty in geothermometer equations and analytical errors may result in an uncertainty in the estimate of pressure of more than a few hundred bars (Ohmoto 1986).

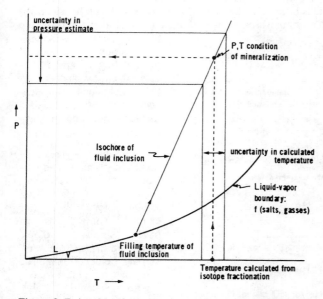

Figure 2. Estimation of pressure conditions of mineralization from combined fluid inclusion and isotopic data (from Ohmoto 1986).

A substantial amount of fluid inclusion data pertaining to hydrothermal U deposits exists in the literature (Campbell 1955; Robinson 1955; Harlass and Schützel 1965; Kranz 1968; Kotov et al. 1968; Leroy and Poty 1969; Naumov and Mironova 1969; Tugarinov and Namov 1969; Gornitz and Kerr 1970; Kotov et al. 1970; Barsukov et al. 1971; Naumov et al. 1971; Regova et al. 1971; Badham et al. 1972; Sassano et al. 1972; Robinson and Ohmoto 1973; Cuney 1974; Knipping 1974; Little 1974; Arnold and Cuney 1974; Oparysheva et al. 1974; Poty et al. 1974; Shegeliski and Scott 1975; Pagel et al. 1980, Wilde et al. 1989; Kotzer and Kyser 1995). However, much of the fluid inclusion data are of limited value for determining pressure of U mineral precipitation because homogenization temperatures are often reported without estimates of fluid composition, and the minerals that contain the fluid inclusions are not associated with U mineral precipitation. Fluid inclusion and stable isotopic studies of hydrothermal U deposits of the Massif Central in France suggest that uraninite precipitated from CO_2-bearing, low salinity fluids at temperatures of 340-350°C and pressures of 700-800 bars (Cuney 1974; Poty et al. 1974), whereas unconformity-type U deposits from Saskatch-ewan, Canada formed from high salinity, low CO_2-bearing fluids at temperatures of 200°C and pressures >1 kbar (Pagel et al. 1980; Kotzer and Kyser 1995).

ISOTOPIC COMPOSITION OF HYDROTHERMAL SYSTEMS

Identifying the source of the fluids that transport the elements that comprise mineral deposits is of great interest to the mining and exploration industry and is required to develop appropriate genetic models of mineral deposition. The source is generally identified by comparing the stable isotopic composition of well-characterized fluid reservoirs (see below) with the isotopic composition of the mineral precipitating fluids, as determined by analysis of fluid inclusions or calculated from stable isotopic measurements and geothermometers.

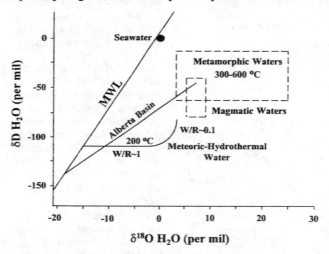

Figure 3. The relationship between the δD and $δ^{18}O$ values for various fluids. The isotopic composition of typical seawater, magmatic and metamorphic fluids, and basinal brines from the Alberta and Gulf coast basins are shown. The isotopic compositions of fluids from which minerals can precipitate and fractionate isotopes include those along the meteoric water line (MWL), fluids in the Canadian Shield (Yellowknife and Sudbury), meteoric-hydrothermal fluids, and geothermal fluids (Geothermal ^{18}O-shifted). The trajectory of evaporative seawater is included (modified from Taylor 1979; reprinted with permission of J. Wiley & Sons, from the book *Hydrothermal Ore Deposits*, p. 244, Fig. 6.4).

Varying degrees of uncertainty can be associated with the characterization of fluids using stable isotopic data. For example, some fluids can have similar isotopic compositions (i.e. the overlapping fields of magmatic and metamorphic waters; Fig. 3). Secondly, analytical uncertainty associated with isotopic measurements and geothermometers can cause large uncertainties in the calculated isotopic compositions of fluids. In addition, to calculate the isotopic composition of a fluid associated with a suite of minerals, using a geothermometer, it is necessary to identify mineral pairs that are in textural equilibrium, with the assumption that they are also in isotopic equilibrium. Thirdly, errors associated with the characterization of fluids may occur when comparing the isotopic composition of ancient mineral precipitating fluids that no longer exist (e.g. ancient seawater) with modern day reservoirs (modern day seawater), and assumptions that they are similar in their isotopic composition. Fourthly, evolved or mixed-source fluids may have a range of possible compositions, making it difficult to characterize fluid using stable isotopes (see below) (Taylor 1987).

Open vs. closed system

Reference to "open" or "closed" systems has created some confusion in the literature because of different definitions of a system. The "closed system" referred to in most studies is analogous to an autoclave that contains water and rock, and is maintained at a constant temperature (Fig. 4a). Any water/rock ratio can be achieved by adjusting the amount of starting material. This is known as Taylor's "closed system" (Taylor 1979). In natural closed systems (Fig. 4b), the water/rock ratio never exceeds 0.02 to 0.1 for the O system, and the isotopic composition of the water is buffered by the rock (Taylor 1979; Ohmoto 1986).

Taylor (1979) also describes an "open system," which is analogous to a flow-through system, where an autoclave with valves on both ends is filled with rock and is heated at a given temperature (Fig. 4c). Water is introduced in the vessel and allowed to equilibrate with the rock, and then is replaced by a new volume of water (Taylor 1979). Taylor's "open system" may not be applicable to natural open systems because the water that interacts with a mass of rock at a given temperature may have undergone isotopic exchange with other rocks at different temperatures and thus may not have retained its original isotopic composition (Ohmoto 1986). In this chapter, "open" and "closed" systems,

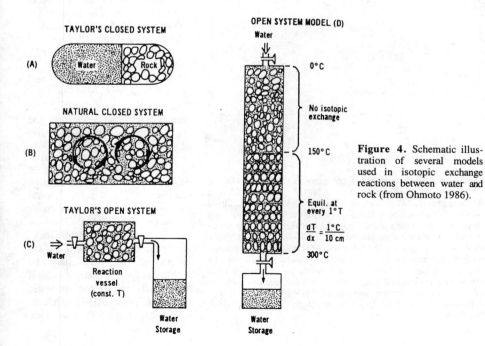

Figure 4. Schematic illustration of several models used in isotopic exchange reactions between water and rock (from Ohmoto 1986).

respectively, are defined as those to which fractional (Rayleigh) and continuous equilibrium processes apply.

Oxygen and hydrogen isotopes

Water is the dominant constituent of mineral precipitating fluids. The isotopic composition of the fluid can be determined by measuring either the H and O isotopic compositions of the water directly from fluid inclusions, or by analyzing hydrous minerals for which the temperatures of formation have been determined. Using mineral-water geothermometers, the isotopic composition of the fluid can then be calculated (Taylor 1974). The O and H isotopic fractionation factors between minerals and water are generally considered to be independent of solution chemistry. However, O isotopic fractionation factors may vary by as much as 3‰ due to salinity effects, as demonstrated by experiments at 600°C (Truesdell 1974; Ohmoto and Oskvareck 1985; Horita 1989a,b; Horita et al. 1993a,b; Horita et al. 1995) and 10‰ at temperatures of 150 to 300°C for H isotopes (Kazahaya and Matsuo 1986; Horita 1989a; Horita et al. 1995).

Kotzer and Kyser (1995) calculated the O and H isotopic composition of fluids in equilibrium with diagenetic clay minerals and quartz in basinal sandstones and metasedimentary rocks of the Athabasca basin, Canada, using their measured δD- and $\delta^{18}O$-values and temperatures of formation. Three isotopically distinct fluids that affected the basin and were associated with U mineral precipitation were characterized: (1) a 200°C basinal brine with δD- and $\delta^{18}O$-values of -60‰ and +4‰, respectively, (2) 200°C basement fluid with δD- and $\delta^{18}O$-values of -60‰ to -15‰ and +2‰ to +9‰, respectively, and (3) late low temperature (50°C) meteoric waters with δD- and $\delta^{18}O$-values of -130‰ and -16‰, respectively (Fig. 5). Mixing between the basin and basement fluids resulted in the precipitation of U minerals and dravite.

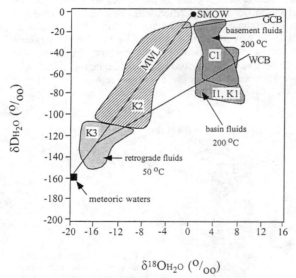

Figure 5. Fields delineating the range in calculated δD and $\delta^{18}O$ values of fluids at 200°C and 50°C in equilibrium with different generations of diagenetic clays (i.e. C1=stage I chlorite, I1-stage 1 illite, K1-stage 1 kaolinite, K2-stage 2 kaolinite, and K3-stage 3 kaolinite). Also shown for reference are the stable isotopic composition of Standard Mean Ocean Water (SMOW); meteoric waters presently within the Athabasca Basin (meteoric waters); the meteoric water line along which most of the meteoric waters plot (MWL); the trend of basinal brines from the Western Canadian Basin (WCB) and Gulf Coast Basin (GCB) (modified from Kotzer and Kyser 1995; reprinted with permission of Elsevier, from *Chem. Geol.*, 120, p. 70, Fig. 10).

Water/rock ratios may be calculated from O and H isotope variations in rocks and minerals. At low water/rock ratios the δD- and $\delta^{18}O$-values of the rock change very little, whereas the δD- and $\delta^{18}O$-values of the fluid may be substantially shifted. At high water/rock ratios the opposite occurs (Taylor 1979). Quantitative estimates of water/rock ratios in areas of hydrothermal alteration are often calculated assuming open-system fluid circulation (Fig. 4c) and the equation of Taylor (1979) for isotope exchange:

$$W/R \text{ (open)} = \ln[(\delta w^i + \Delta - \delta r^i)/\{\delta w^i - (\delta r^f - \Delta)\}] \qquad (12)$$

where i = initial isotopic composition and f = final isotopic composition of the water (w) or rock (r); Δ = mineral-water isotopic fractionation factor. Kotzer and Kyser (1995) used this method to calculate water/rock ratios ranging from 0.2 to 0.8 for unconformity-type U deposits from the Athabasca basin, Canada and much higher water/rock ratios (~10) in reactivated fault zones.

Carbon isotopes

The $\delta^{13}C$ values of carbonates can be used to estimate the $\delta^{13}C$ (H_2CO_3) in fluids where the fractionation factor between the carbonate mineral and the aqueous C species and the temperature of carbonate mineral precipitation are known. Where more than one carbonate species is present in the fluid, an isotopic mass balance must be done (analogous to Eqn. 11) to calculate the $\delta^{13}C_{\Sigma C}$ values (Ohmoto 1972, 1986; Taylor 1987). The concentration of C species in solution depends on the fO_2, pH, and concentration of C in fluids (Langmuir 1997). Many studies have attempted to quantify pH- fO_2 conditions of mineral precipitating fluids based on this approach (Rye and Ohmoto 1974; Ohmoto and Rye 1979).

Table 1. Carbon isotopic data for carbonates from Great Bear Lake and Beaverlodge districts

Locality	Mineral	Paragenesis	$\delta^{13}C_{PDB}$ (‰)
Great Bear Lake District*	dolomite	stage 3 veins	-4.1 to -2.8
Echo Bay mine	calcite	stage 4 veins	-3.9 to -2.3
	dolomite	stage 5 veins	-3.0 to -1.4
Beaverlodge District**	calcite	pre-veins	-6.7 to -2.5
Fay mine/Bolger pit	calcite	stage 1 veins	-8.7 to -1.6
	dolomite	stage 2 veins	-5.2 to +0.6
	calcite	stage 3 veins	-3.8 to -1.6
	calcite	stage 4 veins	-17.0 to -4.3
	calcite	stage 5 veins	-16.9 to -15.9

*Data from Robinson and Ohmoto (1973)
**Data from Sassano et al. (1972)

Carbon isotopic studies of carbonates associated with U minerals at Echo Bay mines, Canada (Table 1) showed that the $\delta^{13}C$ value of the fluids was relatively homogenous (Robinson and Ohmoto 1973). However, Sassano et al. (1972) found that carbonates from the Beaverlodge U deposit, Canada, have a range of $\delta^{13}C_{PDB}$ values from 0 to -17‰ (Table 1). This large variation was interpreted to be the result of a major change in the source of the C during the evolution of the hydrothermal system or because of redox reactions involving graphite and hydrocarbons. Kyser et al. (1988) showed that the $\delta^{13}C_{PDB}$ values of graphite near unconformity-type U deposits from the Athabasca basin, Canada, have a limited range (-25±5‰) and the differences in isotopic composition are not a function of distance from the U mineralization, intensity of alteration or deformation. They suggested that the similarity in the $\delta^{13}C_{PDB}$ values between altered and unaltered graphite was due to oxidation of graphite ($\delta^{13}C \sim -22$‰) producing CO_2.

Sulfur isotopes

Sulfur in many hydrothermal mineral deposits is fixed as either sulfides or sulfates and is thought to have originated either from seawater or igneous sources. The stable isotopic composition of S in sulfides is controlled by a number of factors including fO_2, pH, concentration of S in fluids, presence of sulfate-reducing bacteria, whether closed or open system conditions prevailed during the formation of the S-bearing mineral, and the temperature at which the sulfide mineral formed (Ohmoto 1986; Kyser 1987). Therefore, the $\delta^{34}S$ values of sulfides can be used to estimate the $\delta^{34}S$ (H_2S) in fluids when the fractionation factor between the sulfide mineral and the aqueous S species is known. Similar to C, where more than one S species is present in the fluid, an isotopic mass balance must be carried out (analogous to Eqn. 11) to calculate the $\delta^{34}S_{\Sigma S}$ values (Ohmoto 1972, 1986; Taylor 1987). The concentration of S species in solution depends on the fO_2, pH, and concentration of S in fluids (Fig. 6; Ohmoto 1986).

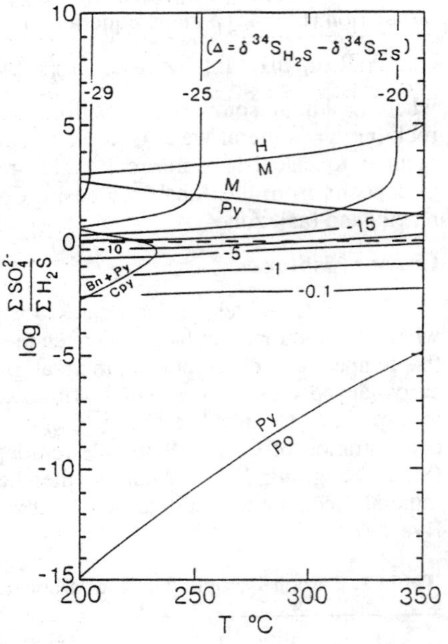

Fractionation of S isotopes between oxidized and reduced forms of S are substantial (i.e. several ‰). Therefore, S-bearing minerals forming in redox conditions tend to have variable $\delta^{34}S$ values. The $^{34}S/^{32}S$ ratio of sulfides and sulfates can be used as an isotopic tracer if the system in which they formed behaved as a closed system with S existing as either reduced or oxidized species (Kyser 1987). If a large number of co-existing sulfides have similar $\delta^{34}S$ values, the oxidation state of S must have been below a certain SO_4/H_2S ratio in a closed system or remained constant in an open system (Ohmoto 1986). If a significant amount of oxidized S is present when sulfides are forming in a closed system, the sulfide minerals will become increasingly enriched in ^{34}S as S is removed from the oxidized reservoir (Kotzer and Kyser 1990a).

Figure 6. Deviation of $\delta^{34}S(H_2S)$ from the isotopic composition of fluid ($\delta^{34}S_{\Sigma S}$) as a function of oxidation (SO_4^{2-}/H_2S ratio) and temperature at neutral pH, $\Sigma S = 0.01$ m, mK^+ = 0.05, mCa^{2+} = 0.5, and mMg^{2+} = 0.25 (from Ohmoto and Rye 1979). Reprinted with permission of J. Wiley & Sons, from the book *Hydrothermal Ore Deposits*, Chapter 10, p. 533, Fig. 10.9.

Uraninite in hydrothermal deposits is generally associated with sulfides such as pyrite and marcasite (Rich et al. 1977). Table 2 lists $\delta^{34}S$ values of sulfides from several U deposits from around the world. Robinson and Ohmoto (1973) have shown that there is a progressive change with time in the isotopic composition of S in sulfides from the Echo Bay U mine, Canada (Table 2). This change in $\delta^{34}S$ values from ~-22‰ in the earliest sulfide to ~+27‰ in the latest sulfide is consistent with a gradual reduction of sulfate to sulfide in solutions with a constant initial $\delta^{34}S$ value of +25±3‰. The progressive reduction of S was attributed to reaction of hydrothermal solutions with ferromagnesian minerals. However, reduction through mixing with reduced gases is not ruled out. Robinson and Ohmoto (1973) suggest that the hydrothermal solutions from this deposit evolved along the path from A to G in Figure 7. Large variations in isotopic compositions

Table 2. Sulfur isotopic data for sulfides from selected U deposits

Locality	Mineral	Paragenesis	$\delta^{34}S_{CTD}$(‰)
Great Bear Lake District[1]	host rock pyrite	pre-vein	+2.4 to +5.1
Echo Bay mine	acanthite	stage 4 veins	-21.5 to -19.7
	marcasite	stage 4 veins	-2.0 to +0.4
	acanthite	stage 5 veins	-4.1 to -0.5
	galena	stage 5 veins	+2.4 to +19.9
	sphalerite	stage 5 veins	+4.1 to +12.6
	chalcopyrite	stage 5 veins	+3.7 to +13.5
	bornite	stage 5 veins	+5.1 to +14.6
	acanthite	stage 6 veins	+21.3
	mckinstryite	stage 7 veins	+23.0 to +27.1
Terra mine[2]	chalcopyrite	pre-vein	-10.2 to +4.0
	pyrite	pre-vein	-9.6 to +5.1
	chalcopyrite	stage 1 veins	+1.0 to +3.9
	marcasite	stage 1 veins	+0.9 to +4.0
	galena	stage 3 veins	-2.0 to +4.2
	pyrite	stage 3 veins	-1.2
	bismithunite	stage 3 veins	+2.7
	matildite	stage 3 veins	+1.4 to +1.9
	chalcopyrite	stage 4 veins	+4.0 to +4.3
	pyrite	stage 4 veins	-26.0
Orphan mine, Arizona[3]	pyrite		-27.2 to -2.8
	chalcopyrite		-2.8
	chalcocite		-18.6
	sphalerite		-19.0
	galena		-21.2
Schwartzwalder mine, Colorado[4]	pyrite		-36.9 to +0.8
	pyrrhotite		-2.4 to -1.3
	chalcopyrite		+1.4
	galena		-5.3
South Alligator River District, Australia[5]	pyrite		-5.6 to +12.3
	galena		-3.4 to +10.4
Shinkolobwe deposit, Zaire[6]	pyrite+siegenite		-1.3 to +6.9

Data from: [1]Robinson and Ohmoto (1973); [2]Robinson and Badham (1974); [3]Gornitz and Kerr (1970); [4]Heyse (1971); [5]Ayers and Eadington (1975); [6]Dechow and Jensen (1965).

of S in sulfides from hydrothermal U deposits are not restricted to Echo Bay. Heyse (1971) also found a wide range in $\delta^{34}S$ values from -37‰ to +4‰ for sulfides from the Schwartzwalder mine, Colorado (Table 2). The $\delta^{34}S$ values of sulfides from several other deposits listed in Table 2 show similar variations.

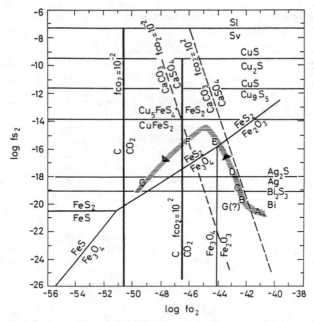

Figure 7. Log fs_2-log fO_2 mineral stability relationship at 150°C. Points A to G correspond to the probable path for the mineral precipitating fluid at Echo Bay mine, Canada. Reprinted with permission of *Economic Geology*, v. 68, p. 635-656, from Robinson and Ohmoto (1973).

Kotzer and Kyser (1990a) used the Pb and S isotopic composition of sulfides and sulfate minerals associated with unconformity-type U deposits from the Athabasca Basin to elucidate movement of late meteoric water along fault zones in the vicinity of the U deposits and to estimate the magnitude of the interaction between late oxidizing meteoric waters and the early-formed U minerals. They showed that sulfides in late fractures associated with the McArthur River U deposit had heavy $\delta^{34}S$ values and radiogenic $^{206}Pb/^{204}Pb$ ratios, indicating that the meteoric waters associated with the sulfides had a high content of radiogenic Pb likely acquired during alteration of uraninite (Fig. 8). However, late sulfides and sulfates from the Key Lake U deposit have $^{206}Pb/^{204}Pb$ ratios that are consistent with a non-radiogenic source and variable $\delta^{34}S$ values (Fig. 8), suggesting that late meteoric waters that precipitated the late sulfides and sulfates did not extensively interact with the U minerals (Kotzer and Kyser 1990a).

STABLE ISOTOPIC COMPOSITION OF PRINCIPAL FLUID RESERVOIRS

The principal fluid reservoirs are seawater, meteoric, juvenile, connate, formation waters, magmatic, metamorphic fluids, and basinal brines. Seawater, meteoric, and juvenile water are considered reference waters because they have uniquely defined, but not necessarily measurable isotopic compositions at their source. The other types of subsurface waters are considered recycled waters because they are evolved mixtures of reference waters (Ohmoto 1986; Sheppard 1986; Taylor 1987). The characterization of these fluid reservoirs is essential so that the fluids that are associated with U deposition can be identified and characterized.

Seawater

Seawater is defined as water of open oceans and in seas having direct access to the oceans (Sheppard 1986). The O and H isotopic composition of modern seawater is relatively constant ($\delta D = 0\pm10‰$ and $\delta^{18}O = 0\pm1‰$) except in areas with high evaporation, where the δD and $\delta^{18}O$ values are higher (i.e. Red Sea) or in areas where seawater has mixed with meteoric water, which decreases δD and $\delta^{18}O$ values. The O and H isotopic composition of ancient seawater, however, is less constrained because its isotopic composition can only be determined by indirect methods such as calculating the effect of melting the ice caps or by analyzing minerals that were formerly in equilibrium with seawater at a known temperature (Sheppard 1986).

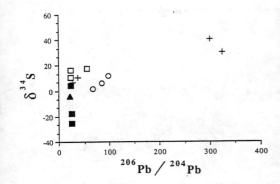

Figure 8. $\delta^{34}S$ values versus $^{206}Pb/^{204}Pb$ ratios of sulfide and sulfate minerals occurring in the Athabasca Basin. (□,○,▲) early sulfides from Key Lake, (■) late sulfides from Key Lake and (+) McArthur River (reprinted with permission of Saskatchewan Geological Society from *Modern Exploration Techniques*, v. 10, p. 125, Fig. 7, Kotzer and Kyser 1990a).

Dissolved C in seawater is mostly in the form of HCO_3^- and has a $\delta^{13}C_{PDB}$ value of ~0‰ but varies slightly with depth in the water column. However, the $\delta^{13}C_{PDB}$ value of dissolved C in pore water, within a few centimeters of the sediment-water interface, can be markedly different (McCorkle et al. 1985). Variations in the $\delta^{13}C_{PDB}$ values of marine carbonates vary from -1 ‰ to +3 ‰, indicating that the $\delta^{13}C_{PDB}$ value of ancient seawater may not have remained constant (Holser 1984). In closed anoxic basins, dissolved C may become enriched in ^{13}C due to bacteriogenic reduction of CO_2 to CH_4 (Stiller and Magaritz 1974), or by evaporation (Stiller et al. 1985). Seawater carbonate represents an important source of C (Taylor 1987).

Sulfur in seawater occurs as sulfate (SO_4). Sulfate has a residence time in the ocean of 7.9 million years and is removed as either sulfide (sedimentary or organically bound sulfide) or sulfate (gypsum) (Holland 1978). The $\delta^{34}S$ value of SO_4 in present-day seawater is ~+20.9‰ (Rees 1978). However, analyses of sulfates in evaporites indicate that there was considerable variation in the $\delta^{34}S$ value of SO_4 in seawater during the Cenozoic and Paleozoic (Holser 1979, 1984; Claypool 1980). Seawater is an important source of S, containing 10% of the S in the crust and ocean system (Ohmoto 1986).

Meteoric water

Meteoric waters are defined as waters of any age that originated as precipitation including snow, rain, ice, lake, river and low-temperature groundwaters. They are derived from ocean water through atmospheric circulation processes (Sheppard 1986).

The mean O and H isotopic composition of meteoric waters at a specific locality is related to temperature, latitude, altitude, distance from the coast, and rate of precipitation (Figs. 9 and 10), and are related by the meteoric water line (MWL) equation:

$$\delta D = 8\,\delta^{18}O + 10 \qquad (13)$$

(Epstein and Mayeda 1953; Friedman 1953; Craig 1961; Dansgaard 1964; Friedman et al. 1964; Yurtsever and Gat 1981; Sheppard 1986). Most meteoric waters that have not

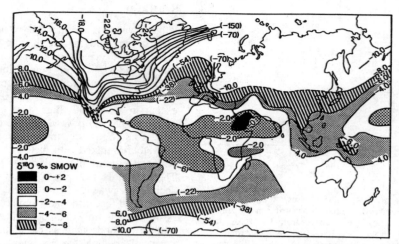

Figure 9. Global distribution of *mean* $\delta^{18}O$ and δD. The δD values are shown in brackets (from Sheppard 1986).

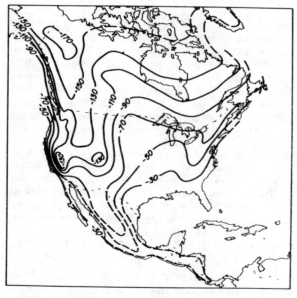

Figure 10. Map of North America showing generalized contours of average δD values of present day meteoric waters (from Taylor 1979). Reprinted with permission of John Wiley & Sons, from the book *Hydrothermal Ore Deposits*, p. 243, Fig. 6.3.

experienced large temperature fluctuations or evaporation plot within ±1‰ of the MWL and have negative δD and $\delta^{18}O$ values (Fig. 11).

Sources of C in meteoric water include atmospheric CO_2, carbonate minerals, soils, and microbial CO_2 (Deines 1980). The $\delta^{13}C_{PDB}$ value of atmospheric CO_2 is -7.0‰, which is lighter than CO_2 from volcanoes (-3 to -5‰) and heavier than CO_2 produced by combustion of fossil fuels (~-80 ‰) (Taylor 1986). Meteoric waters may acquire S during the oxidation of sulfides and sulfate minerals and solution of abiogenic and biogenic gases (Taylor 1987).

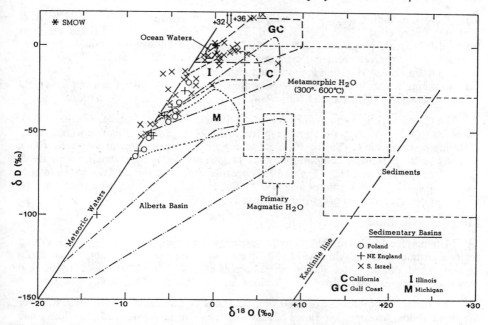

Figure 11. Isotopic compositions of formation waters, ocean waters, meteoric waters, metamorphic waters, magmatic waters, and common sedimentary rocks. The kaolinite weathering line is given for reference (from Sheppard 1986 and from Savin and Epstein 1970). Reprinted with permission of Elsevier, from *Geochim. Cosmochim Acta*, v. 34, p. 25-42.

Magmatic fluids and juvenile waters

Juvenile water is derived from the mantle and has never interacted with the hydrosphere (Sheppard 1986). The existence of such water is difficult to prove. It is defined as water in isotopic equilibrium with mafic magmas at temperatures near 1200°C (Ohmoto 1986). The estimated δD and $\delta^{18}O$ values for juvenile waters are -65±20‰ and +6±1‰, respectively (Taylor 1979; Ohmoto 1986). The contribution of this water to the crustal reservoir and thus to mineral precipitating fluids is probably insignificant.

Magmatic fluids are defined as the volatiles that exsolve from magmas as H_2O, CO_2, SO_2, and H_2S (Taylor 1987). The isotopic composition of these volatiles varies as a function of both source material and isotopic fractionation during degassing (Taylor 1986). The O and H isotopic compositions of magmatic fluids are calculated from the measured isotopic compositions of unaltered igneous rocks or minerals (Taylor and Sheppard 1986) by applying the mineral-H_2O stable isotope geothermometers at temperatures of 700°C to 1200°C (Ohmoto 1986).

The $\delta^{18}O$ values of magmatic fluids are typically in the range of +5.5‰ to +9.5‰ (Taylor 1979; Sheppard 1986). However, magmatic fluids may have $\delta^{18}O$ values that plot outside this range because the O isotopic fractionation factors between minerals and melts or fluids depend strongly on the chemical composition of the melt and fluid, and pressure (Shettle 1978; Ohmoto and Oskvarek 1985). In addition, isotopic exchange between magmatic fluids and wall rocks can occur during cooling and therefore may change the original isotopic compositions of magmatic fluids (Ohmoto 1986).

The δD values of magmatic fluids can vary during degassing and are dependent on many factors, including whether degassing occurs in a closed- or open-system (Nabelek et al. 1983; Taylor et al. 1983; Taylor 1986). The δD values of magmatic fluids exsolved from felsic magmas range from -30‰ to -60‰ (Sheppard 1986). These fluids represent recycled fluids, derived from the melting of hydrous crust (Magaritz and Taylor 1976; Savin and Epstein 1970). Fluids associated with mafic magmas have slightly lower δD values ranging from -50‰ to -70‰ and apparently unaltered MORB has even lower δD values of -80±10‰ (Kyser and O'Neil 1984).

The C isotopic composition of both felsic and mafic magmas is similar (Taylor 1986). Exsolved CO_2 has $\delta^{13}C_{PDB}$ values ranging from -5.5‰ to -3.0‰. This suggests mantle-buffering of magmatic C (Taylor 1986). The S isotopic composition of magmatic fluids depends on the S isotopic composition of the magma and the relative abundance of the S species such as SO_2 and H_2S (Taylor 1987) because SO_2 concentrates ^{34}S relative to H_2S by ~+2‰ to +4‰ at temperatures near 800°C (Thode et al. 1971). Magmatic fluids may be important in the deposition of igneous U deposits.

Metamorphic fluids

Metamorphic fluids are defined as fluids released by dehydration of minerals during regional metamorphism (White 1974; Sheppard 1986). However, it is not certain that a separate fluid phase exists during metamorphism. Pore fluids, meteoric fluids and magmatic fluids may all be involved to various degrees (Rumble 1982; Valley 1986).

The isotopic composition of metamorphic fluids is calculated in a similar way for magmatic fluids. Metamorphic fluids have a wide range in δD values (0‰ to -70‰) and $\delta^{18}O$ values (+3‰ to +20‰; Sheppard 1986). The wide range in isotopic composition of metamorphic fluids is likely due to the variable isotopic composition of the different rock types that are being metamorphosed and their water-rock history (Ohmoto 1986).

Carbon in metamorphic fluids generally occurs as CH_4 and CO_2 (Taylor 1987). CO_2 is the dominant C-bearing species in the absence of graphite, whereas CH_4 occurs a lower temperatures and fO_2 (Ohmoto and Kerrick 1977; Holloway 1984). Decarbonation reactions during metamorphism produce CO_2 that is enriched in ^{13}C and therefore metamorphic fluids with $CO_2/CH_4 >> 1$ can have $\delta^{13}C_{PDB}$ values from -2‰ to +6.0‰. Fluids buffered by graphite can have lower $\delta^{13}C_{PDB}$ values for CH_4 (-34±4‰) and CO_2 (-13±2‰; Bottinga 1968; Ohmoto and Kerrick 1977; Valley and O'Neil 1981).

The S isotopic composition of metamorphic fluids is not well known. The breakdown of pyrrhotite and pyrite during metamorphism may be the source of reduced S in metamorphic fluids (Ripley 1981; Franklin 1986) and therefore the resulting H_2S may be enriched in ^{34}S by +0.5‰ to +0.9‰ at 300 to 500 C (Kajiwara and Krouse 1971).

Connate and formation waters, and basinal brines

Connate waters are waters trapped in the pores of sediments at the time of formation, so that their initial isotopic composition would be identical to those of ancient seawater (Sheppard 1986). Formation waters are also found in the pores of sediments, but there is no implication concerning their age or origin (Sheppard 1986). Basinal brines are very saline (10 to 40 wt % dissolved salts) formation waters.

The changes in $\delta^{18}O$ and δD values of formation waters depend on the type of initial fluid (i.e. seawater vs. meteoric water), the lithology of the rocks or sediments through which the fluids passed, and other geochemical parameters. The O and H isotopic composition of formation waters from different sedimentary basins are summarized in Figure 11. The $\delta^{18}O$ values of formation waters tend to increase with increasing temperature and salinity. For many basins, the δD values also increase with increasing temperature or $\delta^{18}O$ values (Sheppard 1986).

Generally, brines with the lowest temperatures and salinity have the lowest $\delta^{18}O$ and δD values, approaching present-day local meteoric water. High temperature and salinity brines from different basins appear to have similar isotopic compositions ($\delta^{18}O$ = +7±3‰ and δD = -10±30‰; Ohmoto 1986). However, evaporite brines formed from the evaporation of seawater can have $\delta^{18}O$ and δD values as high as +10‰ and +30‰, respectively (Lloyd 1966; Pierre et al. 1984; Knauth and Beeunas 1986; O'Neil et al. 1986; Longstaffe 1987). For further discussion see Longstaffe (1987) and Welhan (1987).

Basinal brines play an important role in the deposition of some of the largest U deposits in the world (e.g. unconformity-type U deposits). Kotzer and Kyser (1995) used the measured $\delta^{18}O$ and δD values of diagenetic minerals such as illite, kaolinite, and quartz in conjunction with stable isotope geothermometers to calculate the temperature and $\delta^{18}O$ and δD values of the diagenetic, high salinity basinal brines associated with the Athabasca basin. They concluded that these 200 °C brines had $\delta^{18}O$ and δD values of +4±2‰ and -60±20‰, respectively. These oxidizing brines transported the U and precipitated uraninite when they mixed with reducing fluids (see below).

STABLE ISOTOPIC STUDIES OF URANIUM DEPOSITS

The world's major deposits of uranium occur in several distinctly different geological environments. Fifteen principal types of U deposits are recognized and are listed in Table 3. More than 40 subtypes and classes can be attributed to the 15 types and are described in detail by Dahlkamp (1993). This classification is largely based on the genetic interpretation and host environment of these deposits. Therefore, it involves some overlap among the forty deposit groups and is open to interpretation. However, it is simple and provides a framework when discussing and comparing U deposits that occur in different geological settings. However, not all deposit types are of economic importance. Approximately 95% of the Earth's U reserve is derived from four principal types of U deposits: (1) unconformity-type, (2) quartz-pebble conglomerate, (3) sandstone, and (4) vein-type (Nash et al. 1981; Dahlkamp 1993).

Table 3. U deposit types*

1. Quartz-pebble conglomerate
2. Unconformity-type
3. Sandstone
4. Volcanic hosted
5. Intrusive hosted
6. Synmetamorphic
7. Epimetamorphic
8. Vein
9. Collapse breccia
10. Breccia complex
11. Surficial
12. Metasomatic
13. Black shale
14. Lignite
15. Phosphorite

*Modified from Dahlkamp (1993)

The geology and processes that are involved in the genesis of these major types of U deposits have been described in numerous studies. Most stable isotopic studies of U deposits use the stable isotopic composition of clay, carbonate and silicate minerals associated with U-bearing minerals to characterize the fluids and mechanisms responsible for U deposition. Few studies report O or H isotopic values for U-bearing minerals. Due to the importance of U as a source of energy and its use in nuclear weapons, some of the data on U deposits are likely unavailable to the general scientific community. Therefore, the following is a review of material that is available to the scientific community.

Unconformity-type uranium deposits

Unconformity-type U deposits are associated with an unconformity that separates crystalline basement rocks that are laterally altered by weathering and overlain by sediments. The largest unconformity-type U deposits occur in two regions; Northern Territory, Australia, and Saskatchewan, Canada.

Canadian deposits. The majority of the Canadian unconformity-type U deposits occur along the eastern margin of the Proterozoic sandstone Athabasca Basin (Fig. 12). The Athabasca Basin consists of a sequence of Helikian poly-cyclic, mature, fluvial to marine quartz sandstones collectively referred to as the Athabasca Group (Ramaekers and Dunn 1977; Ramaekers 1981). These rocks unconformably overly Aphebian metasedimentary and Archean gneisses of the Wollaston Domain of the Trans-Hudson Orogen (Lewry and Sibbald 1977, 1980; Lewry et al. 1985; MacDonald 1987). Diagenesis of the Athabasca sandstones formed basin-wide assemblages of clay, oxide, and silicate minerals (Kotzer and Kyser 1995).

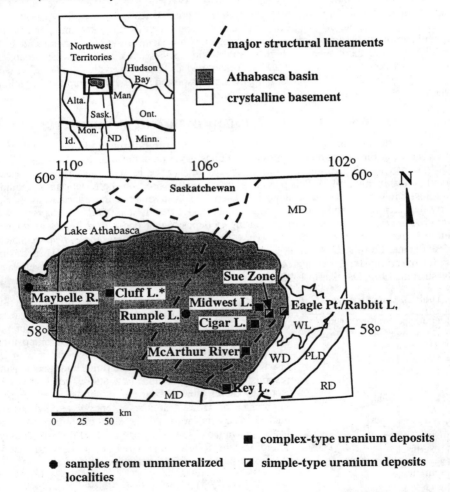

Figure 12. Map showing the extent of the Athabasca Basin, location of the U deposits and major lithostructural domains (modified from Fayek and Kyser 1997). Abreviations: MD: Mudjatik Domain, WD: Wollaston Domain, PLD: Peter Lake Domain, RD: Rottenstone Domain, and WL: Wollaston Lake, R: river, L: lake.

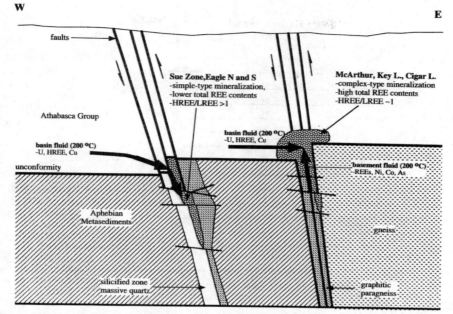

Figure 13. Schematic cross-section of the two major types of unconformity-type U deposits found in the Athabasca Basin, showing a simplified pattern of circulation of basin- and basement-derived fluids (from Fayek and Kyser 1997).

The U deposits occur at or near the intersection between Hudsonian-age faults and the unconformity. The majority of the high-grade U mineralization in these deposits occurs as lenses of massive uraninite, the most common reduced (U^{4+}) U-bearing mineral. On the basis of the spatial association of U minerals with the unconformity and the sulfide mineral assemblages associated with U mineralization, two distinct types of unconformity-type U deposits have been identified in the Athabasca Basin. Uranium deposits formed at the unconformity, hosted partially by sandstone, are referred to as complex-type deposits because Ni-Co-As-Fe-Cu-Pb sulfides and arsenides are associated with U minerals. In contrast, U deposits hosted entirely within fractures in basement rocks are referred to as simple-type deposits, with trace amounts of sulfides and arsenides associated with the U mineralization (Fig. 13).

Intense hydrothermal alteration in the sandstone, basement gneisses and metasediments surrounding the U deposits is characterized by large halos of illitized sandstone and local enrichments of K, Mg, Ca, B, U, Ni, Co, As, Cu, and Fe (e.g. Hoeve et al. 1980; Wallis et al. 1983; Kotzer and Kyser 1990a, 1995). Extensive areas of silicification with dravite, kaolinite, chlorite and graphite are associated with some deposits (Marlatt et al. 1992). Subsequent infiltration of low-temperature meteoric water along reactivated fault zones that host U mineralization precipitated blocky kaolinite, partially altered uraninite to coffinite and uranyl oxide hydrate minerals, and locally remobilized and degraded many of the original high-grade deposits. The fluid-mineral age relationships of various minerals are summarized in Figure 14.

Kotzer and Kyser (1995) measured the δD and $\delta^{18}O$ values of the diagenetic clay and silicate minerals, hydrothermal alteration associated with U-bearing mineral and basement rocks, and blocky kaolinite in reactivated fault zones. Using fluid inclusions from

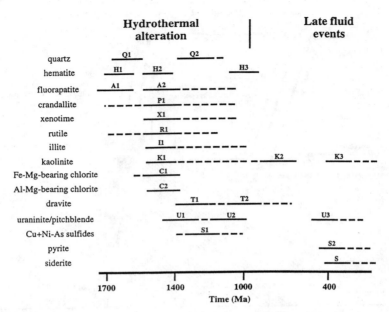

Figure 14. Fluid-mineral-age relationships of various minerals throughout the Athabasca Basin (from Fayek and Kyser 1997). Letters refer to specific generations of minerals.

diagenetic quartz overgrowths and mineral-water geothermometers, Kotzer and Kyser (1995) determined that three isotopically distinct fluids affected the basin and were associated with U mineral precipitation: (1) a 200°C basinal brine with δD and $\delta^{18}O$ values of -60‰ and +4‰, respectively, (2) 200°C basement fluid with δD and $\delta^{18}O$ values of -60‰ to -15‰ and +2‰ to +9‰, respectively, and (3) late, low-temperature (50°C) meteoric waters with δD and $\delta^{18}O$ values of -130‰ and -16‰, respectively (Fig. 5).

The intermediate δD and $\delta^{18}O$ values of hydrothermal minerals associated with U-bearing minerals, relative to the δD and $\delta^{18}O$ values of diagenetic minerals and alteration minerals found in basement rocks, as well as other isotopic measurements (see Chapter 4, this volume), showed that these hydrothermal minerals formed from mixing between the basin and basement fluids. The δD and $\delta^{18}O$ value of the fluid in equilibrium with these minerals was calculated to be -55±25‰ and +4±‰, respectively. Therefore, the U-bearing minerals (uraninite), which are in textural equilibrium with the hydrothermal minerals, likely precipitated from the same fluid (i.e. by mixing between the basin and basement fluids; Kotzer and Kyser 1995).

Recent studies of U minerals from unconformity-type U deposits in Saskatchewan, Canada, indicate that there are three stages (stage 1, stage 2, and stage 3) of U minerals that coincide with the three major fluids events (Fayek and Kyser 1997). High reflectance unaltered stage 1 and 2 uraninites have $\delta^{18}O$ values from -32‰ to -19.5‰ (Hoekstra and Katz 1956; Hattori et al. 1978; Kotzer and Kyser 1993, 1995; Fayek and Kyser 1997). Theoretical and experimental uraninite-water geothermometers (Hattori and Halas 1982; Zheng 1991, 1995; Fayek and Kyser 1998) indicate that uraninite would have been in equilibrium with fluids that had $\delta^{18}O$ values of less than -11‰ at ca. 200°C, the temperature at which the U deposits formed (Kotzer and Kyser 1990b). However, the dominant fluids that equilibrated with silicate and clay minerals associated with primary U mineralization from both simple and complex-type deposits are saline fluids having $\delta^{18}O$ values of 4±3‰ (Kotzer and Kyser 1990b, Rees 1992, Kotzer and Kyser 1993, Percival et al. 1993). Therefore, the uraninite-water geothermometers, in conjunction with the low $\delta^{18}O$ values of ca. -28‰ for uraninite, indicate that the O isotopic composition of the uraninite was completely overprinted by relatively recent, low temperature meteoric fluids with $\delta^{18}O$ values of ca. -18‰ (Hoekstra and Katz 1956; Hattori et al. 1978; Kotzer and Kyser 1990a, 1993), and that the uraninite can exchange O isotopes with fluids, resulting in substantial differences in U-Pb isotope systematics, but only minor disturbances to their chemical composition and texture. However, these fluids must be reducing because U mineral chemistry and solubility is largely a function of fO_2 and uraninite is stable only under very reducing conditions ($fO_2 < 10^{-25}$, Romberger 1985; Cramer 1995).

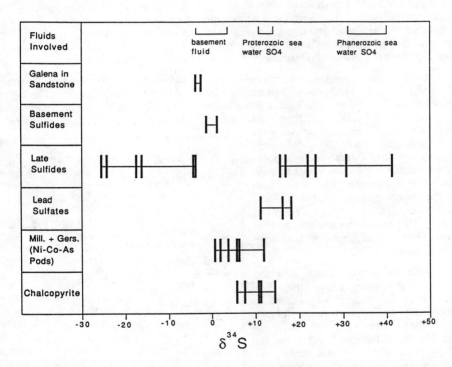

Figure 15. $\delta^{34}S$ values of various sulfides and sulfates associated with U minerals from the Athabasca Basin. The $\delta^{34}S$ values of Millerite (Mill.), gersdorffite (Ger.) and chalcopyrite which are co-evil with uraninite can be modeled by a mixing of basement fluid and basin fluid. Reprinted with permission of Saskatchewan Geological Society from *Modern Exploration Techniques*, v. 10, p.124, Fig. 6, Kotzer and Kyser (1990a).

Sulfur isotopic compositions of sulfides were used to further constrain the fluid compositions during development of U deposits in the Athabasca Basin. The range in $\delta^{34}S$ values for Ni-Co-As sulfides and chalcopyrite associated with U minerals also suggests fluid mixing between two isotopically distinct fluids; basement fluid with $\delta^{34}S$ values near 0‰ and a basin fluid containing a Proterozoic seawater having a $\delta^{34}S$ value near +12‰ (Fig. 15). These sulfides reflect the reducing nature of the hydrothermal system in the vicinity of the U deposits. Pyrite occurring in late fractures in the alteration halo and elsewhere in the basin has highly variable $\delta^{34}S$ values (Fig. 15). The late paragenesis of the pyrite coupled with highly variable $\delta^{34}S$ values indicate their formation during periods of extreme fO_2 variations believed to have developed during the period of re-activation and fracturing of fault zones and subsequent incursion of low-temperature meteoric waters into the Athabasca Basin (Kotzer and Kyser 1990a). Lead sulfate minerals occurring in Pb dispersion haloes in the sandstones above some deposits (i.e. Key Lake) have $\delta^{34}S$ values similar to Devonian seawater sulfate, indicating late-stage fluid movement around the U deposits (Kotzer and Kyser 1990a).

Various types of organic matter have been shown to exist with U deposits (Leventhal et al. 1987; Kyser et al. 1988; Dubessy et al. 1988). Kyser et al. (1988) completed a study of the graphite and bituminous material associated with the U deposits. They showed that hydrocarbon buttons associated with U minerals at Key Lake have low $\delta^{13}C$ values (approximately -50 ‰), whereas similar bituminous materials near U minerals have $\delta^{13}C$ values similar to altered and unaltered graphites. The low $\delta^{13}C$ values of the hydrocarbon buttons are attributed to radiolysis and hydrolysis of graphite and water intimately associated with U minerals (Dubessy et al. 1988; Kotzer and Kyser 1990a).

Australian deposits. Unconformity-type U deposits from the Northern Territory, Australia are located in three areas within the Pine Creek geosyncline (Fig. 16). The largest deposits are located in the Alligator River district (Nash et al. 1981). The Precambrian sequence in the Alligator River area consists of middle Proterozoic sandstones, conglomerates and carbonaceous sediments that unconformably overly Archean to early Proterozoic granitoids and metasediments. The area has been regionally metamorphosed where temperatures reached 600±50°C and pressures of 6 to 8 kbars (Ferguson 1979) However, metamorphism pre-dates U mineral precipitation (Donnelly and Ferguson 1979).

Uranium minerals occur near the unconformity, along shear zones that cut across a variety of sediments (Needham and Stuart-Smith 1979). Uraninite is the most abundant U-bearing mineral (Ewers and Ferguson 1979). Locally, the sediments are characterized by extensive chlorite and the presence of interbedded massive calcite, magnesite and dolomite (Ewers and Ferguson 1979). All the deposits in the region, with the exception of the Nabarlek deposit, are mineralogically similar (Donnelly and Ferguson 1979; Ewers et al. 1983), and U minerals are associated with breccia zones containing sulfides, carbonates, graphite, and chlorite (Ewers and Ferguson 1980). Most deposits are associate with massive bedded carbonates. However, no such carbonates are found at the Nabarlek deposit. In addition, U minerals at the Nabarlek deposit are associated with hematite and sericite, which are absent in the other deposits that contain chlorite, pyrite, chalcopyrite, galena, and sphalerite (Donnelly and Ferguson 1979; Ewers et al. 1983).

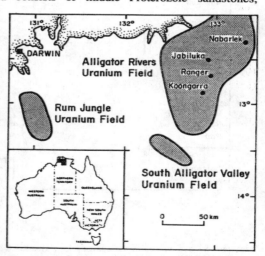

Figure 16. Locality map of the Pine Creek geosyncline showing major U fields. Reprinted with permission of *Economic Geology*, v. 78, p. 823-837, Ewers, Ferguson and Donnelly (1983).

Sulfides occur as disseminated grains throughout the sedimentary strata and in vugs and veins within the brecciated U mineral zones (Donnelly and Ferguson 1979). Disseminated sulfides, mainly pyrite, have $\delta^{34}S$ values near 0‰ (Fig. 17) and are interpreted to have formed in an environment where bacterial reduction was absent (Monster et al. 1979; Cameron 1982; Hoefs 1997). In contrast, vug and vein sulfides are comprised of chalcopyrite, galena, sphalerite, and pyrite (Donnelly and Ferguson 1979). The $\delta^{34}S$ values of these sulfides range from -6‰ to +10‰ (Fig. 18). The relatively $\delta^{34}S$-depleted isotopic composition of the sulfides, in conjunction with fluid inclusion homogenisation temperatures near ~200°C (Ypma and Fuzikawa 1979), suggest that these sulfides formed from

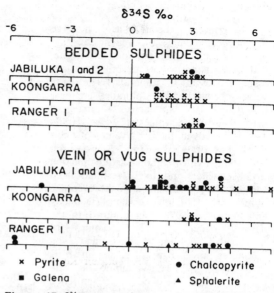

Figure 17. $\delta^{34}S$ values for bedded and vein sulfides (from Donnelly and Ferguson 1979).

hydrothermal fluids with isotopically heavy sulfate (Donnelly and Ferguson 1979; Hoefs 1997) because sulfides precipitated from hydrothermal solutions containing sulfate generally have light $\delta^{34}S$ values (Ohomoto 1972; Ohmoto 1986).

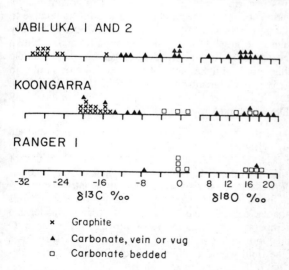

Figure 18. $\delta^{13}C$ and $\delta^{18}O$ values for graphite and bedded and vein carbonates (Donnelly and Ferguson 1979).

Sulfur isotope fractionation between coexisting sulfide mineral pairs (pyrite-chalcopyrite and pyrite-galena) in apparent equilibrium were used in conjunction with the appropriate S isotope geothermometer (Kajiware and Krouse 1971) to calculate temperatures of sulfide precipitation, and therefore to infer temperatures for U mineral deposition. Temperatures of sulfide and U mineral precipitation were calculated to be between 227°C and 315°C, which are slightly higher than fluid inclusion temperatures. The differences between the temperatures from fluid inclusion and temperatures calculated using isotope fractionation factors suggest that sulfide and U minerals precipitated at depths of 500 to 1000 m

(Donnelly and Ferguson 1979).

Carbonates occur as massive interbedded layers of magnesite and dolomite with minor amounts of calcite, and in vugs and veins in breccia zones associated with U minerals. The $\delta^{13}C$ values of interbedded dolomites are near 0‰ (Fig. 19), similar to sedimentary marine carbonate values (0±4‰; Hoefs 1997). However, the $\delta^{18}O$ values of these carbonates of +16‰ (Fig. 18) are not consistent with carbonate precipitation from seawater, and likely resulted from carbonate recrystallization in the presence of relatively ^{18}O-depleted waters (Donnelly and Ferguson 1979). In contrast to the narrow spread of $\delta^{13}C$ values of interbedded dolomites, vein and vug dolomites have a wide range in $\delta^{13}C$ values (Fig. 19). The variability is likely caused by the presence of ^{12}C-enriched organically derived CO_2 ($\delta^{13}C$ = -25±5‰; Eichmann and Schidlowski 1975) during dolomite precipitation (Donnelly and Ferguson 1979). A reasonable correlation exists between the $\delta^{13}C$ values and $\delta^{18}O$ values of dolomites from all the deposits (Fig. 19). However, the majority of the dolomites from the Jabiluka deposits have slightly higher $\delta^{13}C$ values and plot on a different curve relative to dolomites from the deposits (Fig. 19). The difference between the two curves is attributed to greater involvement of organically derived CO_2 during dolomite precipitation at the Koongarra and Ranger deposits relative to the Jabiluka deposits, which would cause carbonates to have lower $\delta^{13}C$ values (Donnelly and Ferguson 1979). The $\delta^{13}C$ values of graphite from the Koongarra and Jabiluka deposits are difficult to interpret because it is difficult to distinguish between multiple generations of organic matter. Graphite from Jabiluka has $\delta^{13}C$ values that are similar to sedimentary organic matter ($\delta^{13}C$ = -25±5‰; Eichmann and Schidlowski 1975). However, graphite from the Koongarra deposit is ^{13}C-enriched (-21 to -14‰) and therefore has likely interacted with ^{13}C-enriched fluids (Donnelly and Ferguson 1979).

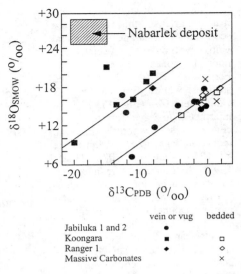

Figure 19. $\delta^{13}C$ values versus $\delta^{18}O$ values for carbonates from the Koongarra and Jabiluka deposits (from Donnelly and Ferguson 1979).

Carbonates from the Nabarlek deposit, which occur as inclusions and veinlets and are associated with uraninite, have a narrow range of $\delta^{13}C$ and $\delta^{18}O$ values from -19.7‰ to -14.8‰ and +25.1 to +28.7‰, respectively (Ewers et al. 1983). The $\delta^{13}C$ values of samples of calcite and dolomite are similar to the $\delta^{13}C$ values of carbonates from the other deposits in the region (see above); however their $\delta^{18}O$ values are significantly more ^{18}O-enriched. These high $\delta^{18}O$ values and low $\delta^{13}C$ values suggest that the hydrothermal fluids that precipitated carbonates at the Nabarlek deposit have interacted with organic matter and the metamorphic host rock at high temperatures. Thus, at low water/rock ratios the $\delta^{18}O$ value of the hydrothermal fluid would approach the $\delta^{18}O$ values of the host rock (Friedman and O'Neil 1977), the $\delta^{13}C$ value of the fluid would be light, and hydrothermal carbonates would have heavy $\delta^{18}O$ values and light $\delta^{13}C$ values (Ewer et al. 1983).

Quartz-pebble conglomerate deposits

Quartz-pebble conglomerate deposits consist of U and metallic elements (i.e. native

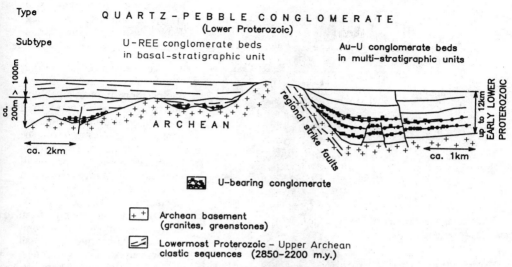

Figure 20. Schematic cross-section through two styles of quartz-pebble conglomerate U deposits (from Dahlkamp 1993). Reprinted with permission of Springer-Verlag from *Uranium Ore Deposits*, 460 p.

Au) and are generally similar throughout the world (Pretorius 1981; Vennemann et al. 1995). However, U/Au ratios differ from deposit to deposit (Vennemann et al. 1995). The largest deposits (containing the world's largest resource of Au and second largest of U) occur within the Late Archean Witswatersrand Supergroup, Witswatesrand region of South Africa and the Early Proterozoic Huronian Supergroup, Elliot Lake, Ontario, Canada (Nash et al. 1981; Dahlkamp 1993). The distribution of deposits and generalized stratigraphy of the Witswatesrand and Huronian Supergroups are shown in Figure 20. Other smaller occurrences are located in Brazil, India, Australia, Russia, and the United States (Houston and Karlstrom 1979; Dahlkamp 1993). A hydrothermal origin was initially considered for these deposits (Davidson 1957); however, it now seems that they are dominantly the product of placer processes with some post depositional redistribution (i.e. modification by diagenetic/metamorphic processes; Liebenberg 1956; Ramdohr 1958; Robertson 1962; Robertson 1976; Pretorius 1976, 1981; Ruzika 1988; Dahlkamp 1993).

Quartz-pebble conglomerate deposits are generally restricted to basal lower Proterozoic units unconformably overlying Archean rocks (Fig. 20; Dahlkamp 1993). Gold- and U-bearing strata are characterized by detrital grains of native Au, U and pyrite (Nash et al. 1981; Dahlkamp 1993). These minerals are concentrated along unconformities and along "leaders" at the Witswatersrand, believed to represent algal mats developed on unconformities (Button 1979). The mineralogy of the conglomerates is typical of mature conglomerates and sandstones except for the presence of uraninite and Au, rounded grains of pyrite, and the absence of magnetite (Nash et al. 1981).

Uraninite and the heavy-metal-bearing minerals were likely liberated from their host rocks by physical weathering and erosional processes prior to development of an anaerobic atmosphere, permitting fluvial transport (Nash et al. 1981; Dahlkamp 1993). The diverse heavy mineral assemblage reflects petrologically different sources (Nash et al. 1981; Dahlkamp 1993). Under these conditions, uraninite and pyrite escaped oxidation during transportation. Post-depositional redistribution and mineral re-crystallization, mainly due to diagenetic processes, led to the formation of minerals such as brannerite, rutile and anatase

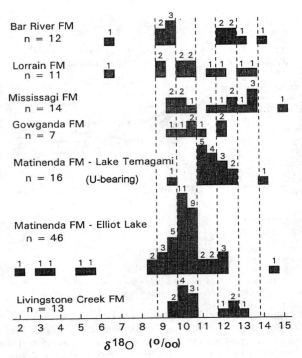

Figure 21. Comparison of the distributions of $\delta^{18}O$ values of quartz pebbles from a number of conglomerates from the Huronian Supergroup, Canada. Reprinted with permission of *Economic Geology*, v. 91:5, p. 322-342, from Vennemann, Kesler, Frederickson, Minter, and Heine (1995).

by reaction of fluids with uraninite, illmenite and magnetite (Nash et al. 1981; Dahlkamp 1993).

Vennemann et al. (1992, 1995) measured the O isotopic composition of quartz pebbles, quartz sands, and cherts from the mineral-bearing quartz-pebble conglomerates of the Witswatersrand and Huronian Supergroups as a means of evaluating the source terrain and post-depositional fluid-rock interaction. Quartz pebbles, quartz sands, and cherts from different lithologies that comprise the U-bearing Huronian Supergroup show a wide range in $\delta^{18}O$ values. Therefore, provenance may have played an important role in determining the Au and U contents of different lithologies. The $\delta^{18}O$ values of quartz pebbles are normally distributed about a mean of 10.2‰ (Fig. 21) and the quartz sands and cherts generally have $\delta^{18}O$ values <11.0‰ (Fig. 22a,b). For the Huronian Supergroup, a crude and empirical relationship exists between Au content and proportion of quartz pebbles with $\delta^{18}O$ values >11.5‰ (Fig. 23; Vennemann et al. 1995). Thus, the source of the pebbles with high $\delta^{18}O$ values may have been Au-bearing hydrothermal quartz veins which generally consist of ^{18}O-enriched quartz (Vennemann et al. 1995).

In contrast, quartz pebbles, quartz sands, and cherts from different lithologies that comprise the U and Au-bearing Witswatersrand Supergroup show much lower variation in their $\delta^{18}O$ values. Quartz pebbles have a mean $\delta^{18}O$ value of 11.4‰ (Fig. 24), whereas quartz sands have $\delta^{18}O$ values <11.0‰, similar to those from the Huronian Supergroup (Fig. 24). However, an empirical relationship does not exist between Au content and $\delta^{18}O$ values of quartz pebbles for the Witswatersrand Supergroup (Fig. 23; Vennemann et al. 1995). The relatively similar $\delta^{18}O$ values of quartz pebbles and quartz sands from different lithologies suggests that they came from a similar source and suggests a more or less uniform source terrane for the Witswatersrand basin as a whole (Vennemann et al. 1995).

A comparison of the distribution of $\delta^{18}O$ values of quartz pebbles and quartz sands between conglomerates of the Witswatesrand and Huronian Supergroups suggests that the most prominent sources of Au are likely the Archean lode Au deposits that may be hosted by granites or greenstones (Fig. 25). The U is most likely derived from the compositionally restricted granitic rocks (Vennemann et al. 1995).

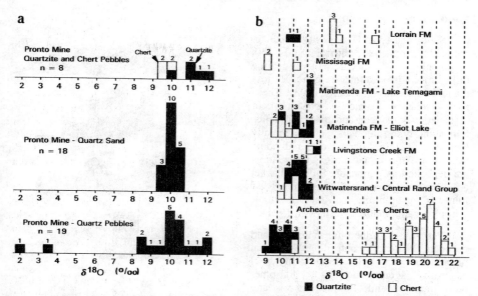

Figure 22. (a) Compilation of the $\delta^{18}O$ values of quartz pebbles, sands, quartzite, and chert pebbles from the Pronto mine, Matinenda Formation, Elliot Lake area, and (b) Compilation of the $\delta^{18}O$ values of chert and quartzite pebbles from several conglomerates from the Huronian Supergroup, Canada (Lorrain, Mississagi, Matinenda, Elliot Lake, and Livingstone Creek), Archean cherts from Wawa and Temagmi banded iron formations, Canada and Central Rand Group, Witswatersrand. Reprinted with permission of *Economic Geology*, v. 91, p. 322-342, from Vennemann, Kesler, Frederickson, Minter and Heine (1995).

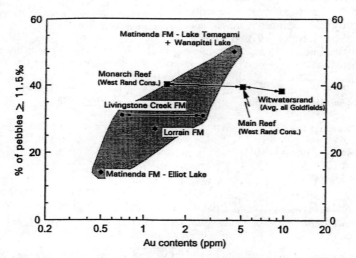

Figure 23. Plot of Au content (in ppm) versus the percentage quartz pebbles with $\delta^{18}O$ values 11.5‰, for conglomerates in the Huronian and Witswatersrand Supergroups (reprinted with permission of *Economic Geology*, v. 91, p. 322-342, from Vennemann, Kesler, Frederickson, Minter and Heine 1995).

Sandstone-type deposits

Sandstone-type U deposits occur in reduced continental fluvial and less commonly in mixed fluvial-marine (arkosic) sandstones that are generally less than 400 Ma old (Dahlkamp 1993). These rock sequences typically contain both diagenetically reduced and oxidized facies. Sandstone-type deposits are generally localized within parts of individual sandstone depositional units. These units are deposited as permeable fluvial channel and bar sandstones. The primary U minerals are uraninite, coffinite and uraniferous organic matter (Nash et al. 1981).

The most important physical and geochemical factors controlling U mineral precipitation are: (1) permeability, (2) adsorptive agents such as coal, other humic material, and Ti-oxide minerals, and (3) reducing agents such as carbonaceous matter, S species related to biogenic sulfate reduction, and limited oxidative destruction of pyrite or marcasite (Nash et al. 1981).

Geochemical data and isotopic ages suggest that sandstone-type deposits formed shortly after sedimentation (Miller and Kulp 1963; Berglof 1970; Lee and Brookins 1978; Ludwig 1978, 1979; Ludwig et al. 1984). Some deposits are enriched in

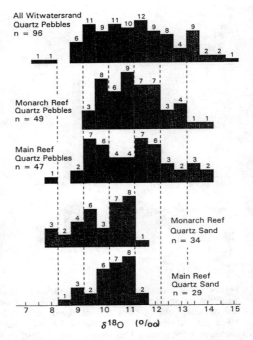

Figure 24. Compilation of the $\delta^{18}O$ values of quartz pebbles and quartz sands from the Main and Monarch reefs of the West Rand Consolidated Au mine and Witswatersrand conglomerates. Reprint permission same as Figure 23.

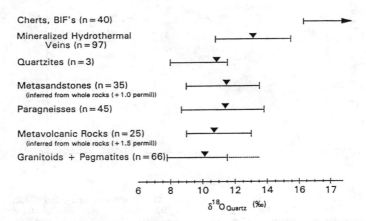

Figure 25. Compilation of the $\delta^{18}O$ values for quartz from Archean granitoids and associated pegmatites, metavolcanic rocks, paragneisses, metasandstones, quartzites, mineralized hydrothermal veins, and cherts. Arrows indicate the mean values. Reprint permission same as Figure 23.

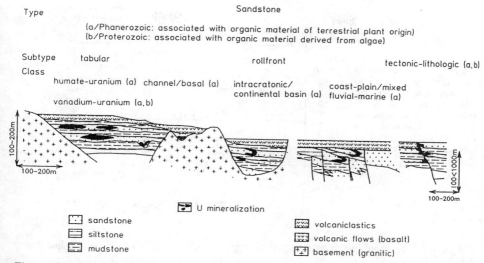

Figure 26. Schematic cross-section of three different types of sandstone-type U deposits including, (1) tabular, (2) roll-front, (3) tectonic-lithologic (from Dahlkamp 1993). Reprinted with permission of Springer-Verlag, from the book *Uranium Ore Deposits*, 460 p.

Se, Mo and V, which occur in discrete zones. The zoning is attributed to deposition across a geochemical gradient produced either at the mixing interface between two fluids (Shawe 1956), or a single solution that is changing due to reaction with the host rock (Rackley et al. 1968; Granger and Warren 1969). The source of U in these deposits is likely volcanic ash in tuffaceous mudstones and granites associated with host sandstones that were leached by saline oxidizing fluids (Denson and Gill 1956; Waters and Granger 1953; Rosholt and Bartel 1969; Stuckless et al. 1977; Granger and Warren 1978; Nash et al. 1981).

Based on the structural environment and elemental associations, sandstone-type U deposits may be divided into three general sub-types: (1) tabular V-U (e.g. Henry Basin, Utah, Oklo, Gabon, Africa), (2) roll front or roll-type (e.g. South Texas Coastal Plains,

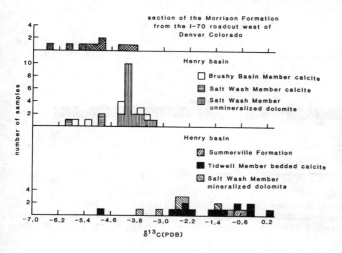

Figure 27. (a) $\delta^{13}C$ values of calcite and dolomite from the Summerville and Morrison Formations.

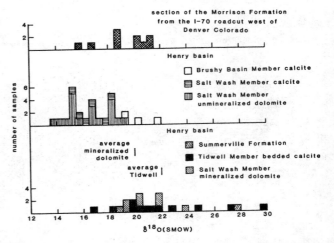

Figure 27, continued.
(b) $\delta^{18}O$ values of calcite and dolomite from the Summerville and Morrison Formations. Reprinted with permission of *Economic Geology*, v. 73, p. 1690-1705, from Goldhaber, Reynold and Rye (1978).

USA), and (3) tectonic-lithologic (e.g. Grants Uranium Region, USA) (Fig. 26; Dahlkamp 1993).

Tabular vanadium-uranium. Tabular V-U deposits in the Henry Basin, Utah occur in fluvial sandstones of the Salt Wash Member of the Morrison Formation of late Jurassic age (Northrop and Goldhaber 1990). Mineralized and weakly mineralized zones are bounded above and below by dolomite-rich cements. Vanadium occurs mostly as V chlorite. Coffinite, the dominant U-bearing mineral, occurs as small (2-5 μm) euhedral to subhedral crystals coating detrital quartz grains and intergrown with pore-lining and pore-filling V chlorite. The C and O isotopic composition of the dolomite cements (Fig. 27a,b) and waters in equilibrium with authigenic clay minerals (Fig. 28) associated with coffinite indicate that U mineral precipitation occurred at the boundary between two isotopically distinct fluids: (1) a brine which precipitated carbonates in rocks below the zone of mineralization and (2) meteoric fluids. The ^{34}S-depleted S isotopic composition of pyrite ($\delta^{34}S$ = -22‰ to -48‰) associated with coffinite indicate that the brine was enriched in

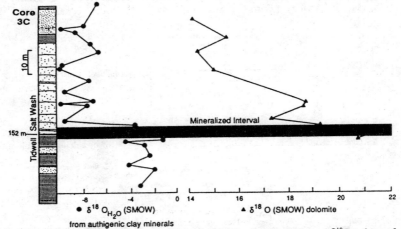

Figure 28. $\delta^{18}O$ values of water calculated from authigenic clay minerals, and $\delta^{18}O$ values of dolomite versus depth from core 3C. Reprinted with permission *of Economic Geology*, v. 85, p. 215-269, from Northrop and Goldhaber (1990).

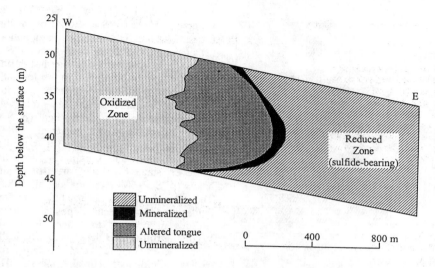

Figure 29. Cross section through a typical roll-type deposit showing relative position of the redox interface. Reprinted with permission of *Economic Geology*, v. 73, p. 1690-1705, from Goldhaber, Reynold and Rye (1978).

sulfate (Northrop and Goldhaber 1990) because sulfides precipitated from hydrothermal solutions containing sulfate generally have depleted $\delta^{34}S$ (Ohmoto 1972; Ohmoto 1986).

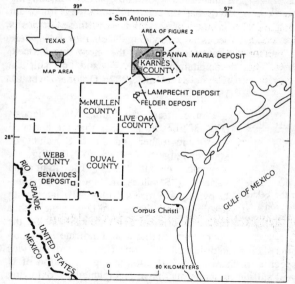

Figure 30. Map of south Texas coastal plain showing locations of the Panna Maria, Benavides, Felder, and Lamprecht U deposits (reprinted with permission of *Economic Geology*, v. 77:1, p. 541-556, from Reynold, Goldhaber and Carpenter (1982).

Roll-type. Roll-type deposits generally involve U mineralization that follows the redox front separating non-oxidized and oxidized sandstone. The mineralization transects the host rock strata, and in cross-section the deposits form crescent-shaped lenses and have altered inner zones or tongues (Fig. 29). Uraninite and coffinite occur as interstitial grains and coating on sand grains and are concentrated along the front part of the redox zone (Dahlkamp 1993).

Roll-type U deposits of the south Texas coastal plain have formed both in host rocks with and without organic matter (Reynolds et al. 1982). The Felder and Lamprecht deposits (Fig. 30) occur in rocks that are devoid of organic matter. Four distinct stages of sulfide minerals

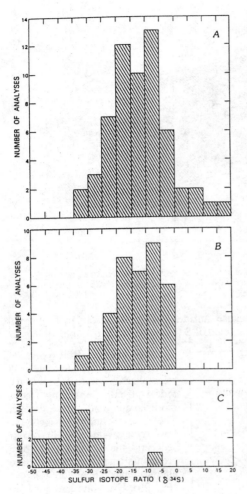

Figure 31. $\delta^{34}S$ values for sulfides from barren (A) and mineralized zones (B) from the Panna Maria deposit and mineralized zone from the Felder deposit (C). Reprinted with permission of *Economic Geology*, v. 77, p. 541-556, from Reynold, Goldhaber and Carpenter (1982).

are recognized: (1) a generation of isotopically light ($\delta^{34}S < -20‰$) pyrite that formed before U mineral precipitation, (2) a generation of isotopically light ($\delta^{34}S < -20‰$) pyrite that formed before and during U mineral precipitation, (3) isotopically heavy ($\delta^{34}S > 0‰$) pyrite that formed after U mineral precipitation, and (4) a relatively late generation of isotopically light ($\delta^{34}S < -20‰$) marcasite (Goldhaber et al. 1983). The Benavides deposit (Fig. 30) is similar to the Felder and Lamprecht deposits; however, stage 1 sulfides are isotopically heavy ($\delta^{34}S > 0‰$) (Reynolds et al. 1982). The altered tongue associated with the Felder and Lamprecht deposits contains sulfides in contrast to other Texas and Wyoming roll-type U deposits that typically have altered tongues that contain iron oxides (Goldhaber et al. 1983). The distribution of U, Mo, and Se are typical of roll-type U deposits that result from reaction between a geochemically reduced rock and an oxygenated U-bearing fluid (Goldhaber et al. 1983). Sulfide bearing fault-leaked solutions from underlying hydrocarbon accumulations were important in the preservation of these deposits because reducing conditions were maintained within the host sandstones, precipitating stage 3 and 4 sulfides (Reynolds et al. 1982; Goldhaber et al. 1983).

In contrast, the Panna Maria deposit (Fig. 30) occurs in rocks that are enriched in organic matter and the $\delta^{34}S$ values of sulfides range from -34‰ to -1‰ (Fig. 31; Reynolds et al. 1982). The light and variable $\delta^{34}S$ values of stage 1 pyrite (pre-U pyrite) and iron sulfides associated with uraninite precipitation are consistent with a biogenic origin. This deposit lacks the characteristics that would indicate formation and preservation involving reductants such as fault-leaked solutions (i.e. lack of sulfides with $\delta^{34}S > 0‰$). However, the specific mechanism for reduction and precipitation of U has not been identified. Uranium and sulfide minerals may have precipitated by oxidizing U-bearing fluids interacting with reducing organic matter, where reduction of U occurred in the presence of H_2S from organic matter or by chelation or adsorption of U onto organic matter. Reduction of U complexes in solution by metastable S species also has not been ruled out (Reynolds et al. 1982).

Vein-type uranium deposits

Vein deposits consist of U minerals in lenses or sheets, fissures, breccias, and stockwork (Dahlkamp 1993). These uraniferous veins can occur within leucogranites and syenites, near the margins of granites, volcanic rocks that are emplaced above a granitic complex, sediments and sedimentary basins, and metamorphosed sediments. The principal U minerals are uraninite and coffinite. Complex vein deposits contain Co, Ni, Bi, Ag, Cu, Pb, Zn, Mo, and Fe sulfides, arsenides and sulfarsenides, whereas these minerals are absent in simple vein deposits (Dahlkamp 1993).

Magmatic and hydrothermal fluids (recirculating meteoric waters) are generally associated with U mineral precipitation (Letian 1986; Wallace and Whelan 1986; Patrier et al. 1997). However, basinal brines may be associated with some complex vein deposits (e.g Beaverlodge, Saskatchewan, Canada and Bernardan U deposit, Western Marche, France; Rees 1992; Kotzer and Kyser 1993; Patrier et al. 1997). For example, fluids in equilibrium with the clay minerals associated with coffinite at the Bernardan U deposit, Western Marche, France had $\delta^{18}O$ values between 5‰ and 10‰ and precipitated coffinite at temperatures near 100°C (Patrier et al. 1997). The isotopic composition of these fluids is consistent with diagenetic fluids (Lee et al. 1989; Kotzer and Kyser 1993; Warren and Smalley 1993; Ziegler et al. 1994; Wilkinson et al. 1994). Kotzer and Kyser (1993) showed that fluids in equilibrium with silicate minerals associated with uraninite from complex vein-type deposits, Beaverlodge, Canada had similar $\delta^{18}O$ values (4±2‰) to the diagenetic fluids that precipitated the large unconformity-type U deposits from the Athabasca Basin, Canada.

Carbonates associated with uraninite from the Schwartzwalder vein U deposit, Colorado, have a $\delta^{18}O$ value of 18.1‰ (Wallace and Whelan 1986). The calculated $\delta^{18}O$ values of fluids associated with these carbonates range from 4.3‰ to 8.2‰. The O isotopic composition of the fluids during uraninite deposition is compatible with a magmatic source, felsic metamorphic rocks or diagenetic fluids (Wallace and Whelan 1986). The carbonates associated with uraninite have $\delta^{13}C$ value of -3.6‰, which is similar to the $\delta^{13}C$ value of the Proterozoic gneiss host-rock carbonates (-3.7‰). Therefore, the source of the C was likely the gneiss host rock. Sulfur isotopic compositions of sulfides associated with uraninite have a restricted range of -5.1‰ to +0.2‰. Therefore, the S could have originated from the metamorphic host rocks, which have $\delta^{34}S$ values near 0.6‰. However, the S values do not preclude a magmatic or sedimentary source (Wallace and Whelan 1986). The light stable isotopic data alone do not distinguish between a magmatic and evolved non-magmatic water. However, these data, in conjunction with geologic and other isotopic data, eliminate a magamtic contribution to the Schwartzwalder vein U. The likely source of the U mineral precipitating fluid is the metamorphic rocks basement (Wallace and Whelan 1986).

Vein-type U deposits from China are associated with granites and volcanic rocks and formed at temperatures near 200°C. Carbonates associated with uraninite have restricted $\delta^{13}C$ values from -7.9‰ to -7.0‰, indicating that the C is from a deep magmatic source (Letian 1986; Tiangang and Zhizhang 1986; Yu 1989). Pyrite associated with uraninite and barren veins have a range of $\delta^{34}S$ values from −15.3‰ to -4.9‰, the youngest veins having pyrite with the heaviest $\delta^{34}S$ values (Tiangang and Zhizhang 1986). This trend in $\delta^{34}S$ values likely reflects deposition from an evolving hydrothermal fluid. The $\delta^{18}O$ values of fluids in equilibrium with silicates associated with uraninite range from -5.5‰ to +3.3‰ and the δD values of fluid inclusions range from -88‰ to -54‰. These data are consistent with the hydrothermal model for these deposits; however, the data suggests that a meteoric water component may be associated with the formation of these U deposits (Letian 1986;

Tiangang and Zhizhang 1986; Yu 1989).

CONCLUSIONS

Stable isotopes have been used successfully to characterize fluids associated with many different types of U deposits. These fluids include magmatic, metamorphic, hydrothermal fluids, and basinal brines. The stable isotopic composition of silicates, oxides and sulfides associated with U minerals, in conjunction with geologic and other isotopic data can be used to develop models for the different types of U deposits. These models may include time, temperature, pressure, source of U, and the mechanism involved in U precipitation. In the future, *in situ* stable isotopic analyses (e.g. by SIMS or laser ablation) of complex phases such as uraninite may help refine these models and permit us to better understand the process involved in the formation and degradation of these deposits.

ACKNOWLEDGMENTS

The authors thank Teresa Mae Lassak for her diligence in obtaining the references used in this manuscript. Helpful comments by Dr. Cécile Engrand on earlier versions of this manuscript were greatly appreciated. The manuscript benefited greatly from the reviews by Drs. Peter C. Burns, Fred J. Longstaffe, and Tom G. Kotzer.

REFERENCES

Arnold M, Cuney M 1974 Une succession anormale de minéraux et ses conséquences sur l'example de la minéralisation uranifère des Bois Noirs-Limouzat (Forez-Massif Central français). Acad Sci Comptes Rendus Ser. D 279:535-538
Ayers DE, Eadington PJ 1975 Uranium mineralization in the South Alligator River Valley. Mineral Depos 10:27-41
Badham JPN, Robinson BW, Morton RD 1972 The geology and genesis of the Great Bear Lake silver deposit. 24th Int'l Geol Congress, Sec 4:541-548
Barsukov VL, Sushchevskaya TM, Malyshev BI 1971 Composition of solution forming pitchblende in a uranium-molybdenum deposit (abst.). Fluid Inclusion Research-Proceedings of COFFI 4:10
Berglof WR 1970 Absolute age relationships in selected Colorado Plateau uranium ores. Unpub PhD Thesis, Columbia Univ 149 p
Bigeleisen J 1965 Chemistry of isotopes. Science 147:463-471
Bigeleisen J, Mayer MG 1947 Calculation of equilibrium constants for isotopic exchange reactions. J Chem Phys 15:261-267
Bottinga Y 1968 Calculation of fractionation factors for carbon and oxygen exchange in the system calcite-carbon dioxide-water. J Phys Chem 72:4338-4340
Bottinga Y, Javoy M 1973 Comments on oxygen isotope geothermometry. Earth Planet Sci Lett 20:250-265.
Button A 1979 Algal concentrations in the Sabi River, Rhodesia: Depositional model for the Wiwatersrand carbon? Econ Geol 74:1876-1882
Campbell DD 1955 Geology of the pitchblende deposits of Port Radium, Great Bear Lake, N.W.T. Unpub PhD thesis, Calif Inst Tech
Cameron EM 1982 Sulfate and sulfate reduction in early Precambrian oceans. Nature 296:145-148
Claypool GE, Holser IR, Sakai H, Zak I 1980 The age curves of sulfur and oxygen isotopes in marine sulfate and their mutual interpretation. Chem Geol 28:199-259
Craig H 1961 Isotopic variations in meteoric waters. Science 133:1702-1703
Cramer JJ 1995 The Cigar Lake U deposit: analogue information for Canada's nuclear fuel waste disposal concept. *AECL Rep.* 11204:32
Cuney M 1974 Le gisement uranifère des Bois-Noirs-Limouzat (Massif Central France)-Relations entre minéraux et fluides. Unpub PhD thesis, Nancy, 174 p
Dahlkamp FJ 1993 Uranium Ore Deposits. Springer-Verlag, Berlin, 460 p
Dansgaard W 1964 Stable isotopes in precipitation. Tellus 16:436-468
Davidson RV 1957 On the occurrence of uranium in ancient conglomerates. Econ Geol 52:668-693
Dechow R, Jensen ML 1965 Sulfur isotopes of some Central African sulfide deposits. Econ Geol 60:894-

941
Deines P 1977 On the oxygen isotopic distribution among mineral triplets in igneous and metamorphic rocks. Geochem Cosmochim Acta 41:1709-1730

Deines P 1980 The isotopic composition of reduced organic carbon. *In* Fritz and Fontes (Eds) Handbook of Environmental Isotope Geochemistry 1: The Terrestrial Environment, 329-406

Denson NM, Gill JR 1956 Uranium-bearing lignite and its relation to volcanic tuff in eastern Montana and North Dakota. U S Geol Surv Prof Paper 300:413-418

Donnelly TH, Ferguson J 1979 A stable isotope study of three deposits in the Alligator River Uranium *In* Field NT, Int'l Uranium Symposium on the Pine Creek Geosyncline, 61-65

Driesner T 1997 The effect of pressure on deuterium-hydrogen fractionation in high-temperature water. Science 277:791-794

Driesner T 1998 Molecular-scale study of hydrogen and oxygen isotope salt and pressure effects (abstr). GSA Annual meeting–Toronto 30-7:A318.

Drummond SE, Ohmoto H 1985 Chemical evolution and mineral deposition in boiling hydrothermal systems. Econ Geol 80:126-147

Dubessy J, Pagel M, Beny J-M, Christensen H, Hickel B, Kosztolanyi C, Poty B 1988 Radiolyses evidence by H_2-O_2 and H_2-bearing fluid inclusions in three uranium deposits. Geochim Cosmochim Acta 52:1155-1167

Epstein S, Mayeda T 1953 Variations of 18O content of waters from natural sources. Geochim Cosmochim Acta 4:213-224

Ewers GR, Ferguson J 1979 Mineralogy of the Jabiluka, Ranger, Koongara and Nabarlek uranium deposits. Int'l Uranium Syposium on the Pine Creek Geosyncline, 67-70

Ewers GR, Ferguson J 1980 Mineralogy of the Jabiluka, Ranger, Koongara and Nabarlek uranium deposits. *In* Ferguson J, Goleby AD (Eds) Uranium in the Pine Creek Geosyncline. I A E A, Vienna, 363-374

Ewers GR, Ferguson J, Donnelly TH 1983 The Nabarlek uranium deposit, Northern Territory, Australia: Some petrologic and geochemical constraints on the genesis. Econ Geol 78:823-837

Eichmann R, Schidlowski M 1975 Isotopic fractionation between coexisting organic carbon-carbonate pairs in Precambrian sediments. Geochim Cosmochim Acta 39:585-595

Fayek M, Kyser TK 1997 Characterization of multiple fluid events and Rare Earth-Element mobility associated with formation of unconformity-type uranium deposits in the Athabasca Basin, Saskatchewan. Can Mineral 35:627-658

Fayek M, Kyser TK 1998 Low temperature oxygen isotopic fractionation in the uraninite-UO_2-CO_2-H_2O system. Geochim Cosmochim Acta (in press)

Ferguson J 1979 Metamorphism in the Pine Creek geosyncline and its bearing on stratigraphic correlations. Int'l Uranium Syposium on the Pine Creek Geosyncline, 71-75

Franklin JM 1986 Volcanic associated massive sulfide deposits-An update. *In* Andrew CJ, Crowe WA, Finlay S, Pennell WM, Pyne JF (Eds) Geology and Genesis of Mineral Deposits in Ireland. Irish Assoc Econ Geol Spec Vol, 49-69

Friedman I 1953 Deuterium content of natural water and other substances. Geochim Cosmochim Acta 4:89-103

Friedman I, O'Neil JR 1977 Compilation of stable isotope fractionation factors of geochemical interest. *In* Friedman I, O'Neil JR (Eds) Data of Geochemistry, 6th ed. Geol Surv Prof Paper 440-KK

Friedman I, Redfield AC, Schoen B, Harris J 1964 The variation in dueterium content of natural waters in the hydrogenic cycle. Rev Geophys 2:177-224

Gornitz V, Kerr PE 1970 Uranium mineralization and alteration, Orphan Mine, Grand Canyon, AZ. Econ Geol 65:751-768

Granger HC, Warren CG 1969 Unstable sulfur componds and the origin of roll type uranium deposits. Econ Geol 64:160-171

Granger HC, Warren CG 1978 Some speculations on the genetic geochemitsry and hydrology of roll-type uranium deposits. Wyoming Geol. Assoc. Guidebook, 30th Ann Field Conf, 349-361

Goldhaber MB, Reynold RL, Rye RO 1978 Origin of the south Texas Roll type uranium deposit: II. Sulfide petrology and sulfur isotope studies. Econ Geol 73:1690-1705

Goldhaber MB, Reynold RL, Rye RO 1983 Role of fluid mixing and fault related sulfide in the origin of the Ray Point uranium district, south Texas. Econ Geol 78:1043-1063

Harlass E, Schützel H 1965 Zur paragenetischen Stellung der Uranpechblende in den hydrothermalen Lagerstätten deswestlichen Erzgebirges. Zeits Angew Geologie 11:569-581

Hattori K, Halas S 1982 Calculation of oxygen isotope fractionation between uranium dioxide, uranium trioxide and water. Geochim Cosmochim Acta 46:1863-1868

Hattori K, Muehlenbachs K, Morton D 1978 Oxygen isotope geochemistry of uraninites. Geol Soc Am-Geol Soc Canada, Prog Abst 10:A417

Heyse JV 1971 Mineralogy and paragenesis of the Schwartzwalder mine uranium ore, Jefferson County,

Colorado. U S Atomic Energy Comm. Rept GJO-912, 91 p
Hoefs J 1997 Stable Isotope Geochemistry (4th ed.). Springer-Verlag, Berlin, 201 p
Hoekstra HR, Katz JJ 1956 Isotope geology of some uranium minerals. Geological Survey Professional Paper 300. Contributions to the geology of uranium and thorium by the United States Geological Survey and Atomic Energy Commission for the United Nations Int'l Conference on Peaceful Uses of Atomic Energy, Geneva, 543-547
Hoeve J, Sibbald TII, Ramaekers P, Lewry JF 1980 Athabasca Basin unconformity-type uranium deposits: A special class of sandstone-type deposits. *In* Fergusom S, Goleby A (Eds) Uranium in the Pine Creek Geosyncline, Vienna, I A E A, 575-594
Holland HD 1978 The Chemistry of the Atmosphere and Oceans. J. Whiley & Sons, New York, 355 p
Holloway JR 1984 Graphite-CH4-H2O-CO2 equilibria at low grade-metamorhic conditions. Geology 12:455-458
Holser WT 1979 Trace elements and isotopes in evaporites. *In* Burns, RG (Ed) Marine Minerals. Rev Mineral 6:295-346
Holser WT 1984 Gradual and abrupt shifts in ocean chemistry during Phanerozoic time. *In* Holland HD, Trendall AF (Eds) Patterns of Change in Earth Evolution. Springer-Verlag, New York, 123-143
Horita J 1989a Analytical aspects of stable isotopes in brines. Chem Geol 79:107-179
Horita J 1989b Stable isotope fractionation factors of water in hydrated saline mineral brine syatems. Earth Planet Sci Lett 95:2797-2817
Horita J, Wesolowski DJ, Cole DR 1993a The activity-composition relationship of oxygen and hydrogen isotopes in aqueous salt solutions. 1. Vapor-liquid water equilibration of single salt solutions from 50°C to 100°C. Geochim Cosmochim 57:2797-2817
Horita J, Cole DR, Wesolowski DJ 1993b The activity-composition relationship of oxygen and hydrogen isotopes in aqueous salt solutions. 2. Vapor-liquid water equilibration of mixed salt solutions from 50°C to 100°C. Geochim Cosmochim Acta 57:4703-4711
Horita J, Cole DR, Wesolowski DJ 1995 The activity-composition relationship of oxygen and hydrogen isotopes in aqueous salt solutions. 3. Vapor-liquid water equilibration of NaCl solutions to 350 °C. Geochim Cosmochim Acta 59:1139-1151
Horita J, Driesner T, Cole DR, Venneman T 1998 Experimental study of pressure effect on D/H fractionation between hydrous minerals and water (abstr.). GSA Annual Meeting–Toronto, 30:A271
Houston RS, Karlstrom KE 1979 Uranium-bearing quartz-pebble conglomerates: Exploration model and United States resource potential: Grand Junction. U S Department of Energy Rept GJBX-1(80), 510 p
Hulston JR 1978 Methods of calculating isotopic fractionation in minerals. *In* Stable isotopes in the earth sciences. DSIR Bull. 220:211-219
Javoy M 1977 Stable isotopes and geothermometry. J Geol Soc 133:609-636
Kazahaya K, Matsuo S 1986 D/H and $^{16}O/^{18}O$ fractionation in NaCl aqueous-vapor systems at elevated temperatures (abstr.). Terra Cognita 6:262
Kajiwara Y, Krouse HR 1971 Sulfur isotope partitioning in metallic sulfide systems. Can J Earth Sci 8:1397-1408
Knauth LP, Beeunas MA 1986 Isotope geochemistry of fluid inclusions in Premian halite with implications for the isotopic history of ocean water and origin of saline formation waters. Geochim Cosmochim Acta 50:419-433
Knipping HD 1974 The concepts of supergene versus hypogene emplacement of uranium at Rabbit Lake, Saskatchewan, Canada. *In* Formation of Uranium Ore Deposits, 531-549, Int'l Atomic Agency, Vienna
Kotov Ye I, Timofeev AV, Khoteev AD 1968 Formation temperatures of some uranium hydrothermal deposits (abst.). Fluid Inclusion Research-Proceedings of COFFI 1:44
Kotov Ye I, Timofeev AV, Khoteev AD 1970 Formation temperatures of minerals of uranium hydrothermal deposits. Fluid Inclusion Research-Proceedings of COFFI 3:38
Kotzer TG, Kyser TK 1990a The use of stable and radiogenic isotopes in the identification of fluids and processes associated with unconformity-type uranium deposits. *In* Modern Exploration Techniques, Sask Geol Soc Spec Publ 10:115-131
Kotzer TG, Kyser TK 1990b Fluid history of the Athabasca Basin and its relation to uranium deposits. *In* Summary of investigations 1990. Sask. Energy and Mines, Sask Geol Surv, Misc Rep 90-4:153-157
Kotzer TG, Kyser TK 1993 O, U and Pb isotopic and chemical variations in uraninite: Implications for determining the temporal and fluid history of ancient terrains. Am Mineral 78:1262-1274
Kotzer TG, Kyser TK 1995 Petrogenesis of the Proterozoic Athabasca Basin, northern Saskatchewan, Canada, and its relation to diagenesis, hydrothermal uranium mineralization and paleohydrogeology. Chem Geol 120:45-89
Kranz RL 1968 Participation of organic compounds in the transport of roe metals in hydrothermal solutions. Inst. Mining and Metallurgy Trans Sect B 77:B26-B36
Kyser TK 1987 Equilibrium fractionation factors for stable isotopes. *In* Kyser TK (Ed) Stable Isotope

Geochemistry of Low Temperature Fluids; Mineralogical Association of Canada Short Course 13:1-84

Kyser TK, O'Neil JR 1984 Hydrogen isotope systematics of submarine basalts. Geochim Cosmochim Acta 48:2123-2133

Kyser TK, Wilson MR, Ruhrmann G 1988 Stable isotopic constraints on the role of graphite in the genesis of unconformity-type uranium deposits. Can J Earth Sci 26:490-498

Langmuir D 1997 Aqueous Environmental Geochemistry. Prentice Hall, New Jersey

Lee MJ, Brookins DG 1978 Rubidium-strontium minimum ages of sedimentation, uranium mineralization, and provenance, Morrison Fromation (Upper Jurassic), Grants mineral belt, New Mexico. Am Assoc Petrol Geol Bull 62:1673-1683

Lee M, Aronson JL, Savin SM 1989 Timing and conditions of Permian Rotliegende sandstone diagenesis, southern North Sea, K-Ar and oxygen isotopic data. AAPG Bull 73:195-215

Leroy J, Poty B 1969 Recherches préliminaires sur les fluides associés à la genese des minéralisations en uranium du Limousin (France). Mineralium Deposita 4:395-400

Letian D 1986 Granite-type uranium deposits of China. In Vein Type Uranium Deposits. I A E A, Vienna, 377-394

Leventhal JS, Grauch RI, Therelkeld CN, Lichte FE 1987 Unusual organic matter associated with uranium from the Claude deposit, Cluff Lake, Canada. Econ Geol 82:1169-1176

Lewry JF, Sibbald TII 1977 Variation in lithology and tectonometamorphic relationships in the Precambrian basement of northern Saskatchewan. Can J Earth Sci 14:1453-1467

Lewry JF, Sibbald TII 1980 Thermotectonic evolution of the Churchill Province in Northern Saskatchewan. Tectonophysics 68:45-82

Lewry JF, Sibbald TII, Scheledewitz DCP 1985 Variation in character of Arcean rocks in the Western Churchill Province and its significance. In Ayers LD, Thurston PC, Card KD, Weber W (Eds) Evolution of Archean supracrustal sequences. Geological Association of Canada, Special Paper 28:239-261

Liebenberg WR 1956 The occurrence and origin of gold and radiocative minerals in the Witwatersrand. Geol Soc South Africa Trans 58:101-254

Little HW 1974 Uranium in Canada. GSC paper 74-1A:137-139

Lloyd RM 1966 Oxygen isotope enrichment of sea water by evaporation. Geochim Cosmochim Acta 30:801-814

Longstaffe FJ 1987 Stable isotope studies of diagenetic processes. In Kyser, TK (Ed) Stable Isotope Geochemistry of Low Temperature Fluids; Mineralogical Association of Canada Short Course 13:187-257

Ludwig KR 1978 Uranium-daughter migration and U/Pb isotope apparent ages of uranium ores, Shirley Basin, Wyoming. Econ Geol 73:29-49

Ludwig KR 1979 Age of uranium mineralization in the Gas Hills and Cooks Gap districts, Wyoming, as indicated by U-Pb isotope apparent ages. Econ Geol 74:1654-1668

Ludwig KR, Simmons KR, Webster JD 1984 U-Pb isotope systematic and apparent ages of uranium ores, Ambrosia Lake and Smith Lake districts, Grants mineral belt, New Mexico. Econ Geol 79:322-337

MacDonald R 1987 Update on the Precambrian geology and domainal classification of northern Saskatchewan. In Summary of investigations 1987. Saskatchewan Geol Surv, Miscellaneous Rept 87-4:87-104

Magaritz M, Taylor HP Jr 1976 Oxygen, hydeogen and carbon isotope studies of the Franciscan formation, Coast Ranges, California. Geochim Cosmochim Acta 40:215-235

Marlatt J, McGill B, Matthews R, Sopuck V, Pollock G 1992 The discovery of the McArthur River uranium deposit, Saskatchewan, Canada. In Homeniuk IA (Ed) The Geological Society of CIM; first annual field conference, 118-127

McCorkle DC, Emerson SR, Quay PD 1985 Stable carbon isotopes in marine porewaters: Earth Planet Sci Lett 74:13-26

Melander L 1960 Isotope Effects on Reaction Rates. Ronald, New York

Miller DS, Kulp JL 1963 Isotopic evidence on the origin of the Colorado Plateau uranium ores. Geol Soc Am Bull 74:609-630

Monster J, Appel PWU, Thode HG, Schidlowski M, Carmichael CM 1979 Sulfur isotope studies in early Archean sediments from the Isna, West Greenland: Implications for the antiquity of bacterial sulfate reduction. Geochim Cosmochim Acta 43:405-413

Nabelek PI, O'Neil JR, Papike JJ 1983 Vapor phase exsolution as a controlling factor in hydrogen isotopic variation in granitic rocks: The Notch Peak granitic stock, Utah. Earth Planet Sci Lett 66:137-150

Nash JT, Granger HC, Adams SS 1981 Geology and concepts of genesis of important types of uranium deposits. In Economic Geology 75th Anniv Vol, 63-117

Naumov GB, Mironova OP 1969 Das Verhalten der Kohlensäure in hydrothermalen Löungen bei der Bildung der Quarz-Nasturan-Kalzit-Gänge des Erzgebirges. Zeitschr Angew Geologie 15:24-241

Naumov GB, Motorina ZM, Naumov VB 1971 Conditions of formation of carbonates in veins of the Pb-Co-Ni-Ag-U type. Geochem Int'l 6:590-598

Needham RS, Stuart-Smith PG 1979 Geology of the Alligator Rivers Uranium Field. Int'l Uranium Syposium on the Pine Creek Geosyncline, 143-147

Northrop HR, Goldhaber MB 1990 Genesis of the tabular-type vanadium uranium deposits of the Henry Basin, Utah. Econ Geol 85:215-216

Ohmoto H 1972 Systematics of sulfur and carbon isotopes in hydrothermal ore deposits. Econ Geol 67:551-579

Ohmoto H 1986 Stable isotope geochemistry of ore deposits. *In* Valley JW, Taylor HP, O'Neil JR (Eds) Stable isotopes in high temperature geological processes. Rev Mineral 16:491-559

Ohmoto H, Rye RO 1974 Hydrogen and oxygen isotopic compositions of fluid inclusions in the Kuroko deposits. Japan Econ Geol 69:947-953

Ohmoto H, Kerrick D 1977 Devolatillization equilibria in graphitic systems. Am J Sci 277:1013-1044

Ohmoto H, Rye RO 1979 Isotopes of sulfur and carbon. *In* Barnes HL (Ed) Geochemistry of Hydrothermal Ore deposits, 2nd edn, J Wiley & Sons, New York, 509-567

Ohmoto H, Oskvareck JD 1985 An experimental study of oxygen isotope exchange reactions in the system Muscovite-CO_2-H_2O-KCl-NaCl at 400 °C and P=5 kbars. (abstr.). Prog Abstracts, 2nd Int'l Symp on Hydrothermal Reactions, Penn State Univ, August, 1985, 47

O'Neil JR 1986 Theoretical and experimental aspects of isotopic fractionation. *In* Valley JW, Taylor HP, O'Neil JR eds. Stable isotopes in high temperature geological processes. Rev Mineral 16:1-40

O'Neil JR, Johnson CM, White LD, Roedder E 1986 The origin of fluids in the salt beds of the Delaware Basin, New Mexico and Texas. Applied Geochem 1:265-271

Oparysheva LG, Shmariovich Ye M, Larkin ED, Shchetochkin V 1974 Characteristics of uranium mineralization in sedimentary blanket and basement granites. Int'l Geol 16:600-609

Pagel M, Poty B, Sheppard SMF 1980 Contributions to some Saskatchewan uranium deposits mainly from fluid inclusions and isotopic data. *In* Uranium in Pine Creek Geosyncline. Ferguson S, Goleby A (Eds) IAEA, Vienna, 639-654

Patrier P, Beaufort D, Bril H, Bonhomme M, Fouillac AM, Aumâitre R 1997 Alteration-mineralization at the Bernardian U deposit (western Marche, France): The contribution of alteration petrology and crystal chemistry of secondary phases to a new genetic model. Econ Geol 92:448-467

Percival JB, Bell K, Torrance JK 1993 Clay mineralogy and isotope geochemistry of the alteration halo at the Cigar Lake uranium deposit. Can J Earth Sci 30:698-704

Pierre C, Ortlieb L, Person A 1984 Suratidal evaporitic dolomite at Ojo de Liebre Lagoon: Mineralogical and isotopic arguments for primary crystallization. J Sed Petrol 54:1049-1061

Polyakov VB 1998 On anharmonic and pressure corrections of the equilibrium isotopic constants for minerals. Geochim Cosmochim Acta 62:3077-3085

Poty BP, Leroy J, Cuney M 1974 Les inclusions fluides dans les minerais des gisements d'uranium intragranitiques du Limousin et du Forez (Massif Central, France). *In* Formation of Uranium Ore Deposits, 569-582, Int'l Atomic Agency, Vienna

Pretorius DA 1981 Gold and uranium in quartz-pebble conglomerates. Econ Geol 75th Anniv Vol, 117-138

Pretorius DA 1976 The nature of the Witwatersrand gold-uranium deposits. *In* Wolf KH (Ed) Handbook of Stratabound and Stratiform Ore Deposits 7, Elsevier, Amsterdam, 29-88

Rackley RI, Shockey PN, Dahill MP 1968 Concepts and methods of uranium exploration. Wyoming Geol Assoc 20th Annual Field Guidebook, 115-124

Ramaekers P 1981 Hudsonian and Helikian basins of the Athabasca Region, northern Saskatchewan. *In* Campbell FHA (Ed) Proterozoic Basins of Canada. Geol Surv Can Paper 81-10:219-233

Ramaekers P, Dunn CD 1977 Geology and geochemistry of the eastern margin of the Athabasca Basin. Sask Geol Soc, Spec Pub 3:297-322

Ramdohr P 1958 New observations on the ores of the Witwatersrand and their genetic significance. Geol Soc South Africa Bull 61

Rees CE 1978 Sulfur isotope measurements using SO_2 and SF_6. Geochim Cosmochim Acta 42:383-390.

Rees MI 1992 History of the fluids associated with the Lode-Gold Deposits, and complex U-PGE-Au vein-type deposits, Goldfields Peninsula, northern Saskatchewan, Canada. Unpub MSc thesis, Univ. of Saskatchewan, 209 p

Rogova VP, Nikitin AA, Naumov, GB 1971 Mineralogic-geochemical conditions of the localization of U-Mo deposits in volcanogenic sedimentary formations (abst.). Fluid Inclusion Research-Proceedings of the COFFI 4:68-69

Reynolds RL, Goldhaber MB, Carpenter DJ 1982 Biogenic and nonbiogenic ore-forming processes in the south Texas uranium district: Evidence from the Panna Maria deposit. Econ Geol 77:541-556

Rich RA, Holland HD, Petersen U 1977 *Hydrothermal Uranium Deposits*. Elsevier Publishing Co., New York, 264 p

Richet P, Bottinga Y, Javoy M 1977 A review of H, C, N, O, S, and Cl stable isotope fractionation among gaseous molecules. Ann Rev Earth Planet Sci 5:65-110

Ripley EM 1981 Sulfur isotope studies of the Dunka Road Cu-Ni deposit, Duluth, Minnesota. Econ Geol 76:610-620

Robertson DS 1962 Thorium and uranium variations in the Blind River ores. Econ Geol 57:1175-1184

Robertson JA 1976 The Blind River uranium deposits: The ores and their setting. Ontario Dept Mines Misc Paper 65, 45 p

Robinson SC 1955 Mineralogy of uranium deposits, Goldfields, Saskatchewan. GSC Bull 31, 128 p

Robinson BW, Badham JPN (1974) Stable isotope geochemistry and the origin of the Great Bear Lake silver deposits, N.W.T., Canada. Can J Earth Sci 11:698-711

Robinson BW, Ohmoto H 1973 Mineralogy, fluid inclusions, and stable isotopes of the Echo Bay U-Ni-Ag-Cu deposits, Northwest Territories, Canada. Econ Geol 68:635-656

Rogova VP, Nikitin AA, Naumov GB 1971 Mineralogic-geochemical conditions of localization of U-Mo deposits in volcanogenic sedimentary formations (abst.). Fluid Inclusion Research-Proceedings of C.O.F.F.I. 4:68-69

Romberger SB 1985 Transport and deposition of uranium in hydrothermal systems of temperatures up to 300 C: Geological implications. *In* Uranium geochemistry, resources: The Institution of Mining and Metallurgy, 12-17

Rosholt JN, Bartel AJ 1969 Uranium, Thorium, and Lead systematics in Granite Nountains, Wyoming. Earth Planet Sci Lett 7:141-147

Rumble D III 1982 Stable isotope fractionation during metamorphic volatilization reactions. *In* Ferry JM (Ed) Characterization of Metamorphism Through Mineral Equilibria. Rev Mineral 10:327-353

Ruzika V 1988 Geology and Genesis of uranium deposits in the early Proterozoic: Blind River-Elliot Lake Basin, Ontario, Canada. *In* Recognition of Uranium Provinces. Proc Tech Comm

Rye RO, Ohmoto H 1974 Sulfur and carbon isotopes and ore genesis: A review. Econ Geol 69:826-842

Sassano GP, Fritz P, Morton RD 1972 Paragenesis and isotopic composition of some gangue minerals from the uranium deposits of Eldorado, Saskatchewan. Can J Earth Sci 9:141-157

Savin SM, Epstein S 1970 The oxygen and hydrogen isotope geochemistry of clay minerals. Geochim Cosmochim Acta 34:25-42

Shawe DR 1956 Significance of roll ore bodies in the genesis of uranium-vanadium deposits on the Colorado Plateau. *In* Page LR, Stocking HE, Smith HB (Eds) Contributions to the geology of uranium and thorium. U S Survey Prof Paper 300:239-241

Shegelski RJ, Scott SD 1975 Geology and mineralogy of the Ag-U-arsenide veins of the Camsell River district, Great Bera Lake, N.W.T. (abstr). Geol Soc America Abstr with Programs 7:857-858

Sheppard SMF 1986 Characterization and isotopic variations in natural waters. *In* Valley JW, Taylor HP, O'Neil JR (Eds) Stable isotopes in high temperature geological processes. Rev Mineral 16:165-183

Shettle DL Jr 1978 Experimental determination of oxygen isotopic fractionation between H_2O and hydrous silicate melts. PhD thesis, Pennsylvania State University, 116 p

Stiller M, Margaritz M 1974 Carbon-13 enriched carbonate in interstitial waters of Lake Kinneret sediments. Limnol Oceanogr 19:849-853

Stiller M, Roynick JS, Shasha S 1985 Extreme carbon-isotope enrichments in evaporating brines. Nature 316:434-435

Stuckless JS, Bunker CM, Bush CA, Doering WP, Scott JH 1977 Geochemical and petrologic studies of a uraniferous granite from the Granite Mountains, Wyoming. U S Geol Surv Jour Res 5:61-81

Taylor BE 1986 Magmatic volatiles: isotopic variation of C, H, and S. *In* Valley JW, Taylor HP Jr, O'Neil JR (Eds) Stable Isotopes in High Temperature Geological Processes. Rev Mineral 16:185-225

Taylor BE 1987 Stable isotope geochemistry of ore forming-fluids. *In* Kyser TK (Ed) Stable Isotope Geochemistry of Low Temperature Fluids; Mineralogical Association of Canada Short Course 13:337-445

Taylor BE, Eichelberger JC, Westrich HR 1983 Hydrogen isotopic eveidence of rhyolitic magma degassing during shallow intrusion and eruption. Nature 306:541-545

Taylor BE, O'Neil JR 1977 Stable isotopes of metasomatic Ca-Fe-Al-Si skarns and associated metamorphic and igneous rocks, Osgood Mountains, Nevada. Contrib Mneral Petrol 63:1-49

Taylor HP 1974 The application of oxygen and hydrogen isotope studies to problems hydrothermal alteration and ore deposition. Econ Geol 69:843-883

Taylor HP 1979 Oxygen and hydrogen isotopic relations in hydrothermal mineral deposits. *In* Geochemistry of Ore Deposits, 2nd edition. Barnes HL (Ed) J. Wiley & Sons, New York, 236-277

Taylor HP 1987 Comparison of hydrothermal systems in layered gabbros and granites, and the origin of low-$\delta^{18}O$ magmas. *In* Magmatic processes: physiochemical properties. Geochem Soc Spec Publ 1:337-357

Taylor HP, Sheppard MF 1986 Igneous Rocks: I. Process of isotopic fractionation and isotope systematics.

In Valley JW, Taylor HP Jr, O'Neil JR (Eds) Stable Isotopes in High Temperature Geological Processes. Rev Mineral 16:227-272

Thode HG, Cragg CB, Hulston JR, Rees CE 1971 Sulfur isotope exchange between sulfur dioxide and hydrogen sulfide. Geochim Cosmochim Acta 35:35-45

Tiangang L, Zhizhang H 1986 Vein uranium deposits in granites of the Xiazhuang ore field. *In* Vein Type Uranium Deposits. I A E A, Vienna, 359-377

Truesdell AH 1974 Oxygenisotopic activities and concentrations in aqueous salt solutions at elevated temperatures-consequences for isotopic geochemistry. Earth Planet Sci Lett 23:378-396

Tugarinov AI, Naumov VB 1969 Thermobaric conditions of formation of hydrothermal uranium deposits. Geochem Int'l 6:89-103

Urey HC 1947 The thermodynamic properties of isotopic substances. J Chem Soc 1947:562

Valley JW 1986 Stable isotope geochemistry of metamorphic rocks. *In* Valley JW, Taylor HP Jr, O'Neil JR (Eds) Stable Isotopes in High Temperature Geological Processes. Rev Mineral 16:445-490

Valley JW, O'Neil JR 1981 $^{13}C/^{12}C$ exchange between calcite and graphite: a possible thermometer in Grenville marbles. Geochim Cosmochim Acta 45:411-419

Vennemann TW, Kesler SE, O'Neil JR 1992 Stable isotope composition of quartz pebbles and their fluid inclusions as tracers of sediment provinance: Implications for gold- and uranium-bearing quartz pebble conglomerates. Geology 20:837-840

Vennemann TW, Kesler SE, Frederickson GC, Minter WEL, Heine RR 1995 Oxygen isotope sedimentology of gold- and uranium-bearing Wiwatersrand and Huronian Supergroup quartz-pebble conglomerates. Econ Geol 91:322-342

Wallace AR, Whelan JF 1986 The Schwartzwalder uranium deposit, III: Alteration, vein mineralization, light stable isotopes, and genesis of the deposit. Econ Geol 81:872-888

Wallis R, Saracoglu N, Brummer J, Golightly J 1983 Geology of the McClean uranium deposits. Geol Surv Canada, Paper 82-11:71-110

Warren EA, Smalley PC 1993 The chemical composition of North Sea formation waters, a review of their heterogeneity and potential applications. *In* Parker JR (Ed) Petroleum Geology of Northwest Europe, Proceedings of the 4th conference. London Geol Soc, 1347-1352

Waters AC, Granger HC 1953 Volcanic debris in uraniferous sandstones and its possible bearing on the origin and precipitation of uranium. U S Geol Surv Circ 224, 26 p

Welhan JA 1987 Stable isotope hydrology. *In* Kyser TK (Ed) Stable Isotope Geochemistry of Low Temperature Fluids; Mineralogical Association of Canada Short Course 13:129-161

Wilde AR, Mernagh TP, Bloom MS, Hoffmann CF 1989 Fluid inclusion evidence on the oriigin of some Australian unconformity-type uranium deposits. Econ Geol 84:1627-1642

Wilkinson M, Fallick AE, Keaney MJ, Haszeldine RS, McHardy WJ 1994 Stable isotopes in illite, the case of meteoric water flushing within the Upper Jurassic Fulmar Formation sandstones, U.K., North Sea. Clay Minerals 29:567-574

White DE 1974 Diverse origins of hydrothermal ore fluids. Econ Geol 69:954-973

Yurtsever Y, Gat JR 1981 Atmospheric waters. *In* Gat JR, Gonfiantini R (Eds) Stable Isotope Hydrology: Dueterium and Oxygen-18 in the Water Cycle. Technical Report Series #210, I A E A, Vienna, 103-142

Ypma PJM, Fuzikawa K 1979 Fluid inclusion studies as a guide to the composition of uranium ore forming solutions of the Nabarlek and Jabiluka deposits, N.T., Australia. Int'l Uranium Symposium on the Pine Creek Geosyncline, 212-215

Yu S 1989 Characteristics and regional geological environments of uranium deposits in Mesozoic volcanics in east China. *In* Uranium Deposits in Magmatic and Metamorphic Rocks. I A E A, Vienna TC-571/6, 77-92

Ziegler K, Sellwood BW, Fallick AE 1994 Radiogenic and stable isotope evidence for age and origin of authigenic illites in the Roliegende, southern North Sea. Clay Minerals 29:555-565

Zheng Y 1991 Calculation of oxygen isotope fractionation in metal oxides. Geochim Cosmochim Acta 55:2299-2307

Zheng Y 1995 Isotope fractionation in uranium oxides. Nucl Sci Tech 6:193-197

5
Environmental Aqueous Geochemistry of Actinides

William M. Murphy
Center for Nuclear Waste Regulatory Analyses
Southwest Research Institute
6220 Culebra Road
San Antonio, Texas 78238

Everett L. Shock
Department of Earth and Planetary Sciences
Campus Box 1169
Washington University
St. Louis, Missouri 63130

INTRODUCTION

The isolation and migration of nuclear wastes; genesis of Th and U mineral deposits; transport of actinides in groundwater, hydrothermal solutions, and subduction fluids; and other geochemical and environmental processes involving actinides depend to a large extent on the aqueous chemistry of these elements. Much of that chemistry can be revealed through use of thermodynamic and kinetic data for the reactions that control the solubility of solids; partitioning of species between aqueous solutions, solids, solid solutions, and mineral surfaces; formation and stability of colloids; uptake of dissolved species by organic matter; and microbial uses as sources of redox energy. However, aqueous actinide chemistry in environmental and geochemical systems is complicated by a variety of phenomena including the coexisting occurrence of dissolved species in multiple valence states and as polynuclear species, disproportionation, radiolysis effects, and colloid stability. All these processes affect the speciation of the actinides in aqueous solution. Variations in temperature, pressure, and solution composition (pH, oxidation-reduction potential, ionic strength, and the presence of complex-forming ligands) mean that scores of aqueous actinide species can be involved in a diverse array of geochemical and environmental processes.

Countless papers are devoted to the aqueous solution chemistry of actinide elements, and numerous reviews are available. However, many aspects of aqueous actinide chemistry remain uncertain or unknown. Among many notable reviews of the aqueous geochemistry of actinides are those by Allard (1982), Allard et al. (1984), Newton and Sullivan (1985), Kim (1986), Katz et al. (1986), Fuger (1992), Seaborg (1993), Kim (1993), and Clark et al. (1995). Recent reviews by international committees on the chemical thermodynamics of U and Am include evaluations of the thermodynamic properties of aqueous species (Grenthe et al. 1992; Silva et al. 1995). The Appendix offers brief summaries of selected reviews to direct readers to more detailed discussions than permitted in this review. Among these reviews, that by Kim (1986), in particular, focuses on natural geochemical systems. Natural occurrences of aqueous U and Th are widely reported, and aqueous concentrations of other actinides, primarily post-nuclear pollutants, have been analyzed in seawater, fresh surface waters, and groundwaters.

Generalities on environmental aqueous actinide chemistry are summarized in this report, and readers are referred to more extensive and specialized reviews for additional details. The primary objectives of this review are to identify complications in the understanding of environmental aqueous actinide chemistry, to guide efforts in prediction of the thermodynamic properties of aqueous actinides, and to identify areas for future research.

GENERALITIES OF ENVIRONMENTAL AQUEOUS ACTINIDE GEOCHEMISTRY

Many characteristics of aqueous actinide geochemistry are well established. General conclusions summarized in the review by Allard et al. (1984) bear repetition, and are followed by a summary of some more recent observations.

- Thorium and U are common in surface and near-surface waters.
- Environmental transuranic species are derived primarily from nuclear technology.
- The aqueous speciation of actinides in low-temperature geochemical solutions is dictated primarily by the oxidation-reduction potential, the pH, and the total dissolved carbonate concentration.
- Hydroxide and carbonate are dominant actinide complex forming agents.
- Fluoride, phosphate, and organic ligands may also contribute to aqueous actinide speciation in particular systems.
- Under reducing conditions aqueous Th, U, and Np occur predominantly in the tetravalent state, Pu occurs predominantly in the trivalent or tetravalent state, and Am occurs in the trivalent state.
- Under oxidizing conditions the common oxidation states of aqueous actinides are tetravalent for Th, hexavalent for U, pentavalent for Np, tetravalent or pentavalent for Pu, and trivalent for Am.
- "Dominating species in solution under most environmental conditions are for An(III): An^{3+}, $AnOH^{2+}$, $AnCO_3^+$, $An(CO_3)_2^-$, and possibly organic complexes; for An(IV): $An(OH)_4$, ($An(OH)_5^-$); for An(V): $AnO_2(CO_3)_2^{3-}$, $AnO_2(CO_3)_3^{5-}$; for An(VI): $AnO_2(CO_3)_2^{2-}$, $AnO_2(CO_3)_3^{4-}$" (Allard et al. 1984).
- Colloidal forms of actinides are common particularly for trivalent and tetravalent species.

In the case of U(VI) species the conclusions of Allard et al. (1984) were substantiated in the review by Grenthe et al. (1992), who also observed that at lower pH (e.g. less than ~5) UO_2^{2+} is the predominant uranyl species in solutions containing carbonate at concentrations of 10^{-4} to 10^{-3} molar. Pabalan and Turner (1997), making use of the Grenthe et al. (1992) data, describe the predominance of $(UO_2)_2CO_3(OH)_3^-$ in near neutral solutions. Turner et al. (1998) conclude that the uncomplexed NpO^+ ion predominates at pH up to 8 in dilute nitrate solutions at equilibrium atmospheric CO_2 pressure and that $NpO_2CO_3^-$ tends to predominate over $NpO_2(CO_3)_2^{3-}$. A variety of spectrometric studies provide evidence for the existence of polynuclear carbonate and hydroxide aqueous species of actinides particularly at elevated actinide concentrations (see reviews by Baes and Mesmer 1976; Grenthe et al. 1992; Plyasunov and Grenthe 1994; Silva et al. 1995). However, the stoichiometries of these species can be uncertain and their relevance to solutions with trace concentrations of actinides is questionable. In synthetic systems at elevated concentrations of carbonate, a large number of aqueous carbonate and hydroxide complexes have been proposed (e.g. Newton and Sullivan 1985). These complexes are unlikely to predominate in natural solutions where constraints from carbonate mineral equilibria limit concentrations of total dissolved carbonate (see below).

Table 1. Redox reactivity of actinides at typical environmental conditions (summarized from Allard 1982; Clark et al. 1995).

Th(IV)	Nonreactive
Pa(IV)	Nonreactive at low oxidation potentials, but easily oxidized
Pa(V)	Nonreactive
U(IV)	Nonreactive, but oxidizes even at low oxidation potentials
U(VI)	Nonreactive
Np(IV)	Nonreactive, but oxidizes in air
Np(V)	Nonreactive
Np(VI)	Nonreactive at high oxygen potentials, but easily reduced
Pu(III)	Nonreactive at low oxygen potentials, but easily oxidized
Pu(IV)	Disproportionates to Pu(III) and Pu(VI)
Pu(V)	Disproportionates to Pu(IV) and Pu(VI)
Pu(VI)	Nonreactive at high oxygen potentials, but easily reduced
Am(III)	Nonreactive
Cm(III)	Nonreactive

Complications in environmental aqueous actinide geochemistry

Numerous challenging aspects of aqueous actinide geochemistry are recognized including the effects of radiolysis, disproportionation, redox disequilibria, colloids, complexes with humic and fulvic acids and other organic compounds, temperature, and the vanishingly low concentrations of transuranic actinides in environmental systems (e.g. Curtis et al. 1999). Although these phenomena have been widely noted, and although aqueous actinides at low concentrations can have important environmental consequences because of radiological effects, in many cases no generally accepted or systematic treatment is available to characterize their effects on aqueous actinide geochemistry. Nevertheless, an attempt is provided in this section to summarize possible implications of each of these phenomena in the context of environmental aqueous actinide chemistry. The general stability characteristics of aqueous actinides of valences occurring in environmental and geochemical systems are summarized in Table 1.

Individual actinide elements can coexist in multiple valence states in a given solution. As an example, the III, IV, V, and VI valence states of Pu were detected in solubility studies by Nitsche et al. (1994) who studied Pu solubilities in dilute, oxidizing groundwater at pH 7 and 8.5 and 25°C and 60°C. At 25°C the IV and V valence states predominated and at 60°C the VI valence state predominated. Some valence states of aqueous actinides are readily oxidized or reduced, and radiolysis effects can cause intermediate oxidation states to disproportionate spontaneously. Again, using Pu as an example of the complications of aqueous actinide redox chemistry, the reduced (III) form is easily oxidized, the IV valence state disproportionates to III and VI, the V state disproportionates to IV and VI, and the VI valence state can be easily reduced (e.g. Allard 1982; Clark et al. 1995). These relations are summarized in Table 1. Radiolytic processes also produce aqueous species from other solutes or from H_2O that promote redox reactions, polymerization, and colloid formation.

Throughout surface and near-surface environments, redox reactions may control or strongly influence the solubility, exchange, and transport of actinides. Many redox reactions are notoriously sluggish at low temperature, including many involving aqueous actinides. As an example, the chemistry of U in environmental and geological fluids is dominated by the difference in behavior of the predominant tetravalent and hexavalent ions. The tetravalent form generally has low solubility (e.g. concentrations are limited by U(IV) phases such as uraninite or coffinite); whereas the hexavalent form is relatively soluble as the uranyl ion (UO_2^{2+}) and its complexes. Even at relatively low equilibrium oxidation potentials uranyl species may dominate aqueous U speciation although uraninite is the stable solid phase. Transport of U is greatly enhanced if it is present in solution as UO_2^{2+} ions and complexes. The predominant form of U(IV), neutral $U(OH)_4$ is not readily sorbed to charged solid surfaces. In contrast, sorption of U by minerals or organic matter may be enhanced if it is present as charged uranyl forms. In fact, sorption of oxidized (charged) uranyl ions by organic matter may precede reduction to much less soluble U(IV) solids (e.g. Zielinski and Meier 1988; Carroll et al. 1992; Barnes and Cochran 1993; Wersin et al. 1994). In contrast, the strong affinity of carbonate ligands for uranyl in aqueous solutions effectively competes with sorption, such that sorption of charged uranyl carbonate species is limited (e.g. Pabalan et al. 1998).

As a result of sluggish kinetics, redox reactions involving U can provide sources of metabolic energy for microorganisms (Lovley et al. 1991; Gorby and Lovley 1992; Lovley and Phillips 1992). Similar microbially-mediated reactions are plausible for other actinides that are present in multiple oxidation states (especially Pu and Np). Expanding thermodynamic models of actinide transport to include microbial processes would help to place those processes in a broader biogeochemical context. Progress of this kind will enable exploration of connections between microbial activity, mineral-fluid reactions, and the redistribution of actinides during low temperature alteration of U ores, weathering of granite and other natural sources of U, deposition and diagenesis of black shales and other sediments, soil development, and geologic isolation of radioactive waste.

Filtering water samples shows that many actinides are present in colloidal form. Either polymeric compounds of hydrolyzed actinides (i.e. "real" colloids) or actinides sorbed on natural colloids (i.e. "pseudocolloids") are commonly observed. Kim (1986) presents data for Am hydroxide solubility in nonsaline water from the Gorleben area of Germany, which show a strong dependence of measured Am concentration on solution filter size down to 1 nm, and attributed this effect largely to the role of pseudocolloids. In addition, Kim (1986) notes a distinction in the colloidal behavior between ^{239}Pu which shows negligible effects of colloids in PuO_2 solubility studies, and ^{238}Pu, which shows appreciable effects of colloids, and attributed this difference to effects of α radiation. Recoil effects of α radiation are concluded to destroy pseudocolloids. In contrast, α recoil effects on actinide hydroxide precipitates are asserted to generate real colloids of the actinide hydroxides. In light of these observations, Kim (1986) offers the discouraging conclusion that "[s]olubility equilibria of transuranic compounds in natural aquatic solutions cannot easily be described thermodynamically, since there is no possibility to quantify the colloid generation involved."

A similarly problematic situation concerns the speciation of aqueous actinides in environmental or geochemical systems that are rich in organic matter. Humic substances (insoluble humin, as well as dissolved humic and fulvic acids) tend to bind trace metals and influence speciation and transport of such elements in surface and ground waters. However, the structures of these compounds are highly variable and measurements are often limited to composite stability constants for poorly-identified clusters of organic functional groups. Partly for this reason, studies reported in the literature on the complexation or chelation of actinides by humic substances are relatively limited (e.g. Kim 1986). Nevertheless, the well-established association of U with organic matter in geochemical processes indicates that ignoring actinide-organic interactions may lead to false impressions of actinide mobility in many environmental applications.

Temperature is also predicted to affect strongly the aqueous speciation of actinides, but few empirical studies are available to assess this effect. As an example, in their comprehensive study of the thermodynamics of Am, Silva et al. (1995) cite only a handful of studies that yield complex association constants at temperatures other than 25°C, and those range from 10 to 55°C. On the other hand, correlation algorithms and equations of state can assist the effort to predict high temperature speciation (see below), especially in conjunction with well-designed and well-characterized experiments.

Aqueous actinides in environmental and geochemical systems

In general, concentrations of actinides are low under natural conditions, and transuranic actinides exist in vanishingly low concentrations generally well below those controlled by solubility limits. Concentrations are commonly expressed in units[1] of µg/L, ppb, or fCi/L.

Uranium occurs at an average of 3 µg/L in seawater (Goldberg et al. 1971, cited in Kim 1986), and continental surface waters contain 0.1 to 500 µg/L U (Rogers and Adams 1970, cited in Kim 1986). Th occurs in the ocean at 6×10^{-4} µg/L (Allard et al. 1984). Concentrations of Pu(III+IV) in 22 filtered (<0.45 µg) lake water samples ranged from 0.06 to 9.6 fCi/L and Pu(V+VI) ranged from less than detection to 0.57 fCi/L in the same samples (Wahlgren and Orlandini 1982, cited in Watters 1983). Pu(239+240) in Lake Michigan surface water ranged from 0.1 fCi/L to 0.8 fCi/L during the period from 1973 to 1977 (Wahlgren, Robbins, and Edgington 1980, cited in Watters 1983). The Lake Michigan Pu concentrations varied in annual cycles with lower values in the late summer and higher values in the winter, which was attributed to coprecipitation of Pu with carbonate in the summer (Watters 1983). Graf (1994) summarized 166 Pu analyses from river waters in the Northern Rio Grande Valley during the period 1974-1988, based on data from Los Alamos National Laboratory. The average ^{238}Pu concentration was found to be below detection, and the average ^{239,240}Pu concentration was 4.1 fCi/L. The highest individual measurements were 90 fCi/L for ^{238}Pu and 130 fCi/L for ^{239,240}Pu. No information was provided by Graf (1994) on filtering techniques.

Carefully sampled but unfiltered groundwater from a coastal plain aquifer collected from a background water chemistry well on the Savannah River Site, South Carolina, had the following actinide concentrations: ^{239}Pu: <0.06 fCi/L; ^{232}Th: 6±1 fCi/L; ^{241}Am: 0.4±0.2 pCi/L; ^{243,244}Cm 0.7±2 pCi/L; ^{238}U: 15±1 pCi/L (Kaplan et al. 1994). In nearby wells, radionuclide contaminated water was collected in a similar fashion, and filtering tests by Kaplan et al. (1994) showed that 54 percent of Pu, 52 percent of U, and 34 percent of Am and Cm was filterable at sizes greater than 0.5 nm. They also showed that the filterable fraction of actinides in groundwater samples varies with Pu > Th > U > Am ≈ Cm among three locations at the Savannah River Site. In addition, Kaplan et al. (1994) document size dependence of the actinide content in colloids, revealing that in many cases a majority of the filterable colloids containing actinides are in the 500 to 3000 molecular weight (MW) range (or between about 0.5 and 1 nm). In one groundwater sample, 100 percent of Th was in the 0.4 µm to 3000 MW range, 100 percent of Pu in the 500 to 3000 MW range, and 100 percent of Am, Cm, and Ra passed through the 500 MW filter (<0.5 nm). Uranium was nearly equally distributed between remaining in solution or being trapped in the 500 to 3000 MW range. (Filter sizes based on MW are used primarily for separation of large organic molecules at physical sizes smaller than 0.4 µm; use of these filters for actinide colloid size distribution does not correspond strictly to colloid mass.)

Recently, Kersting et al. (1999) documented colloidal transport of Pu in groundwater

---

[1] One femtocurie (fCi) equals 1×10^{-3} picocurie (pCi) or 1×10^{-15} Ci, and 1 Ci = 3.7×10^{10} Bq (Becquerel); 1 mole = 1 Bq λ^{-1} N_A^{-1}, where λ stands for the decay constant (sec^{-1}) and N_A represents Avogadro's number.

at the Nevada Test Site. The unique $^{240/239}$Pu ratio from underground weapons tests permitted them to identify the source for filterable Pu in groundwater. As a result they were able to demonstrate that Pu had moved 1.3 km with the groundwater since the test in 1968 (groundwater samples were collected in 1997) at concentrations far in excess of those explained by solubility. The rate of transport of Pu reported in that study falls in the range of estimated velocities for groundwater flow (1-80 m yr^{-1}) at the site. Although the precise composition of the colloids was not identified, fine particulate Pu oxide and/or clays and zeolites with sorbed Pu are likely candidates. Sorption on mineral surfaces is often thought to be a mechanism for retarding the transport of solutes in environmental aqueous solutions, but, as Kersting et al. (1999) show, sorption may be limited if the mineral particles are small enough to be carried by moving groundwater as colloids.

Uranium and other actinides in hot water

Many fossil hydrothermal systems show evidence of U transport by aqueous fluids, sometimes leading to the formation of hydrothermal U deposits (Wenrich 1985; Wallace and Whelan 1986; Wilde et al. 1989; Turpin et al. 1990; George-Aniel et al. 1991; Johnson and McCulloch 1995; Komninou and Sverjensky 1995a; 1995b; 1996; Raffensperger and Garven 1995a; 1995b; Kotzer and Kyser 1995). Magmatic/hydrothermal fluids are called on to fractionate Th from U during the formation of volatile-rich melts (Keppler and Wyllie 1990) and are invoked to explain geochemical signatures of subduction-related rocks. Transport U is indicated by analyses of hot springs and hydrothermal fluids (Michard et al. 1983; Chen and Wasserburg 1986; Sturchio et al. 1987; 1989), as well as analyses of fluid inclusions (Irwin and Roedder 1995). A dramatic example of the interaction of high-temperature aqueous fluids and U is provided by the Oklo Natural Reactor in Gabon where temperatures are estimated to have reached 250°C in the low grade zones and probably exceeded 400°C in the reactor zones (Holliger et al. 1978; Loss et al. 1989; Gauthier-Lafaye et al. 1989; Eberly et al. 1994). It has been suggested that the ability of water to act as a neutron moderator, coupled with variations in the amount of water in the system, may have led to criticality cycles over the lifetime of the natural reactor (Cowan 1976).

Redistribution of U by high-temperature aqueous solutions may alter minerals used for U-lead dating (Kotzer and Kyser 1993; Pan et al. 1993). As a consequence, dates obtained from many zircons, titanites, badellyites, and other minerals can reflect the timing of aqueous alteration during metamorphic and/or hydrothermal processes. The widespread use of U series disequilibrium to date geologic materials is predicated on short-term U exchange between rock units and aqueous fluids (see review by Ivanovich 1994).

The effects of pH and oxidation state on the speciation of aqueous U

Attention paid to the aqueous speciation of U makes it possible to construct oxidation potential-pH diagrams that show relations among different oxidation states of aqueous U and their hydrolysis products. One such diagram at 25°C and 1 bar is shown in Figure 1. There are several ways to describe the oxidation potential appropriate for specific problems, environments, or experimental approaches. Among these, Eh (oxidation potential of a half-cell measured relative to the standard hydrogen half-cell), pe (negative base ten logarithm of the activity of the aqueous electron, e$^-$), fO_2 (fugacity of oxygen), and fH_2 (fugacity of hydrogen) are in common usage. All of these are interchangeable given the right conversion factors, and can be evaluated from values of ΔG_r for the reactions of interest. The variable chosen to represent oxidation potential in Figure 1 is fH_2. The parameter fH_2 is related to fO_2 simply by the equilibrium constant for the reaction $2\ H_2O = 2\ H_2 + O_2$. The advantage of this choice for theoretical work is that, unlike Eh, fH_2 is independent of pH. Solid lines in Figure 1 separate fields of relative predominance of aqueous species. It follows that the lines correspond to values of log fH_2 and pH where pairs of aqueous species are at equal

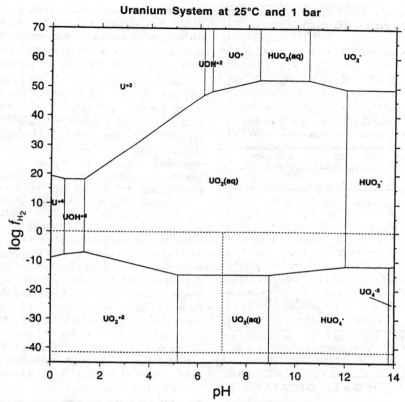

Figure 1. (after Shock et al. 1997b) Oxidation potential-pH diagram for the system U-O-H at 25°C and 1 bar. Solid lines separate fields of relative predominance of aqueous uranium species. Dashed horizontal lines bracket the stability range of H_2O, and the dashed vertical segment indicates neutral pH. Note that all predominant aqueous species stable with H_2O are in either the U(IV) or U(VI) oxidation state. All aqueous species and water are considered to be in their standard state for construction of this diagram.

activities. The horizontal dashed lines in this plot correspond to upper and lower bounds of the range of fH_2 where H_2O is stable. The vertical dashed segment indicates neutral pH.

The critical review by Grenthe et al. (1992) provides values of the standard Gibbs free energy of formation for 11 aqueous species in the U-O-H system. Eight more values were selected or estimated by Shock et al. (1997b) leading to a total of 19 aqueous species representing four oxidation states. Standard state thermodynamic properties for all 19 aqueous U species were used to construct Figure 1, and of these 13 have fields of relative predominance in the diagram. This observation does not mean that the other 6 species are unstable or nonexistent at these conditions, but that they are minor constituents of the total speciation of aqueous U. Note that UO_2^+ and its hydroxide complexes do not exhibit fields of relative predominance, consistent with the disproportionation of UO_2^+ to U^{4+} and UO_2^{2+}. The predominance fields for U^{3+} and its hydroxide complexes fall outside the range of H_2O stability, indicating that U(III) species would be minor constituents of any aqueous U solution at equilibrium. In fact, all of the predominant species in the range of the diagram consistent with the stability of H_2O are in the U(IV) or U(VI) oxidation state. This diagram

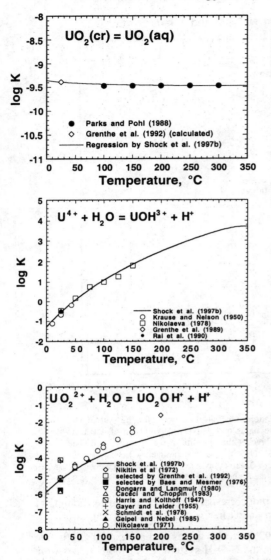

Figure 2. Comparison of experimental (symbols) and calculated equilibrium constants for reactions involving aqueous uranium species as functions of temperature.

can be thought of as a map of the equilibrium surface and changes in this surface with pressure and temperature reveal much about the changing distribution of aqueous U in natural solutions.

Predicting the effects of temperature and pressure on diagrams of this type, or on the aqueous speciation of U in natural solutions where many inorganic and organic complexes can form, requires standard state thermodynamic values for individual aqueous U species or for reactions involving them. In general, experimental data of this type at elevated temperatures and pressures are in short supply. A few examples of equilibrium constants are shown in Figure 2 for reactions involving aqueous U species. The upper plot shows the mild temperature dependence of a reaction expressing the solubility of uraninite where the aqueous species, $UO_2(aq)$, is a nonconventional form of the neutral fourth hydroxide complex of U(IV): $U(OH)_4$. The first hydrolysis constants of the U^{4+} cation are shown in the middle plot where it can be seen that there is close agreement between experimental (symbols) and theoretical (curve) values. This close agreement is the case for reactions involving the reduced U(IV) species, and there is less agreement for the oxidized U(VI) species. An indication of this is given in the bottom plot in Figure 2 which shows a comparison between experimental data and theoretical predictions of the first hydrolysis constant for UO_2^{2+}. Possible sources for disagreement include ambiguities in predictive methods when they are applied to oxycations (see Shock et al. 1997a; 1997b), uncertainties surrounding the oxidation states of high temperature experiments, and lack of demonstration of reversible equilibrium in the experiments.

One case in which there is close agreement between theoretical estimates and experimental data is illustrated in Figure 3. The symbols in the upper plots represent uraninite solubility results from experiments conducted at elevated temperatures and high total pressures of H_2 (500 bars) such that U occurs in tetravalent valence. The curves in these figures are calculated, and the speciation that corresponds to these total solubility curves is illustrated in the lower plots. The pH-independent ranges of these and other data from this study were used to define the symbols

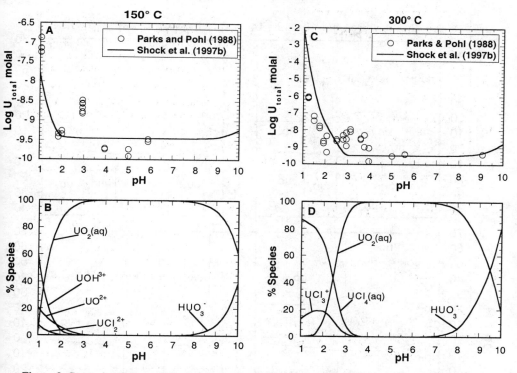

Figure 3. Comparison of experimental data on the solubility of uraninite at 150°C (A and B) and 300°C (C and D) and high pressures of H_2, (500 bars, Parks and Pohl 1988) as a function of pH. Upper plots show total concentrations of U in solution from the experiments (symbols) and that calculated from thermodynamic data (curves). The lower plots illustrate the predominant speciation of U that combine to yield the curves in the upper figures. Note that chloride complexes appear to become increasingly important at low pH as temperature increases.

in the upper plot of Figure 2, and were used in the regression to obtain standard state data for $UO_2(aq)$. The other thermodynamic properties used in the speciation calculations are estimates, and it can be seen that these estimates give a reasonable explanation of the strong pH dependence at low pH. At lower temperatures (like the example for 150°C) enhanced solubility of uraninite can be explained by the formation of charged hydrolysis products of U^{4+}. At higher temperatures, the enhanced solubility of uraninite in these experiments is better explained by the formation of chloride complexes of U^{4+} (HCl was used to obtain the low pH values in all of these experiments). Further details about the regression of uraninite experimental data, estimation of standard state thermodynamic properties for U ions and complexes and prediction of hydrolysis constants are given by Shock et al. (1997b).

Hydrolysis constants and redox relations among aqueous U species developed by Shock et al. (1997b) can be used to generate oxidation potential-pH diagrams at other temperatures and pressures. Diagrams at vapor-liquid saturation pressures (P_{SAT}) for the system H_2O and 100°, 150°, 200°, 250°, 300°, and 350°C are shown in Figure 4. The axes limits of these plots are all identical which helps in visualizing the changes in the speciation of U as temperature increases at P_{SAT}. Comparison of the 100°C plot in Figure 4 with the 25°C plot in Figure 1 shows that one species no longer has a predominance field (U^{4+}) and that three new species have appeared in the acidic pH range (UO^{2+}, UO_2^+ and UO_2OH^+),

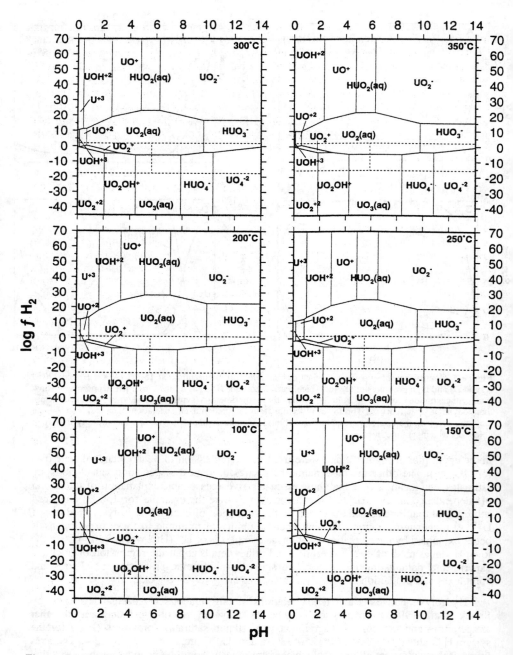

Figure 4. (after Shock et al. 1997b) Oxidation potential-pH diagram for the system U-O-H at elevated temperatures and P_{SAT}. Solid lines separate fields of relative predominance of aqueous uranium species. Dashed horizontal lines bracket the stability range of H_2O, and the dashed vertical segments indicates neutral pH at each temperature. A comparison among these plots and that in Figure 1 is given in the text.

for a total of 15 fields. The log fH_2 range of the predominance field for $UO_2(aq)$ has shrunk from both top and bottom with increasing temperature suggesting that both the reduced, U(III), and oxidized, U(VI), forms of U are stabilized by increasing temperature. Note also that there is an extremely narrow field of relative predominance of UO_2^+ in the 100°C diagram, again suggesting slightly lower overall stability of the U(IV) species. As temperature increases, the $UO_2(aq)$ field continues to shrink owing to the shift in the stability ranges of hydroxide complexes to lower pH values (at least up to 250°C). In addition, the UO_2^+ field broadens slightly in its fH_2 range. All plots in Figure 4 show that the same subset of 15 species predominates, although some are conscripted to predominance in narrow ranges of fH_2 and pH, and others are nearly driven from the diagram at 350°C (U^{3+}, UOH^{3+}).

CALCULATION OF STANDARD STATE PROPERTIES FOR AQUEOUS SPECIES AS FUNCTIONS OF TEMPERATURE AND PRESSURE

Thermodynamic and kinetic data are the foundation for quantitative models of aqueous actinide transport, as well as the dissolution, precipitation and alteration of solids which contain actinides, or to which actinides are adsorbed. There are abundant thermodynamic data at 25°C and 1 bar, as well as at elevated temperatures and pressures, for many actinide-bearing solids (see reviews by Cordfunke and O'Hare 1978; Lemire and Tremaine 1980; Morss 1982; Fuger 1982; Fuger et al. 1983; Mikheev and Myasoedov 1985; David 1986; Grenthe et al. 1992; Silva et al. 1995; among others). In contrast, as temperature increases, the availability of experimentally determined thermodynamic data for aqueous actinide species decreases dramatically. Estimates of high-temperature thermodynamic properties for aqueous U and Pu species by Barner and Scheuerman (1978) and Lemire and Tremaine (1980) employed methods from the 1960s (Criss and Cobble 1964a,b; Helgeson 1967). These estimation methods are superseded by more accurate techniques that allow predictions of standard state thermodynamic properties at high temperatures and pressures for aqueous ions, electrolytes, and acids, as well as inorganic and organic complexes (Shock and Helgeson 1988; 1990; Shock et al. 1989; 1992; 1997a; Shock and Koretsky 1993; 1995; Shock 1995; Sverjensky et al. 1997; Prapaipong et al. 1999). Recently, these methods were used to make predictions for aqueous U species (Shock et al. 1997b) and aqueous actinide ions (Shock et al. 2000).

Calculating the chemical speciation of aqueous solutions and the solubilities of solids in environmental aqueous solutions is facilitated with values of the apparent standard molar Gibbs free energies $\Delta \bar{G}_{P,T}^{\circ}$ of the solids and aqueous species at the temperature and pressure of interest. Values of $\Delta \bar{G}_{P,T}^{\circ}$ can be evaluated from

$$\Delta \bar{G}_{P,T}^{\circ} = \Delta \bar{G}_f^{\circ} - \bar{S}_{Pr,Tr}^{\circ}(T - Tr) + \int_{Tr}^{T} \bar{C}_{Pr}^{\circ} dT - T \int_{Tr}^{T} \bar{C}_{Pr}^{\circ} \ln T + \int_{Pr}^{P} \bar{V}_T^{\circ} dP \qquad (1)$$

where $\bar{G}_f^{\circ}$ stands for the standard partial molar Gibbs free energy of formation from the elements at the reference pressure (Pr) and temperature (Tr) of 25°C and 1 bar, $\bar{S}_{Pr,Tr}^{\circ}$ represents the standard partial molar entropy at the reference conditions, $\bar{C}_P^{\circ}$ and $\bar{V}^{\circ}$ designate the standard partial molar isobaric heat capacity and standard partial molar volume, respectively, and T and P stand for the temperature and pressure of interest.[2]

---

[2] Standard states generally adopted in geochemistry are unit activity of the pure solvent at any temperature and pressure, unit fugacity of a pure gas at any temperature and 1 bar, unit activity of a stoichiometric component of a solid in the corresponding pure phase at any temperature and pressure, and, for aqueous species, unit activity of a hypothetical one molal aqueous solution referenced to infinite dilution at any temperature and pressure.

Equations of state for the temperature and pressure dependence of $\overline{V}°$ and $\overline{C_p}°$ of solids, gases and aqueous species allow evaluation of the integrals in Equation (1). Reviews by Cordfunke and O'Hare (1978), Lemire and Tremaine (1980), Grønvold et al. (1984), Holley et al. (1984), Fuger et al. (1983), Flotow et al. (1984), Hildenbrand et al. (1985), Grenthe et al. (1992), and Silva et al. (1995), among others, provide equation-of-state parameters and thermodynamic data for actinide gases, minerals, and other solids. Equations and parameters for calculating $\overline{C_p}°$ of U-bearing solids are given by Grenthe et al. (1992), and corresponding information for Am-bearing solids is summarized by Silva et al. (1995). For aqueous species, the revised Helgeson-Kirkham-Flowers (HKF) equation of state (Helgeson et al. 1981; Tanger and Helgeson 1988; Shock et al. 1992) allows evaluation of the integrals in Equation (1) and calculation of values of $\Delta \overline{G}°_{P,T}$ to 1000°C and 5 kbar.

The revised-HKF equation of state takes as its foundation the Born equation for ion solvation, and accounts for structural and solvation contributions to the standard partial molar properties of aqueous species. There are seven species-dependent equation-of-state parameters for aqueous species, all of which can be obtained through regression of experimental data. In the absence of experimental data at elevated temperatures and pressures, the equation-of-state parameters can be estimated with correlation algorithms designed to minimize uncertainty (Shock and Helgeson 1988; 1990; Shock et al. 1989; 1997a; Shock 1995; Amend and Helgeson 1997). In addition, several strategies developed in conjunction with the revised-HKF equations are useful for estimating standard state thermodynamic properties of aqueous species at 25°C and 1 bar when they are not available from experimental data (Shock and Koretsky 1993; 1995; Sverjensky et al. 1997; Shock et al. 1997a; Prapaipong et al. 1999). The following section demonstrates the use of these correlation methods to make estimates of equilibrium constants for Am(III) complex association at high temperatures. Americium was selected because it is the predominant form of Am at all redox conditions encountered in environmental systems, because the recently reviewed thermodynamic data set provided by Silva et al. (1995) supplies values of standard state association constants at 25°C based on experimental measurements, and because of the general lack of data for the thermodynamic properties of aqueous Am species at elevated temperatures.

Estimating the temperature dependence for equilibrium association constants

Application of Equation (1) to calculate equilibrium aqueous association constants (β) follows from the relations

$$-2.303 \, RT \log \beta = \Delta \overline{G}°_r \qquad (2)$$

and

$$\Delta \overline{G}°_r = \Sigma \, v_{i,r} \, \Delta \overline{G}°_{P,T,i} \qquad (3)$$

where $\Delta \overline{G}°_r$ stands for the standard Gibbs free energy of reaction, R represents the gas constant, $v_{i,r}$ refers to the stoichiometric reaction coefficient of the ith species in the rth reaction (which is positive for products and negative for reactants), and $\Delta \overline{G}°_{P,T,i}$ designates the apparent standard Gibbs free energy of formation of the ith species in the reaction. Once values of the apparent standard Gibbs free energy for individual aqueous species at high temperatures and pressures are estimated with the revised-HKF equations, they can be combined to yield values of log β using Equations (2) and (3).

Silva et al. (1995) critiqued available thermodynamic data for Am solids and aqueous species, including complexes. They conclude that log β at 25°C and 1 bar for the reaction

$$Am^{3+} + Cl^- = AmCl^{2+} \qquad (4)$$

is 1.05±0.06. Suppose the problem of interest is at 200°C and 100 bars, and an estimate of log β is required despite the absence of experimental data. Referring to Equation (1), values are needed for the standard partial molar entropies, heat capacities, and volumes of Am^{3+}, Cl^-, and $AmCl^{2+}$. These data for Cl^- are well established from experimental measurements and are summarized by Tanger and Helgeson (1988), Shock and Helgeson (1988), and Shock et al. (1997a), who also provide revised-HKF parameters for Cl^-. In contrast, there are no experimental measurements leading to standard state entropies, volumes, or heat capacities for Am^{3+} or $AmCl^{2+}$. Recent estimates of the thermodynamic properties for Am^{3+} are given by Shock et al. (2000). Their value of $\bar{S}°$ is estimated from the ionic radius of Am^{3+} using methods developed by Sassani and Shock (1992; 1994), which are based on experimental data for other trivalent cations and are in close agreement with similarly-derived estimates of $\bar{S}°$ by Morss (1986) and Fuger and Oetting (1976). The value of $\Delta \bar{G}_f°$ adopted by Shock et al. (2000) is calculated from the experimentally-determined standard state value of the enthalpy of formation (ΔH_f) and the estimated value of $\bar{S}°$. Estimates of $\bar{V}°$ and $\overline{C_p}°$ for Am^{3+} were made by Shock et al. (2000) using correlations of $\bar{V}°$ vs. $\bar{S}°$ and $\overline{C_p}°$ vs. $\bar{S}°$ and the assumption that correlations for trivalent lanthanides provide useful estimates for trivalent actinides.

The value of $\Delta \bar{G}_f°$ for $AmCl^{2+}$ used in this review was calculated from the log β value for Reaction (4) at 25°C and 1 bar from Silva et al. (1995). For estimates of other standard state properties, methods provided by Sverjensky et al. (1997) can help. The crux of the argument for using the estimation methods outlined by Sverjensky et al. (1997) is that the predictions will be consistent with data for many other inorganic complex association reactions that have been studied experimentally at high temperatures and pressures. This rationale is especially true in the case of chloride complexes; Sverjensky et al. (1997) employ high-temperature data for 27 chloride complexes in constructing their estimation methods. These methods are built on correlations between the standard state properties of association reactions at 25°C and 1 bar and the absolute standard partial molar properties of the aqueous cations at the same temperature and pressure.[3] It follows that $\bar{S}°$ for $AmCl^{2+}$ can be calculated from an estimate of $\Delta \bar{S}°$ for the association reaction (Eqn. 4). Sverjensky et al. (1997) give a number of correlation expressions for estimating $\Delta \bar{S}°$ leading to a general form (their Eqn. 73), which can be written for the first chloride complex of Am^{3+} as

$$\Delta \bar{S}° = 0.19228\ \bar{S}°^{abs}_{Am^{3+}} + 25.686\ (cal\ mol^{-1}\ K^{-1}) \tag{5}$$

where $\bar{S}°^{abs}$ refers to the absolute standard partial molar entropy, which is related to its conventional counterpart by

$$\bar{S}°^{abs} = \bar{S}° + Z_j\ \bar{S}°^{abs}_{H^+} \tag{6}$$

where Z_j stands for the charge of the cation and $\bar{S}°^{abs}_{H^+}$ has the value -5.0 cal mol^{-1} K^{-1}. Equation (5) is built on several correlations involving values of $\Delta \bar{S}°$ for chloride association reactions obtained by regression of many sets of high-temperature experimental data. Correlations for $\Delta \bar{V}°$ and $\Delta \overline{C_p}°$ are analogous in form to the $\Delta \bar{S}°$ correlations, and are also based on regression of experimental data at elevated pressures and temperatures combined with values for $\bar{V}°$ and $\overline{C_p}°$ for ions. The resulting values of $\bar{V}°$ and $\overline{C_p}°$, together with other standard state properties and revised-HKF equation-of-state parameters, are given in Table 2 for $AmCl^{2+}$ and other Am(III) complexes for which Silva et al. (1995) provide log β values at 25°C and 1 bar.

---

[3] It should be noted that absolute entropies are used in the correlations described by Sverjensky et al. (1997; their equations 66 through 73) despite how the text of that paper may read. For best results, use the PRONSPREP97 code available from the GEOPIG web site (http://zonvark.wustl.edu/geopig/) for making estimates of monovalent ligand complexes other than hydroxide (for hydroxide complexes see Shock et al. 1997a).

Table 2. Summary of standard state thermodynamic data for Am^{3+} (from Shock et al. 2000) and inorganic complexes based on log β values at 25°C and 1 bar selected by Silva et al. (1995), together with revised-HKF equation-of-state parameters needed to calculate the same data at high temperatures and pressures.

Species	$\Delta \bar{G}_f^{\circ\,a}$	$\Delta \bar{H}_f^{\circ\,a}$	$\bar{S}^{\circ\,b}$	$\overline{C}_p^{\circ\,b}$	$\bar{V}^{\circ\,c}$	$a_1 \times 10^d$	$a_2 \times 10^{-2\,a}$	a_3^{e}	$a_4 \times 10^{-4\,f}$	c_1^{b}	$c_2 \times 10^{-4\,f}$	$\omega^a \times 10^{-5}$
Am^{3+}	-142900.	-147400.	-48.9	-35.8	-40.2	-2.9531	-14.9848	11.6223	-2.1594	6.5235	-10.3270	2.3161
$AmOH^{2+}$	-190845.[i]	-186550.[i]	-2.5[g]	-34.4[g]	4.5[g]	2.7959	-0.9559	6.1280	-2.7394	-4.8573	-10.0418	1.0929
AmO^+	-180340.[i]	-171250.[i]	13.6[g]	-76.3[g]	7.1[g]	2.8674	-0.7813	6.0593	-2.7466	-35.6706	-18.5768	0.3444
$HAmO_2$	-221190.[i]	-206620.[i]	48.7[g]	-135.[g]	23.9[g]	5.0241	4.4881	3.9810	-2.9644	-73.2025	-30.5340	-0.0300
AmF^{2+}	-214880.[h]	-220920.[i]	-14.2[j]	2.8[j]	-37.6[j]	-2.8891	-14.8330	11.5730	-2.1658	19.5257	-2.4823	1.2776
AmF_2^+	-285490.[h]	-302130.[i]	-10.[j]	6.1[j]	-34.[j]	-2.6218	-14.1804	11.3165	-2.1927	16.1668	-1.8012	0.7003
$AmCl^{2+}$	-175710.[h]	-184780.[i]	-22.[j]	-1.2[j]	-18.5[j]	-0.2357	-8.3539	9.0265	-2.4336	18.1703	-3.3178	1.3914
$AmNO_3^{2+}$	-171220.[h]	-204070.[i]	-32.[j]	23.[j]	-7.3[j]	1.3476	-4.4881	7.5070	-2.5934	33.6975	1.6122	1.5372
$AmNO_2^{2+}$	-153470.[h]	-178190.[i]	-29.3[j]	10.2[j]	-11.3[j]	0.7899	-5.8466	8.0341	-2.5372	25.9108	-0.9964	1.5067
$AmH_2PO_4^{2+}$	-417130.[h]	-460840.[i]	-25.7[j]	40.4[j]	-5.0[j]	1.6320	-3.7971	7.2436	-2.6219	43.0978	5.1660	1.4477
$AmSCN^{2+}$	-122510.[h]	-136050.[i]	-31.7[j]	35.6[j]	-0.6[j]	2.2644	-2.2529	6.6360	-2.6858	41.0941	4.1830	1.5372
$AmCO_3^+$	-279730.[h]	-280130.[i]	103.7[j]	-61.9[m]	-22.7[j]	-1.7286	-11.9966	10.4522	-2.2830	-38.6375	-15.6435	-1.0199
$Am(CO_3)_2^-$	-412070.[h]	-370270.[i]	84.7[k]	88.0[k]	–	–	–	–	–	60.6334	14.8910	0.3461
$AmSO_4^+$	-326080.[h]	-314020.[i]	-12.4[j]	-56.8[m]	-3.7[j]	1.5392	-4.0158	7.3119	-2.6129	-20.9484	-14.6047	0.7384
$Am(SO_4)_2^-$	-506130.[h]	-485990.[i]	-4.5[k]	-77.8[k]	16.6[k]	4.6821	3.6494	4.3193	-2.9298	-25.2191	-18.8824	1.6997

Notes: a) cal mol^{-1}. b) cal mol^{-1} K^{-1}. c) cm^3 mol^{-1}. d) cal mol^{-1} bar^{-1}. e) cal K mol^{-1} bar^{-1}. f) cal K mol^{-1}. g) estimated with methods described by Shock et al. (1997a) using values for Am^{3+} in the table. h) calculated from log β taken from Silva et al. (1995), $\Delta \bar{G}_f^{\circ}$ for Am^{3+} from the table and for anions from Shock et al. (1997a). i) calculated from values of $\Delta \bar{G}_f^{\circ}$ and $\bar{S}^{\circ}$ in the table using S° for elemental Am from Silva et al. (1995). j) estimated with methods described by Sverjensky et al. (1997) using values for Am^{3+} in the table. k) estimated with methods developed in this study based on those described by Sverjensky et al. (1997) using values for Am^{3+} in the table. m) estimated with methods deduced from Sverjensky et al. (1997) using values for Am^{3+} in the table.

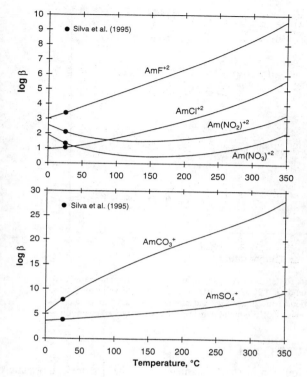

Figure 5. Association constants for aqueous Am(III) complexes as functions of temperature at P_{SAT}. Symbols represent experimental data at 25°C and 1 bar as reviewed by Silva et al. (1995), and curves correspond to predictions made in this study with the revised-HKF equations using estimation procedures outlined by Sverjensky et al. (1997) (see text). Note that temperature changes have highly variable effects on the extent of association of these complexes, and that the carbonate complex exhibits enormous increases in stability as temperature increases.

With these data and parameters it is now possible to estimate $\log \beta_r$ at elevated temperatures and pressures. Examples are shown in Figure 5 for Reaction (4) and other association reactions. Note that the formation of $AmCl^{2+}$, AmF^{2+}, $AmCO_3^+$, and $AmSO_4^+$ complexes are favored by an increase in temperature, but that $AmNO_3^{2+}$ and $AmNO_2^{2+}$ are less associated as temperature increases, at least below about 150 to 200°C.

The estimated equation-of-state parameters for hydroxide complexes of Am(III) listed in Table 2 were obtained with methods described by Shock et al. (1997a) and require comment. The formalism adopted in that study requires that hydroxide complexes be recast into nonconventional species with formulas that never contain H_2O. Thus, AmO^+ corresponds to $Am(OH)_2^+$, and $HAmO_2(aq)$ to $Am(OH)_3(aq)$, and the values of $\Delta \overline{G}_f^\circ$ of the nonconventional species differ from those of the conventional species by amounts equivalent to $\Delta \overline{G}_f^\circ$ of the appropriate stoichiometry of H_2O. In these nonconventional forms, the correlations obtained by Shock et al. (1997a) can be used directly to estimate $\overline{S}^\circ$, $\overline{V}^\circ$ and $\overline{Cp}^\circ$ of the nonconventional complexes. Resulting estimates of $\log \beta$ for the hydrolysis reactions

$$Am^{3+} + H_2O = AmOH^{2+} + H^+ \tag{7}$$
$$Am^{3+} + H_2O = AmOH_2^+ + 2 H^+ \tag{8}$$

and

$$Am^{3+} + 2 H_2O = HAmO_2(aq) + 3 H^+ \tag{9}$$

are shown in Figure 6. On incorporating these data in speciation calculations, the distinction between conventional and nonconventional hydroxide complexes vanishes.

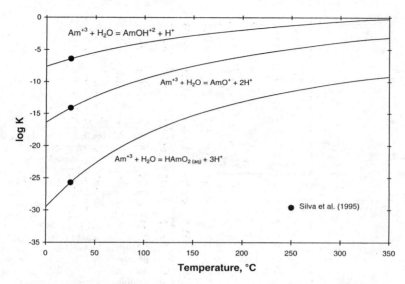

Figure 6. Hydrolysis constants for aqueous Am(III) as functions of temperature at P_{SAT}. Symbols represent experimental data at 25°C and 1 bar as reviewed by Silva et al. (1995), recalculated to conventions adopted in this study, and curves correspond to predictions made in this study with the revised-HKF equations using estimation procedures outlined by Shock et al. (1997a) (see text).

Estimating values of log β at 25°C and 1 bar

Thermodynamic data and equation-of-state parameters for the fourteen inorganic complexes in Table 2 provide a means to extrapolate 25°C log β values, critically reviewed by Silva et al. (1995), to high temperature. The resulting predictions form a starting point for estimating the speciation of Am^{3+} in aqueous solutions at high temperatures and pressures. Nevertheless, a more complete view of Am^{3+} speciation could be obtained if data for other inorganic and organic complexes were included. Many methods for estimating log β values have been proposed, and some emphasize predictions for actinide complexes (Brown and Wanner 1987; Brown and Silva 1987). In the following sections we use methods proposed by Sassani and Shock (2000) to estimate log β values at 25°C and 1 bar for inorganic complexes, and methods proposed by Shock and Koretsky (1993; 1995) and Prapaipong et al. (1999) to estimate log β values for several organic complexes. Corresponding values of $\Delta \overline{G}_f^°$ are given in Tables 3 and 4, together with other data and parameters estimated for these complexes.[4]

Inorganic complexes

The critical review by Silva et al. (1995) provides a strong foundation for estimating missing values of log β at 25°C and 1 bar using procedures developed by Sassani and Shock (2000). These methods are of two types: those for estimating the first association constant and those permitting estimates of higher-order association constants. Values of $\Delta \overline{G}_f^°$ for the first association reaction (and from it log $β_1$) can be estimated from values of $\Delta \overline{G}_f^°$ for cations using a series of charge-dependent linear correlations, which have uncertainties that are generally ≤1 kcal mol^{-1} (Shock and Koretsky 1995; Prapaipong et al.

---

[4]All thermodynamic data and equation-of-state parameters for Am complexes in Tables 2-4 are available in a format for use with the SUPCRT92 code (Johnson et al. 1992) from the GEOPIG web site.

Table 3. Summary of standard state thermodynamic data for inorganic complexes of Am^{3+} estimated in this study. Values of log β at 25°C and 1 bar were estimated with methods from Sassani and Shock (2000). Other standard state data and revised-HKF equation-of-state parameters estimated with methods from Sverjensky et al. (1997) and Shock and Helgeson (1988).

Species	log β	$\Delta\overline{G}_f^a$	$\Delta\overline{H}_f^{o\,a}$	$\overline{S}^{o\,b}$	$\overline{Cp}^{o\,b}$	$\overline{V}^{o\,c}$	$a_1^d \times 10$	$a_2^e \times 10^{-2}$	a_3^e	$a_4^f \times 10^{-4}$	c_1^b	$c_2^f \times 10^{-4}$	$\omega^a \times 10^{-5}$
$AmCl_2^+$	1.52	-207730.	-224330.	-5.	-3.7	6.4	2.8832	-0.7385	6.0333	-2.7484	9.7834	-3.7964	0.6305
$AmCl_3$	0.50	-237720.	-264600.	2.8	-42.9	34.1	6.4188	7.8944	2.6402	-3.1053	-19.2952	-11.7628	-0.0300
$AmCl_4^-$	0.45	-268230.	-308380.	0.5	-118.5	64.2	11.1715	19.4992	-1.9210	-3.5850	-48.4478	-27.2169	1.6238
AmF_3	7.41	-355030.	-386410.	-19.6	-25.8	-30.	-2.3562	-13.5317	11.0616	-2.2196	-9.2860	-8.2838	-0.0300
AmF_4^-	7.89	-423020.	-474140.	-46.	-92.5	-26.5	-0.9839	-10.1809	9.7446	-2.3581	-26.7871	-21.9300	2.3239
$Am(NO_3)_2^+$	1.77	-198330.	-256390.	-4.6	54.1	30.1	6.1236	7.1736	2.9235	-3.0755	43.5785	7.9761	0.6223
$Am(SCN)_2^+$	1.35	-100420.	-120030.	-4.5	84.3	44.3	8.0619	11.9065	1.0633	-3.2712	61.2410	14.1151	0.6223
$Am(SCN)_3$	1.51	-78480.	-104000.	23.2	110.5	94.2	14.6427	27.9751	-5.2524	-3.9354	70.5618	19.4692	-0.0300
$AmHCO_3^{2+}$	2.40	-286460.	-315940.	-26.6	37.7	-11.7	0.7185	-6.0241	8.1108	-2.5299	41.5986	4.6140	1.4574
$AmClO_4^{2+}$	1.82	-147420.	-189870.	-35.8	42.3	7.9	3.4489	0.6429	5.4904	-2.8055	45.5927	5.5441	1.6005

Notes: a) cal mol^{-1}. b) cal mol^{-1} K^{-1}. c) cm^3 mol^{-1}. d) cal mol^{-1} bar^{-1}. e) cal K mol^{-1} bar^{-1}. f) cal K mol^{-1}. g) calculated from values of $\Delta\overline{G}_f$ and $\overline{S}^o$ in the table using $S°$ for elemental Am from Silva et al. (1995).

1999; Sassani and Shock 2000). In addition, values of log β_2, log β_3, and log β_4 (where applicable) can be estimated from log β_1 using a second series of linear correlations whose slopes are derived from the coordination of the cation in aqueous solution. Examples of these correlations are given in Figure 7, where it can be seen that the scatter of points around these correlations leads to uncertainties that are generally on the order of 0.5 unit of log β. As a result, even in cases where no experimental data are available, provisional estimates can be made that are consistent with the majority of available standard state data. Estimates of log β for inorganic Am complexes at 25°C and 1 bar made in this study, together with estimated thermodynamic values and equation-of-state parameters, are listed in Table 3.

The predicted temperature dependence of association constants for inorganic complexes calculated with data and parameters from Tables 2 and 3 are shown in Figure 8. Among the chloride complexes, $AmCl_2^+$ and $AmCl^{2+}$ are estimated to be the most highly associated at temperatures ≤ 200°C. Above this temperature the predicted association constants for $AmCl_3(aq)$ are more positive than those of $AmCl^{2+}$ and approach the value for $AmCl_2^+$ at 350°C. At all temperatures $AmCl_4^-$ is less associated than the other chloride complexes. In contrast, AmF_4^- is predicted to be the most highly associated fluoride complex at temperatures ≤ 50°C. AmF^{2+} is the least associated of the fluoride complexes at all temperatures, and AmF_2^+ and $AmF_3(aq)$ are the most highly associated complexes at elevated temperatures. Equilibrium constants for $AmHCO_3^{2+}$ and $AmClO_4^{2+}$ association predicted in this study exhibit increases with increasing temperature that are much milder than the increases shown by chloride and fluoride complexes. Predictions for $AmH_2PO_4^{2+}$ indicate that this complex is less stable at elevated temperatures than it is at 25°C and 1 bar. This behavior is similar to but more strongly pronounced than the destabilization of nitrate and nitrite complexes with temperature. The log β values for $AmNO_3^{2+}$ and $AmNO_2^{2+}$ shown in Figure 8 are the same as those shown in Figure 5. In addition, estimated values for $Am(NO_3)_2^+$ are shown in Figure 8 where it can be seen that elevated temperature greatly enhances the stability of this complex relative to $AmNO_3^{2+}$. Although concentrations of SCN^- are not commonly reported in natural waters, association constants for thiocyanide complexes of Am^{3+} estimated in this study are shown in Figure 8 to minimize at around 100°C and increase in stability with increasing temperature. No minima are predicted for the temperature dependence of log β for $Am(SO_4)_2^-$ or $Am(CO_3)_2^-$, but the increase in stability of these complexes with temperature is considerably less dramatic than for $AmSO_4^+$ and $AmCO_3^+$ as shown in Figure 8 (note that the curves for the latter complexes are the same as in Fig. 5).

Organic complexes

Estimation methods used in this study for organic complexes are conceptually similar to those for inorganic complexes developed by Sverjensky et al. (1997) and Sassani and Shock (2000). A series of linear correlations allows estimates of $\Delta \overline{G}_r^\circ$ for the first association reaction, log β_2 from log β_1, entropies and heat capacities of association, and, in the case of complexes of monovalent organic ligands, volumes of association. Correlations for monovalent ligands used in this study come from Shock and Koretsky (1995) and those for oxalate from Prapaipong et al. (1999). There are fewer data for organic complexes of trivalent cations than for inorganic complexes. Partly for this reason, Shock and Koretsky (1995) proposed a single correlation to estimate $\Delta \overline{G}_r^\circ$ for monovalent-ligand complexes of trivalent cations. This correlation was used to estimate values of $\Delta \overline{G}_r^\circ$ (and therefore log β_1 and $\Delta \overline{G}_f^\circ$) for formate, acetate, propanoate, butanoate, lactate, and glycolate complexes listed in Table 4. Prapaipong et al. (1999) found that there were even fewer data for divalent-ligand complexes of trivalent cations, precluding construction of an analogous estimation method. The value of log β for the association of Am^{3+} with oxalate^{2-}

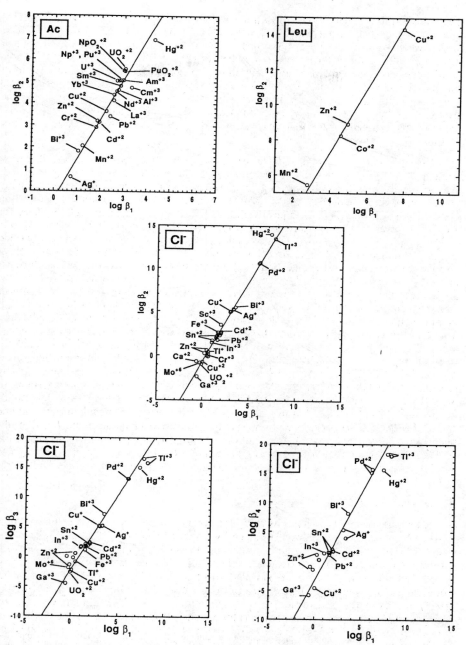

Figure 7. Correlations among association constants at 25°C at 1 bar based on experimental measurements. Those for acetic acid (Ac) and leucine (Leu) are consistent with relations discussed by Shock and Koretsky (1995), and those for chloride show the changes in slope as the order of association increases as discussed by Sassani and Shock (2000).

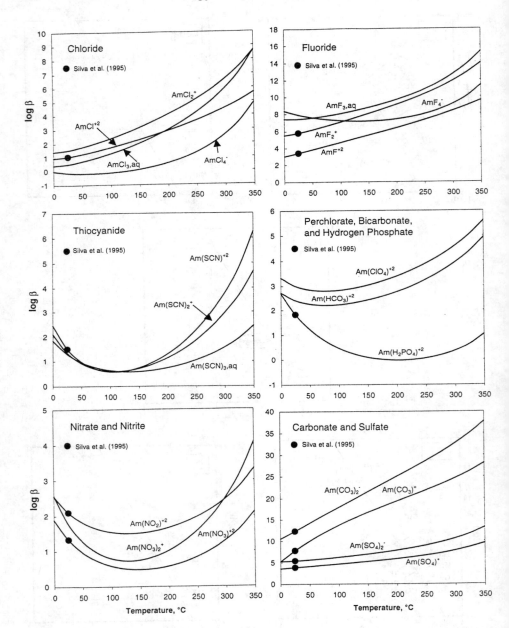

Figure 8. Association constants for aqueous inorganic complexes of Am(III) as functions of temperature at P_{SAT}. Symbols represent experimental data at 25°C and 1 bar as reviewed by Silva et al. (1995), and curve correspond to predictions made in this study with the revised-HKF equations using estimation procedures outlined by Sverjensky et al. (1997) (see text). Additional values of log β at 25°C and 1 bar were estimated in this study using procedures discussed by Sassani and Shock (2000).

Table 4. Summary of standard state thermodynamic data for organic complexes of Am^{3+} estimated in this study. Values of log β and other standard state data at 25°C and 1 bar were estimated with methods from Shock and Koretsky (1995) for monovalent ligands, and Prapaipong et al. (1999) for oxalate, unless otherwise specified. Revised-HKF equation-of-state parameters were estimated with the algorithm from Shock and Helgeson (1988).

Species	log β	$\Delta \bar{G}_f^{\circ a}$	$\Delta \bar{H}_f^{\circ a}$	$\bar{S}^{\circ b}$	$\bar{C}_p^{\circ b}$	$\bar{V}^{\circ c}$	$a_1^d \times 10$	$a_2^a \times 10^{-2}$	a_3^e	$a_4^f \times 10^{-4}$	c_1^b	$c_2^f \times 10^{-4}$	$\omega^a \times 10^{-5}$
$AmAc^{+2}$	2.97[g]	-235220.	-266700.	-25.2	64.9	4.2	2.8878	-0.7273	6.0289	-2.7489	57.3678	10.1565	1.4382
$Am(Ac)_2^+$	5.07[g]	-326360.	-385560.	-4.1	154.3	54.4	9.4478	15.2906	-0.2668	-3.4111	102.2060	28.3793	0.6143
$Am(Ac)_3$	6.54[g]	-416630.	-505490.	10.6	232.6	110.3	16.8468	33.3569	-7.3677	-4.1579	142.1213	44.3415	-0.0300
$AmFor^{+2}$	2.76	-230530.	-252410.	-25.8	12.6	-10.1	0.9287	-5.5108	7.9090	-2.5511	26.7819	-0.5050	1.4477
$Am(For)_2^+$	5.2	-317720.	-357420.	-4.	29.3	24.1	5.2992	5.1607	3.7147	-2.9923	28.9586	2.9204	0.6143
$AmProp^{+2}$	2.76	-233450.	-275460.	-28.	110.8	18.7	4.8782	4.1327	4.1187	-2.9498	84.5913	19.4948	1.4769
$Am(Prop)_2^+$	4.41	-322480.	-401070.	-3.9	263.7	85.	13.6283	25.4982	-4.2788	-3.8330	166.3624	50.6784	0.6143
$AmBut^{+2}$	2.76	-231320.	-283770.	-30.4	136.	34.	6.9922	9.2945	2.0899	-3.1632	99.7521	24.6365	1.5168
$Am(But)_2^+$	4.65	-318540.	-416630.	-4.2	324.	117.4	18.0692	36.3415	-8.5407	-4.2813	201.6874	62.9565	0.6143
$AmGlyc^{+2}$	2.76	-267840.	-307400.	-27.8	65.5	3.6	2.8119	-0.9126	6.1017	-2.7412	58.0507	10.2699	1.4769
$Am(Glyc)_2^+$	4.58	-391490.	-465260.	-3.9	155.6	53.1	9.2598	14.8314	-0.0864	-3.3921	102.9852	28.6502	0.6143
$AmLac^{+2}$	2.72[h]	-269140.	-319200.	-30.5	103.	19.9	5.0696	4.6000	3.9350	-2.9691	80.3905	17.9069	1.5168
$Am(Lac)_2^+$	6.23[h]	-396460.	-489730.	-4.2	245.1	87.8	14.0044	26.4165	-4.6398	-3.8710	155.4532	46.8867	0.6143
$AmOx^{+2}$	7.3[i]	-313960.	-342090.	-38.7	-189.1	–	–	–	–	–	-95.1410	-41.5542	1.1383

Notes: a) cal mol^{-1}. b) cal mol^{-1} K^{-1}. c) cm^3 mol^{-1}. d) cal mol^{-1} bar^{-1}. e) cal K mol^{-1} bar^{-1}. f) cal K mol^{-1}.

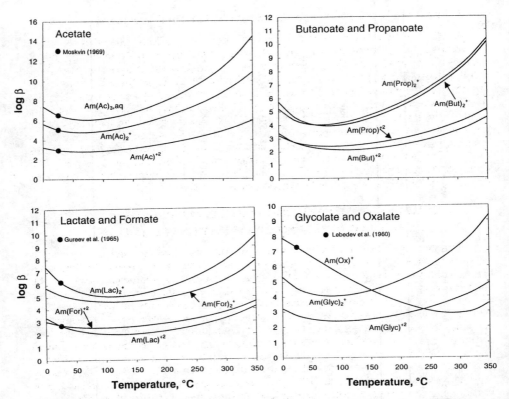

Figure 9. Association constants for aqueous complexes between Am(III) and several organic ligands as functions of temperature at P_{SAT}. Symbols represent experimental data at 25°C and 1 bar, and curves correspond to predictions made in this study with the revised-HKF equations using estimation procedures outlined by Shock and Koretsky (1995) and Prapaipong et al. (1999) (see text). Additional values of log β at 25°C and 1 bar were estimated in this study using procedures discussed by Shock and Koretsky (1995) and Sassani and Shock (2000).

selected by Prapaipong et al. (1999) is listed in Table 4. Note that values of $\overline{V}°$, and therefore the a_1, a_2, a_3, and a_4 parameters in the revised-HKF equation-of-state (see: Shock et al. 1997a for equations), cannot be estimated for organic divalent-ligand complexes (see: Prapaipong et al. 1999).

Predicted values of log β for Am-organic complexes as functions of temperature using data and parameters from Table 4 are shown in Figure 9. In each case the monovalent organic acid ligands show similar behavior, as illustrated by the acetate complex predictions. Higher-order complexes (those with more ligands) are more highly associated than lower-order complexes, and the relative stabilities of the complexes are maintained as temperature increases. Note that the predicted behavior of lactate and formate complexes are quite similar, as are those of propanoate, butanoate, and glycolate. All, like the acetate complex association constants, exhibit minima in the range 50 to 100°C. In contrast, association constants for the Am-oxalate complex are predicted to decrease with increasing temperature and minimize at 300°C. If this trend is typical for dicarboxylate ligands, then it can be expected that monocarboxylate organic ligands will account for more of the Am^{3+} in high-temperature organic-rich solutions than their dicarboxylate counterparts.

Activity coefficients for aqueous actinide ions and complexes

Activity coefficients are required to perform calculations for solutions that are not at the standard state. Because the standard state for aqueous species is hypothetical, activity coefficients are needed for all calculations involving geochemical analyses, or for model solutions designed to probe or illustrate the behavior of geologic fluids. In this study activity coefficients were estimated with the approach advocated by Helgeson (1969) for treating trace ions and charged complexes in geologic fluids. Activity coefficients for neutral complexes are assumed to be equal to one. Other methods to evaluate activity coefficients are in use. For example, see Grenthe et al. (1992, Appendix B) for details on use of the specific ion theory in interpreting data for U species. At elevated ionic strengths (e.g. >1 molal) ion interaction type activity-composition relations are required. For example, the thermodynamics of Np(V) and U(VI) speciation in concentrated solutions are described in Neck et al. (1995) and Pokrovsky et al. (1998).

EXAMPLES OF SPECIATION CALCULATIONS FOR AM(III) IN GEOLOGICAL AND ENVIRONMENTAL FLUIDS

Equilibrium constants for the wide variety of Am complexes considered here make it possible to predict the speciation of Am^{3+} in surface water, groundwater, hydrothermal fluids, sedimentary basin brines, and other geochemical and environmental fluids. In the following discussion speciation of Am^{3+} is predicted for six natural aqueous solutions: the Missouri River water, groundwater from the J-13 well in the vicinity of Yucca Mountain, Nevada, seawater, Gorga hot spring in western Sicily, an oilfield brine from the San Joaquin valley of California (24x-11), and a hydrothermal fluid venting along the East Pacific Rise at 21°N (Table 5). There are no analytical data for Am^{3+} in any of these natural fluids, and concentrations of Am^{3+} are likely to be well below detection in all. Nevertheless, the forms that Am^{3+} would take can be explored for each of these solutions by introducing a trace concentration (10^{-12} ppm) in each speciation model.[5] Comparison of results for these six speciation calculations leads to some generalizations about the aqueous speciation of Am^{3+} in environmental solutions.

Temperatures and fluid compositions for the speciation calculations are given in Table 5. In each case, pH values are for the measured temperature of solution and were recalculated based on reported values of pH at 25°C. Oxidation states are represented by Eh where values are available from analyses, or by values of the fugacity of oxygen (fO_2) when calculated from mineral equilibria (EPR 21°N) or metastable equilibria among organic compounds and minerals (24x-11 brine). Concentrations are listed in units of mg/L, and total carbonate is listed as bicarbonate. Note that temperatures of these natural solutions range from 23.5° to 350°C, pressures correspond to 1 bar at temperatures <100°C or vapor-liquid saturation pressures of H_2O for temperatures >100°C, pH ranges from acidic to slightly alkaline, and redox potential from reduced to oxidized.

Results of speciation calculations are shown as bar graphs in Figure 10 where the heights of the bars correspond to the percentage of Am^{3+} accounted for by each species. The aqueous species are in the same order in each plot to aid comparison. At the low temperature and slightly alkaline conditions of the Missouri River, Am^{3+} would be primarily present as carbonate complexes with about 15% accounted for by hydrolyzed species. Groundwater from the J-13 well is slightly less alkaline than the Missouri River and contains about 25% less total carbonate. These differences are enough to shift the

---

[5] All speciation calculations in this study were conducted with the EQ3NR code (Wolery 1992) using a data file consistent with SUPCRT92 and all subsequent publications from GEOPIG through 1998 and the estimates made here for Am^{3+} complexes.

Table 5. Natural water compositions used in speciation calculations described in text. All concentrations are in mg/L.

	Missouri River	J-13 Well Water	Sea water*	Gorga hotspring	Greeley brine**	EPR21°N vent fluid†
	(a)	(b)	(c)	(d)	(e)	(f)
Temp (°C)	23.5	25	25	49.5	138	350
pH	8.2	7.41	8.22	6.72	5.23	4.32
Eh	0.4	0.34	0.5	0.2	—	—
Log $f\,O_2$	—	—	—	—	−53.0	−30.24
Na^+	68	45.8	10768	353	10392	9931.6
Ca^{+2}	57	13	412.3	211	147	625.22
Mg^{+2}	21	2.01	1291.8	73	44.4	2.43×10^{-7}
K^+	5.04	5.04	399.1	23	100	907.08
Cl^-	15	7.14	19353	608	13044	17443
Br^-	—	—	67.3	2.397	88	6.41×10^{-2}
Sr^{+2}	—	0.04	8.14	—	31	—
Ba^{+2}	—	—	0.02	—	8.5	1.0986
Li^+	—	0.048	0.181	—	2.43	—
I^-	—	—	0.062	—	24.4	—
F^-	0.4	2.18	1.39	3.798	—	—
Cd^{+2}	1.0×10^{-12}	—	0.0001	—	—	—
Cu^{+2}	1.0×10^{-12}	—	0.0007	—	—	2.2241
Zn^{+2}	1.0×10^{-12}	—	0.0049	—	—	6.9313
Mn^{+2}	3.0×10^{-4}	0.012	0.0002	—	0.186	52.734
Fe^{+2}	0.001	0.04	0.002	—	3.88×10^{-2}	92.929
Al^{+3}	1.0×10^{-10}	0.008	0.002	1.63×10^{-4}	8.47×10^{-3}	0.1403
Pb^{+2}	0.002	—	5.0×10^{-5}	—	—	6.382×10^{-2}
SiO_2,aq	—	60.97	4.2	29	70	1057.5
$B(OH)_3$,aq	—	0.766	26.381	—	519	—
HCO_3^-	194	140.4	123.38	244	15045	353.29
SO_4^{-2}	220	18.4	2712	500	52	4.803×10^{-12}
HPO_4^{-2}	—	0.12	6.29×10^{-2}	—	—	—
NO_3^-	0.46	8.78	0.29	0.22	—	—
O_2,aq	—	5.6	6.6	—	—	—

Notes: (a) According to USGS, St. Joseph, MO (1997); (b) Data from Harrer et. al. (1990); (c) Recalculated from Nordstrom, et al. (1979); (d) Data from Favara et al. (1998); (e) Data from Fisher and Boles (1990); (f) Data from Von Damm (1990) and McCollom and Shock (1997).
* also contains: Cs^+, 0.0004; Co^{+2}, 0.00005; CrO_4^{-2}, 6.692 x 10^{-4}; $H_2AsO_4^-$, 0.0075244; Hg^{+2}, 0.00003; MoO_4^{-2}, 0.008335; Ni^{+2}, 0.0017; NO_2^-, 0.02; NH_4^+, 0.03; and Rb^+, 0.117
** also contains: formic acid, 14; acetic acid, 3825; propanoic acid, 626; and butanoic acid, 50
† also contains: HS^-, 241.41; and H_2,aq, 3.434

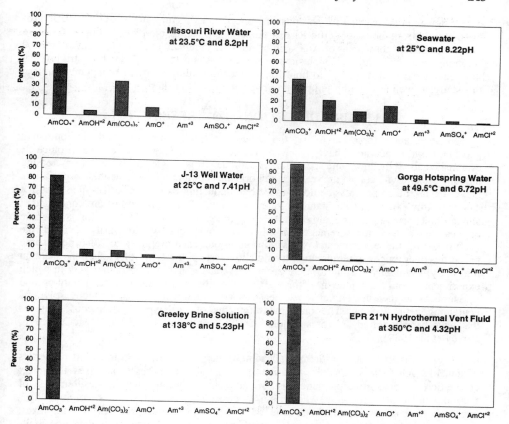

Figure 10. Examples of predicted speciation of Am(III) in natural waters calculated with the equilibrium constants estimated in this study. Also included in these calculations are equilibrium constants for hundreds of other inorganic and organic complexes of major, minor, and trace elements present in the solutions based on the analytical data and other constraints presented in Table 5.

speciation such that >80% of the Am^{3+} is predicted to be present as $AmCO_3^+$. Note that trace percentages of the unassociated Am^{3+} ion and $AmSO_4^+$ are also predicted to be present. Carbonate complexes and hydrolyzed species dominate the calculated speciation of Am^{3+} in seawater, and the sum of Am^{3+} ion, $AmSO_4^+$ and $AmCl^{2+}$ account for <10% of the inorganic complexes. It should be kept in mind that no provisions are included in this speciation model for association between Am^{3+} and dissolved organic compounds, particulate organic matter, or colloids, which may influence or control the speciation of Am^{3+}.

As shown in Figure 8, the influence of temperature is to increase the stability of carbonate complexes of Am^{3+}. This result is reflected in the results for Gorga hot spring shown in Figure 6 where the calculated speciation is >90% $AmCO_3^+$. Similarly, the predicted speciation of Am^{3+} in the 114°C oilfield brine is completely dominated by the $AmCO_3^+$ complex despite the presence of elevated concentrations of chloride and organic acids in this brine. At their natural concentrations, none of the chloride or organic ligands would influence significantly the speciation of Am^{3+} if it were present in this brine. Similar conclusions can be reached for the possible speciation of Am^{3+} in the submarine hydrothermal fluid from the East Pacific Rise. Comparison of all of the plots in Figure 6 shows that the carbonate complexes of Am^{3+} will be dominant in many geologic and environmental solutions, in accord with the conclusions reached by Allard et al. (1984). It is striking that this generalization also appears to hold for the medium-salinity oilfield brine considered in this study, as well as the high-temperature hydrothermal vent fluid.

FUTURE DIRECTIONS FOR RESEARCH

Many research topics involving the aqueous solution chemistry and environmental geochemistry of actinides have received attention for decades. As a result, much is established about the inorganic speciation of actinides in geologic fluids, and an increasing amount is known about the effects of organic compounds on actinide speciation. Nevertheless, much progress could be made by conducting experiments that yield data that can be readily extrapolated to standard state conditions. Numerous studies are conducted at single ionic strengths, and although this approach may seem useful for a specific application, a few measurements at other concentrations would vastly expand the usefulness of the experimental work. In addition, measurements of thermodynamic properties over ranges of temperature would allow tests of the types of predictions made in this study, eventually leading to more accurate predictive methods. Another way in which experiments would compliment efforts to make predictions at elevated temperatures and pressures is to measure partial molar heat capacities and volumes of aqueous solutions containing actinide ions. The only measurements of this type for aqueous actinides that have been reported are for solutions of UO_2^{2+} and Th^{4+} (Morss and McCue 1976; Hovey et al. 1989).

A great diversity of measurements leading to thermodynamic and transport properties of aqueous actinides is possible. One area leading to breakthroughs involves the direct connection of spectroscopic analysis of aqueous solutions with measurements of thermodynamic properties (see chapter on spectroscopic methods in this volume). This approach may be useful to obtain data on mixed ligand and polynuclear complexes. Another approach to clarifying speciation may be to use the extremely low detection limits in the high-mass range provided by sector-field inductively-coupled plasma mass spectrometry coupled with mixing, diffusion and/or adsorption experiments. Applying these methods to the study of colloids may be particularly appropriate for studies of actinide availability and transport in groundwaters. In turn, low detection limits for ICP-MS make it possible to use actinides in tracer studies of transport phenomena. In addition, it seems likely that organisms known to metabolize various oxidation states of U (Lovley et al. 1991; Gorby and Lovley 1992; Lovley and Phillips 1992) could also gain energy from other actinide systems that are out of redox equilibrium. So far, little is known about the rates of microbially-mediated transformations of actinides.

ACKNOWLEDGMENTS

Thanks are due to Justin Glessner and Gavin Chan for technical assistance. A constructive and thorough review by Roberto T. Pabalan contributed substantially to this presentation. This paper documents work performed for the U.S. Nuclear Regulatory Commission (NRC) by the Center for Nuclear Waste Regulatory Analyses (CNWRA) under contract NRC-02-97-009. The work was conducted on behalf of the NRC Office of Nuclear Material Safety and Safeguards, Division of Waste Management. The report is an independent product of the CNWRA and does not necessarily reflect the views or regulatory position of the NRC.

APPENDIX

Allard (1982): A mixture of empirical and estimated thermodynamic data were used to construct a data base for a model actinide in III, IV,V, and VI valences, for its inorganic complexes with OH^-, CO_3^{2-}, HPO_4^{2-}, $H_2PO_4^-$, F^-, and SO_4^{2-}, and for the solubilities of oxide, hydroxide, carbonate, phosphate, and fluoride solids. On the basis of these values it was concluded that hydroxide and carbonate complexes would dominate environmental aqueous actinide chemistry. In general, complex strength increases in the order $AnO_2^+ < An^{3+} \leq AnO_2^{2+} < An^{4+}$. Using these data, aqueous speciation and solubilities were calculated for actinides of fixed valence for various constraints on aqueous carbonate. Introducing standard potentials for redox reactions among species of U, Np, Pu, and Am, aqueous speciation and solubilities were presented as a function of pH for oxidation potentials roughly corresponding to atmospheric O_2 pressure and to equilibrium between ferrous and ferric minerals.

Newton and Sullivan (1985): Empirical reduction potentials for aqueous actinide species are summarized for 1 M carbonate and perchloric acid solutions demonstrating generally that carbonate complexes stabilize dissolved actinides of valences that form strong complexes.

Kim (1986): The quantities of transuranic species in the environment are summarized. General aspects of solubility, complexation, redox reactions, and colloid generation for transuranic species are presented. The role of colloids in natural environments is emphasized leading to the conclusion that thermodynamic treatments of aqueous actinide concentrations are challenging. Solubility and sorption characteristics of transuranic species are a strong functions of colloid size in the nanometer scale. Hence, environmental aqueous actinide chemistry depends on the filtering capacity of the geochemical medium.

Kim (1993): Controls on the aqueous chemistry of transuranium elements are summarized. Kim (1993) emphasizes that the chemistry of transuranic elements in natural aquatic systems can be substantially different than in laboratory or industrial systems, noting in particular hydrolysis under environmental aquatic conditions, multicomponent competition of dissolved species in reactions, aqueous complexation with organic and inorganic ligands, colloidal effects, and sluggish redox transitions. Examples are drawn primarily from studies of groundwaters from the Gorleben site in Germany, which are variously saline and nonsaline.

Fuger (1992): A critical survey is presented of the literature on the thermodynamic properties of hydroxide and carbonate complex association reactions for aqueous actinides noting often large discrepancies in reported data and absent data. A critically reviewed selection of thermodynamic data is presented. For uranium the selection coincides with values selected also by Fuger et al. (1992) or Grenthe et al. (1992). Fuger (1992) offers an alternate to the limit for the thermodynamic properties for the $UO_2(OH)_2$ species reported in the review by Grenthe et al. (1992).

Grenthe et al. (1992): The first in a series of critical reviews of the thermodynamic properties of selected elements of radiological significance, this book offers a careful fairly comprehensive review of experimentally determined thermodynamic properties of uranium species and compounds. Data review led to recommendations for a set of selected standard state data for many aqueous uranium species, which have been incorporated in data bases such as that supporting the EQ3/6 geochemical modeling software. A notable feature of the selected data is that only minimum limits are given for the standard state free energies of $UO_2(OH)_2$ and $U(OH)_5$. Stabilities of these species are less than represented by the minimum free energy values. Also, the second book in this series (Silva et al. 1995) contains an extensive commentary on the data and analyses in Grenthe et al. (1992) including numerous errata.

Clark et al. (1995): A thorough review of carbonate and hydroxide complexes of actinides is provided with an extensive bibliography and a focus on spectroscopic and other modern techniques for characterizing aqueous complexes. Aqueous complex structures are interpreted in the context of actinide coordination in crystalline structures. For U, Np, and Pu in the hexavalent state spectroscopic evidence points to the predominance of $AnO_2(CO_3)$,

$AnO_2(CO_3)_2^{2-}$, $AnO_2(CO_3)_3^{4-}$, and $(AnO_2)_3(CO_3)_6^{6-}$. Evidence exists also for the occurrence of mixed uranyl carbonate or bicarbonate hydroxide complexes. For pentavalent Np, Pu, and Am the carbonate complexes $AnO_2(CO_3)_n^{(2n-1)}$ (for $n = 1, 2,$ or 3) have been observed. Data for tetravalent aqueous actinide species are equivocal. Either mixed hydroxide carbonate or carbonate species predominate. For trivalent Am and Cm complexes with 1, 2, and 3 carbonate species have been observed, but uncertainty exists with regard to the occurrence of mixed carbonate hydroxide species.

Silva et al. (1995): A critical review of the thermodynamic properties of Am compounds and species is presented culminating in presentation of a set of selected standard state thermodynamic data. Notably no standard state enthalpies are selected for aqueous complexes of Am, limiting use of these data to 25°C. Estimation procedures provided in this review offer a means to extend the usefulness of this data set.

Shock et al. (1997b): This paper takes as its foundation the summary and critique of thermodynamic data for uranium by Grenthe et al. (1992) to build equilibrium oxidation potential-pH and distribution-of-species diagrams in the U-O-H system at various temperature and pressure. Regression of experimental data and several correlation methods provide the thermodynamic data and revised-HKF parameters to estimate equilibrium constants for redox and hydrolysis reactions involving 19 aqueous species. The diagrams show that U(IV) and U(VI) species predominate in aqueous solution, that increasing temperature stabilizes U(VI) and U(III) species relative to U(IV) species, but that the latter dominate at oxidation states consistent with mineral buffer assemblages at near-neutral pH. U(VI) species can be stabilized relative to U(IV) species at low pH. The results suggest that hydrothermal transport of uranium requires low pH or potent complexes of U(IV).

Shock et al. (2000): The main purpose of this paper is to provide a set of standard state thermodynamic data and revised-HKF parameters for actinide ions. To reach that goal the authors examined many sources of data and distinguished values in the literature that are based closely on experimental measurements from estimated values. Using the experimentally-derived values, they construct several correlations to refine values of estimated data, including relations among crystallographic radii, standard state entropies, volumes, and heat capacities, standard Gibbs free energies of formation, and standard potentials for redox reactions. Comparisons are made with other estimation methods, and extensive tables summarize the experimental data adopted in the study, estimates made with the new correlations, and estimates made by other investigators. The results provide a thermodynamically-consistent starting point for calculating actinide speciation in aqueous solutions over wide ranges of temperature and pressure.

REFERENCES

Allard B 1982 Solubilities of actinides in neutral or basis solutions. *In* Actinides in Perspective, NM Dedlstien (Ed) Pergamon, p 553-580

Allard B, Olofsson U, Torstenfelt B 1984) Environmental Actinide Chemistry. Inorg Chim Acta 94:205-221

Amend JP, Helgeson HC 1997) Group additivity equations of state for calculating the standard molal thermodynamic properties of aqueous organic species at elevated temperatures and pressures. Geochim Cosmochim Acta 61:11-46

Baes CF Jr, Mesmer RE 1976 The Hydrolysis of Cations. John Wiley & Sons, Inc.

Barner HE, Scheuerman RV 1978 Handbook of Thermodynamic Data for Compounds and Aqueous Species. Wiley, New York, 156 p

Barnes CE, Cochran JK 1993 Uranium geochemistry in estuarine sediments: Controls on removal and release processes. Geochim Cosmochim Acta 57:555-569

Brown PL, Sylva RN 1987 Unified theory of metal ion complex formation constants. J Chem Res (S) 4-5 (M):0110-0181

Brown PL, Wanner H 1987 Predicted formation constants using the unified theory of metal ion complexation. Organization for Economic Co-operation and Development, Nuclear Energy Agency, Paris, 102 p

Caceci MS, Choppin GR 1983 The first hydrolysis constant of uranium (VI). Radiochim Acta 33:207-212

Carroll SA, Bruno J, Petit J-C, Dran J-C 1992 Interactions of U(VI), Nd, and Th(IV) at the calcite-solution interface. Radiochim Acta 58/59:245-252

Chen JH, Wasserburg GJ 1986 The U-Th-Pb systematics in hot springs on the East Pacific Rise at 21°N and Guayamas Basin. Geochim Cosmochim Acta 50:2467-2479

Clark DL, Hobart DE, Neu MP 1995 Actinide carbonate complexes and their importance in actinide environmental chemistry. Chem Reviews 95:25-48

Cordfunke EHP, O'Hara PAG 1978 The Chemical Thermodynamics of Actinide Elements and Compounds, Part 3. Miscellaneous Actinide Compounds. International Atomic Energy Agency, Vienna

Cowan GA 1976 A natural fission reactor. Scientific American 253:36-47

Criss CM, Cobble JW 1964a The thermodynamic properties of high temperature aqueous solution. IV. Entropies of the ions up to 200° and the correspondence principle. J Am Chem Soc 86:5385-5390

Criss CM, Cobble JW 1964b The thermodynamic properties of high temperature aqueous solution. V. The calculation of ionic heat capacities up to 200°. Entropies and heat capacities above 200°. J Am Chem Soc 86:5390-5393

Curtis D, Fabryka-Martin J, Dixon P, Cramer J 1999 Nature's uncommon elements: Plutonium and technetium. Geochim Cosmochim Acta 63:275-285

David F 1986 Oxidation reduction and thermodynamic properties of curium and heavier actinide elements. *In* Handbook on the Physics and Chemistry of the Actinides. Freeman AJ, Keller C (Eds), North-Holland, Amsterdam, p 97-128

Dongarra G, Langmuir DL 1980 The stability of UO_2OH^+ and $UO_2[HPO_4]_2^{2-}$ complexes at 25°C. Geochim Cosmochim Acta 44:1747-1751

Eberly P, Janeczek J, Ewing RC 1994) Petrographic analysis of samples from the uranium deposit at Oklo, Republic of Gabon. Radiochim Acta 66/67:455-461

Favara R, Grassa F, Inguaggiato S, D'Amore F 1998 Geochemical and hydrogeological characterization of thermal springs in Western Sicily, Italy. J Volcanol Geotherm Res 84:125-141

Fisher JB, Boles JR 1990 Water-rock interactions in Tertiary sandstones, San Joaquin Basin, California, U.S.A.: Diagenetic controls on water composition. Chem Geol 82:83-101

Flotow HE, Huschke JM, Yamauchi S 1984) The Chemical Thermodynamics of Actinide Elements and Compounds. Part 9: The Actinide Hydrides. International Atomic Energy Agency, Vienna

Fuger J 1992 Thermodynamic properties of actinide aqueous species relevant to geochemical problems. Radiochim Acta 58/59:81-91

Fuger J 1982 Thermodynamic properties of the actinides: Current perspectives, *In* Actinides In Perspective, Proceedings of the Actinides-1981 Conference. Pergamon Press, New York. p 409-431

Fuger J, Oetting FL 1976 The Chemical Thermodynamics of Actinide Elements and Compounds. Part 2. The Actinide Aqueous Ions. International Atomic Energy Agency, Vienna

Fuger J, Parker VB, Hubbard WN, Oetting FL 1983 The Chemical Thermodynamics of Actinide Elements and compounds. Part 8: The Actinide Halides. International Atomic Energy Agency, Vienna.

Gauthier-Lafaye F, Weber F, Ohmoto H 1989 Natural fission reactors of Oklo. Econ Geol 84:2286-2295

Gayer KH, Leider H 1955 The solubility of uranium trioxide, UO_3H_2O, in solutions of sodium hydroxide and perchloric acid at 25°. J Am Chem Soc 77:1448-1450

Geipel G, Nebel D 1985 Hydrolysis of uranyl ions in solutions of decreased water concentration. Isotopenpraxis 21:253-254 (in German)

George-Aniel B, Leroy JL, Poty B 1991 Volcanogenic uranium mineralization in the Sierra Peña Blanca district, Chihuahua, Mexico: Three genetic models. Econ Geol 86:233-248

Gorby YA, Lovley DR 1992 Enzymatic uranium precipitation. Environ Sci Tech 26:205-207

Graf WL 1994) Plutonium and the Rio Grande Environmental Change and Contamination in the Nuclear Age. Oxford University Press, New York, 329 p

Grenthe I, Bidoglio G, Omenetto N 1989 Use of thermal lensing spectrophotometry (TLS) for the study of mononuclear hydrolysis of uranium (IV). Inorg Chem 28:71-74

Grenthe I, Fuger J, Konings RJM, Lemire RJ Muller AB, Nguyen-Trung C, Wanner H 1992 Chemical Thermodynamics of Uranium. Nuclear Energy Agency. North-Holland, 715 p

Grønvold F, Drowart J, Westrum EF Jr 1984) The Chemical Thermodynamics of Actinide Elements and Compounds. Part 4: The Actinide Chalcogenides (Excluding Oxides). International Atomic Energy Agency, Vienna

Gureev ES, Dedov VB, Karpacheva FM, Lededev IA, Rizhov MN, Trukhlyaev PS, Shvetsov IK, Yakovlev GN 1965 Method of extraction and some chemical properties of trans-plutonium elements. *In* Proc Third Internat Conf Peaceful Uses Atom Energy, United Nations, New York, p 553-560.

Harrar JE, Carley JF, Isherwood WF, Raber E 1990 Report of the committee to review the jus of J-13 well water in Nevada nuclear waste storage investigations. Lawrence Livermore National Laboratory, UCID-21867, Livermore, CA

Harris, Kolthoff 1947 The polarography of uranium. III. Polarography in very weakly acid, neutral or basic solution. J Am Chem Soc 69:446-451

Helgeson HC 1967 Thermodynamics of complex dissociation in aqueous solution at elevated temperatures. Jour Phys Chem 71:3121-3136

Helgeson HC 1969 Thermodynamics of hydrothermal systems at elevated temperatures and pressures. Am J Sci 267:729-804

Helgeson HC, Kirkham DH, Flowers GC 1981 Theoretical prediction of the thermodynamic behavior of aqueous electrolytes at high pressures and temperatures. IV. Calculation of activity coefficients, osmotic coefficients, and apparent molal and standard and relative partial molal properties to 5 kb and 600°C. Am J Sci 281:1241-1516

Hildenbrand DL, Gurvich LV, Yungman VS 1985 The Chemical Thermodynamics of Actinide Elements and Compounds, part 3: The Gaseous Actinide Ions. International Atomic Energy Agency, Vienna

Holley CE, Rand MH, Storms EK 1984) The Chemical Thermodynamics of Actinide Elements and Compounds Part 6: The Actinide Carbides. International Atomic Energy Agency, Vienna.

Holliger P, Devillers D, Retali G 1978 Evaluation des temperatures neutroniques dans les zones de reaction d'Oklo par l'etude des rapports isotopiques. In Natural Fission Reactors. IAEA, Vienna, p 553-568

Hovey JK, Nguyen-Trung C, Tremaine PR 1989 Thermodynamics of aqueous uranyl ion: Apparent and partial molar heat capacities and volumes of aqueous uranyl perchlorate from 10 C to 55 C. Geochim Cosmochim Acta 53:1503-1509

Irwin JJ, Roedder E 1995 Diverse origins of fluid in magmatic inclusions at Bingham (Utah, USA), Butte (Montana, USA), St. Austell (Cornwall, UK), and Ascension Island (mid-Atlantic, UK), indicated by laser microprobe analysis of Cl, K, Br, I, Ba + Te, U, Ar, Kr, and Xe. Geochim Cosmochim Acta 59:295-312

Ivanovich M 1994) Uranium series disequilibrium: Concepts and applications. Radiochim Acta 64:81-94

Johnson JP, McCulloch MT 1995 Sources of mineralising fluids for the Olympic Dam deposit (South Australia): Sm-Nd isotopic constraints. Chem Geol 121:177-199

Johnson JW, Oelkers EH, Helgeson HC 1992 SUPCRT92: A software package for calculating the standard molal thermodynamic properties of minerals, gases, aqueous species, and reactions from 1 to 5000 bar and 0 to 1000°C. Computers Geosci 18:899-947

Kaplan DL, Bertsch PM, Adriano DC, Orlandini KA 1994) Actinide association with groundwater colloids in a coastal plain aquifer. Radiochim Acta 66/67:181-187

Katz JJ, Seaborg GT, Morss LR 1986 The chemistry of the actinide elements. Chapman and Hall, London, 2 volumes

Kersting AB, Efurd DW, Finnegan DL, Rokop DJ, Smith DK, Thompson JL 1999 Migration of plutonium in ground water at the Nevada Test Site. Nature 397:56-59

Keppler H, Wyllie PJ 1990 Role of fluids in transport and fractionation of uranium and thorium in magmatic processes. Nature 348:531-533

Kim JI 1986 Chemical behavior of transuranic elements in natural aquatic systems. In Handbook on the Physics and Chemistry of the Actinides. Freeman AJ, Keller C (Eds) Elsevier Science Publishers, Amsterdam:413-455

Kim JI 1993 The chemical behavior of tranuranium elements and barrier functions in natural aquifer systems. Mater Res Soc Symp Proc 294:3-21

Komninou A, Sverjensky DA 1995a Hydrothermal alteration and the chemistry of ore-forming fluids in an unconformity-type uranium deposit. Geochim Cosmochim Acta 59:2709-2723

Komninou A, Sverjensky DA 1995b Pre-ore hydrothermal alteration in an unconformity-type uranium deposit. Contrib Mineral Petrol 121:99-114

Komninou A, Sverjensky DA 1996 Geochemical modeling of the formation of an unconformity-type uranium deposit. Econ Geol 91:590-606

Kotzer TG, Kyser TK 1993 O, U, and Pb isotopic and chemical variations in uraninite: Implications for determining the temporal and fluid history of ancient terrains. Am Mineral 78:1262-1274

Kotzer TG, Kyser TK 1995 Petrogenesis of the Proterozoic Athabasca Basin, northern Saskatchewan, Canada, and its relation to diagenesis, hydrothermal uranium mineralization and paleohydrogeology. Chem Geol 120:45-89

Kraus KA, Nelson F 1950 Hydrolytic behavior of metal ions. I. The acid constants of uranium(IV) and plutonium(IV). J Am Chem Soc 72:3901-3906

Lebedev IA, Pirozhkov SV, Yakovlev GN 1960 Determination of the composition and the instability constants of oxalate, nitrate and sulfate complexes of Am^{III} and Cm^{III} by the method of ion exchange. Radiokhimiya 2:549-558 (in Russian)

Lemire RJ, Tremaine RM 1980 Uranium and plutonium equilibria in aqueous solutions to 200°C. J Chem Eng Data 25:361-370

Loss RD, Rosman KJR, De Laeter JR, Curtis DB, Benjamin TM, Gancarz AJ, Maeck WJ, Delmore JE 1989 Fission-product retentivity in peripheral rocks at the Oklo natural fission reactors, Gabon. Chem Geol 76:71-84

Lovley DR, Phillips EJP 1992 Reduction of uranium by Desulfovibrio desulfuricans. Appl Environ Microbiol 58:850-856

Lovley DR, Phillips EJP, Gorby YA, Landa ER 1991 Microbial reduction of uranium. Nature 350:413-416

McCollom TM, Shock EL 1997 Geochemical constraints on chemolithoautotrophic metabolism by microorganisms in seafloor hydrothermal systems. Geochim Cosmochim Acta 61:4375-4391

Michard A, Albarede F, Michard G, Minster JF, Charlou JL 1983 Rare-earth elements and uranium in high-temperature solutions from East Pacific Rise hydrothermal vent field (13°N). Nature 303:795-797

Mikheev NB, Myasoedov BF 1985 Lower and higher oxidation states of transplutonium elements in solutions and melts, In Handbook on the Physics and Chemistry of The Actinides, Volume 3. Freeman AJ, Keller C (Eds) Elsevier Science Publishers BV, North-Holland, p 347-386

Morss LR 1982 Complex oxide systems of the actinides. In Actinides In Perspective, Proceedings of the Actinides-1981 Conference. Pergamon Press, New York. p 381-407

Morss LR 1986 Thermodynamic Properties. In The Chemistry of the Actinides, Vol 2, Katz SS, Seaborg GT, Morss LR (Eds) Chapman and Hall, New York, p 1278-1360

Morss LR, McCue MC 1976 Partial molal entropy and heat capacity of the aqueous thorium(IV) ion. Thermochemistry of thorium nitrate pentahydrate. J Chem Eng Data 21:337-341

Moskvin AI 1969 Complex formation of the actinides with anions of acids in aqueous solutions. Sov Radiochem 11:447-449

Neck V, Fanghanel Th, Rudolph G, Kim JI 1995 Thermodynamics of neptunium(V) in concentrated salt solutions: Chloride complexation and ion interaction (Pitzer) parameters for the NpO_2^{2+} ion. Radiochim Acta 69:39-47

Newton TW, Sullivan JC 1985 Actinide carbonate complexes in aqueous solution. In Handbook of the Physics and Chemistry of the Actinides. Freeman AJ, Keller C (Eds) Elsevier, p 387-406

Nikitin AA, Sergeyeva EI, Khodakovsky IL, Naumov GB 1972 Hydrolysis of uranyl in the hydrothermal region. Geokhimiya 3:297-307 (in Russian)

Nikolaeva NM 1971 The study of hydrolysis and complexing of uranyl ions in sulphate solutions at elevated temperatures. Izv Sib Otd Akad Nauk SSSR 62-66 (in Russian)

Nikolaeva NM 1978 The hydrolysis of U^{4+} ions at elevated temperatures. Izv Sib Otd Akad Nauk SSSR 91-95 (in Russian)

Nitsche H, Roberts K, Prussin T, Müller A, Becraft K, Keeney D, Carpenter SA, Gatti RC 1994) Measured solubilities and speciations from oversaturation experiments of neptunium, plutonium, and americium in UE-25p#1 well water from the Yucca Mountain region. LA-12563-MS. Los Alamos National Laboratory, Los Alamos, NM

Nordstrom DK, Plummer LN, Wigley TML, et al. 1979 A comparison of computerized chemical models for equilibrium calculations in aqueous systems. In Chemical Modeling in Aqueous Systems. Jenne EA (Ed) Am Chem Soc Symp Ser 93:857-892

Pabalan RT, Turner DR 1997 Uranium(6+) sorption on montmorillonite: Experimental and surface complexation modeling study. Aquatic Geochem 2:203-226

Pabalan RT, Turner DR, Bertetti FP, Prikryl JD 1998 Uranium^{VI} sorption onto selected mineral surfaces. In Adsorption of Metals by Geomedia. Jenne EA (Ed) Academic Press, San Diego, p 99-130

Pan Y, Fleet ME, MacRae ND 1993 Late alteration in titanite ($CaTiSiO_5$): Redistribution and remobilization of rare earth elements and implications for U/Pb and Th/Pb geochronology and nuclear waste disposal. Geochim Cosmochim Acta 57:355-367

Parks GA, Pohl DC 1988 Hydrothermal solubility of uraninite. Geochim Cosmochim Acta 52:863-875

Pokrovsky OS, Bronikowski MG, Moore RC, Choppin GR 1998 Interaction of neptunyl(V) and uranyl(VI) with EDTA in NaCl media: Experimental study and Pitzer modeling. Radiochim Acta 80:23-29

Prapaipong P, Shock EL, Koretsky CM 1999 Metal-organic complexes in geochemical processes: Temperature dependence of standard partial molal thermodynamic properties of aqueous complexes between metal cations and dicarboxylate ligands. Geochim Cosmochim Acta (in press)

Plyasunov AV, Grenthe I 1994) The temperature dependence of stability constants for the formation of polynuclear cationic complexes. Geochim Cosmochim Acta 58:3561-3582

Raffensperger JP, Garven G 1995a The formation of unconformity-type uranium ore deposits 1. Coupled groundwater flow and heat transport modeling. Am J Sci 295:581-636

Raffensperger JP, Garven G 1995b The formation of unconformity-type uranium ore deposits 2. Coupled hydrochemical modeling. Am J Sci 295:639-696

Rai D, Felmy AR, Ryan JL 1990 Uranium (IV) hydrolysis constants and solubility product of $UO_2 \cdot xH_2O$ (am). Inorg Chem 29:260-264

Sassani DC, Shock EL 1992 Estimation of standard partial molal entropies of aqueous ions at 25°C and 1 bar. Geochim Cosmochim Acta 56:3895-3908
Sassani DC, Shock EL 1994) Errata to Sassani and Shock 1992 Geochim Cosmochim Acta 58:2756-2758
Sassani DC, Shock EL (2000 Estimation of standard state association constants for aqueous complexes at 25 C and 1 bar (in preparation)
Seaborg GT 1993 Overview of the actinide and lanthanide (the f) elements. Radiochim Acta 61:115-122
Schmidt KH, Sullivan JC, Gordon S, Thompson RC 1978 Determination of hydrolysis constants of metal cations by a transient conductivity method. Inorg Nucl Chem Lett 14:429-434
Shock EL 1995 Organic acids in hydrothermal solutions: Standard molal thermodynamic properties of carboxylic acids, and estimates of dissociation constants at high temperatures and pressures. Am J Sci 295:496-580.
Shock EL, Helgeson HC 1988 Calculation of the thermodynamic and transport properties of aqueous species at high pressures and temperatures: Correlation algorithms for ionic species and equation of state predictions to 5 kb and 1000°C. Geochim Cosmochim Acta 52:2009-2036
Shock EL, Helgeson HC 1990 Calculation of the thermodynamic and transport properties of aqueous species at high pressures and temperatures: Standard partial molal properties of organic species. Geochim Cosmochim Acta 54:915-945
Shock EL, Koretsky CM 1993 Metal-organic complexes in geochemical processes: Calculation of standard partial molal thermodynamic properties of aqueous acetate complexes at high pressures and temperatures. Geochim Cosmochim Acta 57:4899-4922
Shock EL, Koretsky CM 1995 Metal-organic complexes in geochemical processes: Estimation of standard partial molal thermodynamic properties of aqueous complexes between metal cations and monovalent organic acid ligands at high pressures and temperatures. Geochim Cosmochim Acta 59:1497-1532
Shock EL, Helgeson HC, Sverjensky DA 1989 Calculation of the thermodynamic and transport properties of aqueous species at high pressures and temperatures: Standard partial molal properties of inorganic neutral species. Geochim Cosmochim Acta 53:2157-2183.
Shock EL, Oelkers EH, Johnson JW, Sverjensky DA, Helgeson HC 1992 Calculation of the thermodynamic properties of aqueous species at high pressures and temperatures: Effective electrostatic radii, dissociation constants, and standard partial molal properties to 1000°C and 5 kb. J Chem Soc Faraday Trans 88:803-826
Shock EL, Sassani DC, Willis M, Sverjensky DA 1997a Inorganic species in geologic fluids: Correlations among standard molal thermodynamic properties of aqueous ions and hydroxide complexes. Geochim Cosmochim Acta 61:907-950
Shock EL, Sassani DC, Betz H 1997b Uranium in geologic fluids: Estimates of standard partial molal properties, oxidation potentials and hydrolysis constants at high temperatures and pressures. Geochim Cosmochim Acta 61:4245-4266
Shock EL, Sassani DC, Betz H. (2000 Actinide ions in aqueous solution: Summary of standard state properties and estimates at high temperatures and pressures. Appl Geochem (in press)
Silva RJ, Bidoglio G, Rand MH, Robouch PB, Wanner H, Puigdomenech I. 1995 Chemical Thermodynamics of Americium. Nuclear Energy Agency. Elsevier, Amsterdam, 374 p
Sturchio NC, Binz CM, Lewis CH III 1987 Thorium-uranium disequilibrium in a geothermal discharge zone at Yellowstone. Geochim Cosmochim Acta 51:2025-2034
Sturchio NC, Böhlke JK, Binz CM 1989 Radium-thorium disequilibrium and zeolite-water in exchange in a Yellowstone hydrothermal environment. Geochim Cosmochim Acta 53:1025-1034
Sverjensky DA, Shock EL, Helgeson HC 1997 Prediction of the thermodynamic properties of aqueous metal complexes to 1000°C and 5 kb. Geochim Cosmochim Acta 61:1359-1412
Tanger JC IV, Helgeson HC 1988 Calculation of the thermodynamic and transport properties of aqueous species at high pressures and temperatures. Revised equations of state for the standard partial molal properties of ions and electrolytes. Am J Sci 288:19-98
Turner DR, Pabalan RT, Bertetti FP 1998 Neptunium(V) sorption on montmorillonite: An experimental and surface complexation modeling study. Clays Clay Mineral 46:256-269
Turpin L, Leroy JL, Sheppard SMF 1990 Isotopic systematics (O, H, Sr, Nd) of superimposed barren and U-bearing hydrothermal systems in a Hercynian granite, Massif Central, France. Chem Geol 88:85-98
Von Damm KL 1990 Seafloor hydrothermal activity: Black smoker chemistry and chimneys. Ann Rev Earth Plant Sci 18:173-204
Wallace AR, Whelan JF 1986 The Schwartzwalder Uranium Deposit, III: Alteration, vein mineralization, light stable isotopes, and genesis of the deposit. Econ Geol 81:872-888
Watters RL 1983 Aquatic chemistry of plutonium. *In* Plutonium Chemistry. Carnall WT, Choppin GR (Eds) ACS Symposium Series 216. American Chemical Society, Washington, DC, p 297-315
Wenrich KJ 1985 Mineralization of breccia pipes in northern Arizona. Econ Geol 80:1722-1735

Wersin P, Hochella MF Jr, Persson P, Redden G, Leckie JO, Harris DW 1994) Interaction between aqueous uranium (VI) and sulfide minerals: Spectroscopic evidence for sorption and reduction. Geochim Cosmochim Acta 58:2829-2843

Wilde AR, Mernagh TP, Bloom MS, Hoffman CF 1989 Fluid inclusion evidence on the origin of some Australian unconformity-related uranium deposits. Econ Geol 84:1627-1642

Wolery TJ 1992 EQ3/6, A software package for geochemical modeling of aqueous systems: Package overview and installation guide. UCRL-MA-110662-Pt 1, Lawrence Livermore National Laboratory, Livermore, CA

Zielinski RA, Meier A 1988 The association of uranium with organic matter in Holocene peat: An experimental leaching study. Appl Geochem 3:631-643

6

Uranium Ore Deposits—
Products of the Radioactive Earth

Jane A. Plant, Peter R. Simpson and Barry Smith
British Geological Survey
Kingsley Dunham Centre
Keyworth, Nottingham, NG12 5GG, United Kingdom

Brian F. Windley
Department of Geology
The University of Leicester
University Road
Leicester, LE1 7RH, United Kingdom

INTRODUCTION

Uranium ore deposits are of intrinsic economic importance as a source of fissionable ^{235}U and ^{238}U, as well as ^{226}Ra and other radioactive species. They are also important as natural analogues of nuclear waste repositories, and they provide fundamental information on Earth processes through time. The genesis of early Proterozoic quartz pebble conglomerate uranium deposits is central to the debate about the evolution of the early atmosphere and the time at which it became oxidizin and hence about the beginnings of life on Earth.

Prior to 1942, uranium was used mainly for coloring glass and ceramic glazes. Following the pioneering work of Marie Curie, uranium ores were also utilised as a source of the daughter product ^{226}Ra for the treatment of cancer, but in 1942 controlled nuclear fission of ^{235}U was demonstrated, which led to the development of its use in nuclear energy and weaponry. Nuclear fuel is produced by the enrichment of natural ^{235}U from about 0.7% to about 3.5%. The residue of depleted uranium, which contains about 0.2% ^{235}U, is used chiefly in armor piercing shells and for counterweights because of its pyrophoric properties on impact and its high density. World production of uranium in 1996 was approximately 36,000 tonnes U (OECD/NEA 1998).

Until the 1940s U and ^{226}Ra were obtained mainly from small, high-grade veins associated with the High Heat Production (HHP) leucogranites of South West England and the Erzgebirge in Germany. In contrast, modern uranium mining is increasingly concerned with large high-grade deposits such as those of the Athabasca basin in Saskatchewan, Canada, and at Pine Creek, Northern Territory, Australia. In this study we consider all of the types of uranium mineralization, that have been economically important during the period of a little over 50 years since the beginning of the nuclear industry.

We first consider the physico-chemical properties of uranium; and the complex interplay between its geochemistry, radioactive decay and heat production and the evolution of the Earth with particular reference to continental crust formation and uranium ore deposition. Particular attention is paid to the late Archaean-early Proterozoic, the middle and late Proterozoic, the early Permian and the Mesozoic-Tertiary periods when many economically important deposits formed. Recent classifications of uranium ore deposits are briefly considered before reviewing the main ore deposit types and metallogenic models in relation to their palaeo-tectonic settings. Preservation, which is particularly important in natural analogue studies aimed at developing safe nuclear waste repositories, is considered briefly.

Throughout the chapter, emphasis is given to the re-evaluation of well-documented classical ore bodies in relation to modern concepts of continental lithospheric convergence and extension, and the evolution of these processes through time. The development of uranium provinces in which uranium is enriched in different lithologies of varying age and, in some cases, where several types of economic uranium deposits occur is also considered. The main thesis of the chapter is that many important uranium deposits and provinces are ultimately related to evolved felsic igneous rocks emplaced at a high level in the crust during the final stages of orogenesis or anorogenically. The chapter is designed to complement detailed comprehensive studies of uranium deposits, for example that of Dahlkamp (1993) and economic classifications of uranium deposits such as that of the IAEA (1996) and OECD/NEA (1998). Comprehensive accounts of the mineralogy and aqueous behavior of uranium are discussed in Chapters 3 and 5 of this volume.

SOME CHARACTERISTICS OF URANIUM

Uranium, Th and K are the main elements contributing to natural terrestrial radioactivity. Uranium has two primary isotopes, ^{238}U and ^{235}U, that occur at the present time in the proportion 99.3% ^{238}U to 0.7% ^{235}U. Both have long and complex decay series (Table 1), in contrast to Th which has only one relatively simple decay chain, from ^{232}Th. Of the three naturally-occuring isotopes of K (^{39}K, ^{40}K and ^{41}K), only ^{40}K is radioactive; the respective isotopic abundances are 93.08, 0.012 and 6.9%. Uranium, Th and K are lithophile elements with Pauling electronegativities of 1.7, 1.3 and 0.8 respectively, and they are strongly partitioned into the continental crust.

The isotopes ^{238}U, ^{235}U, ^{232}Th and ^{40}K decay with half lives so long that significant amounts remain in the Earth, providing a continuing source of heat. The present day heat production of ^{238}U is 9.4×10^{-5} W Kg^{-1}, of ^{235}U 5.7×10^{-4} W Kg^{-1}, of ^{232}Th 2.7×10^{-5} W Kg^{-1}, and of ^{40}K 2.8×10^{-5} W Kg^{-1}; these values particularly for ^{235}U were higher in the past. The slow decay of these isotopes provides the basis for radiometric age dating and isotopic modelling of the evolution of the Earth and its crust. Precise estimates have been made of the age of the Earth and events in its history, initially using the radioactive decay of ^{238}U to ^{206}Pb and ^{235}U to ^{207}Pb, and more recently the decay of ^{40}K to ^{40}Ar, ^{87}Rb to ^{87}Sr, and ^{147}Sm to ^{143}Nd. Ratios of radiogenic to non-radiogenic isotopes eg $^{87}Sr/^{86}Sr$ and $^{143}Nd/^{144}Nd$, also provide information on processes operating during the formation of crust and mantle.

RADIOACTIVITY AND EARTH PROCESSES

The age of the Earth is now generally accepted to be about 4550 Ma, based on U/Pb ratios and Rb-Sr isochron ages of chondritic meteorites thought to represent undifferentiated samples of the early solar system (Patterson 1956; Faure 1986). Chondritic meteorites also provide information on the starting composition of the Earth (Table 2). Chemical segregation in the primitive Earth began when melting occurred as temperatures rose as a result of accretionary impacts, gravitational energy released by separation of an Fe-Ni(S) core, and radioactive decay. Initially, heat production from U, Th and K was about five times greater than at the present day (Fig. 1). In addition, there is evidence that heat production in the early Earth was increased by the decay of short-lived radio-isotopes, such as ^{129}I and ^{244}Pu and possibly ^{26}Al (Brown and Mussett 1993). Formation of the core is thought to have been complete within about 100 Ma (Brown and Mussett 1993), but development of the Earth's crust and mantle has continued to the present time. Models for crust formation link the nature and the rate of the processes involved to the decline of

Table 1a. ^{235}U decay series

Isotope	Half-life	Principal decay mode
^{235}U	7.0×10^8 a	alpha
^{231}Th	26 h	beta
^{231}Pa	3.3×10^4 a	alpha
^{227}Ac	22 a	beta
^{227}Th	19 d	alpha
^{223}Ra	11.4 d	alpha
^{219}Rn	4.0 s	alpha
^{215}Po	1.8 ms	alpha
^{211}Pb	36.1 m	beta
^{211}Bi	2.2 m	alpha
^{207}Tl	4.8 m	beta
^{207}Pb	stable	—

Table 1b. ^{238}U decay series

Isotope	Half-life	Principal decay mode
^{238}U	4.5×10^9 a	alpha
^{234}Th	24 d	beta
^{234}Pa	6.8 h	beta
^{234}U	2.4×10^5 a	alpha
^{230}Th	7.3×10^3 a	alpha
^{226}Ra	1.6×10^3 a	alpha
^{222}Rn	3.8 d	alpha
^{218}Po	3.1 m	alpha
^{214}Pb	27 m	beta
^{214}Bi	20 m	beta
^{214}Po	160 µs	alpha
^{210}Pb	22 a	beta
^{210}Bi	5.0 d	beta
^{210}Po	138 d	alpha
^{206}Pb	stable	—

Table 1c. ^{232}Th decay series

Isotope	Half-life	Principal decay mode
^{232}Th	1.4×10^{10} a	alpha
^{228}Ra	5.8 a	beta
^{228}Ac	6.1 h	beta
^{228}Th	1.9 a	alpha
^{224}Ra	3.7 d	alpha
^{220}Rn	56 s	alpha
^{216}Po	0.15 s	alpha
^{212}Pb	11 h	beta
^{212}Bi	61 m	beta
^{212}Po	0.30 µs	alpha
^{208}Pb	stable	—

Data from Kaye and Laby (1986).

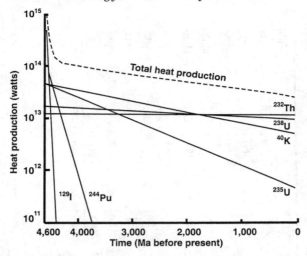

Figure 1. Heat production from the important radioactive isotopes incorporated at the Earth's formation. K, U and Th values are extrapolated back from present-day average abundances in the crust and mantle. Values for short-lived radio-isotopes are estimated from nucleosynthesis theory and decay products in the Allende meteorite. The curvature in the Total heat production line during the first 500 Ma is an estimate reflecting the heat sources available at that time; note that the vertical axis is logarithmic. Used by permission of Kluwer, from Brown and Mussett (1993), Fig. 11.1, p. 213.

energy generated by radioactive decay (Windley 1995).

The Earth is concentrically layered. The core, which is predominantly iron, comprises about 31% of the Earth's mass and 16% of its volume. The temperature at the center of the Earth is thought to be about 8000°C, falling to 3000°-4500°C at the core-mantle boundary.

The mantle, which comprises 68% of the mass of the Earth and 84% of its volume, contains a far greater range of chemical elements in major amounts than the core, and these occur in a variety of minerals depending on pressure and temperature. The mantle undergoes basal heating from the core, as well as internal, radioactive heating although little is known about radio-element abundances, especially in the lower mantle. Heat is transported by thermal convection associated with deformation by solid-state creep. Heat and mass transfer within the mantle may be retarded by at least one major discontinuity, the most pronounced of which is at 670 km; in which case, the lower and upper mantle must be strongly decoupled (Ringwood and Irifune 1988). Successive episodes of crust formation would then rapidly deplete the upper mantle in incompatible elements (including K, Th and U) and there would be only minor replenishment from the less-depleted lower mantle. Alternatively, the 670 km discontinuity may act as only a weak barrier to heat and mass transfer (Christensen 1989).

The mantle is overlain by a thin (~6 km thick) basaltic (Fe-Mg silicate) crust beneath the oceans and a thicker (average 35 km), lighter crust of intermediate composition beneath the continents. The radioactive heat generation of the principal layers of the Earth is shown in Table 3. It is apparent that the Earth's mantle remains the main generator of radioactive heat, despite the high heat productivity of the continental crust since this constitutes only about 0.7% of the Earth by volume. Heat transfer in the Earth is primarily

Table 2. Distribution of radioactive elements (ppm) in the Earth and their heat production

	K	U	Th	Th/U	K/U	Total heat production	
						($W \cdot kg^{-1}$)	($W \cdot m^{-3}$)
CI carbonaceous chondrites	545	0.0074	0.029	3.92	73649	1.54×10^{-12}	4.00×10^{-9}
Primitive mantle	250	0.021	0.085	4.05	11905	4.37×10^{-12}	1.14×10^{-8}
Bulk silicate earth	240	0.023	0.0795	3.46	10435	4.42×10^{-12}	1.15×10^{-8}
Continental crust							
Upper	27500	2.5	10.5	4.20	11000	5.30×10^{-10}	1.38×10^{-6}
Middle (Archaean)	17500	2.2	8.4	3.82	7954	4.44×10^{-10}	1.15×10^{-6}
Lower (Archaean)	8333	0.05	0.42	8.40	166660	1.67×10^{-11}	4.33×10^{-8}
Lower and Middle (post-Archaean)	20000	1.25	6	4.80	16000	2.86×10^{-10}	7.43×10^{-7}
Average (1)	17500	1.3	5.7	4.38	13461	2.82×10^{-10}	7.34×10^{-7}
Average (2)	12500	1.25	4.8	3.84	10000	2.53×10^{-10}	6.58×10^{-7}
Oceanic crust							
Normal Mid-ocean ridge basalt	600	0.047	0.12	2.55	12766	7.88×10^{-12}	2.05×10^{-8}
Ocean island basalt	12000	1.02	4	3.92	11765	2.09×10^{-10}	5.43×10^{-7}

Average (1): upper crust, Weaver and Tarney (1984), Average (2): upper crust, Taylor and McLennan (1981). From McDonough and Sun (1995), Plant and Saunders (1996), Sun and McDonough (1989), Taylor and McLennan (1981), Weaver and Tarney (1984). Heat-production values are calculated by assuming the following present-day heat production: ^{238}U: 9.4 x 10^{-5} W·kg^{-1}; ^{235}U: 5.7 x 10^{-4} W·kg^{-1}; ^{232}Th: 2.7 x10^{-5} W·kg^{-1}. ^{40}K: 2.8 x10^{-5} W·kg^{-1}. We also assume the present-day ^{238}U/^{235}U ratio is 137.88, and that ^{40}K is 0.00118 % of total K. (Used by permission of the editor of *Radiation Protection Dosimetry*, from Plant and Saunders (1996), Table 3, p.27.)

Table 3. Radioactive heat generation as a percentage of present total heat outflow

Continental crust (roughly equal to granodiorite and granulite) ~ 0.7% of Earth's volume	~10%
Oceanic crust (broadly gabbro) ~ 0.2% of Earth's volume	~0.15%
Mantle (undepleted peridotite) ~ 84% of Earth's volume	~30%
Core (iron) ~ 16% of Earth's volume	negligible*
Total	~40%

*Assuming negligible radioactivity in the core (data from Brown and Mussett 1993). Used by permission of the editor of *Radiation Protection Dosimetry*, from Plant and Saunders (1996), Table 4, p. 28.

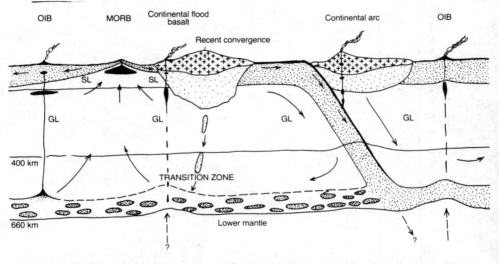

Figure 2. Cartoon illustration of the upper circa 800 km of the Earth, indicating possible sources for the ocean island basalt, mid-ocean ridge basalt, continental flood basalts and arc andesites. Arrows indicate the directions of relative lithospheric plate motion and (simplistically) the convective regime in the upper mantle. Possible regions of material exchange between the upper and lower mantle are indicated by broken arrows. Zones containing a small amount of melt are shown solid; sub-continental mantle lithosphere is highly stippled; sub- oceanic mantle lithosphere is heavily stippled. GL = garnet lherzolite: SL = spinel lherzolite. After Davies et al. (1989) and Floyd (1991). Used by permission of Kluwer, from Brown and Mussett (1993), Fig. 9.5, p 184.

through convection and, in the lithosphere, by conduction.

According to plate tectonic theory, the lithosphere (the crust and outer part of the mantle) is divided into huge plates, moving in a complex inter-relationship with mantle convection as heat is dissipated from the Earth (Fig. 2). New oceanic crust of broadly basaltic composition is fractionated from decompressed mantle material beneath oceanic spreading centers such as the mid-Atlantic Ridge. As plates move away from ridges they cool, increase in density and are finally destroyed as they are subducted back into the mantle.

Where oceanic lithosphere is subducted beneath another oceanic plate an island arc forms characterized by high seismic activity, high heat flow and active volcanoes bordered by a subterranean trench. Initially, magmatism is dominated by tholeiites evolving to calc-alkaline and subsequently more alkaline leucogranite magmas enriched in U and other incompatible elements temporally and away from the trench as in Japan (Taira et al. 1992).

Cordilleran orogens, such as those of the western Americas, reflect subduction of oceanic lithosphere beneath continental plates. They are characterized by a trench with turbiditic sediments, high-heat flow, calc-alkaline (mainly andesite-rhyolitic) volcanism, and by monzonitic-granodioritic intrusions. Extensional collapse of thickened parts of such orogens can give rise to molasse basins. Many allochthonous oceanic and continental fragments may be accreted. Again, the latest most·evolved igneous rocks emplaced furthest away from the trench are generally the most enriched in U.

Continent-continent collision is associated with an initial compressional stage with thrusting, crustal thickening and the generation of upper crustal melt (sedimentary protolith) granites followed by uplift and erosion. Subsequently gravitational collapse of the overthickened crust is associated with extensional tectonics, basin formation and emplacement of late orogenic granites as a result of rapid advective thinning of a reduced thermal boundary conduction layer (Dewey 1988). Models involving extension following the Cretaceous Laramide orogeny have been used to explain the Basin and Range terrane of western North America (Platt and England 1994) and extension following the Late Palaeozoic Variscan orogeny, has also been invoked to explain the formation of the French Massif Central (Malavieille 1993). Such settings are generally characterized by abundant late-post orogenic uraniferous leucogranites classified here as evolved I-type granites similar in composition to those formed in extensional anorogenic crustal rift systems (A-type granite) like those of Nigeria (Bowden and Kinnaird 1984a).

Heat outflow from oceanic lithosphere by conduction and convection of sea water, especially at spreading centers, accounts for approximately 70% of the total outflow of 4.2×10^{13} W from the Earth. The remaining 30% of the heat being lost through the continents (Table 4).

Table 4. Heat outflow as percentage of total outflow

Oceanic heat flow (by conduction)	60%
Oceanic hydrothermal convection	10%
Continental heat flow	30%
Total	100% = 4.2×10^{13} W

Data from Sclater et al. (1980). Used by permission of the editor of *Radiation Protection Dosimetry*, from Plant and Saunders (1996), Table 5, p. 28.

In addition to plate tectonic activity, plumes rising through the mantle may carry heat from the Earth's interior (Morgan 1971; Davies 1988; Sleep 1990). Plumes of hot mantle material from below the 670 km discontinuity or the 2900 km core-mantle boundary have been the primary agents responsible for the fragmentation of supercontinents such as Pangaea (Windley 1995). An intra-continental rift, such as the present day East African Rift Valley, or Benue Trough represent the first stage in the process. Extension of the lithosphere in the rift is commonly associated with alkaline magmatism, alkali flood basalts, phonolites and trachytes, or plutonic ring complexes of carbonatites, syenites and nephelinites. The magmas are characterized by enrichment of volatiles and alkalis, and locally of uranium. In Africa for example, uranium deposits are associated with the doming, rifting and sedimentary stages of evolution of plume-generated rifts (Olade 1980): Uranium, tin and niobium minerals are enriched in alkali granites and syenites in ring complexes and uranium, niobium and rare earth elements are enriched in carbonatites. Stratiform uranium minerals with copper, cobalt and manganese occur in later red-bed sedimentary rocks.

Computer simulation suggests that large continental plates insulate the underlying mantle, by providing a radiothermal blanket in which the cooling effects of sea-floor spreading and subduction are lacking. This causes convective mantle upwelling, continental uplift and extensive magma generation (Fig. 3) (Gurnis 1986).

SEPARATION OF RADIOTHERMAL CONTINENTAL CRUST

Some of the oldest dates for the continental crust are for the Acasta gneiss of the Slave Province, Canada (4017 Ma), the Nulliak volcanics and sedimentary rocks of

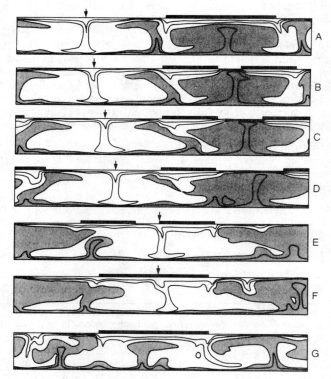

Figure 3. Computer simulation model showing how the upwelling of hot mantle (grey) and downwelling of cold mantle (unshaded) causes fragmentation of a supercontinent (black rectangle) in stages A-C. (Simulation published by Gurnis (1986). The continents may reassemble, perhaps 100-200 Ma later (stage F) above the main one of downwelling (arrow). But after 400-500 Ma the new supercontinent has created its own upwelling (stage G) and the fragmentation process is repeated. Used by permission of Kluwer, from Brown and Mussett (1993), Fig. 11.13, p 228.

Table 5. Concentrations of some radio-elements in Earth's crust and mantle (ppm).

	Undepleted mantle	Ocean Crust (N-MORB)*	Continental crust	Continental enrichment
Rb	0.7	1	32	× 46
Th	0.08	0.15	3.5	× 44
U	0.02	0.05	0.91	× 46
Sr	21	150	260	× 12
Nd	0.97	10	16	× 17
Sm	0.32	3.1	3.5	× 11

N-MORB: Normal mid-ocean ridge basalt. Data from Taylor and McLennan (1985), O'Nions (1987), Sun and McDonough (1989), and Brown and Musset (1993). (Used by permission of the editor of *Radiation Protection Dosimetry*, from Plant and Saunders (1996), Table 6, p 29.)

Labrador (>3800 Ma), and the Isua supracrustals and pre-Amitsoq association of Greenland (>3700 Ma) (Windley 1995). These ages are considerably younger than the age of the Earth, since the first lithosphere was probably thin, unstable and easily reabsorbed into the mantle. The additional heat available in the Archaean is thought to have escaped by rapid convection and the generation and recycling of ocean crust (Burke et al. 1976; Sleep 1979).

Continental crust is thought to separate as a result of a complex series of fractionation events, beginning with the separation of basaltic magma comprising a 10-15% partial melt at ocean-ridge spreading centers (Fig. 2). During subduction, dehydration reactions in the descending slab and partial melting in the hydrated mantle wedge overlying subduction zones give rise to more fractionated basaltic-andesite magma, which initially may become underplated onto older continental crust (Saunders et al. 1988, 1991). Within the crust, the mobilization, assimilation, storage and homogenization processes generate mainly granitoid magmas with base-level isotopic and chemical values established by blending subcrustal and deep-crustal magmas at the mantle-crust transition. Scavenging of mid- to upper-crustal silicic-alkalic/alkaline melts, and intracrustal assimilation-fractionation-crystallization processes can subsequently modify ascending magmas (Hildreth and Moorbath 1988).

Uranium and Th undergo a series of complex fractionation events, resulting in highly variable levels in different rock types (Table 2). This begins with the preferential incorporation of U, in hydrothermally altered ocean crust (Michard and Albarede 1985). The elements are further fractionated at subduction zones, where substantial amounts of U, K and Th appear to be preferentially removed from the slab in hydrous fluids and other volatile streams, and transported into the crust. Some U may be retained in the slab in minor phases such as rutile in eclogite, and transported into the deep mantle.

Uranium, Th, K and Rb are large-ion lithophile elements (LILEs) that do not fit readily into the structures of the main high-temperature rock-forming minerals (Rogers and Adams 1969a). During the series of fractionation events that led to the formation of continental crust, LILEs are partitioned preferentially into small-volume, low-temperature melts, and become progressively more concentrated, so that certain types of highly evolved granite, rhyolite, and alkaline complexes contain exeptionally high concentrations of radioactive elements (Tables 2 and 5).

PRIMARY ENRICHMENT OF URANIUM IN FELSIC IGNEOUS ROCKS

The association of uranium with acid igneous rocks is central to understanding uranium ore deposits, but it is particularly controversial.

Granites have been classified into I (Igneous precursor), S (Sedimentary protolith) and A (Anorogenic)-types, based on studies in the Lachlan Foldbelt, Australia (Chappell and White 1974; Chappell and Stephens 1988; Collins et al. 1982). However, in the Lachlan Foldbelt the three types of intrusions have similar multi-element patterns close to the post-Archaean average shale composition (Plant and Tarney 1994). Elsewhere the much greater spectrum of granitoid compositions has led to many re- and sub-classifications of granite types, with confusion, particularly about the importance of upper and lower-crust and mantle components in petrogenesis. Here three main granite endmembers are distinguished based mainly on tectonic setting and geochemistry. In addition, there are small-volume S-type granitoids, which generally have levels of U close to the average crustal abundance of 2-3 ppm (Plant et al. 1985).

Tonalite-trondhjemite granitoids

The tonalite-trondhjemite granitoids (TTG) that are typical of Archaean granite-gneiss terrains, commonly contain mafic-ultramafic inclusions, that may be relics of their source material. The more silicic granitoids show marked HREE depletion with positive Eu anomalies (Plant and Tarney 1994), features that distinguish them from evolved I- or A-type granitoids. Their spidergram patterns are distinctive (Fig. 4), with marked negative Nb, P, Ti, and Y anomalies, and positive or no Sr anomalies, even where enriched in K. The TTG are most readily explained by hydrous (up to 15%) melting of a mafic source with hornblende or garnet in the residue (Weaver and Tarney 1980). Formation as a result of high degrees of partial melting under conditions of high pH_2O also means that such granitoids are underplated at deep levels in the crust and are almost always associated with strong deformation. Comparison of the chemistry of the TTG suite at granulite facies with those at amphibolite facies (Weaver and Tarney 1980, 1981; Rollinson and Fowler 1987) shows marked depletion in U, Th and Rb, and to less extent K and Pb.

"I-type" calc-alkaline orogenic granitoids

Spidergrams for I-type calc-alkaline granitoids are similar to the Archaean TTG suite (Plant and Tarney 1994), although they have slight positive P anomalies, compared with the negative P of Archaean granitoids. Also they lack the extreme isotopic compositions typical of Archaean gneiss. Their composition is similar worldwide (Plant and Tarney 1994). The more evolved intrusions of the suite may be enriched in U, Th, K and other incompatible elements, such as Mo, probably as a result of fractionation processes (Atherton and Plant 1985).

Calc-alkaline magmas are generally emplaced early in the plutonic cycle, and closest to subduction zones reflecting the breakdown of hornblende and formation of hydrous magmas (Saunders et al. 1991). Hornblende generally has a high K/Rb ratio, and on thermal breakdown produces garnet-bearing residues which retain HREEs leading to the HREE depletion patterns typical of such granitoids. Hornblende has a limited range of stability in the mantle wedge and is replaced by phlogopite at greater depths (cf. Wyllie 1983) and distal to the subduction zone. The stabilities of both minerals are substantially enhanced if fluorine replaces hydroxyl in the structures (Foley 1991), supporting models that require pargasitic hornblende, K-richterite and phlogopite to have a strong controlling influence on magma compositions (e.g. Sudo and Tatsumi 1990), so that F increases in the latest, drier magmas in which U tends to be most enriched.

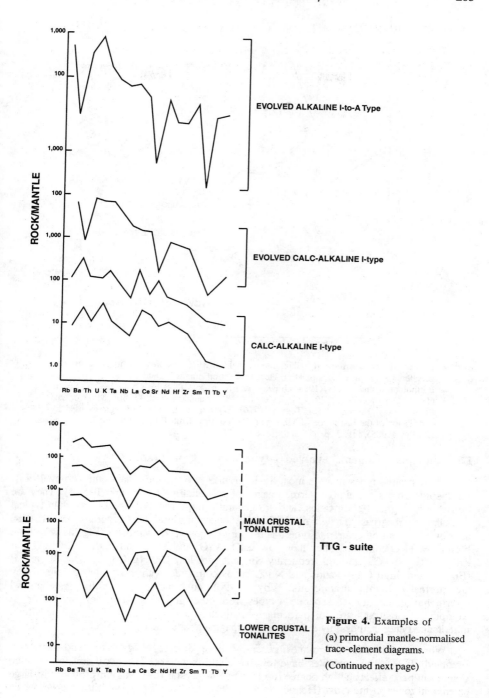

Figure 4. Examples of (a) primordial mantle-normalised trace-element diagrams. (Continued next page)

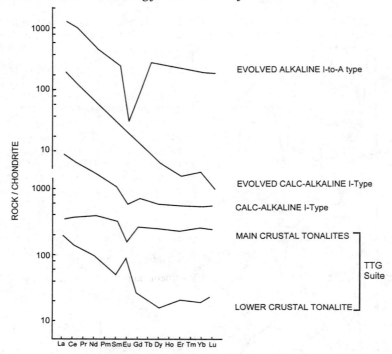

Figure 4, continued. (b) Chondrite-normalised REE diagrams to show the primary geochemical characteristics of High Heat Production granites. Data for (a) and (b) from Bartholomew and Tarney (1984); Bowden and Kinnaird (1984a,b); Chatterjee and Strong (1984); Garson et al., (1975); O'Brien et al. (1985); Saunders et al. (1980); Simpson and Plant (1984); Tauson (1982); Teggin (1976); Watson et al. (1984). Diagrams used by permission of the Institution of Mining and Metallurgy, from Plant, O'Brien, Tarney and Hurdley (1985), Fig. 5, p. 275 and Fig. 6, p. 276.

Late and post-orogenic evolved I-type and A-type granites

These granite types include most Sn-U granites worldwide (Plant and Tarney 1994). Their spidergrams are almost mirror images of calc-alkaline intrusions (Fig. 4). They are Fe-rich with higher Fe/Mg ratios, and are generally more reduced (lower fO_2) than typical calc-alkaline magmas. They have high REE concentrations but generally flatter REE distributions than other granite types, and a large negative Eu anomaly (Fig. 4) with high F and low H_2O contents. They have low CaO, Al_2O_3 and MgO and high K_2O/Na_2O and Zr, Nb, Ta, Th, Ga, Zn and frequently Sn, but very low Cr, Ni, Co, Sc, Ba, Sr, and P (Fig. 4) with high Ga/Al ratios and Nb/Zr ratios compared with most other granites; they are normally highly uraniferous. Eby (1990) constructed numerous geochemical discrimination diagrams to separate A-types from other granitoids, but their Sr, Nd or Pb isotopic compositions are not particularly distinctive and they have only simple zircon inheritance patterns, possibly reflecting high magma temperatures (>900°C).

Whereas calc-alkaline granites generally have dioritic or gabbroic end-members, "evolved" I- or A-type granites have few basic end-members. The geochemistry of such granites in part reflects a high degree (up to 50%) of crystal fractionation probably in large extension zones in the crust (Henney et al. 1994). Evolved I-type granites are emplaced late in orgenesis, frequently post-orogenically in extensional collapse orogens. Many granites of this type were emplaced across Europe, following the extensional collapse of the Variscan orogenic belt.

A-type granite magmas are anorogenic, anhydrous, and frequently alkaline. Many Sn granites have A-type characteristics so that they are unlikely to be higher temperature remelts of sources that had earlier yielded I-type granites (Collins et al. 1982) since all granite magmas are undersaturated with respect to Sn (Taylor and Wall 1992), which would be removed with first stage melts. An explanation involving thermal breakdown of (fluor-) phlogopite-rich assemblages is preferred to explain the higher temperatures of magma generation and their high U, K, Th, Rb, Sn and F contents. Hot dry magmas can rise to higher levels in the crust than hydrous magmas (Brown and Fyfe 1970), where they can initiate hydrothermal convective systems. Emplacement of A-type granites from such deep levels is possible only under extensional conditions (associated with uprise of hot asthenosphere) which in turn creates the fracture systems in the upper crust that facilitate hydrothermal activity.

There was a great deal of anorogenic granite magmatism in the mid-Proterozoic, associated with some of the world's largest mineral deposits, including Mount Isa, Broken Hill and the uraniferous Olympic Dam deposit in Australia (Sawkins 1989) following assembly of a mid-Proterozoic supercontinent, and uprise of hot mantle (Barley and Groves 1992). This may reflect superplumes impinging on mature phlogopite-rich sub-continental lithosphere.

In summary, granitoids resulting from hydrous melting of underplated thick warm mafic crust in the Archaean, or subducted oceanic crust in more recent times, contain low levels of uranium. Calc-alkaline I-type intrusions generally have characteristics reflecting melting of hornblendic mantle wedge and contain only moderate levels of uranium. Granitoids formed by melting thick sedimentary piles or mixed sediment-amphibolite sequences in accretionary tectonic environments also have low uranium contents. Evolved I- and A-type magmas generally contain high levels of uranium reflecting melting of phlogopitic sub-continental lithosphere in anhydrous conditions. They are emplaced late in orogenesis especially during extensional collapse of orogens or in anorogenic settings.

Evolved I- and A-type granites have high heat-production (HHP). In addition to hosting penecontemporaneous U and Sn mineralization, they can provide thermal anomalies in the crust capable of focusing hydrothermal solutions long after magmatic cooling has occurred and following unroofing of the intrusions and/or the deposition of cover sediments (Simpson et al. 1979, Plant et al. 1983, Brown et al. 1987). Fehn et al. (1978) quantitatively modelled hydrothermal convection associated with such radioactive granites (4% K, 15 ppm U and 57 ppm Th) with heat productivity about 3.5 times that of average granite. Such intrusions were shown to be capable of maintaining convection for a few million years after emplacement, but the heat generated by radioactive decay was shown to be capable of regenerating low temperature hydrothermal convection during periods of tectonism or increased heat flow at intervals over hundreds of millions of years. Isotopic age dates from such granites are likely to give ages of geothermal activity rather than magmatic events. Erosion of evolved I- and A-type granitoids particularly when hydrothermally altered or when minerals have become metamict, may act as a source of U for sedimentary deposits (Finch 1996).

CRUSTAL EVOLUTION

There are various models for crustal growth, but the most favored suggest progressive continental growth throughout the last 4000 Ma at a rate declining with the

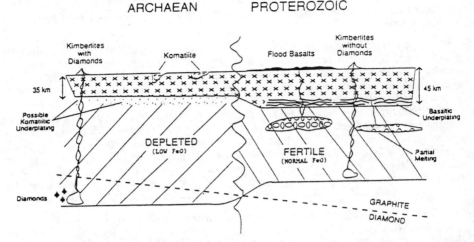

Figure 5. Model for Archaean and Proterozoic crustal evolution emphasizing differences in chemical properties of uppermost mantle. Proterozoic crust develops above fertile mantle that is source of basaltic composition. Crustal underplating leads to development of thicker crust. Archaean crust develops above initially hotter mantle that is depleted in Fe through eruption of komatiite lavas. Magmatic underplating, if present, is ultramafic and is seismically indistinguishable from normal mantle. Cold lithospheric keel may act as thermal boundary against crustal underplating by basaltic magmas from asthenosphere. (Used by permission of The Geological Society of America, from Durrheim and Mooney (1991), Fig. 3, p 608.)

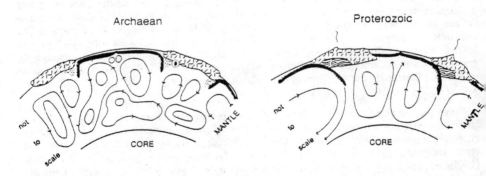

Figure 6. Model to explain Archaean and Proterozoic crustal evolution emphasizing turbulent versus stable convection patterns. Hotter Archaean mantle lacks stable zones of upwelling and particularly downwelling that are needed to develop long-lived subduction zones and associated magmatic arcs. Proterozoic subduction zones are long lived, and overriding crust is underplated (wavy pattern) and thickened at seaward-migrating subduction zones. Archaean magmatic products may have been ultramafic and thus seismically indistinguishable from normal upper mantle, and so the Archaean crust remains thin. Used by permission, from Durrheim and Mooney (1991), Fig. 4, p. 609.

exponential decay of long-lived radioisotopes, a progressive decrease in plume activity and a corresponding increase in plate-tectonic movement (Windley 1995). Production of oceanic and continental crust has led to the depletion of incompatible elements, including K, U and Th, in the upper mantle (Gast 1968; O'Nions and McKenzie 1993). Estimated abundances and the degree of crustal enrichment of radioactive elements in the main crust and mantle layers are given in Table 5.

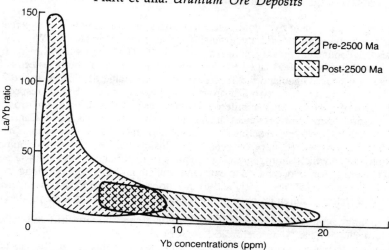

Figure 7. Slope of the rare-earth element profile (light- heavy as given by the La/Yb ratio) plotted against Yb concentration in 644 granite (sensu lato) samples, 319 from the Archaean and 325 from Proterozoic and Phanerozoic complexes. Over 95% of each sample group plots in the field shown—for original data see Martin (1986). Data are chondrite-normalized, which is the conventional way of expressing rare-earth data. Note: La = Lanthanum; Yb = Ytterbium. Used by permission of Kluwer, from Brown and Mussett (1993), Fig. 11.8, p. 222.

From 3800 to about 2500 Ma ago, the Earth's crust is thought to have been highly mobile (Figs. 5 and 6), reflecting the high overall radiogenic heat flow, with heat loss dominated by mantle-plume activity. Archaean greenstone belts have been suggested to reflect emplacement of hot, high Mg komatiite magmas, rarely found in Phanerozoic successions, and basalts over plumes rising through hot Archaean mantle (Campbell et al. 1989). Granite batholiths were more tonalitic (sodic) in composition with highly fractionated REE patterns (Fig. 7); and volcanism was generally bimodal rather than andesitic (Windley 1995). Early Archaean rocks tend to have less U, Th and K than younger, chemically equivalent rocks.

Between about 3200 and 2600 Ma ago there was a major surge of continental growth at arcs and Cordilleran-type margins as a result of accretion followed by continent-continent collision (Windley 1995). Granite compositions changed from Na-rich, light REE-enriched patterns typical of the Archaean to compositions typical of the Proterozoic and later (Fig. 7), although the composition of tonalite remained the same (Condie 1989). Some of the first HHP granites, which include those of the Kaapvaal and Kalahari cratons in Africa, were emplaced in extensional collapse orogens. Deposition of the first known uranium ores in quartz pebble conglomerates overlapped in time with emplacement of these granites (see below).

The shift in geological and tectonic conditions across the Archaean-Proterozoic boundary, although diachronous, was profound, possibly reflecting the assembly of a supercontinent. Emplacement of uraniferous granites was widespread at this time and important uranium provinces were initiated—for example in the Yilgarn block, Australia, the Superior Province, Canada and the Sao Francisco Craton, Brazil. By the early Proterozoic the continental lithosphere had attained a degree of rigidity comparable to that of the present day with the start of modern-style plate tectonic processes (Figs. 5 and 6).

Kerr (1989) has suggested that Proterozoic geology reflects cycles of continental aggregation with accretion over cold subduction zones, and of continental dispersal related to plume activity accompanied by extensive anorogenic magmatism. Windley (1995) proposes that continental breakup between 2400 and 2000 Ma ago was followed by major crustal growth between 2000 and 1600 Ma ago in the central to southwestern USA, the Ketilidian of western Greenland the Svecofennian of the Baltic Shield and in the 2100 Ma old Birimmian and Trans-Amazonian orogens of West Africa and Brazil, respectively. A further phase of post-orogenic uraniferous granite magmatism in collision orogens gave rise to important uranium provinces including the Aphebian basins of Canada (Finch 1996), the Pine Creek 'geosyncline,' Australia and the Franceville basin, Gabon. By 1600-1500 Ma ago accretionary orogens and collision had led to the development of a supercontinent in which anorogenic magmatism was important. Between 1400 and 1000 Ma ago closure of earlier oceans led to extensive collisional orogens that belong to a single Grenvillian system (Hoffman 1991) and to the formation of the Rodinian supercontinent that included Baltica, Laurentia, most of Gondwana and Siberia soon after 1000 Ma ago.

Beginning about 1000 Ma ago a further stage of fragmentation, drift and collision led to the formation of the Neoproterozoic Pan African orogenic belts and the consequent formation of the Pannotia Supercontinent. The break-up of this led to the Iapetus ocean and ultimately to the Caledonian and Variscan orogens in the Lower and Upper Palaeozoic respectively. Later events included the formation of the Alpine-Himalayan mountain belt when southern Europe, Africa and India collided with the Eurasian continental plate and the Tethyan oceanic plate was consumed.

The importance of different types of uranium deposits have been shown to vary through time reflecting episodes of emplacement of uraniferous leucogranites, rhyolites and alkaline complexes at high levels in the continental crust (Dahlkamp 1993). Uranium provinces, which comprise clusters of substantial uranium concentrations including economic deposits, occur in geologically and tectonically distinct regions. In many cases the source rocks for U deposits are of Archaean or Proterozoic age although the deposits may have formed in much younger sediments or metasediments considerable distances away in geologically different settings. In North America uranium provinces have been related to the tectonic evolution of the western active margin of the North American plate with igneous rocks providing both direct and indirect sources of uranium (Finch 1996). The most important periods of uranium mineralization there were Early Proterozoic, Late Triassic to Early Jurassic, Early Cretaceous, Oligocene and Miocene; the Tertiary mineralization which occurs throughout much of western North America and the southeastern coastal region being the most extensive (Finch 1996). Worldwide important uranium provinces were initiated during the late Archaean to early Proterozoic, Middle Proterozoic, early Permian (Fig. 8), Jurassic, Tertiary and to a lesser extent the late Proterozoic.

DISTRIBUTION OF RADIOACTIVITY IN THE CONTINENTAL CRUST

The average abundance of U, Th and K in the crust is 2.6 ppm, 10 ppm and 1% respectively (Taylor 1964). Assuming the continental crust comprises an upper sedimentary-metamorphic complex, a middle layer approximating to granodiorite-tonalite composition and a lower layer of essentially granulite (Windley 1995) levels of radionuclides should increase progressively to higher levels (Table 2).

Recently, the availability of radiometric and geochemical survey data has provided detailed information on the distribution of radioactive elements over large areas of the Earth's continental crust. Radiometric survey data are collected by surface gamma-ray spectrometric measurements normally carried out in the field or from the air; geochemical

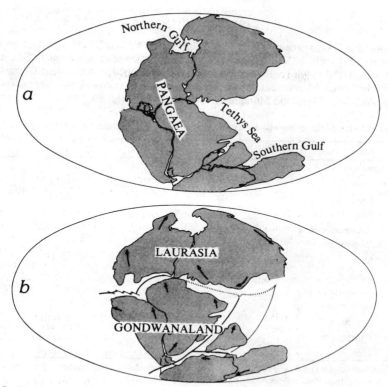

Figure 8. The distribution of modern continents (a) 200 Ma ago and (b) 180 Ma ago, just after fragmentation started to break up the supercontinent Pangaea into the northern and southern landmasses, Laurasia and Gondwanaland (Windley, 1995). Arrows in (b) indicate the direction of continental movements. Evidence is based on a combination of palaeoclimatic and palaeomagnetic data. Used by permission of Kluwer, from Brown and Mussett (1993), Fig. 11.17, p. 232.

surveys are based on the systematic analysis of samples such as rocks, soils, stream sediments and ground and surface waters (Darnley et al. 1995).

In the crust, U, Th, K and other minor radioactive isotopes are concentrated preferentially in acid relative to intermediate, basic and ultrabasic igneous rocks that have extremely low levels of radionuclides (Table 6), and in alkaline relative to calc-alkaline or tholeiitic rocks. For example, alkali basalts have up to 2 ppm U whereas tholeiitic basalts average 0.1-0.5 ppm U and alkaline granites commonly contain two to ten times the U content of calc-alkaline granites, which average 2-6 ppm.

In igneous rocks, K generally occurs in major silicates such as feldspars and micas, whereas U and Th occur mainly in accessory minerals such as thorite, monazite and zircon. Some minerals with high contents of U and Th lose their crystallinity as a result of internal radiation damage and become metamict, and these minerals are more readily altered and hydrated.

In the case of metamorphic rocks the U content commonly reflects that of their protolith. High-grade granulite facies rocks are frequently depleted in U compared to their unmetamorphosed equivalents however (Dostal and Capedri 1978; Heier 1979). This is probably the result of dehydration reactions whereby uranium is removed as a result of metamorphic dewatering processes (Weaver and Tarney 1984).

Uranium has four oxidation states U(III), U(IV), U(V) and U(VI) the most important of which are U(IV) and U(VI). Thorium occurs only in the tetravalent state. Uranium of valency U^{4+} and Th^{4+} form solid solutions with elements such as tetravalent Ce and Zr in minerals, which may be resistant to weathering and commonly occur as detrital grains in alluvial deposits (see Finch and Murakami, Chapter 3, this volume).

Table 6. Mean contents and typical ranges of U in various rock types (ppm)

Rock type	Mean	Range
Igneous		
Mafic	0.8	0.1–3.5
Diorite and quartz diorite	2.5	0.5–12
Silicic	4.0	1.0–22
Alkaline intrusive		0.04–20
Sedimentary		
Shale	3.0	1–15
Black shale		3–1250
Sandstone	1.5	0.5–4
Orthoquartzite	0.5	0.2–0.6
Carbonate	1.6	0.1–10
Phosphorite		50–2500
Lignite		10–2500

Data from Rogers and Adams (1969b) (Used by permission of the editor of *Radiation Protection Dosimetry*, from Plant and Saunders (1996), Table 8, p 31.)

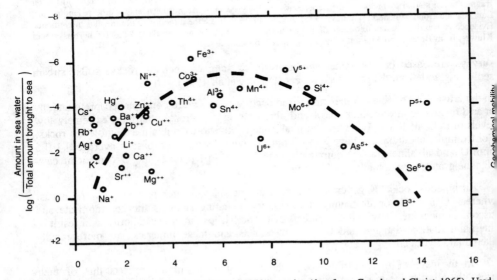

Figure 9. Anomalous geochemical mobility of U(VI) species (data from Garrels and Christ 1965). Used by permission of the Institution of Mining and Metallurgy, from Dall'Aglio (1972), Fig. 2, p. 123.

In the surface environment, U(IV) species are readily oxidized to uranyl U(VI) species that have anomalously high solubility (Fig. 9); and in oxidized red-bed sediments U levels are commonly less than 1 ppm (Table 6). Uranium is readily precipitated in reducing environments, for example in sandstones rich in organic matter or Fe sulfides. Uranium can also be enriched in shales, in which it is frequently concentrated in organic matter to levels of 30-60 ppm U or more. Phosphate-rich sediments, derive U from sea water through the ability of U to form phosphate complexes, such as $(UO_2HPO_4)^{2-}$ at high pH, and phosphorite can average 10-60 ppm U (Rogers and Adams 1969b). Some lignite and coal are also enriched in U; some lignite contains as much as 0.25% U (Table 6).

Table 7. Progressive enrichment of U in the formation of U deposits

	Average ppm U	Cumulative enrichment factor
Bulk earth	2×10^{-2}	1
Continental crust	1	50
High Heat Production Granite	10	500
High-grade U deposits	$10^4 - 10^5$	$5 \times 10^5 - 5 \times 10^6$

(Used by permission of the editor of *Radiation Protection Dosimetry*, from Plant and Saunders (1996), Table 9, p 33.)

The crust-building cycle leads to progressive enrichment of U from levels of 0.02 ppm in bulk Earth to levels approximately double that in normal mid-ocean ridge basalt oceanic crust and approximately 50 times that in continental crust. Some acid igneous rocks contain 150-200 times the concentration of uranium in bulk Earth (Table 7). The subsequent remobilization and concentration of uranium in sedimentary rocks to form high-grade uranium ore deposits can involve cumulative enrichment factors of up to 5×10^6 (Table 7). The processes with the greatest potential to produce significant concentrations of uranium are thus crustal. They include (1) concentration from fluids generated by dehydration/partial melting reactions of lower crustal rocks leaving behind depleted granulite complexes. (2) crystal fractionation of alkaline complexes in continental rift systems. (3) high to low temperature fluid rock interaction associated with the emplacement of highly evolved leucogranites and rhyolites. (4) ground and surface water flow in relation to the evolution of certain types of sedimentary basins with the concentration of uranium at redox boundaries or complexation with ligands such as phosphate; and the (5) concentration of uranium in in surficial deposits such as calcrete as a result of weathering, some of which may reflect the development and preservation of ancient landscapes.

CLASSIFICATIONS OF URANIUM DEPOSITS

Several classifications of uranium deposits have been proposed, for example by Nash et al. (1981), Barthel et al. (1986) Dahlkamp (1993) and OECD/NEA (1994). Uranium deposits are also included in general ore deposit classifications, for example by Cox and Singer (1986), Plant and Tarney (1994), Ekstrand (1984) and Eckstrand, Sinclair and Thorpe (1995).

The OECD/NEA Red Book (1998) has the latest uranium deposit classification and this is published every two years. In 1996 the IAEA published a Guidebook to accompany the IAEA Map of World Distribution of Uranium Deposits (IAEA 1996). On the basis of geological setting and in order of economic importance the classes of uranium deposits are (1) Unconformity related, (2) Sandstone, (3) Quartz-pebble conglomerate, (4) Veins,

(5) Breccia complex (Olympic Dam), (6) Intrusive, (7) Phosphorite, (8) Collapse breccia, (9) Volcanic, (10) Surficial, (11) Metasomatite, (12) Metamorphic, (13) Lignite, and (14) Black shale. The book contains a description of nearly 650 deposits of which 582 are located on a world geological map. In this chapter it is possible to discuss only the main types of uranium ore deposits, and a relatively simple scheme is used based on the deposit types listed by the OECD/NEA 1998. The deposit types are discussed in an order based on their geological setting rather than economic importance however. Thus some deposits of relatively little economic importance are discussed before major deposits such as unconformity uranium deposits. The aim is to link deposit formation to the overall cycle of uranium enrichment in the Earth and to suggest new exploration criteria for uranium ore deposits at the recconnaissance to regional scale. Two main groups of deposits are recognized, those of igneous plutonic or volcanic association including metamorphic deposits and those of sediment/sedimentary basin association (Table 8). The main departure from the IAEA classification is that different vein-type uranium deposits are considered in relation to their geological setting rather than as a single economic class.

URANIUM DEPOSITS OF IGNEOUS PLUTONIC AND VOLCANIC ASSOCIATIONS

This group of uranium deposits, which includes some vein-type uranium deposits, formed as a result of the redistribution of uranium from refractory accessory minerals, such as zircon in igneous or metamorphic rocks, into uranium ore minerals, such as uraninite. This occurs as a result of high to low temperature fluid-rock interaction processes in the crust, involving fluids which have equilibrated with magma, metamorphic, formational, hydrothermal and/or meteoric aqueous fluids. Primary uranium minerals formed at high temperature (and pressure) in silicate systems are replaced by phases that are stable in aqueous conditions at low pressure and progressively lower temperature. The redistribution of elements involves interaction between rock and solutions that contain variable concentrations of acid ligands (F^-, Cl^-, CO_3^{2-}, S^{2-}), strong bases (Na^+, K^+, Ca^{2+}) and complexes of amphoteric elements (Si, Al, Zr, Sn, W) resulting in the replacement of the primary silicate and accessory mineral assemblages by secondary silicate, gangue and uranium ore minerals.

The general behavior of elements in water can be deduced from their ionic potential (Gordon et al. 1958) (Fig. 10). Thorium and tetravalent U are of intermediate ionic potential and hence are relatively immobile in aqueous solutions. Hydrothermal systems associated with granites involve more complex reactions than simple hydrolysis and the neutral to acid pH range of such systems preclude OH complexes. Experimental (e.g. Eugster 1984) and fluid inclusion studies (Haapala and Kinnuen 1979; Kelly and Rye 1979; Patterson et al. 1981; Collins 1981; So et al. 1983) indicate that Cl^- is the main anion in such ore-forming systems, with deposition temperatures ranging from 500-200°C, although F^- is a particularly important ligand in the mobilization of U (Beus and Grigorian 1977; Romberger 1984).

Examples of reactions capable of releasing ore forming elements including U and Sn and W from primary minerals, have been summarized by Eugster (1984). The most important, which takes place in acid reducing conditions, is congruent dissolution, mainly of oxides (such as magnetite) to form metal halide species and water. Highly uraniferous granites commonly have extremely low Ca contents favoring low pH conditions and high fluoride activity. Other reactions such as incongruent dissolution of aluminosilicates such as K feldspar and biotite to form muscovite and melt-fluid reactions are less important for large highly charged cation such as U^{4+} or Th release (Eugster 1984).

Deposition of U ore minerals occurs as a result of the neutralization of acid solutions and the conversion of metal fluorides and bicarbonates to oxide. In granites and schists neutralization involves the conversion of feldspar and biotite to muscovite (Hemley and Jones 1964). Rocks that contain Ca, for example limestones and basic or ultrabasic rocks, also neutralize acid (Kwak and Askins 1981) and reduce fluoride activity.

Hydrothermal systems generally evolve from high temperature acid-reducing conditions to neutral oxidizing conditions, reflecting the influx of meteoric fluids. In these conditions release reactions are mainly exchange reactions involving bivalent octahedral cations such as ferrous Fe and alkaline elements, the release of most large highly charged cations such as Th, ceasing as the pH and f_{O_2} increase and T decreases (Eugster 1984).

Table 8. Classification of uranium deposits

A. IGNEOUS PLUTONIC AND VOLCANIC ASSOCIATION
  1 Igneous Plutonic Association
    1.1 *Magmatic uranium deposits – formed by differentiation of evolved uraniferous magmas*
      1.1.1 Alkaline complex deposits
    1.2 *Formed as result of high-to-low temperature hydrothermal activity associated with high-level granite magmatism*
      1.2.1 Granite associated deposits including vein-type deposits
      1.2.2 'Perigranitic' vein deposits
      1.2.3 Metasomatite deposits
  2 Igneous Plutonic and Volcanic Association
    2.1 *Deposits associated with granite magmatism and acid volcanic and volcaniclastic sequences in anorogenic settings*
      2.1.1 Breccia complex deposits
  3 Igneous Volcanic Association
    3.1 *Formed as a result of high-to-low temperature hydrothermal activity associated with high-level mainly felsic volcanics*
      3.1.1 Volcanic deposits

B. METAMORPHIC ASSOCIATION
  1 Formed by metamorphic fluids probably derived from igneous or sedimentary rocks previously enriched in uranium
    1.1 Synmetamorphic deposits
    1.2 Vein deposits in metamorphic rocks

C. SEDIMENT/SEDIMENTARY BASIN ASSOCIATION
  1 Continental
    1.1 *Associated with late post-orogenic sedimentary basins having mainly clastic fill – formed or modified in some cases by intra-basinal fluid flow*
      1.1.1 Quartz-pebble conglomerate deposits
      1.1.2 Unconformity-related deposits
      1.1.3 Sandstone deposits
      1.1.4 Sediment-hosted vein deposits
      1.1.5 Collapse breccia deposits
      1.1.6 Lignite deposits
    1.2 *Penecontemporaneous with sedimentation or formed by surface weathering*
      1.2.1 Surficial deposits
  2 Marine
    2.1 *Oceanic*
      2.1.1 Phosphorite deposits
    2.2 *Epicontinental*
      2.2.1 Black shale deposits

Main types are based on classes used by the OECD/NEA (1998).

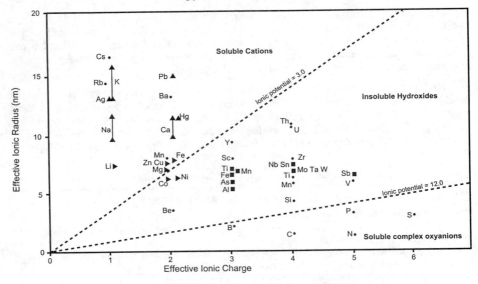

Figure 10. Effective ionic charge plotted against ionic radius for some common ions (after Gordon et al. 1958). Diagram used by permission of the Institution of Mining and Metallurgy, from Plant, O'Brien, Tarney and Hurdley (1985), Fig. 2, p. 267.

The results of experimental studies of hydrothermal systems are consistent with observations of geothermal areas where high to low pressure steam contains high chloride and fluoride levels, e.g. Ellis and Mahon (1967). With decreasing temperature such systems evolve into acid sulfate waters. Post-magmatic circulation involves neutral chloride waters containing Li^+, Na^+, K^+, Rb^+, Cs^+, NH_4^+, Br^-, F^-, $HCO3^-$, SO_4^{2-}, $HAsO_4^{2-}$, $H_2BO_3^-$. Studies of the Na/K and H^+ ion activities in deep waters from hydrothermal areas (Brown and Ellis 1970) in relation to experimental data (Ellis and Mahon 1967) indicate that the high temperature (200-359°C) water compositions in many rock types correspond to an equilibrium assemblage of quartz-K feldspar-albite-muscovite (Ellis 1973). Throughout the chapter high temperature water-rock interaction refers to these conditions in which a secondary granite mineral assemblage can form. Kaolinite becomes the stable aluminosilicate at lower temperature as the $H^+/(Na^++K^+)$ ratio increases. In such conditions uranium is easily remobilised because of the number and stability of oxidized uranium complexes (Langmuir 1978).

Evidence that primary silicate or accessory minerals have reacted extensively with aqueous fluids identifies granites from which U is most likely to have been concentrated into veins or other deposit types (Plant 1986; Govett and Atherden 1988).

Changes associated with water-rock interaction in granites

Evidence that high temperature (>200°C) water-rock interaction has occurred include: the presence of greisens formed by high temperature acid leaching; sub-solidus K feldspar in megacrystic and "feather textured" granite as a result of high temperature K metasomatism; albitized rocks reflecting high temperature Na metasomatism; and the occurrence of veined or metasomatized breccia pipes which contain fragments of granite or country rock in tourmaline or quartz-rich matrices. Biotite is replaced by muscovite or tourmaline, and primary accessory minerals such as zircon and sphene are destroyed in areas of intense greisenization. Sericitization, chloritization, kaolinization and hematization are related to lower temperature (<200°C) water-rock interaction (Simpson et al. 1979; Taylor 1979). Lexan plastic fission track studies show that U occurs in primary accessory minerals often in biotite crystals in unaltered granite. It is redistributed into ore minerals such as uraninite in secondary feldspar or quartz in granites affected by high temperature alteration. At lower temperatures it occurs adsorbed on clay minerals or hematite (Simpson et al. 1979). The minerals in which ore-forming elements occur in altered granites are compared with those of unaltered granites, in which the primary assemblage is magmatic, in Table 9. Changes in major element chemistry as a result of alteration are shown in Table 10 (Stemprok and Skvor 1974).

Geochemically one of the simplest criteria indicative of fluid-rock alteration associated with granites is the increased variability of trace and major element contents (Beus and Grigorian 1977; Plant and Tarney 1994) although the mean whole rock contents may remain unchanged. Hydrothermal activity is also evident in variation diagrams for example for Rb/Sr, Rb/Ba, K/Rb, Sr/Ba and U/Th and Mg/Li. Such diagrams show simple trends reflecting diadochic substitution in rocks formed mainly by magmatic processes but the elements are dispersed and scattered in mineralized intrusions, although plots of high field strength elements such as V/TiO_2 retain simple magmatic trends (Plant et al. 1985).

Evidence of hydrothermal alteration can also be obtained by comparing plots of alkaline elements or their ratios with those of the relatively immobile high field strength elements or their ratios, for example, Ti to V (Plant et al. 1985), Zr to Sn, V to Nb, since they show igneous fractionation trends and are resistant to alteration. Increased B/Ga and B/SiO_2 ratios have been shown to be indicators of B metasomatism as a result of high temperature interaction of granites with shale-derived fluids (Lister 1979; Simpson et al. 1979). Chondrite normalized REE patterns generally reflect the magmatic history of intrusions, but around areas where acid reducing fluids are present, REE patterns may be affected (Alderton et al. 1980). Dispersion of the Ga/Al ratio, which is normally constant in igneous rocks (Nockolds and Allen 1958) may also occur with the deposition of high Al/Ga minerals such as topaz from highly acid solutions (Beus and Grigorian 1977).

The quantity and number of generations of fluid inclusions of different compositions increase in altered granites reflecting the extent and duration of high to low temperature water-rock interaction (Roedder 1977; Rankin and Alderton 1982). Stable isotopes (particularly ^{18}O, ^{16}O) also provide evidence of water-rock interaction. Greisenization and kaolinization involve meteoric water depleted in ^{18}O (Taylor 1977), but ^{18}O enrichment of granites may occur as a result of interaction of magma with pelitic or shale-derived fluids at high temperature, or as a result of boiling of hydrothermal fluids (Coleman 1979). The Rb-Sr, K-Ar and U-Pb isotopic systems may also be disturbed in mineralized granites, since these elements are highly mobile in hydrothermal systems. Pervasive Sr isotope contamination by volume diffusion occurs at temperatures in excess of 500°C in feldspar (Dickin et al. 1980). Mass transfer may occur as a result of reactions involving dissolution or exchange of Sr at lower temperatures and Sr may continue to be redistributed as a result of hydrolysis of plagioclase in low-temperature (<100°C) hydrothermal systems. (Edmunds et al. 1984) Studies of thermal metamorphic aureoles around granites provide further evidence of high-temperature water-rock interaction. It was shown for the Easky adamellite in Ireland that the formation of a metamorphic aureole involved flow of H_2O from the host rocks into a water-undersaturated granite magma. Desilication and formation of corundum-bearing assemblages occurred whilst garnet and biotite were oxidized to

Table 9. Principal host minerals for trace elements in mineralized and unmineralized acid-intermediate igneous rocks.

Element	Mineralized Rocks	Unmineralized Rocks	
		Intermediate Rocks	Acid Rocks
Li	lithium micas (pegmatites), spodumene, amblygonite	hornblende, biotite, feldspar	biotite, feldspar
Rb	secondary micas and clay minerals, low-temperature K-feldspar (adularia)	biotite and other micas, feldspars	micas, K feldspar
Be	beryl, bertrandite, phenacite, chrysoberyl	plagioclase, muscovite	plagioclase, muscovite
F	fluorite, topaz, villiaumite, apatite	amphiboles, micas	micas
Cu	sulfides	pyroxenes, amphiboles, biotite, magnetite	biotite in Cu-Mo porphyries
Zn	sulfides	amphiboles, biotite, magnetite	biotite
Pb	sulfides	feldspars, biotite, various U- & Th-rich accessory minerals	feldspars, various U- & Th-rich accessory minerals
Nb	ferrocolumbite, pyrochlore, loparite, Nb,Ta oxides	ilmenite, zircon, titanite, biotite	ilmenite, zircon, titanite, biotite
Ta	tantalite, microlite, Nb,Ta oxides	ilmenite, zircon, titanite, biotite	ilmenite, zircon, titanite, biotite
Mo	molybdenite, wolfenite	magnetite, ilmenite, titanite, plagioclase	titanite, K-feldspar, plagioclase
Sn	cassiterite, tourmaline	biotite, muscovite, ilmenite	biotite, muscovite, ilmenite
Cs	...	biotite, muscovite	biotite, muscovite
W	tungstates	biotite	biotite
Th	monazite, thorite	zircon, titanite, biotite, monazite, apatite, xenotime, allanite	zircon, titanite, monazite, apatite, biotite, xenotime, allanite
U	uraninite, phosphates, hematite, Nb,Ta oxides	zircon, titanite, biotite, monazite, apatite, xenotime, allanite	zircon, titanite, monazite, apatite, biotite, xenotime, allanite
REE	complex (REE) silicates, monazite, xenotime, allanite, Nb,Ta oxides	biotite, plagioclase (Eu), monazite, allanite, titanite	apatite, zircon, micas (LREE)
Ba	barite, celsian, cymrite	micas, plagioclase	micas, K-feldspar

Table after Beus and Grigorian (1977).

Table 10. Some effects of secondary alteration of granites
(after Stemprok and Skvor 1974)

Process	Change in modal composition	Change in chemical composition
albitization	− K feldspar − Ca plagioclase + albite	+ Na_2O − K_2O
muscovitization	− K feldspar − plagioclase − biotite + muscovite	− K_2O
greisenization	− K feldspar − plagioclase − biotite + quartz + mica	− Na_2O + SiO_2 + Al_2O_3
silicification	− K feldspar − biotite	+ SiO_2
sericitization	− K feldspar − plagioclase	+ K_2O − CaO
kaolinization	− K feldspar − plagioclase	− K_2O − Na_2O

magnetite in the aureole (Yardley and Long 1981). Also, the iron ore mineral suite of the Ardara granite, Ireland, was shown by Atkin (1979) to be a function of host rock lithology as a result of oxidation-reduction reactions.

Conditions for water-rock interaction. There is a relation between granite magmatism, tectonic lineaments and metalliferous mineralization. Magma intrusion in the Andean batholith for example, was focussed along a linear structure (70 Ma) (Pitcher 1978). The super-units were undersaturated with water for most, if not all, of their crystallization history, and are thought to have solidified at a sub-volcanic level, as a result of interaction with groundwater (Pitcher 1979). The pervasive hydrothermal activity and mineralization in the Arequipa segment of the batholith has also been attributed to quenching of the magma as it hit groundwater (Atherton et al. 1979).

The emplacement of granites in seismically active zones provides a powerful means of initiating and maintaining hydrothermal centers in response to changes in permeability (which vary by an order of 10^{12} between highly permeable and impermeable rocks) and temperature gradients. Fractured granites can have permeabilities several orders of magnitude greater, even under higher pressure (Pratt et al. 1977), than unfractured granites (Davies 1969). Unfractured granites have very low permeabilities of the order of 10^{-3} millidarcies in surface conditions, decreasing by two orders of magnitude at 6 km depth (Brace et al. 1968).

Hydration of granites in the crust could involve mass transfer, whereby lower density hydrous phases formed by interaction between granite and host rocks along contacts and in xenolith swarms are carried convectively in the advancing magma. Bartlett (1969) claims that convection occurs in plutons even if the viscosity is $>10^8$ poise.

Radioactive intrusions and volcanics that have not undergone fluid rock interaction, such as the Dido and Dunbeno granites, Queensland (Govett and Atherden 1988), or some of the granites in the Arabian Shield (Jackson and Ramsay 1986) are likely to retain U in refractory mineral phases. They may be important as large volume sources of U for sediment-associated deposits, especially Precambrian intrusions in which uraniferous minerals have become metamict. Nevertheless, granites in which U has been redeposited by water-rock interaction, are the most likely to contain vein-type deposits of U and to be important sources of uranium for sedimentary ore deposits.

Several types of uranium deposits can be related to this general model, including granite-associated vein type, metasomatite, volcanic and, to some extent, breccia complex uranium deposits.

Magmatic uranium deposits

Deposits included in this type by the OECD/NEA (1998) are those associated with intrusive or anatectic rocks of different chemical composition (alaskite, granite, monzonite, peralkaline syenite, carbonatite and pegmatite).

Uranium deposits in alkaline complexes are associated with (1) peralkaline syenite, for example the Ilimaussaq complex, Greenland (although these deposits are of minor economic importance); (2) carbonatites, for example Palabora, South Africa; and (3) granites and associated igneous rocks including alaskite, monzonite and pegmatite. Type examples of the latter include the Rössing deposit, Namibia, and deposits associated with the Limousin granites in the Massif Central, France. Uranium occurrences associated with pegmatites in the Bancroft area, Canada are further examples of this type of deposit (OECD/NEA 1998).

Anorogenic alkaline intrusions, such as the Illimaussaq nepheline syenite complex in Greenland, are generally the most enriched in U. The Illimaussaq complex averages about 60 ppm U with concentrations locally reaching 300 ppm in coarse-grained differentiates (Bailey et al. 1981; Kunzendorf et al. 1982). Baddeleyite and zircon bearing veins in the alkaline Pocos de Caldas intrusion in Brazil also contain several hundred ppm U with enrichment also in Th, Zr, Nb and REE in nepheline syenites (Waber, Schorscher and Peters 1992). Uranium in such complexes is generally bound in refractory minerals such as zircon. Carbonatite complexes contain concentrations of several tens of ppm U also mostly bound in refractory minerals such as pyrochlore. The Palabora complex of South Africa for example averages 40 ppm U (Dahlkamp 1993). Alkaline and carbonatite intrusions are generally believed to reflect anhydrous, lower crustal and upper mantle magmatic processes in the presence of high pCO_2 and pF_2 (Bailey et al. 1981). Such intrusions were emplaced following the assembly of continents or supercontinents and may have been generated by rising mantle plumes impinging on mature continental lithosphere. The Illimausaq complex was emplaced in the Gardar rift about 1150 m years ago following the middle Proterozoic Ketilidian orogenic event in West Greenland, and there are similar uranium occurrences in comparable settings from late Archaean to Recent times.

Vein-type uranium deposits associated with granite. The Rössing deposit in western-central Namibia has total reserves of 64,900 mt in the $80/kg U Reasonably Assured Resources (RAR) category (OECD/NEA 1990) grading 0.025-0.035% U of which about 46,640 mt have been recovered until 1991 (Dahlkamp 1993). The deposit, which is hosted by a late to post-orogenic alaskite granite complex, is emplaced in steeply dipping, brittle concordant and discordant structures cutting an amphibolite-grade metamorphic complex affected by polyphase deformation. The development of feldspar metablastites in the host schists, coupled with geochemical and fluid inclusion studies (Cuney 1980), indicate extensive fluid-rock interaction during emplacement. There is considerable controversy about the genesis of the deposits. Smith (1965) suggested an origin as a result of partial melting of uraniferous and fluid-rich metasediments, whereas Toens and Corner (1980) suggested a source in older Precambrian rocks and overlying metasediments followed by fractional crystallization to produce a fluid-rich pegmatite granite. If these models were correct, the alaskites, which are the lowest melt fraction, should have been the first intrusions to have been emplaced, whereas isotopic age dating confirms that diorites and granites predate alaskite emplacement (Berning 1986). Moreover, as with all such models for granite-associated uranium deposits, it is difficult to understand how the U and ore forming fluids could have been retained during the main tectonic events associated with prograde metamorphism. Comprehensive trace element data for HFS elements such as REE, have not been published for the Rössing granites, although Th contents range up to 98 ppm suggesting that the granites are evolved I-type intrusions. Their emplacement through late structures undergoing active deformation would have facilitated pervasive water-rock interaction. The extremely low Th/U ratios of the intrusions and their highly variable contents of U (38-1120 ppm) and K (2-8 percent) are consistent with water-rock interaction but not with magmatic fractionation. Pyrite, chalcopyrite, bornite, molybdenite, arsenopyrite, magnetite, hematite and fluorite minerals are also suggested to reflect hydrothermal alteration together with uraninite which occurs in inclusions and intergranular spaces and microfractures in quartz, feldspar and biotite (Berning 1986; Brynard and Andreoli 1988). The presence of uraninite inclusions in feldspar suggests that water-rock interaction may have occurred initially at high temperatures > 200°C to 350°C (Ellis and Mahon 1977).

A similar model can be applied to the uranium deposits associated with the Variscan Saint Sylvestre granite in the Limousin region of the Massif Central in France. Most of the orebodies are small with the entire district having estimated reserves only of the order of 34,000 mt U grading 0.1-0.5% U (OECD/NEA 1990; Dahlkamp 1993). The deposits are in veins including those of 'episyenite' in a highly evolved facies of the post-tectonic Saint Sylvestre granite which contains 71-74% SiO_2 with high concentrations of U, Th, Be, F, Sn and W (Friedrich and Cuney 1989) and which is cut by synmagmatic mylonite zones. The granites have been affected by greisenization, quartz dissolution, muscovitization, albitization, chloritization and argillitization. The ore mineralogy consists mainly of uraninite associated with pyrite, marcasite, quartz, chalcedony, fluorite and baryte. The favorable leucogranites also contain Sn, W, Li, Nb and Ta and Mo minerals. The granites associated with ore deposits are emplaced into narrow regional lineaments, several hundred kilometres long marked by elongated gravity lows (Dardel et al. 1979). As in the case of the Rössing alaskites the granites were emplaced during the last phase of intense regional thrusting, through narrow active lineaments, but in the case of the Saint Sylvestre intrusion the lineament was in a major synmagmatic ductile deformation zone (Friedrich et al. 1987). No data for large highly charged elements are available other than for Th, for which levels are similar to those of Rössing and HHP granites generally (Friedrich 1981). Genesis of the uraniferous magmas is ascribed here to anhydrous melting of lower crust/upper mantle under conditions of high pCO_2 and pF_2 following extensional collapse of the Variscan orogen. Crystal fractionation, which by comparison with other Variscan granites could have been up to 50% (Henney et al. 1994), further enriched uranium. The remobilization of U from granite into ore deposits is thought to have been initiated by high temperature interaction between magma and fluid during its emplacement in an active deformation zone. This is reflected by high and variable $^{86/87}Sr$ ratios and ^{18}O values and features such as greisenization. Further enrichment of U occurred as a result of hydrothermal fluids flowing through regional fracture systems giving rise to structurally controlled, U enriched mica-dominant episyenite in which feldspar replaces nearly all pre-existing minerals except quartz (Fig. 11). Leroy (1984) suggests the temperatures at which fluid-rock interaction

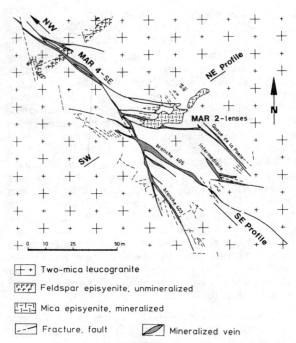

Figure 11. La Crouzille, Margnac mine. Plan view of mineralization at intersection of Margnac 4 veins with Margnac 2 mica-episyenite. After de Fraipont et al. (1982); diagram used by permission of Springer-Verlag, from Dahlkamp (1993), Fig. 5.31, p. 207.

occurred were in the region of 260-370°C associated with the loss of 32% SiO_2 and up to 2.5% Na and Sr, and an increase of up to 3% K_2O and of Rb. Subsequently kaolinisation occurred as a result of the influx of meteoric fluids into the hydrothermal system.

The model proposed here is similar to that of Poty et al. (1986). However the granite is suggested to be a uraniferous evolved I-type intrusion. Also many of the features attributed by Poty et al. (1986) to late stage magmatic processes are suggested to reflect high temperature water-rock interaction during emplacement of the granite through a deformation zone into the crust.

'Perigranite' vein-type uranium deposits. The classical uranium and polymetallic (Ag, Co, Ni) Jachymov orefield in the western Erzgebirge in central Europe, where mining ceased in 1963, contained total uranium resources of 8,500 mt U ranging between 0.1 to 0.85% U (OECD/NEA 1998; Dahlkamp 1993). The deposits, which are structurally controlled, are associated with a highly evolved Variscan granite complex emplaced into metasediments; with features similar to those interpreted as indicative of water-rock interaction in the Limousin region. A similar model is suggested for Jachymov, although with a geothermal field extending into the host metasediments. This produced a different ore assemblage to that formed by predominantly intra-granitic water-rock interaction at Limousin.

The same model can be used to explain similar uranium deposits elsewhere, for example in the contact metamorphic rocks of the Variscan granites of the Iberian Meseta where evolved Variscan granites enriched in U and Th are emplaced into high-grade, intensely fractured pelites and sammites. Another example is the perigranite uranium vein deposits of the Pribram District, Czech Republic (Dahlkamp 1993).

Small concentrations of uranium resulting from supergene processes are know in several ore districts associated with altered uraniferous granites. Models that involved significant ore deposition as a result of oxidizing meteoric fluids descending into radiothermal intrusions (Moreau 1977; Barbier 1974) are no longer accepted by most economic geologists, except some minor deposits in arid and semi-arid environments (see below).

Metasomatite uranium deposits

These deposits are included in a separate group by Dahlkamp (1993) and the IAEA (1996) and OECD/NEA (1998), despite their close association with uraniferous granites enriched in Th, Nb and REE. The principal ore minerals are Th-rich uraninite and U-Th oxides. Metasomatite uranium deposits are characterized by disseminated uranium in structures affected by alkali (Na-dominant) metasomatism. There are two main types:

(1) Those in which alteration is predominantly albitization. This is associated with desilicification and where there is insufficient Al to form albite, aegirine and/or arfvedsonite accompanied by enrichment of Na_2O and depletion of K_2O and SiO_2. K-feldspar and Ca-rich feldspar and quartz are replaced by albite.

Uranium is associated with metasomatized metasediments in the Krivorozhsky-Zheltye Vody district, Krivoy Rog, Ukraine (Batashov et al. 1984; Kazansky and Laverov 1977). At Zheltye Vody the age of the uranium deposit is about 1770 Ma old. The ore is associated with intense metasomatism of Middle and Upper Proterozoic metasediments along lineaments into which microcline granites are emplaced. There are four main phases of alteration: early Fe-Mg-metasomatism that generated stratiform iron ore bodies, Na-metasomatism, carbonate-metasomatism, and finally silicification to form secondary quartzites (Batashov et al. 1984; Kazansky and Laverov 1977). Uranium minerals include uraninite, coffinite and brannerite. The principal U-bearing mineral assemblages are (1) apatite, zircon (uraniferous apatite, zircon and titanite); (2) amphibole-nenadkevite (nenadkevite, brannerite); (3) carbonate-uraninite and aegirine; and (4) sulfide minerals with coffinite, and uraninite (Zhukova 1980). The metasomatites hosting the U are tens of metres wide and may extend up to 1500 m in length, dying out at depth below about 1000 m. The Zheltye Vody uranium deposit is associated with a large iron ore deposit with similarities to Olympic Dam (W. Finch, pers. comm.).

(2) Those in which alteration is mainly weak argillitisation of feldspars including secondary albite, and locally intense hematitization (Collot 1981). At Ross Adams, Bokan Mountain, USA, for example, hematite is pervasive in microfractured albite and coats uraninite grains. The uranium minerals occur where the granite is enriched in Na_2O, Al_2O_3, U, Th and REE (Collot 1981). Uraninite, thorium rich uraninite, uranothorianite, uranothorite, thorite, some brannerite, and coffinite are the main ore minerals with accessory allanite, bastnaesite, monazite, zenotime, galena, pyrite, magnetite, zircon, apatite, fluorite, and calcite and locally abundant hematite (Collot 1981).

Individual deposits may contain a few tens up to ~720 mt U averaging from less than 0.1 to more than 1%, up to 2.5% U (Dahlkamp 1993).

Other possible examples include Radium Hill, Willyama Block, South Australia; Gunnar, Sakatchewan, Canada; Lagoa Real, Bahia, Brazil; Kitongo, Cameroun; Mosquito Gulch, Nonacho, North West Territories, Canada; Espinharas, Paraiba, Brazil; and the Kouvervaara, Kuusamo Schist Belt, Finland (Dahlkamp 1993).

The classification of deposits as metasomatites derives from Russian work e.g.

Kazansky et al. 1968, Smirnov 1977. It is suggested here that at least some metasomatites represent Na-rich fluid-rock interaction with uraniferous evolved I- or A-type granites.

Breccia-complex uranium deposits

This group, which includes the relatively recently discovered Olympic Dam deposit in South Australia, occurs mainly in middle-Proterozoic rocks formed during a period characterized by anorogenic magmatism and extension of continental lithosphere. Host rocks include brecciated or altered granite, felsic volcaniclastic and sedimentary sequences. Uranium deposits, where they occur, immediately overly granitoid complexes. The type example, Olympic Dam in the Stuart Shelf region of South Australia, contains 2000 Mt of ore that contains 0.05% U 1.6% Cu, 0.6 g/t Au and 3.5 g/t Ag (Roberts and Hudson 1983; Scott 1987, OECD/NEA 1990). Proved ore reserves of 49,500 mt U at a grade of 0.068% U were recently reported (Uranium Information Centre 1999). Other deposits which have been suggested by Oreskes and Hitzman (1993) to belong to this group of Proterozoic iron-oxide deposits (Cu-U-Au-Ag-REE), include the Redbank district, Australia, the Great Bear Lake deposit, Canada, and the south-east Missouri iron province, USA. Most of the deposits are discordant like Olympic Dam but the Bayan Obo deposit, China, and the Kiruna deposit, Sweden are concordant with interstratified sedimentary or volcaniclastic host rocks. One characteristic of the breccia-complex group of deposits is their enormous size. Many contain >100 Mt of mineralized rock although not all of the deposits contain significant uranium concentrations.

In addition to their common middle-Proterozoic age and tectonic setting, the deposits are characterized by their high contents of hematite or magnetite and extensive wall-rock iron metasomatism. In addition, most deposits show a sequence from sodic alteration at depth through potassic to sericitic alteration and silicification near the surface. Genetic hypotheses have included liquid immiscibility of an iron-oxide melt, exhalation of iron-rich fluids, and hydrothermal replacement of alumino-silicate wall rocks by iron-rich fluids (see Oreskes and Hitzman 1993). The most favored model, based on the Olympic Dam deposit, suggests that breccia complex deposits form primarily by shallow hydrothermal processes, related to major regional structures with Olympic Dam representing a near-surface end-member of the group.

The Roxby Downs Granite, which hosts the Olympic Dam deposit, is an undeformed Middle Proterozoic granite that intrudes Proterozoic and Archaean basement (Creaser 1989; Reeve et al. 1990). The Olympic Dam deposit, which extends over an area of approximately 7 km by 3 km, consists of several coalescing, steeply dipping hematitic bodies within brecciated and altered granite, which becomes progressively less disrupted away from the deposit (Fig. 12). There are three main breccia types: brecciated granite, heterolithic breccia, and hematite breccia. In addition, volcanic conglomerates and tuffaceous sediments are exposed in a fault-bounded block near the center of the deposit (Oreskes and Einaudi 1990). The breccia bodies generally strike north-northwest, parallel to regional photo-lineaments (O'Driscoll 1985) and faults that predate or overlap with the period of ore formation (Oreskes and Einaudi 1990).

The ore minerals consist primarily of Cu-Fe sulfides and uraninite, subordinate coffinite and locally minor brannerite, disseminated in breccia. The highest ore grades are associated with bornite in breccia (Reeve et al. 1990). Brecciated Roxby Down granite typically contains low-grade pyrite-chalcopyrite zones that grade laterally into higher-grade bornite-chalcocite zones in heterolithic breccia. Locally, high-grade uranium zones also occur in hematite breccia, but the hematite-breccia core of the deposit and of many breccia units are almost entirely barren of sulfides. Uraninite is closely associated with Cu-Fe sulfides throughout the deposit, and, in general, high Cu grades correlate with high U

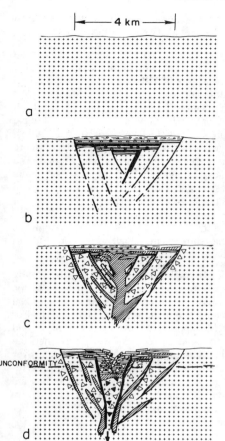

Figure 12. Origin of the Olympic Dam deposit. Hypothetical sequence of events in forming the Olympic Dam deposit. (A) Roxby Downs Granite is unroofed, and exposed, shortly after formation at 1600 Ma. (B) Extensional faulting forms localized basin; volcano-clastic sediments (Gawler Range equivalents?) accumulate. Possible early phases of hydrothermal activity. (C) Breccia complex develops as a result of hydraulic fracturing (open triangles) and hydrothermal activity and intense metasomatism (diagonal rules) along precursor faults. Possible exhalative activity and (or) infiltration of iron-rich fluids into pre-existing sedimentary pile. (D) Rebrecciation of early breccias (closed triangles). Continued, extensional faulting may rotate early-formed subvertical breccia bodies into less steeply dipping orientations. Late-stage collapse of central column entrains sediments in the centre of the breccia complex. Finally, deposit is eroded to level marked unconformity, approximately 300 m below the present surface (from Oreskes and Einaudi 1990). Diagram used by permission of the Geological Association of Canada, from Oreskes and Hitzman (1993), Fig. 4, p. 621.

grades; low-grade Au and Ag zones (averaging 0.6 g/t and 3.5 g/t respectively) are associated with Cu-U zones, although the patterns are more complex. REEs occur in monazite intergrown with sericite, hematite and/or Cu-Fe sulfides, and, unlike Cu and U, they are concentrated in the most hematitic breccias in the center of the deposit (Oreskes and Einaudi 1990; Johnson and Cross 1991). Minerals in cross-cutting veins parallel to major regional structures consist mainly of chalcocite and bornite with subordinate uranium and minor amounts of arsenides of Cu, Ni, Co, native Au and Ag, and abundant hematite. The main gangue minerals are fluorite and quartz with some sericite and chlorite. The evidence of REE patterns and Nd isotope systematics indicate that the Roxby Downs Granite is not the source of REEs in the deposit. Oreskes and Einaud (1990) and Reeve et al. (1990) suggest that the mineralization was broadly contemporaneous with formation of the breccias,

At Olympic Dam hydrothermal alteration, which extends about 1 km from the mineralized breccia complex into pervasively fractured host granite, is dominated by hematite and sericite, with locally abundant silica. Nearer to the breccias, sericite becomes more abundant, replacing plagioclase and rimming K feldspar. Mafic minerals are converted to chlorite and epidote ± hematite. Close to the deposit, hematite replaces sericite, and hematite alteration grades into intense Fe metasomatism. Locally, particularly in the faulted eastern side of the deposit, intense hematitization may be associated with silicification and higher than average gold levels (Reeve et al. 1990).

Thus the Olympic Dam uranium deposit is associated with a huge, hydrothermal breccia complex. The morphology, zonation and mineral textures of the breccia complex suggest intense mechanical brecciation and iron metasomatism of the host granite along a pre-existing north-northwest-trending fault system, and by hydraulic fracturing and/or explosive release of volatiles. Iron metasomatism occurred by precipitation of hematite in veins, fissures, and vugs and by replacement of precursor minerals. Although Olympic Dam is no longer regarded as a sediment-hosted ore deposit (Oreskes and Hitzman 1993), hematitic siltstone fragments in the breccias, suggest it formed near to the surface where hydrothermal fluids were vented and hydrothermal breccias were exposed and eroded. The distribution of economic Cu-U-Au-Ag-REE zones at Olympic Dam suggests an initial sulfur-bearing and a later, supergene event (Oreskes and Hitzman 1993). According to Oreskes and Einaudi (1990), some high-grade chalcocite zones formed partly as a result of supergene enrichment where the orebody was exposed at the Proterozoic unconformity, which is a conclusion supported by evidence of disruption of Nd-isotope systematics (Johnson and Cross 1991). It is at this stage that sulfides may have been converted to hematite. The close association of U with Cu probably reflects redox conditions.

There is clear spatial relationship between the Olympic Dam deposit and the Roxby Downs Granite although the latter appears to have been eroded and unroofed prior to the main mineralizing events. Rapid unroofing may have occurred in response to regional extension in an area of high-heat flow and continuing magmatic activity. Hydrothermal or supergene events between 1500 Ma and 1300 Ma appear to have remobilized U and disturbed Rb-Sr systematics. Hence Rb-Sr dates for igneous rocks of the Stuart Shelf are consistently younger than U-Pb zircon dates for the same samples. Also ^{18}O systematics for quartz-feldspar pairs from the Roxby Downs Granite show evidence of subsolidus re-equilibration; and some biotites give ages older than the whole rocks from which they are separated. The evidence is consistent with pervasive disturbance of the isotopic systems by a major low-moderate-temperature hydrothermal event approximately 80-100 m.y. after emplacement of the Roxby Downs Granite (Creaser 1989).

The main components of the model proposed by Oreskes and Hitzman (1993) involve a shallow environment of formation; no particular host rock; a weak association with contemporaneous igneous activity; strong evidence of regional structural control; and abundant evidence of intense hydrothermal alteration and replacement. It is suggested here that the deposits are associated with anorogenic granite in major continental lithospheric extension zones reflecting mantle plume activity rather than subduction-related tectonism as suggested by Oreskes and Hitzman (1993) (Fig. 13). Radiothermal A-type granite intrusions would thus have been central to the development of breccia complex uranium deposits, including mineralization events, which post-dated granite unroofing.

Volcanic uranium deposits

Uranium deposits of this type, commonly associated with Mo, F, Th and REE, occur stratabound and/or in tectonic structures in felsic volcanic rocks and associated sediments. As in the case of intrusive-associated uranium deposits, the volcanic rocks are formed mainly in late to post-orogenic or continental anorogenic extensional tectonic settings of post-Archaean age (Gandhi and Bell 1995). The volcanic rocks are most commonly subalkaline to peralkaline, although occasionally uranium can be concentrated in potassic-alkaline intermediate rocks. The host rocks are commonly albitized, hematitized and/or altered to carbonate. Type examples include (1) the rhyolite-hosted Michelin deposits in Labrador, Canada, (2) the trachyte-hosted Rexspar deposit, British Columbia, Canada,

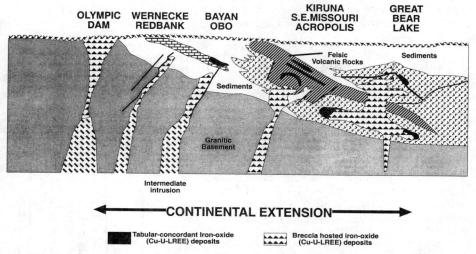

Figure 13. Summary of regional settings of Proterozoic iron-oxide deposits. Schematic representation of the tectonic setting and host-rock sequence for Proterozoic iron-oxide (copper-uranium-REE-gold) deposits. Most occur in continental regimes in areas of extension or rifting. The deposits form in a variety of host-rock characteristic of this setting. (Adapted from Hitzman et al. 1992) Diagram used by permission of the Geological Association of Canada, from Oreskes and Hitzman (1993), Fig. 6, p. 630.

(3) the ignimbrite-hosted Pena de Blanca deposits, Chihuahua, Mexico, (4) the Xiangshan deposit, China, and (5) McDermitt, USA. The rhyolite-hosted Pleutajokk deposit, Sweden, is included in this group by Dahlkamp (1993) although according to the IAEA (1996) it is a metasomatite deposit.

Uranium deposits occur in all volcanic lithofacies from proximal to distal settings (Fig. 14), including volcaniclastic sediments in central calderas, lava flows and ash flow fields, volcano-sedimentary basins, domes, and breccias (including diatremes). Ore zones are stratabound away from volcanic vents as in the Michelin and Rexspar deposits (Gandhi and Bell 1995). In proximal deposits, ore occurs primarily in discordant veins and to a lesser extent disseminated in breccia fillings. The uranium deposits may be synvolcanic in volcanic and proximal exhalative vents. Alternatively they may be epigenetic, and have formed as a result of high to low temperature circulation of fluids including heated meteoric water or groundwaters enriched in uranium by weathering processes. Volcanic uranium type deposits are not known from andesitic volcanic suites (Gandhi and Bell 1995; Ruzicka 1995b).

Uranium deposits of this type commonly range between 500 and 10,000 t U grading from 0.04 to 4% U (Cox and Singer 1986, p. 162-164). The Anderson deposit in the USA contains 20 000 t U. A cluster of deposits in the Streltsov district in eastern Russia contains up to 100,000 t of U. The Michelin deposit in Labrador contains about 7 000 t U at an average grade of 0.1% U, and the Rexspar deposit in British Columbia has about 700 t U at an average grade of 0.075% U.

Volcanic uranium deposits in Labrador and Sweden are hosted by extensive sequences of flows, tuffs, and volcaniclastic sediments erupted during the late- and post-tectonic stages of the Makkovikian-Svecofennian orogeny 1900-1700 Ma (Gandhi 1978, 1986; Lindroos and Smellie 1979; Gustafsson 1981; Halenius et al. 1986, Schärer et al. 1988). Several uranium occurrences are also found in felsic volcanic rocks of the Great Bear

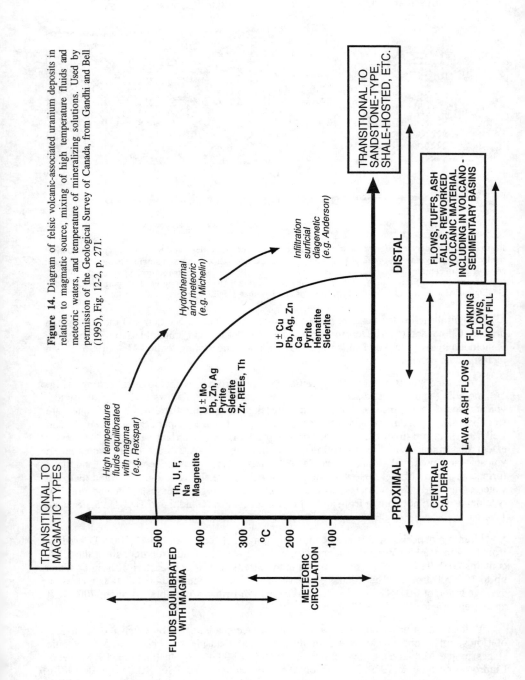

Figure 14. Diagram of felsic volcanic-associated uranium deposits in relation to magmatic source, mixing of high temperature fluids and meteoric waters, and temperature of mineralizing solutions. Used by permission of the Geological Survey of Canada, from Gandhi and Bell (1995), Fig. 12-2, p. 271.

magmatic zone, developed on the west side of the Wopmay orogen 30 to 60 million years after the orogenic peak about 1900 Ma (Hoffman 1980; Hildebrand et al. 1987).

Epigenetic mineralization of this type ranges in age from close to that of the host rocks to significantly younger. Most of the deposits in the Labrador uranium district give isotopic ages in the range 1800-1750 Ma, while their felsic volcanic host rocks, gives ages between 1860 and 1805 Ma (Gandhi 1978, 1986; Schärer et al. 1988). In Sweden, the Pleutajokk and associated deposits, and their host rocks are essentially contemporaneous with those in Labrador (Gustafsson 1981; Hålenius et al. 1986) and formed part of the same early Proterozoic terrane prior to the opening of the Atlantic Ocean (Gower et al. 1990; Gandhi and Bell 1993). Fracture-controlled occurrences in the Baker Lake basin, Canada give discordant uraninite ages between 1800 and 1700 Ma (Miller and LeCheminant 1985). The Rexspar deposit is Palaeozoic, while most of the other examples are Phanerozoic.

Complex deposits such as Rexspar have high Th/U ratios and are enriched in REEs, Mo, F, Pb, and Zn although the Michelin deposit contains only traces of these elements. In some cases proximal vein deposits with low Th/U ratios are also enriched in Mo, F, Pb, and Zn. Uranium occurs mainly as uraninite, and to a lesser extent as coffinite in Th-poor occurrences and many high-grade veins and disseminated deposits. In the Th-rich Rexspar deposit, the dominant U minerals are uraninite, uranoan thorite, and uranothorite (Bell 1985), but a great variety of other minerals include pyrite, fluorophlogopite, apatite, fluorite, celestite, galena, sphalerite, molybdenite, scheelite, siderite, calcite, barite, quartz, albitic plagioclase, bastnaesite, and monazite (Preto 1978). Fluorite, celestite, and minerals containing REE, such as monazite, are potentially economic.

Younger volcanic uranium deposits occur in the Basin and Range province of southwestern USA, mainly Arizona and New Mexico, and the Rio Grande Rift zone in Mexico which is of Tertiary age (Sherborne et al. 1979; Wenrich et al. 1990). The zone represents an extensional regime in which uplift was accompanied by mylonitization, the development of detachment zones, and synsedimentary faults, coupled with intrusion of deep-seated alkali-rich melts (Zoback et al. 1981; Jones et al. 1992). This gave rise to fault-bounded, terrestrial volcaniclastic basins at the surface, in some cases U mineralization evolving into sandstone-type deposits. The late Tertiary volcanic rocks have been related to back-arc extension and regional extensional tectonics that produced the basin and range topography (Zoback and Thompson 1978). Major uranium deposits occur in Tertiary volcanic and related sedimentary lacustrine rocks associated with calderas (Bagby 1986). Deposits in the Province include McDermitt, Nevada (Dayvault et al. 1985); Marysvale, Utah (Cunningham et al. 1982); Date Creek Basin, Arizona (Otton et al. 1990); Pena Blanca, Chihuahua, Mexico (Goodell 1981, 1985; Cardenas-Flores 1985; Sallas and Nieto 1991). The deposits are volcanic, composite hydrothermal veins and tabular lacustrine sandstone uranium deposits in volcanic settings and contain 425 to 17,000 tons of U, grading 0.04 to 0.085 percent U (Finch 1996). Ore minerals are uraninite, coffinite, metatyuyamunite, carnotite, uranophane, jordisite, fluorite, molybdenite and cinnabar. At McDermitt uranium mineralization is attributed to hydrothermal hot spring and meteoric alteration of mainly peralkaline, ash flow tuffs and domes. At Pena Blanca, geothermal activity involving convection of groundwater, deposited uranium in rhyolitic ignimbrites. In the Date Creek Basin diagenetic alkaline carbonate-bearing pore water deposited uranium in carbonaceous tuffaceous mudstone (Otton et al. 1990).

The Pocos de Caldas caldera in the State of Minas Gerais, southeastern Brazil is a ring structure composed of alkaline, volcanic and plutonic rocks, mainly phonolites and nepheline syenites of Mesozoic age. Postmagmatic hydrothermal events have led to widespread argillitization and zeolite formation. The latest, more localised, hydrothermal event is thought to have formed the uranium vein deposits in the complex. These include the Osamu Utsumi mine, which comprises mainly uranium minerals with subsidiary thorium and REE enrichment (Miller et al. 1994; Waber, Schorscher and Peters 1992). The principal uranium minerals are uraninite and brannerite. Secondary uraninite with pyrite occurs at the redox front beneath laterite and saprolite layers.

In volcanic uranium deposits uranium is thought to be enriched, initially in the felsic magma, and subsequently by high-low temperature hydrothermal fluids circulating through the volcanic pile. The main debate concerns whether or not the primary magmatic enrichment of U and associated incompatible elements is of mantle origin (Locardi 1985; Treuil 1985) or of metamorphic-anatectic origin from the middle or lower crust (Chen and Fang 1985). We favor a lower crustal origin possibly by melting of previously underplated igneous rocks (Wyborn et al. 1987; Wyborn 1988). The magma is suggested to have been enriched in uranium and incompatible elements as a result of partial melting in anhydrous conditions with high pCO_2 and pF_2 followed by assimilation-fractionation-crystallization processes in the crust as suggested for evolved I- and A-type granite magmas (see above).

Volcanic-associated uranium mineralization is suggested to reflect a similar sequence of high-low temperature fluid-rock interaction to that of granite-associated uranium deposits. In some cases, where there is adequate thickness of felsic volcanic rock or where there is a buried HHP granite, several episodes of post-magmatic mineralization could occur. Arsenide U-Ag vein deposits with isotopic ages of 1775-1500 Ma related to granite in the Great Bear Lake area in the Northwest Territories, Canada, are hosted by volcanic and volcaniclastic rocks with isotopic ages of 1870-1840 Ma (Miller 1982). They may have formed by such a process.

METAMORPHIC URANIUM DEPOSITS

Uranium-enriched fluids formed by metamorphic processes, including dehydration and partial melting are generated in conditions of low water:rock ratios. The chemistry of such fluids is strongly influenced by reactions with the host rock (Criss and Taylor 1986) indicated by fluid inclusions with high CO_2 contents and enriched in ^{18}O. Depending on the temperature and salinity of the fluid, CO_2 and H_2O can remain miscible (Bowers and Helgeson 1983). In the absence of near-surface oxidation however, CO_2 effervesces with falling temperature and pressure giving rise to fluids of intermediate pH. At 300-400°C fluid chemistry approaches equilibrium with the typical greenschist assemblages of quartz, albite or oligoclase, chlorite, muscovite, carbonate, pyrite ± epidote, graphite, K-feldspar depending on the rock type. In such conditions, magnetite and biotite buffer the pH to ±1 of neutrality and the fo_2 to the stability field of pyrite (Barnes 1979; Ellis and Mahon 1977). Such fluids differ from those associated with acid igneous rocks, which are generally acid and reducing, evolving to neutral to alkaline oxidizing systems in which ^{18}O values approach those of meteoric water. Also metamorphic fluids are likely to be focussed along specific structural zones of increased permeability, giving rise to uranium deposits of limited lateral extent, compared to those associated with igneous systems.

Synmetamorphic uranium deposits

Synmetamorphic uranium mineral occurrences are most commonly of low grade (0.00085-0.085% U and contain subeconomic (kg to 850 mt U) resources (Dahlkamp and Scivetti 1981). The Forstau district, Austria may contain uranium minerals of this type. The process is thought to be important in enriching uranium which is subsequently remobilised to form other types of ore deposit (Dahlkamp 1993). Other uranium deposits classified as synmetamorphic by Dahlkamp (1993) include Ambindra Kemba, Fort Dauphin, Madagascar and possibly Mary Kathleen, Queensland, Australia.

Uranium vein deposits in metamorphic rocks

Deposits of this type occur in lenses and veinlets in mylonite or shear zones associated with major regional fault systems in metamorphic complexes. Type examples occur in the Beaverlodge area, Saskatchewan, Canada and at Rozna-Olsi and Zadni Chodov in the Bohemian massif, Czech Republic (Ruzicka 1971, 1995b). Average ore grades are in the range 0.2% to about 0.45% U. The largest metal tonnage in the Fay orebody, Beaverlodge was >8000 t of U grading 0.25%, with the highest grade of 0.5% in the Habore body which contained >500 t U (Ward 1984). The Rozna-Olsi deposit has produced more than 10,000 t U (OECD/NEA 1998), and has know resources of 2,800 mt U recoverable at costs below $80/kg U (OECD/NEA 1994).

In the Beaverlodge area, Saskatchewan, numerous orebodies are hosted by mylonitized feldspathic quartzite, brecciated and mylonitized granitic gneiss, altered argillite, and brecciated feldspar-carbonate rocks. The host rocks comprise a Lower Proterozoic metasedimentary and a Middle Proterozoic sedimentary and volcanic suite deformed at about 2600 Ma (Stockwell 1982) and about 1700 Ma (Stockwell 1982). In the late Archaean, gneiss domes and U-bearing pegmatite dykes were formed. The Hudsonian Orogeny caused mylonitization of Lower Proterozoic rocks and reactivation of major fault systems, providing the main structural control for the deposits. Most deposits are spatially related to the St. Louis Fault and its crossfaults. The main mineralization event occurred during the 1,700 Ma event, although uraninite also gives ages of 1,100 to 900 Ma, corresponding with a Grenvillian event and of about 200 Ma (Ward 1984).

The Beaverlodge deposit is vertically zoned (Ruzicka 1989). The upper parts consist mainly of uraninite, with brannerite increasing at depth. Locally the veins contain coffinite, uranium-rich organic matter, and oxides, as well as Fe, Cu, Pb and Zn selenides, and Pb sulfides.

The principal gangue minerals are carbonates with minor quartz, chlorite, and albite. Fluid inclusion studies of quartz and carbonates (Sassano 1972) indicate that the uranium-bearing fluids were derived from host rocks through exchange reactions initially at temperatures of 500_oC, falling to 80_oC. Oxygen isotopes suggest this occurred in an essentially closed system. Albitization occurred early in the mineralization process and wall rocks adjacent to the orebodies are commonly hematitized, feldspathized, chloritized, and carbonatized. Locally oligoclase and quartz cement wall-rock breccias (Ruzicka 1989).

The deposits in the Beaverlodge area, Canada are considered to represent the last stage in a cycle whereby U was introduced with granitoid plutons at about 2,600 Ma before being further concentrated in sedimentary rocks that were subsequently mylonitized and fractured during the Hudsonian Orogeny. Uranium was remobilised into fracture fillings, stockworks, and disseminations.

The Rozná Olsí and Zadní Chodov uraninite-coffinite deposits in the Czech Republic are associated with shear and fault zones containing chlorite, carbonate and, locally, graphite, quartz, and albite gangue. The mineralized zones of the Rozná Olsí deposits, which are several tens of kilometres long, are associated with a major fault system, which contains other smaller deposits. The Zadní Chodov deposit also follows a 60 km long regional fault system (Ruzicka 1971).

In the Bohemian Massif, U in mixed volcanic and sedimentary rocks is thought to have been mobilised during metamorphic, tectonic, and igneous events and deposited in mylonitic and fault zones (Ruzicka 1971, 1995b). Uranium ore deposition is accompanied by chloritization, carbonatization, and locally, albitization of the host rocks.

URANIUM DEPOSITS OF SEDIMENT AND SEDIMENTARY BASIN ASSOCIATION

The formation of uranium deposits in sedimentary rocks and sedimentary basins partly reflects the distribution of evolved felsic igneous rocks and hence plate tectonic evolution following the end of the Archaean. The change in the type and abundance of uranium deposits also reflects the evolution of the biosphere, atmosphere and hydrosphere. For example, the appearance of land plants during the Devonian led to the formation of new types of deposits such as those associated with lignite and more recently peat. Since U solubility is particularly affected by redox conditions, changes in the nature of some types of sedimentary uranium ore deposits through time have been linked to increases in the amount of oxygen in the atmosphere, e.g. Holland (1962) and Roscoe (1969). For example, Lower Proterozoic quartz-pebble conglomerates, the earliest known uranium ore deposits, have been suggested to reflect the presence of a reducing atmosphere (Cloud 1968). Subsequently, Middle Proterozoic unconformity-type deposits formed as a result of oxidizing formational fluids precipitating U at redox fronts (Ruzicka 1995a); and younger sandstone deposits were formed by the interaction of highly oxidizing uraniferous meteoric fluids with reducing agents, such as humic detritus, at depth in sedimentary basins (Finch 1996).

The atmosphere and oceans were produced by volcanic outgassing and cooling with the result that the early atmosphere contained large amounts of CO_2, N_2 and H_2O. Early organisms developed about 3500 Ma ago in oceans shielded from ultraviolet radiation. The first oxygen is thought to have been produced by photochemical dissociation of water and then by the photosynthesis of simple prokaryotes, such as blue-green algae. Although a few microfossils have been found in Archaean sediments as old as 3300 Ma ago they are much more common in Proterozoic rocks. The Proterozoic landscape is thought to have been one of barren, eroding continents with shallow continental seas and lagoons in which thick sandstone sequences, algal limestones, and in the early Proterozoic banded iron formation were deposited (Brown and Mussett 1993). Approximately twenty percent of the surface, as today, comprised deeper ocean basins (Brown and Mussett 1993). By the Middle Proterozoic, red beds, such as those of the Athabasca and Kombolgie formations of Saskatchewan, Canada and Northern Territory, Australia became more abundant as well as sedimentary sulfates and limestone formations.

There is considerable evidence for the widespread deposition of sulfate evaporites from 3500 Ma ago. Baryte deposits have been found in the 3450-3500 Ma Fig Tree Group of the Barberton Belt and the 3500 Ma parts of the Eastern Pilbara succession (Windley et al. 1984). In the Early Proterozoic, there are also beds of evaporite and halite, gypsum and anhydrite pseudomorphs indicative of sulfate deposition by brines in a supratidal sabkha environment (Windley et al. 1984). There is thus evidence of a sulfur cycle involving oxidation-reduction extending back at least to 3500 Ma ago. The presence of abundant sulfate in shallow water in the Archaean and early Proterozoic may indicate free oxygen derived from the oxidation of reduced sulfur by microprobes using photo-synthetic oxygen (Maissoneuve 1982), although photochemical reactions in the atmosphere and surface oceans could also have yielded sulfate by oxidation of volcanic hydrogen sulfide.

Throughout the Proterozoic, photosynthetic organisms increased oxygen levels and several authors have proposed an 'oxyatmoconversion' between 2400 and 2200 Ma ago (Roscoe 1969) or 1900 or 2000 Ma ago (Holland 1962). However according to Schidlowski et al. (1975) carbon isotope evidence from limestones and dolomites indicates that about 80% of the oxygen in the present atmosphere was released prior to 3000 Ma ago, and Windley et al. (1984) also provide evidence that the atmosphere was oxidizing by this time.

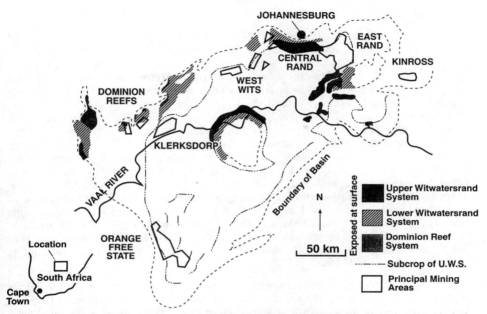

Figure 15. Map of the Witwatersrand Basin illustrating sub-Ventersdorp geology and showing principal mining areas. Used by permission of The Royal Society, from Simpson and Bowles (1977), Fig. 1, p. 529.

By 1500 Ma ago, prokaryotic life-forms had been joined by eukaryotes, simple cellular micro-organisms with a nucleus. Soft-bodied ancestors of a wide range of metazoan animals that started secreting skeletons around the Cambrian-Precambrian boundary have also been found in late Proterozoic rocks from central Australia. Diverse eukaryotes, which consume and produce oxygen, have regulated atmospheric gases since the late Proterozoic producing an atmosphere like that of the present day.

Changes in the biosphere and in the proportions of dissolved and atmospheric gas have affected the nature of uranium ore deposits. The relationship between these factors and the tectonic and thermal evolution of the Earth is considered further below.

Quartz-pebble conglomerate uranium deposits

Known quartz-pebble conglomerate uranium deposits are restricted to basal Lower Proterozoic beds unconformable on Archaean basement, which consists of metamorphic complexes into which granite magmas were emplaced. Magmatism commonly continued during basin filling. The principal type areas for economic deposits are in the Witwatersrand basin, South Africa (Fig. 15), and the Elliot Lake region, Ontario, Canada. Sub-economic occurrences have been reported in western Brazil, where the Jacobina Group unconformably overlies granites of the Archaean Sao Francisco Craton (Gama 1982), at Karnataka, India, and in the USA, Australia, Finland and Russia (see Dahlkamp 1993; Roscoe and Minter 1993).

The Witwatersrand and Elliot Lake deposits are large low-grade resources with mining grade varying between 0.013 and 0.11% U (Dahlkamp 1993). In the case of the Witwatersrand, uranium is a by-product of gold production. Resources of the Elliot Lake District (Agnew Lake, Denison, Pronto and Stanleigh properties) have been estimated at cluster size and grade ranges of 2500-203,000 mt U at 0.03-0.10% U (Finch 1996), with the potential for REE by-products.

The most commonly used model for this type of deposit is the modified placer model (Pretorius 1974; Roscoe and Minter 1993), whereby quartz-pebble gravels are considered to have been deposited in proximal sedimentary facies on the margins of ensialic rift basins. The Witwatersrand placers typically occur above angular unconformities whereas the Elliot Lake sediments occur in the basal beds of fluvial fans deposited where stream gradients decreased (Roscoe 1959). The Elliot Lake sediments are younger (about 2450 Ma old) according to Roscoe and Minter (1993) or 2250 Ma according to Finch (1996) than those of the Witwatersrand (2950-2710 Ma old) and the Dominion Reef Group in South Africa; (which are older than 3060 Ma) (Rundle and Snelling 1977). The Elliot Lake sediments, which occur in a Huronian pericratonic sedimentary basin at the southern margin of the Canadian Shield, are generally thicker and are associated with felspathic quartzite beds rather than quartz arenite beds. Kerogen that is enriched in U and Au is abundant in some of the South African ores, but uraniferous kerogen laminae are rarer at Elliot Lake and gold contents are lower (Roscoe and Minter 1993). A pre-2400 Ma atmosphere containing little free oxygen is invoked to explain the deposits on the basis that uraninite oxidizes readily in surface water and that no pyritic palaeoplacers have been found in rocks younger than 2400 Ma old (although minimum ages for some placer deposits are constrained only by dates as young as 2200 and 1900 Ma).

Some of the principal problems with the modified placer model are (1) the higher content of gold in the Witwatersrand deposits which greatly exceeds that found in the Elliot Lake district, where the basement is known to be more auriferous (Roscoe and Minter 1993); (2) the need for a source of sulfur for authigenic and allogenic pyrite (Dimroth and Kimberley 1976; Roscoe and Minter 1993); (3) evidence that a significant proportion of the pyrite, gold, low-thorium uraninite, and other uranium minerals are authigenic; Lexan plastic fission track method have shown that much of the uranium in the Witwatersrand occurs in phyllosilicates and organic matter[1] and appears to have been deposited from solution (Simpson and Bowles 1977, 1981). (4) the presence of abundant altered detrital titanium minerals. Rutile is replaced by ilmenite in magnetite-ilmenite intergrowths and magnetite is replaced by quartz. Also, brannerite has replaced rutile and anatase and rims the detrital grains (Roscoe and Minter 1993). (5) the detrital uraninite typical of the Witwatersrand and Elliot Lake deposits is high-Th uraninite, which is similar to that found at the present time in the Indus river (Simpson and Bowles 1977).

Recently arguments about detrital placer (Minter 1976; Hallbauer 1975), modified placer (Roscoe and Minter 1993), and hydrothermal (Davidson 1960; Schidlowski 1981) origins for the deposits have been reopened by Phillips et al. (1987, 1989), and Barnicoat et al. (1997). Barnicoat et al. (1997) argue that the gold and associated uranium minerals in the Witwatersrand are all epigenetic hydrothermal in origin and formed as a result of stratabound acid metasomation with fluid flow concentrated along lithological boundaries. No information is provided on the ultimate source of the uranium and gold, however, and the model fails to account for the presence of thorian uraninite in the Dominion Reef and Witwatersrand conglomerates of the same age (3050 ± 60 Ma) as those of granites around the edge of the basin (Rundle and Snelling 1977). At Elliot Lake, Canada, the early Proterozoic rocks, which host the uranium deposits, have well-preserved detrital grains of uraninite. These grains have survived sub-aerial transport from Pre-Huronian source areas

---

[1] The idea that all the U in the basin is in detrital grains; e.g. Liebenberg (1958), Ramdohr (1974) and Roscoe (1969), predates the introduction of fission-track techniques capable of detecting dispersed U (autoradiography has a much higher detection limit).

and also give pre-depositional isotopic ages (Roscoe and Minter 1993). According to Barnicoat et al. (1997), Au is precipitated as a result of interactions of the hydrothermal fluid with shale-derived hydrocarbons in the basin. Some authors favor an abiotic origin for the kerogen (see e.g. Landais et al. 1990; and Barnicoat, et al. 1997). Others invoke micro-organic deposition of the hydrocarbon (e.g. Pretorius 1975; Simpson and Bowles 1977, 1981; Minter 1981; Smits 1990). Barnicoat at al. (1997) argue that the occurrence of residual mesophase pyrobitumen in post-primary porosity fractures renders untenable an origin as in situ algal material. However, they examined only rocks of the Central Rand (the stratigraphically youngest Witwatersrand sequence). The main Carbon Leader host for gold and uranium occurs in the older West Rand sequence.

As noted by Fyfe (1974), bacterial sulfate reduction is important in regions where hydrogen-bearing geothermal waters rise through sediments. Moreover the genesis of kerogen by micro-organism activity at hot springs best explains the presence of laterally extensive stratabound, hydrocarbon layers between thick oligomictic quartz-pebble conglomerate horizons in the Witwatersrand basin. Bacteria are known to promote iron mineral reactions, including sulfate reduction to pyrite. Maximum reduction of U(VI) occurs at temperatures, pH and redox conditions optimal for anaerobic growth of bacteria (Zhang 1998). Ancient prokaryotic communities may also have affected gold dissolution and precipitation (Mossman and Dyer 1985). Biofilms can also account for abrupt changes in mineralogy that are difficult to account for thermodynamically. Micro-organism activity at hot springs in a sedimentary basin containing placer thorian uraninite and gold best explains the ore mineral assemblage of the Carbon Leader in the Witwatersrand. In the Devonian Orcadian basin of Scotland, extensive stromatolite horizons also developed where uraniferous fluids migrated from the incipient North Sea oilfield basin (U. McL. Michie, UK, NIREX, pers. comm.).

According to Simpson and Bowles (1977, 1981) and Barnicoat et al. (1997) sulfidation occurred as a result of diagenesis and later hydrothermal events. Simpson and Bowles (1977, 1981) also identify one or more phases of reworking of the sulfide rich pebbles and detritus variously enriched in uranium prior to sulfide deposition. They present textural evidence that sulfidation affected pebbles and detritus of widely different composition including abundant fragments of pyrophyllite. These authors suggest that the pyrophyllite detritus was derived from hydrothermally altered granites around the edge of basin. This is supported by detailed fluid inclusion studies of pebbles (Shepherd 1977) which also indicate major high temperature predepositional uranium mineralization with minor remobilization as a result of relatively cool circulating basinal fluids. In contrast, Barnicoat et al. (1997) interpret pyrophyllite in the Central Rand Group as the product of in situ acid hydrothermal alteration, although its distribution was mapped mainly using the PIMA instrument, which would be unable to distinguish between detrital and in situ hydrothermal pyrophyllite.

Simpson and Bowles (1977) related some of the gold and pyrite minerals in the basin to hydrothermal remobilization as a result of discordant intrusion of Ventersdorp-age dolerite dykes that cut the basin. New evidence based on Scanning Electron Microscopy and cathodoluminescence studies (Barnicoat et al. 1997) shows that uraninite, hydrocarbon, quartz, kaolinite, pyrite, pyrrhotite, gersdorffite, arsenopyrite, galena and gold are associated with extensive subhorizontal fractures.

According to de Wit. (1992) the late stages (3100-2600 Ma) of the evolution of the Kaapvaal craton, in which the Witwatersrand basin occurs involved accretion of crustal fragments at Cordilleran-type margins. Finally Himalayan-type continent-continent collision occurred between the Kaapvaal and Zimbabwe cratons in Southern Africa with the formation at 2680 Ma of the Limpopo orogen, which has a symmetrical thrust structure similar to that of modern extensional orogens (Treloar et al. 1992). Large-scale crustal thickening and underplating led to rapid uplift and denudation and the deposition of thick clastic sequences in the Witwatersrand foreland rift basin (Burke et al. 1986; Coward et al. 1995). The position of the Central Rand basin within the Kaapvaal craton is thought to reflect oblique collision between the Kaapvaal and Zimbabwe cratons and the formation of the NE-SW-trending Witwatersrand basin on the overriding (Kaapvaal craton) plate prior to thermal subsidence (Clendenin et al. 1987, 1988). Hence, basin-filling reflects collision followed by extensional collapse of the Limpopo belt accompanied by the development of fluvial systems comparable to those of the present-day Indus river, draining the Himalayas.

Late and post orogenic HHP granites that were formed by rapid advective thinning of a reduced thermal boundary conduction layer during extensional collapse of the Limpopo orogen are suggested to have been intensely hydrothermally altered during emplacement. They provided a source for detrital quartz, pyrophyllite, thorian uraninite, uraniferous phyllosilicates and dissolved uranium. The 3000 Ma old granites were shown by Simpson and Plant (1984) to have all of the characteristics of HHP granites, but their spidergram patterns suggest higher degrees of fractionation than younger examples.

Major granite magmatism continued around the basin margins during basin filling between 3100-2700 Ma ago (Buck and Minter 1987). Uranium-Pb ages on detrital zircon grains record continued erosion of the hinterland with unroofing of successively younger granites (Robb and Meyer 1985; Barton et al. 1989). A thick sequence of Ventersdorp lavas was erupted at 2700 Ma.

In conclusion, the earliest uranium deposits were formed in the Witwatersrand basin in a regime of intense tectonic, magmatic and sedimentological activity following extensional collapse of the Limpopo collision orogen. Rapid basin filling by rivers similar to those draining the present day Himalayas deposited placers containing Th-uraninite and gold. Subsequently diagenesis and geothermal activity within the basin further enriched uranium and gold to form the important deposits of the Carbon Leader. Later hydrothermal circulation was associated with regional tectonism.

The (often strongly reducing) conditions under which burial diagenesis below the sediment water interface and mineralization occurred in the Witwatersrand and similar Lower Proterozoic basins means that it is not possible to extrapolate from sedimentary to atmospheric conditions. Throughout the Phanerozoic almost all of the oxidized iron that entered the sedimentary column has been reduced by organic matter except in rare circumstances (Curtis 1977). The composition of detrital uraninite, which has a high thorium content, makes it unnecessary to invoke a reducing atmosphere to explain the formation of quartz-pebble conglomerate uranium deposits.

Unconformity-related uranium deposits

These deposits include the largest known high-grade deposits of uranium. They comprise a major portion of the world's low cost resources of uranium and more than one half of the annual uranium production of Canada and Australia (OECD/NEA 1998). The most important are spatially associated with basement complexes unconformably overlain by Middle Proterozoic clastic red-bed sedimentary cover sequences in intracratonic basins. The basement complexes consist of Archaean to Early Proterozoic granitoid domes, flanked by Lower Proterozoic metasedimentary rocks that include graphitic pelites (Derry 1973). The deposits are mostly associated with faults and fracture zones cutting the unconformities with mineralization apparently related to tectonic/thermal events affecting the basins (Ruzicka 1993). Such deposits formed mainly in the period 1800-800 million years ago,

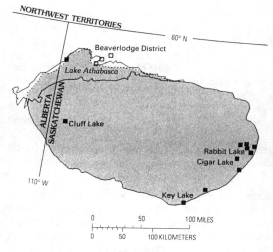

Figure 16. Details of the Athabasca Basin Uranium Province, Saskatchewan and Alberta, Canada. Used by permission of United States Geological Survey, from Finch (1996), Fig. 1, p. 5.

but small low-grade deposits also formed during Phanerozoic times. The most important known deposits are in the Athabasca Basin in Saskatchewan, Canada (Fig. 16), where examples include the Cigar Lake, Key Lake and Rabbit Lake deposits in the east of the Basin, and Cluff Lake in the west, all associated with the Athabasca unconformity. Important deposits such as Nabarlek, Koongarra, Jabiluka 1 and 2 and Ranger 1 and 2 also occur beneath the middle Proterozoic Kombolgie unconformity in the Alligator River or Pine Creek 'geosyncline' area of Northern Territory, Australia.

Total reserves and production of the Athabasca Basin region are in excess of 370 000 mt U with average grades varying between 0.11% and 12.2% U (Dahlkamp 1993) but most major deposits have grades of more than 0.85% U. The district includes Key Lake (75,000 mt U, average grade ca. 2.5% U) the largest high-grade, low-cost (open pit) uranium mine in the world, and Cigar Lake (108,000 mt U, average grade 12.2% U) one of the largest high-grade deposits in the world (OECD/NEA 1998). The Cigar Lake deposit has been studied as a Natural Analogue for a radioactive waste repository. The Australian deposits are generally of lower grades but have a larger range of tonnage (Battey et al. 1987). By 1987 the total resource discovered in the Pine Creek region was 300,000t U (Battey et al. 1987).

The Athabaska Basin uranium Province in a deep localised basin formed on the early Proterozoic Laurentian Craton (Hoffman 1988) at the western margin of the Canadian Shield (Finch 1996). Northeast-trending faults divide the main basin into sub-basins in which the Athabaska Group was deposited (Ramaekers 1981; Sibbald 1988; Sibbald et al. 1991). The important Cluff Lake group of deposits occurs in the Palaeozoic Carswell impact structure with a single cluster of deposits at the northern margin of the western part of the basin with numerous clusters along the southeastern boundary (Finch 1996).

The Athabasca U deposits are hosted by various lithologies below, at, and above the unconformity between Lower Proterozoic basement and Middle Proterozoic cover sediments; they occur as veins, breccias, and open-space fillings commonly associated with reverse and normal faulting. Most deposits show evidence of at least three different episodes of alteration. First retrograde metamorphism of the basement amphibolites (Ruzicka 1993), followed by weathering and erosion to form a regolith (Macdonald 1985), and subsequently hydrothermal alteration. This commonly includes argillitic alteration

(Sopuck et al. 1983) and can also include dolomitization, for example at the Rabbit Lake deposit (Heine 1986). Fracture-bound deposits, such as the Eagle Point deposit, are commonly monometallic and of medium grade (0.085-0.25% U) while deposits in overlying cover sediments, such as Cigar Lake are of high grade (0.085-12%) U and are generally polymetallic (U+Ni+Co+As) associated with bitumen and clays. The Ni, Co, and As occur mainly as sulfides. Uranium minerals consist mainly of uraninite with coffinite and brannerite. The deposits have fluid-inclusion and isotopic signatures consistent with high salinity ore-forming fluids, that ranged in temperature from 160 to 2000°C (similar to diagenetic fluids) (Kyser et al. 1989; Kotzer and Kyser 1990, 1992) and had initial mineralization ages mainly between 1,330 and 1,360 Ma, which are about 100 Ma younger than the depositional age of the Middle Proterozoic cover sediments (Cumming and Krstic 1992). These authors also record evidence of periods of rejuvenation of mineralization in the Athabasca basin at 1,070, 550 and 225 Ma ago.

There are some significant differences between deposits of the Athabasca and Pine Creek regions. For example, the known deposits in the Pine Creek region occur in either altered breccia zones in Lower Proterozoic graphitic metapelites or carbonate rocks below the Middle Proterozoic unconformity. Isotopic age dates for several deposits suggest that deposition post-dated that of the Kombolgie formation although some dates, eg for the Ranger deposit, suggest a pre-Kombolgie age. There is evidence of much later phases of mineralization or re-equilibration of isotopic ages in the district, eg Ewers et al. (1984) and Kotzer and Kyzer (1993). There is also less evidence of regolith development than in the Athabasca basin.

Several models have been proposed to explain unconformity-related uranium deposits. The supergene model originally proposed by Knipping (1974) involved leaching of uranium and other metals from Lower Proterozoic basement rock by groundwaters with precipitation at redox boundaries, such as carbonaceous metasediments, prior to deposition of the red-bed sediments. Such a model fails to account for the extensive mineralized zones in the red-bed cover sequence, that isotopic age dating shows to post-date red-bed deposition (Bray et al. 1987) and which overprints and alters any regolith present. 'Metamorphic hydrothermal' (von Pechmann 1981), and 'magmatic hydrothermal' models (Binns et al. 1980) have also been proposed to account for the genesis of this type of deposit. In the case of Jabiluka deposit in Northern Australia, Hegge and Rowntree (1978) suggest that ore fluids were generated during metamorphism of uraniferous granites and metasediments in the basement, whereas Binns et al. (1980) invoked convection driven by post-tectonic radiothermal granites. Other authors (e.g. Dahlkamp 1978, 1982) have proposed a complex multiphase model. In the Athabasca basin this involved syn-sedimentary concentration of uranium between 2200-1900 Ma, and remobilization and concentration during the Hudsonian orogeny at 1900-1800 Ma; further enrichment during regolith formation between 1800-1350 Ma, followed by diagenetic enrichment at 1350 Ma and finally further remobilization between 1000-200 Ma (Dahlkamp 1978,1982). The principal difficulty with this model is the many isotopic age dates, that clearly demonstrate that ore deposition postdates deposition of the Athabasca sediments (Kotzer and Kyzer 1993).

The diagenetic model proposed by Hoeve et al. (1980) is now one of the most commonly used models to account for the deposits. According to Hoeve the ores formed as a result of the interaction of oxidizing metalliferous fluids, heated to temperatures of 180-220°C (Pagel 1977) during their migration through the basin, with a redox front. This was associated with carbonaceous fluid or methane that ascended through fractures in the basement beneath the unconformity (Fig. 17). Wallis et al. (1986) support this model and propose a "two-fluid" system (involving a reductant plume) to account for redox variations.

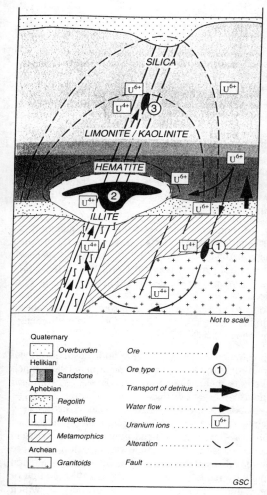

Figure 17. Conceptual model of unconformity-associated uranium deposits. A generalized vertical cross-section. Arrows indicate flow paths of oxidized and reduced convective waters. Circled numbers indicate locations of various styles of mineralization: (1) high grade polymetallic mineralization at the unconformity, (2) medium grade monometallic, mineralization below the unconformity, (3) low grade monometallic mineralization in sedimentary cover rocks above the unconformity. Used by permission of the Geological Survey of Canada, from Ruzicka (1995a), Fig. 7-7, p. 207.

Hoeve et al. (1980) and Sibbald and Quirt (1987) consider the Athabasca sediments as a possible source of uranium and other metals although airborne gamma-ray spectrometry and regional geochemistry indicate extremely low-surface radioactivity over the basin (Cameron 1983). The basin is surrounded on two and a half sides by extensive areas of older rocks enriched in U and Th, however (Darnley 1981).

Although the diagenetic model accounts for many features of the deposits, there is evidence to indicate that the formation of unconformity-type uranium deposits was preceded by several cycles of uranium enrichment following emplacement of large Archaean-Lower Proterozoic radiothermal granites. These highly evolved granites were enriched in incompatible elements such as Rb, Th, Pb, Sn and Nb (Binns et al. 1980) and are probably evolved I- or A-type granites. As discussed above, such granites can cause hydrothermal circulation long after magmatic cooling, when changes in tectonic or thermal conditions regenerate hydrothermal convection cells. The suggestion of Binns et al. (1980) that radiothermal granites could have been a major source of uranium and heat in the Pine Creek 'geosyncline' is supported by evidence that a zone of uranium enrichment associated with Lower Proterozoic radioactive granitoids exists beneath the Athabasca basin in Canada (Darnley 1981). Like Binns, Darnley considered that the radioelement-enriched granites could have provided heat sources capable of regenerating convection cells at intervals over hundreds of millions of years. Such a modification to the basic diagenetic model helps explain how uranium minerals can precipitate episodically over a long period of time. Hoeve et al. (1980) postulated deposition during periods of basic dyke emplacement, but radiothermal granites are likely to have been more important, although the thermal and tectonic perturbations accompanying dyke emplacement are likely to have regenerated granite-associated convection.

Sandstone uranium deposits

Sandstone uranium deposits are the second most important economic deposits of uranium (OECD/NEA 1998). The most significant deposits generally occur in permeable, medium- to coarse-grained and poorly sorted arkosic and quartzitic sandstones. The host sandstones are of fluvial or marginal marine facies and to a lesser extent, of lacustrine and aeolian facies, and occur in intracratonic basins. The host sediments are commonly overlain by, or interbedded with, acid volcaniclastic beds. Uranium minerals are generally associated with organic detritus, which in post-Devonian hosted examples comprises organic land-plant detritus and in Proterozoic hosted deposits comprises marine algal matter. Pyrite and/or marcasite is common, in most sandstone uranium deposits. Unoxidized deposits consist of uraninite and/or coffinite. Sandstone ores associated with vanadium typically oxidize to uranium vanadates, such as carnotite. Those associated with copper minerals oxidize to copper-uranium minerals such as metatorbernite and cuprosklodowskite, respectively. Non-vanadiferous sandstone ores oxidize to hydrous uranium oxides, and uranophane, a uranium silicate, is common.

The deposits are commonly of low- to medium-grade (0.04-0.4% U) generally as much as up to 50,000 mt U but some districts with several deposits have several hundred thousand mt U (Finch 1996). Important examples are in Permian, Triassic, Jurassic, Cretaceous and Tertiary host sandstone formations in the Western Cordillera of the USA. Similar deposits occur in Cretaceous and Permian sandstone formations in Argentina; in Carboniferous deltaic sandstone formations in Nigeria; Permian lacustrine sediments in France and Permian sandstone formations in the Alps. The deposits in Precambrian marginal marine sandstone in Gabon including those that formed the Oklo natural reactors are also considered as sandstone-type deposits by e.g. Dahlkamp (1993).

Tabular uranium sandstone deposits hosted by Upper Palaeozoic and Mesozoic fluvial sediments are typical of the Colorado Uranium Province (Granger and Finch 1988; Finch 1991, Finch 1996). The main mineral districts include the San Juan Basin, Uravan and Lisbon valley mineral belts as well as areas of New Mexico, Colorado and Utah. The deposits were formed in three main episodes of mineralization. Late Triassic-Early Jurassic, Late Jurassic, and Early Cretaceous. Many ores, for example, tabular deposits in the San Juan Basin were redistributed into roll-front deposits and in veins along faults in Late Cretaceous and early Tertiary times as a result of Laramide deformation. The ore mineralogy is mainly uraninite, and coffinite variously associated with Cu and V. Alteration mainly comprises bleaching of the host rocks by reduction of iron minerals. The source of the uranium in the deposits has been suggested by Finch (1996) to be silicic ash, which occurs in thick fine-grained units lying above the main Triassic and Jurassic host sandstone layers in the intracratonic sedimentary basins. The source of the ash is suggested to have been volcanic arcs to the west and south of the edge of the North America plate during the Late Triassic and Late Jurassic. The uranium ores are suggested by Finch (1996) to have been precipitated by reduction between uranium bearing groundwater and an underlying saline brine.

The Oklo deposits in the Franceville basin in southeastern Gabon contain the only known example of natural fission reactors (IAEA 1975a,b; 1978, European Commmission 1998). The ^{235}U content in the ore is depleted from the normal 0.72 percent to an average of 0.62 percent (Miller et al. 1994). The host rocks are Lower Proterozoic sandstones, mudstones and felsic volcaniclastic rocks. The deposits containing the natural fission reactors are in the basal sandstone-conglomerate unit. The reactors became critical about 2000 Ma ago and operated intermittently for about 10^5 to 10^6 years afterwards. About six to twelve tonnes of ^{235}U underwent fission producing about four tons of plutonium. (Jakubick and Church 1986). Factors favoring fission included the high concentration of uranium which in places exceeded forty two percent U, the ratio of ^{235}U to ^{238}U which, in the Early Proterozoic (older than 1,800 Ma) approached that of nuclear fuel, the low content of elements such as boron and vanadium which are neutron absorbers (Gauthier-Lafaye 1997) and the presence of water that acted as a moderator.

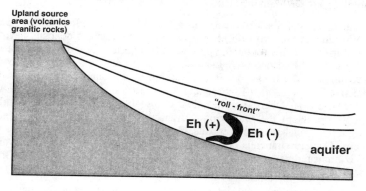

Figure 18. Model for roll front type uranium deposits (modified after Kissin 1988). Diagram used by permission of Elsevier Science, from Plant and Tarney (1994), Fig. 2-8, p. 23.

Roll front deposits which are typical of the Rocky Mountain and Intermontane Basin Uranium Province of the USA are thought to reflect dynamic fluid flow with mass transport of uranium in groundwater moving down dip within a confined aquifer in response to a hydraulic gradient (Fig. 18). Uranium minerals precipitate at redox fronts that may be abrupt or diffuse depending on the nature, abundance and distribution of reducing agents. The typical crescent-shaped cross-section of roll front deposits is thought to reflect the more rapid migration of groundwater through the middle layers of permeable sandstone units than in the case of tabular deposits. Roll-front deposits are considered to reflect a single major cycle of uplift, erosion, sedimentation and uranium deposition although other cycles may cause mineralization of older host rocks as a result of renewed regional uplift (Harshman and Adams 1981).

The Rocky Mountain and Intermontane Basin Uranium Province forms part of the Rocky Mountain structural province, which is a reactivated craton characterized by basement uplifts associated with reverse faulting (Bayer 1983). The uranium province is essentially defined by the extent of Laramide uplifts and basins (Finch 1996). Roll front uranium deposits formed in the basins and vein uranium deposits formed in fractured Precambrian metamorphic and Lower Palaeozoic sedimentary rocks. Vein deposits also formed during the Tertiary when uranium was derived from Proterozoic volcanic rocks deposited in a back-arc basin. The basins in which roll-front deposits formed are latest Cretaceous to Eocene (Gries 1983). The main sandstone deposit districts include the Powder River Basin, Gas Hills, Shirley Basin, Crooks Gap and the Hullett Creek districts, Wyoming (Chenoweth 1991) and the Tallahassee Creek District, Colorado. The host rocks are Cretaceous and Tertiary continental fluvial sandstones in which iron minerals are oxidized up dip of the roll-front deposits and reduced down dip along the advancing redox interface. The ore minerals are uraninite, coffinite and carnotite associated in some deposits with V. The roll-front sandstone deposits developed between uplifts in Tertiary times. They were formed by oxidizing uranium bearing groundwater that entered the host sandstones from the edges of the basin. Two possible sources for the uranium were (1) uraniferous

Precambrian granite (2,900-2,600 Ma) that provided sediment for the host sandstone and (2) overlying Oligocene volcanic ash sediments. Uranium mineralization in the Province began with initiation of Laramide uplift (70 Ma) and peaked in the Oligocene. This was associated with crustal extension between New Mexico in the south and Montana and North and South Dakota in the north (Dickinson 1981).

In the coastal plains of South Texas USA and north Mexico (Goodell 1985; Salas and Nieto 1991, Worrall and Snelson 1989) roll-front sandstone uranium deposits have formed in nearshore marine sediments containing pyrite and marcasite. The host rocks are Eocene, Oligocene, Miocene and Pliocene fluvial sandstone that grade towards the Gulf of Mexico into marine sediments. The host units are stacked but rarely mineralized with uranium so that orebodies overlie one another (Adams and Smith 1981). A major source of volcanic ash was derived from an Oligocene volcanic arc probably in West Texas (Clark et al. 1982). All the deposits in the Gulf Coast Uranium Province, USA and Mexico, are marginal marine roll-front sandstone deposits. Carbonaceous plant material is sparse in the host formations. Contemporaneous down-to-coast growth faults are considered to be the source of the H_2S gas reductant for most of the uranium deposits (Reynolds and Goldhaber 1983). Deposits contain 425-8,500 tons of U grading 0.034-0.25 percent U.

Other sandstone uranium deposits include those of Australia including the Frome Embayment, Westmoreland / Pandanus Creek, and the Ngalia, Amadeus, Carnarvon and Officer Basins and in the far west of China, the Yili basi (OECD/NEA 1998).

Sediment-hosted vein-type uranium deposits

Vein-type uranium deposits occur in various sedimentary rocks in the type area of the Shinkolobwe mine, Shaba district of the Katanga Copper Belt, Democratic Republic of the Congo (Derriks and Vaes 1956; Derricks and Oosterbosch 1958). The Pitch Mine, Colorado, USA (Nash 1988) is another example.

At Shinkolobwe, the host rocks are dominantly siliceous dolomite and dolomitic and carbonaceous shales, partly affected by Mg-metasomatism with magnesite replacing dolomite. The deposit occurs in heavily fractured rocks in a zone bounded by two major faults about 200 m apart and overlain by flat-lying wedges of impermeable strata in a fold belt close to where the fold axis changes direction interpreted by de Magnee and Francois (1985) as being on the flank of an anticlinal, breccia-filled structure. The breccia is composed of numerous, occassionally very large fragments containing Cu-Co minerals of Kupferschiefer type. No granitoids or other igneous intrusions are known in the area. At Shinkolobwe, uraninite and oxidized uranium minerals are associated with Co-Ni sulfides and selenides and Fe, Cu, Mo, Pb and Zn sulfides, trace amounts of precious metals, and phosphates (monazite). Gangue minerals include magnesite, dolomite, quartz and chlorite. The uranium, which occurs in veins, stockworks, along bedding planes, and in breccia matrices, is highly variable in grade and distribution. The Shinkolobwe mine closed in 1960 having produced 19,080 mt U (CEA 1969). Grades ranged from 0.085 to >0.85% U (CEA 1969). Elsewhere in the belt districts may contain up to several 10,000 mt U.

The host sediments were deposited 1300-620 Ma ago and deformed during the Luffilian orogeny at 840-710, 670-620 Ma (Cahen 1970), see also Finch et al. (1995). Isotopic ages of uraninite of ca. 706 Ma and 670 Ma (Cahen et al. 1971) suggest that mineralization was associated with the Luffilian orogeny to the south.

At Shinkolobwe and other deposits such as Kalongwe and Swambo in the Katanga copper belt of the Democratic Republic of the Congo and Zambia, uranium veins always occur in beds underlying cupriferous strata, which locally contain high levels of uranium at the base of a thick pile of sediments of shallow marine origin. Maucher (1962) reports values of up to 50 to 100 ppm U in the basal marine sediments. Most early workers (e.g. Derriks et al. 1956, 1958) favor a magmatic hydrothermal origin. Meneghel (1979) suggested a synsedimentary origin for the uranium, which was later redistrubuted into structures to form veins. Whereas de Magnee and Francois (1985) considered that circulating pressurized hot brines formed the Shinkolobwe and other deposits in the district.

A model involving intrabasinal fluid flow in the Katanga belt driven by regional tectonism is favored here. Oxidizing formational basinal brines probably deposited uranium and subseqently copper at a reducing front formed by shallow marine sedimentary sequences, which also acted as an aquiclude. Ngongo-Kashisha (1975) established a regional north-south mineral zonation in the Shaba district of the Katanga belt. Uranium, Cu, Ni, Co, Mo associated with magnesite and monazite, predominate in the south where Shinkolobwe and other uranium deposits occur. Further north U, Ni, magnesite and monazite decrease as Cu and Co increase.

Collapse-breccia pipe uranium deposits

Deposits of this type occur in roughly circular (10-300 m diameter), vertical pipes up to 1000 m deep filled with collapse breccia in which uranium is concentrated in the matrix and arcuate fracture zones around the pipe. The only known examples of this type of deposit occur in the Arizona Strip, Arizona, USA (Weinrich 1985, 1986; Wenrich and Sutphin 1989). They are a low-cost high-grade resource ranging from a few tonnes to about 2,100 mt grading between 0.25 and 0.85% U. Known resources of the Arizona Strip region are about 11,900 mt U (Dahlkamp 1993); some Au, Ag and Cu have also been recovered. The principal uranium mineral is uraninite and, to a lesser extent coffinite, and their alteration products, commonly accompanied by Cu, Ag and other ore and gangue minerals, such as baryte and anhydrite. Weak local alteration includes bleaching of red beds, calcification, local dolomitization, kaolinitization, and some sulfidation of infill and immediately adjacent wall rocks. Pyrite zones 3-15m thick enriched in (up to 15%) pyrite and marcasite cap the uranium deposits.

The pipes developed in Pennsylvanian and Permian limestone, sandstone, and shale formations, and were mineralized with high-grade uranium mainly in the late Triassic. Uranium-Pb dating of two of the deposits suggest that they were formed in the Permian (260-254 Ma) (Ludwig and Simmons 1992).

The source of uranium, as in the case of tabular sandstone uranium deposits, which also occur in the Colorado Uranium Province, USA is suggested to have been overlying volcaniclastic sediments (Finch 1996). According to Dahlkamp (1993) other elements may have been derived from sediments surrounding the deposits or from depth.

Lignite uranium deposits

Uranium associated with lignite or immediately adjacent clay and or sandstone, is absorbed onto carbonaceous matter or occurs as uranyl humates on sub-bituminous-grade coalified land plant detritus (IAEA 1996). The uranium often occurs in stratiform concentrations below intraformational unconformities. Large low-grade deposits (0.01 percent) uranium occur in the Williston Basin, North and South Dakota, USA (Denson and Gill 1965). Uraniferous tuffs are interbedded with the lignite bearing succession and the origin of the lignite deposits is suggested to be closely related to that of tabular sandstone deposits (Finch 1996). Other lignite uranium deposits occur in the Sokolov and Trutnov basins, Bohemia, Czech Republic, in the Serres and Ketili basins, Greece, at Freital,

Saxony and Stockheim, Barvaria, Germany and the Springbok Flats-Karoo Sub-basin, Transvaal, South Africa.

The uranium concentrated in lignites in these regions could be derived from uraniferous granites or volcanics bordering basins or from interbedded or overlying pyroclastic rocks containing high concentrations of uranium. Uranium could have been transported, by surface and groundwater, with precipitation by absorption on organic matter, which also acts as a reductant. Enrichment could also have occurred diagenetically or later as a result of groundwater flow. Such deposits, which are generally too small or low grade to be economic appeared in post-Devonian rocks after the emergence of land plants.

Surficial uranium deposits

These deposits occur in predominantly unconsolidated near-surface sediments commonly of Tertiary to Recent age which have not been subjected to deep burial and may or may not have been calcified to some degree. Uranium deposits, associated with calcrete, which occur in Australia, Namibia and Somalia in semi-arid areas where water movement is mainly subterranean are included in this type (OECD/NEA 1998). Other environments for surficial uranium deposits include peat and bog, karst caverns and pedogenic and structural fills (OECD/NEA 1998). These uranium deposits are mostly small and low grade. The Yeelirrie deposit, Australia, is potentially economic (Butt et al. 1984). Uranium occurs almost exclusively as U(VI) species or adsorbed onto minerals or organic matter. The Group includes duricrusted shallow sediments or soils in arid to semi-arid areas of Australia, Namibia and Somalia, and peat bogs or clay hollows in cool temperate climates and sedimentary detritus in karst caverns. Examples of shallow sedimentary deposits such as Yeelirrie, Australia, occur in fluvial, channel fill or in playa lake facies rocks for example the Lake Maitland deposit, Australia, which is cemented or replaced by calcrete, dolocrete, silcrete or gypcrete (Cavaney 1984).

Calcrete deposits generally contain <8.5 to 4240 mt U grading 0.025 to 0.068% U although Yeelirrie contains > 44,000 mt U grading about 0.13% U. Daybreak Mine in the USA, where uranium occurs in a palaeochannel near-surface structure (W. Finch pers. com.), produced 21 mt at a grade of 0.25% U. Ore recovered from karst caverns was mostly a few tonnes U up to 68 mt U averaging 0.425-0.75% U.

In the case of deposits in arid and semi-arid environments, the uranium is precipitated from migrating groundwater enriched in uranium and other elements or as a result of fluctuations in the water table, in internal drainage basins. Uranium is precipitated by processes such as reduction, complexation and adsorption. The uranium occurs in these types of deposits and karst filling sediments as vanadates. Elsewhere uranium occurs in surficial deposits predominantly as phosphate minerals, associated with calcrete, gypcrete or laterite on joints and fracture surfaces or crusts eg at Daybreak in the USA,

Surficial deposits are commonly located in regions containing uraniferous igneous rocks. For example, in the Yilgarn Block, Australia, late Archaean granites contain up to 80 ppm U (Gamble 1984) and in the Namib Desert deposits occur in regions containing uraniferous Proterozoic granites and pegmatites. Boyle (1984) suggested that the potential for economic surficial uranium deposits, decreases from hot arid to semi-arid, and temperate to tropical conditions.

Peat bog deposits which form in cool temperate wet conditions can contain up to 42 mt averaging 85 ppm rarely up to 0.17% U (Dahlkamp 1993). Uranium occurs as urano-organic complexes.

In peat bog and karst cavern deposits, the disequilibrium between U and its daughter isotopes indicates a relatively recent age of formation. Such data are unavailable for duricrust varieties. Parts of the Yilgarn craton of Western Australia where Yeelirrie and Lake Maitland are situated, may have been exposed to subaerial conditions since the middle and late Proterozoic (Daniels 1975; Morris 1980; Butt et al. 1984) although uranium mineralization is continuing at the present day (R. Finch, pers. com.).

Phosphorite uranium deposits

Low concentrations of uranium predominantly in fine-grained apatite and fluorapatite are associated with sedimentary phosphates. Examples include nodular phosphorite, locally reworked to pebbles interbedded with shallow marine beds.

The Florida Phosphorite Uranium Province occupies most of the State of Florida in the USA and extends a little way into Georgia (Cathcart et al. 1984). The resources of >850,000 tons of U are the largest of any in North America (Cathcart et al. 1984).

Uranium is recovered as a by-product of phosphoric acid manufacture. Bedded phosphorites occur at Montpelier, Idaho, USA and in North Africa and the Middle East. Uranium occurs in these rocks in cryptocrystalline fluor-carbonate apatite in which uranium substitutes for Ca. There is no alteration, other than weathering, of the deposits and uranium contents are generally negatively correlated with carbonate. One of the most notable features of the deposits is their extent of (up to several 1000s of km^2), but they are generally of low grade <20-300 ppm; and uranium is recovered mainly as a by-product of phosphoric acid production.

The most important deposits represent synsedimentary concentrations of uranium in shallow marine phosphates on the trailing edge of continental passive margins. They have U concentrations up to 6500 ppm averaging 60-200 ppm (OECD/NEA 1990). They are thought to represent sites where there was upwelling of deep marine waters saturated in P with respect to apatite. Slow rates of sedimentation permit precipitation of U from seawater replacing Ca in apatite, and low activities of CO^{2+} allow U to remain in solution (Dahlkamp 1993). Near-shore phosphates and nodular phosphorites contain lower levels of uranium (20-80 ppm). Major Phanerozoic phosphorites formed in warm climates at low latitudes (Cook and McElhinny 1979). According to Cook and Shergold (1986), the development of significant phosphorites in the early Proterozoic coincided with the formation of the first extensive continental plates.

Uraniferous black shale

These deposits which comprise low concentrations of uranium in carbonaceous marine black shales occur mainly in Lower Palaeozoic rocks such as the uraniferous middle Cambrian Alum Shale, Ranstad, Sweden (Bell 1978). Other deposits include the Devonian-Carboniferous Chatanooga Shale, USA (Mutschler et al. 1976), the Chanziping deposit in the Guanxi region, China (Simpson and Yu 1989). The shales typically contain very high contents of organic matter with pyrite/marcastic and thin phosphatic intercalations. Such deposits can extend over 100s to 10,000s of km^2 and are thought to have formed by adsorption of uranium onto organic matter or phosphatic nodules from seawater. Uranium is adsorbed on organic and clay particles with concentrations ranging from 50 to 400 ppm U; they constitute only a potential resource perhaps as a by-product of oil or other metal (Co, Cr, Mo, Mn, REE, V and P) recovery.

CONCLUSIONS

The age of the Earth is generally accepted to be about 4,500 Ma. Initially, heat production from the long-lived isotopes of U, Th and K was about five times greater than at the present day, and the first lithosphere was probably thin, unstable and easily incorporated into the mantle. Early Archaean rocks have less U, Th and K than younger chemically equivalent rocks and no economic U deposits are known from the Archaean.

Marked changes in continental structure and geological processes occurred diachronously across the Archaean-Proterozoic boundary, coinciding with the assembly of the first supercontinent between 2,600 and 2,400 Ma ago by processes comparable to modern plate tectonics. These processes culminated in continent-continent collision followed locally by extensional collapse and erosion of overthickened crust in a regime of intense tectono-magmatic activity. Over the same interval of time there was a marked change in the composition of granitoids from Na rich and light REE enriched, to more potassic sub-alkaline to alkaline and enriched in U and other incompatible elements.

The initiation of plate tectonics began the process whereby radioelements were progressively enriched in continental crust. The first stage of the process occurs when basalt is fractionated from decompressed mantle at constructive plate margins, such as the present day mid-Atlantic ridge. Radioelements are further enriched in the lower crust and hydrated mantle wedge above subduction zones, although felsic magmas, which include the tonalite-trondhjemite granitoid suite, most calc-alkaline granites, and andesites generated in such hydrous conditions above subduction zones contain only low to average levels of radioelements. Further enrichment of radioelements occurs as a result of the assimilation-fractionation-crystallization of late- to post-tectonic or anorogenic felsic magmas. Such magmas are generated in anhydrous conditions during extensional collapse of overthickened crust in orogenic belts or as a result of extension related to later mantle plumes impinging on mature subcontinental lithosphere. The development of plume activity reflects the accumulation of heat beneath radiothermal continental crust, especially beneath supercontinents.

Uranium provinces mainly reflect the presence of highly fractionated anhydrous late- to post-orogenic or anorogenic sub- to peralkaline felsic intrusions or volcanic rocks in the crust. Radiothermal granites are capable of reinitiating convective fluid flow long after magmatic cooling has occurred and erosion and unroofing have taken place. Their presence explains most uranium provinces. Intrusive, metasomatite, volcanic-associated and several vein-type uranium deposits are related to high-to low-temperature fluid-rock interaction during emplacement of such granites. Some of the uranium deposits may reflect later hydrothermal activity associated with the HHP granites, reflected by disturbances to isotopic systems. Breccia complex uranium deposits are also related to radiothermal A-type granites in basement rocks underlying felsic volcanic sequences associated with regional, extensional fracture systems.

There is a continuum between uranium deposits associated with igneous rocks and deposits formed in late- to post-orogenic continental sedimentary basins with predominantly clastic fill. Hence, the early Proterozoic quartz-pebble conglomerate deposits reflect rapid erosion of collisional orogens containing highly radioactive late- to post-tectonic granites, that had been hydrothermally altered. Geothermal activity associated with hot springs and intense microbiological activity in foreland basins, such as the Witwatersrand, further enriched U, with remobilization during later tectonic events. The presence of such sequences in the geological record indicates only that conditions of diagenesis were reducing, which is commonly the case at the present day and cannot be used to provide evidence on the composition of the ancient atmosphere. Unconformity-related uranium deposits may reflect the presence of HHP granites in the basement. The uranium in sandstone deposits is also derived from acid igneous rocks. Volcanic uranium deposits generally occur in calderas and distally in fault-bounded terrestrial basins containing felsic volcaniclastic beds enriched in uranium. Uranium deposits in the extensional Basin and Range Province in the SE USA are associated with acid igneous activity.

Uranium deposits also reflect changes in the biosphere and to some extent the hydrosphere and atmosphere through time. For example, the organic material with which U ore is frequently associated in sandstone-type deposits changed from marine algal sapropelic debris in the Proterozoic Oklo and Gabon deposits to humic detritus from land plants in deposits formed after the Devonian such as those of the western USA. Uranium deposits associated with lignite and coal also occur only in post-Devonian sediments.

In the oceans and on the continental shelves the development of the first economically significant phosphorite deposits followed the initiation of modern-style plate tectonics in the Lower Proterozoic. The main period of uranium enrichment of black shales occurred in the Palaeozoic when marine black shales were deposited in shallow partially closed epicontinental basins which remained tectonically stable over significant periods of time reflected by low sedimentation rates, strongly reducing anaerobic conditions and abundant sapropelic or humic detritus.

The nature of uranium deposits thus records changes in the Earth's tectono-magmatic evolution through time, partly as a result of declining heat generation from U, Th and K, and in the case of sediment-associated deposits as a result of interactions with the evolving biosphere, hydrosphere and atmosphere. The largest, most enriched uranium deposits, which are of unconformity type, are preserved in relatively poorly consolidated sedimentary basin sequences on cratons formed following the large-scale transfer of radioelements from the mantle to the crust, leaving behind a refractory keel of relatively stable sub-continental lithosphere.

ACKNOWLEDGMENTS

The authors thank Drs. M.J Atherton, C.J. Morrisey and P.M. Stone for their comments on the manuscript. They especially thank Warren I. Finch for his many helpful comments on an early draft of the chapter. The chapter is published by permission of the Director of BGS (NERC).

REFERENCES

Adams SS, Smith RB 1981 Geology and recognition criteria for sandstone uranium deposits in mixed fluvial shallow marine sedimentary sequences, South Texas-final report. U.S. Department of Energy, Grand Junction Office, Colorado, Bendix Field Engineering Corporation, GJBX-4 (81) 146 p

Alderton DHM, Pearce JA, Potts RJ 1980 Rare earth element mobility during granite alteration: evidence from southwest England. Earth Planet Sci Lett 49:149-165

Atherton MP, McCourt WJ, Sanderson LM, Taylor WP 1979 The geochemical character of the segmented Peruvian Coastal Batholith and associated volcanics. *In* Origin of Granite Batholiths. Atherton MP, Tarney J (Eds) p 45-64

Atherton MP, Plant JA 1985 High heat production granites and the evolution of the Andean and Caledonian continental margins. *In* High Heat Production (HHP) Granites, Hydrothermal Circulation and Ore Genesis. Inst Min Metall, London, p 263-285

Atkin BP 1979 Oxide-sulphide-silicate mineral phase relations in some thermally metamorphosed granitic pelites from Co. Donegal, Republic of Ireland. *In* The Caledonides of the British Isles—Reviewed. Harris AL, Holland CH, Leake BE (Eds) Spec Publ Geol Soc London 8:375-386

Bagby WC 1986 Descriptive mode of volcanic U. *In* Mineral Deposit Models, Cox DP, Singer DA (Eds) U S Geol Surv Bull 1693, 162 p

Bailey JC, Rose-Hansen J, Lovborg L, Sorensen H 1981 Evolution of Th and U whole-rock contents in the Ilimaussaq intrusion. Rapp Gronlands Geol Unders 103, 87 p

Barbier J 1974 Continental weathering as a possible origin of vein-type uranium deposits. Mineral Deposita 9:271-288

Barley ME, Groves DI 1992 Supercontinent cycles and the distribution of metal deposits through time. Geology 20:291-294

Barnes HL 1979 Solubilities of ore minerals. *In* Geochemistry of Hydrothermal Ore Deposits. Barnes HL (Ed) Wiley, New York, p 334-381

Barnicoat IC, Henderson IHC, Knipe RJ, Yardley BWD, Napier RW, Fox NPC, Kenyon AK, Muntingh DJ, Strydom D, Winkler KS, Lawrence SR, Cornford C 1997 Hydrothermal gold mineralization in the Witwatersrand basin. Nature 386:820-824

Barthel F, Dahlkamp FJ, Fuchs H, Gatzweiler R 1986 Kernenergierohstoffe. *In* Angewandte Geowissenschaften. Bender F (Ed) Ferd Enke Verlag, Stuttgart, p 268-298

Bartholomew DS, Tarney J 1984 Geochemical characteristics of magmatism in the southern Andes (45-60°S). *In* Andean Magmatism, Chemical and isotopic constraints. Harmon RS, Barreiro BA (Eds) Shiva Publications, Orpington, Kent, UK, p 220-250

Bartlett RW 1969 Magma convection, temperature distribution and differentiation. Am J Sci 267:1067-1082

Barton ES, Compston W, Williams IS, Bristom JW, Hallbauer DK, Smith CB 1989 Provenance ages for the Witwatersrand Supergroup and the Ventersdorp Contact Reef: constraints from ion microprobe U-Pb ages of detrital zircons. Econ Geol 84:2012-2019

Batashov BG, Onoprienko ME, Kudlaev AR 1984 The Zheltorechenskoe uranium deposit. 27th Int'l Geol Congr Abstracts Vol IX, Part 1, Moscow, p 330-332

Battey GC, Miezitis Y, McKay AD 1987 Australian Uranium Resources. Bureau of Mineral Resources, Geology and Geophysics, Canberra, Resource Report 1, 69 p

Bayer KC 1983 Generalized structural, lithologic, and physiographic provinces in the fold and thrust belts of the United States (exclusive of Alaska and Hawaii). U S Geol Surv Special Map, scale 1:2,500,000

Bell RT 1978 Uranium in black shales: a review. *In* Short Course in Uranium Deposits: Their Mineralogy and Origin. Kimberley MM (Ed) Mineral Assoc Can, Short Course Handbook 3:207-329

Bell K 1985 Geochronology of the Carswell area, Northern Saskatchewan. *In* The Carswell Structure Uranium Deposits, Saskatchewan. Laine R, Alonso D, Svab M (Eds) Geol Assoc Can Spec Paper 29:33-46

Berning J 1986 The Rössing uranium deposit, South West Africa/Namibia. *In* Mineral Deposits of Southern Africa. Anhaeusser CR, Maske S (Eds) Geol Soc S Africa 2:1819-1832

Beus AA, Grigorian SV 1977 Geochemical Exploration Methods for Mineral Deposits. Applied Publishing Ltd, Illinois, USA, 287 p

Binns RA, McAndrew J, Sun SS 1980 Origin of uranium mineralization at Jabiluka. *In* Uranium in the Pine Creek Geosyncline. Ferguson J, Goleby AB (Eds) Proc Int'l Uranium Symp on the Pine Creek Geosyncline, Sydney, Australia 1979. IAEA, Vienna, p 543-562

Bowden P, Kinnaird JA 1984a Geology and mineralization of the Nigerian Anorogenic Ring Complexes, with a geological map at the scale 1:500 000, Geol Jb Hannover B56:3-65

Bowden P, Kinnaird JA 1984b Petrological and geochemical criteria for the identification of (potential) ore bearing Nigerian granitoids. Proc 27th Int'l Geol Congress, Moscow, VNU Science Press, Utrecht, Netherlands, 9:85-119

Bowers TS, Helgeson HC 1983 Calculations of the thermodynamic and geochemical consequences on non-ideal mixing in the system H_2O-CO_2-NaCl on phase relations in geologic systems. Equations of state for H_2O-CO_2-NaCl fluids at high pressures and temperatures. Geochim Cosmochim Acta 47:1247-1275

Boyle DR 1984 The genesis of surficial uranium deposits. *In* Surficial Uranium Deposits. TECDOC-322. IAEA, Vienna, p 45-52

Brace WF, Walsh JB, Frangos WT 1968 Permeability of granite under high pressure. J Geophys Res 73:2225-2236

Bray CJ, Spooner ETC, Hall CM, York D, Bills TM, Krueger HW 1987 Laser probe $^{40}Ar/^{39}Ar$ and conventional K/Ar dating of illites associated with the McClean unconformity-related uranium deposits, north Saskatchewan, Canada. Can J Earth Sci 24:10-23

Brown PRL, Ellis AJ 1970 The Ohoki-Broadlands hydrothermal area, New Zealand: mineralogy and related geochemistry. Am J Sci 269:97-113

Brown GC, Fyfe WS 1970 The production of granite melts during ultra-metamorphism. Contrib Mineral Petrol 28:310-318

Brown GC, Mussett AE 1993 The Inaccessible Earth, An Integrated View of Its Structure and Composition. 2nd Edn. Chapman & Hall, London, 276 p

Brown GC, Ixer RA, Plant JA, Webb PC 1987 The geochemistry of granites beneath the North Pennines and their role in orefield mineralisation. Trans Inst Min Metall, Sect B, Appl Earth Sci 96:65-76

Brynard HJ, Andreoli MAG 1988 The overview of the regional, geological and structural setting of the uraniferous granites of the Damara Orogen, Namibia. *In* Recognition of Uranium Provinces. IAEA, Vienna, p 195-212

Buck SG, Minter WEL 1987 Paleocurrent and lithological facies control of uranium and gold mineralisation in the Witwatersrand Carbon Leader Placer, Carletonville Goldfield, South Africa. *In* Uranium Deposits in Proterozoic Quartz-Pebble Conglomerates. IAEA-TECDOC-427. IAEA, Vienna, 335-353

Burke K, Dewey JF, Kidd WSF 1976 Dominance of horizontal movements, arc and microcontinental collisions during the later permobile regime. *In* The Early History of the Earth. Windley BF (Ed) Wiley-Interscience, London, p 113-129

Burke K, Kidd WSF, Kusky T 1985 Is the Ventersdorp rift system of southern Africa related to a continental collision between the Kaapvaal and Zimbabwe cratons at 2.64 Ga ago? Tectonophysics 115:1-24

Burke K, Kidd WSF, Kusky T 1986 Archean foreland basin tectonics in the Witwatersrand, South Africa. Tectonics 5:439-456

Butt CRM, Mann AW, Horwitz RC 1984 Regional setting, distribution and genesis of surficial uranium deposits in calcretes and associated sediments in Western Australia. *In* Surficial Uranium Deposits. TECDOC-322. IAEA, Vienna, p 121-127

Cahen L 1970 Etat actuel de la geochronologie du Katangien, Mus Royal Afrique Centre, Annales, Ser 80, Sci Geol 65:7-14

Cahen L, Francois A, Ledent D 1971 Sur l'age des uraninites de Kambove ouest et de Kamoto principal et revision des connaissances relatives aux mineralisations uraniferes du Katanga et du Copperbelt de Zambia. Soc Geol Belgique Annales 94:185-198

Cathcart JB, Sheldon RP, Gulbrandsen RA 1984 Phosphate rock resources of the United States U S Geol Surv Circ 888, 48 p

Cameron EM (Ed) 1983 Uranium exploration in Athabasca Basin. Geological Survey of Canada Paper 82-11, 310 p

Campbell IH, Griffiths RW, Hill RI 1989 Melting in an Archaean mantle plume: heads it's baskets, tails its komatiites. Nature 339:697-699

Cardenas-Flores D 1985 Volcanic stratigraphy and U-Mo mineralization of the Sierra de Pena Blanca district, Chihuahua, Mexico. *In* Uranium Deposits in Volcanic Rocks. Proc Tech Comm Mtg in Uranium in Volcanic Rocks. IAEA, El Paso, Texas, 2-5 April 1984, IAEA, Vienna, Austria, IAEA-TC-490/31:125-136

Cavaney RJ 1984 Lake Maitland uranium deposit. *In* Surficial Uranium Deposits, TECDOC-322, IAEA, Vienna, p 137-140

CEA (Commissariat a l'Energie Atomic) 1969 The French Uranium Mining Industry. CEA, Paris, 123p

Chappell BW, Stephens WE 1988 Origin of infracrustal (I-type) granite magmas. Trans Royal Soc Edinburgh, Earth Sci 79:71-86

Chappell BW, White AJR 1974 Two contrasting granite types. Pacific Geology 8:173-174

Chatterjee AK, Strong DF 1984 Rare earth and other element variations in greisens and granites associated with the East Kemptville tin deposit, Nova Scotia, Canada. Trans Inst Min Metall, Sect B, Appl Earth Sci 93:59-70

Chen Zhaobo, Fang Xiheng 1985 Main characteristics and genesis of Phanerozoic vein-type uranium deposits, *In* Uranium Deposits in Volcanic Rocks. IAEA, Vienna, p 69-82

Chenoweth WL 1991 A summary of uranium production in Wyoming. *In* Mineral Resources of Wyoming. Frost BR, Roberts S (Eds) Wyoming Geol Assoc 42nd Field Conf Guidebook, Laramie, Wyoming, September 14-18, 1991, p 169-179

Christensen UR 1989 Models of mantle convection: one or several layers. Phil Trans Royal Soc London 328:417-424

Clark KF, Foster CT, Damon PE 1982 Cenozoic mineral deposits and subduction-related magmatic arcs in Mexico. Geol Soc Am Bull 93:533-544

Clendenin CW, Charlesworth EG, Maske S 1987 Tectonic style and mechanism of Early Proterozoic successor basin development, southern Africa. Econ Geol Res Unit, Univ Witwatersrand, Johannesburg, Information Circ 197, 26 p

Clendenin CW, Charlesworth EG, Maske S 1988 An early Proterozoic three-stage rift system, South Africa. Tectonophysics 145:73-86

Cloud PE 1968 Atmospheric and hydrospheric evolution in the primitive earth. Science 160:729-736

Coleman ML 1979 Isotopic analysis of trace sulphur from some 'S' and 'I' type granites: heredity or environment. *In* Origin of Granite Batholiths. Atherton MP, Tarney J (Eds) Shiva Publications, Orpington, Kent, UK, p 129-135

Collins PLF 1981 The geology and genesis of the Cleveland tin deposit, Western Tasmania: fluid inclusion and stable isotope studies. Econ Geol 76:365-392

Collins WJ, Beams SD, White AJR, Chappell BW 1982 Nature and origin of A-type granites with particular reference to southeastern Australia. Contrib Mineral Petrol 80:189-220

Collot B 1981 Le granite albitique hyperalcalin de Bokan Mountain (SE Alaska) et ses mineralisations U-Th. Sa place dans la Cordillere Canadienne, 3° cycle thesis, USTL, Montpellier, 238 p

Condie KC 1989 Geochemical changes in basalts and andesites across the Archean-Proterozoic boundary: identification and significance. Lithos 23:1-18

Cook PJ, McElhinny MW 1979 A re-evaluation of the spatial and temporal distribution of sedimentary phosphate deposits in the light of plate tectonics. Econ Geol 74:315-330

Cook PJ, Shergold JH (Eds) 1986 Phosphate Deposits of the World, Vol. 1—Proterozoic and Cambrian Phosphorites. Cambridge Univ Press, Cambridge, UK

Coward MP, Spencer RM, Spencer CE 1995 Early Precambrian processes. Geol Soc London, Spec Publ 95:243-269

Cox DP, Singer DA 1986 Mineral Deposit Models. U S Geol Surv Bull 1693

Creaser RA 1989 The Geology and Petrology of Middle Proterozoic Felsic Magmatism of the Stuart Shelf, South Australia. PhD Thesis, Latrobe Univ, Melbourne, Australia, 434 p

Criss RE, Taylor HP, Jr 1986 Meteoric-hydrothermal systems. Rev Mineral 16:373-424

Cumming GL, Krstic D 1992 The age of unconformity-related uranium mineralization in the Athabasca Basin, northern Saskatchewan, Can J Earth Sci 29:1623-1639

Cuney M 1980 Preliminary results on the petrology and fluid inclusions of the Rössing uraniferous alaskites. Trans Geol S Africa 83:39-45

Cunningham CG, Ludwig KR, Naeser CW, Weiland EK, Mehnert HH, Steven TA, Rasmussen JD 1982 Geochronology of hydrothermal uranium deposits and associated igneous rocks in the eastern source area of the Mount Belknap Volcanics, Marysvale, Utah. Econ Geol 77:453-463

Curtis CD 1977 Sedimentary geochemistry: environments and processes dominated by involvement of an aqueous phase. Phil Trans Royal Soc London A286:353-372

Dahlkamp FJ 1978 Geologic appraisal of the Key Lake, U-Ni deposits, northern Saskatchewan, Econ Geol 73:1430-1449

Dahlkamp FJ 1982 Genesis of uranium deposits. An appraisal of the present state of knowledge. In Ore Genesis: The State of the Art. Amstutz GG, El Goresy A, Frenzel G, Kluth C, Moh G, Wauschkuhn A, Zimmermann RA (Eds) Springer, Berlin, p 644-654

Dahlkamp FJ 1993 Uranium Ore Deposits. Springer-Verlag, Berlin, 460 p

Dahlkamp FJ, Scivetti N 1981 IUREP/International uranium resources evaluation project: Austria, OECD/NEA, Paris, 87 p

Daniels JL 1975 Palaeogeographic development of Western Australia-Precambrian. In Geology of Western Australia. Memoir No. 2, Geol Surv Western Australia, p 437-445

Dall'Aglio M 1972 Planning and interpretation criteria in hydrogeochemical prospecting for uranium. In Uranium Prospecting Handbook. Bowie SHU, Davis M, Ostle D (Eds) Inst Mining Metallm, p 121-134

Dardel J, Peinador Fernandes A, Jamet P Le Caignee R, Moreau M, Serrano JR, Ziegler V 1979 Gisements d'uranium dans les schistes peribatholitiques (type iberique) et dans les leucogranites (type Limousin) (Portugal, Espagne, France), livret guide d'excursion, Sci Terre, Nancy XXIII, 47 p

Darnley AG 1981 The relationship between uranium distribution and some major crustal features in Canada. Mineral Mag 44:425-436

Darnley AG, Bjorklund A, Bolviken B, Gustavsson N, Koval PV, Plant JA, Steenfelt A, Tauchid M, Xie Xuejing, Garrett RG, Hall GEM 1995 A global geochemical database for environmental and resource management. Recommendations for International Geochemical Mapping. Final Report IGCP Project 259, 122 p

Davidson CF 1960 The present state of the Witwatersrand controversy. Mining Mag 102:84-95, 149-159, 222-229

Davies SN 1969 Porosity and permeability of natural materials. In Flow Through Porous Metals. DiWiest DSM (Ed), Academic Press, New York, p 54-89

Davies GF 1988 Ocean bathymetry and mantle convection 1. Large-scale flow and hotspots. J Geophys Res 93:467-480

Davies GF, Norry MJ, Gerlach DC, Cliff RA 1989 A combined chemical and Pb-Sr-Nd isotope study of the Azores and Cape Verde hot spots: the geodynamical implications. In Magmatism in the Ocean Basins. Saunders AD, Norry M.J (Eds) Geol Soc London Spec Publ 42:231-255

Dayvault RD, Caster SB, Berry MR 1985 Uranium associated with volcanic rocks of the McDermitt caldera, Nevada and Oregon, In Uranium Deposits in Volcanic Rocks, Proc Tech Comm Mtg in Uranium in Volcanic Rocks. IAEA, El Paso, Texas, 2-5 April 1984, IAEA, Vienna, Austria, IAEA-TC-490/31:379-409

de Fraipont P, Marquaire CH, Sierak JP. Horrenberg JCC, Ruhland M 1982 Réseau de fractures et repartition des corps tectoniques elementaires de fracturation : Recherche sur le controle tectonique de la minéralisation uranifere du Secteur de Pény (La Crouzille, France). *In* Vein-type and Similar Uranium Deposits in Rocks Younger than Proterozoic. IAEA, Vienna, p 87-102

de Magnee I, François A 1985 Genesis of the Kipushi (Cu, Zn, etc...) and the Shinkolobwe (U, Ni, Co) deposits (Shaba, Zaire) in direct relation with Proterozoic salt diapirs (abstr). Fortschr Mineral 63:140

Denson NM, Gill JR 1965 Uranium-bearing lignite and carbonaceous shale in the southwestern part of the Williston Basin. U S Geol Surv Prof Paper 463, 75 p

Derriks JJ, Oosterbosch R 1958 The Swambo and Kalongwe deposits compared to Shinkolobwe: contribution to the study of Katanga uranium. *In* UN, Int'l Conf Peaceful Uses of Atomic Energy, 2nd, Geneva, Proc 2, Survey Raw Material Resources, p 663-695

Derriks JJ, Vaes JF 1956 The Shinkolobwe uranium deposit. Current status of our geologic and metallogenic knowledge. *In* UN, Int'l Conf Peaceful Uses of Atomic Energy, 1st, Geneva, Proc 6, Geology of Uranium and Thorium, p 94-128

Derry DR 1973 Ore deposition and contemporaneous surfaces. Econ Geol 68:1374-1380

de Wit MJ, Roering C, Hart RJ, Armstrong RA, de Ronde CEJ, Green RWE, Tredoux M, Peberdy E, Hart RA 1992 Formation of an Archaean continent. Nature 357:553-562

Dewey J 1988 Extensional collapse of orogens. Tectonics 7:1123-1139

Dickin AP, Exley RA, Smith BM 1980 Isotopic measurement of Sr and O exchange between meteoric-hydrothermal fluid and the Coire Uaigneich granophyre, Isle of Skye, NW Scotland. Earth Planet. Sci Lett 51:58-70

Dickinson WR 1981 Plate tectonic evolution of the southern Cordillera, *In* Relations of tectonics to ore deposits in the southern Cordillera. Dickinson WR, Payne WD (Eds) Arizona Geol Soc Digest 14:113-135

Dimroth, E, Kimberley MM 1976 Precambrian atmospheric oxygen; evidence in sedimentary distributions of carbon, sulfur, uranium and iron. Can J Earth Sci 13:1161-1185

Dostal J, Capedri S 1978 Uranium in metamorphic rocks. Contrib Mineral Petrol 66:409-414

Durrheim RJ, Mooney WD 1991 Archean and Proterozoic crustal evolution: evidence from crustal seismology. Geology 19:606-609

Eby GN 1990 The A-type granitoids: a review of their occurrence and chemical characteristics and speculations on their petrogenesis. Lithos 26:115-134

Eckstrand OR (Ed) 1984 Canadian mineral deposit types: a geological synopsis. Geol Surv Can, Econ Geol Rep 36, 86 p

Eckstrand OR, Sinclair, WD, Thorpe RI 1995 Geology of Canadian Mineral Deposit Types. Geol Surv Can, Geology of Canada 8, 640 p

Edmunds WM, Andrews JN, Burgess WG, Kay LF, Lee DJ 1984 The evolution of saline and thermal groundwaters in the Carnmenellis granite. Mineral Mag 48:407-424

Ellis AJ 1973 IAGC Symp Tokyo 1970. Ingerson E (Ed) 1:1-22

Ellis AJ, Mahon WAJ 1967 Natural hydrothermal systems and experimental hot water/rock interactions. Geochim Cosmochim Acta 31:519-538

Ellis AJ, Mahon WAJ 1977 Chemistry and Geothermal Systems, Academic Press, New York, 392 p

Eugster HP 1984 Granites and hydrothermal ore deposits: a geochemical framework. Mineral Mag 49:7-23

European Commission 1998 Oklo working group. Proc First Joint EC-CEA workshop on the Oklo-natural analogue Phase II project held in Sitges, Spain, 18-20 June 1997. Louvat D, Davies C (Eds) Report, EUR 18314 EN, 328 p

Ewers GR, Ferguson J, Needham RS, Donnelly TH 1984 Pine Creek Geosyncline NT. *In* Proterozoic Unconformity and Stratabound Uranium Deposits. TECDOC-315, IAEA, Vienna, p 135-206

Faure G 1986 Principles of Isotope Geology. 2nd Edn. Wiley, New York, 589

Fehn U, Cathles LM, Holland H D 1978 Hydrothermal convection and uranium deposits in abnormally radioactive plutons. Econ Geol 83:1556-1566

Finch RJ, Suksi J, Rasilainen K, Ewing RC 1995 U-series dating of secondary uranium minerals; the age of oxidative alteration at the Shinkolobwe mine, Shaba, southern Zaire. Geol Assoc Can, Mineral Assoc Can, Can Geophys Soc, Joint Ann Mtg, Abstracts 20:31

Finch WI 1991 Maps showing distribution of uranium deposits in the Colorado Plateau uranium province—A cluster analysis. U S Geol Surv Misc Field Studies Map MF-2080, scale 1:2,500,000

Finch WI 1996 Uranium Provinces of North America—Their Definition, Distribution, and Models. U S Geol Surv Bull 2141, 18 p

Floyd PA 1991 Ocean islands and seamounts. *In* Oceanic Basalts. Floyd PA (Ed) Blackie, Glasgow, p 174-218

Foley S 1991 High pressure stability of the fluor-and hydroxy-end members of pargasite and K-richterite. Geochim Cosmochim Acta 55:2689-2694

Friedrich M 1981 Petrographie et geochimie des granites de St Jouvent et de la Brame—Relations avec la distribution de l'uraninite, Massif de St Sylvestre, France, CREGU (Cent. Rech. Geol Uranium), Vandoeuvre-les-Nancy, Rep 81-2, 50 p

Friedrich M, Cuney M 1989 Uranium enrichment processes in peraluminous magmatism. In Uranium Deposits in Magmatic and Metamorphic Rocks. IAEA, Vienna, p 11-36

Friedrich, M, Cuney M, Poty B 1987 Uranium geochemistry in peraluminous leocogranites. In Concentration mechanisms of uranium in geological environments—A conference report. Uranium 3:353-385

Fyfe WS 1974 Geochemistry. Clarendon Press, Oxford, 103 p

Gama HM 1982 Serra de Jacobina gold-bearing metasedimentary sequence paleoplacer gold deposits. In Excursions Guidebook, Int'l Symp on Archean and Early Proterozoic Geologic Evolution and Metallogenesis. Landim PMB (Ed) p 119-133

Gamble D 1984 The Lake Raeside uranium deposit. In Surficial Uranium Deposits. TECDOC-322, IAEA, Vienna, p 141-148

Gandhi SS 1978 Geological setting and genetic aspects of uranium occurrences in the Kaipokok Bay-Big River area, Labrador. Econ Geol 73:1492-1522

Gandhi SS 1986 Uranium in Early Proterozoic Aillik Group, Labrador. In Uranium Deposits of Canada. Evans EL (Ed) Spec Vol 33:70-82

Gandhi SS, Bell RT 1993 Metallogenic concepts to aid exploration for the giant Olympic Dam-type deposits and their derivatives; In Proc 8th Quadrennial IAGOD Symposium, Ottawa, Canada, August 1990. Int'l Assoc Genesis of Ore Deposits, Maurice YT (Ed) E Schweizerbart'sche Verlagsbuchhandlung, Stuttgart, p 787-802

Gandhi SS, Bell RT 1995 Volcanic-associated Uranium. In Geology of Canadian Mineral Deposit Types. Eckstrand OR, Sinclair WD, Thorpe RI (Eds) Geol Surv Can, Geology of Canada 8:269-276

Garrels RM, Christ CL 1965 Solutions, Minerals and Equilibria. Harper and Row, New York NY, 450 p

Garson MS, Young B, Mitchell AHG, Tait BAR 1975 The geology of the tin belt in peninsular Thailand around Phuket, Phangnga and Takua Pa. IGS Overseas Memoir 1

Gast PW 1968 Upper mantle chemistry and evolution of the Earth's crust. Princeton Univ Press, Princeton, NJ, p 15-27

Gauthier-Lafaye F 1997 The last natural nuclear fission reactor. Nature 387:337

Goodell PC 1981 Geology of the Pena Blanca uranium deposits, Chihuahua, Mexico. In Uranium in volcanic and volcaniclastic rocks. Goodell PC, Waters AC (Eds) Am Assoc Petrol Geol, Studies in Geology 13:275-291

Goodell PC 1985 Chihuahua City uranium province, Chihuahua, Mexico. In Uranium deposits in volcanic rocks. Proc Tech Comm Mtg, El Paso, Texas, 2-5 April 1984, IAEA, Vienna, IAEA-TC-490/19:97-124

Gordon M, Tracey JI, Ellis MW 1958 Geology of the Arkansas Bauxite Region. U S Geol Surv Prof Paper 299

Govett GJS, Atherden PR 1988 Applications of rock geochemistry to productive plutons and volcanic sequences. J Geochem Expl 30:223-242

Gower CF, Ryan B, Rivers T 1990 Mid-Proterozoic Laurentia-Baltica: an overview of its geological evolution and a summary of contributions made in this volume. In Mid-Proterozoic Laurentia-Baltica. Gower CF, Ryan B, Rivers T (Eds) Geol Assoc Can Spec Paper 32:1-20

Granger HC, Finch WI 1988 The Colorado Plateau Uranium Province USA. In Recognition of Uranium Provinces, Proc Tech Comm Mtg on Recognition of Uranium Provinces, London, England, 18-20 September 1985. IAEA, Vienna, p 157-193

Gries RR 1983 North-south compression of Rocky Mountain foreland structure. In Rocky Mountain foreland basin and uplifts. Lowell JD (Ed) Rocky Mountain Assoc Geologists 1983 p 9-32

Gurnis M 1986 Large-scale mantle convection and the aggregation and dispersal of supercontinents. Nature 322:695-9

Gustafsson B 1981 Uranium exploration in the N. Vasterbotten-S. Norrbotten province, northern Sweden. In Uranium Exploration Case Histories, IAEA, Vienna, p 333-352

Haapala I, Kinnuen K 1979 Fluid inclusions in cassiterite and beryl in greisen veins in the Evrajoki stock, southwestern Finland. Econ Geol 74:1231-1238

Halenius U, Smellie JAT, Wilson MR 1986 Uranium genesis within the Arjeplog-Arvidsjaur-Sorsele uranium province, northern Sweden. In Vein-Type Uranium Deposits. IAEA, Vienna, TECDOC-361:21-42

Hallbauer DK 1975 The plant origin of the Witwatersrand "carbon." Minerals Sci Eng 7:111-131

Harshman EN, Adams SS 1981 Geology and recognition criteria for roll-type uranium deposits in continental sandstones, US-DOE, GJBX-1 81, 185 p

Hegge MR, Rowntree JC 1978 Geologic setting and concepts on the origin of uranium deposits in the East Alligator River region, NT, Australia. Econ Geol 73:1420-1429

Heier KS 1979 The movement of uranium during higher grade metamorphic processes. Phil Trans Royal Soc London A291:413-421

Heine TH 1986 The geology of the Rabbit Lake uranium deposit, Saskatchewan, *In* Uranium Deposits of Canada. Evans EL (Ed) CIM Spec Vol 33:134-143

Hemley JJ, Jones WR 1964 Chemical aspects of hydrothermal alteration with emphasis on hydrogen metasomatism. Econ Geol 59:538-569

Henney PJ, Jones CE, Plant JA, Tarney J 1994 Cairngorm. *In* Granites, metallogeny, lineaments and rock-fluid interactions. Plant J.A. and 24 others, British Geol Surv Res Report SP/94/1, 266 p

Hildebrand RS, Hoffman PF, Bowring SA 1987 Tectono-magmatic evolution of the 1.9Ga Great Bear magmatic zone, Wopmay orogen, northwestern Canada. J Volcan Geotherm Res 32:99-118

Hildreth W, Moorbath S 1988 Crustal Contributions to ARC Magmatism in the Andes of Central Chile. Contrib Mineral Petrol 98:455-489

Hitzman MW, Oreskes N, Einaudi MT 1992 Geological characteristics and tectonic setting of Proterozoic iron oxide (Cu-U-Au-REE) deposits. Precamb Res 58:241-287

Hoeve J, Sibbald TII, Ramaekers P, Lewry JF 1980 Athabasca Basin unconformity-type uranium deposits: a special class of sandstone-type deposits. *In* Uranium in the Pine Creek Geosyncline. Ferguson J, Goleby AB (Eds) IAEA, Vienna, p 575-594

Hoffman PF 1980 Wopmay Orogen: a Wilson cycle of Early Proterozoic age in the northwest of the Canadian Shield. *In* Continental Crust and its Mineral Deposits. Strangway DW (Ed) Geol Assoc Can Spec Paper 20:523-549

Hoffman PF 1988 United plates of America, the birth of a craton; Early Proterozoic assembly and growth of Laurentia. Ann Rev Earth Planet Sci 16:543-603

Hoffman PF 1991 Did the breakout of Laurentia turn Gondwanaland inside-out? Science 252:1409-1412

Holland HD 1962 Model for the evolution of the Earth's atmosphere. *In* Petrologic Studies; a Volume to Honour AF Buddington. Geol Soc Am, p 447-477

IAEA 1975a Proceedings of a Symposium on the Oklo Phenomena. IAEA Technical Report, STI/PUB/405. IAEA, Vienna

IAEA 1975b Proceedings of the Technical Committee Meeting on Natural Fission Reactors. IAEA Technical Report, STI/PUB/475. IAEA, Vienna

IAEA 1996 Guidebook to Accompany IAEA Map of World Distribution of Uranium Deposits. IAEA, Vienna

Jackson NJ, Ramsay CR 1986 Post-orogenic felsic plutonism, mineralization and chemical specialization in the Arabian Shield. Trans Inst Min Metall; Sect B, Appl Earth Sci 95:83-93

Jakubick AT, Church W 1986 Oklo natural reactors: geological and geochemical conditions—A review, Atomic Energy Board of Canada, Research Report, INFO-0179, Ottawa, Canada

Johnson JP, Cross KC 1991 Geochronological and Sm-Nd constraints on the genesis of the Olympic Dam Cu-U-Au-Ag deposit, South Australia. *In* Source, Transport and Deposition of Metals. Pagel M, Leroy JL (Eds) Proc 25 Years' SGA Anniv Mtg, Nancy, Aug 30-Sept 3, 1991, AA Balkema, Rotterdam, p 395-400

Jones CH, Wernicke BP, Farmer GL, Walker JD, Coleman DS, McKenna LW Perry FV 1992 Variations across and along a major continental rift: an interdisciplinary study of the Basin and Range Province, western USA. Tectonophysics 213:57-96

Kaye GWC, Laby TH 1986 Tables of Physical and Chemical Constants and Some Mathematical Functions, 15th Edn. Longman Group, London, Bath Press, Avon, UK, 447 p

Kazansky VI, Krupenikov VA, Omel'Yanenko BI, Pruss AK 1968 Structural and petrological conditions of the formation of uraniferous albitites. Geol Rudn Mestorozhd, Dokl Akad SSSR 10:3-16 (in Russian)

Kazansky VI, Laverov NP 1977 Deposits of uranium. *In* Ore Deposits of the USSR. Smirnov VI (Ed) Vol. II, Pitman Publ, London, San Francisco, Melbourne, 349-424

Kelly WC, Rye RD 1979 Geologic fluid inclusions and stable isotope studies of the tin-tungsten deposits of Panasqueira, Portugal. Econ Geol 74:1721-1819

Kerr RA 1989 Another movement in the dance of the plates. Science 244:529-30

Kissin SA 1988 Nickel-cobalt-native silver (five element) veins: a rift-related ore type. *In* Proc Volume, North American Conference on Tectonic Control of Ore Deposits and the Vertical and Horizontal Extent of Ore Systems, Kisvarsanyi G, Grant SK (Eds) Univ Missouri Press, Rolla, p 268-279

Knipping HD 1974 The concepts of supergene versus hypogene emplacement of uranium at Rabbit Lake, Saskatchewan, Canada. *In* Formation of Uranium Ore Deposits. IAEA, Vienna, p 531-548

Kotzer TG, Kyser TK 1990 The use of stable and radiogenic isotopes associated with unconformity-type uranium deposits. *In* Modern Exploration Techniques, Beck LS, Sibbald TII (Eds) Saskatchewan Geol Soc Spec Publ 9:115-131

Kotzer TG, Kyser TK 1993 O, U, and Pb isotopic and chemical variations in uraninite: Implications for determining the temporal and fluid history of ancient terrains. Am Mineral 78:1262-1274

Kotzer TG, Kyser TK 1992 Isotopic, mineralogic and chemical evidence for multiple episodes of fluid movement during prograde and retrograde diagenesis in a Proterozoic basin. In Water-Rock Interaction. Kharaka Y, Maest N (Eds) Balkema, Rotterdam, p 1177-1181

Kunzendorf H, Nyegaard P, Nielsen BL 1982 Distribution of characteristic elements in the radioactive rocks of the northern part of Kvanefjeld, Ilimaussaq Intrusion, South Greenland. Geol Surv Greenland, Copenhagen Rep 109:1-32

Kwak TAP, Askins PW 1981 Geology of the Fe-Sn-W (-Be-Zn) Skarn (Wrigglite) at Moina, Tasmania. Econ Geol 76: 439-467

Kyser TK, Kotzer TG, Wilson MR 1989 Isotopic constraints on the genesis of unconformity-type uranium deposits: Geol Assoc Can-Mineral Assoc Can Abstracts 14:A120

Langmuir D 1978 Uranium solution-mineral equilibria at low temperatures with applications to sedimentary ore deposits. Geochim Cosmochim Acta 42:547-569

Landais P, Duessy J, Robb LJ, Nouel C 1990 Preliminary chemical analyses and Raman spectroscopy on selected samples of Witwatersrand kerogen. Econ Geol Res Unit, Univ Witwatersrand, Johannesburg, Information Circ 222, 8 p

Leroy J 1984 Episyenitization dans le gisement d'uranium du Bernardan (Marche). Comparaison avec des gisements similaires du Nord Ouest du Massif Central francais, Mineralum Deposita 19:26-35

Liebenberg WR 1958 The mode of occurrence and origin of gold and radioactive minerals in the Witwatersrand System, the Dominion reef, the Ventersdorp contact reef and the Black reef. Geol Soc S Africa Trans 58:101-254

Lindroos H, Smellie JAT 1979 A stratabound uranium occurrence within Middle Precambrian ignimbrites at Duobblon, northern Sweden. Econ Geol 74:1118-1130

Lister CJ 1979 Quartz-cored tourmaline from Cape Cornwall and other localities. Proc Ussher Soc 4:402-418

Locardi E 1985 Uranium in acidic volcanic environments. In Uranium Deposits in Volcanic Rocks. IAEA, Vienna, 17-28

Ludwig KR, Simmons KR 1992 U-Pb dating of uranium deposits in collapse breccia pipes of the Grand Canyon region. Econ Geol 87:1747-1765

Macdonald CC 1985 Mineralogy and geochemistry of the sub-Athabasca regolith near Wollaston Lake. In Geology of Uranium Deposits. Sibbald TII, Petruk W (Eds) CIM Spec Vol 32:155-163

McDonough WF, Sun SS 1995 The composition of the Earth. Chem.Geol 120:223-253

Maissoneuve J 1982 The composition of Precambrian ocean waters. Sed Geol 31:1-11

Malavieille J 1993 Late orogenic extension in mountain belts: insights from the Basin and Range and the late Paleozoic Variscan belt. Tectonics 12:1115-1130

Martin H 1986 Effect of steeper Archean geothermal gradient on geochemistry of subduction-zone magmas. Geology 14:753-756

Maucher A 1962 Die Lagerstatten des Urans. Vieweg Verlag, Braunschweig, 162 p

Meneghel L 1979 Uranium occurrence in the Katanga System of north-western Zambia, In Uranium deposits in Africa, geology and exploration. IAEA, Vienna, p 97-122

Michard A, Albarede F 1985 Hydrothermal uranium uptake at ridge crests. Nature 317:244-246

Michie U McL, UK, NIREX 1996 Personal communication to JAP

Miller AR, LeCheminant AN 1985 Geology and uranium metallogeny of Proterozoic supracrustal successions central District of Keewatin NWT. with comparisons to northern Sakatchewan. In Geology of Uranium Deposits. Sibbald TII, Petruk W (Eds) CIM Spec Vol 32:167-185

Miller RG 1982 The geochronology of uranium deposits in the Great Bear batholith, Northwest Territories. Can J Earth Sci 19:1428-1448

Miller W, Alexander R, Chapman N, McKinley I, Smellie J 1994 Natural analogue studies in the geological disposal of radioactive wastes. Studies in Environmental Science 57, Amsterdam, Elsevier, 395 p

Minter WEL 1976 Detrital gold, uranium, and pyrite concentrations related to sedimentology in the Precambrian Vaal Reef placer, Witwatersrand, South Africa. Econ Geol 71:157-176

Minter WEL 1981 The distribution and sedimentary arrangement of carbon in South African Proterozoic placer deposits. In Genesis of Uranium- and Gold-Bearing Precambrian Quartz-Pebble Conglomerates. Armstrong FC (Ed) Geol Surv Prof Paper 1161-A-BB, p 1-4

Moreau M 1977 L'uranium et les granitoides: essai d'interpretation. In Geology, Mining and Extractive Processing of Uranium. Jones MJ (Ed) Inst Min Metall London, p 83-102

Morgan WJ 1971 Convection plumes in the lower mantle. Nature 230:42-43

Mossman DJ, Dyer BD 1985 The geochemistry of Witwatersrand-type gold deposits and the possible influence of ancient prokaryotic communities on gold dissolution and precipitation. Precamb Res 30:303-319

Mutschler PH, Hill JJ, Williams BB 1976 Uranium from the Chattanooga Shale. U S Bur Mines Inform Circ 8700, 85 p
Nash JT 1988 Geology and geochemistry of the Pitch uranium mine area, Saguache County, Colorado. U S Geol Surv Prof Paper 1797, 38 p
Nash JT, Granger HC, Adams SS 1981 Geology and concepts of genesis of important types of uranium deposits. Econ Geol 75th Anniversary Volume 63-116
Ngongo-Kashisha 1975 Sur la similitude entre les gisements uraniferes (type Shinkolobwe) et les gisements cupriferes (type Kamoto) au Shaba, Zaire. Ann Soc Geol Belg 98:449-462
Nockolds SR, Allen R 1958 Geochemical exploration. Foreign Literature Publ House, Moscow, 176 p
Morris RC 1980 A textural and mineralogical study of the relationship of iron to banded iron-formation in the Hamersley Iron Province of Western Australia. Econ Geol 75:184-209
OECD Nuclear Energy Agency 1990 Uranium, Resources, Production and Demand 1989, OECD, Paris
OECD Nuclear Energy Agency 1994 Uranium, Resources, Production and Demand 1993, OECD, Paris
OECD Nuclear Energy Agency 1998 Uranium, Resources, Production and Demand 1997, OECD, Paris
O'Brien C, Plant JA, Simpson PR, Tarney J 1985 The Geochemistry, Metasomatism and Petrogenesis of the Lake District Granites. J Geol Soc London 142:1139-1157
O'Brien C 1985 The Petrogenesis and Geochemistry of British Caledonian Granites, with special reference to Mineralized Intrusions. PhD Thesis, Univ Leicester (unpublished)
O'Driscoll EST 1985 The application of lineament tectonics in the discovery of the Olympic Dam Cu-U-Au deposit at Roxby Downs, South Australia. Global Tectonics and Metallogeny 3:43-57
Olade MA 1980 Plate tectonics and metallogeny of intra-continental rifts and aulacogens in Africa—A review. In Proc 5th Quad IAGOD Symp, Schweizerbart, Stuttgart, p 91-111
O'Nions, RK 1987 Relationships between chemical and convective layering in the Earth. J Geol Soc London 144:259-74
O'Nions RK, McKenzie D 1993 Estimates of mantle thorium/uranium ratios from Th, U and Pb isotope abundances in basaltic melts. Phil Trans Royal Soc London A 342:65-77
Oreskes N, Einaudi MT 1990 Origin of REE-enriched hematite breccias at the Olympic Dam Cu-U-Au-Ag deposit, Roxby Downs, South Australia. Econ Geol 85:1-28
Oreskes N, Hitzman MW 1993 A Model for the Origin of Olympic Dam-type Deposits. In Mineral Deposit Modeling. Kirkham RV, Sinclair WD. Thorpe RI, Duke JM (Eds) Geol Assoc Can Spec Paper 40:615-633
Otton JK, Bradbury JP, Forester RM, Hanley JH 1990 Paleontological analysis of a lacustrine carbonaceous uranium deposit in the Anderson mine, Date Creek basin, west-central Arizona (USA). Ore Geology Reviews 5:541-552
Pagel M 1977 Microthermometry and chemical analysis of fluid inclusions from the Rabbit Lake uranium deposit, Sakatchewan, Canada, Abstract, Trans Inst Min Metall Sect B Appl Earth Sci 86:157
Patterson CC 1956 Age of meteorites and the Earth. Geochim Cosmochim Acta 10:230-237
Patterson DJ, Ohmoto H, Soloman M 1981 Geologic setting and genesis of cassiterite-sulphide mineralisation at Renison Bell, Western Tasmania. Econ Geol 76:393-438
Phillips GN, Myers RE, Palmer JA, 1987 Problems with the placer model for Witwatersrand gold. Geology 15:1027-1030
Phillips GN, Myers RE 1989 The Witwatersrand gold fields; Part II, an origin for Witwatersrand gold during metamorphism and associated alteration. In The Geology of Gold Deposits: The Perspective in 1988. Reid R, Keays W, Ramsay RH, Groves DI (Eds) Econ Geol Monogr 6:598-608
Pitcher WS 1978 The anatomy of a batholith J Geol Soc London 135:157-182
Pitcher WS 1979 Comments on the geological environments of granites. In Origin of Granite Batholiths. Atherton MP, Tarney J (Eds) p 1-8
Plant JA 1986 Models for granites and their mineralizing systems in the British and Irish Caledonides. In Geology and Genesis of Mineral Deposits in Ireland. Andrew CJ, Crowe RWA, Finlay S, Pennell WM, Pyne JF (Eds) Irish Assoc Econ Geol 121-156
Plant J A, O'Brien C, Tarney J, Hurdley J 1985 Geochemical criteria for the recognition of High Heat Production Granites. In High Heat Production (HHP) Granites, Hydrothermal Circulation and Ore Genesis. Inst Min Metall, London, p 263-285
Plant JA, Saunders AD 1996 The Radioactive Earth. Radiation Protection Dosimetry 68:25-36
Plant JA, Tarney J 1994 Mineral deposit models and primary rock geochemical characteristics. In Handbook of Exploration Geochemistry, Drainage Geochemistry 6:11-72. Hale M, Plant JA (Eds) Elsevier, Amsterdam
Plant JA, Watson JV, Simpson PR, Green PM, Fowler MB 1983 Metalliferous and mineralised Caledonian granites in relation to regional metamorphism and fracture systems in northern Scotland. Trans Inst Min Metall, Sect B, Appl Earth Sci 92:33-42

Platt JP, England PC 1994 Convective removal of lithosphere beneath mountain belts: thermal and mechanical consequences. Am J Sci 294:307-336
Poty B, Leroy J, Cathelineau M, Cuney M, Friedrich M, Lespinasse M, Turpin L 1986 Uranium deposits spatially related to granites in the French part of the Hercynian Orogen. *In* Vein type Uranium Deposits. TECDOC-361. IAEA, Vienna, p 215-246
Pratt HR, Swolfs HS, Brace WF, Black AD, Hardin JW 1977 Elastic and transport properties of an in situ jointed granite. Int'l J Rock Mechanics Mining Sci 14:35-45
Preto VA 1978 Setting and genesis of uranium mineralization at Rexspar. Can Inst Mining Metall Bull 71:82-88
Pretorius DA 1974 The nature of the Witwatersrand gold-uranium deposits. Econ Geol Res Unit, Univ Witwatersrand, Johannesburg, Information Circular 206, 43 p
Pretorius DA 1975 The depositional environment of the Witwatersrand goldfields: A chronological review of speculations and observations. Mineral Sci Eng 7:18-47
Ramaekers P 1981 Hudsonian and Helikian Basins of the Athabaska region, northern Saskatchewan. *In* Proterozoic basins of Canada. Campbell FHA (Ed) Geol Surv Can Paper 81-10:219-233
Ramdohr P 1974 The uranium and gold deposits of Witwatersrand; Blind River District; Dominion Reef; Serra de Jacobina—Microscopic analyses and a geological comparison. *In* Geochemistry and the Origin of Life. Kvenvolden KA (Ed) Dowden, Hutchison and Ross, Stroudsburg, 14:188-194. [Reprinted from Abhandlungen der Deutschen Akademie der Wissenschaften zu Berlin, Klasse für Geologie, 3 (1958).]
Rankin AH, Alderton DHM 1982 Fluid inclusion evidence for the evolution of hydrothermal fluids in southwest England. J Geol Soc London 141:447-452 (abstr)
Reeve JS, Cross KC, Smith RN, Oreskes N 1990 The Olympic Dam Copper-uranium-gold-silver deposit, South Australia. *In* Geology of Mineral Deposits of Australia and Papua New Guinea. Hughes F, (Ed), Australasian Inst Mining Metall Monograph 14:1009-1035
Reynolds RL, Goldhaber MB 1983 Iron disulfide minerals and the genesis of roll-type uranium deposits. Econ Geol 78:105-120
Ringwood AE, Irifune T 1988 Nature of the 650 km seismic discontinuity: implications for mantle dynamics and differentiation. Nature 331:131-137
Robb LJ, Meyer M 1985 The nature of the Witwatersrand hinterland. Econ Geol Res Unit, Univ Witwatersrand, Johannesburg, Information Circular 175, 25 p
Roberts DE, Hudson GRT 1983 The Olympic Dam Copper-uranium-gold deposit, Roxby Downs, South Australia. Econ Geol 78:99-822
Roedder E 1977 Fluid inclusions as tools of mineral exploration. Econ Geol 72:03-525
Rogers JJW, Adams JAS 1969a Geochemistry of uranium. *In* Handbook of Geochemistry. Wedepohl KH (Ed) 2, 5, Springer-Verlag, Berlin, p 92-E1 to 92-E5
Rogers JJW, Adams JAS 1969b Uranium. *In* Handbook of Geochemistry. Wedepohl KH (Ed) 2, 4, Springer, Berlin, p 92-B to 92-O
Rollinson HR, Fowler MB 1987 The magmatic evolution of the Scourian complex at Gruinard Bay. *In* Evolution of the Lewisian and Comparable High Grade Terrains. Park RG, Tarney J (Eds) Geol Soc London Spec Pub 27:7-71
Romberger SB 1984 Transport and deposition of uranium in hydrothermal systems at temperatures up to 300°C: geological implications. *In* Uranium Geochemistry, Mineralogy, Geology, Exploration and Resources. De Vivo B, Ippolito F, Capaldi G, Simpson PR (Eds) Inst Min Metall, p 12-17
Roscoe SM 1959 On thorium-uranium ratios in conglomerate and associated rocks near Blind River, Ontario. Econ Geol 54:511-512
Roscoe SM 1969 Huronian rocks and uraniferous conglomerates in the Canadian Shield. Geol Surv Can Paper 68-40, 205 p
Roscoe SM, Minter WEL 1993 Pyritic Paleoplacer Gold and Uranium Deposits. *In* Mineral Deposit Modeling, Kirkham RV, Sinclair WD. Thorpe RI, Duke JM (Eds) Geol Assoc Can Spec Paper 40:103-124
Rundle CC, Snelling NJ 1977 The geochronology of uraniferous minerals in the Witwatersrand Triad; an interpretation of new and existing U-Pb age data on rocks and minerals from the Dominion Reef, Witwatersrand and Ventersdorp Supergroups. Phil Trans Royal Soc London A286:567-583
Ruzicka V 1971 Geological comparison between East European and Canadian uranium deposits. Geol Surv. Can Paper 70-48 196 p
Ruzicka V 1989 Conceptual genetic models for important types of uranium deposits and areas favourable for their occurrence in Canada. *In* Uranium Resources and Geology of North America. TECDOC-500. IAEA, Vienna, p 49-79
Ruzicka V 1993 Unconformity-type uranium deposits. *In* Mineral Deposit Modelling, Kirkham RV, Sinclair WD, Thorpe RI, Duke JM (Eds) Geol Assoc Can Spec Paper 40:125-149

Ruzicka V 1995a Unconformity-Associated Uranium. *In* Geology of Canadian Mineral Deposit Types. Eckstrand OR, Sinclair, WD, Thorpe RI (Eds) Geol Surv Can, Geology of Canada 8:197-210

Ruzicka V 1995b Vein Uranium. *In* Geology of Canadian Mineral Deposit Types. Eckstrand OR, Sinclair, WD, Thorpe RI (Eds) Geol Surv Can, Geology of Canada 8:277-285

Salas GP, Nieto FC 1991 Geology of uranium deposits in Mexico. *In* Economic Geology, Mexico. Salas GP, (Ed), The Geology of North America, Geol Soc Am, P-3:161-165

Sassano GP 1972 The nature and origin of the uranium mineralization at the Fay mine, Eldorado, Saskatchewan, Canada. PhD Thesis, Univ Alberta, Edmonton (unpubl)

Saunders AD, Norry MJ, Tarney J 1988 Origin of MORB and chemically-depleted mantle reservoirs: trace element constraints. *In* Oceanic and Continental Lithosphere: Similarities and Differences. Menzies MA, Cox KG (Eds) Journal of Petrology, Special Lithosphere Issue. Oxford Univ Press, Oxford, UK, p 415-445

Saunders AD, Norry MJ, Tarney J 1991 Fluid influence on the trace element compositions of subduction zone magmas. Phil Trans Royal Soc London, A 335:377-392

Saunders AD, Tarney J, Weaver SD 1980 Transverse geochemical variations across the Antarctic Peninsula: implications for the genesis of calc-alkaline magmas. Earth Planet Sci Lett 46:344-360

Sawkins FJ 1989 Anorogenic felsic magmatism, rift sedimentation, and giant Proterozoic Pb-Zn deposits. Geology 17:657-660

Scharer U, Krogh TE, Wardle RJ. Ryan AB, Gandhi SS 1988 U-Pb ages of Lower and Middle Proterozoic volcanism and metamorphism in the Makkovik Orogen, Labrador. Can J Earth Sci 25:1087-1107

Schidlowski M 1981 Uraniferous constituents of the Witwatersrand conglomerates—Ore microscopic observations and implications for the Witwatersrand metallogeny. *In* Genesis of Uranium- and Gold-Bearing Precambrian Quartz-Pebble Conglomerates. Armstrong FC (Ed) Geological Survey Professional Paper 1161-A-BB, N1-S29

Schidlowski M, Eichmann R, Junge CE 1975 Precambrian sedimentary carbonates: Carbon and oxygen isotope geochemistry and implications for the terrestrial oxygen budget. Precamb Res 2:1-69

Sclater JG, Jaupart C, Galson D 1980 The heat flow through oceanic and continental crust and the heat loss of the earth. Rev Geophys Space Sci 18:269-311

Scott IR 1987 The development of an ore reserve methodology for the Olympic Dam copper-uranium-gold deposit. Resources and Reserves Symposium, Australasian Inst Mining Metall 99-103

Sherborne JE Jr, Buckovic WA, Dewitt DB, Hellinger TS, Pavlak SJ 1979 Major uranium discovery in volcaniclastic sediments, basin and range province, Yavapai County, Arizona, Am Assoc Petrol Geol Bull 63:621-646

Shepherd TJ 1977 Fluid inclusion study of the Witwatersrand gold-uranium ores. Phil Trans Royal Soc London A 286:549-565

Sibbald TII 1988 Geology and genesis of the Athabasca Basin uranium deposits. *In* Recognition of Uranium Provinces, Proc Tech Comm Meeting of Uranium Provinces, IAEA, London 18-20 September 1985. IAEA, Vienna, Austria, p 61-105

Sibbald TII, Quirt D 1987 Uranium deposits of the Athabasca Basin. Saskatchewan Res Council Publ R-855-1-G-87, 72 p

Sibbald TII, Quirt DH, Gracie AJ 1991 Uranium deposits of the Athabasca Basin, Saskatchewan (Field Trip 11), Geol Surv Canada Open File 2166, 56 p

Simpson PR, Bowles JFW 1977 Uranium mineralisation of the Witwatersrand and Dominion Reef systems. Phil Trans Royal Soc London A 286:527-548

Simpson PR, Bowles JFW 1981 Detrital uraninite and pyrite: Are they evidence for a reducing atmosphere? *In* Genesis of Uranium- and Gold-Bearing Precambrian Quartz-Pebble Conglomerates, Armstrong FC (Ed) U S Geol Surv Prof Paper 1161-A-BB, p S1-S12

Simpson PR, Brown GC, Plant JA, Ostle D 1979 Uranium mineralization and granite magmatism in the British Isles. Phil Trans Royal Soc London A 291:385-412

Simpson PR, Plant JA 1984 Role of heat production granites in uranium province formation. *In* Uranium Geochemistry, Mineralogy, Geology, Exploration and Resources. De Vivo B, Ippolito F, Capaldi G, Simpson PR (Eds) Inst Min Metall, p 167-178

Simpson PR, Yu Shiqing 1989 Uranium Metallogeny, Magmatism and Structure in Southeast China. *In* Metallogenesis of Uranium Deposits. Proc Tech Comm Mtg, 9-12 March 1987, IAEA, Vienna, p 357-367

Sleep NH 1979 Thermal history and degassing of the earth: some simple calculations. J Geol 87:671-686

Sleep NH 1990 Hotspots and mantle plumes: some Phenomology. J Geophys Res 95:6715-6736

Smirnov VI 1977 Ore Deposits of the USSR. Pitman, London, 1258 p

Smith DAM 1965 The geology of the area around the Khan and Swakop Rivers in South West Africa. Mem Geol Surv S Africa 3 (SWA Series) 113 p

Smits G 1990 The geochemical history of the sedimentary rocks of the Witwatersrand as reflected in the mineralogy of the heavy mineral assemblage of the uranium-bearing reefs of the Central Rand Group. PhD Thesis, Potchefstroom Univ, Potchefstroom, Republic of South Africa, 194 p (unpubl)

So CS, Shelton KL, Rye DM 1983 Geologic, sulphur isotopic and fluid inclusion study of the Ssang Jeon tungsten mine, Republic of Korea. Econ Geol 78:157-163

Sopuck VJ, De Carle A, Wray EM, Cooper B 1983 Application of lithogeochemistry to the search for unconformity-type uranium deposits in the Athabasca Basin, In Uranium Exploration in Athabasca Basin. Cameron EM (Ed) Geol Surv Can Paper 82-11:192-205

Stemprok M, Skvor P 1974 Composition of tin-bearing granites from the Krusne Hory metallogenic province of Czechoslovakia. Sb Geol Ved Losiskova Geol SU 16

Stockwell CH 1982 Proposals for time classification and correlation of Precambrian rocks and events in Canada and adjacent areas of the Canadian Shield, Part 1: a time classification of Precambrian rocks and events. Geol Surv Can Paper 80-19, 135 p

Sudo A, Tatsumi Y 1990 Phlogopite and K-amphibole in the upper mantle: implications for magma genesis in subduction zones. Geophys Res Lett 17:29-32

Sun SS, McDonough WF 1989 Chemical and Isotopic Systematics of Ocean Basalts: Implications for Mantle Composition and Processes. In Magmatism in the Ocean Basins, Saunders AD, Norry MJ (Eds) Geol Soc Lond Spec Publ 42:313-45

Taira A, Pickering KT, Windley BF, Soh W 1992 Accretion of Japanese island arcs and implications for the origin of Archean greenstone belts. Tectonics 11:1224-1244

Tauson LV 1982 Geokhimia i metallogenia latitovykh. Serii Geol rudn Mestogozhdenii 3:3-14 (in Russian)

Taylor SR 1964 Abundance of chemical elements in the continental crust: a new table. Geochim Cosmochim Acta 28:1273-1284

Taylor SR, McLennan SM 1981 The composition and evolution of the continental crust; rare earth element evidence from sedimentary rocks. Phil Trans Royal Soc London A 301:381-399

Taylor SR, McLennan SM 1985 The Continental Crust; its Composition and Evolution. Blackwell, Oxford

Taylor HP 1977 Water/rock interaction and the origin of H_2O in granite batholiths. J Geol Soc London 133:509-558

Taylor HP 1979 Oxygen and hydrogen isotope relationships in hydrothermal mineral deposits. In Geochemistry of Hydrothermal Ore Deposits. Barnes HL (Ed) John Wiley & Sons, New York, p 236-277

Taylor JR, Wall VJ 1992 The behavior of tin in granitoid magmas. Econ Geol 87:403-420

Teggin DE 1976 Rubidium-strontium whole rock ages of granites from northern Thailand. In Proceedings of the Seminar on isotopic dating. Li CY (Ed) UNDP, UN ESCAP, Bangkok

Toens PD, Corner B 1980 Uraniferous alaskitic granites with special reference to the Damara orogenic belt. S Africa Atomic Energy Board Res PER 55, 18 p

Treloar PJ, Coward MP, Harris NBW 1992 Himalayan-Tibetan analogies for the evolution of the Zimbabwean Craton and Limpopo belt. Precamb Res 55:571-587

Treuil M 1985 A global geochemical model of uranium distribution and concentration in volcanic rock series. In Uranium Deposits in Volcanic Rocks. IAEA, Vienna, 53-68

Uranium Information Centre, Melbourne 1997

von Pechmann E 1981 Mineralogy of the Key Lake U-Ni orebodies, Saskatchewan, Canada: Evidence for their formation by hypogene hydrothermal processes. In Geology of Uranium Deposits. Sibbald TII, Petruk W (Eds) CIM Spec Vol 32:27-37

Waber N, Schorscher HD, Peters T 1992 Hydrothermal and supergene uranium mineralization at the Osamu Utsumi mine, Pocos de Caldas, Minas Gerais, Brazil. J Geochem Expl 45:53-112

Wallis RH, Saracoglu N, Brummer JJ, Golightly JP 1986 The geology of the McClean uranium deposits, northern Sakatchewan. In Uranium Deposits of Canada. Evans EL (Ed) CIM Spec Vol 33:193-217

Ward DM 1984 Uranium geology, Beaverlodge area. In Proterozoic Unconformity and Stratabound Uranium Deposits. TECDOC-315, IAEA, Vienna, p 269-284

Watson JV, Fowler MB, Plant JA, Simpson PR 1984 Variscan-Caledonian comparisons: Late orogenic granites. Proc Ussher Soc 62-12

Weaver BL, Tarney J 1980 Rare-earth geochemistry of Lewisian granulite-facies gneisses, north west Scotland: implications for the petrogenesis of the Archaean lower continental crust. Earth Planet. Sci Lett 51:279-296

Weaver BL, Tarney J 1981 Lewisian gneiss geochemistry and Archaean crustal development models. Earth Planet. Sci Lett 55:171-180

Weaver BL, Tarney J 1984 Empirical approach to estimating the composition of the continental crust. Nature 310:575-577

Wenrich KJ 1985 Mineralization of breccia pipes in northern Arizona. Econ Geol 80:1722-1735
Wenrich KJ 1986 Uranium mineralization of collapse breccia pipes in northern Arizona, western United States. *In* Vein-type Uranium Deposits, TECDOC-361. IAEA, Vienna, p 395-414
Wenrich KJ, Sutphin HB 1989 Lithotectonic setting necessary for formation of uranium-rich solution-collapse breccia-pipe province, Grand Canyon region, Arizona, *In* Metallogenesis of Uranium Deposits. IAEA, Vienna, p 307-344
Wenrich KJ, Chenoweth WL, Finch WI, Scarborough RB 1990 Uranium in Arizona. *In* Geologic evolution of Arizona, Jenney JP, Reynolds SJ. (Eds) Arizona Geol Soc Digest 17:759-794
Windley BF 1984 The Evolving Continents. 2nd Edn. Wiley, London
Windley BF 1995 The Evolving Continents. 3rd Edn. Wiley, London, 526 p
Windley BF, Simpson PR, Muir MD 1984 The role of atmospheric evolution in Precambrian metallogenesis. Fortschritte der Mineralogie 62:253-267
Worrall DM, Snelson S 1989 Evolution of the northern Gulf of Mexico, with emphasis on Cenozoic growth faulting and the role of salt. *In* The geology of North America—An overview. Bally AW, Palmer AR (Eds) Geol Soc Am, Boulder, Colorado, The Geology of North America A97-138
Wyborn LAI 1988 Petrology, geochemistry and origin of a major Australian 1880-1840 Ma felsic volcano-plutonic suite: a model for continental felsic magma generation. Precamb Res 40/41:37-60
Wyborn LAI, Page RW, Parker AJ 1987 Geochemical and geochronological signatures in Australian Proterozoic igneous rocks. *In* Geochemistry and Mineralization of Proterozoic Volcanic Suites. Pharaoh TC, Beckinsale RD, Rickard D (Eds) Geol Soc London Spec Publ 33:377-394
Wyllie PJ 1983 Experimental and thermal constraints on the deep-seated parentage of some granitoid magmas in subduction zones. *In* Migmatites, Melting and Metamorphism. Atherton MP, Gribble CD (Eds) Shiva Publications, Nantwich, UK, p 37-51
Yardley BWD, Long CB 1981 Contact metamorphism and fluid movement around the Easky adamellite, Ox Mountains, Ireland. Mineral Mag 44:125-131
Zhang C 1998 Microbial uranium reduction and biomineralization: implication for immobilization of toxic metals and radionuclides. Int'l Mineral Assoc Abstracts, Toronto, A46
Zhukova VY 1980 Mineralogy and primary zoning of hydrothermal metasomatic uranium deposits in Precambrian iron formations. *In* Albitized Uranium Deposits. Six articles translated from Russian literature. Avrashov A (Translator) US-DOE, GJBX-193 (80):91-114
Zoback ML, Thompson GA 1978 Basin and range rifting in northern Nevada—Clues from a mid-Miocene rift and its subsequent offsets. Geology 6:111-116
Zoback ML, Anderson RE, Thompson GA 1981 Cainozoic evolution of the state of stress and style of tectonism of the Basin and Range province of the western United States. Phil Trans Royal Soc London 300A:407-434

7 Mineralogy and Geochemistry of Natural Fission Reactors in Gabon

Janusz Janeczek

Faculty of Earth Sciences
University of Silesia
Bedzinska 60, 41-200 Sosnowiec, Poland

INTRODUCTION: NATURAL FISSION REACTORS

The ratio between two most abundant U isotopes, ^{235}U (0.72%) and ^{238}U (99.28%), is remarkably constant and only small deviations from the present-day value of 0.0072 have been observed in nature. There is, however, one place on Earth that we know, where the $^{235}U/^{238}U$ ratio deviates dramatically and can be as low as 0.0038. This is a group of U deposits located in the southeastern part of Gabon, Africa (Fig. 1). Sustained fission chain reactions occurred in these U deposits approximately 2 billion years ago.

The discovery of natural U depleted in fissile ^{235}U was made in 1972 in the U enrichment plant at Pierrelatte in France (Bodu et al. 1972). A small deficiency in ^{235}U of only 0.0031 atomic % was observed during a routine check of isotopic abundances in UF_6 prepared from U ore shipped from Gabon. Subsequent investigation indicated the U deposit at Oklo was the source of the isotopic anomaly. Isotopic analyses of U in the ore and the discovery of fission products provided firm evidence that spontaneous nuclear chain reactions had indeed occurred at Oklo (Neuilly et al. 1972). Interestingly, the Oklo deposit had been mined for two years before this unique isotopic anomaly was recognized.

The serendipitous discovery of natural fission reactors came as a surprise but it was not entirely unexpected. The possibility that large-scale nuclear reactions may have occurred in the Earth's crust was first proposed by Noetzlin in 1939 and Odagiri in 1940 (cf. Kuroda 1975). In 1953 Wetherill and Inghram found that both spontaneous fission of ^{238}U and neutron-induced fission of ^{235}U contributed to the production of stable xenon isotopes in U ore. They estimated that the U deposit they studied was 25% of the way to becoming nuclear reactor (Kuroda 1975). In 1956 Kuroda calculated factors necessary to start chain fission reactions in rich U ore from the Johangeorgenstadt deposit in Saxony and came to the conclusion that self-sustained fission reactions might have occurred there had the ore been about 2 Ga old. In fact, the U deposit at Johangeorgenstadt is some 1.7 Ga younger.

Soon after the discovery of the natural fission reactors at Oklo an effort was undertaken to define and model the conditions required to initiate and maintain fission reactions in nature. An impressive amount of data on geology, geochemistry, geochronology and physics of the natural reactors was gathered and published in two conference proceedings volumes: The Oklo Phenomenon (1975) and Natural Fission Reactors (1978).

During the 1980s, interest in the Oklo phenomenon waned, though a few important papers on the geochemistry and geochronology of the natural reactors were published. The revival of interest in natural fission reactors was triggered by the recognition that the reactors could be studied as natural analogues for some of the important processes that may occur in a nuclear-waste repository. While the usefulness of the Oklo deposit in this respect was recognized early (Walton and Cowan 1975; Cowan 1978; Hagemann and Roth 1978; Leachman and Bishop 1978) it was not until late eighties that a formal research project on

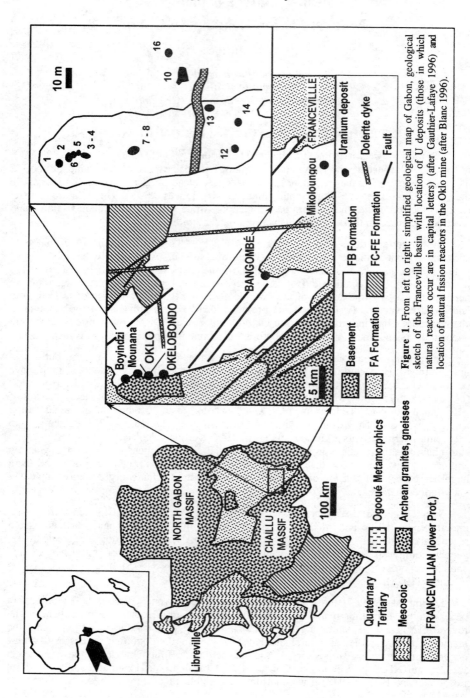

Figure 1. From left to right: simplified geological map of Gabon, geological sketch of the Franceville basin with location of U deposits (those in which natural reactors occur are in capital letters) (after Gauthier-Lafaye 1996) and location of natural fission reactors in the Oklo mine (after Blanc 1996).

natural fission reactors as natural analogues was initiated by the French organizations responsible for the nuclear waste management (Blanc 1996). In 1989 an international and interdisciplinary working group was organized in order to thoroughly characterize the natural fission reactors for the purpose of understanding the release and migration of radionuclides in a radioactive-waste repository. Emphasis has been on migration and retention of fission products and modeling radionuclide transport through the rocks around the reactors. Some of the results obtained during the early stage of that project were recently summarized in a review paper by Gauthier-Lafaye et al. (1996). New interesting observations on geochemistry and mineralogy of the natural fission reactors can be expected as an outcome of this research project, which ends in 1999. Unfortunately for the scientific community further investigation of the natural fission reactors may be hindered, because the mining operation at the Oklo-Okelobondo deposit was terminated in late 1997.

The literature on natural fission reactors is voluminous. Unfortunately many contributions have appeared in technical reports, which are not readily available. The present review offers a mineralogical and geochemical perspective of the Oklo phenomenon, particularly as it applies to the neutronics of the reactors and the fate of fission products and actinides. This review summarizes our knowledge of the mineralogy and geochemistry of natural fission reactors as of mid 1998.

Geological setting

Natural fission reactors occur in the Franceville basin in SE Gabon (Fig. 1). The Palaeoproterozoic Franceville basin consists of uraniferous conglomerates and sandstones, arkosic sandstones, shales locally intercalated with chert, manganese-rich rocks and volcanic rocks (Gauthier-Lafaye 1986; Gauhtier-Lafaye and Weber 1989). This 2000 to 2500 m thick series is collectivelly called the Francevillian and it lies unconformably on the Archean (2.9-2.6 Ga) crystalline basement. A fragment of the crystalline basement, known as the Mouanana horst, crops out within the Franceville basin. The Francevillian is divided into five formations coded FA to FE. The basal FA formation consists of fluviatile and deltaic conglomerates and sandstones. All U deposits that occur along the Mounana horst (Oklo-Okelobondo, Mounana, Boyindzi) and at Bangombé occur at the top of FA formation. In Oklo, at the top of FA formation lies a uraniferous sandstone, named by the miners the C1 layer (Gauthier-Lafaye, pers. comm.). The fission reactors, occur within the C1 layer. The FA formation is overlain by black shales of the FB formation.

The age of deposition for the Francevillian as determined by Rb-Sr and K-Ar methods is 2143±143 Ma before present (b.p.) (Bonhomme et al. 1982). The Sm-Nd dates of minerals indicate that early diagenesis occurred 2099±115 Ma and 2036±79 Ma b.p. (Bros et al. 1992). Despite an old age, the Francevillian of the Franceville basin is practically unmetamorphosed and undeformed except for zones of early fracturing where U is concentrated (Gauthier-Lafaye and Weber 1989). However, locally intense folding occurred in the U deposists along the basin margins. Illite-1M and chlorite help define the degree of metamorphism at the top of the FA formation. The fluid inclusions studies in calcite and quartz overgrowths determined the maximum temperature reached in the Franceville basin during diagenesis at 200°C at pressure of 400 bars (Gauthier-Lafaye and Weber 1989).

The Franceville basin is cross-cut by a numerous large dolerite dykes (Fig. 1). The age of the dolerite intrusion is uncertain, with age determinations ranging from 977 and 981±27 Ma, reported by Bonhomme et al. (1982) for feldspars from dolerite, to the currently accepted age of 780 Ma for the dolerite intrusion at Oklo (Michaud 1998).

There are six U deposits in the Franceville basin (Fig. 1). Four of them, Boyindzi,

Mouanana, Oklo, and Okelobondo were economically important, with total U resources estimated at 40,000 tons of U (Gauthier-Lafaye and Weber 1989). All of these deposits occur along the eastern slope of the Mouanana horst (Fig. 1). At Oklo and Okelobondo U mineralization is confined to the 5 to 8 m thick C1 layer, which forms a broad monoclinical fold diping to the East. At Bangombé U mineralization occurs at the top of a wide anticline. The southernmost U deposit at Mikouloungou, unlike the other deposits, is located in the lower FA formation and is associated with a major E-W fault.

There are two types of U deposits in the Franceville basin: a low-grade deposit with the U content between 0.1 and 1%, and a high-grade deposit with the U concentrations up to 10% (Gauthier-Lafaye and Weber 1989). It is the latter type of U deposit in which fission chain reactions occurred 2 Ga ago.

The origin of U deposits in the Franceville basin

The generally accepted model for the origin of the U deposits in the Franceville basin have been presented by Gauthier-Lafaye (1986) and Gauthier-Lafaye and Weber (1981 1989). According to this model the primary source of U was detrital uraniferous thorite deposited in fluviatile conglomerate, now at the bottom of the FA formation. The thorite was probably from nearby granites and gneisses of the Chaillu massif. The concentration of U in the uraniferous thorite in red conglomerates ranges from 0.3 to 5.9 wt %. The average U concentration in the conglomerates is 9.6 ppm, with a maximum value of 25 ppm-U. These are the highest U concentrations known in the Franceville basin. Apparently, U was preferentially removed from thorite by oxidizing fluids. The origin of the oxidizing fluids is uncertain, but they may have been related to water trapped in closed porosity during deposition of conglomerates, or they may be meteoric waters that descended along fractures and faults during uplift of the Franceville basin (Fig. 2). These waters percolated through the U- and Th-bearing conglomerate and FA sandstones, dissolving sulfate and carbonate cement. As a result, oxidized and highly saline U(VI)-bearing fluids migrated upwards due to the convective circulation in the basin. Fluid-inclusion studies of calcite associated with the primary U mineralization show that fluid temperatures were in the range of 130-150°C, salinities were high (> 20 wt % eq. NaCl), and fluids were enriched in ^{18}O ($\delta^{18}O$ = +6-9‰ SMOW) (Michaud 1998).

Uranium-bearing conglomerates were subsequently covered by thick deltaic and marine deposists, which now form the upper part of the FA formation and shales of the FB

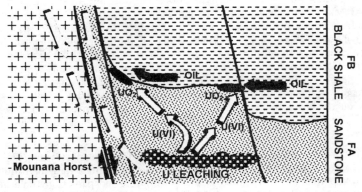

Figure 2. The conceptual model of Gauthier-Lafaye and Weber (1989) for the origin of U deposits in the Franceville basin. Empty arrows represent flow direction of meteoric waters, stipled arrows are U(VI)-bearing fluids, and black arrows labelled "oil" show flow direction of reducing hydrocarbon-bearing fluids.

formation. The FB shales are rich in marine organic matter and reached pressures and temperatures sufficient to reach the "oil window" during burial of the Franceville series at depth up to 4 km. The petroleum that was produced during burial migrated into the FA sandstones and accumulated in numerous structural traps. Uranium mineralization occurred at 2000±50 Ma, when oxidized U(VI)-bearing fluids encountered reduced hydrocarbon-bearing fluids (Fig. 2). Uranium precipitated as uraninite in pores, as well as in hydraulically induced fractures that form an extensive network in the sandstones (Gauthier-Lafaye 1986). Soon after precipitation of uraninite, neutron induced chain fission reactions began in the richest U ore, resulting in the formation of the nuclear reactor zones.

Reactor zones

Natural fission reactors are zoned bodies that consist of high-grade U ore (the reactor core) enveloped by a mantle of clay minerals (clays of the reactor) composed mainly of chlorite or illite (Fig. 3). Uranium concentrations in reactor cores range from 20 to 87 wt %. In clays of the reactors, concentrations of U vary from the ppm level to 3 wt % (Gauthier-Lafaye et al. 1996). A reactor core and the clay mantle together form the reactor zone (RZ). The boundary between reactor zone and the underlying sandstone is rather sharp and is marked by the thin layer of hematite and sometimes by the concentration of uraniferous organic matter and uraninite (Fig. 4). Quartz in sandstone adjacent to the reactor zone is corroded.

Fifteen natural fission reactors have been recognized in the Franceville basin since 1972. Most are located in the Oklo-Okelobondo deposit, although one reactor was discovered in 1985 at Bangombé some 20 km south of Oklo (Fig. 1), hence it is known as the Bangombé reactor. Consecutive numbers were assigned to the reactors following their discovery. Some numbers (11, 12 and 14) were given to U ore bodies based on single determinations of the $^{235}U/^{238}U$ ratio, but have not been confirmed as true fission reactors (Blanc 1996). The latest reactor (number 16) was discovered at Oklo in spring of 1991. Reactors RZ-1 to RZ-9 and RZ-15 were mined out (Gauthier-Lafaye et al. 1996). However, a small fragment of RZ-2, which played a crucial role in understanding the Oklo phenomenon, has been preserved in the open pit mine as the Gabonese National Monument (Blanc 1996).

Not all reactors have been studied to the same extent. The best characterized are RZ-2 and RZ-10 at Oklo and the Bangombé reactor. Reactor zones occur at various depths (Fig. 5), due to the dip of the mineralized layer caused by tectonic deformation some 1950 Ga ago. The deepest known reactor is at Okelobondo, and it was recently sampled and investigated (Jensen and Ewing, submitted). The Bangombé reactor occurs at the depth of only 12 m below ground level The fifteen natural fission reactors are not identical. They differ significantly in size, geometry, U concentration, mineral composition, content of organic matter, and degree of depletion of ^{235}U (Table 1). Some of these differences were original, others were caused by processes related to nuclear reactions and post-reactional geological events.

The primary indicator of neutron-induced fission reactions in a U deposit is the depletion in fissile ^{235}U. Ranges of $^{235}U/^{238}U$ ratios reported from reactor zones are shown in Figure 6. The concentration of ^{235}U in the U minerals not only varies among reactors, but also varies within a single reactor zone. Usually the central portions of reactor cores are more deficient in ^{235}U than outer parts of the reactor (Fig. 7). Depleted U outside reactor zones provides strong evidence for migration of actinides out of reactor zones.

The minimum value of the $^{235}U/^{238}U$ ratio given in Table 1 may not be representative of the entire reactor. The actual amount of highly depleted U in a reactor can be rather

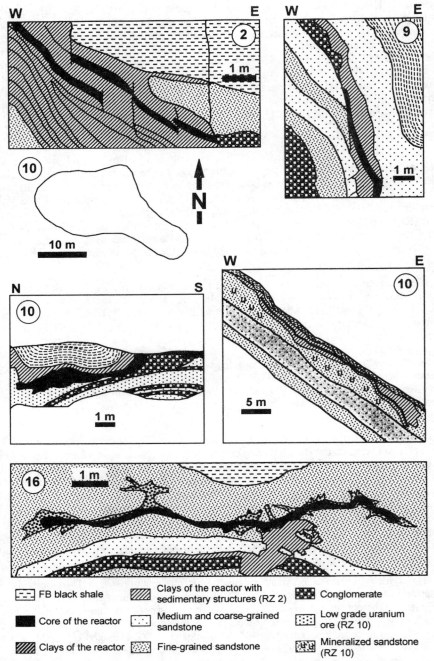

Figure 3. Cross-sections through reactor zones at Oklo. Numbers in circles refer to the numbers of reactor zones. Plan view of RZ-10 is also shown (RZ-2 and RZ-9 from Gauthier-Lafaye 1986; RZ-10 and RZ-16 from Holliger and Gauthier-Lafaye 1996).

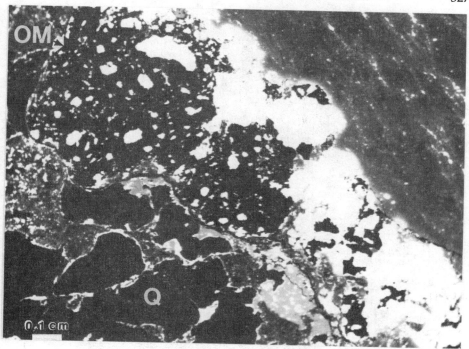

Figure 4. Back scattered electron image of the border between Mg-chlorite of RZ-10 at Oklo (upper right) with FA sandstone. The border is marked by abundant uraninite (white) and blebs of uraninite-bearing bitumen (OM). Quartz (Q) in sandstone is corroded and partially replaced by bitumen. Matrix in sandstone consists of diagenetic Fe-chlorite. Note uraninite interleaved with chlorite in the reactor zone at upper right.

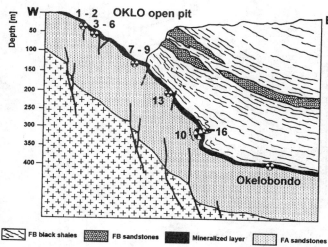

Figure 5. Cross-section of Oklo-Okelobondo U deposit showing the dip of the mineralized layer and the depth of the occurrence of reactor zones. Modified from Gauthier-Lafaye (1996).

328 URANIUM: Mineralogy, Geochemistry and the Environment

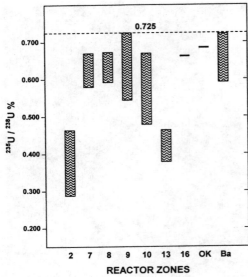

Figure 6. Ranges of values of the $^{235}U/^{238}U$ ratio reported for various natural fission reactors. Only single values were available for RZ-16 and Okelobondo (OK).

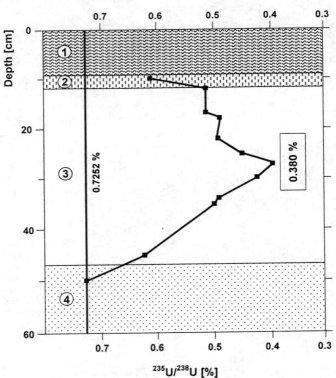

Figure 7. Distribution of U depleted in ^{235}U in RZ-13 and in its immediate vicinity (modified from Hidaka and Holliger 1998). 1—FB black shale; 2—transition zone; 3—reactor core (massive U minerals); 4—FA sandstone.

Table 1. Selected characteristics of natural fission reactors. Note: Ba = Bangombé

Reactor zone	1	2	7-8-9	10	13	Ba
Size (length × width) [m]	40 × 18	12 × 18	Total length 30 (RZ-9: 7)	27 × 30 × 15	10 × 6	1.5 km²
Thickness of the core [cm]	20 - 30	20 - 50, locally 100	a few	a few to 100	20-30	5 - 15
Maximum UO_2 [%]	40	50-60	25-30	65	87	54
Minimum $^{235}U/^{238}U$ [%]	0.409	0.292	0.543 RZ 9 0.579 RZ 7	0.479	0.380	0.590
Organic matter	Absent	Absent	Abundant	Abundant	Absent	Abundant
Clay mantle	Abundant	Abundant	Poorly developed	Well developed	Patchy, from none to thick	30 cm at the top of the reactor core
	chlorite > illite	chlorite > illite	chlorite > illite	chlorite		illite > chlorite

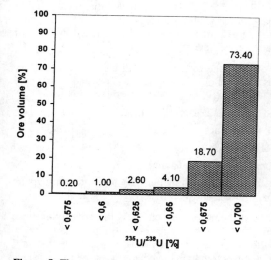

Figure 8. The percentage of the volume of U minerals with various $^{235}U/^{238}U$ ratios in RZ-10. Drawn from data in Holliger and Gauthier-Lafaye (1996).

small. For instance, in RZ-10 at Oklo the volume occupied by U minerals with the lowest values of $^{235}U/^{238}U$ (between 0.575 and 0.550%) is only 0.2% of the total volume of the reactor zone (about 7500 m³); whereas, only slightly depleted U occupies about 73% of this volume (Fig. 8).

Natural fission reactors differ in the development of clay mantle (Table 1). For example, clay mantle is a prominent feature of RZ-2, RZ-10 and Okelobondo but the clay mantle is almost non-existent in RZ-13. The origin of the clay mantle is related to the desilification of sandstone and its replacement by clay minerals. The desilification of sandstone was caused by hydrothermal fluids generated by heat released during fission chain reactions (Gauthier-Lafaye et al. 1989). Silica released from the dissolved sandstone in the reactor zones migrated and may have precipitated in reactor cores again upon their cooling. For instance, the lowest $^{235}U/^{238}U$ ratio in the core of RZ-10 is 0.00386 and it was observed in uraninite associated with the secondary (neogenic) quartz (Gauthier-Lafaye, pers. comm.).

Physics of natural fission reactors

The principles of how the natural fission reactors functioned and the conditions necessary to initialize and maintain self-sustaining fission chain reactions were determined and modeled from the knowledge of fission reactions in man-made nuclear reactors (Keller

1978; Naudet 1991; Oversby 1996).

Nuclear fission is the splitting of a nucleus into two or more lighter nuclei, fission products. Fission of a nucleus that occurs without external excitation is called the spontaneous fission. The most important of the naturally occurring nuclides that can undergo spontaneous fission is ^{238}U. During spontaneous fission of ^{238}U, 2.2 neutrons are produced per fission event. Neutrons are classified on the basis of their energies into fast (>0.1 MeV), slow (about 1 eV) and thermal (about 0.025 eV). Fast neutrons have such high velocities that the probability of a fast neutron colliding with ^{235}U nucleus and causing neutron-induced fission is small. Fast neutrons lose their energies to become thermal neutrons upon collision with masses close to 1 amu; e.g. ^{1}H. Hence, water can "thermalize" fast neutrons. Substances used to slow fast neutrons in man-made reactors are called moderators. Thermal neutrons can induce fission in ^{235}U. Neutrons produced from a fission event in ^{235}U may cause other ^{235}U nuclei to fission, which in turn produces additional neutrons, creating a chain reaction. However, ^{238}U, which is the principal isotope of natural U, captures too many neutrons to allow induced chain reactions in ^{235}U unless there is sufficient amount of ^{235}U. The minimum ^{235}U/^{238}U ratio required for operation of a natural reactor is approximately 1%.

Uranium-235 decays faster than ^{238}U because half-life of ^{235}U (7.1×10^8 a) is an order of magnitude shorter than the half-life of ^{238}U (4.5×10^9 a). This means that the ^{235}U/^{238}U ratio was higher in the geological past than today. Two billion years ago the ^{235}U/^{238}U ratio was 3.5%, which is in the range of ^{235}U/^{238}U ratios in fuels artificially enriched in ^{235}U for man-made reactors using normal (light) water as a moderator.

Apart from the high concentrations of ^{235}U, the following conditions were necessary for self-sustaining chain fission reactions to occur in U deposits at Oklo and Bangombé (Naudet 1991):

1. total U concentrations of at least 10% in a 2 meter thick layer.
2. U ore seams at least 0.5 m thick (ideally at least 63 cm).
3. a water (moderator) to U ratio of about 6%
4. the presence of neutron reflectors (quartz in sandstone)
5. low concentrations of neutron absorbers; i.e. elements with high neutron-capture cross sections, such as B, Li, Mn, HREE and V.

Some of these constraints were not as stringent because at Oklo the conditions were especially favorable. For instance, the amount of water may have been higher than 6 wt % because the porosity of the sediment 2 billion years ago was 40% as compared with 1 to 5 vol % in the sandstones today. Therefore, the reactors may have even been overmoderated.

Most of the reactors are thinner that the minimum thickness of 50 cm required by models of fission reactions (Table 1). At the time of criticality, the U orebodies were surrounded by sandstone. The Si and O that comprise quartz in sandstone have small neutron-capture cross sections, so that sandstones acted as neutron reflectors, returning neutrons back to reactor cores. Naudet (1991) estimated that in the presence of neutron reflectors the actual thickness of the U orebody necessary to achieve criticality was less than 50 cm.

The low abundance of non-fissile neutron absorbers was also fortiutous. For instance, vanadium, an efficient neutron absorber, is abundant element in the nearby U deposit at Mouanana but it is essentially absent at Oklo and Bangombé. In this respect, it is interesting to note that Parnell (1996) reported 1 wt % V in clays surrounding the unspecified reactor.

Table 2. Relative contribution of fissile nuclides to fission-product inventory in reactor zones (RZ)

RZ	Fission contribution (%) from:		
	^{235}U	^{238}U	^{239}Pu
9*	80.5 - 93.5	5.1 - 11.7	1.7 – 9.6
10**	89 - 92	4 - 7	3 – 4
13**	75	18	7
Ba**	92	4	4

Notes: * data from Loss et al. (1988
** data from Hidaka and Holliger (1998)

Thermal neutron-induced fission of ^{235}U was the principal, but not the only fission process in the natural reactors. Fission of ^{238}U by high-energy neutrons and of ^{239}Pu by thermal neutrons also occurred in natural reactors. The proportion of a nuclide produced by fission is called the fission yield. Contributions of each fission process to the fission product inventory can be determined by analysing the isotopic composition of an element having isotopes with distinctly different fission yields for the fission of ^{235}U, ^{238}U, and ^{239}Pu (Curtis et al. 1989). One element suitable for such estimates is palladium. The relative contributions of fissile nuclides to the fission product inventory for various reactor zones are given in Table 2.

To understand how and for how long the natural fission reactors operated, the neutronic parameters have to be known. Estimates of nuclear parameters in the natural fission reactors are based on the isotopic abundances of U and fission products. In particular, some of REE isotopes, such as ^{143}Nd, ^{149}Sm, ^{155}Gd and ^{176}Lu, have large neutron capture cross-sections and, therefore, have large fission yields. Their abundances can be used to determine neutron physics of the natural reactors. The following parameters define the operation of natural fission reactors:

- Neutron fluence (τ)—the time integrated neutron (n) flux expressed in units of n/cm^2.
- Restitution factor (C)—the fraction of ^{235}U produced from ^{239}Pu by alpha-decay relative to ^{235}U consumed by fission reactions.
- Spectrum index (R)—the fraction of non-thermal neutrons in the neutron energy spectrum.
- Number of fissions due to ^{235}U, ^{238}U, and ^{239}Pu.
- Duration of the fission reactions (Δt).

The nuclear parameters of natural fission reactors have been estimated by numerous authors (Naudet 1991, and literature therein; Gauthier-Lafaye et al. 1996; Holliger and Gauthier-Lafaye 1996; Hidaka and Holliger 1998) and are summarized in Table 3. From data in Table 3 it is obvious that RZ-2 is unique among the natural fission reactors in terms of neutron fluence, which in that reactor was an order of magnitude higher than in other reactor zones due to the highest fission rate. The reactors differ also in duration of sustained fission reactions (Table 3).

The lowest values of the restitution factor were calculated for RZ-13 (Table 3), which in conjunction with the high rate of fast neutrons induced fission of ^{238}U, indicate that the neutrons in RZ-13 were not well thermalized (Hidaka and Holliger 1998).

Table 3. Average nuclear parameters of natural fission reactors

Reactor zone	2	9	10	13	OK	Ba
τ (10^{20} n/cm^2)	13.62	1.05 - 7.58	5.25 – 7.98	7.8	3.3	3.08
C	0.490	0.471 - 0.688	0.304 – 0.382	0.111	0.70	0.484
R	0.183	0.119 - 0.932	0.111 – 0.181	0.241	0.14	0.196
Δt (10^5 a)	0.62	0.64 - 5.56	1.43 – 1.98	0.24	2.70	1.10

τ—neutron fluence, C—restitution factor, R—spectrum index, Δt—duration of criticality. Note: Data for RZ-2 and Okelobondo reactor (OK) from Holliger and Gauthier-Lafaye (1996), for RZ 9 from Loss et al. (1988), for RZ-10, RZ-13 and the Bangombé reactor (Ba) from Hidaka and Holliger (1998).

Estimates of nuclear parameters for natural fission reactors rely on isotopic abundances of elements and are, therefore, sensitive to losses of elements from the system. If certain isotopes were partially released from reactors, then the calculated nuclear parameters are meaningless. The wide range of nuclear parameters estimated for RZ-9 (Table 3) can be explained by preferential loss of Nd during dissolution of uraninite (Curtis et al. 1989).

The ^{176}Lu/^{175}Lu ratio can be used to evaluate the average equilibrium temperature of neutrons during criticality because the neutron-capture cross section of ^{176}Lu is highly dependent on the neutron energy (Holliger and Devillers 1981). Temperatures calculated from the ^{176}Lu/^{175}Lu ratio vary from sample to sample, ranging from 220°C to 360°C in RZ-2 to RZ-5, to 380°C in RZ-10. Some calculated temperatures exceeded 1000°C (Hidaka and Holliger 1998); however, there is no mineralogical or geological evidence for such a high temperature in the reactors zones. Hidaka and Holliger (1998) attribute these questionable estimates to uncertaintities in measuring the neutron-capture cross section of ^{176}Lu.

Fluid inclusions in quartz overgrowths in sandstone adjacent to the reactor zones showed temperatures in the range of 400 to 500°C, which may represent temperatures of fluids during the operation of the reactors (Gauthier-Lafaye et al. 1996). Before criticality, the temperature of sediments buried at a depth of 1.5 km was about 160°C. A steep thermal gradient of 100°C m^{-1} between reactor cores and host rocks is suggested by oxygen isotopes in reactor clays (Gauthier-Lafaye et al. 1989; Pourcelot and Gauthier-Lafaye 1998a). Heat generated by nuclear reactions probably dissipated in rocks within 20 m from the reactor zones (Gérard et al. 1998).

Heat from nuclear reactions turned ambient groundwaters into hydrothermal fluids capable of dissolving quartz in sandstone adjacent to uraninite. These hydrothermal fluids dissolved sandstone, which was replaced by clay minerals (Gauthier-Lafaye 1986). Dissolution of both quartz and uraninite and reprecipitation of U in the reactors may have further enriched the U ore. Consumption of water due to argilitization of sandstone caused the reactors to shut down periodically as there was not enough moderator for slowing fast neutrons. After the reactors cooled and circulating groundwaters penetrated the reactors again, fission reactions began anew. Thus, at least some reactors operated intermittently.

The amount of U consumed in fission reactions varies among reactors. Approximately 3600 kg of ^{235}U was fissioned in RZ-1 and RZ-2 (Naudet 1991) as compared to 480 kg in RZ-7 to RZ-9 and 650 kg in RZ-10 (Gauthier-Lafaye et al. 1996). The amount of energy produced by RZ-1 to RZ-10 was about 20 000 MW per year. The most energy was produced by RZ-1 to RZ-6 (16 500 MW yr^{-1}). Neutron fluxes were on the order of 10^7 to 10^8 n cm^{-2} s^{-1}; i.e. much lower than in typical man made nuclear-power reactors (10^{13} to

10^{14} n cm^{-2} s^{-1}), but comparable to some research reactors (Celinski 1997). The maximum burnup of ^{235}U reached in natural reactors during the several thousands of years of their operation was slightly less than burn-ups of nuclear fuel achieved in modern power reactors after 3 years.

The major difference betwen conditions under which fission occurred in the natural fission reactors and the conditions of fission reactions in man-made reactors is in the role of water. In the natural reactors, hydrothermal fluids directly contacted uraninite during the fission reactions. Hot water in man-made reactors does not contact UO_2 fuel pellets, so that fission reactions occur under dry conditions (Janeczek et al. 1996).

MINERALOGY OF NATURAL FISSION REACTORS

The major minerals in the reactor zones are uraninite, illite, chlorite, coffinite, and galena. Accessory minerals are either neogenic or are relicts of detrital material (e.g. zircon). Neogenic minerals are of special interest because they commonly incorporated fission products into their structures. Moreover, trace-element compositions of neogenic minerals can be used as geochemical indicators.

The mineral assemblages of reactor cores and their associated clay mantles are not significantly different from each other. They differ in the uraninite/clay ratio, which is highest in reactor cores. It was recognized early that samples from the Oklo U deposits are extremely heterogeneous with respect to most of their physical, chemical and mineralogical properties (Dran et al. 1975).

Uranium minerals

Uraninite

Uraninite is the major U mineral in all U deposits in the Franceville basin. Petrographic descriptions of uraninite that occurs outside the reactor zones are given by Geffroy (1975). He noticed that fine-grained, massive uraninite ("pitchblende") and tiny xenomorphic grains of uraninite occurred in sandstones; whereas, well-crystallized uraninite is typical of the reactor zones. "Pitchblende" and xenomorphic uraninite in sandstone are dispersed within pores and are embedded in solid bitumen. Uraninite veinlets in sandstone commonly observed.

There are two major types of uraninite occurrences in the reactor cores: (1) massive uraninite and (2) euhedral, angular or rounded grains dispersed in both a clay matrix and solid bitumen (Janeczek and Ewing 1995). The massive uraninite consists of densely packed cubic, cubo-octahedral, and octahedral crystals and large irregular grains embedded in a clay matrix (Fig. 9). Some crystals display oscillatory zoning (Holliger 1992; Gauthier-Lafaye et al. 1996). While the oscillatory zoning is a crystal growth feature, there is additional zoning caused by alteration processes (Janeczek and Ewing 1992, 1995, 1996a). Some uraninite specimens, particularly in RZ-10, have highly reflectant cores surrounded by lower contrast rims (Fig. 10). The boundary between the core and rim is highly irregular and patchy. The core of the crystal is rich in Pb and represents the original uraninite. Its age was determined at 1968 ± 50 Ma (Holliger 1992).

Grain sizes of uraninite range from a few microns to a few millimeters. Some samples display a "porphyritic" texture due to relatively large grains of uraninite within a "mush" of fine-grained uraninite. Geffroy (1975) described uraninite aggregates that consisted of subhedral, slightly rounded crystals of uraninite arranged in a "cog-wheel" manner. A peculiar "fibrous" habit of uraninite can be observed in some samples from RZ-10 and RZ-Bangombé (Fig. 11). The "fibrous" appearance is an artifact of the intersection of uraninite

334 URANIUM: Mineralogy, Geochemistry and the Environment

Figure 9. Reflected light optical micrograph of uraninite aggregates from the Bangombé reactor zone. Xenomorphic grains of uraninite (bright) occur in the illite matrix.

Figure 10. BSE image of zoned idiomorphic uraninite crystals from RZ-10. The high-contrast core of the crystals is Pb-rich. The core is a relict of the original 1.97 Ga old uraninite crystal. The low-contrast zone surrounding the core is Pb-depleted and enriched in Si and Ca. The crystals have both embayed grain boundaries indicative of dissolution and outgrowths indicative of recrystallization. Black background is Mg-chlorite.

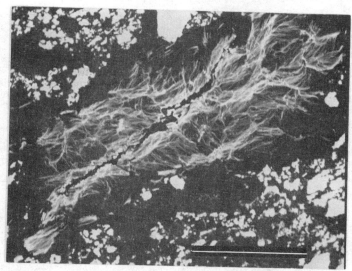

Figure 11. BSE image of "fibrous" aggregates of uraninite in Al-rich clay minerals. Het Al.-rich clay minerals are surrounded by illite in which only angular grains of uraninite occur. Scale bar 100μm.

platlets that have precipitated on the cleavage planes of clay minerals.

Analysis of grain morphologies and textural observations suggest that there were at least three stages of uraninite crystallization (Dymkov et al. 1979; Eberly et al. 1994). In some cases, partially resorbed crystals of uraninite are rimmed by outgrowths, which are indicative of recrystallization (Fig. 10).

Uraninite is often brecciated and fractured. There are a variety of mineralized fractures filled with calcite, apatite, galena, and pyrite. Cracks in uraninite developed predominantly due to tectonic shear and extensile stresses (Eberly et al. 1994). Chemical alteration and replacement by coffinite, seen in back-scattered electrons (BSE) images as low contrast regions, occur along cracks, fractures and grain margins (Janeczek and Ewing 1992).

Uraninite commonly exhibits various stages of alteration and dissolution (Fig. 12). Features indicative of uraninite dissolution in natural reactors include embayed and fractal grain boundaries, relicts of undissolved material, and replacement of dissolved grains by coffinite and other minerals (Janeczek and Ewing 1992, 1995; Bros et al. 1995; Jensen et al. 1997; Jensen and Ewing, in press). Commonly, interiors of uraninite grains were partly or entirely dissolved, forming the "atoll-like" textures (Fig. 12). Holes in the centers of uraninite grains are filled with clay minerals or with mixtures of coffinite, Ti-oxide, and galena. Some uraninite grains are also partly or entirely replaced by coffinite.

X-ray powder diffraction patterns of uraninite from the natural reactors reveal uraninite speciemnts with a wide range of values for the unit cell parameter (Table 4). Sunder et al. (1995) noted that uraninite from RZ-13 gave an X-ray powder diffraction pattern with symmetric peaks, indicative of single-phase uraninite. In contrast, an X-ray powder diffraction pattern for uraninite from RZ-10 showed broad, slightly asymmetric peaks due to overlapping peaks from two cubic phases with different unit-cell volumes (Janeczek and Ewing 1992, 1996a). Unit-cell parameters for this uraninite sample were a_1 = 5.495 Å and a_2 = 5.455 Å. The largest unit-cell parameters are for Pb-rich uraninite with an average of 17.2 wt % Pb, which occurs as relicts in Pb-depleted (6.95 wt % Pb) uraninite with *opposite page*

Figure 12. BSE images of uraninite grains from RZ-10 showing various stages of alteration and dissolution. The lower contrast of the core of the polycrystalline aggregate in image (A) is caused by an increased content of Si and depletion in Pb due to initial alteration. A portion of the core was unaltered. At more advanced stages of alteration (B), the core of the uraninite grains dissolved and replaced by Mg-chlorite and illite. Note the enhanced contrast at the edges of the grain due to elevated concentration of Pb. Extensive fracturing initiated dissolution of uraninite grains along microfratures (C), resulting in a "web-like" pattern of altered regions. Further brecciation of grains and dissolution along fractures (D) caused disintegration of uraninite grains and, in some cases, their disappearance. There are altered (low contrast) regions in uraninite at the interface with clay (D). From Janeczek and Ewing (1995).

Table 4. Unit-cell parameters for uraninite from the natural fission reactors in Gabon

	RZ-2	RZ-9	RZ-10	RZ-13	RZ-16	OK	Ba	Unspec
	5.4346[a]	5.4444[a]	5.495[c]	5.4524[a]	5.4523[a]	5.4565[a]	5.4217[a]	5.49[e]
a, Å	5.4310[a]		5.455[c]	5.446[d]				5.44[e]
	5.4269[a]		5.4521[a]					5.39[e]
	5.44[b]		5.46[d]					5.44[f]
	5.40[b]		5.44[d]					5.43[f]

[a] Pourcelot and Gauthier-Lafaye (1998); [b] Simpson and Bowles (1978); [c] Janeczek and Ewing (1992); [d] Sunder et al. (1995); [e] Kruglov et al. (1975); [f] Weber et al. (1975). Unspec = samples from the unspecified reactor zones, most probably from RZ-2. OK = Okelobondo; Ba = Bangombé

smaller unit-cell parameter (Janeczek and Ewing 1996a). Pourcelot and Gauthier-Lafaye (1998) tried to relate reactor depth with unit-cell parameter and the full width at half-maximum intensity (FWHM) of the 111 diffraction peak of uraninite. They concluded that uraninite close to the surface is poorly crystalline as compared to deep-seated reactors. However, the FWHM of uraninite from the shallow Bangombé reactor (FWHM = 0.388) located at 12 m below ground is the same as for RZ-10 situated 450 m below ground (Pourcelot and Gauthier-Lafaye 1998). The difference between the largest uraninite unit-cell parameter (5.4565 Å from the Okelobondo reactor) and the lowest (5.4217 Å from the Bangombé reactor) observed by Pourcelot and Gauhtier-Lafaye (1998) is, in fact, smaller than the difference between the largest and the smallest unit-cell parameters measured by Janeczek and Ewing (1992, 1996a) from a sample from RZ-10 (0.040 Å) and much smaller than the difference between the largest and smallest unit-cell parameters (0.10 Å) observed by Kruglov et al. (1975) in uraninite from an unspecified reactor. Oklo uraninite with small unit-cell parameters have lower concentrations of interstitial Pb and, perhaps, a higher ratio of U^{6+}/U^{4+}. Unit-cell expansion of relicts of unaltered uraninite is caused mainly by the accumulation of radiogenic Pb, most probably in interstitial sites in uraninite.

Despite the large neutron flux and high doses of alpha-decay events, (i.e. recoils and alpha-particles) experienced by uraninite in the natural reactors, uraninite remains crystalline due to rapid annealing of radiation-induced defects. As an example, an alpha-decay event dose, of $D_\alpha = 1.4 \times 10^{18}$ α-decay events/mg, was calculated for uraninite from RZ-10 with 66.24 wt % U and a "chemical" age of 1800 m.y. (determined from U and Pb concentrations). TEM observations of this uraninite revealed only a few dislocation loops in otherwise perfectly crystalline material (Janeczek and Ewing 1996a). These are, perhaps, extended defects formed in the process of coalescence of migrating point defects. High-resolution transmission electron microscopy (HRTEM) of uraninite from RZ-10 reveal polycrystalline aggregates of nm to μm sized crystallites within single grains (Fig. 13) (Jensen and Ewing, in press). Crystallites are at high angles one to another and display slightly bent lattice fringes. The d-spacings of lattice fringes correspond to uraninite with

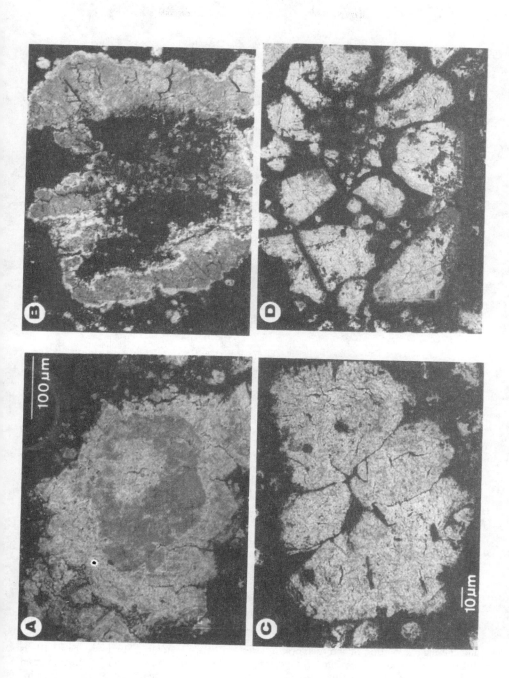

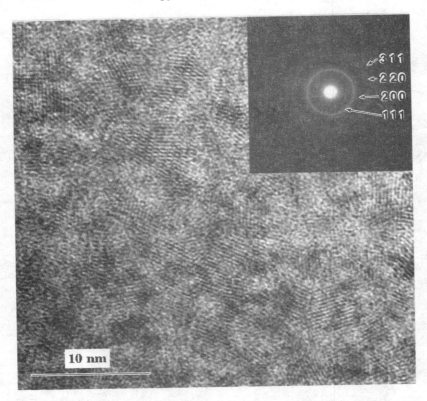

Figure 13. High resolution TEM lattice image and electron diffraction pattern of uraninite from the core of RZ-10. Courtesy of K. Jensen.

the unit-cell parameter of 5.46(6) Å (Jensen, pers. comm.). Low angle grain boundaries and distinct crystallites with high angle grain boundaries are microstructural features indicative of high radiation damage in uraninite (Casas et al. 1998) Polycrystalline regions in uraninite from natural reactors may also be attributed to the structural response of the annealing process of radiation damage or may be due to recrystallization of the uraninite during cooling of the reactor after termination of criticality. Close to the phyllosilicate matrix, the crystallite diameters in the uraninite usually decreases to 5 to 10 nm (Jensen and Ewing, in press).

Electron microprobe analyses of uraninite from the natural fission reactors are given by Simpson and Bowles (1978), Janeczek and Ewing (1992, 1993, 1995), Gauthier-Lafaye et al. (1996), Jensen et al. (1997), Savary and Pagel (1997), Jensen and Ewing (1998, in press) and Pourcelot and Gauhtier-Lafaye (1998). All microprobe analyses reveal elements incompatible with the uraninite structure; i.e. Ti, Si and Fe (Table 5), which are here collectively termed incompatible impurity elements (IIE), in contrast to compatible impurity elements (CIE). Lead is a major IIE in uraninite, but because of high Pb concentrations, its radiogenic origin and its important role in understanding the physical and chemical evolution of the reactors, it is treated here separately. The presence of IIE is due to the alteration of uraninite. As the result of alteration, uraninite and its alteration and replacement products are often intimately intergrown on a microscopic or even sub-

Table 5. Ranges of major and minor elements average concentrations in uraninite from natural fission reactors in Gabon based on electron microprobe analyses (wt %)

RZ:	2	8	9	10	13	16	Ba	OK
UO_2	80.5-89.1	89.1	87.2-89.2	72.3-81.6	86.1-89.9	86.4-90.5	87.0-90.0	87.4-90.2
ThO_2	<0.1-0.2	n.d.	<0.1	0.0-0.3	<0.1-0.2	0.0-0.1	0.0-0.2	<0.1
SiO_2	0.5-0.9	0.5	0.4-0.8	0.0-0.9	0.5-0.9	0.3-0.9	0.1-0.8	0.6-0.7
TiO_2	<0.1-0.4	0.0	0.1-0.3	<0.1-0.4	0.0-0.5	0.0-0.4	<0.1-0.7	<0.1-0.2
ZrO_2	0.1-0.3	n.d.	0.1	0.1-0.3	0.1	n.d.	0.1-0.3	<0.1
PbO	6.0-12.4	4.8	4.8-5.9	11.7-21.9	5.5-6.3	5.3-6.8	4.7-6.4	4.9-6.43
CaO	0.8-1.6	0.9	1.0-1.7	0.3-1.6	1.3-2.0	1.2-2.3	1.1-1.8	0.9-1.5
FeO	0.3-0.5	0.3	0.4-0.6	0.0-0.4	0.2-0.6	0.3-0.5	0.5-0.9	0.5-0.6
RE_2O_3	0.46		0.52	0.22-0.79	0.20	0.14	0.10-0.56	0.11

Note: Numbers in parantheses refer to the number of analyses reported in the literature: Janeczek and Ewing (1992, 1995), Jensen and Ewing (in press) (RZ-2, RZ-9, RZ-10, RZ-13, RZ-16, Bangombé, Okelobondo); Gauthier-Lafaye et al. (1997) (RZ-8, RZ-10, RZ-13); Jensen and Ewing (1998) (Bangombé); Savary and Pagel (1997) (RZ-10 and RZ-16); Pourcelot and Gauhtier-Lafaye (1998) (RZ-9, RZ-10, RZ-13, RZ-16). Only analyses in which concentrations of each incompatible impurity element was below 1.0 wt % were considered for the preparation of the table.

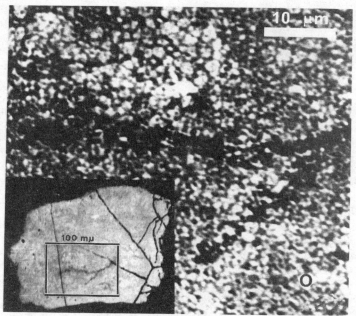

Figure 14. BSE image of the interior of uraninite grain (shown in the inset) within bitumen from the Okelobondo reactor zone. The uraninite is highly altered and partially replaced by clay and coffinite. Circle at lower right represents a nominal electron-beam size of 1 μm, which illustrates the difficulty of obtaining accurate EMP analyses of single-phase material from altered uraninite.

microscopic scale. Therefore, even with a nominal electron-beam diameter of 1 μm there is always a possibility that elements from adjacent material are included in analyses (Fig. 14). High Ti contents in uraninite analyses from some of the reactors (RZ-2 and Bangombé) is due to both Ti-oxides finely dispersed in the uraninite matrix and Ti-oxide coatings on the uraninite grains (Janeczek and Ewing 1995). Only analyses with IIE less than 1 wt % (as oxide) are included in Table 5.

Other impurity elements comonly reported in electron microprobe analyses of uraninite from fission reactors include Ca, Th and rare earths elements (REE) (Table 5). These are compatible impurity elements, as their ionic radii and crystal-chemical properties enable them to proxy for U in the fluorite-type structure. However, contributions from peak-overlaps of Ca, REE and U have to be considered and appropriate correction procedures must be undertaken during the electron microprobe analyses of uraninite to avoid false results. A strong positive correlation between Ca and U was noted in chemical analyses of whole-rock samples from RZ-2 (Branche et al. 1975), which, together with ion microprobe data and BSE images (Dran et al. 1975; Havette et al. 1975), suggested that Ca was in uraninite, not in inclusions of Ca minerals. However, a negative correlation between U + Pb (primordial elements) and Ca + Si was found in uraninite from RZ-10 and Ca distribution in altered crystals followed the same pattern as Si (Janeczek and Ewing 1995). This is particularly evident in the distribution of Si and Ca in zoned crystals from RZ-10 (Fig. 15). Whereas the Si content is low in the Pb-rich core of the crystal, the amount of Si increases by an order of magnitude (from less than 1 wt % to over 1 wt %) in the outer, Pb-depleted alteration zone. Calcium distribution follows the same pattern. Observations by HRTEM revealed nanometer-sized inclusions of calcite adjacent to uraninite grains (Jensen, pers. comm.). Obviously such nanoinclusions and microveins may contribute Ca to the electron microprobe analyses of uraninite. On the other hand, there are uraninite specimens with low concentrations of IIE and relatively high concentrations of Ca (Fig. 16). This is especially true for relicts of unaltered uraninite. Therefore, it is possible that, though Ca can occupy structural sites in uraninite, some Ca reported in microprobe analyses of uraninite is due to microscopic inclusions of other phases or is an artifact of the data collection problems.

Concentrations of REE determined from electron microprobe analyses of uraninite from natural reactors usually are below 0.5 wt %, with Ce being dominant (Janeczek and Ewing 1992; Jensen and Ewing 1998). The highest concentrations of REE (Ce_2O_3 = 1.04 wt % and Nd_2O_3 = 0.90 wt %) were reported for uraninite from sandstone in the Oklo open pit, 326 m from the closest reactor zone (Jensen and Ewing, in press).

Based on major element (U, Pb, Ca, Si) chemistry, uraninite specimens from the natural reactors can be grouped into four categories (Janeczek and Ewing 1995):

1. Uraninite crystals with Pb-rich (up to 20 wt % PbO) cores and Pb-depleted (7 wt % PbO), Si- and Ca-enriched altered regions. These uraninite crystals occur only in RZ-2 and RZ-10.
2. Unzoned, Pb-rich uraninite (PbO > 10 wt %; up to 19 wt % PbO) with low silica contents (SiO_2 < 0.5 wt %). This uraninite occurs only in RZ-2 and RZ-10.
3. Uraninite with relatively low Pb contents (average of 5.2 wt % PbO) and low Si contents.
4. Uraninite with relatively low Pb contents (average of 5.2 wt % PbO) and high silica contents (SiO_2 > 0.5 wt %).

The distinction between low and high Si contents is arbitrary. As seen in Figure 16, when impurity elements are considered there are two major groups of chemical

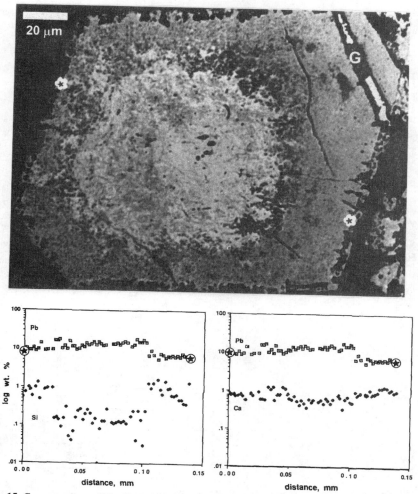

Figure 15. Concentrations of Pb, Si and Ca across a crystal of zoned uraninite from RZ-10. Asterisks on the crystal mark the extreme points of an electron microprobe traverse across the crystal. G = galena. Modified from Janeczek and Ewing (1995).

compositions of uraninite: those in which Pb/(Pb+Ca+IIE) > 85% and those for which this ratio is less than 85%.

There is a striking difference in Pb concentrations between uraninite from RZ-10 and uraninite from other reactor zones (Table 5 and Fig. 17). Analyses of uraninite from RZ-10 cover a wide range of Pb concentrations from over 19 wt % to 5.2 wt %; whereas, the greatest difference between the highest and the lowest Pb concentrations in uraninite from other reactors is less than 2 wt % (Fig. 17). As a result, Pb contents in uraninite from all reactor zones (except for RZ-10 and one sample from RZ-2) are similar and cluster around a mean value of 5.15±0.40 wt % Pb. Values for the standard deviations for PbO, and, in most cases, for UO_2, are an order of magnitude higher for uraninite from RZ-10 than for uraninite from other reactors, a result of chemical heterogeneity among uraninite specimens

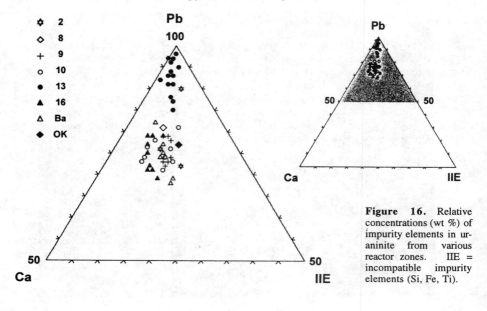

Figure 16. Relative concentrations (wt %) of impurity elements in uraninite from various reactor zones. IIE = incompatible impurity elements (Si, Fe, Ti).

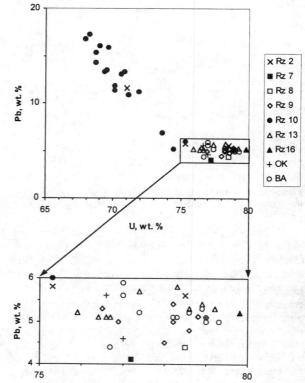

Figure 17. Concentrations of Pb vs. those of U in uraninite from natural fission reactors. The boxed area contains data for uraninite from reactor zones other than RZ-10.

in RZ-10 (Janeczek and Ewing 1995).

Low analytical totals for some analyses in Table 5 are due, in part, to analytical errors, but they may also be caused by hexavalent U. Values of the U^{6+}/U^{4+} ratio determined from XPS spectra of bulk samples from RZ-10 and RZ-13 range from 0.10±0.03 to 0.19±0.05 (Sunder et al. 1992, 1995). The ratio of U^{6+}/U^{4+} = 0.19 gives x = 0.16 in the idealized formula of uraninite, UO_{2+x}, if no other cations are considered. Identical value of 0.19 for the U^{6+}/U^{4+} ratio was obtained for uraninite from RZ-Bangombé based on stoichiometric considerations (Janeczek and Ewing 1995). These values are rather high and it is not clear if they are primary or are caused by partial oxidation of the uraninite.

Table 6. Chemical compositions (wt %) of LWR spent fuel and uraninite from two reactor zones.

	LWR[+]	RZ-16	RZ-10
UO_2	95.4	90.5	77.4
PuO_2	0.9	Utr	Utr
AcO_2	0.1	–	–
F.P.	3.6	?	0.3 – 0.5
PbO	–	5.6	18.7
ThO_2	–	0.1	0.1
CaO	–	1.2	0.3
FeO	–	0.5	0.1
SiO_2	–	0.3	0.3
U^{6+}/U^{4+}	0.001	?	0.10 – 0.19

Note: [+]for burn-up of 35 Mwd/kg U, 10 years after discharge from the reactor. Ac. = Np and Am; F.P. = fission products.

Uraninite in natural fission reactors is a natural spent nuclear "fuel," but its properties are the result not only of nuclear reactions but of geological processes, the later having the greatest impact. The chemical composition of uraninite from natural reactors is significantly different from irradiated UO_2 in man-made spent nuclear fuel (Table 6).

Recently Pourcelot and Gauthier-Lafaye (1998) attempted to measure O isotopes in uraninite from natural reactors. Oxygen was extracted from ~10 mg of uraninite concentrates. Uraninite samples were highly depleted in ^{18}O and δ ^{18}O values ranged from -0.2‰ to -13.2‰ SMOW. These data are of limited use only, because the material analyzed was contaminated with additional minerals that contributed to the O yield. Nevertheless, an important conclusion from the study by Pourcelot and Gauthier-Lafaye (1998) is that there is no significant difference in δ ^{18}O between near-surface reactors (RZ-2, Bangombé); i.e. those exposed to oxidizing groundwater and deep-seated reactors at Oklo and Okelobondo.

Coffinite

Coffinite is the second most abundant uraniferous mineral in and around all natural fission reactors. Coffinite was recognized by early investigators of the Oklo deposit (Geffroy and Lissillour 1974; Geffroy 1975). Yet, there are only a few good quality chemical analyses with the coffinite stoichiometry reported in the literature (Table 7). This is due to the fine-grained masses of coffinite and its intimate intergrowths with uraninite and clay minerals. Therefore, even electron microprobe analyses performed with an electron beam size of 1 μm are not always free from impurities. Chemical analyses with a few weight percent SiO_2 may reflect mixtures of uraninite, coffinite and clay minerals in various proportions. There are not enough reliable chemical analyses of coffinite from Oklo and Bangombé to make general statements about the chemistry of coffinite among the various reactors. Coffinite from sandstone beneath the Bangombé reactor is enriched in phosphorous and REE to the extent that it forms a solid solution with ningyoite, $(U,Ca,Ce)(PO_4)_2 \cdot 1-2H_2O$ (Janeczek and Ewing 1996b); whereas, coffinite from the core

Table 7. Electron microprobe analyses of coffinite from natural fission reactors and their vicinity.

Wt % oxide	$USiO_4$	RZ 10*	RZ 10#	RZ Ba+	RZ Ba+	RZ Ba^i	RZBa^i	RZBa^i
SiO_2	16.41	15.02	16.22	7.94	8.18	5.01	4.88	6.31
UO_2	73.75	66.26	71.42	68.72	69.75	69.28	71.25	66.04
ThO_4			0.04	0.18		0.42	0.56	0.27
TiO_2		0.24	0.41	0.29	0.54	0.63	0.48	0.35
ZrO_2						0.12	0.65	0.47
Al_2O_3			0.91	0.32	0.46	0.51	0.26	0.42
La_2O_3		0.15		0.00	0.00			
Ce_2O_3		0.28	0.24	0.74	0.60	0.70	0.66	0.75
Pr_2O_3				0.24	0.15			
Nd_2O_3				1.24	0.87	1.86	1.38	1.82
Sm_2O_3				0.34	0.24			
FeO				0.20	0.23	0.55	0.49	1.39
PbO		1.34	2.79	0.79	1.08	1.56	1.70	0.82
CaO		2.30	1.04	1.93	1.77	1.06	0.78	0.95
P_2O_5		0.50		8.68	7.67	8.00	7.19	7.93
SO_3						2.50	3.77	3.17
Total	100.00	86.23	94.31	90.69	91.54	92.24	94.10	90.71
Cations on the basis of 4 oxygens								
U	1.00	0.92	0.90	0.88	0.90	0.89	0.90	0.82
Ti		0.01	0.02	0.01	0.02	0.03	0.02	0.01
Ce		0.01		0.02	0.01	0.01	0.01	0.01
Nd				0.025	0.02	0.04	0.03	0.04
Fe			0.01	0.01	0.01	0.03	0.02	0.06
Pb		0.02	0.04	0.01	0.02	0.02	0.03	0.01
Ca		0.15	0.06	0.12	0.11	0.07	0.05	0.06
Si	1.00	0.94	0.92	0.46	0.48	0.29	0.28	0.35
Al				0.02	0.03	0.03	0.02	0.03
P		0.03		0.42	0.38	0.39	0.35	0.37
S						0.11	0.16	0.13

Note: *Janeczek and Ewing 1992, #Janeczek and Ewing 1995, +Janeczek and Ewing 1996c; ^{i}Jensen and Ewing, 1998

of RZ-10 has much lower concentrations of phosphorous and rare earths (Table 7). The coupled substitution, $Si^{4+} + U^{4+} = P^{5+} + (LREE)^{3+}$, may explain P- and LREE-enriched coffinite from Bangombé. Jensen and Ewing (1998) described P- and S-rich coffinite enriched in LREE from sandstone beneath the Bangombé reactor (Table 7) and suggested that S substituted for Si in those coffinite specimens. In RZ-10, uraninite in fractures and fissure veins is commonly intimately associated with or embedded in a matrix of a U-Zr-silicate. The U-Zr-silicate is locally the major phase in the core of RZ-10 (Jensen and Ewing, in press). It is possible that the U-Zr-silicate is Zr-rich coffinite, which forms limited solid-solution with zircon.

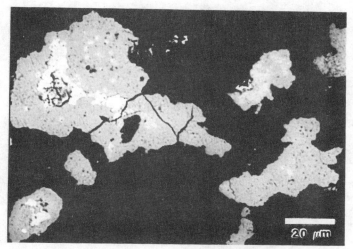

Figure 18. A cluster of uraninite (white) grains replaced by P-rich coffinite (grey) in sandstone 20 cm below the core of the Bangombé reactor. BSE image. Black background is quartz. Uranium in both uraninite and coffinite is isotopically depleted ($^{235}U/^{238}U = 0.0061$).

Low analytical totals in electron microprobe analyses of coffinite from Oklo and Bangombé (Table 7) suggest the presence of water. Whether it occurs as molecular water or as structurally bonded OH groups, or both, is not readily determined. Low analytical totals may also result from U^{6+}.

Paragenetic relationships among uraninite and coffinite (Fig. 18) indicate that coffinite replaced uraninite (Janeczek and Ewing 1992, 1995, 1996c; Jensen and Ewing 1998, in press). This process was particularly pervasive along fractures and grain boundaries (Fig. 19). Geffroy (1975) postulated that some "pitchblende" in the surroundings of a natural reactor formed at the expense of coffinite.

Uranyl minerals

Uranyl minerals in the natural fission reactors are rare and their occurrences are limited to weathered zones and fractures accessible to oxidizing groundwater. Uranyl minerals are abundant on surfaces of mined or drilled samples, on exposed walls in the open pit mine at Oklo, and in tunnels and shafts. In some cases uranyl minerals are also found in sandstones beneath the reactor zones. Uranyl minerals identified in and around natural reactors include: alpha-uranotile, torbernite, fourmarierite, rutherfordine, and wölsendorfite (Geffroy and Lissillour 1974; Branche et al. 1975; Geffroy 1975). These minerals were found in normal U ore exposed in an unspecified part of the open pit mine at Oklo. Bourrel and Pfiffelman (1972) identified francevillite in the weathered zone at Oklo. Francevillite, $(Ba,Pb)(UO_2)_2(V_2O_8) \cdot 5H_2O$, and vanuralite, $Al(UO_2)_2(V_2O_8) \cdot 11H_2O$, were originally described from the Mounana deposit north of Oklo (Geffroy et al. 1964). Both minerals occur at the surface and are associated with "pitchblende," minor coffinite and other vanadium minerals including, karelianite, montroseite and roscoelite. The Mounana deposit has yielded two more new uranyl minerals: curienite, $Pb(UO_2)_2(V_2O_8) \cdot 5H_2O$, and metavanuralite, $Al(UO_2)_2(V_2O_8)(OH) \cdot 8H_2O$, (Gauthier-Lafaye and Weber 1989). Vanadium minerals have not been reported from the natural reactors.

Uranyl minerals described recently from sandstone beneath the Bangombé reactor

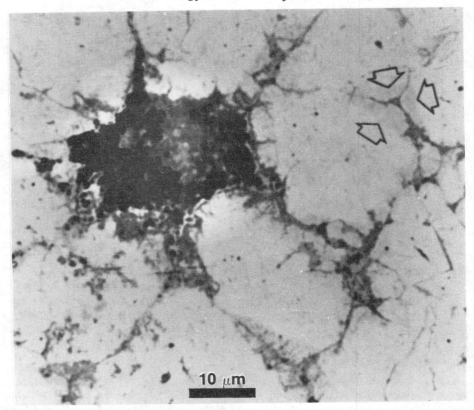

Figure 19. BSE image showing various stages of coffinitization of uraninite from RZ-10. Arrows point to incipient coffinite (light grey) at the triple point of the uraninite grain boundary. Coffinite (grey) and chlorite (black) entirely replaced uraninite in the central part of the image. Scale bar 10 μm. From Janeczek and Ewing (1992).

include françoisite-(Nd), $RE(UO_2)_3O(OH)(PO_4)_2 \cdot 6H_2O$, (Janeczek and Ewing 1996b), uranopilite or zippeite, schoepite, unspecified uranyl orthophosphate (Jensen et al. 1997), iron uranyl phosphate hydrous with a composition of $Al_{0.06}Fe_{0.89}(UO_2)_3(PO_4)_3 \cdot nH_2O$ and uranyl sulfate hydroxide hydrate with a composition between $(UO_2)_9(SO_4)(OH)_{16} \cdot nH_2O$ and $(UO_2)(SO_4)_2(OH)_{30} \cdot nH_2O$ (Jensen and Ewing 1998).

Elongated, bladed crystals of françoisite-(Nd) have been observed in the secondary porosity of the quartz arenite some 20 cm outside the reactor core. Only light REE have been detected in that mineral. The Ce/Nd ratio is equal to one. Françoisite-(Nd) from Bangombé is enriched in radiogenic Pb (2.55 wt % PbO). Analyses by SHRIMP have shown that U in the françoisite-(Nd) is depleted in ^{235}U (Hidaka, pers. comm.).

A hydrated iron uranyl phosphate and a hydrated uranyl sulfate occur within dissolution voids in uraninite. Both minerals also occur in microfractures and, in some cases, rim uraninite. The iron uranyl phosphate also occurs in voids of the silicified breccia zones in association with pyrite. Some of the uranyl sulfates observed by Jensen and Ewing (1998) formed during sample preparation after cutting the sample in water. This author has observed clusters of uranopilite crystals on a goethite-coated fracture surface

Figure 20. Scanning electron microscope image of an aggregate of platy crystals of uranopilite on a goethite-lined fracture surface beneath the Bangombé reactor. Scale bar: 5 μm.

from sandstone adjacent to the Bangombé reactor (Fig. 20). Recently, torbernite, dated at 26500 y, was found in a fracture 1.5 m beneath the Bangombé reactor (Bros et al. 1998).

In RZ-2, RZ-9 and at Okelobondo, minor amounts of unidentified Pb-uranyl sulfate hydroxides were observed in uraninite corrosion rims and at the interface between altered uraninite and corroded galena (Jensen and Ewing, in press). Calcium-uranyl sulfate hydroxide hydrates associated with a Cu-uranyl sulfate hydroxide hydrate occur in coatings on the sandstone walls in galleries accessible from the Oklo open pit (Jensen and Ewing, in press).

Uranyl phosphates and uranyl sulfates are the most abundant uranyl minerals in and around natural reactors. Their formation is apparently a product of water-rock interactions. Detrital monazite, florencite and apatite, all of which occur in the sandstone matrix, may have been the source of phosphorous for uranyl phosphates formation. The abundance of uranyl sulfates formed under ambient conditions suggests that oxidation of pyrite is a significant source of oxidized sulfur.

Major associated minerals

Clay minerals

Illite and chlorite are the major clay minerals in the reactor zones. They predominate over uraninite in the clay mantle of the reactors and constitute a significant portion of reactor cores. Illite is abundant in most of the reactor cores, except for RZ-10, which is dominated by chlorite. Chlorites are a major constituent of clays of the reactor except for the Bangombé reactor, in which illite predominates over chlorite. In RZ-1 and RZ-2 clay minerals envelope U ore (Fig. 3), whereas in other zones argillaceous rocks are interbedded with sandstone of the C1 layer (Gauthier-Lafaye et al. 1989).

Chemical analyses of illite from various natural fission reactors are reported by Gauthier-Lafaye et al. (1989, 1996) and Eberly et al. (1996). Some chemical analyses of illite show elevated concentrations of trivalent Fe, perhaps due to inclusions of fine-grained hematite. Illite polytypes reveal thermal histories of reactor zones. The associations of illite polytypes with each unit in and around reactor zones (illite polytypes in parentheses) are as follows (Gauthier-Lafaye et al. 1989): core (1M); reactor clays (1M and 2M); U-bearing sandstone (1M); barren sandstone (1Md).

The high-temperature 2M illite polytype in the clay mantle of reactors (mainly in RZ-2) is related to the heat generated by nuclear reactions. Analyses of oxygen isotopes in illite indicate temperatures between 200° and 225°C, consistent with the 1M to $2M_1$ transition in illite (Gauthier-Lafaye et al. 1989). The low-temperature polytype of illite in reactor cores is explained by its crystallization during cooling of the reactors, in the temperature range of 150° to 125°C. This conclusion is supported by certain growth features of illite in reactor cores. Illite crystallized perpendicular to the faces of uraninite crystals (Gauthier-Lafaye and Weber 1978) and filled dissolution vugs in uraninite (Janeczek and Ewing 1992). Crystallite sizes of illite from the core of the Bangombé reactor and those of some illite specimens from the clay mantle are smaller (45 to 52 nm) than for illite from the clay mantle (63 to 67 nm) and FB shale (69 to 79 nm) (Eberly et al. 1996). Small crystallite sizes may be caused by radiation damage in illite adjacent to uraninite. Numerous elongated voids with longer axes parallel to the (001) cleavage planes of illite have been observed by this author under TEM in a sample from the core of RZ-10.

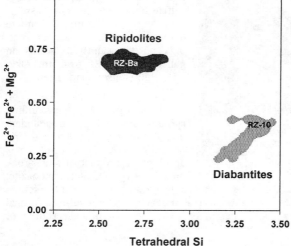

Figure 21. Discrimination diagram for chlorites from RZ-10 and Bangombé reactor. Modified from Eberly et al. (1995).

The chemical compositions of chlorites from natural reactors differ from those of chlorites from the sandstone matrix. The latter are Fe-rich, whereas chlorites from the reactor zones is Mg-chlorites, Fe-Mg-chlorites and Al-chlorites. Chlorites from various reactors also differ in composition. For example, chlorites from the Bangombé reactor are enriched in Fe and contain more tetrahedral Al than chlorite from RZ-10 (Fig. 21). Dioctahedral Al-chlorites (sudoite and donbassite) have been found in RZ-10, Bangombé and Okelobondo (Eberly et al. 1995, 1996; Gauthier-Lafaye et al. 1996; Jensen and Ewing, submitted). Vermiculitized Al-rich chlorites occur in RZ-9, a result recent weathering (Pourcelot and Gauthier-Lafaye 1998b). Mg-chlorites and Al-chlorites in RZ-10 are of hydrothermal origin. Based on oxygen isotopic compositions two genetic groups of chlorites can be distinguished at RZ-10: chlorites located 70 cm from the reactor core formed during criticality; whereas, chlorites closer to the reactor core crystallized during the reactor's cooling stage, as for illite-1M (Pourcelot and Gauthier-Lafaye 1998b). Values of $\delta^{18}O$ for Mg-chlorites and sudoite range from 0 to 15‰ SMOW. Oxidation of Fe in chlorites at the border of RZ-10 may have been caused by water radiolysis (Pourcelot and Gauthier-Lafaye 1998b). Chlorites played important role in trapping migrating radionuclides. Observations of, and experiments on samples from natural reactors both suggest that chlorites are more efficient scavengrs of actinides than is illite (Eberly et al. 1995, 1996).

Kaolinite occurs in reactor zones that have been affected by weathering (Bros et al. 1995; Gauthier-Lafaye et al. 1996). Kaolinite-group minerals are among major constituents of the weathered FB shales.

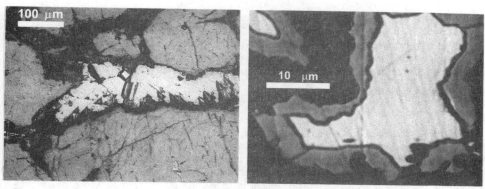

Figure 22. BSE images of galena with radiogenic Pb crystallized among uraninite grains; note elongated galena grains (left image, RZ-9) and coffinite grain from RZ-10 with center entirely occupied by galena.

Galena

Galena is ubiquitous in all reactor zones. Galena occurs within uraninite grains as spherical and lensoidal inclusions and veinlets, as well as within the clay matrix and in solid bitumen (Fig. 22). Galena crystals that grew freely in clay minerals are commonly euhedral, up to 1 mm across. In some uraninite crystals from RZ-10 and RZ-13, single grains of galena account for 40 vol % of the uraninite (Fig. 22). In numerous samples, the number of galena inclusions depends on grain size. Inclusions are abundant in uraninite grains 10 μm across and larger. Smaller uraninite grains are free of galena. In some uraninite crystals, galena has precipitated along grain boundaries, subgrain boundaries, and in fractures, leaving the interiors of uraninite grains free of inclusions (Fig. 23). In other uraninite crystals, interiors are filled with tiny spherical galena particles, whereas rims are free of galena precipitates (Fig. 23).

Trace elements have been analysed in galena from clays of RZ-10 and RZ-13, sulfide veins and in galena from both mineralized and barren sandstone (Raimbault 1992; Peycelon

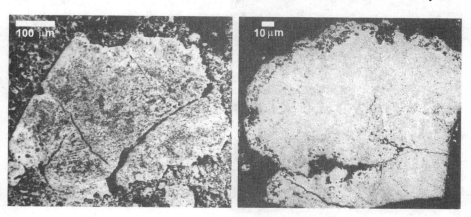

Figure 23. BSE images of uraninite from RZ-9 showing two habits of galena. A "peppery" appearance of the cubo-octahedral crystal of uraninite (left image) is due to abundant tiny galena grains that crystallized within the uraninite at point defects. The annihilation of point defects at the crystal's perimeter resulted in a precipitate-free rim. The oposite effect is evident in the right image that shows galena exsolved along grain boundaries, leaving the interior of the uraninite grain free of precipitates.

and Raimbault cf. Holliger and Gauthier-Lafaye 1996). Galena from reactor zones contains less Ag than galena from either U-bearing sandstone or sulfide veins in which it is associated with chalcopyrite. Galena from the U-bearing sandstone is relatively rich in Se (400 ppm) compared to barren sandstone (0.5 ppm Se). Galena from reactor zones contains ^{98}Mo, which is practically absent in galena from the host rocks.

Accessory minerals

Native elements

Native Pb occurs in RZ-10 (Guthier-Lafaye et al. 1993, 1996; Savary and Pagel 1997) and in solid bitumen from an unspecified reactor zone (Parnell 1996). In RZ-10, native Pb occurs in microfractures in clays of the reactor, as isolated crystals, and as 5-μm thick overgrowths on galena in the reactor's core. Veinlets of native Cu, 5- to 10-μm thick, occur in association with Fe-chlorite cement in sandstone 20 m from RZ-10 (Savary and Pagel 1997).

Ru-, Rh-, and Pd-bearing minerals

Curtis et al. (1989) suggested that, as in the case of anthropogenic spent nuclear fuel, such products of nuclear reactions at Oklo as Ru, Pd, Mo and Tc may have formed mineral assemblages. In many commercial spent UO_2 fuels, Ru, Rh, Pd, together with Mo and Tc, form an alloy (the so called epsilon phase) that precipitates as metal particles along grain boundaries and other defects in UO_2. Curtis et al. (1989) predictions were confirmed by findings of arsenides and sulf-arsenides of Ru, Rh, and Pd, in RZ-10 and RZ-13 (Holliger 1992) (Fig. 24). These minerals contain fissiogenic Ru, Rh, Te, and Pd, together with radiogenic Pb and normal isotopes of S and As (Holliger 1992; Hidaka et al. 1994). Ru-bearing minerals commonly occur at the uraninite grain-boundaries in the core of the Okelobondo reactor zone (Jensen and Ewing, submitted). Grains of those minerals are up to 4 μm in diameter. EDS-analysis revealed Ru and As as major constituents occassionally with minor Co, Ni, and S. Another type of Ru-minerals occur in clays of the Okelobondo reactor. In addition to Ru and As, they contain Pb, Co, and Ni as major elements (Jensen and Ewing, submitted). Ru-bearing minerals have also been found in sandstone at Bangombé some 20 cm below the core of the reactor zone (Janeczek and Ewing 1996b). The Ru-minerals occur as inclusions, about 1.5 μm in diameter, in uraninite altered to coffinite (Fig. 25). Ru-bearing grains, less than 5 μm across, occur in highly silicified sandstone near RZ-10 (Savary and Pagel 1997).

Figure 24. BSE image of an aggregate of fissiogenic Ru-bearing arsenides (arrow) within altered uraninite from RZ-10. White xenomorphic grains and crystals are galena.

Ruthenium-bearing minerals occur as polyphase inclusions in or adjacent to uraninite (Fig. 24) and are termed

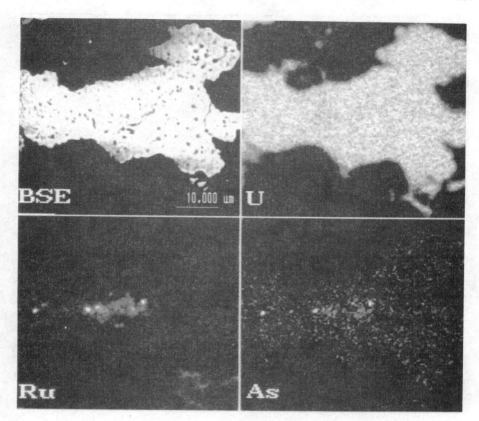

Figure 25. BSE image of coffinite (shown also in Fig. 17) with Ru-bearing arsenide and X-ray maps of U, Ru, and As. Sample from the sandstone in the vicinity of the Bangombé reactor zone.

metallic aggregates by analogy with Ru-bearing metallic segregations in spent nuclear fuel from commercial reactors (Holliger 1992; Hidaka et al. 1994; Gauthier-Lafaye et al. 1996, Savary and Pagel 1997). This term, however, is misPbing because, unlike Ru-rich metals in spent nuclear fuel, Ru, Rh, and Pd in natural reactors are bound to As or S (or both) (Table 8).

Table 8. Compositional ranges for aggregates of PGE minerals from RZ-10 & RZ-13 (at %)

	Ru	*Rh*	*Pd*	*Pb*	*Te*	*As*	*S*	*Se*
RZ 10*	18.12-31.79	3.13 - 6.64	n.r.	14.16 - 28.22	0.30 - 1.13	25.11 - 34.08	3.94 – 19.44	n.r.
RZ 13*	16.94 - 43.63	1.72 - 6.61	n.r.	21.03 - 28.22	1.65 - 4.20	3.60 - 13.64	5.63 – 11.11	n.r.
AGM1**	23.12	4.57	3.83	22.03	1.00	31.63	12.87	0.31
AGM2**	31.79	6.64	3.21	14.16	0.47	35.20	3.94	0.04

Note: *Data from Holliger and Gauthier-Lafaye (1996); ** single analyses from Holliger (1992); n.r. = not reported

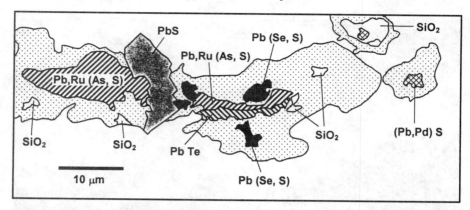

Figure 26. A distribution map of Ru- and Pb-bearing minerals within Zr-rich coffinite (stippled area) in RZ-10. Simplified from Holliger and Gauthier-Lafaye (1996).

Sizes of Ru-mineral inclusions in uraninite from RZ-10 and RZ-13 range from a few microns to 100 μm across. Grains of Ru-minerals at Bangombé and Okelobondo are closer in size to those found in commercial reactor fuels and may represent segregations that occurred during the original operation of the natural fission reactors (Janeczek et al. 1996).

Electron microprobe and ion microprobe studies of Ru-bearing aggregates have shown that they consist of several fine-grained phases intergrown with galena and U minerals (Fig. 26). As a result, chemical data can vary significantly among analyses, precluding unambiguous mineralogical identification of Ru minerals (Table 8). However, elemental and isotopic mapping of Ru-bearing aggregates by Holliger and Gauthier-Lafaye (1996) allows an approximation of their mineral composition (Fig. 26). For instance, grains containing Pb and Te are most probably altaite, and those with Pb, S, and Se may represent a solid solution between galena and claustahalite.

Most, if not all Pb, reported from electron microprobe analyses of Ru minerals is in galena (Holliger and Gauthier-Lafaye 1996; Gauthier-Lafaye et al. 1996). Galena is always associated with Ru-bearing aggregates and may be intimatelly intergrown with them (Figs. 24 and 26). Calculations of mineral formulae from data provided by Holliger (1992), Hidaka et al. (1994), and Holliger and Gauthier-Lafaye (1996), assuming that all Pb and Se are in galena and all Te is in altaite, reveal (for some analyses) stoichiometries close to those of ruthenarsenite, RuAs, and ruarsite, RuAsS. For instance, the ratio of (Ru+Rh+Pd) to As, calculated from analysis AGM2 in Table 8, is 1:1.15, which results in a formula identical to that of ruthenarsenite: $(Ru_{0.77}Rh_{0.16}Pd_{0.07})As$. Palladium reported by Holliger (1992) may be in Ru arsenides and sulf-arsenides or in some unknown mineral. It is possible that some phases encountered in Ru-bearing aggregates are not stoichiometric. Some exotic or even new minerals may remain to be discovered.

Pyrite and other sulfides

Pyrite is fairly abundant in some reactors. Pyrite is replaced by hematite in quartz-illite veins in the core of RZ-16 (Savary and Pagel 1997). Abundant pyrite is dispersed in illite within portions of the reactors clays at Bangombé (Janeczek and Ewing 1996c). However, pyrite is more abundant in host rocks of reactor zones. Trace elements in pyrite from sandstones surrounding the reactor zones were used to determine chemistry and identify generations of fluids circulating around reactors (Raimbault et al. 1996). Based on trace-element compositions, pyrite adjacent to reactor zones can be grouped into three categories

(Raimbault 1993):
1. Pyrite with high Sb and Ag and moderate As concentrations, with a ratio of Ni/Co below 0.3.
2. Pyrite rich in Mo (up to 240 ppm) regardless of the amount of As and Sb.
3. Mo-poor pyrite, low in As and Sb and a ratio of Ni/Co approximately one.

Mo-rich pyrite is restricted to fracture-fillings; whereas pyrite crystals disseminated in the sandstone matrix are slightly enriched in Sb (Raimbault and Peycelon 1995). Pyrite found in the dolerite dike at Oklo is distinct from the above three categories, being poor in all four trace elements. Pyrite from clays of RZ-13 has low concentrations of As and Sb and a wide range of values for Ni/Co (from 0.2 to 6).

A few platy grains of molybdenite, up to 100 nm across, adjacent to uraninite were observed by TEM in a sample from RZ-10 (Janeczek and Ewing 1995). Selected-area electron-diffraction patterns confirm molybdenite in addition to galena. Molybdenum is among the fission products released from uraninite at Oklo in substantial amounts (Curtis et al. 1989; Brookins 1990).

Other sulfides found in small amounts in the reactor zones include chalcopyrite, covellite, digenite, bornite and sphalerite.

Oxides

Hematite, Ti-oxides, minium, and goethite are among accessory oxides and oxyhydroxides identified in the reactor zones. Euhedral grains of Ti-oxides (probably anatase) are abundant in clays of the Bangombé reactor. They are often rimmed by secondary uraninite and coffinite (Fig. 27).

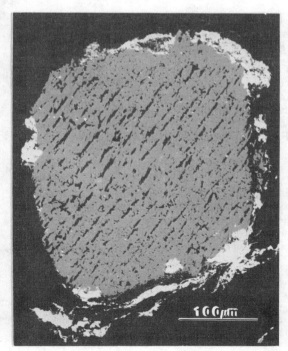

Figure 27. BSE image of an anatase crystal rimmed by Zr-rich coffinite and secondary uraninite. Sample from the clays of the Bangombé reactor.

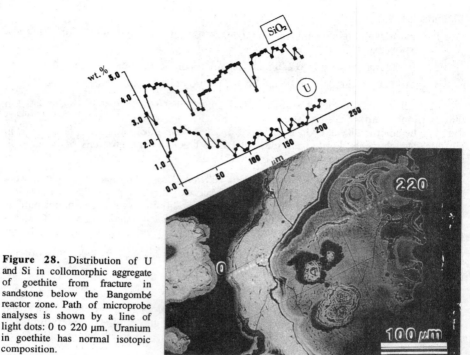

Figure 28. Distribution of U and Si in collomorphic aggregate of goethite from fracture in sandstone below the Bangombé reactor zone. Path of microprobe analyses is shown by a line of light dots: 0 to 220 μm. Uranium in goethite has normal isotopic composition.

Minium (Pb_3O_4) was identified by Raman microspectroscopy in chlorite aggregates surrounding uraninite at the border of RZ-10 (Savary and Pagel 1997). Minium occurs as elongated, rounded or globular aggregates 10 to 50 μm across.

Iron oxyhydroxides occur in reactors affected by weathering. Collomorphic aggregates of uraniferous goethite occurs in fractures below the Bangombé reactor (Fig. 28). Electron microprobe analyses of these aggregates reveal U, Si, Al., Pb and P as impurity elements. The maximum concentration of U is 2 wt % U. A zig-zag pattern of the U distribution in some of the growth-bands in goethite suggests oscillatory adsorption of U on the growing iron oxyhydroxide (Fig. 28). Uranium in the goethite has normal isotopic composition.

Calcite

Fibrous calcite occurs in veins in both units of RZ-10. White, pink and black calcite has been observed. Pink calcite is associated with hematite and iron oxyhydroxides. The fibrous nature of fracture-filling calcite indicates that it crystallized during dilation of veins (Fig. 29). Veins of solid bitumen cut across fibrous calcite veins in RZ-10 (Eberly et al. 1994). Calcite with minor Pb is associated with uraninite and minium at the border of RZ-10 (Savary and Pagel 1997).

The isotopic composition of calcite can be used to distinguish between different generations of calcite or different environments during calcite crystallization (Holliger and Gauthier-Lafaye 1996). Similar C and O isotopic ratios for calcite associated with the dolerite dike and those from RZ-13, which is close to the dike, suggest that calcite in RZ-13 was affected by hydrothermal fluids generated by the dolerite magma. Calcite from RZ

Figure 29. Optical transmitted-light micrograph of fibrous calcite vein from RZ-10. Progressive dilation of fracture is suggested by the orientation of calcite fibers, which have grown perpendicular to the walls of the calcite vein. Modified from Eberly et al. (1994).

10 has a constant C isotopic ratio in the range -5 to -15 ‰ $\delta^{13}C$, whereas calcite from RZ 13 has a wider range of isotopic ratios (Holliger and Gauthier-Lafaye 1996).

Phosphates

Apatite is a common accessory mineral in RZ-10 and RZ-16 and has recently been studied extensively, because it is considered a good geochemical indicator (Carpéna and Sere 1994; Hidaka et al. 1994; Raimbault et al. 1995; Bros et al. 1996; Savary and Pagel 1997). Euhedral zoned apatite crystals a few millimeters across occur in RZ-10 and RZ-16. The cores of apatite crystals in RZ-16 consist of fluorapatite surrounded by an inclusion-rich region. Fluorine in this region was partially substituted by Cl, and Si proxied for P. Fluid inclusions and solid inclusions occur along growth zones in apatite crystals. Uraninite, hematite, calcite and an unidentified LREE-rich mineral were found in the solid inclusions. The outer rims of apatite crystals consist of fluorapatite with minor Al and Mg. Some crystals are rimmed by a few micrometers of calcite. Surfaces of apatite crystals are corroded where in contact with aluminous chlorites. Apatite crystals from RZ-10 have two zones, one with a composition close to $(Ca,Sr,Zr)_{10}(PO_4)_6F_2$, the other, an inclusion-rich zone, has a composition with the approximate formula: $(Ca,Sr,Zr)_{10}(PO_4,SiO_4)_6(F,Cl)_2$ (Carpéna and Sere 1994). The latter apatite zone is enriched in LREE, with La, Ce, and Nd concentrations of 800, 2000, and 4000 ppm, respectively. REE patterns in apatite from reactor zones are strikingly different from apatite in surrounding sandstones (Raimbault et al. 1996).

In addition to the aforementioned crystals, veins of massive, almost pure fluorapatite separate aggregates of uraninite and the clay matrix in RZ-10 (Janeczek and Ewing 1993).

Crandallite group

Minerals of the crandallite group were recognized in illite within and beneath the Bangombé reactor zone (Janeczek and Ewing 1996c), and in the core of RZ-13 (Gauthier-Lafaye et al. 1996; Dymkov et al. 1997). Based on chemical compositions, florencite-(La) occurs in the Bangombé reactor zone and florencite-(Ce) occurs in sandstone (Table 9). The compositions of these two minerals can be written $Cra_3Goy_{17}Flo_{80}$ and $Cra_9Goy_{10}Flo_{81}$, respectively, in which coefficients correspond to percentages of end-members crandallite-(Ca), goyazite-(Sr), and florencite-(REE) (Fig. 30). Aluminous phosphates from RZ-13 show compositional ranges from $Cra_{38}Goy_{38}Flo_{24}$ to $Cra_{32}Goy_{36}Flo_{32}$ (Table 9 and Fig. 30).

Table 9. Electron microprobe analyses of crandallite-group minerals from reactor zones (wt %)

RZ	P_2O_5	SiO_2	Al_2O_3	Fe_2O_3	CaO	SrO	La_2O_3	Ce_2O_3	Pr_2O_3	Nd_2O_3	SO_3	F	Cl
BaS	26.30	0.07	29.94	1.36	0.78	1.98	8.29	13.54	n.d.	3.00	0.58	b.d.l.	b.d.l.
Ba	28.17	b.d.l.	30.12	0.36	0.39	3.45	12.38	11.31	n.d.	2.02	0.37	b.d.l.	b.d.l.
13[a]	18.97	1.18	27.04	n.d.	4.13	7.60	5.68	0.95	0.66	0.37	0.03	1.47	0.30
13[b]	19.51	0.92	29.48	3.26	3.91	8.00	7.08	2.81	0.81	0.39	n.d.	1.51	0.32

Note: BaS and Ba are florencite-(Ce) and florencite-(La) from sandstone and from the illite unit of the Bangombé reactor zone (Janeczek and Ewing, 1996); 13[a] and 13[b] are phosphates from the core of RZ-13 analysed by Gauthier-Lafaye et al. (1996) and Dymkov et al. (1997), respectively. BaS contains also 0.80 wt % ThO_2 and 13[b] has 3.24 wt % SO_3. b.d.l. = below detection limit; n.d. = not determined.

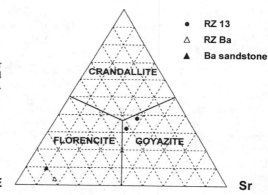

Figure 30. Compositional diagram for crandallite-group minerals from RZ-10 and the Bangombé reactor, and from FA sandstone in Bangombé.

Florencite outside the Bangombé reactor zone occurr in the matrix of coarse-grained sandstone together with monazite-(Ce), apatite, TiO_2 (probably anatase), zircon, and illite. Grains of florencite contain numerous inclusions of thorite, chalcopyrite and galena. Florencite-(La) is embedded in illite and occurs as overgrowths of elongate rhombohedral crystals up to 110 µm long and 50 µm wide. Some florencite grains contain inclusion of galena and U- and Ti-bearing phases. Secondary uraninite and coffinite precipitated on the surfaces of florencite and replaced it along fractures. Aluminous hydroxy phosphates in RZ-13 occur as euhedral grains, tens to hundreds of microns across, within massive uraninite that is veined by calcite, and within the chlorite matrix (Dymkov et al. 1997). Grains of Al hydroxy phosphate are commonly replaced by chlorites, iron sulfides and galena. Unit-cell parameters of the Al hydroxy phosphate, determined by electron diffraction are: $a = 6.98(3)$ Å and $c = 16.35(4)$ Å (space group $P3m$).

Only light REE are detected in Al hydroxy phosphates. La predominates over Ce (La/Ce = 7.0 and 7.7) in aluminous phosphates from RZ-13; whereas La only slightly exceeds Ce in florencite from the Bangombé reactor (La/Ce = 1.09). Ce exceeds La in florencite from the Bangombé sandstone (La/Ce = 0.62).

Zircon

Detrital zircon is an abundant accessory mineral in both FA sandstones and reactor

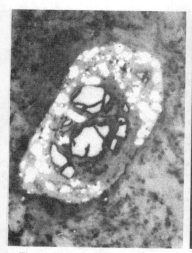 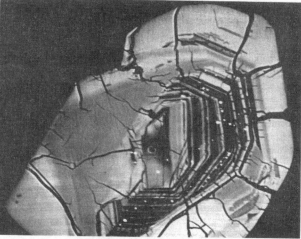

Figure 31. BSE images of detrital zircon crystals from clays of the Bangombé reactor. The crystal shown at left has a partially dissolved metamict core surrounded by altered zircon with numerous inclusions of galena. The crystal shown at right consists of an oscillatory zoned core enveloped and replaced by neogenic zircon.

zones, mainly in the clays of reactors (Fig. 31). Neogenic zircons occur in reactor zones, commonly as overgrowths on detrital zircons. Zoned zircon crystals from RZ-10 have cores of detrital zircon surrounded by inclusion-rich zircon overgrowths (Sere et al. 1994). Zirconium, Si, Hf, U, Al, P, Ca, Fe, and Y were detected in that zone as major components. Solid inclusions, 5 to 10 µm across, consist of galena and minor uraninite. The outermost zones of zircon crystals are rich in inclusions. Zircon crystals with abundant galena inclusions also occur within clays of the Bangombé reactor (Fig. 31). Sub-micron inclusions of U-rich and Hf-poor zircon occur within black calcite in sandstone around RZ-10 (Raimbault et al. 1996), and this zircon is enriched in HREE relative to most zircons in the FA sandstone.

Anomalous isotopic compositions of minerals in the reactor zones

Geffroy and Lissilour (1974) were the first to recognize that mineral assemblages in natural fission reactors are not significantly different from those in other sedimentary U deposits. Ruthenium minerals in reactor zones are the notable exception. Chemical and physical properties of uraninite and other minerals from natural fission reactors are not exceptional. What really makes uraninite and some other minerals from natural fission reactors truly unique are their isotopic compositions.

Minerals containing depleted U and those which incorporated major amounts of fission products have distinctly different isotopic compositions from their counterparts formed elsewhere in the Earth's crust. We may call this isotopic composition anomalous because it deviates from normal isotopic compositions. Minerals with anomalous isotopic compositions include uraninite, coffinite, Ru minerals, and crandallite-group minerals. Additionally, apatite incorporated some fission products and Pu as trace elements. To fully describe minerals in and around reactor zones, a knowledge of their isotopic composition is necessary. Because minerals with anomalous isotopic compositions do not differ in their physical or chemical properties from species with normal isotopic compositions, secondary ion mass spectroscopy (SIMS) is indispensable for studying minerals from natural reactors.

Ion microprobe analyses of samples from natural fission reactors at Oklo were pioneered by Havette et al. (1975), who showed qualitative relations between U, C, Ca, Ti and fission products. A major breakthrough in the age determination of uraninite from fission reactors was made by Holliger (1992). He used SIMS to measure isotopic compositions of Pb and U within relicts of unaltered uraninite in altered single uraninite crystals. From those data he obtained a precise concordant age of 1968±50 Ma. Perhaps an even more important application of SIMS was the discovery of Ru-bearing minerals with 100% fissiogenic ^{99}Ru (Holliger 1992), a decay product of now-extinct ^{99}Tc. In one sample from RZ-13, ^{99}Ru is the major isotope of Ru (Table 10). Arsenic and S found in Ru minerals have normal isotopic compositions. Ruthenium arsenides in natural reactors have Ru isotopic compositions that are dramatically different from normal Ru (Fig. 32). Assuming that at least some Ru minerals are ruthenarsenite, the formula of ruthenarsenite in natural reactors may be written Ru_fAs, where subscript f stands for fissiogenic, to distinguish fission-product-containing ruthenarsenite from ruthenarsenite with normal isotopic compositions. Tellurium in PbTe found in association with Ru minerals is probably also 100% fissiogenic.

Uranium in uraninite that experienced fission chain reactions is depleted in fissile ^{235}U. Individual grains of uraninite are commonly more depleted in ^{235}U than bulk samples. This is illustrated by a sample from the core of RZ-9 (sample GL.3535), in which the ^{235}U/^{238}U ratio was measured in both the bulk rock and from single grains of uraninite by ion microprobe (Holliger and Gauthier-Lafaye 1996). The ^{235}U/^{238}U ratio of 0.697 in the bulk sample exceeds the mean value of 0.674 obtained from eight ion-microprobe measurements of uraninite grains (beam diameter of 30μm). The ^{235}U/^{238}U ratios in uraninite grains range from 0.645 to 0.702.

Table 10. Average isotopic composition of Ru in Ru minerals from RZ-10 and RZ-13 (at %)

	^{96}Ru	^{98}Ru	^{99}Ru	^{100}Ru	^{101}Ru	^{102}Ru	^{104}Ru
RZ-10	< 0.01	< 0.01	23.76	1.18	30.48	27.13	17.43
RZ-13	< 0.01	< 0.01	42.54	3.14	22.33	20.08	11.91
STD	5.54	1.86	12.74	12.60	17.05	31.57	18.66

After Hidaka and Holliger (1998)

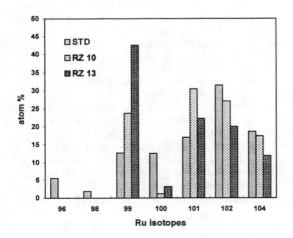

Figure 32. Abundances of Ru isotopes in Ru-bearing arsenides from RZ-10 and RZ-13 compared to the terrestrial standard (STD). Drawn from data by Hidaka and Holliger (1998).

Duffy (1975) attempted to find a zoned distribution of $^{235}U/^{238}U$ ratios in single uraninite grains. Although he failed to detect any isotopic zoning, the possibility of isotope zoning in uraninite from natural reactors cannot be dismissed. Isotopic zoning seems confirmed by SIMS data provided by Holliger and Gauthier-Lafaye (1996), who measured $^{235}U/^{238}U$ ratios in numerous single uraninite grains from RZ-9, RZ-10 and RZ-13, and from sandstone adjacent to RZ-10. Areas in uraninite grains analysed by Holliger and Gauthier-Lafaye (1996) ranged from 0.01 to 0.02 mm^2 (estimated from individual beam sizes [20-30 µm] and the number of spots analysed). Values of the $^{235}U/^{238}U$ ratio varied greatly within analysed areas in individual grains, except for uraninite with normal U compositions. The difference between the maximum and minimum values of $^{235}U/^{238}U$ ratio ($\Delta^{235}U/^{238}U$) in each grain is a measure of isotopic heterogeneity of the uraninite. There is a strong negative correlation between $\Delta^{235}U/^{238}U$ and mean values of $^{235}U/^{238}U$ ratios for each grain (Fig. 33). The less uniform the distribution of ^{235}U, the higher the values of $\Delta^{235}U/^{238}U$. Several interpretations of these results are possible. It could be that ^{235}U was fissioned unevenly within individual uraninite grains, that this isotope was redistributed after criticality, or that the uneven distribution of ^{235}U reflects the distribution of Pu, which formed from neutron capture by ^{238}U. Data of Holliger and Gauthier-Lafaye (1996) cannot be used to distinguish isotopic zoning from uneven, though random distributions of ^{235}U, as expected from alteration of uraninite. Clearly, suitable samples and analytical spot sizes smaller than 10 µm are required to help answer this question.

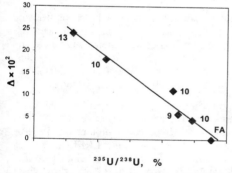

Figure 33. A plot of $^{235}U/^{238}U$ ratios in single uraninite grains against the difference between the maximum and minimum values of $^{235}U/^{238}U$ ratio ($\Delta^{235}U/^{238}U$) in each grain. Numbers refer to reactor zones; FA indicates sandstone. Calculated and drawn from data by Holliger and Gauthier-Lafaye (1996).

Another mineral with an anomalous isotopic composition is coffinite. Uranium in P-rich coffinite from sandstone close to the Bangombé reactor is deficient in ^{235}U. Coffinite also contains fissiogenic LREE. According to Hidaka (pers. comm.) coffinite that hosts Ru-bearing minerals (Fig. 25) has a mean $^{235}U/^{238}U$ ratio of 0.0061. A lower value of 0.0059 was detected in coffinite from approximately 30-50 cm from the Bangombé reactor.

Depleted U ($^{235}U/^{238}U$ = 0.00635; Hidaka, pers. comm.) was found in françoisite-(Nd), from within secondary pores in sandstone some 20 cm beneath the core of the Bangombé reactor. REE in this mineral may also have anomalous isotopic compositions, although no data are currently available. Janeczek and Ewing (1996c) postulated that U and LREE in françoisite-(Nd) were derived from nearby P-rich coffinite, which contains depleted U and fissiogenic LREE.

Secondary-ion mass spectrometry revealed that more than 98% of Ce and Nd and 100% of Sm are fissiogenic in crandallite-group minerals from RZ-13 (Dymkov et al. 1997). Between 27 and 30% of Nd and 67 to 71% of Sm in florencite-(La) from RZ-Bangombé are fissiogenic (Janeczek and Ewing 1996c). Contributions for La in both cases could not been calculated because ^{139}La strongly dominates ^{138}La. Isotopic spectra for Sr and Ba are natural. Dymkov et al. (1997) observed a relation between depletion of LREE in uraninite adjacent to aluminous phosphates and the enrichment in LREE of Al-rich phosphates. At Bangombé, abundances of fissiogenic Nd and Sm are higher than

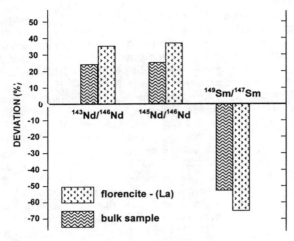

Figure 34. Deviation (%) of $^{143}Nd/^{146}Nd$, $^{145}Nd/^{146}Nd$, and $^{149}Sm/^{147}Sm$ ratios from terrestrial values in florencite-(La) and bulk samples from clays of the Bangombé reactor. Drawn from data by Janeczek and Ewing (1996).

abundances of fissiogenic Nd (19.9%) and Sm (52.6%) observed by Bros et al. (1993) in a bulk sample from the clay mantle (Fig. 34). Relative to the clay mantle, enrichments in fissiogenic Nd and Sm in florencite suggest that LREE were preferentially incorporated into florencite during the influx of fissiogenic LREE, which were released from uraninite.

Crandalite in general and florencite in particular may have played a more significant role in fixing migrating fissiogenic REE than other minerals. Whereas abundances of fissiogenic REE in apatite from the Oklo reactor zones are higher than their abundances in florencite from Bangombé, the highest concentrations of fissiogenic Nd and Sm in apatite from RZ-10 at Oklo were only 501 and 60 ppm, respectively (Hidaka et al. 1994), as compared with 3.1% of Nd and 0.2% Sm in florencite (Janeczek and Ewing 1996c).

Apatite in reactor zones incorporated several hundred ppm of depleted U and fissiogenic Nd and Sm (Hidaka et al. 1994; Bros et al. 1996; Raimbault et al. 1996). The amount of fissiogenic LREE in apatite varies widely. Estimates by Hidaka et al. (1994) and by Bros et al. (1996) show that between 73 and 89% of Ce, 37 and 93% of Nd, and 34 and 96% of Sm are fissiogenic in apatite. At least two types of hydrothermal apatite in and around RZ-16 was suggested on the basis of isotopic compositions (Bros et al. 1996). One type occurs in the clays of the reactor, and is characterized by a high $^{235}U/^{238}U$ ratio (up to 0.00804) and a ratio of fissiogenic Nd (Nd_f) to U greater than unity (Fig. 35). The second type of hydrothermal apatite, found in the core of RZ-16, is depleted in ^{235}U and has Nd_f/U less than one (Fig. 35). The unusually high $^{235}U/^{238}U$ ratio in apatite from the reactor's clay results from the incorporation of ^{239}Pu that migrated out of the reactor and later decayed to ^{235}U.

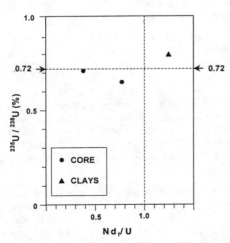

Figure 35. The amount of fissiogenic Nd normalized to U against the $^{235}U/^{238}U$ ratio in apatite from RZ-10 and RZ-13. Modified from Bros et al. (1996).

Preferential loss of U from the reactor zone is indicated by high Nd_f/U ratios (Bros et al. 1996).

An isotopic anomaly for Eu was found in apatite from RZ-10 and RZ-16 (Raimbault et al. 1996). Apatite $^{153}Eu/^{151}Eu$ ratios cluster around 2.8 in RZ-10 and 11 in RZ-16. These ratios are completely different than for fissiogenic isotopes (0.37) and the terrestrial standard (1.093). Raimbault et al. (1996) explain the unusual $^{153}Eu/^{151}Eu$ ratios by chemical fractionation between ^{151}Sm, the precursor to ^{151}Eu and Eu^{2+} cations. The half-life of ^{151}Sm is 88.73 years, so that fixation of fissiogenic LREE in apatite must have occurred during fission reactions, which lasted no longer than 2×10^5 years in RZ-10. Traces of fissiogenic Rb and Ba were found in apatite from RZ-10 (Hidaka et al. 1994). Estimated concentrations of fissiogenic Ba range from 0.3% to 1.76% of total Ba (94 to 480 ppm).

Analyses of isotopic compositions of single mineral grains are of outmost importance for studies of geochemical migration and retention of actinides and fission products. While determinations of isotopes in bulk samples provide valuable information about general trends during migration, the determination of isotopes in single minerals distinguishes minerals with normal isotopic compositions from minerals that contain nuclear-reaction products. Isotopic compositions of bulk samples are averages, which depend on the number of minerals with anomalous isotopic compositions. To illustrate this we may consider observations made by this author on samples of sandstone from the vicinity of the Bangombé reactor. Uraninite that occurs in sandstones has two possible geneses. Uraninite may be primary with normal (unfissioned) U or it may be secondary, formed from U depleted in ^{235}U that migrated out of the reactor zones. Both primary and secondary uraninite occurs close to each other and have identical chemical compositions; i.e. similar concentrations of radiogenic Pb and other impurity elements. The only way to recognize their distinct origins is to measure their $^{235}U/^{238}U$ ratios *in situ*.

Recently, very precise isotopic data for Nd and Sm in uraninite from the natural reactors have been obtained by high mass resolution using a sensitive high resolution ion microprobe (SHRIMP) (Hidaka 1998).

URANIFEROUS ORGANIC MATTER

Organic matter played an important role in both the origin of U deposits in the Franceville basin and in subsequent containment of actinides and fission products in some natural reactors. Many natural fission reactors contain abundant organic matter, with the exceptions of reactors RZ-1 through RZ-6 (Naudet 1991). Organic matter in natural fission reactors occurs in the form of solid bitumen. According to Mossman et al. (1993) and Parnell (1996) bitumen is an allochtonous organic solid with a negligible solubility in organic solvents and a reflectance above 2%. Originally deposited as a viscous liquid, bitumen formed from a polymerized organic solid, kerogen, which did not migrate after deposition.

Analyses of total organic carbon (TOC) in the stratigraphic column of the Francevillian show black shales of the FB, FC, and FD formations to be highly enriched in organic carbon: up to 20% (Cortial et al. 1990). Such concentrations of organic carbon in black shales are comparable to concentrations in major oil-source rocks. The fact that organic matter became concentrated here some 2 Ga ago is relevant to our understanding of the evolution of life on Earth. Biological precursors of organic matter in the Francevillian were algae, and relicts from those organisms occur as alga cyanophyceae and bacterial forms in FB black shales (Cortial et al. 1990). Laminated algal and altered cyanobacterial remains both occur in reactors RZ-7 through RZ-9 (Mossman et al. 1993). Kerogen that is isolated from black shales contains particles similar to those of sapropelic "coal." Organic matter in

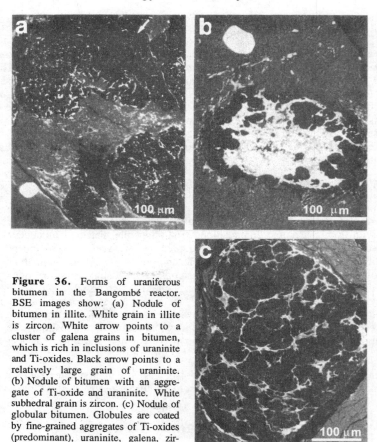

Figure 36. Forms of uraniferous bitumen in the Bangombé reactor. BSE images show: (a) Nodule of bitumen in illite. White grain in illite is zircon. White arrow points to a cluster of galena grains in bitumen, which is rich in inclusions of uraninite and Ti-oxides. Black arrow points to a relatively large grain of uraninite. (b) Nodule of bitumen with an aggregate of Ti-oxide and uraninite. White subhedral grain is zircon. (c) Nodule of globular bitumen. Globules are coated by fine-grained aggregates of Ti-oxides (predominant), uraninite, galena, zircon and illite.

the Francevillian achieved a high thermal maturity during diagenesis, extending into the coke field (Vandenbroucke et al. 1978; Cortial et al. 1990). However, atomic ratios H/C and O/C in organic matter did not reach the anthracite stage and organic matter is, therefore, unmetamorphosed. Hydrothermal reactions during diagenesis generated liquid bitumen and petroleum from kerogen in black shales of FB and FA formations. Migration of liquid organic matter (petroleum) is evident in black shales, and the FB formation black shales were the most probable sources of hydrocarbons that migrated into FA sandstones (Gauthier-Lafaye 1986; Gauthier-Lafaye and Weber 1989, 1997; Mossman et al. 1998). Liquid bitumen migrated into the reservoir FA sandstone along fractures and solidified there. Boundaries between bitumen and quartz in the FA sandstone are irregular due to penetrative replacement of quartz by bitumen (Parnell 1996) (Fig. 4).

There are two types of bitumen in the FA formation: uraniferous and barren. Barren bitumen occurs as angular aggregates in primary pores in sandstone. Uranium concentrations in barren bitumen are less than 30 ppm (Cortial et al. 1990). Uraniferous bitumen occurs as nodules in secondary pores and fractures in sandstone and conglomerate (Fig. 36), and diameters of bitumen masses may reach a few centimeters. Nodules of uraniferous solid bitumen contain abundant uraninite and galena (Fig. 36). Based on U isotopic compositions of uraninite inclusions, uraniferous bitumen can be further subdivided into bitumen with fissioned ^{235}U and bitumen with isotopically normal U (normal U mineralization). Both sub-types of uraniferous bitumen may occur in a single reactor zone. Bitumen with normal U mineralization is common outside reactor zones and other U deposits in the Franceville basin. Gamma-ray spectroscopy of uraniferous bitumen reveals ^{235}U, ^{234}Th, ^{231}Th, ^{230}Th, ^{227}Ac, ^{226}Ra, ^{224}Ra, ^{214}Pb and ^{214}Bi (Rigali and Nagy 1997).

Elementary analyses for total C show that solid bitumen in fission reactors contains between 6% and 89% organic carbon (Nagy et al. 1993). The composition of bitumen is heterogeneous even on a micro-scale.

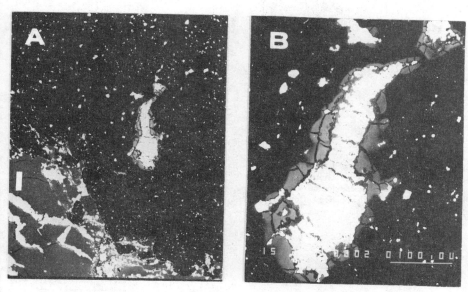

Figure 37. BSE image of a contact between a nodule of bitumen and sandstone from the vicinity of the Bangombé reactor. Dispersed uraninite and galena occur in the nodule. White veins within quartz grains (grey) are chalcopyrite. The BSE image on the right shows an enlargement of elongated uraninite grain embedded in bitumen. The bright rind around the grain contains a mixture of organic matter and uraninite. The uraninite is partially replaced by illite (dark gray textured regions).

Uraninite inclusions in bitumen have diameters that vary by several hundreds of microns and are often elongated and preferentially oriented (Fig. 37). Prefered orientation reflects flow paths of viscous bitumen that developed during agglomeration into nodules. From this observation one may conclude that precipitation of uraninite on viscous bitumen occurred at a stage of thermal maturity corresponding to the "oil window" (Cortial et al. 1990). Large uraninite inclusions exhibit halos with increased reflectance and high contrast in BSE images (Fig. 37). These halos are thought to be regions of radiation damage (Parnell 1996). Halos differ chemically from adjacent bitumen in that they contain dispersed U and Pb and, in some cases, increased oxygen contents. There are no significant differences in chemical composition between uraninite from bitumen and uraninite embedded in clay minerals, regardless of reactor zone. In particular, concentrations of Pb and other imurity elements are of the same order of magnitude as elsewhere both within and outside reactor zones (Janeczek and Ewing 1995).

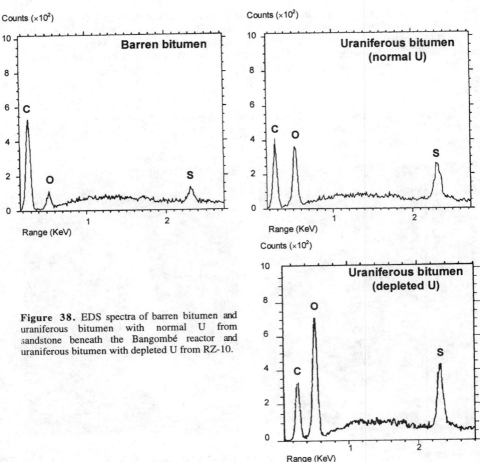

Figure 38. EDS spectra of barren bitumen and uraniferous bitumen with normal U from sandstone beneath the Bangombé reactor and uraniferous bitumen with depleted U from RZ-10.

In addition to uraninite and galena inclusions, nodules of bitumen may contain clay minerals between coalesced bitumen masses, on surfaces of large uraninite crystals, and on native Pb. Rarely, a U-titanium phase, with a composition similar to brannerite, occurs within the bitumen (Parnell 1996). Galena, native Pb and clay minerals form mineral annuli around uraninite grains. Thin veins of galena, uraninite and native Pb commonly extend from bitumen into the host rock (Parnell 1996). Galena sometimes predominates over uraninite in nodules of solid bitumen, as indicated by nodules of bitumen from RZ-9 with unusually high reflectance (Janeczek and Ewing 1995). High reflectance results from abundant and finely dispersed submicron-sized galena.

Where uraninite inclusions cannot be resolved at scales of a few Ångstroms, as for some aggregates of uraniferous bitumen (as shown by EDS spectra), U is thought to occur as organo-uraniferous complexes (Cortial et al. 1990).

Uraniferous bitumen is oxidized (Vandenbroucke et al. 1978), and, therefore, has elevated O/C ratios (Fig. 38). Moreover, uraniferous bitumen from fission reactors is slightly more oxidized than bitumen with isotopically normal uraninite inclusions (Fig. 38). Increased O/C ratios in bitumen from reactors is associated with increased H/C ratios.

Oxygenation of solid bitumen can be explained as a result of reaction with free radicals of O_2 and H_2 produced by the radiolysis of water (Holliger et al. 1997; Savary and Pagel 1997). Alternatively, high H/C and O/C ratios in some samples may result, in part, from abundant clay minerals (Rigali and Nagy 1997). Barren bitumen and uraniferous bitumen differ also in their respective concentrations of S. The highest concentrations of S are found in bitumen with depleted U (Fig. 38).

Microfocused-laser Raman spectra showed that bitumen in and near reactors RZ-7 to RZ-9 consist of highly condensed aromatic hydrocarbons with abundant cryptocrystalline graphite in domains smaller than one micron across. Cryptocrystalline graphite domains are connected by short aliphatic, aromatic and alkyl-aromatic hydrocarbon bridges within the macromolecular matrix of solid bitumen (Nagy et al. 1993). Graphite remote from natural fission reactors occurs as domains one micron across and display higher crystallinities than graphite in reactors (Nagy et al. 1991; Mossman et al. 1993). Organic matter from reactors is less mature than organic matter found elswhere in the Francevillian. The lower maturity is probably due to the combined effects of radiation damage, hydrothermal reactions and late addition of younger, organic matter that migrated into the natural fission reactors (Leventhal et al. 1989; Nagy et al. 1991; Nagy et al. 1993; Mossman et al. 1993). The degree of aromacity of bitumen in the reactor zones correlates with the $^{235}U/^{238}U$ ratio in uraninite inclusions (Holliger et al. 1997). The aromacity of bitumen decreases with decrease in the $^{235}U/^{238}U$ ratio. This observation is in accord with earlier reports on the increase of aromacity of bitumen with the increasing distance from the reactor zones (Vandenbroucke et al. 1978; Nagy et al. 1991). The decrease in aromacity of bitumen is explained by hydrogenation of the aromatic rings caused by water radiolysis. Bitumen in the reactor zones is enriched in ^{13}C, which indicates both thermal alteration and radiolysis of organic matter (Holliger et al. 1997)

High concentrations of organic free radicals have been detected in the natural reactors at Oklo by electron spin resonance (ESR) spectroscopy (Rigali and Nagy 1997). Organic free radicals are organic compounds or C moieties having one or more unpaired electrons. These radicals may have helped reduce U(VI) during initial U deposition (Nagy et al. 1991). Organic free radicals may have been generated during neutron and ionizing irradiation of viscous bitumen Pbing to its solidification via free-radical polymerization (Rigali and Nagy 1997). Concentrations of organic free radicals in natural reactors are on the order of 10^{19}-10^{20}/g of organic carbon and are similar to concentrations in 3 Ga old uraniferous solid organic matter in Witwatersrand (Rigali and Nagy 1997).

Bitumen in and near natural fission reactors appears to have been an effective physical barrier to the migration of actinides and fission products from the enclosed uraninite (Nagy et al. 1991). Results of ion microprobe-mass spectrometry showed that ^{235}U and various fission products have been retained in uraninite inclusions and did not migrate into the surrounding bitumen (Nagy et al. 1993). However, some short-range migration of U and Pb has been noted in bitumen from RZ-9, within a 20-50 µm perimeter around uraninite inclusions (Parnell 1996). Access of aqueous solutions that might have dissolved uraninite inclusions and released radionuclides was apparently hindered by small pore sizes, which are only a few tens of Ångstroms across, and by the inability of the bitumen to be wetted (Rigali and Nagy 1997).

GEOCHEMISTRY OF NATURAL FISSION REACTORS

Actinides, fissiogenic and non-fissiogenic elements

Elements in and around the natural fission reactors can be grouped into actinides, fissiogenic and non-fissiogenic. The latter may have been present before criticality or been

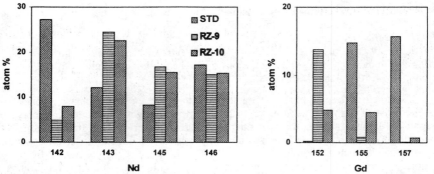

Figure 39. Bar chart showing isotope abundances of Nd and Gd in samples from RZ-9 and RZ-10 compared to normal terrestrial abundances (STD). Drawn from data by Hidaka et al. (1992a).

introduced during or after criticality, but they are unrelated to fission processes. Non-fissiogenic elements include radiogenic Pb and Bi, and elements introduced into the reactors by circulating fluids. Elements formed by neutron capture during criticality are not fission products. Therefore, total concentrations of elements in natural fission reactors are sums of concentrations of primordial elements, fissiogenic elements, and post-reaction elements. Total elemental concentrations are modified because some primordial isotopes were anihilated by neutron-capture reactions during reactor operations. Neutron-capture cross sections of primordial elements are major factors controlling present-day concentrations. For example, some REE isotopes, such as ^{142}Nd, ^{149}Sm, ^{155}Gd, and ^{157}Gd, have such large neutron-capture cross sections that, due to neutron capture, their abundances in natural reactors are much smaller than their abundances in terrestrial standards (Fig. 39).

To distinguish between the abundances of fission products from their primordial counterparts the measured abundance of an exclusively primordial isotope is normalized to the abundance of the same isotope in pure (standard) element. In practice, only isotopes with small neutron-capture cross sections and low fission yields are considered; e.g. ^{92}Mo, ^{106}Cd, ^{116}Sn, ^{124}Te and ^{142}Nd (Hidaka et al. 1988; Curtis et al. 1989). By comparing their abundances with natural abundances, the proportion of fissiogenic elements can be evaluated. The observed isotopic abundance of an element E (E_o) includes the abundances of primordial isotopes (E_p) and the abundances of fissiogenic isotopes (E_f); therefore,

$$E_o = xE_p + yE_f$$

where x and y are the proportions of primordial and fissiogenic isotopes, respectively, of the total. Because $x = 1 - y$, and $E_o = (1 - y)E_p + yE_f$, the proportion of a fissiogenic isotope can be calculated from the equation,

$$y = (E_o - E_p)/(E_f - E_p).$$

Some elements in reactor zones show little evidence of a primordial component (Ru, Pd and Te); whereas, others (Mo, Cd, Sn and Nd) have significant primordial contributions (Loss et al. 1988).

Fissiogenic isotopes consist of a mixture of fission products from ^{235}U, ^{238}U, and ^{239}Pu. Therefore, the total abundance of an element E in a natural reactor is expressed as follows:

$$(E) = (E)_{primordial} + (E)_{fission} = \{(E)_{primordial} + [(E_{fission})^{235}U + (E_{fission})^{238}U + (E_{fission})^{239}Pu]\}.$$

The procedure for calculating the fissiogenic component from the isotopic abundance of an element can be found in Hidaka et al. (1994).

Natural fission reactors provide a unique opportunity to study the behaviors of fission products and some transU elements in geologic systems. This is relevant to the potential behaviors of those same elements in a geologic repository for radioactive waste. Major goals of geochemical investigations of natural fission reactors are to determine:
- retentivities of radionuclides and fission products in the reactors
- migration and retardation mechanisms
- groundwater compositions and other geochemical parameters affecting reactor stabilities and radionuclide migration.

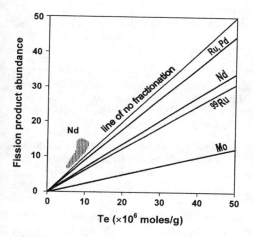

Figure 40. A diagram of the abundances (represented by regression lines) of Mo, Nd, Pd, Ru and ^{99}Ru (now as ^{99}Tc) relative to Te in samples from RZ-9. The area of the enrichment relative to Te is to the left of the no fractionation line, and area of deficiency is to the right. Shadowed region shows a cluster of points for Nd. Nd is both enriched and depleted relative to Te. Modified from Curtis et al. (1989).

Several approaches have been used to estimate the mobilities of selected nuclides. One is to use Eh-pH diagrams to evaluate possible behaviors of elements in aqueous solutions (Brookins 1978). A more direct approach is to determine the degree to which one fission product is retained relative to another. This is the relative retentivity (R_E) of a fissiogenic element. Curtis et al. (1989) used Te as the reference fission product because it was the most abundant one in samples they analyzed from RZ-9. They calculated R_E by dividing the sum of abundances of each element E in a sample by the total fissiogenic Te in each sample:

$$R_E = \Sigma \, C_E / \Sigma \, C_{Te}$$

When elements are present in the same proportions as their ideal fission yields, $R_E = 1$ and $\Sigma C_E = \Sigma C_{Te}$, so that a plot of ΣC_E vs. ΣC_{Te} has a slope of one and a zero intercept (line of no fractionation) (Fig. 40). Deviations from the line of no fractionation indicates enrichment or depletion in an element relative to Te. Curtis et al. (1989) determined that enrichment in a fission product relative to Te was a result of secondary retention.

Hidaka and Holliger (1998) chose ^{143}Nd as a reference fissiogenic isotope and calculated the relative retentivity of each fissiogenic element, E, from the following equation:

$$R_E = [(C_E/C_{Nd}) \times (M_{Nd}/M_E) \times (Y_{Nd}/Y_E)]$$

where C_E, C_{Nd}, M_E, M_{Nd}, Y_E and Y_{Nd} are concentrations (C), mass numbers (M) and fission yields (Y) of fissiogenic element E and ^{143}Nd, respectively.

Retentivities of fissiogenic elements can be estimated from fission-product yield curves by comparing observed abundances of fissiogenic isotopes with calculated abundances (Fréjacques et al. 1975; Hidaka et al. 1992; Hidaka and Holliger 1998). The fission product pairs from a neutron-induced fission event have approximately equal masses, and a plot of fission-product yield against mass number is symmetric, with two

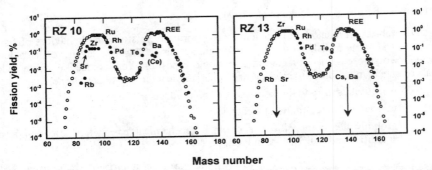

Figure 41. Fission product yield curves for samples from RZ-10 and RZ-13. Open circles show calculated fission product yield, solid circles are measured concentrations of isotopes. Arrows indicate the complete loss of fission products from a sample. Modified from Hidaka and Holliger (1998).

maxima (at mass numbers 95 and 139) and a minimum (Fig. 41). For example, a sample from RZ-10 has fission-yields for Zr smaller than expected from the calculated fission-product yield curve (Fig. 41), suggesting that some fissiogenic Zr was lost from this sample. On the other hand, in a sample from RZ-13, Zr data plot on the fission-product yield curve, which means that Zr was entirely retained in the sample.

Elements related to nuclear processes in natural fission reactors include actinides, actinide daughters, as well as a wide range of fission products and their daughters. Their abundances and behaviors during and after criticality are discussed briefly below.

Actinides

Our disscussion of the actinides begins with the most abundant in the natural fission reactors. Whereas U and Th can be measured directly in samples from the reactors, there is only indirect evidence for extinct Pu and Np, which were relatively abundant during criticality. Ultratrace concentrations of recently formed Pu, however, occur in uraninite from reactor zones. Some Am certainly formed in the natural fission reactors; however, Am yields were too small to document the geochemical history of this actinide (Brookins 1990).

Uranium

Uraninite in natural reactors was affected by numerous processes and events that caused alteration and dissolution, with subsequent redistribution of U and its daughters (Fig. 42). Estimates of U release, based on the modal analysis of a sample from the core of RZ-10, suggest that as much as 6.3 g of U may have been released from each cubic centimeter of the core due to partial dissolution of uraninite during or shortly after reactor operation (Janeczek and Ewing 1992). Curtis and Gancarz (1983) estimated that 10% of U, representing about 100 tons of U may have been dissolved and transported from reactor zones at Oklo. The thickness of the core of the Bangombé reactor (15 cm) is too small to account for neutron-induced-fission chain reactions. A substantial mass of U minerals at Bangombé dissolved. Textural observations of U minerals show that the dissolution of uraninite was related to the formation of clay minerals in the reactor zones.

The coffinitization of uraninite was another process responsible for the release of U. Changes in chemical composition during coffinitization of uraninite from RZ-10 are shown in Figure 43. The simplified reaction for the replacement of uraninite by coffinite is:

$$UO_2 + H_4SiO_2 = USiO_4 + 2 H_2O$$

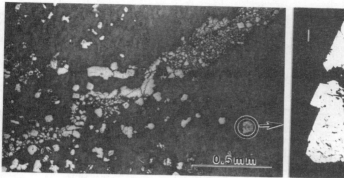

Figure 42. Reflected light optical micrograph of a vein-like aggregate of uraninite crystals in Mg-chlorite from RZ-10. BSE image (on the right) of the encircled crystal shows an atoll-like dissolution feature.

Assuming densities of UO_2 and $USiO_4$ as 10 and 7.4 gcm^{-3}, respectively, simple calculations give 3.5 g of U released from 1 cm^3 of UO_2.

The distance U migrated from the reactor zones can be estimated by measuring $^{235}U/^{238}U$ ratios in the vicinity of the reactors. Bros et al. (1995) found depleted U ($^{235}U/^{238}U = 0.007113$) in a sample of black shale at 6.52 m above the core of the Bangombé reactor zone. The concentration of U in that sample was low (3.8 ppm). A sample of chlorite with 3.3 ppm U that is significantly enriched in ^{235}U was found in argillaceous rock above RZ-10 (Bros et al. 1993), with a $^{235}U/^{238}U$ ratio of 0.00768 in the bulk sample. After dissolving the sample with 1N HCl, the insoluble residue was even more enriched, with a maximum $^{235}U/^{238}U$ ratio of 0.01051. The leachate was strongly depleted, with a $^{235}U/^{238}U$ ratio of 0.006805. Such an enrichment in ^{235}U was interpreted by Bros et al. (1993) as evidence for ^{239}Pu migration and incorporation of ^{239}Pu into chlorite, where it decayed to ^{235}U (see also §9.4.2.3). However, the incorporation of Pu into the chlorite structure is highly unlikely. Perhaps, ^{239}Pu was sorbed or precipitated on the cleavage planes of chlorite.

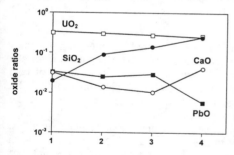

Figure 43. Variation in oxide concentrations (shown as oxide ratios) due to coffinitization of uraninite from RZ-10. 1—uraninite; 2-3—partially altered uraninite (mix of uraninite and coffinite); 4—coffinite.

Uranium depleted in ^{235}U also migrated downwards, into underlying sandstones (Fig. 44). Increased concentrations of slightly depleted U ($^{235}U/^{238}U = 0.00717$) and fissiogenic Nd have been found in the clays of RZ-10 at the boundary with adjacent sandstone (Hemond et al. 1992; Menet et al. 1992). This observation provides evidence for the retention of migrating radionuclides at the interface between the reactor zone and sandstone due to both physical and chemical barriers to migration (Hemond et al. 1992). Migration paths in sandstone included fractures (Fig. 44), grain boundaries, and secondary porosity. Depleted U occurs within fractures in sandstone 60 cm from RZ-10; whereas the $^{235}U/^{238}U$ ratio in the sandstone matrix is normal. Secondary uraninite with depleted U occurs in fractures and secondary pores in sandstone 20 to 40 cm beneath the Bangombé reactor.

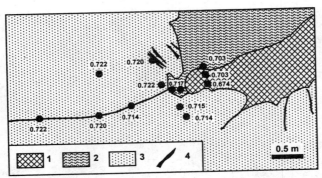

Figure 44. Distribution of $^{235}U/^{238}U$ ratios in and around RZ-10. 1—reactor core, 2—clays of the reactor, 3—sandstone, 4—fractures and groups of fractures. A composite drawing from Menet et al. (1992) and Hemond et al. (1992).

Regional geological events mobilized not only U in natural reactors, but isotopically normal U in low-grade U deposits was also mobilized. This is evidenced by normal $^{235}U/^{238}U$ ratios in U-bearing goethite, which occurs as coatings on fractures in sandstone beneath the Bangombé reactor. Therefore, migration patterns of U in the vicinity of reactor zones can be complex.

Retardation mechanisms include precipitation of secondary uraninite and sorption of U onto existing minerals. Chlorite and anatase were particularly effective at sorbing U (Figs. 27 and 45). Secondary uraninite, interleaved with ripidolite occurs within the clays of the Bangombé reactor; whereas illite surrounding the reactor is U free (Eberly et al. 1995) (Fig. 45).

The behavior of U during the last 1 Ma can be evaluated by studying U-series disequilibria. Bros et al. (1998) measured activity ratios of ^{238}U daughters from the Bangombé reactor zones. At the begining of the ^{238}U-decay sequence two stages of beta decay Pbs to ^{234}U ($t_{1/2} = 2.48 \times 10^5$ a), which, in turn, decays by alpha emission to ^{230}Th ($t_{1/2} = 7.52 \times 10^4$ a). The ratios $^{230}Th/^{238}U$ and $^{234}U/^{238}U$ provide information on the geochemical behavior of U. The central datum point in the diagram of $^{230}Th/^{238}U$ vs. $^{234}U/^{238}U$ (Fig. 46) represents secular equilibrium between parent and daughter nuclides, for which activity ratios of daughters are equal to that of the parent. The diagonal line in Figure 46 represents secular equilibrium between ^{230}Th and ^{234}U. The horizontal line represents secular equilibrium between the parent ^{238}U and its daughter ^{234}U. If both ratios $^{230}Th/^{238}U$ and $^{234}U/^{238}U$ are less than unity, then ^{234}U was lost from the sample; whereas, ratios greater than unity indicate accummulation of ^{234}U.

Uranyl minerals from the Bangombé reactor zone and its vicinity reflect the relatively recent mobilization of U by oxidizing groundwaters. Bros et al. (1998) determined the age of torbernite from a fracture in sandstone by the $^{230}Th/^{238}U$ method determining a formation age of 63 500 years. Bros et al. (1998) did not determine isotopic composition of U in their torbernite sample, and it is, therefore, unclear if torbernite represents migration of depleted U or whether it formed from normal U mobilized from uraniferous sandstone. Because of their high solubilities in rainwater, uranyl sulphates on fracture surfaces may represent very recent U precipitation.

Uranium and its daughters in most samples from the Bangombé reactor zone are not in secular equilibrium, with the exception of one sample from the reactor core, which plots

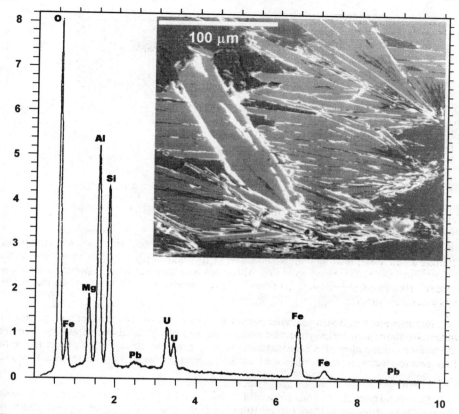

Figure 45. EDS spectrum and BSE image of a vein of Mg-Fe-chlorite, with inclusions of secondary uraninite (white), from within the illite matrix. Sample from the clay zone of the Bangombé reactor.

close to the center of the diagram in Figure 46. Therefore, the Bangombé reactor zone has not been a closed geochemical system with respect to U for at least 1 Ma (Bros et al. 1998). Most data in Figure 46 plot within a narrow range of values (±0.5) close to $^{234}U/^{238}U$ equilibrium, which may reflect preferential removal of ^{234}U from clay minerals, and subsequent precipitation of U that is enriched in ^{234}U. Preferential loss of ^{234}U is caused explained by the relatively easy removal of ^{234}U from along alpha-recoil damage tracks near mineral surfaces. Dissolved U also decays to ^{234}U, which can be adsorbed by clays or other minerals.

Present-day migration of U is measured directly from U concentrations in groundwaters percolating through reactor zones. Uranium concentrations in groundwaters around reactor zones never exceed 10^{-7} mol L^{-1} (Toulhoat et al. 1996). Even relatively oxidized groundwaters that penetrate near the surface of the Bangombé reactor extract little U from U minerals. Low U concentrations in groundwater (5.6×10^{-8} to 8.8×10^{-9} mol L^{-1} U) are explained by the influence of organic matter in the Bangombé reactor, which lowers the redox potential of incoming groundwaters (Toulhoat et al. 1996). This may be true, however, reduction must be accomplished by a form of organic matter other than bitumen in the reactor, because it is difficult to explain how a hydrophobic solid can decrease the redox potential of water. In this author's opinion, pyrite and other base-metal sulfides (e.g.

chalcopyrite, covellite) that are abundant in the reactor zone play the major role in buffering the redox potential.

Some U in groundwaters within and close to the reactor zones is depleted in ^{235}U. The lowest knwon ^{235}U/^{238}U ratio of 0.00687 was measured in groundwater from the Okelobondo reactor, which contained only 1.45×10^{-9} mol L^{-1} U (Toulhoat et al. 1996).

Thorium

Primordial Th in reactor zones occurs mainly in detrital zircon and monazite. Thorite inclusions are common in florencite-(Ce) from sandstone near the Bangombé reactor (Janeczek and Ewing 1996c). The Th content in uraninite from the natural reactors is rather low (Table 5). Thorium in natural reactors is practically all ^{232}Th, most of which was produced by alpha decay of both ^{240}Pu and ^{236}U (Frejacques et al. 1975; Hidaka and Holliger 1998). U-236 is a product of neutron capture by ^{235}U. High concentrations of Th (3588 ppm) in the core of RZ-13 suggests high production rates of ^{240}Pu and ^{236}U due to high neutron fluences in that reactor (7.8×10^{20} n/cm^2) (Hidaka and Holliger 1998).

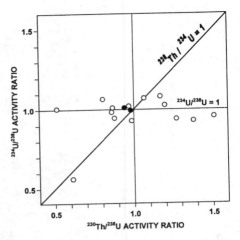

Figure 46. ^{230}Th/^{238}U and vs. ^{234}U/^{238}U activity ratios for samples from the core (solid circles) and clay minerals (solid circles) of the Bangombé reactor and from fractures in sandstone (open circles). Modified from Bros et al. (1998).

Thorium is stable in reactor zones because it substitutes extensively for U in uraninite. For instance, Th concentrations in the Bangombé reactor decrease dramatically from 1400 ppm in the reactor core to more typical values of 10-15 ppm in the adjacent FB shale (Bros et al. 1995), suggesting that Th is confined by uraninite and clays of the reactor. Hence, Th is released as uraninite dissolves. Thorium is less mobile than U in most groundwaters. Therefore, enrichment of Th relative to U in RZ-2 (Fréjacques et al. 1975), and in some samples from RZ-10, is explained by preferential migration of U caused by local increases in redox potential. Th may have behaved similar to Pu, if Pu and U fractionated early. Enrichment of Th observed in a sandstone sample from near RZ-10 may have resulted from the decay of ^{240}Pu that migrated from RZ-10 (Hidaka and Holliger 1998).

Plutonium

Natural Pu is monoisotopic ^{239}Pu ($t_{1/2}$ = 24,100 years), produced from neutron capture by ^{238}U: ^{238}U(n,γ)^{239}Np (β$^-$) → ^{239}Pu. Plutonium-239 decays by alpha-decay to ^{235}U. Total masses of ^{239}Pu measured in U minerals from RZ-9 are on the order of 10^{-12} g, corresponding to concentrations in the range 13×10^8 to 48×10^8 atoms/g U (Curtis et al. 1992). These are extremely small quantities, similar to those in other rich U deposits. Higher concentrations of Pu, up to 63×10^8 atoms/g-U, have been observed only from the Shinkolobwe deposit, Zaire (Curtis et al. 1992). However, during the operation of the natural fission reactors the amounts of Pu produced was much higher than it is today. Hidaka and Holliger (1998) estimated that the contribution of thermal neutron-induced fission of ^{239}Pu to the total neutron-induced-fission events in RZ-10 was about 4%. The amount of ^{239}Pu formed by ^{238}U capture of neutrons can be inferred from the restitution factor (Table 2) because ^{239}Pu decays to ^{235}U. A significant portion of present-day ^{235}U in natural fission reactors formed by alpha decay of ^{239}Pu, with the noticeable exception of RZ-13, for which the restitution factor is unusually small (Table 2).

The short-lived Pu isotope, ^{240}Pu (half-life 6.57 × 10^3 yr), was produced at Oklo from neutron capture by ^{239}Pu (Hidaka and Holliger 1998). Plutonium-240 decays to ^{232}Th, which is readily measured. However, ^{232}Th also forms by alpha decay of ^{236}U that was produced from neutron capture by ^{235}U.

The ionic radius of Pu^{4+} (1.04 Å) is close to that of U^{4+} (1.06 Å), which explains extensive PuO$_2$–UO$_2$ solid solution. Therefore, the geochemical behavior of Pu at Oklo was probably controlled by the geochemical behavior of uraninite. Based on an Eh-pH diagram for solid and dissolved Pu species in water, Brookins (1978) suggested that Pu should have been retained in natural reactors. However, some Pu apparently migrated out of reactor cores as suggested by the discovery, some 30 cm from the core of RZ-10, of chlorite with U enriched in ^{235}U (^{235}U/^{238}U = 0.007682) from (Bros et al. 1993). The only explaination for this enrichment is to assume that some ^{235}U is formed by alpha decay of ^{239}Pu. An even higher ^{235}U/^{238}U ratio of 0.00804 was found in apatite from clays of RZ-16 (Bros et al. 1996). Bros et al. (1996) used that ^{235}U/^{238}U ratio to calculate a ^{239}Pu/^{238}U ratio of 0.013 at the time of nuclear reactions, much higher than the value calculated from neutronic parameters only. To explain this apparent discrepancy, Bros et al. (1996) suggested the preferential loss of U over Pu at the edge of RZ-10, Pbing to enrichment of Pu that was incorporated into apatite.

Dissolved Pu was sorbed by chlorite. The relative efficacy of chlorite from RZ-Bangombé to sorb Pu was demonstrated by a simple experiment in which petrographic thin sections of sandstone and the reactor's clay mantle were exposed to asolutions with 10^{-6} M ^{239}Pu. The distribution of Pu on thin sections was determined by autoradiography, which showed that Pu sorption in all thin sections was much more effective on chlorite than on illite or kaolinite (Eberly et al. 1996).

Neptunium

Neptunium was produced in the natural reactors by the following neutron-capture reactions involving ^{235}U and ^{238}U (Hidaka and Holliger 1998):

^{235}U(n, γ)^{236}U, ^{236}U(n,γβ)^{237}Np

^{238}U(n, 2nβ)^{237}Np

^{238}U(n, γ2β)^{239}Pu, ^{239}Pu(n, γ)^{240}Pu, ^{240}Pu(n, γβα)^{237}Np

Neptunium-237 has a half-life of 2.1 × 10^6 years and unltimately decays to stable ^{209}Bi. All Bi in reactors is considered to have formed by decay of ^{237}Np (Hidaka and Holliger 1998).

The ionic radius of Np^{4+} (1.10 Å) is similar to the that of U^{4+} (1.06 Å), so there is a complete solid solution between NpO$_2$ and UO$_2$ (Brookins 1990). An Eh-pH diagram for Np shows that NpO$_2$ is the stable Np-bearing solid at the E_H and pH conditions present in most reactors, suggesting that Np would be retained in uraninite (Brookins 1978). However, Fréjacques et al. (1975) suggested ^{237}Np migrated from RZ-2. Based on a discrepancy between calculated and observed ratios of ^{209}Bi/^{238}U, Fréjacques et al. (1975) estimated that up to 30% of ^{209}Bi was lost from the core of RZ-2. A large difference between calculated and measured abundances of ^{209}Bi was also observed in the core of RZ-10 (Hidaka and Holliger 1998). However, it is possible that some or all ^{209}Bi migrated only after complete decay of Np. Therefore, redistribution of ^{209}Bi may reflect mobility of Bi, not Np. The rocks around the reactor zones are not enriched in Bi, suggesting that Bi, though redistributed, has been confined to reactor zones.

The highest concentration of Bi (463 ppm) measured is from RZ-13 (Hidaka and Holliger 1998), which suggests that RZ-13 had the highest neutron flux and, therefore, the highest production rate of ^{237}Np.

Actinide daughters: Bi and Pb

The behavior of Bi in natural fission reactors is discussed in the section devoted to Np, because essentially all Bi in natural reactors is a decay product of ^{237}Np. The chemical properties of Bi are similar to those of Pb, and the geochemical behavior of Bi can be deduced by tracking the behavior of more abundant Pb. By analogy with Pb, Hidaka and Holliger (1998) suggest that Bi was redistributed within the core of RZ-10 but did not migrate outside reactor zones.

Lead is a major impurity element in uraninite from natural reactors (Table 5). Lead in natural reactors is almost entirely radiogenic (Havette et al. 1975). Concentrations of common Pb, ^{204}Pb, are no more than 1% of all Pb. Ratios of ^{204}Pb/^{206}Pb in Pb-bearing minerals from various reactors range from 1.4×10^{-5} to 2.5×10^{-3}. In uraninite, ^{204}Pb/^{206}Pb ratios span a narrow range, from 1×10^{-5} to 6×10^{-5}, reflecting extreme depletion in radiogenic Pb.

Lead does not fit readily into the uraninite structure because of its large ionic radius (1.37 Å for eight-fold coordination). Like in other U deposits, mobilization of Pb was a common process in natural reactors. The widespread redistribution of radiogenic Pb affected geochronological studies at Oklo; reported U-Pb ages of U minerals from Oklo range from 1780 to 2050 Ma (Lancelot et al. 1975; Devillers and Menes 1978; Gancarz 1978). Discordant ages as low as 375 Ma (Gancarz 1978) and 330 Ma (Nagy et al. 1991) result from the loss of Pb from U minerals. Cowan (1978), Gancarz (1978) and Gancarz et al. (1980) postulated continuous volume diffusion for 2 Ga as the mechanism for removing Pb from uraninite at Oklo. However, Janeczek and Ewing (1995) showed that several mechanisms, rather than a single process, controlled the loss of Pb from uraninite in the reactors. Release mechanisms included mineral dissolution, grain boundary diffusion, exsolution via continuous precipitation, and volume diffusion. As a result of these processes, Pb contents in uraninite from all reactor zones except RZ-2 and RZ-10 are similar and vary around a mean value of 5.2 wt % PbO (Fig. 17). All of these processes were thermally activated and episodic. The main episode of Pb loss was related to a regional increase in igneous activity in the Franceville basin, which produced numerous large dolerite dikes some 1200 Ma after operation of natural reactors.

Ratios of SiO_2/UO_2 to PbO/UO_2 in uraninite can be used to distinguish dissolution-dominated Pb loss from volume diffusion of Pb (Janeczek and Ewing 1995). The vertical distribution of data points in Figure 47; i.e. constant PbO/UO_2 values for uraninite from reactor zones, and increasing SiO_2/UO_2 ratios reflect the increasing dominance of dissolution-dominated processes relative to volume or grain-boundary diffusion processes. Figure 9-47 shows that uraninite from RZ-2 and RZ-13, as well as some from RZ-10, higher values of SiO_2/UO_2, is affected by dissolution more than uraninite from Bangombé and RZ-16.

Replacement of uraninite by coffinite was another process that led to Pb release (Janeczek and Ewing 1992) (Fig. 43). Based on chemical compositions of uraninite and coffinite from RZ-10, Janeczek and Ewing (1995) estimated that 75 mol % of the Pb was released from one uraninite crystal upon replacement by coffinite. In fact, all of the Pb may have been lost from the original uraninite during its replacement, because the Pb in coffinite may have formed entirely after coffinite had precipitated

Lead released from uraninite did not migrate long distances. High S^{2-} activities and reducing conditions in the natural reactors caused virtually all Pb to precipitate as galena. Native Pb formed in places where concentration of Pb exceeded available S^{2-} (Parnell 1996; Savary and Pagel 1997). Local increases in redox potentials due to radiolysis resulted in formation of Pb oxides (Savary and Pagel 1997). According to Gancarz et al. (1980) Pb lost from uraninite in RZ-9 migrated 3 to 5 m into the basal FA conglomerate.

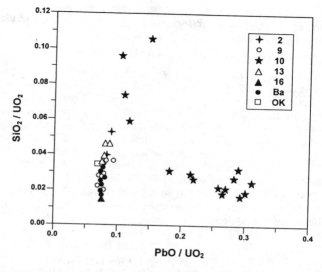

Figure 47. Plot of PbO against SiO_2 normalized to UO_2 plotted against PbO/UO_2 (molecular proportions) for uraninite from natural fission reactors showing the distinction between the dissolution-dominated mechanism of Pb loss and volume diffusion of Pb.

Fission products

Based on their geochemical behaviors, fission products from natural reactors can be grouped into: (1) gaseous elements (Kr and Xe), (2) non-gaseous volatile elements (Cs), (3) elements insoluble in uraninite (Sr, Rb and Ba), (4) elements compatible with the uraninite structure (Zr, Y and lanthanides) and (5) chalcophile metals (strong affinities for S and As; e.g. PGE and Tc). Of those fission products in reactor zones, the least is known about noble gases, Kr and Xe, which were apparently lost from the reactor zones (Naudet 1991). The same fate may apply to halides. Fissiogenic ^{129}I, with half-life 2×10^7 years, decays to ^{129}Xe, which escaped from reactor zones (Hagemann and Roth 1978).

Alkali and alkaline earth elements (Rb, Sr, Cs, and Ba)

Most Cs, Sr and Rb solids are highly soluble in groundwaters and these elements are incompatible in the uraninite structure. These elements were not well retained in uraninite and could migrate far from reactor zones, as shown by early geochemical studies at Oklo (Hagemann et al. 1974; Brookins et al. 1975). Barium-135 and ^{137}Ba are stable decay products of ^{135}Cs and ^{137}Cs, respectively, and the natural ratio of ^{135}Ba to ^{137}Ba is 0.5835. Any significant increase above that value is an indicator of fissiogenic Cs. Natural Cs is monoisotopic (^{133}Cs), so that the behavior of fissiogenic Cs can be determined only indirectly from $^{135}Ba/^{137}Ba$ ratios.

Brookins (1981) found small amounts of fissiogenic Rb and Sr in clay minerals from one unspecified reactor at Oklo, and those clay minerals showed elevated $^{135}Ba/^{137}Ba$ ratios. Based on correlations between fissiogenic Rb and Sr with $^{135}Ba/^{137}Ba$ and Sr with Rb, Brookins suggested that possibly 100% of the alkali and alkaline earth elements were fixed in clays of reactors after these elements were released from uraninite. Recently, Hidaka et al. (1992) found indirect evidence that some fissiogenic Cs was retained in RZ-10. They observed deviations in abundances of Ba isotopes, when normalized to abundances of ^{136}Ba, a primordial Ba isotope unaffected by nuclear reactions. Hidaka et al. (1992) concluded that the chemical behaviors of ^{135}Ba and ^{137}Ba are closer to that of Cs than to ^{138}Ba. The retentivity of ^{138}Ba correlates best with concentrations of Ba, rather than with Cs concentrations. Hidaka et al. also assumed that most Cs (^{133}Cs) in RZ-10 was generated by fission. Chemical fractionation between ^{137}Cs and its stable daughter, ^{137}Ba, must be assumed in order to explain the isotopic anomalies (Hidaka et al. 1992).

The geochemical behavior of ^{90}Sr can be traced by studying ^{90}Zr, because the latter is a stable daughter of relatively short-lived ^{90}Sr. As for other fission products, the Zr isotopic composition is a mixture of primordial and fissiogenic isotopes (Hidaka et al. 1994b). A comparison of isotopic ratios for fissiogenic Zr (estimated from empirical fission yields) with those on the core of RZ-10 reveals a large dicrepancy between expected and observed values of ^{90}Zr/^{91}Zr (Hidaka et al. 1994b). Isotopic ratios for other Zr isotopes do not show significant differences between observed and expected values. Substantial enrichment in ^{90}Zr results from enrichment of its precursor, ^{90}Sr. Apparently, chemical fractionation between Zr and Sr occurred before ^{90}Sr decayed to ^{90}Zr. A positive isotopic anomaly for ^{90}Zr indicates retention of ^{90}Sr in the core of RZ-10, but this conclusion does not apply to other natural reactors. Samples from RZ-2 do not show significant deviation from expected abundances of fissiogenic ^{90}Zr (Hidaka et al. 1994b). Retentivity of ^{90}Zr relative relative to Nd is as high as 22.9% in the core of RZ-10, whereas in other reactor zones it is negligible (Hidaka et al. 1994b). The retentivities of fissiogenic Sr, Cs and Ba did not exceed 5% of their original abundances, and the retentivity of Rb was even lower, in agreement with observations from other reactors. Hidaka et al. (1994b) explain the low retentivity of Rb by thermally enhanced diffusion caused by high temperatures during criticality.

Rare Earth Elements

The isotopic compositions and geochemical behaviors of REE in natural fission reactors have been intensively studied by numerous investigators (Ruffenach et al. 1975; Loubet and Allegre 1977; Brookins 1978; Holliger and Devillers 1981; Hidaka and Masuda 1988; Hidaka et al. 1988; Loss et al. 1988; Curtis et al. 1989; Hidaka et al. 1992a,b; Hemond et al. 1992; Menet et al. 1992; Bros et al. 1995; Hidaka and Holliger 1998). Anomalous isotopic compositions of REE, especially of Nd, were used as a proof that self-sustained fission chain reactions had occurred in uraninite at Oklo (Neuilly et al. 1972). Rare earth elements are sensitive indicators of nuclear reactions because light REE have high fission yields and many REE isotopes have large neutron-capture cross sections. As a result, fissiogenic and non-fissiogenic REE have markedly different isotopic compositions (Fig. 39).

Examples of measured REE concentrations in and around some reactor zones are given in Table 11. In comparison with most U deposits, natural reactors are highly enriched in light REE (Fig. 48). A sample from clays of RZ-10 analysed by Menet et al. (1992) is exceptional in its relative enrichment in heavy REE (Table 11). Heavy REE enrihment can arise if the analyzed sample contains elevated concentrations of detrital zircon, which is known to concentrate heavy REE. Detrital monazite-Ce, florencite-(Ce) and, possibly, zircon and apatite are major sources of primordial REE in FA sandstone hosting nuclear reactors.

The initial fractionation of REE in the natural fission reactors was fortituitous, because heavy REE have large neutron cross sections, and act as neutron poisons to prevent nuclear chain reactions. Primordial heavy REE, which were present in small amounts in U minerals, absorbed neutrons and were transmuted to other elements. The amount of heavy REE generated during fission was small. The combined effects of transmutation of primordial HREE and low production rates fissiogenic HREE caused isotopic abundances of HREE to be much lower than those of terrestrial REE standards (Hidaka and Masuda

1988). This is especially true for ^{155}Gd and ^{157}Gd (Fig. 39).

Neodynium and Ce are the most abundant REE in and around the natural reactors (Table 11). The highest concentrations of Nd are in the Bangombé natural reactor (Bros et al. 1995).

Rare earth elements in reactor zones consist of primordial and fissiogenic isotopes. Estimates of the proportion of fissiogenic REE suggest that REE heavier than Gd are entirely primordial (Hidaka et al. 1988).

Similar ionic radii of U^{4+} and REE^{3+} (1.26 to 1.05 Å for eight coordination) make REE potentially compatible in the uraninite structure, provided charge-balance mechanisms are available. Inspection of fission-product yield curves for uraninite from several reactor zones (Fig. 41) suggests the high retentivity of REE in uraninite (Hidaka and Holliger 1998). However, mobilization of REE, particularly Nd, in and around reactor zones has been noted (Curtis et al. 1989; Hemond et al. 1992; Menet et al. 1992). Bros et al. (1995) found fissiogenic Nd, Sm and Eu in FB shale as far as 5.5 m above the core of the Bangombé reactor. They found that chemical fractionation of migrating REE resulted in a Sm/Nd ratio 13-times larger than that in the reactor's core. Uranium, Nd, and Sm also

Table 11. Concentrations of REE (ppm), U (wt %) and values of $^{235}U/^{238}U$ (%) in natural fission reactors and their surroundings.

	RZ-7 RZ-8 RZ-9	FB shale near RZ-10	Core RZ-10	Core RZ-10	Clays of RZ-10	Sandstone near RZ-10	RZ-2	Ba
La	100	28	210	119	39.6	199	75.3	n.r.
Ce	341	107	453	313	163.1	225	226	n.r.
Pr	87	20.3	170	110	n.r.	38.8	n.r.	n.r.
Nd	355	98.8	575	382	96.1	224	239	819.8
Sm	73.6	18.9	112	67.6	38.8	38.9	40.7	191.8
Eu	11.6	4.2	9.6	6.5	8.5	2.4	4.8	5.4
Gd	15.9	8.6	18.9	15.3	59.9	42.8	8.8	n.r.
Tb	1.7	1.0	2.0	1.5	n.r.	3.9	n.r.	n.r.
Dy	5.0	4.4	9.3	5.6	34.5	15.8	4.8	n.r.
Ho	2.0	0.8	2.5	2.0	n.r.	2.78		n.r.
Er	3.8	2.2	5.8	4.3	29.6	5.7	1.8	n.r.
Tm	0.4	0.2	0.6	0.4	n.r.	0.7		n.r.
Yb	2.9	1.3	4.0	2.4	16.6	3.9	1.4	n.r.
Lu	0.5	0.2	1.7	1.2	4.0	0.6		n.r.
U	51.1	5.86	24.3	14.9	14.12	2.99	66.2 ppm	50.37
$^{235}U/^{238}U$	0.6812	0.6301	0.5793	0.5649	0.717	0.7254	0.6024	0.5902

Note: data for RZ-7 to RZ-9 and FB shale from Hidaka et al. (1992); for RZ-10 from Hidaka and Holliger (1998); for clays of RZ-10 from Hemond et al. (1992) and Menet et al. (1992); for Bangombé reactor from Bros et al. (1995). n.r. = not reported.

fractionated in the core of RZ-10 (Gauthier-Lafaye et al. 1996). Mineralogical evidence for the redistribution and migration of fissiogenic REE includes florencite-(La) within clays of the Bangombé reactor (Janeczek and Ewing 1996a) and REE-bearing crandalite in RZ-13 (Gauthier-Lafaye et al. 1996; Dymkov et al. 1997). Coffinite rich in fissiogenic REE and REE-phosphates occur in sandstone some 20 cm beneath the Bangombé reactor (Janeczek and Ewing 1996b). Apatite in RZ-10 and RZ-16 (Bros et al. 1996) contains fissiogenic REE, indicating redistribution of fissiogenic REE within the reactor zones.

Samariun-147, the daughter of ^{147}Pm ($t_{1/2}$ = 2.62 yr), is relatively abundant in the Oklo reactor zones, and this is regarded as proof that Pm was more abundant at Oklo two-billion years ago than it is in most U deposits today (Kuroda 1982).

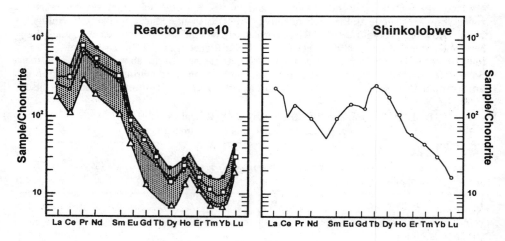

Figure 48. Chondrite-normal-ized plot of REE abundances in RZ-10 and in U minerals from the Shinkolobwe mine, Zaire. Modified from Hidaka et al. (1992b).

Technetium

Technetium is of major concern in radioactive-waste disposal because the fission-generated isotope, ^{99}Tc, with half-life 2.13×10^5 years, remains highly radioactive for periods longer than those of regulatory concern for most proposed underground repositories for high-level waste (10,000 years). Under oxidizing conditions Tc is potentially mobile, which, due to both radioactivity and toxicity, makes it a biologically significant radionuclide in high level waste. Naturally occurring ^{99}Tc forms in highly concentrated U deposits by spontaneous fission of ^{238}U (Curtis et al. 1994). Whereas present-day concentrations of ^{99}Tc in most U deposits are on the order of a few picograms per gram of rock (10^{-11} wt %), Tc was produced in large quantities in natural reactors by neutron-induced fission of ^{238}U. Technetium-99 decays to ^{99}Ru, and a small portion of ^{99}Tc was also converted to ^{100}Ru by neutron capture. Ruthenium-99 and ^{100}Ru are, therefore, indicators of ^{99}Tc behavior generated in natural reactors. Within approximately 10^6 years after the end of fission reactions, essentially all ^{99}Tc produced in the reactors had decayed to ^{99}Ru, so that for a long time, the fission-produced isobars at mass 99 reflected two elements: Tc and Ru (Loss et al. 1988). Analyses of isotopic abundances in uraninite from RZ-9 show that a significant amount of Tc was lost from the reactor; whereas most Ru was retained in uraninite (Loss et al. 1988). Gancarz et al. (1980) found that samples from above certain reactor zones were enriched in ^{99}Tc (now ^{99}Ru) relative to U. They proposed that ^{99}Tc was released from uraninite, transported by hot aqueous fluids and deposited approximately 10 m above the reactors. All samples contain ^{99}Ru in proportions different than those produced by fission (Curtis et al. 1989), which means that a significant portion of ^{99}Ru was either lost or gained. Most dissolved Tc was retained within the first few meters of sandstone adjacent to the core of RZ-9 (Loss et al. 1989) (Fig. 49). Technetium-99 was chemically fractionated from Ru within 1.2 million years after criticality ended (Gancarz et al. 1980; Loss et al. 1989). Hidaka and Holliger (1993) and Hidaka et al. (1999) use the parameter

$$\Delta^{99}Ru = [(^{99}Ru/^{101}Ru)_{obs}/(^{99}Ru/^{101}Ru)_{calc} - 1] \times 100\%$$

to estimate the degree to which Ru and Tc were chemically fractionated in the reactors. Positive values of $\Delta^{99}Ru$ indicate a gain of Tc; whereas negative values indicate losses. Although isotopic abundances measured in reactors RZ-1 to RZ-9 give only negative values of $\Delta^{99}Ru$, the $\Delta^{99}Ru$ values for RZ-10 and RZ-13 span a range of -5.15 to +11.6 (Hidaka and Holliger 1993). The highest values of $\Delta^{99}Ru$ (up to +118) were observed in Ru minerals from RZ-13. This is taken as an evidence for preferential release from uraninite of Tc (relative to Ru), followed by migration and precipitation of Tc under reducing conditions. Precipitation of Tc led to Ru minerals being enriched in ^{99}Tc (now ^{99}Ru) relative to Ru. Non-fissiogenic ^{100}Ru in Ru minerals is explained by neutron capture by ^{99}Tc. Thus ^{99}Tc must have been released from uraninite during criticality or within 10^6 years after criticality ended.

Platinum group elements (PGE): Ru, Rh and Pd

Ruthenium isotopes that originated through the decay of, or neutron capture by ^{99}Tc are discussed above. Three other Ru isotopes are also found at Oklo: ^{101}Ru, ^{102}Ru and ^{104}Ru (Fréjacques et al. 1975; Curtis et al. 1989; Hidaka et al. 1993). All are entirely fissiogenic, produced by radioactive decay of short-lived fission products. Rhodium is monoisotopic, making it is impossible to distinguish between primordial and fissiogenic components. Hidaka and Holliger (1998) assumed that all Rh found in natural reactors is fissiogenic, by analogy with other PGE.

The retention of Ru and Pd relative to Te in RZ-9 was estimated to be 91% and 88%, respectively (Curtis et al. 1989) (Fig. 40). Data for Ru, Rh and Pd in RZ-10 and RZ-13 plot on the calculated relative fission-yield curve (Fig. 41), which is evidence for the high

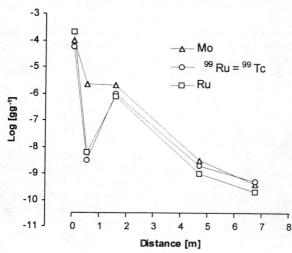

Figure 49. Concentrations of Mo, Ru, and ^{99}Ru (now as ^{99}Tc) in sandstone adjacent to RZ-9. Drawn from data by Loss et al. (1989).

retentivity of these elements. The small amounts of Ru and Pd that were released from uraninite (Fig. 49) reprecipitated in host sandstone close to RZ-9 (Loss et al. 1989). PGE were redistributed within the reactor zones, as evidenced by the precipitation of PGE-bearing minerals.

Low-yield mass region for fission products: Ag, Cd and Te

Silver, Cd and Te occur on the symmetric valley and peak sides of the fission-yield curve (Pd also belongs to this group but is discussed with other PGE) (Fig. 41). Inventories and behaviors of Ag, Cd and Te in natural fission reactors have been studied by numerous authors (Brookins 1978; De Laeter et al. 1980; Loss et al. 1984, 1988, 1989; Curtis et al. 1989; Hidaka et al. 1992; Hidaka and Holliger 1998). Tellurium in the reactor zones is entierly fissiogenic, whereas substantial amounts of primordial Ag and Cd occur in the reactors. Two fissiogenic isotopes of Te, ^{125}Te and ^{126}Te, may be daughters of ^{125}Sb and ^{126}Sn. Large discrepancies between measured and calculated abundances of ^{126}Te in RZ-10 and RZ-13 were explained by chemical fractionation of Te from Sn (Hidaka and Holliger 1998). This fractionation must have occurred during, or shortly after operations of the reactors because ^{126}Sn has a short half life (1×10^5 years). A comparison between abundances of ^{126}Te and Te in RZ-9 revealed both enrichment and depletion in ^{126}Te. This was taken as evidence for the redistribution of ^{126}Sn, which resulted in the secondary retention of ^{126}Te (Curtis et al. 1989).

Tellurium (and ^{126}Sn) was well retained in the reactor zones. Small amounts of Te migrated into adjacent sandstones and was confined within the first few meters of reactor zones. The high retentivity of Te, despite incompatibility in the uraninite structure, was explained by Brookins (1978, 1990) as a result of the wide stability field of native Te apparent from Eh-pH calculations. Although native Te has not been found in natural reactors, Te is closely associated with PGE minerals. An inclusion of (most probably) altaite (PbTe) was observed in a uraninite from RZ-10 (Fig. 26).

In contrast to the behavior of Te, both Ag and Cd were highly mobile in natural fission reactors (De Laeter et al. 1980; Loss et al. 1984; Curtis et al. 1989). Almost all fissiogenic Cd was lost from the host uraninite in RZ-9, and the concentration of Cd in the surrounding sandstone is low, suggesting that a substantial amount of fissiogenic Cd migrated from the reactor zone and was not well retained by the surrounding sandstone (Loss et al. 1989).

There are deficiencies of Ag in reactor zones, which suggests that some fissiogenic Ag migrated from the reactor zones. However, fissiogenic Ag cannot be distingushed from its primordial counterpart, so that the degree of Ag retentivity cannot be determined (Curtis et al. 1989). Silver appears to have been retained in sandstone in the vicinity of reactors (Loss et al. 1989).

Molybdenum

Both primordial and fissiogenic Mo occur in natural fission reactors. In RZ-9, the concentration of fissiogenic Mo was as high as 72% of total Mo (Curtis et al. 1989). Relative to Nd, between 80% and 90% of fissionogenic Mo migrated from RZ-9 (Figs. 9-40), but the majority of released Mo was retained within the first few meters of the reactor zone (Fig. 49). Migrating Mo was incorporated into secondary sulfides such as pyrite. Molybdenite, presumably containing fissiogenic Mo, was observed by TEM in illite adjacent to uraninite from RZ-10 (Janeczek and Ewing 1995), which implies that Mo was at least partially retained close to uraninite in the cores of natural reactors.

Table 12. Evolution of fluids in and around the natural fission reactors.

Event	Age, Ma	T, °C	Salinity, wt % eq. NaCl	Remarks
Primary U mineralization	2000 ± 50	100-150	> 20	$\delta^{18}O = +6.9‰$ SMOW
Fission chain reactions	1970 ± 50	350-600	1.5 - 10	1st mobilization of elements in NFR
Cooling of the reactors	< 2	From 400 to 150-100	From 5 to > 23	Possible continuation of elements mobilization
Steady-state regime	1900	< 160	?	Precipitation of pyrite
Late Ca-diagenesis/basin uplift (?)	1740-1700	150-170	20	2nd mobilization of elements in NFR $\delta^{18}O = 4-9‰$ SMOW
Large-scale circulation	830-790	170-200	0.8 - 5.2	Cu-Fe-Pb-sulfides
Intrusion of dolerite dikes	1000-750	310	18-20	3rd mobilization of elements in NFR $\delta^{18}O = 22-24‰$ SMOW
Post-Middle Jurassic fault tectonics		< 70		
Weathering	0.0765±0.0068	< 70	low	4th mobilization of elements in shallow NFR $\delta^{18}O$ low

Based on data in Michaud (1998)

Conditions for migration and retention of radionuclides in natural fission reactors

Eleven types of fluids circulating in the Franceville basin at various stages of its evolution have been recognized, all of which were related to major geological events in the basin (Mathieu and Cuney 1998). Fluid characteristics (temperature, pressure, and chemistry) were determined from numerous studies of fluid inclusions in various minerals associated with the reactors (e.g. Openshaw et al. 1978; Dubessy et al. 1988; Gauthier-Lafaye and Weber 1989; Gauthier-Lafaye et al. 1989; Savary and Pagel 1997; Mathieu and Cuney 1998). Trace element and isotopic analyses of accessory minerals such as apatite, calcite, pyrite and galena, have also significantly improved our understanding of the nature and roles of fluids that interacted with natural reactors (Raimbault 1993; Raimbault et al. 1996).

Reactor zones experienced at least three major stages of hydrothermal alteration (Table 12). The first stage of dissolution and alteration of uraninite in reactor zones occurred during reactor operations. Heat generated by fission turned groundwaters into high-temperature (250 to 450°C) and moderately saline (3-10 wt % eq. NaCl) fluids. There may have been local increases in redox potentials caused by radiolysis of fluids in the highly radioactive environments of active reactors.

Cooling of the reactors may also have been a period of dissolution and alteration of uraninite in nuclear reactors. The cooling of reactors lasted for 0.6 to 1 Ma, during which time fluids changed from high temperature and moderately saline to low temperature (100°C) and highly saline (> 23 wt % eq. NaCl). High salinities of hydrothermal fluids enhance dissolution of UO_2, even under reducing conditions (Janeczek and Ewing 1992, and references therein). It was during the cooling stage that low-temperature illite crystallized on uraninite surfaces in the reactor cores (Gauthier-Lafaye et al. 1989). Circulation of fluids during fission chain reactions and during the subsequent cooling stage were presumably local events, because heat generated by nuclear reactions dissipated in surrounding rocks within 20 m of reactor zones (Gérard et al. 1998).

A second stage of hydrothermal alteration of reactor zones was caused by basin-wide circulations of groundwaters in the Franceville basin between 1600 and 1400 Ma, causing mobilization of Pb in reactor zones (Raimbault et al. 1996). Evidence for this second hydrothermal stage is revealed by trace-element analyses of pyrite crystals that are enriched in Sb, Ag, Se and Au. These pyrite crystals are associated with Sb-rich galena from reactor zones, an association that has been interpretted as evidence for the mobilization of radiogenic Pb by large-scale circulation of groundwaters.

A third stage of hydrothermal reactor-zone alteration was related to the intrusion of dolerite dikes in the Franceville basin. Hydrothermal fluids generated by the heat of intrusions were highly saline (18-20 wt % eq. NaCl), with temperatures in the range 110° to 260°C. Ages of Pb loss from uraninite in reactor zones suggest that the major episode of the Pb loss was associated with this third hydrothermal stage. The loss of Pb from uraninite was almost complete in all reactor zones, with the exceptions of RZ-10 and of some uraninite specimens in RZ-2. The age of the intrusion of dolerite at Oklo (780 Ma) coincides with the ages (830-790 Ma) of large-scale circulations of fluids from which Pb-Cu-Fe-sulfides precipitated thoughout the Franceville basin (Raimbault et al. 1996). High activities of S^{2-} in the Franceville basin at that time caused precipitation of virtually all Pb released from uraninite, limiting large-scale migration of Pb and, perhaps, other metals.

The fact that more than 90% of U minerals have remained in the natural fission reactors despite the actions of fluids, has been attributed to reducing conditions prevailing in the reactors. However, Curtis and Gancarz (1983) suggested that radiolysis of water during reactor operations resulted in locally oxidizing conditions within the reactor zones. This suggestion was supported by findings of H_2 in fluid inclusions in quartz from clays of RZ-9 and in quartz surrounding reactors RZ-1 through RZ-6 (Dubessy et al. 1988). Recently, H_2 and O_2 were identified in fluid inclusions in quartz at the boundary between RZ-10 and sandstone (Savary and Pagel 1997). Fluid inclusions occur along healed fractures in detrital quartz grains. Concentrations of H_2 in fluid inclusions range from 13 to 60.4 mol %, and concentrations of O_2 range from 40 to 86.6 mol %. Oxygen commonly predominates over H_2. Wide ranges of homogenization temperatures are reported for fluid inclusions, with maximum values of 440°C in RZ-9, 490°C in quartz above RZ-3, and 412°C in quartz from RZ-10. Homogenization temperatures ranging from 320° to 410°C were reported for fluid inclusions containing O_2 and H_2, with salinities ranging from 17.8 to 0.2 wt % eq.-NaCl (Savary and Pagel 1997).

Fluid inclusions containing CH_4 (Savary and Pagel 1997), and some with CO_2, O_2, and CH_4 (Mathieu and Cuney 1998) have also been reported. These inclusions may be related to radiolysis of organic matter in and around natural reactors. As already mentioned, the O/C ratio is higher in reactors zones than in surrounding host rocks. The reaction of organic matter with O_2 and H_2, as well as with free radicals, led to the consumption of dissolved oxygen. Savary and Pagel (1997) explain the lack of O_2 in fluid inclusions from

RZ-9 by the oxidation of abundant organic matter. Dissolved O_2 may have also been consumed by the oxidation of pyrite. Pyrite relicts within hematite occur in small calcite veins in the oxidized portion of sandstone adjacent to RZ-10 (Savary and Pagel 1997). The range of radiolysis due to gamma radiation is limited to water that is within 40 µm of uraninite surfaces.

High redox potentials (+300 to +400 mV) measured recently in deep and chemically confined groundwaters at Okelobondo (Toulhoat et al. 1996) may be explained by current radiolysis of water that is in contact with uraninite, especially in regions depleted in both organic matter and pyrite.

Despite local increases in redox potentials caused by radiolysis, overall redox conditions in and around natural reactors remained reducing for a long time. The predominantly reducing environment is demonstrated by the existence of abundant galena and coffinite. Also, Ru arsenides and related minerals must have formed under reducing conditions. Native Pb in some reactors is further evidence of reducing conditions and, perhaps, local decreases in S^{2-} activities.

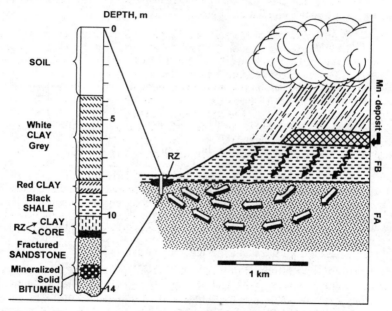

Figure 50. Cross-section through the Bangombé area showing flow paths of meteoric water and oxidized groundwater (modified from Smellie et al. 1993). Geologic profile through the Bangombé reactor zone at left.

Oxidizing conditions in shallow reactors, as indicated by existing uranyl minerals, are recent phenomena related to weathering; i.e. downward percolation of meteoric waters (Fig. 50). Reactor zones that are currently near the surface (RZ-1-9 at Oklo and Bangombé) have been affected by weathering much more than deep-seated reactors. The age of torbernite (63,500 yr) from beneath the Bangombé reactor zone (Bros et al. 1998) suggests a minimum age for the onset of weathering in shallow reactors. Oxidizing waters migrated along numerous fractures in sandstone beneath reactors. Redox fronts developed along these fractures, and advanced via matrix diffusion of groundwaters into unfractured portions of the sandstone. A redox front is marked by red staining of sandstone due to fine-

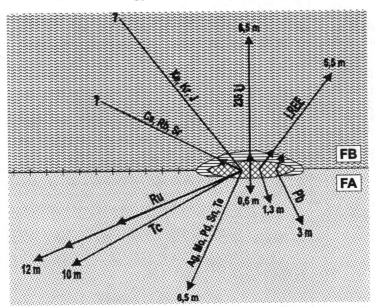

Figure 51. Schematic model summarizing reported minimum distances for migration of U and fission products around the reactor zones.

grained Fe oxides (hematite, goethite, etc.) dispersed within the rock matrix. Some U mobilized by oxidizing fluids either precipitated as uranyl minerals or was sorbed on goethite (Fig. 28). As already discussed, both normal U and depleted U were mobilized during weathering reactions. The uraniferous goethite shown in Figure 28 contains U with a normal isotopic composition, which means that U was derived from isotopically normal U minerals, despite the proximity of this sample to the core of the Bangombé reactor.

Not only has U been oxidized, but bitumen may also have reacted with oxidizing groundwaters. Pourcelot and Gauthier-Lafaye (1998) observed decreases in bitumen oxidation with depth, although bitumen from the deep-seated RZ-13 (250 m) is even more oxidized than bitumen from the shallow Bangombé reactor (12 m). Clearly, the oxidation of bitumen reflects multiple processes, including radiolysis and weathering.

Actinides and fission products that were mobilized during various stages of uraninite dissolution and alteration either reprecipitated within, or migrated out of reactor zones. As a result reactor zones are surrounded by halos of dispersed actinides and fission products released from uraninite. The extent of migration is rather poorly known for gaseous and volatile elements (noble gases, I, Cs, Sr, and Rb), but these elements were almost completely lost from the reactors and their immediate vicinities. Minimum migration distances for other fission products has been determined from the analyses of fission products or their stable daughters in the surrounding rocks. Many geochemical studies suggest that most fission products were confined within about 10 m of reactor zones (Fig. 51). Sorption and mineral precipitation were the most important retardation processes, limiting the mobilities of many dissolved actinides and fission products under the reducing conditions prevailing in and around the reactors. Table 13 summarizes the major mechanisms relevant to releases of actinides and fission products and mechanisms that limited mobilities. Only elements for which retardation mechanisms at natural reactors have been determined are listed in Table 13. Behaviors of other elements are either not well

Table 13. Mobilization and retardation mechanismsms of actinides, their daughters and fission products in natural reactors.

Element(s)	Release from uraninite by	Immobilization by ...	
		Precipitation	Sorption
U	dissolution of uraninite, coffinitization	secondary uraninite, uranyl minerals	chlorite, anatase, florencite, goethite
Pu	dissolution of uraninite	apatite	chlorite
REE	dissolution of uraninite	crandallite-group minerals, apatite, coffinite, REE uranyl phosphate	
Cs, Rb, Sr, Ba	diffusion-related processes		clay minerals (limited)
Tc, PGE	diffusion-related processes	arsenides, sulphides	
Te	diffusion-related processes	altaite (?)	
Pb	diffusion-related processes, coffinitization	aalena, native Pb, minium, arsenides, altaite(?)	

known or have been estimated from theoretical considerations (e.g. Eh-Ph diagrams and crystal chemistries) but have not yet been confirmed by observation.

CONCLUDING REMARKS

The natural fission reactors in Gabon are unique physical phenomena in the Earth's crust. Despite great efforts to identify other natural reactors, nowhere else has evidence for large-scale chain fission reactions been found. Although deviations within 0.01% from the normal $^{235}U/^{238}U$ ratio have been reported for uraninite from various U deposits, it is uncertain if these isotopic anomalies were caused by chemical fractionation or nuclear fission (Apt et al. 1978). It appears that the combination of geological and geochemical conditions required for self-sustained fission chain reactions in highly concentrated U minerals has been extremely rare in nature. Spontaneous fission of ^{238}U was common in ancient, highly concentrated U deposits and was associated with small-scale neutron-induced fission of ^{235}U. However, perhaps even one unfavorable factor was sufficient to have prevented large-scale spontaneous fission chain reactions in most U deposits. Ancient (1.6-1.7 Ga), highly concentrated U deposits from the Alligator River Uranium Field in Australia are good examples of restriction imposed by such unfavorable factors. Despite sufficiently high mass of U and concentrations of fissile ^{235}U, criticality was never achieved in any of the Alligator River deposisits. The maximum thermal neutron fluences derived from Gd and Sm isotope anomalies in those deposits are lower by a factor of 10^4 compared with those reported from natural fission reactors in Gabon (Maas and McCulloch 1990). An important difference between the Oklo and Alligator River U deposits are the higher concentrations of heavy REE in the latter; these elements can act as neutron poisons to prevent sustained nuclear-fission chain reactions.

Despite the apparent uniqueness of the Gabon natural reactors, there is no a single geological feature of the natural fission reactors that is not found in other ancient sedimentary-hosted, U-rich deposits (perhaps with the exception of the occurrence of Ru minerals). For example, the clay mantles around reactor cores is not unusual, because similar clay mantles are known at many U deposits, including those in the Athabasca basin, Canada. Clay mantles in the Oklo and Athabasca Basin deposits resulted from interactions of host sandstones with hydrothermal fluids; however, the origins of those fluids were different. Hydrothermal fluids associated with the natural reactors were generated from the heat of nuclear-fission reactions, whereas in the Athabasca Basin hydrothermal fluids unrelated to fission reactions were generated by deep-seated metamorphic reactions in rocks underlying the Athabasca U deposits.

The uniqueness of the natural fission reactors lies in their isotope geochemistry and isotope mineralogy. Natural fission reactors in Gabon are the only places on Earth where some minerals are, at least in part, composed of elements with non-primordial isotopic abundances. These minerals have isotopic signatures distinct from other terrestrial materials, although the chemical and physical properties of such minerals are the same as their isotopically normal counterparts. Anomalous isotopic compositions enable one to distinguish minerals unrelated to fission reactions from those derived from fissionogenic isotopes.

Hawthorne (1993) listed isotopic order in crystals among significant and difficult problems in Mineralogy that have to be solved. He presumed that isotopes displayed ordering over nonequivalent sites in a structure and that this ordering affects isotope fractionation. Undoubtedly, minerals from the fission reactors are ideal for this kind of mineralogical study because of their unusual isotopic compositions. It was demonstrated in section 9.2.4 that some unknown process (or processes) fractionated U isotopes within uraninite at the natural reactors. Isotopic ordering may be responsible for uneven distributions of U isotopes, but whether heterogeneous distributions of ^{235}U in uraninite resulted from primary zoning or alteration-related zoning is unknown and a matter for further study.

Natural fission reactors provide the rare opportunities to study geochemical behaviors of elements that do not occur naturally in any significant quantities; e.g. Pu and Tc. Their geochemical behavior is not only of academic interest but it has practical applications because these elements are of concern in radioactive waste disposal. Alteration processes in uraninite from the natural fission reactors may be useful in understanding the long term behavior of spent nuclear fuel under repository disposal conditions. It has already been shown that uraninite can be considered analogous to UO_2 in spent nuclear fuel, despite their different thermal histories and irradiation conditions (Janeczek et al. 1996). It is therefore not surprising that the natural fission reactors have been used as "natural analogues" for the disposal of high-level radioactive waste in the lithosphere. Recently, research emphasis has been on more quantitatively estimating the processes of radionuclide migration and retardation within and around the natural reactors (Louvat and von Maravic 1998). For this kind of study a detailed knowledge of mineral assemblages, including their isotopic compositions, is necessary. One of the outcomes of the mineralogical studies in the natural reactors is the illustration of the potential for various minerals (e.g. crandallite, chlorite, and apatite) to act as geochemical barriers that may help limit the migration of certain radionuclides.

Natural fission reactors offer opportunities to study various aspects of nuclear geochemistry, chemical fractionation of isotopes, retardation mechanisms of radionuclides, and perhaps more. In most geological and geochemical studies, isotopic analyses are used for dating, petrogenesis or identifying the sources of certain materials (e.g. origin of reacting fluids). In studying the natural reactors, the usual geologic questions are enlarged by issues related to the neutronics of the system (isotopic abundances, nuclear reactions, neutron fluence and radiation effects). These natural reactors are unique in that one is required to use, not only mineralogy and geochemistry, but also nuclear physics and nuclear chemistry to understand them. Relevant intensive thermodynamic parameters (temperature and pressure) are only part of the story. Neutron fluence must be deduced from isotopic data, which must be done on a grain-by-grain basis. The combination of mineralogy and geochemistry with the study of neutronics of a fission reactor is, perhaps, the most exciting part of studying the Oklo phenomenon.

ACKNOWLEDGMENTS

I am grateful to the editors, Robert Finch and Peter Burns, for inviting me to write this review. Otherwise, I could hardly find an excuse for venturing into the fields of so many experts. Special thanks go to Rodney C. Ewing, who introduced me into the Oklo phenomenon and to many scientific issues associated with natural-analogue studies; this paper could not have been completed without his help, discussions and support. I am indebted Edgar Buck, Rodney C. Ewing and Robert Finch for thoughtful and constructive reviews that considerably improved the final version of this manuscript. I thank Hiroshi Hidaka for reviewing sections on isotope geochemistry. I am grateful to Francois Gauthier-Lafaye for informative discussions and for sharing with me his profound knowledge about the natural reactors. Keld Jensen kindly provided me with the HRTEM image of uraninite and some of his unpublished data; this I greatly appreciate. I also thank Ewa Teper for technical assistance in preparing most of the illustrations. Final thanks must go to Ewa, my wife, for her patient encouragment. In preparation of this chapter, I benefitted from association with the Oklo Working Group organized by the Commission of the European Communities and the Commissariat à l'Energie Atomique of France. This work was financially supported by the KBN (Polish Research Committee) grant 6P04 D 007 14.

REFERENCES

Apt KE, Balagna JP, Bryant EA, Cowan GA, Daniels WR, Vidale RJ, Brookins DG 1978 Search for other natural fission reactors. *In* Natural Fission Reactors, IAEA-TC-119/37 IAEA, Vienna, p 677-689

Blanc P-L 1996 Oklo—Natural Analogue for a Radioactive Waste Repository. Vol. 1. Acquirements of the Project. Nuclear Science and Technology Report EUR 16857/1 EN. 123 p

Bodu R, Bouzigues H, Morin N. and Pfiffelmann JP 1972 Sur l'existence d'anomalies isotopiques rencontrées dans l'uranium du Gabon. CR Acad Sci Paris, 275. Serie D:1731-1732

Bonhomme MF, Gauthier-Lafaye F, Weber F 1982 An example of lower proterozoic sediments: the Francevillian in Gabon. Precambrian Res 18:87-102

Bourrel J, Pfiffelmann JP 1972 La province uranifére du bassin de Franceville (République Gabonaise). Mineral Deposita 7:323-336

Branche G, Chantret F, Guillemaut R, Pouget R 1975 Données chimiques et minéralogiques sur le gisement d'Oklo. *In* The Oklo Phenomenon, IAEA-SM-204/17. IAEA, Vienna, p 119-132

Brookins DG 1978 Application of Eh-pH diagrams to problems of retention and/or migration of fissiogenic elements at Oklo. *In* Natural Fission Reactors, IAEA-TC-119/33. IAEA, Vienna, p 243-254

Brookins DG 1981 Alkali and alkaline earth studies at Oklo. *In* Scientific Basis for Nuclear Waste Management III. Moore JG (Ed) Plenum Press, New York, p 275-282

Brookins DG 1990 Radionuclide behavior at the Oklo nuclear reactor, Gabon. Waste Management 10:285-296

Brookins DG, Lee MJ, Mukhopadhyay B, Bolivar SL 1975 Search for fission-produced Rb, Sr, Cs and Ba at Oklo. *In* The Oklo Phenomenon. IAEA-SM-204/3. IAEA, Vienna, p 401-414

Bros R, Turpin L, Gauthier-Lafaye F, Holliger Ph, Stille P 1993 Occurrence of naturally enriched ^{235}U: Implications for plutonium behaviour in natural environments. Geochim Cosmochim Acta 57:1351-1356

Bros R, Gauthier-Lafaye F, Larque P, Samuel J, Stille P 1995 Mobility of U, Th and lanthanides around the Bangombé natural nuclear reactor (Gabon). Mater Res Soc Symp Proc 353:1187-1194

Bros R, Carpena J, Sere V, Beltritti A 1996 Occurrence of plutonium and fissiogenic REE in hydrothermal apatites from the fossil nuclear reactor 16 at Oklo (Gabon). Proc. of the Fifth Internat Conf, Saint-Malo (France) 1995. Radiochimica Acta 74:277-282

Bros R, Anderson P, Roos P, Claesson S, Holm E, Smellie J 1998 Swedish investigations on the Bangombé reactor zone, U-series disequilibrium and time-scale of radionuclide mobilization processes. *In* Proc 1st Joint EC-CEA Workshop on the OKLO-Natural Analogue Phase II project. Louvat D, Davies C (Eds) Report EUR 18314. p 187-195

Carpéna J, Sere V 1994 Les apatites neoformees lors des reactions nucleaires dans les zones 10 et 16 d'Oklo. Réun Ann Sci Terre, Nancy (abstract)

Casas I., De Pablo J, Giménez J, Torrero ME, Bruno J, Cera E, Finch R Ewing RC 1998 The role of pe, pH, and carbonate on the solubility of UO_2 and uraninite under nominally reducing conditions. Geochim Cosmochim Acta 62:2223-2231

Celi_ski Z 1997 Natural fission reactor at Oklo. PTJ (Advances in Nuclear Technology) 40:11-13 (in Polish)

Cortial F, Gauhtier-Lafaye F, Lacrampe-Couloume G, Oberlin A, Weber F 1990 Characterization of organic matter associated with uranium deposits in the Francevillian formation of Gabon (Lower Proterozoic). Org Geochem 15(1):73-85

Cowan GA 1978 Migration paths for Oklo reactor products and applications to the problem of geological storage of nuclear wastes. *In* Natural Fission Reactors. IAEA-TC-119/26. IAEA, Vienna, p 693-699

Curtis D 1985 The chemical coherence of natural spent fuel at the Oklo nuclear reactors. SKB Technical Report 85-04, Stockholm

Curtis DB, Gancarz AJ 1983 Radiolysis in nature: Evidence from the Oklo nuclear reactors. SKBF/KBS Technical Report 83-10, Stockholm

Curtis D, Benjamin T, Gancarz A, Loss R, Rosman R, Delaeter J, Delmore JE, Maeck WJ 1989 Fission product retention in the Oklo natural fission reactors. Appl Geochem 4:49-62

Curtis D, Fabryka-Martin J, Aguilar R, Attrep M, Roensch F 1992 Plutonium in uranium deposits: Natural analogues of geologic repositories for Pu-bearing nuclear wastes. Proc 3rd Internat Conf High Level Radioactive Waste Management, Las Vegas, Nevada, April 12-16 1992, vol. 1. p 338-344

Curtis D, Fabryka-Martin J, Dixon P, Aguilar R, Rokop D 1994 Radionuclide release rates from natural analogues of spent nuclear fuel. Proc Behaviour of Actinides and Fission Products in the Geosphere. Charleston, South Caroline, USA, p 551-557

De Laeter JR, Rosman KJR, Smith CL 1980 The Oklo natural reactor: cumulative fission yields and retentivity of the symmetric mass region fission products. Earth Planet Sci Lett 50:238-246

Devillers C, Menes J 1975 Contribution du plomb et du Th a l'histoire des reacteurs d'Oklo. *In* The Oklo Phenomenon. IAEA-TC-119/18. IAEA, Vienna, p 495-511

Dran JC, Maurette JC, Petit JC, Drozd R, Hohenberg C, Duraud JP, Le Gressus C, Massignon D 1975 A multidisciplinary analysis of the Oklo ores. *In* The Oklo Phenomenon. IAEA-SM-204/12. IAEA, Vienna, p 223-233

Dubessy J, Pagel M, Beny JM, Christensen H, Hickel B, Kosztolanyi C, Poty B 1988 Radiolysis evidenced by H_2-O_2 and H_2-bearing fluid inclusions in three uranium deposits. Geochim Cosmochim Acta 52:1155-1167

Duffy CJ 1975 Uranium solubilities in the Oklo reactor zones. *In* The Oklo Phenomenon. IAEA-TC-119/36. IAEA, Vienna, p 229-233

Dymkov YuM, Pavlov YeG, Zav'yalov YeN 1979 Phase composition and morphogenetic features of uraninite from the Oklo reactor, Gabon. Geokhimiya 2:217-228 (in Russian)

Dymkov YuM, Holliger P, Pagel M, Gorshkov A, Artyukhina A 1997 Characterization of a La-Ce-Sr-Ca aluminous hydroxy phosphate in nuclear zone 13 in the Oklo uranium deposit (Gabon). Mineralium Deposita 32:617-620

Eberly P, Janeczek J, Ewing RC 1994 Petrographic analysis of samples from the uranium deposit at Oklo, Republic of Gabon. Radiochim Acta 66/67:455-461

Eberly P, Janeczek J, Ewing RC 1995 Precipitation of uraninite in chlorite-bearing veins of the hydrothermal alteration zone (Argile de pile) of the natural nuclear reactor at Bangombé, Republic of Gabon. Mater Res Soc Symp Proc 353:1195-1202

Eberly P, Ewing RC, Janeczek J, Furlano A 1996 Clays at the natural nuclear reactor at Bangombé, Gabon: Migration of actinides. Radiochim Acta 74:271-275

Fréjacques C, Blain C, Devillers C, Hagemann R, Ruffenach J-C 1975 Conclusions tirées de l'étude de la migration des produits de fission. *In* The Oklo Phenomenon. IAEA-SM-204/24. IAEA, Vienna, p 509-524

Gancarz AJ 1978 U-Pb age (2.05×10^9 years) of the Oklo uranium deposit: Natural fission reactors. *In* Natural Fission Reactors, IAEA-TC119/40. IAEA, Vienna, p 513-520

Gancarz A, Cowan G, Curtis D, Maeck W 1980 ^{99}Tc, Pb and Ru migration around the Oklo natural fission reactors. Scientific Basis for Nuclear Waste Management II. p 601-608

Gauthier-Lafaye F 1986 Les gisements d'uranium du Gabon et les réacteurs d'Oklo. Modèle métallogénique de gîtes a fortes teneurs du Protérozoïque inférieur. Sci Géol Mèm 78:206 p

Gauthier-Lafaye F 1996 Introduction to the Oklo problematic. *In* Oklo Working Group. Proc 4th Joint EC-CEA progress and final meeting in Saclay, France, June 1995. von Maravic H (Ed). EUR 16704EN. p 5-16

Gauthier-Lafaye F, Weber F 1978 Études minéralogiques et pétrographiques effectuées a Strasbourg sur les réacteurs naturels d'Oklo. *In* Natural Fission Reactors. IAEA-TC-119/8. IAEA, Vienna, p 199-227

Gauthier-Lafaye F, Weber F, Naudet R, Pfiffelmann JP, Chauvet R, Michel B, Reboul JC 1980 Le gisement d'Oklo et ses reacteurs de fission naturels. Proc Internat Uranium Symp on the Pine Creek Geosyncline. IAEA 663-673 Vienna

Gauthier-Lafaye F, Weber F 1989 The Francevillian (Lower Proterozoic) uranium deposit of Gabon. Econ Geol 84:2267-2285

Gauthier-Lafaye F, Weber F 1993 Uranium-hydrocarbon association in Francevillian uranium ore deposits, Lower Proterozoic of Gabon. *In* Bitumens in Ore Deposits. Parnell J, Kucha H, Landais P (Eds). Springer-Verlag, Berlin, p 276-286

Gauthier-Lafaye F, Weber F, Ohmoto H 1989 Natural fission reactors of Oklo. Econ Geol 84:2286-2295

Gauthier-Lafaye F, Holliger P, Blanc P-L 1996 Natural fission reactors in the Franceville basin, Gabon: A review of the conditions and results of "critical event" in a geologic system. Geochim Cosmochim Acta 60:4831-4852

Geffroy J 1975 Etude microscopique des minerais uraniféres d'Oklo. *In* The Oklo Phenomenon. IAEA-SM-204/18. IAEA, Vienna, p 133-151

Geffroy J, Lissilour J 1974 Etude microscopique des minerais d'Oklo. Bull Inf Sci Tech 193:132-157

Geffroy J, Cesbron F, Lafforgue P 1964 Données préliminaires sur les constituants profonds des minerais uraniféres et vanadiféres de Mounana Gabon. CR Acad Sci Paris 259:601-603

Gérard B, Royer JJ, Le Carlier de Veslud C, Pagel M, Scius H, Gauthier-Lafaye F 1998 3D modelling of heat and mass transfers around a natural fission reactor (Oklo, Gabon). Proc the 1st Annual Progress Meeting of the OKLO-Natural Analogue Phase II Project. Sitges, Spain, June 1997. p 301-307

Hageman R, Roth E 1978 Relevance of the studies of the Oklo natural nuclear reactors to the storage of radioactive wastes. Radiochim Acta 35:241-247

Hagemann R, Lucas M, Nief G, Roth E 1974 Mesures isotopiques du rubidium et du strontium et essais de mesure de l'age de la mineralisation de l'uranium du reacteur naturel fossile d'Oklo. Earth Planet Sci Lett 23:170-176

Havette A, Naudet R 1978 Etude locale par analyse ionique de nouveaux echantillions d'Oklo: Natural fission reactors. *In* Natural Fission Reactors. IAEA-TC-119/13. IAEA, Vienna, p 397-405

Havette A, Naudet R, Slodzian G 1975 Etude par analyse de la répartition et des proportions isotopiques de certain éléments dans échantillons d'Oklo. *In* The Oklo Phenomenon. IAEA-SM-204. IAEA, Vienna, p 463-478

Hawthorne FC 1993 Minerals, mineralogy and mineralogists: Past, present and future. Can Mineral 31:253-296

Hemond C, Menet C, Ménager MT 1992 U and Nd isotopes from the new Oklo reactor 10 (Gabon): Evidence for radioelements migration. Mater Res Soc Symp Proc 257:489-496

Hidaka H 1998 Isotopic analyses by using a sensitive high resolution ion micro-probe (SHRIMP) for natural analogue study of Oklo and Bangombé natural fission reactors. Radiochim Acta 82:327-330

Hidaka H, Masuda A 1988 Nuclide analyses of rare earth elements of the Oklo uranium minerals samples: a new method to estimate the neutron fluence. Earth and Planet Sci Lett 88:330-336

Hidaka H, Masuda A, Fuji I 1988 Abundance of fissiogenic and pre-reactor rare-earth elements in a uranium minerals sample from Oklo. Geochem Journ 22:47-54

Hidaka H, Holliger P, Shimizu H, Masuda A 1992a Lanthanide tetrad effect observed in the Oklo and ordinary uraninites and its implication for their forming processes. Geochem J 26:337-346

Hidaka H, Konishi T, Masuda A 1992b Reconstruction of cumulative fission yield curve and geochemical behaviors of fissiogenic nuclides in the Oklo natural reactors. Geochem J 26:227-239

Hidaka H, Holliger P, Masuda A 1993a Evidence of fissiogenic Cs estimated from Ba isotopic deviations in an Oklo natural reactor zone. Earth Planet Sci Lett 114 391-396

Hidaka H, Shinotsuka K, Holliger P 1993b) Geochemical behaviour of ^{99}Tc in the Oklo natural fission reactors. Radiochim Acta 63:19-22

Hidaka H, Takahashi K, Holliger P 1994a Migration of fission products into micro-minerals of the Oklo natural reactors. Radiochim Acta 66/67:463-468

Hidaka H, Sugiyama T, Ebihara M, Holliger P 1994b Isotopic evidence for the retention of ^{90}Sr inferred from excess ^{90}Zr in the Oklo natural fission reactors: Implication for geochemical behaviour of fissiogenic Rb, Sr, Cs and Ba. Earth Planet Sci Lett 122:173-182

Hidaka H, Holliger P 1998 Geochemical and neutronic characteristics of the natural fossil fission reactors at Oklo and Bangombé, Gabon. Geochim Cosmochim Acta 62:89-108

Hidaka H, Holliger P, Gauthier-Lafaye F 1999 Tc/Ru fractionation in the Oklo and Bangombé natural fission reactors, Gabon. Chem Geol 155:323-333

Holliger P 1992 Les nouvelles zones de reaction d'Oklo: Datation U-Pb et caracterisation in-situ des produits de fission a l'analyseur ionique. CEREM, Note technique DEM N° 01/92, Grenoble. 42 p

Holliger P 1993 Geochemical and isotopic characterization of the reaction zones (uranium, transuranium, Pb and fission products). In Oklo Working Group Meeting. Proc 2nd Joint CEC-CEA Progress Mtg. von Maravic H (Ed) Report EUR 14877. p 27-37

Holliger P, Devillers C 1981 Contribution r la température dans les réacteurs fossiles d'Oklo par la mesure du rapport isotopique du lutetium. Earth and Planet Sci Lett 52:76-84

Holliger P, Gauthier-Lafaye F 1996 Oklo, analogue naturel de stockage de déchets radioactifs (phase 1). Volume 2—Réacteurs de fission et systemes géochimiques anciens. Rapport EUR 16857/2 Commission européenne, Luxembourg. 344 p

Holliger P, Louvat D, Landais P, Kruge M, Ruau O 1997 Organic matter and uraninite from the Oklo natural fission reactors: Natural analogue of radioactive waste containing bitumen and UO_2 irradiated fuel. Nuclear Science and Technology Report EUR 17614:63 p Luxembourg

Janeczek J, Ewing RC 1992 Dissolution and alteration of uraninite under reducing conditions. J Nucl Mater 190:157-173

Janeczek J, Ewing RC 1993 Heterogeneity and alteration of uraninite from the natural fission reactor 10 at Oklo, Gabon. In Oklo Working Group Meeting. Proc 2nd Joint CEC-CEA Progress Mtg. von Maravic H (Ed) Report EUR 14877. p 177-188

Janeczek J, Ewing RC 1995 Mechanisms of Pb release from uraninite in the natural fission reactors in Gabon. Geochim. Cosmochim. Acta 59:1917-1931

Janeczek J, Ewing RC 1996a Uraninite from natural fission reactors in Gabon. In Oklo Working Group. Proc 4th Joint EC-CEA progress and final meeting in Saclay, France, June 1995. EUR 16704EN. p 29-42

Janeczek J, Ewing RC 1996b Phosphatian coffinite with rare earth elements and Ce-rich françoisite-(Nd) from sandstone beneath a natural fission reactor at Bangombé, Gabon. Mineral Mag 60:665-669

Janeczek J, Ewing RC 1996c Florencite-(La) with fissiogenic REEs from a natural fission reactor at Bangombé, Gabon. Am Mineral 81:1263-1269

Janeczek J, Ewing RC, Oversby V, Werme LO 1996 Uraninite and UO_2 in spent nuclear fuel: a comparison. J Nucl Mater 238:121-130

Jensen KA, Ewing RC 1998 Petrography and chemistry of the uraninites and uraninite alteration phases from the uranium minerals-deposit at Bangombé. In Proc 1st annual progress meeting of the OKLO-Natural Analogue Phase II Project. Sitges, Spain, June 1997, Louvat D, Davies C (Eds). pp139-159

Jensen KA, Ewing RC (in press) Microtexture and chemistry of "unaltered" uraninite in the Oklo, Okélobondo, and Bangombé natural fission reactors. Nuclear Science and Technology. Oklo Working Group. Proc Second EC-CEA Workshop on the Oklo-Natural Analogue Phase II project held in Helsinki, Finland, from 16-18 June. Nuclear Science and Technology. Report EUR

Jensen KA, Ewing RC (submitted) The Okelobondo natural fission reactor, southeast Gabon: Geology, mineralogy and retardation of nuclear reaction products. GSA Bulletin

Jensen KA, Ewing RC, Gauthier-Lafaye F 1997 Uraninite: A 2 Ga spent nuclear fuel from the natural fission reactor at Bangombé in Gabon, West Africa. Mater Res Soc Symp Proc 465:1209-1218

Keller C 1978 Das Oklo – Phänomen. Kernbrennstoffkreislauf. Dr. Alfred Hüthig Verlag, Heidelberg

Kruglov AK, Pchelkin VA, Sviderskii MF, Dymkov YuA, Moshanskaya GN, Tchernetsov OK 1975 Preliminary investigations of samples of uranium minerals from the Oklo natural reactor in Gabon. In The Oklo Phenomenon. IAEA-SM-204/8. IAEA, Vienna, p 303-317 (in Russian)

Kuroda PK 1956 On the nuclear physical stability of the uranium minerals. J Chem Phys 25:781-782

Kuroda PK 1977 Fossil nuclear reactor and Pu-244 in the early history of the solar system. IAEA-SM-204/4:479-487 Vienna

Kuroda PK 1982 The Origin of the Chemical Elements and the Oklo Phenomenon. Springer-Verlag. Berlin, Heidelberg, New York, 165 p

Kuroda PK 1993 Technetium in nature and age of the solar system. Radiochim Acta 63:9-18

Lancelot JR, Vitrac A, Allegre CJ 1975 The Oklo natural reactor: age and evolution studies by U-Pb and Rb-Sr systematics. Earth and Planet Sci Lett 25:189-196

Leachman RB, Bishop WP 1978 Relevance of natural retention experience to nuclear waste management. In Natural Fission Reactors. IAEA-TC-119/30. IAEA, Vienna, p 700-707

Loss RD, Rosman KJR, de Laeter JR 1984 Transport of symmetric region fission products at the Oklo natural reactors. Earth and Planet Sci Lett 68:240-248

Loss RD, De Laeter JR, Rosman KJR, Benjamin TM, Curtis DB, Gancarz AJ, Delmore JE, Maeck WJ 1988 The Oklo natural reactors: cumulative fission yields and nuclear characteristics of Reactor Zone 9. Earth and Planet Sci Lett 89:193-206

Loss RD, Rosman KJR, De Laeter JR, Curtis DB, Benjamin TM, Gancarz AJ, Maeck WJ, Delmore JE 1989 Fission product retentivity in peripheral rocks at the Oklo natural fission reactors. Chem Geol 76 71-84

Loubet M, Allegre CJ 1977 Behavior of the rare earth elements in the Oklo natural reactor. Geochim Cosmochim Acta 41- 1539-1548

Louvat D, von Maravic H 1998 Foreword. Nuclear Science and Technology. Oklo Working Group. *In* Proc First EC-CEA Workshop on the Oklo-Natural Analogue Phase II project held in Sitjes, Spain, 18-20 June 1997, Louvat D, Davies C (Eds). Nuclear Science and Technology. Report EUR 18314

Maas R, McCulloch MT 1990 A search for fossil nuclear reactors in the Alligator River Uranium Field, Australia: Constraints from Sm, Gd and Nd isotopic studies. Chem Geol 88:301-315

Mathieu R, Cuney M 1998 Space-time evolution of fluid circulations in the Franceville basin and around Oklo-Okelobondo and Bangombé nuclear reactor zones (Gabon). *In* Oklo Working Group Proc 1st Joint EC-CEA Workshop on the OKLO-Natural Analogue Phase II project. Louvat D, Davies C (Eds). EUR 18314. p 111-121

Menet C, Ménager M-T, Petit J-C 1992 Migration of radioelements around the new nuclear reactors at Oklo: Analogies with a high-level waste repository. Radiochim Acta 58:395-401

Michaud V 1998 Synthesis of existing information on the Oklo basin geology: Questions and prospects. *In* Oklo Working Group Proc 1st Joint EC-CEA Workshop on the OKLO-Natural Analogue Phase II project. Louvat D, Davies C (Eds) EUR 18314. p 27-48

Mossman DJ, Nagy B, Rigali MJ, Gauthier-Lafaye F, Holliger P 1993 Petrography and paragenesis of organic matter associated with the natural fission reactors at Oklo, Republic of Gabon: a preliminary report. Internat J Coal Geol 24:179-194

Mossman DJ, Gauthier-Lafaye F, Nagy B, Rigali MJ 1998 Geochemistry of organic-rich black shales overlying the natural nuclear fission reactors of Oklo, Republic of Gabon. Energy Sources 20:521-539

Nagy B, Gauhtier-Lafaye F, Holliger P, Davis DW, Mossman DJ, Leventhal JS, Rigali MJ, Parnell J 1991 Organic matter and containment of uranium and fissiogenic isotopes at the Oklo natural reactors. Nature 354:472-475

Nagy B, Gauhtier-Lafaye F, Holliger P, Mossman DJ, Leventhal JS, Rigali MJ 1993 Role of organic matter in the Proterozoic Oklo natural fission reactors, Gabon, Africa. Geology 21:655-658

Natural Fission Reactors (Les Réacteurs de Fission Naturels) 1978 International Atomic Energy Agency, Vienna, 754 p

Naudet R 1991 Oklo: Des Réacteurs Nucléaires Fossiles. Étude Physique. Eyrolles, Paris, 685 p

Neuilly M, Bussac J, Fréjacques C, Nief G, Vendryes G, Yvon J 1972 Sur l'existence dans un passé reculé d'une réaction en chaîne naturelle de fissions, dans le gisement d'uranium d'Oklo (Gabon). CR Acad Sci Paris, 275, Série D:1847-1849

Openshaw R, Pagel M, Poty B 1978 Phases fluides contemporaines de la diagenesé des grés, des mouvements tectoniques et du fonctionnement des réacteurs nucléaires d'Oklo (Gabon). *In* Natural Fission Reactors. IAEA-TC-119/9. IAEA, Vienna, p 267-296

Oversby V 1996 Criticality in a high level waste repository. A review of some important factors and an assessment of the lessons that can be learned from the Oklo reactors. SKB Technical Report 96-07. Stockholm, 41 p

Parnell J 1996 Petrographic relationships between mineral phases and bitumen in the Oklo Proterozoic natural fission reactors, Gabon. Mineral Mag 60:581-593

Pourcelot L, Gauthier-Lafaye F 1998a Mineralogical, chemical and O-isotopic data on uraninites from natural fission reactors (Gabon): effects of weathering conditions. CR Acad Sci, Paris, Sciences terre planét 326:485-492

Pourcelot L, Gauthier-Lafaye F 1998b Impact of surface weathering conditions at the Oklo natural reactors on the behaviour of fission products. *In* Oklo Working Group Proc 1st Joint EC-CEA Workshop on the OKLO-Natural Analogue Phase II project. Louvat D, Davies C (Eds) Report EUR 18314, p 181-186

Raimbault L 1992 Geochemistry of apatites, pyrites, and galenas in near- and far-field veins and sandstones around the Oklo fossil reactors (Gabon): Identification of ancient hydrothermal circulations. *In* Oklo Working Group Meeting. Proc 2nd Joint CEC-CEA Progress Mtg. von Maravic H (Ed) Report EUR 14877. p 47-57

Raimbault L, Peycelon H 1995 From near to far field: An attempt to reconstruct ancient fluid circulations around Oklo fossil reactors. Proc 3rd Joint CEC-CEA Oklo Working Group meeting, Brussels, 11-12 October 1993, EUR 16098. p 77-89

Raimbault L, Peycelon H, Blanc JP 1996 Characterization of near- to far-field ancient migrations around Oklo reaction zones (Gabon) using minerals as geochemical tracers. Radiochim Acta 74:283-287

Rigali MJ, Nagy B 1997 Organic free radicals and micropores in solid graphitic carbonaceous matter at the Oklo natural fission reactors, Gabon. Geochim Cosmochim Acta 61:357-368

Ruffenach JC, Menes J, Lucas M, Hagemann R, Nief G 1975 Analyses isotopiques fines des produits de fission et determination des principaux parametres des réactions nucléaires. *In* The Oklo Phenomenon. IAEA-SM-204. IAEA, Vienna, p 371-384

Savary V, Pagel M 1997 The effects of water radiolysis on local redox conditions in the Oklo, Gabon, natural fission reactors 10 and 16. Geochim Cosmochim Acta 61:4479-4494

Sere V, Carpéna J, Kienast J-R 1994 Du zircon neoforme lors des reactions nucléaires d'Oklo. Réun Ann Sci Terre, Nancy (abstract)

Simpson PR, Bowles JFW 1978 Mineralogical evidence for the mode of deposition and metamorphism of reaction zone samples from Oklo. *In* Natural Fission Reactors. IAEA-TC-119/28. IAEA, Vienna, p 297-306

Smellie JAT, Winberg A, Karlsson F 1993 Swedish activities in the Oklo Natural Analogue project. *In* Oklo Working Group. Proc 2nd Joint CEC-CEA Progress Mtg. Report EUR 14877. p 137-148

Sunder S, Sargent FP, Miller NH 1993 X-ray photoelectron spectroscopic study of a sample from a natural fission reactor at Oklo, Gabon. *In* Oklo Working Group. Proc 2nd Joint CEC-CEA Progress Mtg. Report EUR 14877. p 153-166

Sunder S, Miller NH, Duclos AM 1995 Geochemistry of samples from the natural fission reactors in the Oklo uranium deposists: An XPS and XRD study. *In* Proc 3rd Joint CEC-CEA progress meeting. p 153-161

The Oklo Phenomenon (Le Phénoméne d'Oklo) 1975 International Atomic Energy Agency, Vienna, 647 p

Toulhoat P, Gallien JP, Louvat D, Moulin V, l'Henoret P, Guerin R, Ledoux E, Gurban I, Smellie JAT, Winberg A 1996 Preliminary studies of groundwater flow and migration of uranium isotopes around the Oklo natural reactors (Gabon). J Contam Hydrol 21:3-17

Vanderbroucke M, Rouzaud JN, Oberlin A 1978 Etude géochimique de la matiere organique insoluble (kerogene) du minerais uranifere d'Oklo et des schistes apparentes du Francevillien. *In* Natural Fission Reactors. IAEA-TC-119/10. IAEA, Vienna, p 307-332

Walton RD, Cowan GA 1975 Relevance of nuclide migration at Oklo to the problem of geologic storage of radioactive waste. *In* The Oklo Phenomenon. IAEA-SM-204/1. IAEA, Vienna, p 499-507

Weber F, Geffroy J, Le Mercier M 1975 Synthese des etudes minéralogiques et petrographiques des minerais d'Oklo, de leurs gangues et des roches encaissantes. *In* The Oklo Phenomenon. IAEA-SM-204/14. IAEA, Vienna, p 173-192

8 Geomicrobiology of Uranium

Yohey Suzuki and Jillian F. Banfield

Department of Geology and Geophysics
University of Wisconsin—Madison
Madison, Wisconsin 53706

INTRODUCTION

Geomicrobiology is an interdisciplinary field that incorporates geological, chemical, microbiological, and ecological approaches to understand the interdependence of microbial activity and geochemical phenomena. Interest in this field has been stimulated by the recognition that microorganisms exert fundamental controls on the behavior of the majority of elements near Earth's surface. This chapter explores the ways that microorganisms affect the speciation, distribution, and fate of uranium (and other actinides) in near-surface geologic settings.

Microorganisms are virtually ubiquitous near Earth's surface. In excess of 10^8 microorganisms can inhabit one gram of soil, and cell densities of at least one cell per square micron of surface area can occur on some minerals. Microorganisms may directly and indirectly affect reactions at mineral surfaces and in aqueous solutions, and cells may comprise a significant fraction of reactive surface area exposed to fluid in diverse environments. Some microorganisms thrive at temperatures in excess of 100°C (extreme thermophiles). Others occur in Antarctic ice (psychrophiles), concentrated acidic solutions (acidophiles), subsurface aquifers several kilometers below Earth's surface, in saline brines (osmophiles), and at high pressure in the ocean depths.

In addition to an enormous diversity of prokaryotes (bacteria and archaea), eukaryotic microorganisms occur in abundance in many natural environments. For example, Hawksworth (1991) estimated that there are ~1.5 million species of fungi in fresh and ocean waters and soil. Fungi are key constituents of a variety of microbial communities, including the rhizosphere symbioses that are of critical importance to nutrition of higher plants, and lichen communities that colonize and degrade newly exposed rock surfaces. Algae also play critical roles in geochemical cycles, especially through photosynthetic carbon fixation and biomineralization of carbonates and sulfates. Algae most commonly reside in freshwater or marine environments, but they also occur in snow, hot springs, on rock surfaces, and in desert soils (Lee 1980).

The uranium cycle near Earth's surface is mainly controlled by redox reactions. Following reduction of U^{6+} to U^{4+}, uranium is immobilized by precipitation of uraninite (UO_2), the most abundant mineral associated with uranium ore deposits. Conversely, uraninite dissolves via oxidation to form uranyl aqueous complexes and a "corrosion rind" on the surface of uraninite that consists of secondary uranyl minerals (Finch and Ewing 1992). In natural settings, the mobility of uranyl aqueous complexes is controlled in part by adsorption onto solid surfaces and the formation of secondary uranyl minerals. Microorganisms can affect uranium distribution through catalysis of redox reactions or via indirect effects, and may have played a role in uranium ore deposit formation.

Interactions between uranium and microorganisms have been intensively studied for their technological applications, and much has been learned about mechanisms of metal-cell interactions. For economic reasons, bioleaching was an early target for research efforts. *In situ* bioleaching is now an important method for uranium extraction, and fundamental research on microbial abilities to limit the release of uranium and other metals has become increasingly important (Wood 1996).

Uranium biomineralization is of current importance to environmental remediation and materials processing industries. There are significant environmental problems that arise because considerable quantities of industrial uranium-bearing wastes were disposed of in near-surface environments. To prevent the contamination of surface and ground waters with uranium from geochemically reactive waste forms, it is critical that uranium is immobilized in a geochemically inert form until the site can be decontaminated. Modern technological processes also generate vast quantities of metal-contaminated byproducts that must be treated before they can be disposed of in the environment. Uptake and precipitation of uranium by cells may be a cost-effective method to extract and concentrate dissolved uranium from solutions. Similarly, biological sorbents (e.g. waste biomass from fermentation industries) may prove to be inexpensive materials for a variety of heavy metal decontamination processes.

Assuming indefinite, perfect containment is unachievable, another important environmental challenge is to develop models that accurately predict the long-term risks associated with radioactive isotopes released from geological nuclear waste or nuclear weapons repositories. Detailed understanding of microbial uptake of uranium and other actinides and colloidal migration are needed to develop models that account for all the important processes.

In this chapter, we review the results of investigations mostly directed toward understanding microbe-actinide interactions. We focus heavily on the relatively well-studied behavior of uranium, but note that the less well understood phenomena involving microorganisms and other actinides may be predicted based on the geochemical similarities between uranium and these elements. We conclude by drawing attention to some possible areas for future investigations.

BIOLOGICAL CLASSIFICATION OF METAL IONS

Biologically, metal ions are classified by their binding preferences, specifically, whether they seek out O-, N- or S-containing ligands. Pearson (1963) separated metal ions into hard acids (O-seeking) and soft acids (N- or S-seeking). In addition, Nieboer and Richardson (1980) proposed a third class (borderline) defined by behavior intermediate between these groups. Alkaline metals, alkaline earth metals, lanthanides and actinides [including uranium as $(UO_2)^{2+}$] are categorized as hard acids. Relatively few metals (e.g. Hg^{2+}, Au^+, and Ag^+) are classified as soft acids. The borderline metal ions, consisting of the first row transition metals, show increasing N- or S-seeking character in the order: $Mn^{2+} < Zn^{2+} < Ni^{2+} < Fe^{2+}\ Co^{2+} < Cd^{2+} < Cu^{2+} < Pb^{2+}$.

As illustrated in Table 1, O binding sites in biomolecules sought by hard acids are carboxylate, carbonyl, alcohol, phosphate, and phosphodiester groups. Functional groups sought by soft acids include sulphydryl, disulphide, thioether, and amino groups (Blundell and Johnson 1976; Blundell and Jenkins 1977).

THE MICROBIAL CELL—SITES FOR METAL BINDING

In order to familiarize readers with possible metal binding site locations, as well as terms that appear throughout this chapter, the following sections review the structure, composition, and organization of components of microbial cells. The discussion is

Table 1. Functional groups in biomolecules
Source: Blundell and Johnson (1976), Blundell and Jenkins (1977)

Functional groups sought by Hard acids	Functional groups sought by Soft acids
Carboxylate: $R-\overset{\overset{O}{\parallel}}{C}-O^-$	Sulphydryl: $-SH$
Carbonyl: $R-\overset{\overset{O}{\parallel}}{C}-OR,\ R-N\overset{\overset{O}{\parallel}}{H}C-R$	Disulphide: $-S-S-$
Alcohol: $R-\overset{\mid}{\underset{\mid}{C}}-OH$	Thioether: $-SR$
Phosphate: $R-OPO_3^{2-}$	Amino: $-NH_2$
Phosphodiester: $R-O-\overset{\overset{O}{\parallel}}{\underset{\underset{O}{\parallel}}{P}}-O-OR$	

subdivided to allow consideration of different types of organisms, where appropriate. In general, important structural differences are associated with phylogenetically-based groups. Typically, this necessitates separate consideration of organisms from the different domains (bacteria vs. archaea vs. eukaryotes). However, some cell structural details vary considerably between microbial groups within domains, so more detailed subdivisions (e.g. Gram-positive vs. Gram-negative bacteria) will be required in some cases. Special attention is given to identifying anionic sites in cell polymers (especially O-bearing sites, see above), as these play key roles in binding metal cations of particular interest here.

A diagram of a prokayrotic microbial cell is shown in Figure 1. A microbial cell typically consists of the cytoplasm, cytoplasmic membrane, cell wall or outer membrane, and extracellular polymers.

The cytoplasm and cytoplasmic membrane

In the cytoplasm (Fig. 1), all microorganisms have genetic molecules such as

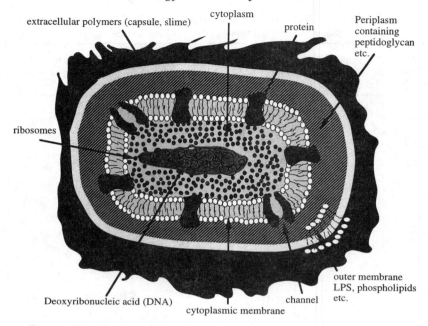

Figure 1. Schematic diagram illustrating the basic components of a prokaryotic cell.

DNA, RNA and plasmids (smaller pieces of DNA that, for instance, can confer toxic metal resistance). In addition, eukaryotes have organelles such as mitochondria (for the citric acid cycle, fatty acid oxidation, etc.), nuclei, chloroplasts (for photosynthesis) and vacuoles. Vacuoles play a variety of roles in degradation of macromolecules, storage of metabolites such as inorganic phosphates, and control of ion concentrations in the cytoplasm (Wiemken et al. 1979; Davis 1986; Wada et al. 1987; Jones and Gadd 1990).

The cytoplasm is surrounded by a highly selective barrier known as the cytoplasmic membrane. The cytoplasmic membrane is essentially a phospholipid bilayer (Fig. 1). The cytoplasmic membrane contains electron transport systems that, for instance, catalyze redox reactions of metals, and membrane transport proteins. The phosphate groups are the main anionic sites in these bilayers.

Cell walls and associated surfaces

Most, but not all microorganisms have a rigid cell wall, with or without loosely attached polymers (exceptions include the bacterial mycoplasma and archaeal thermoplasmales). The diverse cell wall constituents display different reactivities with respect to metal ions and the cell wall constituents vary significantly among microorganisms. The cell wall and/or extracellular polymers are the first regions that interact with surrounding environments. Therefore, it is particularly important to understand the structure of the cell wall and associated molecules involved in metal binding.

Bacteria. Historically, bacteria were subdivided into Gram-positive and Gram-negative groups, based on the structure and composition of the cell wall. Peptidoglycan (Pg) comprises up to ~90% of the cell wall of Gram-positive bacteria, whereas only about 10% of Gram-negative bacterial cell walls consist of Pg (i.e. a few Pg layers; Madigan et al. 1997). The structure of Pg is complex. However, because this compound is a dominant constituent of many bacterial cell walls, and because metal binding is commonly associated with cell walls, some basic details are provided. Pg consists of polysaccharides cross-linked by peptide chains. The polysaccharides are N-acetyl glucosamine (NAG) and N-acetyl muramic acid (NAM). Carboxylate groups at the terminus of individual polysaccharide strands provide the bulk of the anionic character of Pg (Beveridge and Murray 1980).

Teichoic acids (TAs) and teichuronic acids (TUAs) are acidic polysaccharides that are frequently associated with bacterial cell walls. Phosphate and carboxylate residues are anionic sites of TAs and TUAs (McLean et al. 1996). These negatively charged polymers contribute net negative charge of the cell surface, thus can significantly affect passage of ions through the cell wall (Brock et al. 1994).

Another category of reactive biomolecules associated with certain cell walls is lipopolysaccharides (LPS). LPS, unique to Gram-negative bacteria, are major constituents of the outer half of what is essentially a second lipid bilayer (Nikaido and Vaara 1987). LPS consists of three chemically distinct regions (given in order of moving outward): lipid A, the core polysaccharide, and O-polysaccharide side chains. The core polysaccharide is anionic due to its phosphate groups.

The inner half of the outer membrane of Gram-negative bacteria is comprised of phospholipids. The entire outer membrane is relatively permeable due to porin membrane channels that often contain water, and can allow non-specific passage of small molecules and many ions (Madigan et al. 1997).

Fungi. Fungi are classified into five groups: Chytridimycota, Zygomycota, Ascomycota, Basydomycota and Oomycota. Ascomycota is the largest group. The fungal cell wall is mainly composed of a variety of polysaccharides and other polymers, but the combination of polymers varies dramatically between different taxonomic groups. However, ultrastructural studies have shown a general similarity in construction of the fungal cell wall (Bracker 1967; Troy and Koffler 1969; Hunsley and Byrnett 1970; Burnett 1976; Wessel and Sietsma 1981).

The cell wall consists of an inner layer and an outer layer. In fungi other than oomycetes and some yeasts, the inner layer contains a skeleton of crystalline fibrils made up of chitin, $(1 \rightarrow 4)$-β-D-Glycosaminoglycans (Gadd 1993; Wessel and Sietsma 1981). Oomycetes have a skeleton made up of cellulose, $(1 \rightarrow 4)$-β-D-Glucan (Peberdy 1990). Some yeasts have glucans with $(1 \rightarrow 3)$-β and $(1 \rightarrow 6)$-β linkages as the skeletal component (Bartnicki-Garcia 1973). The outer fungal cell wall layer is comprised of matrix components normally containing $(1 \rightarrow 3)$-α-D-Glucan, mannan, β-1,4-mannopyranose and glycoprotein. The matrix component is also present in the micropores of the skeletal component (Wessel and Sietsma 1981). Metal interaction with fungal cell walls may depend upon the cell wall composition (Gadd 1993). Potential functional groups involved in metal binding include carboxyl, amine, hydroxyl, phosphate and sulphydryl groups, although their relative significance is usually difficult to resolve (Strandberg et al. 1981).

Melanins are important fungal pigments that enhance the survival of many species exposed to environmental stress (Bell and Wheeler 1986). A variety of heavy metals can induce or accelerate melanin production in fungi. Melanized cell forms, e.g. chlamydospores, can have high capacities to adsorb metals (Mowll and Gadd 1984). Fungal melanins contain phenolic units, peptide, carbohydrates, aliphatic hydrocarbons and fatty acids, and therefore posses many potential metal-binding sites (Saiz-Jimenez and Shafizadeh 1984; Senesi et al. 1987; Sakaguchi and Nakajima 1987). Oxygen-containing groups in these substances, including carboxyl, phenolic and alcoholic hydroxyl, carbonyl, etc., may be particularly important in metal binding.

Algae. The major algal groups, according to Bold et al. (1980), are Chlorophycophyta (green algae), Charophyta (stoneworts), Euglenophycophyta (euglenids), Phaeophycophyta (brown algae), Chrysophycophyta (golden algae, including diatoms), Pyrrophycophyta (dinoflagellates) and Rhodophycophyta (red algae). Algal cell walls generally consist of an "amorphous" polysaccharide component associated with a skeletal fibrillar material, often a network of cellulose. Alginic acids are common components of the polysaccharide fraction. These are polymers in which D-mannuronic and L-guluronic acids are combined in various proportions (Siegel and Siegel 1973). The mannuronic and guluronic acids contain carboxyl functional groups that may participate in metal binding. The mannuronic acid polymers are fairly linear, but the guluronic acid polymers have a crenellated configuration that allows strong, perhaps multidentate, binding of metals. In some algae, the wall is additionally strengthened by the deposition of calcium carbonate. These forms are often called calcareous or coralline algae (Madigan et al. 1997).

The chemistry of the algal cell wall varies with the algal division, genera, and species (Greene and Darnall 1990). However, carboxylate groups in alginic acids and sulfate groups can be major metal-binding sites (Crist et al. 1992). As the algal cell wall is permeable to molecules with molecular weights up to ~15,000, it can easily be penetrated by metal ions (Madigan et al. 1997).

Lichens. Lichens are associations of a fungus and a photosynthetic symbiont (alga or bacteria). A wide variety of fungi are able to form lichen associations: about 98% are *Ascomycetes*, and 46% of the described *Ascomycetes* are lichenized (Nash 1996; Hawksworth and Hill 1984). The diversity of the photosynthetic partner is much smaller, and many different kinds of lichen may have the same component. The most frequently encountered photosynthetic symbionts are *Trebouxia*, *Trentepohlia* and *Nostoc*. The genera *Trebouxia* and *Trentepohlia* are of an eukaryotic structure and belong to the green algae (phylum *Chlorophyta*). The genus *Nostoc* belongs to the oxygenic photosynthetic bacteria (cyanobacteria; Friedl and Budel 1996). Consequently, lichens contain cell walls of fungi and either of bacteria or algae.

Extracellular polymers

Bacteria typically have extracellular polymers associated with their cell walls. The extracellular polymers, mainly polysaccharides with a repeating sequence of two to six sugar subunits (Sutherland 1972), occur in the form of capsules (tightly associated with the cell surface) or slimes (more loosely associated with the cell surface; Beveridge 1989). Capsule layers extend 25 nm to 1 µm into external environments and slime layers can extend considerably further. Capsule polymers are tightly bound, and can exclude particles such as india ink. However, metal ions are small enough to easily penetrate the capsule layer. Because slimes are 99% water, they are also readily penetrated by metal ions. Carboxylate groups in uronic acid subunits and hydroxyl groups in neutral sugar units are anionic sites in capsule and slime layers. Considerable amounts of protein are also often recovered from capsule material (Corpe 1964, 1975).

Fungi and algae also produce capsular materials. Fungal capsules are formed outside the fungal hypae and are often made up of simple glucans (Wessel and Sietsma 1981). Extracellular mucilage associated with algae sometimes consists of alginic acids that differ from bacterial alginic acids in that they are not acetylated (Percival and McDowell 1981).

MICROBIAL URANIUM ACCUMULATION

Uranium accumulation by microorganisms has been extensively studied in the laboratory. Mechanisms of microbial uranium accumulation are subdivided into those that are metabolism-dependent and those that are metabolism-independent (Gadd 1988, 1993, 1996). Metabolism-dependent accumulation consists of extracellular precipitation with metabolically produced ligands, complexation arising from excreted metabolites, or precipitation due to enzyme-mediated changes in redox state. Metabolism-dependent accumulation can also involve active pumping of metals, leading to intracellular accumulation. Metabolism-independent accumulation involves a physio-chemical interaction between cationic uranium species and negatively charged sites in microorganisms, and can occur whether the cells are living or non-living (even in cell wall fragments).

Metabolism-dependent metal uptake

Enzymatic precipitation of uranyl phosphates. One example of a metabolism-dependent process is uranium mineral precipitation due to the activity of enzymes such as phosphatases. Phosphatases can liberate HPO_4^{2-} from organic compounds, leading to saturation with respect to a metal phosphate solid. For example, a *Citrobacter* sp. isolated from lead-contaminated soil was shown to produce a metal-resistant phosphatase that released phosphate from glycerol 2-phosphate, leading to precipitation of cell bound metal phosphates (Macaskie and Dean 1985; Michel et al. 1986; Macaskie et al. 1988). By this process, phosphatase activity can confer resistance to toxic metals. Precipitation of Cd^{2+}, for example, avoids binding of cadmium to sensitive sites.

Cells that overproduce phosphatase can bind very large quantities of metals. For example, a copper tolerant mutant (dc5c) of the *Citrobacter* sp. that produced three times more phosphatase than the parent strain (N14) accumulated 2500 mg U/g dry cell weight (Macaskie et al. 1988; Macaskie and Dean 1985). The uranyl phosphate mineral was identified by X-ray diffraction (XRD) analysis to be hydrogen autunite, $HUO_2PO_4 \cdot 4H_2O$ (Macaskie et al. 1992; Young and Macaskie 1995). The location of the phosphatase within the cell was examined by electron microscopy of sections of immunogold-labeled cells. Most of the enzyme was distributed in the periplasmic space or associated with the outer membrane (Jeong et al. 1997).

During the initial stage of deposition in response to enzymatic phosphate liberation, the accumulated uranium was visible as electron dense deposits bound to the outer membrane and the cell membrane of the dc5c mutant. It was suggested that membrane phospholipids (located inside the outer membrane and outside the cell membrane) function as sites for initial complexation of the uranyl ion and subsequent nucleation and crystal growth (Macaskie et al. 1994; Jeong et al. 1997). Following the initial stage, more extensive uranium uptake occurred throughout the cell wall matrix. The precipitate then became detached from cells (Jeong et al. 1997). Phosphatase extracted from cells can be inactivated rapidly by $(UO_2)^{2+}$ (Jeong et al. 1997). However, whole cells retained phosphatase activity (resulting in uranyl removal from solution) over many weeks (Macaskie 1990). Thus, it was proposed that the enzyme held in the more protected periplasmic space is able to generate a protective localized layer of inorganic phosphate before penetration of $(UO_2)^{2+}$ across the outer membrane (Jeong et al. 1997). Cells killed by radioactivity can retain their ability to accumulate metals (Strachan et al. 1991) as long as the phosphatase remains active, nucleation sites are retained, and a barrier exists to protect phosphatase activity (Jeong et al. 1997).

Formation of chelating agents. There is some evidence to suggest that cells produce chelating organic molecules in response to metal stress, and these molecules can significantly modify uranium bioavailability. Siderophores are hydroxymate or catechol-bearing chelating molecules that are highly specific for Fe^{3+} and analogous compounds. The biological behavior of uranium, thorium and plutonium resembles that of ferric iron (Hodge et al. 1973) and some siderophores can complex U^{6+} (Macaskie and Dean 1990). *Pseudomonas aeruginosa* grown in culture media containing 10 and 100 ppm of uranium showed an extended lag period before active growth, and growth was significantly inhibited at concentrations of 1000 ppm or higher (Premuzic et al. 1985). During the growth in the media containing 100 ppm of uranium and thorium, *P. aeruginosa* produced fluorescent phenol/catechol pigments and pigments with amino and hydroxamate functional groups. These siderophore-like chelating agents may enhance the mobility and bioavailabiltiy of thorium and uranium by increasing uranium solubility of their respective phases through aqueous complexation (Premuzic et al. 1985).

Enzymatic uranium reduction. Enzymatic reduction of U^{6+} by microbes results in uranium precipitation. This is regarded as metabolism-dependent accumulation of uranium.

Dissimilatory Fe^{3+}-reducing microorganisms are known to generate energy for growth by coupling the oxidation of acetate to carbon dioxide and oxidation of hydrogen to protons with the reduction of ferric iron in anaerobic environments. *Geobacter metallireducens* and *Shewanella putrefaciens* also enzymatically reduce U^{6+} and can grow using uranium as the sole electron acceptor via the following reactions (Lovely et al. 1991):

$$CH_3COO^- + 4\ U^{6+} + 4\ H_2O \rightarrow 4\ U^{4+} + 2\ HCO_3^- + 9\ H^+$$

$$H_2 + U^{6+} \rightarrow U^{4+} + 2\ H^+$$

Thermodynamic calculations have indicated that acetate oxidation coupled to U^{6+} reduction has the potential to generate more than twice as much energy as is obtained from Fe^{3+} reduction per electron transferred (Lovely et al. 1991). The estuarine Fe^{3+}-reducing microoganism, strain BrY (closely related to *S. putrefaciens*) can also reduce U^{6+} (Caccavo et al. 1992).

When *G. metallireducens* was incubated in solutions with uranyl actetate, a black precipitate accumulated. The black precipitate was identified by XRD as uraninite (UO_2). TEM observations of uranium-loaded cells confirmed the precipitate was located outside the cells. No intracellular uranium was detected (Gorby and Lovely 1992). No uranium was reduced in the presence of heat-killed cells of *Geobacter metallireducens* and *Shewanella putrefaciens*, due to destruction of cell wall proteins, and *G. metalireducens* reduced no uranium when metabolically inactive at 4°C (Lovely et al. 1991; Gorby and Lovely 1992).

The sulfate-reducing *Desulfovibrio* species (*D. desulfuricans*, *D. vulgaris*, and *D. baculatum*) are able to reduce U^{6+} (Lovely and Phillips 1992a; Lovely et al. 1993a,c). *D. desulficans* rapidly precipitated 11 g U/ g dry cell weight extracellularly as uraninite (Lovely and Phillips 1992a,b). However, uranium reduction by *D. desulfuricans* was inhibited between 17 mM and 24 mM U^{6+}, probably due to uranium toxicity (Lovely and Phillips 1992a). The reduction of U^{6+} by *D. desulficans* was not inhibited by the addition of 100 μM Mn^{2+}, Fe^{2+}, Mg^{2+}, Zn^{2+}, or Co^{2+}. However, when the 100 μM of copper was added to the incubation solution, U^{6+} reduction was completely inhibited due to loss of activity of the cells (Lovely and Phillips 1992a).

Attempts to grow the U^{6+}-reducing sulfate reducers with U^{6+} as sole electron acceptor mostly have been unsuccessful. However, a recent study by Pietzsch et al. (1998) reports that a sulfate reducing bacterium of genus *Desulfovibrio* isolated from a uranium mining waste pile can grow with U^{6+} as the sole electron acceptor.

Biochemistry of U reduction. The uranium reducing species *G. metallireducens* is a δ-*Proteobacterium*. Its closest known relatives are Gram-negative sulfur- and sulfate-reducers such as *Desulfuromonas acetoxidans* (Lovely et al. 1993a). *G. metallireducens* possessed cytochromes of the c-type (electron carriers in oxidation-reduction reactions), which are predominant in Gram-negative sulfate- and sulfur-reducers as components of the respiratory metabolism (LeGall and Fauque 1988). The c-type cytochromes were oxidized when Fe^{3+}, U^{6+} or nitrate was added to whole cell suspensions. Thus, it was suggested that c-type cytochrome may be a component of the electron transport chain and involved in electron transport to U^{6+} (Lovely et al. 1993a).

The soluble fraction of *D. vulgaris* cells in sodium phosphate buffer (pH 7) accounted for 95% of the H_2-dependent U^{6+} reduction capacity. The membrane-free soluble fraction contained a c-type cytochorome that was identified as cytochrome c3 (Lovely et al. 1993b). The ability of cytochrome c3 to participate in redox reactions involving uranium was demonstrated by showing that when the dithionite-reduced cytochrome c3 was added to a U^{6+} solution, cytochrome c3 was oxidized (Lovely et al. 1993b). When hydrogenase, the typical electron donor for cytochrome c3 in the cell (LeGall and Fauque 1988), was added to a U^{6+} solution with cytochrome c3 and H_2, reduction of U^{6+} occurred and uraninite accumulated (Lovely et al. 1993b). The uraninite crystals formed were smaller than those formed by the whole cells, possibly due to the lack of cell surface nucleation sites (Lovely et al. 1993b).

Enzymatic Tc reduction. Fe^{3+}-reducing bacteria (*S. putrefaciens* and *G. metallireducens*) and a sulfate reducing bacterium (*D. desulfuricans*) also reduce soluble Tc^{7+} into an insoluble low-valence oxide. Reduction is coupled to oxidation of a range of electron donors such as hydrogen, pyruvate and formate (Lloyd et al. 1996; Lloyd et al. 1999). Anaerobically grown cells of *Escherichia coli* also can reduce Tc^{7+}, and hydrogenase is responsible for Tc^{7+} reduction (Lloyd et al. 1997a; Lloyd et al. 1997b). Lloyd et al. (1999) demonstrated that *D. desulfuricans* precipitated the low valence oxide of Tc in the periplasmic space, and a periplasmic hydrogenase rather than a cytoplasmic hydrogenase was involved in Tc^{7+} reduction.

Metabolism-independent uptake of uranium

Intracellular uranium accumulation. Many microorganisms can accumulate uranium intracellularly. In some cases, TEM observations indicate that the uranium deposits are crystalline. Unlike metabolically essential metals such as Cu, Zn, Co, and Mn, which are accumulated intracellularly via energy-dependent transport systems (metabolism dependent), uranium has no essential biological function and is transported into microbial cells only due to increased membrane permeability (e.g. resulting from uranium toxicity). Therefore, intracellular accumulation of uranium is considered as metabolism-independent accumulation.

Uranium that penetrates cell membranes is often immobilized by binding to anionic sites in the cytoplasm or by precipitation due to the change in solution chemistry (e.g. pH, phosphate concentration, etc.). For example, uranium formed dense intracellular deposits in the Gram-negative bacterium *P. aeruginosa* exposed to a 100 ppm uranium solution. Substantial intracellular accumulation occurred within 10 seconds and was not affected by metabolic inhibitors, metal competitors, or pH. Therefore, it was concluded that this intracellular uptake occurs via a passive transport mechanism (Strandberg et al. 1981).

TEM examination of *Pseudomonas sp.* EPS-5028 showed that uranium accumulated intracellularly as needle-like fibrils from a uranyl nitrate solution (Marques et al. 1991). At pH 3, 92% of uranium was taken up from a 50 ppm U solution with a cell density of 2.5 mg/ml in 5 min. Uranium accumulated by this biomass was almost completely removed by three washes with 0.1 M of EDTA and Na_2CO_3. 0.1M potassium oxalate, sodium citrate and nitric acid desorbed 80, 74.4 and 41.4% of the bound uranium, respectively.

Synechococcus elongatus cells were ruptured and individual fractions of the cell were separated to determine the roles of various components in uranium adsorption. Of the uranium absorbed from a 1 ppm U solution, 61%, 32% and 7% were found in the intracellular soluble fraction, the intracellular particulate fraction, and the cell wall fraction, respectively. These data demonstrate that significant uranium accumulates in the cytoplasm of these cells (Horikoshi et al. 1979b). However, it is unclear whether these values reflect equilibrium uptake.

Arthrobacter sp. from a uranium ore deposit accumulated 600 mg U/g dry cell weight intracellularly from a 125 ppm U solution with a cell density of 75 µg/ml at pH 5.8. Uranium uptake was almost complete within 3 minutes. This uranium accumulation was easily removed using 0.1 M of sodium hydrogencarbonate (Sakaguchi 1996). The above studies all indicate that intracellular uranium deposits are often easily removed by strongly complexing ligands or bicarbonate.

Basic approach to studies of biosorption of uranium. Numerous studies have shown that most uranium uptake by microbes is metabolism-independent rather than metabolism-dependent. Biosorption is the term used to describe such non-directed, physio-chemical processes (Volesky and Holan 1995). An electrostatic interaction between anionic sites in cells and positively charged uranyl species is generally regarded as a rapid process, and is considered to be the primary mechanism for uranium biosorption (Haas et al. 1998). Biosorption can involve adsorption, absorption, ion exchange, or precipitation.

It has long been known that uranium binds strongly to biological tissues. In fact, uranium is commonly used as a stain to obtain contrast for visualization of organic materials by TEM (Beveridge 1978). Thus, TEM is a very effective technique for imaging the location of uranium adsorbed by cells to reveal details of the uranium biosorption mechanism.

Uranium accumulation by microorganisms has been extensively studied using batch experiments consisting of solutions containing known cell numbers and uranium concentrations. The ability of microorganisms to bind metals is generally measured on the basis of the amount of uranium bound to 1 g of dry cells (mg U/g dry cells or mM U/g dry cells).

Uranium uptake by microbes is affected by various factors, including pH, contact time, uranium concentrations, cell concentrations, coexisting cations and anions, the physiological state of cells, and pretreatment of cells. Most studies have tried to optimize conditions for uranium uptake by microorganisms by evaluating and controlling these factors.

Examples of uranium uptake by microorganisms. Representative values of microbial uranium uptake are listed in Table 2. These results show that, although the comparison among microbial species in terms of uranium uptake capacity has relative

Table 2. Uranium uptake capacity of bacteria, fungi, alage, and lichens.

Species	U uptake*	pH#	Ref	Species	U uptake*	pH#	Ref
Gram-negative bacteria							
Pseudomonas sp. EPS-5028	55	3	1	Dersxia gummaosa	45.2	4.1	5
Pseudomonas sp. EPS-5028	80	not shown	1	Pseudomonas saccharophilia	87.1	4.6	6
Pseudomonas aeruginosa	100	2.5-4.5	2	Pseudomonas stutzeri	87.1	4.6	6
Pseudomonas aeruginosa	100-150	4	3	Zoogloea ramigera	71.9	4.6	6
Thiobacillus ferrooxidans	102.65^	1.5	4	Zoogloea ramigera	400	3.5	7
Azotobacter vinelandii	59.5	4.1	5				
Gram-positive bacteria							
Mycobacterium smegmatis	187	1**	8	Actinomyces levoris	120	6	13
Arthrobacter sp.	600	5.8	9	Streptmyces viridochromogenes	130	6	13
Bacillus subtilis	615	5	10	Actinomyces flavoviridis	78.1	4.6	6
Bacillus subtilis	84.9	4.6	6	Streptmyces albus	87.3	4.6	6
Micrococcus luteus	74.7	4.6	6	Streptmyces echinatus	81.6	4.6	6
Streptmyces sp.	250	5-6	11	Streptmyces obiraceus	75.4	4.6	6
Streptmyces longwoodensis	440	5	12				
Fungi							
Rhizopus arrhizus	180	4	14	Tricholoma conglobatum	49.2	6	19
Rhizopus arrhizus	195	4	15	Favolus arcularius	45.25	6	19
Rhizopus olgosporus	250	3.7-3.9	16	Inonotus mikadoi	45.75	6	19
Rhizopus oryzea	260	3.7-3.9	16	Saccharomyces cervisiae	100-150	3-4	3
Rhizopus javanicus	250	3.7-3.9	16	Saccharomyces cervisiae	141	4-5	20
Rhizopus formosaensis	200	3.7-3.9	16	Neurospora sitophila	90.7	4.6	6
Rhizopus chinensis	190	3.7-3.9	16	Penicillium chrysogenum	170	3	21
Rhizopus japonicus	200	3.7-3.9	16	Penicillium chrysogenum	165	4.5	14
Rhisopus stolonifer	170	3.7-3.9	16	Penicillium chrysogenum	72.4	4.6	6
Talaromyces emersonii	280	5	17	Penicillium lilacinum	80	4.6	6
Kluyveromyces marxianus	120	5-7	18				
Algae							
Chlorella regularis	67.2	8##	22	Scenedesmus quadricauda	170	7	23
Chlorella regularis	15.6	8##	22				
Lichens							
Cladonia rangiferina	14.5	3.3**	24	Peltigera membranacea	42.2	4.5	25
Cladonia rangiferina	11.6	3.3**	24				

Notes for Table 2:

References:1 Marques et al. (1991), 2 Hu et al. (1996), 3 Strandberg et al. (1981), 4 DiSpirito et al (1983), 5 Cotoras et al. (1992), 6 Nakajima & Sakaguchi (1986), 7 Norberg & Persson (1984), 8 Andres et al. (1993), 9 Sakaguchi (1996), 10 Sakaguchi (1998), 11 Ferris & Myers-Keith (1985), 12 Golab & Orlowska (1990), 13 Horkoshi et al. (1981), 14 Tobin et al. (1984), 15 Tsezos & Bolesky (1981), 16 Teen-Sears et al. (1984), 17 Bengtsson et al. (1995), 18 Bastard et al. (1997), 19 Nakajima & Sakaguchi (1993), 20 Volesky & May-Phillips (1995), 21 Jilek et al. (1978), 22 Horikoshi et al. (1979a), 23 Pribil & Marvan (1976), 24 Boileau et al. (1985), 25 Haas et al. (1998)
* mg U/g dry cell weight, # optimum pH otherwise stated, ^mg U/g dry protein weight,
** optimum pH is unkonwn, ## Optimum pH is 6

meaning, uranium uptake capacity differs greatly with groups, types, and species of microorganisms.

Gram-positive bacteria such as *Bacillus subtilis, Arthrobacter* sp., *Streptomyces longwoodensis* can accumulate exceptionally large amounts of uranium. At pH 5, *Bacillus subtilis* isolated from a uranium mine accumulated 615 mg U/g dry cell weight in the cell wall from a 120 ppm U solution with a cell density of 75 µg/ml. Peptydoglycan in the cell wall was considered to be the primary uranium-binding site. Uranium uptake was almost complete within 3 minutes, but accumulations were also easily removed using 0.1 M of sodium hydrogen carbonate (Sakaguchi 1998). In contrast, *Pseudomonas fluorescence* accumulated uranium along the cytoplasmic and outer membranes. Uranium-crystals bound to the cell surfaces of *P. fluorescens* were investigated by small-angle X-ray scattering (SAXA) and TEM. TEM data showed that platy uranium-crystals, which ranged in size from 10 nm to 1 µm, were aligned along the outer membrane and plasma membrane in the periplasm (Krueger et al. 1993).

TEM observation by Andres et al. (1994) revealed that uranium absorbed by the Gram-positive *Mycobacterium smegmatis* (a lipid-rich and acid-resistant organism) from a uranyl nitrate solution at pH 1, was located in the cytoplasm and the cell wall but was not taken up by either the cytoplasmic membrane or extracellular polymers (Andres et al. 1994). There was no significant difference in the degree of adsorption between living cells and cells inactivated by ultraviolet irradiation, indicating that uptake is independent of cellular metabolic functions (Andres et al. 1993). The ^{31}P-NMR spectra of phospholipid fractions isolated from uranium-loaded biomass showed that uranium was bound to the phosphate groups. Treatment of the isolated peptidoglycan (Pg) fraction with 1mM uranyl nitrate at pH 2 led to marked uranium uptake (1.01 mg U/mg of PG dry weight). Phosphate groups and carboxylate groups of Pg are probably the important uranium binding sites (Andres et al. 1993, 1994).

Zoogloea ramigera, which is a Gram-negative bacterium common in sewage treatment plants, can accumulate up to 400 mg U/g dry cell weight. This organism produces vast amounts of extracellular polysaccharides, and uranium uptake is attributed to complexation between uranyl species and anionic sites in the polysaccharides (Norberg and Persson 1984). *Azotobacter vienlandii*, which also produces anionic polysaccharides, showed higher biosorption of uranium than *Rhizobium trifoli*, an organism that produces neutral polysaccharides as well as anionic polymers (Sutherland 1985; Cotoras et al. 1992). The importance of extracellular polysaccharides in uranium uptake has also been demonstrated by showing that more uranium is taken by polysaccharides extracted from *Pseudomonas sp.* strain EPS-5028 (96 mg U/g polymer) from a 200 ppm U solution at pH 5.7 (Marques et al. 1990). Again, carboxyl groups are inferred to be significant in uranium binding (Congregado et al. 1985).

The fungi *Rhisopus* species and *Talaromyces emersonii* can take up large amounts of uranium. TEM observation of non-living *R. arrhizus* cells exposed to a 100 ppm uranium solution at pH 4 revealed electron dense layers throughout the fungal cell wall, but no electron dense materials anywhere inside the cell walls. Infrared spectra of isolated and purified *R. arrhizus* cell walls were recorded before and after uranium biosorption. The IR absorbance bands at 908 cm^{-1} and 372 cm^{-1} appear after uranium biosorption and can be assigned to the UO_2 v3 and the uranium-nitrogen bond stretch vibrations, respectively. It was suggested that the sequestered uranium is, to a certain extent, associated with the nitrogen of the chitin monomer N-acetylated-D-glucosamine (NAG). Sorption experiments with pure chitin indicated uranium uptake of 6 mg/g. Pure, water-insoluble NAG was also tested in a uranium solution at pH 4. The contact resulted in the production of a fine precipitate. The IR spectra of the precipitate indicated the presence of uranyl hydroxide (all amide bands were absent). From these results, the mechanism of uranium uptake by *R. arrhizus* was attributed to three processes taking place in the cell wall. Initially, uranium was complexed with nitrogen at chitin amine binding sites. Additionally, uranium was adsorbed within the chitin matrix. Subsequently, hydrolysis of the uranium-nitrogen complex occurred, leading to deposition of uranyl hydroxide (Tsezos and Volesky 1982).

There is some doubt about the mechanism proposed above, and the role of chitin in uranium binding has been questioned (Muraleedharan and Venkobachar 1990; Scharer and Byerley 1989; Sakaguchi and Nakajima 1981). Muraleedharan and Venkobachar (1990) pointed out that the uranium uptake of pure chitin accounts for only 3.3% (6 mg /g) of that of the whole cell of *R. arrhizus* (180 mg/g). Sakaguchi et al. (1981) also reported that chitin was a poor uranium sorbent.

Strandberg et al. (1981) showed that the fungus *Saccharomyces cervisiae* accumulated a ~0.2 μm thick layer of needle-like uranium fibrils on cell surfaces after exposure to a 100 ppm U solution at pH 4 for 2 h. Desorption experiments using 0.1 M of nitric acid, EDTA, and $(NH_4)_2CO_3$ for 16 h showed 59.3, 72.3 and 83.5% removal, respectively. Volesky and May-Phillips (1995) used TEM to record a time sequence for uranium accumulation (15 min, 1 h, 5 h, 15 h and 24 h) by *S. cervisiae*. They showed uranium accumulation both extracellularly (on the cell wall surfaces) and intracellularly (throughout the cytoplasm) at pH 4. Uranium microfibrils appeared on the cell wall and plasma membrane after 1 h and increased in abundance after 5 h of contact. Uranium slowly penetrated into the cell interior. As more uranium entered the cell, the cell began to change shape. This was considered a toxic effect. Cells exposed to U solution for 24 h deposited 2 μm long crystalline microfibrils on the outer surface of the cell wall. In the last 15-24 h of contact, a vacuole occupied most of the cytoplasm. Since both living and non-living *S. cervisiae* cells showed a similar tendency to take up uranium extracellularly and intracellularly, the biosorption of uranium might not be associated directly with vital metabolic functions (Volesky and May-Philips 1995).

The cell walls of *S. cervisiae* are composed of glucan and mannan, with a trace of chitin (Volesky and May-Philips 1995). Purified mannan can bind a greater amount of uranium compared with glucan and chitin (Chmielowski et al. 1994). Phosphorylated polysaccharides (phosphomannans) comprise a portion of the cell wall in several yeast species, including *S. cervisiae* (Farkas 1979). Uranium complexation by phosphomannans purified from two *Hansenula* yeast species was related to their phosphate contents (Standberg et al. 1981). Hydroxyl and substituted phosphate groups of mannan may be involved in uranium complexation and subsequent crystallization (Rothstein and Meier 1951; Chmielowski et al. 1994).

Relatively few algal species have been studied, but it appears that *Chlorella regularis* has a high capacity for uranium uptake (Horikoshi et al. 1979a). *Ankistrodesmus* sp. precipitated extracellular uranium-bearing phases with cubic habit. EDS analysis only showed peaks from uranium in the phases (Mann and Fyfe 1985).

Uranium uptake by lichens has been characterized in natural samples and studied experimentally. *Trapelia involuta* growing directly on uranyl phosphate and uranyl arsenate minerals, metatorbernite and metazeunerite, respectively, accumulate high concentrations of uranium. The uptake site is associated with melanin-like pigments within the outer fruiting body walls (McLean et al. 1998).

Barker et al. (1998) used TEM to show that uranium uptake by *P. membranacea*

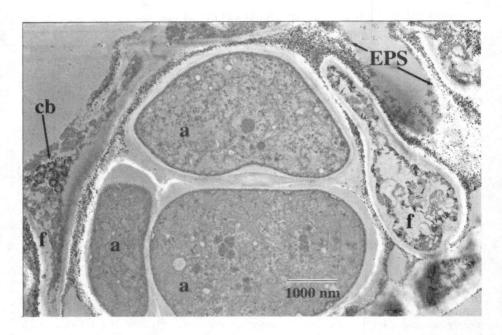

Figure 2. TEM image of *Peltigera membranacea* after 75-minute uptake. a: cyanobacterium, EPS: extracellular polysaccharides, f: fungus, cb: concentric bodies.

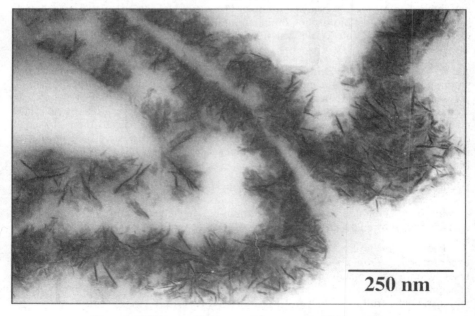

Figure 3. TEM image of *Peltigera membranacea* after 24-hour uptake.

(samples of Haas et al. 1998) is not accomplished by surface complexation alone. TEM images of lichen after contact with 100 ppm of uranium at pH 4 for 75 minutes (Fig. 2, Fig. 4) and 24 hours (Fig. 3) show that uranium is accumulated in crystals. The frequency with which uranium-rich crystals nucleate and grow differs with location within the cells, presumably reflecting differences in organic composition. In fungal cells, uranium accumulated both intracellularly and in extracellular polymers. Sprays of large uranium crystallites formed on the interior cytoplasmic membrane and on some organelle membranes. In cyanobacterial cells, uranium was accumulated in the cell and along thylakoid membranes, which are associated with pigments (chlorophyll a) necessary for photosynthesis (Jensen 1993). A number of small uranium-bearing crystals accumulated in the outermost extracellular layer of both organisms. These small crystals grew into dense fibrils after 24 hours (Fig. 3).

Barker et al. (1998) also demonstrated uranium taken up by *P. membranacea* is strongly concentrated in dense clusters of crystals associated with subcellular organelles (Fig. 4). The organelles, referred to as "concentric bodies" (Peveling 1969), occur in most lichenized fungi (Griffiths and Greenwood 1972). Galun et al. (1974) reported that concentric bodies consist partly of proteins.

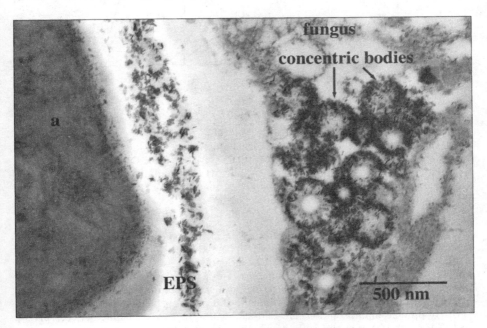

Figure 4. TEM image of concentric bodies after 75-minute uptake.
a: cyanobacterium, EPS: Extracellular polysaccharides.

pH effects on U speciation and U uptake. Solution pH profoundly impacts microbial uranium uptake due to its influence on uranyl speciation in aqueous solutions and on protonation and deprotonation of reactive groups of cell walls, organic molecules, etc. The optimum pH values for uranium uptake by the various types of microorganisms are mostly between pH 3 and pH 6, as shown in Table 2. Maximum uptake is the most frequently observed over the pH 4 to 5 range.

In most studies designed to quantify uranium uptake by microorganisms, experimental solutions were prepared by dissolving uranium nitrate hexahydrates, $UO_2(NO_3)_2 \cdot 6H_2O$, in distilled, deionized water, with the pH adjusted with either NaOH, HCl or HNO_3. The solutions were equilibrated with atmospheric CO_2 pressure ($P_{CO_2} = 10^{-3.5}$). The distribution of uranium species in these bulk solutions as a function of pH can be calculated (we used PHREEQC of Parkhurst 1995) (Fig. 5). The uranium database for these calculations is based on Grenthe et al. (1992). The thermodynamic data for $UO_2(OH)_2$ comes from Choppin and Mathur (1991), as recommended by Langmuir (1997).

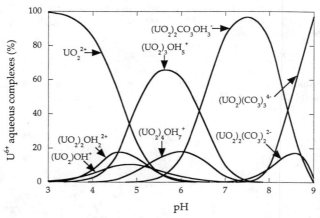

Figure 5. Distribution of U^{6+} aqueous complexes at 25°C and 1 mM ionic strength for 100 ppm U^{6+} and $P_{CO_2} = 10^{-3.5}$.

Below pH of 4.5 the uranyl ion is the predominant uranium species in solutions containing 100 ppm of uranium dissolved from $UO_2(NO_3)_2 \cdot 6H_2O$. Between pH 4.8 and 6.4, $UO_3(OH)_5^+$ is the dominant species. Above pH 6.4, solutions become dominated by anionic uranyl-hydroxyl carbonate or uranyl carbonate species. It should be noted that the pH at which uranyl-hydroxyl carbonate and uranyl carbonate species become predominant depends on the relative concentrations of U and dissolved CO_2.

Hu et al. (1996) reported significant abiotic removal of uranium from 100 ppm uranium solutions between pH 5 and pH 9. The solubility of a uranyl oxide hydrate phase, schoepite, is known to be quite low around pH 6, so it is difficult to distinguish whether uranium removal in experiments is due to microbial effects or chemical precipitation. To understand the role of chemical precipitation of uranium in microbial uptake experiments, the solubility of schoepite at different pH values was calculated (Fig. 6). At 10 ppm uranium, solutions are saturated with respect to schoepite between pH 5.2 and pH 7.4. At 100 ppm and 1000 ppm, solutions are saturated with schoepite between pH 4.6 and pH 8.2 and between pH 4.1 and pH 8.7, respectively.

Fein et al. (1997) and Daughney et al. (1998) carried out titration experiments using *B. subtilis* and *Bacillus licheniformis* to quantify the numbers and identities of the proton active functional groups involved in metal binding and to calculate deprotonation constants for each site. Titration curves fit better to a three-site model, rather than one- or two-site models. This was interpreted to indicate that three different sites participate in metal binding on the cell surfaces. Estimated log dissociation constants of these sites are 4.8, 6.9 and 9.4 for *B. subtilis* and 5.2, 7.5 and 10.2 for *B. licheniformis*.

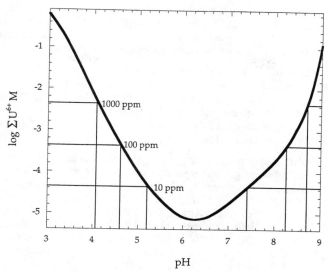

Figure 6. Solubility of schoepite 25°C for $P_{CO_2} = 10^{-3.5}$.

The cell wall of *B. subtilis* contains abundant peptydoglycan and teichoic acids, whereas that of *B. licheniformis* contains less peptydoglycan, and more teichoic and teichuronic acids (Beveridge 1989). The pKa values of 4.8, 6.9 and 9.4 for *B. subtilis* and the pKa values of 5.2, 7.5 and 10.2 for *B. licheniformis* may correspond to carboxylate, phosphate and hydroxl groups, respectively (Fein et al. 1997; Daughney et al. 1998).

Plette et al. (1995) reported that the bacterium *Rhodococcus erythopolis* has 3 functional groups with pKa values of 4.6, 7.8 and 9.96 that participate in metal binding. They attributed the pKa values of 4.6 and 7.8 to carboxylate and phosphate, as did Fein et al. (1997) and Daughney et al. (1998). However, they attribute the pKa value of 9.96 to amine groups rather than hydroxyl groups. Goncalves et al. (1987) and Xue et al (1988) adopted two pKa and one pKa models, respectively, to determine the values of anionic sites in *Klebsiella pneumonia* (pKa values of 4.7 and 7.8) and *Chlamydomonas rheinhardii* (pKa value of 5.8 to 7.6), respectively, However, they did not specify the functional groups that correspond to these sites. The similar pKa values reported in several studies suggest that the deprotonation constants of carboxylate and phosphate groups are relatively constant among the different types of microorganisms studied.

At low pH, positively charged uranyl ions do not interact with protonated sites on the cell surface (Nakajima et al. 1979; Galun et al. 1987). Around pH 5, the ratio of deprotonated to protonated sites is just over 1.0 for carboxylate groups and 0.1 (or less) for the phosphate groups. Positively charged uranyl ions and uranyl hydroxide complexes interact with the deprotonated carboxylate groups. The ratio of deprotonated to protonated sites increases for phosphate groups at neutral and alkaline pHs. However, neutral and anionic aqueous complexes such as uranyl carbonate and uranyl-hydroxide carbonate species do not interact with the deprotonated sites and provide a chemical driving force for the desorption of $(UO_2)^{2+}$ from the bacterium.

The simultaneous importance of surface charge and uranium speciation also has been observed to control adsorption of uranium onto ferric iron minerals, although the pH (ZPC) for cell walls is much lower than that of the ferric iron minerals (Hsi and Langmuir

1985; Waite et al. 1994). Hsi and Langmuir (1985) suggested that UO_2OH^+ and $(UO_2)_3(OH)_5^+$ were adsorbed by hydrous ferric oxide, goethite, and hematite between pH 3 and pH 8. Waite et al. (1994) demonstrated that uranium desorption at alkaline pH is due to weak bonding of $UO_2CO_3^{2-}$ species onto adsorption sites in hydrous ferric oxide.

Effects of coexisting cations on uranium uptake. The capacities of microorganisms to take up uranium can be decreased by coexisting cations due to competition for anionic sites. However, the degree of interference varies with cation concentration and cation type. The ability of a metal to complex with functional groups on the surface of a cell is associated with hydration sphere of the ion which is related with the charge / size ratio (Mullen et al. 1989; Tobin et al. 1984).

Experiments containing microorganisms and similar concentrations of uranium and other divalent metals reveal general trends in competitive uptake. Typically, uranium uptake is not inhibited by the presence of monovalent cations. Divalent cations have a limited ability to compete with U^{6+} and some trivalent and tetravalent cations can outcompete U for binding sites (e.g. Nakajima and Sakaguchi 1993; Tobin et al. 1984; Nakajima and Sakaguchi 1986; Cotoras et al. 1992). For example, Cotoras et al. (1992) showed that K^+, Mg^{2+} and Ca^{2+} at 75-fold molar excess had no effect on uranium uptake at pH 4.1. Hu et al. (1996) compared the relative uptake of U to a selection of divalent and trivalent cations by *Pseudomonas aeruginosa* from solutions containing 0.4 mmol of U and one of 0.4 mmol of the divalent and trivalent cations at pH 2.5 and found that the order of interference with uranium binding was Fe^{3+} (84.9%) > Fe^{2+} (58.7%) > Al^{3+} (54.1%) > Pb^{2+} (34.9%) > Cu^{2+} (26%) > Cr^{3+} (24.4%) > Cd^{2+} (6.6%), Mn^{2+} (4.9%), Ba^{2+} (4.8%), Co^{2+} (4.4%). However, it should be noted that these values may change dramatically as pH is increased. Uranium uptake by *Arthrobacter sp.*, (at pH 6), *Pseudomonas aeruginosa* (at pH 4) and *Pseudomonas sp.* EPS-5028 (at pH 4) was not affected by coexisting cations such as Cu^{2+}, Pb^{2+}, Hg^{2+}, or Cr^{2+} (Shumate 1985; Marques et al. 1991; Sakaguchi 1996). Nakajima and Sakaguchi (1986) showed that 72 species of micro-organisms accumulated more uranium than divalent cations such as Mn^{2+}, Co^{2+}, Ni^{2+}, Cu^{2+}, Zn^{2+}, Cd^{2+}, and Pb^{2+} from a solution containing 4×10^{-5} M of U and 4×10^{-5} M of the other divalent cations at pH 4.6 (however, 11 species accumulated more Hg^{2+} than uranium). The order of interference with uranium-binding was Hg^{2+} > Pb^{2+} > Cu^{2+} > other metals. Other authors have confirmed that ferric ions out-compete uranyl ions for binding sites (Galun et al. 1987; Byerley and Scharer 1987). In fact, when the Fe^{3+}/U^{6+} ratio is greater than approximately 6, the U^{6+} binding of *Pseudomonas sp.* EPS-5028 is completely inhibited at pH 2.5 (Hu et al. 1996).

Aluminum can also affect uranium uptake, but the effect is pH and concentration dependent. At pH 4, increasing the concentration of aluminum in solution results in a progressive decrease in the equilibrium uranium uptake capacity of *R. arrhizus*. At pH 2, aluminum did not have an appreciable effect on the uranium equilibrium uptake capacity, regardless of the aluminum solution concentration. At pH 5 aluminum solubility is low, and aluminum precipitates removing uranium with it (Tsezos et al. 1997).

Th^{4+} can compete effectively with uranium for binding sites. It has been shown that uptake of uranium by *Mycobacterium smegmatis* from a solution containing thorium at the same molar concentration as uranium at pH 1 was reduced to 47% of that in the solution containing uranium only (Andres et al. 1993). Similarly decreased uranium uptake due to the presence of equimolar thorium in solutions was observed in 11 species of fungi (Nakajima and Sakaguchi 1993).

Effects of coexisting anions. Coexisting anions in solutions tend to decrease uranium uptake by forming neutral and anionic complexes with U. These interact weakly with the cell surface.

As noted previously, carbonate can form strong complexes with U. Neutral and negatively charged uranyl carbonate and uranyl-hydroxyl carbonate species are important in solutions with pH values >6. Uranyl-carbonate complexes interact poorly with microbial surfaces because the complexes are neutral or negatively charged, and cell surface sites are negatively charged.

The uranyl ion complexes strongly with phosphate species, forming $UO_2(H_2PO_4)^{3-}$ or $UO_2(HPO_4)_2^{2-}$ (Finch and Ewing 1992). Uranium uptake by *Chlorella* cells decreased with increasing concentration of phosphate ion in solutions at pH 5. Furthermore, 1 mM dihydrogenphosphate reduced uranium uptake to 4.2% of that in the solution without phosphate ion, suggesting that uranyl-phosphate complexes can not be taken up by microbial cells or that uranium phosphate is precipitating (Nakajima et al. 1979).

Uranyl sulfate complexes do not impact upon uranium uptake as strongly as phosphate or carbonate complexes. 60 mM sulfate ion decreased uranium uptake by *S. levoris* by 18% at pH 4 (Byerley and Scharer 1987) and a 10-fold excess of sulfate ion reduced uranium uptake by *Rhizopus arrhizus* by 16% at pH 4 (Tobin et al. 1986). In the case of *P. aeruginosa*, 0.625 M of sulfate ion did not inhibit uranium uptake significantly at pH 2.5 (Hu et al. 1996). Therefore, uranyl sulfate complexes compete poorly for anionic sites.

Nitrate and chlorine form weak complexes with uranyl ions and have little effect on uranium speciation (Tobin et al. 1984; Treen-Sears et al. 1984). Uranium uptake is almost always unaffected by nitrate and chloride (Sakaguchi 1996; Nakajima et al. 1979; Tobin et al. 1986).

Other factors affecting uranium biosorption. The amount of uranium accumulated by microbes increases with the concentration of uranium and cell numbers in solution. However, at high concentrations of uranium (due to surface site saturation) or cell numbers (due to complete uptake from solution), uranium is no longer taken up by cells (Marques et al. 1991; Horikoshi et al. 1981; Friis and Myers-Keith 1986).

Cell walls of microorganisms are very sensitive to the growth medium composition, conditions of growth (e.g. aerobic or anaerobic), and incubation time (Treen-Sears et al. 1984; Volesky and May-Philips 1995). The uranium binding capacity of *S. longwoodensis* was found to increase with culture age in parallel to the increase in total phosphorus content of *S. longwoodensis* cells. *Azotobacter vienlandii* also accumulated more uranium with increased culture age. This species produces bacterial alginate acids extracellularly, and increased age corresponded to more extensive production of extracellular alginate acids (Cotoras et al. 1992).

Pretreatment of cells, for example heating and freeze-drying to kill the microorganisms often modifies uranium uptake. The accumulation of uranium by heat-killed cells is typically greater than accumulation by living cells (Tsezos and Volesky 1981; Galun et al. 1983; Nakajima and Sakaguchi 1986; Nakajima et al. 1979). The biosorbent properties of dead cells are increased because more surface area becomes available due to lysis of the cells or denaturing of proteins (Hu et al. 1996; Rome and Gadd 1987).

XRD studies of heat-dried biomass loaded with uranium and thorium showed that, although thorium loaded samples are amorphous, uranyl loaded samples contained crystalline ammonia uranyl phosphate (Andres et al. 1995). This phase also occurs as the natural mineral, uramphite (Markovic et al. 1988). The ammonia ion could be formed intracellularly by heat-induced transformation of cellular amino acids during drying (Bonner and Castro 1965). Phosphate could be liberated from a source such as adenosine triphosphates, sugar phosphates and polyphosphate granules (Gale and Mclain 1963). However, it is uncertain whether the uranyl phosphate mineral is formed during the biosorption experiment or during drying.

Study of the accumulation of $(UO_2)^{2+}$ and Pb^{2+} by a *Streptomyces* sp. strain revealed that the dead cells accumulated more $(UO_2)^{2+}$ and Pb^{2+} than living cells. This was attributed to a greater degree of exposure of more functional groups to the uranium solution. While Pb^{2+} uptake by dead cells is connected with metal ions binding to cell surface structures, $(UO_2)^{2+}$ uptake by dead cells is due not only to binding to cell wall sites but also to cytoplasmic accumulation (Golab et al. 1991).

Models for uranium biosorption

The Freundlich and Langmuir adsorption isotherm models (Langmuir 1918; Freundlich 1926) have been applied to study uranium biosorption by various microorganisms. For example, it has been shown that uranium adsorption onto *Actinomyces levoris* and *R. arrhizus* cells can be modeled using a Freundlich isotherm (Horikoshi et al. 1979a; Tsezos and Volesky 1981). Uptake of uranium by *Pseudomonas* sp. EPS-5028, *A. vinelandii*, *D erxia gummosa*, *Arthrobacter* sp., *B. subtilles*, *K. marxians*, and *T. emersonni* demonstrates Langumuirian adsorption behavior (Marques et al. 1991; Cotoras et al. 1992; Sakaguchi 1998, 1996; Bustard et al. 1997; Bengtsson et al. 1995). Although the Brunauer-Emmett-Teller (BET) has been less frequently applied, biosorption by *Mycobacterium smegmatis* only followed the BET model (Brunauer et al. 1938; Andres et al. 1993).

Adsorption models such as these are useful for obtaining values for adsorption capacities and for describing equilibrium conditions for adsorption as a function of the sorbate concentration in solution (Volesky and Holan 1995; Rome and Gadd 1987). However, they are too simplistic when applied to complex biological systems. The adsorption models shed no light on the mechanism of adsorption and have little physical meaning. The model isotherms are significantly affected by changing pH, ionic strength, or solution composition. Consequently, all approaches using these models are system-specific, and can not be used to predict adsorption capacities in other systems (Volesky and Holan 1995; Langmuir 1997; Fowle and Fein 1999b).

An alternative approach is to use surface complexation models. Surface complexation models describe adsorption in terms of chemical reactions between surface functional groups and dissolved chemical species (Davis and Kent 1990). Surface complexation models differ from adsorption isotherm models in that the surface complexation models are based on mass law equations and take into account chemical speciation of the surfaces, the activity of sorbing ions, aqueous complexation and the pH of system (Davis and Kent 1990).

The activity of an ion bound to the surface compared to the ion in to the bulk solution declines as follows:

$$(X_s^z) = (X^z)[e^{-\psi F/RT}]^z$$

where z is the charge of ion X, (X_s^z) is the activity of ion X of charge z near the surface, (X^z) is the corresponding activity of X in the bulk solution, $e^{-\psi F/RT}$ is the Bolzman factor, ψ is the potential in volts at the plane of adsorption, and F, R and T are the Faraday constant, ideal gas constant and absolute temperature, respectively (Langmuir 1997).

Using the surface complexation model, Fein et al. (1997), Daughney et al. (1998), and Fowle and Fein (1999b) determined deprotonation constants (mentioned previously) and metal-bacteria stability constants. They adopted the constant capacitance model to account for the electrostatic interaction. The constant capacitance model which assumes all adsorbed ions occupy an adsorption plane immediately adjacent to the solid surface and that the magnitude of the potential declines linearly with distance from the surface (Daughney and Fein 1998). The capacitance of a bacterial surface (C_1) is calculated from the electric potential at the zero plane (ψ_0) and the electric charge (σ_0):

$$C_1 = \sigma_0 / \psi_0$$

Fein et al. (1997) and Daughney et al. (1998) determined the capacitances of the surfaces of *B. subtilis* (8.0 F m^{-2}) and *B. licheniformis* (3.0 F m^{-2}) and the equilibrium constants for the adsorption of Cd^{2+}, Pb^{2+}, and Cu^{2+} onto carboxyl and phosphate groups of *B. subtilis* and *B. licheniformis*. Fowle and Fein (1999b) also measured the stability constant of Ca^{2+} onto the carboxyl group of *B. subtilis* and $PbOH^+$, $CuOH^+$ onto the carboxyl groups of *B. subtilis* and *B. licheniformis*. The equilibrium constants are listed in Table 3.

Table 3. Metal-bacteria stability constants
[References: 1. Fein et al. (1997), 2. Daughney et al. (1998), 3. Fowle and Fein (1996b).]

Metal species	Bacteria*	Adsorption sites	logK	Ref
Cd^{2+}	BS	Carboxyl	3.9	1
Cd^{2+}	BS	Phosphate	3.4	1
Cd^{2+}	BL	Carboxyl	4.4	2
Cd^{2+}	BL	Phosphate	5.4	2
Cu^{2+}	BS	Carboxyl	4.9	1
Cu^{2+}	BS	Phosphate	4.4	1
Cu^{2+}	BL	Carboxyl	6.0	2
Pb^{2+}	BS	Carboxyl	4.7	1
Pb^{2+}	BS	Phosphate	4.2	1
Pb^{2+}	BL	Carboxyl	5.7	2
Pb^{2+}	BL	Phosphate	5.6	2
Al^{3+}	BS	Carboxyl	5.8	1
Al^{3+}	BS	Carboxyl	5.0	1
PbOH	BS	Carboxyl	5.8	3
PbOH	BL	Carboxyl	6.6	3
CuOH	BS	Carboxyl	6.4	3
CuOH	BL	Carboxyl	6.1	3
Ca^{2+}	BS	Carboxyl	2.8	3

It was shown using *B. subtilis* and *B. licheniformis* that the carboxyl- and phosphate-metal complexes have a 1:1 stoichiometry rather than being bidentate or tridentate (Fein et al. 1997; Daughney et al. 1998), and that intracellular uptake is not the mechanism of metal sorption for these species (Fowle and Fein 1999b). Underlying the above analyses are some model assumptions: (1) The reactive surface area (140 m^2/g) can be estimated from the rod-shape of cells (5.0 µm long and 1 µm wide). (2) Cell density in a suspension (2.5 g/L) is calculated assuming 10^{10} cells/ml. (3) A single pKa value can be used for individual functional groups. (4) Negligible involvement of extracellular polymers occurs owing to a wash protocol.

Daughney and Fein (1998) showed the stability constants obtained by the constant capacitance model varied with ionic strength (0.1 M and 0.01 $NaNO_3$), which indicates these constants are not independent of ionic strength. Fowle and Fein (1999a) extended the experimental system with a single metal and a single species of bacteria, either *B. subtilis* or *B. licheniformis*, to a system containing up to 3 types of metals and 2 species of bacteria (*B. subtilis* and *B. licheniformis*). They verified that the surface complexation model, using equilibrium constants from single metal and single bacteria experiments, could be applied to multi-component systems.

Langmuir (1979) compared the stability constants for oxalate and carbonate complexes with the same array of divalent cations and found a linear correlation. Using this type of correlation, measured stability constants for the complexation of metal with carboxyl sites on *B. subtilis* and *B. licheniformis* and known stability constants for the complexation of metals with carboxyl sites of organic acids (oxalate and tiron) were compared. An excellent linear correlation was obtained (Fein et al. 1997; Daughney et al. 1998). Fowle and Fein (1999b) estimated from known stability constants of uranyl-oxalate and uranyl-tiron complexes (Martell and Smith 1977) that the log stability constant of the uranyl-carboxyl complex of *B. subtillus* was 4.4.

Fowle and Fein (1999b) calculated the uranium distribution in a simplified bacteria-water-rock system. The system consisted of a solution containing 10^{-6} M uranium with an ionic strength of 0.1 M, 5g /L of *B. subtilis* cells, and 10^{-4} M ferrihydrite. A solution pH of 5 was selected so that the uranyl ion is dominant. The equilibrium constants are based on the uranyl-carboxyl stability constant and data obtained by Waite et al. (1994) and Dzombak and Morel (1990) for ferrihydrite. The result was that 47% and 39% of uranium were bound to the surfaces of bacteria and ferrihydrite, respectively, with 14% of uranium left in the aqueous phase. Although the system is too simplified and does not take into account intracellular uptake, biofilm formation, extracellular polymers, biomineralization, etc., this example illustrates that quantitative predictions for uranium distribution and transport affected by microbes in natural systems can be derived from the experimental data.

URANIUM TOXICITY, TOLERANCE AND BIOAVAILABILITY

The degree of toxicity of a metal ion (measured by the damage the metal ion can do to the cell) varies with the type of ion, its chemical speciation, and concentration. Elements such as copper and nickel play important roles in diverse metabolic processes, but become toxic to cells when concentrations are high. However, uranium has no known biological function and it is poisonous to cells even in low concentrations. Studies of the degree of resistance of 10 different isolates of *Thiobacillus ferrooxidans* to the metals, copper, nickel, uranium, and thorium also showed that uranium is 20 to 40 times more toxic than either copper or nickel at pH 2.1 (Leduc et al. 1997). Similarly, uranium exhibited a stronger inhibitory effect on the growth of *P. aeruginosa* and *Citrobacter* sp. than thorium (Premuzic et al. 1985; Plummer and Macaskie 1990). The uranyl ion is known to affect *Thiobacillus ferrooxidans* by inhibiting iron oxidation and carbon dioxide fixation (Touvinen and Kelly 1974a,b). However, toxicity may occur in different ways in different organisms.

The degree to which a metal ion will interfere with cellular function will depend upon the ability of the organism to affect the local concentration of the ion. Cells may differ in their ability to prevent passage of toxic ions through the cytoplasmic membrane, to pump ions out of the cytoplasm, and to sequester ions from solution by adsorption and precipitation (i.e. the ability of the organism to affect the ion's bioavailability may vary).

In the following section we consider toxicity responses of organisms because these responses can dramatically affect actinide speciation and mobility in the environment.

Toxicity mechanisms in microorganisms

The toxicity of uranium is derived from its chemical properties and radioactivity. The radioactivity associated with uranium is not lethal to microorganisms due to the long half-life of uranium and the short life cycle of microorganisms (Ehrlich 1996). The chemical properties of uranium cause significant toxic effects, similar to those induced by other metal ions (including heavy metals). The mechanisms of metal-ion toxicity include blocking biologically essential pathways (e.g. transport systems for nutrients and ions), displacing or substituting the essential metal ion from biomolecules and functional cellular units, conformational modification, disruption of cellular and organellar membrane integrity, and denaturation and inactivation of enzymes (Ochiai 1977, 1987). For example, uranium is often taken up more than certain metals such as Cu^{2+}, Zn^{2+}, and Ni^{2+} that have essential functions in some enzymes. Uranium can replace those metals and destroy the enzyme function. Metal ions maximize their toxicity when metal ions enter into the cytoplasm because biologically important molecules such as DNA, RNA, and abundant proteins are located in the cytoplasm.

Factors affecting bioavailability and toxicity of uranium to microorganisms

A subset of the factors affecting the toxicity and bioavailability of uranium are similar to the factors (pH, anions, cations, etc., see above) affecting uranium biosorption (DiSprito et al. 1983; Tuovinen and Kelly 1974a,b; DiSpirito and Tuovinen 1982). In addition, solution redox state, sorption onto inorganic and organic compounds, and complexation by organics will be important.

Solution redox potential will control the valence state of metal ions, affecting both the toxicity and the bioavailablity of the ions (Gadd 1993). Chromium occurs in less toxic trivalent and more toxic hexavalent states. The hexavalent state is very damaging because this ion is highly soluble at physiological pH. Chromium in the trivalent state is less problematic, partly because it is less soluble at physiological pH (Ehrlich 1996). Much less is known about the toxicity of uranium ions. However, because U^{6+} is soluble and U^{4+} is relatively insoluble, U^{4+} is probably less bioavailable. However, U^{4+} binds more strongly to biological molecules so it is perhaps more toxic than U^{6+}.

Ferric oxyhydroxides and organic materials may be very important potential sorbents of U^{6+}. Both ferric oxyhyroxides and organics commonly occur in sediments and soils with high capacity for U^{6+} sorption (Hsi and Langmuir 1985; Tripathi 1983; Wood 1996; Langmuir 1997) and can reduce bioavailability of U^{6+} (Gadd and Griffiths 1978; Babich and Stotzky 1980; Gadd 1993).

Natural organic ligands such as citrate and oxalate produced by microorganisms and higher plants form relatively stable complexes with uranium (Francis et al. 1992). Complexation with organic molecules present in the environment can reduce the bioavailability of uranium. This was shown for *Citrobacter* sp. (Young and Macaskie 1995).

Resistance and tolerance to uranium toxicity

Microorganisms can develop resistance mechanisms to toxic metals that are abundant in the environment by genetic adaptation and/or physiological adaptation (Gadd 1990). The ability to grow in the presence of elevated metal concentrations is found in a wide range of microbial groups and species. Because this ability is often associated with microbial species from unpolluted sites, adaptation is not always necessary (Gadd 1990).

Organisms isolated from contaminant sites tend to be resistant to toxic metal concentrations. For example, Doelman (1979) showed 75% of the bacteria isolated from Pb-polluted soils were able to grow on a medium containing 30 µg Pb ml^{-1}, whereas only 5% of bacteria from less polluted soils were able to grow on the same medium. The number of resistant bacteria in soils polluted by Cd, Cu, Ni, and Pb was approximately 15 times greater than that of bacteria in soils less polluted by the same metals (Duxbury and Bicknell 1983).

Microorganisms from highly uranium-contaminated sites accumulate far larger quantities of uranium than organisms from uncontaminated sites. We infer that these organisms are more tolerant than those isolated from uncontaminated sites. Organisms isolated from uranium mines show especially high abilities to accumulate uranium. In fact, the highest uranium biosorption reported to date was associated with *Bacillus subtilis* isolated from the Ranger uranium mine in northern Australia (615 mg U/g cells was taken up within 1 h at pH of 5; Sakaguchi 1998). The second highest microbial uranium biosorption value reported to date is associated with an *Arthrobacter* isolated from another uranium mine. This species accumulated 600 mg U/g cells intracellularly (Sakaguchi 1996). The sulfate-reducing bacterium *D. desulfuricans* isolated from a uranium mining waste pile was much more effective at reducing U^{6+} over a wider range of pH conditions than the same bacterium isolated from a non-uranium contaminated soil near a gas main (Panak et al. 1998).

Metal tolerance may result from the possession of (or ability to produce) polymers with high capability for metal adsorption, or possession of an impermeable cell wall (Gadd 1990). Tolerance may also be achieved by coincidental change in physico-chemical factors that result in reduced metal toxicity. Microbial resistance mechanisms to toxic metals consist of mainly intracellular or extracellular sequestration by adsorption or precipitation, extracellular binding, enzymatic redox reactions, formation of complexing or chelating agents, or specific metal efflux systems (Silver 1997; Collins and Stotzky 1989; Premuzic et al. 1985).

Genetically encoded metal efflux systems. Most commonly, genetically encoded metal resistance systems (typically efflux pumps) are found on small, circular DNA molecules known as plasmids (see Silver 1997). These plasmids are mobile, i.e. they can spread quickly from cell to cell to allow rapid introduction of resistance to a microbial population. In other cases, the machinery for metal resistance is coded by chromosomal genes. Plasmid-encoded systems tend to be highly metal specific (Silver 1997).

As described above, many uranium biosorption studies show that uranium penetrates the cytoplasm of a variety of microorganisms. To date, no plasmid-encoded efflux system for uranium has been reported. In the absence of such systems, microorganisms living in highly uranium contaminated sites must rely upon other detoxification mechanisms (DiSpirito et al. 1983).

Sequestration into vacuoles. In fungi, the majority of Cd^{2+}, Co^{2+}, Mn^{2+}, Mg^{2+}, Zn^{2+}, K^+, and Th^{4+} taken up is deposited in vacuoles. Within the vacuoles, the metals bind to low molecular weight polyphosphates, thus reducing binding to other organelles and macromolecules (Okorokov et al. 1980; White and Gadd 1987; Gadd and White 1989; Volesky et al. 1993). For example, *S. cervisiae* and *S. uvarum* in a solution containing $MnCl_2$ precipitated crystalline $MnHPO_4$ in vacuoles (Kihn et al. 1988). *Saccharomyces cervisiae* also accumulated uranium in vacuoles (Volesky and May-Philips 1995), possibly for detoxication purposes.

Production of metal binding proteins. A common response to metal toxicity is the synthesis of intracellular metal-binding proteins such as metallothionein, which has a cysteine-rich peptide that may function in detoxification through storage and potentially activation of other detoxification pathways (Mehra and Winge 1991).

Polyphosphates. In cells lacking vacuoles, polyphosphate granules may be involved in metal ion sequestration (Roomans 1980).

MICROBIAL ECOLOGY IN CONTAMINATED SITES

Microbial ecology is a scientific discipline that examines the relationships between various microbial populations and their biotic and abiotic environments. Because of the above mentioned abilities of specific microorganisms to tolerate, adsorb, reduce (or oxidize), and precipitate metals, microbial populations have the potential to greatly modify trace metal geochemistry. Conversely, microbial population structure (species and abundance) may be largely controlled by the metal chemistry. Thus, the geochemical and microbial characteristics of contaminated environments will be interdependent.

Study of microbial ecology is essential in order to understand the interactions between microorganisms and toxic metal ions in contaminated sites. Because only a subset of the organisms present in pristine sites have the ability to adapt to environments highly modified by metal pollution, microbial community structure may depend very heavily on the specific metals present (Gadd 1993).

In general terms, elevated levels of toxic metals are believed to alter microbial communities by reducing cell abundance and species diversity and by selecting for a resistant population (Duxbury 1985; Babich and Stotzky 1985). General reductions in the numbers of microorganisms, including actinomycetes and fungi, often have been recorded in soils polluted with Cu, Cd, Pb, As and Zn (Babich and Stotzky 1985). For example, along a steep gradient in soil Cu and Zn concentrations approaching a brass mill, fungal biomass decreased by about 75% (Nordgren et al. 1983). Similar reductions in numbers of aquatic cyanobacteria and algae in response to metal pollution have been noted (Rai et al. 1981).

Many studies of microorganisms isolated from metal contaminated sites show that prokaryotes (e.g. bacteria such as actinomycetes) are more sensitive than eukaryotes (fungi) to heavy metal pollution of soil. This may be attributed to the higher surface area per volume of prokaryotic cells compared to eukaryotes or it may be due to extra organelles involved in detoxification in eukaryotes (Bhat et al. 1979; Doelman 1979; Williams et al. 1977). Gram-negative bacteria are often more tolerant than Gram-positive bacteria (Babich and Stotzky 1985; Duxbury and Bicknell 1983; Doelman 1985).

The effects of toxic metals on microbial ecology vary with the type and quantity of resources (nutrients) available and the physico-chemical conditions of a habitat, e.g. temperature, pH, salinity and the availability of water, light and oxygen. Water content of soils determines the mobility and availability of toxic metals to microorganisms. Surface area of soils also affects metal toxicity to microbial communities. In most cases, the lowest concentrations of metals causing measurable effects to microbial communities were found in sandy soils and the highest in clay and organic soils (Doelman 1985). This may result from the low reactive mineral surface area available for metal adsorption in sandy materials. However, it can be difficult to separate effects of metals from effects of other environmental components. In polluted sites, a variety of toxins may be present, environments may be nutrient-limited, and pore water or ground waters may have very unusual chemistries (Gadd 1990; Gadd and Griffiths 1978).

Microbial ecology in surfical uranium-contaminated sites and groundwaters

Microbial ecology in high uranium environments (e.g. uranium ore deposits and disposal sites for uranium-bearing wastes) has been studied using traditional microbiological and molecular biological approaches. Questions such as the mechanisms of uranium bioleaching (Isabel de Siloniz et al. 1991; Berthelot et al. 1997) and sulfide mineral oxidation plus toxic metal mobilization in uranium mine mill wastes (Silver 1987; Schippers et al. 1995) have been explored. Potential microbial impacts on the stability of radioactive wastes disposed in geological respositories (West et al. 1992; Francis et al. 1994; Pedersen et al. 1996) and uranium immobilization by microorganisms (Barton et al. 1996) have also been considered. Below, we review results of studies of microbial ecology in uranium-contaminated sites, and discuss their relevance to understanding factors that control uranium in the environment.

Culture-based approaches. Using culture-based methods, Updegraff and Douros (1972) studied microbial ecology at three uranium deposits and revealed that samples of uranium ore and associated sediments contained remarkably few microorganisms, and very few different kinds of microorganisms. 72% of the pure cultures belonged to the genus *Arthrobacter*. Other bacterial genera identified were *Bacillus* and *Streptomyces*. According to Miller et al. (1987), the percentage of *Arthrobacter* was high in the sandy uranium tailings where the organic carbon content was low, whereas the percentage of *Bacillus* and *Streptomyces* sp. was high in organic-rich regions. uranium-mill slimes are predominantly populated with *Bacillus* or fungi in the top soils with relatively high organic carbon contents. A predominance of *Arthtobacter* (from 50-100%) was also found in uraniferous peat samples from the Flodelle Creek mine (Otto and Zielinski 1985). Since *Arthrobacter, Bacillus* and *Streptmyces* sp. are known to take up high amounts of uranium, it has been suggested that these organisms may influence uranium mobility in nature (Sakaguchi 1996, 1998; Friis and Myers-Keith 1986; Nakajima and Sakaguchi 1986).

The microbial ecology of the high-grade uranium deposit at Cigar Lake, Canada was investigated using microscopic observations and culturing techniques (Francis et al. 1994). Uranium concentrations ranged from 0.3 ppm in pH 6 groundwater to 8 ppm in pH 9 groundwater collected from the ore body and host rocks. The total viable plate counts of aerobic bacteria in water samples from the ore body and host rocks were 10^4 to 10^7 colony forming units (CFU) ml^{-1}. Plate counts in ore samples showed 10^5 CFU g^{-1} of aerobic bacteria. Denitrifiers (nitrate reducing bacteria) were the predominant group of microorganisms in all the water samples tested, and these were also detected in the ore samples. As denitrification is limited by the availability of nitrate, Fe^{3+} or U^{6+} may serve as alternate electron acceptors (Francis et al. 1994). Fermenters and sulfate reducers were present in the water samples, but their numbers were 2 to 4 orders of magnitude lower than those of the denitrifiers. The low number of CFU observed in water samples from the ore zone was attributed to radiation effects as well as high pH (Francis et al. 1994).

West et al. (1992) studied the abundance of aerobic and anaerobic heterotrophic bacteria and sulfate reducing bacteria in ore samples and groundwater samples collected from Osamu Utsumi uranium mine. The sample collection ranged from near surface to 50 m depth using culturing and microscopic methods. The total cell counts and CFU's for heterotorphs were variable among the samples, and no correlation between the numbers and depth was observed. Sulfate reducing bacteria were detected in most groundwater samples.

Molecular biological approaches to microbial ecology. It is well established that culture-based approaches provide an extremely incomplete picture of microbial populations. Culturing favors a few species that thrive in media used and obscures the often extremely high diversity of organisms in natural environments. Over the past decade or so, molecular methods based on ribosomal RNA analysis have become widely used. These techniques can be applied directly to environmental samples (without culturing) to identify organisms, as well as to establish the phylogenetic relationships of new organisms to known species (see Barns and Nierzwicki-Bauer 1997 for details). Related RNA probing methods utilize DNA sequence information to label individual cells within natural assemblages (with species, group, or domain-level specificity; see Amman et al. 1995) and allow quantification of the make-up of microbial populations as a function of environmental conditions (e.g. see Schrenk et al. 1998). However, polymerase chain reaction (PCR)-based methods are not without biases. Many of these can be overcome by the appropriate combination of PCR-based and *in situ* hybridizations.

The diversity and abundance of bacteria in groundwaters at the Bangombe reactor in the Oklo uranium deposit were studied by analysis of 16S-rRNA gene sequence information, and by counting microbial cells by microscopy (Pedersen et al. 1996). 16S rRNA gene analysis showed 50% of total clones sequenced belonged to 20 different groups within the Gram-negative bacteria. *Sphaerotilus, Rhodocyclus, Zoogloea* and *Acinetobacter*-like sequences dominated at some sampling sites. Notably, *Arthrobacter, Bacillus* and *Streptmyces* species reported from many other uranium sites using culture-based methods were not abundant/detected. Three sequences distantly related to known species were also predominant (Pedersen et al. 1996). The total number of bacteria ranged from 5×10^4 to 6×10^5 ml^{-1}, and cell counts correlated with total organic carbon (TOC).

16S-rRNA gene sequencing methods were used to analyze bacterial diversity in a soil sample from a depth of 4 to 5 m at a uranium mining waste pile associated with relative acidic (pH 4.5) surface waters with elevated uranium concentrations up to 58mg/L (Puers and Selenska-Pobell 1998). Results revealed organisms related to *Proteobacteria* (24%), Green-non-sulfur bacteria (41%) and bacteria of the *Fibrobacter, Acidobacterium* subdivision (19%). Only 5.5% of the clones belonged to the Gram-positive phylum. The majority of the *Proteobacteria* belonged to the γ-subdivision (15.4%). Within the Gram-positive group, three of the five clones were High-G+C-subdivision members. These results show the predominance of Gram-negative bacteria over Gram-positive bacteria. *Actinomycetes*, which are major High-G+C-subdivision members often isolated from U-contaminated sites, might also be important in this high uranium environment.

IMPLICATIONS FOR UNDERSTANDING GEOLOGICAL PROCESSES

Microbial interactions with uranium may have been of long term geological importance. Although not yet well understood, microorganisms may have played roles in ore formation, cell fossilization, and the uranium geochemical cycle.

Microbial roles in the formation of uranium ore deposits

Seventy percent of economically important uranium ore deposits around the world are classified into three types: quartz-pebble conglomerates, unconformity-type deposits, and sandstone deposits (Nash et al. 1981). In each of these, uranium is generally in the reduced U^{4+} form and microorganisms may have played a role in the formation of these deposits.

The formation of quartz-pebble conglomerate deposits has been attributed to detrital transport and deposition of uraninite under anoxic conditions ~2.6 to 2.2 billion years ago (Holland 1962). Uranium ores originally formed in strata containing abundant organic material, mostly kerogen (thucholite), and uranium mineralization is closely associated with kerogen-rich regions (Willingham et al. 1985). Willingham et al. (1985) suggested that the stratiform kerogen originated from cyanobacteria. Hallbauer and Van Warmelo (1974) observed that the morphology of thucolite was similar to filamentous and branched cells. Their proposal that thucolite is derived from primitive plants, most likely lichen, remains controversial. It is still uncertain how organic material and/or microorganisms affected uranium mineralization.

Unconformity-type deposits formed ~2.2 to 0.4 billion years ago. It has been proposed that the ore formation mechanism involves movement of oxidizing solutions up through faults. Aqueous U^{6+} was reduced and precipitated in carbonaceous marginal marine sediments. Many of the ore bodies occur within graphitic layers. Because the estimated temperature of ore-formation is ~100° to 250°C, a significant role for living microorganisms in U^{6+} reduction is considered unlikely.

Sandstone deposits are the most abundant uranium reservoirs. The sandstone-hosted deposits formed below 50°C by reduction of U^{6+} to insoluble U^{4+}. It is believed that this occurred soon after sandstone sedimentation, when the permeability was high enough to allow passage of ore-forming fluids (Nash et al. 1981). Uranium in these deposits was derived from leaching of granite or tuff. Uranium was transported in neutral to alkaline, oxidizing solutions as uranyl-carbonate complexes (Hostetler and Garrels 1962; Langmuir 1997), reduced, and precipitated as insoluble U^{4+} minerals such as uraninite and coffinite. As uranium ore bodies are associated with organic matter and pyrite, precipitation has typically been attributed to inorganic U^{6+} reduction by aqueous humic acids, hydrogen sulfide derived by sulfate-reducing bacteria, or H_2S-rich natural gas (Nash et al. 1981). However, because sulfate-reducing bacteria can reduce U^{6+} enzymatically, and grow using U^{6+} as the sole electron acceptor, microorganisms may have played a key role in formation of these deposits.

Lovely et al. (1991) reported almost no U^{6+} reduction by sulfide at neutral pH and 30°C after 24 hours. However, ferric iron-reducing bacteria rapidly reduced most of the U^{6+} in the same bicarbonate solutions at the same temperature. Lovely and Phillips (1992a) inoculated *D. desulfuricans* into the bicarbonate buffer containing 0.35 mM U^{6+} and 1 mM sulfide over the temperature range 10 to 80°C. They demonstrated much more U^{6+} was reduced at 30°C than at 80°C. At 80°C, enzymatic reduction did not occur because cells were unable to grow at this temperature. Non-enzymatic reduction of U^{6+} by sulfide was more favorable at 80°C than at 30°C. Based on the slow rate of U^{6+} reduction by organic matter below 120°C (Nakashima et al. 1984), it has been suggested that enzymatic reduction of U^{6+} by sulfate reducing bacteria is a more plausible mechanism for ore precipitation than indirect nonenzymatic reduction by sulfide produced by sulfate reducing bacteria (Lovely et al. 1991; Lovely and Phillips 1992a; Lovely et al. 1993b).

Fossilization of microbial cells

Microorganisms produce specific molecules (lytic enzymes) to degrade cell wall compounds. After cell death, lytic enzymes from decomposing microorganisms break down cell walls. Consequently, cell wall components are typically not preserved in sediments.

Well-preserved microfossils are frequently found in cherts, organic-rich sediments and shales. Metals are associated with these microfossils, and these give rise to contrast for TEM observations (Degens and Ittekkot 1982; Degens et al. 1970). Gloubic and Barghoorn (1977) and Degens and Ittekkot (1982) attributed the preservation of microfossils to the binding of metals by microbial cells.

Ferris et al. (1988) conducted experiments under geothermal conditions to investigate the role of silicification in the preservation of microbial cells. The lytic enzymes degraded the cells with no pretreatment. However, they found that iron-loaded cells remained intact for 150 days. Beveridge et al. (1983) also carried out similar experiments under diagenetic conditions and showed that uranium-loaded *B. subtilis* cells were well preserved, and crystalline uranyl phosphate minerals developed on cell surfaces. Milodowski et al. (1990) obtained natural evidence that fossilization of microbial cells (with filamentous structures analogous to *Actinomycetes*) in uranium-rich environments was associated with formation of platy calcium and uranium and/or vanadium minerals.

Subsurface transport of uranium

Microbes can enhance or retard uranium migration. Suspended colloids (particles with diameters less than 10 µm) can migrate in solutions through porous and fractured subsurface rocks. Microorganisms are considered to be colloids due to their size. Thus, microbial colloidal transport can facilitate movement of adsorbed uranium if cells are not attached to large mineral particles (McCarthy and Zachara 1989). For example, Keswick et al. (1982) reported bacterial migration 920 m through geological media.

Many natural environments, including rocks and soils, are nutrient poor (oligotrophic). Under these conditions, microorganisms often attach to solid surfaces and frequently form biofilms. Pedersen (1996) indicated that bacteria attached to solid surfaces can retard uranium transport in geological settings. However, when prevailing conditions become too harsh for microbes to survive (due to low availability of nutrients or extremes of temperature, pH, etc.), microbes adopt starvation responses by reducing their cell size and/or adhesion abilities (Lappin-Scott and Costerton 1990; Bar-Or 1990). These responses enhance the migration of microbes and uranium in soils and subsurface strata.

In addition to bacterial cells, small mineral particles can be transported as colloids. As discussed in the above section on the results of biosorption experiments, laboratory studies suggest that microorganims can form colloidal-sized uranium precipitates. If these precipitates are detached from microbial cell surfaces, they can be transported in ground water (McCarthy and Zachara 1989).

Microbial impacts on the surface uranium cycle

The most important sink for uranium near Earth's surface is mineral precipitation associated with reduction of U^{6+} to U^{4+} in anoxic marine sediments (Veeh 1967; Anderson et al. 1989; Klinkhammer and Palmer 1991; Lovely et al. 1993c). Marine sediments are typically stratified in terms of redox reactions. U^{6+} is reduced within the Fe^{3+} reduction zone (Cochran et al. 1986; Lovely et al. 1991). Thus, Fe^{3+}-reducing microorganisms, which can grow using U^{6+} as their sole electron acceptor, may play important roles in the uranium cycle in marine sediments (Lovely et al. 1991; Lovely and Phillips 1992a; Lovely et al. 1993c). In the case of estuarine sediments, U^{6+} reduction occurs in the sulfate-reducing zone (Barnes and Cochran 1993). Barnes and Cochran (1993) suggested that in near-shore and estuarine sediments where organic matter is relatively abundant, bacteria may control U^{6+} reduction.

BIOREMEDIATION OF URANIUM— AND OTHER ACTINIDES—CONTAMINATED SITES

Radioactive wastes are classified as high-level wastes (spent fuel etc.), transuranic wastes (wastes from spent fuel reprocessing, nuclear weapons), and low- and intermediate-level wastes, which are generated from a variety of activities (Department of Energy 1979). In addition, in the United States alone, there are more than 230 million tons uranium mill tailings (residues from uranium mining and milling operations; US Department of Energy 1993, 1994). Long term containment of these wastes is of great environmental concern (Landa and Gray 1995).

Low-level wastes and uranium mill tailings have either been disposed of in shallow burial sites or remain at the surface. Radioactive metals are leached out of the wastes and contaminate surrounding ground waters and soils. To protect the environment, these metals should be immobilized. Bioremediation is increasingly attracting attention as a cost effective option for site treatment.

Barton et al. (1996) proposed in-situ bioremediation of U^{6+}-contaminated sites. Their concept is that selective addition of nutrients to stimulate bacterial growth and activity in groundwater will lead to removal of U^{6+} in insoluble forms. Francis (1995) and Phillips et al. (1995) suggested that uranium in contaminated soils could be extracted by addition of bicarbonate or citric acids to form complexes with U^{6+}. The extracted U^{6+} is removed from process solutions by reductive precipitation by U^{6+}-reducing microorganisms.

IMPLICATIONS FOR MIGRATION OF OTHER ACTINIDES

Uranium has many similarities with other actinides in terms of chemical properties. For examples, $(UO_2)^{2+}$ is similar to $(PuO_2)^{2+}$. $(PuO_2)^{2+}$ is less stable and more easily reduced than $(UO_2)^{2+}$. U^{4+} has similar chemical behavior to Th^{4+} and Pu^{4+}. Therefore, it is possible to extrapolate the data for microbial uranium accumulation to other actinides (Macaskie 1991). However, some of other actinides have much higher radioactivity, and this can severely inactivate microbes. Despite this, many radiation-resistant strains of bacteria have been isolated from soil and other environments with high radioactivity (Yoshinaka et al. 1973; Yano et al. 1975; Kiselov et al. 1961; Matsuyama et al. 1973). These microorganisms may have high potential to immobilize actinides in natural settings.

SUGGESTIONS FOR FUTURE WORK

Results of experimental investigations of uranium accumulation and studies of microbial ecology in uranium-rich environments strongly imply microorganisms can play significant roles in uranium immobilization. However, no direct observations of living microorganisms, other than lichen, immobilizing uranium have been published. Despite the fact that each microbial cell is small, cell numbers in uranium-contaminated sites can be considerable. If one gram of typical uranium-contaminated soil (1-10 ppm U) contains ~10^7 cells, using typical uptake values of 100 mg U/ g dry weight cells (see above), we estimate that cells can account for ~1-10 ppm U. Thus, the majority of uranium in a typical uranium contaminated soil could be associated with microorganisms. This remains to be tested. However, if there are ~10^4 cells/g, the majority of uranium is probably associated with other phases. If uranium phases do not dissolve after cells die and if the cell walls do not degrade rapidly, very large quantities of uranium could be sequestered into solids by biomineralization. The real contribution of microorganisms to uranium immobilization requires detailed characterization of microbial populations in contaminated environments, the mechanisms of uranium precipitation and their associated uranium phases.

To build models that quantify the microbial impacts on the fate of uranium and other actinides, we have to precisely understand the mechanisms by which microbes interact with uranium in nature. The thermodynamic dataset for uranyl species, and understanding of functional groups associated with the cell wall of microorganisms, are extremely limited. Experiments designed to measure the stability constants for complexation between various uranyl species and functional groups for various microbes are required.

Based on laboratory studies, it is clear that both intracellular and a variety of forms of extracellular uptake of uranium by microorganisms can be important. Besides, precipitation of U-bearing crystals at cell surfaces is also important. These phenomena represent challenges for development of quantitative models. In addition to the surface complexation models, other approaches should be applied. For instance, the surface precipitation model (Farley et al. 1985) can be used to quantify sorption and subsequent precipitation of uranium at microbial surfaces (Warren and Ferris 1998).

Crystals formed by microorganisms are typically of nanometer scale and quite hydrous, therefore identification is difficult using conventional mineralogical techniques such as XRD and TEM. Understanding of uranium biomineralization requires accurate phase identification. Other techniques, e.g. cryo-TEM, combined with ultra-sensitive image recording materials to minimize beam damage, or synchrotron-based spectroscopic methods, should be helpful in efforts to identify these phases.

The mechanisms of biomineralization are currently quite unclear. Some progress has been made in identifying cell surface molecules involved in initial uranium binding. However, it is clear that uranium interacts with a large number of organic molecules. Uranium toxicity probably arises from uranium binding to proteins, and proteins are implicated to be important in uranium uptake by concentric bodies of lichenized fungi. The details of such interactions may be of fundamental significance, so understanding uranium binding by amino acids and proteins, as well as other organic molecules, may be an important future research priority.

Although many microorganisms accumulate U^{6+} intracellularly, no intracellular detoxification of U^{6+}, other than vacuoles, has been reported. Additional research is needed on microbial uranium detoxification mechanisms in nature.

Uranium toxicity is moderated by various environmental as well as biological factors. Studies of microbial ecology of uranium-contaminated sites should be conducted in parallel with geochemical investigations. 16S-rRNA sequence analysis has revolutionized our ability to study microbial populations, including species that are currently unculturable. The few studies that have used these approaches have revealed the importance of new species of bacteria whose interaction with uranium is still unknown. Further investigations into uranium uptake by microorganisms that are present, rather than just easily cultivated, are critical for understanding of the microbial impacts on natural uranium distributions.

Investigations of microbial interactions with uranium are relevant to understanding of the past and present surface uranium cycle in nature. They are also critical for development of strategies for remediation of actinide-contaminated sites and for prediction of the fate of high-level wastes disposed in geological settings. However, the topic of microbial interactions with heavy metals is comparatively unexplored. This field promises the opportunity for tremendously exciting and important new insights, of relevance to our understanding of the Earth and to human welfare.

ACKNOWLEDGMENTS

The authors thank Dr. Bill Barker for his contributions to this paper, Dr. Simcha Stroes-Gascoyne for loan of her reprint collection, and Dr. David Fowle for his comments on the manuscript. We thank Dr. Jeremy Fein and Dr. Simcha Stroes-Gascoyne for their valuable reviews. YS thanks his colleagues at Japan Atomic Energy Research Institute for their collaboration and is grateful to Yoshida Scholarship Foundation for financial support. The research involved in preparation of this chapter was partially supported by a grant from the U.S. Department of Energy, Basic Energy Sciences Program (Grant # DEFG02-93ER14328).

REFERENCES

Amman RI, Ludwig W, Schleifer K-H 1995 Physiologenetic identificaiton and in situ detection of individual microbial cells without cultivation. Microbiol Rev 59:0-32

Anderson RF, LeHuray AP, Fleisher MQ, Murray JW 1989 Uranium deposition in Saanich Inlet sediments, Vancouver island. Geochm Cosmochim Acta 53:2205-2213

Andres Y, MacCordick J, Hurbert JC 1993 Adsorption of several actinide (Th, U) and lanthanide (La, Eu, Yb) ions by *Mycobacterium smegmatis*. Appl Microbiol Biotechnol 39:413-417

Andres Y, MacCordick J, Hubert JC 1994 Binding sites of sorbed uranyl ion in the cell wall of *Mycobacterium smegmatis*. FEMS Microbiol Lett 115:27-32

Andres Y, MacCordick J, Hurbert JC 1995 Selective biosorption of thorium ions by an immobilized mycobacterial biomass. Appl Microbiol Biotechnol 44:271-276

Babich H, Stotzky G 1980 Environmental factors that influence the toxicity of heavy metal and gaseous pollutants to microorganisms. Critic Rev Microbiol 8:99-145

Babich H, Stotzky G 1985 Heavy metal toxicity to microbe-mediated ecologic processes: a review and potential application to regulatory policies. Environ Res 36:111-137.

Barker WW, Haas JR, Suzuki Y, Banfield JF 1998 U-phosphate biomineralization as a mechanism of U fixation by lichen. Geol Soc Am Abst: 205

Barnes CE, Cochran JK 1993 Uranium geochemistry in estaurine sediments: Controls on removal and release processes. Geochim Cosmochim Acta 57:555-569

Barns SM, Nierzwicki-Bauer SA 1997 Microbial diversity in ocean, surface and subsurface environments. *In* Geomicrobiology. Banfield JF, Nealson KH (Eds) Rev Mineral 35:35-79

Bar-Or Y 1990 The effect of adhesion on survival and growth of microorganisms. Experimentia 46:823-826

Bartnicki-Garcia S 1973 Fundamental aspects of hyphal morphogenesis. *In* Microbial Differentiation. 23rd Symp Soc Gen Microbiol. Cambridge Univ Press, Cambridge, UK, p 245-267

Barton LL, Choudhury K, Thomson BM, Steenhoudt K, Groffman AR 1996 Bacterial reduction of soluble uranium: The first step of in situ immobilization of uranium. Radioactive Waste Mgt Environ Restoration 20:141-151

Bell AA, Wheeler MH 1986 Biosynthesis and function of fungal melanins. Ann Rev Phytopathol 24:411-451

Bengtsson L, Johansson B, Hackett TJ, McHale L, McHale AP 1995 Studies on the biosorption of uranium by *Talaromyces emersonii CBS 814.70* biomass. Appl Microbiol Biotechnol 42:807-811

Berthelot D, Leuduc LG, Ferroni GD 1997 Iron-oxidizing autotrophs and acidophilic heterotrophs from uranium mine environments. Geomicrobiol J 14:317-324

Beveridge TJ 1978 The response of cell walls of *Bacillus subtilis* to metals and to electron-microscopic stains. Can J Microbiol 24:89-104

Beveridge TJ 1989 The structure of bacteria. *In* Bacteria in Nature. JS Poindexter, ER Leadbetter (Eds) Plenum Publ, New York, p 1-65

Beveridge TJ, Meloche JD, Fyfe WS Murray GE 1983 Diagenesis of metals chemically complexed to bacteria: Laboratory formation of metal phosphates, sulfides, and organic condensates in artificial sediments. Appl Environ Microbiol 45:1094-1108

Beveridge TJ, Murray RGE 1980 Sites of metals deposition in the cell wall of *Bacillus subtilis*. J Bacteriol 141:876-887

Bhat PK, Upadhyaya SD, Dagar JC, Singh VP 1979 Assessment of heavy metal toxicity. I Effect on microbial population, mineralization and soil respiration. Cur Sci 48:571-573

Blundell TL, Jenkins JA 1977 The binding of heavy metals to proteins. Chem Soc Rev 6:139-171

Blundell TL, Johnson LN 1976 Protein Crystallography. Academic Press, London, p 199-231

Boileau LJR, Nieboer E, Richardson DHS 1985 Uranium accumulation in the lichen *Cladonia rangiferina*. Part I. Uptake of cationic, neutral, and anionic forms of the uranyl ion. Can J Bot 63:384-389

Bold HC, Alexopoulus CS, Delevoryas T 1980 Morphology of Plants and Fungi, 4th edn. Harper and Row, New York.

Bonner WA, Castro AJ 1965 Natural porducts. *In* Essentials of Modern Organic Chemistry. WA Bonner, AJ Castro (Eds) Reinhold, New York, p 1-540

Bracker CE 1967 Ultrastructure of fungi. Ann Rev Phytopathol 5:343-374

Brock TD, Madigan MT, Martinko JM, Parker J 1994 Biology of microorganisms. Prentice Hall, Englewood Cliffs, New Jersey

Brunauer S, Emmett PH, Teller E 1938 Adsorption of gases in multimolecular layers. J Am Chem Soc 60:309-319

Burnett JH 1976 Fundamentals of Mycology, 2nd edn. Edward Arnold, London

Bustrad M, Donnellan N, Rollan A, MacHale AP 1997a Studies on the biosorption of uranium by a thermotolerant, ethanol-producing strain of Kluyveromyces marxianus. Biopro Eng 17:45-50

Bustard M, Donnellan A, Mchale AP 1997b Studies on the biosorption of uranium by a thermotolerant, ethanol-producing strain of *Kluyvermyces marxianus*. Biopro Eng 17:45-50

Byerley JJ, Scharer JM 1987 uranium (VI) biosorption from process solutions. Chem Eng J 36:B49-B59

Caccavo F Jr, Blakemore RP, Lovely DRE 1992 A hydrogen-oxidizing, Fe(3+)-reducing microorganism from the Great Bay Estuary, New Hampshire. Appl Environ Microbiol 58:3211-3216

Chmielowski J, Worznica A, Klapcinska B 1994 Binding of uranium by yeast cell wall polysaccharides. Bull Polish Acad Sci Biol Sci 42:147-149

Choppin GR, Mathur JN 1991 Hydrolysis of actinyl (VI) cations. Radiochim Acta 52/53:25-28

Cochran JK, Carey AE, Sholkovitz ER, Supernant LD 1986 The geochemistry of uranium and thorium in coastal marine sediments and sediment porewaters. Geochim Cosomochim Acta 50:663-680

Collins YE, Stozky G. 1989 Factors affecting the toxicity of heavy metals to microbes. *In* Metal Ions and Bacteria. TJ Beveridge, RJ Doyle (Eds) Wiley, New York, p 31-90

Congregado F, Estanol I, Espny MJ, Fuste MC, Manresa MA, Marques AM, Guinea J, Simon-Pujol MD 1985 Preliminary studies on the production and composition of the extracellular polysaccharide synthesized by *Pseudomonsa sp. EPS-5028*. Biotechnol Lett 7:883-888

Corpe W 1964 Factors influencing growth and polysaccharide formation by strain of *Chromobacterium violaceum*. J Bacteriol 88:1433-1437

Corpe WA 1975 metal-binding properties of surface materials from marine bacteria. Dev Ind Microbiol 16:249-255

Cotoras D, Millar M, Viedma, P, Pementel J, Mestre A 1992 Biosorption of metal ions by *Azotobacter vinelandii*. World J Microbiol Biotechnol 8:319-323

Crist RH, Oberholser K, McGarriry J, Crist DR, Johnson JK, Brittsan JM 1992 Interaction of metals and protons with algae. 3. Marine algae, with emphasis on lead and aluminum. Environ Sci Technol 26:496-502

Daughney CJ, Fein JB 1998 The effect of ionic strength on the adsorption of H^+, Cd^{2+}, Pb^{2+}, and Cu^{2+} by *Bacillus subtilis* and *Bacillus licheniformis*: A surface complexation model. J Coll Inter Sci 195:53-77

Daughney CJ, Fein JB, Yee N 1998 A comparison of the thermodynamics of metal adsorption onto two common bacteria. Chem Geol 144:161-176

Davis RH 1986 Compartmental and regulatory mechanismsin the arginine pathways of *Neurospora crassa* and *Saccharomyces cerevisiae*. Microbiol Rev 50:280-313

Davis JA, Kent DB 1990 Surface complexiation modeling in aqueous geochemistry. Rev Mineral 23:177-248

Degens ET, Watson SW, Remsen CC 1970 Fossil membranes and cell wall fragment from 7000-year old Black Sea sediments. Science 168:1207

Degens ET, Ittekkot V 1982 *In situ* metal-staining of biological membranes in sediments. Nature 298:262-264

Department of Energy 1979 report to the president by the interagency review group on nuclear waste management. TID-29442, DOE, Washington, DC

DiSpirito AA, Talnagi Jr JW, Tuovinen OH 1983 Accumulation and cellular distribution of uranium in *Thiobacillus ferrooxidans*. Arch Microbiol 135:250-253

DiSpirito AA, Tuovinen OH 1982 Uranous ion oxidation and carbon dioxide fixation by *Thiobacillus ferrooxidans*. Arch Microbiol 133:33-37

Doelman P 1979 Effects of lead pollution on soil microflora. PhD thesis. Verwey Wageningen.

Doelman p 1985 Resistance of soil microbial communities to heavy metals. *In* Microbial Communities in Soil. V Jensen, A Kjoller, LH Sorensen (Eds) Elsevier Sci Publ, New York, p 369-384

Duxbury T, Bicknell B 1983 Metal-tolerant bacterial populations from natural and metal-polluted soils. Soil Biol Biochem 15:243-250

Duxbury T 1985 Ecological aspects of heavy metal responses in microorganisms, In Advances in Microbial Ecology 8: 185-235, Marshall KC (Ed) Plenum Press, New York

Dzomabak DA, Morel EMM 1990 Surface complexation modeling: Hydrous ferric oxide. Wiley, New York

Ehrlich HL 1996 Geomicrobiology. Marcel Dekker, New York

Farkas V 1979 Biosynthesis of cell walls of fungi. Microbiol Rev 43:117-144

Farley KJ, Dzombak DA, Morel FMM 1985 A surface precipitation model for the sorption of cations on metal oxides. J Coll Inter Sci 106:226-242

Fein JB, Daughney J, Yee N, Davis TA 1997 A chemical equilibrium model for metal adsorption onto bacterial surfaces. Geochim Cosmochim Acta 61:3319-3328

Ferris FG, Fyfe WS, Beveridge TJ 1988 Metallic ion binding by *Bacillus subtilis*: Implication for the fossilization of microorganisms. Geology 16:149-152

Finch RJ, Ewing RC 1992 The corrosion of uraninite under oxidizing conditions. J Nuclear Mater 190:133-167

Fowle DA, Fein JB 1999a Competitive adsorption of metal cations onto two Gram-positive bacteria: Testing the cehmical equilibrium model. Geochim Cosmochim Acta (in press)

Fowle DA, Fein JB 1999b Experimental measurements of the reversibility of metal-bacteria adsorption reactions. Chem Geol (in press)

Francis AJ 1995 Microbiological treatment of radioactive wates. In Chemical Pretreatment of Nuclear Waste for Disposal. Schulz WW, Horwitz EP (Eds) Plenum Press, New York, p 115-131

Francis AJ, Dodge CJ, Gillow JB 1992 Biodegradation of metal citrate complexes and implications for toxic-metal mobility. Nature 356:140-142

Francis AJ, Joshi-Tope G, Gillow JB, Dodge CJ 1994 Enumeration and characterization of microorganisms associated with the uranium ore deposit at Cigar Lake, Canada. BNL–49737, Department of Energy, Washington, DC, p 1-42

Freundlich H 1926 Colloid and Capillary Chemistry. Methuen, London

Friedl T, Budel B 1996 Photobionts. In Lichen Biology. Nash TH (Ed) Cambridge Univ Press, New York, p 8-23

Friis N, Myers-Keith P 1986 Biosorprion of uranium and lead by *Streptmyces longwoodensis*. Biotechnol Bioeng 28:21-28

Gadd GM 1988 Accumulation of metals by microorgansims and algae. In Biotechnology—a Comprehensive Treatise (Special microbial processes) 6b:401-433. H-J Rehm (Ed) VCH Verlagsgesellschaft, Weinheim

Gadd GM 1990 Metal tolerance. In Microbiology of Extreme Environments. Edwards C (Ed) Milton Keynes, Open Univ Press, p 59-88.

Gadd GM 1993 Transley review No. 47, Interactions of fungi with toxic metals. New Phytol 124:25-60

Gadd GM 1996 Influence of microorganisms on the environmental fate of radionuclides. Endeavour 20:150-156

Gadd GM, Griffiths AJ 1978 Microorganisms and heavy metal toxicity. Microbial Ecology 4:303-317

Gadd GM, White C 1989 Uptake and intracellular compartmentation of thorium in *Saccharomyces cervisiae*. Environ Poll 61:187-197

Gale GR, Mclain HH 1963 Effect of ethanbutol on cytology of *Mycobacterium smegmatis*. J Bacteriol 86:749-756

Galun M, Behr L, ben-Shaul Y 1974 Evidence for protein content in concentric bodies of lichenized fungi. J Microscopie 19:193-196

Galun M, Keller P, Malki D, Galun E, Siegel SM, Siegel BZ 1983 Removal of uranium (VI) from solution by fungal biomass and fungall wall-related biopolymers. Science 219:285-286

Galun M, Galun E, Siegel BZ, Keller P, Lehr H, Siegel SM 1987 Removal of metal ions from aqueous solutions by *Penicillium* biomass: Kinetic and uptake parameters. Water Air Soil Poll 33:359-371

Gloubic S, Barghoorn ES 1977 Interpretation of microbila fossils with special reference to the Precambrian. I. Fossil Algae. E Fluge (Ed) Springer-Verlag, Berlin, p 1-14

Golab Z, Orlowska B, Smith RW 1991 Biosorption of lead and uranium by *Streptmyces sp*. Water, Air and Soil Pollution 60:99-106

Goncalves MLS, Sigg L, Reutlinger M, Stumm W 1987 Metal ion binding by biological surfaces: Voltammetric assessment in the presence of bacteria. Sci Total Environ 60:105-119

Gorby YA, Lovely DR 1992 Enzymatic uranium precipitation. Environ Sci Technol 26:205-207

Greene B, Darnall DW 1990 Microbial oxygenic photoautotrophs (cyanobacteria and algae) for metal ion binding. In Microbial Metal Recovery. HL Ehrlich, C Brierley (Eds) McGraw-Hill, New York, p 185-222

Grenthe I, Fuger J, Konings RJM, Lemire RJ, Muler AB, Nguyen-Trung C, Wanner H 1992 Chemical thermodynamics of uranium. Chem Thermo Series, Elsevier, Amsterdam

Griffiths HB, Greenwood AD 1972 The concentric bodies of lichenized fungi. Arch Mikrobiol 87:285-302
Haas JR, Bailey EH, Purvis OW 1998 Bioaccumulation of metals by lichens: Uptake of aqueous uranium by *Peltigera membranacea* as a function of time and pH. Am Mineral 83:1494-4502
Hallbauer DK, Van Warmelo KT 1974 Fossilized plants in thucholite from Precambrian rocks of the Witwatersrand, South Africa. Precambrian Res 1:199-212
Hawksworth DL, Hill DJ 1984 The lichen-forming fungi. Blackie, Glasgow
Hawksworth DL 1991 The fungal dimension of biodiversity: magnitude, significance, and conservation. Mycol Res 95:641-655
Hsi C-KD, Langmuir D 1985 Adsorption of uranyl onto ferric oxides: Application of surface complexation site-binding model. Geochim Cosmochim Acta 49:1931-1941
Hodge HC, Stannard JN, Hursh JB 1973 Uranium, plutonium, and transuranic elements: Handbook of experimental pharmacology XXXVI. Springer-Verlag, New York
Holland HD 1962 Model for evolution of the earth's atmosphere. *In* Petrologic Studies, a Volume in Honor of AF Buddington. Engel AEJ, James HL, Leonard BF (Eds) Geol Soc Am. p 447-477
Horikoshi T, Nakajima A, Sakaguchi T 1979a Uptake of uranium by *Chlorella regularis*. Agric Biol Chem 443:617-623
Horikoshi T, Nakajima A, Sakaguchi T 1979b Uranium uptake by *Synechococcus elongatus*. J Ferment Technol 57:191-194.
Horikoshi T, Nakajima A, Sakaguchi T 1981 Studies on the accumulation of heavy metal elements in biological systems XIX. Accumulation of uranium by microorganisms. Eur J Appl Microbiol Biotechnol 12:90-96
Hostetler PB, Garrels RM 1962 Transportation and precipitation of uranium and vanadium at low temperatures, with special reference to sandstone-type uranium deposits. Econ Geol 57:137-167
Hu MZ-C, Norman JM, Faison BD 1996 Biosorption of uranium by *Pseudomonas aeruginosa* Strain CSU: Characterization and comparison studies. Biotechnol Bioeng 51:237-247
Hunsley D, Burnett JH 1970 The ultrastructural architecture of the of the walls of some fungi. J Gen Microbiol 62:203-218
Isabel de Siloniz M, Lorezo P, Perera J 1991 Distribution of oxidizing bacterial activities and characterization of bioleaching-related microorganisms in a uranium mineral heaps. Microbiologia 7:82-89
Jensen ET 1993 Cyanobacterial ultrastructure. *In* Ultrastructure of Microalgae. T Berner (Ed) CRC Press, Boca Raton, p 7-51
Jeong BC, Hawes C, Bonthrone KM, Macaskie LE 1997 Localization of enzymatically enhanced heavy metal accumulation *Citrobacter sp*. And metal accumulation in vitro by loposomes containing entrapped enzyme. Microbiol 43:2497-2507
Jones RP, Gadd GM 1990 Ionic nutrition of yeast—the physiological mechanisms involved and applications for biotechnology. Enzyme Microb Tech 12:402-418.
Keswick, BH, Wang DA, Gerba CP 1982 The use of microorganisms as ground water tracers. Ground Water 20:142-149
Kihn JC, Dassargue CM, Mestdagh MM 1988 Preliminary ESR study of Mn(II) retention by the yeast *Saccharomyces*. Can J Microbiol 34:1230-1234
Kiselov PN, Kashkin KP, Boltaks B, Vitovskaya GA 1961 Acquisition of radio resistant by microbe cells inhabiting media with increases levels of natural radiation. Microbiol (English transl) 30:197-198
Klinkhammer GP, Palmer MR 1991 Uranium in the oceans: where it goes and why. Geochim Cosmochim Acta 55:1799-1806
Krueger S, Olson GJ, Johnsonbaugh D, Beveridge, TJ 1993 Characterization of binding of gallium, platinum, and uranium to *Pseudomonas fluorescences* by small-angle x-ray scattering and transmission electron microscopy. Appl Environ Microbiol 59:4056-4064
Landa ER, Gray JR 1995 US Geological survey research on the environmental fate of uranium mining and milling wastes. Environ Geol 26:19-31
Langmuir I 1918 The adsorption of gases on plane surfaces of galss, mica, and platinum. J Am Chem Soc 40:1361-1403
Langmuir D 1979 Techniques of estimating thermodynamic properties for some aqueous complexes of geochemical interst. *In* Chemical Modeling in Aqueous Systems. EA Jenne (Ed) Am Chem Soc, p 353-387
Langmuir D 1997 Aqueous Environmental Goechemistry. Prentice-Hall, Engelwood Cliffs, New Jersey.
Lappin-Scott HM, Costerton JW 1990 Starvation and penetration of bacteria in soils and rocks. Experimentia 46:807-812
Leduc LG, Ferroni GD, Trevors JT 1997 Resistance to heavy metals in different stains of *Thiobacillus ferooxidans*. World J Microbiol Biotechnol 13:453-455
Lee RE 1980 Phycology. Cambridge Unversity Press, New York

LeGall J, Fauque G 1988 Dissimilatory reduction of sulfur compounds. *In* Biology of Anaerobic Microorganisms. AJB Zehnder (Ed) John Wiley & Sons, New York, p 587-639

Lloyd JR, Young LE, Macaskie LE 1996 A novel PhophorLmager-based technique for monitoring the microbial reduction of technetium. Appl Environ Microbiol 62:578-582

Lloyd JR, Cole JA, Macaskie LE 1997a Reduction and removal of heptavalent technetium from solution by *Escherichia coli*. J Bacteriol 179:2014-2021

Lloyd JR, Harding CL, Macaskie LE 1997b Tc(VII) reduction and accumulation by immobilized cells of *Escherichia coli*. Biotechnol Bioeng 55:505-510

Lloyd JR, Ridley J, Khizniak T, Lyalikova N, Macaskie LE 1999 Reduction of Technetium by Desulfovibrio desulfuricans: Biocatalyst characterization and use in a flowthrough bioreactor. Appl Environ Microbiol 65:2691-2696

Lovely DRE, Phillips EJP 1992a Reduction of uranium by *Desulfovibrio desulfuricans*. Appl Environ Microbiol 58:850-856

Lovely DR, Giovannon SJ, White DC, Champine JE, Phillips EJP, Gorby YA and Goodwin S 1993a Geobacter metallireducers gen. Nov. sp., a microorganism capable of coupling the complete oxidation of organic compounds to the reduction ofiron and other metals. Arch Microbiol 159:336-344

Lovely DR, Phillips EJP, Gorby YA, Landa ER 1991 Microbial reduction of uranium. Nature 350:413-416

Lovely DR, Phillips JP 1992b Bioremediation of uranium contamination with enzymatic uranium reduction. Environ Sci Technol 26:2228-2234

Lovely DR, Roden EE, Phillips EJP, Woodward JC 1993c Enzymatic iron and uranium reduction by sulfate-reducing bacteria. Marine Geol 113:41-53

Lovely DR, Widman PK, Woodward JC, Phillips EJP 1993b Reduction of uranium by Cytochrome c3 of *Desulfovibrio desulfuricans*. Appl Environ Microbiol 59:3572-3576

Macaskie LE 1990 An immobilized cell bioprocess for the removal of heavy metals from aqueous flows. J Chem Technol Biotechnol 49:357-379

Macaskie LE 1991 The application of biotechnology to the treatment of wates produced from the nuclear fuel cycle: Biodegradation and bioaccumulation as a means of treating radionuclide-containing streams. Crit Rev Biotechnol 11:41-112

Macaskie LE, Blackmore JD, Empson RM 1988 Phosphatase overproduction and enhanced uranium accumulation by a stable mutant of a *Citrobacter sp.* Isolated by a novel method. FEMS Microbil Lett 55:157-162

Macaskie LE, Dean ACR 1985 Uranium accumulation by immobilized cells of *Citrobacter sp.* Biotechnol Lett 7:457-462

Macaskie LE, Dean ACR 1990 Uranium accumulation by immobilized biofilms of a *Citrobacter sp.* *In* Biosorption of Heavy Metals. Volesky B (Ed) CRC Press, Boca Raton, Florida, p 199-248

Macaskie LE, Empson RM, Cheetham AK, Grey CPA, Skarnuli J 1992 Uranium Bioaccumulation by a *Citrobacter sp.* As a Result of Enzymatically Mediated Growth of Polycrystalline HUO_2PO_4. Science 257:782-784

Macaskie LE, Blackmore JD, Empson RM 1988 Phosphate overproduction and enhanced uranium accumulation by a stable mutant of a Citrobacter sp. isolated by a novel method. FEMS Microbiol Lett 55:157-132

Macaskie LE, Jeong BC, Tolley MR 1994 Enzymatically accelerated biomineralization of heavy metals: Application to the removal of americium and plutonium from aqueous flows. FEMS Microbiol Rev 14:351-368

Madigan MT, Martinko JM, Parker J 1997 Biology of microorganisms. Prentice Hall, Englewood Cliffs, New Jersey

Mann H, Fyfe WS 1985 Uranium uptake by algae: experimental and natural environments. Can J Earth Sci 22:1899-1903

Marokovic M, Parkovic N, Pavkovic ND 1988 Precipitation of ammonium uranyl phosphate trihydrate-solubility and structural comparison with alkali uranyl (2+) phophates. J Res Natl Bureau Stand USA 93:557-563

Marques AM, Bonet R, Simon-Pujol MD, Fuste MC, Congregado F 1990 Removal of uranium by an exopolysaccharide from *Psedomonas sp*. Appl Microbiol Biotechnol 34:429-431

Marques AM, Roca X, Simon-Pujol MD, Fuste MC, Congregado F 1991 Uranium Accumulation by *Pseudomonas sp.* EPS-5028, Appl Microbiol Biotechnol 35:406-410.

Martell AE, Smith RM 1977 Critical stability constants. III: Other oraganic ligands. Plenum Press, New York

Matsuyama A, Okazawa Y, Arai H, Mijune M 1973 Isolation of radio resistant vegetative bacteria form the high-background radio active area. Sci Pap Inst Phys Chem Pes (Jap) 67:34-40

McCarthy JF, Zachara JM 1989 Subsurface transport of contaminats. Environ Sci Technol 23:496-502

McLean RJC, Frotin D, Brown DA 1996 Microbial metal-binding mechanisms and their relation to nuclear waste disposal. Can J Microbiol 42:392-400

McLean L, Purvis OW, Williamson BJ, Bailey EH 1998 Role fro lichen melanins in uranium remediation. Nature 391:649-650

Mehra RK, Winge DR 1991 Metal ion resistance in fungi. Molecular mechanisms and their regulated expression. J Cell Biochem 45:30-40

Michel LJ, Macaskie LE, Dean ACR 1986 Cadmium accumulation by immobilized cells of a Citrobacter sp. using various phosphate donors. Biotechnol Bioeng 28:1358-1365

Miller CL, Landa ER, Updegraff DM 1987 Ecological Aspects of Microorganisms Inhabiting Uranium Mill Tailings. Microb Ecol 14:141-155.

Milodowski AE, West JM, Pearce JM, Hyslop EK, Basham IR, Hooker PJ 1990 Uranium-mineralized microorganims associated with uraniferous hydrocarbons in southwest Scotland. Nature 347:465-467

Mowll JL, Gadd GM 1984 Cadmium uptake by *Aureobasidium pullulans*. J Gen Microbiol 130:279-284

Mullen MD, Wolf DC, Ferris FG, Beveridge TJ, Flemming CA, Bailey GW 1989 Bacterial sorption of heavy metals. Appl Environ Microbiol 55:3143-3149

Muraleedharan TR, Venkobachar C 1990 Mechanism of biosorption of Copper (II) by *Ganoderma lucidum*. Biotechnol Bioeng 35:320-325

Nakashima S, Disna JR, Perruchot A, Trichert J 1984 Eqperimental study of mechansoims of fixation and reduction of uranium by sedimentary organic matter under diagenesis or hydrothermal conditions. Geochem Cosmochem Acta 48:2321-2329

Nakajima A, Sakaguchi T 1986 Selective accumulation of heavy metals by microorganisms. Appl Microbiol Biotechnol 24:59-64

Nakajima A, Sakaguchi T 1993 Accumulation of uranium by basidomycetes. Appl Microbiol Biotechnol 38:574-578

Nakajima A, Horikoshi T, Sakaguchi T 1979 Studies on the accumulation of heavy metal elements in biological systems V. Ion effects on the uptake of uranium by *Chlorella regularis*. Agric Biol Chem 43:1461-1466

Nash JT, Granger HC, Adams SS 1981 Geology and concepts of genesis of important types of uranium deposits. Econ Geol 75:63-116

Nash TH 1996 Introduction. *In* Lichen Biology. Nash TH (Ed) Cambridge Univ Press, New York, p 1-7

Nieboer E, Richardson DHS 1980 The replacement of the nondesc ript term: heavy metals by a biologically and chemically significant classfication of metal ions. Environ Pollut (Ser B) 1:3-26

Nikaido H, Vaara M 1987 Outer membrane. In *Escherichia coil* and *Salmonella typhimurium* Cellular and Molecular Biology. FC Neidhardt (Eds) Am Soc Microbiol, Washington, DC, p 7-22

Norberg A, Persson H 1984 Accumulation of heavy-metals ions by *Zoogloea ramigera*. Biotech Bioeng 26:239-246

Nordgren A, Baath E, Soderstrom B 1983 microfungi and microbial activity along a heavy metal gradient. Appl Environ Microbiol 45:1829-1837

Ochiai EI 1977 Bioinorganic Chemistry. Allyn and Bacon, Boston

Ochiai EI 1987 General Principles of Biochemistry of the Elements. Plenum Press, New York

Okorokov LA, Lichko LP, Kulaev IS 1980 Vascuoles: main compartments of pottasium, magnesium and phosphate ions in *Saccharomyces carlsbergensis* cells. J Bacteriol 144:661-665

Otto JA, Zielinski RA 1985 Movement and concentration of uranium in young, oraganicrich sediments, stevens country, Washinton. *In* Concentration Mechanisms of Uranium in Geologic Environments. Int'l Meeting Oct 2-5, 1985, Soc fr Mineral Cristallogr, Center for Research on the Geology of Uranium, Nancy, France, p 49-52

Panak P, Kutschke S, Selenska-Pobell S, Geipel G, Bernhard G, Nitsche H 1998 Bacteria from uranium mining wate pile: Interaction with U(VI). Abstracts Euroconference on Bacteria-Metal/Radionuclide Interaction: Basic Research and Bioremediation, p 87-89

Parkhurst DL 1995 users guide to PHREEQC-A computer program for speciation, reaction-path, advective-transport, and inverse geochemical calculations. U S Geol Survey Water Resources Inv Rept 80-96

Pearson RG 1963 Hard and soft acids bases. J Am Chem Soc 84:3533-3539

Peberdy JF 1990 Fungal cell wall—a review. *In* Biochemistry of cell walls and membranes in fungi. Huhn PJ, Trinci APJ, Jung MJ, Coosey MW, Copping LE (Eds) Springer-Verlag, Berlin, p 5-30.

Pedersen K 1996 Investigations of subterranean bacteria in deep crystalline bedrock and their importance for the disposal of nuclear waste. Can J Microbiol 42:382-400

Pedersen K, Arlinger J, hallbeck L, Pettersson C 1996 Diversity and distribution of subterranean bacteria in groundwater at Oklo in Gabon, Africa, as determined by 16S rRNA gene sequencing. Molecul Ecol 5:427-436

Percival E, Mcdowell RH 1967 Chemstry and enzymology of marine algal polysacharides. Academic Press, New York.

Percival E, Mcdowell RH 1981 Algal walls—Composition and biosynthesis. *In* Plant Carbohydrates I. Tanner W, Loewus FA (Eds) Springer-Verlag Berlin, p 277-316

Peveling E 1969 Elektronenoptische Untersuchungen an Flechten: IV. Die Fenstruktur einer Flechten mit Cyanophyceen-Phycobioten. Protoplasma (Wien) 68:209-222

Phillips EJP, Lovely DR, Landa ER 1995 Remediation of uranium contaminated soils with bicarbonate extraction and microbial U(VI) reduction. J Ind Microbiol 14:203-207

Pietzsch K, Hard BC, Babel W 1998 A simple method for determination of the ability of sulphate reducing bacteria to reduce U(VI). Abstracts of Eurocongerence on Bacteria-Metal/Radionuclide Interaction: Basic Research and Bioremediation, p 91-92

Plette ACC, Van Riemsdijk WH, Benedetti MF, van der Wal A 1995 pH dependent charging behavior of isolated cell walls of a gram-positive soil bacterium. J Coll Inter Sci 173:354-363

Plummer EJ, Macaskie LE 1990 Actinide and Lanthanum toxicity towards a *Citrobacter sp.*: Uptake of lanthanum and a strategy for the biological treatment of liquid wastes containing plutonium. Bull Environ Contam Toxicol 44:173-180

Premuzic ET, Francis AJ, Lin M, Schubert J 1985 Induced formationn of chelating agents by *Psudomonas aeruginosa* grown in the presence of thorium and uranium Arch Environ Contam Toxicol 14:759-768

Pribil S, Marvan P 1976 Accumulation of uranium by the chlorococcal alga *Scenedesmum quadricauda*. Arch Hydrobiol Suppl 49 Algological Studies 14:214-225

Puers C, Selenska-Pobell S 1998 Investigation of bacterial diversity in a soil sample of a depleted uranium mining area nearby Johanngeorgenstadt, Saxonia, via 16S-RDNA-Sequence Analysis. Abstracts of Euroconference on Bacteria-Metal/ Radionuclide Interaction: Basic Research and Bioremediation, p 85

Rai LC, Gaur JP, Kumar HD 1981 Phycology and heavy metal pollution. Biol Rev Camb Philos Soc 56:99-151

Rome LD, Gadd GM 1987 Copper adsorption by *Rhizopus arrhizus*, *Cladosporium resinae* and *Penicillium italicum*. Appl Microbiol Biotechnol 26:84-90

Roomans GM 1980 Localization of divalent cations in phosphate-rich cytoplasmic granules in yeast. Phsiol Plant 48:47-50

Rothstein A, Meier R 1951 Relationship of the cell surface to metabolism VI. The chemical nature of uranium-complexing groups of the cell surface. J Cell Comp Physiol 38:245-270

Saiz-Jimenez C, Shafizadeh F 1984 Iron and copper binding by fungal phenolic polymers: An electron spin resonance study. Curr Microbiol 10:281-286

Sakaguchi T, Nakajima A 1981 Adsorption of uranium by chitin phosphate and chitosan phosphate. Agric Biol Chem 45:2191-2195

Sakaguchi T, Nakajima A 1987 Accumulation of uranium by biopigments. J Chem Technol Biotechnol 40:133-141

Sakaguchi T 1996 Bioaccumulation of Uranium. Kyushu Univ Press, Hukuoka, Japan, p 61-95

Sakaguchi T 1998 Removal and recovery of uranium by using microorganimsms. Abstracts of Euroconference on Bacteria-Metal/Radionuclide Interaction: Basic research and Bioremediation; 7-9

Scharer JM, Byerley JJ 1989 Aspects of uranium adsorption by microorganisms. Hydrometallurgy. 21:319-329

Schippers A, Hallmann R, Wentzien S, Sand W 1995 Microbial deversity in uranium mine waste heaps. Appl Environ Microbiol 61:2930-2935

Schrenk MO, Edwards KJ, Goodman RM, Harmers RJ, Banfield JF 1998 Distribution of *Thiobacillus ferooxidans* and *leptospirillum ferrooxidans*: Implication for generation of acid mice dranage. Science 279:1519-1522

Senesi N, Sposito G, Martin JP 1987 Copper (II) and iron (III) complexication by humic acid-like polymers (melanins) from soil fungi. Sci Total Environ 62:241-252

Siegel BZ and Siegel SM 1973 The cehmical compostion of algal cell walls, CRC Critical reviews in Microbiology, Chemical Rubber Company, Cleveland, p 1-26.

Silver M 1987 Distribution of iron-oxidizing bacteria in the Nordic Uranium Tailings Deposit, Elliot Lake, Ontario, Canada. Appl Environ Microbiol 53:8486-852

Silver S 1997 The bacterial view of the periodic table: Specific functions for all elements. *In* Geomicrobiology: Interactions Between Microbes and Minerals. JF Banfield Nealson KH (Eds) Rev Mineral 35:345-360

Strachan LF, Jeong BC, Macaskie LE 1991 Radiotolerance of phosphatases of a *Citrobacter sp.*: potential for the use of this organism in the treatment of wates containing radiotoxic actinides. Proc 21st Mtg Am Chemical Soc Symposium: Biotechnology for Wastewater Treatment 31:128-131

Strandberg GW, Shumate SE, Parrott JR 1981 Microbial Cells as Biosorbents for Heavy Metals: Accumulation of Uranium by *Saccharomyces cerevisiae* and *Pseudomonas aeruginosa*. Appl Environ Microbiol 41:237-245

Sutherland JW 1972 Biosynthesis of microbial exopolysaccharides. Adv Microb Physiol 23:79

Sutherland IW 1985 biosynthesis and composition of Gram-negative bacterial extracellular and wall polysaccharides. Ann Rev Microbiol 39:243-270

Tobin JM, Cooper DG, Neufeld RJ 1984 Uptake of metal ions by *Rhizopus arrizus* biomass. Appl Environ Microbiol 47: 821-824.

Tobin JM, Cooper DG, Neufeld RJ 1986 Influence of aions on metal adsorption by *Rhizopus arrizus* biomass. Biotechnol Bioeng 30:882-886

Treen-Sears ME, Volesky B, Neufeld RJ 1984 Ion exchange/complexation of the uranyl ion by *Rhizopus* biosorbent. Biotechnol Bioeng 26:1323-1329

Tripathi VS 1983 Uranium(VI) transporttion modelling: geochemical data and submodels. PhD Dissertation, Stanford Univ, Palo Alto, California, p 1-291

Troy FA, Koffler H 1969 The chemistry and molecular architecture of the cell walls of *Penicillium chrysogenum*. J Biol Chem 244:5563-5576

Tsezos M, Georgousis Z, Remoudaki E 1997 Mechanism of Aluminum Interference on uranium biosorption by *Rhizopus arrhizus*. Biotechnol Bioeng 55:16-27

Tsezos M, Volesky B 1981 Biosorption of Uranium and Thorium. Biotechnol Bioeng 23:583-604.

Tsezos M, Volesky B 1982 The mechanisms of uranium biosorption by *Rhizopus arrhizus*. Biotechnol Bioeng 24:385-401

Tsezos M, Georgousis Z, Remoudaki E 1997 Mechanism of aluminum interference on uranium biosorption by *Rhizopus arrhizus*. Biotechnol Bioeng 55:16-27

Tuovinen OH, Kelly DP 1974a Studies on the growth of *Thiobacillus ferrooxidans* II. Toxicity of uranium to growing cultures and tolerance conferred by mutation, other metal cations nad EDTA. Archiv Microbiol 95:153-164

Tuovinen OH, Kelly DP 1974b Studies on the growth of *Thiobacillus ferrooxidans* III. Influence of uranium, other metal ions and 2:4-dinitrophenol on ferrous iron oxidation and carbon dioxide fixation by cell suspensions. Archiv Microbiol 95:165-180

Updegraff, D.M. and Douros, J.D. 1972 The relationship of microorganisms to uranium deposits. Dev Ind Microbiol 13:76-90

US Department of Energy 1993 Annual status report on the uranim mill tailings remedial action project. US Department of Energy Report DOE/RW-0006. Rev. 9

US Department of Energy 1994 Integrated data base for 1993. US spent fuel and radioactive waste inventories, projections, and characteristics. US Department of Energy Report DOE/RW-0006. Rev. 9

Veeh HH 1967 Deposition of uranium from the ocean. Earth Planet Sci Lett 3:145-150

Volesky B, May H, Holan ZR 1993 Communications to the Editor Cadmium biosorption by *Saccharomyces*. Biotechnol Bioeng 41:826-829

Volesky B, May-Philips HA 1995 Biosorption of heavy metals by *Saccharomyces cervisiae*. Appl Microbiol Biotechnol 42:797-806

Volesky B, Holan ZR 1995 Biosorption of heavy metals. Biotechnol Prog 11:235-250

Wada Y, Ohsumi Y, Tanifuji M, Kasai M, Anraku Y 1987 Vacuolar channel of the yeast *Saccharomyces cerevisiae*. J Biol Chem 262:17460-17263

Waite TD, Davis JA, Payne, TE, Waychunas G, Xu N 1994 Uranium (VI) adsorption to ferrihydrite: Application of a surface complexation model. Geochim Cosmochim Acta 58:5465-5478

Warren LA, Ferris FG 1998 Continuum between sorption and precipitation of Fe (III) on microbial surfaces. Environ Sci Technol 32:2331-2337

Wessel JGH, Sietsma JH 1981 Fungal cell walls: A survey. In Plant Carbohydrates II: Extracellular carbohydrates. W Tanner, FA Loewus (Eds) Springer-Verlag, Berlin, p 352-394

West JM, McKinley IG, Vialta A 1992 Microbiological analysis at the Pocos de Caldas natural analogue study sites. J Geochem Explor 45:439-449

White C, Gadd GM 1987 The uptake and cellular distribution of zinc in *Saccharomyces cerevisiae*. J Gen Microbiol 133:727-737

Wiemken A, SchellenbergM, Urech K 1979 Vacuoles: the sole compartments of digestive enzymes in yeast (*Saccharomyces cerevisiae*). Archiv Microbiol 123:23-35

Williams ST, Mcneilly T, Wellington EMH 1977 The decomposition of vegetation growing on metal mine waste. Soil Biol Biochem 9:271-275

Willingham TO, Nagy B, Nagy LA, Krinsley DH, Mossman DJ 1985 Uranium-bearing stratiform organic matter in paleoplacers of the lower Huronian Supergroup, Eliot Lake-Blind River region, Canada. Can J Earth Sci 22:1930-1944

Wood SA 1996 The role of humic substances in the transport and fixation of metals of economic interest (Au, Pt, Pd, U, V). Ore Geol Rev 11:1-31

Xue H-B, Stumm W, Sigg L 1988 The binding of heavy metals to algal surfaces. Wat Res 22:917-926

Yano K, Cho B, Yoshinaka T, Yamaguchi H 1975 Survey on radiation resistant asporogenic bacteriain natural environments. In Radiation for a Clean Environment. IAEA-SM-194/202, Vienna, p 85-98

Young P, Macaskie LE 1995 Role of citrate as a complexing ligand which permits enzymatically-mediated uranyl ion bioaccumulation. Bull Environ Contam Toxicol 54:892-899

Yoshinaka T, Yano K, Yamaguchi H 1973 isolation of a highly radio resistant bacteria, *Arthrobacter radiotolerans nov. sp.* Agric Biol Chem 37:2269-2275

9

Uranium Contamination in the Subsurface: Characterization and Remediation

Abdesselam Abdelouas
Center for Radioactive Waste Management
The University of New Mexico
Albuquerque, New Mexico 87131

Werner Lutze and H. Eric Nuttall
Department of Chemical and Nuclear Engineering
The University of New Mexico
Albuquerque, New Mexico 87131

INTRODUCTION

The purpose of U mining is to produce fuel for civilian nuclear power plants and for military programs. Enrichment from pitchblende ores, *in situ* solution mining, and processing of phosphate and V mining byproducts are important procedures used to recover U from minerals. Extraction of U ore from Earth's crust, milling and chemical processing to prepare a U concentrate, known as yellow cake (U_3O_8), are accompanied by the production of large amounts of solid and liquid residues from which heavy metals are often readily leachable by rain, with subsequent contamination of groundwater. These elements, including U and daughter products of U decay chains, are hazardous and may occur in the groundwater in concentrations exceeding groundwater and drinking water protection standards.

Most U mining and subsequent milling of ores resulted in large volumes of radioactive sand-like residues called mill tailings. These tailings contain radioactive elements such as U, Th, Ra, Rn, and non-radioactive heavy metals in low concentrations. Mill tailings typically contain about 85% of the radioactivity present in unprocessed ore. If not properly disposed of, U mill tailings constitute a potential hazard to public health for very long periods of time. Recently, concerns about the hazards related to U mill tailings have been reiterated at an international conference on U mill tailings hosted by the United States Department of Energy (D.O.E. actual month, 1997). Argentina, Australia, Brazil, Canada, Germany, Republic of South Africa, and the United States are wrestling with clean-up and environmental restoration issues associated with U mining and milling. For example, in the United States more than 230 million tons of U mill tailings are stored at mill sites (Morrisson and Spangler 1992). There are 24 inactive U mill tailings sites located in 10 states (Smythe et al. 1995). In Canada, the total tailings accumulation will reach 300 million tons by the end of the century (Al-Hachimi 1992).

In the United States, besides mining and milling of U, processes related to commercial fuel manufacture and the production of nuclear weapons such as reprocessing of spent fuel, fuel fabrication, reactor operations, component fabrication and weapons testing have led to release of radioactive and other hazardous materials into the environment, including groundwater (Linking Legacies 1997). However, the current review focuses on wastes from mining and milling of U only.

Inappropriate conditioning and disposal of tailings waste permit the contaminants to spread into the air, soil, sediment, surface water, and groundwater. Under aerobic and anaerobic conditions, dissolution or immobilization of U is affected or can be affected by one or more of the following processes (Francis 1990; Francis et al. 1991):

- changes in pH and E_H that result in changes of the speciation and/or oxidation state;
- complexation such as chelation by siderophore and other microbial products, and by inorganic species such as carbonate and phosphate;
- bioaccumulation, movement and release due to remineralization elsewhere in the environment.

The extent and direction of U migration are determined by factors such as the rate and direction of groundwater flow, and the extent to which mineral assemblages of host rocks adsorb U from solution (Yanase et al. 1995; Sato et al. 1997).

Uranium deposits can be weathered naturally by oxidizing groundwater flowing through fractured rock and infiltrating from the surface (e.g. Otton et al. 1989; Yanase et al. 1995). Otton et al. (1989) studied uraniferous Holocene sediments from the Caeson range of Nevada and California (U.S.A.) and showed that water associated with these sediments can reach concentrations up to 177 µg/L U. These waters affect private and public water supply systems. Yanase et al. (1995) showed that at the Koongarra U deposit in Australia, U had migrated over distances of approximately 200 m over a time period of 1-1.5 million years.

Francis et al. (1991) reported that the predominant mechanism of dissolution of U from ores is oxidation. The oxidant Fe(III) is produced from Fe(II) in the ore by *Thiobacillus ferrooxidans* (Tuovinen 1986; Muñoz et al. 1995a). Ferric Fe oxidizes UO_2 to UO_2^{2+}. A review of the role of microorganisms on sulfide mineral oxidation, e.g. pyrite, can be found in Nordstrom and Southam (1997). The bacterial catalysis of Fe(II) oxidation depends on several environmental conditions including temperature, pH, total dissolved solids, O concentration, ecology of commensal or predatory species, and hydrologic conditions. Along with other acidophilic Fe- or S-oxidizers, these bacteria commonly occur on surfaces of exposed ores and in acid solutions in U and other metal mines (Schippers et al. 1995; Pizarro et al. 1996).

Garcia Júnior (1993) investigated the potential of applying a bacterial leaching process to leach U from an ore from Figueira-PR, Brazil, under laboratory (shake flasks, column) and pilot scale (column and heap leaching) conditions. Laboratory experiments showed that 60% of U was leached in samples inoculated with *T. ferrooxidans* isolated from the Figueira mine waters, against only 30% in sterilized controls. In the larger scale experiments (column and heap leaching) with *T. ferrooxidans*, 50% of the initial U was leached. Muñoz et al. (1995b) conducted column experiments to leach U ore in the presence of microorganisms. The authors showed that *T. ferrooxidans* played a major role in ore disintegration and U dissolution. Bhatti et al. (1998) studied the oxidative dissolution of U from a uraninite-containing rock in acid sulfate solutions under test conditions relevant to U leach mines and acid mine waters. They showed that addition of *T. ferrooxidans* and $FeSO_4$ enhanced the oxidative dissolution of uraninite. $FeSO_4$ was added to support the redox cycle between U^{IV}/U^{VI} and Fe^{II}/Fe^{III} couples.

Nordstrom and Southam (1997) indicated that mine waste environments and acid mine waters support a wide diversity of autotrophic and heterotrophic bacteria. Autotrophic bacteria such as *T. ferrooxidans* can leak or excrete organic compounds, e.g. pyruvate, that may be utilized by heterotrophic bacteria (Shnaitman and Lundgren 1965; Dugan et al. 1970a,b; Johnson et al. 1979; Ledin and Pederson 1996). Also, dead autotrophic cells will provide more organic sustenance for heterotrophs. However

Nordstrom and Southam (1997) pointed out that the microbial ecology in acid mine waters generated by oxidation of sulfides is still poorly understood. Other types of bacteria are involved in the dissolution of U from rock such as granite, where U is generally present as oxide (Bloomfield and Kelso 1973; Berthlin and Munier-Lamy 1983). The solubilization of U from rocks (syenite and aplite) is due to the formation of soluble organo-uranyl compounds, and simple carboxylic and phenolic acids (Munier-Lamy and Berthlin 1987).

Because of the large volume of groundwater and soil contaminated with U, innovative, economically attractive remediation technologies are needed. To date, pump-and-treat is the most widely used technology. Frequently, this technology has been ineffective in permanently lowering contaminant concentrations in groundwater (Travis and Doty 1990). Moreover, the costs of pump-and-treat can be prohibitive. A recent study by Quinton et al. (1997) showed that groundwater cleanup technologies such as pump-and-treat, permeable reactive barriers with zero-valent Fe, and bio-barriers, are more expensive than intrinsic (*in situ*) bioremediation.

In countries where the annual rate of precipitation is higher than the evaporation rate, e.g. Germany and Canada, U tailings are frequently close to groundwater that connects with creeks, rivers and lakes. Leaching of U can contaminate large volumes of water. In countries with arid to semi-arid weather, e.g. Australia and the western United States, groundwater contamination is a serious problem as well because of the limited quantities of water available to water animals, for irrigation and as drinking water.

We begin this chapter with a brief description of remediation activities for mill tailings in the United States and Germany. Then we report on a case study of leaching of contaminants and the radiological impact of U mill tailings on the environment in Canada. This is followed by a brief review of technologies used in the past, and more recently, to cleanup groundwater contaminated with U. We then review literature on U reduction by microorganisms with applications to groundwater cleanup. If the process is applied *in situ* the term "*in situ* bioremediation" is used. The process involves a multitude of bio-geochemical reactions under anaerobic conditions in groundwater and with the host rock. In addition to U, other dissolved component species, e.g. O, nitrate, and sulfate, and solids in the host rock, e.g. Fe and Mn participate in oxidation-reduction reactions (e.g. Froelich et al. 1979; Luther 1995; Murray et al. 1995). *In situ* bioremediation of U is the primary emerging technology and will be dealt with in fair detail.

LEACHING OF CONTAMINANTS FROM U MILL TAILINGS

This section deals with U leaching from mill tailings into surface and groundwater. Tailings have different chemical compositions depending on the leachants used to extract U from the ore matrix. Depending on mineralogical and economic considerations, two processes, acid leaching with sulfuric or nitric acid, and alkaline leaching with carbonate, have been used.

Tailings are composed of two fractions: solid and aqueous. Proper waste management would keep the solid and liquid portions separate and confined in structures without access to groundwater. The liquid portion could be stored in lined ponds, but breakdown of dams, liners or flooding in the rainy seasons may lead to escape of contaminants. In this case, the vadose zone will be contaminated with hazardous elements, which can reach the water table eventually.

The solid portion of the tailings requires special treatment aimed at long-term confinement of contaminants. Generally, the host gangues of the tailings are not durable enough and weathering mobilizes residual U and other heavy metals. Erosion by wind is another pathway for transporting contaminated material away from the source. Appropriate coverage with layers of soil and rocks can eliminate this pathway and prevent exhalation of radioactive Rn. Deposition of tailings on impermeable strata such as clay, or the use of an engineered liner may limit seepage into the aquifer. At many milling and tailings sites, liners have not been used, and migration of contaminants into the groundwater has occurred.

After storage of tailings, chemical reactions will take place within the tailings pile. Sulfide oxidation and generation of sulfuric acid are the most important reactions. The intensity of acid generation depends on several parameters, including sulfide content, ferric Fe concentration, the population density of sulfide-oxidizing bacteria, and the availability of O. Iron sulfide (e.g. pyrite, FeS_2) and other sulfide materials (CuS, ZnS, PbS) undergo weathering to generate acid solutions, Fe and other dissolved metals. Pyrite can be oxidized by O or by ferric Fe. Luther (1987), Luther et al. (1991), Singer and Stumm (1970), Moses et al. (1987), and Moses and Herman (1991) showed that the most efficient environmental oxidant for pyrite is dissolved Fe(III), not O. The solubility of Fe(III) is very low in neutral and alkaline solutions; the concentrations of Fe(III) are very low and pyrite oxidation by Fe(III) can be neglected. However, the Fe(III) concentration increases with decreasing pH and is frequently high enough in mine and tailings waters to oxidize sulfide. The oxidation of pyrite by O and ferric Fe can be described by the following equations (Singer and Stumm 1970; Moses et al. 1987):

$$FeS_2 + 14\ Fe^{3+} + 8\ H_2O \rightarrow 15\ Fe^{2+} + 2\ SO_4^{2-} + 16\ H^+ \tag{1}$$

$$FeS_2 + 3.5\ O_2 + H_2O \rightarrow Fe^{2+} + 2\ SO_4^{2-} + 2\ H^+ \tag{2}$$

$$14\ Fe^{2+} + 3.5\ O_2 + 14\ H^+ \rightarrow 14\ Fe^{3+} + 7\ H_2O \tag{3}$$

Equations (1) and (2) produce acid. The oxidation of 1 mole of pyrite by Fe(III) produces 16 moles of H. The rate of pyrite oxidation is limited by the rate at which dissolved O oxidizes Fe(II) (Moses and Herman 1991). These authors showed that hydrated Fe(II) is first adsorbed on pyrite, and then oxidized to Fe(III) by dissolved O. Fe(III) is then reduced by S from pyrite. The oxidation rate of pyrite by Fe(III) may increase by 5 to 6 orders of magnitude due to catalytic activity of microorganisms such as *Thiobacillus ferrooxidans* (Environment Canada 1987). These microorganisms can oxidize Fe(II) to Fe(III) as well as ferrous and non-ferrous sulfides. Pyrite oxidation produces acid and thus enhances leaching of U and other hazardous elements such as Cd, Cr, and Zn from the tailings. The net acidity and thus Fe solubility are limited by the chemical properties of gangue minerals that may consume acid by ion-exchange processes and dissolution of minerals such as Ca and Mg carbonate. Conversely, an excess of carbonate in solution can increase the pH beyond the range of optimal growth of bacteria and reduce the catalytic efficiency.

Though most of the U has been extracted during processing of the ore, residual U can be dissolved by rain or groundwater. Junghans and Helling (1998) analyzed samples of U tailings in Germany using X-ray fluorescence and found U concentrations up to 275 mg/kg of tailings. This is much higher than in average uncontaminated soil (1 mg/kg). Junghans and Helling (1998) reported data on the chemical composition of groundwater flowing into the tailings, groundwater within the tailings, and water leaving the tailings (Table 1). Uranium concentrations in the water increased by more than two orders of magnitude as it passed through the tailings. An increase of the pH was explained by dissolution of carbonate-rich minerals, as suggested by an increase in bicarbonate concentration.

Table 1. Chemical composition of groundwaters in the vicinity of a U mill tailings in Germany. Data from Junghans and Helling (1998).

	Inflow groundwater	Tailings groundwater	Outflow groundwater
pH	5.2-5.7	6.8	7.6-7.8
E_H (mV)	446-594	279	374-459
Conductivity (μS/cm)	75-137	1120	827-1009
Bicarbonate (mg/l)	1.2-13.0	139.4	45.1-170.6
Sulfate (mg/L)	17.0-20.0	406.0	236.0-350.0
Iron (mg/L)	0.06	1.8	<0.01-0.08
U_{total} (μg/L)	0.6	707.1	259.8
^{226}Ra (Bq/L)	0.13	0.3	<0.001

Delaney et al. (1998) analyzed solid U tailings and tailings porewater in Canada and found concentrations of U up to 1.4 g/Kg and 0.5 mg/L, respectively. Brookins et al. (1993) studied the geochemical behavior of leachates of U mill tailings at Maybell, CO (USA). Uranium had been leached with sulfuric acid from the host rock. The authors showed that leachates from the tailings deteriorated the quality of the groundwater nearby. Analysis of tailings pore water showed high concentration of U (0.44-1 mg/L), Se (0.95-1.27 mg/L), Fe (380-560 mg/L), Mn (710-870 mg/L) and nitrate (400-420 mg/L). Total dissolved salts amounted to about 11.5-15.3 g/L. The sulfate concentration was high (8.1-11.1 g/L) and constituted at least 70% of the total dissolved salts. The pH was as low as 2.87-3.26. E_H ranged between 480-550 mV indicating oxidizing conditions.

Brookins et al. (1993) conducted column leach experiments to assess the mobility of elements on the surface of the tailings particles. Cations and anions in the water from the column were released in the following order of decreasing concentraion: SO_4 > NH_4 > Al > Mn > NO_3 > U > Fe > Se > PO_4 > Ni > As > Cd. Average concentrations of some species in the leachate of the column were: U = 2.72 mg/L, Se = 0.7 mg/L, Fe = 0.94 mg/L, Mn = 43.3 mg/L, NO_3 = 5.9 mg/L, SO_4 = 2.5 g/L. Al-Hachimi et al. (1993) studied the leachability of hazardous elements from U tailings using batch and column experiments. They showed that dissolution of pyrite from the tailings caused the pH of leaching water to drop to as low as 2.0. Under these acidic conditions, U as well as Th, V, As, Sb, and Ce were solubilized. The concentration of U in the eluate of the column was 3.0 to 8.0 mg/L. Concentrations of other hazardous elements were also elevated; that of Th reached 33 mg/L.

Fernandes et al. (1998) analyzed some pollutants in a acid drainage at Poços de Caldas, Brazil. The average concentrations of pollutants were: ^{238}U = 71 Bq/L, ^{226}Ra = 0.14 Bq/L, Al = 61 mg/L, and Mn = 6.6 mg/L; the pH was 2.9 because of pyrite oxidation. The authors suggested that, given the high concentration of pollutants including U, Al, Mn, and SO_4^{2-} and the low pH, remedial measures are necessary.

U(IV) is much less mobile than U(VI) in groundwater. However, U(IV) is readily oxidized to U(VI) if O is available as gaseous or dissolved species. Oxygen can reach the tailings either by diffusion or by infiltration with rain and groundwater. Oxidation of U(IV) by O is described by Equation (4)

$$UO_2 + 2 H^+ + 1/2 O_2 \rightarrow UO_2^{2+} + H_2O \tag{4}$$

with the reaction rate depending on pH and on complexing species concentrations, such as sulfate (SO_4^{2-}), bicarbonate/carbonate (HCO_3^-/CO_3^{2-}), phosphate ($H_2PO_4^-$), and organics (humic acid). The effect of pH is small between 5 and 10 (Parks and Pohl 1988), but in more acidic solutions (pH<5), the U(IV) dissolution reaction appears to be first order with respect to H concentration.

In the case of U solution mining with sulfuric acid, sulfate complexation will take place. The $UO_2SO_4^0$ species can be significant up to pH = 7 (Langmuir 1978). If the waters become alkaline as a result of ion exchange with the rock matrix or by carbonate-rich mineral dissolution, U carbonate complexes such as $UO_2(CO_3)_2^{2-}$ will dominate. Uranium sulfate and carbonate complexes are soluble and migrate with the groundwater. Brookins et al. (1993) showed that U(VI) can be complexed by humic acids associated with sphagnum peat under acidic to alkaline pH conditions. This may retard U(VI) migration by adsorption of the complex U(VI)-humic acid onto aquifer material. More recently, Bernhard et al. (1998) studied U speciation by laser spectroscopy in three U-related waters from Saxony (Germany). These authors detected the following U species: (1) $Ca_2UO_2(CO_3)_3$(aq) at pH = 7.1 in carbonate- and Ca-containing mine water from Schelma; (2) $UO_2(CO_3)_3^{4-}$ at pH = 9.8 in carbonate-containing and Ca-poor tailings water from Helmsdorf; (3) UO_2SO_4(aq) at pH = 2.6 in sulfate-rich mine water from Konigstein.

Leaching and erosion of mine and mill tailings can cause contamination of groundwater as well as nearby rivers and lakes. This can lead to contamination of fish, or even plants if the water is used for irrigation. Air pollution may result from wind erosion. There are potentially multiple sources of hazard to human health. Uranium is a suspected human carcinogen and a known kidney chemotoxin (Cothern and Lappenbusch 1983). More data on U release and migration from the tailings are reported in the next section.

ENVIRONMENTAL IMPACT AND HEALTH RISKS OF U MILL TAILINGS

Environmental impact and health risks related to U mill tailings occur by two main processes. One is surface soil/water contamination by erosion and wind dispersion of radioactive material and air pollution by Rn emission. The other is contamination of the subsurface including groundwater due to leaching and leaking of radioactive and hazardous metals (Cd, Cu, Pb, Zn) from the tailings.

Suppression of erosion and Rn emission may be accomplished by covering the tailings with clays and/or synthetic materials. Radon is retarded long enough for the Rn to decay before it can reach the atmosphere. This type of environmental protection is technically less challenging than dealing with contaminated groundwater. Tailing and liquids associated with the U extraction process were usually dumped on the ground without construction of liners for prevention of seepage of contaminants into the ground.

In the next two subsections, we briefly describe work conducted in the framework of two major remediation projects carried out in the United States and in Germany. In the U.S. project, surface remediation of mill tailing sites in the western part of the country has just been completed, and subsurface remediation is now being addressed. In the Germany project, surface and subsurface remediation are being attempted at the same time. In the last part of this section, we will report on a project completed in Canada to characterize surface water contamination and related potential health risks.

Remediation work in the western United States: The UMTRA program

In the United States, the Uranium Mill Tailings Radiation Control Act (UMTRCA) of 1978 gave the U.S. Department of Energy (DOE) the responsibility for remediating 24 U mill tailings piles, and associated "vicinity properties" in ten states. The "vicinity properties" are structures located on tailings or built by using tailings, and open land contaminated with tailings. In 1988, the scope of DOE's U mill tailings remediation (UMTRA) project expanded to include groundwater remediation at these sites, to ensure that concentrations of residual radioactive materials in the groundwater meet any one of these three criteria:

- Background levels: defined by concentrations of constituents in nearby groundwater that was not contaminated by processing activities;
- Maximum concentration limits (MCL). Values for inorganic constituents applied to the UMTRA sites are given in Table 2 (Federal Register 1995).
- Alternate concentration limits (ACL) for hazardous constituents that do not pose a substantial hazard to human health and the environment, as long as the limit is not exceeded.

Table 2. Maximum concentrations of inorganic constituents for groundwater protection at UMTRA Project sites (Federal Register, 1995)

Constituents	Maximum concentration mg/L
As	0.05
Ba	1.0
Cd	0.01
Cr	0.05
Pb	0.05
Hg	0.002
Mo	0.1
NO_3	44.0
Se	0.01
Ag	0.05
^{226}Ra and ^{228}Ra	5 pCi/L
^{234}U and ^{238}U	0.044 mg/L (≈30 pCi/L)

Stabilization of U mill tailings and remediation of structures built with them have been completed. Some tailings piles were stabilized in place. Others were relocated to more favorable areas, and then stabilized. In all cases, the tailings are in disposal cells, with clay covers, resembling large municipal landfills. The UMTRA project tailings piles range up to about 4 million cubic meters in volume. As an example, we describe the surface remedial action that was initiated at the Tuba City site, Arizona in 1988 by D.O.E. At this site, the U mill tailings and other associated materials were removed and stabilized in an engineered disposal cell. The remedial action was completed in April 1990. A total of 1,100,000 m³ of contaminated materials was stabilized in a disposal cell covering 20 ha within the 59 ha disposal site. The tailings were disposed of at the Tuba City site by stabilization in place, meaning that the tailings were essentially left undisturbed instead of being moved to a new site. Surrounding contaminated soils due to windblown materials were also placed in the disposal cell. The tailings pile was shaped to allow the placement of progressively less contaminated materials into the final pile configuration. After placing all the contaminated materials into a pile, a clayey sand layer was compacted over the entire pile to reduce Rn emissions. This layer also reduces the infiltration of water and helps to protect groundwater from further contamination. A graded layer of durable rock was then placed over the pile to protect against erosion. Because the tailings were stabilized in place, no liner could be placed on the bottom of the cell to retard leakage. As the wet tailings consolidated under their own weight and additional overburden, contaminated pore water may have been forced out through the bottom of the cell, a phenomena termed transient drainage (D.O.E. 1994).

The flow rate of groundwater is very low at the Tuba City site, 15 m/yr. However, eventually the groundwater at Tuba City will reach the nearby Moenkopi Wash that is used for irrigation. Several groundwater remediation technologies have been evaluated for potential application. We will report on research and development work for a bioremediation process below.

Remediation work in Germany by WISMUT

By the time of cessation of U mining in the former East Germany in 1990, production totaled 220,000 tons of U, making the WISMUT company rank third among the world's U producers since 1946. These intensive mining operations left behind a number of environmental stresses of varying risk potential in the States of Saxony, Thuringia and Saxony-Anhalt (Morrison and Cahn 1991; Holdway 1992; Thein et al. 1993; Knoch-Weber and Mehner 1997). Visible landmarks left behind by past operations include a multitude of mines, waste dumps, tailings ponds, and a huge excavated open pit mine on the outskirts of Ronneburg in Thuringia. These pose various risk potentials and impacts on the environment. WISMUT was assigned the task of carrying out extensive environmental rehabilitation and cleanup of abandoned facilities and wastes. As the areas concerned are densely populated, the actual cleanup and remediation are preceded by comprehensive investigations of suspect areas and background surveys. The work program implemented in 1991 was based on the "Concept for the cleanup, closure, plugging, rehabilitation and recultivation of mining and milling facilities." WISMUT's work is focused on eliminating or mitigating contamination without impeding any long-term remediation option.

Environmental monitoring by WISMUT is organized by the company's fields of operation, which were previously in charge of mining and processing of U ores. The environmental monitoring includes: (1) releases of solid, fluid, gaseous and aerosol components (emissions). It also includes the monitoring of ambient air for radioactive material such as Rn and long-lived alpha emitters in breathing-zone air, dust, and of radionuclides in the recipient streams; (2) discharge of non-radioactive waters into recipient streams. A central environmental database was established in 1990. Each year some 200,000 water data and 700,000 air data are added.

A total surface area of 3,100 ha spread over five facilities is being remediated. This includes more than 2,100 ha of mine spoils and mill tailings. In addition, remediation includes safeguarding and plugging of approximately 1,400 km of open mine workings, the filling of more than 50 shafts and 30 other major outlets (e.g. adits, wells) as well as cleanup of contaminated plant areas, buildings, structures and transportation routes.

Backfilling of acid-generating waste is common practice in mine closure. Objectives and advantages of backfilling include:

- Consolidating multiple rock piles in a single location;

- Providing a presumably secure repository for acid generating mill tailings;
- Preventing development of acidic pit lakes.

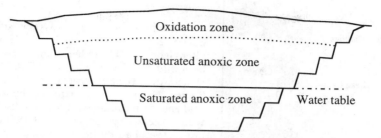

Figure 1. Conceptual zonation of backfilled open pit.

Within an open pit, three potential deposition zones can be identified in terms of O abundance (Fig. 1). From the bottom to the top: a saturated anoxic zone below the water table, an unsaturated and anoxic zone in the middle, and an oxidation zone above. The objective is to place the waste with the highest acid-generating capacity in the saturated anoxic zone to prevent contact of O with sulfide-rich minerals. The waste would be mixed with CaO to neutralize the acid and to reduce metal release. The oxidation zone in the top layer will receive waste with low acid-generating capacity and high neutralization capacity, i.e. high carbonate concentration (calcite and dolomite). The unsaturated anoxic zone in the middle will receive waste with moderate acid-generating capacity.

In 1990 the timeframe for complete remediation was estimated between 10 and 15 years. However, by 1997, this timeframe proved to be too short for conducting the remediation to its conclusion. Unequaled throughout the world in terms of scale, this environmental challenge will require a greater operational and regulatory time span than that estimated in 1990. The duration of the remedial period is now estimated to be 20 years, and will end in the year 2010. An estimated 7 billion EURO have been assigned to accomplish the remediation task.

Long-term remediation strategies will address the problem of subsurface cleanup, especially the groundwater, which up to now has not been addressed. However, evaluation of potential technologies, including for groundwater cleanup, are under way.

Mill tailings in Canada: A case study

In this subsection we report on a study by Veska and Eaton (1991) on environmental aspects and the radiological impact of radionuclide contamination from an abandoned U mill tailings site in the Northwest Territories of Canada. The authors studied the surface water and groundwater movement and the leachability of radionuclides from the tailings. The study includes an evaluation of potential health risks related to the site by estimating external and internal radiation exposure.

General environment and mining activities. The site studied by Veska and Eaton (1991) is located at Rayrock on the Sherman Lake-Marian River system in the Northwest Territories of Canada. The area comprises a large number of lakes, rocky outcroppings, and bogs. Poplar, birch, and coniferous trees are abundant. Though the area has no resident population, the mine buildings have been used periodically by the Dene Indian Nation as hunting and fishing shelters. The climate is subarctic and semi-arid. Ice melting starts in late May and contributes to surface and groundwater recharge. A schematic of

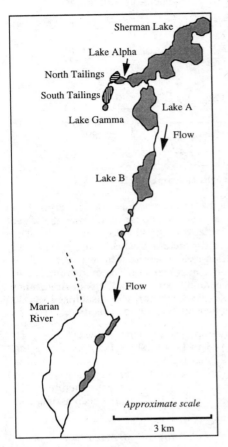

Figure 2. Site location (adapted from Veska and Eaton 1991).

the site is given in Figure 2. The figure shows that the drains of the tailings piles can reach the Sherman Lake through connected natural water systems.

Uranium ore in the Rayrock site was mined between 1957 and 1959. The U ore was pitchblende (UO_2). Other minerals such as hematite (Fe_2O_3), pyrite (FeS_2), and chalcopyrite ($CuFeS_2$) occurred in the ore bodies. Uranium was extracted by sulfuric acid and the resulting tailings were discharged without neutralization. About 70,000 tons of ore were milled during the 2-year period of operation. Two piles of tailings (north and south) were created and then abandoned, covering a surface area of about 62,000 m^2 (Fig. 2).

Piezometers were installed in the north tailings pile and in the surrounding area to measure water table elevations and to estimate direction and velocity of groundwater flow. Laboratory experiments included chemical analyses of water, tailings, soil, and leaching of tailings columns. The column experiments were conducted to evaluate the leachability of radionuclides from the tailings under acidic and neutral conditions.

Water and soil. Hydrological studies of the site showed that the recharge waters entered the northern tailings from the west, south, and east, converging toward the middle of the tailings, and discharge into lake Alpha shown in Figure 3. pH, U and ^{226}Ra concentrations of surface and groundwater are given in Table 3. The groundwater sampled in the tailings had pH values between 3.9 and 7.1. Concentrations of U ranged between 81 and 2800 µg/L inside the tailings but were only 19 µg/L in water located outside the tailings pile (piezometer E). The ^{226}Ra concentrations ranged between 1.3 and 14.0 Bq/l. The concentration in the piezometer at the edge of Lake Alpha and in the lake were close, 3.4 and 3.6 Bq/l, respectively.

Surface water in Alpha Creek contained 30 µg/L U. The U concentration decreased downstream and was only 4 µg/L for Lake Alpha, 5 µg/L for Sherman Lake, and insignificant in the Marian River (0.6 µg/L). The highest ^{226}Ra concentration was observed in Alpha Creek (20 Bq/l), and decreased drastically in the water further downstream (Table 3).

Four stratigraphical layers were identified in the tailings from top to bottom: Coarse tailings, fine tailings, peat, and clay. The thickness of the tailings layers were each about 50 cm. Chemical analyses showed that all types of soil contained small amounts of sulfide. The coarse and fine tailings contained 0.06 wt % and 0.03 wt % sulfide,

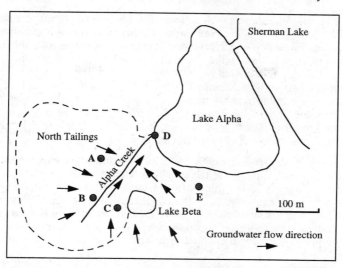

Figure 3. Water table map of the north tailings area (adapted from Veska and Eaton 1991).

Respectively. Though the sulfide concentration was low, its dissolution probably caused an increase of acidity at least locally, as suggested by a blue coloring resulting from the oxidation of $CuFeS_2$. The coarse and fine tailings also contained 65 and 67 mg/kg of U, and 10 Bq/g and 49 Bq/g of ^{226}Ra, respectively. Radionuclide concentration in the underlying clay were low, 11 mg/kg U, 0.3 Bq/g ^{226}Ra. This suggested that the peat layer (containing 180 mg/kg of U and 33 Bq/g of ^{226}Ra) located between the tailings and clay is very efficient in retarding radionuclide migration downward.

Table 3. U and ^{226}Ra concentrations and pH in surface and groundwaters in the proximity of the Rayrock site. Data from Veska and Eaton (1991).

Water sample	pH	U (μg/L)	^{226}Ra (Bq/L)
Groundwater			
A (on tailings)	4.7	81	7.6
B (on tailings)	3.9	2800	7.0
C (on tailings)	7.1	110	14.0
E (edge of tailings and Lake Alpha)	6.2	86	3.4
D (off tailings)	5.0	19	0.3
Surface Water			
Alpha Creek	7.6	30	20
Lake Alpha	7.1	4	3.6
Sherman Lake	7.1	5	0.23
Lake A*	n.r.	≤0.5	0.005
Marian River**	n.r.	0.5-0.8	≤0.4

*Soniassay et al. (1986); **Kalin (1984); n.r. not reported

Leaching of tailings columns with deionized water or a 10^{-3} N sulfuric acid solution showed significant differences between the coarse and fine tailings. Leachate from the coarse tailings was lower in pH and had less ^{226}Ra than the leachate from the fine tailings. The ^{226}Ra concentration in the leachate from the coarse tailings was similar to that in the water samples collected outside but near the tailings site (0.03-0.1 Bq/l). The pH of the leachate from the fine tailings, leached with deionized water, did not change. ^{226}Ra concentrations were similar to those measured in groundwater collected on the tailings site (Table 3). The high surface area of the fine tailings relative to the coarse tailings probably offered more sites with Ra to be leached in the experiments. Furthermore, the fine tailings contained more ^{226}Ra (49 Bq/l) than the coarse (10 Bq/l).

Evaluation of health risks. External and internal pathways of exposure were used to assess the risks from potential use of the site. The external exposures to γ radiation were determined from a survey of the mine site, the tailings and the surrounding area. The internal exposures were evaluated from the concentrations of radionuclides in drinking water and in fish. The calculations were conducted for an individual who spends from 1 day to 1 month on or near the tailings mine site, who eats fish from the Sherman Lake as the only supply of meat for the whole year, and who drinks water from the Marian River.

The calculated effective whole-body dose-equivalent rates at various sites at Rayrock ranged from 0.6 to 73 mSv/y. As expected, the dose rate was very high on the tailings and ranged between 42 and 73 mSv/y. Residence on the tailings for 1 day would result in a whole-body dose of 0.1-0.2 mSv. The dose during continuous residence for one month would be 3.4-6 mSv, about twice the annual dose from average natural radiation in Canada.

The measured concentrations of U, ^{226}Ra, and ^{210}Pb in the water were used to estimate the internal dose from consumption of water. Furthermore, these radionuclides were measured in fish from the lakes surrounding the site. The internal dose from water intake was determined assuming a drinking water intake of 2 L/d from the Rayrock site for 1 day and for 1 month, and for a year for the Marian River. For example, an individual could accumulate a dose exceeding the limit established for public exposure in Canada only by drinking 2 L/d of water for a whole year from Alpha Creek, where ^{226}Ra contributes the most to γ radiation. Consumption of 260 g fish by an individual per day would increase the dose by 12% over that from drinking water from Alpha Creek.

Veska and Eaton (1991) concluded that groundwater is flowing from the tailings to the nearby lake-river system. Though the peat located between the tailings and the underlying clay immobilized a high fraction of contaminants leached from the tailings, a fraction of these elements was transported into the nearby lakes through surface and groundwater. Groundwater flowing through the tailings was contaminated with U with concentrations exceeding the levels for groundwater protection in Canada (100 µg/L) and in the U.S. (44 µg/L). Local acidification probably enhanced the leaching of the tailings. Measurements of radionuclides in the surrounding water system including streams, lakes and rivers showed that concentrations were significantly elevated only in the Alpha Creek. Drinking of 2 L/d from the Alpha Creek would be necessary to acquire a dose exceeding acceptable values in Canada.

No toxicity testing was conducted by Veska and Eaton (1991). However, Rippon and Riley (1996) studied the environmental impact of tailings, located in the Northern Territory, Australia, using toxicity testing protocols. Rippon and Riley (1996) simulated the effect of biological toxicity of tailings in the laboratory. They used washed tailings fines alone and a mixture of washed tailings and natural floodplain sediment in the toxicity tests. Background water was added to tailings/floodplain sediment mixtures with and without aeration to simulate diverse conditions of deposition and mixing of tailings with water. Rippon and Riley (1996) concluded that the non-radiological toxicity of the tailings is low. Experiments with tailings/floodplain sediment with a ratio of 0.40 under anoxic conditions showed an increase of Mn concentrations in solution. Elevated Mn concentrations were correlated with the low population growth of certain organisms (*Hydra viridissima*) and the reduction of the brood size of others (*Moinodaphnia macleayi*). A mixture of tailings/floodplain sediment with low ratio (i.e. 0.10) led to increase of pH in the water ($\approx$ 7) and neutralization of acidity from the floodplain sediment with an initial pH = 4.8. The increase of pH resulted in an increase in population growth of *H. viridissima* relative to the experiment using only the floodplain sediment.

Summary. There is risk of health hazard from U tailings. Radioisotopes of U, Ra, Pb, and other hazardous metals and chemical species such as Mn, Fe, Cu, and nitrate can reach the human body through various pathways, in particular surface and groundwater. The extent of leaching of tailings and dispersion of hazardous materials depends on the origin, treatment, and disposal of tailings (nature of ore, acid and/or alkaline treatment, post-treatment such as neutralization), and also on the climate (temperature, annual precipitation, etc.). In many cases, e.g. in the vicinity of U mill tailings in the western U.S., one or more toxic species are found to exceed concentrations set by groundwater protection standards (e.g. 44 µg/L for U). Values of up to 1 mg of U per liter are not unusual (D.O.E. 1992).

TECHNIQUES FOR U REMEDIATION

Various methods have been suggested for the removal of U from contaminated sites. New technologies are being developed and tested as better, faster and cheaper methods are sought. Combined chemical and biological treatment, as well as phyto-remediation, have been suggested. *In situ* methods have received increased attention compared with pump-and-treat technology. Avoidance or minimization of secondary waste is one way to make technologies economically more attractive and environmentally more compatible.

Innovative or emerging processes

There are several new processes that combine chemical and biological treatment, such as the process developed at Brookhaven National Laboratory by Francis and Dodge (1998). In this process, U is extracted from the soil followed by treatment of the waste stream by bacteria. Francis and Dodge (U.S. patent) have a process for extracting U from soil using water-soluble metal citrate complexes followed by biodegradation and photodegradation. The metals and/or radionuclides are separated by photochemical treatment of the solution in the presence of air. More than 99% of the U extracted from soil was recovered as stabilized $UO_3 \cdot xH_2O$. The technology has been successfully tested in the laboratory to treat contaminated soils and sludge from two different Department of Energy sites. A pilot scale field demonstration for treating incinerator bottom residues is currently under way.

Cornish et al. (1995) suggested the use of selected plants that hyperaccumulate heavy metals 100 times that of conventional plants. This may lead to a low-cost technology for remediation of extensive areas with lightly to moderately contaminated soils. This technology, known as phytoremediation, is particularly suited in areas of land or in water where contaminants of low concentration are widespread. In 1994, a screening test for U-accumulating plants was started at the D.O.E.'s Fernald, Ohio, site and at an abandoned U mine near Clancy, Montana. Modest uptake of U (up to 10 mg U/Kg leaf and stalk biomass) was observed using red clover at Fernald and tansy mustard at Clancy. Application of phytoremediation could be expanded to treat contaminated water using aquatic plants. A wide range of aquatic plants is being considered for treatment of contaminated water, including submerged rooted aquatic plants, floating aquatic plants, and microalgae. Phytoremediaton is still at an early stage of development, but preliminary results are encouraging.

Another emerging technology tested for U removal from groundwater is the application of permeable barriers containing zero-valent Fe (Fe^0). Permeable barriers can be implemented by digging a trench in the flow path of a contaminant plume and backfilling the trench with Fe^0, or injecting a suspension of Fe^0 colloids to create a reducing zone (Cantrell et al. 1995; Kaplan et al. 1996; Cantrell and Kaplan 1997; Cantrell et al. 1997; Gu et al. 1998; Abdelouas et al. 1999a). In a bench-scale test, metallic Fe was found to remove U(VI) effectively by a multitude of mechanisms including reduction/precipitation and adsorption (Cantrell et al. 1995). Gu et al. (1998) showed that nearly 100% of U was removed through reaction with Fe^0 at an initial concentration up to 76 mM and only a small percentage (<4%) of UO_2^{2+} appeared to be adsorbed on the corrosion products of Fe^0 (Fe^{3+}-oxyhydroxides). Abdelouas et al. (1999a) showed that the pH (between 4 and 9) did not significantly affect the reduction rate of U by Fe^0. They found that the U precipitate was poorly crystalline hydrated UO_2 (Fig. 4). The U.S. Environment Protection Agency, in cooperation with D.O.E., are evaluating this technology at the Rocky Flats site in Colorado. There, the groundwater contains both nitrate and U. The technology is also being tested at another D.O.E. site at Oak Ridge. There, a permeable reactive trench and a funnel-and-gate reactive barrier medium are being evaluated. A treatment train with removable cassettes has been designed, constructed, and installed in the funnel and gate, with Fe^0 as the primary treatment medium. Fe^0 is also capable of degrading chlorinated hydrocarbons, MoO_4^-, TcO_4^-, and CrO_4^{2-} (Cantrell et al. 1995), and NO_3^- (Abdelouas et al., unpublished).

Morrison and Spangler (1992) tested 24 industrial materials for use in chemical barriers for U mill tailings remediation. These materials included Fe oxyhydroxides, hydrated lime, fly ash, zeolite, phosphate, Ba chloride, and many others. Some materials such as Ba chloride or Ca phosphate were effective in removing 99% of U in solution with 30 mg/L U, while others such as ferrous sulfate, ferric oxyhydroxide, hematite, Ti oxide, or Ba chloride extracted more than

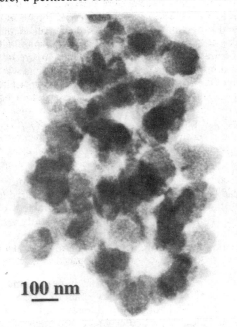

Figure 4. Uraninite formed upon the reduction of U(VI) in the presence of metal iron particles (Fe^0).

96% of the Mo (initial concentration of 8.9 mg/L). Removal from solution was caused by both precipitation (as Ca uranate, Ca molybdate, ferrous molybdate, or Ba molybdate) and sorption (on ferric oxyhydroxide, hematite, Ca phosphate, or Ti oxide). The authors suggested the use of a combination of two or more materials as the chemical barriers to remove effectively both U and Mo.

Another technology being developed by D.O.E. is *in situ* redox manipulation (ISRM), which relies on the fact that unconfined aquifers are usually oxidizing environments. Many contaminants are mobile only under oxidizing conditions. Reducing an aquifer's redox potential allows a variety of redox-sensitive contaminants to be treated. The goal of ISRM methods is to create a permeable treatment zone in the subsurface to remediate these redox-sensitive contaminants. Redox sensitive contaminants migrating through the manipulated zone are immobilized (precipitated) or chemically destroyed. The ISRM approach may be applied to immobilize inorganic contaminants and destroy organic contaminants in groundwater that can not be readily treated by other methods because the contaminants are dispersed over large areas or located tens of meters below the surface. The permeable treatment zone is created downstream in the contaminant plume by injecting appropriate reagents or microbes to reduce Fe^{3+} to Fe^{2+} in the clay of the soil. This reducing zone can also be created by the injection of colloidal Fe or reduced colloidal clays.

The ISRM technology was tested at the Hanford 100H area (Fruchter et al. 1996; Scott et al. 1998). During testing, twenty-one thousand gallons of buffered Na dithionite solution were injected into the unconfined aquifer at the Hanford 100H Area ISRM site. There was no plugging of the well screen or the formation during the test. Dithionite was detected in monitoring wells at least 7.5 m from the injection point. Approximately ninety percent of the reaction product and unreacted reagent were then withdrawn from the aquifer. Preliminary core data showed that, as predicted, 60 to 100% of the Fe in the clays was reduced by dithionite. The reductive capacity was measured by O consumption on cores recovered after the reagent was injected. The measured reductive capacity of the treated soil would be equivalent to 51-85 pore volumes of contaminated water to be transported through the treated zone within 7 to 12 years. These estimates assumed groundwater containing 1 mg/L hexavalent Cr and 9 mg/L dissolved O flowing at a rate of 0.3 m per day. If additional treatment capacity were required to meet target cleanup levels, the treatment zone could be extended or reinjected with dithionite once the available treatment capacity had been expended. The lifetime of the barrier is determined by the accessible, i. E., reducible Fe and the efficiency of the emplacement. Groundwater in the injection zone remained anoxic 18 months after the injection. Hexavalent Cr concentration remained below detection limits, and total Cr levels declined to near detection limits. The authors suggested that the technology could be applied to U remediation. Scott et al. (1998) estimated that their technology is at most half the cost of pump-and-treat.

Conventional technologies: Pump-and-treat

Pump-and-treat processing of U-contaminated groundwater involves extraction of contaminated water, followed by separation process on the surface. Separation processes include but are not limited to:

- ion exchange,
- reverse osmosis,
- bioremediation (reduction vs. adsorption and bioaccumulation),
- reductive precipitation of U (chemical precipitation).

Uranium can readily be removed from solution by ion exchange; in fact this method is a part of the process used in the extraction of U from ore (Merritt 1971). The exchange process usually uses macroporous or gel-type cation-exchange resins (Rastogi et al. 1997; Clifford and Zhang 1994). However, it is difficult to reduce U to drinking water limits (on the order of 0.044 mg/L) using ion exchange. Reverse osmosis has become a common water treatment technology, and is often used in home units to provide high-quality drinking water. Reverse osmosis uses a semi-permeable membrane and pressure to separate water molecules from U(VI) and other contaminants. Large-scale processes have been developed to desalinate seawater using this technique, and a few cities in the world receive all of their drinking water from this process, which is also referred to as hyperfiltration and ultrafiltration (Hiemenz 1986).

Biological processes for removing U from water include bio-sorption, bio-accumulation and bio-reduction. Biosorption can be defined as uptake of U by microorganisms by physico-chemical mechanisms, such as adsorption or ion exchange. The role of microorganisms in the environmental fate of radionuclides and metals is reviewed by Brierley (1991) and Gadd (1997). Bio-sorption technologies have been developed during the last decade, mainly to remove U and other metals (Cd, Cu, Zn, Fe, Pb, Ag) from waste streams and for above-ground treatments (Tsezos et al. 1989, 1997; Golab et al. 1991; Guibal et al. 1992; Volesky 1994; Hu et al. 1996; Bouby et al. 1996; Hu and Reeves 1997; Bustard and Mchale 1998; Bustard et al. 1998; Vecchio et al. 1998). Contaminants can be adsorbed onto the microorganism cell wall surfaces and cell envelopes. Biomass, such as generated as by-products of industrial fermentations and certain metal-binding algae found in large quantities in the sea could be immobilized in polymers or other matrices and used to remediate water contaminated with metals, including U. One of the advantages of biosorption is that it does not require living cells, and thus problems such as contaminants toxicity can be avoided. However, biosorption is not suitable for *in situ* application where clogging of pores with biomass can occur. Uranium can also be accumulated on biomass (e.g. *Citrobacter sp.*) as a result of precipitation with enzymatically liberated inorganic phosphate (HUO_2PO_4) (Macaskie et al. 1992, 1996, 1997; Roig et al. 1995, 1997; Basnakova et al. 1998; Finlay et al. 1999). This process is also called bio-accumulation. *Citrobacter* species were immobil-ized in a polyacrylamide gel or colonized on glass helices and supplied with citrate and glycerol 2-phosphate. 1 mM of uranyl nitrate $[UO_2(NO_3)_2]$ was added to the immobilized bacteria. Up to 90-99% of the U was removed.

Smith et al. (1994) stated that biological treatment technologies for metals remediation are in their infancy. A similar statement could also be made for the application of phytoremediation to the treatment of metals. Today there is emphasis on *in situ* processes because they are cheaper, as demonstrated by Quinton et al. (1997).

The trend in development of remediation technologies is away from pump-and-treat processes, in favor of *in situ* applications. These processes tend to be cheaper and environmentally more compatible. Travis and Doty (1990) pointed out that contaminated aquifers can not be cleaned sufficiently by the pump-and-treat method. The effectiveness of groundwater pumping can be predicted by groundwater transport models (Hall 1988; Mackay 1989; Mackay and Cherry 1989). These authors calculated that continuous pumping for as long as 100 to 200 years may be needed in order to lower contaminant concentrations by a factor of 100, assuming the ideal condition of a totally dissolved contaminant in a homogenous aquifer. Recent experience with pumping contaminated aquifers at 19 sites for 10 years (EPA 1989) showed significant mass removal of contam-inants but little success in reducing concentrations to the target levels. The typical effect of pumping is an initial drop in concentrations by a factor of 2 to 10, followed by little or no further decline. Once the pumps are turned off, contaminant concentrations rise.

MICROBIAL REDUCTION OF URANIUM

Bioremediation of heavy metals and radioelements such as U is receiving increasing interest worldwide. A simplified diagram of the U cycle in nature, including human activities, is given in Figure 5. *In situ* bioremediation of U fits into this cycle by enhancing the reduction of excess U(VI) created by man.

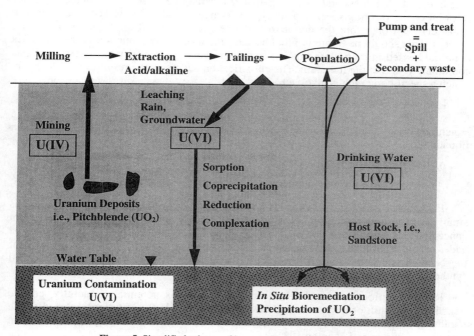

Figure 5. Simplified scheme of human effect on U cycle in nature.

Reduction of uranium

The predominant valence states of U in nature are U(VI) and U(IV) (Langmuir 1978). U(VI) is readily soluble in water under various conditions, with solubility enhanced by aqueous complexation with carbonate, phosphate, organic, and other species. Reduction of U(VI) in anoxic marine sediments is the most important sink of dissolved U (Cochran et al. 1986; Klinkhammer and Palmer 1991). Reduction of U(VI) in the subsurface environment led to the formation of some U ore deposits (Jensen 1958; Hosteler and Garrels 1962; Taylor 1979; Maynard 1983). U(VI) can also be reduced by microorganisms in U waste (Francis et al. 1991; Dodge and Francis 1997; Francis 1998). In nature, the activity of microorganisms can control removal or release of U in pore water (Barnes and Cochran 1993). U(IV) forms an insoluble oxide, uraninite, that is stable under reducing conditions. The solubility is on the order of 10^{-6} g/L between pH 2 and pH 10 (Parks and Pohl 1988). Uraninite nominally UO_2, is the principal ore mineral in many deposits (Rich et al. 1977; Kimberley 1979).

Reduction of uranium by microorganisms

Redox reactions and the role of microbes. The function of microbes in mediating redox processes is analogous to that of a catalyst as the transfer of electrons between an acceptor and a donor is facilitated, increasing reaction rates by several orders of magnitude. Therefore, microbes or their enzymes do not reduce or oxidize redox-sensitive elements. They are not a reaction product of a redox process. Rather, as catalysts, microbes affect the reaction rate, not the free energy of a reaction. However, microbes use some of the free energy of reaction for their metabolic processes. Chemical reactions that are not spontaneous (negative ΔG) cannot be catalyzed by microbes. After sorption of the reacting species, a specific enzyme(s) is activated to accept and release (transfer) electrons from the donor to the acceptor, in case of an oxidation-reduction process. This should be borne in mind because the mechanistic function of the microbes is not addressed in their classification as sulfate reducers, denitrifying bacteria, etc., as commonly found in the literature. The classification addresses the result of the process in which a specific microbe is involved.

Redox reactions are not catalyzed in the absence of certain essential elements such as C, N and P. These elements are needed for cell synthesis. As an example, the stoichiometry of denitrification with acetic acid is shown together with cell synthesis in Equation (5).

$$0.712\ CH_3COOH + NO_3^- =$$
$$0.485\ N_2 + 0.03\ C_5H_7O_2N + 0.273\ CO_2 + HCO_3^- + 0.817\ H_2O \quad (5)$$

Phosphorus is not shown and was not measured because the stoichiometric factor is small compared with that of N. Phosphorus and N must be present in sufficient concentrations in the water that contains the metal, e.g. U, to be reduced, or they must be added, together with an organic substrate (reductant). Equation (5) shows that 3 mol % of the N in nitrate and about 10 mol % of the C from the substrate are used for cell synthesis. These percent values depend on substrate.

Redox reactions are conditioned on a certain redox potential. Generally, pϵ (the negative logarithm of a hypothetical electron concentration in solution), is used as the parameter to quantify the redox intensity or potential range in which a given reaction can take place (Stumm and Morgan 1981).

A species of concern for remediation, e.g. $UO_2(CO_3)_2^{2-}$, is rarely the only redox-sensitive constituent in a contaminated aquifer. Other redox reactions may be catalyzed by microbes as well. These reactions are consecutive or simultaneous, and may be interdependent. They may be catalyzed by the same or different microbes. Separation of certain desirable redox reactions from others is not always possible. As an example, reduction of U in a solution that contains sulfate, leads to formation of H_2S or HS^-, depending on pH. The sequence of reactions with decreasing pϵ is given below.

Sequence of microbially catalyzed reduction reactions (p$\epsilon°$; 25°C and pH = 7):

$1/4\ O_2\ (g) + H^+ + e^- = 1/2\ H_2O$ $\qquad$ p$\epsilon°$ = 13.75

$1/5\ NO_3^- + 6/5\ H^+ + e^- = 1/10\ N_2\ (g) + 3/5\ H_2O$ $\qquad$ p$\epsilon°$ = 12.65

$1/2\ MnO_2\ (s) + 1/2\ HCO_3^-\ (10^{-3}) + 3/2\ H^+ + e^- = 1/2\ MnCO_3(s) + H_2O$ $\qquad$ p$\epsilon°$ = 8.9

$1/2\ UO_2\ (CO_3)_2^{2-} + H^+ + e^- = 1/2\ UO_2(s) + HCO_3^-\ (10^{-3})$ $\qquad$ p$\epsilon°$ = 4.9

$FeOOH(s) + HCO_3^-(10^{-3}) + 2\ H^+ + e^- = FeCO_3(s) + 2\ H_2O$ $\qquad$ p$\epsilon°$ = -0.8

$1/8\ SO_4^{2-} + 9/8\ H^+ + e^- = 1/8\ HS^- + 1/2\ H_2O$ $p\varepsilon° = -3.75$

Convenient organic substrates for reaction are acetate, ethanol, and lactate. Using ethanol, the number of electrons needed per reaction in the above list is provided by the following oxidation reaction, Equation (6):

$$1/12\ C_2H_5OH + 1/4\ H_2O = 1/6\ CO_2 + H^+ + e^- \qquad (6)$$

When oxidizing conditions prevail in a closed system that contains some or all of the redox-sensitive species on the left side of the equations, and the appropriate microbial activity, after amendment with an organic substrate, dissolved O is reduced first, followed by denitrification, etc., until sulfate is reduced to sulfide. The reactions are catalyzed by different microbial enzymes. There is a succession of microbial activity from denitrifiers, to metal reducers, to sulfate reducers. Denitrification must be complete before sulfate reducers are activated. At the end of denitrification the $p\varepsilon$ value is about −2, low enough for sulfate reduction to begin.

A literature review. Microorganisms can reduce U indirectly by producing H_2S or H_2 in the course of other processes (abiotic reduction) or directly using their enzymes (enzymatic reduction). The first microorganisms identified to enzymatically reduce U(VI) were the dissimilatory Fe(III)-reducing microorganisms, *Geobacter metallireducens* and *Shewanella putrefaciens* (Lovley et al. 1991). These microorganisms used U as an electron acceptor and H_2 or acetate as an electron donor to support growth, and tolerated U(VI) concentrations as high as 8 mM. Several authors studied the enzymatic reduction of U(VI) by various pure or mixed cultures of microorganisms, including metal- and sulfate-reducing bacteria under a variety of conditions (Kauffman et al. 1986; Lovley et al. 1991; Francis et al. 1991; Gorby and Lovley 1992; Lovley and Phillips 1992a,b; Barnes and Cochran 1993; Lovley et al. 1993a,b; Francis et al. 1994; Lovley 1995a,b; Phillips et al. 1995; Barton et al. 1996; Tucker et al. 1996, 1998a,b; Uhrie et al. 1996; Hard and Babel 1997; Ganesh et al. 1997; Abdelouas et al. 1998a, b).

Kauffman et al. (1986) studied U(VI) reduction in the presence of indigenous sulfate-reducing bacteria of the genus *Desulfovibrio*. The water and soil samples originated from a U mine in the Ambrosia Lake region in New Mexico. After addition of sucrose to the groundwater/soil, U concentration decreased from 1 mg/L to less than 0.1 mg/L. The decrease of U concentration was accompanied by a decrease in sulfate concentration and the production of H_2S.

Lovley et al. (1991) showed for the first time the enzymatic reduction of U(VI) in the presence of Fe-reducing bacteria (strain GS-15 or *Geobacter metallireducens* and *Shewanella putrefaciens*). The authors showed that these microorganisms can grow by coupling the oxidation of acetate, formate, lactate, pyruvate or H_2 to the reduction of U(VI). They concluded that enzymatic reduction of U(VI) can be much faster than commonly cited abiological mechanisms for U(VI) reduction (e.g. reduction by sulfide). Lovley et al. (1991) suggested the use of bacteria as a method for remediation of environments contaminated with U.

Francis et al. (1991) studied the reduction of U in sediment samples contaminated with U processing wastes. The sediments were contaminated with U and toxic metals such as As, Cd, Cr, Hg, and Zn. The sediment was amended with glucose and ammonium chloride dissolved in deionized water in serum bottles and incubated at 24°C. No bacteria were inoculated in the bottles. The microbial activity was demonstrated by gas production (CO_2, H_2, and CH_4) and by the reduction of U concentration in the solution over 50 days. The concentration of U (as uranyl carbonate) in the sediments decreased as well. Francis et al. (1991) concluded that the reduction of U was due to the activity of indigenous microbes under anaerobic conditions. While U(VI) was reduced and precipitated, Fe(III) and Mn(IV) were reduced and dissolved.

Gorby and Lovley (1992) extended the work of Lovley et al. (1991) on U reduction by microorganisms by identifying the U precipitate formed. These authors used uncontaminated groundwater spiked with uranyl acetate (0.4-1 mM) and Na bicarbonate (30 mM). Acetate was the electron donor. The groundwater was inoculated with the Fe(III)-reducing bacteria (GS-15 or *G. metallireducens*), from an enriched medium. As expected, U(VI) was reduced by GS-15 within a few hours. The precipitate was identified as uraninite (UO_2) by X-ray diffraction and transmission electron microscopy. The authors concluded that the enzymatic reduction of U(VI) by microorganisms can be used as a technique for U(VI) removal from groundwater.

Lovley and Phillips (1992a,b) used sulfate-reducing bacteria to reduce U(VI). They also tested the effect of toxic metals on the growth of bacteria and reduction of U(VI). A pure culture of sulfate-reducing bacteria, *Desulfovibrio desulfuricans*, was used to couple the oxidation of lactate to the reduction of U(VI) with concentrations up to 24 mM. Among a variety of anions (SO_4^{2-}, MoO_4^{2-}, NO_3^-) and toxic metals (Cu^{2+}, Zn^{2+}, Co^{2+}, Mn^{2+}), only high concentrations of Cu (> 100 mM) inhibited U(VI) reduction. Lovley and Phillips (1992a) also showed that U(VI) was reduced by *D. desulfuricans* in mine drainage waters at pH = 4.0. The authors concluded that enzymatic reduction of U is more efficient than biosorption. The authors suggested that enzymatic reduction of U(VI) could be used to remove dissolved U from surface water, groundwater, and waste streams, and the process could offer advantages over others.

Lovley et al. (1993a) investigated the reduction of U(VI) by several strains of sulfate-reducing bacteria. They found that only *Desulfovibrio* species reduced U(VI) to U(IV), and led to precipitation of uraninite. *Desulfovibrio* species also reduced Fe(III) from a 10 mM solution. Addition of 1 or 10 mM of molybdate did not inhibit the reduction of Fe(III) by *D. desulfuricans*. However, none of these species can conserve enough energy from U(VI) and Fe(III) reduction to support growth.

Lovley et al. (1993b) investigated the mechanism for U(VI) reduction by *Desulfovibrio vulgaris* (Hildenborough). They showed that cytochrome *c3*, extracted from this bacterium, reduced U(VI) in the presence of excess hydrogenase and H_2. The cytochrome reductase, which is located in the soluble fraction of the periplasmic region, catalyzed the reductive precipitation of U. Lovley et al. (1993b) also showed that cytochrome *c3* can be reduced chemically by dithionite ($S_2O_4^{2-}$), and the reduced cytochrome *c3* can reduce U(VI) in the absence of hydrogenase and H_2, with a high efficiency. These results suggest that developing a remediation strategy for U and other metals that is based upon the separation and use of enzymes is quite possible, and deserves further investigation.

Francis et al. (1994) extended research on microbial U reduction by studying the oxidation state of U reduced by *Clostridium sp.* The authors detected U(III) but concluded that this valence state is unstable and was rapidly oxidized to U(IV) in an aqueous medium. They also reported that higher concentration of U(VI) (240 µM) inhibited the growth of *Clostridium sp.*

Phillips et al. (1995) studied the reduction of U(VI) in contaminated soils collected from acid and alkaline tailings, as well as from a site contaminated with U as the result of test firing of artillery shells made of depleted U. Prior to reduction, U was dissolved with Na bicarbonate or nitric acid. U(VI) was then reduced in the leachate by *D. desulfuricans* using lactate as the electron donor. However, the reduction rate of U(VI) decreased at higher concentrations of carbonate (100 mM). A carbonate concentration of 30 mM was efficient in extracting most of the U from the soils, and still low enough to observe fast reduction of U(VI).

Lovley (1995a) reviewed studies concerning microbial reduction of U(VI) and concluded that it offers several advantages over currently applied technologies for U removal such as ion exchange and biosorption. The advantages include: 1) high removal rate of U per unit of biomass; 2) reduction and removal of highly soluble U(VI)-carbonate complexes; 3) precipitation of a highly concentrated waste form (UO_2). This concentrate may have a commercial value or, if disposed of, requires little storage space; 4) the potential of treating mixed waste with organic contaminants as electron donors for the reduction of U(VI); and 5) the potential for *in situ* remediation of groundwater.

Barton et al. (1996) investigated aspects of transformation of U(VI) to U(IV) by anaerobic bacteria important for U mill tailings. The authors used sulfate-reducing bacteria (*D. baculatus, D. gigas,* and *D. vulgaris*), and nitrate-reducing bacteria (*Pseudomonas sp.* and *P. putida*). They also investigated the reduction of U(VI) by mixed cultures of anaerobic bacteria in a sludge sample collected from the Albuquerque Municipal Sewage Treatment Facility. Groundwater samples from four U mill tailings sites in the U.S. were also tested for reduction of U(VI) by indigenous bacteria. Barton et al. (1996) found that all types of bacteria tested reduced U(VI) to U(IV), which precipitated as uraninite. More than 90% of the initial 1 mM of U(VI) was reduced with both pure and mixed cultures. Addition of selenate and vanadate had little effect on U(VI) reduction, whereas addition of arsenate and molybdate at 1 mM inhibited U(VI) reduction. The authors suggested the use of anaerobic bacteria as a technique for *in situ* removal of U from groundwater.

Uhrie et al. (1996) studied the removal of U, Se, As and V from aqueous solution by mixed populations of sulfate-reducing bacteria present in a sulfidogenically active sediment obtained from the Laramie River (Albany County, Wyoming). The water was amended with sulfate and lactate and other trace metals. Initial U concentrations ranged between 10 and 40 mg/L. The authors showed that U was completely reduced within 3 days and suggested the use of sulfate-reducing bacteria to remove U, As, Se and V from mine waters. In the case of sulfate-poor waters, sulfate-reducing bacteria can be easily activated by adding sulfate and organic C into the aquifer surrounding the mine zone.

Hard and Babel (1997) isolated several strains of sulfate-reducing bacteria from a waste water pond in Germany and a Cu mine in Norway. They tested reduction of sulfate at different pH and metal (U, Al, Cd, Cr, Cu, Pb, Ni, Mn, Ag) concentrations to simulate the composition of mine waters. All strains were found to use methanol as the electron donor and C source, but were also able to utilize lactate, pyruvate and acetate. One of the isolates grew at pH = 4.0, but all other strains did not grow below pH = 6.0. All strains were found to be tolerant of U at a concentration of 0.01 mM. Of the other metals, only high concentrations of Cd (10 mM) inhibited the growth of bacteria. The authors suggested the use of the isolated sulfate-reducing bacteria for the decontamination of acid U mine waters high in sulfates and metals.

Ganesh et al. (1997) investigated the effect of complexation of U(VI) with organic ligands on its reduction by a sulfate-reducing bacterium (*D. desulfuricans*) and an Fe-reducing bacterium (*Shewanella alga*). The reduction rate of U(VI) depended on the nature of the organic complex and the type of bacteria. The reduction by *D. desulfuricans* of U(VI) complexed with malonate, oxalate, and citrate was slower than that of U(VI) complexed with acetate or 4,5-dehydroxy-1,3-benzene disulfonic acid. The results with *S.*

alga were different than those with *D. desulfuricans*. In fact, *S. alga* reduced U(VI) more rapidly from malonate, oxalate, and citrate complexes than from acetate and the aromatic complexes. The authors concluded that the selection of bacteria for rapid U reduction must depend upon the organic composition of the waste stream.

Tucker et al. (1998a,b) studied the reduction of U(VI) by *Desulfovibrio desulfuricans* immobilized in polyacrylamide gels and in column reactors. Formate, lactate or H_2 served as the electron donor. Uranium removal efficiencies of 86-99% were achieved in the presence of high concentrations (~1 mM) of Mo, Se, and Cr. The authors suggested that the enzymatic reduction of U by immobilized cells of *D. desulfuricans* may be a practical method for removing these metals from solution in a biological reactor.

In summary, the literature on microbially mediated reduction of U provides the following information:

- U(VI) can be reduced to U(IV) by enzymatic activity of microorganisms including metal- and sulfate-reducing bacteria;
- U(VI) can be reduced either by pure cultures or by mixed cultures from waste sludge or from groundwater from U mill tailings sites;
- the reduced U precipitates as uraninite (UO_2);
- complexation of U(VI) with organic and inorganic ligands can inhibit its reduction by microorganisms. Furthermore, complexation of the reduced U(IV) may inhibit its precipitation;
- reduction of U(VI) by microorganisms appears to be more efficient than other techniques of U removal, although further understanding of the process is needed.

A laboratory study on bioremediation of uranium in groundwater from a mill tailings site

This section summarizes recent work by the chapter authors on U removal from groundwater using indigenous bacteria from a mill tailings site in the state of Arizona, U.S.A. Laboratory work was done to provide a database necessary to develop an *in situ* bioremediation process that took the local conditions into account. More details of this work can be found in a series of publications by Abdelouas et al. (1998a,b,c; 1999b).

The site. The site is located near Tuba City, Arizona (Fig. 6). A U mill was operated at the site by Rare Metals Corporation of America from 1956 until 1966. The mill processed approximately 800,000 tons of ore during the 10-year period. Between 1956 and 1962, sulfuric and nitric acid were used in an acid-leach process. From 1962 to 1966, Na carbonate was used in an alkaline-leach process. Mill tailings slurries were placed in three contiguous piles at the site, covering about 25 acres. Water used in processing was discharged into unlined ponds (Anon. 1993 and 1994). It is estimated that 6×10^5 m^3 of water contaminated with tailings metals seeped into the ground and reached the watertable. High concentrations of sulfate and nitrate are present in the groundwater in the vicinity of the site, and the concentration of U is up to 20 times the maximum concentration for groundwater protection in the United States (Federal Register, 1995).

The site is approximately 1550 m above sea level on a gently sloping terrace composed of unconsolidated dune sand and sediment gravel. The limited and highly variable supply of surface water in the semi-arid climate makes groundwater an important resource in the area. Two points of shallow groundwater withdrawal from the underlying Navajo Sandstone exist within 3.2 km of the site, including a low-yield domestic well and a spring that feeds into the Moenkopi Wash (Fig. 6), an intermittent stream that drains to

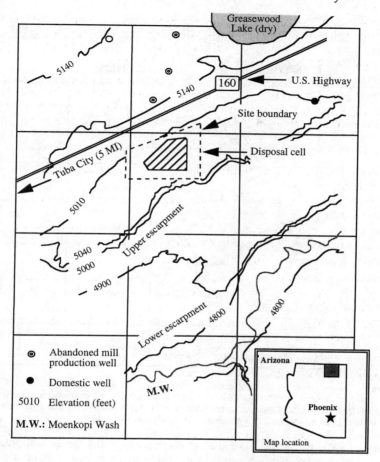

Figure 6. The Tuba City site location (adapted from D.O.E 1994).

The southwest into the Little Colorado River. Water from the Moenkopi Wash is used for stock watering and agricultural diversions by the Navajo and Hopi tribes in the vicinity of the site.

Groundwater. Groundwater for laboratory experiments was collected from several monitoring wells, located within the area of contaminated groundwater and outside the plume (Fig. 7). Water was pumped into plastic bottles that were placed in a nitrogen-flushed glove box in the field. Figures 8 to 10 show that the concentrations of U, nitrate, and sulfate in uncontaminated groundwater (well #915) remained practically constant with time, whereas the respective concentrations in the contaminated groundwater in the heart of the plume increased with time (well #906).

Contaminated groundwater was taken from well #926 and uncontaminated water was obtained from well #948. The *in situ* water temperature was 16°C and did not vary seasonally. The concentration of dissolved O, 3.1 mg/L in well #926, and 7.0 mg/L in well #948, was measured *in situ* using a YSI 5739 field probe. The pH, measured in the glove box, was 6.6 in well #926 and 7.7 in well #948. In the laboratory, the water was

456   URANIUM: Mineralogy, Geochemistry and the Environment

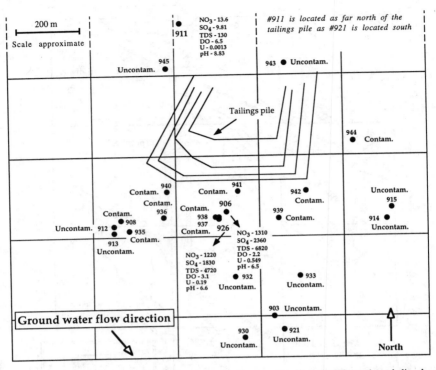

Figure 7. Concentrations in mg/L of certain contaminants, dissolved O (DO) and total dissolved salts (TDS) in groundwater at Tuba City site. Wells are identified by number; their locations are indicated by black dots. Contam.: contaminated; Uncontam. Uncontaminated. Reprinted by permission of Elsevier Science B.V. (Abdelouas et al. 1998c, Fig. 1, p. 345).

stored at 4°C. The pH of the water did not change during storage. In well #926, the U concentration exceeded the U.S. groundwater protection standard by more than 5 times (Table 4). Dissolved O concentrations indicated oxidizing conditions.

Indigenous bacteria. Cultures grown during oxidation-reduction experiments were analyzed. Two denitrifying bacterial species were identified; *Pseudomonas aeruginosa*

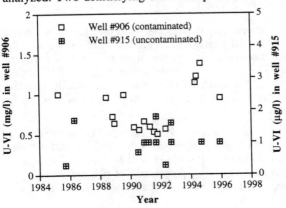

Figure 8. Variation of U concentration with time in well #906 (contaminated) and well #915 (uncontaminated).

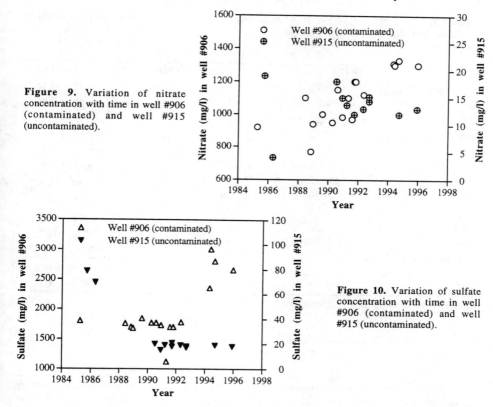

Figure 9. Variation of nitrate concentration with time in well #906 (contaminated) and well #915 (uncontaminated).

Figure 10. Variation of sulfate concentration with time in well #906 (contaminated) and well #915 (uncontaminated).

and *Pseudomonas stutzeri*, and one sulfate-reducing bacterium, *Shewanella putrefaciens*. *Shewanella putrefaciens* were reported by Lovley (1995b) to catalyze reduction of U(VI) in the presence of H_2. The initial concentration of sulfate-reducing bacteria in the groundwater was very low (average 1 cell/100 mL) compared to that of denitrifiers (450 cells/100 mL).

Host rock. The host rock collected at the Tuba City site was Navajo sandstone, a pure and well-sorted eolian quartz sand of Middle Jurassic age. A stratigraphic study of the Navajo sandstone was published by Peterson and Pipiringos (1979). Samples were taken from drilled cores in the groundwater-saturated zone and from the surface. The median grain size diameter (d_{50}) of the sand particles was 110 μm and the porosity ranges from 24 to 28%. Several samples from the contaminated (wells #911 and 932) and uncontaminated (wells #906, 937, and 940) area were analyzed mineralogically and geochemically. Quartz was the dominant phase. Microcline was present but was strongly altered. Clay minerals, including smectite and illite, formed a coating 10 μm thick on the quartz grains. Scanning electron microscope (SEM) images, in conjunction with energy dispersive X-ray (EDX) analyses, showed occasional Ca, probably as calcite, enriched in the clay coating. Iron oxide rims that imparted an orange-brown hue were observed on the quartz under the optical microscope. No differences in the phase composition of the contaminated and uncontaminated samples were seen.

Atomic absorption and inductively-coupled plasma atomic emission spectrometry

Table 4. Chemical composition of groundwaters from the mill tailings site near Tuba City, AZ (mg/L)

	Well #926	Well #948
U	0.250	0.0014
Alkalinity ($CaCO_3$)	572	88
Ca	681	28.2
Mg	329	6.2
Na	172	14
Sr	7.04	0.65
Fe	0.05	0.34
NO_3^-	1220	11.85
SO_4^{2-}	1830	9.56
Cl	88.4	6.9
TOC	4.0	5.0
TDS	4720	130
Temperature (°C)	16.8	17.1
Dissolved Oxygen	3.1	7.0
pH	6.63	7.70
U.S. ground water protection standards* (mg/L)		
U	0.044	
NO_3^-	44	

*Federal Register (1995)

Table 5. Chemical composition of Navajo sandstone

Oxide	(wt %)
SiO_2	95.18 ± 0.82
Al_2O_3	1.82 ± 0.44
MgO	0.23 ± 0.05
CaO	< 0.2
Fe_2O_3	0.05 ± 0.01
TiO_2	0.05 ± 0.01
MnO	0.05 ± 0.003
Na_2O	0.06 ± 0.01
K_2O	1.09 ± 0.22
P_2O_5	< 0.1
Loss	0.66 ± 0.13
Total	99.12 ± 0.72
	(mg/kg)
U	0.44 ± 0.08

(ICP-AES) were used to determine the chemical composition of the sandstone after digestion of the whole rock in nitric acid. Uranium was determined by ICP-mass spectrometry. The pristine sandstone and that in contact with the contaminated groundwater were analyzed for 45 trace elements. Average trace element concentrations were on the order of 1 mg/kg (Table 5).

Iron, presumably present as hematite and oxyhydroxide, was determined by selective leaching of the sandstone grains in a mixture of hydroxylamine-hydrochloride and acetic acid followed by analysis of the leachate. For a variety of samples from the contaminated and uncontaminated area (wells #906, 911, 932, 937, and 940; Fig. 7), the results for Fe_2O_3 ranged between 0.024 and 0.086 wt %, with an average of 0.040 wt % (Table 6). It is important to identify the Fe source because Fe(III) from sandstone can be reduced and dissolved by microorganisms (Lovley and Phillips 1992c; Frederickson and Gorby 1996; Ernsten et al. 1998).

The U concentration in the sandstone was measured in samples from 5 different locations (wells #906, 911, 932, 937, and 940; Fig. 7), with an average concentration of 0.47 mg/kg U. The concentration of U in the samples from the contaminated well #906 and the uncontaminated well #911 are 0.56 mg/kg and 0.55 mg/kg, respectively. The analytical error of the U analyses did not exceed ±10%. Hence, there was no significant variation in the U concentration of the sandstone between locations. Using the value of 0.47 mg/kg of sandstone, the inventory of primary U is 3,000 kg in the sandstone within the 6×10^5 m^3 contaminated groundwater. Using an average concentration of 0.5 mg/L, the inventory in the groundwater is 300 kg of U. Table 6 shows that there is no relationship between the U and Fe content in the sandstone, suggesting that there was no

Table 6. Concentration of U and iron given in mg/kg of sandstone from different wells. Total iron expressed as Fe_2O_3.

Wells # and sandstone color	U	Fe_2O_3
906 (white)	0.56	438
911 (reddish)	0.55	455
932 (white)	0.39	492
937 (white + red spots)	0.45	861
940 (reddish)	0.43	240

adsorption of U from the plume onto the sandstone, and that U is free to migrate with the groundwater. To support this hypothesis the following work was completed: U speciation in the groundwater; acid leaching of sandstone from contaminated and uncontaminated area; sorption of U onto the sandstone. Results of these experiments will be described briefly in the following sections.

Organic and inorganic additives to the groundwater/sandstone. Dissolved C (3 mg/L) and P (0.1 mg/L) concentrations in the groundwater were too low to sustain microbial growth. Indigenous bacteria had to be activated by adding C and P sources. Glucose, methanol, ethanol, lactate, and acetate were used. Ethanol or lactate yielded the highest reduction rate of U and acetate the highest rate of nitrate reduction. Several phosphate sources were tried, with trimetaphosphate (TMP) selected. TMP dissolves slowly in water and does not cause immediate precipitation of Ca phosphate, which removes phosphate from solution. A concentration of TMP of 20 mg/L was enough to support bacterial growth.

Batch and column experiments. The investigation of reduction of aqueous species in contaminated groundwater requires experiments with groundwater in the presence of host rock. This is because the host rock can participate in many ways in reduction-oxidation and other chemical processes. The results would be incomplete if experiments were conducted only with groundwater.

Experiments designed to determine optimal conditions for microbially mediated reduction-oxidation processes are best performed in serum bottles. These experiments are simple and many experiments can be run in parallel, e.g. to identify the best amendments (organic C compound, P source) in terms of high yield and fast reaction. Contaminated groundwater from well #926 and sand were used in batch experiments. Resazurin solution was added as a redox potential indicator. Resazurin becomes colorless at $E_H <$ -100 mV and, depending on pH, pink or blue under more oxidizing conditions.

After determination of optimal operating conditions for reduction of U, semi-dynamic experiments were conducted in columns (Plexiglas tubes, 5 cm wide, 20 cm long) filled with Navajo sandstone. Water was pumped into the column and replaced by Ar in the reservoir, with the effluent collected in an Ar-filled serum bottle. Between sampling, the column was sealed with wax to prevent penetration by O. In some batch and column experiments uranyl nitrate, $UO_2(NO_3)_2 \cdot 6H_2O$, was added to obtain enough reaction product for identification.

Results

Leaching of sandstone. Navajo sandstone samples from the contaminated (well #906) and uncontaminated (well #911) area were investigated to determine the leachability of primary U. Samples were leached at room temperature in 1N HCl.

Table 7. Chemical composition of sandstone before and after leaching with 1N HCl.

(wt %)	Before leaching		After leaching	
	Well #906	Well #911	Well #906	Well #911
SiO_2	95.7	94.0	98.2	95.6
Al_2O_3	1.3	1.9	1.1	2.4
MgO	0.17	0.22	0.10	0.21
CaO	< 0.2	< 0.2	< 0.2	< 0.2
Fe_2O_3	< 0.1	< 0.1	< 0.1	< 0.1
TiO_2	< 0.02	0.05	0.02	0.09
MnO	0.019	0.007	0.001	0.005
Na_2O	< 0.05	0.06	< 0.05	0.07
K_2O	0.98	1.24	0.66	1.19
P_2O_5	< 0.1	< 0.1	< 0.1	< 0.1
Weight loss at 1000°C for 1h	0.50	0.66	0.34	0.61
Total	98.67	98.08	100.44	100.49
(mg/kg)				
U	0.56	0.55	0.38	0.34

The results of the leaching experiments are shown in Table 7, columns 4 and 5. After leaching, the U concentration in the sandstone was lowered by 32% (well #906) and 38% (well #911), respectively. The difference between 32% and 38% was judged not to be significant. The leached fraction of U was probably adsorbed on clay minerals, Fe oxide, and probably on carbonates. The residual U appears to be strongly bonded in other accessory minerals such as zircon. The results in Table 7 do not support adsorption of U on the sandstone from the contaminated groundwater, nor can the contrary be concluded in light of the scattering of U concentrations in the sandstone and the errors associated with the leaching experiments.

Uranium speciation, adsorption and coprecipitation. Results of adsorption experiments of U on sandstone are shown in Table 8. R_D values are defined as in Berry

Table 8. Results of U adsorption experiments with sandstone and groundwater from well #926 at 16°C.

	Initial concentration of U (mg/L)	Initial pH	Final concentration of U (mg/L)	Final pH	Rock /Water (1g in 10 cm^3)	R_D/$cm^3 g^{-1}$
sol. 1:	0.19	6.6	0.19	6.7	1:10	0
sol. 1*:	0.19	6.6	0.02	7.4	1:10	85*
sol. 2:	0.095	7.5	0.089	7.5	1:10	<1
sol. 3:	0.039	7.5	0.027	7.6	1:10	5
sol. 4:	0.027	7.5	0.006	7.6	1:10	31
sol. 5:	0.019	7.6	0.002	7.7	1:10	63

* Experiment conducted at 24°C without sandstone. In this experiment aragonite precipitated on the plastic bottle wall.

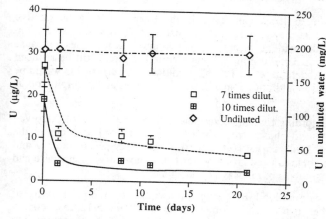

Figure 11. Uranium adsorption on sandstone from well #926 in diluted groundwater at 16°C. Reprinted by permission of Elsevier Science B.V. (Abdelouas et al. 1998c, Fig. 2, p. 352).

et al. (1988) as $R_D = V/M \times (C_o - C)/C$, where R_D is the distribution ratio, V is the volume of solution, M is the mass of solid, C_o is the initial and C the equilibrium concentration of U. Table 8 shows that no change of the initial concentration was noticed in undiluted water from well #926. However, adsorption of U on the Navajo sandstone was significant in the highly diluted groundwater. These results are shown in Figure 11. In the experiment with undiluted groundwater, adsorption of U on the sand particles was inhibited by the high carbonate concentration of 7×10^{-3} M. Calculations of U speciation with the composition of the undiluted groundwater using the EQ3NR code (Wolery 1992) indicated that 98% of the U was complexed by carbonates. The database used to calculate U speciation was taken from Grenthe et al. (1992). The U species were: $UO_2(CO_3)_2^{2-}$ (55%), $UO_2(CO_3)_3^{4-}$ (40%), $UO_2CO_3(aq)$ (3%). The negatively charged species are not adsorbed on the negatively charged SiO_2 surface with a zeta potential of -12 mV. After dilution, the calculated yield of the $UO_2(OH)(aq)$ species was 55%. $UO_2(OH)_2(aq)$ dissociated into H^+ and $UO_2(OH)^+$, and U was, evidently, adsorbed on the sand particles.

Adsorption may be accompanied by coprecipitation, depending upon the experimental conditions and on groundwater composition. Experiments are needed to distinguish between the two processes. Adsorption experiments at 24°C showed a drastic decrease of U concentration, whereas those at 16°C did not. After 44 days at 24°C, the U concentration decreased to 0.02 mg/L though the carbonate concentration is high enough to complex most of the U. A white precipitate had formed, identified by X-ray diffraction and electron microscopy as an aragonite/calcite mixture. Uranium coprecipitated (0.4 g per kg of aragonite/calcite), which explained the depletion of U in solution (Abdelouas et al. 1998c). The saturation index log (Q/K) (Q = ion activity product, K = equilibrium constant) of aragonite, calculated with the help of the EQ3NR code, increased from −0.3 at 16°C (pH = 6.6) to 0.7 at 24°C (pH = 7.4), a value high enough to indicate supersaturation of the solution with respect to aragonite.

The results described above are of practical significance when selecting a remediation process. Ex situ technologies such as pump and treat to remove U will affect the speciation of U and change the temperature of the groundwater. Pumping the contaminated groundwater to the surface to remove U could be accompanied by *in situ* mixing of the carbonate-rich water with uncontaminated groundwater. The result would be that U adsorption on the sandstone increases, leaving a fraction of U in the ground. This fraction would be difficult to quantify. Desorption kinetics and equilibrium would determine when, and to what extent, the water would be recontaminated by U. Storage of groundwater on the surface prior to U removal, especially at the Tuba City site in the summer with temperatures up to 40°C, can result in the formation of a radioactive precipitate, aragonite/calcite, that may require special treatment. *In situ* bioremediation, where water is pumped, amended with an organic nutrient, and reinjected, does not cause dilution or temperature changes. Uranium in carbonate complexes can still be reduced quickly (Phillips et al. 1995; Abdelouas et al. 1998a). Additionally, adsorbed U would be accessible to the microbes and would not be spared from reduction and precipitation as UO_2.

Reduction of nitrate. According to their redox potential, the reduction of dissolved species in groundwater and solid phases from sandstone is expected to occur in the following order: $O_2 > NO_3^- > Mn(IV) > U(VI) > Fe(III) > SO_4$. Oxygen, nitrate, U and sulfate were contained in the contaminated groundwater from Tuba City. Manganese and Fe were present as minor constituents of the Navajo sandstone.

Acetate and ethanol were added to the groundwater to obtain a ratio of C source/nitrate slightly higher than the stoichiometric one of 5/8 and 5/12, respectively (Eqns. 7 and 8). Oxygen was neglected because its concentration is too low compared to nitrate.

$$8 NO_3^- + 5 CH_3COO^- + 3 H^+ = 4 N_2 + 10 HCO_3^- + 4 H_2O \qquad (7)$$

$$12 NO_3^- + 5 C_2H_5OH + 2 H^+ = 6 N_2 + 10 HCO_3^- + 11 H_2O \qquad (8)$$

At 16°C and 24°C, biological denitrification in the columns was complete within 10 and 3 days, respectively. The nitrate concentration decreased from 1.2 g/L to below the detection limit of the ion chromatograph (0.1 mg/L). The results for 24°C are shown in Figure 12. Nitrate reduction was followed by an accumulation of nitrite, which was then reduced to N. In the presence of sufficient host rock material, pH changes in solution were negligible. This may not always be the case. A granitic host rock may have less buffering capacity than a clay or even a sandstone, which usually contains some clay.

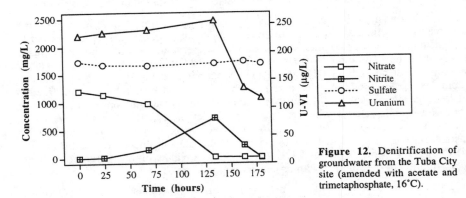

Figure 12. Denitrification of groundwater from the Tuba City site (amended with acetate and trimetaphosphate, 16°C).

The sulfate concentration remained constant during denitrification, in agreement with the sequential nature of the process. At the end of denitrification, the E_H was measured electrochemically and a value of −70 to -100 mV was obtained.

During denitrification the U concentration decreased significantly. Nuttall et al. (1996) showed that this is due to adsorption of U on denitrifying bacteria. Additionally, some U(VI) may have coprecipitated with calcite. Microbially catalyzed reduction of high concentrations of nitrate leads to the generation of proportional amounts of CO_2. This increases the carbonate concentration in solution and supports formation of aragonite/calcite. During denitrification the Ca concentration in the groundwater decreased by 66%. There was no precipitation of U phosphate.

Natural denitrification at a mill tailings was reported by Brookins et al. (1993). These authors showed that nitrate concentrations in the leachate of the U mill tailings site at Maybell, CO decreased by a factor of one thousand beneath the tailings impoundment. They attributed the decrease in nitrate concentration to denitrification by denitrifying bacteria (*Pseudomonas* and *Flavobacterium*) indigenous to the site (Longmire and Thomson 1992). These bacteria used the organic C from groundwater, whose concentration was as high as 212.4 mgC/L (Longmire 1991).

Reduction of U. The redox potential at the end of denitrification (between -70 and -100 mV) was low enough to stimulate growth of sulfate-reducing bacteria (Pfennig et al. 1981; Postgate 1984). The activation of sulfate-reducing bacteria was necessary for U reduction. The U concentration was too low to support the growth of bacteria capable of reducing this metal. To activate *Shewanella putrefaciens*, more ethanol or lactate needed to be added to reach a molar ratio with sulfate of 2/3 (reactions 9 and 10).

$$3\ SO_4^{2-} + 2\ C_2H_5OH = 4\ HCO_3^- + 3\ HS^- + H^+ + 2\ H_2O \tag{9}$$

$$3\ SO_4^{2-} + 2\ C_3H_5O_3 = 6\ HCO_3^- + 3\ HS^- + H^+ \tag{10}$$

After several weeks of column experiments, black spots appeared on the sand. They were identified as mackinawite, $FeS_{0.9}$, indicating that sulfate had been reduced to sulfide and that Fe(III) from the sandstone had been reduced to Fe(II) and then precipitated as sulfide. The U(VI) concentration in the effluent of the column decreased to values as low as 0.5 µg/L. The chemical composition of groundwater from well #926 after denitrification and partial reduction of sulfate in a column experiment is given in Table 9. The concentration of Ca did not change during sulfate reduction, indicating that no calcite precipitated. Hence, U was reduced in the course of sulfate reduction.

Table 9. Chemical composition of groundwater from well #926 before and after bioremediation. The reaction time was 21 d.

	Well #926, before bioremediation	Well #926, after bioremediation
NO_3	1220	<0.1
SO_4	1830	1250
U	0.250	0.014
Mn	0.02	0.76
Fe	0.05	0.59
Ca	681	234*
Mg	329	264
Sr	7.04	3.47

* reached after denitrification and remained constant during reduction of U.

Table 9 shows that the sulfate concentration decreased by 600 mg/L. Thermodynamically, U(VI) is not stable in the presence of sulfide (H_2S/HS^-) and should be reduced to U(IV). Therefore, the hypothesis had to be examined experimentally that U was reduced by sulfide and not by the catalytic activity of bacteria. These experiments are described by Abdelouas et al. (1998a). The authors concluded that reduction of U was

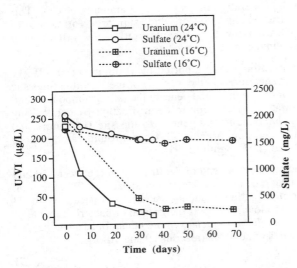

Figure 13. Reduction of U in groundwater from the Tuba City site (amended with ethanol and trimetaphosphate, 16° and 24°C). The initial U(VI) concentration was 250 µg/L. Reprinted by permission of Elsevier Science B.V. (Abdelouas et al. 1998a, Fig. 6, p. 229).

mediated by enzymatic activity of sulfate-reducing bacteria. Competing reactions such as biosorption on biomass, coprecipitation with aragonite/calcite, or abiotic reduction by sulfide produced during the bioreactions were either too slow or too inefficient to explain depletion of U to concentrations of 1 µg/L. This concentration is more than one order of magnitude lower than that mandated by U.S. standard for groundwater protection (44 µg/L).

The rate of U reduction was slower at 16°C than at 24°C (Fig. 13). The U(VI) concentration decreased from 231 to 1.8 µg/L in 36 days at 24°C and from 250 to 15 µg/L in 69 days at 16°C. These reduction rates are slow compared to those reported by Lovley et al. (1991). The difference can be explained by the low initial concentration of sulfate-reducing bacteria (on the order of 1 cell/100 mL) in the groundwater and by the lower reaction temperatures in the work by Abdelouas et al. (1998a).

Experiments with high concentrations of U, up to 235 mg/L added to the groundwater, showed that after denitrification a black precipitate formed (mackinawite and presumably uraninite). Figure 14 shows that U was removed from solution. Final concentrations of U(VI) after 90 days ranged between 0.1 mg/L and 0.2 mg/L. Figure 14 shows that U reduction accompanies sulfate reduction. Obviously, at these higher U concentrations, the final U concentrations do not satisfy the maximum permissible concentration limit of 44 µg/L. Typical U concentrations in contaminated groundwater are much lower than these increased concentrations.

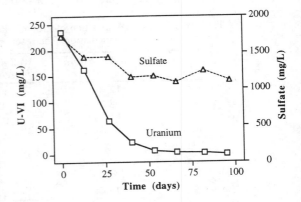

Figure 14. Reduction of U in groundwater from the Tuba City site (amended with ethanol and trimetaphosphate, 24°C). U(VI) concentration was artificially increased by adding uranyl nitrate. Reprinted by permission of Elsevier Science B.V. (Abdelouas et al. 1999b, Fig. 4, p. 362).

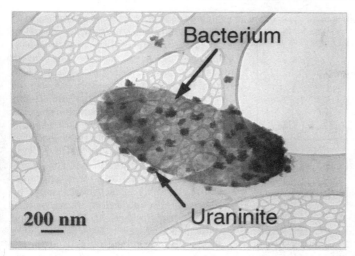

Figure 15. Bacterium with uraninite particle.

Solid reaction products. Microbially mediated reduction-oxidation reactions must be conducted in the presence of the aqueous and solid phase to understand the processes where applied *in situ*. Solid phases form due to interaction between reduction processes in solution and the host minerals. As an example, the formation of mackinawite, $FeS_{0.9}$, was described in the previous section. The sulfide ions formed in solution, and Fe(II) came from Fe(III) in hematite and oxyhydroxides. Uranium was found to be involved in four different processes: complexation with carbonate, coprecipitation, adsorption on biomass, and reduction. Reduction of U(VI) produces U(IV) which precipitates as UO_2. To verify this, the dark precipitate from batch experiments with high U concentrations was studied using transmission electron microscopy. Figure 15 shows that most of this material was attached to cell walls. Microdiffraction revealed that the particles were poorly crystallized uraninite (Fig. 16) that contained Ca (Fig. 17). Macinawite and its electron diffraction pattern are shown in Figure 18.

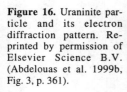

Figure 16. Uraninite particle and its electron diffraction pattern. Reprinted by permission of Elsevier Science B.V. (Abdelouas et al. 1999b, Fig. 3, p. 361).

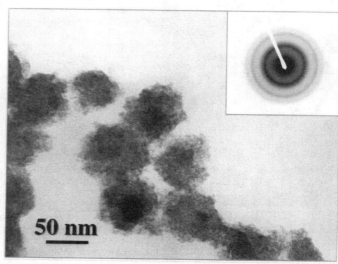

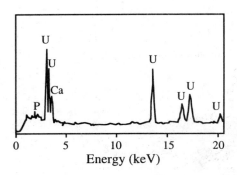

Figure 17. EDS spectrum of uraninite.

Chemical durability of uraninite. If U were removed from groundwater as part of an *in situ* bioremediation process, U(VI) would be reduced *in situ* and would precipitate as uraninite in the host rock. Hence, the water would be free of U but the U was not removed from the site. If this were acceptable in principal, the question would then be whether UO_2 would redissolve and contaminate the water after the recharge of the site with O-rich uncontaminated groundwater. To answer this question, the long-term chemical durability of UO_2 precipitates must be known. After completion of an *in situ* bioremediation campaign, reducing conditions will eventually be lost. Oxidation and dissolution of uraninite by O-rich flowing groundwater may occur. Experiments with freshly bio-precipitated uraninite showed that the solid is dissolved quickly in O-rich water. However, the chemistry in the presence of the host rock (Navajo sandstone) is very different, as a lot of mackinawite was precipitated together with uraninite. To simulate ingress of fresh uncontaminated water into the site (after assumed bioremediation), sandstone columns were flushed with O-rich uncontaminated groundwater from well

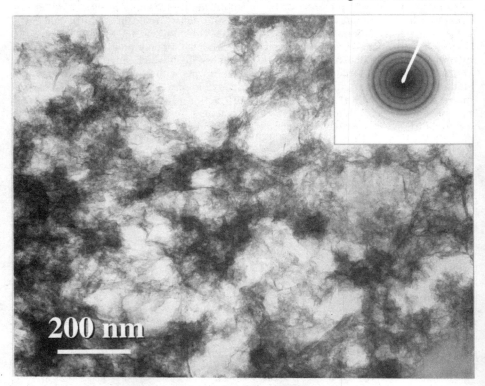

Figure 18. Mackinawite and its electron diffraction pattern.

#948 (Table 4). This groundwater contained only 1.4 µg/L of U. Many pore volumes were pumped through the column at both slow and fast flow rates. All velocities were high, 75-1000 m/yr., compared with the actual velocity of 15 m/yr. of groundwater at the Tuba City site. The results of two leaching experiments are shown in Figure 19. The figure shows the U concentration in the leachate as a function of the volume of uncontaminated water, expressed as pore volumes, passing through the column. After one pore volume, the U(VI) concentration stayed constant with

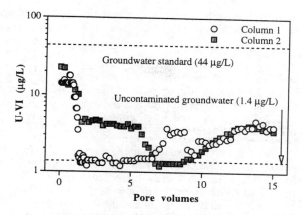

Figure 19. Leaching of uraninite from sandstone columns with uncontaminated groundwater (DO = 7 mg/L). Column 1 (velocity = 75 m/yr.), column 2 (velocity = 1000 m/yr.). U(IV) inventory in the column is 26 µg. Reprinted by permission of Elsevier Science B.V. (Abdelouas et al. 1999b, Fig. 6, p. 364).

an average value of a few µg/L. If the O introduced had oxidized U, the U(VI) concentration should have increased. To explain the result, one must realize that mackinawite was precipitated in large excess compared with uraninite. The inventory of Fe(III) in the host rock is much higher than the small amount of U contained in the water. Mackinawite constitutes an efficient redox buffer, by consuming the dissolved O before it oxidizes U. Mackinawite is oxidized to goethite and sulfate:

$$FeS_{0.9} + 1.8 O_2 \rightarrow 0.9 SO_4^{2-} + Fe^{2+} \tag{11}$$

$$Fe^{2+} + 1/4 O_2 + 1.5 H_2O \rightarrow FeOOH + 2 H^+ \tag{12}$$

The sulfate concentration in the outlet of the column increased from background (9.6 mg/L) to 21 mg/L, indicating continuous oxidation of mackinawite. Another redox buffer is provided by the presence of living denitrifying bacteria. Uncontaminated groundwater contained 11.85 mg/L, which was lowered to 7 mg/L at the outlet. We calculated that before the complete oxidation of mackinawite, uraninite would have been completely oxidized at a very slow rate. A more detailed description of the O balance and additional experiments are described by Abdelouas et al. (1999b).

SUMMARY AND CONCLUSIONS

The presentation and discussion of experimental data show what is needed to acquire a site-specific database that can be used to develop an *in situ* groundwater bioremediation process for U. It was assumed that U would be precipitated from the aqueous phase but would not be removed from the site. This strategy would be advantageous because no secondary waste and no other byproducts are produced. It was further assumed that such a process would consist of an array of boreholes configured in a way that provided optimal underground remediation using indigenous bacteria. Leaving U in the ground requires long-term durability of the precipitate. It was demonstrated that a redox buffer may be generated, along with reduction of U to provide sufficient stability of uraninite and to keep the concentration of U in the groundwater after remediation significantly below acceptable limits, and near the background U concentration. The protective effect of Fe sulfide with respect to uraninite dissolution is a common process in natural U ore deposits (Cross et al. 1991; Dewynne et al. 1993).

Our discussion further showed that it is mandatory to conduct the microbially mediated reduction-oxidation experiments in a system that contains both the groundwater and the host rock material. The chemistry of the system can only be described completely in the presence of all components. Temperature changes, dilution and other processes may have an impact on the kind and the rate of chemical reactions in the system. This must be borne in mind when conducting experiments in the laboratory with material collected in the field.

Whatever the details of an engineered process for U remediation may be, there are two critical issues: one is mixing of the contaminated water with the necessary additives to stimulate bacterial growth underground, and the second is the expectation that indigenous bacteria are ubiquitous so that the process can be applied everywhere. The first issue is strictly technical and there are solutions to this problem. The second issue is uncertain. Though it may be speculated that finding the right bacteria in one place indicates that they are ubiquitous, experiments were conducted by the authors with groundwater and soil from various sites, including four sites in the United States and four sites in Germany. These samples include water with widely varying chemical composition, including pH and temperature. The composition of the soil varied from site to site. If a certain constituent was deficient, e.g. sulfate, it was added to obtain the desired reactions with Fe and U. In all cases, the same results were obtained by stimulating bacterial growth and without specifically identifying the strains catalyzing the chemical reactions. In all cases, uraninite precipitated together with mackinawite. Hence, it is feasible that *in situ* bioremediation can be performed on the assumption that the microbial functionality is ubiquitous without knowing or expecting that a specific bacterial strain is present.

We close this chapter by emphasizing the fact that though significant progress was made in the field of U bioremediation, *in situ* tests are needed to confirm the laboratory results. Our team conducted an *in situ* test but only for biological denitrification in the groundwater from a nitrate-contaminated site in Albuquerque, New Mexico. As expected, the optimal conditions for bacterial growth and nitrate reduction were encountered *in situ* in the field. Nitrate and its products were reduced within 5 days (Abdelouas et al. 1999c).

ACKNOWLEDGMENTS

We are grateful to Dr. Weiliang Gong for his help with TEM work and Neil Butt for his assistance in experiments conducted by the authors. We thank Dr. J.-L. Crovisier for chemical analyses of sandstone samples conducted at Centre de Géochimie de la Surface, CNRS (Strasbourg, France), and P.C. Burns, B.A. Strietelmeier and A.J. Francis for helpful reviews of this chapter.

REFERENCES

Abdelouas A, Lu Y, Lutze W, Nuttall HE 1998a Reduction of U(VI) to U(IV) by indigenous bacteria in contaminated ground water. J Contam Hydrol 35:217-233
Abdelouas A, Lutze W, Nuttall E 1998b Reduction of nitrate and uranium by indigenous bacteria. C R Acad Sci Paris 327:25-29
Abdelouas A, Lutze W, Nuttall E 1998c Chemical reactions of uranium in ground water at a mill tailings site. J Contam Hydrol 34:343-361
Abdelouas A, Lutze W, Nuttall HE, Gong W 1999a Remediation of U(VI)-contaminated water using zero-valent iron. C R Acad Sci Paris 328:315-319

Abdelouas A, Lutze W, Nuttall HE 1999b Oxidative dissolution of uraninite precipitated on Navajo sandstone. J Contam Hydrol 36:353-375

Abdelouas A, Deng L, Nuttall HE, Lutze W, Fritz B, Crovisier JL 1999c In situ biological denitrification of groundwater. C R Acad Sci Paris 328:161-166

Al-Hachimi A 1992 Uranium tailings disposal: Review of current technology. Int'l J Environ Studies 42:53-62

Al-Hachimi A, Evans GJ, Cox B 1993 Leachability of hazardous elements from uranium tailings. Int'l J Environ Studies 46:59-68

Anonymous 1993 Programmatic Environmental Impact Statement for the Uranium Mill Tailings Remedial Action Water Project, DOE/EIS-0198, Oct. 1993

Anonymous 1994. Site Observational Work Plan for the UMTRA Project at Tuba City, Arizona. DOE/AL/62350-160

Barnes CE, Cochran JK 1993 Uranium geochemistry in estuarine sediments: Controls on removal and release processes. Geochim Cosmochim Acta 57:555-569

Barton LL, Choudhury K, Thomson BM, Steenhoudt K, Groffman AR 1996 Bacterial reduction of soluble uranium: The first step of *in situ* immobilization of uranium. Radioac Waste Manag Environ Restor 20:141-151

Basnakova G, Stephens ER, Thaller MC, Rossolini GM, Macaskie LE 1998 The use of *Escherichia Coli* bearing a phon gene for the removal of uranium and nickel from aqueous flows. Appl Microbiol Biotechnol 50:266-272

Bernhard G, Geipel G, Brendler V, Nitsche H 1998 Uranium speciation in waters of different uranium mining areas. J Alloys Compounds 271:201-205

Berry JA, Bourke PJ, Coates HA, Green A, Jefferies NL, Littleboy AK, Hooper AJ 1988 Sorption of radionuclides on sandstones and mudstones. Radiochim Acta 44/45:135-144

Berthelin J, Munier-Lamy C 1983 Microbial mobilization and preconcentration of uranium from various rock materials by fungi. *In* Environmental Geochemistry. R Hallberg (Ed) Ecol Bull (Stockholm) 35:395-401

Bhatti TM, Vuorinen A, Lehtinen M, Tuovinen OH 1998 Dissolution of uraninite in acid solutions. J Chem Technol Biotechnol 73:259-236

Bloomfield C, Kelso WI 1973 The mobilization and fixation of molybdenum, vanadium and uranium by decomposing plant matter. J Soil Sci 24:368-379

Bouby M, MacCordick HJ, Billard I 1996 Biosorption and retention of several actinide and fission-product elements by biomass from *Mycobacterium phlei*. J Radioanal Nucl Chem 210:161-169

Brierley CL 1991 Bioremediation of metal-contaminated surface and groundwaters. Geomacrobiol J 8:201-223

Brookins DG, Thomson BM, Longmire PA, Eller PG 1993 Geochemical behavior of uranium mill tailings leachate in the subsurface. Radioactive Waste Management Nuclear Fuel Cycle 17:269-287

Bustard M, Mchale AP 1998 Biosorption of heavy-metals by distillery-derived biomass. Bioprocess Engineer 19:351-353

Bustard M, Rollan A, Mchale AP 1998 The effect of pulse voltage and capacitance on biosorption of uranium by biomass derived from whiskey distillery spent wash. Bioprocess Engineer 18:59-62

Cantrell KJ, Kaplan DI 1997 Zero-valent iron colloid emplacement in sand columns. J Environ Engrg, ASCE 123:499-505

Cantrell KJ, Kaplan DI, Gilmore TJ 1997 Injection of colloidal Fe^0 particles in sand with shear-thinning fluids. J Environ Engrg, ASCE 123:786-791

Cantrell KJ, Kaplan DI, Wietsma TW 1995 Zero-valent iron for the *in situ* remediation of selected metals in groundwater. J Hazard Mater 42:201-212

Clifford D, Zhang Z 1994 Modifying ion exchange for combined removal of uranium and radium. J AWWA, April, 214-227

Cochran JK, Carey AE, Sholkovitz ER, Surprenant LD 1986 The geochemistry of uranium and thorium in coastal marine sediments and sediment pore waters. Geochim Cosmochim Acta 50:663-680

Cornish JE, Goldberg WC, Levine RS, Benemann JR 1995 Phytoremediation of soils contaminated with toxic elements and radionuclides. Bioremediation of Inorganics 3:55-62

Cothern CR, Lappenbusch WL 1983 Occurrence of uranium in drinking water in the United States. Health Phys 45:89-99

Cross JE, Haworth A, Neretnieks I, Sharland SM, Tweed CJ 1991 Modelling of redox front and uranium movement in a uranium mine at Poços de Caldas. Radiochim Acta 52/53:445-451

Delaney TA, Hockley DE, Chapman JT, Holl NC 1998 Geochemical characterization of tailings at the McArthur River Mine, Saskatchewan. Proc Tailings and Mine Waste '98, Balkema, Rotterdam, p 571-578

Dewynne JN, Fowler AC, Hagan PS 1993 Multiple reaction fronts in the oxidation-reduction of iron-rich uranium ores. SIAM J Appl Math 53:971-989
Dodge CJ, Francis AJ 1997 Biotransformation of binary and ternary citric acid complexes of iron and uranium. Environ Sci Technol 31:3062-3067
D.O.E. This Month 1997 In the aftermath of the Cold War, cleaning up uranium mill tailings has become a worldwide problem. August 1, 1997, vol 20, p 11
D.O.E. 1992 UMTRA groundwater program plan, May 1992, prepared by the U S Department of Energy, UMTRA Project Office, Albuquerque Operations Office, Albuquerque, New Mexico
D.O.E. 1994 Site observational work plan for the UMTRA Project Site at Tuba City, Arizona, prepared by the U S Department of Energy, UMTRA Project Office, Albuquerque Operations Office, Albuquerque, New Mexico
Dugan PR, MacMillan CB, Pfister RM 1970a Aerobic heterotrophic bacteria indigenous to pH 2.8 acid mine water: I. Microscopic examination of acid streamers. J Bacteriol 101:973-981
Dugan PR, MacMillan CB, Pfister RM 1970b Aerobic heterotrophic bacteria indigenous to pH 2.8 acid mine water: II. Predominant slime-producing bacteria in acid streamers. J Bacteriol 101:982-988
Environment Canada 1987 Conservation and protection. Proc Acid Mine Drainage Seminar/Workshop, Environment Canada 23-26 March
E.P.A. 1989 Evaluation of groundwater extraction remedies. U S Environmental Protection Agency. Office of Solid Waste and Emergency Response; EPA/504/0289/054; Washington, DC
Ernstsen V, Gates WP, Stucki J 1998 Microbial reduction of structural iron in clays-A renewable source of reduction capacity. J Environ Qual 27:761-766
Federal Register 1995 Environmental Protection Agency (EPA), CFR 40 Part 192, Groundwater Standards for Remedial Actions at Inactive Uranium Processing Sites, Table 1, p 2866, November 1995
Fernandes HM, Franklin MR, Veiga LH 1998 Acid rock drainage and radiological environmental impacts. A study case of the uranium mining and milling facilities at Poços de Caldas. Waste Management 18:169-181
Finlay JA, Allan VJM, Conner A, Callow ME, Basnakova G, Macaskie LE 1999 Phosphate release and heavy metal accumulation by biofilm-immobilized and chemically-coupled cells of a *Citrobacter* sp. pre-grown in continuous culture. Biotechnol Bioengineer 63:87-97
Francis AJ 1990 Microbial dissolution and stabilization of toxic metals and radionuclides in mixed wastes. Experientia 46:840-851
Francis AJ 1998 Biotransformation of uranium and other actinides in radioactive-wastes. J Alloys Compounds 271:78-84
Francis AJ, Dodge CJ 1998 Remediation of soils and wastes contaminated with uranium and toxic metals. Environ Sci Technol 32:3993-3998
Francis AJ, Dodge CJ, Gillow JB, Cline JE 1991 Microbial transformations of uranium in wastes. Radiochim Acta 52-3:311-316
Francis AJ, Dodge CJ, Lu F, Halada GP, Clayton CR 1994 XPS and XANES studies of uranium reduction by *Clostridium* sp. Environ Sci Technol 28:636-639
Francis AJ, Dodge CJ. Contaminated site reclamation with biochemical recovery of radionuclides and/or toxic metals, U S Patent Numbers 5,047,152 / 5,292,456 / 58,354,688
Frederickson JK, Gorby YA 1996 Environmental processes mediated by iron-reducing bacteria. Current Opinion Biotechnol 7:287-294
Froelich PN, Klinkhammer GP, Bender ML, Luedtke NA, Heath GR, Cullen D, Dauphin P, Hammond D, Hartman B, Maynard V 1979 Early oxidation of organic matter in pelagic sediments of the Eastern Equatorial Atlantic: Suboxic diagenesis. Geochim Cosmochim Acta 43:1075-1090
Fruchter JS, Amonette JE, Cole CR, Gorby YA, Humphrey MD, Istok JD, Olsen KB, Spane FA, Szecsody JE, Teel SS, Vermeul VR, Williams MD, Yabusaki SB 1996 *In situ* redox manipulation field injection test report—Hanford 100 H area, PNNL-11372/UC-602, Pacific Northwest National Laboratory, Richland, Washington
Gadd GM 1997 Roles of micro-organisms in the environmental fate of radionuclides. *In* Health impacts of large releases of radionuclides. Lake JV, Bock GR, Cardew G (Eds) Ciba Foundation Symp 203:94-104 John Wiley & Sons, New York
Ganesh R, Robinson KG, Reed GD, Sayler GS 1997 Reduction of hexavalent uranium from organic complexes by sulfate- and iron-reducing bacteria. Appl Environ Microbiol 63:4385-4391
Garcia Júnior O 1993 Bacterial leaching of uranium ore from Figueira-PR, Brazil, at laboratory and pilot scale. FEMS Microbiol Rev 11:237-242
Golab Z, Orlowska B, Smith RW 1991 Biosorption of lead and uranium by *Streptomyces* sp. Water Air Soil Pollution 60:99-106
Gorby YA, Lovley DR 1992 Enzymatic uranium precipitation. Environ Sci Technol 26:205-207

Grenthe I, Fuger J, Lemire RJ, Muller AB, Nguyen-Trung C, Wanner H (Eds) 1992 Chemical thermodynamics of uranium: Elsevier Science Publishing Company, New York
Gu B, Liang L, Dickey MJ, Yin X, Dai S 1998 Reductive precipitation of uranium(VI) by zero-valent iron. Environ Sci Technol 32:3366-3373
Guibal E, Roulph C, Le Cloirec P 1992 Uranium biosorption by a filamentous fungus *Mucor miehei:* pH effect on mechanisms and performances of uptake. Water Res 26:1139-1145
Hall CW 1988 Pratical limits to pump and treat technology for aquifer remediation. Groundwater quality protection preconference workshop proceedings; water pollution control federation: Dallas, p 7-12
Hard BC, Babel FW 1997 Bioremediation of acid mine water using facultatively methylotrophic metal-tolerant sulfate-reducing bacteria. Microbiol Res 152:65-73
Hiemenz PC 1986 Principles of colloid and surface chemistry, 2nd edn. Marcel Dekker, New York
Holdway DA 1992 Uranium mining in relation to toxicological impacts in inland waters. Ecotoxicol 1:75-88
Hostetler PB, Garrels RM 1962 Transportation and precipitation of uranium and vanadium at low temperatures with special reference to sandstone-type uranium. Econ Geol 57:137-167
Hu Z-C, Norman JM, Faison BD 1996 Biosorption of uranium by *Pseudomonas aeruginosa* strain CSU: Characterization and comparison studies. Biotechnol Bioeng 51:237-247
Hu Z-C, Reeves M 1997 Biosorption of uranium by *Pseudomonas aeruginosa* strain CSU immobilized in a novel matrix. *Biotechnol Prog* 13:60-70
Jensen ML 1958 Sulfur isotopes and the origin of sandstone-type uranium deposits. Econ Geol 53:598-616
Johnson DB, Kelso WI, Jenkins DA 1979 Bacterial streamer growth in a disused pyrite mine. Environ Pollu 18:107-118
Junghans M, Helling C 1998 Historical mining, uranium tailings and waste disposal at one site: Can it be managed? A hydrological analysis. Proc Tailings and Mine Waste '98, Balkema, Rotterdam, 117-126
Kalin M 1984 Rayrock, Northwest Territories, uranium tailings and environment. Toronto, Ontario, Canada: University of Toronto Institute for Environment Studies
Kaplan DI, Cantrell KJ, Wietsma TW, Potter MA 1996 Retention of zero-valent iron colloids by sand columns: Application to chemical barrier formation. J Environ Qual 25:1086-1094
Kauffman JW, Laughlin WC, Baldwin RA 1986 Microbiological treatment of uranium mine waters. Eviron Sci Technol 20:243-248
Kimberley MM 1979 Uranium Deposits, Their Mineralogy and Origin. Mineral Assoc Canada, Univ Toronto Press, 521 p
Klinkhammer GP, Palmer MR 1991 Uranium in the oceans: Where it goes and why? Geochim Cosmochim Acta 55:1799-1806
Knoch-Weber J, Mehner HC 1997 Environmental monitoring by WISMUT rehabilitation units. Kerntechnik 62:187-193
Langmuir D 1978 Uranium-solution mineral equilibria at low temperature with applications to sedimentary ore deposits. Geochim Cosmochim Acta 42:547-569
Ledin M, Pedersen K 1996 The environmental impact of mine wastes - roles of microorganisms and their significance in treatment of mine wastes. Earth-Science Reviews 41:67-108
Linking Legacies 1997 Connecting the Cold War Nuclear Weapons Production Processes to Their Environmental Consequences. U S Department of Energy, Office of Environmental Management, January 1997, DOE/EM-0319, 230 p
Longmire PA 1991 Hydrochemical investigations at a uranium mill tailings site, Maybell, Colorado. PhD Thesis, Dept Geology, Univ New Mexico, Albuquerque, NM, 671 p
Longmire PA, Thomson BM 1992 Evidence for denitrification at a uranium mill tailings site, Maybell, CO. Proc Water and Rock Conference, Park City, Utah, 6 p
Lovley DR 1995a Bioremediation of organic and metal contaminants with dissimilatory metal reduction. J Industr Microbiol 14:85-93
Lovley DR 1995b Microbial reduction of iron, manganese, and other metals, *In* L Donald, L Sparks (Eds) Advances in Agronomy, 54:175-231, Academic Press, New York
Lovley DR, Phillips EJP 1992a Bioremediation of uranium contamination with enzymatic uranium reduction. Appl Environ Microbiol 58:850-856
Lovley DR, Phillips EJP 1992b Reduction of uranium by *Desulfovibrio desulfuricans.* Environ Sci Technol 26:2228-2234
Lovley DR, Phillips EJP 1992c Novel mode of microbial energy metabolism: Organic carbon oxidation coupled to dissimilatory reduction of iron and manganese. Appl Environ Microbiol 54:1472-1480
Lovley DR, Phillips EJP, Gorby Y, Landa E 1991 Microbial reduction of uranium. Nature 350:413-416
Lovley DR, Roden EE, Phillips EJP, Woodward JC 1993a Enzymatic iron and uranium reduction by sulfate-reducing bacteria. Marine Geol 113:41-53

Lovley DR, Widman PK, Woodward JC, Phillips EJP 1993b Reduction of uranium by cytochrome *c3* of *Desulfovibrio vulgaris*. Appl Environ Microbiol 59:3572-3576

Luther, III GW 1987 Pyrite oxidation and reduction: Molecular orbital theory considerations. Geochim Cosmochim Acta 51:3193-3199

Luther, III GW 1995 Trace metal chemistry in porewaters. *In* Metal Contaminated Aquatic Sediments. Allen HE (Ed) Ann Arbor Press, Ann Arbor, Michigan, p 65-80

Luther, III GW, Fredelman TG, Kostka TG, Tsamakis EJ, Church TM 1991 Temporal and spatial variability of reduced sulfur species (FeS_2, $S_2O_3^{2-}$) and porewater parameters in salt marsh sediments. Biogeochem 14:57-88

Macaskie LE, Empson RM, Cheetham AK, Grey CP, Skarnulis J 1992 Uranium bioaccumulation by a *Citrobacter* sp. as a result of enzymically mediated growth of polycrystalline HUO_2PO_4. Science 257:782-748

Macaskie LE, Lloyd JR, Thomas RAP, Tolley MR 1996 The use of micro-organisms for the remediation of solutions contaminated with actinide elements, other radionuclides, and organic contaminants generated by nuclear fuel cycle activities. Nuclear Energy 35:257-271

Macaskie LE, Yong P, Doyle TC, Roig MG, Diaz M, Manzano T 1997 Bioremediation of uranium-bearing wastewater: Bioremediation and chemical factors influencing bioprocess application. Biotechnol Bioeng 53:100-109

Mackay DM 1989 Characterization of the distribution and bahavior of contaminants in the subsurface. Proc National Research Council Water Science and Technology Board Colloq, Washington, DC

Mackay DM, Cherry JA 1989 Groundwater contamination: Pump-and-treat remediation. 2. Environ Sci Technol 23:630-636

Maynard JB 1983 Geochemistry of Sedimentary Ore deposits. Springer-Verlag, New York

Merritt RC 1971 The Extractive Metallurgy of Uranium. Atomic Energy Commission

Morrison SJ, Spangler RR 1992 Extraction of uranium and molybdenum from aqueous solutions: A survey of industrial materials for use in chemical barriers for uranium mill tailings remediation. Environ Sci Technol 26:1922-1931

Morrison SJ, Cahn LS 1991 Mineralogical residence of alpha-emitting contamination and implications for mobilization from mill tailings. J Contam Hydrol 8:1-21

Moses CO, Herman JS 1991 Pyrite oxidation at circumneutral pH. Geochim Cosmochim Acta 55:471-482

Moses CO, Nordstrom DK, Herman JS, Mills AL 1987 Aqueous pyrite oxidation by dissolved oxygen and ferric iron. Geochim Cosmochim Acta 51:1561-1571

Munier-Lamy C, Berthlin J 1987 Formation of polyelectrolyte complexes with the major elements Fe and Al and the trace elements U and Cu during heterotrophic microbial leaching of rocks. Geomicrobiol J 5:119-147

Muñoz JA, González F, Blázquez ML, Ballester A 1995a A study of a Spanish uranium ore. 1: A review of the bacterial leaching in the treatment of uranium ores. Hydrometallurgy 39:39-57

Muñoz JA, Blázquez ML, Ballester A, González F 1995b A study of the bioleaching of a spanish uranium ore. 3. Column experiments. Hydrometallurgy 38:79-97

Murray JW, Codispoti LA, Friederich GE 1995 Oxidation-reduction environments. The suboxic zone in the Black Sea. *In* Aquatic Chemistry. Interfacial and Interspecies Processes. CP Huang, CR O'Melia, JJ Morgan (Eds) Advances Chem Ser 244:157-194

Nordstrom DK, Southam G 1997 Geomicrobiology of sulfide mineral oxidation. Rev Mineral 35:361-390

Nuttall E, Lutze W, Barton L, Choudhury K 1996 Preliminary screening results for *in situ* bioremediation at an UMTRA site, Waste Management '96, Session 45, paper 6, Tucson, Arizona, February 1996

Otton JK, Zielinsky RA, Been JM 1989 Uranium in Holocene valley-fill sediments, and uranium, radon, and helium in waters, Lake Tahoe-Carson Range area, Nevada and California, USA. Environ Geol Water Sci 13:15-28

Parks GA, Pohl DC 1988 Hydrothermal solubility of uraninite. Geochim Cosmochim Acta 52:863-875

Peterson F, Pipiringos GN 1979 Strastigraphic relations of the Navajo Sandstone to middle Jurassic formations, Southern Utah and Northern Arizona. US Geol Surv Prof Paper 1035-B, 43 p

Pfennig N, Widdel F, Trüper HG 1981 The dissimilatory sulfate-reducing bacteria. *In* Starr MP, Stolp H, Trüper HG, Balows A, Schlegel HG (Eds) The Prokaryotes, Vol. 1, p 926-940. Springer-Verlag, Heidelberg

Phillips EJP, Landa ER, Lovley DR 1995 Remediation of uranium contaminated soils with bicarbonate extraction and microbial U(VI) reduction. J Industr Micrbiol 14:203-207

Pizarro J, Jedlicki E, Orellana O, Romero J, Espejo RE 1996 Bacterial populations in samples of bioleached copper ore as revealed by analysis of DNA obtained before and after cultivation. Appl Environ Microbiol 62:1323-1328

Postgate JR 1984 The Sulfate-Reducing Bacteria, 2nd Edn. Cambridge Univ Press, Cambridge, UK, 208 p

Quinton GE, Buchanan RJ, Ellis DE, Shoemaker SH 1997 A method to compare groundwater cleanup technologies. Remediation, Autumn 1997, p 7-16
Rastogi RK, Mahajan MA, Chaudhuri NK 1997 Separation of thorium and uranium product at the tail end of thorium fuel reprocessing using macroporous cation-exchange resin. Sep Sci Technol 32:1711-1723
Rich A, Holland HD, Petersen U 1977 Hydrothermal Uranium Deposits. Elsevier, New York, 264 p
Rippon GD, Riley SJ 1996 Environmental impact assessment of tailings disposal from a uranium mine using toxicity testing protocols. Water Res Bull 32:1167-1175
Roig MG, Manzano T, Diaz M 1997 Biochemical process for the removal of uranium from acid-mine drainages. Water Research 31:2073-2083
Roig MG, Manzano T, Diaz M, Pascual MJ, Paterson M, Kennedy JF 1995 Enzymatically-enhanced extraction of uranium from biologically leached solutions. Int'l Biodeterioration Biodegradation 35:93-127
Sato T, Murakami T, Yanase N, Isobe H, Payne TE, Airy P 1997 Iron nodules scavenging uranium from groundwater. Environ Sci Technol 31:2854-2858
Schippers A, Hallmann R, Wentzien S, Sand W 1995 Microbial diversity in uranium mine waste heaps. Appl Environ Microbiol 61:2930-2935
Schnaitman CA, Lundgren DG 1965 Organic compounds in the spent medium of *Ferrobacillus ferroxidans*. Can J Microbiol 11:23-27
Scott MJ, Metting FB, Fruchter JS, Wildung RE 1998 Surface barrier technology results in cost savings showing that research investment pays off. Soil and Groundwater Cleanup, October 1998, 6-13
Singer PC, Stumm W 1970 Acid mine drainage—The rate limiting step. Science 167:1161-1123
Smith LA, Alleman BC, Copley-Graves L 1994 Biological Treatment Options, Emerging Technology for Bioremediation of Metals, CRC Press, p 1-12
Smythe C, Bierley D, Bradshaw M 1995 The US regulatory framework for long-term management of uranium mill tailings. J Mine Ventil Soci South Africa, October 1995, 246-253
Soniassay R, Ladouceur E, MacDonald A 1986 Monitoring of the Rayrock abandoned mine site, Rayrock, NWT. Yellowknife, Northwest Territories: Indian and Northern Affairs Canada
Stumm W, Morgan JJ 1981 Aquatic Chemistry, John Wiley & Sons, New York
Taylor GH 1979 Biogeochemistry of uranium minerals. In Biogeochemical Cycling of Mineral-Forming Elements. PA Trudinger, DJ Swaine (Eds) p 485-514. Elsevier, New York
Thein J, Hähne R, Klapperich H, Gross U 1993 Problems and approaches related to the cleaning and safe storage of preparation residues from uranium ore mining in eastern Germany. In Contaminated Soil '93. Arendt F, Anokkee GJ, Bosman R, van den Brink WJ (Eds) Kluwer Academic Publishers, Dordrecht, The Netherlands, 999-1001
Travis CC, Doty CB 1990 Can contaminated aquifers at superfund sites be remediated? Eviron Sci Technol 24:1464-1466
Tsezos M, Georgousis Z, Remoudaki E 1997 Mechanism of aluminum interference on uranium biosorption by *Rhizopus arrhizus*. Biotechnol Bioeng 55:16-27
Tsezos M, McCready RGL, Bell JP 1989 The continuous recovery of uranium from biologically leached solutions using immobilized biomass. Biotechnol Bioeng 34:10-17
Tucker MD, Barton LL, Thomson BM 1996 Kinetic coefficients for simultaneous reduction of sulfate and uranium by *Desulfovibrio desulfuricans*. Appl Microbiol Biotechnol 46:74-77
Tucker MD, Barton LL, Thomson BM 1998a Removal of U and Mo from water by immobilized *Desulfovibrio desulfuricans* in column reactors. Biotechnol Bioeng 60:88-96
Tucker MD, Barton LL, Thomson BM 1998b Reduction of Cr, Mo, Se and U by *Desulfovibrio desulfuricans* immobilized in polyacrylamide gels. J Industr Microbiol Biotechnol 20:13-19
Tuovinen OH 1986 Acid-leaching of uranium ore materials with microbial catalysis. Biotechnol Bioengin Symp 13:65-72
Uhrie JL, Drever JI, Colberg PJS, Nesbitt CC 1996 *In situ* immobilization of heavy metals associated with uranium leach mines by bacterial sulfate reduction. Hydrometallurgy 43:231-239
Vecchio A, Finoli C, Disimine D, Andreoni V 1998 Heavy-metal biosorption by bacterial-cells. Fresenuis J Anal Chem 361:338-342
Veska E, Eaton RS 1991 Abandoned Rayrock Uranium mill tailings in the Northwest Territories: Environmental conditions and radiological impact. Health Phys 60:399-409
Volesky B 1994 Advances in biosorption of metals: Selection of biomass types. FEMS Microbiol Rev 14:291-302
Wolery TJ 1992 EQ3NR, A computer program for geochemical aqueous speciation-solubility calculations: Theoretical manual, user's guide, and related documentation (version 7.0): UCRL-MA-110662-PT-IV, Lawrence Livermore Nat'l Lab, Livermore, California
Yanase N, Payne TE, Sekine K 1995 Groundwater geochemistry in the Koongarra ore deposit, Australia (I): Implications for uranium migration. Geochem J 29:1-29

10 Uranium Mineralogy and the Geologic Disposal of Spent Nuclear Fuel

David J. Wronkiewicz
University of Missouri-Rolla
Department of Geology and Geophysics
Rolla, Missouri 65409

Edgar C. Buck
Argonne National Laboratory
Chemical Technology Division
9700 South Cass Avenue
Argonne, Illinois 60439

INTRODUCTION

The safe disposal of nuclear waste represents one of the most daunting tasks facing modern society. This task will require the development of a disposal system that ensures the adequate protection of the health and safety of the public from the release of radionuclides. It is the policy of the government of the United States to pursue a disposal option for these wastes in an underground mined geologic repository. The Yucca Mountain site, in Nevada, is currently being evaluated to determine its suitability as a high-level nuclear waste disposal facility.

The National Research Council indicates that an assessment period for high-level waste disposal should apply for the time period when the greatest health risk occurs, within the limits imposed by the long-term stability of the geologic environment (estimated to be approximately 10^6 years for Yucca Mountain; NRC 1995). A multi-barrier disposal system is being investigated in the United States towards this purpose. This disposal system incorporates engineered barrier components that may include, but are not necessarily limited to, the waste form, internal stabilizers, the canister in which the waste is emplaced, and backfill material located between the canister and adjacent host rock. Such a system would be designed to limit water contact with the waste forms, thereby limiting the long-term release of radionuclides to levels that will not significantly increase the radiological health risks to the general public. The sorptive capacity of the geologic host rock will also be an important barrier to the migration of many radionuclides (SCP 1988; TSPA-VA 1998).

There are numerous types of nuclear waste materials awaiting disposal in many countries. These wastes represent a wide array of materials with variable compositions and radionuclide contents. High-level nuclear waste forms may include spent nuclear fuel, crystalline ceramics, cement, and vitrified glass produced from a molten mixture of glass frit, radioactive liquid and radioactive salt wastes. The types and quantities of each waste form will vary from country to country depending on the policies of the individual governments and the types of radioactive processing facilities present within the respective national boundaries.

Spent nuclear fuel waste in the United States originates from several sources. The majority is derived from the operation of commercial power reactors, and commercial spent fuel represents the largest curie content inventory of any waste form awaiting disposal. This large radionuclide inventory results from both the large volume of spent fuel awaiting disposal and the high concentration of radionuclides associated with the spent fuel rods

0275-0279/99/0038-0010$05.00

(SCP 1988). Nearly 30,000 metric tons of commercial spent fuel were in storage at commercial power reactors by 1995, with future projections suggesting that this quantity will more than double by the anticipated opening of the potential Yucca Mountain high-level nuclear waste repository site in 2010 (Integrated Data Base Report 1995). The proposed Yucca Mountain repository site is conceptually designed to hold 70,000 metric tons of high-level radioactive waste. Under current policy, about 63,000 tons of this total would be spent commercial fuel. Spent fuel originating from former weapons production facilities at several Department of Energy laboratory sites and U.S. Navy nuclear reactors represent additional sources of waste. These fuel types represent only a small percentage of the total U.S. inventory of spent fuel and are much more varied in composition that those derived from commercial power plants (Fillmore and Bulmahn 1994; Louthan et al. 1996). Some of these fuels, in particular those containing pyrophoric metal or corrosion components, would be classified as a RCRA hazardous waste and would therefore be precluded from direct disposal in a repository system (NRC 1998).

The UO_2 in spent nuclear fuel is not stable under oxidizing conditions and will be converted to the U^{6+} state when exposed to an oxidizing atmosphere. In the presence of aerated solutions, the uranium solid will tend to be dissolved as the uranyl (UO_2^{2+}) ion, or be complexed with a variety of inorganic and organic compounds (e.g. CO_3^{2-}, PO_4^{3-}) to form soluble compounds (Langmuir 1978; Shoesmith and Sunder 1992). The uranium may also combine with various precipitants in solution to form secondary uranyl solid phases. These alteration phases may influence the corrosion rate of the spent fuel matrix (or oxidized portions thereof), convection of fluids and oxidants to and from the reacting spent fuel surface, and the solubility of released spent fuel components (including uranium, transuranic species, fission products, and stable daughter decay products). An accurate appraisal of alteration phase development will therefore play an important role in evaluating the corrosion process for spent fuel. The development of a suite of alteration phases during corrosion testing, which are representative of those expected following long-term repository disposal, is often limited by slow reaction kinetics and relatively short test intervals. This effect may be especially relevant when experiments are conducted at low sample surface area/leachant volume ratios. The relatively high fluid content used in these tests effectively dilutes the concentration of components released from the samples (often intentionally), thereby inhibiting the nucleation of alteration phases. Alternatively, analogue studies of the weathering of uraninite (ideally UO_2) can provide information on long-term reaction trends, but the environmental conditions under which these materials were altered may have changed over time and are often not known with any great degree of certainty. A correlation between the alteration processes occurring in experimental and natural analogue reactions is an essential component for evaluating models that predict the effects of alteration phases on the dissolution behavior of spent fuel and the release of radionuclides.

This paper provides a literature summary of recent efforts to evaluate the reaction of spent fuel and identifies potential secondary phases that may occur during spent fuel alteration in an oxidizing environment. These efforts include experimental investigations into the dissolution of unirradiated UO_2, natural occurrences of uranium, spent fuel, and simulated spent fuel (SIMFUEL). The effort here focuses on the alteration of uranium oxide in an oxidizing environment, and thus is directly applicable to the disposal of commercial fuel in the environment expected at the Yucca Mountain site in Nevada.

THE STRUCTURE AND COMPOSITION OF SPENT NUCLEAR FUEL

A typical commercial nuclear reactor fuel assembly consists of vertically stacked UO_2 pellets that are clad in a zirconium metal jacket (Zircaloy-2 or -4; Johnson and Shoesmith 1988; Ewing et al. 1995). Nuclear fuel that is destined for use in commercial reactors in the

United States is enriched in ^{235}U (~3%) in order to achieve criticality during reactor operation. The fuel pellets are fabricated from granules of uranium oxide that have been hot-isostatically pressed. This process produces a pellet with grain sizes that range from two to ten μm and 95% maximum theoretical density.

Spent nuclear fuel that is retired from commercial power service will typically contain between 95 and 99% UO_2 (Johnson and Shoesmith 1988). The remaining fraction is composed of fission products and transuranic elements that formed during nuclear power plant operation. The relative proportions of these inventories will change over time, varying as a function of their original concentration in the spent fuel, differences in half-lives of the various isotopes present, and differences in the production of intermediate decay daughter isotopes. Weber and Roberts (1983) have determined the cumulative radiation dose for commercial high-level waste that has been removed from reactor service for more than ten years. Early radiation doses from these wastes are largely due to the presence of ^{137}Cs and ^{90}Sr. The ^{137}Cs isotope has a half-life of 30.17 years and emits both γ and β radiation during decay to ^{137}Ba. The ^{90}Sr isotope has a half-life of 29.0 years and emits only β radiation during its decay to ^{90}Y, subsequently followed by γ and β emissions during the transformation of ^{90}Y to the stable ^{90}Zr isotope. In general, β-decay is the primary source of radiation during the first 500 years of storage, as it originates primarily from the shorter-lived fission products (Weber et al. 1998).

The relative number of α-decay events arising from the actinide elements in spent fuel will increase in importance over time relative to other types of decay processes. The α-decay events are expected to be the dominant source of radioactivity after approximately 500 years (Weber and Roberts 1983; Weber et al. 1998). The transuranic element inventory in commercial spent fuel is dominated by plutonium, with the total inventory in spent fuel rods expected to be on the order of one percent (Johnson and Shoesmith 1988). The proportion of total decay events arising from the actinide inventory, however, will be dominated by the decay of americium until ~3 × 10^5 years, followed by neptunium, and then finally uranium (Weber et al. 1982). Despite its relative abundance, plutonium will not account for more than 20% of the total dose at any time.

The elevated temperatures encountered during nuclear reactor operation (up to 1700°C) may promote grain growth and diffusive segregation of fission products that are incompatible within the UO_2 crystalline structure. A portion of these fission products will migrate out of the UO_2 matrix under the influence of a temperature gradient and accumulate in the grain boundaries and the gap region between the fuel and Zircaloy cladding. This redistribution of fission products is dependent on the burnup history, with fission product migration increasing with percent burnup (Oversby 1994). Cesium, iodine, and several gas-phase radionuclides are most often noted to be preferentially partitioned into the "gap and grain-boundary" regions (Johnson et al. 1983). Gray and coworkers (Gray and Strachan 1991; Gray and Wilson 1995; Gray 1999) determined that the gap and grain-boundary inventory of ^{137}Cs usually accounted for less than one-to-two percent of the total amount present in light water reactor fuel. This fraction of ^{137}Cs was also determined to be approximately equal to one-third that of the fission gas, whereas ^{129}I contents were nearly equal those of the fission gas fraction (Gray 1999). Numerous studies have noted that these gap and grain-boundary radionuclides may be rapidly dissolved and mobilized during aqueous corrosion (Johnson et al. 1983; Gray and Strachan 1991; Forsyth and Werme 1992; Gray and Wilson 1995; Stroes-Gascoyne et al. 1995). This rapid release is attributed to the location of these radionuclides along easily accessible cladding-gap regions of the spent fuel assembly and their high solubilities. The rate of strontium release from grain-boundary regions of CANDU (CANada Deuterium Uranium) fuels is also noted to be slightly faster than uranium, but still generally more subdued than cesium or iodine (Stroes-

Gascoyne et al. 1993). Gray and Wilson (1995) have determined that the Sr residing in the gap and grain-boundary regions represents less than about 0.2% of the total inventory in spent fuel.

The segregation of fission products into different regions of the spent fuel can influence the patterns of their subsequent release from the waste package during aqueous corrosion. Temporal variations in radionuclide release rates would likely affect the patterns of their subsequent adsorption and/or incorporation in uranyl alteration phases and other solids. For example, the rapid release of Cs associated with the initial exposure of the gap and grain boundary inventory to aqueous fluids would likely lead to an increase in the aqueous activity of Cs ions in solution and possibly the formation of Cs-enriched phases.

Most transuranic and lanthanide element radionuclides occur in solid solution within the UO_2 matrix (Johnson and Shoesmith 1988). However, some of these radionuclides (e.g. plutonium), are preferentially enriched along a narrow band (<200 µm) at the outer edge of the fuel pellet rims (Thomas et al. 1992; Forsyth and Werme 1992). The rims are also characterized by an increase in porosity (due to fission gas accumulation and associated formation of vesicles), higher alpha dose rates, higher fission product concentrations, a finer grained structure, reduction in hardness, and increased fracture toughness (Thomas et al. 1992; Matzke and Speno 1996). The unique characteristics of the spent fuel rims had developed as a result of the self-shielded capture of epithermal neutrons and a resultant enhancement in plutonium production. This process led to a restructuring of the fuel pellets due to loss or refinement of the original UO_2 grains and the formation of micrometer-sized pores (Thomas et al. 1992).

Small metallic inclusions of solid solution alloys containing radionuclides of Ru, Mo, Pd, Tc, Rh and other trace radionuclides (the five-metal ε-phase), have been noted in examinations of sectioned spent fuel samples. The sizes of these inclusions range from ten nanometers to the micrometer size range, with the largest concentrations occurring in regions of the spent fuel samples closest to the grain boundaries and at the fuel pellet rims (Thomas et al. 1989; Forsyth and Werme 1992). Thomas and Charlot (1990) also noted a spatial correlation of xenon and krypton fission gas "particles" closely associated with the five metal ε-phase. The xenon and krypton were reported to have formed as internally pressured condensed-phase particles. The ε-phase alloy inclusions located along grain boundaries and near the pellet rims would be easily accessed by water penetration in these relatively permeable areas. Release of more soluble metal components, such as technetium, from exposed ε-phase inclusions would likely be controlled by the oxidative corrosion rate of the metal alloy.

Spent nuclear fuel and other nuclear waste forms will be enclosed in a disposal container prior to emplacement into a nuclear waste repository. Although a final container reference design for U.S. high-level nuclear waste has not yet been selected, it is likely that the container will be metallic, with an outer layer of corrosion-allowance and an inner layer of corrosion-resistant material (Cantlon et al. 1995; Sagüés 1999). When these two metals are contacted by groundwater containing electrolytes, the outer layer will corrode first, thereby inhibiting corrosion of the inner layer. The outer wall of the container will also provide mechanical strength against handling stresses and impacts from rock falls.

THE REPOSITORY ENVIRONMENT
An unsaturated setting

The Yucca Mountain site is currently being evaluated to determine its suitability for disposal of high-level nuclear wastes in the United States. It is in southern Nevada, ~160 km (100 miles) northwest of Las Vegas (TSPA-VA 1998), just north of the Amargosa

Desert. It is characterized by a semiarid climate with limited seasonal rainfall. The proposed waste emplacement horizon is located in a densely welded and devitrified zone of Miocene aged volcanic tuffs, ~300 m below the crest of Yucca Mountain and 250 m above the local water table. The tuff units at this site are in the unsaturated or vadose zone, with the pore space in the rock being partially saturated (~90%) with liquid water and the remainder being filled with a vadose-zone gas phase. This gas phase will have a composition similar to air, except that its CO_2 content will be increased through interaction with the soil above the repository, the geologic host rock, and engineered barrier system components (Thorstenson et al. 1989; Murphy 1993; TSPA-VA 1998).

The geologic environment in a nuclear waste repository will be subjected to changing temperatures and levels of radioactivity over time (TSPA-VA 1998). The emplacement of radioactive wastes in the repository will result in an increase in the temperature of the host rock adjacent to the waste packages. This increase will arise through the energy released due to the decay of radioactive waste form components. The spent fuel waste packages are expected to heat the surrounding repository environment to higher temperatures than any other types of waste forms due to the greater levels of radioactivity associated with the spent fuel. The repository hydrologic and atmospheric systems are likely to be perturbed by this thermal event. Variations in water flux rates and convection patterns are likely to occur, as are potential changes in groundwater composition as the hydrologic system adjusts to this thermal pulse. The composition of the repository atmosphere may also vary due to changes in the rate of gaseous exchange with host rock components and changes in air and groundwater circulation patterns. A prolonged cooling period will follow the initial heating period, as the radioactivity levels of the various waste packages decrease over time.

The proposed waste emplacement strategy for the Yucca Mountain site will result in a thermal event that heats the repository host rock immediately surrounding the waste packages to a temperature above the local boiling point of water at ~96°C (Buscheck et al. 1996). Water contact with the waste packages will only occur after the temperature of the surrounding host rock falls below the boiling point. This thermal period will thus prevent the onset of liquid water reaching the waste packages for approximately 1000 years following emplacement (TSPA-VA 1998). The actual length of time between waste emplacement and the decrease in repository temperature below the boiling point will depend upon the thermal energy released from the individual waste packages, emplacement geometry, thermal conductivity of the host rock and/or backfill, and hydrologic characteristics of the repository system (Buscheck et al. 1996; TSPA-VA 1998).

The repository atmosphere, solubility limits of various elements, and the dissolution kinetics of various minerals present in the tuff host-rock will influence the composition of groundwater at the Yucca Mountain site. The groundwater composition is also expected to change during thermal fluctuations that result from waste emplacement into the repository. An averaged analysis of water samples collected from well J-13, located at the Nevada Test Site adjacent to the proposed Yucca Mountain repository, indicates a water composition dominated by sodium, silicon, and bicarbonate (Table 1; Harrar et al. 1990). Several other alkali, alkaline earth, and anion elements are present in significant amounts, with many of these constituents representing potential components for the precipitation of uranium mineral phases (Wronkiewicz et al., 1992, 1996, 1997a). The J-13 well water was also reacted with the Topopah Spring Member repository host rock from the Yucca Mountain site at 90°C, producing a fluid (EJ-13) with a significantly elevated silicon concentration, and slight variations in most other components (Table 1).

Effects of radiation on the repository environment

Calculated radioactivity level trends for aged CANDU spent fuel indicate that most of the γ and β source radionuclides will have decayed to near background levels after 300 to 500 years (Sunder et al., 1995). The long-term radiation doses from aged spent fuel will thus be dominated by α radiation fields arising from the decay of long-lived actinides, rather than the γ and β fields that dominate relatively youthful spent fuel. The composition of the spent fuel will also change over time due to the decay of fission products and actinides, and the ingrowth of stable daughter isotopes.

Table 1. Average composition of major components in groundwater at the Yucca Mountain site (mg/L)

Component	J-13^1	EJ-13^2
Si	28.5	45.4
Na	45.8	54.1
K	5.0	8.1
Mg	2.0	1.0
Ca	13.0	8.8
HCO_3^-	128.9	135.0
F	2.2	2.4
Cl	7.1	7.2
NO_3^-	8.8	7.6
SO_4^{2-}	18.4	17.3
pH	7.4	8.2

1. Saturated zone groundwater (Harrar et al. 1990); 2. Tuff rock reacted with J-13 water for 14 days at 90°C (Wronkiewicz et al. 1996).

The high level of radioactivity associated with the decay of fission products and transuranic elements may influence the corrosion processes of spent fuel. Water molecules exposed to radiation energy may be transformed into a variety of radiolysis products, including oxidizing molecules (H_2O_2, O_3) and short-lived radicals (O_2^-, ·OH) that can increase the oxidation potential of the solution contacting the waste form. These species, particularly the ·OH radical, may accelerate the oxidative-dis-solution process of the spent fuel matrix (Sunder et al. 1992, 1995). Oxidative dissolution of spent fuel may also occur in a repository sited in a reducing environment when the redox potential adjacent to the surface of spent fuel is increased by the formation of water radiolysis products. Such a scenario has been evaluated in the proposed Canadian high-level waste disposal system which involves the disposal of CANDU fuel (Shoesmith and Sunder 1992; Sunder et al. 1995). The point at which the threshold dissolution rate, below which oxidative dissolution processes due to the formation of gamma-radiolysis products, becomes insignificant in comparison with chemical dissolution rates in a non-oxidizing environment, is predicted to occur at ~200 years in this system. Similarly, the oxidative dissolution enhancement by formation of α plus β radiolysis products in water will only be important for 600 years or less.

The formation of radiolytic acids (e.g. HNO_3, H_2CO_3, and $H_2C_2O_4$; Wronkiewicz et al. 1992, 1997b) may increase the acidity of the solution in contact with the waste in a moist air or air-water system. The stability of the secondary uranyl phases may be either enhanced or reduced in the presence of these acids, depending on the pH of the initial solution and the stability field of the particular phase(s). Carbonic and carboxylic acid species may also complex with uranium and other actinide elements, thereby increasing the solubility of uranium, enhancing the dissolution rate of the spent fuel, and increasing the mobility of radionuclides.

OXIDATION AND DISSOLUTION PROCESSES FOR URANIUM SOLIDS

The dissolution rate of UO_2 or spent fuel in water will depend upon the redox conditions of the environment to which it is exposed and the presence of water. Although uranium is sparingly soluble in a reducing environment, its solubility may increase many orders of magnitude under oxidizing conditions (Langmuir 1978; Shoesmith and Sunder 1992; Ewing et al. 1995). The oxidation of UO_2 leads to the formation of progressively higher oxidation states of uranium and an oxidized layer at the surface with a composition of UO_{2+x} (where $0 < x \leq 1$). The kinetics of this oxidation step may be further enhanced when the reaction at the sample surface takes place in a thin film of water that is exposed to an oxidizing atmosphere (Posey-Dowdy et al. 1987; Ahn 1996). McGillivray et al. (1994) have noted a similar effect in the reaction of uranium metal, whereby an increase in oxidation rate varied as a function of water vapor pressure, with both $H_2O_{(vap)}$ and O_2 gas forming reactive species (OH^- and O^{2-}, respectively) that diffuse through the uranium lattice structure. The oxidation potential adjacent to a waste form surface may also be increased through the interaction of radiation energy emitted by the waste forms with water (Sunder et al. 1992; Shoesmith and Sunder 1992). The unsaturated geologic medium at the Yucca Mountain site represents such an environment, where an oxidizing atmosphere, limited amounts of liquid water, and radiation energy will be present. This environment, combined with the elevated temperatures associated with waste emplacement, may accelerate oxidative dissolution process for spent fuel once the fuel container and Zircaloy cladding have been breached.

The oxidation process for spent fuel will initiate along the grain boundaries and then progress inwards from the grain boundaries to the cores (Taylor et al. 1980; Thomas et al. 1989; Einziger et al. 1992). A slight shrinkage of the unit-cell parameters of the crystals will occur as uraninite oxidizes from UO_2 to U_4O_9 (Johnson and Shoesmith 1988). This shrinkage can weaken the grain-boundary regions, leading to microfracturing of the samples. The microfracturing, in turn, may allow water to penetrate into the grain boundaries, thus increasing the surface area of the spent fuel pellets exposed to reactive fluids. This process will also expose accumulated gap and grain-boundary fission products (e.g. radionuclides of cesium and iodine) and potentially lead to more rapid release rates (Ahn 1996). The occurrence of microfractures has been noted to induce a friable nature to spent fuel samples, where light crushing or polishing will cause the fuel to separate into individual grains (Gray and Strachan 1991). This process may also be expected to lead to disaggregation of fuel pellets upon exposure to water.

Oxidation beyond U_4O_9 involves a conversion from the cubic uraninite structure into a tetragonal (distorted cubic) structure of the U_3O_7 state. Electrochemical investigations into this process have noted the development of a five-to-ten nanometer thick U_3O_7 surface layer (Johnson and Shoesmith 1988). Continued oxidation to U_3O_8 will be accompanied by extensive recrystallization to an orthorhombic structure (Johnson and Shoesmith 1988; Finch and Ewing 1992). Where only a slight expansion of the unit-cell parameters occurs as uraninite oxidizes from U_4O_9 to U_3O_7, a large degree of expansion will occur upon continued oxidation to a U_2O_5 or U_3O_8 state.

Extended exposure of spent fuel to hot and dry atmospheric conditions, accompanied by U_3O_8 formation, could potentially result in volume expansion, spallation of fuel particles, release of fission gases, and rupture of the enveloping Zircaloy metal cladding (Einziger et al. 1992). However, oxidation tests conducted with spent fuel samples at moderate temperatures (175 to 195°C), and time periods approaching three years in length, appear to reach a maximum oxygen-to-metal ratio endpoint of 2.4. The burnup dependence on the activation energy for oxidation is noted to have a dominating effect on suppressing the rate of U_4O_9-to-U_3O_8 conversion in spent fuel (Choi et al. 1996; Kansa et al. 1999). This inhibiting effect has been attributed to the presence of fission products and higher actinides in the spent fuel matrix. Significant U_3O_8 formation due to dry-air oxidation is not expected to occur in a geological repository for spent fuel, provided the maximum exposure temperature of the fuel to air does not exceed 150°C (Kansa et al. 1999).

The attainment of an oxidation state beyond U_3O_7 in the presence of water or moist air is usually precluded by oxidative dissolution (Johnson and Shoesmith 1988; Shoesmith and Sunder 1992). The U_3O_7 composition thus appears to be a limiting threshold, beyond which the process of oxidative dissolution involving electrochemical rate parameters will become dominant. This corrosion potential threshold occurs at approximately -50 to -100 mV for a fluid immediately in contact with the surface of the uranium solid (Shoesmith and Sunder 1992). The dissolution kinetics of UO_{2+x} may, however, be hindered when only limited amounts of water are present. In tests conducted with UO_2 samples at 225°C, both dehydrated schoepite, $(UO_2)O_{0.25-x}(OH)_{1.5+2x}$ (where $0 \le x \le 0.25$), and U_3O_8 were found as alteration products on samples exposed to aerated steam (Taylor et al. 1995). Dehydrated schoepite occurred as isolated surface patches at relative humidity values of approximately 30 to 70%, possibly reflecting limited adsorption of water along microcracks, pores, and surface impurities. A more uniform film of dehydrated schoepite was noted at relative humidities of >70%, suggesting that the adsorbed film of water completely enveloped the sample and was thick enough to dissolve and mobilize uranium over the entire surface.

The effect of oxidative dissolution processes on the alteration of spent nuclear fuel has been investigated by examining cross-sectioned fragments of altered spent fuel (Finch et al. 1999). These samples had been reacted in the presence of condensed steam or dripping EJ-13 groundwater at 90°C. The samples were progressively altered inwards along a somewhat uniform reaction front, with the original spent fuel material being replaced by precipitated uranyl alteration phases (Fig. 1a). Oxidative dissolution was also noted to occur along the grain boundaries and was enhanced in tests where the samples were contacted by the largest volumes of simulated groundwater. The zone of grain-boundary corrosion also penetrated well in advance of the uniform reaction front, with penetration along grain boundaries occurring to depths exceeding 200 µm after five years of reaction. Dissolution patterns in these regions were characterized by open gaps between the grains and curvilinear voids that penetrate approximately one or two micrometers from the grain boundaries into the cores. This grain-boundary dissolution is expected to increase the overall exposed surface area of the samples and also enhance the potential for the disaggregation of the spent fuel pellets along the weakened grain contacts.

The corrosion progress of unirradiated UO_2 samples was monitored by examining the surficial and cross-sectioned alteration features (Wronkiewicz et al. 1997a). The cross-sectioned UO_2 samples displayed evidence for preferential dissolution along the grain boundaries, resulting in the formation of a polygonal network of open channels (Fig. 1b). Grain-boundary corrosion and polygonal fracturing has also been observed in spent fuel samples as described above (Finch et al. 1999). The corrosion of the grain boundaries also led to the loosening of near surface UO_{2+x} grains and their spallation from the sample surface (Wronkiewicz et al. 1992; Taylor et al. 1995). The intergrain channels were characteristically 0.3 to 1 µm wide, and penetrated into the pellets to an average depth of two-to-five grain boundaries (~15 to 20 µm). The penetration depths, however, were variable, ranging from regions with essentially no visible corrosion of the grain boundaries to regions where penetration occurred to a depth of approximately 50 µm.

Studies of spent fuel microstructures suggest that the association of closely spaced fission gas bubbles along the grain boundaries may provide a preferential pathway for oxygen and fluid penetration, thereby enhancing the internal oxidative dissolution of the fuel along grain boundaries (Thomas et al. 1989). The preferential corrosion of grain boundaries in unirradiated UO_2 specimens (Wronkiewicz et al. 1997a), where fission gases do not exist, suggests that the cause for the rapid intergrain-boundary corrosion may be (at least partially) attributed to localized structural defects in the UO_2 grains. These defects may

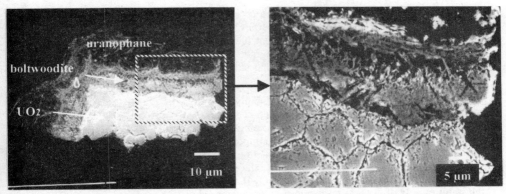

Figure 1. Scanning electron photomicrographs of cross-sectioned spent nuclear fuel and UO_2 samples following reaction with dripping EJ-13 groundwater solution at 90°C. (a) spent fuel after 3.7 years of reaction displaying, from top to bottom, mounting epoxy (black), porous mat of surface alteration phases (Na-boltwoodite), and spent fuel matrix displaying corrosion of the grain boundary regions (unpublished data of R.J. Finch), and (b) UO_2 pellet after eight years of reaction displaying, from top to bottom, mounting epoxy (black), layer of alteration phases (compreignacite), alteration phase development within grain boundary corroded region, and unaltered UO_2. [Used by permission of the editor of the Materials Research Society, from Wronkiewicz et al. (1997a) Fig. 2, p. 522].

have been induced during the hot-isostatic pressing of the samples. Such defects would almost certainly accumulate along the intergrain-boundary regions of the samples, where the dimensions of the grains had changed in response to the hot pressing process. Strain accumulation in the intergrain-boundary regions would lead to higher energy potentials for corrosion relative to grain cores, and hence, faster dissolution rates.

Changes in temperature may affect the dissolution rate of UO_2 and spent fuel, with the release rates of uranium plus other waste form components generally increasing with temperature. For example, the rate of dissolution of UO_2 in deionized and mildly oxidizing water has been shown to increase in accordance with temperature over range of 25 to 90°C (Thomas and Till 1984; de Pablo et al. 1997; Steward and Mones 1997). A change in reaction rate patterns, from a square root to a linear dependence on the carbonate concentration has also been observed between 25 and 40°C and correlated to different reaction mechanisms for UO_2 dissolution (de Pablo et al. 1997).

The effects of temperature and activities of various uranyl complexants (e.g. bicarbonate ions) on the dissolution rates of UO_{2+x} solids may be quite complex. Variations in carbonate content are reported to have a greater effect on uranium release rates than temperature, with UO_2 samples displaying five- to twelve-fold variation in dissolution rates between carbonate concentration levels of 2×10^{-4} and 2×10^{-2} molar (Steward and Mones 1997). The effect of increasing carbonate concentration was even more enhanced when the dissolving solid was dehydrated schoepite, where a rate increase of 25- to 50-fold was noted. Other studies suggest an inverse temperature dependence for uranium dissolution rates when carbonate ions are present. For example, Thomas and Till (1984) demonstrated that the dissolution rate of UO_2 pellets decreased in granitic groundwater as temperatures were increased from 30 to 90°C, with the decreased rate being attributed to the deposition of passivating uranyl phases on the sample surfaces at higher temperatures and alkalinity. This observed inverse temperature dependence is consistent with models suggesting that aqueous uranium complexes may have reaction quotients that vary inversely with temperature (Plyasunov and Grenthe 1994). This trend is also supported by thermodynamic calculations suggesting that uranyl phases such as schoepite,

$[(UO_2)_8O_2(OH)_{12}](H_2O)_{12}$, will have decreased equilibrium constants as a function of increasing temperature over a range of 25 to 100°C (Murphy 1997). Thus, the solubilities of uranyl phases that have formed during spent fuel alteration may increase as the repository system cools down over time (i.e. retrograde solubility). This effect may be further augmented by the increasing thermodynamic stability of some carbonate complexes with decreasing temperature.

Temperatures may also control the stability of individual phase "polymorphs" (e.g. schoepite vs. dehydrated schoepite; Vochten et al. 1990; Taylor et al. 1993; Finch et al. 1998), thereby indirectly influencing the solubility of uranium. The stability regime of the various uranium alteration phases and the composition of the fluid contacting those phases would determine the influence of phase transitions on the alteration pattern of UO_{2+x} solids. Taylor et al. (1993) also demonstrated that the stability fields for the oxidation products $\gamma\text{-}UO_3$ and dehydrated schoepite in a $UO_2\text{-}O_2\text{-}H_2O$ system would be diminished relative to the stability field of U_3O_8 as temperatures were increased over the range of 25 to 200°C.

EFFECTS OF SECONDARY URANIUM MINERALS

Influence on the dissolution of uranium solids

Uranium oxide is thermodynamically unstable in the presence of moisture and an oxidizing environment. The uranium oxide will dissolve under such conditions with the rate determining step for dissolution being the oxidative dissolution of uranium and its release as uranyl ions (UO_2^{2+}) or possibly other aqueous uranyl complexes (e.g. $UO_2(CO_3)_2^{2-}$; Shoesmith and Sunder 1992). The majority of the radionuclides present in spent fuel are contained either as substitution impurities within the uranium oxide crystal structure or as isolated inclusions surrounded by the uranium oxide matrix. The rate of UO_2 matrix dissolution will thus exert a dominating control on the release rate of many radionuclides during the oxidative dissolution of spent fuel.

The uranium released from spent fuel or UO_2 may precipitate as secondary uranyl phases when the oxidative dissolution rate exceeds the capacity of a system to transport dissolved uranyl species away from the reacted surface. The preservation of significant quantities of uranyl minerals as alteration layers around primary uraninite in many natural uranium deposits confirms that the rate of primary uraninite oxidative dissolution in natural systems can be rapid relative to the rate of transport (Murphy and Pearcy 1992). Environments with such a limited transport capacity are typified by desert-like settings like the Yucca Mountain site, where limited rainfall and groundwater recharge have resulted in a relatively deep groundwater table and a slow vertical flux of water through the vadose zone rock column. The accumulation of uranyl phases as alteration products has also been observed in experimental corrosion studies with unirradiated UO_2, SIMFUEL, and spent fuel (Wronkiewicz et al. 1992; Diaz-Arocas and Garcia-Serrano 1997; Finch et al. 1999).

The previously described relationships between secondary phase growth and uranium release suggest that the alteration phases may limit the aqueous transport or diffusion of components near the interface between the reacting solid and the leachant solution. The formation of these phases has been postulated to create a protective barrier that may limit the diffusion of dissolved components, oxidants, and other reactants to or from the sample surface (Thomas and Till 1984; Shoesmith et al. 1996). Such phases could act as a protective layer on the samples, limiting the dissolution rate of the solid. Microscopic evidence supporting such a process, however, is not readily apparent on reacted UO_2 and spent fuel samples (Wronkiewicz et al. 1992, 1997a; Finch et al. 1999, respectively). Detailed examinations of these samples revealed the presence of a porous and fractured alteration layer that appears to be permeable to fluids. The alteration layers may, however, inhibit the contact of fluid with a portion of the UO_{2+x} surface by their physical attachment, or by limiting the convection of fresh leachant solution to the sample surface. The presence of an alteration layer has also been noted to diminish the release of solid UO_{2+x} granules from the sample surface by acting as a cohesive mat that holds the loosened surface particles in place (Wronkiewicz et al. 1997a).

Comparison of corrosion products between spent fuel and UO_2

It is impossible to fully simulate the expected repository corrosion behavior of spent nuclear fuel because there currently are no materials available that replicate a spent fuel assembly that has aged for hundreds or thousands of years. Potential variations in corrosion behavior arise due to differences in radioactivity levels, types of radiation emitted (α, β, and γ), thermal output as a function of radiation dose, temperature history, composition, and distribution of contaminants (e.g. fission products, decay products, and transuranic elements). Commercial spent fuel samples, even those pulled from reactor service nearly 50 years ago, are not representative of an "aged fuel" sample because they emit a much stronger radiation dose and have much higher ($\gamma + \beta)/\alpha$ radiation fields than would be expected for aged spent fuel. The high levels of radioactivity associated with relatively youthful spent fuel may influence its alteration behavior by causing the formation of a variety of radiolysis products (Sunder et al. 1992; Shoesmith and Sunder 1992). Conversely, UO_2 pellets do not contain any of the long-lived fission products, transuranic elements, ingrown daughter decay products, and degradation features (fuel pin cracking, gap and grain-boundary inventories, radiation damage, rim effects, etc.) that are characteristic of spent fuel rods.

Unirradiated UO_2 and spent fuel dissolution reactions were examined in a series of parallel tests conducted at Argonne National Laboratory (Wronkiewicz et al. 1992, 1996, 1997a; Finn et al. 1994, 1995; Finch et al. 1999). An investigation of the alteration progress from eight-year tests with UO_2 indicates that approximately 95% of the uranium that was released during the corrosion of the samples had subsequently precipitated back on the surfaces of the samples, test vessel components, and within the corroded intergrain-boundary regions (Wronkiewicz et al. 1997a; Fig. 1b). Most of this uranium occurred in the form of dehydrated schoepite, a phase comprised of only the uranium released from the sample, oxygen, and water. Additional uranium had also combined with Si, Na, K, Ca, and Mg from the simulated groundwater leachant to form a variety of uranyl alteration phases. The end result of this process was the development of the following paragenetic sequence of alteration phases on the top surfaces of the samples within the first 3.5 years of reaction (Wronkiewicz et al. 1996; phase compositions given in Table 2):

uraninite (UO_{2+x}) → dehydrated schoepite → becquerelite + compreignacite
→soddyite → uranophane + boltwoodite → Na-boltwoodite

This trend apparently was controlled by the nucleation kinetics of the individual phases since uranophane, $Ca[(UO_2)(SiO_3OH)]_2(H_2O)_5$, boltwoodite, $K[(UO_2)(SiO_3OH)](H_2O)_{1.5}$, and Na-boltwoodite, $(Na,K)[(UO_2)(SiO_3OH)](H_2O)_{1.5}$, would be the phases predicted to form based on equilibrium constraints. Bruno et al. (1995) have also noted the formation of schoepite in oxic bentonite-granitic groundwater systems where this phase is thermodynamically unstable with respect to other uranyl phases. These authors suggest that schoepite formation is kinetically favored in this system and that the reaction pathway will eventually evolve towards the formation of uranophane as the end product of alteration. Replacement textures, dissolution pits, and phase overgrowths were also noted to occur during the UO_2 corrosion tests, with these features being used to help define the paragenetic sequence for UO_2 alteration (Fig. 2; Wronkiewicz et al. 1992, 1996). Similar alteration trends have been noted during the weathering of naturally occurring uraninite (Frondel 1956; Finch and Ewing 1991, 1992; Ewing 1993).

Variations in the UO_2 alteration phase sequence were noted along the sides and bottoms of the UO_2 samples (Wronkiewicz et al. 1996). In these areas, a reversed paragenetic trend was noted relative to the sequence previously described for the top

Table 2. Potential uranium minerals for the Yucca Mountain Repository Site

Mixed U^{4+} - U^{6+} phases
Ianthinite $[U_2^{4+}(UO_2)_4O_6(OH)_4(H_2O)_4](H_2O)_5$

Uranyl Oxide Hydrates
Schoepite $[(UO_2)_8O_2(OH)_{12}](H_2O)_{12}$
  Metaschoepite $[(UO_2)_8O_2(OH)_{12}](H_2O)_{10}$
  Dehydrated Schoepite $(UO_2)O_{0.25-x}(OH)_{1.5+2x}$ ($0 \leq x \leq 0.25$)
Studtite $UO_4(H_2O)_4$
  Metastudtite $UO_4(H_2O)_2$

Alkali and Alkaline Earth Uranyl Oxide Hydrates
Agrinierite $(K_2,Ca,Sr)U_3O_{10}(H_2O)_4$
Bauranoite $BaU_2O_7(H_2O)_{4.5}$
Becquerelite $Ca[(UO_2)_3O_2(OH)_3]_2(H_2O)_8$
Billietite $Ba[(UO_2)_3O_2(OH)_3]_2(H_2O)_4$
Calciouranoite $(Ca,Ba,Pb)U_2O_7(H_2O)_5$
  Meta.Calciouranoite $(Ca,Ba,Pb)U_2O_7(H_2O)_2$
Clarkeite $(Na,1/2Ca,1/2Pb)[(UO_2)O(OH)](H_2O)_{0-1}$
Compreignacite $K_2[(UO_2)_3O_2(OH)_3]_2(H_2O)_7$
Na-Compreignacite $(Na,K)_2[(UO_2)_3O_2(OH)_3]_2(H_2O)_7$
Protasite $Ba[(UO_2)_3O_3(OH)_2](H_2O)_3$
Rameauite $K_2CaU_6O_{20}(H_2O)_9$
Wölsendorfite $(Pb,Ba,Ca)_{6.5}[(UO_2)_{14}O_{19}(OH)_4](H_2O)_{12}$

Uranyl Silicates
Uranophane (α and β polymorphs) $Ca[(UO_2)(SiO_3OH)]_2(H_2O)_5$
Boltwoodite $K[(UO_2)(SiO_3OH)](H_2O)_{1.5}$
Na-Boltwoodite $(K,Na)[(UO_2)(SiO_3OH)](H_2O)_{1.5}$
Haiweeite $Ca[(UO_2)_2Si_5O_{12}(OH)_2](H_2O)_3$
Oursinite $(Co,Mg)(UO_2)_2Si_2O_7(H_2O)_6$
Sklodowskite $Mg[(UO_2)(SiO_3OH)]_2(H_2O)_6$
Soddyite $(UO_2)_2(SiO_4)(H_2O)_2$
Swamboite $UH_6(UO_2)_6(SiO_4)_6(H_2O)_{30}$
Uranosilite USi_7O_{17}
Weeksite $K_2(UO_2)_2(Si_5O_{13})(H_2O)_3$

Uranyl Selenates
Guilleminite $Ba[(UO_2)_3(SeO_3)_2O_2](H_2O)_3$

Uranyl Molybdates
Calcurmolite $Ca(UO_2)_3(MoO_4)_3(OH)_2(H_2O)_{11}$
Cousinite $Mg(UO_2)_2(MoO_4)_2(OH)_2(H_2O)_5$
Iriginite $U(MoO_4)_2(OH)_2(H_2O)_3$
Moluranite $H_4U^{4+}(UO_2)_3(MoO_4)_7$
Tengchongite $CaU_6Mo_2O_{25}(H_2O)_{12}$
Umohoite $(UO_2)(MoO_2)(OH)_4(H_2O)_2$

Synthetic Uranium Phases
Protasite-type (Buck et al., 1997) $(Cs_{0.8}Ba_{0.6})(UO_2)_5(MoO_2)O_4(OH)_6 \cdot nH_2O$
Sr-Curite (Burns and Hill, 1999) $Sr_3[(UO_2)_4O_4(OH)_3]_2(H_2O)_2$

Figure 2. Scanning electron microscope photomicrographs of alteration phases formed during the aqueous alteration of UO_2 at 90°C. (a) dehydrated schoepite crystals with fine-grained UO_{2+x} surface particles after 1.5 years of reaction, (b) stellar-shaped accumulation of dehydrated schoepite after 2.25 years of reaction, (c) dehydrated schoepite crystals showing dissolution pits after 8 years of reaction, (d) tabular books of schoepite after 3.5 years of reaction, (e) large tabular becquerelite crystal showing dissolution pits and the precipitation of acicular uranophane crystals, some directly within the pits formed on the becquerelite (3.5 years of reaction), (f) bladed soddyite crystals after 3.5 years of reaction, (g) very fine-grained uranyl silicate (soddyite?) after 3.5 years of reaction, (h) U-Si coating formed over unidentified tabular phase that has been completely dissolved away (3.5 years of reaction), and (i) dense mat of Na-boltwoodite crystals after 8 years of reaction. [Used by permission of the editor of *Journal of Nuclear Materials*, from Wronkiewicz et al. (1992) Fig. 6, p. 117 and Wronkiewicz et al. (1996) Fig. 6, p. 90].

surfaces of the samples. This reversed trend relates to the position of the phases on the sample surface rather than their occurrence as a function of time. As the reacting solution migrated across the UO_2 sample surface it was successively depleted in silicon, alkali, and alkaline earth elements by the precipitation of alteration phases. Thus, soddyite, $(UO_2)_2(SiO_4)(H_2O)_2$, would have formed where Ca, Na, and K activities were sufficiently lowered by the "upstream" precipitation of uranophane and Na-boltwoodite. Dehydrated schoepite eventually reappeared as the dominant alteration phase as the activity of silicic acid was further depleted by the precipitation of soddyite. This reverse paragenetic trend also correlates directly with predictive models evaluating the formation of alteration phases on spent fuel (Bruton and Shaw 1988) and uraninite (Pickett and Murphy 1999) where replenishment of alkali, alkaline earths, and silicon is limited by restricted water flow.

Solid solution compositional changes were also noted in some alkali and alkaline earth uranyl oxide hydrate phases as a function of their position on the surface of the reacted UO_2 samples (Wronkiewicz et al. 1996). For example, K/Na molar ratios of Na-compreignacite, $(Na,K)_2[(UO_2)_3O_2(OH)_3]_2(H_2O)_7$, progressively decreased from ~2.0 on the top surface of the eight year sample to a ratio of ~0.8 on the sample sides. Such a pattern indicates a preference for potassium ions relative to sodium ions in compreignacite, $K_2[(UO_2)_3O_2(OH)_3]_2(H_2O)_7$ (the K/Na ratio of the EJ-13 leachant is 0.17), as well as a progressive decrease of the potassium activity of the solutions as they migrate across the samples.

The formation of secondary uranyl phases may also influence the dissolution rate of UO_{2+x} by controlling the activity of uranyl ions in solution. The uranyl ion activity would progressively increase during the dissolution of the oxidized UO_{2+x} surface until secondary uranyl phases were nucleated. The precipitation of a succession of alteration phases in paragenetic sequence would sequentially lower the solution activity for the uranyl ion. Each successive decrease in solution activity could, in turn, lead to the dissolution of the uranyl phases that had formed earlier. An example of this process was noted in UO_2 corrosion tests where the precipitation of uranophane was correlated to the formation of dissolution pits in earlier formed becquerelite, $Ca[(UO_2)_3O_2(OH)_3]_2(H_2O)_8$ (Fig. 2e; Wronkiewicz et al. 1992). A second example was noted where the precipitation of becquerelite had also led to the formation of dissolution pits in earlier formed dehydrated schoepite (Fig. 2c; Wronkiewicz et al. 1996). A similar behavior is observed in experiments where schoepite is intentionally converted into compreignacite or becquerelite by reaction with potassium- or calcium-bearing solutions, respectively (Vochten and Van Haverbeke 1990; Sandino and Grambow 1994).

Spent nuclear fuel samples were reacted in a manner similar to the previously described UO_2 tests (Finn et al. 1994, 1995). A detailed examination of the reacted samples has revealed the presence of a number of uranyl phases that occur in alteration layers formed on the sample surfaces after several years of reaction at 90°C (Buck et al. 1997, 1998; Finch et al. 1999). The types and amounts of phases present depended on the composition and volume of water coming in contact with the spent fuel. When the samples were reacted in condensed deionized water, the alteration phase assemblage was limited to dehydrated schoepite, metaschoepite, $(UO_2)_8O_2(OH)_{12}](H_2O)_{10}$, and a Cs-Ba-Mo uranyl phase. This relatively simple phase assemblage reflects the limited variety of components present in the reacting system, with all components being derived only from water and spent fuel. A more diverse array of alteration products was noted when the spent fuel samples were reacted in the presence of dripping EJ-13 water (Fig. 1a; Finch et al. 1999). Alteration phases occurred in a relatively dense mat of needle-like uranyl phases that included Na-boltwoodite, β-uranophane, soddyite, Na-compreignacite, the Cs-Ba-Mo phase, and unidentified U-Zr and U-Zr-Pu oxides. The distribution of the various phases in these tests depended on the flow rate of water, with Na-boltwoodite and β-uranophane dominating the alteration phase sequence when relatively high drip rates (1.5 milliliters per week) were used.

Comparisons with natural analogues

Natural uraninite deposits that have been exposed to surface weathering fluids provide an attractive analogue from which to evaluate the long-term corrosion of spent fuel. The study of such deposits can provide fundamental data on reaction processes, mineral paragenesis, phase longevity, and element mobility. Such data can be used to confirm the validity of reaction processes that have been extrapolated from relatively short-term laboratory tests (Weber et al. 1998).

About 200 uranium minerals exist. Such a large number of phases would seemingly complicate attempts to use natural analogues to aid in determining potential reaction pathways for spent fuel. However, a large proportion of these uranium minerals contain either vanadium, phosphorus, or arsenic, three elements that are not expected to be present in the proposed Yucca Mountain repository environment in any significant concentration.

An additional 20 phases belong to the group of uranyl carbonates. None of the carbonate phases have been reported from corrosion tests with spent nuclear fuel or appropriate surrogate materials. Some of these phases may occur, however, in the presence of elevated CO_2 levels, especially during the early thermal pulse following waste emplacement in the repository (TSPA-VA 1998). Lead is also an essential component of many uranium minerals in natural systems, arising as a result of the radioactive decay of parent uranium isotopes. A significant accumulation of lead will not occur during any extended period of regulatory concern for waste disposal due to the long half-lives of the ^{238}U and ^{235}U isotopes (4.47×10^9 and 7.04×10^8 years, respectively). Less than 0.1% of the ^{235}U and <0.02% of the ^{238}U in the spent fuel will have undergone decay and transformation into stable daughter lead isotopes after 10^6 years of disposal. With the removal of V, P, As, Pb, and CO_3^{2-} as possible components, the total number of potential uranium phases is reduced to approximately forty (Table 2). Many of the phases listed in Table 2 would be expected to occur only in trace amounts, if at all, due to the limited mass of some phase-forming components in the spent fuel and/or groundwater.

An interesting natural analogue deposit for UO_2 alteration at Yucca Mountain has been described in a series of papers resulting from a study of the Nopal I uranium deposit located in the Peña Blanca Mining District, Chihuahua, Mexico (Murphy and Pearcy 1992; Cesbron et al. 1993; Leslie et al. 1993; Pearcy et al. 1994; Murphy 1995). This deposit is hosted in a sequence of volcanic rocks that have similar hydrologic, mineralogic, and geochemical characteristics as those of Yucca Mountain, Nevada. Furthermore, the relative youthfulness of uranium ores in the deposit (deposition occurred 8±5 my ago) has precluded the accumulation of significant amounts of radiogenic lead. As a result, the primary uraninite (UO_{2+x}) at the deposit has a composition similar to that expected for commercial spent nuclear fuel, except that elements representative of some of the fission product, transuranic, and daughter decay elements are missing.

The deposition of the primary uraninite mineralization at Nopal occurred under reducing conditions. The uraninite was subsequently exposed to oxidizing groundwater after regional tectonic forces elevated the deposit above the local water table (Murphy 1995). Uranophane crystals from the Nopal I deposit have been isotopically dated at 3.2 million years, with this date reflecting the minimum age for the first exposure of the Nopal uranium minerals to surface weathering processes (Pickett and Murphy 1997).

The primary uraninite at Nopal was progressively altered in oxidizing groundwater, first to ianthinite, $U_2^{4+}(UO_2)_4O_6(OH)_4(H_2O)_4](H_2O)_5$, followed by uranyl oxide hydrates (schoepite, dehydrated schoepite, and becquerelite), and finally uranyl silicates (soddyite, uranophane, weeksite, $K_2(UO_2)_2(Si_5O_{13})(H_2O)_3$, and boltwoodite). Weeksite and boltwoodite also appear to be the latest formed silicate phases (Leslie et al. 1993; Pearcy et al. 1994). A comparison of the reaction paragenesis at Nopal I with UO_2 dissolution under simulated Yucca Mountain test conditions (Wronkiewicz et al. 1992, 1996) indicates a close correlation between the two reaction trends (Fig. 3). Whatever variations occur between the depositional patterns of the two systems can be attributed to differences in the availability of certain cations in the leachant solutions. For example, when Si, Na, Ca, and K were the major components of the EJ-13 solution in the UO_2 tests, the reaction trends were characterized by the formation of alkali and alkaline earth uranyl-bearing oxide hydrates and silicates. At Nopal, the reaction trend is similar, with only minor variations being noted in the trace mineral assemblages. For example, the presence of barium and arsenic has possibly resulted in the formation of trace amounts of billietite, $Ba[(UO_2)_3O_2(OH)_3]_2(H_2O)_4$, and abernathyite, $K[(UO_2)(AsO_4)](H_2O)_3$. The close similarity of major alteration phase assemblages between the UO_2 experiments and Nopal suggests that the likely compositional ranges of dominant spent fuel alteration products in the Yucca

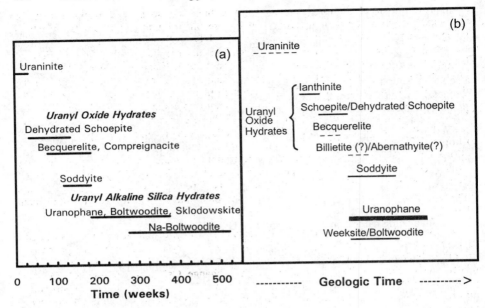

Figure 3. Comparative reaction paragenetic sequences for uranium alteration phases: (a) Experimental sequence developed on UO₂ samples after 8 years of corrosion (Wronkiewicz et al. 1997a), and (b) sequence developed in the Nopal natural uranium deposit, Mexico (Leslie et al. 1993). Weight of lines indicates relative abundance of phase: dashed = minor, thin = abundant, and thick = very abundant. [Used by permission of the editor of *Journal of Nuclear Materials*, from Wronkiewicz et al. (1996) Fig. 7, p. 92].

Mountain environment may be relatively limited and insensitive to small variations in system conditions (Pearcy et al. 1994). This correlation also indicates that the laboratory tests have replicated an environment representative of that occurring during the oxidative weathering of uraninite.

The predominance of Ca-bearing α-uranophane at Nopal (Leslie et al. 1993; Cesbron et al. 1993; Pearcy et al. 1994), relative to the Na-boltwoodite observed in the UO₂ tests (Wronkiewicz 1996), may reflect major differences in cation compositions for the respective systems. Reaction path models for the gas-water-rock geochemical system at Yucca Mountain predict that the aqueous calcium/sodium ratio will progressively decrease with increasing reaction progress (Murphy 1993). The α-uranophane at Nopal may thus reflect the presence of fluids that are less evolved (i.e. have higher Ca/Na ratios) than the Na-enriched EJ-13 solution used in the UO₂ tests. Conversely, the predominance of Na-boltwoodite in the UO₂ tests reflects the more-evolved sodium-rich nature of the EJ-13 leachant (Table 1). For spent fuel alteration, the occurrence of one specific uranyl phase over another may well reflect such changes in fluid composition. These changes may be induced by the reaction of groundwater with the Yucca Mountain host-rock, repository engineered barrier components (waste container, backfill, repository drift liners, etc.), and the spent fuel waste form (fission product and transuranic elements).

Influence of alteration phases on the migration of radionuclides

The dissolution behavior of spent fuel, and the rates and mechanisms of radionuclide release, are critical parameters that need to be determined before the performance of a nuclear waste repository containing spent fuel can be assessed.

The migration of fission products from a breached spent fuel package may be significantly retarded by their incorporation into secondary uranyl phases. Both natural analogue examples and experimental results with UO_2 pellets suggest that the uranium released during the dissolution of uraninite will readily combine with a wide variety of elements to form uranium alteration phases (Frondel 1956; Finch and Ewing 1991, 1992; Wronkiewicz et al. 1992; Leslie et al. 1993). Elements that may be incorporated in the uranyl oxide hydrate and mixed U^{4+}–U^{6+} alteration phases include U, Ca, K, Na, Sr, Pb, Ba, Mo, Bi, Ti, C, Nb, and Ta. Uranyl silicates may also contain U, Ca, K, Na, Cu, Pb, Mg, C, Co, Si, and rare earth elements. Many of these elements also occur as radioactive fission product isotopes and intermediate actinide decay-daughters in spent fuel. For example, ^{14}C, ^{87}Sr, ^{93}Mo, ^{94}Nb, ^{135}Cs, and ^{151}Sm are radionuclides that have long enough half-lives or are present in high enough concentrations in nuclear waste materials that they would present a concern with respect to long-term isolation (>300 years; Ahn 1994; Wronkiewicz et al. 1996; TSPA-VA 1998). It is expected that many of these fission products would become incorporated in alteration phases after their release from the spent fuel.

Burns and Hill (1999) experimentally formed a synthetic analogue of the lead uranyl hydrate curite, with strontium replacing lead in the curite crystal structure. This phase was orthorhombic and had a structural formula approximating $Sr_3[(UO_2)_4O_4(OH)_3]_2(H_2O)_2$. In a related study, Burns (1999) was able to ion exchange Cs^+ ions into the K^+ and Na^+ positions in boltwoodite. Substitutions of both Sr and Cs ions into these mineral structures may have important implications for the migration of radionuclides following the corrosion of spent fuel.

Studies conducted with commercial spent fuel have confirmed that the uranyl oxide hydrates and uranyl silicates will be present as alteration phases following oxidative dissolution and that some of these phases will incorporate fission products in their crystalline structures (Buck et al. 1997). In this study, cesium, barium, and molybdenum were found as primary constituents of an orthorhombic uranyl phase with a structure related to the alkaline earth uranyl oxide hydrates of the protasite group. All three elements are present as radioactive fission products in spent fuel, with additional barium being generated as a stable decay daughter product from radioactive cesium (e.g. $^{137}Cs \rightarrow \,^{137}Ba$). The structural formula determined for this phase is (where $n \approx 6$):
$(Cs_{0.8}Ba_{0.6})(UO_2)_5(MoO_2)O_4(OH)_6 \cdot nH_2O$.

Uranophane has also been detected in corrosion tests with spent fuel (Wilson 1990; Finch et al. 1999). Finch et al. had noted that both Ba and Sr were incorporated in the β-uranophane phase examined in their tests, where they likely replaced Ca in the interlayer sites of the crystal structure. Ruthenium and Tc were also detected as trace constituents. These authors also postulated that these two elements may possibly replace silicon in the structure of β-uranophane, but the authors were unable to confirm this theory. The potential incorporation of Tc in uranyl phases is especially noteworthy because the heptavalent oxidation state expected for this element in an oxidizing environment such as that of Yucca Mountain site will result in the formation of the highly soluble pertechnetate anion (TcO_4^-). The high solubility of this anion complex, combined with the long half-life of ^{99}Tc (213,000 years), results in this radioisotope being among those of most concern for the long-term performance of a repository system.

Transuranic elements represent a unique concern for radioactive waste isolation due to the long half-lives of many isotopes (e.g. $T_{1/2}$ for ^{239}Pu = 24,110 years), the ability to spontaneously fission, and the potential for their accumulation into a configuration that could produce a critical mass. The transuranic species of predominant concern under an oxidizing environment are expected to be $Pu^{3+,4+,5+,6+}$, $Np^{4+,5+}$, Am^{3+}, and Cm^{3+}. The possible incorporation of transuranic elements into uranyl alteration phases was first suggested by Finch and Ewing (1994). Burns et al. (1997) and Miller et al. (1997) further evaluated the potential for such a substitution process in the structures of a variety of uranyl and uranyl silicate phases. Many of these structures contain infinitely extending polyhedral sheets, with cations of higher bond-valence (e.g. U^{6+}) being located within the sheets, whereas mono- or divalent cations and/or H_2O molecules are located between the sheet structures (Stohl and Smith 1981; Burns et al. 1997). Substitutions of Pu^{6+}, Pu^{5+}, or Np^{5+} for U^{6+} were determined to be possible in these uranyl phases, provided that appropriate structural modifications and charge-balance substitutions could be made. Trivalent actinides, such as Pu^{3+} and Am^{3+}, may occur in the sheets of the structure of α-uranophane and boltwoodite, and possibly in the interlayer sites of other uranyl phases. The quadravalent actinides (e.g. Pu^{4+}) might also substitute into the uranyl polyhedral sheets in schoepite, becquerelite, compreignacite, α-uranophane, and boltwoodite. This substitution would be limited, however, due to the valence charge difference between Pu^{4+} and U^{6+} ions. Charge balanced coupled substitution mechanisms have been proposed for this process, with potential mechanisms including the substitution of OH^- for O^{2-}, H_2O for O^{2-}, and the incorporation of an additional low valence cation (e.g. Li^+ for Ba^{2+}).

Buck et al. (1998) provided the first experimental evidence confirming the presence of transuranic elements in uranyl phases. In an examination of the alteration phases formed during the corrosion of spent fuel, these authors noted that crystals of dehydrated schoepite had incorporated Np^{5+} into their structures at a concentration of approximately 550 ppm. The resulting Np:U ratio in this phase varied from between 0.003 and 0.006, representing an enrichment relative to the spent fuel source material with a Np:U ratio of 0.0005.

Kim and Wronkiewicz (1999) have experimentally examined the potential for actinide substitution in uranyl oxide hydrate phases by using Ce^{4+} as a surrogate for Pu^{4+}. Preliminary data obtained from dehydrated schoepite crystals precipitated in the presence of a Ce^{4+} bearing solution suggest a Ce concentration of approximately 25 to 10 ppm, with the concentration progressively decreasing with increasing reaction time and crystal size.

Ianthinite is a mixed valence uranium phase noted as an alteration product in natural uranium deposits (Leslie et al. 1993; Pearcy et al. 1994; Finch and Ewing 1994). The potential for transuranic substitution into this phase is particularly noteworthy because it contains both U^{4+} and U^{6+} in its crystal structure (Burns et al. 1997). Quadravalent actinide substitution (e.g. Pu^{4+} for U^{4+}) can potentially occur directly in this phase without the need for any charge balancing substitution mechanism. Burns has also proposed that billietite may similarly show a high potential for incorporating quadravalent transuranic elements into its structure, though a charge balancing mechanism would be required to accommodate the substitution of a quadravalent transuranic into the U^{6+} position.

The longevity of alteration phases is an important consideration with respect to their potential for retarding the migration of radionuclides. The longevity of a particular phase should significantly exceed the half-life of a component radionuclide if that mineral phase is to be an effective retardant against radionuclide migration. Ideally, the nucleation and dissolution kinetics involving these phases would allow for their formation and persistence in a repository system until all radionuclides of primary concern have decayed to insignificant levels. In natural geologic systems, uranyl oxide hydrate and alkali + alkaline earth uranyl silicate phases may persist for hundreds of thousands of years or longer as metastable alteration phases, thereby acting as a long-term sinks for uranium and other contained radionuclides (Finch et al. 1995). Supporting evidence for this determination can be found in rate constant and dissolution rate calculations that indicate a lifetime for a "one millimeter crystal" of uranophane at 282,000 years (Bruno et al. 1995). By contrast, a crystal of schoepite with similar dimensions would only be expected to last 1570 years based upon the same calculation method. Both the calculated lifetime, and the persistence of these phases in natural systems (Finch et al. 1995; Bruno et al. 1995), occur in contrast to the relatively short existence (several years) of uranyl oxide hydrate phases in 90°C dissolution tests with UO_2 pellets (Figs. 2c,e,h; Wronkiewicz et al. 1992, 1996). In the UO_2 study, becquerelite was found to be an intermediate phase in the paragenetic sequence, replacing earlier formed dehydrated schoepite, only later to be replaced, in turn, by β-uranophane. The large differences in the phase longevity determined in these different studies attests to the complexity of the processes involved and the effect that various parameters (e.g. temperature, solution composition, etc.) may have on kinetics of phase nucleation and dissolution.

The differences in observed longevity between the natural and experimentally produced becquerelite may, of course, be at least partially attributable to the differences in reaction temperatures and solution compositions in the two environments. Other factors may also be important. For example, studies suggest that naturally occurring becquerelite may be less soluble than synthetically produced becquerelite (Finch et al. 1995). This latter comparison is in good agreement with the observation that well-ordered crystalline uranyl oxide hydrate compounds are generally more stable than poorly crystalline (i.e. those with many defect sites) and amorphous materials (Sandino and Grambow 1994; Bruno and Sandino 1989). If the effect of uranyl phases on fission product migration is to be included in performance assessment models for the proposed Yucca Mountain repository site, then we will need to have a greater understanding of the potential pathways for trace element incorporation in secondary uranium phases. Additional studies investigating the compositional variations of these phases, their stability at various temperatures and fluid compositions, their stability as a function of the degree of trace-element substitution, and the distribution coefficients that govern the incorporation of fission products and transuranic elements into their structures may be warranted.

CONCLUSIONS

The overall reaction of UO_2 in a moist and oxidizing environment is controlled by a combination of factors, including oxidative-dissolution of uranium, temperature, precipitation kinetics of uranyl alteration phases, and leachant composition. Results from a number of studies indicate that the alteration rates for UO_2 and spent fuel may be quite rapid in an unsaturated geologic setting at elevated temperatures (Wronkiewicz et al. 1992, 1996; Finn et al. 1995; Finch et al. 1999). The rate at which the UO_2 reactions occur in test environments suggests that an advanced stage of uranyl alteration phase development can develop rapidly within the time constraints of geologic disposal in a repository, once the primary container and Zircaloy cladding that house the spent fuel have been breached.

The sample reaction trends observed in tests with unirradiated UO_2 have closely replicated those that occur in natural geologic systems, such as the Nopal I deposit in Mexico. Both the natural and experimental systems display a paragenetic sequence of mineral phases that is characterized by the following trend: uraninite (UO_{2+x}) → uranyl oxide hydrates (including schoepite-type and alkali + alkaline earth uranyl oxide hydrates) → uranyl silicates → alkali + alkaline earth uranyl silicates (Wronkiewicz et al. 1992, 1996; Leslie et al. 1993; Pearcy et al. 1994). This similarity indicates that the UO_2 in the experiments has reacted by the same mechanism as uraninite in the natural deposits, and thus the experimental and natural analogue reactions may be used to simulate the reaction progress expected for spent fuel at the proposed Yucca Mountain repository. Uranyl silicates have also been observed following extended testing with spent fuel, although the reaction paragenesis for these samples has not yet been determined (Finch et al. 1999). Alkali and alkaline earth uranyl silicates are expected to be the long-term solubility-limiting phases for uranyl ions during the corrosion of spent fuel at the proposed Yucca Mountain repository, provided that a sufficient flux of groundwater components are available for phase formation (Bruton and Shaw 1988; Nguyen et al. 1992; Wronkiewicz et al. 1992; Pickett and Murphy 1999).

The migration of fission products from a breached spent fuel package may be significantly retarded by the formation of a variety of secondary uranyl phases. These phases can incorporate silicon, carbon, alkali, alkaline earth, transition metal, rare earth, and actinide elements into their crystalline structures. A greater understanding of the compositional variations of these phases, their stability at various temperatures and fluid compositions, the distribution coefficient ratios that govern the incorporation of fission products and transuranic elements into their structures, and their nucleation-dissolution kinetics will be required if the effect that uranyl phases have on radionuclide migration is to be included in performance assessment models for a nuclear waste repository.

ACKNOWLEDGMENTS

The authors gratefully acknowledge reviews of a previous version of this manuscript by W.M. Murphy, P.C. Burns, and an unidentified reviewer. This research has been supported in part by the U.S. Department of Energy under contract DE-FG07-97ER14820.

REFERENCES

Ahn TM 1994 Long-term C-14 source term for a high-level waste repository. Waste Mgmt 14:393-408
Ahn TM 1996 Dry oxidation and fracture of LWR spent fuels. U.S. Nuclear Regulatory Commission Report NUREG-1565, 29 p
Bruno J, Sandino A 1989 The solubility of amorphous and crystalline schoepite in neutral to alkaline aqueous solutions. *In* Scientific Basis for Nuclear Waste Management XII. Lutze W, Ewing RC (Eds) Mater Res Soc Symp Proc 127:871-878
Bruno J, Casas I, Cera E, Ewing RC, Finch RJ, Werme LO 1995 The assessment of the long-term evolution of the spent nuclear fuel matrix by kinetic, thermodynamic and spectroscopic studies of uranium minerals. *In* Scientific Basis for Nuclear Waste Management XVIII, Part 1. Murakami R, Ewing RC (Eds) Mater Res Soc Symp Proc 353:633-639
Bruton CJ, Shaw HF 1988 Geochemical simulation of reaction between spent fuel waste form and J-13 water at 25°C and 90°C. *In* Scientific Basis for Nuclear Waste Management XI. Apted MJ, Westerman RE (Eds) Mater Res Soc Symp Proc 112:485-494
Buck EC, Wronkiewicz DJ, Finn PA, Bates JK 1997 A new uranyl oxide hydrate phase derived from spent fuel alteration. J Nucl Mater 249:70-76
Buck EC, Finch RJ, Finn PA, Bates JK 1998 Retention of neptunium in uranyl alteration phases formed during spent fuel corrosion. *In* Scientific Basis for Nuclear Waste Management XXI. McKinley IG, McCombie C (Eds) Mater Res Soc Symp Proc 506:87-94
Burns PC, Ewing RC, Miller ML 1997 Incorporation mechanisms of actinide elements into the structures of U^{6+} phases formed during the oxidation of spent nuclear fuel. J Nucl Mater 245:1-9
Burns PC, Hill FC (1999) Implications of the synthesis and structure of the Sr analogue of curite. Can Mineral (in press)
Burns PC 1999 Cs boltwoodite obtained by ion exchange from single crystals: Implications for the radionuclide release in a nuclear repository. J Nucl Mater 265:218-223
Buscheck TA, Nitao JJ, Ramspott LD 1996 Localized dryout: An approach for managing the thermal-hydrological effects of decay heat at Yucca Mountain. *In* Scientific Basis for Nuclear Waste Management XIX. Murphy WM, Knecht DA (Eds) Mater Res Soc Symp Proc 412:715-722
Cantlon JE, Allen CR, Arendt JW, Brewer GD, Cohon JL, Cording EJ, Domenico PA, Langmuir D, McKetta JJ Jr, Price DL, Verink ED Jr, Wong JJ 1995 U.S. Nuclear Waste Technical Review Board, Report to the U.S. Congress and the Secretary of Energy, 1995 Findings and Recommendations, 53 p
Cesbron F, Ildefonse P, Sichere MC 1993 New mineralogical data on uranophane and β-uranophane;

synthesis of uranophane. Mineral Mag 57:301-308
Choi J-W, McEachern RJ, Taylor P, Wood DD 1996 The effect of fission products on the rate of U_3O_8 formation in SIMFUEL oxidized in air at 250°C. J Nucl Mater 230:250-258
de Pablo J, Casis I, Giménez J, Molera M, Torrero ME 1997 Effect of temperature and bicarbonate concentrations on the kinetics of $UO_2(s)$ dissolution under oxidizing conditions. In Scientific Basis for Nuclear Waste Management XX. Gray WJ, Triay IR (Eds) Mater Res Soc Symp Proc 465:535-542
Diaz-Arocas P, Garcia-Serrano J 1997 Coprecipitation studies from SIMFUEL solutions in 5 m NaCl media. In Scientific Basis for Nuclear Waste Management XX. Gray WJ, Triay IR (Eds) Mater Res Soc Symp Proc 456:543-548
Einziger RE, Thomas LE, Buchanan HC, Stout RB 1992 Oxidation of spent fuel in air at 175 to 195°C. In High Level Radioactive Waste Management, Proceedings of the Third International Conference. Am Nucl Soc, LaGrange Park, Illinois, p 1449-1457
Ewing RC 1993 The long-term performance of nuclear waste forms: Natural materials – three case studies. In Scientific Basis for Nuclear Waste Management XVI. Interrante CG, Pabalan RT (Eds) Mater Res Soc Symp Proc 294:559-568
Ewing RC, Weber WJ, Clinard FW Jr 1995 Radiation effects in nuclear waste forms for high-level radioactive waste. Progress in Nucl Energy 29:63-127
Fillmore DL, Bulmahn 1994 Characteristics of Department of Energy spent nuclear fuel. In Proc Topical Meeting on DOE Spent Nuclear Fuel, 1994, Am Nucl Soc, LaGrange Park, Illinois, p 313ff
Finch RJ, Ewing RC 1991 Alteration of natural UO_2 under oxidizing conditions from Shinkolobwe, Katanga, Zaire: A natural analogue for the corrosion of spent fuel. Radiochim Acta 52/53:395-401
Finch RJ, Ewing RC 1992 The corrosion of uraninite under oxidizing conditions. J Nucl Mater 190:133-156
Finch RJ, Ewing RC 1994 Formation, oxidation and alteration of ianthinite. In Scientific Basis for Nuclear Waste Management XVII. A Barkatt, RA Van Konynenburg (Eds) Mater Res Soc Symp Proc 333:625-630
Finch RJ, Suksi J, Rasilainen K, Ewing RC 1995 The long-term stability of becquerelite. In Scientific Basis for Nuclear Waste Management XVIII, Part 1. Murakami T, Ewing RC (Eds) Mater Res Soc Symp Proc 353:647-652
Finch RJ, Hawthorne FC, Ewing RC 1998 Structural relations among schoepite, metaschoepite, and "dehydrated schoepite." Can Mineral 36:831-845
Finch RJ, Buck EC, Finn PA, Bates JK 1999 Oxidative corrosion of spent UO_2 fuel in vapor and dripping groundwater at 90°C. In Scientific Basis for Nuclear Waste Management XXII. Wronkiewicz DJ, Lee JH (Eds) Mater Res Soc Symp Proc 556, (in press)
Finn PA, Buck EC, Gong M, Hoh JC, Hafenrichter LD, Bates JK 1994 Colloidal products and actinide species in leachate from spent nuclear fuel. Radiochim Acta 66/67:197-203
Finn PA, Buck EC, Hoh JC, Bates JK 1995 Spent fuel's behavior under dynamic drip tests. Global 1995, International Conference on Evaluating Emerging Nuclear Fuel Cycle Systems. Am Nucl Soc Topical Meeting. 241-248
Forsyth RS, Werme LO 1992 Spent fuel corrosion and dissolution. J Nucl Mater 190:3-19
Frondel C 1956 Mineral composition of gummite. Am Mineral 41:539-568
Gray WJ, Strachan DM 1991 UO_2 matrix dissolution rates and grain boundary inventories of Cs, Sr, and Tc in spent LWR fuel. In Scientific Basis for Nuclear Waste Management XIV. Abrajano TA Jr, Johnson LH (Eds) Mater Res Soc Symp Proc 212:205-212
Gray WJ, Wilson CN 1995 Spent fuel dissolution studies FY1991 to FY 1994. Pacific Northwest Laboratory Report PNL-10540/UC-802, December 1995
Gray WJ 1999 Inventories of iodine-129 and cesium-137 in the gaps and grain boundaries of LWR spent fuels. In Scientific Basis for Nuclear Waste Management XXII. Wronkiewicz DJ, Lee LH (Eds) Mater Res Soc Symp Proc 556, (in press)
Harrar JE, Carley JF, Isherwood WF, Raber E 1990 Report of the committee to review the use of J-13 well water in Nevada nuclear waste storage investigations. Lawrence Livermore National Laboratory Report UCID-21867
Integrated Data Base Report—1994 (1995) US Department of Energy Report DOE/RW-0006, Rev. 11
Johnson LH, Burns KI, Joling HH, Moore CJ 1983 Leaching of ^{137}Cs, ^{134}Cs, and ^{129}I from irradiated UO_2 fuel. Nucl Tech 63:470-475
Johnson LH, Shoesmith, DW 1988 Spent Fuel, Chapter 11, In Radioactive Waste Forms for the Future. Lutze W, Ewing RC (Eds) Elsevier, p. 635-698
Kansa EJ, Hanson BD, Stout RB 1999 Grain size and burnup dependence of spent fuel oxidation: Geological repository impact. In Scientific Basis for Nuclear Waste Management XXII. Wronkiewicz DJ, Lee JH (Eds) Mater Res Soc Symp Proc 556, (in press)
Kim CW, Wronkiewicz DJ 1999 Incorporation of radionuclides in the alteration phases of spent nuclear

fuel. Migration '99, abstract (accepted)

Langmuir D 1978 Uranium solution-mineral equilibria at low temperatures with applications to sedimentary ore deposits. Geochim Cosmochim Acta 42:547-569

Leslie BW, Pearcy EC, Prikryl JD 1993 Oxidative alteration of uraninite at the Nopal I deposit, Mexico: Possible contaminant transport and source term constraints for the proposed repository at Yucca Mountain. In Scientific Basis for Nuclear Waste Management XVI. Interrante CG, Pabalan RT (Eds) Mater Res Soc Symp Proc 294:505-512

Louthan MR Jr, Iyer NC, Sindelar RL, Peacock HB Jr 1996 Disposal of aluminum based spent fuels in a repository. In Scientific Basis for Nuclear Waste Management XIX. Murphy WM, Knecht DA (Eds) Mater Res Soc Symp Proc 412:91-98

Matzke HJ, Spino J 1996 Formation of the rim structure in high burnup fuel. In Proc Int'l Workshop "Interfacial Effects in Quantum Engineering Systems." IEQES 96, Proc J Nucl Mater 18 p

McGillivray GW, Geeson DA, Greenwood RC 1994 Studies of the kinetics and mechanisms of the oxidation of uranium by dry and moist air. A model for determining the oxidation rate over a wide range of temperatures and water pressures. J Nucl Mater 208:81-97

Miller ML, Burns PC, Finch RJ, Ewing RC 1997 Transuranium element incorporation into the β-U_3O_8 uranyl sheet. In Scientific Basis for Nuclear Waste Management XX. Gray WJ, Triay IR (Eds) Mater Res Soc Symp Proc 465:581-588

Murphy WM, Pearcy EC 1992 Source-term constraints for the proposed repository at Yucca Mountain, Nevada, derived from the natural analog at Peña Blanca, Mexico. In Scientific Basis for Nuclear Waste Management XV. Sombret CG (Ed) Mater Res Soc Symp Proc 257:521-527

Murphy WM 1993 Geochemical models for gas-water-rock interactions in a proposed nuclear waste repository at Yucca Mountain, Nevada. In Site Characterization and Model Validation Focus "93 Topical Meeting, Am Nucl Soc 115-121

Murphy WM 1995 Natural analogs for Yucca Mountain. Radwaste 2:44-50

Murphy WM 1997 Retrograde solubilities of source term phases. In Scientific Basis for Nuclear Waste Management XX Gray WJ, Triay IR (Eds) Mater Res Soc Symp Proc 465:713-720

NRC 1995 Technical bases for Yucca Mountain standards. Committee on Technical Bases for Yucca Mountain Standards, National Research Council, National Academy Press, Washington DC, 222 p

NRC 1998 Electrometallurgical Techniques for DOE Spent Fuel Treatment: Spring 1998 Status Report on Argonne National Laboratory's R&D Activity. Committee on Electrometallurgical Techniques for DOE Spent Fuel Treatment, National Research Council, National Academy Press, Washington D.C. 78 p

Nguyen SN, Silva RJ, Weed HC, Andrews JRJ 1992 Standard Gibbs free energies of formation at 30°C of four uranyl silicates: Soddyite, uranophane, sodium boltwoodite and sodium weeksite. Chem Therm 24:259-276

Oversby VM 1994 Nuclear waste materials. In Materials Sciences and Technology A Comprehensive Treatment. Chan RW, Haasen P, Kramer EJ (Eds) VCH Verlagsgesellschaft, Germany

Pearcy EC, Prikryl JD, Murphy WM, Leslie BW 1994 Alteration of uraninite from the Nopal I deposit, Peña Blanca District, Chihuahua, Mexico, compared to degradation of spent nuclear fuel in the proposed U.S. high-level nuclear waste repository at Yucca Mountain, Nevada. Appl Geochem 9:713-732

Pickett DA, Murphy WM 1997 Isotopic constraints on radionuclide transport at Peña Blanca. In 7th CEC Natural Analogue Working Group Meeting Proc. von Maravic H, Smellie J (Eds) EUR 17851 EN, European Commission, Luxembourg, 113-122

Pickett DA, Murphy WM 1999 Unsaturated zone waters from the Nopal I natural analog, Chihuahua, Mexico – Implications for radionuclide mobility at Yucca Mountain. In Scientific Basis for Nuclear Waste Management XXII. Wronkiewicz DJ, Lee JH (Eds) Mater Res Soc Symp Proc 556, (in press)

Plyasunov AV, Grenthe I 1994 The temperature dependence of stability constants for the formation of polynuclear cationic complexes. Geochim Cosmochim Acta 58:3561-3582

Posey-Dowty J, Axtmann E, Crerar D, Borscik M, Ronk A, Woods W 1987 Dissolution rate of uraninite and uranium roll-front ores. Econ Geol 82:184-194

Sagüés, AA 1999 Nuclear waste package corrosion behavior in the proposed Yucca Mountain repository. In Scientific Basis for Nuclear Waste Management XXII Wronkiewicz DJ, Lee LH (Eds) Mater Res Soc Symp Proc 556 (in press)

Sandino A, Grambow B 1994 Solubility equilibria in the U(VI)-Ca-K-Cl-H_2O System: Transformation of schoepite into becquerelite and compreignacite. Radiochim Acta 66/67:37-43

SCP 1988—Site Characterization Plan, U.S. Department of Energy, Office of Civilian Radioactive Waste Management, Report DOE/RW-0199

Shoesmith DW, Sunder S 1992 The prediction of nuclear fuel (UO_2) dissolution rates under waste disposal conditions. J Nucl Mater 190:20-35

Shoesmith DW, Sunder S, Bailey MG, Miller NH 1996 Corrosion of used nuclear fuel in aqueous perchlorate and carbonate solutions. J Nucl Mater 227:287-299

Steward SA, Mones ET 1997 Comparison and modeling of aqueous dissolution rates of various uranium oxides. *In* Scientific Basis for Nuclear Waste Management XX Gray WJ, Triay IR (Eds) Mater Res Soc Symp Proc 465:535-542

Stohl FV, Smith DK 1981 The crystal chemistry of the uranyl silicate minerals. Am Mineral 66:610-625

Stroes-Gascoyne S, Tait JC, Porth RJ, McConnel JL, Barnsdale TR, Wartson S 1993 Measurements of grain-boundary inventories of ^{137}Cs, ^{90}Sr, and ^{99}Tc in used CANDU fuels. *In* Scientific Basis for Nuclear Waste Management XVI. Interrante CG, Pabalan RT (Eds) Mater Res Soc Symp Proc 294:41-46

Stroes-Gascoyne S, Moir DL, Kolar M, Porth RJ, McConnel JL, Kerr AH 1995 Measurement of gap and grain-boundary inventories of ^{129}I in used CANDU fuels. *In* Scientific Basis for Nuclear Waste Management XVIII, Part 1. Murakami T, Ewing RC (Eds) Mater Res Soc Symp Proc 353:625-631

Sunder S, Shoesmith DW, Christensen H, Miller NH 1992 Oxidation of UO_2 fuel by the products of gamma radiolysis of water. J Nucl Mater 190:78-86

Sunder S, Shoesmith DW, Miller NH 1995 Prediction of the oxidative dissolution rates of used nuclear fuel in a geological disposal vault due to the alpha radiolysis of water. *In* Scientific Basis for Nuclear Waste Management XVIII, Part 1 Murakami T, Ewing RC eds.) Mater Res Soc Symp Proc 353:617-624

Taylor P, Burgess EA, Owen DG 1980 An x-ray diffraction study of the formation of β-$UO_{2.33}$ on UO_2 pellet surfaces in air at 229 to 275°C. J Nucl Mater 88:153-160

Taylor P, Lemire RJ, Wood DD 1993 The influence of moisture on air oxidation of UO_2: Calculations and observations. Nucl Tech 104:164-170

Taylor P, Wood DW, Owen DG 1995 Microstructures of corrosion films on UO_2 fuel oxidized in air-steam mixtures at 225°C. J Nucl Mater 223:316-320

Thomas GF, Till G 1984 The dissolution of unirradiated UO_2 fuel pellets under simulated disposal conditions. Nucl Chem Waste Mgmt 5:141-147

Thomas LE, Einziger RE, Woodley RE 1989 Microstructural examination of oxidized spent fuel PWR fuel by transmission electron microscopy. J Nucl Mater 166:243-251

Thomas LE, Charlot LA 1990 Analytical electron microscopy of light-water reactor fuels. Ceramic Trans 9:397-407

Thomas LE, Beyer CE, Charlot LA 1992 Microstructural analysis of LWR spent fuels at high burnup. J Nucl Mater 188:80-89

Thorstenson DC, Weeks EP, Haas H, Woodward JC 1989 Physical and chemical characteristics of topographically affected airflow in an open borehole at Yucca Mountain, Nevada, Nucl Waste Isol Unsat Zone. Focus '89, Am Nucl Soc 256-270

TSPA-VA 1998 Viability assessment of a repository at Yucca Mountain, Vol 3: Total system performance assessment, December 1998. U.S. Department of Energy Office of Civilian Radioactive Waste Management Report DOE/RW-0508/V3 (http://www.ymp.gov/vadoc/v3)

Vochten R, De Grave D, Lauwers H 1990 Transformation of synthetic U_3O_8 into different uranium oxide hydrates. Mineral Petrol 41:247-255

Vochten R, van Haverbeke L 1990 Transformation of schoepite into the uranyl oxide hydrates: becquerelite, billietite and wölsendorfite. Mineral Petrol 43:65-72

Weber WJ, Turcotte RP, Roberts FP 1982 Radiation damage from alpha decay in ceramic nuclear waste forms. Radioactive Waste Mgmt 2:295-319

Weber WJ, Roberts FP 1983 A review of radiation effects in solid nuclear waste forms. Nucl Tech 60:178-198

Weber WJ, Ewing RC, Catlow CRA, Diaz de la Rubia T, Hobbs LW, Kinoshita C, Matzka Hj, Motta AT, Nastasi M, Salje EKH, Vance ER, Zinkle SJ 1998 Radiation effects in crystalline ceramics for the immobilization of high-level nuclear waste and plutonium. J Mater Res 13:1434-1484

Wilson CN 1990 Results from NNWSI series 3 spent fuel tests. Pacific Northwest Laboratory Report PNL-7170/UC-802

Wronkiewicz DJ, Bates JK, Gerding TJ, Veleckis E, Tani BS 1992 Uranium release and secondary phase formation during the unsaturated testing of UO_2 at 90°C. J Nucl Mater 190:107-127

Wronkiewicz DJ, Bates JK, Wolf SF, Buck EC 1996 Ten-year results from unsaturated drip tests with UO_2 at 90°C: Implications for the corrosion of spent nuclear fuel. J Nucl Mater 238:78-95

Wronkiewicz DJ, Buck EC, Bates JK 1997a Grain boundary corrosion and alteration phase formation during the oxidative dissolution of UO_2 pellets. *In* Scientific Basis for Nuclear Waste Management XX. Gray WJ, Triay IR (Eds) Mater Res Soc Symp Proc 465:519-526

Wronkiewicz DJ, Bates JK, Buck EC, Hoh J, Emery J, Wang LM 1997b Radiation effects on moist-air systems and the influence of radiolytic product formation on nuclear waste glass corrosion. Argonne National Laboratory Report, ANL-97/15, 238 p

11 Spectroscopic Techniques Applied to Uranium in Minerals

John M. Hanchar
Department of Geology
The George Washington University
Washington, DC 20052

INTRODUCTION

One area of research that may be used to characterize U minerals and materials that contain U involves techniques from many disciplines grouped under the general term of spectroscopy. Spectroscopic tools covering a wide range in energy of the electromagnetic spectrum, from gamma rays to microwaves, and of a variety of experimental configurations, play an important role in contemporary mineral studies. Various spectroscopic tools are used to identify cations, determine elemental concentrations, oxidation states, structural sites, and local environments of ions in materials.

In this chapter I review spectroscopic properties of U in minerals. In particular, I will discuss optical (e.g. absorption and luminescence) and X-ray (e.g. X-ray absorption and X-ray photoelectron) spectroscopic techniques. These are the areas in which the most research has been done on U in minerals. These spectroscopic tools allow for a better understanding of the physical and chemical properties of U in minerals, which is necessary for the study of U minerals in the environment.

This review discusses U in minerals, and U minerals; with U primarily in the U^{4+} or U^{6+} oxidation states both of which are important environmentally. Hexavalent U occurs in minerals as the uranyl ion, UO_2^{2+}, and U^{4+} occurs in minerals as the U^{4+} ion. All of the techniques used to acquire the results presented in this paper have been discussed in detail in the *Reviews in Mineralogy*, Volume 18, "Spectroscopic Methods in Mineralogy and Geology" and in the many references found therein (Hawthorne 1988). My objective for this chapter is to briefly describe each technique, discuss how a particular technique has been used to characterize the chemical and physical properties of U in the material investigated, and to present examples of studies and discuss selected results.

OPTICAL SPECTROSCOPIC METHODS

The optical spectroscopic properties of U in minerals are directly related to the chemical state of U as it occurs in those minerals. In nature and in synthetic materials (including synthetic minerals) U may occur as U ions (e.g. U^{3+}, U^{4+}, U^{5+}, and U^{6+}) with U^{4+} and U^{6+} the dominant oxidation states in minerals (Marfunin 1979a). Many U^{6+} minerals are bright yellow, green, or orange in hand sample. It has often been speculated that the cause of these colors, and the green fluorescence emission observed in some UO_2^{2+} minerals, is due to U^{6+} in the form of the UO_2^{2+} ion (Gleason 1972; Warren et al. 1995).

These emission colors are, in fact, directly related to the intrinsic crystal chemical properties that the structures of U minerals, and other UO_2^{2+}-bearing compounds, exert on the UO_2^{2+} ion (Marfunin 1979a,b; Brittain 1990; Blasse and Grabmaier 1994; Baird and Kemp 1997). Unlike other actinides, and the lower oxidation states of U, U^{6+} has no f electrons, so any luminescence emission from U^{6+}-containing materials is not a $5f$

Table 1. Electron configuration of the uranium atom and uranium ions

Atom or ion	Core electron configuration	Electron configuration
U^0	Rn	$5f^3 6d^1 7s^2$
U^{2+}	Rn	$5f^3 7s^1$
U^{3+}	Rn	$5f^3$
U^{4+}	Rn	$5f^2$
U^{5+}	Rn	$5f^1$
U^{6+}	Rn	$5f^0$

electron-related transition. In other words, the U^{6+} ion is a $5f^0$ ion (Marfunin 1979a; Blasse and Grabmaier 1994). Hence, the optical transitions for UO_2^{2+} are of the charge-transfer type (Blasse and Grabmaier 1994). Table 1 shows the different electron configurations for the neutral U atom and for various U ions.

When making any type of optical spectroscopic measurement, three factors must be taken into account. These are the source of the incident electromagnetic radiation (laser, high-energy electrons, incandescent lamp, ultraviolet lamp, etc.), a device to disperse the transmitted or emitted energy (spectrograph, or spectrometer), and an apparatus to record the light (photomultiplier tube, diode array detector, or charge-coupled device [CCD] detector). The design of these devices has changed considerably over the past 30 years or so, but the experimental principles remain the same.

In the field of luminescence spectroscopy many terms are used for different specialized techniques leading to potential confusion among workers new to the field. These include luminescence, fluorescence, cathodoluminescence, photoluminescence, thermoluminescence, etc. It is this author's position to consider all these different techniques as luminescence. Where appropriate, I elaborate on the different techniques in this review.

In absorption and luminescence spectroscopies, electromagnetic radiation is absorbed, emitted (or both) from a material after an initial bombardment with some form of electromagnetic radiation. The absorption or emission is dispersed and simultaneously recorded with a detector. It is the nature of the source and emitted radiations that leads to the differences in the names of these techniques. Another potential point of confusion is the use of various spectral units in the presentation of luminescence spectra. Table 2 provides the conversion of the different units commonly used, and the spectral colors at different wavelengths (see also Hawthorne 1988).

Absorption spectroscopy

An absorption spectrum of a mineral yields information about the wavelengths of electromagnetic radiation absorbed by that mineral. The spectrum is measured by varying the wavelength of the incident radiation and recording the intensity of the transmitted beam. When light is absorbed by a material, the energy of the absorbed light excites a transition between different electronic energy levels in the material starting from an electronic ground state. The positions of higher energy levels (excited states) can be obtained from the absorption spectrum (Meijerink 1998). This information from the absorption spectrum reveals which oxidation state(s) of U are present, and absorption spectra can be used to characterize, or "fingerprint" materials.

Light transmitted through a sample is detected by a spectrometer, dispersed by a diffraction grating, and the light intensity at different wavelengths is measured. Typically,

Table 2. Conversion of spectral units used in luminescence spectroscopy

Wavelength (nm)	Frequency ν (10^{14} s^{-1})	Wavenumber $\bar{\nu}$ (10^3 cm^{-1})	Energy (eV)	Color
1000	3.0	10.0	1.24	infrared
700	4.3	14.3	1.77	red
620	4.8	16.1	2.00	orange
580	5.2	17.2	2.14	yellow
530	5.7	18.9	2.34	green
470	6.4	21.3	2.64	blue
420	7.1	23.8	2.95	violet
300	10.0	33.3	4.15	near ultra-violet
200	15.0	50.0	6.20	far ultra-violet

a dual-beam configuration is used and the beam of light is passed alternately through a reference material by the use of a chopper. The transmission of the light through the sample is determined by dividing the intensity of the sample beam by the reference beam (Meijerink 1998).

The light source used for absorption spectroscopy is commonly a deuterium lamp for ultraviolet studies (180-350 nm) or a tungsten lamp for near ultraviolet, visible, and infrared investigations (350-3000 nm). A diffraction grating in a spectrometer is then used to disperse the light. A diffraction grating is chosen for the wavelength region of interest. The dispersed light emission is detected using either a photomultiplier tube (180 to 800 nm range) or a PbS detector (800-3000 nm range) (Meijerink 1998).

Absorption spectroscopy of uranium in minerals. The absorption properties of U^{4+} and UO_2^{2+} in a variety of materials are well known, and the ranges in energy over which absorption takes place for each U ion do not overlap (Bénard et al. 1994). The main absorption lines for U^{4+} are from about 450 nm up to about 1.3 μm, whereas the UO_2^{2+} ion absorbs only below 450 nm (Bénard et al. 1994). Richman et al. (1967) and Kisliuk et al. (1967), in related studies, used the absorption spectra of oriented single crystals of synthetic zircon to determine the electronic energy levels of U^{4+}. Zircon was chosen as the host mineral for these studies because of the high symmetry (D_{2d}) of the Zr^{4+} site that is occupied by the U^{4+} ion in zircon, and because no charge compensating species would have to be added to the system had other minerals been used (e.g. CaF_2 with U^{4+} substituted for Ca^{2+}). Polarized absorption spectra were obtained over a range of temperatures from 4 to 300 K, and a wavelength range of 400 to 2500 nm.

The general features of the absorption spectrum for zircon are similar to those of U^{4+} in other materials (Richman et al. 1967), and these features can be used to identify U^{4+} in studies in which one of the objectives is to determine the oxidation state of U. If for a given ion and site symmetry, the energy positions and crystal field levels are already known, assignment of the absorption bands in terms of the corresponding multiplets is possible (Bérnard et al. 1994). By using polarization selection rules, Richman et al. (1967) and Kisliuk et al. (1967) determined that electronic energy levels of U^{4+} in zircon could be identified according to their irreducible representations. Richman et al. (1967) and Kisliuk et al. (1967) determined 30 energy levels for U^{4+} up to 24,000 cm^{-1}. The experimental results were compared with theoretical energy-level calculations. The results obtained by Richman et al. (1967) suggest that the sort of success that had been achieved by Wybourne (1965) in fitting the absorption spectra of lanthanides in crystals could also be achieved for the tetravalent actinides in crystals.

In another study of U in zircon, Mackey et al. (1975) measured absorption spectra and calculated crystal field energy levels of U^{4+}. They used synthetic zircon crystals doped with U^{4+} and natural zircon crystals. In their study, Mackey et al. (1975) attempted, and failed, to reproduce the crystal-field calculation results of Richman et al. (1967). Mackey et al. (1975) experimentally confirmed the absorption spectra results from U^{4+} reported by Richman et al. (1967), however, and found errors in the calculations by Richman et al. (1967) of about 60 cm^{-1} for the zero crystal field. Using a fitting of ten parameters, Mackey et al. (1975) reported an energy level fit that has a standard deviation of 112 cm^{-1}.

In a study of U in zircon, nine natural samples were investigated by Vance and Mackey (1974). The presence and absorption spectroscopic properties of U^{4+} and U^{5+} were investigated. Vance and Mackey (1974) made a series of measurements from 300 to 2500 nm over a temperature range from 15 to 290 K. In addition to the absorption peaks assigned to U^{4+} in zircon (Richman et al. 1967), the authors also identified a set of peaks in the infrared in the 1000 to 1500 nm range.

The assignment of U^{4+} peaks by Vance and Mackey (1974) is in good agreement with the absorption spectra discussed by Richman et al. (1967), except in cases where there was a low concentration of U, or line broadening due to radiation damage (i.e. metamictization) in the zircon crystals. Vance and Mackey (1974) assigned the infrared peaks in their zircon absorption spectra to U^{5+}. In some of their measurements, the infrared peaks dominated the absorption spectrum. Figure 1 is an example of an adsorption spectrum for a brown "high" zircon from Burma (present name Myanmar). The term "high" is from the terminology of Anderson (1963) and refers to the density of the material which is directly related to the degree of metamictization.

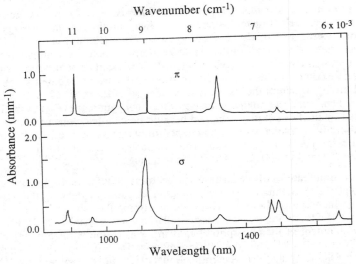

Figure 1. Infrared π and σ absorption spectrum for natural "high" Burma zircon (after Vance and Mackey 1975) at 15 K.

Vance and Mackey (1974) concluded that U^{5+} may be stabilized in zircon by trivalent (e.g. rare earth elements [REE]) and possibly divalent impurity elements, and does not appear to be the result of a valance change caused by radiation damage in the zircon crystals. Vance and Mackey (1974) concluded that the only stable oxidation states of U in natural zircon are U^{4+} and U^{5+}, and they reported that there was no effect on the oxidation state of U even after extreme doses of radiation.

In a continuation study of U^{5+} in zircon, Vance and Mackey (1975) investigated the σ-polarized zero-phonon lines (i.e. peaks in the absorption spectrum) at 1107 and 1493 nm (9030 and 6700 cm^{-1}, respectively). Some weaker lines were also observed in zircon samples containing ~0.1 wt % U^{5+}. The initial study by Vance and Mackey (1974) was primarily to identify U^{5+} in zircon, whereas the latter study was undertaken to further characterize U^{5+} in zircon. Initial measurements were done on synthetic zircon crystals doped with UO_2 (i.e. U^{4+}) and Y_2O_3 at a UO_2 / Y_2O_3 molar ratio of approximately 7:1. The Y^{3+} was added for charge compensation.

Absorption spectroscopic measurements confirmed that all of the U in the synthetic zircon crystals was U^{5+} and present in the different samples at 0.1, 0.03, and 0.01 wt % U^{5+}. Vance and Mackey (1975) noted, however, that U concentrations in selected crystals from a given batch varied by at least a factor of two (although no actual values were reported in that paper). Vance and Mackey (1975) attributed this variation in U content to chemical heterogeneity of U in the crystals. From the results of Zeeman measurements, Vance and Mackey (1975) also concluded that a ground-state near degeneracy exists for remotely charge-compensated U^{5+} in zircon, with ground-states being degenerate to less than 1 cm^{-1}. The two strong lines, at 9030 and 6700 cm^{-1}, appear to be single-ion transitions of U^{5+}.

In a later study, Vance and Mackey (1978) addressed some minor discrepancies between calculated and experimentally determined energy-level calculations for U^{4+} in zircon. In this study, Vance and Mackey (1978) investigated U^{4+} and U^{5+} in the isostructural zircon-group minerals hafnon ($HfSiO_4$) and thorite ($ThSiO_4$) in an attempt to determine the uniqueness of U absorption features observed in zircon. Given that Zr^{4+} and Hf^{4+} have similar ionic radii (0.84 and 0.83 Å, respectively, in eight coordination [Shannon 1976]), and similar electronic configurations, it is not surprising that the absorption spectrum for U^{4+} in hafnon is similar to that of U^{4+} in zircon. The relative intensities of the U absorption peaks are similar in both materials, and the shifts in energy are only a few tens of cm^{-1}. Tetravalent thorium, however, is considerably larger (1.050 Å in eight coordination [Shannon 1976]) than Zr^{4+}, and consequently the absorption spectrum for U^{4+} reported for thorite is significantly different from that for U^{4+} in zircon or hafnon. In thorite, the σ-polarized transitions were shifted a few hundred cm^{-1}. Also, the relative intensities of the U^{4+} peaks were different than those observed in zircon or hafnon. There were also differences in the π absorption spectrum for the thorite crystals.

Vance and Mackey (1978) concluded in their study that, although differences were observed for U^{4+} in the different minerals, no confirmation could be made regarding the uniqueness of the previously calculated energy levels for U^{4+} in zircon. Also, the energy levels for U^{4+} in thorite were not observed at sufficient intensities to be able to derive a meaningful set of fitting parameters for their crystal field model.

Vance (1974) investigated the anomalous absorption spectrum in "low" (in the terminology of Anderson 1963) metamict zircon in the infrared region. Low-temperature polarization studies have allowed for a clear distinction in the zircon absorption spectrum between lines intrinsic to U^{4+} (so-called normal features) (Richmann et al. 1967) and the anomalous peaks in the absorption spectrum. Vance (1974) concluded that anomalous features in the spectrum are due to U^{4+} incorporated in polycrystalline zirconia (ZrO_2) within the metamict zircon. Upon annealing (1200°C) the zircon crystals used in their study, Vance (1974) determined that the normal region of the zircon absorption spectrum is enhanced at the expense of the anomalous region. Vance (1974) concluded that absorption features in the normal and anomalous regions of the spectrum are all due to U.

Fielding (1970) investigated the U and REE contents, and the nature of color centers in a natural zircon crystal. The sample studied was collected from alluvium presumably derived from a pegmatite in the late Permian New England Batholith in New South Wales, Australia. In that study, Fielding (1970) attempted to correlate features in the absorption spectrum, over a spectral range of 1400 to 40000 cm^{-1}, with the distribution of trace impurities such as Fe, the REE, and U. Fielding (1970) suggested that absorption maxima in the region 15000 to 16000 cm^{-1} and 6000 to 8000 cm^{-1} belong to transitions characteristic of U, Fe, and the REE.

In addition to the numerous absorption spectroscopy studies of U in zircon, there been a few other studies involving the occurrence U in other minerals. Hargreaves (1970) investigated the energy levels of U ions in U^{2+}, U^{3+} and U^{4+} in synthetic fluorite (CaF_2) crystals at 20 K. The crystals were grown under high vacuum to eliminate all water and O_2 contamination, and in the presence of abundant F^-. This eliminated residual O and established F^- charge compensation sites in the CaF_2 crystals. The crystals were initially grown with U^{4+} and were subsequently reduced, after crystallization, to the U^{3+} and U^{2+} oxidation states using an intense ultraviolet light source. The conversion from U^{4+} to the lower oxidation states was monitored by measuring the decrease in absorption of the characteristic U^{4+} peaks for fluorite in the infrared.

The reduced U^{3+} and U^{2+} ions did not produce sharp f to f transitions in the infrared region, apparently as a result of selection-rule prohibitions (Hargreaves 1970). The synthetic CaF_2 crystals had the following sample colors. The CaF_2: U^{2+} crystals were apple-green, the CaF_2: U^{3+} crystals were ruby-red, and the CaF_2: U^{4+} crystals were yellow. Hargreaves (1970) also reported CaF_2 crystals of other colors such as cerise or brown, which he attributed to mixed oxidation states or contamination from oxygen. For U^{4+} in CaF_2, Hargreaves (1970) found that the absorption spectra in the infrared, visible, and ultraviolet regions revealed crystal-field-split energy levels of the $5f^2$ electrons of tetravalent U ions in C_{3v} symmetry sites in CaF_2. The absorption spectra for U^{2+} in CaF_2 are probably related to the $5f^37s^1$ configuration, and the U^{3+} in CaF_2 can be attributed to the $5f^3$ configuration of U^{3+}. The results for U^{2+} in CaF_2 are the least conclusive of the study.

In a study of laser-induced fluorescence of U, deNeufville et al. (1981) investigated the optical characteristics of the UO_2^{2+} ion, including absorption and emission (luminescence) spectra in a variety of geological materials. The absorption spectra were acquired from 3200 to 7500 Å. The absorption spectra for three UO_2^{2+} minerals (metatorbernite, $Cu(UO_2)_2(PO_4)_2 \cdot 8H_2O$; meta-autunite, $Ca(UO_2)_2(PO_4)_2 \cdot 8H_2O$; and uranophane, $Ca(UO_2)_2(SiO_3(OH))_2 \cdot 5H_2O$) are displayed in Figure 2.

The absorption spectra of these three UO_2^{2+} minerals reveal a characteristic set of evenly spaced absorption peaks (Fig. 2) which have been observed in absorption spectra of other crystals and solutions containing the UO_2^{2+} ion (Rabinowitch and Belford 1964). These peaks are attributed to the excitation of the O-U-O ion symmetric stretch vibration in the electronically excited states (deNeufville et al. 1981). The broad feature at ~6500 Å in the metatorbernite spectra (Fig. 2) is attributed to Cu^{2+} in that sample. Aside from these results, the absorption spectra of metatorbernite and meta-autinite are nearly identical, reflecting their similar crystal structures. Corresponding features in the metatorbernite absorption spectrum, however, are shifted to slightly longer wavelengths by approximately 40 Å with respect to meta-autinite. The uranophane absorption spectrum has the same general features as the metatorbernite and meta-autinite, although the UO_2^{2+} vibrational bands are less well defined (deNeufville et al. 1981).

Luminescence spectroscopy

Luminescence in minerals involves the bombardment of a mineral surface with a stream of electrons or some form of electromagnetic radiation. To observe luminescence, a short-wave ($250 < \lambda < 300$ nm) or long-wave ($350 < \lambda < 400$ nm) ultraviolet radiation source, high energy electrons (cathodoluminescence), laser (laser fluorescence), protons (ionoluminescence), or X-rays (radioluminescence), are used as the radiation source (Meijerink 1998). Upon absorbing the incident radiation, some light is emitted from the mineral having a wavelength that is longer (lower energy) than the incident radiation that was initially absorbed (Brittain 1990), and greater than that produced by thermal blackbody radiation (Waychunas 1988). The majority of minerals that emit luminescence contain either

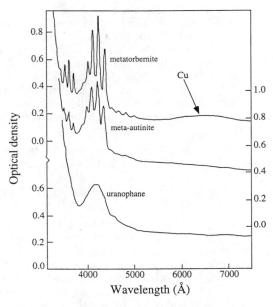

Figure 2. Absorption spectra of natrural metatorbernite, meta-autunite, and uranophane (after deNeufville et al. 1981).

certain transition metals (e.g. Cr^{3+}, Mn^{2+}, or Fe^{3+}), lanthanides (e.g. Sm^{3+}, Eu^{2+}, Eu^{3+}, Tb^{3+}, Dy^{3+}), or actinides (e.g. UO_2^{2+}), or emit luminescence from some type of defect-related phenomena.

Many different types of luminescence phenomena have been investigated and a full discussion can be found in Waychunas (1988) or in Vij (1998). The luminescence spectrum of a sample provides information about the spectral distribution of the light emitted from the sample. This information can be used to identify the element(s) causing the luminescence emission, and in some cases to characterize and "fingerprint" the luminescent material (Marfunin 1979a,b). Luminescence spectroscopy is an extremely useful technique because it can be used on small particles (e.g. <50 microns). During a survey of a polyphase sample, materials with different luminescence emissions may be readily identified, and used to further characterize the sample with other techniques such as Raman, X-ray absorption spectroscopies, electron microprobe, X-ray diffraction, etc. It is a relatively nondestructive technique; however, depending on the incident radiation source, sample heating and damage may occur.

Luminescence spectroscopy of uranium in minerals. Some of the earliest work in the field of luminescence microscopy and spectroscopy was done by Stokes and Becquerel in the 19th century during their studies of U salts as part of a larger investigation of radioactive materials (Pringsheim 1949). The bright colors in hand specimens and the luminescent properties of UO_2^{2+} minerals led to the identification of these minerals in field surveys for U deposits (Marfunin 1979a). Many classic studies on luminescence of the UO_2^{2+} ion in minerals are published in Russian and not available as English translations. Cejka and Urbanec (1990) discuss the key points of several of the Russian studies, and a brief summary from their discussion is presented here.

When the UO_2^{2+} ion is present in minerals, a transition occurs between the free UO_2^{2+} ion and the UO_2^{2+} ion bonded in a crystal structure. This results in a decrease in site symmetry of the UO_2^{2+} ion from D_h to D_{6h} or lower symmetry (Cejka and Urbanec 1990). This change in symmetry leads to a rearrangement in molecular orbitals. The features of the absorption and luminescence spectra of UO_2^{2+} minerals is a function of transitions between different energy levels. These differences are equal to the vibrational energy of the UO_2^{2+} ion and are characteristic properties of that material (Cejka and Urbanec 1990). Each of the broad bands of electronic transitions of the UO_2^{2+} ion represent one series of vibrational transitions. The absorption and luminescence spectra of UO_2^{2+} minerals are characteristic.

Many studies in the 1950s and 1960s assisted in our understanding of the luminescent properties of the UO_2^{2+} ion in minerals within the framework of crystal field and molecular orbital theories as discussed in Marfunin (1979a,b), Fred (1986), and Carnall and Crosswhite (1986). Denning (1992) discusses the results of an extensive study of UO_2^{2+} spectroscopy in synthetic materials.

The cause of the intense green to yellowish-green luminescence commonly observed in UO_2^{2+} minerals when excited by ultraviolet radiation is the UO_2^{2+} ion (Brittain 1990; Blasse and Grabmaier 1994). The UO_2^{2+} ion is a nearly linear O-U-O molecular ion which remains intact upon inclusion into a mineral lattice. The two oxygen atoms remain in axial positions, whereas any ligands bond along the equatorial plane (Cejka and Urbanec 1990). When only oxygen atoms construct the remaining configurational environment, it is possible for the U^{6+} as the central ion to achieve either octahedral or trigonal configuration (UO_6^{6-}), or tetrahedral (UO_4^{2-}) symmetry (Brittain 1990), however, these ions are unknown in minerals (Burns et al. 1997). Also, these ions are generally not luminescent except under cryogenic conditions. The luminescent color of the latter configuration in materials is red (Blasse and Grabmaier 1994).

A characteristic property of UO_2^{2+} luminescence emission is that the luminescence bands are narrower than transition-metal activators but considerably broader than the sharp line emissions characteristic of trivalent REE (Marfunin 1979a,b; Brittain 1990). The energy levels of UO_2^{2+} can be described by using molecular orbital theory; however, a complete theoretical description of these energy is not yet available (Brittain 1990).

The luminescence emission spectra of most UO_2^{2+} minerals are generally similar. Figure 3 is an example of the UO_2^{2+} luminescence emission in synthetic autunite, $Ca[UO_2]_2(PO_4)_2 \cdot 10H_2O$. The UO_2^{2+} emission consists of a single pure electronic transition that is accompanied by a single vibronic progression (Brittain 1990). The UO_2^{2+} spectrum consists of a sequence of bands having peaks at 490, 512, 535,

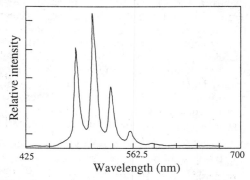

Figure 3. Luminescence spectra of synthetic autunite at room temperature (after Brittain 1990).

561, and 588 nm. This pattern is interpreted as a pure electronic luminescence transition at 490 nm followed by a coupling of the vibrational electronic band (Brittain 1990) resulting in the other peaks in the emission spectrum.

Variations in vibrational spacings of UO_2^{2+} emission spectra in different materials reflect differences in the energies of the total symmetric UO_2^{2+} stretching vibrational modes in the ground and excited states (Brittain 1990). In research on absorption and luminescence of minerals containing UO_2^{2+}, Tarashchan et al. (1974), and Tarashchan (1978) found that UO_2^{2+} luminescence spectra in most UO_2^{2+} minerals are generally similar with a greenish emission. As is the case with many minerals that possess REE induced luminescence, Fe, Cu, Mn, Pb and Bi may act to quench the luminescence of the UO_2^{2+} ion in minerals (Marfunin 1979a,b).

deNeufville et al. (1981) investigated the optical characteristics of the UO_2^{2+} ion absorption and emission in a variety of geological materials. The fluorescence emission spectra were acquired from 4000 to 6600 Å. The fluorescence spectra from three UO_2^{2+} minerals (liebigite, $Ca_2(UO_2)(CO_3)_3 \cdot 10H_2O$; andersonite, $Na_2Ca(UO_2)(CO_3)_3 \cdot 6H_2O$; and schröckingerite, $NaCa_3(UO_2)(CO_3)_3(SO_4)F \cdot 10H_2O$) are presented in Figure 4. The spectra in Figure 4 correspond to those of UO_2^{2+} compounds having UO_2^{2+} with D_{2h} and D_{3h} point symmetry (deNeufville et al. 1981) and are characteristic.

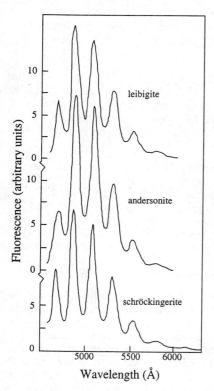

Figure 4. Laser fluorescence (excitation wavelength 3300 Å) spectra of natural leibigite, andersonite, and schröckingerite (after deNeufville et al. 1981).

In nature, some soddyite samples exhibit fluorescence whereas others do not (Frondel 1958). In a study of synthetic soddyite, $(UO_2)_2SiO_4 \cdot 2H_2O$, Vochten et al. (1995) reported the fluorescence spectrum, using an excitation energy wavelength of 430 nm at 298 and 77 K. Vochten et al. (1995) found that at both temperatures, no phosphorescence could be detected at a sampling interval of 2 ms. At 298 K, no fluorescence was detected, but at 77 K a well defined fluorescence spectrum was recorded. They attributed this temperature dependence of the fluorescence emission to the fact that the quantum efficiency of fluorescence of the UO_2^{2+} ion decreases with increasing temperature. The fluorescence spectrum in Figure 5 is characterized by three maxima. The most intense peak is at 547 nm, and two peaks of lower intensity occur at 526 and 568 nm. Based on band gap energy calculations, Vochten et al. (1995) concluded that soddyite at 77 K should be considered an insulator.

In a study of the sorption of the UO_2^{2+} ion on a reference 2:1 smectite clay from Cheto, Arizona, Morris et al. (1994) used luminescence and Raman spectroscopies to determine which UO_2^{2+} solution ions are present and absorbed onto the clay, to determine if multiple sorption sites can be identified for the UO_2^{2+} ion, and to characterize the structure and stability of the UO_2^{2+} ion at absorption sites. The UO_2^{2+}-bearing clay samples were immersed in aqueous solutions of UO_2^{2+} nitrate. The samples were prepared at a high initial ionic strength (~0.1 to 0.3 M), and the UO_2^{2+} loading levels in the samples ranged from ~0.1 to ~53% of the reported cation exchange capacity (e.g. ~1.2 meq/g). The excitation source used for the measurements was either an Ar ion laser ($\lambda = 3638$ Å) or monochromatized ($\lambda = 4000$ Å) output from a Xe arc lamp.

The fluorescence emission spectra vary significantly in intensity and band shape as a function of the UO_2^{2+} concentration, and the equilibrium pH of the solutions from which the clays were prepared. Morris et al. (1994) identified four distinct UO_2^{2+} sorption complexes over a broad range of surface coverage and equilibrium pH values (from ~2.5 to 7). At low and moderate concentrations, Morris et al. (1994) suggest that that the smectite samples contain primarily amphoteric edge-site sorption complexes. For higher concentration levels, Morris et al. (1994) suggest that the sample coverage greatly exceeds the number of available edge sites, and that the species identified for those samples are thought to be exchange complexes in the fixed-charge sites. All the available spectroscopic data indicate that the four sorption complexes consist of monomeric UO_2^{2+} ions.

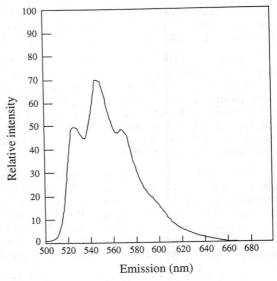

Figure 5. Fluorescence spectra (excitation wavelength 470 nm) of synthetic soddyite at 77 K (after Vochten et al. 1995).

In a study combining X-ray absorption techniques with luminescence, Raman spectroscopy, and other ancillary techniques (e.g. scanning electron microscopy [SEM] with energy dispersive spectroscopy [EDS] and powder X-ray diffraction [XRD]), Morris et al. (1996) investigated the chemical state of U in contaminated soils from the U.S. Department of Energy (DOE) former U processing facility at Fernald, Ohio. X-ray absorption studies on those contaminated soils have shown that ~75 to 95% of the U in bulk samples is U^{6+}. The use of the complimentary spectroscopic techniques allowed Morris, et al. (1996) to identify specific hosts for much of the U in addition to determining the oxidation state of the U with X-ray absorption spectroscopic techniques.

By using luminescence imaging and spectroscopy Morris et al. (1996) found that the U in the contaminated soil samples occurs primarily in tabular, platey, grains which range in size from 10 to 100 microns. Some of the UO_2^{2+}-related luminescence observed by Morris et al. (1996) was green and some was orange-red. Scanning electron microscopy-EDS analyses confirmed the presence of U in both of these luminescent materials. The green luminescent material was determined using powder XRD to be consistent with schoepite, $[(UO_2)_8O_2(OH)_{12}](H_2O)_{12}$, although other UO_2^{2+} minerals that emit green luminescence may also be present. In the orange-red luminescent material, the SEM-EDS analyses revealed that this material contained little silicon, no phosphorus, and only small amounts of calcium and magnesium. The only cations measured by Morris et al. (1996) using SEM-EDS were potassium and aluminum, and trace amounts of Fe. Based on these analyses, Morris et al. (1996) determined that the orange-red luminescent material does not consist of UO_2^{2+} phosphates (including organophosphates) or UO_2^{2+} silicates. The exact identification of the orange luminescent materials, however, was problematic because of inherent variations in the luminescence spectra from sample to sample, and these samples do not yield Raman data because of sample decomposition.

Nicholas (1967) investigated the green luminescence of U^{6+} in synthetic CaF_2 and SrF_2. Spanning a temperature range from 5 to 300 K, Nicholas identified two types of spectra that were attributed to different luminescent centers. Nicholas (1967) also investigated the substitution mechanism of U^{6+} substituting for Ca, or Sr, with four oxygen ions replacing four fluorine ions around the center to provide the necessary charge compensation.

The luminescence spectrum of CaF_2 was interpreted by Nicholas (1967) to be a non-phonon line emission with vibrational structures on the sides. Nicholas (1967) assigned the observed lines and bands on the basis of lattice vibrations, internal local modes of the center, and a combination of these factors. The results obtained by Nicholas (1967) are in good agreement with theoretical calculations. The fluorescence emission of U^{6+} in the SrF_2 is similar to that of CaF_2; however, there are other electronic transitions evident in the luminescence spectrum. Nicholas also concluded that the fluorescence intensity of U in CaF_2 and SrF_2 is strongly dependent on the dopant level of the U.

X-RAY ABSORPTION SPECTROSCOPY

X-ray absorption spectroscopy (XAS) techniques, including extended X-ray absorption fine structure (EXAFS) and X-ray absorption near-edge structure (XANES) involve the irradiation of a sample with a tunable source of monochromatic X-rays from a synchrotron radiation facility (Antonio 1992). These two techniques are collectively referred to as X-ray absorption fine structure (XAFS), and can provide fine-scale chemical and structural information about a specific X-ray absorbing element in a material (Brown et al. 1988; Koningsberger and Prins 1988).

The elements probed may be major elements in the material of interest, trace constituents, dissolved ions in a liquid, or a surface-sorbed component on a solid (Schulze and Bertsch 1995). For elements with a Z greater than 20, XAS can be used on solid materials, liquids, wet samples, and materials suspended in solutions down to the 100 parts per million (ppm) concentration range (Schulze and Bertsch 1995). This is, in part, why many environmental studies have used XAS techniques, and with the availability of higher energy X-rays at third-generation synchrotron facilities, such as the Advanced Photon Source (APS) at Argonne National Laboratory, there will most likely be more studies of this nature in the near future.

Traditionally, diffraction techniques (e.g. X-ray, neutron, electron) have been the primary tools for determining structural information about crystalline and noncrystalline materials. X-ray absorption spectroscopy techniques offer an alternative way of investigating atomic structures of materials based on the absorption of X-rays by individual atoms in the material being probed (Antonio 1992). For poorly ordered minerals, the information obtained using XAS generally cannot be obtained using other techniques (Schulze and Bertsch 1995). There are several excellent reviews of XAS principles and applications applied to biological, chemical, materials, and earth sciences (e.g. Sayers et. al. 1971; Stern 1974; Brown and Donaich 1980, Calas et al. 1987; Brown et al. 1988; Koningsberger and Prins 1988; Hasnain 1991; Charlet and Manceau 1993; Fendorf et al. 1994; Allen et al. 1996). Much of the synchrotron-based research that has been done on minerals of environmental importance has used XAS techniques, and those are the techniques on which I focus here. I discuss several examples in which XAS ha been used to study a structural or chemical aspect of U in minerals.

In an XAFS measurement, the X-ray energy is scanned from just below to above the binding energy of a core-shell electron (e.g. K or L) of an element of interest (Antonio 1992). When the energy of the incident X-rays is equal to the electron binding energy, X-ray absorption occurs and a steeply rising absorption edge is observed. For energies greater than the binding energy, oscillations of the absorption with incident X-ray energy (XAFS) are observed (Brown et al. 1988). This technique is useful for elements from Li to U, and in certain experimental configurations XAFS is nondestructive (Schulze and Bertsch 1995). As the atomic number increases across the periodic table (e.g. from Li to U), the energy required to probe those elements also increases. This means that the absorption edge probed is not necessarily the same one for all elements (e.g. K versus M edges). The main use of XAFS is to determine structural information about the local environment (e.g. ~ 5 Å from the X-ray absorbing element) of elements in materials. This techniques yields information about the interatomic distances and numbers of neighboring atoms (also called coordination number) and the degree of disorder and identity of atoms in the immediate vicinity of the X-ray absorbing atom (Antonio 1992).

In a XANES measurement, the energy region of interest is that which is beyond the pre-edge to approximately 50 eV above the absorption edge of an element and is referred to as the X-ray near edge structure (Brown et al. 1988; Koningsberger and Prins 1988). The exact energy of the pre-edge and the absorption edge are related to the chemical environment of the X-ray absorbing element (Schulze and Bertsch 1995). Hence, one primary application of a XANES experiment is to probe the site symmetry and oxidation state of the X-ray absorbing element. The energy of the pre-edge and the absorption edge are typically increased by ~1-3 eV for each electron removed from the valence shell (Schulze and Bertsch 1995). This change in energy is related to a decrease in shielding of the core electrons (increase in valence) which results in an increase in the binding energy of the core levels. In certain cases, the edge position (energy) varies systematically with oxidation state and so semi-quantitative determinations of the valence of metals in mixed valence compounds can be determined (Schulze and Bertsch 1995).

XAS spectroscopy of uranium in minerals

Much of the environmental research on U using XAS techniques involves determining the chemical environment of U in the mineral of interest. One area of particular interest is in the sediments around DOE U processing facilities such as Savannah River, South Carolina, and Fernald, Ohio. These sediments generally contain U in the UO_2^{2+} form, but some U^{4+} is usually also present (Bertsch et al. 1994). In that the U contamination in these facilities is caused by anthropogenic activities, the chemical and physical state of the U is not always obvious.

Allen et al. (1994) used XAS and luminescence spectroscopy to study the adsorbtion of U on contaminated soil particles at the DOE Feed Materials Production Center in Fernald, Ohio. This facility served for over 30 years as a U processing facility in the DOE nuclear weapons program. Uranium was deposited around the facility in some areas in concentrations as high as 1000 ppm. Allen et al. (1994) determined using XANES spectroscopy that ~80 to 90% of the U at the site is present in the UO_2^{2+} form. Based in part on the XAS results obtained by Allen et al. (1994) plans for the treatment and removal of U-contaminated soil at Fernald have been initiated.

Using a combination of micro-synchrotron X-ray fluorescence (SXRF) and micro-XANES (in this context micro implies an *in situ* spot analysis on the scale of 50 to 200 μm in spatial resolution, as opposed to a bulk sample analysis), Bertsch et al. (1994) studied soil and sediment solids which had received waste from two DOE U processing facilities. The measurements were done on soil and sediment samples which were sprinkled lightly on adhesive Kapton tape. Bertsch et al. (1994) determined that the U in the soil and sediments is primarily U^{6+} and is often associated with Mn and Fe and possibly on the surface of oxyhydroxide minerals. The limits of detection for U in this study was ~1 ppm with a relative accuracy of 5%. Bertsch et al. (1994) also determined that the U^{4+}, and possibly some of the U^{6+}, is associated with Cu perhaps in a Cu-U containing phase. The reduced U is thought to have originated in aeolian particulates rather than from a biologically-mediated reduction process (Bertsch et al 1994).

The differences in the oxidation states of U in the XANES spectrum are due to shifts in energy and intensity of the absorption edge for different oxidation states. Qualitative structural information about the U bonding environments can also be obtained from the XANES measurements from the multiple scattering resonances on the high-energy side of the main absorption feature. These features are related to the local environment around the central absorbing atom (Bertsch et al. 1994).

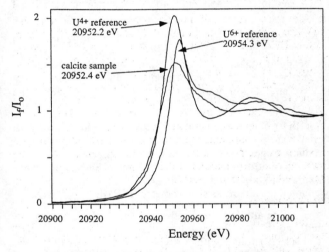

Figure 6. XANES spectra for well characterized reference compounds of
U^{4+} : ($[U^{4+}P_5W_{30}O_{110}]^{11-}$) and
U^{6+} : ($[U^{6+}O_2]^{2+}$)
in 1 M $HClO_4$ media, and XANES spectrum for natural calcite sample (after Sturchio et al. 1998).

The multiple scattering resonance for the high-energy side of the main absorption feature in the U^{6+} spectrum in Figure 6 is distinct for U^{6+}-containing phases in which the multiple scattering is related to the U-O bonds in the UO_2^{2+} (Bertsch et al. 1994). The center position of the multiple resonance feature in the XANES spectrum is inversely related to the U-O bond distance (Bertsch et al. 1994). These two observations in the XANES spectrum help to characterize the "fingerprint" of different valence states of U in earth materials (Farges et al. 1992; Calas et al. 1987).

Hunter and Bertsch (1998) in a subsequent study of sediments contaminated by U at DOE facilities, and using a novel sample preparation technique (a non-reactive Si-polymer), used XANES and SRXF techniques to make elemental and oxidation state distribution maps on individual soil particles prior to extraction for chemical analysis. The XANES measurements determined the relative proportions of U^{6+} and U^{4+} in the sediments, and revealed that greater than 85% of the U in the soil particles is U^{6+}. The complimentary SXRF analyses revealed that the U^{6+} is contained primarily in UO_2^{2+} hydroxide phases, possibly schoepite.

In a study of evaporation pond sediments from the San Joaquin Valley, California, Duff et al. (1997) used SXRF and XANES to characterize the oxidation/reduction conditions in the sediments. The evaporation ponds are used for irrigation drainage and are known to contain elevated levels of U. Knowing the oxidation state of the U in the sediments and in the surrounding hydrologic environment is crucial for understanding the solubility and transport phenomena of the U. Duff et al. (1997) addressed the following issues regarding the U enriched sediments: (1) to determine U^{6+}/U^{4+} ratios in waters in contact with the pond sediments; (2) to determine the oxidation state of U in solids in the reducing pond sediments as the sediments underwent man-induced "in situ" oxidation; and (3) to determine the oxidation state of U in solids with respect to depth in pond sediment surface layers.

The XANES results indicate that the U-containing solid phases in the sediments are 25% U^{4+} and 75% U^{6+}, even in the presence of naturally occurring volatile sulfides. U in these highly reducing pond sediments may occur as a mixed-valence U compound, such as ianthinite, $U^{4+}{}_2(UO_2)_4(OH)_4(H_2O)_9$, wyartite, $CaU^{5+}(UO_2)O(OH)_4(CO_3)(H_2O)_7$, or as a mixture of U^{4+} and $UO_2{}^{2+}$ compounds. Some U may also be present as sorbed species. The oxidation state of the U increased in U^{4+} component with depth, suggesting a gradual reduction of U^{6+} to U^{4+} with time.

To better understand the mechanism(s) involved in reducing U in pond sediments in the previous study from U^{6+}, to U^{4+}, Duff et al. (1999) used the XANES technique to investigate the effects of native pond algae (*Chlorella*), and potential man-induced reducing agents (e.g. acetate, sucrose, and alfalfa shoot), on U redox chemistry in the same pond sediments from the San Joaquin Valley, California, as discussed in the previous study.

In this study by Duff et al. (1999), samples of pond sediments were equilibrated with natural and synthetic pond inlet waters containing approximately 10 mg L^{-1} of U^{6+} to which the above mentioned reducing agents were added. The equilibrations were done under oxidizing and O_2-limiting conditions (Duff et al. 1999). Sediments were investigated for changes in oxidation state using XANES and U concentrations were determined using inductively coupled plasma-mass spectrometry (ICPMS).

Duff et al. (1999) found that in the alfalfa treated sediments, 95% of the U was lost to solution, sulfides precipitated, and the XANES results indicate a reduction of the U^{6+} to U^{4+}. Additionally, Duff et al. (1999) found with XANES measurements that upon exposure to air, the U^{4+} readily oxidized. Much less U^{6+} was reduced in the other two reduction treatments, and the U in the solid phase in the sediments consisted primarily of U^{6+} similar to the natural pond sediments. Duff et al. (1999) concluded that U reduction occurs primarily by electron transfer between sulfide and $UO_2{}^{2+}$ ions, and that biosorption by algae and bacteria is the dominant mechanism for depositing U in the pond sediments.

Dent et al. (1992) used EXAFS to investigate the sorption of $UO_2{}^{2+}$ ions onto montmorillonite. EXAFS spectroscopy allows for this determination directly by studying the surface structure of sorbed species. Dent et al. (1992) determined that in their study, changes in the surface complex structure vary as a function of pH. At low pH (< 3), the data suggest the presence of the $UO_2{}^{2+}$ ion with 5 to 6 equatorial oxygens in a monomeric aquo-complex. At higher pH values (e.g. 4-5) and degree of hydrolysis, a reduction was observed in the EXAFS amplitude of the equatorial oxygen shell and is possibly related to an increase in structural disorder. At a pH of 4.3, the EXAFS data suggest a multi-nuclear species. The results of $UO_2{}^{2+}$ sorption at a pH of 5 indicate a similarity to results obtained in hydrolyzed solutions. Those results are consistent with the uptake of solution species by an outer-sphere, ion exchange mechanism. The results of this study indicate that the structure of the UO_2^{2+} ion is retained on sorption at the montmorillonite/water interface.

In another application of using XAS to determine speciation of U, Chisholm-Brause (1994), investigated the adsorbtion of U on the clay mineral montmorillonite. Chisholm-Brause et al. (1994) studied how the structurally distinct UO_2^{2+} sorption complexes form on montmorillonite as a function of surface coverage of U. Their data suggest that over a twenty-fold increase in surface coverage, and there were systematic changes in the coordination numbers and bond lengths for the oxygen atoms in the equatorial planes of the sorbed UO_2^{2+} species. Chisholm-Brause et al. (1994) determined that these changes may be attributed to the existence of at least three structurally distinct sites on the montmorillonite that interacted with the UO_2^{2+} ions over the concentration range of their study.

Thompson et al. (1998) studied the sorption of U on kaolinite using EXAFS spectroscopy. The samples in their experimental study consisted of aqueous U^{6+} equilibrated with kaolinite. Their EXAFS results provide unique structural and compositional information about U speciation at the kaolinite-water interface, and indicate that the U sorption complex on kaolinite always contains a UO_2^{2+} component surrounded by five equatorial oxygen atoms. They also detected contributions in the UO_2^{2+} EXAFS spectra from Si and/or Al atoms, located at a second nearest neighbor position, 3.3 Å from U. Thompson et al. (1998) propose that this indicates an inner-sphere complexation of U by kaolinite. At lowest pH values (e.g. 7.0 > pH 6.0) and in the presence of air, small mononuclear U species dominate. At higher values of pH (7.5 > pH 7.0) in air, small multinuclear U species dominate. In the absence of CO_2, and at higher pH (e.g. 8.0 pH 7.1), Thompson et al. (1998) determined that the dominant U species contains at least three U atoms which they interpret to mean the formation of multinuclear sorption complexes or surface precipitates (i.e. crystallization of a separate phase). In either situation, kaolinite seems to enhance clustering of U atoms. These results indicate that the use of kaolinite as a sorbent for removal of U from contaminated ponds is promising.

Sturchio et al., (1998) reported the results of an EXAFS and XANES investigation of the local environment and oxidation state of U in a natural calcite ($CaCO_3$) sample from a Mississippi Valley-type (MVT) zinc ore deposit in Tennessee. The XANES results indicate that the U is primarily tetravalent (Fig. 6). The EXAFS data indicate that U substitutes for divalent Ca in calcite. These results indicate that U^{4+} can reside in calcite under reducing conditions, and help validate U-series and U/Pb, dating methods for ancient calcites. In addition, calcite may provide a sink for U in anoxic aqueous environments. It is not clear, however, which element(s) help maintain charge neutrality for U substitution in calcite, and the effect this may have on the extent of U substitution in calcite.

In a study of U^{6+} minerals, including rutherfordine (UO_2CO_3), uranophane ($Ca[UO_2]_2[SiO_3OH]_2 \cdot 5H_2O$), meta-autinite ($Ca[UO_2]_2[PO_4]_2 \cdot 6H_2O$), and meta-ankoleite ($K_2[UO_2]_2[PO_4]_2 \cdot 6H_2O$), and solution model compounds including aqueous UO_2^{2+} ion (0.05 M $UO_2[NO_3]_2$), UO_2^{2+} diacetate ($UO_2[CH_3CO_2]_2$), and UO_2^{2+} nitrate ($UO_2[NO_3]_2 \cdot 6H_2O$), Thompson et al. (1997) used EXAFS spectroscopy to determine the coordination environment of U^{6+} at the solid-water interface. Using the EXAFS computer program FEFF 6, Thompson et al. (1997) did "ab initio" calculations of phase-shift and amplitude functions for fitting the experimental EXAFS data. The "ab initio" calculations reproduced their experimental results, and helped determine phase-shift and amplitude functions for neighboring atoms whose spectral contributions are difficult to isolate from experimental data due to overlaps of Fourier-transform features.

At ambient temperature, Thompson et al. (1997) were able to fit spectral contributions from axial oxygen (1.8 Å), equatorial oxygen (2.2-2.5 Å), nitrogen (2.9 Å), carbon (2.9 Å), silicon (3.2 Å), phosphorus (3.6 Å), distant oxygen (4.3 Å), and U (4.0, 4.3, 4.9, and 5.2 Å) atoms. The spectral contributions from nitrogen, carbon, silicon, and phosphorus, distant oxygen, and distant U (e.g. 4.9 and 5.2 Å) are weak, and could easily go undetected in samples of unknown composition. At a temperature of 10 K, the detection of distant U neighbors extends to 7.0 Å. Thompson et al. (1997) suggest that the ability to detect these atoms indicates that EXAFS may be useful for determining inner-sphere U sorption at solid aluminosilicate-water interfaces. They noted that even for samples that are of unknown composition their results indicate that EXAFS is useful for determining nearest neighbors of U to within 4.3 Å. Additionally, Thompson et al. (1997) propose that these findings shed light on clustering of U and formation of precipitates.

In an investigation of radiation damage in zircon and thorite, Farges and Calas (1991) used EXAFS to examine the effects of metamictization in those minerals. Farges and Calas used EXAFS to probe the Zr-K and the Hf, Th and U L_{III} absorption edges. In a natural metamict zircon sample, Farges and Calas (1991) determined that the U is U^{4+} and in six-coordination with oxygen. This is unlike Hf and Th that are in eight-coordination with oxygen in zircon, as is Zr.

Farges and Calas (1991) were unable to determine the exact location of the U polyhedra in metamict and annealed zircon samples. The lack of a U-U component in the EXAFS spectra, in both metamict and annealed samples, suggests the absence of U-bearing phases as inclusions in the zircon such as uraninite or coffinite. The likely substitution of U for Zr, however, cannot be proven due to the absence of U-Zr contributions in the EXAFS spectra for zircon, as was determined for U-Th in thorite (Farges and Calas 1991).

In a comparison between the EXAFS results and qualitative optical absorption spectroscopy measurements, Farges and Calas (1991) found good agreement with the two complimentary techniques for the oxidation state and local environment for metamict zircon crystals from Sri Lanka and the Ural Mountains.

In metamict thorite, EXAFS results indicate that the U occurs as the UO_2^{2+} ion (with usually two axial O atoms at 1.78 Å and 4-5 equatorial O atoms at 2.2-2.3 Å). Farges and Calas (1991) suggest that the presence of U in the UO_2^{2+} form in metamict thorite can be attributed to weathering enhanced by accumulated radiation damage. In the metamict thorite samples, no medium-range order was detected, so Farges and Calas (1991) concluded that the UO_2^{2+} groups in metamict thorite are in amorphous domains rather than as inclusions.

In a study of radiation damage in natural zirconolite ($[Ca,Th]ZrTiO_7$), Farges et al. (1993) investigated the structural environment of Zr, Th, and U using EXAFS spectroscopy. The oxidation state of U in the metamict zirconolite samples was determined to be U^{4+}, with no indication of higher oxidation states. This is consistent with the lack of physical or chemical alteration in the samples. Comparison with other metamict Zr-Th-U- bearing phases such as zircon or thorite suggest that Zr^{4+}, Th^{4+} and U^{4+} prefer 7, 8, and 6 coordinated sites, respectively, in the metamict zirconolite samples at ambient temperatures and pressures.

X-RAY PHOTOELECTRON SPECTROSCOPY

In X-ray photoelectron spectroscopy (XPS), monoenergetic soft X-rays (e.g. AlK_α or MgK_α) irradiate a sample resulting in the ejection of electrons from the sample. The electrons ejected are from core levels of the atom of the material probed. The identity of the elements present in a sample can be directly determined from the kinetic energies (binding energies) of the ejected photoelectrons which are characteristic of a particular element. On a finer scale, XPS allows for the determination of the oxidation state of the elements present from small variations in the measured kinetic energies and other structural effects. Also, the relative concentrations of the elements present can be determined from the measured photoelectron intensities. In a solid material, the kinetic energies of the ejected photoelectrons leaving the sample are so low that, in practice, XPS probes only the top 1 to 5 nm, depending on the material, the energy of the photoelectrons of interest, and the angle between the sample and detector.

X-ray photoelectron spectroscopy of uranium in minerals

Few studies have been done using XPS to investigate the chemical properties of U in minerals. In a study of U and Th oxide compounds, Veal and Lam (1974) used XPS to investigate the valance-band structures. Although these are synthetic materials, they are discussed here to illustrate the utility of this technique in studying different oxidation states of materials, and because the distribution of U^{4+} and U^{6+} in minerals is a major part of this review. Their XPS results indicate that the electronic structures of UO_2 and ThO_2 are similar apart from the presence of $5f$ electrons in UO_2. These $5f$ electrons contribute to the differences in physical properties of these two materials including magnetism, electrical conductivity, and color (Veal and Lam 1974).

Veal and Lam (1980) also used XPS to investigate the role of $5f$ electrons in bonding in UO_2, U_4O_9 ($UO_{2.25}$), U_3O_8 ($UO_{2.67}$), and β-UO_3. Veal and Lam (1980) found a systematic dependence of the localized $5f$ electron peak intensity in the degree of oxidation in those U compounds. The primary difference in the XPS spectra between UO_2 and U_3O_8 is a significant change in height in of the $5f$ peak (Fig. 7). This change in intensity of the $5f$ peak indicates the transfer of the metal-oxygen bonding electrons back into the localized $5f$ electron states as oxygen is removed from U_3O_8 during oxidation. There is, however, little change in the shape of the bonding band in the XPS spectra for the different U compounds (Fig. 7).

Sunder et al. (1996) examined uraninite-bearing samples from the Cigar lake U deposit in northern Saskatchewan, Canada, and found that U^{6+}/U^{4+} ratios for most samples were less that 0.5, considered an upper threshold for the stability of synthetic UO_{2+x} prior to dissolution (x ~0.33). Reported U^{6+}/U^{4+} ratios ranged from 0.09 to 0.76. Higher U^{6+}/U^{4+} ratios in some samples were attributed to partial oxidation after collection or during drilling, although penetration of oxidizing surface waters along fractures may have resulted in some samples being oxidized *in situ*. Fayek et al. (1997) reported (unidentified) UO_2^{2+} minerals along micrometer-sized veins transecting uraninite from Cigar Lake, the presence of which would significantly enhance U^{6+}/U^{4+} ratios measured by XPS. However, Sunder et al. (1996) do not report UO_2^{2+} minerals in their samples. Sunder et al. (1994) examined uraninite-bearing samples from the Oklo U deposits in Gabon, and found U^{6+}/U^{4+} ratios comparable to, or lower than those reported from Cigar Lake, with values in the range 0.10 to 0.19.

Francis et al. (1994) used XANES and XPS to investigate the reduction of U by the bacteria *Clostridium* sp. Comparisons of the XPS spectra for a control sample, and the sample inoculated with *Clostridium* sp indicated that U^{6+} was effectively reduced to U^{4+}.

XANES analysis of the U M_v absorption edge for U metal, UO_2, and the sample inoculated with *Clostridium* sp also confirmed the reduction of U to U^{4+} and also possibly U^{3+}. Organic acid metabolites, the extracellular components of the cell-culture medium and cells killed by heat treatment all failed to reduce the U under anaerobic conditions. One implication of this study is these results further suggest that the reduction of U only occurred in the presence of growing or resting cells. This indicates that reduction and subsequent stabilization of U can occur through enzymatic action of anaerobic bacteria.

Wersin et al. (1994) used solution analysis and several different spectroscopic techniques, including XPS, to investigate the interaction between UO_2^{2+} and galena and pyrite surfaces under anoxic conditions. The experimental run products examined using XPS indicate that on the reacted mineral surfaces, the concomitant formation of polysulfide materials and a U oxide compound with a U^{6+}/U^{4+} of ~2. They also observed that, at values of pH greater than 6, an increase in the concentrations of oxidized sulfur and iron, and higher relative amounts of UO_2^{2+} on mineral surfaces. Wersin et al. (1994) suggested that XPS is an effective tool for studying the interaction of UO_2^{2+} and sulfide minerals because it allows for the identification of surface reaction products.

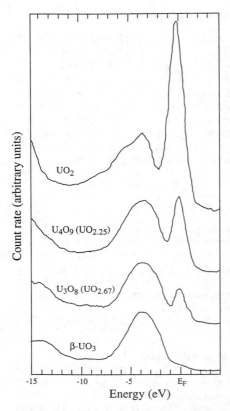

Figure 7. XPS valence spectra for several uranium oxide compounds (after Veal and Lam 1980).

CONCLUDING COMMENTS

There has been surprisingly little work done on spectroscopy of U in minerals. It is this author's position that this situation will change with the current interest in environmental mineralogy and, especially, the cleanup of U contaminated soils and the disposal and long-term containment of radioactive waste. The techniques discussed in this review are complimentary to many other techniques used to obtain chemical and structural information about materials. These techniques and allow for a more thorough characterization of U in minerals.

ACKNOWLEDGMENTS

Thanks to Dorothea Hanchar for her comments on the manuscript, to Steve Wolf and Robert Finch for their official reviews of the manuscript, and especially to Robert Finch for his Herculean editorial efforts and unfailing good nature .

REFERENCES

Allen PG, Berg JM, Chisolm-Brause CJ, Conradson SD, Donohoe RJ, Morris DE, Musgrave JA, Tait CE 1994 Determining uranium speciation in contaminated soils by molecular spectroscopic methods: Examples from the uranium in soils integrated demonstration. Waste Management '94 Meeting Proc, Tucson, Arizona

Allen PG, Bucher JJ, Denecke MA, Edelstein NM, Kaltsoyannis N, Nitsche H, Reich T, and Shuh DK 1996 Application of synchrotron radiation techniques to chemical issues in the environmental sciences. *In* KL D'Amico, L J Terminello, DK Shuh, Eds. Synchrotron Radiation Techniques in Industrial, Chemical, and Materials Science. Plenum Press, New York, p 169-185

Anderson BW 1963 Absorption spectra and properties of metamict zircons. J Gemmology 9:1-6

Antonio MR 1992 Extended X-ray absorption fine structure. *In* CC Brundle, J Charles A Evans, S Wilson (Eds) Encyclopedia of Materials Characterization. Butterworth-Heinemann, Boston, p 214-226

Baird CP, Kemp TJ 1997 Luminescence, spectroscopy, lifetimes and quenching mechanisms of excited states of uranyl and other actinide ions. Prog Reaction Kinetics 22:87-139

Bénard P, Louër D, Dacheux N, Brandel V, Genet M 1994 $U(UO_2)(PO_4)_2$, a new mixed-valence uranium orthophosphate: *Ab initio* structure determination from powder diffraction data and optical and X-ray photoelectron spectra. Chem Mater 6:1049-1058

Bertsch PM, Hunter DB, Sutton SR, Bajt S, Rivers ML 1994 *In situ* chemical speciation of uranium in soils and sediments by micro X-ray absorption spectroscopy. Env Sci Tech 28:980-984

Blasse G, Grabmaier BC 1994 Luminescent Materials. Springer-Verlag, New York

Brittain HG 1990 Luminescence spectroscopy. *In* DL Perry (Ed) Instrumental Surface Analysis of Geologic Materials, p. 283-309. VCH Publishers, Inc., New York

Brown Jr GE, Calas, G, Waychunas GA, and Petaiu J 1988 X-ray absorption spectroscopy and it applications in mineralogy and geochemistry. Rev Mineral 18:431-512

Brown GS, Doniach S 1980 The principles of X-ray absorption spectroscopy. *In* H Winick, S Doniach, Eds. Synchrotron Radiation Research. Plenum Press, New York, p 353-385

Burns PC, Ewing RC, Hawthorne FC 1997 The crystal chemistry of hexavalent uranium; polyhedron geometries, bond-valence parameters, and polymerization of polyhedra. Can Mineral 35:1551-1570

Calas G, Brown Jr GE, Waychunas GA, Petaiu J 1987 X-ray absorption spectroscopic studies of silicate glasses and minerals. Phys Chem Minerals 15:19-29

Carnall WT, Crosswhite HM 1986 Optical spectra and electronic structure of actinide ions in compounds and in solution. *In* JL Katz, GT Seaborg, LR Morss (Eds) The Chemistry of the Actinide Elements, Vol 2. Chapman and Hall, New York, p 1235-1276

Cejka J, Urbanec Z 1990 Secondary uranium minerals. Academia, Praha, 99 p

Charlet L, Manceau AA 1993 Structure, formation, and reactivity of hydrous oxide particles: Insights from X-ray absorption spectroscopy. *In* J Buffle, HP van Leeuwen (Eds) Environmental Particles, Vol 2. Lewis Publishing, Ann Arbor, MI, p 117-164

Chisholm-Brause CJ, Conradson SD, Buscher CT, Eller PG, Morris DE 1994 Speciation of uranyl sorbed at multiple binding sites on montmorillonite. Geochim Cosmochim Acta, 58:3625-3631

deNeufville JP, Kasdan A, Chimenti RJL 1981 Selective detection of uranium by laser-induced fluorescence: a potential remote-sensing technique. 1: Optical characteristics of uranyl geologic targets. Appl Optics 20:1279-1296

Denning RG 1992 AmElectronic-structure and bonding in actinyl ions. Structure and Bonding 79:215-276

Dent AJ, Ramsay JDF, Swanton SW 1992 An EXAFS study of uranyl ion in solutions and sorbed onto silica and montmorillonite clay colloids. J Coll Inter Sci 150:45-60

Duff MC, Amrhein C, Bertsch PM, Hunter DB 1997 The chemistry of uranium in evaporation pond sediment in the San Joaquin Valley, California, USA, using X-ray fluorescence and XANES techniques. Geochim Cosmochim Acta 61:73-81

Duff MC, Hunter DB, Bertsch PM, Arnhein C 1999 Factors influencing uranium reduction and solubility in evaporation pond sediments. Biogeochemistry 45:95-114

Farges F, Calas G 1991 Structural analysis of radiation damage in zircon and thorite; An X-ray absorption spectroscopic study. Am Mineral 76:60-73

Farges F, Ponader CW, Calas G, Brown Jr GE 1992 Structural environments in silicate glass/melt systems: U^{IV}, U^V, and U^{VI}. Geochim Cosmochim Acta 56:4205-4220

Farges F, Ewing RC, Brown Jr GE 1993 The structure of aperiodic, metamict, $(Ca,Th)ZrTiO_7$ (zirconolite): An EXAFS study of the Zr, Th, and U sites. J Mater Res 8:1983-1995

Fayek M, Janeczek J, Ewing RC 1997 Mineral chemistry and oxygen isotopic analysis of uraninite, pitchblende and uranium alteration minerals from the Cigar Lake deposit, Saskatchewan, Canada. Appl Geochem 12:549-565

Fendorf SE, Sparks DL, Lamble GM, Kelley MJ 1994 Applications of X-ray absorption fine structure spectroscopy to soils. Soil Sci Soc Am J 58: 1583-1595

Fielding PE 1970 The distribution of uranium, rare earths, and color centers in a crystal of natural zircon. Am Mineral 55:428-440

Francis AJ, Dodge CJ, Lu FL, Halada GP, Clayton CR 1994 XPS and XANES studies of uranium reduction by *Clostridium* sp. Env Sci Tech 28:636-639

Fred MS 1986 Spectra and electronic structure of free actinide atoms and ions. *In* JL Katz, GT Seaborg LR Morss (Eds) The Chemistry of the Actinide Elements, Vol 2. Chapman and Hall, New York, p 1196-1234

Frondel C 1958 Systematic mineralogy of uranium and thorium. U S Geol Surv Bull 1064, 400 p

Gleason S 1972 Ultraviolet Guide to Minerals. Ultraviolet Products, Inc., San Gabriel, CA

Hargreaves WA 1970 Energy levels of uranium ions in calcium fluoride crystals. Phys Rev B 2:2273-2284

Hasnain SS 1991 X-ray Absorption Fine Structure. Ellis Horwood, New York

Hawthorne FC 1988 Spectroscopic Methods in Mineralogy and Geology. Reviews in Mineralogy, Vol 18. Mineralogical Society of America, Washington, DC

Hunter DB, Bertsch PM 1998 *In situ* examination of uranium contaminated soil particles by micro-X-ray absorption and micro-fluorescence spectroscopies. J Radioanalyt Nuclear Chem 234:237-242

Kisliuk P, Richman I, Wong, EY 1967 Absorption spectrum of U^{4+} in zircon ($ZrSiO_4$). *In* HM Crosswhite, and HW Moos 9Eds) Optical Properties of Ions in Crystals, p 537. Interscience Publishers, New York

Koningsberger DC, Prins R 1988 X-ray Absorption: Principles, Applications, Techniques of EXAFS, SEXAFS, and XANES. John Wiley & Sons, New York

Mackey DJ, Runciman WA, Vance ER 1975 Crystal-field calculations for energy levels of U^{4+} in $ZrSiO_4$. Phys Rev B 11:211-218

Marfunin AS 1979a Spectroscopy, Luminescence, and Radiation Centers in Minerals. Translated from Russian by VV Schiffer. Springer-Verlag, New York

Marfunin AS 1979b Physics of Minerals and Inorganic Materials. Translated from the Russian by NG Egorova and AG Mishchenko. Springer-Verlag, New York

Meijerink A 1998 Experimental Tecnhiques. *In* DR Vij (Ed) Luminescence of Solids. Plenum Press, New York, p 45-94

Morris DE, Chisholm-Brause CJ, Barr ME, Conradson SD, Eller PG 1994 Optical spectroscopic studies of the sorption of $(UO_2)^{2+}$ species on a reference smectite. Geochim Cosmochim Acta 58:3613-3623

Morris DE, Allen PG, Berg JM, Chisholm-Brause CJ, Conradson SD, Donohoe RJ, Hess, NJ, Musgrave JA, Tait CD 1996 Speciation of uranium in Fernald soils by molecular spectroscopic methods: Characterization of untreated soils. Environ Sci Technol 30:2322-2331

Nicholas JV 1967) Luminescence of hexavalent uranium in CaF_2 and SrF_2 powders. Phys Rev 155:151-155

Pringsheim P 1949) Fluorescence and Phosphorescence. Interscience, New York

Rabinowitch E, Belford RL 1964 Spectroscopy and Photochenistry of Uranyl Compounds. Macmillian, New York

Richman I, Kisliuk P, Wong EY 1967 Absorption spectrum of U^{4+} in zircon ($ZrSiO_4$). Phys Rev 155: 262-267

Sayers DE, Stern EA, Lytle FW 1971 New technique for investigating noncrystalline structures: Fourier analysis of the extended x-ray absorption fine structure. Phys Rev Lett 27:1204-1207

Shannon RD 1976 Revised effective ionic radii and systematic studies of interatomic distances in halides and chalcogenides. Acta Crystallogr A32: 751-767

Schulze DG, Bertsch PM 1995 Synchrotron X-ray techniques in soil, plant, and environmental research. Advan Agron 55:1-66

Stern EA 1974 Theory of extended x-ray absorption fine structure. Phys Rev B10:3027-3037

Sturchio NC, Antonio MR, Soderholm L, Sutton SR, Brannon JC 1998 Tetravelent uranium in calcite. Science 281:971-973

Sunder S, Cramer JJ, Miller NH 1996 Geochemistry of the Cigar Lake deposit: XPS studies. Radiochim Acta 74:303-307

Sunder S, Miller NH, Duclos AM 1994 XPS and XRD studies of samples from the natural fission reactors in the Oklo uranium deposits. Mater Res Soc Symp Proc 333:631-638

Tarashchan AN 1978 Luminescence of Minerals. Naukova Dumka, Kiev

Tarashchan AN, Krasilshchikova OA, Platonov AN, Povarennych AS 1974 Luminescence of uranyl minerals. Constitution and Properties of Minerals 8:78-85

Thompson HA, Brown GE, Jr, Parks GA 1997 XAFS spectroscopic study of uranyl in solids coordination and aqueous solution. Am Mineral 82:483-496

Thompson HA, Parks GA, Brown GE, Jr 1998 Structure and composition of uranium^{VI} sorption complexes at the kaolinite-water interface. *In* EA Jenne (Ed) Adsorption of Metals by Geomedia, Academic Press, San Diego, p 349-370

Vance ER 1974 The anomalous optical absorption spectrum of low zircon. Mineral Mag 39:709-714
Vance ER, Mackey, DJ 1974 Optical study or U^{5+} in zircon. J Phys C Solid State Phys 7:1898-1908
Vance ER, Mackey DJ 1975 Further studies of the optical absorption spectrum of U^{5+} in zircon. J Phys C Solid State Phys 8:3439-3447
Vance ER, Mackey DJ 1978 Optical spectra of U^{4+} and U^{5+} in zircon, hafnon, and thorite. Phys Rev B 18:185-189
Veal BW, Lam DJ 1974 X-ray photoelectron studies of thorium, uranium, and their dioxides. Phys Rev B 10:4902-4908
Veal BW, Lam DJ 1980 Photoelectron spectra of actinide compounds. *In* N Edelstein (Ed) Lanthanide and Actinide Chemistry and Spectroscopy, 131, American Chemical Society, p 427-441
Vij DR 1998 Luminescence of Solids. Plenum Press, New York
Vochten R, Van Haverbeke L, Van Springel K, De Grave E 1995 Soddyite: Synthesis under elevated temperature and pressure, and study of some physicochemical characteristics. N Jb Miner Mh 1995:470-480
Warren TS, Gleason S, Bostwick RC, Berbeek ER 1995 Ultraviolet Light and Fluorescent Minerals. Thomas J. Warren, Rio, WV
Waychunas GA 1988 Luminescence, X-ray emission and new spectroscopies. *In* FC Hawthorne (Ed) Spectroscopic Methods in Mineralogy and Geology. Rev Mineral 18:639-698
Wersin P, Hochella MF Jr, Persson P, Redden G, Leckie JO, Harris, DW 1994 Interaction between aqueous uranium(VI) and sulfide minerals: Spectroscopic evidence for sorption and reduction. Geochim Cosmochim Acta 58:2829-2843
Wybourne BG 1965 Spectroscopic Properties of Rare Earths. Interscience Publishers, New York

Infrared Spectroscopy and Thermal Analysis of the Uranyl Minerals

Jiří Čejka

The Natural History Museum
The National Museum
Václavské náměstí 68
CZ-115 79 Praha 1, Czech Republic

INFRARED SPECTROSCOPY OF THE URANYL MINERALS

Introduction

In crystal structures of minerals and synthetic compounds that contain hexavalent U, the U^{6+} cation is almost always present as part of the stable, approximately linear uranyl ion: UO_2^{2+}. The uranyl ion is coordinated by four, five or six equatorial anions in an approximately planar arrangement, perpendicylar to the axis of the uranyl ion, giving tetragonal, pentagonal, and hexagonal dipyramids, respectively, with uranyl oxygens at the apices. In most minerals, the equatorial anions are not strictly coplanar but are shifted by small distances above and below the equatorial plane. Equatorial oxygen atoms belong to various oxyanions, including hydroxyl ions, carbonates, sulfates, silicates, phosphates, and arsenates. Uranyl oxygens are more strongly bonded to the uranium atom than are oxygen atoms shared with oxyanions. The alignment of uranyl groups perpendicular to the equatorial oxyanions commonly results in the formation of polyhedral sheets that are important structural components in most uranyl minerals; if bonds between uranium and equatorial atoms are preferred in one direction, chain structures result. Polyhedral layers or chains are connected by hydrogen bonds between hydroxyl protons in one layer and oxygens (either uranyl or oxyanion) in adjacent layers, or by hydrogen bonds with protons of H_2O groups in interlayer sites. In most minerals, interlayer cations also participate through electrostatic forces and compensate the negative charge of polyhedral layers (Urbanec and Cejka 1979a; Cejka and Urbanec 1990; Burns et al. 1996).

A variety of instrumental methods have been used to determine the structures, thermodynamics, and physical and chemical properties of uranyl minerals. Spectroscopic methods are an indispensable tool in such research. The theoretical, instrumental and experimental base of many spectroscopic methods have been discussed in *Reviews in Mineralogy*, Volume 18, "Spectroscopic Methods in Mineralogy and Geology" (Hawthorne 1988). Vibrational spectroscopy in the mineral sciences has also been discussed by McMillan (1985). The topics of the present contribution are the infrared spectroscopy and thermal analysis of uranyl minerals. The primary emphasis of this paper is to present the current state of practical knowledge and information on measured infrared spectra of uranyl minerals.

Infrared spectroscopy (IR) can improve understanding of the crystal chemistry of uranyl minerals, especially when complemented by an X-ray structural analysis of the material being investigated. Wilkins (1971a,b) recommends using IR in cases in which X-ray analyses or measurements of optical properties fail to produce a unique solution to the structure being investigated. IR spectra of structurally and chemically well-characterized compounds also aid interpretations of IR spectra from unknowns.

The intensity of diffracted X-rays increases as a function of the number of electrons in a crystalline material. Therefore, the structure of the uranium sublattice may be readily determined by X-ray diffraction, whereas light atoms such as oxygen and hydrogen are not so readily located. At best, the positions of protons may be inferred from geometrical and bond-valence considerations. Neutron diffraction can identify H atom positions if sufficient pure material is available for analysis, but this is commonly difficult to acheive for natural uranium minerals, and to date, not all uranyl minerals have been successfully synthesized.

The physical laws that pertain to IR spectroscopy are well understood. Proton vibrations are extremely sensitive to the nature of the chemical bonds by which the protons are connected to adjoining atoms. Vibrations of oxyanions are identified by their characteristic wavenumbers in an IR spectrum. Absorptions corresponding to uranyl vibrations also have well-defined locations in an IR spectrum (Ross 1972, Urbanec and Cejka 1979a; Nakamoto 1986). An important factor to consider is that the number of IR-active vibrations depends upon the site symmetry of an oscillating polyhedron.

The symmetry of the crystal lattice, local site symmetries, and the location of polyhedra in the structure all affect the number and positions of vibration bands in and IR spectrum. Site- and factor-group analyses are commonly used to interpret IR spectra (Nakamoto 1986). This has been done, for example, with natural layered uranyl phosphates (Cejka et al. 1984a). Correlations with X-ray structure analyses, if available, help elucidate details of a material's structure; however, Raman spectra are often necessary for complete interpretation of vibrational–structural relationships. Measurements of single crystals are recommended if available. An advantage of IR spectroscopy is the ability to analyze small samples (i.e. several milligrams or a few tens of micrograms). This is important in mineralogy when a very rare specimen is analyzed. Uranyl minerals (i.e. most secondary uranium minerals) commonly originate as intimate mixtures with other minerals (uranium and non-uranium), and often in only small quantities. Care must be taken to ensure that material chosen for various analyses is one compound and that all material comes from the same small region of the specimen. Under ideal circumstances, large crystals should be selected and broken so that pieces of the same crystal can be used for various analyses. Without taking such care, ambiguous or erroneous results may be obtained (Pagoaga 1983; Cejka and Urbanec 1990).

The character of IR spectra can, in some cases, be influenced by sample preparation (Russell 1974; Gevorkyan et al. 1979, 1981; Gevorkyan and Ilchenko 1985, Cejka et al. 1988c; Cejka and Urbanec 1990). Artifacts of sample preparation are especially common when KBr disks or micropellets are used. Minerals or synthetic compounds possessing structures with narrow stability ranges (e.g. island-like, chain or layer structures, such as uranyl sulfates) may be altered or even fully destroyed during sample preparation, so that recorded IR spectra may not represent the minerals analyzed. Such processes as dehydration (partial or total), ion-exchange reactions, redox-reactions, etc., have been observed when KBr reacts with $CuSO_4 \cdot xH_2O$ and synthetic johannite during grinding or disk preparation (Cejka 1988). In some cases, it is necessary to record IR spectra with a Fourier Transform Infrared Spectrophotometer (FTIR) using Nujol (parafin oil) suspensions, KBr disks, and the Diffusive Reflectance Infrared Fourier Transform technique (DRIFT). Care must be taken when comparing these results with other data sets because of potential problems with sample preparation.

IR spectra are reported for uranyl minerals by Moenke (1962, 1966), Farmer (1974), Gadsden (1975), and Boldyrev (1976). The first study concerning primarily IR spectroscopy of uranium minerals are in two papers by Wilkins (1971a,b). IR spectra of uranyl minerals were published later by Urbanec and Cejka (1979), Wang et al. (1981), Cejka and Urbanec (1990), and Jones and Jackson (1993). However, information about IR spectra of individual uranyl minerals are dispersed throughout the literature, data are commonly reported only as graphs, and in some cases only partly interpreted data are provided. Spectroscopy of synthetic uranium compounds have been summarized by Dieke and Duncan (1949), Rabinowitch and Belford (1964), Volodko et al. (1981), and Novitskiy et al. (1981). Relatively few studies have reported Raman spectra of uranium minerals (e.g. Biwer et al. 1990). Surprisingly, the recent Dana's New Mineralogy (Gaines et al. 1997) contains little information concerning IR spectra for, not only uranium minerals, but non-uranium minerals as well.

In this contribution, some IR spectra and TG curves of well-defined uranyl minerals are included that have not yet been published elsewhere. The reason to include such IR spectra and TG curves is to cover all available information on IR spectra and TA analysis of uranyl minerals, especially from the point of view of practical application. Some uranyl minerals are mentioned only briefly, because detailed data are not available. Detailed information is included especially for some minerals or their synthetic analogues studied by our group. This enables us to suggest possible interpretation of IR spectra and thermal analyses to aid in determining crystallo-chemical relations in minerals. All uranyl minerals mentioned in this contribution are presented with references used for interpretation of their IR spectra and TG and DTA curves. Additional papers related to specific topics are listed in the References section.

Uranyl, UO_2^{2+}

As noted above, the U^{6+} cation is almost always part of the nearly linear uranyl ion, UO_2^{2+}, that is an integral part of tetragonal, pentagonal, and hexagonal dipyramids in mineral structures. Based on 105 well-refined crystal structures of minerals and synthetic uranyl compounds, the following bond information can be inferred (Burns 1999a; Burns et al. 1996, 1997a). The $U-O_I$ (uranyl oxygen) bond length is independent of the equatorial anions that make up the polyhedra: average U-O bond lengths for uranyl ions are 1.79(3) Å (tetragonal dipyramids); 1.79(4) Å (pentagonal dipyramids); and 1.78(3) Å (hexagonal dipyramids). Whereas pentagonal and hexagonal dipyramids always contain a uranyl ion, tetragonal dipyramids do not always contain uranyl ions. There is a continuous series of coordination polyhedra from tetragonal dipyramids with uranyl ions to holosymmetric octahedral geometry. In contrast to uranyl-ion bond lengths, equatorial $U-O_{ligand}$ (O, OH) bond lengths in uranyl polyhedra depend upon coordination number. Average bond $U-O_{ligand}$ lengths for all polyhedra are 2.28(5) Å, 2.37(9) Å, and 2.47(12) Å, respectively (Burns 1999a; Burns et al. 1996, 1997a). Uranyl polyhedra may polymerize with each other or with cation-oxyanion coordination polyhedra to form complex structures. The hierarchy of these crystal structures has been discussed by Burns et al. (1996).

Uranyl vibrational modes (fundamental vibrations)

A free uranyl, UO_2^{2+}, point symmetry $D_{\infty h}$, should exhibit three fundamental modes (Fig. 1):

symmetric stretching vibration v_1;

bending vibration v_2 (δ); and

antisymmetric stretching vibration v_3.

The bending mode is doubly degenerate since it can occur in two mutually perpendicular planes. It can split into its two components when the uranyl ion is placed in an external force field (Hoekstra 1982). Thus, the linear uranyl group, point symmetry $D_{\infty h}$, has four normal vibrations, but only three fundamentals.

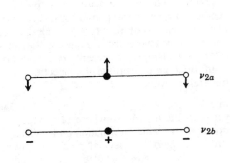

Figure 1. Normal modes of vibration in uranyl, UO_2^{2+} (adapted from Nakamoto 1986).

Uranyl stretching vibrations, v_1 UO_2^{2+} and v_3 UO_2^{2+}

In linearly symmetric uranyl ions belonging to the $D_{\infty h}$ point group, the symmetric stretching vibration, v_1 UO_2^{2+}, exhibits in the region 900–750 cm^{-1} and is Raman active, whereas, v_1 UO_2^{2+} appears in the IR spectrum only in the case of substantial symmetry lowering. The antisymmetric stretching vibration, v_3 UO_2^{2+} (1000–850 cm^{-1}), is active in the infrared and inactive in Raman and is sensitive to isotopic exchange. A lowering of symmetry ($D_{\infty h} \rightarrow C_{\infty v}$, C_{2v} or C_s) causes both the activation of all three fundamentals in the infrared and Raman spectra and the activation of their overtones and combination vibrations (Bist and Pant 1964) (Tables 1 and 2).

Table 1. Correlation between symmetry types of the three vibrations for point groups C_s, C_{2v}, $C_{\infty v}$, $D_{\infty h}$

Assignment	C_s	C_{2v}	$C_{\infty v}$	$D_{\infty h}$
v_1	A′ R, IR	A_1 R, IR	$A_1 = \Sigma^+$ R, IR	Σ_g^+ R
v_2	A′ R, IR	A_1 R, IR	$E_1 = \Pi$ R, IR	Π_u IR
v_3	A′ R, IR	B_1 R, IR	$A_1 = \Sigma^+$ R, IR	Σ_u^+ IR

R, Raman active; IR, infrared active.

Table 2. Correlation between the symmetry types of the two vibrations v_1 and v_3 for point groups D_{nh}

Assignment	D_{2h}	D_{3h}	D_{4h}	D_{5h}	D_{6h}
v_1	A_g R	A_1' R	A_{1g} R	A_1' R	A_{1g} R
v_3	B_{1u} IR	A_2'' IR	A_{2u} IR	A_2'' IR	A_{2u} IR

It is customary to designate one-dimensional species by the letters A or B, two-dimensional species by E, and three-dimensional species by F. A and B denote non-degenerate species. A represents the symmetric species (character = +1) with respect to rotation about the principal axis (chosen as z axis), whereas B represents the antisymmetric species (character -1) with respect to rotation about the principal axis; E and F denote doubly and triply degenerate species, respectively. If two species in the same point group differ in the character of C (other than the principal axis), they are distinguished by subscripts 1, 2, 3, etc. If two species differ in the character of σ (a plane of symmetry) other than $σ_v$ (a vertical plane of symmetry), they are distinguished by ´ and ?. If two species differ in the character of i (a center of symmetry), they are distinguished by subscripts g and u (German: gerade and ungerade). If several different labels are allowed, g and u take precendence over 1, 2, 3, etc. which in turn take precendence over ´ and ?. In the case of linear molecules (point groups $C_{∞v}$ and $D_{∞h}$) these symbols are usually replaced by a system derived from the term symbol for the electronic states of diatomic and other linear molecules. A capital Greek letter Σ, Π, Δ, Φ, etc., is used, corresponding to l = 0, 1, 2, 3, etc., where l is the quantum number for rotation about the molecular axis. These symbols are equivalent to A_1, E_1, E_2, E_3, etc., respectively. For Σ species a superscript + or - is added to indicate the symmetry with respect to a plane which contains the molecular axis. Species of the point group $D_{∞h}$ are further distinguished by the subscripts g or u which specify the symmetry with respect to the inversion (Wilson et al. 1955; Turrell 1972; Nakamoto 1986).

If the uranyl group is linear, but not symmetric, all three vibrational modes $ν_1$, $ν_2$ and $ν_3$ UO_2^{2+} show vibrational activity in Raman as well as in IR spectra. In this case the intensities of $ν_2$ and $ν_3$ UO_2^{2+} in Raman spectra are lower than that for $ν_1$ UO_2^{2+}, whereas in infrared spectrum the intensities of $ν_2$ and $ν_3$ UO_2^{2+} are expected to be somewhat stronger than that of $ν_1$ UO_2^{2+}(Prins 1973). If the uranyl ion is linear, then the frequencies $2ν_3$, $4ν_3$, $6ν_3$, etc., $3ν_1$, $5ν_1$, $7ν_1$, etc. should be IR-forbidden but Raman–allowed whereas $3ν_3$, $5ν_3$, $7ν_3$, etc. and $2ν_1$, $4ν_1$, $6ν_1$, etc. should be IR-allowed but Raman-forbidden (Herzberg 1945; Bullock 1969). A combination band $ν_1 + ν_3$ is also IR-allowed (Bullock 1969). In contrast to this, in the case of $D_{∞h}$ symmetry, only those overtones and combinations should be IR-active which make the resultant vibration effectively antisymmetric (e.g. ($ν_1 + ν_3$), ($2ν_2 + ν_3$), $3ν_2$, $3ν_3$, etc.). The site symmetry for the symmetric linear ion is likely to be D_{nh} or lower (n is the number of ligands in the equatorial plane), (2 ≤ n ≤ 6), depending upon the number and manner in which the ligands (L) are coordinated to the uranyl ion. Under these circumstances, the UO_2L_n will have a number of vibrations of various symmetry types. Of these, uranyl vibrations $ν_1$ and $ν_3$ will have correlations with the symmetry types shown in Table 2 (Bist and Pant 1964). From Table 2, it should be clear that the selection rules for Raman and IR spectra will not change with equatorial ligation. According to Bist and Pant (1964) it appears, therefore, that, although IR spectra of most uranyl compounds indicate that the uranyl ion maintains its linearity, wherever $ν_1$ also occurs, the uranyl ion is interpretted as belonging to point groups $C_{∞v}$, C_{2v}, or even C_s, or else the site symmetry is distorted due to non-planar or asymmetric ligation. The presence of the $ν_1$ vibration in the IR spectrum suggests that the symmetry is reduced, the loss of the $C_∞$ axis.

Several bands attributed to the $ν_1$ and $ν_3$ UO_2^{2+} vibrations and correlated with crystal structure data are interpretted as indicating either several structurally nonequivalent uranyl groups in the unit cell or the co-existence of different crystal modifications in the mineral studied.

Relations between a free UO_2^{2+} group stretching and bending vibrations and their corresponding force constants can be calculated using the following equations (Weidlein et al. 1988):

$$v_1 = 1303 \ [(f+f')/m_o]^{1/2} \ cm^{-1} \tag{1}$$

$$v_2 = 1303 \ [2d \ (2/m_u + 1/m_o)]^{1/2} \ cm^{-1} \tag{2}$$

$$v_3 = 1303 \ [(f-f') \ (2/m_u + 1/m_o)]^{1/2} \ cm^{-1} \tag{3}$$

where m_u and m_o are expressed in atomic mass units (amu), f (stretching f_{uo}), f' (interaction $f_{uo,uo'}$)
and d (bending) force constants in N cm^{-1}. In a simple valence force field, and assuming harmonic vibrations characteristic of a linear ion, the f' interaction force constant can be omitted and the relation between v_3 and v_1 is

$$v_3 = v_1 \ (2m_o/m_u)^{1/2} \tag{4}$$

Under these conditions, the v_3/v_1 ratio is ~ 1.065 (Rabinowitch and Belford 1964). If the f' interaction force constant is respected, the v_3/v_1 may decrease or increase depending upon the sign of f' (Hoekstra 1982).

If the f' interaction force constant is ignored in Equations (1) and (3), the relationship $v_1 = 0.939 \ v_3$ is obtained. Thus, a plot of v_1 against v_3 should lead to a straight line of slope 0.939 with zero intercept. Bullock (1969) applied least-squares to fit data for uranyl nitrato complexes and derived a line of slope 0.89 and intercept 30.8 cm^{-1} on the v_3 axis. Using the same procedure, McGlynn et al. (1961) obtained the expression, $v_1 = 21 + 0.89$ cm^{-1}, whereas Bagnall and Wakerley (1975) obtained, $v_1 = 0.912 \ v_3 = 1.04$ cm^{-1}. Thus, there are four expressions available to be applied for calculating $v_1 UO_2^{2+}$ and $v_3 \ UO_2^{2+}$ relations:

$$v_1 = 0.939 \ v_3 \ (cm^{-1}) \ (\text{McGlynn et al. 1961}) \tag{5}$$

$$v_1 = 0.89 \ v_3 + 21 \ (cm^{-1}) \ (\text{McGlynn et al. 1961}) \tag{6}$$

$$v_1 = 0.89 \ v_3 + 30.8 \ (cm^{-1}) \ (\text{Bullock 1969}) \tag{7}$$

$$v_1 = 0.912 \ v_3 - 1.04 \ (cm^{-1}) \ (\text{Bagnall and Wakerley 1975}) \tag{8}$$

It seems necessary to note that these relations need further study (see, e.g. Seryozhkin et al. 1983). The v_1/v_3 relation (or better, $v_1 = f(v_3)$) may be influenced by the character of the oxyanions, structural details, hydrogen bonding, external force fields around uranyl ions caused by other factors, etc. The use of well-defined mineral and synthetic phases and precise measurements of wavenumbers are indispensable. This was not always the case with older data, many of which are difficult to interpret due to the lack of well characterized natural and synthetic materials. The same is valid also for empirical equations inferred on the basis of Badger's rule, which will be discussed later.

The intensity of the antisymmetric stretching vibration $v_3 \ UO_2^{2+}$ is usually strong or very strong. Therefore, this vibration is easy to observe and correctly assign. the intensity of the symmetric stretching vibration $v_1 \ UO_2^{2+}$ (if IR active) is usually low. The $v_1 \ UO_2^{2+}$ vibration is located in the region of oxyanion skeletal vibrations or δ UOH vibrations, and some coincidences are possible. This may cause a wrong tentative assignment of the vibration. For this reason, an approximate calculation of $v_1 \ UO_2^{2+}$ from $v_3 \ UO_2^{2+}$ using Equations (5)–(8) may be helpful.

The character of IR spectra of uranyl minerals (wavenumbers of $v_1 \ UO_2^{2+}$ and $v_3 \ UO_2^{2+}$ vibrations) may be influenced by conditions under which the minerals formed and the stability of their crystal structures.

Uranyl bending vibration v_2 (δ) UO_2^{2+}

The doubly degenerate bending vibration, v_2 UO_2^{2+} (δ UO_2^{2+}), (approximately 300–200 cm^{-1}) is IR active, and a decrease of symmetry can cause splitting of this vibration into two IR and Raman active components. Coincidences between uranyl bending vibrations and U-O$_{ligand}$ vibrations are possible.

Unfortunately, older papers do not contain data on the uranyl bending vibration. According to Hoekstra (1982), very little information was available on uranyl bending vibrations until 1965, because commercial infrared instruments could not reach the low frequencies required to make such measurements.

Estimating the uranyl (U-O_I) bond length from its stretching vibrations

A general relationship exists between internuclear distance and bond-force constant

$$R_{ij} = \beta f_{ij}^{-1/3} + d_{ij} \tag{9}$$

where R_{ij} is the bond length, f is the bond force constant, β and d are constants characteristic of the bonding atoms (Badger 1934, 1935; see Hoekstra 1982).

An empirical relationship between stretching force constant (f) and the U-O$_I$ (uranyl) bond length R_{U-O_I} was published by Jones (1959), who used X-ray structure data for $K_3UO_2F_5$:

$$R_{U-O_I} = 1.08 \, f^{1/3} + 1.17 \text{ Å} \tag{10}$$

where R_{U-O_I} is the U-O$_I$ (uranyl) bond length in Å and f is the bond force constant in mdyn·Å^{-1} or N·cm^{-1}. Some other equations were inferred using X-ray structure data available for a few uranyl compounds (for details see Hoekstra 1982). Veal et al. (1975) plotted v_3 UO_2^{2+} and R_{U-O_I} (from X-ray diffraction data) for 20 uranyl compounds including uranyl salts and metal uranates and derived, according to Hoekstra (1982), a best fit variant of the Badger's relation

$$R_{U-O_I} = 81.2 \, v_3^{-2/3} + 0.895 \text{ Å} \tag{11}$$

Glebov (1981—30 uranyl compounds included ; 1983—12 uranates and 19 uranyl coordination compounds included; 1985—55 compounds inclusive uranates and uranyl coordination compounds; 1989—64 compounds include 8 uranates, 3 uranyl oxides and hydroxides, and 53 uranyl coordination compounds) published a set of empirical equations. Glebov proved that it is necessary to separate not only uranates, but also different groups of coordination compounds for deriving empirical equations based on the R_{U-O_I} / v_3 UO_2^{2+} relation, with special regard to the possible existence of Badger's, but also anti–Badger's dependence in some coordination compounds. However, this conclusion was not accepted by Seryozhkin and Seryozhkina (1984).

In his last paper, Glebov (1989) discussed seven empirical equations, from which the following ones may be useful:

$$R_{U-O_I} = 72.98 \, v_3^{-2/3} + 1.006 \text{ Å} \tag{12}$$

suitable especially for uranates, uranyl oxides, and uranyl hydroxides,

$$R_{U-O_I} = 162.4 \, v_3^{-2/3} + 0.03 \text{ Å} \tag{13}$$

and

$$R_{U-O_I} = 68.2 \, v_3^{-2/3} + 1.05 \text{ Å} \tag{14}$$

suitable for coordination compounds as uranyl carbonates (eq. 13), and uranyl sulfates, molybdates, and silicates (Eqn. 14).

The newest empirical equation was inferred by Rodriguez S. and Martinez Quiroz (1996a,b) who used seven uranyl compounds for inferring the empirical equation

$$R_{U-O_I} = 52.2476 \, v_3^{-2/3} + 1.17 \text{ Å} \tag{15}$$

Hoekstra (1982) points out that, although there appear to be substantial differencies among the equations relating R_{U-O_I} and v_3 UO_2^{2+} or f, the relatively restricted range in v_3 (approximately 850–1000 cm^{-1}) observed for uranyl salts, together with uncertainties in bond-length measurements from X-ray and neutron diffraction data, make it difficult to select one equation in preference to others.

Equatorial ligand effect on v_3 UO_2^{2+}

Hoekstra (1963, 1982) concluded that the influence of equatorial ligands on uranyl vibrations is often overshadowed by other effects, including ionic bonding of metal cations or hydrogen bonding to the uranyl oxygens.

Anion vibrational modes (fundamental vibrations)

Stretching vibration of the peroxidic group v O_2^{2-} is observed approximately in the region of 880 cm^{-1} (Rocchiccioli 1966; Weidlein et al. 1988). The pyramidal groups SeO_3^{2-} and TeO_3^{2-} with point symmetry C_{3v} exhibit six normal vibrations with four fundamentals (Nakamoto 1986).

The symmetric stretching vibrations v_1, the bending vibrations v_2, the antisymmetric stretching vibrations v_3 and the bending vibrations v_4 (both doubly degenerate) are all active in Raman and IR spectra (Table 3). The decrease of symmetry $C_{3v} - C_s$ is mostly followed by splitting of the doubly degenerate vibrations v_3 and v_4. Although $v_2 > v_4$ holds in all cases, the order of two stretching vibrations v_1 and v_3 depends on the nature of the central metal (Nakamoto 1986).

In the spectrum of planar CO_3^{2-} group with D_{3h} point symmetry and six normal vibrations four fundamentals appear:

v_1 (A_1'), symmetric stretching vibration (Raman-active) 1115–1050 cm^{-1},

v_2 (A_2'), out-of plane bending vibration (IR-active) 880–835 cm^{-1},

v_3 (E'), antisymmetric stretching vibration (IR- and Raman-active) 1610–1250 cm^{-1},

v_4 (E'), in-plane bending vibration (IR- and Raman-active), 770–670 cm^{-1}.

Vibrations v_3 and v_4 CO_3^{2-} are both doubly degenerate. When symmetry decrease takes place ($D_{3h} \rightarrow C_{2v}$, C_s or C_1), the v_1 vibration becomes IR-active and the v_3 and v_4 vibrations are both split (Table 4). The CO_3^{2-} anion behaves as monodentate or bidentate ligand and can be coordinated to U^{6+} mono- or bi-dentately, which should be inferred from the difference between wavenumbers of the split v_3 CO_3^{2-} vibration. Although the number of IR-active fundamentals is the same for C_{2v} and C_s symmetries, the splitting of degenerate vibrations is larger in the bidentate than in the monodentate complex. According to Jolivet et al. (1980), for large f-transition metals, the v_3 splitting can be calculated to about 50-60 cm^{-1} in the monodentate, and 160-190 cm^{-1} in the bidentate structures.

Most uranyl minerals contain anions with tetrahedral symmetry. In the case of a free ion of T_d symmetry, there are 9 normal vibrations characterized by four fundamental distiguishable modes of vibration, two of which are IR-active (Table 5).

In the case of symmetry lowering, $T_d \rightarrow C_{3v}$, C_{2v}, C_1, the symmetric stretching vibration v_1 is activated and can appear in the IR spectrum. A decrease of symmetry is also the cause of IR activation and splitting of the doubly degenerate bending vibration v_2.

Table 3. Vibrational spectra of pyramidal anions

	SeO_3^{2-} (a) [cm^{-1}]	TeO_3^{2-} (b) [cm^{-1}]
ν_1 (A$_1$)	790	740
ν_2 (A$_2$)	452	400
ν_3 (E)	730	665
ν_4 (E)	390	345(?)

(a) Na$_2$SeO$_3$ (Khandelwal and Verma 1976).
(b) UO$_2$TeO$_3$ (Botto 1984).

Table 4. Correlation tables for D_{3h}, C_{2v}, and C_s (Bist and Pant 1964)

Point group	ν_1	ν_2	ν_3	ν_4
D_{3h}	A_1'(R)	A_2?(IR)	E'(IR,R)	E'(IR,R)
C_{2v}	A_1(IR,R)	B_1(IR,R)	A_1(IR,R) + B_2(IR,R)	A_1(IR,R) + B_2(IR,R)
C_s	A'(IR,R)	A?(IR,R)	A'(IR,R) + A'(IR,R)	A'(IR,R) + A'(IR,R)

Table 5. Vibrational spectra of tetrahedral anions

T_d symmetry (free ion)	ν_1(A$_1$) [cm^{-1}] R	ν_2(E) [cm^{-1}] R	ν_3(F$_2$) [cm^{-1}] R, IR	ν_4(F$_2$) [cm^{-1}] R, IR
SiO_4^{4-}	819	340	956	527
PO_4^{3-}	938	420	1017	567
AsO_4^{3-}	837	349	878	463
SO_4^{2-}	983	450	1105	611
VO_4^{3-}	826	336	804	336
MoO_4^{2-}	897	317	837	317
WO_4^{2-}	931	325	838	325

(Ross 1972, 1974a,b; Nakamoto 1986).

Table 6. Correlation tables for T_d, C_{3v}, C_{2v}, and C_1

Point group	ν_1	ν_2	ν_3	ν_4
T_d	A_1(R)	E(R)	F_2(IR,R)	F_2(IR,R)
C_{3v}	A_1(IR,R)	E(IR,R)	A_1(IR,R) + E(IR,R)	A_1(IR,R) + E(IR,R)
C_{2v}	A_1(IR,R)	A_1(IR,R) + A_2(R)	A_1(IR,R) + B_1(IR,R) + B_2(IR,R)	A_1(IR,R) + B_1(IR,R) + B_2(IR,R)
C_1	A(IR,R)	2 A(IR,R)	3 A(IR,R)	3 A(IR,R)

Simultaneously the triply degenerate antisymmetric stretching vibration v_3 and the bending vibration v_4, both active in the IR spectrum, can split into two or three components. From a theoretical point of view, the number of bands assigned to the vibrations v_3 and v_4 should be the same (Table 6; Nakamoto 1986).

It seems necessary to add that $V_2O_8^{6-}$ groups occur in the structures of "uranyl vanadates" rather than VO_4^{3-}. Wilkins (1971a) assumes that condensed vanadate ions $V_2O_8^{6-}$ may be consistent with strong vanadate infrared absorptions well outside the range expected for isolated VO_4^{3-} tetrahedra. The tentative assignments of IR spectra for metatyuyamunite and carnotite were made with VO_3 and V_2O_2 groups vibrations (Botto et al. 1989, Baran and Botto 1976). The symmetry decrease in uranyl coordination compounds and minerals is significant and can be characterized by the following examples (M = metal cation) (Fig. 2) (Nakamoto 1986). The lowering of symmetry caused by coordination is different for the monodentate and bidentate complexes. If more than the expected vibrations of oxyanions are observed, the cause may be in structurally nonequivalent oxyanion groups present in the unit cell of studied phases.

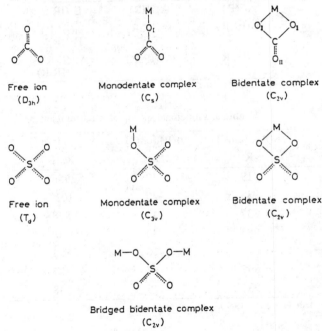

Figure 2. D_{3h} and T_d symmetry lowering in monodentate and bidentate complexes (from Nakamoto 1986).

Molecular water, hydroxyl, hydroxonium and $H_3O^+ \cdot nH_2O$ vibrational modes (fundamental vibrations), hydrogen bonding and deuteroanalogues

Molecular water, hydroxyl and hydroxonium ions play important roles in the crystal structures of minerals (Hawthorne 1992). Most uranyl minerals contain molecular water (H_2O groups), and many contain hydroxyl ions. Many inorganic compounds contain hydroxonium ions (H_3O^+) and their crystal structures have been determined by using X-ray or neutron diffraction. In the case of uranyl minerals, the existence of hydroxonium ions in

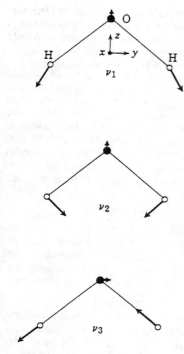

Figure 3. Normal modes of vibration in H₂O molecule (from Nakamoto 1986).

crystal structures of some of these minerals remains an unsettled issue. As is customary, cation-anion charge balance in chemical formulas based on chemical analyses of minerals is often accomplished by adding H^+ or H_3O^+ to the formula, replacing partly or wholly a missing cation. The presence of isolated H^+ is difficult to establish, and determination of H_3O^+ in mineral crystal structures is difficult. Up to the present, the hydroxonium ions was experimentally demonstrated only in synthetic chernikovite, $(H_3O^+)_2(UO_2)_2(PO_4)_2 \cdot 6H_2O$; however, hydroxonium ions are believed to constitute part of the crystal structures of phosphuranylite, sabugalite, uranospathite, vanuranylite, and other uranyl minerals. Chemical formulas for these minerals, including H_3O^+ ions or its hydrates, have been proposed, but unambiguous evidence for such compositions is lacking. The hydroxonium problems in mineralogy have been discussed in several papers (e.g. Melnikov and Melnik 1969; Kukovskiy 1971; Wilkins et al. 1974; Kubisz 1966, 1968; Arkhipenko and Bokiy 1981).

Molecular water (point symmetry C_{2v}) is characterized by three fundamentals, symmetric stretching vibration v_1 (A_1) OH, and antisymmetric stretching vibration v_3 (B_1) OH, both in the range of 3600–2900 cm⁻¹, and bending vibration v_2 (A_1) H_2O (δ H_2O) 1700–1590 cm⁻¹ (Fig. 3) (For H_2O (g) v_1 (A_1) 3656.65 cm⁻¹, v_2 (A_1) 1594.78 cm⁻¹, v_3 (B_1) 3755.79 cm⁻¹, see Ryskin 1974). All vibrations are IR active, but the intensity of the v_1 vibration is very small in comparison with that of the v_3 vibration. In the range of approximately 900–300 cm⁻¹, three types of libration modes can occur: τ—twisting, ω—wagging, and ρ—rocking (Fig. 4). These are due to rotational oscillations of the water molecule, restricted by interactions with neighboring atoms and are classified into three types, depending upon the direction of the principal axis of rotation (Nakamoto 1986; Dothée 1980; Dothée and Camelot 1982; Dothée et al. 1982). Molecular water libration modes are observed only if H_2O groups participate in hydrogen bonds, a fact that may be used as evidence for hydrogen bonds in the crystal structure of the phases under study.

If OH⁻ ions (point symmetry $C_{\infty v}$) are present, they are commonly indicated by sharp bands between 3700 and 3450 cm⁻¹, but sometimes at lower wavenumbers if any

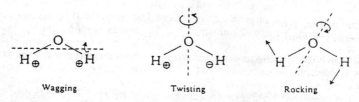

Figure 4. The three librational modes of water in the solid state (from Nakamoto 1986).

appreciable amount of hydrogen bonding is involved. The restricted rotational or libration motion of this ion usually occurs with a wavenumber in the 600–300 cm^{-1} range. Participation of MOH groups in hydrogen bonding manifests itself as the out-of-plane bending vibration, γ MOH. Plyusnina (1977) assumes that stretching vibrations for H$_2$O, OH$^-$ and H$_3$O$^+$, are located in the same region (3750–2000 cm^{-1}) and that the character of an IR spectrum is strongly influenced by hydrogen bonding. In contrast, bending vibrations seem to be more easy detected: δ H$_2$O (1700–1590 cm^{-1}), v_4 (δ) H$_3$O$^+$ (1800–1680 cm^{-1}) and δ MOH (1500 cm^{-1} and over wide range below this value).

Hydroxonium, H$_3$O$^+$ (C$_{3v}$), has 6 normal vibrations that exhibit four fundamentals: v_1 (A$_1$) symmetric stretching vibration 3160 cm^{-1}; v_2 (A$_2$) bending vibration 1140 cm^{-1}; v_3 (E) doubly degenerate antisymmetric stretching vibration 3320 cm^{-1}; and v_4 (E) doubly degenerate bending vibration 1730 cm^{-1}. All four vibrations are Raman and IR active (Arkhipenko and Kovaleva 1978). Additional data can be found in the literature (e.g. Rozière 1973; Wilkins et al. 1974).

Interestingly the range of wavenumbers of δ H$_3$O$^+$, as well as δ H$_2$O vibrations is about 100 cm^{-1}, whereas that of δ MOH is greater than 1000 cm^{-1}. The reason for this range is that the force constant for the angle ∠ROH is determined by the electronic state of oxygen. This state is influenced by the nature of the ionic group, R, and the character of the R-O bond, which in the case of compounds containing OH$^-$ groups, is more heterogeneous than the electronic state of oxygen in H$_2$O groups and hydroxonium ions. In those two molecular groups all oxygen atoms are bonded to the same hydrogen atoms.

In IR spectra of hydroxo compounds, a number of ν OH vibrations and especially δ MOH (δ UOH) vibrations can be observed, the cause of which may be more MOH bonds that differ in their M-OH distances. Increasing the covalency of the M ⇐ OH bond, the ∠MOH angle, and the strength of hydrogen bonding, shifts the δ MOH (in-plane bending) and γ MOH (out-of-plane bending) vibrations to higher wavenumbers. In those cases where groupings of atoms that are bonded to hydroxyl ions remain constant, the interval of this vibration sharply narrows. Bending vibrations of OH groups at the apices of silicate tetrahedra (silanols ≡ SiOH) lie approximately in the range 1470–1400 cm^{-1} independent of the degree of tetrahedra bonding.

Combination bands δ H$_2$O + L H$_2$O (libration modes) in the range 2400–2100 cm^{-1} may also be observed in the IR spectra. Stretching vibrations ν OH are located in the same wavenumber region (3750–2000 cm^{-1}). The distribution of vibration bands in this region are not influenced by the nature of OH groupings, but rather by the forces of the hydrogen bonding network. The relationship between ν OH vibrations (the antisymmetric stretching vibration v_3 H$_2$O) or OH vibration shift ($v_0 - v$, $v_0 = 3756$ cm^{-1}) and OH ··· O distance has been presented elsewhere (e.g. Plyusnina 1977; White 1971). This relationship between OH vibration shift and O-H...O distance enables one to calculate the distance between H-bonded oxygens and to infer the stregnth of the hydrogen bond. Resolution between coordination and H-bonds can be difficult in the region of stretching vibrations ν OH, because both act in the same direction (i.e. by lowering the wavenumber of the stretching vibrations for H$_2$O). The stronger the covalent bond between the H$_2$O oxygen atom and the cation, the more strongly the electron density of OH bonds is redistributed. This is related to an increase of positive charge on the hydrogen atoms of coordinated H$_2$O, and the stronger H-bonds with hydrogen atoms form such molecular groups near proton-acceptor groups.

Absorption bands associated with D$_2$O vibrations occur in the range 2700–2450 cm^{-1} (ν OD), 1250–1180 cm^{-1} (δ D$_2$O), and 1450–1400 cm^{-1} (δ HDO). The isotopic shifting ratios ν OH / ν OD and δ H$_2$O / δ D$_2$O are approximately 1.35, and that of δ H$_2$O / δ HDO is approximately 1.15 (Ross 1972). Isotopic shifting H → D is useful in the study of IR spectra of uranyl minerals, because, in some cases, it enables more precise assignment of observed bands, especially those related to the H$_2$O and D$_2$O vibrations where partly or fully deuterated analogues are available (e.g. Muck et al. 1986a). However, we should point out that crystal structures of rehydrated or redeuterated phases prepared from partly or fully dehydrated minerals or synthetic phases (if such processes are possible) may not be identical to the original compounds. Direct synthesis of deuteroanalogues using D$_2$O (vapor or liquid), or deuterated synthetic compounds, may be the most effective way of synthesizing structurally equivalent compounds; however, in many cases, this is not easily done. It is clear that preparation methods for deuteroanalogues (e.g. temperature applied), may strongly influence the character of IR spectra.

Nakamoto (1986) classified water in inorganic salts as "lattice-related" or coordinated, with no sharp distinction between the two. Lattice water denotes weakly bonded H$_2$O groups in a crystal structure, held by weak hydrogen bonds to an anion, by weak ionic bonds to a metal cation, or both. Coordinated water denotes H$_2$O groups bonded to a metal cation through partially covalent bonds. It should be noted that, according to Nakamoto, the spectra of H$_2$O groups are highly sensitive to their surroundings. IR spectroscopy of water is discussed in detail by Yukhnevich (1974); the nature of water in organic and inorganic compounds is discussed by Karyakin and Kriventsova (1973); and water in minerals is discussed by Rozhkova et al. (1971). Vibrations of protons in minerals (hydroxyl, molecular water, and ammonium) are reviewed by Ryskin (1974). Numerous papers on IR spectra of water in minerals are mentioned by Rossman (1988). Some recent papers on IR spectra of molecular water and hydroxyl ions in minerals are listed at a Caltech web page (http:www.minerals.gps.caltech.edu/).

In minerals, three fundamental types of water may be distinguished: adsorbed, crystal-hydration (lattice or coordination), and constitution water (pimarily hydroxyl ions). We will not discuss so-called "zeolitic water,"the definition of which is not unambiguous. The role of molecular water in minerals is more important than has commonly been acknowledged (for a modern detailed classification of OH groupings in minerals see Hawthorne 1992).

Adsorbed molecular water is usually characterized by relatively narrow absorption bands in the range of 3700–3550 cm^{-1}. Crystal-hydration water can be coordinated by a cation or statistically distributed in various "free" positions in interlayer sites. Several crystal-chemically non-equivalent types of molecular water are commonly evident from several absorption bands in the region of the bending vibration δ H$_2$O associated with doubling of the rocking and wagging libration modes in H-bonded systems.

It is difficult to differentiate the types of water (species) in the region of the stretching vibrations ν OH, because two effects—coordination and hydrogen bonds—act in the same direction; namely, to decrease the wavenumbers of ν OH stretching vibrations. Molecular water that is weakly polarized by cations absorbs in the range of 3650–3400 cm^{-1} (weak hydrogen bonds). More strongly polarized H$_2$O groups (strong hydrogen bonds) are commonly manifested as broad diffuse absorption bands below 3400 cm^{-1}. The wavenumbers of the H$_2$O stretching vibration ν OH are therefore shifted to the long-wavelength region compared with those of vapor-phase water molecules. By contrast, wavenumbers of bending vibrations δ H$_2$O increase toward 1700 cm^{-1}. Constitution water—essentially hydroxyl ions—is released (as H$_2$O) by heating hydroxocompounds. This process (dehydroxylation) takes place at temperatures above those associated with the loss of molecular water (dehydration). However, both processes (dehydroxylation and dehydration) may overlap each other.

Absorption bands assigned to the bending vibration of molecular water, δ H_2O (1700–1590 cm^{-1}), are lacking in the IR spectra of anhydrous hydroxocompounds. The stretching vibrations of OH⁻ groups uninfluenced by hydrogen bond formation are located above 3500 cm^{-1}. If strong H-bonds are formed, the wavenumbers of stretching vibrations for OH⁻ and H_2O are shifted to lower values. Delta M–O–H can be observed in the region below 1500 cm^{-1}.

Evidence for the hydroxyl ion, OH⁻ (e.g. in schoepite) can be investigated by NMR (nuclear magnetic resonance). Strong H-bonds between hydroxyl ions in the structural sheets and H_2O groups in interlayer sites can cause effects associated with the presence of OH⁻ ions that may not be evident in infrared spectra or DTA (differential thermal analysis) curves (Sidorenko et al. 1979).

Dry grinding a mineral sample can lead to structural modification and, in some cases, to substantial loss of crystallinity. To overcome this problem the sample can be moistened; e.g. with an inert, volatile, organic liquid (such as isopropanol). Some interactions between sample and KBr occur even at room temperature during preparation of disks (and micropellets) containing mineral samples that have exchangeable cations, or are soluble in water, or contain water in their structures (Russell 1974). During preparation of KBr disks, a great amount of energy acts on minerals when they are milled and pressed with KBr, and sample treatment may predetermine the character of IR spectra. Some thermo-labile minerals may be subject to chemical reactions, such as hydration or dehydration (Horák and Vítek 1980). Dispersion of some uranyl minerals (e.g. uranyl sulfates) in KBr, and the preparation of disks may cause pressure or temperature increases that can cause structural changes, not only to the interlayer with cations and H_2O groups, but also to the structural units that contain uranyl ions (see Fig. 27: 1,2,3 below) (Cejka et al. 1988a). DRIFT spectra or spectra measured from Nujol suspensions are not influenced by this preparation and may differ from IR spectra measured in KBr disks. DRIFT spectra are usually from powdered material, but results are difficult to interpret because they are partly due to transmission and partly due to reflection, and also depend on particle size (Russell 1974, Hrebicík and Volka 1996). Broad diffuse bands related to stretching vibrations ν OH or bending vibrations δ H_2O (DRIFT, Nujol) serve to characterize IR spectra and correspond beteer to the ideal shapes of IR spectral bands than does the combination of broad and narrow absorption bands commonly observed in KBr disks (Gevorkyan and Ilchenko 1985).

Hydrogen-bearing minerals will presumably possess H-bonds in their structures, and within a H-bonding network, oxygens of H_2O groups and hydroxyl ions may act as H donors. Uranyl oxygens, anion oxygens, oxygens of H_2O groups and hydroxyl ions, that is, all oxygens with a lone electron pair, may act as H-bond acceptors. In the SiO_3OH^{3-} anion three oxygens may act as H-donors, whereas the fourth, part of the silanol group, may act either as H-donor or H-acceptor. Hygroscopic KBr may contact adsorbed water molecules (weakly bonded H_2O groups) and affect, not only weakly bonded, but strongly H-bonded water molecules, which may weaken or disrupt these H-bonds. This means that broad or diffuse bands (especially in DRIFT spectra) may be observed as narrow bands in IR spectra from KBr disks. However, an assignment of IR spectra from KBr disks may actually be easier than assignments in DRIFT spectra, so that KBr disks are commonly used, even if the spectra may not fully correspond to the structures of the minerals studied. An assignment of IR spectra measured in Nujol does not commonly cause problems. If the preparation of a KBr disk is too drastic (the human factor cannot be eliminated), the combined effects of hygroscopic KBr, pressure and temperature, may cause partial or full destruction of mineral structures with H-bearing groups, and the H-bonding network in such minerals may be partly or fully changed or destroyed. Naturally, some rearrangement of H-bonds is assumed to occur during dehydration of H-bearing minerals (e.g. Gevorkyan et al. 1979, 1981; Gevorkyan and Ilchenko 1985). Similar problems were described, e.g. in the case of natural uranyl sulfates, especially uranopilite (Cejka et al. 1984a). Significant changes occur to the IR spectrum of uranophane, following partial destruction of its crystal structure by "rubbing down," with absorption bands that could be assigned to δ OH (1420 and 1360 cm^{-1}) or attributed to SiO$_3$OH^{3-} ions (Gevorkyan et al. 1979, 1981; Gevorkyan, Ilchenko 1985; Plyusnina 1977). During rubbing down, degradation of the uranophane structure continues until, after approximately 30 minutes, one observes absorption bands in the IR spectrum that can be assigned to H$_3$O$^+$ vibrations (Gevorkyan et al. 1979, 1981; Gevorkyan and Ilchenko 1985).

When the crystal structure of a mineral is unstable (e.g. uranyl sulfates) the preparation of KBr disks or micropellets for infrared spectra recording may cause such structure changes that the infrared spectra measured in KBr are not, as noted above, comparable with those in Nujol. The interpretation of these data may lead to erroneous results. In such cases, the infrared spectrum measured in Nujol is the only acceptable representation of the mineral investigated. The presence of vibration bands related to Nujol needs to be evaluated in such cases.

In studying the IR spectrum of synthetic compreignacite, Dothée (Dothée 1980; Dothée and Camelot 1982; Dothée et al. 1982) summarizes the influence of hydrogen bonds on ν OH vibrations observed in IR spectra available for uranyl compounds:

(a) ν OH vibrations narrow enough with wavenumbers higher than 3400 cm^{-1} together with bands below 700 cm^{-1} (H$_2$O libration modes) indicate the presence of crystalline H$_2$O;

(b) ν OH vibrations near 3350 cm^{-1} (lower than in (a)), together with δ UOH (in-plane bending vibration) bands at 900 and 1020 cm^{-1} and γ UOH (out-of-plane bending vibration) at 790 and 880 cm^{-1} (observed, e.g. in UO$_3$·0.8H$_2$O) are related to the OH group bridging three UO$_2^{2+}$ groups;

(c) ν OH bands between 3000 and 3350 cm^{-1}, broader than (a) and (b), together with bands near 600 and 700 cm^{-1} can be assigned to H$_2$O libration modes, and may be related to coordination water;

(d) fine ν OH bands near 3500 cm^{-1}, associated with δ UOH between 800–900 cm^{-1} and γ UOH between 600–700 cm^{-1}, are characteristic of OH groups bridging two UO$_2^{2+}$ groups.

(e) from this point of view, bands associated with δ H$_2$O (bending vibration) in the region of 1600 cm^{-1} are not directly useful, however, they do confirm the presence of H$_2$O groups in the crystal structure. The value of the wavenumber of the δ H$_2$O vibration may only indicate the presence of hydrogen bonds. Dothée's tentative assignment may be useful, especially if IR spectra of uranyl minerals that contain both hydroxyl ions and H$_2$O groups are being studied.

Infrared spectra of the uranyl minerals

Uranyl oxide hydrates, alkali, alkaline-earth and other uranyl oxide hydrates

Ianthinite, [U$_2^{4+}$(UO$_2$)$_4$O$_6$(OH)$_4$(H$_2$O)$_4$](H$_2$O)$_5$ (Fig. 5) (Urbanec and Cejka 1979a)

Ianthinite was thought to be structurally similar in its layers to schoepite (Urbanec and Cejka 1979a). Some uranium, however, is present as U^{4+}. X-ray powder diffraction data of the material studied corroborates that ianthinite alteration is followed by formation of schoepite (Urbanec and Cejka 1979a). No assignment of the IR spectrum of ianthinite was

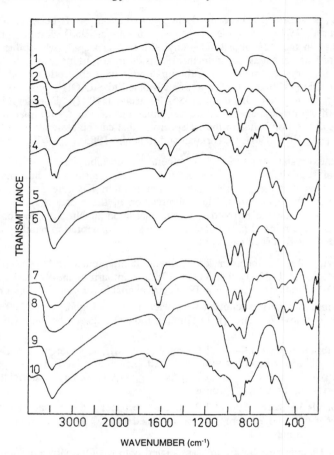

Figure 5. Infrared spectra of some uranyl hydroxide and uranyl silicate minerals: 1. Schoepite (KBr disk); 2. Ianthinite (KBr micropellet); 3. Becquerelite (KBr micropellet); 4. Vandenbrandeite (KBr disk); 5. Curite (KBr disk); 6. Uranophane (KBr micropellet); 7. Sklodowskite (KBr disk); 8. Cuprosklodowskite (KBr micropellet); 9. Kasolite (KBr micropellet); 10. Soddyite (KBr micropellet) (from Cejka and Urbanec 1990).

reported. Crystal structure of ianthinite was recently published by Burns et al. (1997b).

Schoepite, $[(UO_2)_8O_2(OH)_{12}](H_2O)_{12}$ (Fig. 5) (Urbanec and Cejka 1979b)

The character of the IR spectrum of studied schoepite is analogous to that of synthetic $UO_3 \cdot 2H_2O$ (Urbanec and Cejka 1979b). However, it could be probably metaschoepite with regard to the crystal structure study of schoepite by Finch et al. (1996a). There is a difference in the antisymmetric stretching vibration v_3 UO_2^{2+} whose lower wavenumber 930 cm^{-1} in schoepite (958 cm^{-1} in $UO_3 \cdot 2H_2O$, Hoekstra and Siegel 1973) can result from differences in formation conditions between nature and laboratory and also from the experimental conditions of sample preparation for the IR spectra measurement (Urbanec and Cejka 1979b). However, Sobry (1972, 1973) assigned bands at 926 cm^{-1} in α-$UO_3 \cdot 2H_2O$, and at 929 cm^{-1} in β-$UO_3 \cdot 2H_2O$ to v_3 UO_2^{2+}. The existence of two $UO_3 \cdot 2H_2O$ polymorphs appears doubtful (Hoekstra and Siegel 1973). Allen et al. (1996) observed the

IR spectrum of schoepite to be nearly identical to that reported by Hoekstra and Siegel (1973), apart from slight differences in the resolution of the respective bands. Urbanec and Cejka (1979b) assigned the vibrations at 930 cm^{-1} to v_3 UO_2^{2+}, 841 cm^{-1} to v_1 UO_2^{2+}, 1072 and 1018 cm^{-1} to δ UOH, 545 and 520 cm^{-1} (broad) to $v(U-O_{II})$, 3440 and 3160 cm^{-1} to v OH, and 1624 cm^{-1} to δ H_2O. Raman spectrum of schoepite (Maya and Begun 1981) reveals the bands at 840 and 860 cm^{-1} which can be assigned to v_1 UO_2^{2+}. According to Maya and Begun, observation of two well defined bands indicates the uranyl moiety in two different environments within the crystal structure. Allen et al. (1996) assigned bands at 885 and 930 cm^{-1} to v_1 and v_3 UO_2^{2+}, respectively. They discuss the pH influence on the structure of schoepite-like phases formed. It is clear that with increasing pH the structure of schoepite-like phases continuously changes to that of "uranates" and that the v_3 UO_2^{2+} decreases.

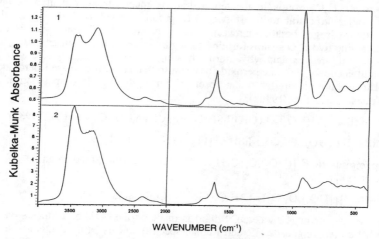

Figure 6. Infrared spectrum of studtite: 1. KBr; 2. DRIFT (from Cejka et al. 1996a).

Studtite, $UO_4 \cdot 4H_2O$ (Fig. 6) (Cejka et al. 1996a)

The IR spectrum of studtite is interpreted with regard to that of its synthetic analogue $UO_4 \cdot 4H_2O$, and synthetic $UO_4 \cdot 2H_2O$ (meta-studtite) (e.g. Deane 1961; Sato 1961a,b; 1976; Rocchiccioli 1966). A difference was observed between the diffusion reflectance spectrum of studtite and that in the KBr disk. It seems probable that preparation of a studtite sample in KBr disk (and hygroscopy of KBr) may alter the hydrogen bonding network in the studtite structure, and partial dehydration can also proceed. A band at 918 cm^{-1} (924 cm^{-1} in KBr) was assigned to antisymmetric stretching vibration v_3 UO_2^{2+}. Symmetric stretching vibration v_1 UO_2^{2+} was observed only in KBr disk as shoulder band at 820 cm^{-1}, which was somewhat lower than its calculated value 841 cm^{-1} using an empirical equation $v_1 = 0.89$ $v_3 + 21$. Weak vibrations in the region 570–400 cm^{-1} can be assigned to the stretching vibrations of weakly bonded secondary (equatorial) oxygen atoms v $U-O_{(2)}$ (Hoekstra and Siegel 1973). The bands found in the range 3500–3000 cm^{-1} are related to the stretching vibration v OH, those at 1700–1620 cm^{-1} to the bending vibrations δ H_2O, while those at 760–600 cm^{-1} to libration modes of water molecules. A shoulder at 880 cm^{-1} (875 cm^{-1}) was assigned to the stretching vibration of the peroxidic group, v O_2^{2-}. Weak vibrations 2377 and 2180 cm^{-1} could correspond to combination bands (e.g. δ H_2O + L H_2O), those in the range 1530–1400 cm^{-1} probably also to combination bands or overtones. Peroxogroup, O_2^{2-}, is typical ligand forming with uranyl ion a set of complex compounds. According to its displacing ability, the peroxogroup can be set in series $CO_3^{2-} \geq O_2^{2-} > OH^- > F^-$ (Chernyayev 1964). As a bidentate ligand O^{2-} forms a strong bond (donor-acceptor coordination) with uranyl in its equatorial plane. Water molecules in studtite are structurally non-equivalent to each other and bonded with different forces. This is confirmed by stepwise dehydration of synthetic $UO_4 \cdot 4H_2O$. The lowering of the stretching vibrations ν OH and the increase of the bending vibrations δ H_2O suggest the formation of hydrogen bonds in the uranyl peroxide hydrates. Water molecules form hydrogen bonds with adjacent uranyl oxygen and probably also peroxide oxygens. Uranyl peroxide tetrahydrate probably forms a characteristic H-bonding network differing from that of uranyl peroxide dihydrate (meta-studtite). It is difficult to establish the geometry of the uranyl polyhedron in uranyl peroxide hydrates. It might be assumed, however, that some H_2O groups may coordinate the uranyl ion in its equatorial plane. Mutual hydrogen bonding between H_2O groups in the tetrahydrate is much stronger than between H_2O groups and other structural units. If two H_2O groups are released during heating or evacuation, the hydrogen bonding network is partly destroyed and a new one is created, with the remaining two H_2O groups forming strong hydrogen bonds to uranyl oxygens and to each other. This may explain why, during heating, the two remaining H_2O groups are released simultaneously with one peroxide oxygen, so that anhydrous uranyl peroxide does not exist. This behavior is similar to that observed during thermal decomposition of uranyl nitrate hydrates (Cejka et al. 1996a).

Becquerelite, $Ca[(UO_2)_6O_4(OH)_6] \cdot 8H_2O$ (Figs. 5 and 7) (Cejka et al. 1998a)

Billietite, $Ba[(UO_2)_6O_4(OH)_6] \cdot 8H_2O$ (Fig. 7) (Cejka et al. 1998a)

Compreignacite, $K_2[(UO_2)_6O_4(OH)_6] \cdot 8H_2O$ (Dothée 1980; Dothée and Camelot 1982; Dothée et al. 1982)

Protasite, $Ba[UO_2)_3O_3(OH)_2] \cdot 3H_2O$ (Fig. 7) (Cejka et al. 1998a)

Tentative assignment (Cejka et al. 1998a) is made on the basis of Dothée's papers, in which synthetic compreignacite (IR and Raman spectra) and its partly deuterated analogue were studied. Perrin's (1976) studies of inorganic uranyl hydroxo-anion compounds are also discussed. The vibration in the range 924–850 cm^{-1} may be assigned to the antisymmetric stretching vibration ν_3 UO_2^{2+} for synthetic becquerelite, billietite and protasite. Two vibrations were observed on the case of becquerelite (924, 891 cm^{-1}), billietite (911, 860 cm^{-1} (sh)), and compreignacite (905, 860-870 cm^{-1} (sh)), only one in synthetic protasite (913 cm^{-1}). Vibrations at 810 (becquerelite), 807 (billietite), 807 (protasite), and 805 cm^{-1} (compreignacite) are assigned to the symmetric stretching

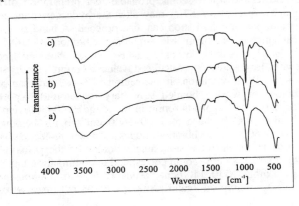

Figure 7. Infrared spectra (KBr disk) of synthetic (a) protasite, (b) billietite, (c) becquerelite (from Cejka et al. 1998a).

vibration v_1 UO_2^{2+}. According to the assignment by Perrin (1976), Dothée (1980), Dothée and Camelot (1982) and Dothée et al. (1982), the following tentative interpretation was presented: v U_3O (bridge elongation) 600–400 cm^{-1}, H_2O libration and γ UOH (out-of-plane bending), 750–620 cm^{-1}, δ UOH (in-plane bending) 1150–980 cm^{-1}, overtones or combination bands or δ UOH 1560–1220 cm^{-1}, δ H_2O bending vibration 1650–1600 cm^{-1}, a broad shoulder at 1715 cm^{-1} in the case of compreignacite. Some coincidences of δ UOH may be possible especially in the region of v_3 UO_2^{2+} and v_1 UO_2^{2+}. Stretching vibrations v OH are observed in the range 3572–2850 cm^{-1}; in becquerelite (synthetic 3497 VS, 3239, 3158, 2980, 2950, 2870 cm^{-1}-shoulders), natural 3572–3463 VSb, 3130, 2950 and 2870 cm^{-1}-shoulders), protasite (synthetic–3553 VS, 3503 VS, 3380, 3256, 2920, 2860 cm^{-1}-shoulders), billietite (synthetic—3531 VSb, 3510 VS, 3265 b, 2910, 2850 cm^{-1}-shoulders, natural—3550 VSb, 3490 VSb, 3278, 2930, 2870 cm^{-1}-shoulders), compreignacite (synthetic—3530 VS, 3475 S, 3300 cm^{-1} (a shoulder).

In the IR spectra measured in Nujol in the range 600–50 cm^{-1}, vibrations corresponding to v_2 UO_2^{2+} were observed at 262 and 251 cm^{-1} (becquerelite), 268 and 245 cm^{-1} (billietite), 266 and 251 cm^{-1} (protasite), and 265 and 245 cm^{-1} (compreignacite). The other vibrations observed in the IR spectra measured in Nujol suspensions especially those at wavenumbers lower than 470 cm^{-1} have been tentatively assigned according to Dothée and Camelot (1982), Dothée et al. (1982) as v U-O$_{(2)}$ and can be separated in v U_3O (bridge elongation, 470–406 cm^{-1}),v $U_3(OH)_3$ (group elongation) ~356 cm^{-1}, γ U_3O (out-of-plane bending) 330–295 cm^{-1}, and γ $U_3(OH)_3$ (out-of-plane bending) or δ $U_3(OH)_3$ (in-plane bending) ~210 cm^{-1}. The lower vibrations are assigned to molecular deformation and lattice modes (Bullock and Parett 1970) and can be related to UO_2^{2+} translations (185–150 cm^{-1}), UO_2^{2+} rotations (140–86 cm^{-1}) and probably also cation translations (90–60 cm^{-1}) (Cejka et al. 1998a).

Infrared absorption spectra measured in KBr disks display more vibrations, especially in the region of v OH, and small shifts of v_3 UO_2^{2+} and v_1 UO_2^{2+} wavenumbers. Differences observed in infrared spectra measured in KBr disks and diffusion reflectance spectra suggest a partly destroyed H-bonding network, structural distortions, or reactions of synthetic phases with hygroscopic KBr (cation exchange?), probably artifacts of sample preparation (Cejka et al. 1998a).

Wölsendorfite (Pb,Ca)$U_2O_7 \cdot 2H_2O$, $PbU_2O_7 \cdot 2H_2O$ (Sejkora et al. 1997; Sobry 1973)

v_3 UO_2^{2+} was observed as a dominant band in the range 840–860 cm^{-1}, v_1 UO_2^{2+} as a shoulder in 750–770 cm^{-1}. An intense band at 757 cm^{-1} in the Raman spectrum of a synthetic Pb-phase was assigned to v_1 UO_2^{2+} in the IR spectra. Vibrations observed below 600 cm^{-1} (575, 513, 447 cm^{-1} in natural phase, and 557–578, 529–532, 467–475, and 409–446 cm^{-1} in synthetic phases) can be assigned to v (U-O$_{II}$), (i.e. to the vibrations U-O$_{ligand}$) (Cejka 1994; Sejkora et al. 1997; Sobry 1972, 1973). However, according to Dothée, these vibrations could be related to the structural motive U_3O. Stretching vibrations v OH were found in the range 3533–2820 cm^{-1} for both natural and synthetic phases. Broad bands indicate a H-bonding network in the crystal structure of wölsendorfite-like phases, and wavenumbers of δ H_2O (1618–1638 cm^{-1}) support this conclusion. A shoulder at 1585 cm^{-1} observed in the IR spectrum of wölsendorfite suggests two structurally non-equivalent types of molecular water; however, this shoulder was not observed in IR spectra from synthetic phases. In the IR spectra of some synthetic phases (Cejka 1994) vibrations in the range 646–721 cm^{-1} were observed and assigned to δ UOH. Infrared spectra of wölsendorfite and its synthetic analogues indicate that the crystal structure of this mineral is neither a pure uranyl oxide hydrate nor a pure uranate structure type, but an intermediate between these two structure types. Baran and Unzeitig (1991) proposed for wölsendorfite the crystal chemical formula $Pb_6[(UO_2)_{12}O_{16}(OH)_4]\cdot 10H_2O$, related to the α-U_3O_8 structure type. The structure of a barian wölsendorfite, $Pb_{6.16}Ba_{0.36}[(UO_2)_{14}O_{19}(OH)_4](H_2O)_{12}$, is composed of slabs of α-U_3O_8 and β-U_3O_8-type sheets with Pb and Ba cations and H_2O groups in interlayer positions (Burns 1999b 1999c).

Masuyites, $PbO\cdot 3UO_3\cdot 4H_2O$ and $4PbO\cdot 9UO_3\cdot 10H_2O$

The composition of synthetic masuyite-like phases vary to a certain extent. However, the average value corresponds to $PbO\cdot 3UO_3\cdot 4H_2O$ (Cejka 1994). Using Baran and Unzeitig's rule (1991), Cejka proposes (on the basis of 13 analyzed synthetic phases) the following crystal chemical formulas for masuyite respecting some changes in the Pb/U molar ratio:

$Pb_3[(UO_2)_8O_9(OH)_4]\cdot 8H_2O$ ($\alpha + \beta$ -U_3O_8 structure type)

$Pb_3[(UO_2)_9O_9(OH)_6]\cdot 9H_2O$ (α or $\alpha + \beta$ -U_3O_8 structure type)

$Pb_3[(UO_2)_{10}O_{10}(OH)_6]\cdot 11H_2O$ ($\alpha + \beta$ -U_3O_8 structure type)

All three structures should be relatively stable from the crystal chemical point of view (Baran and Unzeitig 1991). Recently, Deliens and Piret (1996) described "grooved" masuyites, $PbO\cdot 3UO_3\cdot 4H_2O$, (according to their composition, this may be a Pb analogue of protasite) and "type" masuyites, $4PbO\cdot 9UO_3\cdot 10H_2O$. According to older papers, the composition of masuyite may vary in the range $(1-4)PbO\cdot (3-9)UO_3\cdot (4-10)\cdot H_2O$ (Deliens et al. 1984). Infrared spectra of synthetic masuyite-like phases (Cejka 1994) are similar to those of wölsendorfite-like phases. The main difference is that, for masuyites, an intense band appears in the range 900–911 cm^{-1} with a shoulder at 950 cm^{-1} to which ν_3 UO_2^{2+} can be assigned. ν_1 UO_2^{2+} was located in the range 770–781 cm^{-1} by this author; however, this value seems too low. Two vibrations assigned to ν_3 UO_2^{2+} may be associated with two structurally non-equivalent uranyl groups in the elementary cell. Vibrations below 600 cm^{-1} (410–591 cm^{-1}) correspond to ν (U-O$_{II}$) (U-equatorial oxygens vibrations) and may be related to the structural motive U_3O according to Dothée's assignment. The maximum of a broad band at 3450 cm^{-1} corresponds to ν OH vibrations that occur in the range 2860–3460 cm^{-1}. A shoulder or a weak band at 3530–3550 cm^{-1} may be associated with weakly H-bonded H_2O groups or hydroxyl ions. Dothée assigned bands near 3500 cm^{-1} to the stretching vibrations of hydroxyl bonding two uranyls. Bending vibration δ H_2O was observed in the range 1608–1619 cm^{-1}. Very weak absorptions between 730 and 630 cm^{-1} probably correspond to H_2O libration modes. Bending vibrations δ UOH lie in the range 1025–1180 cm^{-1}.

On the basis of IR spectra of synthetic masuyite-like phases, it can be inferred that these compounds are structurally related to the protasite-becquerelite group. The structure of the crystal of a masuyite crystal studied by Burns and Hanchar (1999), $Pb[(UO_2)_3O_3(OH)_2](H_2O)_3$, also demonstrates that this mineral is closely related to the protasite-becquerelite group.

Vandendriesscheite, $Pb_{1.57}[(UO_2)_{10}O_6(OH)_{11}](H_2O)_{11}$

Cejka (1994) prepared two phases that may be structurally related to vandendriesscheite and proposed the chemical formulas $Pb[(UO_2)_6O_5(OH)_4]\cdot 8H_2O$ and $Pb[(UO_2)_{10}O_7(OH)_8]\cdot 14H_2O$. Both formulas differ from that inferred by Burns (1997) from a single-crystal X-ray structure analysis of vandendriesscheite: $Pb_{1.57}[(UO_2)_{10}O_6(OH)_{11}](H_2O)_{11}$.

Here, we present IR spectra for both synthetic phases. They are similar to IR spectra of synthetic $UO_3 \cdot 2H_2O$ (Hoekstra and Siegel 1973). Interactions between uranyl oxygens and interlayer Pb^{2+} ions and H_2O groups coordinated to Pb^{2+} weaken bond strengths between uranium and uranyl oxygens. This phenomenon is manifested as definite shoulders at 960 and 940 cm^{-1}in the IR spectrum of the sample with U/Pb = 6. The most intense band is at 907 cm^{-1}. All three vibrations are assigned to v_3 UO_2^{2+}. In the sample with U/Pb = 10, there are two bands at 958 and 921 cm^{-1} with a shoulder at 902 cm^{-1} assigned to v_3 UO_2^{2+}. Shoulders near 880 and 843 cm^{-1} can be assigned to v_1 UO_2^{2+}, which agrees with 870 and 846 cm^{-1} in the Raman spectrum of $UO_3 \cdot 2H_2O$ (Hoekstra and Siegel 1973). There may be a coincidence of v_3 UO_2^{2+} and v_1 UO_2^{2+} for various uranyl groups near 846 cm^{-1}. This band might be assigned to the v_3 UO_2^{2+} of the uranyl that is strongly influenced by interlayer Pb^{2+} ions and H_2O groups coordinated to Pb^{2+}. If this assignment is correct, a weak band at 768 cm^{-1}, observed in the first synthetic phase (U/Pb = 6), may be associated with v_1 UO_2^{2+}. Bands below 600 cm^{-1} were assigned to v (U-O$_{II}$) (i.e. U-O$_{equatorial}$ vibrations).

Stretching vibrations v OH in the range 2860–3553 cm^{-1} are observed in IR spectra for both synthetic phases, and sharp maxima are observed at 3553 and 3480 cm^{-1}, with a shoulder near 3349 cm^{-1} (the second synthetic phase). The band at 3553 cm^{-1} may be related to H_2O groups or hydroxyl ions that are only weakly H-bonded in the structure. Intense bands at 1613 and 1617 cm^{-1}, respectively, related to δ H_2O indicate that no strong H-bonds are formed. The extensive network of hydrogen bonds that link adjacent uranyl oxo-hydroxo sheets was derived on the basis of geometrical and crystal-chemical arguements in natural vandendriesscheite by Burns (1997). Libration modes of H_2O groups are located at 717 and 727 cm^{-1} respectively. δ UOH were found in the range 1024–1160 cm^{-1} in both studied phases.

When comparing IR spectra of vandendriesscheite-like phases, masuyite-like phases, wölsendorfite, and wölsendorfite-like phases, a progressive lowering of v_3 UO_2^{2+} vibration is observed. With regard to Baran's (1992) observation, it can be inferred that vandendriesscheite- and masuyite-like phases are structurally related to uranyl oxide hydrate structure types, whereas wölsendorfite and wölsendorfite-like phases form intermediates between the uranyl oxide hydrate and uranate structure types, as discussed above for wölsendorfite.

Curite, $[Pb_{8-x}(OH)_{4-x}(H_2O)_{2x}][(UO_2)_8O_8(OH)_6]_2$ $(0 < x \leq 2)$ or $[U_8O_x(OH)_{30-x}]Pb_3(OH)_{24-x}\cdot(x-21)H_2O$ $(24 \geq x > 21)$ (Fig. 5) (Cejka et al. 1998b)

Infrared bands at 959 and 923 cm^{-1} with shoulders at 932 cm^{-1} and (probably) at 843 or 868 cm^{-1} can be assigned to v_3 UO_2^{2+}, and those at 868, 843 (shoulder) and 762 (shoulder) cm^{-1} to v_1 UO_2^{2+}. There may also be some coincidences with δ UOH to which vibrations, especially at 815 cm^{-1} (shoulder) and 802 cm^{-1}, are assigned. Vibrations at 637 and 621 cm^{-1} may be related to H_2O libration modes or out-of-plane bending γ UOH. Bands observed in the region 3534–2860 cm^{-1} are assigned to v OH. A sharp intense band at 3534 cm^{-1} shows weakly bonded H_2O groups or hydroxyl ions. The band close to 1600 cm^{-1} corresponds to that at 3534 cm^{-1}, and is related to weakly H-bonded H_2O groups. The wavenumber is lower than those of wölsendorfite-, masuyite-, and vandendriesscheite–like synthetic phases. It seems necessary to add that, in the IR spectra of some other curite samples studied in KBr disks, two bands at 1595 and 1633 cm^{-1} are observed, indicating two structurally non-equivalent types of H_2O groups (in agreement with structure and thermal analyses), whereas DRIFT shows only one band at 1598 cm^{-1}. A weak band at 1313 cm^{-1} with a shoulder at 1550 cm^{-1} is probably associated with overtones or combination bands. Bands below 600 cm^{-1} were assigned to bridging bending v UOH (519 cm^{-1}), bridge elongation v U_3O (470–410 cm^{-1}), elongation v $U_3(OH)_3$ (375–360 cm^{-1}), out-of-plane bending γ U$_3$O (315–310 cm^{-1}) and bending ν_2 (δ) UO$_2^{2+}$ (260–250 cm^{-1}), in accordance with Dothée's assignments.

Changes in ν OH, splitting of δ H$_2$O, in intensities of ν_3 UO$_2^{2+}$ vibrations in IR spectra measured in KBr disks may strongly influence the character of the spectra measured. IR-affecting changes may include partial destruction of the H-bond network in, partial dehydration of, and structural modifications to curite during sample preparation (Cejka et al. 1998b).

Contrary to the porposal by Mereiter (1979) that OH ions replace H$_2$O groupps in interlayer sites in synthetic curite, a recent crystal structure study of natural curites (Li and Burns 1999) does not support hydroxyl in the interlayer. Instread, minor variations in the number of O atoms and hydroxyl ions in the sheet of uranyl polyhedra are responsible for maintaining charge balance. According to Li and Burns (1999), the general formula for curite may be written as Pb$_{1.5+x}$[(UO$_2$)$_4$O$_{4+2x}$(OH)$_{3-2x}$](H$_2$O).

Clarkeite, Na[(UO$_2$)O(OH)](H$_2$O)$_{0-1}$ (Sidorenko 1978; Sidorenko et al. 1985)

New chemical and structural data of clarkeite were recently published by Finch and Ewing (1997). Sidorenko et al. (1985) published the IR spectra of synthetic clarkeite, calciouranoite, metacalciouranoite, bauranoite, wölsendorfite, curite, schoepite, Na$_2$U$_2$O$_7$, K$_2$U$_2$O$_7$, CaU$_2$O$_7$, and PbU$_2$O$_7$. The resolution of the spectra is not good and their tentative interpretation seems weak. The spectra are therefore not reproduced here; however, a shifting of ν_3 UO$_2^{2+}$ to lower wavenumbers is observed in some of their data compared with wavenumbers of ν_3 UO$_2^{2+}$ in the becquerelite-protasite group. This observation agrees with the Baran's conclusions (Baran and Tympl 1966; Baran et al. 1985; Baran 1992) that ν_3 UO$_2^{2+}$ decreases from 958 cm^{-1} (UO$_3$·2H$_2$O) with increasing Na/U ratio to 870 cm^{-1} (Na$_2$U$_2$O$_6$(OH)$_2$), 849 cm^{-1} (Na$_2$U$_2$O$_7$), 775 cm^{-1} (Na$_2$UO$_4$) and 590 cm^{-1} (Na$_4$UO$_5$). The extrapolated value for Na$_6$UO$_6$ should be close to 400 cm^{-1}. Ideal clarkeite, Na[(UO$_2$)O(OH)](H$_2$O)$_{0-1}$ (Finch and Ewing 1997) may correspond to Na$_2$U$_2$O$_6$(OH)$_2$. It seems probable that a hypothetical series exists from uranyl oxide hydrates to alkali and alkaline earth uranyl oxide hydrates via some intermediates to anhydrous uranates, associated with crystal structure transformations and transitions.

Vandenbrandeite, CuUO$_2$(OH)$_4$ (Fig. 5) (Cejka 1994a)

In synthetic vandenbrandeite the absorption band at 890–900 cm^{-1} was assigned to ν_3 UO$_2^{2+}$. Some small deviations were observed with wavenumbers probably caused by using different measuring techniques. In the case of natural vandenbrandeite dispersed in KBr disk, the reaction of CuUO$_2$(OH)$_4$ with KBr cannot be excluded (ν_3 UO$_2^{2+}$ 878–888 cm^{-1}). Because of relatively large dispersion of ν_3 UO$_2^{2+}$ values for all measured samples (878–900 cm^{-1}), two possible assignments for ν_1 UO$_2^{2+}$ may be found (i.e. 799–805 cm^{-1}), and less probably 830–844 cm^{-1}. Other weak and medium bands or shoulders in the region 800–1539 cm^{-1} were assigned to δ UOH. However, combination bands or overtones may be located over 1200 cm^{-1}. Bands in the region 538–600 cm^{-1} are related to γ UOH, and those in the region 451–511 cm^{-1} to ν UOH. Bands observed in the region 2810–3511 cm^{-1} with two sharp maxima near 3510 and 3420 cm^{-1} were assigned to ν OH vibration, suggesting a relatively free character of the related hydroxyl ions, while some of those below 2362 cm^{-1} could probably correspond to overtones or combination bands. Weak bands observed in some IR spectra near 1600 cm^{-1} (together with some ν OH vibrations) may indicate small amounts of adsorbed water. The XOH vibrations may not be related only to UOH, but also to CuOH. Weak hydrogen bonds might be expected with regard to the character of ν OH vibrations (Cejka 1994a).

Uranosphärite, $Bi_2U_2O_9 \cdot 3H_2O$ or $BiO(UO_2)(OH)_3$ (Sejkora et al. 1994)

The band at 899 cm^{-1} is associated with v_3 UO_2^{2+}. Bands in the range 3416–2858 cm^{-1} are associated with v OH. According to Sejkora et al. (1994), bands at 1034, 795 and 471 probably correspond to Bi-O vibrations. No other assignment was reported.

Uranyl silicates

Before discussing the ir spectra of the uranyl silicate minerals, some conclusions are presented from the crystal chemical relations of the uranophane mineral group with regard to their ir spectra reported by various authors using a variety of techniques (Gevorkyan and Ilchenko 1985; Gevorkyan et al. 1979, 1981; Cejka 1993a).

(a) Natural uranyl silicates from the uranophane group may contain "mobile" protons. These protons may occur in uranyl silicate sheets (thus forming hydroxyl ions in SiO_3OH^{3-} groups) and also in the interlayer containing H_2O groups (thus forming hydroxonium ions), always bonded to the lone electron pair of oxygen atoms (Gevorkyan and Ilchenko 1985; Gevorkyan et al. 1979, 1981; Cejka 1993a)

(b) Proton "mobility" may be caused by mechanical influences (e.g. by intensive pulverizing). Under such conditions "mobile" protons may become bonded to interlayer H_2O groups, forming hydroxonium ions. However, deformation (distortion) of H_2O groups caused by strong hydrogen bonds may simulate hydroxonium ions in IR spectra. Such distortions may occur when strong hydrogen bonds are formed between hydroxyl ions of the SiO_3OH tetrahedra and neighboring (adjacent) H_2O groups (Gevorkyan et al. 1979, 1981).

(c) "Mobile" protons in layers (structural sheets) are commmonly bonded to apical oxygens of silicate tetrahedra. H_2O groups located near the apical hydroxyl ions form strong hydrogen bonds with the hydroxyl ions. Hydroxyl ions interact strongly with neighbouring H_2O groups. Determining hydroxyl ions in uranyl silicates seems difficult. They could be determined with more confidence if hydrogen bonding is partially or fully destroyed (e.g. by partial dehydration).

(d) The phenomena described in (a)-(c) are observed especially in the case of uranophane and β-uranophane, and, in the main, also of cuprosklodowskite and sklodowskite; i.e. in uranyl silicates with relatively high water contents (2.5-3.5 H_2O per UO_2^{2+}), in boltwoodite, and, with lower probability, in kasolite.

(e) In the structures of uranyl silicates of the uranophane group, various hydrogen-oxygen groupings may form: H_2O groups, hydroxyl ions, and hydroxonium ions. Each of them may originate under specific conditions, and mutual interactions are believed to occur (especially hydrogen bonding). "Mobile" protons are probably the main cause of all such interactions and, therefore, explain disparate results (and interpretations) reported by various researchers using a variety of experimental methods. The inferred chemical formulas for minerals studied commonly differ as a result. The reason for such apparent discrepancies need not be due to analytical errors, or incorect interpretations, but may be caused by real differences in the actual minerals structures which may depend on sample preparation and the method used for structure determination.

Such phenomena (processes) probably also occur in other uranyl mineral groups. It is well known that mobile protons occur in interlayer spaces of some uranyl phosphates and arsenates, especially $HUO_2PO_4 \cdot 4H_2O$ (or better $(H_3O^+)UO_2PO_4 \cdot 3H_2O$), the synthetic analogue of chernikovite, and its arsenate derivative, hydrogen-uranospinite, $HUO_2AsO_4 \cdot 4H_2O$ [$(H_3O^+)UO_2PO_4 \cdot 3H_2O$]. It is not currently clear whether such processes may also occur in, e.g. uranyl oxide hydrates.

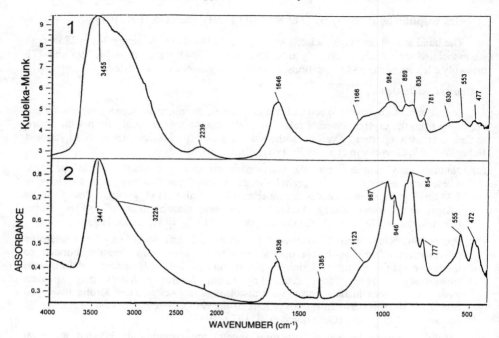

Figure 8. Infrared spectrum of β-uranophane: 1. DRIFT; 2. KBr dis

If diffusion reflectance FTIR spectra (DRIFT) of uranyl silicates are to be interpreted, it is not easy to assign especially absorption bands to ν_3 UO_2^{2+}. The bands observed in this region are of low intensity and are very broad. In contrast, absorption spectra measured in KBr disks show sharp bands that enable tentative assignments; however, the possibility for destruction of the H-bond network in the crystal structure of the minerals being studied must also be considered when preparing samples.

Uranophane, $Ca(UO_2)_2(SiO_3OH)_2 \cdot 5H_2O$ (Fig. 5) (this study)

ν_3 UO_2^{2+} 859 cm⁻¹, ν_1 UO_2^{2+} 790 (a shoulder = sh) cm⁻¹, ν_1 SiO_4^{4-}—possible coincidence with ν_3 UO_2^{2+}, ν_2 SiO_4^{4-} 460 and 485 cm⁻¹ ν_3 SiO_4^{4-} 942 and 1006 cm⁻¹, ν_4 SiO_4^{4-} 560 and 609 cm⁻¹, ν OH 3437 cm⁻¹, 3200 (sh b) cm⁻¹, 2920 and 2850 cm⁻¹, δ H_2O 1636 cm⁻¹, a shoulder at 650 cm⁻¹ may be related to H_2O libration mode, a weak broad band at 1541 cm⁻¹ to δ SiOH. A broad band at ~2200 cm⁻¹ could correspond to overtones or combination bands.

Broad bands in the region of ν OH vibrations and the relatively high wavenumbers of δ H_2O confirm the existence of hydrogen bonds in uranophane crystal structure. A shoulder related to the δ H_2O band observed in some IR spectra of uranophane indicates structurally nonequivalent H_2O groups in the unit cell. This fact was also confirmed with thermal analysis. Weak bands observed in the range 1200–1550 cm⁻¹ may be assigned to δ SiOH. Bands near 250 and 270 cm⁻¹ indicate the ν_2 (δ) UO_2^{2+} vibration.

β-uranophane, $Ca(UO_2)_2(SiO_3OH)_2 \cdot 5H_2O$ (Fig. 8) (this study)

ν_3 UO_2^{2+} 854 cm⁻¹, ν_1 UO_2^{2+} 777 cm⁻¹, ν_1 SiO_4^{4-} 880 (sh) cm⁻¹, ν_2 SiO_4^{4-} 472 and 460 (sh) cm⁻¹, ν_3 SiO_4^{4-} 930 (sh), 946, and 987 cm⁻¹, ν_4 SiO_4^{4-} 555 cm⁻¹; ν OH 3447 and 3225 (sh b) cm⁻¹, δ H_2O 1636 and 1665 (sh) cm^{-1}, δ SiOH ~1450 (w b) cm^{-1}. Structurally unequivalent H_2O groups, bonded with different forces in the crystal structure (confirmed with thermal analysis), and a H-bond network were inferred from the wavenumbers of ν OH and δ H_2O. A band near 299 cm^{-1} is associated with v_2 (δ) UO_2^{2+}.

Cuprosklodowskite, $Cu(UO_2)_2(SiO_3OH)_2 \cdot 6H_2O$ (Fig. 5) (this study)

v_3 UO_2^{2+} 871 cm^{-1}, v_1 UO_2^{2+} 777 cm^{-1}, v_1 SiO_4^{4-}—possible coincidence with v_3 UO_2^{2+}, v_2 SiO_4^{4-} 442 and 490 cm^{-1}, v_3 SiO_4^{4-} 933 and 980 cm^{-1}, v_4 SiO_4^{4-} 557 cm^{-1}, ν OH 3422 (b) cm^{-1}, 3200 (b) cm^{-1}, δ H_2O 1624 cm^{-1}, the shoulder at 1402 cm^{-1} may be associated with δ SiOH vibration. The δ SiOH assignment near 1450 cm^{-1} was given by Akhmanova et al. (1963). The ν OH vibrations wavenumbers decrease and the wavenumber of δ H_2O confirm H-bonds in the structure. v_2 (δ) UO_2^{2+} lies in the range 222–298 cm^{-1}.

Sklodowskite, $Mg(UO_2)_2(SiO_3OH)_2 \cdot 6H_2O$ (Fig. 5) (this study)

v_3 UO_2^{2+} 864 cm^{-1}, v_1 UO_2^{2+} 763 cm^{-1}, v_1 SiO_4^{4-} 850 (sh) cm^{-1}, v_2 SiO_4^{4-} 441 and 502 cm^{-1}, v_3 SiO_4^{4-} 934, 989, 1163 cm^{-1}(?), v_4 SiO_4^{4-} 570 cm^{-1}, ν OH 3490 (b) cm^{-1}, 3317, 3200 (sh b), 2900, 2850 cm^{-1}, δ H_2O 1653 and 1670 (sh) cm^{-1}, δ SiOH 1450 (sh), 1410 (vw) cm^{-1}. A lowering of wavenumbers of the OH grouping vibrations and a vibration with a shoulder assigned to δH_2O confirms H-bonds and structurally nonequivalent H_2O groups in the structure. v_2 (δ) UO_2^{2+} was found at 222 and 262 cm^{-1}.

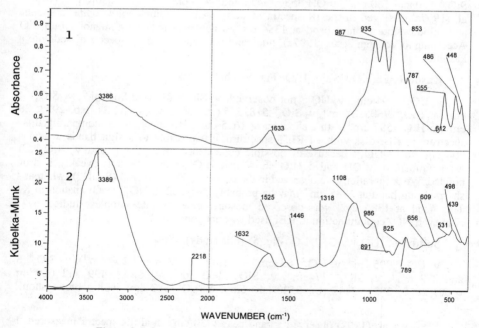

Figure 9. Infrared spectrum of boltwoodite: 1. KBr disk; 2. DRIFT.

Boltwoodite, $K(UO_2)(SiO_3OH) \cdot H_2O$ (Fig. 9) (this study)

v_3 UO_2^{2+} 853 cm^{-1}, v_1 UO_2^{2+} 787 cm^{-1}, (v_2 (δ) UO_2^{2+} 280 cm^{-1}—Plesko 1981), v_1 SiO_4^{4-} 875 (sh) cm^{-1}, v_2 SiO_4^{4-} 448 and 486 cm^{-1}, v_3 SiO_4^{4-} 935 and 987 cm^{-1}, v_4 SiO_4^{4-} 555 and 612 cm^{-1}. ν OH are observed as broad bands in the region 2800–3500 cm^{-1} with maxima near 3150 and 3386 cm^{-1} (3190 b, 3390, and 3465 (sharp) cm^{-1}—Plesko et al. 1992), ~3200, 3377 and 3458 (sharp) cm^{-1}—Vochten et al. (1997a), and 3246, 3401 and 3469 (sharp) cm^{-1}—another sample studied in this contribution). Broad bands near 1620–1640 cm^{-1} are assigned to δ H$_2$O. A shoulder at 1675 cm^{-1}, or a band at 1684 cm^{-1} (this study), or at 1730 cm^{-1} (Plesko 1981; Plesko et al. 1992) shows a strong hydrogen bonding network in the crystal structure of boltwoodite; however, this may indicate H$_3$O ions, in agreement with Stohl and Smith (1981). Some bands are found in the region 1200–1550 cm^{-1} (e.g. 1384 cm^{-1}—Vochten et al. (1997)—asynthetic boltwoodite) in available IR spectra of boltwoodite. These bands may indicate a SiO$_3$OH^{3-} group in the structure and correspond to δ SiOH (see e.g. Vochten et al. 1997; Burns 1998a).

IR spectra of synthetic boltwoodite, synthetic sodium boltwoodite and natural boltwoodite are similar but not identical, especially in the region of ν OH vibration (Vochten et al. 1997a). Vochten et al. proposed for these phases ideal formula K(UO$_2$)(SiO$_3$OH)·H$_2$O and Na(UO$_2$)(SiO$_3$OH)·H$_2$O which is supported by a recent structure determination of boltwoodite, (K$_{0.56}$Na$_{0.42}$)[(UO$_2$)(SiO$_3$OH)](H$_2$O)$_{1.5}$ (Burns 1998a).

Sodium-boltwoodite, (Na$_{0.7}$K$_{0.3}$)(UO$_2$)(SiO$_3$OH)·H$_2$O (Gevorkyan et al. 1979)
Na(UO$_2$)(SiO$_3$OH)·H$_2$O—synthetic phase (Vochten et al. 1997a)

Gevorkyan et al. (1979) assigned the observed bands to: ν_3 UO$_2^{2+}$ 880 cm^{-1}, ν_1 UO$_2^{2+}$ 780 cm^{-1}, ν_1 SiO$_4^{4-}$ (not observed), ν_2 SiO$_4^{4-}$ 477 cm^{-1}, ν_3 SiO$_4^{4-}$ 940 and 995 cm^{-1}, SiO$_4^{4-}$ 530 and 560 cm^{-1}, ν OH 3580, 3475 and 3410 cm^{-1}, δ H$_2$O 1630 cm^{-1}. Vochten et al. (1997a) observed in the IR spectra of boltwoodite, sodium boltwoodite, uranophane and sklodowskite a sharp band at 1384 cm^{-1} which may indicate a silanol group SiOH. According to Vochten et al. (1997a), this band is absent in the IR spectra of soddyite and kasolite.

Kasolite, Pb(UO$_2$)(SiO$_4$)·H$_2$O (Fig. 5) (this study)

ν_3 UO$_2^{2+}$ 904 cm^{-1}, ν_1 UO$_2^{2+}$ not observed, ν_1 SiO$_4^{4-}$ 811 and 864 cm^{-1}, ν_2 SiO$_4^{4-}$ 4 cm^{-1}, ν_3 SiO$_4^{4-}$ 960 (sh) cm^{-1}, ν_4 SiO$_4^{4-}$ 518, 545 (sh), 563 cm^{-1}, ν OH 3421 b, 2923, 28 cm^{-1}, δ H$_2$O 1597 cm^{-1} with a shoulder at 1625 cm^{-1}. However, when comparing this spectrum in KBr disk with a DRIFT spectrum, it is possible to assign bands at 864 and 904 cm^{-1} to ν_3 UO$_2^{2+}$ because of possible coincidence of ν_3 UO$_2^{2+}$ and ν_1 SiO$_4^{4-}$. The wavenumbers of ν OH and δ H$_2$O show that H$_2$O groups are only weakly hydrogen bonded. Weak vibrations are observed in the region 1300–1500 cm^{-1} (SiOH?), whereas the sharp strong band at 1385 cm^{-1} may be probably assigned to NO$_3^-$ contamination in KBr. This band at 1385 cm^{-1} was observed in some new kasolite samples studied by us. However, NO$_3^-$ contamination in KBr used was not confirmed.

Soddyite, (UO$_2$)$_2$SiO$_4$·2H$_2$O (Fig. 5) (this study)

ν_3 UO$_2^{2+}$ 905 cm^{-1}, ν_1 UO$_2^{2+}$ 832 cm^{-1}, (ν_2 (δ) UO$_2^{2+}$ 228 and 268 cm^{-1}). ν_1 SiO$_4^{4-}$ 8 and 872 cm^{-1}, ν_2 SiO$_4^{4-}$ 474 cm^{-1}, ν_3 SiO$_4^{4-}$ 960 cm^{-1}, ν_4 SiO$_4^{4-}$ 499 and 619 cm^{-1}, ν OH 2850, 2927, 3447 (b), 3520 (sh) cm^{-1}, δ H$_2$O 1582 cm^{-1} with a broad shoulder at 1637 cm^{-1}.

Plesko et al. (1992) observed a band near 1730 cm^{-1} in all IR spectra measured, in synthetic phases: soddyite, boltwoodite and weeksite. This band may be due to systematic error in the IR spectra. Vochten et al. (1995b) assume a broad complex band with a maximum at 3459 cm^{-1} corresponds with various ν OH stretching of molecular water, and that at 1579 cm^{-1} to δ H$_2$O. They observed six bands between 1101 and 800 cm^{-1}, bands at 956 cm^{-1} and 871 cm^{-1} are related to ν_3 SiO$_4^{4-}$ and ν_1 SiO$_4^{4-}$, and those at 904 and 835 cm^{-1} to ν_3 UO$_2^{2+}$ and ν_1 UO$_2^{2+}$, respectively. The remaining two bands should be H librations modes. The bands at 615 may be assigned to ν_4 SiO_4^{4-} and those around 2366 cm^{-1} to overtones or combination bands. Moll et al. (1995) infer H$_2$O groups are located in the uranium coordination sphere on the basis of the assignment of bands at 3450 cm^{-1} and 1588 cm^{-1} with a shoulder at 1635 cm^{-1} (adsorbed water weakly hydrogen bonded) to ν OH and δ H$_2$O, respectively. According to these authors, sharp absorption bands at 963 cm^{-1} (ν_3 SiO_4^{4-}) and 878 cm^{-1} (ν_1 SiO_4^{4-}) confirm that the SiO_4^{4-} tetrahedra are isolated and unpolymerized. Band at 835 cm^{-1} is assigned to ν_1 UO_2^{2+}, that at 920 cm^{-1} to ν_3 UO_2^{2+}, and that at 278 cm^{-1} to ν_2 (δ) UO_2^{2+}.

Weeksite, $K_2(UO_2)_2(Si_2O_5)_3 \cdot 4H_2O$ (Fig. 10) (this study)

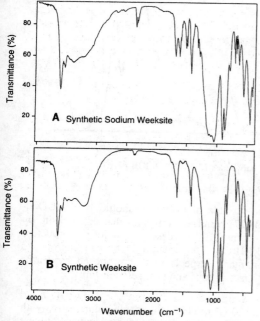

Figure 10. Infrared spectra of (A) synthetic sodium weeksite, (B) synthetic weeksite (from Vochten et al. 1997b).

The chemical composition of weeksite is uncertain. Smith (1984) reports it as $K_2(UO_2)_2Si_6O_{15} \cdot 4H_2O$ (U/Si = 1/3). However, a crystal structure analysis of weeksite-like phase $(K_{0.62}Na_{0.38})_2(UO_2)_2$-$(Si_5O_{13}) \cdot 3H_2O$ (Baturin and Sidorenko 1985), with U/Si ratio 1/2.5 was described and this phase has been accepted as weeksite. According to Baturin and Sidorenko (1985), this phase is one of those studied by Eremenko et al. (1977). Chemical analyses of weeksite by Eremenko et al. are close to those accepted by Smith, i.e. $(K,Na)_2(UO_2)_2(Si_2O_5)_3 \cdot 4H_2O$ Vochten et al. (1997b) published IR spectra for synthetic weeksite and sodium weeksite. The spectra are almost identical to that synthetic weeksite reported by Plesko et al. (1992). The IR spectrum of weeksite described in the range 2100–400 cm^{-1} by Eremenko et al. (1977) is also similar to those of synthetic phases.

ν_1 UO_2^{2+} 865–877 cm^{-1}; ν_2 UO_2^{2+} 265 cm^{-1}; ν_3 UO_2^{2+} 915 cm^{-1}; ν_1 SiO_4^{4-} 789–800 cm^{-1}; ν_2 SiO_4^{4-} 410, 455 cm^{-1}; ν_3 SiO_4^{4-} 970, 1055, 1120(?)cm^{-1}; ν_4 SiO_4^{4-} 545, 640 cm^{-1}; ν OH 2960–3605 cm^{-1}, δ H$_2$O 1625 cm^{-1} with a 1650 cm^{-1} shoulder ; δ SiOH 1385–1400, 1460 cm^{-1}.

Plesko et al. (1992) and Vochten et al. (1997b) suggest sharp bands 3605 (3610), 3540 (3520) and 1625 (1624) cm^{-1} are indicative of H$_2$O molecules located in ordered sites, whereas the broad band between 3000 and 3500 cm^{-1} may be associated with the stretching vibrations of H$_2$O groups having a disordered arrangement (Ryskin 1974; White 1971). It is clear the sharp band at 3605 (3610) cm^{-1} corresponds to H$_2$O groups only weakly hydrogen bonded, whereas the broad bands in the region 3000–3500 cm^{-1} indicate a H-bond network in the structure. δ H$_2$O vibration at 1625 cm^{-1} with a shoulder at 1650 cm^{-1} in the IR spectrum of synthetic weeksite by Vochten et al. (1997b) also indicate hydrogen bonds in the structure. However, bands observed in the region 1200–1500 cm^{-1} may correspond to δ SiOH. Some shoulder below 800 cm^{-1} may be assigned to H$_2$O libration modes. Tentative assignment of polymerized SiO_4 tetrahedra vibrations (e.g. $Si_2O_5^{2-}$—see Nguyen et al. 1992) is not unambiguous on the basis of available data.

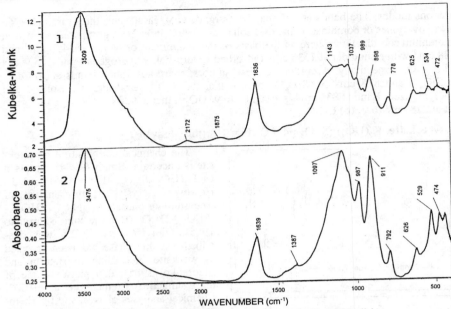

Figure 11. Infrared spectrum of haiweeite: 1. DRIFT; 2. KBr disk.

Haiweeite, $Ca(UO_2)_2(Si_2O_5)_3 \cdot 5H_2O$ (Fig. 11) (this study)

According to the structure analysis, the formula of haiweeite should be $Ca(UO_2)_2$-$[Si_5O_{12}(OH)_2] \cdot 4.5H_2O$ (Rastsvetaeva et al. 1997) which is similar to that of weeksite by Baturin and Sidorenko (1985). Recently, Burns (1999e) described a refined structure analysis of haiweeite, $Ca[(UO_2)_2Si_5O_{12}(OH)_2](H_2O)_3$. The structure of haiweeite proposed by Rastsvetaeva et al. (1997) is similar in many respects to that proposed by Burns (1999e), although the two reported different space groups and unit-cell constants.

ν_3 UO_2^{2+} 911 cm^{-1}, ν_1 UO_2^{2+}—not observed, ν_1 SiO_4^{4-} 792 cm^{-1}, ν_2 SiO_4^{4-} 450 and 474 cm^{-1}, ν_3 SiO_4^{4-} 987, 1050 (sh), 1097, 1130 (sh), 1170 (sh) (some coincidences with δ UOH are possible), ν_4 SiO_4^{4-} 529 cm^{-1} and 626 cm^{-1}, ν OH 3475 cm^{-1} with shoulders 3570, 3250, 2950, 2930 and 2850 cm^{-1}, δ H$_2$O 1639 cm^{-1}, and probably δ SiOH 1387 and 1470 cm^{-1}. Tentative assignment of bands to SiO$_4$-polymers vibrations (such as Si$_2$O$_5^{2-}$) is still uncertain.

Uranyl phosphates and arsenates

Autunite, $Ca(UO_2)_2(PO_4)_2 \cdot (8\text{-}12)H_2O$ (this study)

Meta-autunite, $Ca(UO_2)_2(PO_4)_2 \cdot (6\text{-}8)H_2O$ (Fig. 12) (Cejka Jr. et al. 1984a)

Meta-autunite II, $Ca(UO_2)_2(PO_4)_2 \cdot (0\text{-}2)H_2O$ (Cejka Jr. et al. 1984a)

Meta-torbernite, $Cu(UO_2)_2(PO_4)_2 \cdot (6\text{-}8)H_2O$ (Fig. 13) (Cejka Jr. et al. 1984a)

Meta-uranocircite I, $Ba(UO_2)_2(PO_4)_2 \cdot 8H_2O$ (Cejka Jr. et al. 1984a)

Meta-uranocircite II, $Ba(UO_2)_2(PO_4)_2 \cdot 6H_2O$ (Fig. 14) (Cejka Jr. et al. 1984a)

Sabugalite, $HAl(UO_2)_4(PO_4)_4 \cdot 16H_2O$ (Fig. 15) (Muck et al. 1986b)

Saléeite, $Mg(UO_2)_2(PO_4)_2 \cdot 10H_2O$ (Fig. 15) (Muck et al. 1986b)

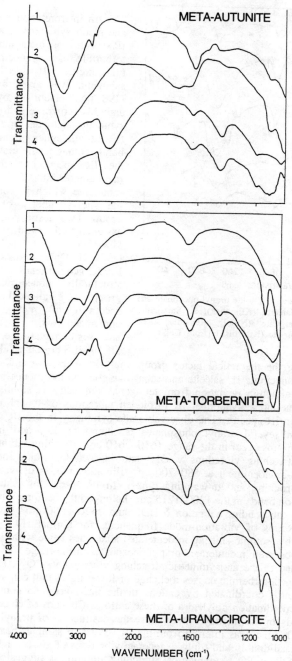

Figures 12, 13, 14. Infrared spectra (KBr disks): of 1. the mineral; 2. a synthetic; 3. mineral-deuteroanalogue; 4. synthetic-deuteroanalogue (from Cejka Jr et al. 1984b, 1984b, and 1985c, respectively).

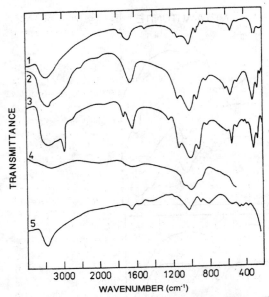

Figure 15. Infrared spectra of some uranyl phosphate and arsenate minerals: 1.Saléeite (KBr disk); 2. Sabugalite (KBr disk); 3. Sabugalite (Nujol); Renardite (KBr micropellet); 5. Walpurgite (KBr disk) (from Cejka and Urbanec 1990).

An interpretation based on site and factor symmetry analysis of the IR spectra of meta-autunite, meta-torbernite, meta-uranocircite I, meta-uranocircite II, saléeite and sabugalite was given by Cejka Jr. (1983) and elaborated by this author and his coworkers together with synthetic analogues and partly deuterated analogues. Four absorption bands belong to the uranyl group in the IR spectra of meta-autunite, meta-uranocircite I, saléeite and sabugalite which is consistent with the site symmetry of the uranyl group. Two absorption bands were assigned to the symmetric stretching vibration v_1 UO_2^{2+} in meta-uranocircite II. Only meta-torbernite may have geometrically distinct and structurally nonequivalent uranyl groups, as is indicated by the Raman spectrum of meta-torbernite, in agreement with X-ray structure analyses of meta-torbernite (Ross et al. 1964; Stergiou et al. 1993), and also formally fitting the theoretical factor group analysis. The newest X-ray structure analyses of meta-uranocircite II, saléeite and sabugalite confirm the monoclinic symmetry of these minerals, although deviation from tetragonal symmetry is small. Splitting of degenerate vibrations and activation of IR forbidden vibrations of the phosphate group were not observed in IR spectra. The observable site symmetry in the IR spectra of the phosphate group in several uranyl phosphates is, therefore, D_{2d} and in others S_4. Vibrations of H_2O groups occur in the range 1630–1650 cm^{-1} (bending vibrations δ H_2O), 3100–3580 cm^{-1} (stretching vibrations v OH) and in meta-torbernite and partially in sabugalite and meta-uranocircite I at 580–700 cm^{-1} (libration modes of water molecules, probably of the ρ-rocking and ω-wagging types). In IR spectra of deuterated uranyl phosphates absorption bands in the 1200–1215 cm^{-1} belong to the bending vibration δ D_2O, 1430–1450 cm^{-1} to the bending vibration δ HDO and 2520–2580 cm^{-1} to the stretching vibrations v OD. Ratios of vibration-mode frequencies for normal H_2O groups to the frequencies caused by isotopic shifting in deuterated analogues agree with published data. Absorption bands observed in deuterated uranyl phosphates in the range 615–645 cm^{-1} can probably be assigned to the antisymmetric stretching vibration v_3 PO_4^{3-}. Dehydration of meta-autunite and meta-torbernite causes a change or lowering of unit cell symmetry. The release of H_2O groups coordinated by cations in the interlayer also contributes to the destruction of the coordination polyhedra of these cations. Changes of electrostatic forces and hydrogen bonds in the interlayer may cause the destruction of uranyl polyhedra in structural sheets. This fact was inferred from the IR spectra of samples heated to 600° C, where a splitting of all doubly and triply degenerate vibrations of UO_2^{2+} and PO_4^{3-} groups and an activation of the v_1 PO_4^{3-} symmetrical stretching vibration is suggested according to the number of absorption bands. From IR spectra of meta-autunite and meta-torbernite dehydrated at 600°C, owing to the splitting of these vibrations, destruction of the crystal structures of these minerals was supposed. On the other hand, the IR spectrum of meta-uranocircite (II), after dehydration, does not change substantially, hence, a relatively stable crystal structure of this mineral is inferred.

A weak absorption band at 810 cm^{-1} is assigned to the symmetric stretching vibration v_1 UO_2^{2+} in the IR spectrum of sabugalite (Cejka Jr. et al. 1984a; Cejka et al. 1988b). The doubly degenerate bending vibration v_2 UO_2^{2+} is split into two components (absorption bands at 254 and 298 cm^{-1}). A sharp absorption band at 915 cm^{-1} is assigned to the antisymmetric stretching vibration v_3 UO_2^{2+}. Activation of the symmetric stretching vibration v_1 PO_4^{3-} and the doubly degenerate bending vibration v_2 PO_4^{3-} was not observed in the IR spectrum measured in a KBr disk. Weak absorption bands at 365 and 397 cm^{-1} and a shoulder at 965 cm^{-1} in the IR spectrum in Nujol, however, may be assigned to these vibrations. Two intense absorption bands at 995 and 1123 cm^{-1} relate to the antisymmetric stretching vibration v_3 PO_4^{3-}. Absorption bands at 470 and 542 cm^{-1} are assigned to the bending vibration v_4 PO_4^{3-}. A weak absorption band at 585 cm^{-1} may probably be assigned to libration modes of water molecules. An absorption band at 1640 cm^{-1} with a shoulder at 1665 cm^{-1} (KBr disk), and that at 1634 cm^{-1} (Nujol), relate to the bending vibration δ H_2O. An intense absorption band at 3400 cm^{-1}, with shoulders at 3100, 3240 and 3580 cm^{-1}, was assigned to the stretching vibration v OH. Kubisz (1969) assumes that $H_9O_4^+$ ions are present in the interlayer of sabugalite. The stretching vibrations (v_1, v_3) H_3O^+ should be observed at approximately 2500–3400 cm^{-1}, and the bending vibrations (v_2, v_4) at aproximately 950–1140 and 1670–1750 cm^{-1}, respectively (Rozi_re 1973). The distinction between H_3O^+ and H_2O entities is difficult in the region of v OH. The bending vibration v_2 H_3O^+ may be partly overlapped by the vibration v_3 PO_4^{3-}. The bending vibration v_4 H_3O^+ is the only one that should be observed unambiguously if H_3O^+ ions are present. The OH stretching band associated with the hydrogen bond between the hydroxonium and an H_2O group is clearly observed at 2760 cm^{-1}. In the infrared spectrum of sabugalite absorption bands (KBr disk) were found neither at 1700–1750 nor at 2700–3000 cm^{-1}. Absorption bands at 1735 and 2965 cm^{-1} and a shoulder at 2720 cm^{-1} in the IR spectrum (Nujol) are consistent with the vibrations v_4, v_1 and v_3 H_3O^+, respectively. The difference between the IR spectra may be caused by the KBr disk preparation, probably associated with structure changes or partial dehydration. The presence of hydroxonium ions or their hydrates in the interlayer of sabugalite can be inferred from these observations. This assumption is supported by the fact that sabugalite (Smolnyvrch deposit, Czech Republic) contains 0.2–0.3 wt % of K$^+$ ions, and the substitution K$^+$ → H_3O^+ is inferred. Sabugalite may therefore be represented by the formula

(H_3O^+)Al(UO_2PO_4)$_4 \cdot$15H_2O or ($H_9O_4^+$)Al(UO_2PO_4)$_4 \cdot$12H_2O.

Uranospathite is in fact fully hydrated sabugalite. Hydroxonium ions may also be expected in this mineral. Finch (1999-written communication), however, suggests that the interlayer may turn out to contain OH$^-$ as in althupite.

The IR spectrum of ferrous substituted saléeite was published by Vochten and Van Springel (1996). Cation H-Ca, H-Mg isomorphism in uranyl phosphates and uranyl arsenates was studied by X-ray diffraction and IR spectroscopy (Sidorenko et al. 1992). The peculiarities of the IR spectra of H_2O groups incorporated into phosphate and arsenate structures, including meta-autunite, meta-torbernite, meta-novácekite and meta-zeunerite, were published by Gevorkyan and Povarennykh (1980).

Bassetite, $Fe^{2+}(UO_2)_2(PO_4)_2 \cdot (8-12)H_2O$ (Fig. 16) (Vochten et al. 1984)

IR spectrum of synthetic bassetite and fully oxidized bassetite was published Vochten et al. 1984). v_1 PO_4^{3-} 940 cm^{-1}, v_2 PO_4^{3-} 420 cm^{-1}, v_3 PO_4^{3-} 1020 cm^{-1}, v_4 PO_4^{3-} 560 cm^{-1}. The bands attributed to the PO_4^{3-} groups are the same for both minerals. H_2O groups give

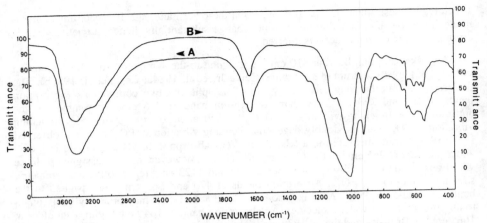

Figure 16. Infrared spectra of (1) synthetic bassetite and (2) fully oxidized synthetic bassetite (from Vochten et al. 1984).

bands in the regions 1600–1650 cm^{-1} (δ H$_2$O) and 2600–3700 cm^{-1} (ν OH). For both compounds, the shape of the band in the δ H$_2$O region is nearly the same, and an extra shoulder is observed in synthetic bassetite. The shape of ν OH is different for the two phases, which is related to changes to the coordinated water molecules. UO$_2^{2+}$ vibrations are not discussed.

Meta-ankoleite, K$_2$(UO$_2$)$_2$(PO$_4$)$_2$·6H$_2$O (Pham Thi et al. 1985a)

ν_1 UO$_2^{2+}$ 820 cm^{-1}, ν_3 UO$_2^{2+}$ 920 cm^{-1}, ν_3 PO$_4^{3-}$ 1000 and 1110, ν_4 PO$_4^{3-}$ 545 cm^{-1}, ν OH 3000–3500 cm^{-1}, δ H$_2$O 1650–1750 cm^{-1}, H$_2$O libration modes 600–700 cm^{-1}.

Uramphite, (NH$_4$)$_2$(UO$_2$)$_2$(PO$_4$)$_2$·(6-8)H$_2$O (Fig. 17) (Baran and Botto 1977)

ν_1 UO$_2^{2+}$ 828 cm^{-1}, ν_3 UO$_2^{2+}$ 922 cm^{-1}, ν_2 UO$_2^{2+}$ 256 cm^{-1}, ν_3 PO$_4^{3-}$ 1000 and 1115 cm^{-1}, ν_4 PO$_4^{3-}$ 543 cm^{-1}, ν OH + ν NH 3200, 3320, 3440 cm^{-1}, δ H$_2$O 1650 cm^{-1}, ν_4 NH$_4^+$ 1430 cm^{-1}, H$_2$O libration modes 465, 618 cm^{-1}, U-PO$_4$ 301 cm^{-1}, combination bands or overtones 1850 and 2090 cm^{-1}.

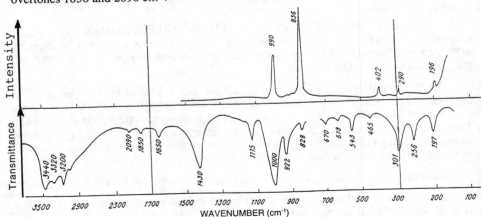

Figure 17. Infrared and Raman spectra of synthetic uramphite (from Novitskiy et al. 1981).

Chernikovite, $(H_3O^+)_2(UO_2)_2(PO_4)_2 \cdot 6H_2O$ (Pham Thi et al 1985b)

ν_1 UO_2^{2+} 820 cm^{-1}, ν_2 UO_2^{2+} 255 cm^{-1}, ν_3 UO_2^{2+} 930 cm^{-1}, ν_3 PO_4^{3-} 980, 1000, 1120 cm^{-1}, ν_4 PO_4^{3-} 545, 620 cm^{-1}, ν OH 3350 cm^{-1}, δ H_2O 1640 (sh) cm^{-1}, ν H_3O^+ 2600 (sh) cm^{-1}, ν_4 (δ) H_3O^+ 1740 cm^{-1}, H_2O libration modes 670 cm^{-1}.

Synthetic chernikovite, $(H_3O^+)_2(UO_2)_2(PO_4)_2 \cdot 6H_2O$ (HUP), is among the best known proton solid conductors and is used in micro-ionic devices. This is why so many publications concern this compound.

Abernathyite, $K_2(UO_2)_2(AsO_4)_2 \cdot (6\text{-}8) H_2O$ (Wilkins et al. 1974)

ν_1 UO_2^{2+} 900 cm^{-1}, ν_3 UO_2^{2+} 953 cm^{-1}, ν_1 AsO_4^{3-} 372 cm^{-1}, ν_3 AsO_4^{3-} 818 cm^{-1}, ν_4 AsO_4^{3-} 818 cm^{-1}, ν_4 AsO_4^{3-} 497 cm^{-1}, ν OH ~3500 cm^{-1}, δ H_2O 1645 cm^{-1}, H_2O libration mode 590 cm^{-1}.

Novácekite, $Mg(UO_2)_2(AsO_4)_2 \cdot 12H_2O$ (Gevorkyan and Povarennykh 1979)

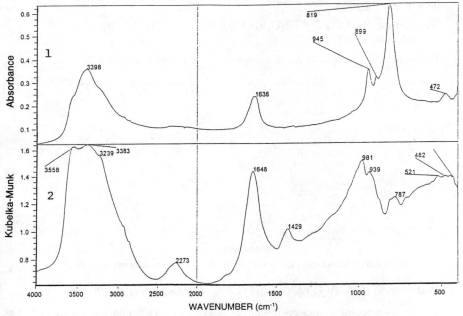

Figure 18. Infrared spectrum of novácekite: 1. KBr disk; 2. DRIFT.

Meta-novácekite, $Mg(UO_2)_2(AsO_4)_2 \cdot (6\text{-}8)H_2O$ (Fig. 18) (this study)

ν_1 UO_2^{2+} 899 cm^{-1}, ν_3 UO_2^{2+} 945 cm^{-1}, ν_3 AsO_4^{3-} 819 cm^{-1}, ν_4 AsO_4^{3-} 472 cm^{-1}, ν OH 3398 cm^{-1} and some shoulders, δ H_2O 1636 cm^{-1}.

Ni(UO$_2$)$_2$(AsO$_4$)$_2 \cdot$7H$_2$O, unnamed mineral (Fig. 19) (this study)

ν_1 UO_2^{2+} 897 cm^{-1}, ν_3 UO_2^{2+} 945 cm^{-1}, ν_3 AsO_4^{3-} 811 cm^{-1}, ν_4 AsO_4^{3-} 477 cm^{-1}, ν OH 3417 cm^{-1} with a shoulder, δ H_2O 1628 cm^{-1} with a shoulder.

Meta-kahlerite, $Fe(UO_2)_2(AsO_4)_2 \cdot 8H_2O$ (Vochten et al. 1986)

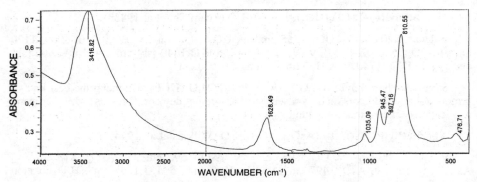

Figure 19. Infrared spectrum of natural nickel uranyl arsenate (KBr disk) (from Ondrus et al. 1997).

IR spectrum of fully oxidized synthetic metakahlerite is available. No detailed assignment was reported.

Vochtenite, $(Fe^{2+},Mg)Fe^{3+}[(UO_2)(PO_4)]_4(OH)\cdot(12-13)H_2O$ (Fig. 20) (Zwaan et al. 1989)

No detailed assignment of the IR spectrum of vochtenite was reported.

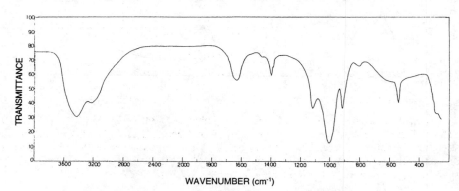

Figure 20. Infrared spectrum of vochtenite (KBr disk) (from Zwaan et al. 1989).

Walpurgite, $(UO_2)Bi_4O_4(AsO_4)_2\cdot 2H_2O$ (Fig. 15) (Sejkora et al. 1994)

v_1 UO_2^{2+}—possible coincidence with v_3 AsO_4^{3-}, v_3 UO_2^{2+} 888 cm^{-1}, v_2 UO_2^{2+} 245 cm^{-1}, v Bi—O 306 cm^{-1}, v_2 AsO_4^{3-} 370 cm^{-1}, v_3 AsO_4^{3-} 778 and 796 cm^{-1}, v_4 AsO_4^{3-} 432 cm^{-1}, v OH 3380, 3520 cm^{-1}, δ H_2O 1604 cm^{-1}, δ MOH 1028 cm^{-1}.

Trögerite, $(UO_2)_3(AsO_4)_2\cdot 12H_2O$ (Shchipanova et al. 1971)

P-trögerite, $(UO_2)_3(PO_4)_2\cdot 5H_2O$ (Sidorenko et al. 1975)

H-uranospinite, $(H_3O^+)_2(UO_2)_2(AsO_4)_2\cdot 6H_2O$ (Wilkins et al. 1974)

H-uranospinite, v_1 UO_2^{2+} 918 cm^{-1}(?), v_3 UO_2^{2+} 950 cm^{-1}, v_2 AsO_4^{3-} 386 cm^{-1}, v_3 AsO_4^{3-} 815 cm^{-1}, v_4 AsO_4^{3-} 486 cm^{-1}, v OH, v H_3O^+ ~3500 cm^{-1}, δ H_2O 1635 cm^{-1}, v_4 H_3O^+ 1740 cm^{-1}, v_2 H_3O^+ 1150 cm^{-1}, H_2O libration mode 610 cm^{-1}, (δ H_2O + libration mode) ~2300 cm^{-1} (Wilkins et al. 1974).

Bands observed in the IR spectrum of trögerite: 480, 550, 810–830, 875–915, 1000–1040, 1650, 3000–3600 cm^{-1} (Shchipanova et al. 1971). The tentative assignment is not given. The IR spectrum of natural H-uranospinite, also given by Shchipanova et al.(1971), differs from that of synthetic phase (e.g. Wilkins et al. 1974; Kreuer et al. 1983a,b).

$(UO_2)_3(PO_4)_2 \cdot 4H_2O$ (synthetic), ν_1 UO_2^{2+} 852 cm^{-1}, ν_3 UO_2^{2+} 933, 950 cm^{-1}, ν_2 UO_2^{2+} 250 cm^{-1}, ν_1 PO_4^{3-} 992, 1000 cm^{-1}, ν_3 PO_4^{3-} 1076, 1123, 1163 cm^{-1}, ν_4 PO_4^{3-} 533, 570, 628 cm^{-1}, ν OH 3225, 3325, 3390, 3598, 3575 cm^{-1}, δ H_2O 1600, 1618, 1631, 1675 cm^{-1}, H_2O libration modes 450, 720 cm^{-1}, ν (U- OPO_3) 286 cm^{-1}, δ (U-OPO_3) 217 cm^{-1} (data taken from Kobets and Umreyko (1983); see also Gmelin 1981; Weigel 1985).

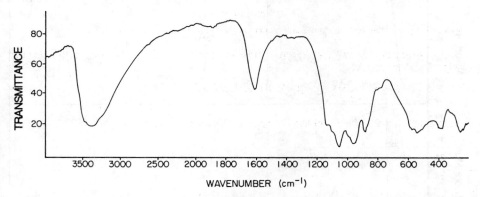

Figure 21. Infrared spectrum of phurcalite (KBr disk) (from Atencio et al. 1991).

Phurcalite, $Ca_2(UO_2)_3O_2(PO_4)_2 \cdot 7H_2O$ (Fig. 21) (Atencio et al. 1991)

The following absorption bands (cm^{-1}) were observed: 3400, 1615, 1140, 1055, 960, 885, 575, 540, 385, and 255. The very broad band centered at 3400 cm^{-1} is due to the stretching vibration of H_2O. The band at 1615 cm^{-1} is a consequence of the bend vibration of H_2O. The peaks observed between 960 and 1140 cm^{-1} are probably those of the stretching vibrations of PO_4 tetrahedra. The band at 885 cm^{-1} is due to ν_3 UO_2^{2+}.

Phosphuranylite, $KCa(H_3O^+)_3(UO_2)_7(PO_4)_4O_4 \cdot 8H_2O$ (Fig. 22) (Sejkora et al. 1994)

Yingjiangite, $(K_{1-x}Ca_x)(UO_2)_3(PO_4)_2(OH)_{1+x} \cdot 4H_2O$ (x = 0.35) or $(K_2,Ca)(UO_2)_7(PO_4)_4(OH)_6 \cdot 6H_2O$ (Fig. 23) (Chen Zangru et al. 1990)

The IR spectrum of yingjiangite indicates that the major uranyl band is about 910 cm^{-1} (ν_3 UO_2^{2+}). PO_4^{3-} bands appear at 540 and 590 cm^{-1} (ν_4 PO_4^{3-}), 990 cm^{-1} (ν_1 PO_4^{3-}), 1040 and 1085 cm^{-1}(ν_3 PO_4^{3-}), ν OH 3200 (b) cm^{-1} and 3500 cm^{-1} and δ H_2O near 1620 cm^{-1}.

The IR spectrum of phosphuranylite: ν_3 UO_2^{2+} 907 and ~895 (sh) cm^{-1}, ν_2 PO_4^{3-} 431, 471 cm^{-1}, ν_3 PO_4^{3-} 1009 cm^{-1} (ν_1 PO_4^{3-}?), 1042, 1090 cm^{-1}, ν_4 PO_4^{3-} 540 cm^{-1}, ν OH 3200 (b) cm^{-1}, 3442 cm^{-1}, δ H_2O 1633 cm^{-1} with a shoulder at ~1650 cm^{-1}. Some weak bands can be assigned to H_2O libration modes in the region 400–800 cm^{-1}. According to the crystal structure of phosphuranylite, H_3O^+ ions should be included in the structure (Demartin et al. 1991). However, the presence of H_3O^+ was not proved in the IR spectrum of the phosphuranylite studied.

Chemical composition, X-ray powder diffraction data and IR spectra of phosphuranylite and yingjiangite are very similar, and the two minerals may be identical.

Renardite, $Pb(UO_2)_3(PO_4)_2(OH)_2 \cdot 6H_2O$(?) (Fig. 15) (Cejka and Urbanec 1990)

No interpretation of the IR spectra of renardite was reported. Deliens et al. (1990) inferred renardite was not a mineral species but a mixture of phosphuranylite and dewindtite.

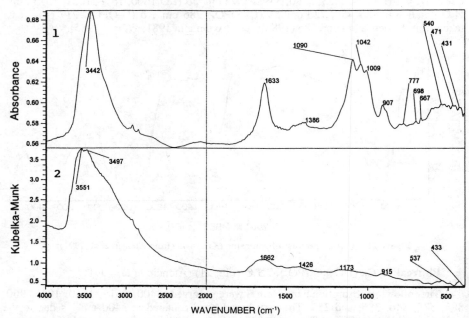

Figure 22. Infrared spectrum of phosphuranylite: 1. KBr disk; 2. DRIFT.

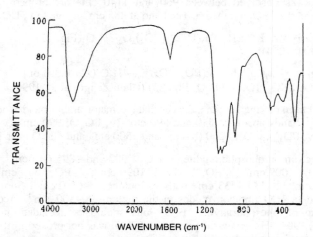

Figure 23. Infrared spectrum of yingjiangite (KBr disk) (from Zhang et al. 1992).

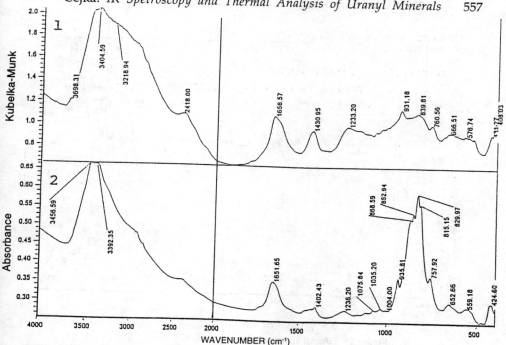

Figure 24. Infrared spectrum of natural $U^{4+}(HAsO_4)_2 \cdot 4H_2O$: 1. DRIFT; 2. KBr disk (Ondrus et al. 1997).

$U^{4+}(HAsO_4)_2 \cdot 4H_2O(?)$, unnamed mineral (Fig. 24) (Ondrus et al. 1997)

The IR spectrum of this unnamed mineral, the composition of which is uncertain (Ondrus et al. 1997), is similar but not identical with that of the synthetic $U^{4+}(HAsO_4)_2 \cdot 4H_2O$ (Chernorukov et al. 1985).

Uranyl vanadates

Carnotite, $K_2(UO_2)_2(V_2O_8) \cdot 3H_2O$ (Fig. 27) (Baran and Botto 1976)

Tyuyamunite, $Ca(UO_2)_2(V_2O_8) \cdot 9H_2O$ (Fig. 27) (Sidorenko et al. 1995)

Meta-tyuyamunite, $Ca(UO_2)_2(V_2O_8) \cdot 3H_2O$ (Botto et al. 1989a; this study)

Strelkinite, $Na_2(UO_2)_2(V_2O_8) \cdot 6H_2O$ (Sidorenko et al. 1985)

Curiénite, $Pb(UO_2)_2(V_2O_8) \cdot 5H_2O$ (Fig. 25) (this study)

Francevillite, $(Ba,Pb)(UO_2)_2(V_2O_8) \cdot 5H_2O$ (Fig. 27)
(Cejka and Urbanec 1990; this study)

Vanuralite, $Al(OH)(UO_2)_2(V_2O_8) \cdot 11H_2O$ (Fig. 26) (this study)

Tl-carnotite, $(K,Tl,Na)_2(UO_2)_2(V_2O_8) \cdot 2H_2O$ (Sidorenko et al. 1985)

$(NH_4)_2(UO_2)_2(V_2O_8) \cdot 5H_2O$, synthetic phase (Baran and Botto 1977)

$Cs_2(UO_2)_2(V_2O_8)$, synthetic phase (Dickens et al. 1992)

$Ag_2(UO_2)_2(V_2O_8)$, synthetic phase (Abraham et al. 1994)

$M^{2+}(UO_2)_2(V_2O_8) \cdot nH_2O$ (M^{2+} = Ni, Zn, Cd) synthetics (Chernorukov et al. 1998a)

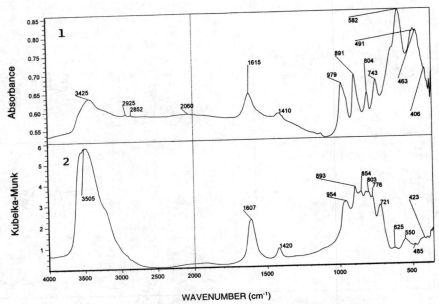

Figure 25. Infrared spectrum of curiénite: 1. KBr disk; 2. DRIFT

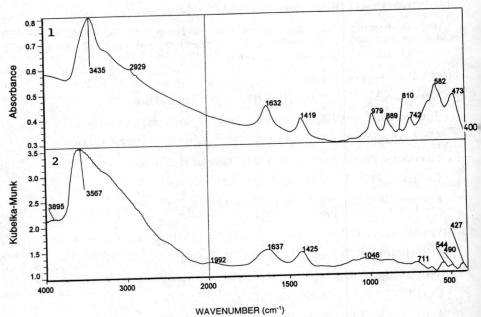

Figure 26. Infrared spectrum of vanuralite: 1. KBr disk; 2. DRIFT.

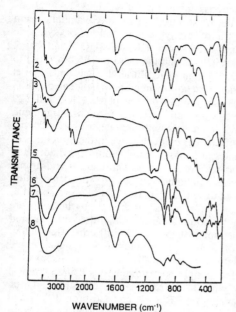

Figure 27. Infrared spectra of some uranyl sulfate and uranyl vanadate minerals: 1. Johannite (Nujol); 2. Johannite (KBr micropellet); Johannite (KBr disk); 4. Johannite (deuteroanalogue-Nujol); 5. Uranopilite (KBr disk); 6. Carnotite (KBr disk); 7. Tyuyamunite (KBr disk); 8. Francevillite (KBr micropellet) (from Cejka and Urbanec 1990).

According to Wilkins (1971a), it would seem that structures with isolated tetrahedra are atypical amongst the uranyl vanadates. The original structure determination on anhydrous carnotite (Sundberg and Sillén 1949) indicates the presence of isolated VO_4^{3-} ions, but a subsequent refinement of the crystal structure (Appleman and Evans 1965) showed $V_2O_8^{6-}$ ions. Condensed vanadate ions are also consistent with the presence of strong vanadate IR absorption bands well outside the range expected for isolated VO_4^{3-} tetrahedra.

All uranyl vanadate mineral and synthetic phases exhibit a layer structure with $[(UO_2)_2V_2O_8]_n^{2n-}$ units. The vanadium atoms are coordinated by five oxygen atoms in the form of square pyramids. Two polyhedra share a common edge, forming $V_2O_8^{-6}$ groups. For the interpretation of IR and Raman spectra of meta-tyuyamunite, Botto et al. (1989) used the following building units in the layers: V_2O_2 double bridges, terminal VO_3 groups, UO_2^{2+} groups, and secondary (equatorial) U-O_{II} bonds in a pentagonal environment. H_2O groups exhibit characteristic stretching, bending, and libration vibrations. The same units were used for tentative assignment of the IR and Raman spectra of $(NH_4)_2(UO_2)_2(V_2O_8) \cdot 5H_2O$ (Botto and Baran 1976), and synthetic carnotite (Baran and Botto 1976). However, some authors (Moenke 1962; Ross 1974; Povarennykh and Gevorkyan 1970; Wang et al. 1981; and, in the main, also Cejka and Urbanec 1990) consider vibrations to be related to those of VO_4^{3-}, whereas Botto et al. suggest that they correspond to VO_3 terminal groups. Tentative assignment of IR vibrations to VO_x groups and distinguishing them from those of UO_2^{2+} therefore does not seem simple. There are discrepancies in the location of corresponding vibrations given by various authors. However, the principal problems are discrepancies in tentative assignments of bands related to stretching vibrations of UO_2^{2+} and VO_3 (VO_4?) groups with regard to papers published by various authors.

Vibrations at approximately 870–894 cm^{-1} may be assigned to v_3 UO_2^{2+}. Some authors (e.g. Povarennykh and Gevorkyan 1970; Ross 1974; Wang et al. 1981; Sidorenko et al. 1995) assigned the band near 980 cm^{-1} to v_3 UO_2^{2+}. According to Botto et al. (1989a), vibrations in the range 870–894 cm^{-1} are related to the antisymmetric stretching vibration v_3 VO_3. Dickens et al (1992) assigned the band at 885 cm^{-1} in the IR spectrum of $Cs_2(UO_2)_2(V_2O_8)$ and Abraham et al. (1994) that at 891 cm^{-1} in the IR spectrum of $Ag_2(UO_2)_2(V_2O_8)$ to v_3 UO_2^{2+}. Recently Chernorukov et al. (1998) assigned the absorption band at 880 cm^{-1} to v_3 UO_2^{2+} and that at 740 cm^{-1} to v_1 UO_2^{2+} in the IR spectra of synthetic $M^{2+}(UO_2)_2(V_2O_8) \cdot nH_2O$ (M^{2+} = Ni, Zn, Cd). With regard to this assignment, v_1 UO_2^{2+} (if IR active) should be observed in the range 777–819 cm^{-1} or 805–843 cm^{-1}, when calculated using v_1 UO_2^{2+} = $f(v_3$ $UO_2^{2+})$ empirical equations. Intensities of v_1 UO_2^{2+} vibrations assigned by Botto et al. (1989a) (748 cm^{-1}) in the IR spectrum of meta-tyuyamunite seem higher than usually observed in the IR spectra of uranyl compounds. For synthetic carnotite, Baran and Botto (1976) assigned the band at 740 cm^{-1} (IR) to v_3 UO_2^{2+} and that at 737 cm^{-1} (Raman) to v_1 UO_2^{2+}; Botto and Baran (1976) assign the band at 735 cm^{-1} (IR) to v_3 UO_2^{2+} and that at 738 cm^{-1} (Raman) to v_1 UO_2^{2+} in ammonium uranyl vanadate; Botto et al. (1989a) suppose v_3 UO_2^{2+} in natural and synthetic meta-tyuyamunite to be located at 803 cm^{-1}, probably at 814 cm^{-1} in synthetic and 808 (sh) cm^{-1} in natural phase (IR), and 829 cm^{-1} (Raman), and v_1 UO_2^{2+} at 748 cm^{-1} (IR) and 747 cm^{-1} (Raman). Such assignment, established on the basis of comparison of uranyl vanadates with uranates (e.g. Baran and Botto 1976), does not seem correct. There are differences evident between the structures of uranyl vanadates and those of anhydrous uranates; e.g. concerning uranyl polyhedra and uranyl oxyanion layer structures. Bond relations with regard to uranium atoms and uranyl oxygens are not the same. Calculated U-O$_1$ bond lengths for uranyl vanadates using "Badger's equations" and wavenumbers assigned to v_3 UO_2^{2+} by Botto et al. correspond to those in uranates. However, U-O bond lengths for "uranyls" in anhydrous uranates are much longer than those observed in uranyl coordination compounds with layered structures. Practically, the same assigned values for v_3 UO_2^{2+} (735–740 cm^{-1}, IR) and v_1 UO_2^{2+} (738–737 cm^{-1}, Raman) in the spectra of synthetic ammonium uranyl vanadate and synthetic carnotite (Baran and Botto 1976; Botto and Baran 1976) therefore seem to be questionable. Uranyl bond lengths calculated for uranyl vanadate minerals using empirical equations and v_3 UO_2^{2+} (870–894 cm^{-1}) agree with those from X-ray structure analyses. Exceptions are found for curiénite (structure analysis by Bor_ne and Cesbron 1971) and francevillite (structure analysis by Shashkin 1975). In the both cases, reported uranyl bond lengths are much shorter than those observed recently for uranyl vanadates (e.g. crystal structure analysis of francevillite by Mereiter 1986). Thus, the assignment of v_3 UO_2^{2+} and v_1 UO_2^{2+} to approximately 880 cm^{-1} and ~800 or ~824 cm^{-1}, respectively, may be likley. Intense bands observed in the range 738–748 cm^{-1} (IR and Raman) are probably related to stretching vibrations of VO$_3$ and not to those of UO_2^{2+}. Respecting the tentative assignment by Botto et al., bands in the range 950–986 cm^{-1} correspond to VO$_3$ symmetric stretch, those at 837–868 cm^{-1} and probably also at 711–776 cm^{-1} to VO$_3$ antisymmetric stretch. There is also another possibility. The lower vibration could be assigned to v V$_2$O$_2$, to which bands observed in the range 612–670 cm^{-1} are related. If tentative assignment given by Botto et al. (1989a) is applied, weak bands in this region could be assigned to libration modes of water molecules; bands observed in the range 447–583 cm^{-1} may be assigned to equatorial v(U-O$_{II}$), those in 400-430 cm^{-1} with δ VO$_3$, those near 350 cm^{-1} to δ V$_2$O$_2$, and bands lower than 300 cm^{-1} to δ UO_2^{2+}.

Absorption bands above 2800 cm^{-1} correspond to v OH stretching vibrations, those in 1600–1637 cm^{-1} to δ H$_2$O bending vibrations. Weak and very weak bands and a shoulder in the range 1000–1500 and 1900–2600 cm^{-1} may be correlated with combination bands or overtones; however, bands in 1000–1425 cm^{-1} observed in the IR spectrum of vanuralite may be assigned to δ UOH.

Uranyl molybdates

matkovskiy et al. (1979) give wavenumbers of v_3 UO_2^{2+} for calcurmolite, CA(UO$_2$)$_3$(MOO$_4$)$_3$(OH)$_2$·7H$_2$O, (905 CM^{-1}), iriginite (788 CM^{-1}) and mourite (883CM^{-1}).

Mourite, UMo$_5$O$_{12}$(OH)$_{10}$(?) (Osipov et al. 1981)

IR spectra of natural mourite (Osipov et al. 1981) and phases produced by heating up to 1000°C were published without detailed interpretation. A band at 940 cm^{-1} was assigned to ν_3 UO_2^{2+} and that at 980 cm^{-1} to MoO_2^{2+} stretching vibration. No further data are available. The ν OH region is not reported. However, a band at ≈1700 cm^{-1} (taken from the graph by Osipov et al. 1981) indicates H-bonded H_2O groups, and is associated with δ H_2O. This observation does not agree with the formula proposed for mourite. The δ H_2O vibration intensity decreases with temperature and this fact is associated with a shifting of the wavenumber to lower values. At 450 °C this band entirely disappears. According to Osipov et al. (1981) the chemical composition and U and Mo valencies are not correctly determined. It seems likely that uranium (U^{4+} and U^{6+}) and molybdenum (Mo^{6+} and Mo^{5+}) valencies may vary. However, if Osipov's mourite is correctly defined, mourite contains molecular water. It is not possible to verify hydroxyl ions because of the incompleteness and low quality of the IR spectrum.

Umohoite, $UO_2MoO_4 \cdot 4H_2O$ (Wang et al. 1981; Seryozhkin et al. 1981)

ν_3 UO_2^{2+} 910 cm^{-1}, ν_1 UO_2^{2+} 818 cm^{-1}, ν_2 MoO_4^{2-} 255 cm^{-1} (possible coincidence with ν_2 UO_2^{2+}), ν_3 MoO_4^{2-} 818 cm^{-1}(?), ν_4 MoO_4^{2-} 360 and 410 cm^{-1}, ν OH 3340–3500 (b) cm^{-1}, δ H_2O 1620 (b) cm^{-1}, 1728 cm^{-1} (Wang et al. 1981).

α-$UO_2MoO_4 \cdot 2H_2O$ (Seryozhkin et al. 1981)

ν_3 UO_2^{2+} 938 and 920 cm^{-1}, ν_1 UO_2^{2+}—not observed, ν_2 UO_2^{2+} 246 cm^{-1}, ν_1 MoO_4^{2-} 952 cm^{-1}, ν_2 MoO_4^{2-} not observed, ν_3 MoO_4^{2-} 823 cm^{-1}, ν_4 MoO_4^{2-} 320 and 364 cm^{-1}, ν OH 3510 and 3440 cm^{-1} (both broad), δ H_2O 1610 cm^{-1}. A band at 715 cm^{-1} may be assigned to H_2O libration modes.

Iriginite, $UO_2Mo_2O_7 \cdot 3H_2O$ or $UO_2(HMoO_4)_2 \cdot 3H_2O$(?), $[(UO_2)(MoO_3OH)_2(H_2O)](H_2O)$ (Zhiltsova et al. 1970)

Tentative assignment of data taken only from the recorded spectrum (Zhiltsova et al. 1970) does not seem possible, because of the low quality of the spectrum. Synthetic so-called iriginite-like phases vary in composition and, for the most part, do not correspond to natural iriginite (Zhiltsova et al. 1970).

Tengchongite, $CaO \cdot 6UO_3 \cdot 2MoO_3 \cdot 12H_2O$ (Fig. 28) (Chen Zhangru et al. 1986)

IR spectrum of tengchongite shows the major uranyl band at about 920 cm^{-1} (ν_3 UO_2^{2+}) and that H_2O groups are present (ν OH at 3430 cm^{-1} and δ H_2O at 1640 cm^{-1}). The molybdate band appears at 780 cm^{-1}.

Deloryite, $Cu_4(UO_2)(MoO_4)_2(OH)_6$ (Sarp and Chiappero 1992)

An IR spectrophotometric study of deloryite shows the presence of absorption bands characteristic of anions MoO_4^{2-} and OH^-. No detailed information was reported.

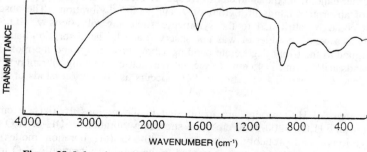

Figure 28. Infrared spectrum of tengchongite (from Chen Zangru et al. 1986).

Uranyl sulfates

The character of IR spectra of uranyl sulphate minerals can be strongly influenced by sample preparation, especially in the case of KBr disks.

Johannite, $Cu(UO_2)_2(OH)_2(SO_4)_2 \cdot 8H_2O$ (Fig. 27) (Cejka et al. 1988b)

The most extensive study on the IR spectrum of johannite, its synthetic analogue and their partly deuterated analogues was published by Cejka et al. (1988a). Space group of johannite is $P\bar{1}$ - C_i^1. Mereiter (1982) supposed monoclinic cell pseudosymmetry, which agrees with Hurlbut's paper (1950). Symmetry analysis of the function groups of johannite was made for the interpretation of the IR spectra (Cejka et al. 1988a). Four normal vibrations of UO_2^{2+} group (point symmetry $D_{\infty h}$) are realized in irreducible representations $\Sigma_g^+ + \Pi_u + \Sigma_u^+$ and nine normal SO_4^{2-} group vibrations (point symmetry T_d) in irreducible representations $A_1 + E + 2F_2$. From the structure analysis of johannite (Mereiter 1982) it is known that both U-O in uranyl are practically the same (1.78 Å) and the angle O-U-O equals 179.6°. C_i or C_1 are the only possible site symmetries of the UO_2^{2+} group for the space group $P\bar{1}$ - C_i^1. An influence of sample preparation (KBr disks and especially micropellets) on the character of the IR spectra of johannite was observed (Cejka et al. 1988c). A reaction between johannite and KBr cannot be excluded. The most reproducible IR spectra were gained in Nujol, which may more accurately represent johannite.

Two absorption bands (216 and 257 cm^{-1}) were assigned to the doubly degenerate bending vibration ν_2 UO_2^{2+}. The splitting of this vibration is in agreement with both site symmetries (2A_u for C_i; 2A for C_1) considered. The medium-strong absorption band at 911 cm^{-1} with a very weak shoulder at 936 cm^{-1} corresponds to the antisymmetric stretching vibration ν_3 UO_2^{2+}. The uranium coordination polyhedron in johannite is an irregular pentagonal dipyramid in which five oxygens (two of OH$^-$ and three of SO_4^{2-} ions) coordinated in the uranyl equatorial plane do not lie exactly in this plane. The presence of the shoulder at 936 cm^{-1} could be caused by this distortion. Three weak absorption bands (780, 821, and 832 cm^{-1}) were observed in the region of the symmetric stretching vibration ν_1 UO_2^{2+}. ν_1 UO_2^{2+} does not become IR active for the site symmetry C_i (A_g) in contrast to the site symmetry C_1 (A). These weak absorption bands can be assigned to the ν_1 UO_2^{2+} or to the bending vibrations of the δ U-O-H type or to the libration modes of water molecules. Coincidence of these vibrations is also possible.

In the case of the SO_4^{2-} anion, C_1 is the only admissible site symmetry. The vibrations ν_1 ($A_1 \rightarrow$ A) and ν_2 (E$\rightarrow$2A) become IR active; simultaneously the doubly degenerate ν_2 and the triply degenerate vibrations ν_3 and ν_4 [$2F_2 \rightarrow 2(3A)$] split. The observed absorption bands assigned to the SO_4^{2-} vibrations do not agree with the space group $P\bar{1}$ - C_i^1, because the site symmetry group must be of the same or lower order as the factor group C_i ($G_S \leq G_F$). On the other hand, from the character of IR spectra recorded one can suppose a certain overlapping of absorption bands in regions of ν_3 and ν_4 SO_4^{2-} vibrations. The absorption band, which could be attributed to the symmetric stretching vibrations ν_1 SO_4^{2-}, was observed only as a weak shoulder or was not observed at all. The absorption band at 619 cm^{-1} was assigned to the triply degenerate bending vibration ν_4 SO_4^{2-} and the most intense absorption bands at 1096 and 1145 cm^{-1} to the antisymmetric stretching vibration ν_3 SO_4^{2-}. The doubly degenerate bending vibration ν_2 SO_4^{2-} occurs as absorption bands at 384 and 422 cm^{-1}.

The vibrations of H_2O groups are manifested in the regions 1600-1660 cm^{-1} (the bending vibration δ H_2O), 3200-3600 cm^{-1} (the stretching vibrations ν OH of H_2O groups and hydroxyl ions), and probably in the range 780-835 (the libration modes). Two absorption bands (1630 and 1655 cm^{-1}) assigned to the bending vibration δ H_2O indi-cate the presence of structurally unequivalent H_2O groups bonded by different forces in the crystal structure of johannite, which agrees with structure and thermal analyses.

IR spectra of deuterated johannites were measured in Nujol suspensions and, owing to the strongly overlapping of absorption bands of Nujol and bending vibration δ HDO ,the latter could not be detected. The absorption bands at 2443, 2592, and 2638 cm^{-1} in the IR spectra of partly deuterated natural johannite were assigned to the stretching vibrations ν OD. From the presence of several absorption bands in the OD region, it can be inferred that H-D isotopic exchange was performed not only in the case of H_2O-D_2O groups but also in structural

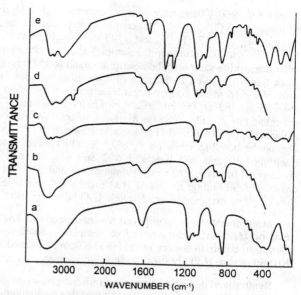

Figure 29. Infrared spectra of jáchymovite (a) Nujol; (b) KBr disk. Uranopilite sample 1: (c) Nujol; (d) KBr disk; uranopilite sample 2: (e) KBr disk (from Cejka et al. 1996b).

hydroxyl ions; however, neither was fully exchanged. The ratio of wavenumbers δ H_2O/δ D_2O = 1.361 agree well with 1.36 (Ross 1972). The intense absorption band at 1206 cm^{-1} may be caused by δ D_2O or by the coincidence of this vibration with ν_3 SO_4^{2-}. Similar overlapping absorption bands of δ D_2O vibration with ν_3 PO_4^{3-} was observed in the IR spectrum of deuterated meta-torbernite (Cejka et al. 1984a). The coincidence of absorption bands for δ U-O-D and ν_4 SO_4^{2-} vibrations is also proposed. Absorption bands related to the UO_2^{2+} and SO_4^{2-} vibrations in IR spectra of deuteroanalogues differ to some extent, however, from those of natural and synthetic johannite, which can be probably attributed to structural modifications related to deuteroanalogue preparation. Dispersion of samples of johannite in KBr and preparation of disks and micropellets may cause pressure or temperature effects (or both), leading to structure changes, not only to the interlayer containing Cu^{2+} and H_2O groups, but to uranyl sulfato hydroxo structural sheets as well (Fig. 27: 1,2,3).

Jáchymovite, $(UO_2)_8(SO_4)(OH)_{14} \cdot 13H_2O$ (Fig. 29) (Cejka et al. 1996b)

Uranopilite, $(UO_2)_6(SO_4)(OH)_{10} \cdot (12-13)H_2O$ (Fig. 29) (Cejka et al. 1996b)

The absorption bands at 904 cm^{-1} (jáchymovite) and 930 cm^{-1} (uranopilite) were assigned to the antisymmetric stretching vibration ν_3 UO_2^{2+}. According to the empirical equation $\nu_1 = 21 + 0.89\nu_3$ (Volodko et al. 1981), the symmetric vibration ν_1 UO_2^{2+} should appear approximately near 826 and 849 cm^{-1}, respectively. Weak absorption at 833 cm^{-1} (jáchymovite) and 840 cm^{-1} (uranopilite) were assigned to this vibration. However, ν_1 UO_2^{2+} vibration need not be observed, especially in the case of uranopilite. All these vibrations may coincide with those related to δ UOH. Doubly degenerate vibration ν_2 UO_2^{2+} is shown at 265 and 256 cm^{-1}, respectively. A weak broad absorption band at 1010 cm^{-1} (observed only in the case of one uranopilite specimen) may be related to the stretching vibration ν_1 SO_4^{2-}, there may be a coincidence with δ UOH. The antisymmetric stretching vibration ν_3 SO_4^{2-}, (triply degenerate), also coinciding with δ UOH, exhibits three absorptions at 1067, 1110, and 1145 cm^{-1} (jáchymovite), nine absorptions in the range 1191-1067 cm^{-1} (one uranopilite sample) and two absorptions at 1139 and 1118 cm^{-1} (the other uranopilite sample). The absorption band at 458 cm^{-1} (jáchymovite) and probably also weak bands observed in the case of uranopilite samples (470 and 476 cm^{-1}, respectively) were assigned to the doubly degenerate bending vibration ν_2 SO_4^{2-}. A moderately intense absorption band as 355 cm^{-1} (jáchymovite) and 362 cm^{-1} (uranopilite), which are probably not related to ν_2 SO_4^{2-}, can be assigned to out-of-plane bending vibration of U-O in the equatorial plane of uranyl (Hoekstra and Siegel 1973; Dothée and Camelot 1982). Triply degenerate bending vibration ν_4 SO_4^{2-} exhibits three bands at 601, 621, and 668 cm^{-1} (jáchymovite) and two bands at 602 and 637 (or 602 and 668) cm^{-1} for uranopilite. Absorption bands at 582 cm^{-1} (jáchymovite) and 550-551 cm^{-1} (two uranopilite samples) are related to libration modes of H_2O groups rather than SO_4^{2-} group vibrations; however, they could be assigned to out-of-plane U-O bending vibrations in the equatorial plane.

Molecular water is manifested by absorptions at 1619 and 1681 cm^{-1} (jáchymovite), 1620 or 1681 cm^{-1} (two uranopilite samples) related to bending vibration δ H_2O, and absorption bands in the region 3550-3110 cm^{-1} assigned to ν OH, and those in the range 581-530 cm^{-1} to H_2O librations as indicated above.

Synthesis of deuteroanalogues of mineral phases in the system UO_3-SO_3-H_2O has not been successful. Therefore, it was not possible to distinguish absorptions related to H_2O or OH$^-$ ions, and to the bending vibrations δ UOH, especially coinciding with vibrations of UO_2^{2+} and SO_4^{2-} groups, and to establish a better assignment (Cejka et al. 1996a).

Zippeites, $M_4^+(UO_2)_6(SO_4)_3(OH)_{10} \cdot 4H_2O$: M^+ = K, Na, NH_4^+ (Fig. 30) (this study)

$M^{2+}(UO_2)_6(SO_4)_3(OH)_{10} \cdot nH_2O$ (M^{2+} = Co, Ni, Mg, Zn, Cd; n = 8 or 16?)

Chemical formulas of zippeites are not clear. Frondel et al (1976) proposed those formulas given above. Haacke and Williams (1979) assumed that the molecular water content in zippeites with divalent cations to be 8 H_2O. Spitsyn et al. (1982) and Kovba

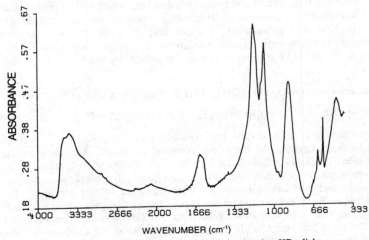

Figure 30. Infrared spectrum of synthetic zippeite: KBr disk.

Table 7. Unit cell parameters of the synthetic zippeites (Cejka and Sejkora 1994)

	a (Å)	b (Å)	c (Å)	β (°)
K	8.698(3)	13.856(8)	17.760(7)	104.17(3)
Na	not indexed	–	–	–
NH₄	8.703(3)	14.167(4)	17.785(6)	104.04(3)
Cd	8.741(3)	14.332(5)	17.742(6)	104.23(2)
Co	8.687(2)	14.224(3)	17.702(4)	104.20(1)
Mg	8.798(3)	14.145(5)	17.688(9)	104.38(3)
Ni	8.756(3)	14.218(7)	17.463(5)	104.39(3)
Zn	8.673(3)	14.236(4)	17.666(8)	104.21(3)

Table 8. ν_3 UO_2^{2+} vibrations in synthetic zippeites (Cejka and Sejkora 1994)

	KBr [cm⁻¹]	DRIFT [cm⁻¹]
Vochten's zippeite	–	924
Vochten's zippeite (this study)	921, 918	913
Zippeite K (a)	872	906
Zippeite K (b)	870	905
Zippeite Na	909, 915	925
Zippeite NH₄⁺	870	873
Zippeite Ni (a)	903	906
Zippeite Ni (b)	906	920
Zippeite Co (a)	880	883
Zippeite Co (b)	883	900
Zippeite Mg	897	912
Zippeite Zn	881	874
Zippeite Cd	890	910

et al. (1982) synthetized a series of phases having a general composition $M^{2+}(UO_2)_2(SO_4)(OH)_4 \cdot 1.5H_2O$, ($M^{2+}$ = Ni, Mg, Co, Zn, Mn) and determined the crystal structure of Zn^{2+} phase, and showed it to be comparable to a synthetic phase $(NH_4)_2SO_4(OH)_4 \cdot 4H_2O$ (Pechurova et al. 1965). All these synthetic phases have been named zippeites. Recently Vochten et al. (1995a) determined the crystal structure of synthetic zippeite, $K(UO_2)_2SO_4(OH)_3 \cdot H_2O$, suggesting that analyses made by Frondel et al. are probably not correct. However, the composition of Frondel's zippeites was confirmed by Haacke and Williams (1979) and O'Brien and Williams (1981) with the exception of water content (8 or 16 H_2O) in synthetic phases containing divalent cations. Unfortunately, new chemical analyses of minerals are lacking.

A series of zippeites was synthetized and charactreized on the basis of X-ray powder diffraction data (Table 7). The IR spectra (DRIFT and KBr disk) were measured (Cejka and Sejkora 1994).

Vochten's synthetic zippeite has unit cell constants, $a = 8.755(3)$, $b = 13.987(7)$, $c = 17.730(7)$ Å, $\beta = 104.13(3)°$. Preparation methods used by Vochten and Cejka are similar but not identical. The cell parameters of synthtetic M^{2+} zippeites are similar to those given by Kovba et al. (1982).

In IR spectra of synthetic phases, ν OH vibrations are located in the region 2834–3606 cm^{-1} (K phase—DRIFT—shows sharp bands near 3600 and 3500 cm^{-1}, Vochten's K phase, kindly supplied by R. Vochten, near 3623 and 3540 cm^{-1} (measured by Cejka), 3625 and 3540 cm^{-1} (reported by Vochten et al. 1995a), and broad bands in the region 3000–3350 cm^{-1} indicating the presence of a complex hydrogen bonding network in the crystal structure, δ H$_2$O near 1600 cm^{-1} and near 1630 cm^{-1}. ν_1 SO$_4^{2-}$ and ν_2 SO$_4^{2-}$ are IR-active and ν_2 SO$_4^{2-}$, ν_3 SO$_4^{2-}$ and ν_4 SO$_4^{2-}$, are split because of symmetry lowering (ν_1 SO$_4^{2-}$ 1000–1010 cm^{-1}, ν_2 SO$_4^{2-}$ 411–470 cm^{-1}, ν_3 SO$_4^{2-}$ 1087–1204 cm^{-1}, ν_4 SO$_4^{2-}$ 580–680 cm^{-1}); some coincidences with δ UOH or H$_2$O libration modes are possible (Fig. 30).

IR spectra of synthetic phases show strong influences on the position of the ν_3 UO$_2^{2+}$ vibration as a function of sample-preparation technique (Table 8). This observation may support the existence of two structural types of natural zippeites, as suggested by Cejka et al. (1985c). According to preliminary results by Cejka and Sejkora (1994), zippeites possess layer structures. Uranyl pentagonal dipyramids share edges to form chains in which hydroxyl groups act as bridging tridentately bonded ligands. Adjacent chains are connected via SO$_4^{2-}$ tetrahedra in uranyl hydroxo sulfate layers. The composition of the layers is probably $[(UO_2)_6(SO_4)_3(OH)_{12-x}]$, where x depends on the cation valency. Cations (M$^+$, M^{2+}), H$_2$O groups and probably some hydroxyl ions occupy interlayer sites. The composition of the interlayer may vary. These conclusions are consistent with crystal structure studies of zippeite-like phases made by Spitsyn et al. (1982), Kovba et al. (1982) and Vochten et al. (1995a).

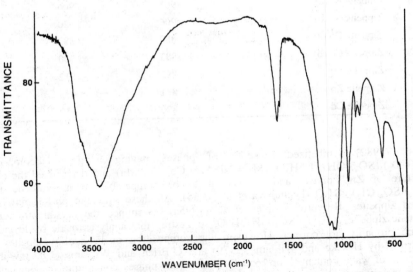

Figure 31. Infrared spectrum of deliensite (KBr disk) (from Vochten et al. 1997c).

Deliensite, Fe^{2+}(UO$_2$)$_2$(SO$_4$)$_2$(OH)$_2$·3H$_2$O (Fig. 32) (Vochten et al. 1997c)

ν_3 UO$_2^{2+}$ 933 cm^{-1}, ν_1 UO$_2^{2+}$ 854–813 cm^{-1}, ν_3 SO$_4^{2-}$ 1120 and 1081 cm^{-1}, ν_4 SO$_4^{2-}$

593 cm^{-1}, ν OH 3409 (b) cm^{-1}, δ H$_2$O 1629 cm^{-1}.

Uranyl carbonates

Interpretating IR spectra of natural and synthetic uranyl carbonates is problematic because of problems with possible coincidences of ν$_2$ CO$_3^{2-}$ and ν$_1$ UO$_2^{2+}$ (if IR-active) and with partial overlapping of ν$_3$ CO$_3^{2-}$ and δ H$_2$O.

Rutherfordine, UO$_2$CO$_3$ (Fig. 32) (Cejka and Urbanec 1988a)

The published wavenumber of ν$_3$ UO$_2^{2+}$ for synthetic uranyl carbonate (Hoekstra 1963) should be 985 cm^{-1} (Hoekstra 1973). Phases prepared in the system UO$_3$-CO$_2$-H$_2$O were characterized by X-ray diffraction and chemical analysis as UO$_2$CO$_3 \cdot x$H$_2$O or as solid solutions UO$_2$(CO$_3$)$_{1-x}$(OH)$_{2x} \cdot n$H$_2$O ($n \geq 0$) (Cejka and Urbanec 1988a). On the basis of IR spectra and thermal analysis, ordered layer (polymeric) structures in hydrothermal phases (ν$_3$ UO$_2^{2+}$ 970–986 cm^{-1}) and "disordered" layer structures in ambient-temperature phases (ν$_3$ UO$_2^{2+}$ 956–965 cm^{-1}) were

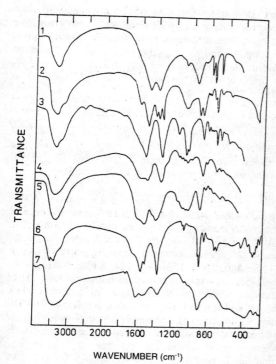

Figure 32. Infrared spectra of some uranyl carbonate minerals: 1. Rutherfordine (KBr micropellet; 2. Sharpite (KBr disk); 3. Schröckingerite (KBr micropellet); 4. Liebigite (KBr disk); 5. Voglite (KBr micropellet); 6. Andersonite (KBr disk); 7. Wyartite (KBr disk) (from Cejka and Urbanec 1990).

inferred. The IR properties of natural uranyl carbonate, rutherfordine (ν$_3$ UO$_2^{2+}$ 985 cm^{-1}) correspond to those of uranyl carbonate phases prepared under hydrothermal conditions (Cejka and Urbanec 1988a). The crystal structure of rutherfordine has been recently refined by Finch et al. (1999).

Tentative assignment of the IR spectrum of rutherfordine (Urbanec and Cejka 1979b; Cejka and Urbanec 1988a 1990) with regard to CO$_3^{2-}$ vibrations calculated by Dik et al. (1989b) is as follows.

ν$_3$ UO$_2^{2+}$ 985 (978, 981, 985) cm^{-1}, ν$_1$ UO$_2^{2+}$ was not observed in the IR spectrum, 889 cm^{-1}—Raman spectrum (Wilkins, 1071b), ν$_2$ UO$_2^{2+}$—not measured in a mineral phase, 255 cm^{-1} (Hoekstra 1973), 257–258 cm^{-1} (Urbanec and Cejka 1979b), 255–260 cm^{-1} (Cejka and Urbanec 1988a). Point symmetry of the CO$_3^{2-}$ ion is close to D$_{3h}$, however, the D$_{3h}$–C$_{2v}$ symmetry lowering was inferred (Cejka and Urbanec 1988a); ν$_1$ CO$_3^{2-}$ 1112 cm^{-1} (calculated 1110 cm^{-1}), ν$_2$ CO$_3^{2-}$ 804 or 806 cm^{-1} (calculated 807 cm^{-1}), ν$_3$ CO$_3^{2-}$ 1415 and 1503, or 1415 and 1510 cm^{-1} (calculated 1433 and 1520 cm^{-1}), ν$_4$ CO$_3^{2-}$ 702 and 781 or 704 and 781 cm^{-1} (calculated 707 and 783 cm^{-1}). Problematic criterion of the splitting values (the wavenumbers of the ν$_3$ CO$_3^{2-}$ vibrations found are 1418–1415 and 1510–1503 cm^{-1}, splitting ~90 cm^{-1}) cannot be succesfully employed to decide whether carbonate groups in rutherfordine structure are bonded monodentately or bidentately. However, both structure reports (Christ et al. 1955; Finch et al. 1999) indicate that CO_3^{2-} groups are both mono- and bi-dentate in natural rutherfordine. Strong broad bands in the region 3000–3600 cm^{-1} and weak band or shoulder near 1600 cm^{-1} may be found in the IR spectrum of some rutherfordine samples. They may indicate that some H-bonded hydroxyl ions or H_2O groups are present in the structure of the rutherfordine sample studied (Urbanec and Cejka 1979b).

Blatonite, $UO_2CO_3 \cdot H_2O$ (Vochten and Deliens 1998)

v_3 UO_2^{2+} 950 cm^{-1}, v_1 CO_3^{2-} 1110 cm^{-1}, v_2 CO_3^{2-} 914 cm^{-1}, v_3 CO_3^{2-} 1367 cm^{-1}, δ H_2O 1630 and 1750 cm^{-1}, v OH 2914, 2935, 3426 (b) cm^{-1}, OH libration 670 cm^{-1}. The antisymmetric stretching vibration v_3 UO_2^{2+} wavenumber (950 cm^{-1}) shows that this mineral may form under ambient conditions in contrast to rutherfordine (v_3 UO_2^{2+} = 985 cm^{-1}).

Sharpite, $Ca(UO_2)_6(CO_3)_5(OH)_4 \cdot 6H_2O$ (Fig. 32) (Cejka et al. 1984b)

v_3 UO_2^{2+} ~914 and ~957 cm^{-1}, v_1 UO_2^{2+} was located at 759–780 cm^{-1} (two medium or weak peaks in this region); however, the coincidence of v_1 UO_2^{2+} and v_2 CO_3^{2-} in the region 811–850 cm^{-1} is more probable, v_2 UO_2^{2+} 252 and 258 cm^{-1} or 254 cm^{-1}, v_1 CO_3^{2-} 1010–1190 cm^{-1}. Some δ UOH vibrations can also be located in this region, v_2 CO_3^{2-} at ~830 and ~848 cm^{-1}, v_3 CO_3^{2-} 1381, 1418, 1445, 1455, 1540 or 1244, 1420, 1449, 1460, 1543 cm^{-1}, v_4 CO_3^{2-} ~690, ~705, ~760, ~775 cm^{-1}; some coincidences with δ UOH or H_2O libration modes are more probable than with v_1 UO_2^{2+}, v OH 3200 (sh), 3425, 3540 (sh) or 3105 (sh), 3250 (sh), 3425, 3540 (sh), or 3445, 3580 (sh), or 2956, 3001, 3437, 3546 cm^{-1}, δ H_2O 1622 or 1640 (b) or 1627 and ~1740 cm^{-1}. The character of the IR spectrum in the v OH region indicates the presence of more structurally nonequivalent (i.e. bonded with different forces) hydroxyl ions or water molecules, and the presence of a hydrogen-bonding network in the structure. Some bands near 1150, 1734–1750, 2500–2650 and 3300 cm^{-1} could correspond to H_3O^+ vibration; however, this assignment is highly speculative. According to the IR spectra assignment, two types of uranium coordination exist, one of them maintains the uranyl nature similar to that in $UO_3 \cdot 2H_2O$ (v_3 UO_2^{2+} = 957 cm^{-1}), whereas in the other (v_3 UO_2^{2+} = 914 cm^{-1}), uranyl oxygen atoms are affected by the hydrogen bonds with H_2O groups and by presence of Ca^{2+} cations and are similar to those in complex hydrated uranyl carbonates. Sharpite contains carbonate groups of two types. If a point symmetry lower than D_{3h} is assumed for CO_3^{2-}, the doubly degenerate modes, the v_4 CO_3^{2-} bending mode, and the v_3 CO_3^{2-} stretching mode are split. From the occurrence of a pair of doublets in the v_4 CO_3^{2-} range and at least four bands in the v_3 CO_3^{2-} range observed in the IR spectrum of sharpite, it may be inferred that sharpite contains two nonequivalent carbonate groups with symmetry lower than D_{3h}.

$Na_4[UO_2(CO_3)_3]$, mineral (Fig. 33) and synthetic phase (Koglin et al. 1979; Ondrus et al. 1997; this study) (the IR spectra prove that these phases are structurally different; see Table 9).

$K_4[UO_2(CO_3)_3]$, synthetic phase (Anderson et al. 1980)

$Rb_4[UO_2(CO_3)_3]$, synthetic phase (Gorbenko-Germanov and Zenkova 1962)

$Cs_4[UO_2(CO_3)_3]$, synthetic phase (Gorbenko-Germanov and Zenkova 1962)

$(NH_4)_4[UO_2(CO_3)_3]$, synthetic phase (Bukalov et al. 1970; Baran 1982)

IR and Raman spectra of sodium, potassium, and ammonium uranyl tricarbonates were published by Novitskiy et al. (1981) without interpretation. For further references concerning IR spectra of anhydrous uranyl tricarbonates see; e.g. Bagnall (1983) and

Cejka: *IR Spetroscopy and Thermal Analysis of Uranyl Minerals*

Volodko et al. (1981) (Table 9).

Table 9. Infrared spectra of anhydrous M^+ uranyl tricarbonates (syn = synthetic)

	$Na\ syn^1$	$Na\ mineral^2$	$K\ syn^3$	$NH_4\ syn^3$	$Rb\ syn^4$	$Cs\ syn^4$
$\nu_3\ UO_2^{2+}$	843	848	881	883	877	877
$\nu_1\ UO_2^{2+}$. (Raman)	809	-	815	831	828[5]	808[5]
$\nu_2\ UO_2^{2+}$	308sh, 316	-	251sh, 258[6] 279, 297sh	231, 276	-	-
$\nu_1\ CO_3^{2-}$	1062	1064	1038–1047	1045	1044	1049
$\nu_2\ CO_3^{2-}$	820–821 843–845 862sh	827, 848	840–850. 855	839	852	-
$\nu_3\ CO_3^{2-}$	1340–1342. 1560	1345,1360sh 1562, 1577	1329-1342 1354-1365 1532-1551 1595-1600	1332 1500-1515	1338 1570	1357 1551
$\nu_4\ CO_3^{2-}$	684-700 732-735	703, 735	714-721	685,715	689 719	691 722

1. Koglin et al. (1979); 2. Ondrus et al. (1997) and this study; 3. Novitskiy et al. (1981);
4. Gorbenko–Germanov and Zenkova (1962); 5. Dik et al. (1989a); 6. Anderson et al. (1980).

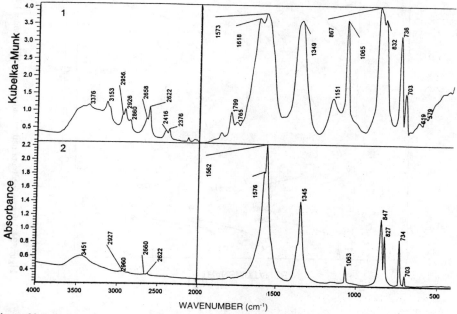

Figure 33. Infrared spectrum of natural $Na_4[UO_2(CO_3)_3]$: 1.DRIFT; 2. KBr disk (from Ondrus et al. 1997).

570 URANIUM: Mineralogy, Geochemistry and the Environment

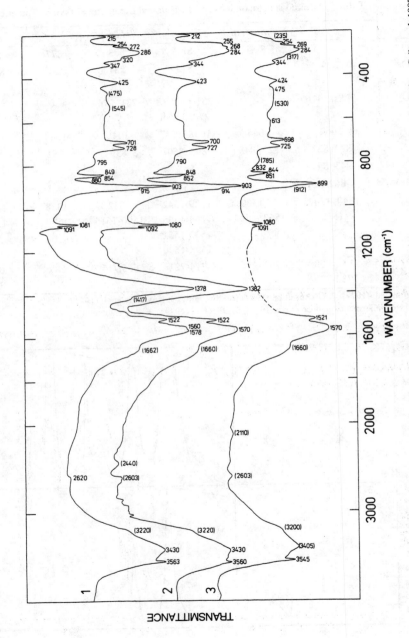

Figure 34. Infrared spectra of natural and synthetic andersonite: 1. mineral (KBr disk); 2. synthetic (KBr disk); synthetic (Nujol) (from Cejka et al. 1987).

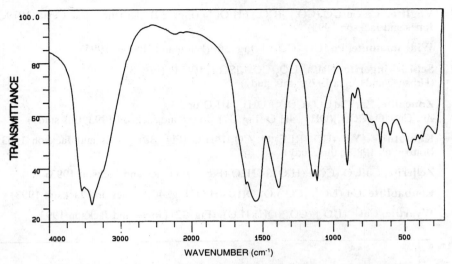

Figure 35. Infrared spectrum of bayleyite (KBr disk) (from Jones and Jackson 1993).

Andersonite, $Na_2Ca[UO_2(CO_3)_3]\cdot xH_2O$ ($x = 6, 5.6, 5.33?$) (Fig. 32, 34) (Cejka et al. 1987)

Andersonite-liebigite intermediate, $Ca_{1.54}Na_{0.63}[UO_2(CO_3)_3]\cdot xH_2O$ ($x = 5.38$) (Vochten et al. 1994)

Bayleyite, $Mg_2[UO_2(CO_3)_3]\cdot 18H_2O$ (Fig. 35) (Jones and Jackson 1993; this study)

Grimselite, $NaK_3[UO_2(CO_3)_3]\cdot H_2O$ (Wang et al. 1981)

Liebigite, $Ca_2[UO_2(CO_3)_3]\cdot 11H_2O$ (Fig. 32) (Urbanec and Cejka 1979c; Jones and Jackson 1993)

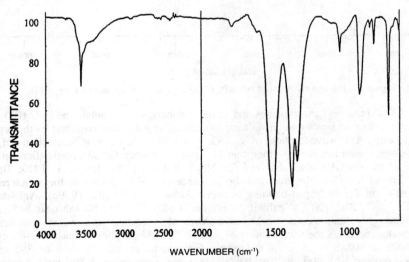

Figure 36. Infrared spectrum of widenmannite (KBr disk) (from Elton and Hooper 1995).

Voglite, $Ca_2Cu[UO_2(CO_3)_3](CO_3) \cdot 6H_2O(?)$ (Fig. 32) (Urbanec and Cejka 1979c; Jones and Jackson 1993)

Widenmannite, $Pb_2[UO_2(CO_3)_3]$ (Fig. 36) (Elton and Hooper 1995)

Schröckingerite, $NaCa_3[UO_2(CO_3)_3]SO_4F \cdot 10H_2O$ (Fig. 32) (Urbanec and Cejka 1979c; this study)

Znucalite, $Zn_{12}Ca[UO_2(CO_3)_3](OH)_{22} \cdot 4H_2O$ or $Zn_{11}Ca[UO_2(CO_3)_3](OH)_{20} \cdot 4H_2O$ (Fig. 37) (Jones and Jackson 1993; this study)

Kamotoite-(Y), $4UO_3 \cdot (REE)_2O_{3.3}CO_2 \cdot 14H_2O$ (Fig. 38) (Jones and Jackson 1993; Botto et al. 1989b; this study)

Zellerite, $Ca[UO_2(CO_3)_2(H_2O)_2] \cdot 3H2O$ (Fig. 39) (Jones and Jackson 1993)

Roubaultite, $Cu_2(UO_2)_3(CO_3)_2O_2(OH)_2 \cdot 4H_2O$ (Fig. 40) (Jones and Jackson 1993)

Wyartite, $CaU^{5+}(UO_2)_2(CO_3)(OH)(H_2O)_7$ (Fig. 32) (Jones and Jackson 1993)

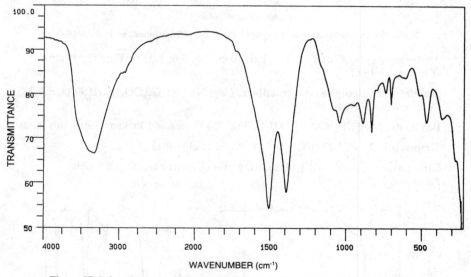

Figure 37. Infrared spectrum of znucalite (KBr disk) (from Jones and Jackson 1993).

v_3 UO_2^{2+} in uranyl tricarbonates and in other minerals, that contain the $[UO_2(CO_3)_3]^{4-}$ complex together with other anions in their structures, appears approximately in the region 926–840 cm^{-1}. The wavenumbers of v_3 UO_2^{2+} may be influenced by interactions of the uranyl oxygen with the cations other than U^{6+} and a stronger ligand coordination on the equator of the uranyl ion as observed for $Na_4[UO_2(CO_3)_3]$ (Koglin et al. 1979; Baran 1982). The v_3 UO_2^{2+} values may also be influenced by H_2O groups in the structure by participating in the hydrogen-bonding network (Urbanec and Cejka 1979c). Anhydrous salts NH_4^+, K^+, Rb^+, and Cs^+ exhibit a vibration at 883-877 cm^{-1}, Na salt near 845 cm^{-1}. Introduction of alkaline-earth metals (and other M^{2+} cations) and water into uranyl tricarbonates increases the wavenumber of v_3 UO_2^{2+}. Liebigite shows it at 893–882 cm^{-1}, andersonite at 905–899 cm^{-1}, an andersonite-liebigite intermediate at 910 or 895 cm^{-1}, schröckingerite at 913–900 cm^{-1}, bayleyite at 903 cm^{-1}, grimselite at 890 cm^{-1}, voglite at

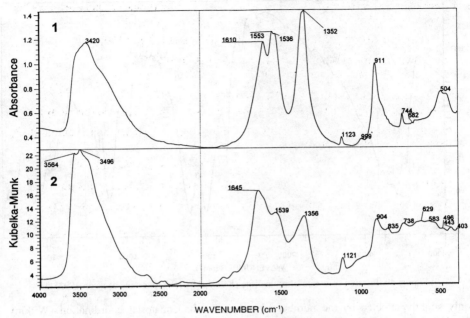

Figure 38. Infrared spectrum of kamotoite-Y: 1. KBr disk; 2. DRIFT.

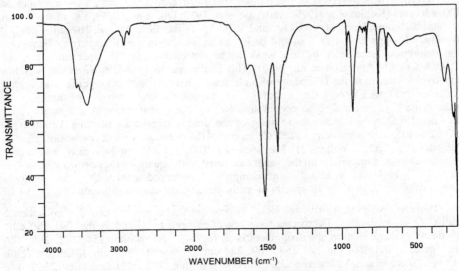

Figure 39. Infrared spectrum of zellerite (KBr disk) (from Jones and Jackson 1993).

904–902 cm^{-1}, widenmannite at 926 cm^{-1}(?), and znucalite 890 cm^{-1}. The uranyl group forms a part of hexagonal dipyramids in uranyl tricarbonates, with the two uranyl oxygen atoms (U-O$_I$—uranyl) at the apexes and six oxygen atoms (U-O$_{II}$, O- equatorial ligand) belonging to bidentate carbonate groups in the equatorial plane. The U-O$_I$ bond lengths are

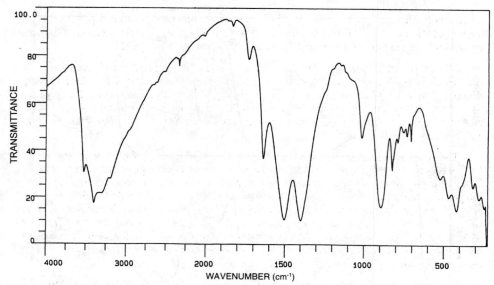

Figure 40. Infrared spectrum of roubaultite (KBr disk) (from Jones and Jackson 1993).

only slightly affected by the number of anions in the equatorial coordination (Wilkins 1971b).

v_1 UO_2^{2+} was observed in the Raman spectra of anhydrous uranyl tricarbonates at 800–830 cm^{-1} (Koglin et al. 1979; Anderson et al. 1980; Dik et al. 1989), i.e. in the region of possible coincidence of v_1 UO_2^{2+} and v_2 CO_3^{2-} in the IR spectrum. Because of $D_{\infty h}$ symmetry lowering, v_1 UO_2^{2+} should be in fact active also in the IR spectrum. However, the absorption band may be too weak to be detected in the IR spectrum, e.g. of $K_4[UO_2(CO_3)_3]$ (Anderson et al. 1980). This is the reason why this vibration was not tentatively assigned in the IR spectra of andersonite, liebigite, schröckingerite, and voglite (Urbanec and Cejka 1979c). Cejka et al. (1987) assume the site symmetry of the UO_2^{2+} in andersonite is C_s, according to crystal structure by X-ray diffraction (Coda et al. 1981; Mereiter 1986), which indicates splitting of the doubly degenerate bending vibration v_2 UO_2^{2+}, and IR activation of v_1 UO_2^{2+}. Cejka et al. (1987) assign a weak absorption band at 795 cm^{-1} to v_1 UO_2^{2+}. Wilkins (1971b) locates v_1 UO_2^{2+} at 836 cm-1 (Raman) which may be more correct, compared with the value calculated with equations expressing v_1 UO_2^{2+} = f(v_3 UO_2^{2+}). However, v_1 UO_2^{2+} seems completely overlapped with v_2 CO_3^{2-} and uniquely correct determination of the IR spectra of many uranyl carbonates is difficult.

Tentative assignment of v_2 (δ) UO_2^{2+} is problematic because of possible coincidence with U-O$_{ligand}$ vibrations. IR spectra have not often been measured in the 180–350 cm^{-1} region. With some caution, the assignment is as follows: andersonite (254, 272, 286, 320 cm^{-1}, or 255, 268, 284 cm^{-1}, or 254, 269, 284 cm^{-1}), bayleyite (285 cm^{-1}), liebigite (286, 316 or 280 cm^{-1}), schröckingerite (252, 286 cm^{-1}), voglite (298 cm^{-1}), znucalite (276 cm^{-1}), kamotoite-(Y) (280 or 270 and 310 cm^{-1}).

Symmetric stretching vibration, v_1 CO_3^{2-}, does not overlap with any UO_2^{2+} vibration. However, it may be overlapped with δ MOH vibrations (if OH$^-$ ions are present, e.g. znucalite) and skeletal vibrations (e.g. v_1 SO_4^{2-} and v_3 SO_4^{2-} stretching vibrations in schröckingerite). In anhydrous uranyl tricarbonates with M$^+$ cations, v_1 CO_3^{2-} lies in the range 1043–1062 cm^{-1}, in hydrated M$^+$, M^{2+} uranyl tricarbonates and related phases it is in the range 989–1151 cm^{-1} (Tables 9 and 10). If structurally nonequivalent carbonate groups are present, more than one band may be observed in the region of v_1 CO$_3^{2-}$.

The out-of-plane bending vibration v_2 CO$_3^{2-}$ lies in the range 820–857 cm^{-1} (Tables 9 and 10). This assignment is difficult, because more bands related to v_2 CO$_3^{2-}$ are possible, if structurally nonequivalent carbonate groups are present, and possible coincidence of v_2 CO$_3^{2-}$ and v_1 UO$_2^{2+}$ (if active in the IR spectrum). The v_1 UO$_2^{2+}$ band is usually weak, while that of v_2 CO$_3^{2-}$ varies from a shoulder (e.g. kamotoite-(Y)) to an intense band (e.g. andersonite).

The doubly degenerate antisymmetric stretching vibration v_3 CO$_3^{2-}$ is in the range 1317–1584 cm^{-1} (Tables 9 and 10). Partial overlap with δ H$_2$O is possible. Splitting of v_3 CO$_3^{2-}$ in most phases studied is greater than that of monodentate carbonato ligands. This observation supports the conclusion that carbonato ligands are bonded bidentately, as inferred from the X-ray structure analysis (e.g. Anderson et al. 1980). More than two v_3 CO$_3^{2-}$ vibrations are split in phases such as the K$^+$ salt, andersonite, liebigite, an andersonite-liebigite intermediate, schröckingerite, znucalite and voglite, which indicates the presence of structurally nonequivalent carbonate groups, a conclusion that is also supported by the doubling of v_1 CO$_3^{2-}$ and v_2 CO$_3^{2-}$ vibrations.

The doubly degenerate in-plane bending vibration, v_4 CO$_3^{2-}$, is in the region 684–750 cm^{-1} (Tables 9 and 10). It is split in Na$^+$, NH$_4^+$, Rb$^+$ and Cs$^+$ salts and andersonite, bayleyite, liebigite, an andersonite-liebigite intermediate, schröckingerite, voglite and znucalite.

Table 10. CO$_3^{2-}$ vibrations in hydrated uranyl tricarbonates and related phases

Mineral	v_1 CO$_3^{2-}$	v_2 CO$_3^{2-}$	v_3 CO$_3^{2-}$	v_4 CO$_3^{2-}$
Andersonite	1080,1092sh	848,851sh	1382,1522,1570b	700,727
	1083,1092sh	847,852	1382,1525,1575	700,728
Liebigite	1074b	849,845	1378,1538,1590	682,737,742
	1070		1379,1515,1549	
Andersonite-liebigite intermediate	1076,1129	(829),846	1364,1524,1563	(673),694,741
Bayleyite	1116,1144	849	1387,1553	668,693,731
Grimselite	1058	850,890	1360,1558,1582	695,720
Schröckingerite	1078,1083sh	820,843	1370,1383sh,1551,1579	706,741
	1082	822,843	1317sh,1377sh,1402b,1553b	708,744
Voglite	1005,1032,1073	833-843	1402-1420,1465sh,1512-1560	685,743
	1030,1053,1073	822,843	1429,1515,1563	675b,743
	1025,1084,1114,1149	838		671,717,746
Widenmannite	1057	830,857	1350,1385,1512	725
Znucalite	992,1042,1104.	832sh	1334sh,1380,1506,1543sh,1584	704,740
	1046,1082		1392,1508	706,742
Kamotoite–(Y)	989sh,1033sh,1121	835	1356b,1534b	738
	999,1123	822sh,833sh,855sh	1357,1367sh,1536,1553	744
	1120	830	1355,1540.	745
	1122,1151	835sh	1364,1537	742

Molecular water is manifested by δ H_2O vibrations at 1594–1705 cm^{-1}, and by stretching vibrations in the range 2850–3564 cm^{-1}. Sharp band at ~3550 cm^{-1} in the IR spectrum of andersonite indicates weakly bonded H_2O groups (related to δ H_2O near 1590 cm^{-1}, which may be overlapped with v_3 CO_3^{2-}) and may correspond to 0.33–0.66 H_2O of the five H_2O per the unit cell (Cejka et al. 1987, 1988b). TG curves of nearly all hydrated uranyl tricarbonates prove that these substances contain two or more types of structurally nonequivalent water molecules. δ H_2O at 1620 cm^{-1} and ν OH at 3457 sh and 3561 cm^{-1} show the likely presence of H_2O in widenmannite, which was originally described as anhydrous (Walenta 1976).

Wavenumbers of δ H_2O vibrations and the character of ν OH vibrations indicate a hydrogen-bonding network in crystal structures of uranyl tricarbonates. Some weak bands observed in the range 433–700 cm^{-1} may be assigned to libration modes of H_2O groups. In znucalite, bands related to δ MOH can be observed. In the case of schröckingerite, vibrational bands near 1095 and 1080 cm^{-1} belong to v_3 SO_4^{2-}, those near 605 and 665 cm^{-1} to v_4 SO_4^{2-} and that near 1015 cm^{-1} to v_1 SO_4^{2-}. Only slight deviations in the series of measured IR spectra of liebigite and schröckingerite reveal the existence of well-defined crystal structure, whereas lower similarity of measured IR spectra for various samples of voglite suggests possible variability in H_2O groups bonding in the structure, in cation kinds or in the number of coordinated carbonate ions in the uranyl equatorial plane (Urbanec and Cejka 1979c; Cejka and Urbanec 1990). Only IR spectra (graphs and peak tables) without any assignment are available for roubaultite, zellerite and wyartite (Jones and Jackson 1993).

Uranyl selenites and tellurites

Haynesite, $(UO_2)_3(OH)_2(SeO_3)_2 \cdot 5H_2O$ (Fig. 41) (Cejka et al. 1999)

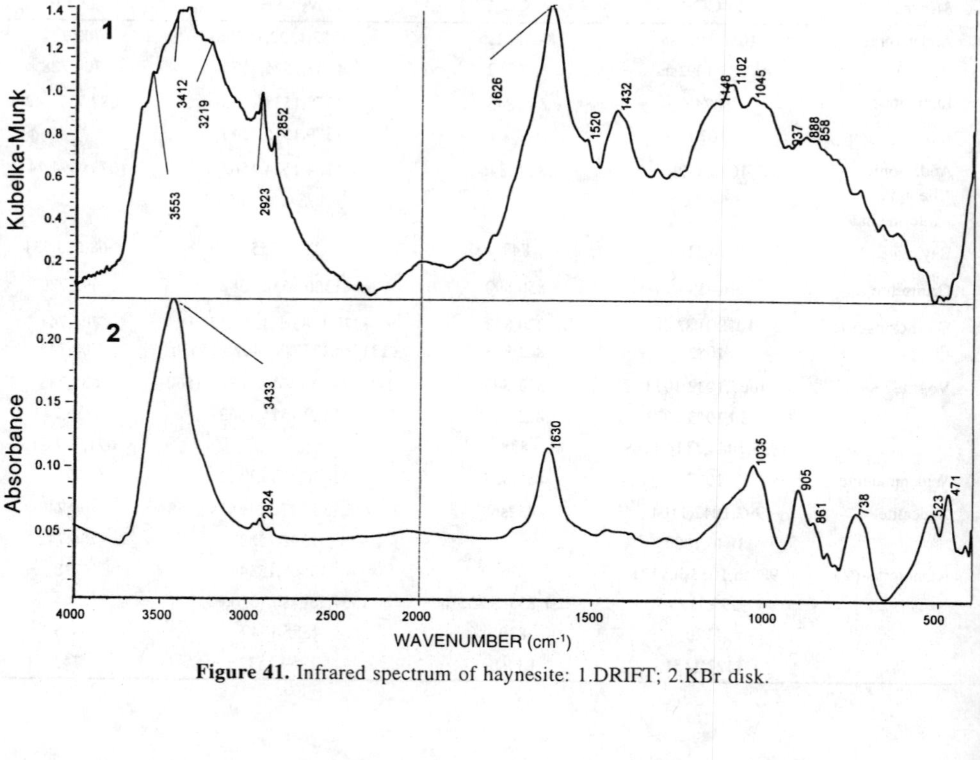

Figure 41. Infrared spectrum of haynesite: 1.DRIFT; 2.KBr disk.

v_1 UO_2^{2+} ~820 and 850 (sh) cm^{-1}, v_3 UO_2^{2+} 905 and 862 cm^{-1}, v_1 SeO_3^{2-} ~820 cm^{-1}, may coincide with v_1 UO_2^{2+} or δ UOH, v_2 SeO_3^{2-} 471 cm^{-1}, v_3 SeO_3^{2-} 738 (b), 800 cm^{-1}, v_4 SeO_3^{2-} not measured, ν OH 3433, 3250 (sh), 2970, 2925, 2850 cm^{-1}, δ UOH 1020 (sh), 1036, 1080 (sh), 1120 (sh), 1150(sh)(?), cm^{-1}, δ H_2O 1630 cm^{-1}; a band at 523 cm^{-1} may be associated with ν (U-O$_{lig}$) vibrations. Weak bands in the region 1200–1500 cm^{-1} may be related to combination bands or overtones (Khandelwal and Verma 1976). Sharp bands observed in DRIFT spectrum of haynesite (3600 (sh), 3554 (sh), 3413, 3350, 3219, 2923, 2853 cm^{-1}), weak bands in the region 1800–2300 cm^{-1}, shoulders in the region 1650–1800 cm^{-1} and complex character of the spectrum in the region 1000–1500 cm^{-1} indicate that several hydroxyl ions are present; however, the presence of H_3O^+ cannot be excluded. Some bands may be related to combination bands or overtones (or both), also the presence of minor CO_3^{2-} in the sample studied cannot be unambiguously excluded (Cejka et al. 1999). The character of the IR spectrum also indicates hydrogen bonds.

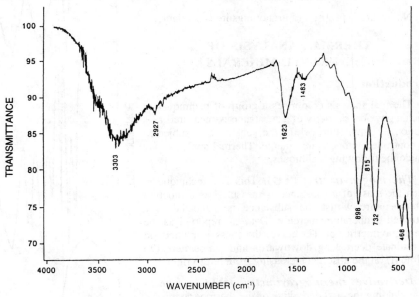

Figure 42. Infrared spectrum of piretite (from Vochten et al. 1996).

Piretite, $Ca(UO_2)_3(SeO_3)_2(OH)_4 \cdot 4H_2O$ (Fig. 42) (Vochten et al. 1996; this study)

The assignment given here is made with regard to that of haynesite (Cejka et al. 1999) and to the interpretation of Vochten et al. (1996).

v_1 UO_2^{2+} 815 cm^{-1} (possible coincidence with v_1 SeO_3^{2-} or δ UOH), v_3 UO_2^{2+} 898 cm^{-1}, v_1 SeO_3^{2-} 815 cm^{-1} (possible coincidence with v_1 UO_2^{2+} or δ UOH), v_2 SeO_3^{2-} 468 cm^{-1}, v_3 SeO_3^{2-} 732 cm^{-1}, v_4 SeO_3^{2-} not observed, ν OH 3303(b), 2927 cm^{-1}, δ H_2O 1623 cm^{-1}, H_2O libration mode or ν (U-O$_{lig}$) 515 cm^{-1}. Weak bands and shoulders in the range 1000–1483 cm^{-1} may be δ UOH or related to combination bands or overtones. The character of the IR spectrum confirms hydrogen bonds.

Schmitterite, UO_2TeO_3 (Fig. 43) (Botto 1984)

U-O bond lengths, which lowers symmetry. The band at 818 cm^{-1} is assigned to v_1 UO_2^{2+} (830 cm^{-1} in Raman spectrum). The bands at 537 and 565 cm^{-1} may be assigned to the secondary U-O stretching vibrations, and the sharp bands at 740 and 655 cm^{-1} correspond to the Te-O stretching vibrations. A band at 400 cm^{-1} may correspond to δ (Te-O$_n$) and four weak bands or shoulders below 400 cm^{-1} may be associated with δ O-Te-O, δ UO_2^{2+} and lattice vibrations.

Cliffordite, UTe_3O_9 (Fig. 43) (Botto 1984)

v_3 UO_2^{2+} 873 cm^{-1}; other assignments are lacking.

Uranyl wolframates

Uranotungstite, $(Fe,Ba,Pb)(UO_2)_2(WO_4)(OH)_4 \cdot 12H_2O$

No infrared spectrum of uranotungstite is available.

THERMAL ANALYSIS OF THE URANYL MINERALS

Introduction

Thermal analysis comprises a group of techniques in which a physical property of a substance is measured as a function of temperature whilst the substance is subjected to a controlled temperature change. Thermal analysis may include the following techniques.

***Thermogravimetry* (TG).** This is a technique in which the mass of a substance is measured as a function of temperature whilst the substance is subjected to a controlled temperature increase. Data are reported as the thermogravimetric, or TG curve; the mass is plotted on the ordinate decreasing downwards and temperature (T) or time (t) on the abscissa increasing from left to right.

***Derivative thermogravimetry* (DTG).** A technique yielding the first derivative of the thermogravimetric curve with respect to either time or temperature. Data are reported as a derivative thermogravimetric, or DTG curve. The derivative is plotted on the ordinate with mass losses downwards and T (temperature)or t (time) on the abscissa, increasing from left to right.

Isobaric mass-change determination. A technique in which the equilibrium mass of a substance at constant partial pressure of the volatile product(s) is measured as a function of temperature. An isobaric mass-change curve is recorded: mass is plotted on the ordinate, decreasing downwards, and T is on the abscissa, increasing from left to right.

***Evolved gas detection* (EGD).** A technique in which the evolution of gas from a substance is measured

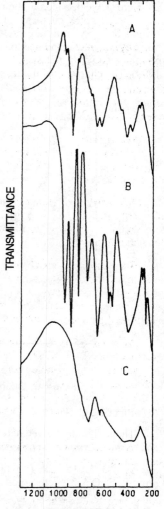

Figure 43. Infrared spectrum of
(A) synthetic cliffordite
(B) schmitterite
(C) U_3O_8
(from Botto 1984).

as a function of temperature.

***Evolved gas analysis* (EGA).** A technique in which the nature or amount of volatile product(s) released by a substance is measured as a function of temperature. The analytical method should always be clearly stated.

Emanation thermal analysis. A technique in which the release of radioactive emanation from a substance is measured as a function of temperature. Introduction of ^{238}Th or ^{224}Ra by coprecipitation provides a source of α-emitting ^{220}Rn. Alternatively, radioactive gases can be directly injected into the samples via ion bombardment (Gallagher 1993).

***Differential thermal analysis* (DTA).** A technique in which the temperature difference between a substance and reference material is measured as a function of temperature whilst both the substance and reference material are subjected to a controlled temperature change. Data are reported as a differential thermal, or DTA curve. The temperature difference (ΔT) is usually plotted on the ordinate with endothermic reactions downwards and temperature or time on the abscissa, increasing from left to right.

***Quantitative differential thermal analysis* (quantitative DTA).** This term covers those uses of DTA where the equipment is designed to produce quantitative results in terms of energy or other physical parameters. Data are plotted in the same manner as for the usual DTA curve.

***Differential scanning calorimetry* (DSC).** A technique in which the difference in energy inputs into a substance and a reference material is measured as a function of temperature whilst the substance and reference material are subjected to a controlled temperature program.

Simultaneous techniques. This term covers the application of two or more techniques to the same sample at the same time- e.g. simultaneous TG and DTA. The names of simultaneous techniques should be separated by *and* or by a hyphen when abbreviated in an acceptable manner; e.g. simultaneous TG-DTA. Unless contrary to established practice, all abbreviations should be written in capital letters without full-stops.

Coupled simultaneous techniques. This term covers the application of two or more techniques to the same sample when all instruments involved are connected through an interface; e.g. simultaneous DTA and mass spectrometry. In coupled simultaneous techniques, as in discontinuous techniques (see below), the first technique to be mentioned is that by which the initial measurement is made; if a DTA instrument and a mass spectrometer are connected through an interface, DTA-MS is the correct form (not MS-DTA).

Discontinuous simultaneous techniques. This term refers to the application to one sample of coupled techniques when sampling for the second technique is discontinuous; e.g. discontinuous simultaneous DTA and gas chromatography, in which discrete portions of evolved volatile(s) are collected from a sample situated in the instrument used for the first technique (DTA).

Some of the principal definitions given here are taken from Lombardi (1980). Only information that is important for the study of uranyl minerals is included. Detailed information concerning individual methods of thermal analysis and relevant terms used can be found in other publications [e.g. Brown (1988), Wendlandt (1986), Wunderlich (1990), Smykatz-Kloss and Warne (1991), and Galagher (1993)].

Theoretical bases of thermal analysis have been given in the book by Seskák (1984). The character of thermal decomposition curves is influenced by many factors, depending on the properties of the compounds being studied and experimental conditions. Progressive standardization and computerization of methods and instruments improve thermal analyses and enhance the reproducibility of results. Sample weights are also reduced, a factor that is critical for studies of very small mineral samples, as is often the case for uranyl minerals.

Methods of thermal analysis, especially DTA, TG, DTG, and DSC are especially useful when combined with X-ray diffraction, gas chromatography, mass spectrometry, and IR spectroscopy. The combination of thermal analysis with high temperature X-ray diffraction techniques (camera or diffractometer) contributes significantly to our knowledge of the mechanisms by which minerals thermally decompose. Infrared spectroscopy can also be used in conjunction with thermal analyses. Analyses of gaseous decomposition products by mass spectrometry or gas chromatography also complements thermal analysis and other analytical methods. Difficulties may occur with very small sample masses. In some cases, it is necessary to correct a recorded TG curve by using a blank run (e.g. Al_2O_3).

TG curves are often used to determine total water content in new minerals because of a common lack of sufficient material to complete detailed chemical or microchemical analyses. The character of TG curves records mass changes, and that of DTG curves reflects the rates at which such changes occur. DTA curves record endothermic reactions observed during heating that may characterize several properties, or changes, in a material, including dehydration, dehydroxylation, structural decomposition, structural transformations, sintering and melting. DTA can also be used to study evaporation or sublimation, and exothermic reactions observed during heating may represent, for example, oxidation (burning of organic inclusions or sulfur, oxidation of Fe^{2+} to Fe^{3+}), and recrystallization of amorphous material that was formed during thermal decomposition (Smykatz-Kloss 1982).

Required sample masses may vary from a few mg to hundreds of mg. The heating rate may vary in the range $0.1-200°C \cdot min^{-1}$. The furnace atmosphere may be static, self-generated, or dynamic-air, inert gases, and vacuum, among others. A typical heating rate is 5 or $10°C \cdot min^{-1}$; air static or dynamic atmosphere ($ml-air \cdot min^{-1}$)- and in some cases nitrogen or argon atmospheres are used (e.g. Vochten and Deliens 1998). Al_2O_3 is commonly used as a reference material (Wendlandt 1986).

With regard to the compositions, structures, properties, conditions of origin and existence of uranyl minerals in nature, these fundamental issues and the following problems can be addressed with thermal analysis in combination with additional methods, especially X-ray diffraction (Cejka and Urbanec 1977, 1990):

(a) Determine dehydration conditions; temperature intervals within which H_2O of various kinds is lost; separate true dehydration (molecular water) from dehydroxylation (so-called "constitution" water; i.e. hydroxyl ions); and define the thermal stability for each dehydration step. Differences among dehydration rates associated with various forms of H_2O in a structure may be of particular interest. Generally, the first water to be lost as temperature is increased are H_2O groups from interlayer or interstitial sites, although if any adsorbed water is present, it is released first. Interlayer H_2O may escape in several steps, as some H_2O groups may be co-ordinated by cations and therefore more strongly bonded. The release of interlayer H_2O is followed by the release of "water" strongly bonded as part of the structural units; e.g. as hydroxyl ions in polyhedral sheets. Földvari et al. (1988) discuss the possibility of using thermal analyses to examine different types of bonding of water in minerals.

(b) Define temperature intervals for the existence of anhydrous phases and the nature of their decomposition. The last dehydration step may be overlapped by the onset of decomposition of a dehydrated phase. Dehydration or decomposition processes in anhydrous phases are commonly associated with the formation of X-ray amorphous phases.

(c) Characterize solid and gaseous decomposition products and reactions between solid decomposition products. According to Sidorenko et al. (1986), various degrees of ordering of a mineral structure and of the H_2O groups in the structure can be inferred from the details of dehydration curves (the release of interlayer water may be stepwise or continuous; and both phenomena were observed while studying various specimens of the same mineral, e.g. synthetic hydrogen meta-autunite).

The course of thermal analysis of uranyl minerals can be characterized by following processes and reactions: dehydration; dehydroxylation (if OH^- ions are present in the mineral structure); decomposition (destruction of the structure) of an anhydrous phase; crystallization of new phases and solid-state reactions between solid decomposition products; oxidation of U^{4+} or Fe^{2+} (if present) and others. These processes may partly overlap one another. Dehydration and dehydroxylation or decomposition of anhydrous phases can be (and mostly are) associated with formation of X-ray amorphous phases.

End products of thermal decomposition of uranyl minerals are commonly $UO_{2.67}$ (especially among minerals without cations besides U and for which anions are lost; e.g. as CO_2 or SO_3 gas), uranates (especially for minerals containing additional cations, such as alkali, alkaline-earth and transition metals, or uranyl phosphates and others, and for uranyl minerals containing anions that do not decompose into gaseous phases). Transition metal uranates (Cu, Co, etc.) commonly decompose, forming $UO_{2.67}$ and the relevant metal oxides or mixtures of $UO_{2.67}$ and metal mono-uranates. This is observed when polyuranates, formed during thermal decomposition processes, decompose further to mono-uranates (Cejka 1987, 1988, 1992; Cejka and Urbanec 1990). Thermal analyses of uranium minerals have been published by Ambartsumyan et al. (1961) and by Cejka and Urbanec (1977, 1990).

Sidorenko et al. (1986) compare DTA curves of various uranyl-mineral groups with respect to ordered and disordered structures, and conclude that endotherms and exotherms lie in approximately the same temperature interval for practically all uranyl-mineral groups (if such thermal effects are produced) (Table 11). According to Sidorenko et al. (1986), the first endotherm is always associated with release of H_2O groups from interlayer sites (in the case of layer compounds) and proceeds approximately in the region 50-200°C. Two types of H_2O release are observed, stepwise and continuous, depending on how a structure or its constituent H_2O groups are ordered. The second endotherm (350-500°C) is usually associated with complete dehydration (and dehydroxylation) and amorphization. Formation of new crystalline phases proceeds approximately in the range 500-800°C, and is characterized by the first exotherm. Solid-state reactions, sintering, melting and other processes may be associated with a third endotherm (800-1000°C).

Unless intermediate and end products (including gaseous decomposition products, if possible) have been identified by independent methods (e.g. X-ray powder diffraction, mass spectrometry, etc.), interpretations of TG and DTA curves, based on mass changes and endothermic or exothermic effects only, must be considered tentative. Unfortunately, intermediate and end products are often unknown, and assignments for TG and DTA curves in the literature are commonly based on incomplete information.

Table 11. DTA characteristics of uranyl minerals (°C)

Mineral group	1st endotherm	2nd endotherm	1st exotherm	3rd endotherm
oxide	95, 135, 170 $^{+)}$	350-400	–	700, 900-1060
hydrates	100-200 $^{x)}$			
silicates	100, 180, 200 $^{+)}$	≈500	650-750	900-1000
phosphates				
autunites	90, 140, 200, 250$^{+)}$	specific	750-800	1000
	100-200 $^{x)}$	specific	750-800	1000
others	100-200			
arsenates				
autunites	75, 115-145 $^{+)}$	300-380	≈ 600	900-1000
	100-200 $^{x)}$	300-380	≈ 600	900-1000
others	200-230	300-380	≈ 600	900-1000
vanadates	70, 90 150 $^{+)}$	350-500	500-600	700-1000
	100-200 $^{x)}$	350-500	500-600	700-1000
sulfates	50, 75, 150, 200 $^{+)}$	400-500	575-600	
	100-200 $^{x)}$	400-500	575-600	
carbonates	50, 75, 150-190 $^{+)}$	350	specific	800-900
	100-200 $^{x)}$	350	specific	800-900
molybdates	100-200	specific	500-600	800-900

Notes :(+) ordered structures, (x) disordered structures.

The aim of this contribution is to review the most important information on the behaviors of uranyl minerals during thermal decomposition. However, only a survey of mostly recent references is provided, although older data are included if recent data are unavailable. TG and DTA curves are presented only as illustrations of processes that occur during thermal decomposition of uranyl minerals. Problems related to chemical kinetics of thermal decomposition processes and their applications have been not included, as these issues do not seem as important as examining more fundamental problems encountered during thermal analysis of uranyl minerals.

Thermal analysis

Uranyl oxide hydrates, alkali, alkaline earth and other hydrates

Thermal analysis of this uranyl-mineral group was reviewed recently by Cejka (1993b). Uranyl oxide hydrates loose H_2O in one or more steps up to approximately 200°C, at which temperature corresponding endotherms are observed. Dehydroxylation up to 350-400°C, associated with formation of amorphous phases, is also commonly characterized by endotherms. Exotherms indicate formation of new crystalline phases (e.g. α-UO_3, $UO_{2.89}$) or U^{4+} oxidation (e.g. ianthinite). Formation of the end product $UO_{2.67}$ may also be associated with an endotherm.

Ianthinite, $[U_2^{4+}(UO_2)_4O_6(OH)_4(H_2O)_4](H_2O)_5$ (Guillemin and Protas 1959)

DTA: endotherms at 110°C (dehydration), 630°C ($UO_3 \rightarrow UO_{2.67}$), exotherm at 320°C ($U^{4+}$ oxidation) (Guillemin and Protas 1959). Some data are available concerning thermal analysis of synthetic phases which may be analogues of ianthinite (Bignand 1955; Cordfunke et al. 1968). A new crystal chemical formula for ianthinite was inferred from a single-crystal structure study (Burns et al. 1997b).

Schoepite, $[(UO_2)_8O_2(OH)_{12}](H_2O)_{12}$ (Protas 1959)

TG [probably metaschoepite (Protas 1959)]: molecular water: 1.25 H_2O (60-125°C); hydroxyl ions: 0.75 H_2O (135-450°C). Protas assigned to his "schoepite specimen" the formula $U_4O_9(OH)_6 \cdot 5H_2O$ which is comparable with the formula reported recently for schoepite (Finch et al. 1996a), except that Protas' specimen had a lower H_2O content. DTA curve of "metaschoepite" corresponds to that of synthetic $UO_3 \cdot 2H_2O$ (endotherms at 130 and 300°C—dehydration and dehydroxylation), at 630 and 700°C (transition of α-UO_3 to $UO_{2.67}$), exotherm at 480°C (α-UO_3 crystalization from amorphous UO_3). Thermal analysis of $UO_3 \cdot 2H_2O$ depends strongly on the chemical history of the substance (see Hoekstra and Siegel 1973). Problems of spontaneous alteration of schoepite into metaschoepite and dehydrated schoepite were discussed recently by Finch et al. (1992, 1996b 1998).

Studtite, $UO_4 \cdot 4H_2O$ (Cejka et al. 1996a)

Metastudtite, $UO_4 \cdot 2H_2O$

TG (Cejka et al. 1996a, and references therein especially concerning synthetic phases): loss of two H_2O up to 145°C, two H_2O (145-300°C). It is not known whether water and oxygen are lost simultaneously during the second dehydration step, with formation of amorphous UO_x (x = 3.0-3.5). The thermal decomposition of studtite can be represented schematically as follows: $UO_4 \cdot 4H_2O \rightarrow UO_4 \cdot 2H_2O \rightarrow UO_x \cdot nH_2O \rightarrow UO_3$ (amorphous) → α-UO_3 (or $UO_{2.89}$) → $UO_{2.67}$ with $3.0 \leq x \leq 3.5$, $0 \leq n \leq 0.5$. TG and DTA curves of synthetic uranyl peroxide tetrahydrate and dihydrate are available (e.g. Rocchiccioli 1966; Sato 1976).

Layer structures are also characteristic of most hydrated uranyl oxides that contain cations in addition to U. Molecular water escapes from interlayer sites in several steps: First, it is released as H_2O not bonded to cations, then as H_2O co-ordinated by cations; both are characterized by endotherms. Dehydration proceeds up to approximately 250°C. Further temperature increases cause destruction of the polyhedral layers that contain uranyl anions, as "hydroxyl" water is lost up to approximately 500°C. This dehydroxylation is characterized by corresponding endotherms. X-ray amorphous phases, formed during dehydroxylation, transform to crystalline phases at higher temperatures (an exotherm—usually very weak—may be observed on the DTA curve). Polyuranates of relevant cations, or mixtures of uranates and $UO_{2.67}$ are end products of decomposition. In the case of curite, PbO is volatilized above 460°C. No information concerning PbO loss during heating of other Pb-bearing uranyl oxide hydrates has been reported.

Agrinierite, $(K_2,Ca,Sr)U_3O_{10} \cdot 4H_2O$ (Cesbron et al. 1972)

DTA: endotherms at 163°C (dehydration), and 1060°C (oxygen release associated with formation of nonstoichiometric uranate phase). TG curve was not reported (dehydration and dehydroxylation proceed up to 500°C) (Cesbron et al. 1972).

Bauranoite, $BaU_2O_7 \cdot (4-5)H_2O$ (Rogova et al. 1973)

DTA: endotherms at 180°C (dehydration) and 790°C, exotherm at 850 °C. TG: mass loss observed up to 700°C is 7.1 wt % and at 800°C 8.0 wt % (Rogova et al. 1972). However, if the composition of bauranoite is correct, the tetrahydrate should contain 9.03 wt % and the pentahydrate 11.05 wt % H_2O.

Becquerelite, $Ca[(UO_2)_6O_4(OH)_6] \cdot 8H_2O$ (natural) (Fig. 44) (Cejka and Urbanec 1983)

DTA: endotherm at 135°C (dehydration). TG: 6 H$_2$O (50-130°C), 1 H$_2$O (130-255°C), 1H$_2$O (255-300°C)—dehydration, 3 H$_2$O (300-470°C)—dehydroxylation partly overlapped with the last dehydration step. A phase of approximate composition CaO·6UO$_{2.92}$ (structurally related to UO$_{2.89}$) is formed in the range 465-600°C. In the range 620-720°C a mixture of CaU$_5$O$_{15.4}$ and UO$_{2.67}$ is produced (Cejka and Urbanec 1983; Cejka et al. 1988e).

Becquerelite (synthetic: 12 H$_2$O or 9 H$_2$O and 6 OH$^-$) (Cejka et al. 1998a)

DTA: endotherms at 226°C (dehydration), 430°C (dehydroxylation), small exotherms at 392 and 600°C indicate the formation of α-UO$_3$ (UO$_{2.89}$), UO$_{2.67}$ and CaUO$_4$. TG: 7 H$_2$O (25-150°C), 2 H$_2$O (150-328°C), 3 H$_2$O = 6 OH$^-$ (328-560°C). Dehydration and dehydroxylation processes may partly overlap with oxygen release up to 710°C. The mechanism of thermal decomposition of the synthetic phase is the same as that of the mineral (Cejka et al. 1998a, and references therein).

Protasite, Ba[(UO$_2$)$_3$O$_3$(OH)$_2$]·3H$_2$O (synthetic: 5 H$_2$O or 4 H$_2$O and 2 OH$^-$) (Cejka et al. 1998a)

DTA: endotherms at 145 and 220°C (dehydration and dehydroxylation), exotherm at 409°C (crystallization of α-UO$_3$). TG: 1 H$_2$O (25-100°C), 3 H$_2$O (100-190°C) (dehydration), 1 H$_2$O= 2 OH$^-$ (190-342°C) (dehydroxylation). Further decomposition associated with oxygen release was observed in two steps up to 864°C and is characterized first by formation of UO$_{2.89}$, UO$_{2.67}$ and uranates in the system BaO-UO$_3$ (first BaUO$_4$, at higher temperatures BaU$_2$O$_7$ and Ba$_2$U$_3$O$_{11}$). A mixture of Ba$_2$U$_3$O$_{11}$ and UO$_{2.67}$ is the end product of thermal decomposition of protasite. Further endotherms at 816, 897 and 987°C partly overlapped by melting or sintering of the oxides can be deduced (Cejka et al. 1998a, and references therein).

Billietite, Ba[(UO$_2$)$_6$O$_4$(OH)$_6$]·8H$_2$O (synthetic: 11 H$_2$O or 8 H$_2$O and 6 OH$^-$) (Cejka et al. 1998a)

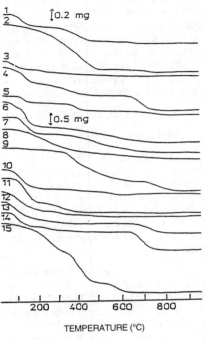

Figure 44. TG curves of some uranyl minerals (Stanton Redcroft Thermobalance TG 750, heating rate 10°C.min^{-1}, dynamic air atmosphere 10 ml min^{-1}): 1. Becquerelite (5.665 mg); 2. Curite (18.415 mg); 3. Cuprosklodowskite (1.731 mg); 4. Jáchymovite (3.829 mg); 5. Andersonite (1.923 mg); 6. Uranophane (15.467 mg); 7. Sklodowskite (13.878 mg); 8. Kasolite (26.522 mg); 9. Soddyite (26.115 mg); 10. Meta-autunite (10.01 mg); 11. Meta-torbernite (9.84 mg); 12. Sabugalite (7.335 mg); 13. Uranopilite (6.945 mg); 14. Johannite (4.975 mg); Sharpite (1.103 mg, mass loss 0.182 mg) (from Cejka and Urbanec 1990).

The report by Pagoaga (1983) and Pagoaga et al. (1987), that billietite contains 4 H$_2$O and 6 OH$^-$, has not been corroborated either for the mineral or its synthetic analogue (Cejka et al. 1998a).

DTA: endotherms at 166 and 237°C (dehydration and dehydroxylation), two exotherms at 398 and 630°C can be assigned to the crystallization of α-UO$_3$ followed by its transition into $UO_{2.89}$ and $UO_{2.67}$ and the formation of $BaUO_4$. Reactions in the $BaO-UO_3$ system are characterized by endotherms at 805, 894 and 985°C. TG: 3 H_2O (25-100°C), 5 H_2O (100-205°C), 3 H_2O = 6 OH⁻ (205-500°C). Oxygen is released in two stages up to 670 and 920°C, respectively, and decomposition continues at higher temperatures. As in the case of protasite, first $UO_{2.89}$ and than $UO_{2.67}$ and $BaUO_4$ are formed, followed by the formation of BaU_2O_7 and $Ba_2U_3O_{11}$. A mixture of $Ba_2U_3O_{11}$ and $UO_{2.67}$ was observed as end products of the thermal decomposition of billietite (Cejka et al. 1998a, and references therein).

Clarkeite, $Na[(UO_2)O(OH)](H_2O)_{0-1}$ (Sidorenko et al. 1985)

According to Sidorenko (1978), an inflection at 100°C and an endotherm at 180°C on the DTA curve for clarkeite and a broad endotherm at 140°C for its synthetic analogue ($Na_2U_2O_7 \cdot xH_2O$, $x \approx 2$) are related to dehydration and dehydroxylation. Crystal structure of clarkeite does not change during heating, and therefore, H2O groups and hydroxyl ions do not play a critical role in the structure of clarkeite (Sidorenko et al. 1985). The composition and structure of clarkeite was published recently by Finch and Ewing (1997).

Compreignacite, $K_2[(UO_2)_6O_4(OH)_6] \cdot 8H_2O$ (Protas 1964)

TG: 8 H_2O (60-200°C) (dehydration), 3 H_2O = 6 OH⁻ (200-480°C) (dehydroxylation)

(Protas 1964). According to Burns (1998b), compreignacite contains only seven H_2O groups and six hydroxyl ions in its structure. Potassium polyuranate is probably formed at higher temperatures.

Curite, $Pb_{8-x}(OH)_{4-2x}(H_2O)_{2x}[(UO_2)_8O_8(OH)_6]_2$ (x = 1.44-2) or $U_8O_x(OH)_{30-x}Pb_3(OH)_{24-x} \cdot (x-21)H_2O$ ($24 \geq x > 21$) (Cejka et al. 1998b)

TG: curite decomposes continuously in the range 20-520°C within which dehydration and dehydroxylation overlap each other. Mass lost up to 520°C corresponds to 3.43 wt %. It is not possible to distinguish dehydration and dehydroxylation. These processes may also partly overlap with PbO release (Taylor et al. 1981; Cejka et al. 1985a). Dehydration and dehydroxylation and PbO loss are associated first with the formation of an amorphous phase, followed by crystallization of phases such as $PbUO_4$ and $Pb_3U_{11}O_{36}$. $Pb_3U_{11}O_{36}$ is the unambiguously defined end product of curite thermal decomposition at 900°C (Cejka et al. 1998b). Water content in curite samples may vary according to locality (Cejka et al 1985a; Cejka et al. 1998b, and references therein).

Fourmariérite, $Pb(UO_2)_4O_3(OH)_4 \cdot 4H_2O$ synthetic phase (Protas 1959):

DTA: endotherms below 100 and below 300°C (dehydration and dehydroxylation, respectively) and 870°C (melting of an anhydrous phase?). TG: dehydration (4 H_2O) and dehydroxylation (2 H_2O = 4 OH⁻) up to 480°C. These data are in excellent agreement with the recent chemical formula of fourmariérite reported by Piret (1985). No PbO loss during heating of synthetic fourmarierite is reported.

Masuyites, $4PbO \cdot 8UO_3 \cdot 10H_2O$ (type masuyites) and $PbO \cdot 3UO_3 \cdot 4H_2O$ (grooved masuyites) synthetic phases

"Curite hydratée" $3PbO \cdot 8UO_3 \cdot 10H_2O$ (Protas 1959)

TG: X-ray powder patterns of the synthetic compounds are similar to those of natural masuyites. Dehydration and dehydroxylation proceed continuously up to 460°C. From the perspective of compositional and XRD data, synthetic masuyite-like phases corresponding to "grooved" masuyites as described by Deliens and Piret (1996) are characterized on the

DTA curve by a broad weak endotherm in the range 20-80°C (adsorbed water), an endotherm at 150°C (dehydration), a broad endotherm at 250-280°C (dehydroxylation) and 950°C (melting). The TG curve indicates release of adsorbed water up to 90°C, 2.7 H_2O (180°C), 1.3 H_2O (440°C), 0.3 H_2O (500°C) (Cejka 1994). The different types of masuyites were described recently by Deliens and Piret (1996). The structure of a masuyite crystal, $Pb[(UO_2)_3O_3(OH)_2](H_2O)_3$, has recently been reported by Burns and Hanchar (1999).

Rameauite, $(K_2,Ca)U_6O_{20}\cdot 9H_2O$ (Cesbron et al. 1972)

DTA: endotherms at 95, 130, and 170°C (dehydration) and 1060°C (oxygen release and $UO_{2.67}$ formation). TG: the main part of water is released up to 130°C, the remaining water up to 500°C (Cesbron et al. 1972).

Uranosphärite, $Bi_2U_2O_9\cdot 3H_2O$ (Ambartsumyan et al. 1961)

Amorphous uranosphärite(?) studied by Ambartsumyan et al. (1961) is characterized on the DTA curve by endotherms at 160°C (an inflection), 220°C (dehydration), 480°C (decomposition of anhydrous phase), 800°C (sintering of the substance which melts at 1000°C without decomposition). TG: 2 H_2O (50-100°C), the remaining water is released in the range 150-300°C. The synthetic phase (Protas 1959) dehydrates between 390 and 530°C (TG curve). Protas (1959) proposed that the structure of uranosphärite contains hydroxyl ions instead of molecular water.

Vandenbrandeite, $CuUO_2(OH)_4$ (Cejka 1994a)

Mineral: DTA: endotherms at 370-395°C (dehydroxylation), 950-960 and 955-980°C (decomposition or phase transition of $CuUO_4$), three endotherms in the range 990-1025°C (decomposition, sintering, or melting?), exotherm at 545-555°C (crystallization of new phases). TG: dehydroxylation (250-380°C- mass loss 9.94 wt % and 9.49 wt %), partial decomposition of $CuUO_4$ (up to 960°C—10.43 wt % and 10.10 wt %).

Synthetic vandenbrandite: DTA shows endotherms at 370-395°C (dehydroxylation), an endotherm at 1070°C (decomposition, sintering, or melting?), and an exotherm at 575°C, (crystallization of new phases—CuU_3O_{10}, $CuUO_4$). TG shows dehydroxylation (250-380°C, corresponding to a mass loss of 10.61 wt %), and partial decomposition of $CuUO_4$ (up to 960°C, with mass loss of 11.52 wt %). CuU_3O_{10} and $CuUO_4$ crystallize during heating (430-500°C) of the dehydroxylated, amorphous compound, which then transform into $CuUO_4$ (600-800°C), followed by partial decomposition and formation of $UO_{2.67}$ (Cejka 1994a).

Vandendriesscheite, $Pb_{1.57}[(UO_2)_{10}O_6(OH)_{11}](H_2O)_{11}$ (Protas 1959; Cejka 1994)

TG: vandendriesscheite dehydrates in three or four steps up to 440°C (mass loss 9.3 wt %) (Protas 1959). There are two synthetic phases structurally related to vandendriesscheite, with proposed formulae, $Pb[(UO_2)_6O_5(OH)_4]\cdot 8H_2O$ and $Pb[(UO_2)_{10}O_7(OH)_8]\cdot 14H_2O$ (Cejka 1994). The first synthetic analogue dehydrates up to 420°C in four steps: adsorbed water (80°C), 1 H_2O (100°C), 6 H_2O (200°C), 3 H_2O (420°C). The second phase also dehydrates in four steps, adsorbed water (60°C), 1 H_2O (100°C), 10 H_2O (180°C), 7 H_2O (470°C). (Cejka 1994). DTA curves of both synthetic samples are characterized by broad endotherms in the range 20-80°C (adsorbed water), 100, 150, 270-280, and 325-330°C (dehydration and dehydroxylation), and 915-925°C (melting and decomposition?). Potential PbO loss during heating was not reported.

Wölsendorfite, $(Pb,Ca)U_2O_7\cdot 2H_2O$, $PbU_2O_7\cdot 2H_2O$ (Protas 1959; Cejka 1994).

DTA: endotherms at 150 and 240°C. TG: dehydration up to 250°C (Potdevin and Brasseur 1958). Synthetic $PbU_2O_7\cdot 2H_2O$ dehydrates continuously up to 300°C (TG). No endotherms were observed on the DTA curve (Protas 1959). DTA curve of wölsendorfite, studied by Belova and Fedorov (1974), indicates two endotherms—130°C (dehydration), and 790°C (decomposition of anhydrous phase?). Synthetic wölsendorfite-like, Pb-only phases dehydrate up to 400°C; TG shows adsorbed water (90°C), 1 H_2O (200°C), 1 H_2O (410°C). The DTA curve indicates very weak endotherms, 20-60°C (broad; adsorbed water), 150°C (dehydration), and 950°C (melting, decomposition?) (Cejka 1994). The structure of a barian wölsendorfite, $Pb_{6.16}Ba_{0.36}[(UO_2)_{14}O_{19}(OH)_4](H_2O)_{12}$, was recently described by Burns (1999b 1999c).

Uranyl silicates

Thermal decomposition of uranyl silicates has been recently reviewed by Cejka (1994b).

Uranophane, $Ca(UO_2)_2(SiO_3OH)_2\cdot 5H_2O$ (Fig. 44) (Urbanec et al. 1985a)

DTA: two endotherms at 160°C (weak) and 215°C (large) are associated with dehydration (Cejka et al. 1986a; Urbanec et al. 1985a). Ambartsumyan et al. (1961) observed three endotherms: 100-120°C, 160-190°C (dehydration), 450-600°C (dehydroxylation), Gevorkyan et al. (1979) at 100 and 180°C (dehydration), and 500°C (dehydroxylation). A weak exotherm at 745°C may reflect formation of crystalline phases (CaU_2O_7 or $UO_{2.67}$) (Ambartsumyan et al. 1961). TG: 1 H_2O ($\rightarrow$ 95°C), 2 H_2O (95-155°C), 1 H_2O (155-480°C), 1.5 H_2O (480-820°C). Amorphous SiO_2 and CaU_2O_7 or $UO_{2.67}$ are probably end products of thermal decomposition of uranophane. Some oxygen may also be released from CaU_2O_7.

β-Uranophane, $Ca(UO_2)_2(SiO_3OH)_2\cdot 5H_2O$ (this study)

DTA: two endotherms at 150-180°C (strong) and 190-200°C (weak). Further decomposition proceeds as for that of uranophane (Ambartsumyan et al. 1961). Galliski and De Upton (1986) observed two endotherms (104 and 600-800°C). The second endotherm may be associated with dehydroxylation. Camargo and DeSousa (1975) reported two endotherms at 165 and 750°C. Our new measurement (TG) shows that β-uranophane loses 3.5 H_2O at 220°C, 1.5 H_2O at 600°C, and approximately 1 H_2O at 950°C(?). The last dehydroxylation stage may partly coincide with some oxygen release from CaU_2O_7 or $UO_{2.67}$ formation.

It is evident that dehydration of both uranophane and β-uranophane proceeds in two steps, which indicates structurally nonequivalent H_2O groups (i.e. bonded with different forces). Dehydroxylation proceeds in one stage above 400°C, and may be related to dehydroxylation of SiO_3OH^{3-}. Any H_3O^+ would be expected to be released at lower temperatures and, if present, may coincide with dehydration.

Cuprosklodowskite, $Cu(UO_2)_2(SiO_3OH)_2\cdot 6H_2O$ (Fig. 44) (Urbanec et al. 1985a)

DTA shows endotherms at 140, 160 (an inflection), 220°C (dehydration), 640°C (dehydroxylation and structural decomposition), and 960°C (melting); an exotherm is observed at 650°C (crystallization of new phases) (Ambartsumyan et al. 1961). In nitrogen atmosphere, cuprosklodowskite loses 4 H_2O (75-150°C), 2 H_2O (200-250°C), and 1 H_2O (300-650°C). No further mass loss was observed up to 800°C (Rosenzweig and Ryan 1975). Heating in air resulted in loses of 4 H_2O up to 145°C, 1 H_2O (145-228°C), and 2 H_2O plus 0.5 O_2 up to 844°C (Urbanec et al. 1985a; Cejka et al. 1986a). Dehydration and dehydroxylation processes may partly overlap. Copper uranates (or $UO_{2.67}$ and amorphous

CuO) and amorphous SiO_2 are end products of the thermal decomposition of cuprosklodowskite.

Sklodowskite, $Mg(UO_2)_2(SiO_3OH)_2 \cdot 6H_2O$ (Fig. 44, 45) (this study)

DTA: endotherms at 218, 350°C (dehydration), 650°C (dehydroxylation and structure destruction), 940°C(?). TG: 2.5 H_2O ($\rightarrow$ 227°C), 2.5 H_2O ($\rightarrow$ 644°C), 0.5 H_2O (824°C?). The sample studied contained only approximately 5.5 H_2O (Cejka et al. 1986a). Recently, our TG curve for another sklodowskite sample shows the loss of 4 H_2O up to 225°C, one H_2O up to 425°C and two H_2O up to 900°C. The first two stages correspond to dehydration and the third one to dehydroxylation (may be partly overlapped by dehydration) associated with formation of Mg uranates and amorphous SiO_2.

As for uranophane and β-uranophane, cuprosklodowskite and sklodowskite contain structurally non-equivalent water molecules. According to our experience, the water content in sklodowskite can vary. Dehydroxylation proceeds at relatively high temperature, in agreement with the existence of SiO_3OH^{3-} in the structures of cupro-sklodowskite and sklodowskite. Whether oxygen is released during dehydroxyl-ation is not known.

Boltwoodite, $K(UO_2)(SiO_3OH) \cdot H_2O$ (this study)

DTA: endotherms at 170°C (dehydration), 708°C (dehydroxylation?) (Honea 1961; Pu Congjian 1990). TG (our recent measurement): 1 H_2O (340°C), 0.5 H_2O(?) (900°C). Initial mass loss is due to dehydration, and the second mass loss may be related to SiO_3OH^{3-} dehydroxylation. An anhydrous phase transforms into crystalline $K_2U_2O_7$ and amorphous SiO_2. A crystal structure study of boltwoodite by Burns (1998) clearly shows the presence of SiO_3OH^{3-} groups in the uranyl silicate layers of this mineral, as first suggested by Vochten et al. (1997a).

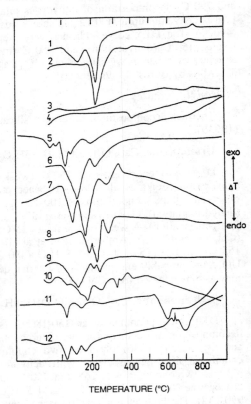

Figure 45. DTA curves of some uranyl minerals (micro-DTA instrument, heating rate 10°C min^{-1}, static air atmosphere. reference sample Al_2O_3): 1. Uranophane (30 mg); 2. Sklodowskite (40 mg); 3. Kasolite (60 mg); 4. Soddyite (40 mg); 5. Meta-autunite (30 mg); 6. Meta-autunite synthetic (30 mg); 7. Meta-torbernite (30 mg); 8. Meta-torbernite synthetic (30 mg); 9. Meta-uranocircite II synthetic (20 mg); 10. Sabugalite (17 mg); 11. Uranopilite (20 mg); 12. Johannite (10 mg) (from Cejka and Urbanec 1990).

Sodium boltwoodite, $(Na_{0.7}K_{0.3})(H_3O)(UO_2)(SiO_4) \cdot H_2O$ or $Na(UO_2)(SiO_3OH) \cdot H_2O$ (Zhiltsova et al. 1976)

DTA: endotherm at 150°C (dehydration), Na- or K-uranates are formed at 600°C (Zhiltsova et al. 1976). Amorphous SiO_2, formed during dehydroxylation, remains unchanged.

Kasolite, $Pb(UO_2)(SiO_4)\cdot H_2O$ (Fig. 44, 45) (Cejka et al. 1986a; Urbanec et al. 1985a)

DTA: endotherms at 100-200°C (dehydration), 650-800°C (destruction of anhydrous lead uranyl silicate and formation of lead uranate (Ambartsumyan et al. 1961)). Ambartsumyan's TG curve for kasolite shows most water is released between 100-200°C; dehydration is finished at 500°C. However, Cejka et al. (1986a) observed a broad endotherm in the range 100-360°C and a continuous mass decrease up to 900°C, corresponding to one $H_2O(?)$; dehydration may be probably partly overlapped with oxygen release during lead uranate decomposition and possibly PbO release, as observed for curite. Lead uranate ($PbUO_4$) and amorphous SiO_2 are formed during thermal decomposition of kasolite (Ambartsumyan et al. 1961; Urbanec et al. 1985a).

Oursinite, $(Co_{0.86}Mg_{0.10}Ni_{0.04})O\cdot 2UO_3\cdot 2SiO_2\cdot 6H_2O$ or $M^{2+}(UO_2)_2(SiO_3OH)_2\cdot 5H_2O$ (Deliens and Piret 1983)

TG: dehydration takes place in several steps: most water is released up to 325°C (~3 $H_2O \rightarrow 100°C$, ~2 $H_2O \rightarrow 325°C$) and ~1 H_2O (325-500°C) (Deliens and Piret 1983). The last dehydration step may correspond to SiO_3OH^{3-} dehydroxylation. However, the temperature related to dehydroxylation is relatively low compared with other uranyl silicates; this may be due to the presence of transition elements, which may decrease the stability field of the dehydrated compound.

Swamboite, $U_{1/3}H_2(UO_2)_2(SiO_4)_2\cdot 10H_2O$ or $U_{1/3}(UO_2)_2(SiO_3OH)_2\cdot 10H_2O$ (Deliens and Piret 1981), or $U^{6+}H_6(UO_2)_6(SiO_4)_6\cdot 30H_2O$ (Mandarino 1999)

Swamboite dehydrates in four steps: 4 H_2O (220°C), 2 H_2O (260°C), 2 H_2O (310°C), 2 H_2O (340°C), as H_2O is lost from the interlayer, and dehydroxylates(?) up to 460°C (1 H_2O) (Deliens and Piret 1981). Dehydroxylation may be related to SiO_3OH^{3-} dehydration.

Uranosilite, $UO_3\cdot 7SiO_2\cdot (0-1)H_2O$ (Walenta 1983)

According to Walenta (1983), the structure of uranosilite appears stable up to 900°C.

Soddyite, $(UO_2)_2SiO_4\cdot 2H_2O$ (Fig. 44, 45) (Urbanec et al. 1985a)

DTA: endotherms at 400 and 724(?)°C, exotherm at 750 °C (Urbanec et al. 1985a; Cejka et al. 1986), at 47, 400 and 692°C—synthetic phase (Moll et al. 1995). The first endotherm is associated with evaporation of adsorbed water, the second one with release of molecular water, and the third one with structural decomposition and formation of $UO_{2.67}$ and amorphous SiO_2. TG: adsorbed water is released approximately up to 320°C, 1 H_2O (320-450°C) and 1 H_2O (450-720°C), 0.3 O_2 (720-960°C) (Urbanec et al. 1985a). TG-DTG-DTA curves (simultaneously recorded) of synthetic soddyite were published by Kuznetsov et al. (1981), who proposed the existence of a poorly crystalline intermediate uranyl silicate, $2UO_3\cdot (1-x)SiO_2$, in a mixture with amorphous SiO_2. This uranyl silicate is stable up to 670°C, at which point it decomposes to $UO_{2.67}$ and more amorphous SiO_2.

Weeksite, $K_2(UO_2)_2(Si_2O_5)_3\cdot 4H_2O$ (Tarkhanova et al. 1975; this study)

According to our recent measurement, the TG curve of weeksite shows a continuous mass loss up to 660°C (~4 H_2O) and a second mass loss up to 800°C. The structure of the

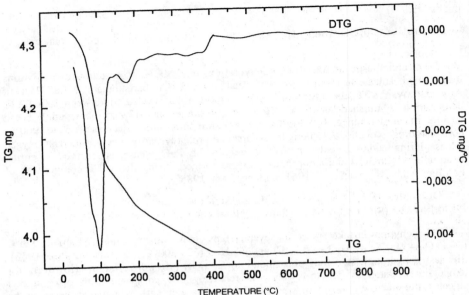

Figure 46. TG-DTG curves of haiweeite: sample weight 3.937 mg, heating rate 10°C min^{-1}, dynamic air atmosphere 10 ml min^{-1}.

anhydrous phase remains crystalline up to 900°C (Tarkhanova et al. 1975). End products of the thermal decomposition of weeksite are $K_2U_2O_7$ and amorphous SiO_2.

Haiweeite, $Ca(UO_2)_2(Si_2O_5)_3 \cdot 5H_2O$ (Fig. 46) (this study)

According to our recent TG curve, haiweeite dehydrates in several steps: 3 H_2O (~85°C), 1 H_2O (135°C), 1 H_2O (240°C), 2 H_2O (615°C); dehydration is followed by formation of CaU_2O_7 or $UO_{2.67}$ and amorphous SiO_2. CaU_2O_7 may loose some oxygen at 900°C. The haiweeite sample studied contained two more H_2O than shown in the IMA-approved formula. The chemical formula of haiweeite should probably be $Ca(UO_2)_2[Si_5O_{12}(OH)_2] \cdot 4.5H_2O$ (Rastsvetaeva et al. 1997). Burns (1999e) refined the crystal structure of haiweeite, reporting the formula $Ca[(UO_2)_2Si_5O_{12}(OH)_2](H_2O)_3$.

Note: Thermal decomposition end products of uranophane, β-uranophane and haiweeite may be calcium diuranate and amorphous SiO_2 or calcium oxide, $UO_{2.67}$ and amorphous SiO_2, which depends on specific experimental conditions.

Uranyl phosphates and arsenates

The existence of several dehydration steps is a characteristic feature for minerals and synthetic compounds of the autunite and meta-autunite groups. Not all fully hydrated phases or phases observed during dehydration have been found in nature. Dehydration is affected by several factors, including mineral formation and stability, and ionic radii of interlayer cations. Some dehydration processes are reversible; others are irreversible. Dehydration takes place in several steps up to approximately 300°C and is usually characterized by several (2 to 4) endotherms. Multiple endotherms indicate that H_2O groups are bonded within the interlayer by various forces, with some H_2O groups co-ordinated by cations. Different arrangements of H_2O groups and cations in interlayer sites may be manifested by different temperature intervals of related endotherms (Muck et al. 1986).

Compounds related to the meta-autunite group do not possess the first endotherm on DTA curves characteristic of fully hydrated phases, i.e. fewer endotherms are assigned to dehydration in the less-hydrated compounds.

Molecular water contents of distinct hydrated phases may vary to some extent, causing the character of TG and DTA curves to vary for differ among specimens of the same mineral. Such complications explain why data in extensive references are not identical. Hoffmann and Weigel (Hoffmann 1972; Weigel 1985; Weigel and Hoffmann 1976) suggest that uniquely defined hydrates of autunites do not exist. Water in autunites is not zeolitic in character, because changes of hydration affect the structures. Thus water is an important constituent of the interlayer and of the crystal structure of these minerals.

The entire dehydration process and transition to anhydrous phases takes place in uranyl arsenates at higher temperatures than those in similar uranyl phosphates. Dehydration of uranyl arsenates is more complicated, making interpretations more difficult. The occurrence of fully hydrated phases of autunites is an exception; most transform to meta-autunites at ambient temperatures. Dehydration at higher temperatures is usually associated with formation of X-ray amorphous anhydrous compounds. However, anhydrous barium uranyl phosphate remain crystalline up to approximately 1000°C. Crystalline phases formed by rehydrating dehydrated compounds may not be identical to the original minerals, and structural arrangements of H_2O groups may be quite different. Walenta (1965b) showed that most autunite-group minerals display two or more hydration steps; however, some phases were prepared only in the laboratory. Loss of molecular water is accompanied by a decrease in the c cell dimension and increase in refraction indices. Exotherms that may correspond to the crystallization of new anhydrous phases is interesting from the point of view that exotherms are not necessarily observed on DTA curves of all specimens of given mineral (Cejka 1983). Also, exotherms are more frequent in uranyl arsenates than in uranyl phosphates. Thermal decomposition of autunites has been studied by many authors (for additional references see, e.g. Cejka and Urbanec 1990).

Autunite, $Ca(UO_2)_2(PO_4)_2 \cdot (10-12)H_2O$ (Leonova 1958)

Meta-autunite I, $Ca(UO_2)_2(PO_4)_2 \cdot (2-6)H_2O$ (Fig. 44, 45) (Cejka et al. 1985a)

Meta-autunite II, $Ca(UO_2)_2(PO_4)_2 \cdot (0-2)$ or $(0-6)H_2O$ (Cejka et al. 1985b)

Autunite-DTA: endotherms 92, 141, 216°C related to dehydration, meta-autunite I-DTA endotherms 140 and 216°C (Ambartsumyan et al. 1961). Autunite dehydrates in three steps, meta-autunite in two steps. TG (Vochten and Deliens 1980): Autunite (10.5 H_2O) ⇔ meta-autunite I (7H_2O)—the reversible process proceeds at 30°C; meta-autunite I → meta-autunite II (5 H_2O)—the irreversible process proceeds in the range 60-85°C. Many papers have been published on the thermal behaviors of autunite and meta-autunite (e.g. Leonova 1958; Leo 1960; Takano 1961; Hoffmann 1972; Weigel and Hoffmann 1976; Cejka 1983; Cejka et al. 1985b; Cejka et al. 1985a). Anhydrous calcium uranyl phosphate (formed at 500°C) melts and decomposes at 1100°C with formation of $UO_{2.67}$.

Bassetite, $Fe^{2+}(UO_2)_2(PO_4)_2 \cdot 8H_2O$ (Ambartsumyan et al. 1961; Vochten et al. 1984)

DTA: (mineral) endotherm 150°C (dehydration), exotherm 300-500°C (Fe^{2+} oxidation); (synthetic) endotherm 120°C (dehydration), exotherm ≈ 270°C (250-400°C). In the case of synthetic phase, an endotherm was observed in the exotherm region (295°C), associated with 1 H_2O release. TG: (mineral) 6 H_2O (25-160°C), 1 H_2O (160-235°C), 1 H_2O (235-400°C); (synthetic) 7 H_2O (25-160°C), 1 H_2O (160-235°C), 1 H_2O (235-400°C) (Ambartsumyan et al. 1961; Vochten et al. 1984; Vochten and Brizzi 1987). After dehydration, the anhydrous phase decomposes. In the range 700-1000°C, an iron uranium pyrophosphate may be formed, which melts at 1075°C with formation of $UO_{2.67}$. Fe^{2+} can easy oxidize, and the fully oxidized phase, $Fe^{3+}(UO_2)_2(PO_4)_2(OH) \cdot 6H_2O$, and its thermal decomposition have also been described (Vochten et al. 1984).

Chernikovite, $(H_3O^+)_2(UO_2)_2(PO_4)_2 \cdot 6H_2O$ (Ambartsumyan et al. 1961; Vochten and Deliens 1980; Hoffmann 1972)

DTA: endotherms at 160 and 220°C (dehydration), and 500-700°C (Ambartsumyan et al. 1961). Synthetic phase: endotherms at 65°C ($6H_2O$) and 125-200°C ($2H_2O$) (Vochten and Deliens 1980). The number of endotherms may vary depending on experimental conditions. TG (isobaric thermal decomposition under constant H_2O vapor pressure): 4 H_2O (65°C), 2 H_2O (97°C), 2 H_2O (163°C) (Hoffmann 1972, Weigel and Hoffmann 1976). Some other intermediate hydrates may be found. Octa- and dihydrate should bee the most stable. Anhydrous phase decomposes to $(UO_2)_2P_2O_7$ (500-600°C) and $U_2O_3P_2O_7$ (800-900°C) (Johnson et al. 1981; Barten 1980; Barten and Cordfunke 1980a 1980b).

Meta-ankoleite, $K_2(UO_2)_2(PO_4)_2 \cdot 6H_2O$ (Gonzales Garcia and Romero Diaz 1959; Johnson et al. 1981)

Synthetic phase: DTA shows endotherms at 62, 120 and 175°C—dehydration (Gonzales Garcia and Romero Diaz 1959). TG: 5 H_2O (40-75°C), 1 H_2O (75-100°C) (Johnson et al. 1981). Existence of hexahydrate, monohydrate and anhydrous phase was confirmed by Hoffmann (Hoffmann 1972; Weigel and Hoffmann 1976).

Lehnerite, synthetic phase, $Mn(UO_2)_2(PO_4)_2 \cdot 8H_2O$ (Vochten et al. 1990)

DTA: three endotherms up to 100°C, endotherm near 180°C and near 240°C, all associated with dehydration. TG: 0.5 H_2O (adsorbed water) (25-40°C), 0.5 H_2O (40-60°C), 4.5 H_2O (60-100°C), 1 H_2O (100-225°C), 0.5 H_2O (225-250°C), 0.5 H_2O (250-450°C) (Vochten 1990). According to Pozas-Tormo et al. (1986), the DTA curve of synthetic lehnerite is characterized by endotherms at 60, 74, 115, 215 and 260°C and the TG curve indicates a stepwise loss of 3 H_2O (25-80°C), 4.5 H_2O (80-150°C), and 2 H_2O (150-300°C).

Sodium meta-autunite, $Na_2(UO_2)_2(PO_4)_2 \cdot 8H_2O$ (Hoffmann 1972; Weigel and Hoffmann 1976)

The mineral seems to be a hexa- rather than octahydrate. Synthetic phase: DTA shows endotherms at 87.5, 137.5 and 200°C due to dehydration (Gonzales Garcia and Romero Diaz 1959). TG: the octahydrate dehydrates as follows: 6 H_2O (60-70°C), 2 H_2O (100-160°C). An anhydrous compound is formed above 190°C (Gmelin 1981). The hexahydrate loses 4 H_2O up to 95°C (3 H_2O at 80°C, 1 H_2O at 95°C), 2 H_2O at 170°C (Hoffmann 1972; Weigel and Hoffmann 1976).

Ranunculite, $(HAl)_2(UO_2)_2(PO_4)_2(OH)_6 \cdot 8H_2O$ (Deliens and Piret 1979b)

TG: 4 H_2O (400°C), 1 H_2O (700°C), 1 H_2O (1000°C?) (Deliens and Piret 1979b).

Sabugalite, $HAl(UO_2)_4(PO_4)_4 \cdot 16H_2O$ (Figs. 44, 45) (Cejka et al. 1988b; Vochten and Pelsmaeker 1983)

DTA: endotherms at 191, 244 and 320°C—dehydration, and a broad one in the range 380-450°C. TG: ~9 H_2O (150°C), ~4.5 H_2O (150-270°C), ~3 H_2O (270-440°C). Decomposition of the anhydrous phase proceeds in the range 440-675°C. The anhydrous phase loses oxygen during decomposition and is transformed into crystalline phases whose compositions depend on temperature. $UO_{2.67}$ and phases in the system UO_3-P_2O_5 and

Al_2O_3-P_2O_5 are formed. The first new crystalline, partly dehydrated phase is isostructural with meta-autunite II. Uranospathite is equivalent to fully hydrated sabugalite; sabugalite is the same as meta-uranospathite I, and partly dehydrated sabugalite (metasabugalite) is equivalent to meta-uranospathite II (Cejka et al. 1988b). Ambartsumyan et al. (1961) observed dehydration at 100°C (7 H_2O), 200°C (4.5 H_2O), and 300°C (4.5 H_2O). Vochten and Pelsmaekers (1983) reported three endotherms on a DSC curve (50, 120 and 175°C) corresponding to water releases: 4 H_2O (65°C), 7 H_2O (65-126°C), and 5 H_2O (126-220°C).

Saléeite, $Mg(UO_2)_2(PO_4)_2 \cdot 10H_2O$ (Piret and Deliens 1980; Vochten and Van Doorselaer 1984)

DTA: endotherms at 70, 95, 270 and 320°C (dehydration), TG: 7.7 H_2O (20-167°C), 3 H_2O (167-700°C) (Piret and Deliens 1980). According to Hoffmann (1972), three region of hydrates should exist (9-10 H_2O, 7-9 H_2O, and 4-6 H_2O). Vochten and Van Doorselaer (1984) observed endotherms at 68, 110, 288 and 354°C corresponding to 3 H_2O (25-91°C), 5 H_2O (91-193°C), 1 H_2O (193-308°C), and 1 H_2O (308-590°C).

Threadgoldite, $Al(UO_2)_2(PO_4)_2(OH) \cdot 8H_2O$ (Deliens and Piret 1979a)

TG: $\approx$3 H_2O (145°C), $\approx$4 H_2O (225°C), $\approx$2 H_2O (700°C) (Deliens and Piret 1979a).

Torbernite, $Cu(UO_2)_2(PO_4)_2 \cdot (8-12)H_2O$ (Pozas-Tormo et al. 1986)

Metatorbernite, $Cu(UO_2)_2(PO_4)_2 \cdot (4-8)H_2O$ (Fig. 44, 45) (Cejka et al. 1985a; Vochten 1983)

Torbernite (synthetic)

DTA: endotherms at 75, 110, 143 and 237°C—dehydration, TG: 3 H_2O (25-80°C), 5 H_2O (80-150°C), 2 H_2O (150-300°C) (Pozas-Tormo et al. 1986).

Metatorbernite

DTA: endotherms at 120, 150 and 255°C (Ambartsumyan et al. 1961), 133, 172 and 265°C (Cejka et al. 1985d). DTA curves of synthetic metatorbernite indicate endotherms at 195, 222 and 295°C (Cejka et al. 1985a) and 120, 150, 150-450°C (a broad one) (Vochten et al. 1979; Vochten 1983). TG curve of synthetic metatorbernite shows dehydration, 4 H_2O ($\rightarrow$ 120°C), 2 H_2O (120-150°C), 2 H_2O (150-450°C) (Vochten 1983). Anhydrous phase transforms into uranium copper pyrophosphate, which is stable up to 800°C. Decomposition proceeds at higher temperature with formation of $UO_{2.67}$.

Uramphite, $(NH_4)_2(UO_2)_2(PO_4)_2 \cdot 6H_2O$ (Ambartsumyan et al. 1961; Kobets and Umreyko 1976)

DTA shows endotherms at 125 and 175°C due to dehydration, 500°C due to NH_3 release. An exotherm at 600°C is due to crystallization of uranium pyrophosphate. TG shows loss of 4 H_2O (30-100°C), 2 H $_2$O (100-200°C) (Ambartsumyan et al. 1961); the synthetic phase shows endotherms at 110°C due to dehydration, 180°C due to pyrophosphate formation, and at 425°C due to NH_3 release; an exotherm at 690°C corresponds to crystallization of amorphous pyrophosphate intermediate, $U_2O_3P_2O_7$ being the principle decomposition product. TG shows loss of 6 H_2O ($\rightarrow$ 150°C) (Kobets and Umreyko 1976; Lepilina and Smirnova 1984), for further data see also Hoffmann (1972; Weigel and Hoffmann 1976), Johnson et al. (1981) and especially Gmelin (1981) and Kobets and Umreyko (1983).

Uranocircite I, $Ba(UO_2)_2(PO_4)_2 \cdot 12H_2O$

Uranocircite II, $Ba(UO_2)_2(PO_4)_2 \cdot 10H_2O$

No thermal decomposition data on these two minerals are available.

Meta-uranocircite I, $Ba(UO_2)_2(PO_4)_2 \cdot 8H_2O$ (Walenta 1963; Weigel and Hoffmann 1976)

DTA: endotherms at 130°C (meta I → meta II transition), 185, 270, 1030 and 1060°C. The first three endotherms correspond to dehydration, the last two to decomposition of anhydrous phase (Walenta 1963). TG (synthetic): 2 H_2O (50-60°C), 1 H_2O (70-80°C), 3 H_2O (85-100°C), 2 H_2O (170-200°C) (Weigel and Hoffmann 1976).

Meta-uranocircite II, $Ba(UO_2)_2(PO_4)_2 \cdot 6H_2O$ (Fig. 45) (Cejka et al. 1985a; Vochten et al. 1992)

DTA (synthetic): endotherms at 165, 220 and 245°C (Cejka et al. 1985a), in nitrogen atmosphere endotherms at approximately 40, 80, 160, 190 and 245°C (broad), and TG showing dehydration 4 H_2O (94°C), 1 H_2O (94-175°C), 0.5 H_2O (175-206°C), 0.5 H_2O (206-300°C) (Vochten et al. 1992). Meta-uranocircite II remains crystalline up to 1000°C, as inferred from X-ray powder diffraction, infrared spectroscopy and thermal analysis (see e.g. Cejka et al. 1985a; Cejka 1983; Cejka et al. 1985b).

Uranospathite, $HAl(UO_2)_4(PO_4)_4 \cdot 40H_2O$ (Walenta 1978)

Arsenuranospathite, $HAl(UO_2)_4(AsO_4)_4 \cdot 40H_2O$ (Walenta 1978)

Uranospathite is unstable, and unless preserved at low temperature or in a humid atmosphere, it converts into a new phase with the loss of 20-24 H_2O. The partly dehydrated phase is identical to sabugalite. Arsenuranospathite is unstable under ambient conditions and converts to a lower hydrate containing about 20 H_2O. No further data concerning thermal decomposition of the two minerals are available (Walenta 1978); also see the description of sabugalite above.

Xiangjiangite, $(Fe^{3+},Al)(UO_2)_4(PO_4)_2(SO_4)_2OH \cdot 22H_2O$ (Hunan 230 Inst. 1978)

DTA: endotherms at 95, 174, 330 and 477°C—dehydration and dehydroxylation(?), and ~754°C (SO_3 release) (Hunan 230 Inst. 1978).

Dumontite, $Pb_2[(UO_2)_3O_2(PO_4)_2] \cdot 5H_2O$ (Frondel 1958)

Dumontite dehydrates up to 300°C, oxygen release is inferred in the range 600-900°C (Frondel 1958).

Furongite, $Al_2(UO_2)(PO_4)_2(OH)_2 \cdot 8H_2O$ (Zhi-xiong Wang 1979)

DTA, TG and DTG curves of furongite were published (Zhi-xiong Wang 1979) [the paper was not seen by this author]. Furongite from Kobokobo, $Al_2(UO_2)_2(PO_4)_3OH \cdot 13.5H_2O$, dehydrates and dehydroxylates in three or four steps (Deliens and Piret 1985a).

Phurcalite, $Ca_2(UO_2)_3(PO_4)_2O_2 \cdot 7H_2O$ (Braithwaite et al. 1989; Atencio 1991)

DTA: endotherm at 150°C, melting at 900°C, TG: dehydration—the main part of water is released up to approximately 200°C; however, the mass loss continues to higher temperatures (above 500°C, mass loss may correspond to slow P_2O_5 volatilization) (Braithwaite et al. 1989; Atencio 1991; Atencio et al. 1991).

Moreauite, $Al_3UO_2(PO_4)_3(OH)_2 \cdot 13H_2O$ (Deliens and Piret 1985b)

TG: the mineral dehydrates and dehydroxylates in two partly overlapping steps up to 500°C. The exact mechanisms of thermal decomposition are not clear (Deliens and Piret 1985b).

Yingjiangite, $(K_{1-x}Ca_x)(UO_2)_3(PO_4)_2(OH)_{1+x} \cdot 4H_2O$ ($x = 0.35$) (Chen Zhangru et al. 1990) or $(K_2,Ca)(UO_2)_7(PO_4)_4(OH)_6 \cdot 6H_2O$ (Zhang Jingyi et al. 1992)

DTA: endotherms at 190 and 269°C (dehydration and dehydroxylation, respectively), at 944°C (due to the oxidation of U^{4+} formed during thermal decomposition of the anhydrous material) (Chen Zhangru et al. 1990); endotherms at 172, 223 (an inflection), 523 (broad), 900 and 960°C. Dehydration (TG) proceeds up to 800°C (Zhang Jingyi et al. 1992).

Parsonsite, $Pb_2(UO_2)(PO_4)_2 \cdot 2H_2O$ (Frondel 1958)

The mineral is anhydrous at ≈300°C (Frondel 1958). A recent structure determination of parsonsite indicates that it is anhydrous (Burns 1999d).

Phosphuranylite, $KCa(H_3O^+)_3(UO_2)_7(PO_4)_4O_4 \cdot 8H_2O$ (this study)

TG shows mass loss (8.15 wt %) at ≈280°C corresponding to ideal water content (8.61 wt %) (Demartin et al. 1991). However, the decomposition proceeds to higher temperature. At ≈620°C 10.70 wt % is lost and at 1000°C 13.25 wt % is lost. The mechanisms of decomposition are not clear (this study).

Dewindtite, $Pb_3[H(UO_2)_3O_2(PO_4)_2]_2 \cdot 12H_2O$ (Frondel 1958)

The mineral loses water up to 300°C, oxygen at 400°C (Frondel 1958).

Abernathyite, $K_2(UO_2)_2(AsO_4)_2 \cdot 8H_2O(?)$ (Garcia Gonzales and Romero Diaz 1959; Hoffmann 1972; Weigel and Hoffmann 1976)

Synthetic phase corresponds to hexahydrate (Hoffmann 1972; Weigel and Hoffmann 1976). DTA shows an inflection at 62°C, endotherms at 120 and 175°C due to dehydration (Gonzales Garcia and Romero Diaz 1959), TG shows loss of 5 H_2O (85°C), 1 H_2O (210°C) (Hoffmann 1972; Weigel and Hoffmann 1976).

Hallimondite, $Pb_2(UO_2)(AsO_4)_2$ (Walenta 1965a)

DTA: endotherm at 940°C due to melting (Walenta 1965a).

Heinrichite, $Ba(UO_2)_2(AsO_4)_2 \cdot 10H_2O$ (Walenta 1965b; Hoffmann 1972)

Heinrichite is unstable and irreversibly transforms to metaheinrichite at ambient temperatures.

Meta-heinrichite, $Ba(UO_2)_2(AsO_4)_2 \cdot 8H_2O$ (Walenta 1965b; Hoffmann 1972)

DTA: endotherms at 145°C (an inflection), 185, 250°C (dehydration), 1080 and 1100°C decomposition of anhydrous barium uranyl arsenate (Walenta 1965b); TG: 6 H_2O (120°C), 2 H_2O (180-220°C) (Walenta 1965b); 5 H_2O (110°C), 2 H_2O (110-160°C) in the case of heptahydrate (Hoffmann 1972; Weigel and Hoffmann 1976).

Kahlerite, $Fe^{2+}(UO_2)_2(AsO_4)_2 \cdot 12H_2O$ (Walenta 1964)

Kahlerite is unstable and irreversibly transforms to metakahlerite at ambient temperatures.

Meta-kahlerite, $Fe^{2+}(UO_2)_2(AsO_4)_2 \cdot 8H_2O$ (Walenta 1964; Vochten et al. 1986)

Synthetic kahlerite: DTA shows endotherms at 185, 230 and 350°C (dehydration), at 935 and 1015°C (decomposition of anhydrous phase), an exotherm at 620°C (crystallization of an anhydrous Fe^{2+} uranyl arsenate or oxidation of Fe^{2+}) (Walenta 1964). DSC (nitrogen atmosphere): endotherms at 50, 110 and 225°C (dehydration) (Vochten et al. 1986). TG: 5 H_2O (125°C), 3 H_2O (125-260°C) (Walenta 1964), in nitrogen atmosphere 1 H_2O (75°C), 4 H_2O (75-195°C), 3 H_2O (195-365°C) (Vochten et al. 1986).

Kirchheimerite, $Co(UO_2)_2(AsO_4)_2 \cdot 12H_2O$ (Walenta 1964)

Meta-kirchheimerite, $Co(UO_2)_2(AsO_4)_2 \cdot 8H_2O$ (Walenta 1964; Vochten and Goeminne 1984)

Kirchheimerite is unstable and irreversibly transforms to metakirchheimerite at ambient temperatures.

Synthetic meta-kirchheimerite: DTA shows endotherms at 115, 195 and 300°C (dehydration), 1005°C (decomposition), and an exotherm at 620°C (crystallization of an anhydrous Co-uranyl arsenate). Two additional weak exotherms apparent in the data at higher temperatures are not discussed (Walenta 1964). DSC (7 H_2O): endotherms at 45, 122 and 275°C (dehydration) (Vochten and Goeminne 1984). TG: 6 H_2O (115°C), 2 H_2O (115-240°C) (Walenta 1964), 1 H_2O (25-85°C), 3.5 H_2O (85-200°C), 2.5 H_2O (200-500°C) (Vochten and Goeminne 1984).

$Ni(UO_2)_2(AsO_4)_2 \cdot 7H_2O$, unnamed mineral phase (Ondrus et al. 1997; Vochten and Goeminne 1984)

Synthetic phase—DSC: endotherms 74, 124 and 302°C—dehydration. TG: 1 H_2O (25-100°C), 3 H_2O (100-150°C), 3 H_2O (150-500°C) (Vochten and Goeminne 1984).

Novácekite I, $Mg(UO_2)_2(AsO_4)_2 \cdot 12H_2O$ (Walenta 1964)

Novácekite II, $Mg(UO_2)_2(AsO_4)_2 \cdot 10H_2O$ (Walenta 1964)

Meta-novácekite, $Mg(UO_2)_2(AsO_4)_2 \cdot 8H_2O$ (Walenta 1964; Hoffmann 1972; Weigel and Hoffmann 1976)

Novácekite I is unstable and reversibly transforms to novácekite II, novácekite II reversibly transforms to meta-novácekite (Walenta 1964).

Synthetic meta-novácekite: DTA shows endotherms at 120, 180, 320, 395°C (dehydration), 1010°C (decomposition). If novácekite II is present in the sample, an endotherm at 95°C is observed, weak exotherm at 485°C (Walenta 1964). DTA curve of the mineral shows endotherms at 119, 187, 231, 239 and 369°C (dehydration), 968°C (decomposition, $UO_{2.67}$ formation and As_2O_5 release) (Skvortsova et al. 1974). TG: 6 H_2O (110°C), 2 H_2O (in two steps 110-240 and 240-300°C) (Walenta 1964); ≈novácekite II (10 H_2O) → 2 H_2O (90°C), 2 H_2O (120°C), 4 H_2O (250°C), 2 H_2O (350°C) (Hoffmann 1972; Weigel and Hoffmann 1976). Leonova (1958) prepared an undecahydrate which transforms into a heptahydrate (60-70°C), and hexahydrate (115-130°C).

Sodium meta-uranospinite, $(Na_2,Ca)(UO_2)_2(AsO_4)_2 \cdot 5H_2O$ (Ambartsumyan et al. 1961)

DTA: endotherms at 140 and 200°C—dehydration (Kopchenova and Skvortsova 1957; Ambartsumyan et al. 1961). TG (unstable synthetic Na phase containing approximately 9 H_2O): 2 H_2O (60°C), 4 H_2O (70°C), 2 H_2O (165°C), 1 H_2O (190°C) (Hoffmann 1972; Weigel and Hoffmann 1976). The TG curve of this compound is not readily interpreted because of some discrepancies between the chemical compositions of the synthetic compound and the mineral.

Trögerite, $(UO_2)_3(AsO_4)_2 \cdot 12H_2O$(?) (Ambartsumyan et al. 1961)

DTA: endotherms at 130 and 180°C—dehydration. Tensimetric measurements showed dehydration in four steps, 8 H_2O (20-100°C), 2 H_2O (100-200°C), 1.4 H_2O (300°C). The remaining water is lost at 400°C. A crystalline anhydrous phase was formed at 400°C, which decomposed with $UO_{2.67}$ formation and As_2O_5 release at 1000°C (Ambartsumyan et al. 1961). Synthetic phase containing 11 H_2O dehydrates in four steps, 5 H_2O (80°C), 2 H_2O (100°C), 3 H_2O (200°C), 1 H_2O (310°C) (Barten and Cordfunke 1980c). Barten (1980) describes thermal decomposition of anhydrous $(UO_2)_3(AsO_4)_2$. The natural and synthetic phases are often confused with H-meta-uranospinite, $(H_3O^+)_2(UO_2)_2(AsO_4)_2 \cdot 6H_2O$.

H-uranospinite (H-meta-uranospinite), $(H_3O^+)_2(UO_2)_2(AsO_4)_2 \cdot 6H_2O$ (Hoffmann 1972; Weigel and Hoffmann 1976)

Chernikovite (H-uranyl phosphate = HUP) and H-meta-uranospinite (H-uranyl arsenate = HUAs) both are proton-conducting solid electrolytes (Howe and Shilton 1980).

Synthetic- DTA:endotherm at ~90°C—dehydration. No further thermal effects were observed on the DTA curve of the synthetic phase (Guillemin 1956). A phase transition was observed in the range 18-28°C (an endotherm at 25°C) (Benyacar and Abeledo 1974). TG: dehydration proceeds in one step up to 330°C (Guillemin 1956). Isobaric thermal analysis curve indicates two dehydration steps, 5-6 H_2O (80°C), 2 H_2O (150-210°C) (Hoffmann 1972; Weigel and Hoffmann 1976). Octa-, tetra-, tri- and dihydrates may exist (Howe and Shilton 1980). Crystalline anhydrous $(UO_2)_2As_2O_7$ was prepared by heating HUAs at 550°C for two days in a stream of dry oxygen (Barten and Cordfunke 1980c).

Uranospinite, $Ca(UO_2)_2(AsO_4)_2 \cdot 12H_2O$(?) (Walenta 1964/1965)

Meta-uranospinite, $Ca(UO_2)_2(AsO_4)_2 \cdot 8H_2O$ (Ambartsumyan et al. 1961)

DTA: 120, 175, 240°C (dehydration), 990°C (melting) (Ambartsumyan et al. 1961). Synthetic phase shows endotherms at 100, 140, 195°C—dehydration and an exotherm at 40°C (Guillemin 1956). TG: 6 H_2O (50-110°C), 2 H_2O (200-300°C) (Ambartsumyan et al. 1961).

Uranospinite (synthetic) $Ca(UO_2)_2(AsO_4)_2 \cdot (10H_2O)$

DTA: endotherms at 160, 215, 25o, 285°C—dehydration, 1030 and 1045°C (decomposition), exotherms at 605 and 1065°C (Walenta 1964/1965).TG: 8 H_2O (130°C), 2 H_2O (185-260°C)(Walenta 1964, 1965). According to Ambartsumyan et al. (1961), the structure of the mineral remains unchanged in the range 285-500°C, decomposition of the anhydrous phase proceeds in the range 500-800°C, a new undefined crystalline phase is formed. This decomposes at 1000°C with $UO_{2.67}$ formation.

Zeunerite, $Cu(UO_2)_2(AsO_4)_2 \cdot (10-16)H_2O$ (Walenta 1964/1965)

Zeunerite (synthetic): DTA: endotherms at 145, 170, 215, 255°C (dehydration), 970 and 1015°C—two-step decomposition of copper uranyl arsenate associated with $UO_{2.67}$ formation, exotherms at 560 and 665°C (crystallization of anhydrous copper uranyl arsenate) (Walenta 1964/1965).

Meta-zeunerite, $Cu(UO_2)_2(AsO_4)_2 \cdot 8H_2O$ (Vochten and Goeminne 1984)

Meta-zeunerite: DTA: endotherms at 100, 130 and 180°C—dehydration (Guillemin 1956), at 100, 140, 180°C—dehydration, 970°C—melting of anhydrous phase (Ambartsumyan et al. 1961). DSC synthetic phase): endotherms at 108, 130, 185 and 230°C (Vochten and Goeminne 1984). TG: 1 H_2O (25-112°C), 4 H_2O (112-165°C), 0.5 H_2O (165-218°C), 2.5 H_2O (218-500°C) (Vochten and Goeminne 1984). Guillemin (1956) observed formation of an intermediate 2.5 hydrate, isostructural with meta-autunite II. Formation of a new compound, probably a crystalline anhydrous copper uranyl arsenate, is assumed to occur in the range 400-900°C (Ambartsumyan et al. 1961).

Vochten et al. (1981) inferred that structural transitions among anhydrous uranyl arsenates occur at higher temperatures than for uranyl phosphates.

Uranyl vanadates

Carnotite, $K_2(UO_2)_2(V_2O_8) \cdot (1-3)H_2O$ (Ambartsumyan et al. 1961)

TG: carnotite dehydrates up to 200°C (2 H_2O 50-100°C, 1 H_2O 100-200°C), which is characterized by a broad endotherm in the range 50-200°C on the DTA curve (Ambartsumyan et al. 1961). Sidorenko (1978) assumes this water loss does not cause any structure change. Synthetic carnotite is anhydrous and not hygroscopic (Abraham et al. 1993).

Strelkinite, $Na_2(UO_2)_2(V_2O_8) \cdot 6H_2O$ (Alekseeva et al. 1974)

TG: Strelkinite containing approximately 4 H_2O dehydrates in two steps, 2 H_2O (50-280°C), 2 H_2O (280-1000°C?) (Alekseeva et al. 1974). Synthetic hydrated $Na_2(UO_2)_2(V_2O_8) \cdot 4H_2O$ dehydrates also in two steps, 2 H_2O (50-100°C), 2 H_2O (100-500°C). Hydration of anhydrous phase to tetrahydrate proceeds at ambient temperature (Abraham et al. 1993).

Anhydrous carnotite and strelkinite both appear stable at high temperatures; however, no detailed information is available.

Curiénite, $Pb(UO_2)_2(V_2O_8) \cdot 5H_2O$ (Fig. 47) (Cesbron 1970a; this study)

DTA: endotherms at 190, 710 and 910°C, synthetic at 190, 725 and 905°C (Cesbron and Morin 1968), synthetic at 116 (an inflection), 141, 714 and 903°C, exotherms at 326 and 476°C (Cesbron 1970a). Endotherms are related to dehydration, fusion of anhydrous phase, and oxygen release. The assignment of exotherms is not clear. $UO_{2.67}$ seems to be one of the end products of curiénite thermal decomposition (Cesbron 1970a). TG: curiénite dehydrates up to 690°C (5 H_2O) (this study).

Francevillite, $(Ba,Pb)(UO_2)_2(V_2O_8) \cdot 5H_2O$ (Cesbron 1970a; Cesbron and Morin 1968)

DTA: endotherms at 112, 156 and 176°C (an inflection)- dehydration, 869°C—melting of anhydrous phase, 1005°C- $UO_{2.67}$ formation and oxygen release. TG: francevillite dehydrates in three steps, 2 H_2O (20-90°C), 2 H_2O (90-120°C), 1 H_2O (120-140°C). (Branche et al. 1957; Cesbron 1970a; Cesbron and Morin 1968).

Both dehydrated curiénite and dehydrated francevillite rehydrate readily.

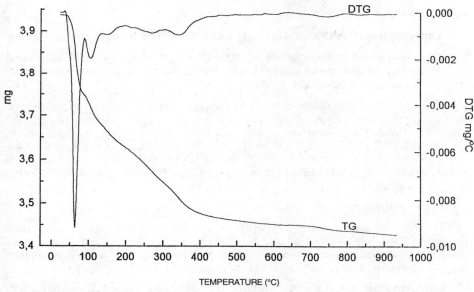

Figure 47. TG-DTG curves of curiénite: sample weight 4.322 mg, heating rate 10°C min^{-1}, dynamic air atmosphere 10 ml min^{-1}.

Sengiérite, $Cu_2[(UO_2)_2(V_2O_8)](OH)_2 \cdot 6H_2O$ (Guillemin 1956)

TG: synthetic phase dehydrates in two steps, 4 H_2O (130-170°C), 2 H_2O (170-250°C), and dehydroxylates at 500°C (Guillemin 1956). Thermal analysis of sengiérite indicates OH$^-$ ions, as confirmed by the structure determination (Piret et al. 1980).

Tyuyamunite, $Ca(UO_2)_2(V_2O_8) \cdot (5-8.5)H_2O$ (Ambartsumyan et al. 1961)

Meta-tyuyamunite, $Ca(UO_2)_2(V_2O_8) \cdot (3-5)H_2O$ (Stern et al. 1956)

An equilibrium between both hydrates (~5, ~8.5 H_2O) was shown to be a function of partial pressure of water vapor at 24°C (Stern et al. 1956). TG: tyuyamunite dehydrates in two steps, 6 H_2O (40-100°C), 2 H_2O (200-300°C), characterized by endotherms at 110-140 and 270-280°C on the DTA curve. An exotherm at 500°C is associated with crystallization of a new phase, an anhydrous calcium uranyl vanadate. A structural change is observed in the range 800-1000°C. A lower hydrate was not found during thermal decomposition of tyuyamunite (Ambartsumyan et al. 1961).

Vanuralite, $Al(UO_2)_2(V_2O_8)(OH) \cdot 11H_2O$ (Cesbron 1970b)

Meta-vanuralite, $Al(UO_2)_2(V_2O_8)(OH) \cdot 8H_2O$ (Cesbron 1970b)

An equilibrium between vanuralite (~11 H_2O), meta-vanuralite (~8 H_2O) and a phase containing 2 H_2O (meta-vanuralite II?) at 20°C is a function of water vapor pressure (Cesbron 1970b). TG: vanuralite dehydrates in several steps up to 500°C. These steps are better characterized on the DTA curve of vanuralite: endotherms at 73, 96, 153, 340, 415°C corresponding to dehydration, endotherm at 519°C to dehydroxylation. Two exotherms at 563 and 602°C are related to recrystallization of anhydrous phase. An endotherm at 802°C indicates oxygen release and formation of $UO_{2.67}$ and the melting of the remainder. The endotherm related to the loss of 3 H_2O at 73°C is not evident in the data for meta-vanuralite

(Cesbron 1970b).

Vanuranylite, $(H_3O^+,Ba,Ca,K)_2(UO_2)_2(V_2O_8) \cdot 4H_2O(?)$ (Buryanova et al. 1976)

DTA: endotherms at ~50(an inflection), 90, ~130°C (an inflection), 220°C correspond to dehydration. An endotherm at 870°C indicates the decomposition and fusion of an anhydrous phase (Buryanova et al. 1965).

Stepwise dehydration observed in uranyl vanadate minerals indicates the presence of H_2O groups bonded by different forces in the interlayer and in part co-ordinated by cations. Most dehydrated uranyl vanadates readily rehydrate. The thermal decomposition of synthetic $M^{2+}(UO_2)_2(V_2O_8) \cdot nH_2O$ (M^{2+}= Ba, Pb, Sr, Mn, Co, Ni) (Cesbron 1970a), $(NH_4)_2(V_2O_8) \cdot nH_2O$ (Botto and Baran 1976), $Ag_2(UO_2)_2(V_2O_8) \cdot 4H_2O$ (Abraham et al. 1994), and $M^{2+}(UO_2)_2(V_2O_8) \cdot nH_2O$ (M^{2+}= Ni, Zn, Cd) (Chernorukov et al. 1998a) have also been studied.

Uranyl molybdates

Thermal analysis of natural molybdates is a complex problem due to polyvalent properties of, not only uranium, but also molybdenum. Poorly characterized single-mineral samples were studied in several cases, making thermal-analysis curves difficult to compare between nominally identical minerals (Dara and Sidorenko 1967).

Calcurmolite, $Ca(UO_2)_3(MoO_4)_3(OH)_2 \cdot 11H_2O$, shows an endotherm at 170°C (dehydration), and an exotherm at 560°C (crystallization of $UO_{2.67}$?). The intermediate is X-ray amorphous in the range 200-400°C, and sinters over 800°C (Dara and Sidorenko 1967). Phases similar to calcurmolite may also contain sodium (Ca, Na uranyl molybdates). Two endotherms corresponding to dehydration were observed on DTA curves of Na-bearing phases. However, thermal analyses of such phases are variable and complex (Dara and Sidorenko 1967; Skvortsova et al. 1969; Zhiltsova et al. 1970).

Iriginite is nominally $UO_2Mo_2O_7 \cdot 3H_2O$, but the composition is ill defined and potentially variable. Dehydration of iriginite probably proceeds in two steps, with endotherms at either 144 and 184°C, 170 and 240°C, or 220 and 260°C (Dara and Sidorenko 1967). According to Trunov and Kovba (1963), a metastable anhydrous dimolybdate is formed, which decomposes to α-UO_2MoO_4 and MoO_3 (exotherm at 450°C). The synthetic phase studied may not correspond to iriginite. An ill-defined compound crystallizes after dehydration. An additional endotherm is observed in the range 600-700°C. An exotherm in the range 500-600°C probably corresponds to formation of $UO_{2.67}$ (Dara and Sidorenko 1967; Karpova et al. 1968; Zhiltsova et al. 1970).

The DTA curve of **moluranite**, $H_4U^{4+}(UO_2)_3(MoO_4)_7 \cdot 18H_2O$, shows an inflection at 90°C, an endotherm at 200°C (dehydration), and an exotherm at 600°C (crystallization of $UO_{2.67}$). Above 700°C, the DTA curve shows some complex behavior (Dara and Sidorenko 1967).

An endotherm (200°C) on the DTA curve of **sedovite**, $U(MoO_4)_2$, probably corresponds to dehydration. However, the mineral is nominally anhydrous. An exotherm approximately at 580°C, is related to the crystallization of $UO_{2.67}$ (Dara and Sidorenko 1967).

Umohoite, $UO_2MoO_4 \cdot 4H_2O$, dehydrates up to 400°C (endotherms at 170 and 370°C), an exotherm at 520°C corresponds to the formation of new crystalline phases (MoO_3 and $UO_{2.67}$) (Trunov and Kovba 1963; Dara and Sidorenko 1967; Sidorenko 1978).

Tengchongite, $CaO \cdot 6UO_3 \cdot 2MoO_3 \cdot 12H_2O$, shows endotherms at 83°C (dehydration) and 853°C (melting of anhydrous phase), an exotherm at 574°C corresponds to the formation of $UO_{2.67}$ (Chen Zhangru et al. 1986).

The most detailed study of a uranyl molybdate was published on the thermal decomposition of **mourite**, $U^{4+}Mo_5^{6+}O_{12}(OH)_{10}(?)$, (Dara and Sidorenko 1967; Osipov et al. 1981). A region of endotherms (40-430°C), exotherms (500-600°C) and a sharp endotherm at 720-730°C were observed on the DTA curve (the last endotherm is not associated with any mass loss). Adsorbed water is released between 40 and 130°C; molecular water from the interlayer escapes in two temperature ranges: 130-180°C and 180-260°C. Structural decomposition takes place between 260 and 430°C, and is related to dehydroxylation and the release of water from interlayer positions. Crystallization of $UMoO_6$, oxidation of U^{4+} and the possible oxidation of Mo^{5+}, proceed in the range 500-600°C. A phase with composition $UMoO_6$ exists in the range of 600-1000°C.

If thermal decompositions of individual uranyl molybdates are compared with each other, it can be inferred that they differ most in the low-temperature region (i.e. dehydration). Dehydration is accompanied by some structural changes. At higher temperatures, mechanisms of decomposition are similar, although crystallization of new phases ($UO_{2.67}$, MoO_3, $UMoO_6$) may depend on the U^{4+}/U^{6+} ratio in the original mineral (Dara and Sidorenko 1967; Sidorenko 1978).

Uranyl sulfates

Jáchymovite, $(UO_2)_8(SO_4)(OH)_{14}\cdot 13H_2O$ (Fig. 44, 45) (Cejka et al. 1996b)

Uranopilite, $(UO_2)_6(SO_4)(OH)_{10}\cdot(12-13)H_2O$ (Fig. 44, 45) (Cejka et al. 1996b)

DTA: dehydration and dehydroxylation of jáchymovite are characterized by a pronounced endotherm at 140°C and a less intense endotherm (an inflection?) at 200°C, whereas dehydration and dehydroxylation of uranopilite is characterized by endotherms at 118, 144 and 176°C, and an inflection(?) at 200°C. Endotherms assigned to the thermal decomposition of anhydrous phases formed were observed at 705°C and 714°C, respectively. SO_3 and O_2 are released and $UO_{2.67}$ is formed. In the case of uranopilite, exotherms at 623, 653 and 668°C correspond to crystallization of new phases (α-UO_3, α-UO_2SO_4) (Cejka et al. 1982b), for jáchymovite this crystallization could be probably derived from the course of the DTA curve at approximately 400°C. Transition of α-UO_2SO_4 in β-UO_2SO_4 was not proved (Cejka et al. 1996a). According to the TG curves, both minerals dehydrate and dehydroxylate in several steps: jáchymovite ~5 H_2O (20-97°C), ~5 H_2O (97-145°C), ~10 H_2O (145-576°C); uranopilite ~7 H_2O (20-66°C), ~10.5 H_2O (66-517°C). Dehydration and dehydroxylation processes partly overlap and are associated with the formation of X-ray amorphous phases.

Meta-uranopilite, $(UO_2)_6(SO_4)(OH)_{10}\cdot 5H_2O(?)$

Meta-uranopilite does not seem to be a product of partial dehydration of uranopilite. This conclusion was inferred from the fact that partially dehydrated uranopilite, the composition of which corresponds to meta-uranopilite, is X-ray amorphous (Cejka et al. 1996b; Ondrus et al. 1997). There is a small difference between the thermal decomposition of "anhydrous" jáchymovite and uranopilite. The mass loss of anhydrous uranopilite in the range 600-660°C (~0.4 wt %) can be explained by the transformation of α-UO_3 to $UO_{2.89}$. The thermal decomposition of the both minerals is finished at 884°C, and is associated with the formation of $UO_{2.67}$ and SO_3 and release of O_2 (Cejka et al. 1996a).

Johannite, $Cu(UO_2)_2(SO_4)_2(OH)_2\cdot 8H_2O$ (Fig. 44, 45) (Cejka et al. 1988a)

The DTA curve of johannite shows endotherms at 135, 165 and 215°C (dehydration), exotherms at 370 and 495°C (crystallization of new phases), and an endotherm at 760°C (decomposition of crystalline phases). From the TG curve it is inferred that dehydration takes place in the range 60-400°C in several steps, to which three endotherms are assigned. Four H_2O groups escape in the range 60-100°C (H_2O groups not co-ordinated by Cu^{2+} cations and is bonded by hydrogen bonds only)(Cejka et al. 1988a), followed by two H_2O (100-150°C), and three H_2O (2 H_2O + 2 OH^-; 150-400°C). The use of a larger sample mass for simultaneously recording TG-DTG-DTA curves of synthetic johannite (Cejka et al. 1988a) enabled the characterization of a more detailed course of dehydration and dehydroxylation: dehydration corresponds to losses of four H_2O (80-140°C; endotherm at 136°C), 2 H_2O (140-185°C; endotherm at 174°C), 2 H_2O (185-290°C), followed by dehydroxylation ($2OH^- \rightarrow 1\ H_2O + O^{2-}$) in the range 290-550°C; endotherm at 360°C. As it was proved by chemical gas analysis, no SO_3 escaped up to 400°C (Cejka et al. 1988a; Sokol and Cejka 1992). Two exotherms appeared on the DTA curves (natural at 370 and 495°C, and synthetic at 448 and 504°C, respectively) and are associated with formation of crystalline phases (β-UO_2SO_4, CuU_3O_{10}) (Cejka et al. 1988a). Further decomposition accompanied by SO_3 release is observed in the range 600-900°C for both natural and synthetic johannites, and is associated with CuU_3O_{10} and $CuUO_4$ formation. However, two steps were apparent in this temperature interval when a larger mass was used for the synthetic sample. These processes are characterized by endotherms at 760°C (natural) and 710 and 880°C (synthetic). A small endotherm at 985°C was observed on the DTA curve of synthetic johannite (Cejka a et al. 1988a). Kovba et al. (1982) ascribed for synthetic johannite endotherms at 150, 190, 360, 770, 840, and 920°C, exotherms at 450 and 510°C. In the case of a small sample mass of natural johannite, dehydration and dehydroxylation processes and the release of SO_3 and O_2 during uranate formation may overlap. Better resolution was observed on studying larger samples of synthetic johannite.

Due to rehydration of partly dehydrated johannite in humid air, observed d-spacings using high temperature X-ray camera are especially useful. At 90°C, d-spacings of johannite may correspond to the tetrahydrate described briefly by Kovba et al. (1982). At 140°C several weak and diffuse lines were observed; at 175°C the intermediate formed during dehydration is X-ray amorphous. β-UO_2SO_4 crystallizes at 380°C; however, according to Cordfunke (1972), α-UO_2SO_4 may be the first crystalline phase formed followed by transformation into the β-phase (α-UO_2SO_4 was admixed with β-UO_2SO_4). A reaction between β-UO_2SO_4 and amorphous CuO, associated with formation of CuU_3O_{10} occurs at 530°C. CuU_3O_{10} was observed in the range 530-960°C. $CuUO_4$ formed in the range 650-960°C. Both copper uranates (CuU_3O_{10} and $CuUO_4$) decompose at 960°C and higher temperatures into Cu_2UO_4, $UO_{2.67}$ and $O_{2\ (g)}$ (Cejka et al. 1988a). Kovba et al. (1982) assume the formation of $CuUO_4$ and U_3O_{8-x} with a composition near to $UO_{2.61}$ in this temperature region (Cejka et al. 1988a, and references therein). The course of thermal decomposition for synthetic johannite was investigated by simultaneous TG-DTA-MS (thermogravimetry- differential thermal analysis-mass spectrometry) (Sokol and Cejka 1992). Dehydration and dehydroxylation processes occur in four stages (60-450°C). Decomposition of the anhydrous compound ($CuO \cdot 2UO_3 \cdot 2SO_3$) is characterized by evolution of SO_3 (in two steps) and oxygen. Small exotherms at 655 and 688°C in argon and air, respectively, mark the onset of SO_3 evolution and are associated with the formation of crystalline CuU_3O_{10} (Sokol and Cejka 1992).

Deliensite, $Fe(UO_2)_2(SO_4)_2(OH)_2 \cdot 3H_2O$ (Vochten et al. 1997c)

The TG curve (dynamic N_2 atmosphere) of deliensite shows that dehydration and dehydroxylation do not take place in distinct steps. Instead, over an interval between 110-

500°C, the dehydration is complete, corresponding to a loss of 8.63 wt % H_2O. Between 500-800°C, decomposition of the sulfate group occurs, with a loss of 17.87 wt % SO_3 (Vochten et al. 1997c).

Zippeites, $M^+_4(UO_2)_6(SO_4)_3(OH)_{10} \cdot 4H_2O$ and $M^{2+}_2(UO_2)_6(SO_4)_3(OH)_{10} \cdot (8 \text{ or } 16)H_2O$ (Haacke and Williams 1979; Kovba et al. 1982; Cejka and Sejkora 1994; Vochten et al. 1995a)

Results of the TG analysis of some M^{2+}-zippeites are given in Table 12. (Haacke and Williams 1979). Chemical composition and crystal structure of zippeite group minerals are not yet definitively proved. However, the molar ratio $UO_2^{2+}/SO_4 = 2$ seems to be accurate (Cejka et al. 1985b). Dehydration and dehydroxylation occur in at least two stages. Dehydroxylation may partly overlap with the last stage of dehydration. SO_3 and O_2 escape approximately in the range 650-900°C. In the case of alkali zippeites, the decomposition continues to higher temperatures. Polyuranates of relevant cations may form intermediates. Some polyuranates, e.g. CoU_3O_{10}, an intermediate of the thermal decomposition of Co-zippeite, decompose into $UO_{2.67}$ and relevant metal oxides. Zn-zippeite, $Zn(UO_2)_2SO_4(OH)_4 \cdot 1.5H_2O$, (TG-DTG-DTA simultaneously recorded) dehydrates up to 180°C, dehydroxylates at 280°C. SO_3 is released up to 760°C with the formation of $ZnU_3O_{10} + ZnO$. Zinc uranate decomposes to ZnO and U_3O_{8-x} at 940°C (Kovba et al. 1982). Combined TG and DTA of synthetic zippeites prove that anhydrous intermediates of alkali zippeites are more stable then those containing divalent cations (Cejka and Sejkora 1994).

Synthetic potassium zippeite, $K(UO_2)_2(SO_4)(OH)_3 \cdot H_2O$, (combined TG and DSC, nitrogen dynamic atmosphere): from the TG curves, a total loss of 6.17 wt % H_2O was deduced. In the range 30-150°C, 1.97 H_2O groups are lost, whereas between 150 and 450°C, 0.41 H_2O groups are lost. The total number of H_2O groups lost over the entire temperature range is therefore 2.38. The loss of SO_3 takes place in the range 650-900°C (Vochten et al. 1995a).

Table 12. Thermogravimetric analysis of synthetic M^{2+} zippeites octahydrates

Compound	Transition	Temperature(°C)	Cumulative weight loss (wt %)	
			calculated	observed
Mg-zippeite	-5 H_2O	73-159	4.0	3.8
	-8 H_2O	190-560	10.3	10.5
	-(3SO_3+O_2)	690-850	22.3	22.0
Co-zippeite	-5 H_2O	66-152	3.9	4.0
	-8 H_2O	160-549	10.0	10.3
	-(3SO_3+O_2)	690-860	21.6	22.0
Ni-zippeite	-5 H_2O	66-146	3.9	4.0
	-8 H_2O	180-485	10.0	10.4
	-(3SO_3+O_2)	680-830	21.6	21.9
Zn-zippeite	-5 H_2O	80-159	3.8	3.9
	-8 H_2O	200-560	10.0	10.0
	-(3SO_3+O_2)	680-880	21.5	21.4

(Haacke and Williams 1979)

Coconinoite, $Fe_2^{3+}Al_2(UO_2)_2(PO_4)_4(SO_4)(OH)_2 \cdot 2H_2O$ (Young et al. 1966)

DTA: a large endotherm with double top between 150-250°C which corresponds to the loss of the large amount of water and a smaller endotherm near 800°C which corresponds to the loss of SO_3. Exotherms below 100 and at 600°C are not reported (Young et al. 1966).

Uranyl carbonates

Rutherfordine, UO_2CO_3 (Fig. 48) (Cejka 1969; Vulpius et al. 1997; Meinrath 1997)

Synthetic uranyl carbonate—DTA: endotherm at ~610°C (decarbonation and oxygen release); TG: mass loss corresponding to CO_2 and O_2 release (~550-640°C). Synthetic uranyl carbonate decomposes practically in one step forming amorphous UO_3, which immediately transforms into $UO_{2.67}$. Both decomposition processes overlap each other. However, a crystalline phase UO_{3-x} ($x \approx 0.1$) is thought to be an intermediate (Cejka 1969; Cejka and Urbanec 1973, 1988a and references therein).

Blatonite, $UO_2CO_3 \cdot H_2O$ (Vochten and Deliens 1998)

Under a constant N_2 flow and heating rate 5°C.min^{-1}, the loss of H_2O and CO_2 is complete at 500°C (Vochten and Deliens 1998).

Joliotite, $UO_2CO_3 \cdot nH_2O$ ($n \leq 2$) (Walenta 1976)

Joliotite dehydrates at 100°C, an X-ray amorphous phase forms up to 300°C. At this temperature anhydrous phase probably may begin to decompose. No TG and DTA curves are available (Walenta 1976).

Sharpite, $Ca(UO_2)_6(CO_3)_5(OH)_4 \cdot 6H_2O$ (Fig. 45) (Cejka et al. 1984b)

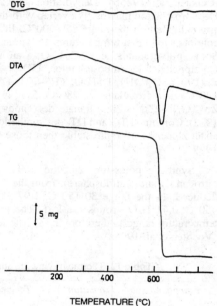

Figure 48. TG-DTG-DTA curves of synthetic rutherfordine simultaneously recorded; MOM Derivatograph OD 101, heating rate 10°C min^{-1}, static air atmosphere, 200 mg (from Cejka and Urbanec 1990).

TG: sharpite dehydrates and dehydroxylates in two steps: 3 H_2O (50-212°C), 5 H_2O (212-330°C—an overlap of dehydration and dehydroxylation). Anhydrous phase decarbonates in two steps: four CO_2 (330-500°C), one CO_2 (500-600°C). Further steps of the thermal decomposition of sharpite correspond to the thermal decomposition of becquerelite (Cejka and Urbanec 1983; Cejka et al. 1984b).

Grimselite, $K_3Na[UO_2(CO_3)_3] \cdot H_2O$ (Walenta 1972)

Synthetic phase—DTA: endotherms at 260°C(?), 405, 475, 730, 835°C (without any assignment). Endotherms may indicate a series of decomposition reactions, polymorphic transformations and other solid state reactions. X-ray powder patterns after heating of grimselite at 600°C indicate an unknown crystalline phase. The decomposition product at 800°C is X-ray amorphous (Walenta 1972).

Andersonite, $Na_2Ca[UO_2(CO_3)_3] \cdot nH_2O$ (n = 5.33—Mereiter 1986; n = 5.6—Coda et al. 1981; n = 6—Axelrod et al. 1951) (Cejka 1969; Cejka et al. 1987)

Synthetic phase—DTA: endotherms at 190°C (dehydration), 360°C (destruction of the structure), 710°C (a monoclinic phase $Na_{0.75}Ca_{0.25}UO_{3.625}$ formation), 870°C (unreacted $CaCO_3$ decomposition, unreacted sodium carbonate melting, hexagonal phase $Na_{0.5}Ca_{0.5}UO_{3.75}$ formation).

Natural phase releases 4 H_2O (50-155°C), 1.6 H_2O (155-315°C) (dehydration), 1.5 CO_2 (315-435°C), the remaining CO_2 is released in two steps (435-690 and 690-830°C). Decarbonation processes overlap. The last dehydration step and the beginning of decarbonation also overlap. According to Cejka (1969), a monoclinic phase $Na_{0.75}Ca_{0.25}UO_{3.625}$ is formed at temperatures close to 710°C. During further heating at higher temperatures (710-800°C), this monoclinic phase reacts with sodium carbonate and calcium carbonate and a hexagonal phase $Na_{0.5}Ca_{0.5}UO_{3.75}$ is formed. Within 800-900°C, when both $Na_{0.75}Ca_{0.25}UO_{3.625}$ and $Na_{0.5}Ca_{0.5}UO_{3.75}$ co-exist, unreacted calcium carbonate begins to decompose, whereas unreacted sodium carbonate undergoes melting. At elevated temperatures, both phases react with each other as well as with the sodium carbonate flux and the calcium oxide, thus being gradually converted to the monoclinic phase, the composition of which is close to $Na_{0.5}Ca_{0.5}UO_{3.625}$. This phase differs from the initial monoclinic phase by small shifts in the X-ray diffraction peaks (Cejka 1969; Cejka and Urbanec 1988b; Cejka et al. 1987). These phases formed during thermal decomposition of andersonite (Cejka 1969) and described in detail by Spitsyn et al. (1962) may be structurally related to clarkeite (Finch and Ewing 1997).

$Na_{0.63}Ca_{1.54}[UO_2(CO_3)_3] \cdot xH_2O$ (x = 5.38), a synthetic liebigite-andersonite intermediate (Vochten et al. 1994)

DSC (dynamic nitrogen atmosphere): four endotherms and an inflection up to 200°C, one endotherm in the range 200-300°C three endotherms (one weak) in the range 300-400°C.

TG: three mass losses related to dehydration, five mass losses indicating decarbonation (DSC and TG curves are not interpreted). Dehydration and decarbonation may partly overlap which can be inferred from the character of the TG curve (Vochten et al. 1994).

$Na_4[UO_2(CO_3)_3]$, unnamed mineral (Ondrus et al. 1997; Cejka 1969; this study)

Synthetic phase—DTA: endotherms at 440°C (decomposition of the uranyl tricarbonate), and 870°C (melting of unreacted Na_2CO_3). TG: synthetic phase decomposes approximately in the range 300-550°C which is associated with the formation of $Na_2U_2O_7$ and Na_2CO_3 and a release of 1.5 CO_2. The sodium diuranate gradually reacts with sodium carbonate to form α-Na_2UO_4 (approximately within 650-750°C, 0.5 CO_2 is lost) and at about 870°C the unreacted sodium carbonate melts. An irreversible polymorphic transformation of α-Na_2UO_4 to β-Na_2UO_4 takes place approximately within the range 900-950°C. Finally, β-Na_2UO_4 reacts with fused sodium carbonate to form Na_4UO_5. This reaction is associated with the release of the last CO_2. The mechanism of the thermal decomposition of the synthetic phase related to the unnamed mineral agrees with that proposed by Cejka in 1969 (Cejka 1969; Ondrus et al. 1997).

Bayleyite, $Mg_2[UO_2(CO_3)_3] \cdot 18H_2O$ (Cherkasov et al. 1968; Mayer and Mereiter 1986)

Synthetic phase: Cherkasov et al. (1968) found two steps of dehydration: 50-200°C (endotherms at 50 and 120°C), 10 H_2O in the range 20-60°C; 7.5 H_2O up to 200°C; the remaining water is associated with the formation of an anhydrous phase at 300°C. The anhydrous phase releases CO_2 in the range 300-800°C. According to Mayer and Mereiter (1986), the TG curve shows the compound dehydrates completely in the temperature range 45-200°C, yielding a yellow X-ray amorphous intermediate. DTG and DTA curves show synthetic bayleyite dehydrates in three apparently overlapping steps (6+6+6 H_2O). An endotherm at 55°C (no mass change) corresponds to incongruent melting of crystals, followed by the first dehydration step. Two further endotherms (90 and 130°C) are associated with dehydration, whereas endotherms at 335 and 530°C are associated with decarbonation (2 + 1 CO_2 up to 600°C). An exotherm at 600°C apparently corresponds to formation of crystalline $MgUO_4$ and MgO (Mayer and Mereiter 1986).

Liebigite, $Ca_2[UO_2(CO_3)_3]\cdot 11H_2O$ (Cejka 1969; Cejka and Urbanec 1979c; Amairy et al. 1997)

DTA: endotherms at 105°C—dehydration, 480°C—destruction of the structure and partial decarbonation. TG: dehydration proceeds in two steps, ~9 H_2O (25-150°C) and the remaining water is lost up to 300°C. From the course of the TG curve, it is inferred that 6 H_2O are released in the range 25-100°C, 3.5 H_2O in the range 100-200°C and the remaining water in the range 200-300°C. Decarbonation proceeds in three or four steps, respectively, 2 CO_2 (300-510°C) or ~0.5 CO_2 (300-450°C) and ~1.5 CO_2 (450-510°C), ~0.5 CO_2 (510-650°C) and 0.5 CO_2 650-900°C. The decomposition steps partly overlap. During thermal decomposition of liebigite, $UO_{2.67}$, $CaUO_4$, Ca_3UO_6 and Ca_2UO_5 are formed. Ca_2UO_5 is the end product of the thermal decomposition of liebigite (900-1200°C) (Cejka 1969; Cejka and Urbanec 1977, 1979c; Ambartsumyan et al. 1961; Amayri et al. 1997).

Roubaultite, $Cu_2(UO_2)_3(CO_3)_2O_2(OH)_2\cdot 4H_2O$ (Cesbron 1970c)

TG: roubaultite dehydrates approximately in the range 200-500°C and dehydroxylates between 520-560°C (Cesbron et al. 1970c). Two steps found on the TG curve by Cesbron et al. in fact correspond to dehydration (5 H_2O) and decarbonation (2 CO_2) (Ginderow and Cesbron 1985). CuO and $UO_{2.67}$ or copper uranates may be formed at higher temperatures.

Schröckingerite, $NaCa_3[UO_2(CO_3)_3]SO_4F\cdot 10H_2O$ (Ambartsumyan et al. 1961; Cejka and Urbanec 1980)

The mineral loses six H_2O at room temperature over H_2SO_4 or when heated in air at 75°C (Hurlbut 1954; Smith 1959). DTA: endotherms at 130°C (dehydration), and 390-400°C (decomposition of anhydrous phase), exotherm at 400-410°C (formation of an ill-defined crystalline cubic phase; Ambartsumyan et al. 1961). TG: 10 H_2O (60-280°C), followed by decarbonation (over 375°C) and solid state reactions. Ca_2UO_5 is one of the end products of the thermal decomposition (Ambartsumyan et al. 1961; Cejka and Urbanec 1977, 1980, 1990).

Lepersonnite, $CaO\cdot(REE)_2O_3\cdot 24UO_3\cdot 8CO_2\cdot 4SiO_2\cdot 60\ H_2O$ (Deliens and Piret 1982)

DTA: endotherms 40, ~85, ~125°C—dehydration(?); TG: 35-290°C (dehydration), 310-500°C (decarbonation) (Deliens and Piret 1982).

Wyartite, $CaU^{5+}(UO_2)_2(CO_3)(OH)(H_2O)_7$ (Guillemin and Protas 1959)

DTA: endotherms at 80 and 160°C (dehydration and dehydroxylation?), 630°C (decarbonation and $UO_{2.67}$ formation), exotherm at 280°C (U^{4+} oxidation) (Guillemin and Protas 1959). Bignand (1955) reports dehydration proceeds up to 260°C and decarbonation up to 650°C. According to Burns and Finch (1999), the structural formula of wyartite corresponds to $CaU^{5+}(UO_2)_2(CO_3)(OH)(H_2O)_7$.

Widenmannite, $Pb_2[UO_2(CO_3)_3] \cdot xH_2O$ (x=?) (Walenta 1976)

Widenmannite was heated at 100 and 200°C without any changes. A color change was observed at 300°C and the structure of widenmannite was destructed at 600°C (Walenta 1976).

Urancalcarite, $Ca(UO_2)_3CO_3(OH)_6 \cdot 3H_2O$ (Deliens and Piret 1984)

TG: water and CO_2 are released up to 450°C (Deliens and Piret 1984).

Kamotoite-(Y), $4UO_3 \cdot [(Y,Nd,Gd,Sm,Dy)_2O_3 \cdot 3CO_2 \cdot 14.5H_2O$ (Botto et al. 1989b)

DTA: endotherms at 60 and 85°C—dehydration, 540°C—decarbonation. TG: dehydration ($\rightarrow$ 85°C), decarbonation ($\rightarrow$ 540°C). The product of pyrolysis of kamotoite is a mixture of Ln_2O_3 and UO_3 detected at 700°C (Botto et al. 1989b).

Znucalite, $Zn_{12}(UO_2)Ca(CO_3)_3(OH)_{22} \cdot 4H_2O$ (Ondrus et al. 1990)

DTA: endotherms at ~ 400°C—dehydration and dehydroxylation, a broad endotherm at 600°C, corresponds to decarbonation. TG: adsorbed water is lost (110°C), dehydration and dehydroxylation (180-450°C- the process proceeds in two steps), decarbonation (450-960°C). Uranates and polyuranates of Zn were found above 500°C (Ondrus et al. 1990).

Uranyl selenites and tellurites

Derriksite, $Cu_4(UO_2)(SeO_3)_2(OH)_6 \cdot H_2O$

Derriksite decomposes in several steps (Cesbron et al. 1971). According to the TG curve, it dehydrates (1 H_2O) up to 272°C, dehydroxylation occurred in the range 272-367°C (3 H_2O) without formation of an anhydrous phase. In the range 367-463 and 463-646°C, two SeO_2 molecules (1+1) are released. The release of the second SeO_2 molecule is probably accompanied by UO_3 decomposition to $UO_{2.67}$. It seems possible that copper uranate, CuO and $UO_{2.67}$ form end products of derriksite thermal decomposition (Cesbron et al. 1971).

Interpretation of the TG curve of **guilleminite** (Pierrot et al. 1965) is difficult to rectify with the structural formula, $Ba(UO_2)_3(SeO_3)_2O_2(H_2O)_3$, reported by Cooper and Hawthorne (1995). Guilleminite dehydrates in two steps: 3 H_2O (50-150°C) and 2 H_2O (150-250°C), the latter being thought to be dehydroxylation; however, no hydroxyl ions are reported from the structure determination. After complete dehydration, SeO_2 is released in several steps up to 960°C. Barium uranate and also $UO_{2.67}$ are probably formed. The temperature range for all the processes indicated above and possible overlapping are difficult to verify. If all SeO_2 is incrementally released, the process should be partly overlapped with that of dehydration. However, the temperature range (below 250°C) for such SeO_2 release is low. On the other hand, if 3 H_2O are accepted as correct, there is a difference between water content determined from crystal structure study (4.2 wt %) and the observed mass decrease up to 250°C (6.0 wt %). The sum of the mass loss at 960°C (21.8 wt %) corresponds with the theoretical value derived from the new formula (21.436 wt %). The observed molecular weight inferred from the TG curve of guilleminite is 1293.39 amu, and that calculated from XRD study is 1287.40 amu

If the water content in **marthozite**, $Cu(UO_2)_3(SeO_3)_3(OH)_{12} \cdot 7H_2O$, is correct (10.61 wt %), it is not possible to interpret the TG curve of marthozite reported by Cesbron et al. (1969). At 100°C, associated with the loss of 3 H_2O, a metaphase containing (7+1) H_2O is formed. At 180°C (all molecular water lost is ~ 7 H_2O), a monoclinic phase is formed. Dehydration and dehydroxylation of marthozite proceed up to 485°C (10.5 wt %, 11 H_2O). If all SeO_2 is released up to 945°C, the observed mass loss is 17.7 wt % (16.3 wt % SeO_2 and 1.4 wt % O_2 corresponding to $UO_{2.67}$ formation) (Cesbron et al., 1969). However, the mass loss related to SeO_2 (16.3 wt %) is less than that determined by chemical analysis (23.9 wt % SeO_2) (Cesbron et al., 1969). New chemical analyses of marthozite seem necessary.

The TG curve of **piretite**, $Ca(UO_2)_3(SeO_3)_2(OH)_4 \cdot 4H_2O$, shows that dehydration and dehydroxylation do not take place in distinct steps (Vochten et al., 1996).

Cliffordite, UTe_3O_9, decomposes above 750°C. This decomposition is associated with formation of **schmitterite**, UO_2TeO_3, and a loss of two molecules TeO_2. Subsequent heating of schmitterite to 950°C leads to $UO_{2.67}$ and TeO_2 and O_2 release (Botto, 1984; Misra et al., 1998).

Uranyl tungstates (wolframates)

Uranotungstite, $(Fe,Ba,Pb)(UO_2)_2(WO_4)(OH)_4 \cdot 12H_2O$, releases most of its water up to 120°C. Dehydration is complete at 300°C. No further mass loss was observed at 350°C (Walenta, 1985). No TG or DTA curves are available.

ACKNOWLEDGMENTS

The author would like to express his thanks to Bob Finch and Peter Burns for inviting him to contribute to this volume of Reviews in Mineralogy, to the reviewers Bob Finch, Janusz Janeczek, and John M. Hanchar for insightful reviews of the manuscript and their suggestions, to Peter Burns, Bob Finch and Rod Ewing for providing him with preprints of their studies submitted for publication, to Bob Finch and John M. Hanchar for careful revision of the English text, Jana Ederová and Miloslava Novotná (both of the Prague Institute of Chemical Technology) for recording some of thermoanalytical curves and measuring some of the infrared spectra, to Jan Holec (National Museum Prague) for his help with the figures in this manuscript, and to his son Jan Cejka and his daughter-in-law Bohunka Cejková for their help with preparation of the manuscript. Financial support for the research on uranyl minerals realized in the National Museum Prague has been provided by the State Department of Culture of the Czech Republic with the grant PK96MO5BP122.

REFERENCES

Abraham F, Dion Ch, Saadi M 1993 Carnotite analogues: synthesis, structure and properties of the $Na_{1-x}K_xUO_2VO_4$ solid solution ($0 \leq x \leq 1$). J Mater Chem 3:459-463

Abraham F, Dion Ch, Tancret N, Saadi M 1994 $Ag_2(UO_2)_2V_2O_8$: a new compound with the carnotite structure. Synthesis, structure and properties. Advanced Mater Research Vols 1-2 Scitec Publications Switzerland, p 511-520

Akhmanova MV, Karyakin AV, Yukhnevich GV 1963 Determination of hydroxyl groups in silicate minerals with IR-spectroscopy. Geokhimiya 1963:581-585 (in Russian)

Alekseeva MA, Chernikov AA, Shashkin DP, Konkova EA, Gavrilova IN 1974 Strelkinite—a new uranyl vanadate. Zap Vses Miner Obshch 103:576-580 (in Russian)

Allen PG, Shuh DK, Bucher JJ, Edelstein NM, Palmer CEA, Silva RJ, Nguyen SN, Marquez LN, Hudson EA 1996 Determinations of uranium structures by EXAFS: Schoepite and other U(VI) oxide precipitates. Radiochim Acta 75:47-53

Amairy S, Geipel G, Matz W, Schuster G, Baraniak L, Bernhard G, Nitsche H 1997 Synthesis and characterization of calcium uranyl carbonate: $Ca_2[UO_2(CO_3)_3]\cdot 10H_2O$ (Liebigite). Annual Report 1997, p 14-15, Inst Radiochem Forschungszentrum, Rossendorf

Ambartsumyan TsL, Basalova GI, Gorzhevskaya SA, Nazarenko NG, Khodzhayeva RP 1961 Thermal investigations of uranium and uranium-bearing minerals. Gosatomizdat, Moscow, 148 p (in Russian)

Anderson A, Chieh Ch, Irish DE, Tong JPK 1980 An X-ray crystallographic, Raman, and infrared spectral study of crystalline potassium uranyl carbonate, $K_4UO_2(CO_3)_3$. Can J Chem 58:1651-1658

Appleman DE, Evans HT Jr 1965 The crystal structure of synthetic anhydrous carnotite, $K_2(UO_2)_2V_2O_8$ and its cesium analogue, $Cs_2(UO_2)_2V_2O_8$. Am Mineral 50:825-842

Arkhipenko DK, Kovaleva LT 1978 Analysis of vibrational spectra of H_3O^+ and $H_3O^+\cdot 3H_2O$ ions. In Rentgenografiya i spektroskopiya mineralov. VS Sobolev (Ed) p 17-26, Nauka, Novosibirsk (in Russian)

Arkhipenko DK, Bokiy GB 1981 Determination of oxonium ions in minerals by means of physicochemical investigation methods. Mineral Zh Kiev 3:27-34 (in Russian)

Atencio D 1991 Furcalita e outros minerais uraníferos secundarios de Perus, Sao Paulo. Thesis, Universidade de Sao Paulo Instituto de Geociencias, Sao Paulo, 147 p

Atencio D, Neumann R, Silva AJGC, Mascarenhas YP 1991 Phurcalite from Perus, Sao Paulo, Brazil, and redetermination of its crystal structure. Can Mineral 29:95-105

Axelrod JM, Grimaldi FS, Miltom Ch, Murata KJ 1951 The uranium minerals fro the Hillside mine, Yavapai County, Arizona. Am Mineral 36:1-22

Badger RM 1934 A relation between the internuclear distances and bond force constants. J Chem Phys 2:128-131

Badger RM 1935 The relation between internuclear distances and force constants of molecules and its application to polyatomic molecules. J Chem Phys 3:710-714

Bagnall KW 1983 Uranium carbonates and carbonato complexes. In Gmelin Handbook of inorganic chemistry 8th edn Supplement Volume C13:1-26, Springer-Verlag, Berlin

Bagnall KW, Wakerley MW 1975 Infrared and Raman spectra of the uranyl ion. J Inorg Nucl Chem 37:329-330

Baran EJ, Botto IL 1976 Das Schwingungsspektrum des synthetischen Carnotits. Monatshefte Chem 107:633-639

Baran EJ, Botto IL 1977 Die IR-Spektren von $NH_4UO_2AsO_4 \cdot 3H_2O$ und $NH_4UO_2PO_4 \cdot 3H_2O$. Monatshefte Chem 108:781-790

Baran V 1982 Ammonium uranyl carbonate—complex compound with variable coordination number? Coll Czechoslov Chem Commun 47:1269-1281

Baran V 1992 Uranium(VI)- oxygen chemistry. Nuclear Research Institute, Rez, 136 p

Baran V, Tympl M 1966 Infrared spectra of sodium uranates. Z anorg allg Chem 347:175-183

Baran V, Unzeitig M 1991 Uranyl oxide hydrate minerals from the standpoint of coordination chemistry. Neues Jb Miner Mh 2:63-75

Baran V, Sourková L, Spalová J 1985 Water-free precipitation uranate with maximum sodium content Na/U=1.0. J Radioanal Nucl Chem Letters 95:331-338

Barten H 1980 Thermochemical investigations on uranyl phosphates and arsenates. Thesis, ECN Report-188, Netherlands Energy Research Foundation, Petten, 145 p

Barten H, Cordfunke EHP 1980a A study on the thermal stability of uranyl phosphates $(UO_2)_3(PO_4)_2$, $(UO_2)_2P_2O_7$, and $UO_2(PO_3)_2$ using a static non-isothermal method. J Inorg Nucl Chem 42:75-78

Barten H, Cordfunke EHP 1980b The formation and stability of hydrated and anhydrous uranyl phosphates. Thermochim Acta 40:357-365

Barten H, Cordfunke EHP 1980c The formation and stability of hydrated and anhydrous uranyl arsenates. Thermochim Acta 40:367-376

Baturin SV, Sidorenko GA 1985 Crystal structure of weeksite $(K_{0.62}Na_{0.38})_2(UO_2)_2(Si_5O_{13})\cdot 3H_2O$. Doklady Akad Nauk SSSR 282:1132-1136 (in Russian)

Belova LN, Fedorov OV 1974 New data on wölsendorfite. Zap Vses Miner Obshch 103: 718-719 (in Russian)

Benyacar MAR De, Abeledo MEJ De 1974 Phase transition in synthetic troegerite at room temperature. Am Mineral 59:763-767

Bignand C 1955 Sur les propriétés et les synthèses de quelques minéraux uranifères. Bull Soc fr Minér Crist 78:1-26

Bist HD, Pant DD 1964 Infrared spectra of uranyl salts. J Pure Appl Phys 2:107-120

Biwer BM, Ebert WL, Bates JK 1990 The Raman spectra of several uranyl-containing minerals using a microprobe. J Nucl Mater 175:188-193

Boldyrev AI 1976 Infrared spectra of minerals. Nedra, Moscow 199 p (in Russian)

Borène J, Cesbron F 1971 Structure cristalline de la curiénite, Pb(UO$_2$)$_2$(VO$_4$)$_2$·5H$_2$O. Bull Soc fr Minér Crist 94:8-14

Botto IL 1984 Spectroscopic characterization of synthetic schmitterite. Acta SudAmericana Quím 4:71-77

Botto IL, Baran EJ 1976 Über Ammonium-Uranyl-Vanadat und die Produkte seiner thermischen Zersetzung. Z anorg allg Chem 426:321-332

Botto IL, Baran EJ 1982 Die IR-Spektren der Phasen $A_{1/2}{}^{3+}B_{1/2}{}^{5+}Te_3O_8$ (A = Fe,In,Sc; B = Nb,Ta) und UTe$_3$O$_9$. Z anorg allg Chem 484:210-214

Botto IL, Baran EJ, Deliens M 1989a Vibrational spectrum of natural and synthetic metatyuyamunite. Neues Jb Miner Mh 5:212-218

Botto IL, Garcia AC, Deliens M 1989b Thermal and IR spectroscopic characterization of kamotoite. Coll Czechoslov Chem Commun 54:1263-1268

Boyko NW, Seryozhkina LB, Seryozhkin VN 1985 Synthesis and physico-chemical investigation of calcium hydroxo chromato uranylates. Radiokhimiya 26:183-186 (in Russian)

Braithwaite RSW, Paar WH, Chisholm JE 1989 Phurcalite from Dartmoor, Southwest England, and its identity with "nisaite" from Portugal. Mineral Mag 53:583-589

Branche G, Ropert ME, Chantret F, Morignat B 1957 La francevillite, nouveau minéral uranifère. C R Acad Sci Paris 245:89-91

Brown ID 1995 Anion-anion repulsion, coordination number, and the asymmetry of hydrogen bonds. Can J Phys 73:676-682

Brown ME 1988 Introduction to thermal analysis. Techniques and applications. Chapmann and Hall, London

Bukalov SS, Vdovenko VM, Ladygin IN, Suglobov DN 1970 Raman spectra of the uranyl anion complexes. Zh Prikl Spektroskopii 12:341-344 (in Russian)

Bullock JI 1969 Raman and infrared spectroscopic studies of the uranyl ion: the symmetry stretching frequency, force constants, and bond lengths. J Chem Soc A1969:781-784

Bullock JI, Parrett FW 1970 The low frequency infrared and Raman spectroscopic studies of some uranyl complexes: the deformation frequency of the uranyl ion. Can J Chem 48:3095-3097

Burns PC 1997 A new uranyl oxide hydrate sheet in vandendriesscheite: Implications for mineral paragenesis and the corrosion of spent nuclear fuel. Am Mineral 82:1176-1186

Burns PC 1998a Implications of solid-solution in the structure of boltwoodite. Can Mineral 36:1069-1075

Burns PC 1998b The structure of compreignacite, K$_2$[(UO$_2$)$_3$O$_2$(OH)$_3$]$_2$(H$_2$O)$_7$. Can Mineral 36:1061-1067

Burns PC 1999a The crystal chemistry of uranium. Chapter 3, this volume

Burns PC 1999b Wölsendorfite: a masterpiece of structural complexity. Geol Assoc Can–Mineral Assoc Can Joint Ann Mtg, Sudbury 1999, Abstracts 24:16

Burns PC 1999c A new complex sheet of uranyl polyhedra in the structure of wölsendorfite. Am Mineral (in press)

Burns PC 1999d A new uranyl phosphate chain in the structure of parsonsite. Am Mineral (submitted)

Burns PC 1999e A new uranyl silicate sheet in the structure of haiweeite and comparison to other uranyl silicates. Can Mineral (submitted)

Burns PC, Finch RJ 1999 Wyartite: Crystallographic evidence for the first pentavalent uranium mineral. Am Mineral (in press)

Burns PC, Hanchar JM 1999 The structure of masuyite, Pb[(UO$_2$)$_3$O$_3$(OH)$_2$](H$_2$O)$_3$. Can Mineral (in press)

Burns PC, Miller ML, Ewing RC 1996 U^{6+} minerals and inorganic phases: a comparison and hierarchy of crystal structures. Can Mineral 34:845-880

Burns PC, Hawthorne FC, Ewing RC 1997a) The crystal chemistry of hexavalent uranium: Polyhedral geometries, bond-valence parameters, and polyhedral polymerization. Can Mineral 35:1551-1570

Burns PC, Finch RJ, Hawthorne FC, Miller ML, Ewing RC 1997b The crystal structure of ianthinite, [U$_2{}^{4+}$(UO$_2$)$_4$O$_6$(OH)$_4$(H$_2$O)$_4$](H$_2$O)$_5$: A possible phase for Pu^{4+} incorporation during the oxidation of spent nuclear fuel. J Nucl Mater 249:199-206; Mater Res Soc Symp Proc 465:1193-1200

Buryanova EZ, Strokova GS, Shitov VA 1965 Vanuranylite—a new mineral. Zap Vses Miner Obshch 94:437-443 (in Russian)

Camargo WGR De 1971 Minerais uraniferos de Perus, SP. Boletim IGA No 2:85-201

Camargo WGR De, Souza IM 1975 Thermal behavior of β-uranophane. Kristall und Technik 10:571-579

Casas I, Perez I, Torrero E, Bruno J, Cera E, Duro L 1997 Dissolution studies of synthetic soddyite and uranophane. SBK Technical Report 97-15, Swedish Nuclear Fuel and Waste Management Co., Stockholm, 36 p

Cejka Jan 1988 Influence of preparation on the character of the IR spectra of inorganic compounds and minerals (the study of the solid state reactions in the system CuSO$_4$.nH$_2$O and KBr). Unpublished report, Prague Inst Chem Tech, 42 p (in Czech)

Cejka Jan 1994 Study of synthetic analogues of the minerals in the PbO-UO$_3$-H$_2$O system. Thesis, Prague Inst Chem Tech, 93 p (in Czech)

Cejka J 1969 To the chemistry of andersonite and thermal decomposition of dioxotri-carbonatouranates. Coll Czechoslov Chem Commun 34:1635-1656

Cejka J 1987 Mechanism of the thermal decomposition of secondary uranium minerals with regard to its technological application. *In* Proc Mineral Seminar: Chemistry and physical properties of minerals and their technological applications. DT CSVTS Ústí nL, p 37-40 (in Czech)

Cejka J 1988 Methods of thermal analysis in mineralogy. *In* Proc Mineral Seminary "Modern methods in applied mineralogy" DT CSVTS Ústí nL, p 102-106 (in Czech)

Cejka a J 1992 Thermal analysis of uranyl minerals I. Introduction. Casopis Národního muzea rada prírodovedná 159:101-104 (in Czech)

Cejka J 1993a To the problem of molecular water, OH⁻ and H_3O^+ ions in the structure of uranyl silicates of the uranophane group. *In* Proc 5th Mineral Seminary Audit VT Ústí nL, p82-84 (in Czech)

Cejka J 1993b Thermal analysis of uranyl minerals II. Uranyl hydroxide hydrates and uranates. Casopis Národního muzea rada prírodovedná 161: 89-96 (in Czech)

Cejka 1994a Vandenbrandeite, $CuUO_2(OH)_4$: Thermal analysis and infrared spectrum Neues Jb Miner Mh 3:112-120

Cejka J 1994b Thermal analysis of uranyl minerals III.Uranyl silicates. Casopis Národního muzea rada prírodovedná 162:93-99 (in Czech)

Cejka J, Urbanec Z 1973 Chemie der Uranylcarbonate I. Das System Rutherfordin-Sharpit- Schoepit. Coll Czechoslov Chem Commun 38:2327-2346

Cejka J, Urbanec Z 1975 Secondary uranium minerals in the collections of the National Museum in Prague III. Sharpite. Casopis Národního muzea rada prírodovedná 144:11-19 (in Czech)

Cejka J, Urbanec Z 1976 Secondary uranium minerals in the collections of the National Museum in Prague IV. Liebigite. Casopis Národního muzea rada prírodovedná 145:26-36 (in Czech)

Cejka J, Urbanec Z 1977 The application of thermal analysis in the scientific reassessment of secondary uranium minerals deposited in the National Museum in Prague. Casopis Národního muzea rada prírodovedná 146:114-125

Cejka J, Urbanec Z 1979a Thermal analysis of natural uranyl sulphates. *In* Proc 8th Conf Thermal Anal Termanal 79´ High Tatras, p 177-180 (in Czech)

Cejka J, Urbanec Z 1979b Secondary uranium minerals in the collections of the National Museum in Prague V. Voglite. Casopis Národního muzea rada prírodovedná 148:69-76 (in Czech)

Cejka J, Urbanec Z 1979c Contribution to the thermal decomposition of liebigite. Casopis Národního muzea rada prírodovedná 146:177-180 (in Czech)

Cejka J, Urbanec Z 1980 Secondary uranium minerals in the collections of the National Museum in Prague VI. Schröckingerite. Casopis Národního muzea rada prírodovedná 142:60-69 (in Czech)

Cejka J, Urbanec Z 1981 Thermal and infrared spectrum analyses of johannite. *In* Proc 2nd Europ Symp Thermal Analysis Aberdeen. D Dollimore (Ed) Heyden, London, p 520-524

Cejka J, Urbanec Z 1983 To the thermal decomposition of mineral becquerelite, $CaO.6UO_3 \cdot (10-11)H_2O$. Casopis Národního muzea rada prírodovedná 152:26-36 (in Czech)

Cejka J, Urbanec Z 1988a Contribution to the hydrothermal origin of rutherfordine, UO_2CO_3. Casopis Národního muzea rada prírodovedná 157:1-10 (and references therein)

Cejka J, Urbanec Z 1988b Thermal and infrared spectrum analyses of natural and synthetic andersonite. J Thermal Anal 33:389-394

Cejka J, Urbanec Z 1990 Secondary uranium minerals. Transactions of the Czechoslovak Academy of Sciences, Math Natur History Series 100:93 p, Academia, Prague

Cejka J, Sejkora J 1994 Zippeite group minerals. *In* Proc 6th Mineral Seminar Audit Ústí nL, p 28-30 (in Czech)

Cejka J, Mrázek Z, Urbanec Z, Vasí_cková S 1982a High temperature X-ray, thermal and infrared spectrum analysis of johannite and its deuteroanalogues. *In* Proc 7th Int'l Conf Thermal ICTA Kingston Vol 1 (B Miller, ed) Wiley-Heyden, Chichester, p 713-718

Cejka J, Mrázek Z, Urbanec Z 1982b Molecular water in uranopilite. *In* Proc 9th Conf Thermal Anal Termanal 82´ High Tatras, p 87-88 (in Czech)

Cejka J, Mrázek Z, Urbanec Z 1984a Secondary uranium minerals in the collections of the National Museum in Prague IX. Uranopilite. Casopis Národního muzea rada prírodovedná 153:172-182 (in Czech)

Cejka J, Mrázek Z, Urbanec Z 1984b New data for sharpite, a calcium uranyl carbonate. Neues Jb Miner Mh 3:109-117

Cejka J, Cejka J Jr, Muck A 1985a Thermal analysis and infrared spectra of some natural and synthetic uranium micas. Thermochim Acta 86:387-390

Cejka J, Mrázek Z, Urbanec Z 1985b Secondary uranium minerals in the collections of the National Museum in Prague X. Zippeites. Casopis Národního muzea rada prírodovedná 154:94-106 (in Czech)

Cejka J, Urbanec Z, Mrázek Z 1986a Secondary uranium minerals in the collections of the National Museum in Prague XII. Uranyl silicates. Casopis Národního muzea rada prírodovedná 155:30-44 (in Czech)
Cejka J, Cejka J Jr, Muck A 1986b Contribution to the structural chemistry of uranium micas V. Metauranocircite I and metauranocircite II. Casopis Národního muzea rada prírodovedná 155:81-90 (in Czech)
Cejka J, Urbanec Z, Cejka J Jr 1987 To the crystal chemistry of andersonite. Neues Jb Miner Mh 11:488-511 (and references therein)
Cejka J, Urbanec Z, Cejka J Jr, Mrázek Z 1988a Contribution to the thermal analysis and crystal chemistry of johannite, $Cu[(UO_2)_2(SO_4)_2(OH)_2] \cdot 8H_2O$. Neues Jb Miner Abh 159:297-309 (and references therein)
Cejka J, Urbanec Z, Cejka J Jr, Ederová J, Muck A 1988b Thermal and infrared spectrum analyses of sabugalite. J Thermal Anal 33:395-399
Cejka J, Cejka J Jr, Urbanec Z 1988c Becquerelite: thermal analysis and infrared spectrum. *In* Proc 11th Conf Thermal Anal Termanal 88´ High Tatras, p 95-96 (in Czech)
Cejka J, Sejkora J, Deliens M 1996a New data on studtite, $UO_4 \cdot 4H_2O$ from Shinkolobwe, Shaba, Zaire. Neues Jb Miner Mh 3:125-134 (and references therein)
Cejka J, Sejkora J, Mrázek Z, Urbanec Z, Jarchovsky T 1996b Jáchymovite, $(UO_2)_8(SO_4)(OH)_{14} \cdot 13H_2O$, a new uranyl mineral from Jáchymov, the Krusné Hory Mts., Czech Republic. Neues Jb Miner Abh 170:155-170 (and references therein)
Cejka J, Sejkora J, Skála R, Cejka Jan, Novotná M, Ederová J 1998a Contribution to the crystal chemistry of synthetic becquerelite, billietite and protasite. Neues Jb Miner Abh 174:159-180 (and references therein)
Cejka J, Cejka Jan, Skála R, Sejkora J, Muck A 1998b New data on curite from Shinkolobwe, Zaire. Neues Jb Miner Mh 9:385-402 (and references therein)
Cejka J, Sejkora J, Deliens M 1999 To the infrared spectrum of haynesite, a hydrated uranyl selenite, and its comparison with other uranyl selenites. Neues Jb Miner Mh 6:241-252
Cejka J Jr 1982 Infrared spectra of natural uranyl phosphates. Unpublished report, Prague Inst Chem Tech, 29 p (in Czech)
Cejka J Jr 1983 Contribution to the structural chemistry of uranium micas. Thesis, Prague Inst Chem Tech, 90 p
Cejka J Jr, Muck A 1985 Contribution to the structural chemistry of uranium micas IV. Metaautunite. Casopis Národního muzea rada prírodovedná 154:183-193 (in Czech)
Cejka J Jr, Muck A, Cejka J 1984a To the infrared spectroscopy of natural uranyl phosphates. Phys Chem Minerals 11:172-177
Cejka J Jr, Muck A, Hájek B 1984b Vibrational spectra of secondary uranium minerals metaautunite and metatorbernite. Sci Papers Prague Inst Chem Tech B29:175-188
Cejka J Jr, Muck A, Urbanec Z, Cejka J 1985a Thermal and infrared spectrum analyses of curite. In Proc 8th Int'l Conf Thermal Anal ICTA Bratislava 2:637-640. Thermochim Acta 93:637-640
Cejka J Jr, Muck A, Cejka J 1985b Infrared spectra and thermal analysis of synthetic uranium micas and their deuteroanalogues. Neues Jb Miner Mh 3:115-126
Cejka J Jr, Muck A, Hájek B, Cejka J 1985c To the crystal chemistry of metauranocircites. Sci Papers Prague Inst Chem Tech B30:67-81
Cejka J Jr, Muck A, Hájek B 1985d Contribution to the structural chemistry of uranium micas II. Metatorbernite. Casopis Národního muzea rada prírodovedná 154:36-44 (in Czech)
Cesbron F 1970a Etude cristallographique et comportement thermique des uranylvanadates de Ba, Pb, Sr, Mn, Co, et Ni. Bull Soc fr Minér Crist 93:320-327
Cesbron F 1970b Nouvelles donées sur la vanuralite. Existence de la méta-vanuralite. Bull Soc fr Minér Crist 93:242-248
Cesbron M, Morin N 1968 Une nouvelle espèce minérale: la curiénite. Etude de la série francevillite-curiénite. Bull Soc fr Minér Crist 91:453-459
Cesbron F, Oosterbosch R, Pierrot R 1969 Une nouvelle espèce minérale: la marthozite. Uranyl sélénite de cuivre hydraté. Bull Soc fr Minér Crist 92:278-283
Cesbron F, Pierrot R, Verbeek T 1970 La roubaultite, $Cu_2(UO_2)_3(OH)_{10} \cdot 5H_2O$, une nouvelle espèce minérale. Bull Soc fr Minér Crist 93:550-554
Cesbron F, Pierrot R, Verbeek T 1971 La derriksite, $Cu_4(UO_2)(SeO_3)_2(OH)_6 \cdot H_2O$, une nouvelle espèce minérale. Bull Soc fr Minér Crist 94:534-537
Cesbron F, Brown WL, Bariand P, Geffroy J 1972 Rameauite and agrinierite, two new complex uranyl oxides from Margnac, France. Mineral Mag 38:781-789
Chen Zhangru, Huang Yuzhu, Gu Xiaofa 1990 A new uranium mineral—yingjiangite. Acta mineral Sinica 10:102-105 (in Chinese)

Chen Zhangru, Luo Keding, Tan Falan, Zhang Yi, Gu Xiaofa 1986 Tengchongite, a new mineral of hydrated calcium uranyl molybdate. Kewue Tongbao 31:396-401

Cherkasov VA, Zhagin BP, Golandskaya ZD 1968 Synthesis of $Mg_2[UO_2(CO_3)_3] \cdot 18H_2O$ (bayleyite). Zh Neorg Khim 13:1205-1206 (in Russian)

Chernikov AA, Sidorenko GA, Valueva AA 1977 New data on minerals of the ursilite-weeksite group. Zap Vses Miner Obshch 96:553-564

Chernorukov NG, Korshunov IA, Voynova LI 1985 New uranium compounds. Uranium(IV) monohydrogenarsenate tetrahydrate. Radiokhimiya 27:676-679 (in Russian)

Chernorukov NG, Suleymanov EV, Knyazev AV, Suchkov AI 1998a Synthesis, structure and properties of $A''(VUO_6)_2 \cdot nH_2O$ (A'' = Ni, Zn,Cd). Zh Neorg Khim 43:1085-1089 (in Russian)

Chernorukov NG, Karyakin NV, Suleymanov EV, Alimzhanov MI 1998b Synthesis and study of uranovanadates of alkaline-earth metals., Russ J Gen Chem 68:839-843

Chernorukov NG, Karyakin NV, Suleymanov EV, Belova YuS 1998c Synthesis and study of compounds $Mg(PUO_6)_2 \cdot nH_2O$ and $Mg(AsUO_6)_2 \cdot nH_2O$. Zh Neorg Khim 43:380-383 (in Russian)

Chernorukov NG, Suleymanov EV, Dzhabarova ST 1998d Synthesis and study of compounds in the systems $Ni(PUO_6)_2-H_2O$ and $Ni(AsUO_6)_2-H_2O$. Zh Neorg Khim 43:1090-1095 (in Russian)

Chernyaev II (Ed) 1964 Uranium complex compounds. Nauka, Moscow, 491 p (in Russian)

Christ CL, Clark JR, Evans Jr HT 1955 Crystal structure of rutherfordine UO_2CO_3. Science 121:472-473

Coda A, Della Giusta A, Tazzoli V 1981 The structure of synthetic andersonite, $Na_2Ca[UO_2(CO_3)_3] \cdot xH_2O$ ($x \cong 5.6$). Acta Crystallogr B37:1496-1500

Colomban Ph, Pham Thi M, Novak A 1985 Vibrational study of phase transitions and conductivity mechanism in $H_3OUO_2PO_4 \cdot 3H_2O$ (HUP). Solid State Commun 53:747-751

Cooper MA, Hawthorne FC 1995 The crystal structure of guilleminite, a hydrated Ba-U-Se sheet structure. Can Mineral 33:1103-1109

Cordfunke EHP 1972 The system uranyl- sulphate-water II. Phase relationships and thermochemical properties of the phases in the system $UO_3-SO_3-H_2O$. J Inorg Nucl Chem 34:1551-1561

Cordfunke EHP, Prins G, Van Vlaanderen P 1968 Preparation and properties of the violet "U_3O_8 hydrate." J Inorg Nucl Chem 30:1745-1750

Dara AD, Sidorenko GA 1967 X-ray thermographic investigation of uranyl molybdates. Atom Energiya 23:126-133 (in Russian)

Deane, AM 1961 The infrared spectra and structure of some hydrated uranium trioxides and ammonium diuranates. J Inorg Nucl Chem 21:238-252

Deliens M, Piret P 1979a Les phosphates d' uranyle et d' aluminium de Kobokobo IV. La threadgoldite, $Al(UO_2)_2(PO_4)_2(OH) \cdot 8H_2O$. Bull Minéral 102:338-341

Deliens M, Piret P 1979b Ranunculite, $AlH(UO_2)(PO_4)(OH)_3 \cdot 4H_2O$, a new mineral. Mineral Mag 43:321-323

Deliens M, Piret P 1981 La swamboite, nouveau silicate d'uranium hydraté du Shaba, Zaire. Can Mineral 19:553-557

Deliens M, Piret P 1982 Bijvoetite et lepersonnite, carbonates hydratés d'uranyle et de terres rares de Shinkolobwe, Zaire. Can Mineral 20:231-238

Deliens M, Piret P 1983 L'oursinite, $(Co_{0.86}Mg_{0.10}Ni_{0.04})O.2UO_3 \cdot 2SiO_2 \cdot 6H_2O$, nouveau minéral de Shinkolobwe, Zaire. Bull Minéral 106:305-308

Deliens M, Piret P 1984 L'urancalcarite, $Ca(UO_2)_3CO_3(OH)_6 \cdot 3H_2O$, nouveau minéral de Shinkolobwe, Zaire. Bull Minéral 107:21-24

Deliens M, Piret P 1985a Les phosphates d'uranyle et d'aluminium de Kobokobo VIII. La furongite. Ann Soc Géol Belg 108:365-368

Deliens M, Piret P 1985b Les phosphates d'uranyle et d'aluminium de Kobokobo VII. La moreauite, $Al_3UO_2(PO_4)_3(OH)_2 \cdot 13H_2O$, nouveau minéral. Bull Minéral 108:9-13

Deliens M, Piret P 1996 Les masuyites de Shinkolobwe (Shaba, Zaire) constituent un groupe formé de deux variétés distinctes par leur composition chimique et leurs propriétés radiocristallographiques. Bull Inst Royal Sci Natur Belg, Sci Terre 66:187-192

Deliens M, Piret P, Van der Meersche E 1990 Les minéraux secondaires du Zaire. Deuxième complement. Inst Royal Sci Natur Belg et Musée Royal de l' Afrique Centrale Tervuren, 39 p

Demarco RE, Richards DE, Collopy TJ, Abbot RC 1959 Evidence for the existence of peroxyuranic acid. J Am Chem Soc 81:4167-4169

Demartin F, Diella V, Donzelli S, Gramaccioli CM, Pilati T 1991 The importance of accurate crystal structure determination of uranium minerals I. Phosphuranylite, $KCa(H_3O)_3(UO_2)_7(PO_4)_4O_4 \cdot 8H_2O$. Acta Crystallogr B47:439-446

Demartin F, Gramaccioli CM, Pilati T 1992 The importance of accurate crystal structure determination of uranium minerals II. Soddyite $(UO_2)_2(SiO_4) \cdot 2H_2O$. Acta Crystallogr. C48:1-4

Dickens PG, Stuttard GP, Ball RGJ, Powell AV, Hull S, Patat S 1992 Powder neutron diffraction study of the mixed uranium-vanadium oxides $Cs_2(UO_2)_2(V_2O_8)$ and UVO_5. J Mater Chem 2:161-166

Dieke G, Duncan ABF 1949 Spectroscopic properties of uranium compounds. McGraw Hill New York

Dik TA, Umreyko DS, Nikanovich MV, Klavsut GN 1989a Coordination influence on spectral characteristic of CO_3^{2-} anion in uranyl carbonate complexes. Koord Khim 15:225-231

Dik TA, Klavsut GN, Nikanovich MV, Umreyko DC 1989b Spectral-structural investigation of UO_2CO_3. Koord Khim 15:1435-1438 (in Russian)

Dorhout PK, Rosenthal GL, Ellis AB 1988 Solid solutions of hydrogen uranyl phosphate and hydrogen uranyl arsenate. A family of luminescent lamellar hosts. Inorg Chem 27:1159-1162

Dothée D 1980 Spectroscopie vibrationelle de l'hexauranate de potassium hydraté en vue de l'étude de phase du système UO_3-KCl-H_2O. Thesis, Université de Franche-Comté, 143 p

Dothée, DG, Camelot MM, 1982 Spectroscopie de vibration de l'hexauranate de potassium hydraté I. Les fréquences attribuables aux mouvements des atomes d'oxygène: hypothèses sur la structure de la couche anionique. Bull Soc Chim France 1982:97-102

Dothée BR, Fahys BR, Camelot MM 1982 Spectroscopie de vibration de l'hexauranate de potassium hydraté II. Les mouvements des atomes d'hydrogène: hypothèses sur la structure de l'eau dans l'uranate. Bull Soc Chim France 1982:103-108

Elton NJ, Hooper JJ 1995 Widenmannite from Cornwall, England: the second world occurrence. Mineral Mag 59:745-749

Emsley J 1981 Very strong hydrogen bonding. Chem Soc Rev 9:91-124

Eremenko GK, Ilmenev ES, Azimi NA 1977 Finding of weeksite group minerals in Afghanistan. Doklady Akad Nauk SSSR 237:1191-1193 (in Russian)

Farmer VC (Ed) 1974 The infrared spectra of minerals. The Mineralogical Society, London, 539 p

Finch RJ, Ewing RC 1992 The corrosion of uraninite under oxidizing conditions. J Nucl Mater 190:133-156

Finch RJ, Ewing RC 1997 Clarkeite: New chemical and structural data. Am Mineral 82:607-619

Finch RJ, Miller ML, Ewing RC 1992 Weathering of natural uranyl oxide hydrates: Schoepite polytypes and dehydration effects. Radiochim Acta 58/59:433-443

Finch RJ, Cooper MA, Hawthorne FC 1996a The crystal structure of schoepite $[(UO_2)_8O_2(OH)_{12}](H_2O)_{12}$. Can Mineral 34:1071-1088

Finch RJ, Hawthorne FC, Ewing RC 1996b Schoepite and dehydrated schoepite. Mater Res Soc Proc 412:831-838. Can Mineral 34:1071-1088

Finch RJ, Hawthorne FC, Ewing RC 1998 Structural relations among schoepite, metaschoepite and dehydrated schoepite. Can Mineral 36:831-845

Finch RJ, Cooper MA, Hawthorne FC, Ewing RC 1998 Refinement of the crystal structure of rutherfordine. Can Mineral 37 (in press)

Földvári M, Paulik F, Paulik J 1988 Possibility of thermal analysis of different types of bonding of water in minerals. J Thermal Anal 33:121-132

Frondel C 1958 Systematic mineralogy of uranium and thorium. U S Geol Survey Bull 1064, 400 p

Frondel C, Ito J, Honea RM, Weeks AM 1976 Mineralogy ot the zippeite group. Can Mineral 14:429-436

Gadsden JA 1975 Infrared spectra of minerals and related inorganic compounds. Butterworth, London, 277 p

Gaines RV, Skinner HCW, Foord EE, Mason B, Rosenzweig A 1997 Dana's new mineralogy. John Wiley & Sons, New York

Gallagher PK 1993 Thermal analysis. In Advances in Analytical Geochemistry 1:211-257. JAI Press Inc

Galliski MA, De Upton IL 1986 Betauranofano primario de la manifestacion nuclear Verde I, Agua del Desierto, Departamento Los Andes, Provincia de Salta, Argentina. Revista Asoc Argent Mineral Petrol Sedimentol 17:55-60

Gevorkyan SV, Povarenykh AS 1980 The peculiarities of IR spectra of water molecules incorporated into phosphate and arsenate structures. Mineral Zh Kiev 2:29:36 (in Russian)

Gevorkyan SV, Ilchenko EA 1985 Applied aspects of IR spectroscopy of minerals. Preprint 85. Institute of geochemistry and physics of minerals Kiev, 40 p (in Russian)

Gevorkyan SV, Povarennykh AS, Rakovich DI 1973 Infrared absorption spectra of uraninite and curite. Konstitutsiya i svoystva mineralov Kiev 7:90-106 (in Russian)

Gevorkyan SV, Matkovskiy AO, Povarennykh AS, Sidorenko GA 1979 Infrared spectroscopic investigation of some uranyl silicates. Mineral Zh Kiev 1:78-85 (in Russian)

Gevorkyan SV, Povarennykh AS, Ignatov SI, Ilchenko EA 1981 Peculiarities of the investigation of IR spectra of minerals. Mineral Zh Kiev 3:3-11 (in Russian)

Ginderow D, Cesbron F 1985 Structure de la roubaultite, $Cu_2(UO_2)_3(CO_3)_2O_2(OH)_2 \cdot 4H_2O$. Acta Crystallogr C41:654-657

Glebov VA 1981 Electronic structure and properties of uranyl compounds. Frequencies of O-U-O stretching vibrations and the Badger's formulas. Koord Khim 7:388-395 (in Russian)

Glebov VA 1983 Electronic structure and properties of uranyl compounds. Energoatomizdat, Moscow, 88 p (in Russian)
Glebov VA 1985 Stretching vibration frequencies and interatomic distances in uranyl compounds. In Proc 3rd All-state Conf Uranium Chemistry, Moscow, p 38. Nauka, Moscow (in Russian)
Glebov VA 1989 Stretching vibration frequencies and interatomic distances in uranyl compounds. In Uranium Chemistry. Laskorin BN, Myasoedov BF (Eds) p 68-75. Nauka, Moscow (in Russian)
Gmelin Handbook of inorganic chemistry (Brown D) 1981 8th ed Uranium Supplement Volume C14, p 72-117. Springer-Verlag, Berlin
Gonzáles García F, Romero Díaz R 1959 Constitucion y propriedades físico-químicas de algunos fosfatos dobles de uranilo y substancias analogas IV. Analisis térmico diferential. Anal Real Soc Esp Fíz Quím 55:419-428
Gorbenko-Germanov DS, Zenkova RA 1962 Vibrational spectra of alkali metal uranyl tricarbonates and uranyl trinitrates. Physical Problems in Spectroscopy I:422-427. Academy of Sciences, Moscow (in Russian)
Gorbenko-Germanov DS, Zenkova RA 1966 On the vibrational structure of basic and excitated level of UO_2^{2+} in $K_4[UO_2(CO_3)_3]$. Optika i Spektroskopiya 20:842-847 (in Russian)
Guillemin C 1956): Contribution à la minéralogie des arséniates, phosphates et vanadates de cuivre. Bull Soc fr Minér Crist I Arséniates de cuivre 79:7-95, II Phosphates et vanadates de cuivre 79:219-275
Guillemin C, Protas J 1959 Ianthinite et wyartite. Bull Soc fr Minér Crist 82:80-86
Haacke DF, Williams PA 1979 The aqueous chemistry of uranium minerals p I Divalent cation zippeites. Mineral Mag 43:539-541
Hammer VMF, Libowitzky E, Rossman GR 1998 Single-crystal IR spectroscopy of very strong hydrogen bonds in pectolite, $NaCa_2[Si_3O_8(OH)]$, and serandiete, $NaMn_2[Si_3O_8(OH)]$. Am Mineral 83:569-576
Hawthorne FC (Ed) 1988 Spectroscopic methods in mineralogy and geology. Reviews in Mineralogy, Vol 18. Mineralogical Society of America, Washington DC, 698 p
Hawthorne FC 1992 The role of OH and H_2O in oxide and oxysalt minerals. Z Kristallogr 201:183-206
Herzberg G 1945 Molecular spectra and molecular structure II. Infrared and Raman spectra of polyatomic molecules. Van Nostrand New York
Hoekstra HR 1963 Uranium-oxygen bond lengths in uranyl salts: uranyl fluoride and uranyl carbonate. Inorg Chem 2:492-495
Hoekstra HR 1973 Personal written communication
Hoekstra HR 1982 Vibrational spectra. In Gmelin Handbook of inorganic chemistry Uranium Supplementum 15:211-240 Springer Verlag, Berlin
Hoekstra HR, Siegel S 1973 The uranium trioxide-water system. J Inorg Nucl Chem 35:761-779
Hoffmann G 1972 Über Phosphate und Arsenate von Uran(VI) und Neptunium(VI). Thesis, Ludwig-Maxmilians-Universität München, 156 p
Honea RM 1959 New data on gastunite, an alkali uranyl silicate. Am Mineral 44:1047-1056
Honea RM 1961 New data on boltwoodite, an alkali uranyl silicate. Am Mineral 46:12-25
Horák M, Vítek A 1980 Processing and interpretation of vibrational spectra. SNTL Praha, 430 p (in Czech)
Howe AT, Shilton MG 1980 Studies on layered uranium(VI) compounds II. Thermal stability of hydrogen uranyl phosphate and hydrogen uranyl arsenate (HUP and HUAs). J Solid State Chem 31:393-399
Hrebi_ík M, Volka K 1996 Diffuse-reflectance infrared spectroscopy. Theory and application. Chem Listy 90:80-92 (in Czech)
Huang CK, Kerr PF 1960 Infrared study of the carbonate minerals. Am Mineral 45:311-324
Hunan team China 1976 A new mineral discovered in China. Furongite, $Al_2(UO_2)(PO_4)_2(OH)_2 \cdot 8H_2O$. Acta Geol Sinica 2:203-204 (in Chinese)
Hunan team China 1978 Xiangjiangite, a new uranium mineral discovered in China. Sc Geol Sinica No 2:183-188 (in Chinese)
Hurlbut CS Jr 1950 Studies of uranium mineral (IV): Johannite. Am Mineral 35:531-535
Hurlbut CS 1954 Studies of uranium minerals (XV): Schroeckingerite from Argentina and Utah. Am Mineral 39:901-907
Johnson ChM, Shilton MG, Howe AT 1981 Studies of layered uranium compounds VI. Ionic conductivities and thermal stabilities of $MUO_2PO_4 \cdot nH_2O$, where M= H, Li, Na, K, NH_4 or 1/2 Ca and where n is between 0 and 4. J Solid State Chem 37:37-43
Jolivet JP, Thomas Y, Taravel B 1980 Vibrational study of coordinated CO_3^{2-} ions. J Mol Struct 60:93-98
Jones GC, Jackson B 1993 Infrared transmission spectra of carbonate minerals. Chapman and Hall, London
Jones LH 1958 Systematics in the vibrational spectra of uranyl complexes. Spectrochim Acta 10:395-403
Jones LH 1959 Determination of U-O bond distance in uranyl complexes from their infrared spectra. Spectrochim Acta 11:409-411
Karpova LN, Zhiltsova IG, Sidorenko GA, Sgibneva AF, Koroleva GI 1968 On conditions of iriginite formation. Geokhimiya 1968:166-172 (in Russian)

Karyakin AV, Kriventsova GA 1973 The nature of water in organic and inorganic compounds. Nauka, Moscow,176 p (in Russian)

Karyakin NV, Chernorukov NG, Suleymanov EV, Mokhalov LA, Alimzhanov MI 1998 Physicochemical properties of $NaVUO_6$ and $NaVUO_6 \cdot 2H_2O$. Russ J Gen Chem 68:504-508

Khandelwal BL, Verma VP 1976 Liquid-liquid distribution, spectrophotometric, thermal, IR and Raman studies on selenito uranylates. J Inorg Nucl Chem 38:763-769.

Kobets LV, Umreyko DS 1976 Investigation of thermal stability of $NH_4UO_2PO_4 \cdot 3H_2O$. Zh Neorg Khim 21:2161-2165 (in Russian)

Kobets LV, Umreyko DS 1983 Uranium phosphates. Uspekhi Khimii 52:897-921 (in Russian)

Koglin E, Schenk HJ, Schwochau K 1979 Vibrational and low temperature optical spectra of the uranyl tricarbonato complex $[UO_2(CO_3)_3]^{4-}$. Spectrochim Acta 35A:641-647

Kopchenova EV, Skvortsova KV 1957 Sodium uranospinite. Doklady Akad Nauk SSSR 114:634-636 (in Russian)

Kovba LM, Tabachenko NV, Seryozhkin VN 1982 Synthesis and physico-chemical investigation of new uranyl hydroxo sulfates. Doklady Akad Nauk SSSR 266:1148-1152 (in Russian)

Kreuer KD, Rabenau A, Messer R 1983a Proton conductivity in the layer compound $H_3OUO_2AsO_4 \cdot 3H_2O$ (HUAs) I. Conductivity in the orthorhombic low-temperature phase and the vehicle mechanism of proton transport in solids. Appl Phys A32:45-53

Kreuer KD, Rabenau A, Messer R 1983b Proton conductivity in the layer compound $H_3OUO_2AsO_4 \cdot 3H_2O$ (HUAs) II. Conductivity in the tetragonal high temperature phase. Appl Phys A32:155-158

Kubisz J 1966 On the existence of hydronium hydrates $H_9O_4^+$ and $H_{15}O_7^+$ in minerals. Mineral Mag 35:1071-1079

Kubisz J 1968 The role of positive hydrogen-oxygen ions in minerals. Prace Mineralogiczne 11, 75 p Polska Akademia Nauk, Warsaw (in Polish)

Kukovskiy EG 1971 To the hydroxonium problem in mineralogy. Mineral Sbornik Lvov 25:283-286 (in Russin)

Kuznetsov LM, Tsvigunov AN, Makarov ES 1981 Hydrothermal synthesis and physicochemical investigation of soddyite synthetic analogue. Geokhimiya 1981:1493-1508 (in Russian)

Legros JP, Legros R, Masdupuy E 1972 Sur un silicate d' uranyle, isomorphe du germanate d'uranyle, et sur les solutions solides correspondantes. Application à l'étude structurale du germanate d'uranyle. Bull Soc Chim France 1972:3051-3060

Lepilina RG, Smirnova NM 1984 Thermograms of inorganic phosphate compounds. Nauka, Leningrad (in Russian), p 166-173

Leo GW 1960 Autunite from Mt Spokane, Washington. Am Mineral 45:99-128

Leonova EN 1958 Synthesis of secondary uranium phosphates and arsenates. Trudy Inst Geol Rudnykh Mestorozhd Petrografii Mineralogii i Geokhimii Akad Nauk SSSR 30:37-55 (in Russian)

Li Jianzhong 1988 A preliminary study on andersonite. Youkuang Dizhi 4:235-239 (in Chinese)

Li Y, Burns PC 1999 A single-crystal X-ray diffraction study of the crystal chemistry of curite. GAC-MAC Joint Ann Meeting Sudbury 1999, Abstracts Vol 24:71

Lombardi G 1980 For better thermal analysis. 2nd ed, Int'l Confederation Thermal Anal ICTA Rome, 46 p

Ludikov VI, Zhiltsova IG, Sidorenko GA, Perlina SA 1979 On the synthesis of β-uranophane. Doklady Akad Nauk SSR 245:212-215 (in Russian)

Lutz HD 1995 Hydroxide ions in condensed materials—correlation of spectroscopic and structural data. Struct Bond 82:85-103

Mandarino JA 1999 Fleischer's Glossary of mineral species 1999. The Mineralogical Record Inc, Tucson, Arizona

Mandirola OB De 1961 Síntesis hidrotérmica de silicatos de uranio y calcio. Anal Asoc Quim Argent 49, 9 p

Mandirola OB De, Silberman E 1961 Espectrometria infrarroja de silicatos de uranio y calcio. Anal Asoc Quim Argent 49, 10 p

Matkovskiy AO, Gevorkyan SV, Povarennykh AS, Sidorenko GA, Tarashchan AN 1979 On the bond characteristics U-O in uranyl minerals by their IR spectroscopy data. Mineral Sbornik Lvov 33:11-22 (in Russian)

Matkovskiy AO, Platonov AN, Tarashchan AN, Sidorenko GA 1980a New data on metaautunite luminescence and spectroscopy. Mineral Zh Kiev 2:46-53 (in Russian)

Matkovskiy AO, Tarashchan AN, Povarennykh AS 1980b Luminescence of schroeckingerite and andersonite at 4.2 K. Doklady Akad Nauk Ukr SSR Ser B 2:22-25 (in Russian)

Maya L, Begun GM 1981 A Raman spectroscopy study of hydroxo and carbonato species of the uranyl(VI) ion. J Inorg Nucl Chem 43:2827-2832

Mayer H, Mereiter K 1986 Synthetic bayleyite, $Mg_2[UO_2(CO_3)_3] \cdot 18H_2O$: Thermochemistry, crystallography and crystal structure. TMPM Tschermaks mineral petrogrMitt 35:133-146

McGlynn SP, Smith JK, Neely WC 1961 Electronic structure, spectra and magnetic properties of oxycations III. Ligation effects on the infrared spectrum of the uranyl ion. J Chem Phys 35:105-116
McMillan P 1985 Vibrational spectroscopy in the mineral sciences. In Microscopic to macroscopic-atomic environments to thermodynamic properties. Rev Mineral 14:9-63
Meinrath G 1997 Speziation des Urans unter hydrogeologischen Aspekten. Wiss Mitt Inst Geol 4, 150 p, Technische Universität, Freiberg
Meinrath G, Kimura T 1993a Behavior of U(VI) solids under conditions of natural aquatic systems. Inorg Chim Acta 204:79-85
Meinrath G, Kimura T 1993b Carbonate complexation of the uranyl(VI) ion. J Alloy Comp 202:89-93
Melnikov VS, Melnik YuM 1969 Oxonium problem in mineralogy. Mineral Sbornik, Lvov 23:235-250 (in Russian)
Mereiter K 1982 Die Kristallstruktur des Johannites, $Cu(UO_2)_2(OH)_2(SO_4)_2 \cdot 8H_2O$. Tschermaks mineral petrogr Mitt 30:47-57
Mereiter K 1986a Crystal structure refinements of two francevillites $(Ba,Pb)[(UO_2)_2V_2O_8] \cdot 5H_2O$. Neues Jb Miner Mh 12:552-560
Mereiter K 1986b Neue kristallographische Daten über das Mineral Andersonit. Anzeiger Österr Akad Wiss Math-Naturwiss Kl 3:39-41
Mishra R, Manboodiri PN, Tripathi SN, Bharadwaj SR, Dharwadkar SR 1998 Vaporization behavior and Gibbs' energy of formatipon of $UTeO_5$ and UTe_3O_9 by transpiration. J Nucl Mater 256:139-144
Moenke H 1962/1966 Mineralspektren I, II. Akademie Verlag, Berlin
Moll H 1997 Zur Wechselwirkung von Uran mit Silicat in wässrigen Systemen. Thesis, Technische Universität, Dresden, 206 p
Moll H, Matz W, Schuster G, Brendler E, Bernhard G, Nitsche H 1995 Synthesis and characterization of uranyl orthosilicate, $(UO_2)_2SiO_4 \cdot 2H_2O$. J Nucl Mater 227:40-49
Moroz IKh, Sidorenko GA 1974 Crystallochemical peculiarities of autunite-metaautunite structure type. Rentgenografiya mineralnogo syrya 10:37-55 (in Russian)
Moroz IKh, Valueva AA, Sidorenko GA, Zhiltsova IG, Karpova LN 1973 Crystal chemistry of uranium micas. Geokhimiya 1973:210-223 (in Russian)
Mrázek Z, Novák M 1984 Secondary uranium minerals from Zálesí and Horní Lostice in the Rychlebské hory Mts., nothern Moravia. Acta Musei Moraviae Sci Natur 67:7-35 (in Czech)
Muck A, Cejka J Jr 1985 Infrared spectra of secondary uranium minerals saléeite and sabugalite. Sci Papers Prague Inst Chem Tech B30:55-66
Muck A, Cejka J Jr, Cejka J 1985 Contribution to the structural chemistry of uranium micas III. Saléeite. Casopis Národního muzea rada prírodovedná 154:45-50 (in Czech)
Muck A, Cejka J Jr, Cejka J, Urbanec Z 1986a Infrared spectroscopic study of H-D isotopic effect in layered uranium micas $M^{2+}(UO_2PO_4)_2 \cdot nX_2O$; ($M^{2+}$ = Cu, Ca, Ba; X = H, D). Sci Papers Prague Inst Chem Tech B31:71-93
Muck A, Cejka J Jr, Cejka J 1986b Contribution to the structural chemistry of uranium micas VI. Sabugalite. Casopis Národního muzea rada prírodovedná 155:119-124 (in Czech)
Nakamoto K 1986 Infrared and Raman spectraa of inorganic and coordination compounds. J Wiley and Sons New York, 484 p
Naumova IS, Valueva AA, Sidorenko GA 1982 On the nature of β-uranotile structure. Mineral Zh Kiev 4:57-61 (in Russian)
Nguyen SN, Silva RJ, Weed HC, Andrew JE Jr 1992 Standard Gibbs free energies of formation at the temperature 303.15 K of four silicates: soddyite, uranophane, sodium boltwoodite and sodium weeksite. J Chem Termodynamics 24:359-376
Nikanovich MV, Kovrikov AB, Popov VG, Sevchenko AN, Umreyko DS 1976 Calculation and investigation of vibrational spectrum of calcium uranyl phosphate hexahydrate. Doklady Akad Nauk SSSR 231:320-323 (in Russian)
Nikanovich MV, Umreyko DS, Sevchenko AN 1980 Vibrational spectra and structure of double uranyl phosphates. Zh Prikl Spektroskopii 32:658-663 (in Russian)
Noe-Spirlet MR, Sobry R 1974 Les uranates hydratés de forment pas une série continue. Bull Soc Royale Sci Liège 43:164-171
Novak A 1974 Hydrogen bonding in solids. Correlation of spectroscopic and crystallographic data. Struct Bond 18:177-216
Novitskiy GG, Umreyko DS, Shamanovskaya RP 1969 Infrared spectra and structure of uranium peroxide hydrates, $UO_4 \cdot xH_2O$. Zh Neorg Khim 14:2313-2315 (in Russian)
Novitskiy GG, Komyak AI, Umreyko DS 1981 Uranyl compounds, Vol 2: Atlas of spectra. Beloruss State University, Minsk, 216 p (in Russian)
Nyfeler D, Armbruster T 1998 Silanol groups in minerals and inorganic compounds. Am Mineral 83: 119-125

Nyfeler D, Hoffmann Ch, Armbruster T, Kunz M, Libowitzky E 1997 Orthorhombic Jahn-Teller distortion and Si-OH in mozartite, $CaM_2^{3+}O[SiO_3OH]$: A single-crystal X-ray, FTIR, and structure modeling study. Am Mineral 82:841-848

O'Brien TJ, Williams PA 1981 The aqueous chemistry of uranium minerals. 3. Monovalent cation zippeites. Inorg Nucl Chem Letters 17:105-107

Ohwada K 1968 Estimation of U-O bond distance in uranyl compounds from their infrared spectra. Spectrochim Acta 24A:595-599

Omori KP, Kerr PF 1963 IR studies of saline sulphate minerals. Bull Geol Soc Am 74:709-734

Ondrus P, Veselovsky F, Rybka R 1990 Znucalite. $Zn_{12}(UO_2)Ca(CO_3)_3(OH)_{22} \cdot 4H_2O$, a new mineral from Príbram, Czechoslovakia. Neues Jb Miner Mh 9:393-400

Ondrus P, Veselovsky F, Skála R, Císarová I, Hlousek J, Fryda J, Vavrín I, Cejka J, Gabasová A 1997 New naturally occurring phases of secondary origin from Jáchymov (Joachimsthal). J Czech Geol Soc 42:77-107

Osipov BS, Nazarenko NG, Shilyakova II, Shchipanova OV, Kazantsev VV 1981 Thermoanalytical investigations of mourite. In Uranium chemistry. BN Laskorin (Ed) p 21-36, Nauka, Moscow (in Russian)

Pagoaga MK 1983 The crystal chemistry of the uranyl oxide hydrate minerals. Thesis, Univ Maryland, College Park, 147 p

Pagoaga MK, Appleman DE, Stewart JM 1987 Crystal structures and crystal chemistry of the uranyl oxide hydrates becquerelite, billietite and protasite. Am Mineral 72:1230-1238

Pechurova NI, Kovba LM, Ippolitova EA 1965 Solid phases formed by the reations of ammonium hydroxide with uranyl nitrate and sulfate. Zh Neorg Khim 10:918-922 (in Russian)

Perrin A 1976 Nouveaux nitrato complexes dinucléaires et halogéno complexes tétranucléaires de l'uranyle: Préparation, études structurales et vibrationnelles. Thesis, Université de Rennes, 208 p

Pham Thi M, Velasco G 1983 Vibrational study of hydrogenated uranyl phosphate (HUP). Solid State Ionics 9/10:1055-1060

Pham Thi M, Colomban Ph 1985a Morphological, X-ray and vibrational study of various uranyl phosphate hydrates. J Less-Common Metals 108:189-216

Pham Thi M, Colomban Ph 1985b Cationic conductivity, water species motions and phase transitions in $H_3OUO_2PO_4 \cdot 3H_2O$ (HUP) and MUP related compounds ($M^+ = Na^+, K^+, Ag^+, Li^+, NH_4^+$). Solid State Ionics 17:295-306

Pham Thi M, Colomban Ph, Novak A 1985a Vibrational study of $H_3O^+UO_2PO_4 \cdot 3H_2O$ (HUP) and related compounds. Phase transitions and conductivity mechanisms:Part I $KUO_2PO_4 \cdot 3H_2O$ (KUP). J Phys Chem Solids 46:493-504

Pham Thi M, Colomban Ph, Novak A 1985b Vibrational study of $H_3O^+UO_2PO_4 \cdot 3H_2O$ (HUP) and related compounds. Phase transitions and conductivity mechanisms: Part II $H_3OUO_2PO_4 \cdot 3H_2O$. J Phys Chem Solids 46:565-578

Pierrot R, Toussaint J, Verbeek T 1965 La guilleminite, une nouvelle espèce minérale. Bull Soc fr Minér Crist 88:132-135

Piret P 1985 Structure cristalline de la fourmariérite, $Pb(UO_2)_4O_3(OH)_4 \cdot 4H_2O$. Bull Minéral 108:659-665

Piret P, Deliens M 1980 Nouvelles donées sur la saléeite holotype de Shinkolobwe. Bull Minéral 103:630-632

Piret P, Declerq JP, Wauters-Stoop D 1980 Structure cristalline de la sengiérite. Bull Minéral 103:176-178

Plesko EP 1981 Phase relations in the quaternary system K_2O-UO_3-SiO_2-H_2O. Thesis, Penn State Univ, University Park, Pennsylvania, 68 p

Plesko EP, Scheetz BE, White WB 1992 Infrared vibrational characterization and synthesis of a family of hydrous alkali uranyl silicates and hydrous uranyl silicate minerals. Am Mineral 77:431-437

Potdevin H, Brasseur H 1958 Etude d' uranates minéraux et synthétiques. Bull Acad Royale Belg Sér 5 44:874-912

Povarennykh AS 1979 IR spectra of some hydroxides and oxyhydrates. Konstitutsiya i svoystva mineralov Kiev 13:78-87 (in Russian)

Povarennykh AS, Gevorkyan SV 1970 Peculiarities of IR spectra of vanadates. Mineral Sbornik Lvov 24:254-260 (in Russian)

Pozas-Tormo R, Moreno-Real L, Martínez-Lara M, Bruque-Gamez S 1986 Layered metal uranyl phosphates. Retention of divalent ions by amine intercalates of uranyl phosphates. Can J Chem 64:30-34

Prins G 1973 Investigations on uranyl chloride, its hydrates, and basic salts. Thesis, Univ Amsterdam, 119 p

Protas J 1959 Contribution à l'étude des oxydes d'uranium hydratés. Bull Soc fr Minér Crist 82:239-272

Protas J 1964 Une nouvelle espèce minérale: la compreignacitee, $K_2O \cdot 6UO_3 \cdot 11H_2O$. Bull Soc fr Minér Crist 87:365-371

Pu Congjian 1990 Boltwoodite discovered for the first time in China. Acta Mineral Sinica 10:157-160 (in Chinese)
Rabinowitch E, Belford RL 1964 Spectroscopy and photochemistry of uranyl compounds. McMillan, New York (Russian translation Atomizdat, Moscow, 1968, 344 p)
Rastsvetaeva RK, Arakcheeva AV, Pushcharovskiy DYu, Atencio D, Menezes Filho LAD 1997 New silicon band in the structure of haiweeite. Kristallografiya 42:1003-1009 (in Russian)
Rocchiccioli C 1966 Etude par thermogravimétrie, analyse thermique différentielle et spectrographie d´absorption infrarouge des hydrates du peroxide d´uranium. C R Acad Sci 263:1061-1063
Rodriguez S, Martínez-Quiroz E 1996a Determinacion por espectroscopia vibracional de la distancia y fuerza de enlace del par U-O en complejos de uranilo. Rev Soc Quím Méx 40:115-121
Rodriguez S, Martínez-Quiroz E 1996b Frecuencia vibracional del ion uranilo con la electronegatividad del uranio. Rev Soc Quím Méx 40:215-219
Rogova VP, Belova LN, Kiziyarov GP, Kuznetsova NN 1973 Bauranoite and metacalciouranoite—new minerals of the group of hydrous uranium oxides. Zap Vses Miner Obshch 102:75-81 (in Russian)
Rosenzweig A, Ryan RR 1975 Refinement of the crystal structure of cuprosklodowskite, $Cu(UO_2)_2(SiO_3OH)_2 \cdot 6H_2O$. Am Mineral 60:448-453
Ross SD 1972 Inorganic infrared and Raman spectra. McGraw-Hill New York, 414 p
Ross SD 1974a Phosphates and other oxy-anions of group V. In The Infrared spectra of minerals. VC Farmer (Ed) p 383-422. The Mineralogical Society, London
Ross SD 1974b Sulphates and other oxy-anions of group VI. In The Infrared Spectra of Minerals. Farmer VC (Ed) p 423-444. The Mineralogical Society, London
Rossman GR 1988 Vibrational spectroscopy of hydrous components. Rev Mineral 18:193-206
Rozhkova EV, Ershova KS, Solntseva LS, Sidorenko GA, Likhonina EV 1971 To the classification of molecular water in minerals. Nedra, Moscow, 80 p (in Russian)
Rozière J 1973 Spectroscopie de vibration des hydrates du proton dans les cristaux. Thesis, Université des Sciences et Techniques du Languedoc, 150 p
Ruchkin ED, Durasova SA 1964 Optical investigation of hydrated uranium peroxide. Izv Sibirskogo Otdel Akad Nauk SSSR Ser Khim Nauk 7:62-66 (in Russian)
Russell JD 1974 Instrumentation and techniques. In The infrared spectra of minerals. VC Farmer (Ed) p 11-35. The Mineralogical Society, London
Ryskin YaI 1979 The vibrations of protons in minerals: hydroxyl, water and ammonium. In The infrared spectra of minerals. Framer VC (Ed) p 137-181. The Mineralogical Society, London
Sarp H, Chiappero PJ 1992 Deloryite, $Cu_4(UO_2)(MoO_4)_2(OH)_6$, a new mineral from the Cap Garonne mine near Le Pradet, Var, France. Neues Jb Miner Mh 2:58-64
Sato T 1961a Uranium peroxide hydrates. Naturwiss 46:668
Sato T 1961b Thermal decomposition of uranium peroxide hydrates. Naturwiss 46:693
Sato T 1976 Thermal decomposition of uranium peroxide hydrates. J Appl Chem 26:207-213
Seidl V, Knop O, Falk M 1969 Infrared studies of water in crystalline hydrates: gypsum, $CaSO_4 \cdot 2H_2O$. Can J Chem 47:1361-1368
Sejkora J, Veselovsky F, Srein V 1994 The supergene mineralization of uranium occurrence Ryzoviste near Harrachov (Krkonose Mts., Czech Republic). Acta Musei Nationalis Pragae, Hist Natur B50:55-91
Sejkora J, Mazuch J, Abert F, Srein V, Novotná M 1997 Supergene mineralization of the uranium deposit Slavkovice, western Moravia. Acta Musei Moraviae, Sci Natur 81:3-24 (in Czech)
Sejkora J, Cejka J, Kotrl_ M, Novotná M 1998 Saléeite from Ryzoviste near Harrachov, Czech Republic. Bull Miner Petrol Odd National Museum Prague 6:217-220 (in Czech)
Seryozhkin VN, Seryozhkina LB 1984 On the use of modified Badger's equations to uranyl coordination compounds. Zh Neorg Khim 29:1529-1532 (in Russian)
Seryozhkin VN, Seryozhkina LB, Fakeeva OA, Olikov NN 1981 On the demonstration of nonplanar uranyl ions by luminescent and vibration spectra of uranyl molybdates and sulfates. Zh Neorg Khim 26:3321-3328 (in Russian)
Seryozhkin VN, Seryozhkina LB, Soldatkina MA, Zolin VF, Lokshin BV 1983 On the correlation of the results of the X-ray structure and spectroscopy investigations of uranyl compounds with tetrahedral oxyanions of VI.group elements. Koord Khim 9:92-96 (in Russian)
Seryozhkina LB, Seryozhkin VN, Soldatkina MA 1982 IR spectroscopic investigation of coordination type of sulfato groups in uranyl compounds. Zh Neog Khim 27:1750-1757 (in Russian)
Sesták J 1984 Thermophysical properties of solids. Academia Prague, 464 p
Shashkin DP 1975 Crystal structure of francevillite, $Ba[(UO_2)_2(VO_4)_2] \cdot 5H_2O$. Doklady Akad Nauk SSSR 220:1410-1413 (in Russian)
Shchipanova OV, Belova LN, Pribytkov PV, Katargina AP 1971 New data on structure and diagnostics of trögerite and hydrogen-uranospinite. Doklady Akad Nauk SSSR 197:178-181 (in Russian)

Shen Caiqing 1988 Experiental researches on the formation of andersonite. Yanshi Kuangwuxie Zashi 7:79-88 (in Chinese)
Sidorenko GA 1978 Crystal chemistry of uranium minerals. Atomizdat, Moscow, 216 p (in Russian)
Sidorenko GA, Zhiltsova IG, Moroz IKh, Valueva AA 1975 Synthesis and crystal chemical investigation of P-analogue of trögerite. Doklady Akad Nauk SSSR 222:444-447 (in Russian)
Sidorenko GA, Matkovskiy AO, Kalinichenko AM 1979 On the type of "OH groupings" in natural uranium hydroxides. Konstitutsiya i svoystva mineralov Kiev 13:15-18 (in Russian)
Sidorenko GA, Valueva AA, Dubinchuk VT, Naumova IS, Yudin RN 1985 To the crystal chemistry of uranyl hydroxides. Kristallokhimiya i strukturnyy tipomorfizm mineralov, p 126-134. Nauka Leningrad (in Russian)
Sidorenko GA, Gorobets BS, Dubinchuk VT 1986 Current methods of the mineralogical analysis of uranium ores. Energoatomizdat, Moscow, 184 p (in Russian)
Sidorenko GA, Zhiltsova IG, Valueva AA, Perlina SA 1992 On calcium-magnesium uranium micas. Mineral Zh Kiev 14:58-67 (in Russian)
Sidorenko GA, Naumova IS, Valueva AA, Skvortsova KV 1995 On Tl-carnotite and isomorphism in uranyl vanadates. Mineral Zh Kiev 17:67-72 (in Russian)
Skvortsova KV, Kopchenova EV, Sidorenko GA, Kuznetsova NN, Dara AD, Rybakova LI 1969 Calcium-sodium uranyl molybdates. Zap Vses Mineral Obshch 108:679-688 (in Russian)
Skvortsova KV, Sidorenko GA, Moroz IKh, Rybakova LI, Zhiltsova IG 1974 First finding of metanovácekite in the USSR. Zap Vses Mineral Obshch 103:606-611 (in Russian)
Smith DK Jr 1959 An X-ray crystallographic study of schroeckingerite and its dehydration product. Am Mineral 44:1020-1025
Smith DK Jr 1984 Uranium mineralogy. In Uranium geochemistry, mineralogy, geology, exploration and resoursces DeVivo et al. (Eds) p 43-88. Institution of Mining and Metallurgy, London
Smykatz-Kloss W 1982 Application of differential thermal analysis in mineralogy. J Thermal Anal 23:15-44
Smykatz-Kloss W, Warne SStJ (Eds) 1991 Thermal analysis in geosciences. Springer-Verlag, Berlin
Smyslova IG 1972 On Tl-carnotite. Zap Vses Mineral Obshch 101:87-90 (in Russian)
Sobry R 1972 Implications structurales de l'étude des propriétés physico-chimiques des "uranates hydratés." Thesis, Université de Liège, 177 p
Sobry R 1973 Etude des "uranates hydratés" II. Examen des propriétés vibrationelles des uranates hydratés de cations bivalents. J Inorg Nucl Chem 35:2753-2768
Sokol F, Cejka J 1992 A thermal and mass spectrometric study of synthetic johannite, $[Cu(UO_2)_2(SO_4)_2(OH)_2]\cdot 8H_2O$. Thermochim Acta 206:235-242
Spitsyn VI, Van Shi-Khua, Kovba LM 1962 Study of mixed uranates of sodium and calcium. Best Moskovskogo Univ Nr 5:60-62
Spitsyn VI, Kovba LM, Tabachenko VV, Tabachenko NV, Mikhaylov YuN 1982 To the investigation of basic uranyl salts and polyuranates. Izv Akad Nauk SSSR, Ser Khim 1982:807-812 (in Russian)
Stergiou AC, Rentzeperis PJ 1993 Refinement of the crystal structure of metatorbernite. Z Kristallogr 205:1-7
Stern TW, Stieff LR, Girhard MN, Meyrowitz R 1956 The occurrence and properties of metatyuyamunite, $Ca(UO_2)_2(VO_4)_2\cdot(3-5)H_2O$. Am Mineral 41:187-201
Sundberg I, Sillén LG 1949 On the crystal structure of KUO_2VO_4 (synthetic anhydrous carnotite) Arkiv Kemi 1:337-351
Takano Y 1961 X-ray study of autunite. Am Mineral 46:812-822
Tarkhanova GA, Sidorenko GA, Moroz IKh, 1975 First finding of a mineral of the weeksite group in the USSR. Zap Vses Mineral Obshch 104:598-603 (in Russian)
Taylor JC, Stuart WI, Mumme IA 1981 The crystal structure of curite. J Inorg Nucl Chem 43:2419-2423
Tridot G 1955 Contribution à l'étude des composés peroxydés de l'uranium et du molybdène. Thesis, Université de Paris, 46 p
Trunov VK, Kovba LM 1963 Investigation of uranyl wolframates and molybdates. Vestnik Moskovskogo Univ No 6:34-35 (in Russian)
Turrell G 1972 Infrared and Raman spectra of crystals. Academic Press, London, 384 p
Urbanec Z, Cejka J 1974 Chemie der Uranylcarbonate II. Beitrag zur Frage der Uranylcarbonatbildung unter normalen Bedingungen. Coll Czechoslov Chem Commun 39:2891-2910
Urbanec Z, Cejka J 1975 Thermal analysis of the UO_3-CO_2-H_2O system. In Proc 4th Int'l Conf Thermal Analysis ICTA I:943-953. I Buzás (Ed) Akademiai Kiadó, Budapest
Urbanec Z, Cejka J 1979a Infrared spectroscopy used for the study and scientific reassessment of secondary uranium minerals from the collections of the National Museum in Prague, Casopis Národního muzea rada prírodovedná 148:16-31

Urbanec Z, Cejka J 1979b Infrared spectra of rutherfordine and sharpite. Coll Czechoslov Chem Commun 44:1-9

Urbanec Z, Cejka J 1979c Infrared spectra of liebigite, andersonite, voglite and schroeckingerite. Coll Czechoslov Chem Commun 44:10-23

Urbanec Z, Cejka J 1979d Thermal analysis of synthetic andersonite: Comparison of different techniques applied. *In* Proc 8th Conf Thermal Anal Termanal '79, High Tatras, p 213-216 (in Czech)

Urbanec Z, Cejka J 1980 Thermal and infrared spectrum analysis of uranopilite. *In* Proc 6th Int'l Conf Thermal Analysis ICTA Thermal Analysis 2:359-364. Hemminger W (Ed) Birkhauser Verlag, Basel

Urbanec Z, Mrázek Z, Cejka J 1985a Thermal and infrared spectrum analysis of some uranyl silicate minerals. *In* Proc 8th Int'l Conf Thermal Analysis ICTA 2:525-528. Thermochim Acta 93:525-528

Urbanec Z, Mrázek Z, Cejka J 1985b Thermal, X-ray and infrared spectrum analysis of a new uranyl sulphate mineral. Thermochim Acta 86:383-386

Van Haverbeke L, Vochten R, Van Springel K 1996 Solubility and spectrochemical characteristics of synthetic chernikovite and meta-ankoleite. Mineral Mag 60:759-766

Veal BW, Lam DJ, Carnall WT, Hoekstra HR 1975 X-ray photoemission spectroscopy study of hexavalent uranium compounds. Phys Rev B12:5651-5663

Vochten R 1983 Formation of secondary uranyl phosphates in the oxidation zone of uranium deposits. Spec Publ Geol Soc South Africa 7:287-293

Vochten R 1990 Transformation of chernikovite and sodium autunite into lehnerite. Am Mineral 75:221-225

Vochten R, Deliens M 1980 Transformation of curite into metaautunite. Paragenesis and electrokinetic properties. Phys Chem Minerals 6:129-143

Vochten R, Goeminne A 1984 Synthesis, crystallographic data, solubility and electrokinetic properties of metazeunerite, metakirchheimerite and nickel-uranylarsenate. Phys Chem Minerals 11:95-100

Vochten R, Van Doorselaer 1984 Secondary uranium minerals of the Cunha Baixa mine. Mineral Record 1984:293-297

Vochten R, Van Springel K 1996 A natural ferrous substituted saléeite from Arcu su Linnarbu, Capoterra, Cagliari, Sardinia. Mineral Mag 60:647-651

Vochten R, Deliens M 1998 Blatonite, $UO_2CO_3 \cdot H_2O$, a new uranyl carbonate monohydrate from San Juan County, Utah. Can Mineral 36:1077-1081

Vochten R, Huybrechts W, Remaut G, Deliens M 1979 Formation of metatorbernite starting from curite: Crystallographic data and electrokinetic properties. Phys Chem Minerals 4:281-290

Vochten R, Piret P, Goeminne A 1981 Synthesis, crystallographic data, solubility and elektrokinetic properties of copper-, nickel-, and cobalt-uranylphosphates. Bull Minéral 104:457-467

Vochten R, De Grave E, Pelsmaekers J 1984 Mineralogical study of bassetite in relation to its oxidation. Am Mineral 69:967-978

Vochten R, De Grave E, Pelsmaekers J 1986 Synthesis, crystallographic and spectroscopic data, solubility and electrokinetic properties of metakahlerite and its Mn analogue. Am Mineral 71:1037-1044

Vochten R, Van Haverbeke L, Van Springel K 1992 Transformation of chernikovite into meta-uranocircite II, $Ba(UO_2)_2(PO_4)_2 \cdot 6H_2O$, and study of its solubility. Mineral Mag 56:367-372

Vochten R, Van Haverbeke L, Van Springel K 1993 Synthesis of liebigite and andersonite, and study of their thermal behavior and luminescence. Can Mineral 31:167-171

Vochten R, Van Haverbeke L, Van Springel K, Blaton N, Peeters OM 1994 The structure and physicochemical characteristics of a synthetic phase compositionally intermediate between liebigite and andersonite. Can Mineral 32:553-561

Vochten R, Van Haverbeke L, Van Springel K, Blaton N, Peeters OM 1995a The structure and physicochemical characteristics of synthetic zippeite. Can Mineral 33:1091-1101

Vochten R, Van Haverbeke L, Van Springel K 1995b Soddyite: Synthesis under elevated temperature and pressure, and study of some physicochemical characteristics. Neues Jb Miner Mh 10:470-480

Vochten R, Blaton N, Peeters O, Deliens M 1996 Piretite, $Ca(UO_2)_3(SeO_3)_2(OH)_2 \cdot 4H_2O$, a new calcium uranyl selenite from Shinkolobwe, Shaba, Zaire. Can Mineral 34:1317-1322

Vochten R, Blaton N, Peeters O, Van Pringel K, Van Haverbeke L 1997a A new method of synthesis of boltwoodite and of formation of sodium boltwoodite, uranophane, sklodowskite and kasolite from boltwoodite. Can Mineral 35:735-741

Vochten R, Blaton N, Peeters O 1997b Synthesis of sodium weeksite and its transformation into weeksite. Neues Jb Miner Mh 12:569-576

Vochten R, Blaton N, Peeters O 1997c Deliensite, $Fe(UO_2)_2(SO_4)_2(OH)_2 \cdot 3H_2O$, a new ferrous uranyl sulfate hydroxyl hydrate from Mas d'Alary, Lodève, Hérault, France. Can Mineral 35:1021-1025

Volodko LV, Komyak AI, Umreyko DS 1981 Uranyl compounds Vol 1. Spectra and structure. Beloruss State University Minsk, 432 p (in Russian)

Vulpius D, Nicolai R, Geipel G, Matz W, Bernhard G, Nitsche H 1997 Preparation and characterization of uranyl carbonate. Annual Report 1997:20-21. Inst Radiochem Forschungszentrum, Rossendorf

Walenta K 1963 Über die Barium-Uranylphosphatmineralien Uranocircit I, Uranocircit II, Meta-Uranocircit I und Meta-Uranocircit II von Menzenschwand im südlichen Schwarzwald. Jh Geol Landesamt Baden-Württemberg 6:113-135

Walenta K 1964,1964/1965 Beiträge zur Kenntnis seltener Arsenatmineralien unter besonderer Berücksichtung von Vorkommen des Schwarzwaldes. Tschermaks mineral petrogr Mitt 9:111-174; 9:252-282

Walenta K 1965a Die Uranglimmergruppe. Chem Erde 24:254-278

Walenta K 1965b Hallimondite, a new uranium mineral from the Michael mine near Reichenbach (Black Forest, Germany). Am Mineral 50:1143-1157

Walenta K 1972 Grimselit, ein neues Kalium-Natrium-Uranylkarbonat aus dem Grimselgebiet (Oberhasli, Kt Bern, Schweiz). Schweiz mineral petrogr Mitt 52:93-108

Walenta K 1976 Widenmannit und Joliotit, zwei neue Uranylkarbonatmineralien aus dem Schwarzwald. Schweiz mineral petrogr Mitt 56:167-185

Walenta K 1978 Uranospathite and arsenuranospathite. Mineral Mag 42:117-128

Walenta K 1979 Über den Hügelit. Tschermaks mineral petrogr 26:11-19

Walenta K 1983 Uranosilit, ein neues Mineral aus der Uranlagerstätte von Menzenschwand im südlichen Schwarzwald. Neues Jb Miner Mh 6:259-269

Walenta K 1985 Uranotungstit, eine neues sekundäres Uranmineral aus dem Schwarzwald. Tschermaks mineral petrogr Mitt 34:25-34

Wang Wenguang, Zhang Shuling, Lu Jun 1981 Infrared spectral characteristics of some common uranium minerals. Sci Geol Sinica 7:235-246 (in Chinese)

Wang Zhi-Xiong 1979 The mineralogical investigation of furongite. Sci Sinica 12:199-206 (in Chinese)

Weidlein J, Müller U, Dehkicke K 1988 Schwingungsspektroskopie. G. Thieme, Stuttgart, 228 p

Weigel F 1985 The carbonates, phosphates and arsenates of the hexavalent and pentavalent actinides. In Handbook of the Physics and Chemistry of the Actinides. Freeman AJ, Keller C (Eds) p 243-288. Elsevier Sci Publishers

Weigel F, Hoffmann G 1976 The phosphates and arsenates of hexavalent actinides P I. Uranium. J Less-Common Metals 44:99-123

Wendlandt WW(1986 Thermal analysis, 3rd ed, Wiley Interscience, New York

White WB 1971 Infrared characteristization of water and hydroxyl in the basic magnesium carbonate minerals. Am Mineral 56:46-53

Wilkins RWT 1971a Infrared spectroscopy in the mineralogical sciences of uranim ores. Neues Jb Miner Mh 10:440-450

Wilkins RWT 1971b U-O bond lengths and force constants in some uranyl minerals. Z Kristallogr 134:285-290

Wilkins RWT, Mateen A, West GW 1974 The spectroscopic study of oxonium ions in minerals. Am Mineral 59:811-819

Wilson Jr EB, Decius JC, Cross PC 1955 Molecular vibrations. The theory of infrared and Raman vibrational spectra. McGraw-Hill, New York, 388 p. Wunderlich B 1990 Thermal analysis. Academic Press, New York

Young EJ, Weeks AD, Meyrowitz R 1966 Coconinoite, a new uranium mineral from Utah and Arizona. Am Mineral 51:651-663

Yukhnewich GV Infrared spectroscopy of water. Nauka, Moscow, 208 p (in Russian)

Zhang Jingyi, Wan Anwa, Gong Wenshu 1992 New data on yingjiangite. Acta Petrol et Mineral 11:178-184 (in Chinese)

Zhiltsova IG, Karpova LN, Sidorenko GA, Valueva AA 1970 Formation of metastable and stable iriginite modifications by treatment of uranium containing solutions on powellite. Geokhimiya 1970:1019-1023 (in Russian)

Zhiltsova IG, Sidorenko GA, Karpova LN, Tarkhanova GA, Valueva AA 1976 On two structural modifications of boltwoodite. Trudy Mineral Muz AE Fersmana 25:35-41 (in Russian)

Zhiltsova IG, Ludikov VI, Sidorenko GA, Valueva AA 1981 On the stability of the uranophane crystal structure. Kristallokhimiya mineralov, p 82-87, Nauka, Leningrad (in Russian)

Zwaan PC, Arps CES, De Grave E 1989 Vochtenite, $(Fe^{2+},Mg)Fe^{3+}[UO_2/PO_4]_4(OH)\cdot(12-13)H_2O$, a new uranyl phosphate mineral from Wheal Basset, Redruth, Cornwall, England. Mineral Mag 53:473-478

13 Analytical Methods for Determination of Uranium in Geological and Environmental Materials

Stephen F. Wolf
Chemical Technology Division
Argonne National Laboratory
Argonne, Illinois 60439

INTRODUCTION

Uranium is ubiquitous in nature, being more abundant in the Earth's crust than Sb, Cd, Hg, or Ag. As a trace element, U is a constituent of igneous, metamorphic, and sedimentary rocks. Uranium can also be found concentrated as a minor or major constituent in numerous geological environments. In the hydrosphere, U is present at trace levels in surface waters, ground waters, and seawaters. While the existence of U has been known for more than two centuries, its use reached a new era with the discovery of nuclear fission in 1938 and subsequent development in the production of atomic weapons and energy. More than fifty years of nuclear weapons production, testing, and energy generation has left a legacy of environmental contamination in 132 sites in 31 states. As strategies for environmental remediation and disposal of U-contaminated materials are developed and implemented, the ability to perform accurate and precise determinations of U in a variety of natural matrices has become increasingly important. The study of U in geological materials particularly should give insight to the long-term disposition of U in the environment. These studies frequently require the use of analytical methods that are capable of determining U concentrations along with a suite of other elements. The modern analytical chemist has a variety of techniques available for the determination of U as a major, minor, and trace component. The purpose of this chapter is to review those techniques and procedures that have been developed for the determination of U in geological and environmental materials. The specific method chosen for the determination of any element is dependent on several factors: sample matrix, sample size, analyte concentration, the need to determine multiple analytes, cost of analysis, time of analysis, equipment available, and the skill and training of the analyst. Any of these factors can potentially influence the selection of an analytical method.

SAMPLE DISSOLUTION METHODS

A majority of the bulk analytical techniques discussed in this review require that a homogeneous, stable, aqueous solution be prepared prior to elemental analysis. While the determination of U in both dissolved and suspended groundwater fractions can provide useful information on U transport mechanisms, (for example, Porcelli et al. 1997; Haraguchi et al. 1998) more frequently, total dissolution of material is desired. The dissolution of suspended solids in water and minerals can be accomplished by two common methods: fluxed decomposition and acid dissolution. Fluxed decompositions are performed at atmospheric pressures in a crucible resistant to the fusion medium. Acid dissolutions are performed with the sample and reagents in a beaker on a hot plate at atmospheric pressure, in a closed digestion bomb at elevated pressures in a convection oven, and in a closed digestion bomb at elevated pressures in a microwave oven. Dolezal et al. (1968) have published a useful compilation of decomposition techniques for inorganic analysis.

Fluxed decomposition

For fluxed decompositions, a 100 to 1000-mg solid sample is mixed with several grams of one or more of the available flux materials, such as sodium hydroxide, sodium peroxide, sodium carbonate, lithium tetraborate, lithium metaborate, or many related compounds. The mixture is heated over a Meker burner or in a furnace until the mixture forms a well-mixed molted fusion of material. After cooling, the fusion can readily be dissolved with a dilute mineral acid such as HCl or HNO_3. The determination of trace U in a flux-decomposed sample can be complicated due to the high total dissolved solids (TDS) of the resulting solution and the potential for contamination caused by impurities in the flux reagents. However, flux decomposition is used for many standard classical wet chemical procedures for the determination of U at major and minor levels.

Acid dissolution

For acid dissolution, an approximately 100-mg solid sample is mixed with several milliliters of one or more concentrated mineral acids and heated either in an open Teflon beaker or a closed dissolution bomb that is lined with Teflon and has a steel jacket. The types of acids selected depend on the chemical nature of the sample, the elements desired for determination, and the method of determination. A hot plate-based digestion in an open-vessel with HNO_3, HCl, and H_2SO_4 can be used for the dissolution of many of the U-containing minerals, such as autunite, carnotite, phosphuranylite, torbernite, tyuyamunite, uraninite, and uranophane (Meites 1963). A HF-HNO_3 mixture is frequently used for the dissolution of silicates. In these cases the HF must be removed by fuming with $HClO_4$ or complexation with boric acid so that insoluble actinide fluorides are converted to a soluble form. A set of standard procedures for both the hot-plate digestion and convection-oven digestion of water and solid samples with HCl and HNO_3 has been published by the American Society for Testing and Materials (ASTM) in ASTM D1971-95 (1995a). Procedures for microwave-heated digestion in a closed vessel with HCl and HNO_3 for the determinations of metals in groundwater are given in ASTM D4309-91 (1991a). These procedures are applicable for digestion of suspended solids in water and leaching elements from most solid materials but would not yield total dissolution of silica-containing materials. The United States (US) Environmental Protection Agency (EPA) has developed Method 3052 (1996) for total dissolution of inorganic constituents using a microwave oven. This method has been validated for U determinations on the Standard Reference Material (SRM) 2704 Buffalo River Sediment. An up to date overview of microwave digestion techniques can be found in Kingston and Haswell (1997). Acid-based digestions and dissolutions result in solutions with low TDS and can be used for direct determination of many elements, including U, with out further chemical treatment. The availability of high-purity mineral acids makes acid digestion and dissolution ideal for determination of trace U contents in water and geological samples.

CLASSICAL WET CHEMICAL TECHNIQUES

By definition, classical chemical techniques are those well-established analytical procedures that set the standard of accuracy by which all other techniques are judged. Typically, these techniques involve the dissolution of a macroscopic quantity (several grams) of the sample followed by separation, purification, and quantitative determination of the analyte. The determination step can be performed with a variety of well-established laboratory techniques, including gravimetric, electrogravimetric, conductometric, colorometric, volumetric, and spectrophotometric methods. Most of these methods can only be used when U is a major or minor component. Spectrophotometric methods are capable of determining U at trace levels. These procedures have been used in the field of ore and mineral analysis. An overview of these classical techniques and procedures for their application to U can be found in Dean (1995) and Ridden and Warf (1950). Many classical wet chemical techniques have been adapted as sample preparation steps for instrumental methods of determination, where the elimination of sample matrix is required for accurate results. The specific set of procedures chosen for U determination is dependent on the sample matrix, U concentration in the sample, and other elements that may need to be determined. Classical chemical techniques are time consuming; however, when they are skillfully performed, extremely high accuracy can be obtained. For this reason, these techniques are frequently used to produce and to certify custom standards for use within a laboratory. Many classical methods do not require special instrumentation and can be performed with the equipment and materials found in many analytical laboratories. The purpose of the following section is to give a few examples of wet chemical procedures that are used for the determination of U in geological materials. The dissolution of the sample, which may be a complex mixture of minerals, the separation from interfering elements, and the determination of U are illustrated.

One classical wet chemical method for the determination of U in silicate minerals is given in Ingamells and Pitard (1986). In this method a large quantity of silicate material is decomposed via sintering with sodium peroxide in a platinum crucible. The fusion is dissolved in HNO_3, and U is precipitated with a mixture of Na_2CO_3 and K_2CO_3 (Sandell 1959). This precipitate is evaporated and redissolved in nitric acid. Uranium can be effectively extracted with ethyl acetate. The extracted U is evaporated and fused with K_2CO_3, Na_2CO_3, and NaF and determined via fluorometry. Fluorometry is discussed further below.

Gravimetric determination

The sulfide-carbonate-hydroxide method can be used for U determination in geological samples. The sample is dissolved with HNO_3 and H_2SO_4. Arsenic is removed with HBr; acid sulfide metals are precipitated from the acidic solution with H_2S; Al, Cr, and Fe are removed with NH_4OH and $(NH_4)_2CO_3$; Co and Ni are precipitated the ammoniacal solution with H_2S. The solution is acidified with HCl, and U is precipitated with NH_4OH. The precipitate is filtered, washed, ignited, and determined gravimetrically as U_3O_8. Alternatively, the sample is decomposed via fluxed fusion, the fusion is dissolved in HCl and U(IV) is precipitated with addition of 6% cupferron solution. The precipitate is filtered, washed, ignited, and determined gravimetrically as U_3O_8. This separation effectively removes Al, Cr, Mn, Zn, and PO_4^{3-} but the presence of Fe, Ti, V(V), and Zr will interfere with the U determination.

Volumetric determination

After acid dissolution, U can be determined volumetrically via titration. One very versatile and thoroughly tested wet chemical method for the determination of U is the modified Davies and Gray technique (Davies and Gray 1964; C1267-94 1994a). In this method, an aliquot of the dissolved U sample is evaporated and redissolved in dilute HNO_3. The U is reduced to U(IV) with the addition of Fe(II) in concentrated H_3PO_4 that contains H_3NO_3S. The Fe(II) is selectively oxidized with the addition of HNO_3 in the presence of a Mo(IV) catalyst. After the addition of a V(IV) solution, the U(IV) is titrated with Cr(VI) to a potentiometric end point. This method is exceptionally accurate but requires milligrams of U in the sample aliquot.

Colorimetric determination

The solid geological sample is dissolved in HNO_3 and HF. Most of the Fe is extracted with $CH_3COOC_2H_5$. Samples containing heavy metals require their separation by treatment with H_2S. Uranium is reduced and precipitated with cupferron, along with Ti and V. The cupferrates are ignited, the U is oxidized, and Ti and V are removed by extraction with cupferron. The U in the resulting solution is determined colorimetrically by comparison with prepared standards.

URANIUM SEPARATIONS

It is frequently necessary to preconcentrate and extract U in order to reduce the volume of solution and increase the concentration of U. This is particularly true for U at the low concentrations it is typically found in nature. There are several separation methods that can accomplish this: coprecipitation, liquid-liquid extraction, and ion-exchange. Several specific examples are described below.

Coprecipitation

A very quick way to separate U from solution is by coprecipitation. Actinides will coprecipitate from a carbonate-free solution with Fe(III). Carbonate is removed upon heating the solution. Coprecipitation is accomplished by adding several milligrams of $FeCl_3$ to an acidified solution and adjusting the solution pH to basic with the addition of NH_4OH. Uranium will coprecipitate with $Fe(OH)_3$. This method was used by Wolf et al. (1997) in extract trace actinides from coral sand.

Liquid-liquid extraction

Liquid-liquid extraction can be accomplished with a wide range of organic acids, ketones, ethers, esters, alcohols, and phosphoric acid derivatives (Lally 1992). Uranium(IV) can be extracted with ethyl acetate after treatment of an acidified solution with $Al(NO_3)_3 \cdot 9H_2O$ (Guest and Zimmerman 1955). Uranium(VI) can be extracted with 25% tributyl phosphate in toluene (Mair and Savage 1986). Methyl isobutyl ketone has also been used in large-scale separation processes. Recently, Eichrom Industries Inc. has developed several actinide specific exchange resins for separation of actinides from acidic matrices such as acid-stabilized groundwaters and dissolved minerals. While these resins function by passing U-containing solution down an exchange column containing the material, the separation is based on liquid-liquid extraction. One of these extraction chromatographic resins, TRU Spec resin, consists of octyl(phenyl)-N,N-diisobutyl-carbamoylmethylphosphine oxide in tributyl phosphate supported by an inert polymeric substrate, Amberlite XAD-7 (Horwitz et al. 1993). The second extraction chromatographic resin, U/TEVA Spec resin, comprised of diamylphosphonate sorbed on Amberlite XAD-7 and possess tetravalent ion specificity.

Ion-exchange

Ion-exchange resins are typically high-molecular-weight organic polymers containing a large number of functional groups that provide the basis for the separation. Cation exchange can be performed on resins possessing $-SO_3H$ functional groups. Cation exchange is not commonly used of a method of U preconcentration and separation due to the lack of selectivity of the UO_2^{2+} cation over other divalent metal ions. Anion exchange can be performed on resins possessing functional groups such as $-NR_3^+$. Uranyl ions form strong anionic sulfate and chloride complexes that can be the basis for U separation from other metals in solution. Uranium separations from a variety of metals can be performed by use of basic anion exchange resins such as Bio-Rad AG1X8.

NUCLEAR METHODS

Radiometric techniques

The natural radioactivity inherent to the isotopes of U can provide the means for their direct analysis by radiometric analytical techniques. In principle, the α-particles, β-particles, and γ-rays emitted by U isotopes can be used both for the qualitative identification and quantitative measurement of U isotopes. The abundances, half-lives ($t_{1/2}$), modes of decay, and thermal neutron-capture cross section (σ_n) in units of barns (1 b = 10^{-24} cm^2) for the long-lived U isotopes, including the naturally occurring U isotopes, are given in Table 1. Since quantitative analysis with radiometric techniques is based on the measurement of an absolute disintegration rate of the isotope of interest, the precision is inversely related to the specific activity, counting time, counting efficiency, and number of atoms of the radionuclide of interest. The long half-lives and, therefore, low specific activities of the U isotopes complicate their direct determination via radiometric techniques. Samples containing low concentrations of U need to be counted for relatively long times on extremely stable spectrometers possessing low backgrounds to obtain precision adequate for quantitation of long-lived radionuclides. While several radiometric techniques are amenable to the determination of U at concentrations typical for those in geological and environmental samples, the most commonly used radiometric ones are high-resolution α- and γ-spectrometry, described briefly below.

Table 1. Nuclear data for long-lived uranium isotopes (Firestone 1998)

Nuclide	Abundance	$t_{1/2}$	Decay Mode(s)	σ_n (b)
^{232}U		68.9 y	α (100%) ^{24}Ne (9x10^{-11}%)	74 76$_f$
^{233}U		1.592x10^5 y	α (100%) SF (<6x10^{-11}%) ^{24}Ne (<9.5x10^{-11}%)	530$_f$ 46
^{234}U	0.0055%	2.455x10^5 y	α (100%) SF (1.64x10^{-9}%) ^{24}Ne (9x10^{-12}%) Mg (1.4x10^{-11}%)	100$_{g+m}$
^{235}U	0.7200%	7.038x10^8 y	α (100%) SF (7.0x10^{-9}%) ^{24}Ne (8x10^{-10}%)	580$_f$ 98
^{236}U		2.342x10^7 y	α (100%) SF (9.4x10^{-8}%)	5.1
^{238}U	99.2745%	4.468x10^9 y	α (100%) SF (5.45x10^{-5}%) β$^-$β$^-$ (2.2x10^{-10}%)	2.7

α-Spectrometry

α-Spectrometry provides the means to identify and quantify individual α-emitting radionuclides based on the measurement of emitted α-particles specific to the decay of the radionuclide of interest. All of the naturally occurring and long-lived non-naturally occurring isotopes of U emit α-particles with energies ranging from ~4 to 5.3 MeV

Table 2. Nuclear data for α-spectrometric determination of uranium isotopes (Firestone 1998)

Nuclide	α-Energy (keV)	α-Yield (%)
^{232}U	5.320	68.15
	5.264	31.55
^{233}U	4.824	84.4
	4.784	13.2
^{234}U	4.775	71.38
	4.722	28.42
^{235}U	4.596	5.0
	4.556	4.2
	4.502	1.7
	4.414	2.1
	4.398	55
	4.366	17
	4.215	5.7
^{236}U	4.494	73.8
	4.445	25.9
^{238}U	4.198	79.0
	4.151	20.9

(Table 2). Despite the short range of α-particles through matter, their detection possesses advantages as compared to the detection of β-particles, and γ-rays, the major advantage being the extremely low backgrounds that are achievable in α-spectrometers. This low background is due, in part, to the short ranges of α-particles allowing effective shielding of detectors from potential background sources. While α-spectrometry can be performed with ionization chambers, magnetic spectrometers, and scintillation detectors, the most commonly used detectors are ion-implanted silicon detectors. A typical α-spectrometer with 450-mm^2 ion-implanted silicon detectors has energy resolutions <20 keV, efficiencies >25% for source-to-detector spacing <1 cm, an energy range from 3 to 8 MeV, and backgrounds <1 × 10^{-4} counts per second. These detectors are typically contained in an integrated spectrometer containing the vacuum chamber, detector, bias supply for the detector, complete amplification system (preamplifier, amplifier, and biased amplifier), and calibration pulser. Data acquisition is controlled by a personal computer equipped with a multichannel analyzer for pulse height analysis along with data acquisition and processing software.

α-Spectrometry requires elaborate chemical preparation procedures before the sample can be counted. Samples need to be decomposed, the element of interest chemically separated and purified, and the chemical yield determined for precise quantitative analysis. Chemical yield is typically determined by spiking the sample with a non-naturally occurring U isotope such as ^{232}U prior to chemical treatment. The purified sample is electrodeposited onto a stainless steel, titanium, or platinum planchet to obtain a uniform thin source. The thin source minimizes sample self-attenuation of α-particles, which degrades the spectrum due to inelastic scattering of the α-particles. Electrodeposition is carried out in a buffered electrolyte matrix such as $(NH_4)_2SO_4/H_2SO_4/NH_4OH$ adjusted to pH 3.5 and at a constant current (0.5 A/cm^2) and low voltage.

For α-spectrometric measurement, samples are placed in a vacuum chamber at a fixed distance from a detector with a known background and counting efficiency. The background is determined by counting an electrodeposited experimental blank. The efficiency of the detector for a given geometry is determined by counting a calibrated source at that geometry. After counting the sample for a time sufficient to attain reasonable statistical error, the α-spectrum is corrected for background and detector efficiency. The chemical yield is calculated from the recovery of the spike as determined from the α-spectrum. The number of counts per unit time of each isotope can then be related to their concentration in the sample. Analysis of a typical U α-spectrum is straightforward due to the complete resolution of the ^{238}U, ^{235}U, ^{234}U, and ^{232}U peaks. A typical α-spectrometry system such as the one described above can be used for the determination of total U abundances and isotopic ratios. The chemical treatment allows preconcentration of the U from individual samples, thus allowing the processing of large samples for U determinations. Typical backgrounds and efficiencies would require a total of 0.4 µg of natural U to achieve 10% counting statistics for a 24-h sample count. Examples of several applications of α-spectrometry to the determination of U and the measurement of U-series disequilibria are given in Ivanovich and Murray (1992). Standard methods for the determination of U in water have been published by the U.S. EPA (EPA-600/7-79-093 1979) and the ASTM (D3972-90 1990; D3084-95 1995b).

Table 3. Nuclear data for γ-spectrometric determination of uranium isotopes (Firestone 1998)

Nuclide	$t_{1/2}$	γ-Energy (keV)	γ-Yield (%)
^{238}U series			
^{234}Th	24.10 d	63.2, 63.6	85
		93.1, 93.5	100
^{234m}Pa	1.175 m	94.66	0.12
		98.44	0.19
		1001	0.59
^{226}Ra	1600 y	186.1	100
^{214}Pb	26.8 m	352	36
		295	
		242	3.7
		53	
		786	
^{214}Bi	19.9 m	2204	6
		1764	17
		1238	6.0
		1120	16
		609.3	47
^{235}U	7.038×10^8 y	185.7	54.0
		143.8	10.5

γ-Spectrometry

γ-Spectrometry provides the means to identify and quantify individual radionuclides based on the measurement of emitted γ-rays specific to nuclear transitions of the radionuclide of interest or one or more of its progeny. The energies and intensities for γ-rays of the U isotopes and progeny that are most frequently used for the determination of U in geological and environmental materials are given in Table 3. γ-Rays are highly penetrating, and their measurement is straightforward since most γ-ray peaks can be individually resolved with semiconductor detectors. The technique requires very little sample preparation, particularly when compared to α-spectrometric analysis. It is non-destructive and multiple radionuclides can be measured simultaneously. Typically, all that is required is that the sample be placed in front of the detector of a calibrated γ-spectrometry system. The detector needs to be calibrated for energy and efficiency at the geometry at which the sample is counted. Ideally, the sample size and matrix of the sample need to be similar to that of the standardization source. This is typically achieved by preparing a mock-up sample that has been spiked with known amounts of radionuclides that emit γ-rays with energies comparable to those of interest. A γ-spectrometry system typical for most modern laboratories utilizes a high-purity germanium (HPGe) detector biased with a power supply, a preamplifier, spectroscopy amplifier, and pulse height analyzer. HPGe detectors are required to be at liquid-nitrogen temperature when they are under high-voltage bias and require connection to Dewar via a cryostat. The detector and sample are normally shielded with alternating layers of Pb, Cu, and a low-Z material such as Lucite. This graded shielding system minimizes the X-rays scattered from the lead shielding and shields the detector from external sources. Data acquisition and spectral analysis can be controlled by a personal computer equipped with data acquisition and processing software for quantitative analysis. A schematic diagram of a typical γ-spectrometry system is illustrated in Figure 1.

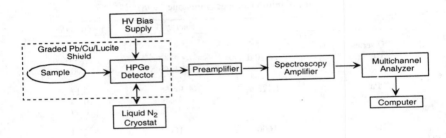

Figure 1. Schematic diagram of a γ-spectrometry system.

Most HPGe detectors are configured as a junction between n- and p-type semiconductor materials and come in a variety of geometries, sizes, and shapes. The selection of a specific detector depends on the types of samples measured. The most common geometry is a coaxial HPGe detector. Typical HPGe detectors used in geochemical and environmental analysis have volumes of 100-200 cm^3, possess efficiencies of approximately 30 to 40% relative to a 7.6 × 7.6-cm NaI(Tl) detector, and have resolutions <2 keV at 1333 keV. These systems are useful for a wide spectrum of applications and are normally calibrated in the range from approximately 50 to 2000 keV. Larger detectors with relative efficiencies >150% are available but at the expense of slightly lower relative resolutions (>2.3 keV at 1333-keV). These high-efficiency detectors are useful for counting larger-volume, low-activity environmental samples. Another approach to high-efficiency γ-spectrometry is to use a well-type HPGe. Well-type detectors offer the ultimate absolute counting efficiency for small samples. The name "well type" is given since this type of detector possesses a cavity that can surround a small sample. A typical well only has an available volume of < 10 cm^3, which precludes the analysis of large volume samples. Finally, thin planar HPGe detectors equipped with a beryllium window offer high resolutions for the measurement of γ-rays with energies between 3 and 300 keV. These detectors are transparent to higher energy γ-rays and thus possess lower background from Compton scattering of high-energy γ-rays.

Due to its long half-life, the quantitation of ^{238}U is usually based on the measurement of short-lived daughters such as ^{234}Th (Harbottle and Evans 1997). Thorium-234 has a half-life of 24.1 d and is essentially always in equilibrium in environmental samples such as minerals and soils unless the material has been recently chemically treated or the U has been recently deposited. Measurement is typically based on a 93 keV or the 63 keV γ-ray doublets. Measurement of the these γ-rays generally can be improved by using a detector sensitive to low-energy photons such as a planar HPGe with thin beryllium windows. A γ-spectrum of a solid UO_2 sample weighing 10 g collected on a high-resolution planar HPGe detector is given in Figure 2. Determination of ^{238}U can also be performed by measurement of the 1001-keV γ-ray from the decay of ^{234m}Pa ($t_{1/2}$ = 1.175 m). The low branching ratio for this γ-ray precludes its use for samples with U concentrations <10 μg/g. A recently published example of the application of γ-spectrometry for the measurement of U is Gu et al. (1997). In a study of the deposition of dust in the Chinese loess plateau, activities of ^{238}U, as well as ^{232}Th and several of their daughter products, were determined non-destructively using a low background well-type HPGe detector.

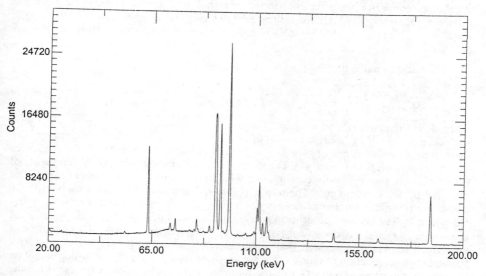

Figure 2. Gamma spectrum of a 10-g solid UO_2 sample counted for one day. The spectrum was collected on high-resolution planar HPGe detector possessing thin beryllium. The spectrum illustrates prominent peaks from ^{234}Th, ^{234m}Pa, and ^{235}U.

In the case of ^{235}U, identification and quantitation can be performed directly via measurement of its strong 185.7-keV γ-ray. A correction for interference from the 186.1-keV γ-ray, ^{226}Ra ($t_{1/2}$ = 1600 y), is required for samples in secular equilibrium. However, this interference can be corrected by indirectly determining the contribution of ^{226}Ra based on measurement of γ-rays from ^{214}Pb and ^{214}Bi. The α-decay of ^{226}Ra produces ^{222}Rn ($t_{1/2}$ = 3.8235 d), which rapidly establishes equilibrium with both ^{214}Pb and ^{214}Bi. The short half-lives (26.8 m and 19.9 m, respectively) and abundance of γ-rays emitted by these two daughter radionuclides makes it possible to correct for ^{226}Ra in samples containing U at levels <1 μg/g when secular equilibrium can be assumed

(Harbottle and Evans 1997). The determination of the concentration ^{235}U can be used to calculate total U if a known isotopic distribution is assumed. Standard methods for the determination of γ-ray emitters in water have been published by the ASTM (D3649-91 1991b).

Neutron activation analysis

Neutron activation analysis (NAA) is based on the transmutation of nuclei to radionuclei or excited nuclear states by capture of neutrons. Qualitative and quantitative analyses can be performed by measurement of the α-particles, β-particles, γ-rays, or delayed neutrons that may be emitted as the excited nuclei or their excited daughter products relax with a calibrated spectrometer. The emitted radiation can be measured during irradiation or after a delay. The most frequently used method of neutron activation analysis of U-containing materials is activation with thermal neutrons. Typically, NAA is performed using neutrons produced in a nuclear reactor or by an isotopic source such as ^{252}Cf and subsequently thermalized (most probable energy of 0.025 eV). Samples and flux monitor standards are typically irradiated simultaneously and analyzed for high-precision quantitative analysis. If no chemical treatment is performed after irradiation, the technique is called instrumental neutron activation analysis (INAA). If chemical separations are performed after irradiation, the technique is called radiochemical neutron activation analysis (RNAA). Activation analysis employing epithermal neutrons, (i.e. neutrons with energies of 0.1-1 eV) is called epithermal neutron activation analysis (ENAA).

The basic equation that governs NAA is as follows

$$R = \frac{m}{M} N \phi \sigma_n \varepsilon (1 - e^{-\lambda t}) \tag{1}$$

where R is the count rate of a sample containing a mass m (g) of a given nuclide with molecular weight M (g^{-1} mol) at the end of irradiation, and N is Avogadro's constant. The thermal neutron flux density (ϕ) is the number of thermal neutrons incident on a unit area of sample per unit time (n cm^{-2} s^{-1}). The thermal neutron cross section (σ_n) is a measure of the probability that the nuclide of interest will capture a neutron and be activated. The units for σ_n are given in cm^2. The term ε is the efficiency of the detector measuring the count rate (absolute number of decays per number of decays counted) for the decaying radionuclide. The final set of terms in parentheses corrects the activity for the duration of irradiation t (s) and the decay constant λ (s^{-1}) of the specific nuclide. This equation is valid for samples thin enough that attenuation of the neutron flux can be ignored. Examination of equation (1) shows that the absolute sensitivity of NAA is dependent on the flux ϕ, the cross section σ_n for the nuclear reaction, the decay constant λ for the nuclide of interest, and the irradiation duration t. A thorough overview of the equations governing NAA can be found in texts such as Friedlander et al. (1981) and Ehman and Vance (1991).

Non-destructive NAA determination of U in geological and environmental materials can be performed by several approaches (Parry 1991). One approach is NAA with subsequent detection of delayed neutrons. This method possesses a high specificity for U and has been used for many years in the exploration of U ores. The method is based on the measurement of delayed neutrons following a short irradiation with thermal neutrons. Uranium-235 undergoes thermal-neutron-induced fission. The resulting neutron-rich fission products undergo β$^-$ decay. While most of the fissiogenic daughter nuclei are in excited states and relax via γ-ray emission, some daughter nuclei possess sufficient energy for relaxation to occur via delayed-neutron emission. The delayed neutrons are emitted rapidly after β decay with the half-life of the β⁻ decaying parent radionuclide. The neutrons are typically detected with polyethylene-moderated BF_3 proportional counter systems. The BF_3 detector is extremely sensitive to neutrons and insensitive to the γ-rays and β-particles emitted by other activated elements in the sample. The technique, therefore, is highly selective with respect to ^{235}U. If the sample is irradiated with a neutron flux possessing an appreciable fast-neutron component, there will be a contribution from fast-neutron fissions of ^{238}U. This condition is not problematic since this nuclide is still U. If the sample contains significant amount of ^{232}Th, a correction for fast-neutron fissions is required for accurate U determination. The dominant ^{235}U fission products that decay via delayed-neutron emission are ^{89}Br ($t_{1/2}$ = 4.4 s), ^{87}Br ($t_{1/2}$ = 55.6 s), ^{94}Rb ($t_{1/2}$ = 2.8 s), ^{137}I ($t_{1/2}$ = 24.5 s), and ^{135}Sb ($t_{1/2}$ = 1.7 s). For example, the decay scheme for ^{87}Br is

$$^{87}_{35}Br \xrightarrow{55.60s} {}^{87}_{36}Kr + {}^{0}_{-1}\beta^- + \bar{\nu} \xrightarrow{76.3m} {}^{86}_{36}Kr + {}^{1}_{0}n + {}^{0}_{-1}\beta^- + \bar{\nu} \qquad (2)$$

The NAA with delayed-neutron detection can be automated for online and unattended determination of U content in solid and liquid samples. A system described by Rosenberg et al. (1977) determines the U content in 10-g samples that have been packed in polyethylene containers and irradiated for 60 s in a flux of 10^{16} n cm^{-2} s^{-1}. Samples are then counted for 60 s after a decay of 20 s. Even though natural U is only 0.720% ^{235}U, and less than 1% of the neutrons released in the fission are from delayed-neutron emitters, the thermal cross section of ^{235}U is so large (582 b) that the technique possesses sufficient sensitivity and selectivity for routine analysis of U-containing materials. The reported method detection limit is 0.6 μg/g. This method can be also be coupled with γ-ray spectrometry for the determination of additional elements in the sample.

Fission track analysis

If a sample is irradiated in close proximity to an inorganic glass or plastic insulating material, the heavy nuclides generated from thermal-neutron-induced fission will produce submicroscopic tracks. The measurement of these tracks after etching is the basis for fission track analysis (FTA). Fission track analysis has been applied to geological dating (Wagner 1966) and, recently, the measurement of uranium in pore water (Barnes and Cochran 1990). The FTA method can be performed with small samples but generally only gives precisions of 6 to 15% (Toole et al. 1984; Barnes and Cochran 1990).

Instrumental neutron activation analysis

Minor and trace quantities of U are typically determined non-destructively by INAA. This method consists of determining the intensity of the 105-keV γ-ray photopeak from the decay of ^{239}Np. Neptunium-239 is formed from thermal-neutron capture of ^{238}U and subsequent decay the following reaction:

$$^{238}_{92}U + {}^{1}_{0}n \longrightarrow {}^{239}_{92}U \xrightarrow{23.5 \min} {}^{239}_{93}Np + {}^{0}_{-1}\beta^- + \bar{\nu} \xrightarrow{2.33d} {}^{239}_{93}Pu + {}^{0}_{-1}\beta^- + \bar{\nu} \qquad (3)$$

The amount of ^{239}U and ^{239}Np produced is directly related to, and therefore a measure of, the amount of ^{238}U present in the original sample. At high concentrations U can be determined by measurement of the 74-keV γ-ray photopeak from the decay of ^{239}U ($t_{1/2}$ = 23.5 m). Detection limits are enhanced with use of a detector sensitive to low-energy photons such as a planar HPGe with a thin beryllium window. Both of these methodologies are enhanced with a higher ratio of epithermal neutrons.

INAA is most frequently used for non-destructive trace determination U along with a suite of elements in bulk geological samples. Routine applications of INAA for the determination of trace elements including U in geological materials in the literature include the analysis of fluorites (Joron et al. 1997) and precious and common opals (McOrist and Smallwood 1997). Joron et al. (1997) reported U detection limits of 6.9 ng/g for a 8-15 h irradiation at a flux of 1.1×10^{13} n cm^{-2} s^{-1} followed by a one-week sample cooling and 3000-s count time.

The high sensitivity of INAA makes this technique ideal for the analysis of minute amounts of material such as individual phases, minerals, and crystals separated from bulk material. Schrauder et al. (1996) used INAA to determine U along with a suite of trace elemental in 13 micro-inclusion-bearing fibrous diamonds from Botswana in order to constrain the composition of fluids in the mantle. Raimbault et al. (1997) developed a technique to extract single crystals and perform multielemental analysis, including U, with INAA. The INAA technique required only a few micrograms of mineral and facilitated the measurement of more than 30 elements in multiple samples from the Oklo U deposit in Gabon. This work also utilized the presence of epithermal neutrons in the reactor flux producing (n, p) reaction products and extending the number of elements that could be quantified. This methodology was able to detect low µg/g concentrations of U in single crystal samples in a variety of minerals.

Radiochemical neutron activation analysis

With respect to NAA techniques, RNAA offers detection limits for samples with U concentrations lower than can be quantified with INAA. The RNAA quantitation of U is based on the radiochemical analysis of one or more fission products produced from the thermal-neutron-induced fission of ^{235}U. Typically, the irradiated sample is decomposed and equilibrated with a milligram amount of the element of interest known as the carrier. The carrier element, along with the radionuclide produced during sample irradiation, is then chemically separated and purified. The amount of radionuclide recovered from the chemical separation is determined gravimetrically from the amount of carrier element recovered. The purified elemental sample can then be radiometrically assayed by γ-spectrometry. RNAA is a destructive technique that is more time-consuming and requires greater analytical and chemical laboratory skills than does INAA. The chemical separation of the element of interest, however, results in a sample possessing a γ-spectrum with virtually no background from other γ-ray emitting species. The technique has, therefore, lower detection limits than INAA. Procedures for the determination of trace U in geological materials via fission ^{132}Te are given in Anders et al. (1988). Using a thermal neutron flux of 1×10^{14} n cm^{-2} s^{-1} this methodology can achieve detection limits in the range of 10-100 pg/g for U.

ATOMIC SPECTROMETRIC TECHNIQUES

Atomic spectrometric methods are based on the absorption and emission of electromagnetic radiation from atoms or ions. Information specific to the electronic structure of individual elements is manifested in the ultraviolet and X-ray regions of the electromagnetic spectrum. Each element possesses a unique electronic structure and, therefore, unique atomic spectrum. The absorption, emission, and fluorescence of electromagnetic radiation of a wavelength specific to an electronic transition of a particular element can serve as the basis for analysis and determination of that element. Uranium is amenable to analysis and determination via several atomic spectrometric-based techniques. A survey of recent analytical and geoanalytical literature reports that very few new developments have been made in the application of atomic spectrometric techniques to geological materials (Lipschutz et al. 1999). This is an indication of the maturity of these techniques and the large pre-existing knowledge base.

Typically, an analytical atomic spectrometric determination involves the atomization of the analyte from solution and subsequent measurement of the absorption, emission, and fluorescence of specific wavelengths of electromagnetic radiation. Atomic absorption occurs when a ground-state atom absorbs energy in the form of specific wavelengths of electromagnetic radiation and is elevated to an excited electronic state. This can be done by interrogating the atomized sample with electromagnetic radiation of a wavelength specific to the electronic transition of interest and measuring the change in transmitted light. The intensity of the absorbed light can be related to the amount of the element of interest. Atomic emission occurs because, when an atom in an excited electronic state relaxes to its ground state, at which point it radiates a characteristic line spectrum. For atomic emission the element of interest must be excited to a higher electronic energy level. This can be achieved via thermal collisions in a flame, electrical discharge, or plasma source. The wavelength of interest is separated from interfering light sources via a monochromator, and the intensity is measured. The intensity of the emitted radiation can be related to the amount of element present. Finally, atomic fluorescence involves the selective excitation of the species of interest with a monochromatic source of electromagnetic radiation followed by measurement of a specific wavelength of electromagnetic radiation produced by the radiative de-excitation of the element. The intensity of the emitted radiation can be related to the amount of element present. The following sections summarize several of the most commonly utilized atomic spectrometric techniques and their application to the determination of U.

Atomic absorption spectrometry

Historically, atomic absorption spectrometry (AAS) has been an essential tool for the analysis of geological and environmental materials. Flame AAS first became widely available in the early 1960s, and it still finds use because of the relative simplicity of its implementation. In flame atomization AAS, a solution of the sample is introduced as a aerosol into a flame by means of a pneumatic nebulizer in combination with an expansion chamber. Nebulization converts the liquid sample to an aerosol with a stable and reproducible size distribution. This aerosol is swept into a flame with a carrier gas. The flame serves to drive off the sample solvent and convert the analyte to gas-phase atoms. The most widely used flame for the determination of U is a mixture of nitrous oxide and acetylene. A nitrous oxide acetylene flame produces a temperature of 3200 K. The primary objective of the flame is to dissociate molecules of analyte into atoms. The atoms in the flame are then exposed to a source of electromagnetic radiation that produces a line wavelength specific to an electronic transition in the element of interest. The most widely used line source for AAS is the hollow cathode lamp. The hollow cathode lamp is a glass envelope filled to a low pressure with an inert gas containing a hollow cathode made of the element of interest. A voltage is applied between the anode and the cathode, resulting in ionization of the inert gas and acceleration of the ions toward the cathode. The inert gas ions sputter metal ions from the cathode. The metal ions are subsequently excited by further collisions with the inert gas ions producing an intense spectrum characteristic of the metal of the cathode. The atoms in the flame absorb this light, and a transition to a higher electronic energy state occurs.

The amount of light absorbed can be related to the concentration of the element of interest via the Beer-Lambert law

$$A = abc \tag{4}$$

where A is the absorbance, a is the absorptivity (L g^{-1} cm^{-1}), b is the atom cell width (cm), and c is the concentration (g L^{-1}) of the element in the unit cell. Measurement of the amount of light absorbed is achieved with a combination of a grating or prism monochromator and a detection device such as a photomultplier tube, photodiode array, or charge transfer device. The monochromator isolates the wavelength of interest from other sources of light such as the radiation source and light emitted by other species in the flame. A schematic diagram of a typical flame AAS is illustrated in Figure 3. The AAS determination of U content is typically performed via the 358.5 nm line (Slavin 1975). Modern flame based AAS instrumentation can only achieve detection limits of 40 µg/mL for U in solution (Dean 1995). This relatively high detection limit is due to the low efficiency for atomizing refractory U at flame temperatures.

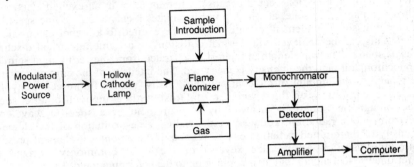

Figure 3. Schematic diagram of a flame AAS.

The relatively poor detection limits of flame AAS can be improved by substituting the flame source with a graphite furnace for sample atomization. This technique is known as electrothermal atomic absorption spectrometry (ETAAS) or graphite furnace atomic absorption spectrometry (GFAAS). A standard practice for measuring trace elements in water by ETAAS is given in ASTM D3919-94a (1994b). Electrothermal atomization is typically performed with a graphite tube platform furnace. A small sample of the solution is inserted onto a platform mounted in the furnace, and the sample is heated through a profile that dries, ashes, and finally atomizes the sample. Chemical modifiers are frequently utilized to increase the volatility of the sample matrix. Background correction is required due to scattering by molecular species, salt particles, and smoke. The electrothermal source produces a transient signal, which requires fast electronics for data handling. Background correction can be performed via a system that employs the continuum radiation from a deuterium lamp in combination with a tuned amplifier. Electrothermal atomic absorption spectrometry offers several advantages over flame AAS, including small sample sizes (microliters), low background, increased sensitivity. However, ETAAS generally is more complex and more susceptible to interferences, and requires greater operator skill.

Electrothermal atomic absorption spectrometry can achieve detection limits of 30 ng/mL with a 20-µL U-containing solution. Both AAS and ETAAS are generally single-element methods. The accuracy of AAS and ETAAS dependents on the complexity of the sample matrix; however, accuracies better than ±1% can be achieved if concentrations are significantly higher than the detection limits. A complete overview of atomic spectrometric instrumentation can be found in Skoog (1985), Slavin (1978), and Haswell (1992).

Atomic emission spectrometry

Atomic emission spectrometry (AES) can be performed by employing a variety of methods to excite the element of interest to a higher electronic energy level: flames, electrical discharge sources such as direct-current arcs or alternating-current arcs, high-voltage alternating–current spark discharge, and an inductively coupled plasma (ICP). As the excited ion relaxes, it emits electromagnetic radiation at a specific wavelength. The emitted radiation is then passed through a grating or prism monochromator to isolate the wavelength of interest. A photodetector measures the intensity of the selected wavelength of light, and the intensity can be directly related to concentration.

Flame emission spectrometry

Atomic emission spectrometry with flames (flame emission spectrometry or FES) is performed on instrumentation similar to AAS. In FES the sample is sprayed into a flame possessing sufficient energy to excite the element to a level at which it will radiate a characteristic line-emission spectrum. The 358.5-nm emission line can be used for the FES determination of U, as well as FAAS. Ionization of U must be suppressed by adding an alkali element such as potassium or cesium to the sample and standards. A nitrous oxide and acetylene flame is required. Flame emission spectrometry can achieve detection limits of 100 ng/mL (Dean 1995).

Inductively coupled plasma-atomic emission spectrometry

The most widely used approach to AES utilizes an ICP source. Since the sensitivity of AES analysis is dependent on the number of atoms promoted to a higher energy electronic state, a more energetic source should result in a sensitivity increase. The main advantage of the ICP source is that its high temperature, up to 8000 K, ensures complete atomization of molecular species and excitation of even the most refractory elements such as U. Inductively coupled plasma-atomic emission spectrometry first became widely available in the middle 1970s and has become a popular instrument for multielemental analysis of solutions. A complete overview of ICP-AES instrumentation can be found in Montaser and Golightly (1992).

In ICP-AES, a solution of the sample is typically introduced as an aerosol into the ICP source by means of a pneumatic nebulizer in combination with an expansion chamber. Nebulization converts the liquid sample to an aerosol with a stable and reproducible size distribution. This aerosol is swept into the ICP with a stream of Ar called the sample carrier gas. A typical ICP source contains a torch which is configured as three concentric quartz tubes through which Ar is flowing. The diameter of the largest tube is approximately 2 cm, and the Ar that flows through it at a rate of approximately 14 L/min is called the cooling gas. The middle tube contains Ar flowing at a rate of 0 to 1 L/min. This auxiliary flow can be used to affect the shape of the plasma and extend the torch life. The aerosol sample and carrier gas is introduced into the center tube with a flow of approximately 1 L/min. The torch is surrounded by a water-cooled induction coil that is powered by a radio-frequency generator. Typical frequencies are 27.5 and 40 MHz with a power of 1.3 kW. An impedence matching circuit is used to maintain a stable plasma. The source is called an inductively couple plasma since the coil acts as an inductor. The ICP atomizes molecular species and excites and ionizes the atoms. As the analyte leaves the plasma, the atoms and ions cool, and electromagnetic radiation is produced by the radiative de-excitation at wavelengths specific to individual elements. The electromagnetic radiation can be measured in a configuration that is either axial or radial with respect to the plasma. The electromagnetic radiation passes through a set of slits, the wavelength of interest is selected by a grating or prism monochromator, and its intensity is measured by a detection device such as a photomultiplier tube or an array detector. The monochromator can be scanned or a polychromater can be employed to simultaneously measure the intensity of multiple wavelengths. The intensity can be directly related to concentration. A schematic diagram of a typical ICP-AES is given in Figure 4. The emission spectrum of U consists of thousands of resolvable lines. The 385.96-nm emission line is typically selected for ICP-AES determination of U detection limits of 20 ng/mL can be achieved (Dean 1995). Accuracies of ±1% can be achieved if concentrations are significantly higher than the detection limits.

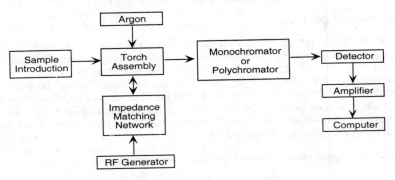

Figure 4. Schematic diagram of an ICP-AES.

The primary advantage of ICP-AES over flame AAS for the determination of U in solution is the much lower detection limits and multielement capability and high sample throughput. Uranium detection limits for ICP-AES are slightly superior to those of ETAAS, and the multielement capability and increased sample throughput make ICP-AES an attractive technique for routine determination of U and other elements in solution. In terms of cost and maintenance ICP-AES instrumentation is more expensive and more complex than AAS, ETAAS, and FES. While all atomic spectrometric techniques generally determine total U concentration without isotopic selectivity, a high-resolution ICP-AES has been developed for the determination of U isotopic ratios (Edelson 1992).

Atomic fluorescence spectrometry

Atomic fluorescence spectrometry (AFS) is also an atomic emission technique. Like AAS and AES, AFS involves the conversion of the sample to gas-phase atoms via a flame, furnace, or a laser. The atomized analyte is then exposed to a source of electromagnetic radiation that produces a line wavelength specific to an electronic transition in the element of interest. Atomic fluorescence spectrometry can generally provide detection limits comparable to ETAAS and ICP-AES; however, AFS has not demonstrated significant advantages that would warrant significant commercial development. As such, the technique is not as widely used.

Fluorometry

Uranium is unique among most elements in that it can be determined directly via fluorometry without the addition of a fluorescent chelating agent. A standard test method for determining traces of U in water by fluorometry is given in ASTM D2907-91 (1991c). In the fluorometric method an aliquot of the liquid sample is pipetted into a platinum disk containing a sodium fluoride-lithium fluoride flux and is evaporated and fused in a furnace. The fused disk can then be directly exposed to an ultraviolet source such as a mercury-arc lamp in combination with a filter or monochromator at 365 nm. The intensity of the fluorescence is measured at 560 nm. The presence of anions such as

Cl⁻ and cations such as Cd^{+2}, Cr^{+3}, Co^{+2}, Cu^{+2}, Fe^{+3}, Mg^{+2}, Mn^{+2}, Ni^{+2}, Pb^{+2}, Pt^{+4}, Th^{+4}, and Zn^{+2} quench U fluorescence and lower the sensitivity of the technique. Niobium and Ta are reported to enhance U fluorescence. In samples containing significant quantities of these elements, analyses can be performed after chemical separation of the U or dilution of the sample if sufficient concentrations of U are present. The fluorometric technique can determine U concentrations as low as 5 ng/mL. A method has recently been developed for the determination of trace U in geological materials by laser fluorometry after low temperature dissolution with HF and HNO_3 for economical high-throughput analyses (Ramdoss et al. 1997).

Phosphorimetry

Lower detection limits can be achieved by utilizing the ability of uranyl ions to phosphoresce when excited to a triplet state. The uranyl ion can be directly determined phosphormetrically in a H_3PO_4 or H_2SO_4 solution. The acid is added to protect the uranyl ion from various intermolecular mechanisms that quench luminescence. The sample is excited at 254 nm, and the intensity of the yellow-green emission can be measured in a fluorimeter and related to concentration. This method has a measurement sensitivity of 100 ng/mL. Use of a pulse-laser excitation source can increase the sensitivity of this method greatly. In the pulsed laser phosphorimetric method, an aliquot of sample is pipetted into a glass vial, mixed with HNO_3 and H_2O_2, and taken to dryness. The residue is dissolved in HNO_3 and a complexant such as H_3PO_4. The pulsed-laser phosphorimetric technique can determine U concentrations as low as 50 pg/mL. A standard test method for determining trace U in water by pulsed-laser phosphorimetry is given in ASTM D5174-91 (1991d). Both the fluorometric and phosphorimetric techniques are single-element methodologies dedicated to the determination of U. Each technique determines total U concentration independent of the specific isotopic composition of the U.

X-ray fluorescence

X-ray fluorescence (XRF) can provide qualitative identification and quantitative determination of U in a variety of matrices. In XRF the sample is typically bombarded with a beam of X-ray energy electromagnetic radiation at wavelengths shorter than the spectral line desired. Samples can be liquid or solid. The excitation removes an electron from an inner shell of the target atom, fluorescence X-rays are emitted as the resulting vacancy is filled, and electrons from the outer shell make the transition to the inner shell as part of the relaxation process. The excitation is typically provided by X-ray tubes. Each element emits characteristic secondary fluorescent X-rays that can be selectively detected by using a diffraction crystal to disperse the X-rays by wavelength (wavelength-dispersive spectrometry) or detected directly with use of a SiLi detector (energy-dispersive spectrometry). The intensity of the detected X-rays can be related to concentration. A schematic diagram of a typical XRF is given in Figure 5. The XRF method is very rapid and can compete with wet-chemical methods in terms of accuracy, but only for major constituents of the sample. Solid samples are typically homogenized and pressed or fused into a pellet for quantitative determinations. Matrix effects require matching of standards to unknown samples for accurate results. Detection limits are typically in the µg/g range for heavy elements such as U, and multielement analysis is readily performed.

A procedure for the determination of U soils can be found in ASTM C1255-93 (1993). The procedure calls for use of a Rh anode X-ray tube with a Mo, Rh, or Ag secondary target to excite a well-homogenized pressed powdered pellet sample. This procedure can be used to determine as little as 20 µg/g of U and could, therefore, be adapted as a rapid method for the analysis of U in geological materials.

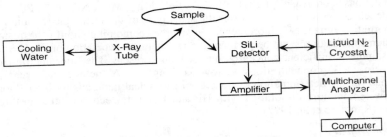

Figure 5. Schematic diagram of an XRF.

It should be mentioned that electron beams, accelerator produced X-rays, and accelerator produced high-energy protons can be used as an excitation source for *in situ* X-ray fluorescence analysis of geological materials with high spacial resolution. Scanning electron microscopy (SEM) and electron microprobe analysis (EMPA) can be distinguished from XRF by use of electrons as an excitation source. Accelerator-produced X-rays and protons for XRF analysis are used in synchrotron X-ray fluorescence microprobe analysis (SXRFM) and micro-proton induced X-ray emission analysis (μ-PIXE), respectively. Both SXRFM and μ-PIXE are extremely useful for determining the micro-distribution of uranium in minerals at major, minor, and trace concentrations. These microprobe techniques generate X-rays by similar mechanisms to that of conventional XRF. The high sensitivity of SXRFM results from the high intensity of the X-rays that can be generated by synchrotron sources. The μ-PIXE technique uses a focussed beam of protons to selectively irradiate different areas of the sample. The collision of the charged proton with the target atom induces electron shell vacancies in the target atom. Fluorescence X-rays are emitted as the resulting vacancies are filled. The μ-PIXE technique is particularly useful for the determination of U due to lower backgrounds at energies typical for heavier elements. Both SXRFM and μ-PIXE are limited by the availability of synchrotron X-ray sources and particle accelerators.

As with most other atomic spectrometric techniques, XRF determines total U concentration independent of the specific isotopic composition of the U. Further discussion on XRF instrumentation and analytical methods can be found in Bertin (1975) and Tertian et al. (1982).

MASS SPECTROMETRY

Mass spectrometric methods employ the separation and detection of gas-phase ions on the basis of the mass-to-charge ratio (m/z). Typically, mass spectrometry involves the atomization and ionization of the analyte from solution or solid sample, where the ion beam is focused and directed with a system of electrostatic ion lenses; after dispersal in space or time these ions are directly detected. The most common mass analyzers used in inorganic mass spectrometry are single-focusing magnetic analyzers, double-focusing analyzers (single-focusing magnetic analyzer and a radial electrostatic field in tandem), time-of-flight analyzers, and quadrupole analyzers. Mass spectrometry provides one of the most sensitive methods for the determination of trace and ultratrace elements in a wide variety of matrices. In contrast to many of the atomic spectrometric techniques outlined in this review, mass spectrometry provides the means for the determination of isotopic ratios of individual elements. As such, mass spectrometric techniques are invaluable for the determination of U in geological and environmental samples.

Solid state mass spectrometry

Inorganic mass spectrometric techniques are defined primarily by the method of sample introduction, atomization, and ionization. Several methods of sample introduction have been adapted for the introduction of solid state inorganic materials for mass spectrometric analysis: the spark source, glow discharge source, secondary ion sputter, laser ionization, and thermal ionization.

Spark source mass spectrometry

In spark source mass spectrometry (SSMS), solid samples are evaporated, atomized, and ionized via electron bombardment between two pin-shaped electrodes across which a high-potential (30 kV), high-frequency plasma is generated. The ions are introduced into a mass spectrometer, separated on the basis of m/z, and detected.

Glow discharge mass spectrometry

Glow discharge mass spectrometry (GDMS) functions when Ar^+ ions, formed at a low pressure, are accelerated to a cathode prepared from a solid sample. The resulting sputtered neutral atoms are ionized in the glow discharge plasma and are introduced into a mass spectrometer. Quantification in both SSMS and GDMS generally proves difficult if no suitable Standard Reference Material (SRM) is available.

Secondary ion mass spectrometry

Secondary ion mass spectrometry (SIMS) utilizes bombardment of ions possessing sufficient energy to penetrate into a solid sample and cause the sputtering of neutral particles and ions. The ions are then introduced into a mass spectrometer, separated on the basis of m/z, and detected. As with other solid-state mass spectrometric techniques quantification have generally proven difficult if no suitable SRM is available. The strength of SIMS is the ability to perform isotopic ratio analysis with high sensitivity and good topographic resolution in both depth and lateral position. In recent years, new SIMS instruments have been established and applied to the *in situ* determination of U, Th, and Pb isotopic compositions. Compston et al. (1984) has demonstrated that U isotopic analysis can be performed as part of U-Pb age determinations by ion probes such as the sensitive high mass resolution ion microprobe (SHRIMP). More recently, Meyer et al. (1996) used SHRIMP to measure the ages and isotopic composition of these elements in lunar zircon crystals from a granophyre.

Laser ablation mass spectrometry

Focused electromagnetic radiation from a laser has been used to sample the analyte with spacial resolution. Laser ablation (LA) has been used to ablate neutral atoms with spacial resolution prior to introduction into various types of mass sepctrometers, such as an inductively coupled plasma mass spectrometers (ICPMS). Laser ablation ICPMS has been developed for the direct analysis of trace elements in solid samples with spatial resolutions as low as 10 μm. The development of LA-ICPMS as a microprobe technique has been the subject of a substantial amount of publications in the last two years (Lipschutz et al. 1999). The analytical capabilities of LA-ICPMS have been described in Longerich et al. (1996). Norman at al. (1996) has reported LA-ICPMS detection limits of 50 ng/g for determination of U in pyroxenes and garnets. Machado and Gauthier (1996) used LA-ICPMS to determine the concentration and isotopic composition of U in zircon and monazite. The LA-ICPMS technique provides a slightly simpler, higher sample throughput, and lower cost alternative to the SHRIMP technique. The main limitation of this technique is the lower precision.

Resonant ionization mass spectrometry

Lasers can also be used to directly ionize the analyte through non-resonant pathways as with laser ionization mass spectrometry (LIMS). Specific nuclides can be ionized prior to mass spectrometric analysis through the use of one or more lasers tuned to specific wavelengths. This technique is resonant ionization mass spectrometry (RIMS). The RIMS method has been used for the determination of $^{235}U/^{238}U$ ratios in several SRMs achieving precisions of 0.4% (Green and Sopchyshyn 1989). The instrumentation, however, is complex and is not well suited for routine analysis of geological and environmental samples.

Thermal ionization mass spectrometry

Thermal ionization mass spectrometry (TIMS) is performed by depositing as a salt the dissolved, chemically purified sample on a filament composed of a refractory metal with a high electron work function. Rhenium, Ta, Pt, and W are typical filament materials. The element is atomized by increasing the temperature of the filament. The temperature of the filament or an adjacent filament causes ionization of the atomized element. The ions are then mass-analyzed. Thermal ionization mass spectrometry utilizing isotope dilution (ID) is one of the most precise and accurate methods for the determination of a single element. Precisions better than 0.01% can be obtained. The primary disadvantage is that extensive sample preparation is required and that only one element can be determined. Procedures and applications of TIMS to U measurements are summarized in Chen et al. (1992) and references therein. The authors developed a methodology for the determination of ^{238}U abundances, $^{238}U/^{235}U$ ratios, and $^{234}U/^{238}U$ ratios in small cosmochemical samples. The method utilized a ^{233}U-^{236}U double spike to correct for instrumental fractionation, high-yield, low-background chemical separation procedures, and filament-loading techniques. This methodology could determine $^{234}U/^{238}U$ ratios with a precision of 0.5% (2σ) for a sample size of 5×10^9 ^{234}U atoms.

Inorganic mass spectrometric techniques have been reviewed recently (Becker and Dietze 1998). While all of these mass spectrometric techniques can potentially be used for U determinations, the technique that has generated the most effort for the determination of U in geological and environmental samples is inductively coupled plasma mass spectrometry (ICPMS).

Inductively coupled plasma mass spectrometry

The ICP source is an effective and efficient atomization and ionization source that allows convenient introduction of solutions. When combined with a quadrupole mass analyzer, the large dynamic range, nominal resolution, high sample throughput and relatively simple spectra make this an extremely useful analytical tool for performing trace elemental analysis. Inductively coupled plasma mass spectrometers became commercially available in the early 1980s and have since gained wide acceptance as a sensitive and accurate method for geoanalytical and environmental applications.

The ICP source used in ICPMS is essentially identical to that used in ICP-AES with a few minor modifications. The ICP produces a population of atomic and molecular ions at high gas temperature and near atmospheric pressure. The most critical component of an ICPMS is the interface that samples these ions and allows their transfer to a lower pressure region without significant bias. This is achieved by means of a sampling cone possessing an orifice approximately 1-mm diameter. Ions flow through this orifice into an expansion chamber that is mechanically pumped to lower the pressure. The sampled material forms a supersonic jet. This jet flows through a skimmer cone also possessing an

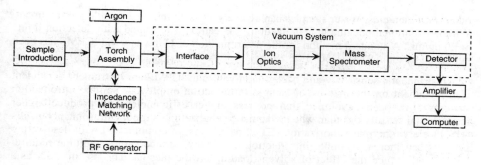

Figure 6. Schematic diagram of an ICPMS.

orifice with an approximately 1 mm in diameter. Ions can then be guided to the mass spectrometer with a series of electrostatic ion lenses. The spectra generated by ICPMS are relatively simple, consisting primarily of singly charged ions with some additional doubly charged ions, oxides, and hydrides. The doubly charged ions, hydrides, and oxides can be minimized to approximately <1% of the parent ion. Other molecular interferences are also possible. Most commercial ICPMS instruments use a fast-scanning quadrupole as a mass filter, resulting in nominal mass resolution. Quadrupole-based ICPMS instruments are capable of scanning the entire useful elemental mass range (from ^{6}Li to ^{238}U) in 100 ms. Transmitted ions are detected with a electron multiplier detector. A schematic diagram of a typical ICPMS instrument is given in Figure 6. A thorough description of the principals of ICPMS operation and instrumentation can be found in Jarvis et al. (1992). Instrument control and data acquisition of most modern commercial ICPMS instrumentation are performed via a personal computer equipped with a user friendly software interface. In terms of cost and maintenance ICPMS instrumentation is more expensive and more complex that that of AAS, ETAAS, FES, and ICP-AES.

Elemental analysis with ICPMS requires the determination of the concentration of at least one isobar-free isotope and known isotopic distribution for the element. Determination of U is straightforward due to the virtual absence of isobaric or molecular interferences in typical environmental and geological samples. A mass spectrum of a 50 ng/mL U-containing solution is given in Figure 7. Quantitation is typically based on the most abundant U isotope, ^{238}U. Uranium is frequently determined along with a suite of

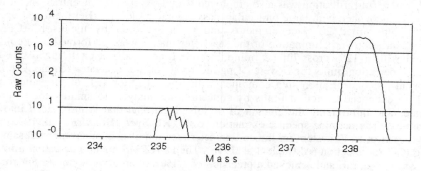

Figure 7. Spectrum of a 50 ng/mL U-containing sample, collected using conventional liquid sample introduction, pneumatic nebulization and a quadrupole-based ICPMS in the author's laboratory. The sample was prepared from NIST SRM 3164. Note that at this concentration the only peaks evident are from ^{235}U and ^{238}U. Generally no isobaric or molecular interferences are present in this m/z range.

other trace elements. Water samples can be analyzed with minimal sample pre-treatment. With conventional pneumatic nebulization, ^{238}U can be determined with detection limits of 20 pg/mL. A standard method for the determination of U in water via ICPMS has been published by the ASTM (D5673-96 1996a).

Instrumental sensitivity and background values in ICPMS are instrument-dependent and dependent on the method of sample introduction employed. Sample introduction systems have been developed that possess higher efficiencies and produce higher instrumental sensitivities than when pneumatic nebulization is used. Ultrasonic nebulizers (USN), electrothermal vaporization (ETV) nebulizers, in combination with desolvating systems can increase sample introduction efficiency, reduce sample volume requirements, and reduce the effect of solvent loading on the plasma. The use of USN as a method of sample introduction increases efficiency and sensitivity by removing aqueous sample matrix, i.e. water. The suitability of USN was evaluated for analysis of U in industrial wastewater by Crain and Mikesell (1992). This work demonstrated detection limits of 0.5 pg/mL. Several methods for sample introduction used in ICPMS are described in Montaser et al. (1998).

The direct analysis of solutions with TDS greater than ~0.2% can be problematic for ICPMS. Potential interferences from polyatomic species and degradation of signal resulting from the buildup of solid deposits on the nebulizer and sampling cones has led to the development and application of alternative strategies for sample pretreatment, sample introduction, and instrumental calibration for the determination of elements in solutions with high TDS. Chemical separation techniques continue to be developed and applied to the preparation of environmental and geological samples prior to ICPMS analysis. Crain et al. (1995) describe a chemical procedure for separation of actinides from soil leachates prior to determination with ICPMS. The method was capable of detection of 8 ng/L of soil using conventional pneumatic nebulization. This limit was an order-of-magnitude lower when ultrasonic nebulization was employed. Fluorinated metal alkoxide glass (MAF-8HQ) was used to immobilize 8-hydroxyquinoline for the separation and preconcentration of 22 trace elements (including U) in seawater (Sohrin et al. 1998). The trace elements were eluted with 0.5 M HNO_3 and analyzed via ICPMS. The method possessed a precision of <10%, and contamination limited detection limits. Haraguchi et al. (1998) developed a method for the determi-nation of trace elements (including U) in seawater and lake water by ion chromatography (IC) coupled with ICPMS. The chelating resin Chelex 100 was used for sample preconcentration. Furthermore, ultrafiltration was also employed to selectively preconcentrate metal ions that were associated with large molecular mass organic molecules. Bettinelli and Spezia (1995) determined 20 trace elements and U in seawater by means IC-ICPMS. Preconcentration and elimination of the sea water matrix were performed on a column packed with a 12% cross-linked iminodiacetate-functionalized chelating resin. The reported detection limits for U were 21 pg/mL for a 100 mL seawater sample. Use of a chemical preconcentration step combined with flow injection (FI) of the sample can result in enhanced detection limits by concentrating analytes to smaller total sample volumes and minimizing the amount of dissolved material that is introduced into the instrument. The actinide-specific extraction resin TRU Spec (Horwitz et al. 1993) was used as part of an FI system for the on-line determination of U in environmental samples using ICPMS detection (Aldstadt et al. 1996). The FI-ICPMS method detection limit for ^{238}U was 0.3 pg/mL and achieved a precision of 1.1% and an accuracy of ±1.8% for the determination of ^{238}U in certified groundwater samples. A standard method for the determination of U in soils via acid dissolution FI-ICPMS has been published by the ASTM (C1310-95 1995c). The estimated detection limits achievable by this procedure for ^{234}U are 30 ng/L of dissolved soil in solution.

Electrochemical methods can be used to selectively separate certain analytes from interfering species. Pretty et al. (1998) report use of an anodized glassy carbon electrode to preconcentrate U prior to analysis with ICPMS. Elimination of 2×10^4 µg/mL of Na was attained with a 90-s cell washout time. This method was used to quantify U in seawater material NASS-4 at several levels of dilution. Recoveries ranged from 100.4 to 119.4%, and results agreed with the certified U concentration within estimated 95% confidence limits.

Calibration methods

Bulk elemental analysis of geological materials by ICPMS involves the dissolution of the material with mineral acids followed by direct analysis. Acid dissolutions are performed so that TDS values do not exceed 0.2%. The most common approach to calibration is external calibration followed by linear regression. Internal standards such as ^{209}Bi and ^{230}Th are commonly used to compensate for variations in instrumental response. This approach was used to determine a suite of more than 30 trace elements (including U) in a study of 12 fine-grained Apollo 12 basalts (Snyder et al. 1997). Shinotsuka and Ebihara (1997) determined concentrations of Th, U, and lanthanides in chondritic meteorites using external standards for linear calibration. Comparisons of results with those from RNAA revealed systematic differences between the two data sets. The authors concluded that the differences were due to difficulties with the RNAA data. A standard method for the determination of U in soils via acid dissolution and ICPMS has been published by the ASTM (C1345-96 1996b). The estimated detection limits achievable by this procedure for U are 500 ng/g. Assuming a sample in mineral acids with TDS of 0.2% and solution detection limits of 20 pg/mL, U concentrations of approximately <40 ng/g can be determined in the solid sample. The current generation of commercial quadrupole-based ICPMS instruments is capable of U detection limits of 100 fg/mL in solution (Brenner et al. 1998).

Eggins et al. (1997) described a new calibration methodology for rapid and precise determination of 40 trace elements in geological samples. The method uses multiple internal standards (both naturally occurring elements and isotopically enriched isotopes) and external standardization to correct for the complex variations in instrumental mass response. Uranium-235 is used as the internal standard for the ^{238}U determination. The external standard is a reference solution that is analyzed periodically during the course of a multiple-sample analysis. Calibration is performed with dissolved reference material BHVO-1. Calibration with a sample possessing a composition similar to the unknown samples is referred to as matrix matching. The accuracy of this analysis and calibration methodology was demonstrated on several reference materials to be better than ±3% and was able to determine U at concentrations <10 ng/g in the solid sample. The Eggins methodology was used to determine trace element abundances in Western Australian basalts and gabbros (Sylvester et al. 1997). Niobium/uranium ratios determined in this work indicate that parts of the Late Archean mantle beneath Western Australia underwent melt extraction and resulted in the formation of continental crust comparable to that seen in the mantle.

Isotope dilution is commonly employed to improve the accuracy of elemental ICPMS determinations. Isotope dilution requires multiple measurements for each sample and is more time consuming than external calibration. Despite the additional time effort required for ID analysis, the high level of accuracy achievable by this method makes this an option, particularly when an exhaustive compositional determination is not required. Nominal mass resolution and the availability of isobar-free nuclides make ID-ICPMS particularly applicable to the determination of U. In ID analysis of U, a known quantity of a spike isotope, typically a minor isotope, is added to the sample. Uranium determinations can be performed using ^{234}U, ^{235}U, or ^{236}U. If chemical separation is required, the spike is chemically homogenized and equilibrated with the sample so that total isotopic exchange takes place between the spike and sample. After the isotope-diluted element is isolated, the isotope ratio R is measured with the mass spectrometer. Another advantage of performing ID analysis is the correction for potential losses during chemical processing. The basic equation for ID analysis is

$$G_S = 1.66 \times 10^{-18} \frac{M}{W_s} N_{Sp} \left(\frac{h_{Sp}^2 - Rh_{Sp}^1}{Rh_S^1 - h_S^2} \right) \tag{5}$$

where G_s is the content of the sample (µg/g), M is the atomic weight of the element of interest (g/mol), W_s is the sample weight (g), N is the number of atoms, h is the isotope abundance (%), the superscripts 1 and 2 indicate the light and heavy isotope, and subscripts S and Sp indicate the sample and the spike, respectively. The suitability of ID-ICPMS analysis was evaluated for the determination of U (Joannon et al. 1997). Uranium concentrations were measured in two separate solutions of six international rock reference materials, and the results differed by 1-3% for concentrations in the range of 10-400 ng/g. Toole et al. (1991) used ID-ICPMS to assay small volumes (0.25 mL) of sediment pore waters for U content using ^{236}U as a spike. The only pretreatment required was dilution by a factor of 20, which gave sufficient volume for three replicate analyses per sample. Accurate results were obtained for samples ranging in concentration from 0.5 to 10 ng/mL detection limits of 2 pg/mL of sample. Isotope dilution ICPMS analysis can yield accuracies <±1% in a wide variety of matrices.

High resolution-inductively coupled plasma mass spectrometry

Perhaps the technique that shows the highest promise for improvements in inorganic elemental and isotopic analysis is high resolution (HR)-ICPMS. This technique combines an ICP source with a high-precision, double-focusing, mass spectrometer. A review of HR-ICPMS instrumentation and applications has been published (Moens and Jakubowski 1998). While introduced in 1988, the availability of commercial second- and third-generation HR-ICPMS instruments has made a significant impact on planetary sciences, earth sciences, and environmental chemistry. Commercial HR-ICPMS instruments, when operated at resolutions >2000, are capable of separating many interfering polyatomic ions from isotopes of interest. However, when operated at lower resolutions, a HR-ICPMS equipped with a single detector is capable of measurement with higher sensitivities, lower backgrounds, and higher precisions than typical quadrupole-based ICPMS instruments. A mass spectrum of a 1 ng/mL U-containing solution measured on a VG Elemental Axiom HR-ICPMS is given in Figure 8. The spectrum was collected by the author using an instrumental resolution setting of 460. The instrument possesses a 2-GHz/ppm sensitivity and a <0.1 cps background. This corresponds to detection limits of <1 fg/mL for U in solution or approximately 2,500,00 atoms of U. This would correspond to approximately 1 pg/g in solid samples with no preconcentration assuming a typical 0.2% TDS sample dissolution. The determination of elements at these low concentrations require increased attention to cleanliness during sample preparation and analysis including working in class 100-type clean rooms. The ability to determine U concentrations at these levels will undoubtedly make an impact on trace actinide studies in the future.

Isotope ratio determination

Nominal mass resolution and the availability of isobar-free nuclides make ICPMS

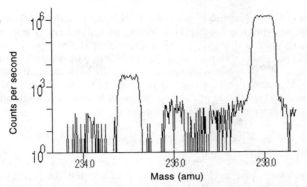

Figure 8. Spectrum of a 1 ng/mL U-containing sample, collected using conventional liquid sample introduction, pneumatic nebulization and a HR-ICPMS in the author's laboratory. The sample was prepared from NIST SRM 3164. Note that even at these low concentrations, peaks from naturally occurring minor U isotopes ^{234}U and non-naturally occurring ^{236}U are evident. The source of the ^{236}U may be contamination. The flat top peaks generated by magnetic sector-based HR-ICPMS allow the determination of isotopic ratios with increased precision compared to quadrupole-based ICPMS.

measurement of U isotope ratios straightforward for both quadrupole-based and high-resolution instruments. Typical precisions for quadrupole-based ICPMS instruments range from RSD of 0.1-1.0% limiting its usefulness for the measurement of many of the geologically important isotopic systems. Despite these limitations, the ease at which ICPMS is able to measure isotope ratios makes it useful for a few select isotopic systems that do not require precisions better than ~0.1%. Ketterer (1998) determined ^{234}U/^{235}U ratios as a proxy for ^{234}U/^{234}U in lake water, seawater, and dissolved corals. Mass discrimination correction was performed via internal standardization with Ir, producing IrAr$^+$ at masses 231 and 233. Precisions of 0.2-0.5% were obtained. Superior precisions can be achieved on HR-ICPMS instrumentation designed in a multi-collector (MC) configuration. When used in the MC configuration, the precision of ICPMS is comparable to TIMS. A review of MC-ICPMS instrumentation and applications including U isotopic analysis has been reported (Halliday et al. 1998). This work reports results of replicate analysis of NIST SRM-906 measured by TIMS and MC-ICPMS. Precisions of approximately 1δ unit (2σ), comparable to the best achievable TIMS precision, were achieved. The external reproducibility of MC-ICPMS was superior to TIMS. Multi-collector ICPMS will undoubtedly be established as the preferred method for U isotopic determinations due to its high precision, ease of sample introduction, high sample throughput, and user friendly interface.

SUMMARY

Many techniques are available for the determination of U in geological and environmental materials. Uranium is unique in that it is one of few naturally occurring radionuclides and can be determined by direct radiometric counting techniques. These techniques, however, are limited in their applicability due to the low specific activity of U and the low concentrations at which it is typically found in nature. While it is possible to determine U as a major or minor constituent in geological samples using wet chemical methods, instrumental techniques are used nearly exclusively for the determination of U at trace levels. Most of the standard geoanalytical techniques such as AAS, ICP-AES, XRF, INAA, and ICPMS have been adapted for U determination. Uranium is unique among most elements in that it can be determined directly via fluorometry and phosphorimetry without the addition of a fluorescent chelating agent. A substantial database of standard analytical methods suitable for U determination has been compiled and is maintained by the American Society for Testing and Materials. A review of trends in techniques used for bulk U determinations indicate that ICPMS is one of the most commonly used methods today. The strength of ICPMS resides in its high sensitivity, multielement capability, and high sample throughput. Further development of applications of ICPMS and the more recently available HR-ICPMS for accurate, ultratrace U determinations in geological and environmental samples should improve our understanding of the behavior of U in the environment.

ACKNOWLEDGMENTS

The author thanks Professor Clive R. Neal and Dr. Elane Streets for technical reviews of this manuscript. Joseph E. Harmon provided valuable editorial review. The author is especially grateful to his wife Heather for support through the many evenings and weekends that were used to produce this work.

This manuscript has been created by the University of Chicago, as Operator of Argonne National Laboratory ("Argonne"), under Contract No. W-31-109-ENG-38 with the U.S. Department of Energy. The U.S. Government retains for itself, and others acting on its behalf, a paid-up, nonexclusive, irrevocable worldwide license in said article to reproduce, prepare derivative works, distribute copies to the public, and perform publicly and display publicly, by or on behalf of the Government.

REFERENCES

Aldstadt JH, Kuo JM, Smith LL, Erickson MD 1996 Determination of uranium by flow injection inductively coupled plasma mass spectrometry. Anal Chim Acta 319:135-143

American Society for Testing and Materials 1990 Standard test method for isotopic uranium in water by radiochemistry. ASTM D3972-90, Annual Book of ASTM Standards, Vol 11.02

American Society for Testing and Materials 1991a Standard practice for sample using closed vessel microwave heating technique for the determination of total recoverable metals in water. ASTM D4309-91, Annual Book of ASTM Standards, Vol 11.01

American Society for Testing and Materials 1991b Standard test method for high-resolution gamma-ray spectrometry of water. ASTM D3649-91, Annual Book of ASTM Standards, Vol 11.02

American Society for Testing and Materials 1991c Standard test methods for microquantities of uranium in water by fluorometry. ASTM D2907-91, Annual Book of ASTM Standards, Vol 11.02

American Society for Testing and Materials 1991d A standard test method for trace uranium in water by pulsed-laser phosphorimetry. ASTM D5174-91, Annual Book of ASTM Standards, Vol 11.02

American Society for Testing and Materials 1993 A standard test method analysis of uranium and thorium in soils by energy dispersive X-ray fluorescence spectroscopy. ASTM C1255-93, Annual Book of ASTM Standards, Vol 12.01

American Society for Testing and Materials 1994a Standard test method for U by iron(II) reduction in phosphoric acid followed by chromium(VI) titration in the presence of vanadium. ASTM C1267-94, Annual Book of ASTM Standards, Vol 12.01

American Society for Testing and Materials 1994b Standard practice for measuring trace elements in water by graphite furnace atomic absorption spectrophotometer. ASTM D3919-94a, Annual Book of ASTM Standards, Vol 11.01

American Society for Testing and Materials 1995a Standard practice for digestion of samples for determination of metals by flame atomic absorption or plasma emission spectroscopy. ASTM D1971-95, Annual Book of ASTM Standards, Vol 11.01

American Society for Testing and Materials 1995b Standard practice for alpha-particle spectrometry of water. ASTM D3084-95, Annual Book of ASTM Standards, Vol 11.02

American Society for Testing and Materials 1995c Standard test methods for determining radionuclides in soils by inductively coupled plasma-mass spectrometry using flow injection preconcentration. ASTM C1310-95, Annual Book of ASTM Standards, Vol 12.01

American Society for Testing and Materials 1996a Standard test method for elements in water by inductively coupled plasma-mass spectrometry. ASTM D5673-96, Annual Book of ASTM Standards, Vol 11.02

American Society for Testing and Materials 1996b Standard test methods for analysis of total and isotopic uranium and total thorium in soils by inductively coupled plasma-mass spectrometry. ASTM C1345-96, Annual Book of ASTM Standards, Vol 12.01

Anders E, Wolf R, Morgan JW, Ebihara M, Woodrow AB, Janssens M-J, Hertogen J 1988 Radiochemical neutron activation analysis for 36 elements in geological material: Au, Ag, Bi, Br, Cd, Cs, Ge, In, Ir, Ni, Os, Pd, Rb, Re, Sb, Se, Sn, Te, Tl, U, and Zn as well as Sc, Y, and REE *In* Nuclear Science Series, Radiochemistry Techniques, USDOE Office of Scientific and Technical Information, NAS-NS-3117

Barnes CE, Cochran JK 1990 Uranium removal in oceanic sediments and oceanic U balance. Earth Planet Sci Lett 97:94-101

Becker JS, Dietze H-J 1998 Inorganic trace analysis by mass spectrometry. Spectrochim Acta 52B:177-187

Bertin EP 1975 Principles and Practices of X-ray Spectrometric Analysis, 2nd edn. Plenum, New York

Bettinelli M, Spezia S 1995 Determination of trace elements in sea water by ion chromatography-inductively coupled plasma mass spectrometry. J Chromatog 7090A:275-281

Brenner IB, Liezers M, Godfrey J, Nelms S, Cantle J 1998 Analytical characteristics of a high efficiency ion transmission interface (S mode) inductively coupled plasma mass spectrometer for trace element determinations in geological and environmental samples. Spectrochim Acta 53B:1087-1107

Chen JH, Edwards RL, Wasserburg GJ 1992 Mass spectrometry and applications to uranium-series disequilibrium: Uranium series disequilibrium applications to earth, marine, and environmental sciences. M Ivanovich, RS Harmon (Eds) Clarendon Press, Oxford, p 174-206

Compston W, Williams IS, Meyer C 1984 U-Pb geochronology of zircons from the lunar breccia 73217 using a sensitive high mass-resolution ion Microprobe. Geochim Cosmochim Acta 89:B525-B534

Crain JS, Mikesell BI 1992 Detection of sub-ng/L actinides in industrial wastewater matrices by inductively coupled plasma-mass spectrometry. Appl Spectrosc 46:1498-1502

Crain JS, Smith LL, Yaeger JS, Alvarado JA 1995 Determination of long-lived actinides in soil leachates by inductively coupled plasma-mass spectrometry. J Radioanal Nucl Chem 194:133-139

Davies W, Gray W 1964 A rapid and specific volumetric method for the precise determination of uranium using ferrous sulfate as a reductant. Talanta p 1203

Dean JA 1995 Analytical Chemistry Handbook. McGraw-Hill, New York

Dolezal J, Povondra P, Sulcek Z 1968 Decomposition techniques in inorganic analysis. London Iliffe Books Ltd, London

Edelson MC 1992 High-resolution plasma spectrometry. *In* Inductively coupled plasmas in analytical atomic spectrometry, 2nd edn. A Montaser DW Golightly (Eds) VCH Publishers, New York, p 341-372

Eggins SM, Woodhead JD, Kinsley LPJ, Mortimer GE, Sylvester P, McCulloch MT, Hergt JM, Handler MR 1997 A simple method for the precise determination of 40 trace elements in geological samples by ICPMS using enriched isotope internal standardization. Chem Geol 134:311-326

Ehmann WD, Vance DE 1991 Radiochemistry and Nuclear Methods of Analysis. John Wiley & Sons, New York

Firestone RB 1998 Table of Isotopes, 8th edn. CM Baglin (Ed) John Wiley & Sons, New York

Friedlander G, Kennedy JW, Macias ES, Miller JM 1981 Nuclear and Radiochemistry. John Wiley & Sons, New York

Green LW, Sopchyshyn FC 1989 Quantification of uranium isotopes by multiphoton ionization time-of-flight mass spectrometry. Int'l J Mass Spectrom Ion Phys 89:81-95

Gu ZY, Lal D, Liu TS, Guo ZT, Southon J, Caffee MW 1997 Weathering histories of Chinese loess deposits based on uranium and thorium series nuclides and cosmogenic ^{10}Be. Geochim Cosmochim Acta 61:5221-5231

Guest RJ, Zimmerman JB 1955 Determination of uranium in uranium concentrates. Anal Chem 27:931-936

Halliday AN, Lee D-C, Christensen JN, Rehkämper M, Yi W, Luo X, Hall CM, Ballentine CJ, Pettke T, Stirling C 1998 Applications of multiple collector-ICPMS to cosmochemistry, geochemistry, and paleoceanography. Geochim Cosmochim Acta 62:919-940

Haraguchi H, Itoh A, Kimata C, Miwa H 1998 Speciation of yttrium and lanthanides in natural water by inductively coupled plasma mass spectrometry after preconcentration by ultrafiltration and with a chelating resin. Analyst 123:773-778

Harbottle G, Evans CV 1997 Gamma-ray methods for determining natural and anthropogenic radionuclides in environmental and soil science. Radioact Radiochem 8:38-46

Haswell SJ 1992 Atomic Absorption Spectroscopy. Elsevier, Amsterdam

Horwitz EP, Chiarizia R, Dietz ML, Diamond H, Essling AM, Graczyk D 1992 Separation and preconcentration of uranium from acidic media by extraction chromatography. Anal Chim Acta 266:25-37

Horwitz EP, Dietz ML, Chiarizia R, Diamond H, Nelson DM 1993 Separation and preconcentration of actinides from acidic media by extraction chromatography. Anal Chim Acta 281:361-372

Ingamells CO, Pitard FF 1986 Applied Geochemical Analysis. John Wiley & Sons, New York
Ivanovich M, Murray A 1992 Spectroscopic methods. *In* Uranium series disequilibrium applications to earth, marine, and environmental sciences. M Ivanovich, RS Harmon (Eds) Clarendon Press, Oxford, p 127-173
Jarvis KE, Gray AL, Houk RS 1992 Handbook of Inductively Coupled Plasma Mass Spectrometry. Chapman and Hall, New York
Joannon S, Telouk P, Pin C 1997 Determination of U and Th at ultra-trace levels by isotope dilution inductively coupled plasma mass spectrometry using a geyser-type ultrasonic nebulizer: Application to geological samples. Spectrochim Acta 52B:1783-1789
Joron JL, Treuil M, Raimbault L 1997 Activation analysis as a geochemical tool: Statement of its capabilities for geochemical trace element studies. Radioanal Nucl Chem 216:229-235
Ketterer ME 1998 High-precision determination of $^{234}U/^{238}U$ activity ratios in natural waters and carbonates by ICPMS. *In* Proceedings of Pittcon 98, New Orleans, LA, March 1-5, 1998, p 993
Kingston HM, Haswell SJ (Eds) 1997 Microwave-enhanced chemistry fundamentals, sample preparation, and applications. American Chemical Society, Washington, DC
Lally AE 1992 Chemical procedures *In* Uranium series disequilibrium applications to earth, marine, and environmental sciences. M Ivanovich, RS Harmon (Eds) Clarendon Press, Oxford, p 94-126
Lipschutz ME, Wolf SF, Hanchar JM, Culp FB 1999 Geochemical and cosmochemical materials. Anal Chem 71:1R-20R
Longerich HP, Jackson SE, Günther D 1996 Laser ablation inductively coupled plasma mass spectrometric transient data acquisition and analyte concentration. J Anal At Spectrom 11:899-904
Machado N, Gaithier G 1996 Determination of 207Pb/206Pb ages on zircon and monazite by laser-ablation ICPMS and application to a study of sedimentary provenance and metamorphism in southeastern Brazil. Geochim Cosmochim Acta 60:5063-5073
Mair MA, Savage DJ 1986 U.K. Atomic Energy Agency Report, ND-R-134
McOrist, GD, Smallwood, A 1997 Trace elements in precious and common opals using neutron activation analysis. Radioanal Nucl Chem 223:9-15
Meites L (Ed) 1963 Handbook of Analytical Chemistry, McGraw-Hill, New York
Meyer C, Williams IS, Compston W 1996 Uranium-lead ages for lunar zircons: Evidence for a prolonged period of granophyre formation from 4.32-3.88 Ga. Meteoritics Planet Sci 31:370-387
Moens L, Jakubowski N 1998 Double focussing mass spectrometers in ICPMS. Anal Chem 70:251A-256A
Montaser A, Golightly DW (Eds) 1992 Inductively Coupled Plasmas in Analytical Atomic Spectrometry, 2nd edn. VCH Publishers, New York
Montaser A, Minnich MG, McLean JA, Liu H, Caruso J, McLeod CW 1998 Sample introduction in ICPMS *In* Inductively Coupled Plasma Mass Spectrometry, 2nd edn. A Montaser (Ed). VCH Publishers, New York, p 83-264
Norman MD, Pearson NL, Sharma A, Griffen WL 1996 Quantitative analysis of trace elements in geological materials by laser ablation ICP-MS: Instrumental operating conditions and calibration values of NIST glasses. Geostand Newsl 20:247-261
Parry SJ, 1991 Geological applications *In* Activation Spectrometry in Chemical Analysis. JD Winefordner, IM Kolthoff (Eds) John Wiley & Sons, New York, p 206-207
Porcelli D, Andersson PS, Wasserburg GJ, Ingri J, Baskaran M 1997 The importance of colloids and mires for the transport of U isotopes through the Kalix River watershed and Baltic Sea. Geochim Cosmochim Acta 61:4095-4113
Pretty JR, Duckworth DC, Van Berkel GJ 1998 Electrochemical sample pretreatment coupled on-line with ICPMS: Analysis of U using an anodically conditioned glassy carbon working electrode. Anal Chem 70:1141-1148
Raimbault L, Peycelon H, Joron JL 1997 Single-crystal trace element analysis in rock-forming minerals by instrumental neutron activation analysis. Radioanal Nucl Chem 216:221-228
Ramdoss K, Gomathy Amma B, Umashankar V, Rangaswamy R 1997 Cold dissolution method for the determination of uranium in various geological materials at trace levels by laser fluorometry. Talanta 44:1095-1098
Ridden CJ, Warf JC 1950 Uranium *In* Analytical Chemistry of the Manhattan Project. CJ Rodden (Ed) McGraw Hill, New York, p 3-159
Rosenberg RJ, Pitkanen V, Sorsa A 1977 An automated uranium analyzer based on delayed neutron counting. J Radioanal Chem 37:169-179
Sandell EB 1959 Colorimetric Determination of Traces of Metals, 3rd edn, Interscience, New York
Schrauder M, Koeberl C, Navon O 1996 Trace element analyses of fluid-bearing diamonds from Jwaneng, Botswana. Geochim Cosmochim Acta 60:4711-4724

Shinotsuka K, Ebihara M 1997 Precise determination of rare earth elements thorium, and uranium in chondritic meteorites by inductively coupled mass spectrometry—a comparative study with radiochemical neutron activation analysis. Anal Chim Acta 338:237-246

Skoog DA 1985 Principles of Instrumental Analysis, 3rd edn. Saunders College Publishing, New York

Slavin M 1978 Atomic absorption spectroscopy, 2nd edn. John Wiley & Sons, New York

Snyder GA, Neal CR, Taylor LA, Halliday AN 1997 Anatexis of lunar cumulate mantle in time and space: Clues from trace-element, strontium, and neodymium isotopic chemistry of parental Apollo 12 basalts. Geochim Cosmochim Acta 61:2731-2747

Sohrin Y, Iwamoto SI, Akiyama S, Fujita T, Kugii T, Obata H, Nakayama E, Goda S, Fujishima Y, Hasegawa H, Ueda K, Matsui M 1998 Determination of trace elements in seawater by fluorinated metal alkoxide glass-immobilized 8-hydroxyquinoline concentration and high-resolution inductively coupled plasma mass spectrometry detection. Anal Chim Acta 363:11-19

Sylvester PJ, Campbell IH, Bowyer DA 1997 Niobium/uranium evidence for early formation of the continental crust. Science 275:521-523

Tertian M, Baker MD, Christie A, Tyson JF 1982 Principles of Quantitative X-ray Fluorescence Analysis. Heyden, London

Toole J, Thomson J, Wilson TRS, Baxter MS 1984 A sampling artefact affecting the uranium content of deep-sea porewaters obtained from cores. Nature 308:263

U.S. Environmental Protection Agency 1979 Radiometric method for the determination of uranium in water. EPA-600/7-79-093

U.S. Environmental Protection Agency 1996 Microwave assisted acid digestion of siliceous and organically based matrices: In Test Methods for Evaluating Solid Waste—Update (III), EPA Method 3052

Wagner GA 1966 Age determination of tektites and other natural glasses by traces of natural ^{238}U fission (fission track method). Z Naturforschg 21A:733-745

Wolf SF, Bates JK, Buck EC, Dietz NL, Fortner JA, Brown NR 1997 Physical and chemical characterization of actinides in soil from Johnston Atoll. Environ Sci Tech 31:467-471

14 Identification of Selected Uranium-bearing Minerals and Inorganic Phases by X-ray Powder Diffraction

Frances C. Hill

Department of Civil Engineering and Geological Sciences
University of Notre Dame
Notre Dame, Indiana 46556

INTRODUCTION

The ability to identify minerals and other solid phases in a natural or synthetic sample is a crucial first step for many geochemical investigations. If the composition and structure of a material is unknown, ideally, a crystal of the phase in question is obtained that is of a suitable size and crystallinity to allow for the unambiguous solution of the structure by single-crystal X-ray diffraction. Because lattice parameters and the positions of atoms in the unit cell are revealed in a structural solution, single-crystal X-ray methods provide not only chemical information about a crystal, but can also differentiate between polymorphs. The availability of increasingly sensitive detectors for single crystal diffractometers, such as the CCD (charge-coupled device) area detector, has revolutionized the field of phase characterization by facilitating the structural solution for very small and/or weakly diffracting crystals. For crystals of insufficient size, crystallinity or both, and for crystals of known materials or materials with a limited range of possible compositions, X-ray powder diffraction can provide important information on structural and compositional variations of the material.

The presence of uranium in a crystalline material can have both positive and negative impact on data collected in an X-ray experiment. The scattering of X-rays by atoms in a crystal is roughly proportional to the number of electrons, and thus the presence of uranium in a crystal will dominate the resulting X-ray diffraction pattern, sometimes producing inaccurate results for lighter atoms in a structure (Glusker and Trueblood 1985). Conversely, uranium is a very efficient scatterer of X-rays, resulting in intense diffraction peaks in most cases, using a minimal amount of sample. Uranium-bearing phases often lack crystals that are suitable for a single-crystal X-ray experiment due to many factors including the presence of exceedingly small crystals of poor crystallinity, and the prevalence of twinned crystals. Absorption of X-rays by uranium-bearing materials can be substantial, requiring the application of specialized absorption corrections to the data. Because of these drawbacks, X-ray powder methods have been used extensively for phase identification. This study provides information on the identification of uranium-bearing phases from powder X-ray diffraction data, and presents a tabulation of calculated powder diffraction data in a concise manner that can serve as a reference for those involved in phase identification.

A BRIEF REVIEW OF X-RAY DIFFRACTION

A crystal may be defined as a solid material that is composed of an array of atoms arranged in a periodic pattern that extends in three directions (Cullity 1978). In general, the distance between atoms in a crystal is in the range of 10^{-10} m, usually reported as 1 Å. X-radiation is the name given to that part of the electromagnetic spectrum for which the wavelength of the radiation varies between approximately 0.5 and 2.5 Å, roughly of the same order of magnitude as the interatomic distance in crystals. Thus, shining X-rays on a crystalline material can result in diffraction of the incident beam by the crystal, which acts as a diffraction grating, resulting in scattered beams for which the directions and intensities are a function of the arrangement and type of atoms in the crystal. A diffracted X-ray beam is produced only when scattering of the incident X-ray beam from the atoms in the crystal is coherent (i. e. in phase), and can only be detected when a detector is in the proper angular position to record the event. The conditions that must be satisfied for X-ray diffraction to occur in a crystal are described by the Laue equations, derived by Max von Laue in a treatise that was awarded a Nobel prize in 1912. In this work, Laue showed that, for a periodic row of atoms spaced a distance a apart, the incident X-ray beam, S_0, is scattered coherently by the atoms only if the path difference between S_0 and the scattered X-ray beam, S, is equal to an integral number of wavelengths (Azaroff 1968). Extending this concept to a three-dimensional periodic array, the Laue equations may be expressed as:

$$\mathbf{a} \cdot (\mathbf{S} - \mathbf{S}_0) = h\lambda \qquad \mathbf{b} \cdot (\mathbf{S} - \mathbf{S}_0) = k\lambda \qquad \mathbf{c} \cdot (\mathbf{S} - \mathbf{S}_0) = l\lambda$$

where $\mathbf{a}$, $\mathbf{b}$ and $\mathbf{c}$ are the primitive lattice vectors and h, k and l are integers. Employing geometrical arguments, an alternative method for visualizing diffraction by a crystal was proposed by W.L. Bragg in 1912. This relationship between X-ray wavelength, lattice spacing and angular position is given by the Bragg equation:

$$n\lambda = 2d\sin\theta$$

where λ is the wavelength of the incident X-radiation, d is a lattice spacing in the crystal, n is an integer and θ is the angle the incident beam makes with a lattice in the crystal. If λ is fixed (as for most modern X-ray experiments), only lattice planes within the crystal for which the interplanar spacing and the diffraction angle satisfy the Bragg equation will produce a coherent diffracted beam of X-rays. It can be shown readily that the Laue equations and the Bragg equation are equivalent expressions for the necessary conditions for X-ray diffraction to occur in a crystal (see Azaroff 1968).

In a powder X-ray diffraction (XRD) experiment, a sample of crystalline material that has been ground to a fine powder is placed in a diffractometer [see Bish and Post (1989) for a discussion of sample preparation techniques], and X-rays of a known wavelength (commonly CuK_α with a wavelength of 1.5418 Å) irradiate the sample as it is slowly rotated through a range of angles. The intensity and direction of the diffracted beam is measured by a detector and recorded. The result is reported as a diffraction pattern in which intensities (commonly converted from incoming X-ray intensity to electronic pulses and displayed as counts per second (cps)) are plotted as a function of the angular position of the detector. The positions and intensities of peaks in the diffraction pattern can then be analyzed to identify the sample. Phase identification is often accomplished by comparing XRD powder patterns of known materials to that of the unknown material through searches of computerized databases of powder diffraction data, such as that provided by the International Centre for Diffraction Data (ICDD), the Inorganic Crystal Structure Database (ICSD) or the Joint Commission on Powder Diffraction Standards (JCPDS).

POWDER DIFFRACTION DATA FOR URANIUM-BEARING PHASES

One of the more comprehensive compilations of powder diffraction data for uranium-bearing phases can be found in Smith (1984). Containing 231 valid mineral species (and a few unnamed minerals), this collection of diffraction data provides a useful reference for naturally occurring uranium phases. A large number of powder diffraction data sets exist for uranium-bearing phases (both mineral and non-mineral) in searchable databases, as part of a larger collection of diffraction data. However, several problems can be encountered with these databases when attempting to identify an unknown phase. For example, several contradictory diffraction patterns can be retrieved for the same phase in many cases. Intensities of diffraction peaks commonly vary widely from pattern to pattern for the same phase due to variations in sample preparation and minor variations in sample chemistry. Additionally, many uranium-bearing phases have similar d-spacings producing the most intense diffraction peaks, so that an unknown sample might be narrowed down to a choice of two or three possibilities at best. This problem is often encountered when the phases in question differ slightly in composition such as in hydration state (Finch et al. 1997). All of these factors can make positive identification of uranium-bearing phases difficult.

Most discrepancies that are encountered in searchable databases can be attributed to varying techniques of sample preparation of the standards. Many uranium-bearing phases are composed of sheetlike structures of uranyl polyhedra that are loosely held together by interlayer cations and H_2O groups (Burns et al. 1996), and yield a perfect one-directional cleavage in a manner that is analogous to clays and micas. This property can produce preferred orientation of a powder sample, and give rise to diffraction peaks with inaccurate intensities that may make identification of the unknown phase difficult. The very small crystals so often encountered when examining uranium-bearing phases can make it difficult for other analytical techniques to be employed for phase characterization, primarily because sample purity becomes a question when crystals are so small. As a result, data exists in searchable databases for which reported structures or formulas (or both) are incorrect. Additionally, as discussed above, uranium atoms dominate X-ray diffraction pattern relative to lighter elements, so that subtle differences between phases may be obscured.

The purpose of this study is to present an additional database to aid in the identification of uranium-bearing phases. In an effort to circumvent some of the problems encountered in more commonly available databases, a set of calculated diffraction data for uranium-bearing phases has been produced. All data reported in this study have been calculated using interactive software for the PC: XPOW (Downs et al. 1993), one of many programs readily available for the calculation of powder diffraction patterns. This software uses lattice and structural parameters, determined from single crystal X-ray experiments, to calculate diffraction patterns. A target X-ray source is chosen and the diffraction pattern is calculated over a range of 2θ values selected by the user. The program assumes an occupancy factor of 1.0 for each atom in the asymmetric unit unless a value is specified by the user. Values for isotropic displacement factors can be included in the calculation if available. The results of this calculation are contained in a file that displays for each observed diffraction peak calculated by the program, a 2θ value, the d-spacing, the relative intensity, and the Miller indices. The results may be plotted as a diffraction pattern using the accompanying software XPOWPLOT (Downs et al. 1993). As with any calculation method of this type, assumptions are made as to features such as the shape of diffraction peaks and the resolution of peak overlaps. The handling of these features in the resulting diffraction profiles provides the greatest discrepancies between experimental diffraction data and calculated diffraction data. The calculated diffraction data presented in this work are intended to augment the experimental diffraction data, and aid in the characterization of uranium-bearing phases.

Tables 1 through 15 (in the Appendix of this chapter) contain calculated d-spacings corresponding to the five most intense diffraction peaks for each uranium-bearing phase. Lattice parameters are also given for each phase along with the space group symbol. The data has been divided according to the chemistry of each of the phases, with Table 1 listing uranyl oxide and oxide hydrate phases, Table 2 listing uranyl oxide and oxide hydrate phases that contain cations besides uranium, and Tables 3–15 listing uranium-bearing phases that also contain other anionic groups such as silicates or carbonates. Most of the minerals and phases in these tables contain uranium in its oxidized form, U^{6+}, though a limited number of phases containing U^{4+} and U^{5+} have been included for structures that were well characterized. The d-spacing for the most intense diffraction peak is given in the column labeled $d^1_{(hkl)}$ where the number 1 indicates the most intense diffraction peak (and 5 the least intense), and the subscript refers to the Miller indices of the diffracting plane.

Most uranium-bearing phases included in this database are from structures solved using single crystal X-ray methods. In a few cases, exceptionally well refined Rietveld analyses are also included. For many of the phases considered, more than one refinement exists for a given structure. A careful examination of experimental results, including reported tolerances for structural parameters and final R-values, was made to decide which refinement would be used for a particular phase. Many uranium-bearing phases are hydrated. Because only a small number of single-crystal refinements include positional parameters for hydrogen atoms (difficult to determine when a heavy atom like uranium is present), hydrogen atoms were removed from data files when reported. Calculated diffraction patterns are virtually indistinguishable when calculated with and without hydrogen atoms. The bulk of uranium-bearing phases included in this study are inorganic, though a few organic uranyl phases are included. The range of 2θ values used for the calculations is 5–70°. The X-ray target source is CuK_α.

COMPARISON OF EXPERIMENTAL AND CALCULATED POWDER X-RAY DIFFRACTOGRAMS

How do calculated powder X-ray diffraction patterns compare with those collected in an X-ray experiment? To illustrate similarities and differences, four examples have been chosen. Calculated diffraction patterns were plotted using XPOWPLOT (Downs et al. 1993). The experimental data were selected from available databases and recast in a form suitable for use with XPOWPLOT, which requires a listing of d-spacings and relative intensities for identified diffraction peaks, then plotted in the same manner as the calculated data. No effort is made with this software to simulate a background, so the resulting plots appear without any background.

Figure 1 shows a comparison of experimental and calculated powder patterns for the mineral boltwoodite, $K(UO_2)(H_2O)(SiO_3OH)$. Both diffraction patterns contained in this figure are based on the data from Stohl and Smith (1981). The crystal structure of boltwoodite was reported by Stohl and Smith (1981), and lattice parameters and atomic coordinates reported by them were used to calculate the diffraction pattern (Fig. 1b). The experimental diffraction pattern for boltwoodite in Figure 1a was recorded for a powder sample of boltwoodite from the same locality as the single crystal used to determine the structure (Stohl and Smith 1981). The experimental pattern was only reported to a value of 47° 2θ, whereas the lower diffraction pattern was calculated to 65° 2θ. The calculated and experimental diffraction patterns show excellent agreement, both for peak positions and relative intensities.

Figure 2 compares experimental and calculated diffraction data for the mineral derriksite, $Cu_4(UO_2)(SeO_3)_2(OH)_6$, a uranium-bearing phase composed of infinite chains of uranyl polyhedra (Burns et al. 1996). The experimental diffraction pattern for derriksite was produced by Cesbron et al. (1971) from a natural specimen from the Shaba Province in the Democratic Republic of the Congo. The diffraction pattern in Figure 2b was calculated from data reported in Ginderow and Cesbron (1983a). Derriksite exhibits

Hill: X-ray Powder Diffraction Identification of U-bearing Phases

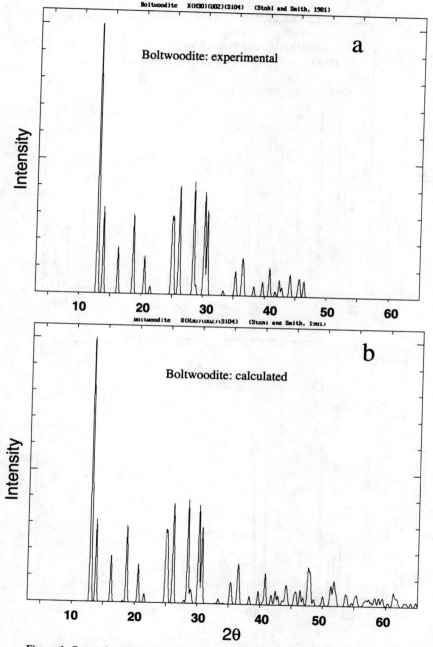

Figure 1. Comparison of (a) the experimental and (b) the calculated diffraction patterns for the mineral boltwoodite.

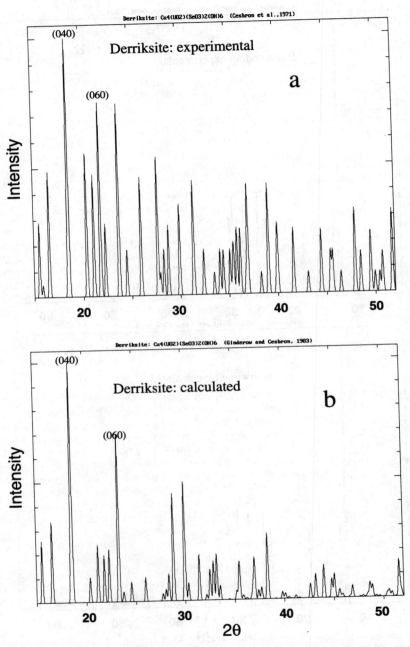

Figure 2. Comparison of (a) the experimental and (b) the calculated diffraction patterns for the mineral derriksite.

very good {010} cleavage. The two most intense diffraction peaks in the diffraction pattern of derriksite [labeled (040) and (060) in Fig. 2] correspond to diffraction from lattice planes in the crystal that parallel this cleavage direction. Consequently, despite generally poor agreement between the peak intensities for the experimental and calculated diffraction patterns, the most intense diffraction peaks match very well, aiding in positive identification. This example illustrates that the presence of sample cleavage, and the potential preferred orientation that can occur during sample preparation, can actually be used to advantage in identifying a phase, a technique that has been used extensively by those studying clays and micas (Klug and Alexander 1974).

In Figure 3, experimental and calculated diffraction patterns for the inorganic phase Sr_3UO_6, a uranium-bearing phase composed of isolated uranyl polyhedra, are compared. The experimental data was taken from Rietveld (1966), whereas the pattern in Figure 3b. was produced from data reported in Loopstra and Rietveld (1969). This material lacks good cleavage, thus preferred orientation should not play a role in the resulting experimental diffraction pattern. Very good agreement exists between the two diffraction patterns, even in the range of 2θ greater than 45°.

Figure 4 compares experimental and calculated diffraction patterns for soddyite, $(UO_2)_2(SiO_4)(H_2O)$, a mineral composed of a framework of uranyl polyhedra. The experimental data are from Stohl and Smith (1981), while the calculated pattern was calculated from data reported in Demartin et al. (1992). Several major diffraction peaks that are not present in the calculated data are evident in the experimental diffraction pattern. Relative intensities also vary greatly between the two diffraction patterns, perhaps reflecting preferred sample orientation caused by the cleavage planes {001} (perfect) and {111} (good) that exist in soddyite. Extraneous diffraction peaks in the experimental pattern may indicate the presence of a minor contaminating phase.

SUMMARY

In general, the agreement between experimental and calculated diffraction patterns tends to be good. The four figures illustrate common situations encountered in the identification of uranium-bearing phases. Calculated diffraction patterns indicate positions of diffracted peaks and their relative intensities without extraneous peaks, variations in diffracted peak intensities that may arise from experimentally derived data, or both. Depending on whether a sample has been prepared properly and on how pure the sample is, a diffraction pattern collected from an unknown phase may closely resemble that of a standard from a database, or it may differ. Ranges of data outside of the experimental range of 2θ values reported can be produced by a calculated diffraction pattern to aid in verification. Chemical variability can be simulated by altering cations that occur on a particular site in a crystal, so that solid-solution series within a mineral group or inorganic system can be simulated and compared with an experimentally determined diffraction pattern. Sample preparation techniques can also be evaluated and improved with the aid of calculated diffraction patterns. Calculated diffraction data, when used in conjunction with experimental data contained in searchable databases, can be a valuable tool for aiding the identification of uranium-bearing phases by X-ray diffraction methods.

ACKNOWLEDGMENTS

This work was funded by the Environmental Management Sciences Program of the United States Department of Energy (DE-FG07-97ER14820) to Peter Burns. The manuscript was greatly improved by thorough reviews by Robert J Finch and Deane K. Smith. R.L. Hill provided invaluable aid in the preparation of many of the data tables.

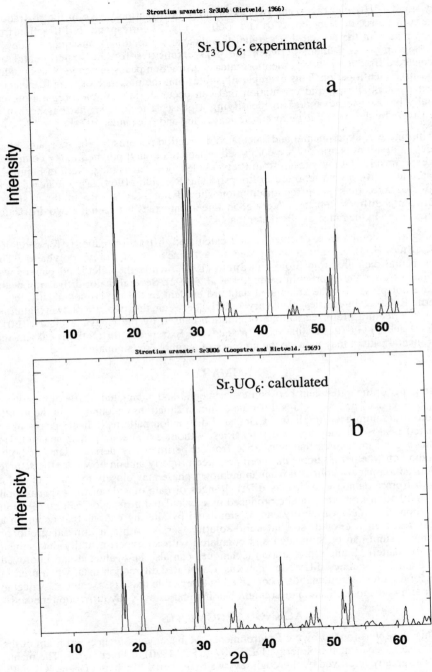

Figure 3. Comparison of (a) the experimental and (b) the calculated diffraction patterns for the phase Sr_3UO_6.

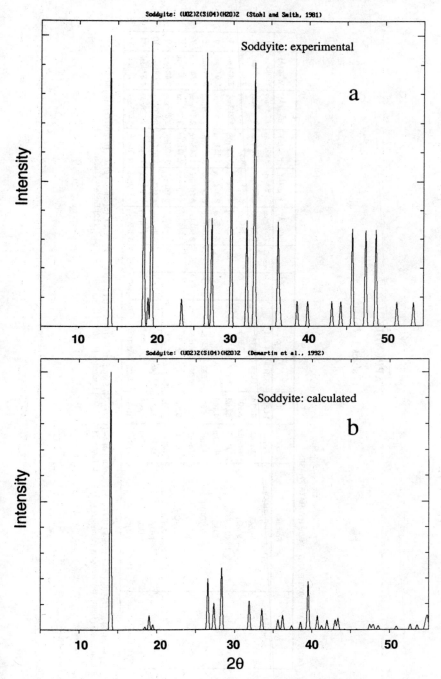

Figure 4. Comparison of (a) the experimental and (b) the calculated diffraction patterns for the mineral soddyite.

APPENDIX
(Tables 1–15)

TABLE 1. URANYL OXIDE AND OXIDE HYDRATE PHASES

Name	Formula	S.G.	a	b	c	α	β	γ	d^1_{hkl} (Å)	d^2_{hkl} (Å)	d^3_{hkl} (Å)	d^4_{hkl} (Å)	d^5_{hkl} (Å)	ref.
schoepite	$[(UO_2)_8O_2(OH)_{12}](H_2O)_{12}$	$P2_1ca$	14.337	16.813	14.731				7.366(002)	3.253(242)	3.626(240)	3.223(402)	3.683(004)	1
	α-U_3O_8	$Amm2$	6.716	11.96	4.147				6.716(100)	2.642(131)	2.874(031)	3.358(200)	2.184(231)	2
	γ-UO_3	$I4_1/amd$	6.901		19.975				4.384(103)	3.451(200)	6.523(101)	2.440(220)	2.637(107)	3
	$[H_2(UO_2)_3O_4]$	P-1	6.802	7.417	5.556	108.5	125.5	88.2	4.207(011)	4.746(110)	3.291(121)	3.212(120)	3.394(201)	4
	δ-UO_3	Pm-$3m$	4.146						4.146(100)	2.932(110)	1.854(210)	1.693(211)	2.073(200)	5
	γ-$[(UO_2)(OH)_2]$	$P2_1/c$	5.56	5.522	6.416		112.7		4.038(011)	5.129(100)	3.678(111)	2.831(112)	3.141(102)	6
	α-UO_3	$Amm2$	3.961	6.86	4.166				3.961(100)	3.561(011)	2.648(111)	2.593(120)	1.789(131)	7
	β-$[(UO_2)(OH)_2]$	$Pbca$	5.644	6.287	9.937				3.869(111)	4.969(002)	2.601(113)	3.144(020)	1.927(131)	8
	α-$[(UO_2)(OH)_2]$	$Cmca$	4.242	10.302	6.868				3.406(111)	5.151(020)	2.488(131)	2.857(121)	3.434(002)	9
	β-U_3O_8	$Cmcm$	7.069	11.445	8.303				3.357(130)	4.152(002)	2.611(132)	3.535(200)	2.691(202)	10
	UO_2	Fm-$3m$	5.458						3.151(111)	1.93(220)	1.646(311)	2.729(200)	1.576(222)	11
	β-UO_3	$P2_1$	10.34	14.33	3.91		99		3.029(4,221)	3.583(040)	4.81(210)	3.075(220)	3.1(121)	12

(1) Finch et al. (1996), (2) Loopstra (1977), (3) Loopstra et al. (1977), (4) Siegel et al. (1972a), (5) Weller et al. (1988), (6) Siegel et al. (1972b), (7) Loopstra & Cordfunke (1966), (8) Taylor & Bannister (1972), (9) Taylor (1971), (10) Loopstra (1970), (11) Rundle et al. (1948), (12) Debets (1966)

Table 2 — next page

TABLE 3. URANIUM-BEARING PHASES CONTAINING As

Name	Formula	S.G.	a	b	c	α	β	γ	d^1_{hkl} (Å)	d^2_{hkl} (Å)	d^3_{hkl} (Å)	d^4_{hkl} (Å)	d^5_{hkl} (Å)	ref.
orthowalpurgite	$Bi_4O_4[(UO_2)(AsO_4)_2](H_2O)_2$	$Pbcm$	5.492	13.324	20.685				10.343(002)	3.207(115)	3.088(133)	3.276(132)	5.601(022)	1
walpurgite	$Bi_4O_4[(UO_2)(AsO_4)_2](H_2O)_2$	P-1	7.135	10.426	5.494	101.5	110.8	88.2	10.207(010)	3.269(021)	3.051(131)	3.109(121)	3.131(211)	2
	$Li[(UO_2)(AsO_4)](D_2O)_4$	$P4/n$	7.097		9.19				9.19(001)	3.549(200)	3.31(201)	2.421(221)	2.509(220)	3
	$NH_4[(UO_2)(AsO_4)](H_2O)_3$	$P4/ncc$	7.189		18.191				9.096(002)	3.343(202)	3.595(200)	2.448(222)	2.542(220)	4
abernathyite	$K[(UO_2)(AsO_4)](H_2O)_3$	$P4/ncc$	7.176		18.126				9.063(002)	3.336(202)	3.588(200)	2.443(222)	2.537(220)	4
	$KH_3O[(UO_2)(AsO_4)]_2(H_2O)_6$	$P4/ncc$	7.171		18.048				9.024(002)	3.332(202)	3.586(200)	2.441(222)	2.535(220)	4
seelite	$Mg_3O[(UO_2)(AsO_4)]_2(H_2O)_4$	$C2/m$	18.207	7.062	6.661		99.7		8.973(200)	3.012(112)	3.286(220)	4.487(400)	3.531(020)	5
	$Mg[(UO_2)(AsO_3)_7(AsO_4)_{3}]_2(H_2O)_7$	$C2/m$	16.194	7.071	6.67		99.7		8.967(200)	4.484(400)	3.016(-112)	3.289(220)	3.536(020)	6
metazeunerite	$Cu[(UO_2)(AsO_4)]_2(H_2O)_8$	$P4_2/nmc$	7.105		17.704		89.9		8.852(002)	3.553(200)	3.297(202)	2.512(220)	2.417(222)	7
	$[(UO_2)D(AsO_4)](D_2O)_4$	P-1	7.164	7.112	17.554	90.2		90	8.777(002)	3.556(002)	3.582(200)	3.729(014)	3.314(-202)	8

(1) Krause et al. (1995), (2) Mereiter (1982b), (3) Fitch et al. (1982a), (4) Ross & Evans (1964), (5) Bachet et al. (1991), (6) Piret & Piret-Meunier (1994), (7) Hanic (1960), (8) Fitch et al. (1982b)

TABLE 4. URANIUM-BEARING PHASES CONTAINING B

Name	Formula	S.G.	a	b	c	α	β	γ	d^1_{hkl} (Å)	d^2_{hkl} (Å)	d^3_{hkl} (Å)	d^4_{hkl} (Å)	d^5_{hkl} (Å)	ref.
	$K_6[(UO_2)_2(B_{16}O_{24}(OH)_8)](H_2O)_{12}$	$P2_1/n$	12.024	26.45	12.543		94.7		11.984(100)	10.916(110)	9.085(021)	8.880(121)	13.225(020)	1
	$Na[(UO_2)(BO_3)]$	$Pcam$	10.712	5.78	6.862				10.712(100)	2.778(311)	3.409(211)	3.268(422)	2.663(021)	2
	$[Mg(UO_2)(B_2O_5)]$	$Pbcm$	9.747	7.315	7.911				9.747(100)	3.956(002)	4.874(200)	3.424(221)	3.665(102)	3
	$Li[(UO_2)(BO_3)]$	$P2_1/c$	5.767	10.574	6.835		105		5.571(100)	3.835(120)	4.127(021)	3.301(002)	3.23(102)	4
	$[(UO_2)_2(B_2O_4)]$	$C2/c$	12.504	4.183	10.453		122.2		4.423(002)	3.85(111)	5.29(200)	4.92(202)	2.934(311)	5
	$[Ca(UO_2)_2(BO_3)_2]$	$C2$	16.512	8.169	6.582		97		4.085(020)	8.195(200)	7.311(110)	3.679(401)	4.829(201)	6

(1) Behm (1985), (2) Gasperin (1988), (3) Gasperin (1987a), (4) Gasperin (1987d), (5) Gasperin (1987e), (6) Gasperin (1987d)

TABLE 2. URANIUM-BEARING OXIDES AND OXIDE HYDRATES

Name	Formula	S.G.	a	b	c	α	β	γ	d^1_{hkl} (Å)	d^2_{hkl} (Å)	d^3_{hkl} (Å)	d^4_{hkl} (Å)	d^5_{hkl} (Å)	ref.
	$K_2U_2O_7$	R-3m	3.96	3.96	19.82	90	120	90	17.165(001)	8.582(002)	5.722(003)	4.291(004)	3.96(010)	1
protasite	$Ba[(UO_2)_3O_3(OH)_2](H_2O)_3$	Pc	12.295	7.221	6.956		90.4		12.295(100)	7.221(010)	6.227(110)	3.464(120)	3.611(020)	2
fourmarierite	$Pb[(UO_2)_4O_3(OH)_4](H_2O)_4$	$Cmc2_1$	13.986	16.4	14.293				8.536(111)	3.17(242)	3.573(004)	4.796(131)	3.141(402)	3
becquerelite	$Ca[(UO_2)_3O_2(OH)_3]_2(H_2O)_8$	$Pna2_1$	13.86	12.3	14.92				7.831(111)	7.46(092)	3.19(232)	3.529(230)	5.261(121)	4
billietite	$Ba[(UO_2)_3O_2(OH)_3]_2(H_2O)_4$	$Pna2_1$	12.072	30.167	7.146				7.542(040)	4.192(231)	3.229(042)	4.558(211)	3.771(080)	2
	$Cs_4[(UO_2)_5O_7]$	Pbcn	18.776	7.07	14.958				7.479(002)	3.668(104)	3.032(512)	3.316(510)	3.196(022)	5
compreignacite	$K_2[(UO_2)_3O_2(OH)_3]_2(H_2O)_7$	Pnnm	14.859	7.175	12.187				7.43(200)	3.192(213)	3.535(013)	3.231(220)	3.715(400)	6
sayrite	$Pb_2[(UO_2)_5O_6(OH)_2](H_2O)_4$	$P2_1/c$	10.704	6.96	14.533		116.8		7.041(102)	3.111(311)	3.52(204)	3.058(113)	3.12(122)	7
	$Na_2(UO_2(O_3))(H_2O)_9$	$P2_1/c$	6.413	17.292	14.186		98.5		7.015(002)	7.361(021)	8.646(020)	5.114(120)	5.448(022)	8
	$K_2[(UO_2)_2O_3]$	$P2_1$	6.931	7.69	6.984		109.7		6.575(011)	3.186(121)	3.313(120)	3.288(002)	3.263(200)	9
Sr-curite	$Sr_{2.84}[(UO_2)_4O_4(OH)_3]_2(H_2O)_2$	Pnam	12.314	12.961	8.405				6.157(200)	3.39(122)	3.134(140)	3.06(202)	3.526(022)	10
curite	$Pb_5[(UO_2)_8(OH)_6](H_2O)_3$	Pnma	12.551	13.003	8.39				6.147(111)	3.073(222)	4.195(002)	3.369(212)	6.276(200)	11
	$K_8[(UO_2)_2O_6]$	P-1	6.474	9.698	6.344	101.2	102.3	109	6.05(110)	5.941(001)	2.626(131)	5.74(011)	2.645(211)	12
	$(Na_2U_2O_7)_{.5}$	R-3m	6.34			36.1			5.92(110)	3.177(110)	3.342(100)	2.701(211)	1.964(011)	13
	$Na[(UO_2)O(OH)](H_2O)_{0-1}$	R-3m	3.954	3.954	17.730				5.910(003)	3.362(101)	3.194(012)	2.955(006)	2.710(104)	14
clarkeite	Li_6UO_6	R-3	8.381						5.487(101)	3.33(021)	4.191(110)	3.629(012)	2.295(122)	15
	K_2LiUO_6	P-3m	6.193		5.388				5.363(006)	3.783(101)	3.1(110)	2.396(021)	5.338(001)	16
	$Na_4[UO_2O_3]$	I4/m	7.557		4.641				5.344(110)	3.955(101)	2.732(121)	2.39(310)	3.779(200)	12
	$α-Na_2UO_4$	Cmmm	9.74	5.72	3.49				4.932(110)	2.837(201)	2.849(111)	4.87(200)	2.824(310)	17
	$Ca_2[UO_2O_3]$	$P2_1/c$	7.914	5.441	11.448		108.8		4.863(011)	3.2(211)	3.01(013)	5.419(002)	3.746(200)	18
	Ca_3UO_6	$P2_1$	5.728	5.958	8.301		90.6		4.84(011)	2.916(112)	2.938(-112)	4.129(110)	4.691(101)	19
	$Cu(UO_2)O_2$	$P2_1/c$	5.475	4.957	6.569		118.9		4.793(020)	3.256(102)	3.755(011)	3.549(111)	2.876(220)	20
vandenbrandeite	$Cu[(UO_2)(OH)_4]$	P-1	7.855	5.449	6.089	91.4	101.9	89.2	4.463(101)	5.262(101)	3.393(111)	3.976(011)	2.963(201)	21
	$Na_{11}U_5O_{16}$	$P4_232$	9.543						4.268(210)	2.386(222)	2.755(222)	5.51(111)	1.687(440)	22
	$[Sr_3(UO_2)_{11}O_{14}]$	Pmmn	28.508	8.381	6.733				4.191(002)	3.367(002)	2.625(022)	2.095(040)	1.832(420)	23
	$[Pb_3(UO_2)_{11}O_{14}]$	Pmmn	28.459	8.379	6.765				4.19(002)	3.383(022)	2.632(022)	14.23(200)	2.095(040)	24
	$α-K_2UO_4$	I4/mmm	4.335		13.13				4.116(101)	6.565(002)	3.08(103)	3.065(110)	1.773(213)	25
	$β-Na_2[(UO_2)_2O_2]$	Pbca	5.808	5.975	11.718				3.924(111)	5.859(002)	2.849(113)	2.988(020)	2.904(200)	26
	Sr_3UO_6	$P2_1/c$	6.013	6.214	8.614		90.2		3.858(020)	3.866(-111)	3.496(102)	3.107(020)	3.007(200)	27
	$Li_4[UO_2O_3]$	I4/m	6.736	6.736	4.457				3.717(101)	4.763(110)	3.368(200)	2.496(211)	2.13(310)	28

Phase	S.G.	a	b	c			d_1(hkl)	d_2(hkl)	d_3(hkl)	d_4(hkl)	d_5(hkl)	Ref
Li$_2$[(UO$_2$)O$_2$]	Pnma	10.547	6.065	5.134			3.673(111)	5.274(200)	2.617(311)	1.837(222)	2.629(220)	29
FeUO$_4$	Pbcn	4.888	11.937	5.11			3.387(111)	3.086(130)	3.882(021)	1.968(132)	2.577(041)	30
α-Cs$_2$[(UO$_2$)$_2$O$_3$]	C2/m	14.528	4.264	7.605	112		3.368(400)	4.065(110)	3.526(002)	3.078(-402)	2.695(-312)	31
Ba[(UO$_2$)$_2$]	Pbcm	5.744	8.136	8.237			3.347(120)	3.32(120)	2.894(022)	2.039(222)	2.872(200)	32
Ba$_2$U$_3$O$_7$	Imma	8.155	11.315	8.193			3.308(220)	3.318(022)	2.89(02)	2.022(242)	2.829(040)	33
MgUO$_4$	Imam	6.52	6.595	6.924			3.298(020)	3.462(002)	2.388(002)	3.26(200)	1.926(222)	34
Pb[(UO$_2$)O$_2$]	Pbcm	5.536	7.968	8.212			3.298(102)	3.234(120)	2.859(202)	1.989(222)	2.768(200)	35
Sr$_2$[(UO$_2$)O$_3$]	P2$_1$/c	8.104	5.661	11.918	109		3.295(211)	3.13(013)	5.059(011)	5.634(002)	2.776(202)	18
Sr[(UO$_2$)O$_2$]	Pbcm	5.49	7.977	8.13			3.267(102)	3.227(120)	2.847(022)	1.976(222)	5.49(100)	18
Bi$_2$[(UO$_2$)O$_2$]	C2	6.872	4.009	9.69	90	90	3.261(111)	3.238(201)	2.817(112)	3.23(003)	1.989(310)	36
α-Sr[(UO$_2$)O$_2$]	R-3m	6.587			35.3		3.24(110)	2.771(211)	6.17(111)	1.997(011)	3.085(222)	37
γ-Sr[(UO$_2$)O$_2$]	R-3m	6.542			35.5		3.233(110)	2.761(211)	6.123(111)	1.994(011)	3.062(222)	37
β-Cs$_2$[(UO$_2$)$_2$O$_3$]	C2/m	14.615	4.32	7.465	113.8		3.181(-311)	3.344(111)	6.83(001)	3.733(-113)	3.343(111)	31
Rb$_2$UO$_4$	I4/mmm	4.345		13.83			3.162(103)	3.072(110)	6.915(002)	1.791(213)	2.173(200)	13
UC$_2$O$_6$	P-3m	4.99		4.622			3.157(101)	4.322(100)	2.495(110)	1.695(-12)	2.038(102)	38
Cs$_2$[U$_2$O$_7$](D$_2$O)$_{0.444}$	C2/m	14.531	4.274	7.601	113		3.156(311)	3.35(111)	3.498(206)	6.997(001)	3.344(400)	39
PbTe[(UO$_2$)O$_3$]	P2$_1$/c	7.813	7.061	13.775	93.7		3.15(121)	3.489(202)	3.09(014)	2.99(212)	3.376(113)	40
Ca[(UO$_2$)O$_2$]	R-3m	6.268			36		3.134(110)	5.855(111)	3.295(100)	2.666(211)	1.937(011)	18
α-Cd[(UO$_2$)O$_2$]	R-3m	6.233			36.1		3.124(110)	2.655(211)	1.931(011)	2.91(222)	5.82(111)	18
BaU$_2$O$_7$	I4$_1$/amd	7.128		11.95			3.08(211)	3.477(103)	3.853(112)	1.926(224)	3.061(202)	41
K$_2$[(UO$_2$)$_3$O$_8$](UO$_2$)$_2$	Pbam	6.945	19.533	7.215			3.079(151)	3.608(200)	3.129(201)	3.405(150)	7.215(001)	42
Pb$_3$UO$_6$	Pnma	13.719	12.351	8.213			3.06(222)	3.388(212)	3.088(040)	2.998(420)	7.047(101)	43
K$_8$BiU$_6$O$_{24}$	Pm-3m	8.631					3.052(220)	2.158(400)	1.762(420)	4.316(200)	1.526(440)	44
Sr$_3$UO$_6$	P2$_1$	5.959	6.179	8.553	90.2		3.032(112)	3.025(112)	3.089(200)	2.98(200)	2.145(220)	45
Moctezumite												
Ba$_2$MgUO$_6$	Fm-3m	8.379					2.962(220)	4.838(111)	1.71(422)	2.095(400)	2.526(311)	46
Ba$_2$FeUO$_6$	Fm-3m	8.361					2.956(220)	1.707(422)	2.09(400)	4.827(111)	1.478(440)	47
Ba$_2$CrUO$_6$	P-3	8.301					2.935(220)	1.694(422)	2.075(400)	4.793(111)	1.467(440)	47
β-Cd[(UO$_2$)O$_2$]	Cmmm	7.023	6.849	3.514			2.856(111)	3.512(200)	3.425(020)	1.874(311)	4.903(110)	41
Mn[(UO$_2$)O$_2$]	Imma	6.647	6.984	6.75			2.811(121)	4.736(101)	3.375(002)	3.324(200)	1.803(321)	48
CoUO$_4$	Imma	6.497	6.952	6.497			2.772(121)	4.594(101)	3.249(202)	1.769(321)	2.297(202)	49

(1) Jove et al. (1988) (2) Pagoaga et al. (1987), (3) Piret (1985), (4) Piret-Meunier & Piret (1982), (5) van Egmond (1976a), (6) Burns (1998), (7) Piret et al. (1983), (8) Alcock (1968), (9) Saine (1989), (10) Burns and Hill (in press), (11) Taylor et al. (1981), (12) Wolf & Hoppe (1986), (13) Kovba et al. (1958), (14) Finch et al. (1997), (15) Wolf & Hoppe (1985), (16) Wolf & Hoppe (1987), (17) Kovba et al. (1961), (18) Loopstra & Rietveld (1969), (19) Rietveld (1966), (20) Siegel & Hoekstra (1968), (21) Rosenzweig & Ryan (1977b), (22) Bartram & Fryxell (1970), (23) Cordfunke et al. (1991), (24) Ijdo (1993a), (25) Kovba (1971), (26) Cordfunke & Ijdo (1995), (27) Ijdo (1993b), (28) Hoekstra & Siegel (1982), (29) Giebert et al. (1978), (30) Bacmann & Bertaut (1967), (31) van Egmond (1976b), (32) Reis et al. (1976), (33) Cordfunke & Ijdo (1988), (34) Zachariasen (1954), (35) Cremers et al. (1986), (36) Koster et al. (1975), (37) Fujino et al. (1977), (38) Collomb et al. (1976), (39) Mijlhoff et al. (1993), (40) Swihart et al. (1993), (41) Yamashita et al. (1981), (42) Allpress (1965), (43) Kovba (1972), (44) Sterns et al. (1986), (45) Gasperin et al. (1991), (46) Padel et al. (1972), (47) Grenet & Michel (1971), (48) Bacmann & Bertaut (1966), (49) Bertaut et al. (1962)

TABLE 5. URANIUM-BEARING PHASES CONTAINING C

Name	Formula	S.G.	a	b	c	α	β	γ	d^1_{hkl} (Å)	d^2_{hkl} (Å)	d^3_{hkl} (Å)	d^4_{hkl} (Å)	d^5_{hkl} (Å)	ref.
andersonite	$Na_2Ca(UO_2)(CO_3)_3(H_2O)_{5.333}$	$R\text{-}3m$	17.904	17.904	23.753	90	90	120	12.984(101)	5.69(211)	7.918(003)	5.256(122)	5.168(300)	1
bayleyite	$Mg_2UO_2(CO_3)_3(H_2O)_{18}$	$P2_1/a$	26.56	15.256	6.505		92.9		8.842(300)	13.225(110)	5.977(011)	13.263(200)	7.628(020)	2
liebigite	$Ca_2(UO_2)(CO_3)_3(H_2O)_{11}$	$Aba2$	16.699	17.557	13.697		99.4		8.779(020)	5.4(022)	8.349(200)	6.849(002)	3.608(142)	3
swartzite - synthetic	$CaMgUO_2(CO_3)_3(H_2O)_{12}$	$P2_1/m$	11.08	14.634	6.439		93.4		8.758(110)	5.495(-111)	7.317(020)	4.797(021)	6.353(001)	4
	$Sr_2(UO_2)(CO_3)_3(H_2O)_8$	$P2_1/c$	11.379	11.446	25.653		92.3		8.063(100)	5.989(-113)	11.359(100)	4.914(114)	3.498(-132)	5
schröckingerite	$NaCa_3[(UO_2)(CO_3)_3](SO_4)F(H_2O)_{10}$	$P\text{-}1$	9.634	9.635	14.391	91.4	96.4	120.3	7.18(002)	8.308(010)	14.36(001)	4.787(003)	5.602(102)	6
	$(NH_4)_4(UO_2)(CO_3)_3$	$C2/c$	10.679	9.373	12.85		95.1		6.357(-111)	5.972(111)	6.385(002)	5.306(200)	3.874(202)	7
	$K_4(UO_2)(CO_3)_3$	$C2/c$	10.24	9.198	12.222		90.9		6.114(-111)	5.811(111)	3.401(113)	2.672(-132)	6.087(002)	8
roubaulite	$Cu_2[(UO_2)_3(CO_3)_2O_2(OH)_2](H_2O)_4$	$P\text{-}1$	7.767	6.924	7.85	92.2	99.5	93.5	5.564(101)	7.512(100)	3.229(120)	3.184(112)	4.144(111)	9
Sr-swartzite	$SrMgUO_2(CO_3)_3(H_2O)_{12}$	$P2_1/m$	11.216	14.739	6.484				5.542(-111)	8.848(110)	7.37(020)	4.83(021)	6.395(001)	4
	$Ca_3Na_{1.5}(H_3O)_{.5}(UO_2)(CO_3)_3)(H_2O)_8$	$Pnnm$	18.15	16.866	18.436				5.132(222)	8.433(020)	6.467(202)	3.996(420)	4.609(004)	10
	$[(UO_2)(CO_3)]$	$Pmmn$	4.845	9.205	4.296				4.603(020)	2.144(220)	2.301(040)	2.423(200)	1.947(022)	11
rutherfordine	$Tl_2UO_2(CO_3)_3$	$C2/c$	10.684	9.309	12.726		94.9		3.532(-113)	3.33(221)	3.272(-311)	2.721(-132)	4.839(-112)	12
	$Cs_4(UO_2)(CO_3)_3(H_2O)_6$	$P2_1/n$	18.723	9.647	11.297		96.8		3.496(-222)	9.296(200)	4.431(021)	5.202(102)	2.943(502)	13

(1) Mereiter (1986c), (2) Mayer & Mereiter (1986), (3) Mereiter (1982a), (4) Mereiter (1986b), (5) Mereiter (1986d), (6) Mereiter (1986a), (7) Serezhkin et al. (1983) (8) Han et al. (1990), (9) Ginderow & Cesbron (1985), (10) Vochten et al. (1994), (11) Christ et al. (1955), (12) Mereiter (1986f), (13) Mereiter (1988)

TABLE 6. URANIUM-BEARING PHASES CONTAINING Cl

Name	Formula	S.G.	a	b	c	α	β	γ	d^1_{hkl} (Å)	d^2_{hkl} (Å)	d^3_{hkl} (Å)	d^4_{hkl} (Å)	d^5_{hkl} (Å)	ref.
	$K_2[(UO_2)_4Cl_4O_2(OH)_2(H_2O)_4](H_2O)_2$	$P\text{-}1$	12.15	12.33	8.026	110.5	96.3	138.7	11.316(110)	5.997(001)	6.397(100)	4.992(111)	3.324(011)	1
	$[(UO_2)_4Cl_2O_2(OH)_2(H_2O)_6](H_2O)_4$	$P2_1/c$	11.645	10.101	10.206		105.8		11.205(100)	7.503(110)	4.899(210)	4.105(102)	7.041(011)	2
	$Rb_2[(UO_2)_2O_2Cl_8(H_2O)_2](H_2O)_2$	$P2_1/c$	8.54	8.096	21.735		111.7		7.794(-102)	3.776(-212)	7.515(011)	5.667(110)	3.108(016)	3
	$[(UO_2)_2Cl_2H_2O]$	$P2_1/m$	5.828	8.534	5.557		97.8		5.774(100)	4.286(101)	5.506(001)	4.267(200)	3.425(111)	4
	$[(UO_2)_2Cl_2(H_2O)_3]$	$Pnma$	12.738	10.495	5.547				5.445(210)	4.904(011)	5.248(020)	4.577(111)	5.086(101)	5
	$[(UO_2)(OH)Cl(H_2O)_2]$	$P2_1/n$	17.743	6.136	10.725		95.5		5.039(210)	5.32(011)	5.338(002)	4.386(020)	3.769(-212)	6
	UO_2Cl_2	$Pnma$	5.725	8.409	8.72				4.786(101)	3.159(121)	3.207(112)	4.205(020)	6.053(011)	7
	$Cs_9[(UO_2)OCl_9]$	$P2_1/m$	8.734	4.118	7.718		105.3		3.17(111)	3.795(-102)	2.929(-211)	6.493(-101)	3.603(011)	8

(1) Perrin & Le Marouille (1977), (2) Åberg (1976), (3) Perrin (1977), (4) Taylor & Wilson (1974), (5) Debets (1968), (6) Åberg (1969), (7) Taylor & Wilson (1973), (8) Allpress & Wadsley (1964)

TABLE 7. URANIUM-BEARING PHASES CONTAINING Mo

Name	Formula	S.G.	a	b	c	α	β	γ	d^1_{hkl} (Å)	d^2_{hkl} (Å)	d^3_{hkl} (Å)	d^4_{hkl} (Å)	d^5_{hkl} (Å)	ref.
	$(NH_4)_2(Er_2UMo_{12}O_{42})(H_2O)_{22}$	$P\text{-}1$	13.094	13.195	16.846	89.7	95.2	93.4	16.777(001)	13.172(010)	13.017(100)	10.36(0-11)	10.36(011)	1
	$(NH_4)_2(Sc_2UMo_{12}O_{42})(H_2O)_{26}$	$P\text{-}1$	12	12.23	13.001	60.1	117.8	114.6	10.293(010)	10.306(100)	10.645(001)	9.87(-111)	9.034(-101)	2
	$CuH_6(UMo_{12}O_{42})(H_2O)_{12}$	$R\text{-}3$	10.554	10.554	10.554	78.5	78.5	78.5	10.198(100)	7.897(110)	3.399(301)	3.075(0-13)	3.463(311)	3
	$Mg(UO_2)_6(MoO_4)_7(H_2O)_{14}$	$C222_1$	11.313	20.163	23.877				10.082(020)	10.082(020)	5.136(024)	9.118(111)	3.569(242)	4
	$(NH_4)_3H(ThUMo_{12}O_{42})(H_2O)_{15}$	$R\text{-}3c$	13.512			87.6			9.866(110)	9.736(110)	3.616(3-12)	3.172(4-11)	3.037(3-31)	5
umohoite	$H_2(Nd_2(H_2O)_{12}UMo_{12}O_{42})(H_2O)_{12}$	$C2/c$	24.225	21.323	10.982				9.352(6-11)	10.614(020)	12.059(200)	6.029(400)	5.672(131)	6
	$[(UO_2)(MoO_4)](H_2O)_4$	$P2_1/c$	6.32	7.5	57.8		94		8.91(111)	14.415(004)	4.805(012)	3.023(126)	3.216(122)	7
iriginite	$K_2[(UO_2)(MoO_4)_2]$	$P2_1/c$	12.269	13.468	12.875		95.1		7.207(008)	3.822(6-311)	3.164(141)	3.645(312)	3.618(230)	8
	$[(UO_2)(MoO_3OH)_2](H_2O)_2$	$Pca2_1$	12.77	6.715	11.53				6.403(002)	3.227(113)	3.126(121)	3.193(400)	5.283(111)	9
	$Ba(UO_2)_3(MoO_4)_4(H_2O)_4$	$Pbca$	17.797	11.975	23.33				6.385(200)	4.171(410)	4.878(204)	3.477(232)	3.888(006)	10
	$\alpha\text{-}(UO_2)_3(MoO_4)(H_2O)_2$	$P2_1/c$	13.612	11.005	10.854		113.1		5.103(122)	6.26(200)	7.394(011)	3.511(221)	4.992(002)	10
	$Mg(UO_2)_3(MoO_4)_4(H_2O)_8$	$Cmc2_1$	17.105	13.786	10.908				4.819(021)	10.734(110)	8.553(200)	3.010(223)	3.365(402)	12
	$Zn(UO_2)_3(MoO_4)_4(H_2O)_8$	$Cmc2_1$	17.056	13.786	10.919				4.816(221)	10.722(110)	8.528(200)	3.011(223)	3.219(023)	12
	$Sr(UO_2)_6(MoO_4)_7(H_2O)_{19}$	$C222_1$	11.166	20.281	24.061				4.812(221)	9.782(011)	5.174(024)	3.583(242)	9.061(111)	4
	$(UO_2)(MoO_4)$	$P2_1/c$	7.195	5.484	13.58		104.6		4.531(222)	3.917(111)	4.308(110)	5.524(-102)	3.343(-113)	13
	$[Ca(UO_2)(MoO_4)_4O_{10}]$	$P\text{-}1$	13.239	6.651	8.236	90	90.4	120.2	4.21(012)	3.31(410)	3.308(210)	3.326(220)	6.71(101)	14
	U_2MoO_8	$P2_12_12$	6.734	23.240	4.115				4.118(002)	3.358(160)	2.602(161)	11.620(020)	3.332(210)	15
deloryite	$Cu_4[(UO_2)(MoO_4)_2](OH)_6$	$C2/m$	19.94	6.116	5.52		104.2		4.115(001)	3.74(311)	4.833(400)	2.484(221)	3.268(510)	16
	$CuUO_2(MoO_4)_2(OH)_6$	$C2/m$	19.839	5.511	6.101			104.5	3.734(3-11)	4.089(1,-11)	4.802(400)	3.251(501)	2.475(212)	17

(1) Tat'yanina et al. (1982a), (2) Samohvalova et al. (1990a), (3) Tat'yanina et al. (1990a), (4) Tabachenko et al. (1984a), (5) Tat'yanina et al. (1982b), (6) Samohvalova et al. (1990b), (7) Makarov & Anikina (1963), (8) Sadikov et al. (1988), (9) Serezhkin et al. (1973), (10) Tabachenko et al. (1984b), (11) Serezhkin et al. (1980a), (12) Tabachenko et al. (1983), (13) Serezhkin et al. (1980b), (14) Lee & Jaulmes (1987), (15) Serezhkin et al. (1972), (16) Pushcharovsky et al. (1996), (17) Tali et al. (1993)

TABLE 8. URANIUM-BEARING PHASES CONTAINING Nb

Name	Formula	S.G.	a	b	c	α	β	γ	d^1_{hkl} (Å)	d^2_{hkl} (Å)	d^3_{hkl} (Å)	d^4_{hkl} (Å)	d^5_{hkl} (Å)	ref.
	K[(UO$_2$)(NbO$_4$)]	Pbca	7.579	11.321	15.259				7.630(022)	3.498(211)	4.546(024)	3.163(024)	2.935(213)	1
	Cs[(UO$_2$)(NbO$_4$)]	P2$_1$/c	7.43	8.7	10.668		105.1		7.174(100)	3.72(120)	3.587(200)	3.222(113)	3.441(112)	2
	[(UO$_2$)Nb$_3$O$_8$]	Fddd	7.38	12.78	15.96				3.196(220)	3.99(004)	2.494(224)	3.195(040)	1.845(260)	3
	[(UO$_2$)TiNb$_2$O$_8$]	Fddd	7.28	12.62	16.02				3.153(220)	4.005(004)	2.477(224)	3.155(040)	4.075(113)	4

(1) Gasperin (1987b), (2) Gasperin (1987c), (3) Chevalier & Gasperin (1968), (4) Chevalier & Gasperin (1969)

TABLE 9. URANIUM-BEARING PHASES CONTAINING N

Name	Formula	S.G.	a	b	c	α	β	γ	d^1_{hkl} (Å)	d^2_{hkl} (Å)	d^3_{hkl} (Å)	d^4_{hkl} (Å)	d^5_{hkl} (Å)	ref.
	[(UO$_2$)$_3$O(OH)$_3$(H$_2$O)$_6$](NO$_3$)(H$_2$O)$_4$	P-1	8.026	11.276	12.346	109.7	99.4	88.6	9.562(011)	7.081(101)	6.246(110)	4.469(121)	4.302(122)	1
	(UO$_2$)(NO$_3$)$_2$(H$_2$O)$_2$	Cm	10.52	5.93	6.95		72		6.61(001)	5.101(110)	4.759(201)	4.387(111)	5.003(200)	2
	UO$_2$(NO$_3$)$_2$(H$_2$O)$_6$	Cmc2$_1$	13.15	8.02	11.42				6.575(200)	5.872(111)	6.847(110)	5.71(002)	4.311(202)	3
	(NH$_4$)$_2$UO$_2$(NO$_3$)$_4$	P2$_1$/c	6.39	7.74	12.87		108		6.542(011)	4.872(-111)	6.077(100)	6.12(002)	4.126(111)	4
	UO$_2$(NO$_3$)$_2$(H$_2$O)$_6$	Cmc2$_1$	13.197	8.035	11.467				5.889(111)	6.599(200)	4.400(112)	6.863(110)	4.328(022)	5
	UO$_2$(NO$_3$)$_2$(H$_2$O)$_2$	P2$_1$/c	14.124	8.432	7.028		108		5.238(011)	4.722(-211)	6.716(200)	3.571(220)	3.717(211)	6
	[(UO$_2$)(OH)(NO$_3$)]$_2$(H$_2$O)$_4$	P-1	8.622	8.628	10.393	109.6	105.6	99.7	5.228(111)	3.719(202)	9.17(001)	7.96(100)	3.48(102)	7
	Rb[(UO$_2$)(NO$_3$)$_3$]	R-3c	9.384		18.899				4.692(110)	4.085(104)	6.162(012)	2.575(214)	2.709(300)	8

(1) Åberg, M. (1978), (2) Vdovenko et al. (1962), (3) Vdovenko et al. (1967), (4) Kapshukov et al. (1971), (5) Taylor & Mueller (1965), (6) Dalley et al. (1971), (7) Perrin (1976), (8) Zalkin et al. (1989)

TABLE 10. URANIUM-BEARING PHASES CONTAINING P

Name	Formula	S.G.	a	b	c	α	β	γ	d^1_{hkl} (Å)	d^2_{hkl} (Å)	d^3_{hkl} (Å)	d^4_{hkl} (Å)	d^5_{hkl} (Å)	ref.
phuralumite	$Al_2[(UO_2)_3(PO_4)_2(OH)_2](OH)_4(H_2O)_{10}$	$P2_1/c$	13.836	20.918	9.428		112.4		10.459(020)	10.913(110)	12.792(100)	6.122(130)	5.23(020)	1
althupite	$AlTh(UO_2)[(UO_2)_3O(OH)(PO_4)_2]_2$	$P-1$	10.953	18.567	13.503	72.6	68.2	84.2	10.169(100)	5.085(200)	3.264(220)	5.808(121)	4.414(123)	2
saléeite	$Mg[(UO_2)_2(PO_4)_2](H_2O)_{10}$	$P2_1/c$	6.951	19.947	9.896		135.2		9.974(020)	4.813(031)	3.46(-151)	3.293(022)	4.433(-122)	3
threadgoldite	$Al[(UO_2)_2(PO_4)_2(OH)](H_2O)_8$	$C2/c$	20.168	9.847	19.719		110.7		9.433(200)	3.466(-224)	5.333(-312)	3.366(024)	4.098(312)	4
	$ND_4[(UO_2)_2(PO_4)_2](D_2O)_3$	$P4/ncc$	7.022		18.091				9.046(002)	3.273(202)	3.511(200)	2.394(222)	2.483(220)	5
	$K[(UO_2)_2(PO_4)_2](H_2O)_3$	$P4/ncc$	6.994		17.784				8.892(002)	3.254(202)	3.497(200)	2.382(222)	2.473(220)	6
	$[(UO_2)H(PO_4)](H_2O)_4$	$P4/ncc$	6.995		17.491				8.746(002)	3.498(200)	3.247(202)	2.380(222)	2.473(220)	7
metatorbernite	$Cu_{92}[(UO_2)(PO_4)]_2(H_2O)_8$	$P4/nmm$	6.972		17.277				8.635(002)	3.486(200)	3.233(202)	2.465(-121)	2.370(222)	8
meta-uranocircite	$Ba[(UO_2)(PO_4)]_2(H_2O)_6$	$P2_1/c$	9.789	9.882	16.688			89.9	8.467(101)	8.434(002)	4.270(121)	4.265(-121)	6.39(102)	9
upalite	$Al[(UO_2)_3(PO_4)_2O(OH)](H_2O)_7$	$P2_1/a$	13.704	16.82	9.332		111.5		8.410(020)	6.247(021)	3.857(202)	4.205(040)	3.506(222)	10
meta-autunite	$Ca[(UO_2)(PO_4)]_2(H_2O)_6$	$P4/nmm$	6.96		8.4				8.400(001)	3.480(200)	3.215(201)	2.461(220)	2.362(221)	11
vanmeersscheite	$U(UO_2)_3(PO_4)_2(OH)_6(H_2O)_4$	$P2_1mn$	17.06	16.76	7.023				8.380(020)	5.978(220)	11.956(110)	3.069(501)	3.512(002)	12
phurcalite	$Ca_2(UO_2)_3O_2(PO_4)_2(H_2O)_7$	$Pbca$	17.415	16.035	13.598				8.018(020)	2.891(522)	3.100(560)	3.13(024)	4.256(312)	13
phosphuranylite	$KCa(H_3O)_3(UO_2)_7(PO_4)_4O_4(H_2O)_8$	$Cmcm$	15.899	13.74	17.3				7.95(200)	3.153(240)	2.88(225)	5.853(202)	3.09(025)	14
françoisite-Nd	$Nd[(UO_2)_3O(OH)(PO_4)_2](H_2O)_6$	$P2_1/c$	9.298	15.605	13.668		112.8		7.803(020)	3.129(124)	5.77(120)	3.901(040)	2.874(322)	15
dewindtite	$Pb_3[H(UO_2)_3O_2(PO_4)_2(H_2O)_{12}$	$Cmcm$	16.031	17.264	13.605				7.289(021)	5.874(220)	8.016(200)	11.747(110)	3.131(204)	16
	$Rb(UO_2(HPO_4)(H_2PO_3))(H_2O)_3$	$P2_1/c$	8.589	10.275	13.23		106.8		6.420(110)	5.911(012)	5.391(012)	3.166(-222)	2.939(-212)	17
ulrichite	$Cu[Ca(UO_2)(PO_4)_2](H_2O)_4$	$C2/m$	12.79	6.85	13.02		91.03		6.394(110)	5.496(-111)	3.197(010)	4.521(020)	3.425(020)	18
	$[(UO_2)(H_3PO_4)_2(H_2O)](H_2O)_2$	$P2_1/c$	11.369	13.899	7.481		113.7		6.144(011)	3.311(-321)	5.002(-121)	4.233(130)	8.332(110)	19
	$Na_{3.5}UO_2)_3(H_2PO_4)(PO_4)_3$	$P1$	6.675	6.922	10.732	84	82.3	89.4	5.513(011)	4.296(020)	4.644(111)	5.299(101)	3.526(003)	20
	$Na_2(UO_2)(P_2O_7)$	$Pna2_1$	13.259	8.127	6.973				5.292(011)	4.136(211)	4.064(020)	2.994(401)	3.487(002)	21
	$(UO_2)H(PO_4)_3$	$P2_1/c$	9.811	20.814	8.695			94.1	4.893(200)	3.882(112)	6.289(111)	5.19(040)	6.666(021)	22
	$U(UO_2)(PO_4)_2$	$Cmcm$	7.088	9.036	12.702				4.191(112)	2.760(114)	6.351(002)	3.544(200)	4.519(020)	23
	$Cs[(UO_2)(PO_3)_3]$	$P2_1/c$	6.988	10.838	13.309		104.3		3.818(122)	5.743(110)	5.542(012)	3.311(122)	3.479(031)	24
dumontite	$Pb_2[(UO_2)_3O_2(PO_4)_2](H_2O)_5$	$P2_1/m$	8.118	16.819	6.983		109		3.742(210)	4.274(031)	3.47(102)	2.997(051)	6.983(110)	25
	$K_4[(UO_2)(PO_4)_2]$	$P4_2/nmc$	6.985		11.865				3.493(002)	5.933(002)	3.01(202)	2.28(012)	1.746(400)	26

(1) Piret et al. (1979), (2) Piret & Deliens (1987), (3) Miller & Taylor (1986), (4) Khosrawan-Sazedj (1982b), (5) Fitch & Fender (1983), (6) Fitch & Cole (1991), (7) Morosin (1978), (8) Stergiou et al. (1993), (9) Khosrawan-Sazedj (1982a), (10) Piret & Declercq (1983), (11) Makarov & Ivanov (1960), (12) Piret & Deliens (1982), (13) Atencio et al. (1991), (14) Demartin et al. (1991), (15) Piret et al. (1988), (16) Piret et al. (1990), (17) Mistryukov & Mikhajlov (1985), (18) Birch et al. (1988), (19) Krogh-Andersen et al. (1985), (20) Gorbunova et al. (1980), (21) Linde et al. (1984), (22) Sarin et al. (1983), (23) Benard et al. (1996), (24) Linde et al. (1978), (25) Piret & Piret-Meunier (1988), (26) Linde et al. (1980)

TABLE 11. URANIUM-BEARING PHASES CONTAINING Se

Name	Formula	S.G.	a	b	c	α	β	γ	d^1_{hkl} (Å)	d^2_{hkl} (Å)	d^3_{hkl} (Å)	d^4_{hkl} (Å)	d^5_{hkl} (Å)	ref.
guilleminite	$Ba[(UO_2)_3(SeO_3)_2O_2](H_2O)_3$	$Pmn2_1$	7.084	7.293	16.881				8.441(002)	6.695(011)	7.293(010)	3.564(021)	5.518(012)	1
	$[(UO_2)(SeO_4)(H_2O)_2](H_2O)_2$	$C2/c$	14.653	10.799	12.664		119.9		7.925(111)	4.845(221)	4.34(221)	3.659(313)	5.399(020)	2
	$(NH_4)[(UO_2)F(SeO_4)](H_2O)$	$Pnma$	8.45	13.483	13.569				6.742(020)	3.462(221)	4.034(201)	2.852(124)	3.432(123)	3
	$UO_2Se_2O_5$	P-1	9.405	11.574	6.698	93	93.7	109.7	5.431(011)	8.393(-110)	3.12(-112)	8.821(100)	3.039(-1-12)	4
demesmaekerite	$Pb_2Cu_5[(UO_2)_2(SeO_3)_3]_2(OH)_6(H_2O)_2$	P-1	11.955	10.039	5.639	89.8	100.4	91.3	5.407(101)	3.336(311)	2.959(301)	4.741(111)	3.722(021)	5
	$[(UO_2)(SeO_3)]$	$P2_1/m$	5.408	9.278	4.254		93.4		5.399(100)	3.518(020)	3.132(021)	2.762(121)	3.245(101)	6
derriksite	$Cu_4[(UO_2)(SeO_3)_2](OH)_6$	$Pn2_1m$	5.57	19.088	5.965				4.772(020)	6.363(030)	3.818(050)	2.983(020)	3.097(141)	7
	$[(UO_2)(HSeO_3)_2(H_2O)]$	$C2/c$	6.354	12.578	9.972		82.3		4.658(111)	6.289(020)	4.941(002)	3.885(022)	3.004(113)	8
	α-$(UO_2)(SeO_4)$	$P2_1/c$	6.909	5.525	13.318		103.8		4.201(120)	4.265(110)	5.335(-102)	3.233(004)	6.71(100)	9

(1) Cooper & Hawthorne (1995), (2) Serezhkin et al. (1981a), (3) Blatov et al. (1989), (4) Trombe et al. (1985), (5) Ginderow & Cesbron (1983b), (6) Loopstra & Brandenburg (1978), (7) Ginderow & Cesbron (1983a), (8) Mistryukov & Michailov (1983), (9) Brandenburg & Loopstra (1978)

TABLE 12. URANIUM-BEARING PHASES CONTAINING Si

Name	Formula	S.G.	a	b	c	α	β	γ	d^1_{hkl} (Å)	d^2_{hkl} (Å)	d^3_{hkl} (Å)	d^4_{hkl} (Å)	d^5_{hkl} (Å)	ref.
β-uranophane	$Ca[(UO_2)(SiO_3OH)]_2(H_2O)_5$	$P2_1/c$	13.966	15.443	6.632		91.4		13.962(100)	7.722(020)	6.757(120)	3.861(040)	3.181(420)	1
haiweeite	$Ca(UO_2)_2[Si_5O_{12}(OH)_2](H_2O)_{4.5}$	$P2_12_12_1$	14.263	17.998	18.395				9.188(002)	7.030(121)	4.599(004)	5.347(221)	3.725(241)	2
sklodowskite	$Mg(UO_2)_2(SiO_3OH)_2(H_2O)_6$	$C2/m$	17.382	7.047	6.61		105.9		8.359(200)	4.179(400)	2.983(112)	4.319(111)	3.247(220)	3
α-uranophane	$Ca[(UO_2)(SiO_3OH)]_2(H_2O)_5$	$P2_1$	15.909	7.002	6.665		97.3		7.89(200)	3.945(400)	4.807(101)	2.982(012)	3.2(220)	4
weeksite	$(K_{.62}Na_{.38})_2(UO_2)_2(Si_5O_{13})(H_2O)_3$	$Cmmm$	7.092	17.888	7.113				7.113(001)	8.944(020)	5.567(021)	2.914(151)	3.174(201)	5
boltwoodite	$K(H_3O)[(UO_2)(SiO_4)]$	$P2_1$	7.073	7.064	6.638		105.8		6.806(100)	3.135(120)	2.91(012)	2.954(112)	4.738(011)	6
cuprosklodowskite	$Cu[(UO_2)(SiO_3OH)]_2(H_2O)_6$	P-1	7.052	9.267	6.655	109.2	89.8	110	6.365(110)	6.084(011)	4.063(121)	4.832(101)	2.973(021)	7
soddyite	$(UO_2)_2(SiO_4)(H_2O)_2$	$Fddd$	8.334	11.212	18.668				6.297(111)	3.148(222)	2.278(226)	3.355(131)	2.803(040)	8
	$Na_2(UO_2)(SiO_4)$	$I4_1/acd$	12.718		13.376				4.608(202)	4.497(220)	2.617(422)	2.683(224)	3.18(400)	9
kasolite	$Pb[(UO_2)(SiO_4)](H_2O)$	$P2_1/c$	6.704	6.932	13.252		104.2		3.058(120)	6.101(011)	3.525(112)	4.19(113)	2.914(014)	10

(1) Viswanathan & Harneit (1986), (2) Rastsvetaeva et al. (1997), (3) Ryan & Rosenzweig (1977), (4) Ginderow (1988), (5) Baturin & Sidorenko (1985), (6) Stohl & Smith (1981), (7) Rosenzweig & Ryan (1975), (8) DeMartin et al. (1992), (9) Shashkin et al. (1974), (10) Rosenzweig & Ryan (1977a)

TABLE 13. URANIUM-BEARING PHASES CONTAINING S

Name	Formula	S.G.	a	b	c	α	β	γ	d^1_{hkl} (Å)	d^2_{hkl} (Å)	d^3_{hkl} (Å)	d^4_{hkl} (Å)	d^5_{hkl} (Å)	ref.
	$[(UO_2)(SO_4)(H_2O)_2](H_2O)_{1.5}$	$P2_1/c$	16.887	12.492	6.735		90.9		16.885(100)	10.042(110)	4.58(021)	6.995(210)	3.367(002)	1
	$Mg[(UO_2)_2(SO_4)_2](H_2O)_{11}$	$C2/c$	11.334	7.715	21.709		102.2		10.609(002)	5.305(004)	6.278(111)	5.029(113)	5.875(112)	2
	$[(UO_2)(SO_4)(H_2O)_2]_2(H_2O)_3$	$Pca2_1$	11.227	6.79	21.186				10.593(002)	5.603(111)	4.96(202)	2.973(313)	5.426(201)	3
	$[(UO_2)(SO_4)(H_2O)_2](H_2O)_{1.5}$	$C2/c$	13.7	10.79	11.91		110.8		8.252(110)	3.874(022)	5.395(020)	5.57(002)	3.97(310)	4
johannite	$Cu[(UO_2)_2(OH)_2(SO_4)_2](H_2O)_8$	$P-1$	8.903	9.499	6.812	109.9	112	100.4	7.722(100)	6.119(011)	3.861(200)	6.144(101)	4.379(120)	5
	$[(UO_2)(SO_4)]_2H_2(H_2O)_5$	$C2/c$	11.008	8.242	15.619		113.7		7.151(002)	5.508(112)	6.381(110)	3.525(114)	3.345(312)	6
zippeite	$K[(UO_2)_2(SO_4)(OH)_3](H_2O)$	$C2/c$	8.755	13.987	17.73		104.1		6.994(002)	3.111(224)	3.119(222)	3.485(204)	3.473(204)	7
	$(NH_4)[(UO_2)_2(SO_4)_2](H_2O)_2$	$P2_1/c$	7.783	7.403	20.918		102.3		6.838(102)	6.961(011)	5.995(012)	3.434(114)	5.331(111)	8
	$Mn[(UO_2)_2(SO_4)_2(H_2O)](H_2O)_4$	$P2_1$	6.506	11.368	8.338		90.8		6.505(100)	3.914(012)	3.45(031)	8.337(001)	4.696(021)	9
	$(NH_4)_2U(SO_4)_2(H_2O)_4$	$P2_1/c$	6.707	19.033	8.83		97.3		6.445(021)	6.28(110)	6.653(100)	4.379(002)	4.756(040)	10
	$K_2[(UO_2)_2(SO_4)_3](H_2O)_2$	$Pnma$	13.806	11.577	7.292				5.633(111)	6.903(200)	4.436(220)	3.689(311)	3.311(131)	11
	$UO_2(OH)(NH_2SO_3)](H_2O)_3$	$P2_1/c$	6.125	17.361	9.054		117.2		4.906(121)	7.305(011)	5.904(021)	3.694(111)	8.681(020)	12
	$U(SO_4)_2(H_2O)_2$	$Pnma$	14.674	11.093	5.688				4.785(111)	6.120(210)	3.083(321)	7.337(200)	4.495(201)	13
	$\beta-(UO_2)(SO_4)$	$P2_1/c$	6.76	5.711	12.824		102.9		4.216(012)	4.316(110)	5.144(-102)	3.331(112)	2.809(-212)	14
	$Cs_2[(UO_2)_2(SO_4)_3]$	$P-42_1m$	9.62		8.13				4.14(201)	3.803(211)	3.489(112)	3.401(220)	5.217(111)	15

(1) van der Putten & Loopstra (1974), (2) Serezhkin et al. (1981b), (3) Zalkin et al. (1978), (4) Brandenburg & Loopstra (1973), (5) Mereiter (1982c), (6) Alcock et al. (1982), (7) Vochten et al. (1995), (8) Niinistö et al. (1978), (9) Tabachenko et al. (1979), (10) Bullock et al. (1980), (11) Niinistö et al. (1979), (12) Toivonen & Laitinen (1984), (13) Kierkegaard (1956), (14) Brandenburg & Loopstra (1978), (15) Ross & Evans (1960)

TABLE 14: URANIUM-BEARING PHASES CONTAINING V

Name	Formula	S.G.	a	b	c	α	β	γ	d^1_{hkl} (Å)	d^2_{hkl} (Å)	d^3_{hkl} (Å)	d^4_{hkl} (Å)	d^5_{hkl} (Å)	ref.
	$Ni[(UO_2)_2(V_2O_8)](H_2O)_4$	$Pnma$	10.6	8.25	15.12				8.68(101)	7.56(022)	3.78(004)	4.125(020)	2.642(321)	1
	$K_2[(UO_2)_2(V_2O_8)]$	$P2_1/c$	6.59	8.403	10.43		104.2		6.389(100)	3.128(013)	3.51(120)	5.056(002)	3.102(113)	2
	$Cs_2[(UO_2)_2(V_2O_8)]$	$P2_1/c$	10.51	8.45	7.32		106.1		5.406(011)	10.098(100)	3.517(022)	3.177(311)	3.366(300)	3
sengierite	$Cu_2[(UO_2)_2(V_2O_8)](OH)_2(H_2O)_6$	$P2_1/c$	10.599	8.093	10.085		103.4		4.905(202)	6.243(011)	3.218(311)	5.022(111)	3.122(022)	4
francevillite	$Ba_{.96}Pb_{.04}[(UO_2)_2(V_2O_8)](H_2O)_5$	$Pbcn$	10.419	8.51	16.763				3.002(312)	4.191(004)	4.255(020)	8.382(002)	5.181(112)	5
curienite	$Pb[(UO_2)_2(V_2O_8)](H_2O)_5$	$Pbcn$	10.4	8.45	16.34				2.986(012)	5.114(112)	4.085(004)	4.225(020)	2.937(024)	6

(1) Borène & Cesbron (1970), (2) Sundberg & Sillen (1948), (3) Applemann & Evans (1965), (4) Piret et al. (1980) (5) Mereiter (1986e), (6) Borène & Cesbron (1971)

TABLE 15. MISCELLANEOUS URANIUM-BEARING PHASES

Name	Formula	S.G.	a	b	c	α	β	γ	d^1_{hkl} (Å)	d^2_{hkl} (Å)	d^3_{hkl} (Å)	d^4_{hkl} (Å)	d^5_{hkl} (Å)	ref.
	$Cs_2(UO_2)_3Br_4$	$P2_1/c$	9.959	9.806	6.415		104.8		9.269(100)	3.210(300)	3.846(021)	5.088(-111)	4.236(111)	1
	$[(UO_2)(CrO_4)(H_2O)_2](H_2O)_{3.5}$	$P2_1/c$	11.179	7.119	26.49		94.2		8.845(102)	6.000(110)	5.508(104)	3.562(214)	5.874(104)	2
	$Pb_2(UO_2)(TeO_3)_3$	$P2_1/n$	11.605	13.389	6.981		91.2		8.768(100)	3.095(022)	3.308(-311)	3.347(040)	3.253(311)	3
	$Cu(H_2O)_4(UO_2)_2(HGeO_4)_2(H_2O)_2$	$C2/m$	17.66	7.148	6.817		112.8		8.14(200)	4.07(400)	3.574(020)	6.284(001)	3.273(220)	4
	$Sr[(UO_2)(CrO_4)(OH)]_2(H_2O)_8$	$P-1$	8.923	9.965	11.602	106.6	99.1		7.575(101)	6.189(102)	3.788(202)	4.459(121)	5.442(002)	5
	$K[(UO_2)(CrO_4)(OH)](H_2O)_{1.5}$	$P2_1/c$	13.292	9.477	13.137		104.1		6.446(200)	4.578(112)	6.36(02)	3.223(400)	5.281(112)	6
	$(UO_2)_2[GeO_4](H_2O)_2$	$Fddd$	8.179	11.515	19.397				6.306(111)	4.642(113)	3.334(220)	2.819(226)	3.42(131)	7
schmitterite	$(UO_2)_2(CuO)$	$P-1$	6.516	7.614	5.615	109.5	125.2	97.3	4.188(001)	3.287(120)	3.258(201)	6.95(010)	2.094(002)	8
	$[(UO_2)(TeO_3)]$	$Pca2_1$	10.161	5.363	7.862				3.688(210)	3.109(202)	3.171(012)	1.982(212)	5.363(010)	9
	UO_2F_2	$R-3m$	5.775	5.775	5.775	42.7			3.458(100)	5.240(111)	3.304(110)	2.671(211)	6.057(001)	10
brannerite	$UT_{i2}O_6$	$C2/m$	9.812	3.77	6.925		119	90	3.452(110)	4.753(-201)	3.357(-202)	2.9(290)	1.709(622)	11
	$(UO_2)(Te_3O_7)$	$Pa-3$	11.335						3.272(222)	2.834(400)	2.00(440)	4.01(220)	1.467(440)	12
cliffordite	Ba_2CrUO_6	$Pa-3$	8.301	8.301	8.301				2.935(220)	1.694(422)	2.075(400)	4.793(111)		13

(1) Mikhailov & Kuznetsov (1971), (2) Serezhkin & Trunov (1981), (3) Branstätter (1981a), (4) Legros & Jeannin (1975a), (5) Serezhkin et al. (1982), (6) Serezhkina et al. (1990), (7) DeMartin et al. (1992), (8) Dickens et al. (1993), (9) Meunier & Galy (1973), (10) Zachariasen (1948), (11) Szymanski & Scott (1982), (12) Branstätter (1981b), (13) Grenet & Michel (1971)

REFERENCES

Åberg M 1969 The crystal structure of [(UO$_2$)$_2$(OH)$_2$Cl$_2$(H$_2$O)$_4$]. Acta Chem Scand 23:791–810
Åberg M 1976 The crystal structure of [(UO$_2$)$_4$Cl$_2$O$_2$(OH)$_2$(H$_2$O)$_6$]·4H$_2$O, a compound containing a tetranuclear aquachlorohydroxooxo complex of uranyl(VI). Acta Chem Scand A30:507–514
Åberg M 1978 The crystal structure of hexaaqua-tri-my-hydroxo-my-3-oxo-triuranyl (VI) nitrate tetrahydrate, ((UO$_2$)$_3$O(OH)$_3$(H$_2$O)$_6$)(NO$_3$)(H$_2$O)$_4$. Acta Chem Scand A32:101–107
Alcock NW 1968 The crystal and molecular structure of sodium uranyl triperoxide. J Chem Soc A 1968:1588–1594
Alcock NW, Roberts MM, Brown D 1982 Actinide structural studies. 3. The crystal and molecular structures of UO$_2$SO$_4$·H$_2$SO$_4$·5H$_2$O and (NpO$_2$SO$_4$)$_2$·H$_2$SO$_4$·4H$_2$O. J Chem Soc, Dalton Trans 1982:869–873
Allpress JG 1965 The crystal structure of barium diuranate BaU$_2$O$_7$. J Inorg Nucl Chem 27:1521–1527
Allpress JG, Wadsley AD 1964 The crystal structure of caesium uranyl oxychloride Cs$_{0.9}$(UO$_2$)OCl$_{0.9}$. Acta Crystallogr A17:41–46
Applemann DE, Evans HT Jr 1965 The crystal structures of synthetic anhydrous carnotite, K$_2$(UO$_2$)$_2$V$_2$O$_8$ and its ceasium analogue, Cs$_2$(UO$_2$)$_2$V$_2$O$_8$. Am Mineral 50:825–842
Atencio D, Neumann R, Silva AJGC, Mascarenhas YP 1991 Phurcalite from Perus, São Paulo, Brazil, and redetermination of its crystal structure. Can Mineral 29:95–105
Azaroff LV 1968 Elements of X-ray Crystallography. New York, MacGraw Hill, 610 p
Bachet B, Brassy C, Cousson A 1991 Structure de Mg((UO$_2$)(AsO$_4$))$_2$·4H$_2$O. Acta Crystallogr C47:2013–2015
Bacmann M, Bertaut EF 1966 Paramètres atomiques et structure magnétique de MnUO$_4$. J Physique 27:726–734
Bacmann M, Bertaut EF 1967 Structure du nouveau compose UFeO$_4$. Bull Soc fr Mineral Crist 90:257–258
Bartram SF, Fryxell RE 1970 Properties and crystal structure of NaUO$_3$ and Na$_{11}$O$_{16}$U$_5$. J Inorg Nucl Chem 32:3701–3706
Baturn SV, Sidorenko GA 1985 Crystal structure of weeksite (K$_{.62}$Na$_{.38}$)$_2$(UO$_2$)$_2$[Si$_5$O$_{13}$]·3H$_2$O. Dokl. Akad. Nauk SSSR 282:1132–1136 (in Russian)
Behm H 1985 Hexpotassium(cyclo-octahydroxotetracosaoxohexadecaborato) dioxouranate(VI) dodecahydrate, K$_6$[UO$_2${B$_{16}$O$_{24}$(OH)$_8$}]·12H$_2$O. Acta Crystallogr C41:642–645
Benard P, Louer D, Dacheux N, Brandel V, Genet M 1996 Synthesis, ab initio structure determination from powder diffraction and spectroscopic properties of a new diuranium oxide phosphate. AQIEF 92:79–87
Bertaut EF, Delapalme A, Fovrat F, Pauthenet R 1962 Etude des uranates de cobalt et de manganese. J Physique Radium 23:477–485
Birch WD, Mumme WG, Segnit ER 1988 Ulrichite: a new copper calcium uranium phosphate from Lake Boga, Victoria, Australia. Austral Mineral 3:125–131
Bish DL, Post JE (Eds) 1989 Modern Powder Diffraction. Reviews in Mineralogy, Vol 20, Mineralogical Soc America, Washington, DC, 369 p
Blatov VA, Serezhkina LB, Serezkhin LB, Trunov VK 1989 Crystal structure of NH$_4$UO$_2$SeO$_4$F·H$_2$O. Zh Neorg Khim 34:162–164 (in Russian)
Borène J, Cesbron F 1970 Structure cristalline de l'uranyl-vanadate de nickel tétrahydraté Ni(UO$_2$)$_2$(VO$_4$)$_2$·4H$_2$O. Bull Soc fr Minéral Cristallogr 93:426–432
Borène J, Cesbron F 1971 Structure cristalline de la curiénite Pb(UO$_2$)$_2$(VO$_4$)$_2$·5H$_2$O. Bull Soc fr Minéral Cristallogr 94:8–14
Brandenburg NP, Loopstra BO 1973 Uranyl sulphate hydrate, UO$_2$SO$_4$·3.5H$_2$O. Cryst. Struct Commun 2:243–246
Brandenburg NP, Loopstra BO 1978 β-uranyl sulphate and uranyl selenate. Acta Crystallogr B34:3734–3736
Branstätter F 1981a Synthesis and crystal structure determination of Pb$_2$(UO$_2$)((TeO$_3$)$_3$). Z Kristallogr 155:193–200
Branstätter F 1981b Non-stoichiometric, hydrothermally synthesized cliffordite. Tschermaks mineral petrogr Mitt 29:1–8
Bullock JI, Ladd MC, Povey DC, Storey AE 1980 The chemistry of the actinoids. Part 7(1). Crystal structure analysis of (NH$_4$)$_4$U(SO$_4$)$_2$(H$_2$O)$_4$ and comments on the structure of U$_2$(SO$_4$)$_3$(H$_2$O)$_9$. Inorg Chim. Acta 43:101–108
Burns PC 1998 The structure of compreignacite, K$_2$[(UO$_2$)$_3$O$_2$(OH)$_3$]$_2$(H$_2$O)$_7$. Can Mineral 36:1061–1067
Burns PC, Hill FC (in press) Implications of the synthesis and structure of the Sr analogue of curite. Can Mineral
Burns PC, Miller ML, Ewing RC 1996 U^{6+} minerals and inorganic phases: a comparison and hierarchy of

crystal structures. Can Mineral 34:845–880

Cesbron F, Pierrot R, Verbeek T 1971 La derriksite, $Cu_4(UO_2)(SeO_3)_2(OH)_6(H_2O)$, une nouvelle espece minerale. Bull Minéral 94:534–537

Chevalier R, Gasperin M 1968 Structure cristalline de l'oxyde double UNb_3O_{10}. CR Acad Sci Paris C267:481–483

Chevalier R, Gasperin M 1969 Synthèse en monocristaux et structure cristalline de l'oxyde $UTiNb_2O_{10}$. CR Acad Sci Paris C268:1426–1428

Christ CL, Clark JR, Evans HT Jr 1955 Crystal structure of rutherfordine, UO_2CO_3. Science 121:472–473

Collomb A, Gondrand M, Lehmann M, Capponi JJ, Joubert JC 1976 Etude par diffraction X et neutronique d'un monocristal de UCr_2UO_6 obtenu par synthèse hydrothermale sous tres haute pression. Determination des structures cristallographique et magnétique. J Solid State Chem 16:41–48

Cooper MA, Hawthorne FC 1995 The crystal structure of guilleminite, a hydrated Ba-U-Se sheet structure. Can Mineral 33:1103–1109

Cordfunke EHP, Ijdo DJW 1988 $Ba_2U_2O_7$: crystal structure and phase relationships. J Phys Chem Solids 49:551–554

Cordfunke EHP, Ijdo DJW 1995 Alpha and beta Na_2UO_4: structural and thermochemical relationships. J Solid State Chem 115:299–304

Cordfunke EHP, Van Vlaandersen P, Onink M and Ijdo DJW 1991 $Sr_3U_{11}O_{36}$: crystal structure and thermal stability. J Solid State Chem 94:12–18

Cremers TL, Eller PG, Larson EM, Rosenzweig A 1986 Single-crystal structure of lead uranate(VI). Acta Crystallogr C42:1684–1685

Cullity BD 1978 Elements of X-ray Diffraction. 2nd edn. Reading, Massachusetts, Addison-Wesley

Dalley NK, Mueller MH, Simonsen SH 1971 A neutron diffraction study of uranyl nitrate dihydrate. Inorg Chem 10:323–328

Debets PC 1966 The structure of β-UO_3. Acta Crystallogr 21:589–593

Debets PC 1968 The structure of uranyl chloride and its hydrates. Acta Crystallogr B24:400–402

Demartin F, Diella V, Donzelli S, Gramaccioli CM, Pilati T 1991 The importance of accurate crystal structure determination of uranium minerals. I. Phosphuranylite $KCa(H_3O)_3(UO_2)_7(PO_4)_4O_4 \cdot 8H_2O$. Acta Crystallogr B47:439–446

Demartin F, Gramaccioli CM, Pilati T 1992 The importance of accurate crystal structure determination of uranium minerals. II. Soddyite $(UO_2)_2(SiO_4) \cdot 2H_2O$. Acta Crystallogr C48:1–4

Dickens PG, Stuttard GP, Patat S 1993 Structure of CuU_3O_{10}. J Mater Chem 3:339–341

Downs RT, Bartelmehs KL, Gibbs GV, Boisen MB Jr 1993 Interactive software for calculating and displaying X-ray or neutron powder diffractometer patterns of crystalline materials. Am Mineral 78:1104–1107

Finch RJ, Cooper MA, Hawthorne FC, Ewing RC 1996 The crystal structure of schoepite, $[(UO_2)_8O_2(OH)_{12}](H_2O)_{12}$. Can Mineral 34:1071–1088

Finch RJ, Ewing RC 1997 Clarkeite: New chemical and structural data. Am Mineral 82:607–619

Finch RJ, Miller ML, Hawthorne FC, Ewing RC 1997 Distinguishing among schoepite, $[(UO_2)_8O_2(OH)_{12}](H_2O)_{12}$, and related minerals by X-ray powder diffraction. Powd Diff 12:230–238

Fitch AN, Cole M 1991 The structure of $KUO_2PO_4 \cdot 3D_2O$ refined from neutron and synchrotron-radiation powder diffraction data. Mater Res. Bull 26:407–414

Fitch AN, Fender BEF 1983 The structure of deuterated ammonium uranyl phosphate trihydrate, $ND_4UO_2PO_4 \cdot 3D_2O$ by powder neutron diffraction. Acta Crystallogr C39:162–166

Fitch AN, Fender BEF, Wright AF 1982a The structure of deuterated lithium uranyl arsenate tetrahydrate $LiUO_2AsO_4 \cdot 4D_2O$ by powder neutron diffraction. Acta Crystallogr B38:1108–1112

Fitch AN, Wright AF, Fender BEF 1982b The structure of $UO_2DAsO_4(H_2O)_4$ at 4 K by powder neutron diffraction. Acta Crystallogr B38:2546–2554

Fujino T, Masaki N, Tagawa H 1977 The crystal structures of α- and γ-$SrUO_4$. Z Kristallogr Kristallgeom Kristallphys Kristallchem 145:299–309

Gasperin M 1987a Synthèse et structure du dibouranate de magnésium MgB_2UO_7. Acta Crystallogr C43:2264–2266

Gasperin M 1987b Synthèse et structure de trois niobouranates d'ions monovalents: $TlNb_2U_2O_{11.5}$, $KNbUO_6$, et $RbNbUO_6$. J Solid State Chem 67:219–224

Gasperin M 1987c Synthèse et structure du niobouranate de césium: $CsNbUO_6$. Acta Crystallogr C43:404–406

Gasperin M 1987d Synthèse et structure du borouranate de calcium: $CaU_2B_2O_{10}$. Acta Crystallogr C43:1247–1250

Gasperin M 1987e Structure du borate d'uranium UB_2O_6. Acta Crystallogr C43:2031–2033

Gasperin M 1988 Synthèse et structure du borouranate de sodium, $NaBUO_5$. Acta Crystallogr C44:415–416

Gasperin M 1990 Synthèse et structure du borouranate de lithium $LiBUO_5$. Acta Crystallogr C46:372–374

Gasperin M, Rebizant J, Dancausse JP, Meyer D, Cousson A 1991 Structure de $K_9BiU_6O_{24}$. Acta Crystallogr C47:2278–2279

Giebert E, Hoekstra HR, Reis AH. Jr, Peterson SW 1978 The crystal structure of lithium uranate. J Inorg Nucl Chem 40:65–68

Ginderow D 1988 Structure de l'uranophane alpha, $Ca(UO_2)_2(SiO_3OH)_2 \cdot 5H_2O$ Acta Crystallogr C44:421–424

Ginderow D, Cesbron F 1983a Structure de la derriksite, $Cu_4(UO_2)(SeO_3)_2(OH)_6$. Acta Crystallogr C39:1605–1607

Ginderow D, Cesbron F 1983b Structure de la demesmaekerite, $Pb_2Cu_5(SeO_3)_6(UO_2)_2(OH)_6 \cdot 2H_2O$. Acta Crystallogr C39:824–827

Ginderow D, Cesbron F 1985 Structure de la roubaultite, $Cu_2(UO_2)_3(CO_3)_2O_2(OH)_2 \cdot 4H_2O$. Acta Crystallogr C41:654–657

Glusker JP, Trueblood KN 1985 Crystal Structure Analysis, a Primer. 2nd edition. New York, Oxford University Press

Gorbunova, YuE, Linde SA, Lavrov AV, Pobedina AB 1980 Synthesis and structure of $Na_{6-x}(UO_2)_3(H_xPO_4)(PO_4)_3$ ($x = 0.5$). Dokl Akad Nauk SSSR 251:385–389 (in Russian)

Grenet JV, Michel A 1971 Etude cristallographique des composes Ba_2FeUO_6 et Ba_2CrUO_6. Ann Chem (Paris) 1971:83–88

Han J-C, Rong, S-B, Chen, S-B, Wu, X-R 1990 The determination of the crystal structure of tetrapotassium uranyl tricarbonate by powder X-ray method. Chinese J Chem 1990:313–318

Hanic F 1960 The crystal structure of meta-zeunerite $Cu(UO_2)_2(AsO_4)_2 \cdot 8H_2O$. Czech. J Phys 10:169–181

Hoekstra H, Siegel S 1964 Structural studies of Li_4UO_5 and Na_4UO_5. J Inorg Nucl Chem 26:693–700

Ijdo DJW 1993a $Pb_3U_{11}O_{36}$, a Rietveld refinement of neutron powder diffraction data. Acta Crystallogr C49:654–656

Ijdo DJW 1993b Redetermination of tristrontium uranate(VI). A Rietveld refinement of neutron powder diffraction data. Acta Crystallogr C49:650–652

Jove J, Cousson A, Gasperin M 1988 Synthesis and crystal structure of $K_2U_2O_7$ and mossbauer (273Np) studies of $K_2Np_2O_7$ and $CaNpO_4$. J Less-Common Metals 139:345–350

Kapshukov II, Volkov YF, Muskvitsev EP, Lebedev IA, Yakovlev GN 1971 Crystal structure of uranyl tetranitrate. Zhur Struk Khim 12:94–98

Khosrawan-Sazedj F 1982a The crystal structure of meta-uranocircite. II. $Ba(UO_2)_2(PO_4)_2(H_2O)_6$. Tschermaks mineral petrogr Mitt 29:193–204

Khosrawan-Sazedj F 1982b On the space group of threadgoldite. Tschermaks mineral petrogr Mitt 30:111–115

Kierkegaard P 1956 The crystal structure of $U(SO_4)_2(H_2O)_4$. Acta Chem Scand 10:599–616

Klug HP, Alexander LE 1974 X-ray Diffraction Procedures for Polycrystalline and Amorphous Materials, 2nd edition. New York, Wiley, 966 p

Koster AS, Renaud JPP, Rieck GD 1975 The crystal structure at 295 and 1275 K of bismuth uranate, Bi_2UO_6. Acta Crystallogr B31:127–131

Kovba LM 1971 The crystal structure of potassium and sodium monourantes. Radiokhim 13:309–311

Kovba LM 1972 Crystal structure of $K_2U_7O_{22}$. J Struct Chem 13:235–238

Kovba LM, Ippolitova EA, Simanov, YuP, Spitsyn VI 1958 The X-ray investigation of alkali elements. Dokl Akad Nauk SSSR 120:1042–1044 (in Russian)

Kovba LM, Ippolitova EA, Simanov, YuP, Spitsyn VI 1961 The crystal structure of uranates. Uranates containing uranyloxygen chains. Zhur Fiz Khim 35:719–722

Krause W, Effenberger H, Brandstätter F 1995 Orthowalpurgite, $(UO_2)Bi_4O_4(AsO_4)_2 \cdot 2H_2O$, a new mineral from the Black Forest, Germany, Eur J Mineral 7:1313–1324

Krogh-Andersen E, Krogh-Andersen IG, Ploug-Soerensen G 1985 Structure determination of a substance alleged to be hendecahydrogen diuranyl pentaphosphate. Solid State Protonic Conduct. Fuel Cells Sens, Eur Workshop "Solid State Mater Low Medium Temp. Fuel Cell Monit, Special Emphasis Proton Conduct." Goodenough JB, Jensen J, Potier A (Eds) Odense Univ Press, Odense, Denmark, p 191–202

Lee MR, Jaulmes S 1987 Nouvelle série d'oxydes dérivés de la structure de α-U_3O_8: M(II) UMo_4O_{16}. J Solid State Chem 67:364–368

Legros JP, Jeannin Y 1975a Coordination de l'uranium par l'ion germanate. I. Structure d'un uranyl germanate de cuivre $Cu(H_2O)_4(UO_2HGeO_4)_2(H_2O)_2$. Acta Crystallogr B31:1133–1139

Legros JP, Jeannin Y 1975b Coordination de l'uranium par l'ion germanate. II. Structure du germanate d'uranyle dihydraté $(UO_2)_2GeO_4(H_2O)_2$. Acta Crystallogr B31:1140–1143

Linde SA, Gorbunova YE, Lavrov AV 1980 Crystal structure of $K_4UO_2(PO_4)_2$. Zh Neorg Khim 25:1992–1994 (in Russian)

Linde SA, Gorbunova YE, Lavrov AV, Kuznetsov VG 1978 Synthesis and structure of $CsUO_2(PO_3)_3$

crystals. Dokl Akad Nauk SSSR 242:1083–1085 (in Russian)
Linde SA, Gorbunova YE, Lavrov AV, Pobedina AB 1984 The synthesis and the structure of crystals of uranyl pyrophosphate $Na_2UO_2P_2O_7$. Zh Neorg Khim 29:1533–1537 (in Russian)
Loopstra BO 1970 The structure of β-U_3O_8. Acta Crystallogr B26:656–657
Loopstra BO 1977 On the crystal structure of α-U_3O_8. J Inorg Nucl Chem 39:1713–1714
Loopstra BO, Brandenburg NP 1978 Uranyl selenite and uranyl tellurite. Acta Crystallogr B34:1335–1337
Loopstra BO, Cordfunke EHP 1966 On the structure of α-UO_3. Rec Trav Chim Pays-Bas 85:135–142
Loopstra BO, Rietveld HM 1969 The structure of some alkaline-earth metal uranates. Acta Crystallogr B25:787–791
Loopstra BO, Taylor JC, Waugh AB 1977 Neutron powder file studies of the gamma uranium trioxide phases. J Solid State Chem 20:9–19
Makarov YS, Anikina LI 1963 Crystal structure of umohoite $(UMo_6(H_2O)_2)\cdot 2H_2O$. Geochemistry 1963:14–21
Makarov YS, Ivanov VI 1960 The crystal structure of meta-autunite, $Ca(UO_2)_2(PO_4)_2\cdot 6H_2O$. Dokl Acad Sci USSR, Earth Sci Sect 132:601–603
Mayer H, Mereiter K 1986 Synthetic bayleyite, $Mg_2[UO_2(CO_3)_3]\cdot 18H_2O$: thermochemistry, crystallogrphy and crystal structure. Tschermaks mineral petrogr Mitt 35:133–146
Mercier R, Pham, Thi M, Colomban P 1985 Structure, vibrational study and conductivity of the trihydrated uranyl bis(dihydrogenphosphate): $UO_2(H_2PO_4)_2(H_2O)_3$. Solid State Ionics 15:113–126
Mereiter K 1982a The crystal structure of liebigite, $Ca_2UO_2(CO_3)_3\cdot\sim 11H_2O$. Tschermaks mineral petrogr Mitt 30:277–288
Mereiter K 1982b The crystal structure of walpurgite, $(UO_2)Bi_4O_4(AsO_4)_2(H_2O)_2$. Tschermaks. mineral petrogr Mitt 30:129–139
Mereiter K 1982c Die Kristallstruktur des Johannits, $Cu(UO_2)_2(OH)_2(SO_4)_2\cdot 8H_2O$. Tschermaks mineral petrogr Mitt 30:47–57
Mereiter K 1986a Crystal structure and crystallographic properties of a schröckingerite from Joachimsthal. Tschermaks mineral petrogr Mitt 35:1–18
Mereiter K 1986b Synthetic swartzite, $CaMg[UO_2(CO_3)_3](H_2O)_{12}$ and its strontium analogue, $SrMg[UO_2(CO_3)_3](H_2O)_{12}$: crystallography and crystal structures. N Jb Mineral Mh 481–492
Mereiter K 1986c Neue kristallographische Daten über das Uranmineral Andersonit. Anzeiger der Oesterreich Akad Wiss, Mathematisch-Naturwissenschaftliche Klasse 3:39–41
Mereiter K 1986d Structure of strontium tricarbonato- dioxouranate(VI) octahydrate. Acta Crystallogr C42:1678–1681
Mereiter K 1986e Crystal structure refinements of two francevillites, $(Ba,Pb)[(UO_2)_2V_2O_8](H_2O)_5$. N Jb Mineral Mh 552–560
Mereiter K 1986f Structure of thallium tricarbonatodioxouranate(VI). Acta Crystallogr C42:1682–1684
Mereiter K 1988 Structure of caesium tricarbonato-dioxouranate(VI) hexahydrate. Acta Crystallogr C44:1175–1178
Meunier G, Galy J 1973 Structure cristalline de la schmitterite synthétique $UTeO_5$. Acta Crystallogr B29:1251–1255
Mijlhoff FC, Ijdo DJW, Cordfunke EHP 1993 The crystal structure of α- and β-$Cs_2U_2O_7$. J Solid State Chem 102:299–305
Mikhailov YN, Kuznetsov VG 1971 The crystal structure of caesium tetrabromouranate, $Cs_2(UO_2)Br_4$. Zhur Neorg Khim 16:2512–2516
Miller SA, Taylor JC 1986 The crystal structure of saléeite, $Mg[UO_2PO_4]_2\cdot 10H_2O$. Z Kristallogr 177:247–253
Mistryukov VE, Michailov YN 1983 The characteristic properties of the structural function of the selenitogroup in the uranyl complex with neutral ligands. Koord Khim 9:97–102 (in Russian)
Mistryukov VE, Michailov YN 1985 Crystal structure of $Rb(UO_2)(HPO_3)(H_2PO_3))\cdot 3H_2O$ and $K_2(UO_2)(HPO_3)_2)\cdot 2H_2O$. Koord Khim 11:1393–1398
Morosin B 1978 Hydrogen uranyl tetrahydrate, a hydrogen ion solid electrolyte. Acta Crystallogr B34:3732–3734
Niinistö L, Toivonen J, Valkonen J 1978 Uranyl (VI) compounds. I. The crystal structure of ammonium uranyl sulfate dihydrate, $(NH_4)_2UO_2(SO_4)_2\cdot 2H_2O$. Acta Chem Scand A32:647–651
Niinistö L, Toivonen J, Valkonen J 1979 Uranyl (VI) compounds. II. The crystal structure of potassium uranyl sulfate dihydrate, $K_2UO_2(SO_4)_2\cdot 2H_2O$. Acta Chem Scand A33:621–624
Padel L Poix P, Michel A 1972 Preparation et etude cristallographique du systeme Ba_2MgUO_6–$Ba_2Fe_{1.333}U_{0.667}O_6$. Rev Chim Mineral 9:337–350
Pagoaga MK, Appleman DE, Stewart JM 1987 Crystal structures and crystal chemistry of the uranyl oxide hydrates becquerelite, billietite, and protasite. Am Mineral 72:1230–1238
Perrin A 1976 Structure cristalline du nitrate de dihydroxo diuranyle tétrahydraté. Acta Crystallogr

B32:1658–1661

Perrin A 1977 Préparation étude structurale et vibrationnelle des complexes $M_2U_2O_5Cl_4 \cdot 2H_2O$ (M = Rb, Cs): mise en evidence d'un anion tétranucléaire $[(UO_2)_4O_2Cl_8(H_2O)_2]^{4+}$. J Inorg Nucl Chem 39:1169–1172

Perrin A, Le Marouille JY 1977 Structure cristalline et moléculaire du complex tétranucléaire $K_2(UO_2)_4O_2(OH)_2Cl_4(H_2O)_6$. Acta Crystallogr B33:2477–2481

Piret P 1985 Structure cristalline de la fourmariérite, $Pb(UO_2)_4O_3(OH)_4 \cdot 4H_2O$. Bull Minéral 108:659–665

Piret P, Declercq J-P 1983 Structure cristalline de l'upalite $Al[(UO_2)_3O(OH)(PO_4)_2] \cdot 7H_2O$. Un exemple de macle mimétique. Bull Minéral 106:383–389

Piret P, Declercq J-P, Wauters-Stoop D 1980 Structure cristalline de la sengiérite. Bull Minéral 103:176–178

Piret P, Deliens M 1982 La vanmeersscheite $U(UO_2)_3(PO_4)_2(OH)_6 \cdot 4H_2O$ et la méta-vanmeersscheite $U(UO_2)_3(PO_4)_2(OH)_6 \cdot 2H_2O$, nouveaux minéraux. Bull Minéral 105:125–128

Piret P, Deliens M 1987 Les phosphates d'uranyle et d'aluminium de Kokobobo. IX. L'althupite $AlTh(UO_2)[(UO_2)_3O(OH)(PO_4)_2]_2(OH)_3 \cdot 15H_2O$, nouveau minéral; propriétés et structure cristalline. Bull Minéral 110:65–72

Piret P, Deliens M, Piret-Meunier J 1988 La françoisite-(Nd), nouveau phosphate d'uranyle et de terres rares; propriétés et structure cristalline. Bull Minéral 110:65–72

Piret P, Deliens M, Piret-Meunier J, Germain G 1983 La sayrite, $Pb[(UO_2)_5O_6(OH)_2] \cdot 4H_2O$. Nouveau minéral; propriétés et structure cristalline. Bull Minéral 106:299–304

Piret P, Piret-Meunier J 1988 Nouvelle détermination de la structure cristalline de la dumontite $Pb_2[(UO_2)_3O_2(PO_4)_2] \cdot 5H_2O$. Bull Minéral 111:439–442

Piret P, Piret-Meunier J 1994 Structure de la seelite de Rabejac (France). Eur J Mineral 6:673–677

Piret P, Piret-Meunier J, Declercq J-P 1979 Structure of phuralumite. Acta Crystallogr B35:1880–1882

Piret P, Piret-Meunier J, Deliens M 1990 Composition chimique et structure cristalline de la dewindtite $Pb_3[H(UO_2)_3O_2(PO_4)_2]_2 \cdot 12H_2O$. Eur J Mineral 2:399–405

Piret-Meunier J, Piret P 1982 Nouvelle determination de la structure cristalline de la becquerelite. Bull Minéral 105:656–610

Pushcharovsky D. Yu, Rastsvetaeva RK, Sarp H 1996 Crystal structure of deloryite $Cu_4(UO_2)(Mo_2O_8)(OH)_6$. J Alloys Compd 239:23–26

Rastsvetaeva RK, Arakcheeva AV, Pushcharovskii DYu, Atencio D, Menezes Filho LAD 1997 New silicon band in the Haiweete structure. Crystallogr Reports 42:1003–1009

Reis AH, Jr, Hoekstra HR, Giebert E, Peterson SW 1976 Redetermination of the crystal structure of barium uranate. J Inorg Nucl Chem 38:1481–1485

Rietveld HM 1966 The crystal structure of some alkaline earth metal uranates of the type M_3UO_6. Acta Crystallogr 20:508–513

Rosenzweig A, Ryan RR 1975 Refinement of the crystal structure of cuprosklodowskite, $Cu[(UO_2)_2(SiO_3OH)_2] \cdot 6H_2O$. Am Mineral 60:448–453

Rosenzweig A, Ryan RR 1977a Kasolite, $Pb(UO_2)(SiO_4) \cdot H_2O$. Cryst Struct Commun 6:617–621

Rosenzweig A, Ryan RR 1977b Vandenbrandeite $Cu(UO_2)(OH)_4$. Cryst Struct Commun 6:53–56

Ross M, Evans HT. Jr 1960 The crystal structure of cesium biuranyl trisulfate, $Cs_2(UO_2)_2(SO_4)_3$. J Inorg Nucl Chem 15:338–351

Ross M, Evans HT. Jr 1964 Studies of the torbernite minerals. I. The crystal structure of abernathyite and the structurally related compounds $NH_4(UO_2AsO_4) \cdot 3H_2O$ and $K(H_3O)(UO_2AsO_4)_2 \cdot 6H_2O$. Am Mineral 49:1578–1602

Rundle RE, Baenziger NC, Wilson AS, McDonald RA 1948 The structures of carbides, nitrides and oxides of uranium. J Am Chem Soc 70:99–105

Ryan RR, Rosenzweig A 1977 Sklodowskite, $MgO \cdot 2UO_3 \cdot 2SiO_2 \cdot 7H_2O$. Cryst Struct Commun 6:611–615

Sadikov GG, Krasovskaya TI, Polyakov YA, Nikolaev VP 1988 Structural and spectral studies on potassium dimolybdatouranylate. Inorg Mater 24:91–96

Saine MC 1989 Synthèse et structure de KU_2O_7 monoclinique. J Less-Common Metals 154:361–365

Samohvalova EP, Molchanov VN, Tat'yanina IV, Torchenkova EA 1990a Crystal structure of $(NH_4)_2Sc_2UMo_{12}O_{42} \cdot 12H_2O$. Koord Khim 16:207–211

Samohvalova EP, Molchanov VN, Tat'yanina IV, Torchenkova EA 1990b Crystal structure of $H_2Nd_2(H_2O)_{12}UMo_{12}O_{42} \cdot 12H_2O$. Koord Khim 16:1277–1282

Sarin VA, Linde SA, Fikin LE, Dudarev VYA, Gorbunova YE 1983 Neutronographic study of $UO_2H(PO_3)_3$ monocrystals. Zh Neorg Khim 28:1538–1541 (in Russian)

Serezhkin VN, Boiko NV, Trunov VK 1982 The crystal structure of $Sr[UO_2(OH)CrO_4]_2 \cdot 8H_2O$. Zh Neorg Khim 23:270–273 (in Russian)

Serezhkin VN, Chuvaev VF, Kovba LM, Trunov VK 1973 The structure of synthetic iriginite. Dokl Akad Nauk SSSR 210:873–876. (in Russian)

Serezhkin VN, Efremov VA, Trunov VK 1980 The crystal structure of $\alpha UO_2Mo_4 \cdot H_2O$. Kristallogr 25:861–865 (in Russian)

Serezhkin VN, Kovba LM, Trunov VK 1972 Structure of uranyl molybdate. Kristallogr 17:1127–1130

Serezhkin VN, Soldatkina MA, Boiko NV 1983 The refinement of the crystal structure $(NH_4)_4[UO_2(CO_3)_3]$. J Struct Chem 24:770–774

Serezhkin VN, Soldatkina MA, Efremov VA 1981a Crystal structure of the uranyl selenate tetrahydrate. J Struct Chem 22:451–454

Serezhkin VN, Soldatkina MA, Efremov VA 1981b Crystal structure of $MgUO_2(SO_4)_2 \cdot 11H_2O$. J Struct Chem 22:454–457

Serezhkin VN, Trunov VK 1981 The crystal structure of $UO_2CrO_4 \cdot 5.5H_2O$. Kristallografiya 26:301–304 (in Russian)

Serezhkin VN, Trunov UK, Makarevich LG 1980b The refined crystal structure of uranyl molybdate, Kristallogr 25:858–860 (in Russian)

Serezhkina LB, Trunov VK, Kholodkovskaya LN, Kuchumova NV 1990 Crystal structure of $K[UO_2CrO_4(OH)] \cdot 1.5H_2O$. Koord Khim 16:1288–1291. (in Russian)

Shashkin DP, Lur'e EA, Belov NV 1974 Crystal structure of $Na_2[(UO_2)SiO_4]$. Kristallografiya 19:958–963 (in Russian)

Siegel S, Hoekstra HR 1968 The crystal structure of copper uranium tetroxide. Acta Crystallogr B24:967–970

Siegel S, Hoekstra HR, Giebert E 1972b The structure of γ-uranyl dihydroxide, $UO_2(OH)_2$. Acta Crystallogr B28:3469–3473

Siegel S, Viste A, Hoekstra HR, Tani BS 1972a The structure of hydrogen triuranate. Acta Crystallogr B28:117–121

Smith DK 1984 Uranium mineralogy. *In* Uranium Geochemistry, Mineralogy, Geology, Exploration and Resources. Ippolito F, DeVero B, Capaldi G (Eds) Institution of Mining and Metallurgy, London, p 43–88

Stergiou AC, Rentzeperis PJ, Sklavounos S 1993 Refinement of the crystal structure of metatorbernite. Z Kristallogr 205:1–7

Sterns M, Parise JB, Howard CJ 1986 Refinement of the structure of trilead(II) uranate(VI) from neutron powder diffraction data. Acta Crystallogr C42:1275–1277

Stohl FV, Smith DK 1981 The crystal chemistry of the uranyl silicate minerals. Am Mineral 66:610–625

Sundberg I, Sillen LG 1948 On the crystal structure of KUO_2VO_4 (synthetic anhydrous carnotite). Arkiv Foer Kemi. 1:337–351

Swihart GH, Gupta PKS, Schlemper EO, Back ME, Gaines RV 1993 The crystal structure of moctezumite $[PbUO_2](TeO_3)_2$. Am Mineral 78:835–839

Szymanski JT, Scott JD 1982 A crystal structure refinement of synthetic brannerite UTi_2O_6 and its bearing on rate of alkaline-carbonate leaching of brannerite in ore. Can Mineral 20:271–280

Tabachenko VV, Balashov VL, Kovba LM, Serezhkin VN 1984b Crystal structure of barium uranyl molybdate $Ba(UO_2)_3(Mo_4)_4 \cdot 4H_2O$. Koord Khim 10:854–857 (in Russian)

Tabachenko VV, Kovba LM, Serezhkin VN 1983 Crystal structure of molybdatouranylates of magnesium and zinc of composition $M(UO_2)_3(Mo_4)_4(H_2O)_8$ (M = Mg, Zn). Koord Khim 9:1568–1571 (in Russian)

Tabachenko VV, Kovba LM, Serezhkin VN 1984a Crystal structures of $Mg(UO_2)_6(Mo_4)_7 \cdot 18H_2O$ and $Sr(UO_2)_6(Mo_4)_7 \cdot 15H_2O$. Koord Khim 10:558–562 (in Russian)

Tabachenko VV, Kovba LM, Serezhkin LB 1979 Crystal structure of manganese sulfatouranylate $MnUO_2(SO_4)_2 \cdot 5H_2O$. Koord Khim 5:1563–1568 (in Russian)

Tali R, Tabachenko VV, Kovba LM 1993 Crystal structure of $Cu_4UO_2(MoO_4)_2(OH)_6$. Zh Neorg Khim 38:1450–1452 (in Russian)

Tat'yanina IV, Chernaya TS, Torchenkova EA, Simonov VI, Spitsyn VI 1979 Crystal structure of the U(IV) heteromolybdate $CuH_6(UMo_{12}O_{42})(H_2O)_{12}$. Dok Akad Nauk SSSR Danka 247:1162–1165

Tat'yanina IV, Fomicheva EB, Molchanov VN, Zovodnik VE, Bel'sky VK, Torchenkova EA 1982a The crystal structure of $(NH_4)_2(Er_2UMo_{12}O_{42})(H_2O)_{22}$. Kristallogrr 27:233–238

Tat'yanina IV, Molchanov VN, Torchenkova EA, Kazanskii LP 1982b The crystal structure of $(NH_4)_3H(ThUMo_{12}O_{42})(H_2O)_{15}$ and its spectroscopic properties. Koord Khim 8:1261–1267

Taylor JC 1971 The structure of the α form of uranyl hydroxide. Acta Crystallogr B27:1088–1091

Taylor JC, Bannister MJ 1972 A neutron diffraction study of the anisotropic thermal expansion of β-uranyl dihydroxide. Acta Crystallogr B28:2995–2999

Taylor JC, Mueller MH 1965 A neutron diffraction study of uranyl nitrate hexahydrate. Acta Crystallogr 19:536–543

Taylor JC, Stuart WI, Mumme IA 1981 The crystal structure of curite. J Inorg Nucl Chem 43:2419–2423

Taylor JC, Wilson PW 1973 The structure of anhydrous uranyl chloride by powder neutron diffraction. Acta Crystallogr B29:1073–1076

Taylor JC, Wilson PW 1974 The structure of uranyl chloride monohydrate by neutron diffraction and the disorder of the water molecule. Acta Crystallogr B30:169–175

Toivonen J, Laitinen R 1984 Uranyl hydroxide sulphamate trihydrate $UO_2(OH)(NH_2SO_3)(H_2O)_3$. Acta Crystallogr C40:7–9

Trombe JC, Gleizes A, Galy J 1985 Structure of a uranyl diselenite $UO_2Se_2O_5$. Acta Crystallogr C41:1571–1573

van der Putten N, Loopstra BO 1974 Uranyl sulphate $2.5H_2O \cdot UO_2SO_4 \cdot 2.5H_2O$. Cryst Struct Commun 3:377–380

van Egmond AB 1976a Investigations on cesium uranates. V. The crystal structures of Cs_2UO_4, $Cs_4U_5O_{17}$, $Cs_2U_7O_{22}$ and $Cs_2U_{15}O_{46}$. J Inorg Nucl Chem 38:1649–1651

van Egmond AB 1976b Investigations on cesium uranates. VI. The crystal structure of $Cs_2U_2O_7$. J Inorg Nucl Chem 38:2105–2107

Vdovenko VM, Stroganov EV, Sokolov AP 1967 About the structure of uranyl nitrate hexahydrate. Radiokhimiya 9:127–130

Vdovenko VM, Stroganov EV, Sokolov AP, Lungu G 1962 The structure of uranyl nitrate dihydrate. Radiokhimiya 4:59–66

Viswanathan K, Harneit O 1986 Refined crystal structure of β-uranophane $Ca(UO_2)_2(SiO_3OH)_2 \cdot 5H_2O$. Am Mineral 71:1489–1493

Vochten R, Van Haverbeke L, Van Springel K, Blaton N, Peeters OM 1994 The structure and physicochemical characteristics of a synthetic phase compositionally intermediate between liebigite and andersonite. Can Mineral 32:553–561

Vochten R, Van Haverbeke L, Van Springel K, Blaton N, Peeters OM 1995 The structure and physicochemical characteristics of synthetic zippeite. Can Mineral 33:1091–1101

Watkin DJ, Denning RG, Prout K 1991 Structure of dicaesium tetrachlorodioxuranium(VI). Acta Crystallogr C47:2517–2519

Weller MT, Dickens PG, Penny DJ 1988 The structure of δ-UO_3. Polyhedron 7:243–244

Wolf R, Hoppe R 1985 Neues ueber oxouranat: Ueber α-Li_6UO_6. Mit einer Bemerkung ueber β-Li_6UO_6. Z Anorgan Allegem Chem 528:129–137

Wolf R, Hoppe R 1986 Neues üeber Oxouranat (VI): Na_4UO_5 and K_4UO_5. Rev Chim Minérale 23:828–848

Wolf R, Hoppe R 1987 Ein neues Oxouranat (VI): $K_2Li_4UO_6$ mit einer Bemerkung ueber $Rb_2Li_4UO_6$ und $Cs_2Li_4UO_6$. Z Anorgan Allegem Chem 534:34–42

Yamashita T, Fujino T, Masaki N, Tagawa H 1981 The crystal structure of α- and β-$CdUO_4$. J Solid State Chem 37:133–139

Zachariasen WH 1948 Crystal chemical studies of 5f-series of elements. I. New structure types, Acta Crystallogr A1:265–268

Zachariasen WH 1954 Crystal chemical studies of 5f-series of elements. XXI. The crystal structure of magnesium orthouranate. Acta Crystallogr A7:788–791

Zalkin A, Ruben H, Templeton DH 1978 Structure of a new uranyl sulphate hydrate, α-$2UO_2SO_4 \cdot 7H_2O$. Inorg Chem 17:3701–3702

Zalkin A, Templeton LK, Templeton DH 1989 Structure of rubidium uranyl(VI) trinitrate. Acta Crystallogr C45:810–811